FRIEDEL-CRAFTS ALKYLATION CHEMISTRY

A CENTURY OF DISCOVERY

ROYSTON M. ROBERTS
The University of Texas at Austin
Austin, Texas

ALI ALI KHALAF
The University of Assiut
Assiut, Egypt

MARCEL DEKKER, INC.   New York and Basel

Library of Congress Cataloging in Publication Data

Roberts, Royston M.
Friedel-Crafts alkylation chemistry.

Includes bibliographical references and index.
1. Friedel-Crafts reaction. I. Khalaf, Ali Ali,
[date] . II. Title.
QD281.A5R63 1984 547'.21 84-4265
ISBN 0-8247-6433-1

MARCEL DEKKER, INC.
270 Madison Avenue, New York, New York 10016

Current printing (last digit):
10 9 8 7 6 5 4 3 2 1

PRINTED IN THE UNITED STATES OF AMERICA

This Book is Dedicated to

PROFESSORS

COSTIN D. NENITZESCU

and

ECATERINA CIORANESCU-NENITZESCU

with

Appreciation and Affection

Preface

A comprehensive treatise, *Friedel-Crafts and Related Reactions*, edited by George A. Olah, appeared in 1963-1965. One of the present authors (R.M.R.), with the collaboration of G. J. Fonken, contributed a chapter to that work, and in 1965 he wrote a short review which appeared in *Chemical & Engineering News* (January 25, 1965) under the title "Friedel-Crafts Chemistry." Shortly after its appearance, Dr. Maurits Dekker suggested that an expansion of that review into a monograph which would represent "the heart of Friedel-Crafts chemistry" would be valuable, because the size of the Olah treatise made it impractical for the working chemist, from the viewpoint of both convenience and expense.

This project was initiated with enthusiasm, and its prospects for success were augmented when Dr. Khalaf agreed to become a collaborator; however, other commitments of both authors delayed its completion. In the meantime, Professor Olah updated portions of his contributions to the comprehensive treatise and combined them with some new material to produce a monograph which borrowed with our blessing the title of the 1965 *Chemical & Engineering News* review, "Friedel-Crafts Chemistry." Owing to the differences in viewpoint and interest, it turned out that there was little duplication between Olah's 1973 monograph and our manuscript in preparation, and we concluded that there was still a need for our contribution. One way of subdividing "Friedel-Crafts chemistry" is in terms of *alkylation* and *acylation* reactions. This book is concerned almost entirely with the first category, the larger and more interesting area of alkylation reactions and the competing processes that accompany them, hence the title *Friedel-Crafts Alkylation Chemistry*.

To the average organic chemist, "the Friedel-Crafts reaction" is most likely to bring to mind the reaction of an alkyl halide with benzene in the presence of aluminum chloride catalyst. He or she is also likely to be aware that the alkylbenzene so produced may be of uncertain structure because of the possibility of isomerization, but may not know exactly what the facts are, or where to learn them. This situation is a reasonable one, even at the present time. Rearrangements accompanying Friedel-Crafts alkylations were recognized at a very early date, but many of the early reports were incorrect or at least incomplete because of the lack of modern analytical techniques, and the resultant errors and misconceptions have been perpetuated throughout most of the reviews and textbooks up to the present time.

Although alkylations with alkyl halides have been central to Friedel-Crafts chemistry since the first observations of Friedel and Crafts, the chapter on this subject in the large Olah treatise was far from complete and contained several errors with regard to rearrangements accompanying alkylations. This chapter was not one of the four brought up to date by Olah for his 1973 monograph. In the last two decades, through research in our own laboratories and those of others, most of the mysteries surrounding the rearrangements and other competing processes accompanying Friedel-Crafts alkylations have been cleared up, so that a definitive picture of their relationship to the rest of Friedel-Crafts chemistry may now be given.

This book begins in Chapter 1 with a brief introduction and overview of the whole area of Friedel-Crafts alkylation chemistry. The complications produced by rearrangements and by competing reactions occurring before, during, and/or after alkylation are outlined briefly here. Chapter 2 treats in detail the most common complication, the rearrangement of an alkylating agent preceding its attachment to an aromatic substrate. This chapter includes a survey of the generation and chemistry of carbocations in superacid media.

Chapter 3 is a comprehensive survey of alkylations of arenes with all types of alkyl halides, in which the errors and inconsistencies of the older literature and even textbooks published in very recent years are sorted out and corrected.

Chapters 4 and 5 will probably be of special interest to industrial and other practicing organic chemists. In Chapter 4 alkylations by alkenes, alcohols, and other industrially useful alkylating agents are cataloged and described in detail and, in Chapter 5, a similar extensive treatment is given to alkylations by di- and polyfunctional alkylating agents.

Cyclialkylations make up a wide and important area of Friedel-Crafts alkylation chemistry. Some of the di- and polyfunctional reagents described in Chapter 5 lead to bicyclic and polycyclic products, so that their reactions may be considered to be cyclialkylations. Chapter 6 is devoted to the additional large group of cyclialkylations that occur through intramolecular ring closures of arylhaloalkanes, arylhydroxyalkanes, and arylalkenes.

The last two chapters describe those major complications to Friedel-Crafts alkylations other than the rearrangements of the alkylating agents that occur before or simultaneously with the alkylation step. In Chapter 7 the transalkylations and reorientations which have long been known to accompany alkylations are described and interpreted in modern mechanistic terms. In Chapter 8 the more recently recognized rearrangements, dealkylations, and fragmentations that arenes undergo in the presence of Friedel-Crafts catalysts are described, so that the effects of these reactions as complications to desired alkylations may be seen.

We are pleased to acknowledge the continuing encouragement given for this project by Dr. Maurits Dekker, without which the book would never have been completed. We are grateful to Terry Parsons Smith for editorial and technical assistance at an early stage of the work, and to the University of Texas Research Institute for a grant which made this assistance possible. We are indebted to Dr. Thomas L. Gibson, Jr., for similar excellent assistance at a later stage, and we thank him and other former graduate students and postdoctoral associates for their contributions to the research reported in this monograph. Their names will be found in association with those of the authors in references throughout the book. One other person who was invaluable to the completion of the project was Susie Pruett, who typed most of the text and drew most of the figures. Her ability to transform illegible and sometimes slightly Arabic script into clear typescript was a major miracle.

Preliminary planning for the book was done during a summer visit at the Philipps University of Marburg and further organization of material was carried out during a semester spent at the Polytechnic Institute of Bucharest. For these opportunities, one of us (R.M.R.), expresses his appreciation to the Fulbright-Hayes program and to the host institutions, and he also wishes especially to thank Professor Karl Dimroth (Marburg) and Professors A. T. Balaban and E. Cioranescu-Nenitzescu (Bucharest) for their hospitality during these visits. A.A.K. is grateful for support from Assiut University (Assiut, Egypt) and from King Abdulaziz University (Jeddah, Saudi Arabia).

Both authors are greatly indebted to the Robert A. Welch Foundation for generous support in the form of fellowships and other financial aid to the research at the University of Texas at Austin. We believe that it may be seen that this research has resulted in significant contributions to the new Friedel-Crafts alkylation chemistry described in this book.

Finally, we wish to thank our wives, Phyllis and Magida, for their unselfish encouragement and their understanding of the neglect they uncomplainingly encountered during the long gestation period of the book.

Royston M. Roberts
Ali Ali Khalaf

Contents

1

Introduction and Scope

I. DISCOVERY OF THE FRIEDEL-CRAFTS REACTION

The reaction of benzene with "amyl chloride" in the presence of aluminum chloride to produce "amylbenzene" was carried out by Charles Friedel and his American collaborator James Mason Crafts in Paris on May 14, 1877, and the results were reported orally to the Chemical Society of France a few days later [1]. This was apparently the first typical Friedel-Crafts alkylation reaction, although Friedel and Crafts had earlier tested the behavior of various combinations of aluminum, aluminum chloride, organic halides, and hydrocarbons. Their oral report contained the statement: "With a mixture of chloride and hydrocarbon, the formation is established, in good yield, of hydrocarbons from the residues of the hydrocarbon less H and from the chloride less Cl. It is thus that ethylbenzene, amylbenzene, benzophenone, etc., are obtained" [1]. Thus in their first communication they described products of both alkylation and acylation. Friedel and Crafts recognized the practical importance of their discovery, as is shown by the fact that they immediately secured patents in both France and England on procedures for preparing hydrocarbons and ketones. Their judgment was accurate; probably no other reaction has been of more practical value. Major processes for the production of high-octane gasoline, synthetic rubber, plastics, and synthetic detergents are applications of *Friedel-Crafts chemistry*. Aside from its commercial importance, this area of organic chemistry encompasses classic examples of some of the most interesting aspects of modern organic theory: electrophilic aromatic substitution and carbocation formation and rearrangement. In the following chapters we examine both the practical and theoretical aspects of Friedel-Crafts chemistry.

II. REARRANGEMENTS ACCOMPANYING ALKYLATION

Within a matter of months after its discovery, one of the limitations to the alkylation reactions as a synthetic method was revealed by the Russian chemist G. Gustavson [2]: the same product, isopropylbenzene, was obtained from either *n*-propyl bromide or isopropyl bromide and benzene and aluminum chloride. This was the first indication that the alkyl group being attached to the aromatic ring may suffer a structural rearrangement. (It is now known that the reaction with *n*-propyl bromide gives some *n*-propylbenzene as well as the major product, isopropylbenzene, but the presence of the normal product went undetected in the early work.) The formation of isopropylbenzene was attributed to the izomerization of *n*-propyl bromide, brought about by the effect of the catalyst prior to alkylation. This isomerization of the alkylating agent was demonstrated separately [3]. *sec*-Butylbenzene was reported to be the alkylation product from *n*-butyl chloride, and *t*-butylbenzene the product from both *t*-butyl and isobutyl chlorides. In the reaction of the latter halide, a gaseous hydrocarbon thought to be isobutylene was observed and this was assumed to be the intermediate for the rearranged product [4]. Interestingly, unchanged isobutyl chloride was recovered from one of the reaction mixtures.

Although these particular early examples of rearrangement have been checked and confirmed by modern analytical techniques, quite a number of incorrect accounts of rearrangements have appeared in the literature. On the basis of both correct and incorrect reports, generalizations and extrapolations have been made that range from misleading to downright incorrect, and some of these fallacies have carried over into even recent publications [5]. Specific cases will be examined in detail in the following chapters, but we shall try to set the record straight briefly at this point.

The generalization that "the usual tendency in rearrangement of the alkyl groups during alkylation is in the direction primary → secondary → tertiary" [5-(c)-(e)] is only partially correct, because although primary groups usually rearrange to secondary, secondary groups do not usually rearrange to tertiary. *n*-Butyl halides lead to *sec*-butyl groups, but not usually to *t*-butyl [5(a),(b)]. *t*-Alkyl halides do not always form *t*-alkylbenzenes [5(c)], but may indeed produce *sec*-alkylbenzenes. Thus it may be seen that the generalization which has perhaps made the strongest impression, "regardless of the configuration of the alkyl halide, the final product contains an alkyl group of the highest possible branching" [5(a)], is nevertheless wrong.

III. TYPES OF ALKYLATING AGENTS

Although the alkylating agents involved in the discovery of the Friedel-Crafts reaction were alkyl halides, and these compounds are most commonly used in textbook examples of the reaction, many other types of alkylating agents have been used. In fact, some of the most important commercial applications employ alkenes, such as the synthesis of styrene from ethylene and cumene from propylene. Alcohols and ethers are also useful types of alkylating agents. Alkylations with alkyl halides are discussed in Chap. 3 and those involving the other alkylating agents are discussed in Chap. 4.

Some of the most interesting and useful applications of the Friedel-Crafts reaction are those in which di- and polyfunctional alkylating agents are em-

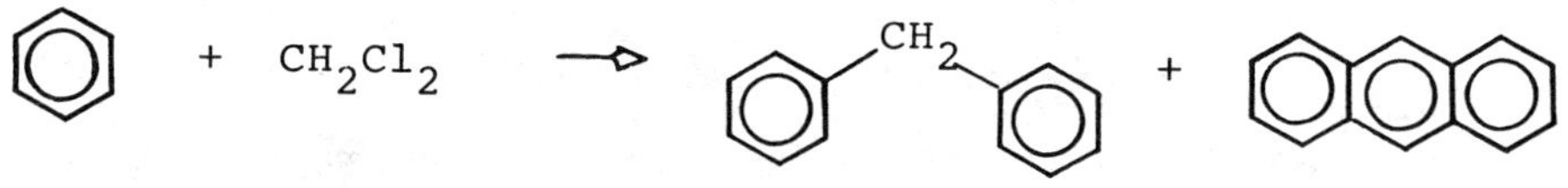

+ CH_2Cl_2 → CH_2 + (1)

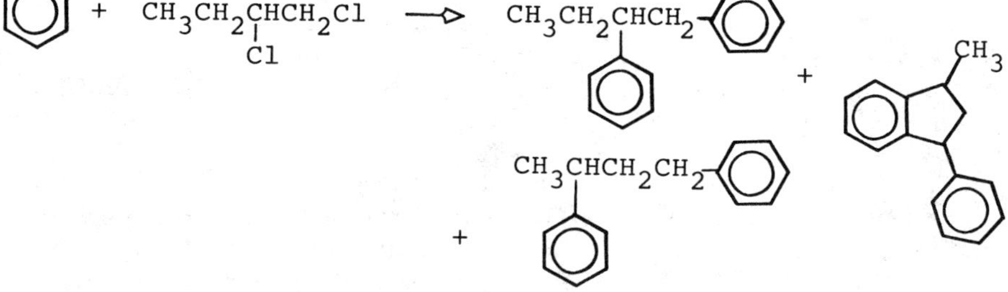

+ $CH_3CH_2CHClCH_2Cl$ → $CH_3CH_2CH(C_6H_5)CH_2$– + CH_3 + $CH_3CH(C_6H_5)CH_2CH_2$– (2)

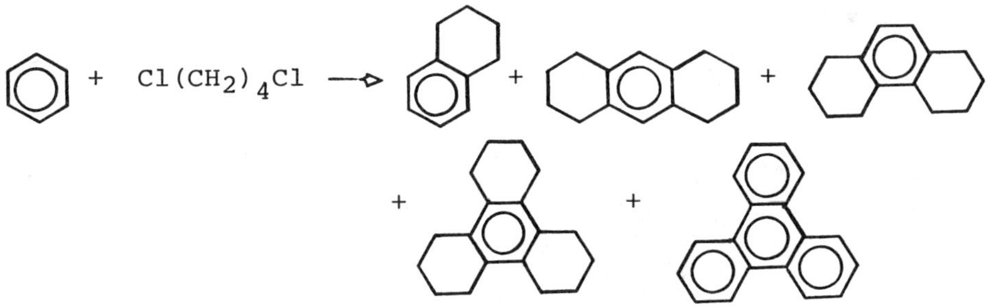

$$C_6H_6 + Cl(CH_2)_4Cl \longrightarrow$$

(3)

ployed with arene substrates. Reactions with dihalides may lead to a wide variety of products, such as diarylalkanes, polycyclic arenes, indans, tetralins, and other alicylic arenes, as shown in Eqs. (1) to (3).

The reaction of an unsaturated alkyl halide with an arene may take place at either or both of the reactive sites, that is, the halide and/or the double bond, depending on the type of catalyst and the conditions. In general, protonic catalysts favor reaction of the double bond and metal halides favor reaction at the halide site:

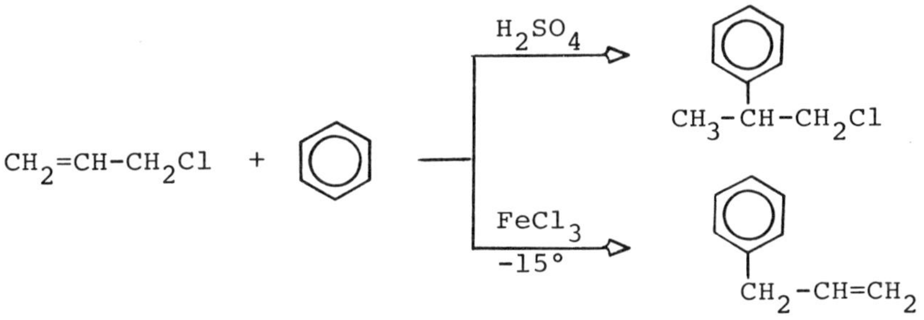

$$CH_2{=}CH{-}CH_2Cl + C_6H_6 \xrightarrow{H_2SO_4} C_6H_5{-}CH(CH_3){-}CH_2Cl \quad (CH_3{-}CH{-}CH_2Cl)$$

$$CH_2{=}CH{-}CH_2Cl + C_6H_6 \xrightarrow[-15°]{FeCl_3} C_6H_5{-}CH_2{-}CH{=}CH_2$$

IV. CYCLIALKYLATIONS

Some of the products in Eqs. (1) to (3) are formed by processes in which the difunctional alkylating agent first reacts with the arene to yield an aryl-substituted monofunctional compound, which then undergoes an intramolecular alkylation to produce the alicyclic ring. Although the term *cyclialkylation* has been applied more generally (to any Friedel-Crafts alkylation of arenes with difunctional compounds such as dichlorides, diols, and dienes in which a new ring is formed), we shall use it in the more limited sense to describe intramolecular ring closures such as that of an arylhaloalkane (Eq. 4).

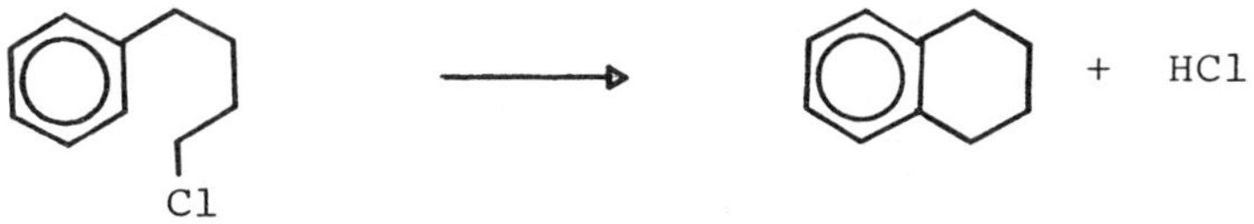

$$C_6H_5(CH_2)_4Cl \longrightarrow \text{tetralin} + HCl \quad (4)$$

Reactions of di- and polyfunctional alkylating agents with arenes (some of which produce cyclizations) are described in Chap. 5 and cyclialkylations are described in Chap. 6.

V. TRANSALKYLATIONS AND REORIENTATIONS

Another feature of Friedel-Crafts alkylation observed from the start was polyalkylation. Alkylation of benzene did not stop with the introduction of one alkyl group onto the ring; di-, tri-, and higher-alkylated benzenes were produced, although their production could be controlled to some extent by using a molar excess of benzene and by variations in other experimental parameters. It was also shown that alkyl groups could be transferred from one aromatic ring to another by the catalytic effect of aluminum chloride. For example, when ethylbenzene was heated with the catalyst, a mixture of benzene, ethylbenzene, diethylbenzene (mainly *m* and *p* isomers), and higher-boiling material was produced [6]. An unusual feature of both polyalkylations and transalkylations (also called disproportionations) was the rather high proportions of *meta*- and 1,3,5-oriented isomers produced. These have been explained in terms of the dealkylation of a 1,2,4-trialkyl derivative (to produce an *m*-dialkylbenzene) or the reorientation of 1,2-, 1,4-, and 1,2,4-derivatives in the presence of an excess of catalyst. As discussed in detail in Chap. 7, the type of Friedel-Crafts rearrangement represented by a reorientation of an *ortho*- or a *para*-dialkylbenzene to a *meta*-dialkylbenzene is intimately related to the transalkylation reaction: the reorientation of most dialkylbenzenes cannot be demonstrated without the concurrent production of transalkylation products. Thus it is possible that some observed reorientations take place by an intermolecular transalkylation process rather than by a true intramolecular reorientation (Eqs. 5 and 6).

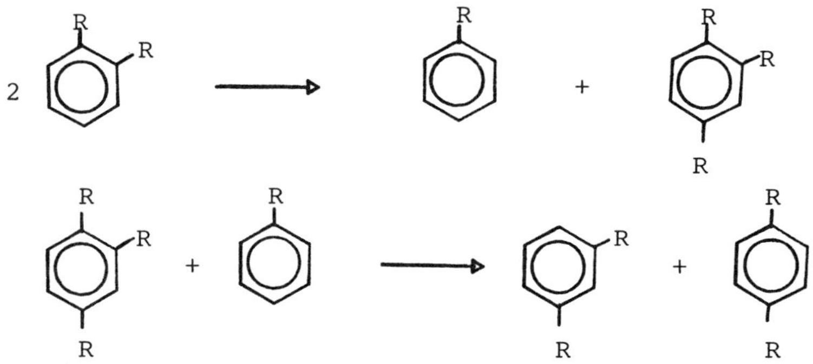

(5)

(6)

VI. REARRANGEMENTS OF ARENES

Although the Friedel-Crafts alkylation reaction has been widely used as a synthetic procedure, as we have already noted there are inherent limitations which have complicated its practical applications. These include polyalkylation, which can be controlled without too much difficulty, the problem of producing the desired orientation of the alkyl group, or groups, when more than one is introduced, and the possible rearrangement of the alkyl group whenever it is larger than ethyl. The latter two complications cannot be controlled in some cases, but if they are correctly recognized, the utility and reliability of Friedel-Crafts alkylation procedures are little impaired.

More recently an additional complication has been discovered, the isomerization of the side chain of an alkylbenzene *after* attachment to the aromatic ring [7]. This new reaction has been called the *alkylbenzene rearrangement*. It is the occurrence of this unexpected rearrangement which is responsible for some of the fallacies in the generalizations mentioned in Sec. II. For example, alkylation of benzene with *t*-pentyl chloride and aluminum chloride gives the secondary alkylbenzene 2-methyl-3-phenylbutane as the major product, because the initially formed *t*-pentylbenzene undergoes a subsequent rearrangement (Eqs. 7 and 8).

$$CH_3\text{-}CH_2\text{-}\underset{Cl}{\overset{CH_3}{C}}\text{-}CH_3 + C_6H_6 \longrightarrow CH_3\text{-}CH_2\text{-}\underset{C_6H_5}{\overset{CH_3}{C}}\text{-}CH_3 + HCl \qquad (7)$$

$$CH_3\text{-}CH_2\text{-}\underset{C_6H_5}{\overset{CH_3}{C}}\text{-}CH_3 \xrightarrow{AlCl_3} CH_3\text{-}\overset{CH_3}{CH}\text{-}\underset{C_6H_5}{CH}\text{-}CH_3 \qquad (8)$$

Details of this and of other alkylbenzene rearrangements are given in Chap. 8.

VII. DEALKYLATIONS AND FRAGMENTATIONS OF ARENES

True reversibility of Friedel-Crafts alkylation implies dissociation of an alkylbenzene into benzene and alkene, or alkyl halide, etc. Under ordinary alkylating conditions the thermodynamic equilibrium is so far on the side of alkylation that the reaction is irreversible for all practical purposes. However, some of the early workers in Friedel-Crafts chemistry observed the evolution of saturated hydrocarbon gases from reaction mixtures in which an alkylbenzene was heated with aluminum chloride. In 1886 Anschutz [6] reported that a saturated gas, free of chlorine, was produced from cymene, and toluene was found among the liquid products of the reaction. The gas was later identified as propane [8]. The evolution of propane from cymene suggests that a transient isopropyl cation is formed and may abstract a hydride ion from some source in a fast step that diverts it from the alkylation equilibrium. This hypothesis was supported by results of experiments in which cycloalkanes (as sources of hydride ions) were added to various alkylbenzenes in the presence of aluminum chloride [9]. Alkanes corresponding to the side chains of the alkylbenzenes were evolved in an order of ease in conformity with carbocation stability: *t*-butyl > *sec*-buytl > isopropyl (ethyl and methyl were not cleaved).

$$C_6H_5\text{-}\underset{CH_3}{\overset{CH_3}{C}}\text{-}CH_3 + C_6H_{12} \xrightarrow{AlCl_3} H\text{-}\underset{CH_3}{\overset{CH_3}{C}}\text{-}CH_3 + C_6H_5\text{-}C_6H_{11} \qquad (9)$$

Actually, secondary and tertiary alkylbenzenes require no added source of hydride ions for dealkylation by aluminum chloride, and even primary alkylbenzenes such as *n*-butylbenzene undergo some (ca. %5) dealkylation at 100°C [10]. The obvious source of the hydride ion necessary for the formation of an alkane after cleavage from the side chain of an alkylbenzene molecule is the side chain of another molecule; the secondary and tertiary benzylic hydrogens should be especially susceptible to abstraction. Other sources of hydride exchange are possible and will be discussed when mechanisms for dealkylation are described in detail in Chap. 8.

Ipatieff and Pines [9] reported that dealkylation of *sec*-butylbenzene produced only *n*-butane. With the added capability of gas chromatography, more recently it was possible to demonstrate that isobutane was also produced from *sec*-butylbenzene, and the dealkylation product from 2-phenylpentane was found to be almost entirely (96%) isopentane. Thus we see here another type of Friedel-Crafts rearrangement—the rearrangement of an alkyl group accompanying the dealkylation reaction.

The development of sensitive and convenient analytical methods in recent years has led to the recognition of one further type of reaction of alkylbenzenes that takes place when they are heated with Friedel-Crafts catalysts, especially at higher temperatures. Whereas dealkylation applies to a reaction producing an alkane corresponding to the side chain of an alkylbenzene, the term *fragmentation* may be used for the process that produces alkanes with fewer carbon atoms than the original alkylbenzene. Only a few scattered observations of such reactions were noted until recently. However, gas chromatography has made it easy to demonstrate that almost a complete spectrum of lower alkanes and alkylbenzenes is produced by heating an alkylbenzene with aluminum chloride at 100°C for a short time [10].

It is probably impossible to study fragmentation as an individual reaction, because the required conditions are more vigorous than those for dealkylation, rearrangement, transalkylation, and reorientation. In fact, there is evidence that common intermediates are involved in several of these reactions. The interrelationship of these reactions is explored in Chap. 8.

REFERENCES

1. Fridel, C., and J. M. Crafts, Bull. Soc. Chim. Fr., (2), *27*, 530 (1877).
2. Gustavson, G., J. Russ. Phys.-Chem. Soc., *10*, 269 (1877); Chem. Ber., *11*, 1251 (1878).
3. Kekule, A., and H. Schrötter, Chem. Ber., *12*, 2279 (1879); Gustavson, G., Chem. Ber., *16*, 958 (1883).
4. Schramm, J., Monatsh. Chem., *9*, 613 (1888).
5. (a) Calloway, N. O., Chem. Rev., *17*, 332, 367 (1935); (b) Gilman, H., and R. M. Meals, J. Org. Chem., *8*, 135 (1943); (c) Drahowzal, F. A., in *Friedel-Crafts and Related Reactions*, (G. A. Olah, ed.), Vol. 2, Part 1, Wiley-Interscience, New York, 1964, p. 446; (d) Olah, G. A., in *Friedel-Crafts and Related Reactions*, Vol. 1, (G. A. Olah, ed.), Wiley-Interscience, New York, 1963, p. 68; (e) Olah, G. A., *Friedel-Crafts Chemistry*, Wiley-Interscience, New York, 1973, p. 68; (f) Hendrickson, J. B., D. J. Cram, and G. S. Hammond, *Organic Chemistry*, 3rd ed., McGraw-Hill, New York, 1970, p. 670.
6. Anschutz, R., Justus Liebigs Ann. Chem., *235*, 177 (1886).
7. (a) Schmerling, L., and J. P. West, J. Am. Chem. Soc., *76*, 1917 (1954); (b) Roberts, R. M., and S. G. Brandenberger, Chem. Ind., 227 (1955); Roberts, R. M., Intra-Sci. Chem. Rep., *6*(2), 89 (1972), and references given there.
8. Schorger, A. W., J. Am. Chem. Soc., *39*, 2671 (1917).
9. Ipatieff, V. N., and H. Pines, J. Am. Chem. Soc., *59*, 56 (1937).
10. Roberts, R. M., E. K. Baylis, and G. J. Fonken, J. Am. Chem. Soc., *85*, 3454 (1963).

2

Rearrangements of Alkylating Agents

The whole field of Friedel-Crafts chemistry began with a report by Friedel and Crafts on the action of aluminum chloride on alkyl chlorides. In 1877, these pioneers made two preliminary oral communications to the Chemical Society of France describing their experiences relating to the action of aluminum chloride on alkyl chlorides particularly "amyl chloride" [1,2]. They stated that the interaction of amyl chloride with aluminum chloride in the cold "gave rise to an intense evolution of hydrogen chloride, to some combustible gases from saturated hydrocarbons, and to some condensed products."

A month later, Friedel and Crafts published a more detailed account of their findings in which they emphasized that the active catalyst in these reactions was the metal chloride [3]. In these reports, however, there is no indication that they were aware of the isomerizing effect of the catalyst on the alkyl chloride.

In fact, the real interest in the study of the rearrangements of alkyl halides by metal halide catalysts arose after Gustavson claimed that the reaction of benzene with either *n*-propyl or isopropyl bromide in the presence of aluminum chloride resulted only in the formation of isopropylbenzene [4]. The first explanation of Gustavson's claim was given in terms of isomerization of *n*-propyl to isopropyl bromide prior to the alkylation process itself. This type of explanation inspired numerous chemists to investigate the behavior of alkylating agents under the influence of Friedel-Crafts catalysts in order to understand the results of alkylations utilizing these agents. More important, in recent years a great deal of effort has been devoted toward elucidating the complex nature of the rearrangements of alkylating agents under Friedel-Crafts conditions; e.g., to resolve the question of whether or not these rearrangements occur concurrent with or consecutive to substrate ionization, and to seek evidence for or against the involvement of intermediates such as alkenes, protonated cyclopropanes, and primary carbocations. The present chapter concentrates on understanding the various isomerization processes observed when Friedel-Crafts alkylating agents such as halides and alcohols are subjected to the action of Friedel-Crafts catalysts.

In discussing these rearrangements we shall distinguish between two types: (1) those induced by metal halides, which will be discussed first; and (2) those induced by strong acids.

I. REARRANGEMENTS OF ALKYLATING AGENTS IN THE PRESENCE OF METAL HALIDE CATALYSTS

A. Rearrangements in Acyclic Systems

1. Rearrangements in linear acyclic systems

a. Ethyl rearrangements

Studies on the isomerization of ethyl halides similar to those of propyl halides are made possible by labeling one of the carbon atoms with carbon 13 or carbon 14. The case of ethyl halide rearrangement is of particular importance because, first, it presents an ideal model to determine whether or not primary to primary carbocation rearrangements may occur, and second, as will be evident later, it gives more insight into the mechanism of Friedel-Crafts alkylation.

The finding that labeled ethyl halide can be rearranged in the presence of aluminum halide was first reported by Roberts et al. [5]. These workers found that ethyl-β-^{14}C chloride was almost completely isomerized by standing over aluminum chloride at room temperature for 1 hr. The same workers also reported that the Friedel-Crafts alkylation of benzene with ethyl-β-^{14}C chloride gave ethylbenzene with no isotope-position rearrangement in the ethyl side chain, if the ethyl-β-^{14}C chloride was added to a mixture of benzene and aluminum chloride.

The authors used the alkylation reaction as a means of determining the extent of isotopic rearrangement produced in ethyl-β-^{14}C chloride by treatment with aluminum chloride in the absence of benzene. Thus the ethyl chloride recovered after standing over aluminum chloride for 1 hr was used to alkylate benzene, the ethylbenzene obtained was oxidized to benzoic acid, and the molar radioactivity of the latter was measured.

A more detailed study of a similar isotopic isomerization was conducted by Sixma and Hendriks [6]. These investigators carried out the isomerization of ethyl-α-^{14}C bromide with and without solvents using aluminum bromide as catalyst. They found that the rate of isomerization was lower in nitrobenzene than in carbon disulfide, and still lower in nitromethane. The isomerization was much faster in the absence of solvents. From measurements of the reaction rate at temperatures between 25 and 50°C, Sixma and Hendriks calculated the apparent activation energy of the reaction as 19.1 kcal/mol. Interterestingly, these authors also pointed out that the isomerization of ethyl bromide by aluminum bromide does not take place via elimination and readdition, since no appreciable amounts of hydrogen are exchanged for deuterium in the presence of deuterium bromide.

The isomerization of ethyl-β-^{14}C iodide in the presence of aluminum chloride was similarly investigated by Lee et al. [7]. Since ethylation of benzene with ethyl-β-^{14}C iodide was also shown to proceed with no detectable rearrangement, the investigators applied the alkylation-oxidation procedure previously used by Roberts et al. [5] as a means of measuring the rearrangement of the ^{14}C-labeled atoms from the β to the α positions of the ethyl iodide.

As in the cases of ethyl chloride and bromide, the degree of isotope-position rearrangement induced by aluminum chloride on ethyl-β-^{14}C iodide was shown to be dependent on temperature, contact time, and ratio of the catalyst to the ethyl iodide.

It is of interest to point out here that analysis of the data obtained by the various groups of workers on the isotopic rearrangement of radioactive ethyl halides [5-7], together with those of Brown and Wallace [8] on the addition compounds of aluminum halides with alkyl halides, reveals that the apparent order of isomerization by aluminum chloride under equivalent conditions is *n*-propylbromide > ethyl chloride > ethyl bromide > ethyl iodide. For example, whereas ethyl-β-^{14}C chloride underwent almost complete isotopic rearrangement during 1 hr of treatment with aluminum chloride at room temperature, the bromide underwent about 58% rearrangement during 2 hr at 50°C, and the iodide less than 8% rearrangement under conditions similar to those used for the chloride. Complete equilibration of the bromide was only achieved during 8 hr at 80°C and that of the iodide after about 42 hr at 40°C.

Among ethyl halides, the order of reactivity above is not surprising and can easily be rationalized in terms of the order of polarization of the carbon-halogen bond. The fact that the rate of isomerization of *n*-propyl to isopropyl halide is considerably faster than that of ethyl halides was explained by Brown and Wallace [8] in terms of simultaneous transfer and participation of the hydride ion in either ionization or shift of the halogen without ionization [see Sec. I.A.1.b(i)].

In an attempt to evaluate the effect of phenyl substitution on the isomerization of ethyl halides, Adema and Sixma [9] investigated the kinetics of the isomerization of 1-bromo-2-phenylethane-1-^{14}C to 1-bromo-2-phenylethane-2-^{14}C under the influence of aluminum bromide in carbon disulfide. With this compound, the reaction proceeded so fast that its rate could only be followed in the temperature range −70 to −100°C. The kinetic results of the present work combined with earlier data [10] from spectrophotometric measurements led the authors to suggest that phenylethyl bromide and aluminum bromide combine under the reaction conditions to form a 1:1 complex in which the bromine atom of the halide is linked to aluminum by a coordinate bond. They visualized the rate-determining step in this isomerization as being the reaction of $C_8H_9Br\text{-}AlBr_3$ with one of the ions formed by the ionization of the complex of β-phenylethyl bromide and aluminum bromide dimer ($C_8H_9Br\text{-}Al_2Br_6$). Since it was not possible in the light of the available kinetic measurements to conclude what ions would be formed, no discussion was given concerning their structures.

In a subsequent paper, Adema and Sixma [11] studied the kinetics of the reaction involving the exchange of bromine atoms between 1-bromo-2-phenylethane-^{82}Br and aluminum bromide in carbon disulfide in the tempera-true range −80 to −100°C. They estimated the energy of activation for this process to be 10.6 kcal/mol. In agreement with their previous findings about the corresponding isomerization reaction, they concluded that the kinetics of bromine exchange could be explained only if nearly all aluminum bromide was present as the complex $C_8H_9Br\text{-}AlBr_3$.

As to the mechanism of isotopic rearrangement of ethyl halides under the influence of aluminum halides, it has been generally accepted that the process proceeds via the transitory existence of primary carbocations according to the following simple mechanism [12–14]:

$$CH_3{}^{14}CH_2Cl \rightleftarrows CH_3{}^{14}CH_2{}^{+}AlCl_4{}^{-} \rightleftarrows {}^{14}CH_3CH_2{}^{+}AlCl_4{}^{-} \rightleftarrows {}^{14}CH_3CH_2CL + AlCl_3$$

$$\underline{1}$$

A modification of this scheme can be given in which the classical equilibrating system 1 is replaced by a bridged nonclassical ethyl cation 2. This is, however, less likely, as recent theoretical calculations have indicated that the classical equilibrating ethyl cation 1 is ca. 9 to 11 kcal/mol more stable than the nonclassical ethyl cation 2 [15-19].

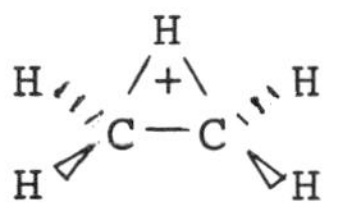

```
       H
 H,,  /+\  ,,H
   'C—C'
 H⊲      ⊳H
       2
```

Support for the involvement of transitory alkyl carbocation in alky halide-Lewis acid systems has been inferred from a number of experimental observations. These include vapor pressure depressions of methyl and ethyl chlorides in the presence of gallium chloride [20], electrical conductivities of aluminum halide-alkyl halide systems [21-25] and of alkyl fluorides in the presence of BF_3 [26], dielectric polarizability of solutions of aluminum halides in ethyl halides [27,28], exchange of halogen between methyl, ethyl, or *n*-propyl halide and aluminum halides [6,29,30] and the recently observed intramolecular and intermolecular scrambling of the hydrogens and carbons of labeled and unlabeled ethyl fluoride in excess SbF_5-SO_2, $FSO_3H(D)$-SbF_5-SO_2, and $H(D)SbF_5$-SO_2, [31,32].

The inclusion of halogen exchange among the numerous kinds of evidence supporting the conversion of the simple alkyl halides into ions under the influence of metal halides has, however, been questioned on various grounds, among which are the following: (1) The observation made by Fairbrother [28] that in benzene solution a reaction mixture of triphenylmethyl bromide and stannic bromide showed an increase in light absorption corresponding to less than 0.1% ionization. Since the tendency for ionization must be far greater in triphenylmethyl halides than in simpler primary, or even ordinary secondary or tertiary, halides, Fairbrother's observation raises serious doubts about the simple ionization of ordinary alkyl halides. (2) The finding of Sixma et al. [33] that halogen exchange between C_2H_5-^{82}Br and $AlBr_3$ followed third-order kinetics. (3) The independent findings by Brezhneva et al. [34] and by Trotters [35] that halogen interchange of ethyl and *n*-propyl halides with aluminum halides proceeded more rapidly than rearrangement of exchange products. These observations, while suggesting that a simple ionization and recombination mechanism is an inadequate explanation for the halogen exchange reaction, cannot be extrapolated to the ^{14}C-position rearrangement, as the two processes were shown to proceed at completely different rates, with the latter being considerably slower [9,11]. This slowness of the isotope-position rearrangement infers the involvement of a highly developed primary carbocation which is capable of undergoing the observed ^{14}C scrambling.

It followed from the work of Brown and co-workers [8,36] on halogen exchange between metal halides and alkyl halides that the initial stage in the reaction between the alkylating agent and the Friedel-Crafts catalyst must be the formation of an addition compound.

$$RX + MX_3 \rightleftarrows RX{:}MX_3$$

In a subsequent stage, this addition compound might or might not undergo ionization depending on a number of factors, such as the nature of the alkylating group R and the experimental conditions. For further information on such addition compounds, see Chap. 3 and references given there.

b. Propyl Rearrangements

(i) *n*-Propyl to isopropyl rearrangements. In 1879, Kekule and Schrötter [37] reported the discovery that *n*-propyl bromide could be isomerized completely to isopropyl bromide in the presence of aluminum chloride. This discovery was shortly confirmed by Gustavson [38-40] using the same reactants and later by other workers using *n*-propyl chloride and either aluminum chloride [23] or aluminum chloride-silica gel mixtures [41] as catalysts. These early isomerization results, together with the announcement of Gustavson [4] that *n*-propyl and isopropyl bromide interact with benzene in the presence of aluminum chloride to form the same substance, isopropylbenzene, were considered by early workers as supporting evidence of an alkylation mechanism involving prior isomerization of *n*-propyl to isopropyl halides [13,37,42,43]. More recent findings, however, do not fit this simple view.

From later investigations, it appears that the degree of isomerization of *n*-propyl to isopropyl halide depends on reaction variables, including the proportion and type of catalyst used, as deactivation of the catalyst is caused by the polymeric by-products [44-46]. Thus, *n*-propyl chloride could be isomerized to isopropyl chloride with aluminum chloride but not with $ZnCl_2$-HCl [47-49] or $AlCl_2HSO_4$ [50]. The influence of reaction conditions on the extent of isomerization can be seen from the following reports. The treatment of *n*-propyl bromide with $AlBr_3$ was claimed to form an equilibrium bromopropane mixture containing 91% of the isopropyl isomer when the reaction was performed at 26°C [51,52]. On the other hand, when the reaction was carried out at reflux temperature, the equilibrated bromopropanc products were shown to contain the *n*-propyl isomer in 1.3% proportion by infrared analysis [53] and 2% by vapor-phase chromatography [54].

In recent years, various other studies have been made on the Lewis acid-induced rearrangements of propyl halides, chiefly with the aim of determining the reaction mechanism. A simple path for this isomerization would be dehydrohalogenation to propene followed by readdition of hydrogen chloride according to Markownikoff's rule to form isopropyl chloride. This possibility was considered by McKenna and Sowa [55,56], who proposed that olefins are involved as intermediates in the boron trifluoride-catalyzed alkylations of benzene with alcohols. Price, however, pointed out that it is by no means necessary to suppose that an olefin is formed as an intermediate in all cases, since benzyl chloride, benzyl alcohol, and benzhydrol, which cannot form olefins, are particularly active alkylating agents [13]. He offered an alternative explanation for the isomerization of alkyl groups under alkylation conditions in terms of an ionic mechanism [12] based on the general theory of carbocation rearrangements outlined by Whitmore [57]. According to this mechanism, a catalyst-alkyl halide complex is first formed, followed by dissociation to form a carbocation that can rearrange by a hydride shift to another (more stable) carbocation before reacting with the arene.

$$CH_3CH_2CH_2Cl + AlCl_3 \longrightarrow CH_3CH_2\overset{+}{C}H_2{-}{-}{-}AlCl_4^-$$

$$\downarrow$$

$$CH_3\overset{+}{C}HCH_3 \xleftarrow{\sim H:^-} CH_3CH_2\overset{+}{C}H_2 + AlCl_4^-$$

$$\downarrow C_6H_6 \qquad\qquad \downarrow C_6H_6$$

$$CH_3CH(C_6H_5)CH_3 \qquad\qquad CH_3CH_2CH_2C_6H_5$$

An elegant test of the theory that an alkene is an intermediate in alkyl halide isomerization was provided by experiments designed by Doering and coworkers [58] using deuterium chloride. They allowed *n*-propyl chloride to isomerize in the presence of aluminum chloride and deuterium chloride at 0°C. They then analyzed the product, isopropyl chloride, for deuterium content by mass spectrometry and found it to be devoid of deuterium. Under essentially identical conditions, propene readily added deuterium chloride to form deuteroisopropyl chloride. Accordingly, Doering concluded that propene could not be an intermediate; instead, he proposed that a carbocation, or its equivalent, was formed, followed by an intramolecular 1,2 hydride shift and recombination with chloride ion from the catalyst complex.

$$CH_3CH_2CH_2Cl + AlCl_3 \longrightarrow CH_3CH_2\overset{+}{C}H_2{---}ClAlCl_3^-$$

$$\downarrow \sim H:^-$$

$$CH_3\underset{\underset{Cl}{|}}{C}HCH_3 + AlCl_3 \longleftarrow CH_3\overset{+}{C}HCH_3{---}ClAl\bar{C}l_3$$

The action of aluminum bromide on *n*-propyl and isopropyl bromide was also investigated by Brown and Wallace [8]. They again established the formation of a 1:1 addition compound between the alkyl halide and the catalyst as a first step in the reaction. Moreover, they found that the rate of *n*-propyl to isopropyl isomerization was considerably faster than the rate of dehydrohalogenation and that the isomerization rate was not affected by the presence of hydrogen bromide. They also observed that hydrogen bromide was produced from *n*-propyl bromide far more rapidly than from ethyl bromide. From these results they concluded that the simple ionization of the addition complex could not be the rate-determining step and that the propyl isomerization was facilitated by simultaneous transfer of a hydride ion of the type proposed by Winstein and his collaborators [59]. Two alternative assisted pathways were, however, suggested for the isomerization of the addition compound, one with ionization as in path A and another without ionization as in path B.

$$CH_3\text{-}\underset{\underset{H}{|}}{\overset{\overset{H}{..}}{C}}\text{-}CH_2{:}\ddot{B}r{:}AlBr_3 \xrightarrow{A} CH_3\text{-}\underset{\underset{H}{|}}{\overset{\overset{H}{+}}{C}}\text{-}\ddot{C}H_2AlBr_4^-$$

$$CH_3\text{-}\underset{\underset{H}{|}}{\overset{\overset{H}{..}}{C}}\text{-}CH_2{:}\ddot{B}r{:}AlBr_3 \xrightarrow{B} CH_3\text{-}\underset{\underset{\underset{AlBr_3}{:\ddot{B}r:}}{}}{\overset{\overset{H}{|}}{C}}\text{-}CH_3$$

The authors stated that either of these two mechanisms could account for their observations, as well as for Doering's results [58] but that their data did not allow them to decide between the alternative mechanisms. However, they pointed out that a comparison of the rate of exchange of the halogen in an *n*-propyl halide with the rate of isomerization should afford conclusive data on which such a decision could be made.

Later, this comparison was made independently by Beers [60] in 1958 and by Trotter [35] in 1963. Beers found that exchange of bromine between bromine-labeled aluminum bromide and *n*-propyl bromide occurred much faster than the formation of isopropyl bromide. Trotter found that halide exchange between *n*-propyl bromide or *n*-propyl iodide and aluminum chloride yielded, initially, an exchange product containing a high percentage of *n*-propyl chloride. Similarly, in the ^{14}C-labeled compounds ethyl bromide [6,33] and phenylethyl bromide [9,11] it was found that in carbon disulfide solution the bromide exchange proceeded faster than the isotope-position rearrangement. These data appear to rule out Brown's mechanism of concerted shifts without ionization.

The elimination-addition mechanism for the isomerization of *n*-propyl to isopropyl halides under Friedel-Crafts conditions has been further discredited by more recent work. In 1964, Douwes and Kooyman [61] performed the aluminum bromide-catalyzed isomerization of *n*-propyl bromide in the presence of deuterium bromide; in accordance with the results of Doering et al. [58] no incorporation of deuterium was observed, indicating that the isomerization did not involve reversible dehydrobromination. In order to obtain more detailed information in regard to the isomerization mechanism on the basis of H/D isotope effects, the same authors conducted a series of experiments starting with 1-bromopropane-2-d and 1-bromopropane-2,2-d_2. The results of these experiments showed significant isotope effect, with the hydrogen migrating 3.4 times faster than deuterium. To account for these results and for the finding that the exchange between organic and inorganic bromine was faster than isomerization to isopropyl bromide, Douwes and Kooyman favored a mechanism involving a fast reversible formation of an intermediate in which the bromine atoms have lost their identity, followed by a slow hydrogen migration within this intermediate leading to isomerization.

$$\text{n AlBr}_3 + n\text{-C}_3\text{H}_7\text{Br} \xrightleftharpoons{\text{fast}} (n\text{-C}_3\text{H}_7^{+}\text{Al}_n\text{Br}_{3n+1}^{-})$$

$$(n\text{-C}_3\text{H}_7^{+}\text{Al}_n\text{Br}_{3n+1}^{-}) \xrightarrow{\text{slow}} \text{CH}_3\text{CHBrCH}_3 + \text{n AlBr}_3$$

It was pointed out that in the equations above, the *n*-propyl cation should not be regarded as a fully developed cation but as an *n*-propyl fragment carrying a significantly greater positive charge than in free *n*-propyl bromide.

Karabatsos and co-workers [45,62] studied the reaction of 1-bromopropane-1, 1-d_2, $-2,2$-d_2, and -1-^{13}C as well as of 2-bromopropane-2-d and -1, 1,1,3,3,3-d_6 with aluminum bromide (5.8:1 mole ratio of 1-BrPr to $AlBr_3$) at 0°C. They found that treatment of either 1-bromopropane or 2-bromopropane under these conditions gave equilibrium mixtures consisting of 6 ± 0.5% 1-bromopropane and 94 ± 0.5% 2-bromopropane. They also observed that the conversion of 1-bromopropane to 2-bromopropane was retarded by the presence of a small amount of diethyl ether and that the conversion of 1-bromopropane-2,2-d_2 to 2-bromopropane was slower than those of the other deuterium-labeled 1-bromopropanes under comparable conditions. The latter observation, noted also by Douwes and Kooyman [61], was attributed to the k_H/k_D isotope effect associated with the 1,2-hydride shift leading to 2-bromopropane. Nuclear magnetic resonance (NMR) and mass spectral analysis of the 2-bromopropane products obtained from the partial reactions of 1-bromopropane-1,1-d_2 and 1-bromopropane-1-^{13}C with aluminum bromide gave the data summarized in Eqs. (1) and (2).

$$CH_3CH_2CD_2Br \longrightarrow CH_3\underset{\displaystyle Br}{C}HCHD_2 + CH_3\underset{\displaystyle Br}{C}DCH_2D + CH_2D\underset{\displaystyle Br}{C}HCH_2D \quad (1)$$

100% d_2

45% conversion	>	98.7%	<	1.3%
79% conversion	>	98.0%	<	2.0%
80% conversion	>	97.0%	<	3.0%

$$CH_3CH_2{}^{13}CH_2Br \longrightarrow CH_3\underset{\displaystyle Br}{C}H{}^{13}CH_3 + CH_3{}^{13}\underset{\displaystyle Br}{C}HCH_3 \quad (2)$$

100% ^{13}C > 99.5% < 0.5%

On the basis of these data, it was concluded that the 2-bromopropanes were mainly isotope-position unrearranged, meaning that they had arisen primarily from an essentially irreversible 1,2-hydride shift process. Similar conclusions about the nature of *n*-propyl to isopropyl rearrangements may also be reached by analyzing the data obtained by Lee and Woodcock [63] from the alkylation of benzene with 1-chloropropane-1,1-d_2 in the presence of aluminum chloride.

In a more recent paper, Lee and Woodcock [46] reported findings about isotopic scrambling during partial isomerization of 1-^{14}-C-1-chloropropane to 2-chloropropane induced by treatment with aluminum chloride at 0°C. They found that the extents of *n*-propyl to isopropyl isomerization and of isotopic scrambling in such partial reactions were very susceptible to changes in reaction conditions. They also found that the 2-chloropropane obtained in their experiments contained small amounts of ^{14}C activity (1.4 to 3.0%) at the C-2 position, whereas Karabatsos et al. [45], found less than 0.5% isotope-position rearrangement. The difference between the results of Karabatos et al. [45], and those of Lee and Woodcock [46], was considered by the latter workers to be practically insignificant, and generally it may be concluded that only relatively small amounts of isotope-position rearrangements in the 2-halopropanes could have occurred in both these two sets of experiments. The same authors also mentioned that in private communications Deno had pointed out that in such partial isomerizations of labeled 1-halopropane to 2-halopropane, some isotope-position rearrangements have taken place in the 1-halopropane during the course of the reaction and hence some of the 2-halopropane could have been derived from isotopically scrambled 1-halopropane. Consequently, the authors attributed the presence of ^{14}C at the C-2 position of the 2-chloropropane to isomerizations involving isotopically labeled 1-chloropropane-2-^{14}C. To explain the conversion of 1- to 2-chloropropane Lee and Woodcock [46] proposed the formation of a propyl cation-$AlCl_4$ anion ion pair or complex in which the positive moiety could perform as a classical carbocation, which can, among other things, undergo an irreversible 1,2 hydride shift to give 2-propyl cation. Results from carefully designed solvolysis [45,64-67] and deamination [68-72] experiments with *n*-propyl derivatives showed conclusively that the *n*-propyl to isopropyl rearrangements in these closely related reactions also occurred by an intramolecular hydrogen transfer. The alternative possibility, involving the formation of propene and its subsequent reaction with solvent, was excluded by the experimental evidence.

In their paper, Lee and Woodcock [46] discussed some interesting results by Frigerio and Shaw [72-74] that led the latter authors to criticize the classical carbocation mechanism for the isomerization of *n*-propyl to isopropyl halides in the presence of aluminum halides. Frigerio and Shaw found that the reaction of 1-halopropanes-1-^{14}C with aluminum halides gave 80 to 90% yields of 2-halopropanes in which the ^{14}C activity remained entirely at C-1. Further, they indicated that reactions of 1-chloropropane-1-^{36}Cl with aluminum chloride, 1-bromopropane-1-^{82}Br with aluminum bromide, 1-chloropropane with aluminum bromide, and 1-bromopropane with aluminum chloride showed little or no halogen exchange.* For example, when 1-chloropropane-1-^{36}Cl was treated with 2.19 mol % of ordinary $AlCl_3$, the distribution of the total ^{36}Cl activity in the resulting products was 34.9, 63.5, and 1.6%, respectively, in the 1-chloropropane, 2-chloropropane, and $AlCl_3$. Frigerio and Shaw thus concluded that the alkyl halogen remained tightly bound to the parent molecule throughout isomerization and suggested that one of the possible mechanisms for *n*-propyl to isopropyl halide isomerization could involve the intermediate formation of halonium ion-AlX_3H^- anion ion pairs rather than the classical 1-propyl and 2-propyl cations.

AlX_3H^- AlX_3H^-

$CH_3CH{-}\overset{*}{C}H_2$ (bridged by X^+) ⇌ $CH_2{-}CH{-}\overset{*}{C}H_3$ (bridged by X^+)

⇅   ⇅

$CH_3CH_2{*}CH_2X$ + $CH_3CHX{*}CH_3$ $CH_3CHX{*}CH_3$ + $XCH_2CH_2{*}CH_3$

The halonium ion mechanism above was disqualified by Lee and Woodcock [46] on the ground that since all halogen exchange between $AlCl_3$ and 1-chloropropane-1-^{36}Cl had not been completely excluded, formation of the propyl cation-$AlCl_4$ anion ion pair or complex could not be eliminated. Moreover, the halonium ion mechanism is not in accord with the recent reports of Karabatsos et al. [45] on the aluminum bromide-induced isomerizations of deuterated propyl bromides.

(ii) Isotopic *n*-propyl rearrangements Before embarking on a detailed discussion of the occurrence of isotopic *n*-propyl rearrangements under Friedel-Crafts conditions, it seems of interest to review briefly their history and their involvement in other closely related carbocation reactions.

This type of rearrangement was first discovered in 1953 by Roberts and Halmann [68] in the course of deaminating 1-aminopropane-1-^{14}C in 35% perchloric acid solution. Degradation and radioassay of the 1-propanol-^{14}C obtained in this reaction indicated that 8.5% of the original carbon-14 had moved to the C-2 and/or C-3 position of the 1-propanol produced. However, in proposing that all of the 1-propanol was formed by the carbocation processes visualized in scheme 1, the authors assumed that the rearranged

*These results are in sharp contradiction with those of Beers [60], which indicated that bromine exchange between bromine-labeled aluminum bromide and *n*-propyl bromide occurred much faster than the formation of isopropyl bromide.

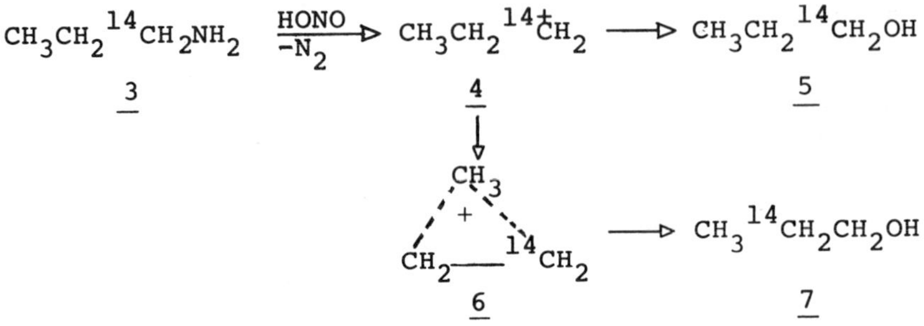

$$CH_3CH_2{}^{14}CH_2NH_2 \xrightarrow[-N_2]{HONO} CH_3CH_2{}^{14}\overset{+}{C}H_2 \longrightarrow CH_3CH_2{}^{14}CH_2OH$$

3 4 5

4 → 6 (bridged ion: CH_3 bridging CH_2—$^{14}CH_2$, +) $\longrightarrow CH_3{}^{14}CH_2CH_2OH$ (7)

Scheme 1

label was located exclusively at C-2. The observed 8.5% of isotope-position rearrangement was explained by having 17% of the 1-propanol originating by way of 6 and 83% from 4.

Since this report by Roberts and Halmann, the deamination of 1-aminopropane has been the subject of extensive studies [70,71,75-78] and there has been disagreement among the various workers about the extent and nature of the rearrangement. For example, while Roberts and Halmann [68] estimated a total rearrangement of 8.5% to the 2 position, Reutov and Shatkina [75,76a] concluded that 8% of the original ^{14}C of the amine had found its way to the C-3 and that none of the label was at C-2. In explaining their results, the latter workers considered two probable mechanisms, one involving a 1,3 hydride shift (Eq. 3) and another involving two consecutive 1,2 hydride shifts (Eq. 4). However, they ruled out the latter possibility on the ground that no 1-propanol was formed in the deamination of 2-aminopropane [79].

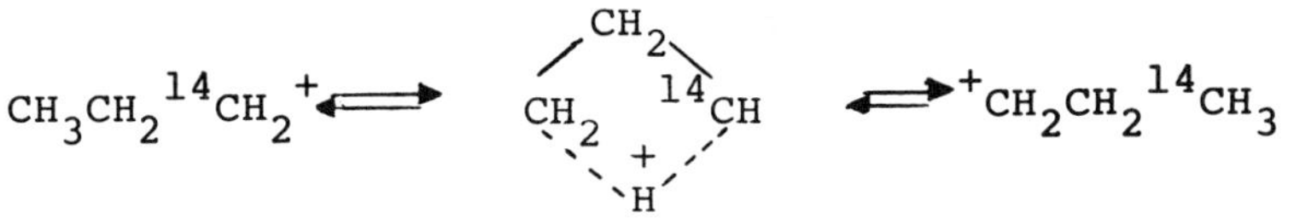

$$CH_3CH_2{}^{14}CH_2{}^{+} \rightleftarrows [CH_2(CH_2)^{14}CH \cdots H]^{+} \rightleftarrows {}^{+}CH_2CH_2{}^{14}CH_3 \quad (3)$$

$$CH_3CH_2{}^{14}CH_2{}^{+} \rightleftarrows CH_3\overset{+}{C}H{}^{14}CH_3 \rightleftarrows {}^{+}CH_2CH_2{}^{14}CH_3 \quad (4)$$

Further work was then designed to distinguish between the two alternative mechanisms. In 1962, Karabatsos and Orzech [69] studied the deamination of 1-aminopropane-1,1,2,2-d_4 (8) and they concluded that the 1-propanol produced consisted of 88% 9 (unrearranged), 12% 10, and no 11. Since 11 would be expected to form by the successive 1,2

CH_3-CD_2-CD_2-NH_2 (8)

CH_3-CD_2-CD_2-OH (9)

CHD_2-CD_2-CH_2OH (10)

CH_3-CHD-CH_2OH (11)

hydride shifts that produce the ^{14}C rearrangements of Eq. (4), the absence of 11 was taken as evidence *against* this mechanism and *for* the 1,3 shift of Equation (3). Further backing for the idea of a 1,3 hydride shift was provided by Reutov [80], who reported that the deamination of 1-aminopropane-2-t gave 1-propanol product in which the label was exclusively at C-2 (1-propanol-2-t) and a 2-propanol product having tritium at both C-1 and C-2.

The development of the idea of protonated cyclopropane intermediates [81-84] and the mounting interest in their application to the carbocation rearrangement of the propyl system inspired a more careful reinvestigation of the previously summarized deamination results. By reexamining the deamination of 1-aminopropane, Lee and Kruger [71,85] obtained the results shown in Eq. (5). Similarly, the deamination of 1-aminopropane-1-t was shown [70] to give the results summarized in Eq. (6).

$$CH_3CH_2{}^{14}CH_2NH_2 \rightarrow \underset{95.9\%}{CH_3CH_2{}^{14}CH_2OH} + \underset{2.2\%}{CH_3{}^{14}CH_2CH_2OH} + \underset{1.9\%}{{}^{14}CH_3CH_2CH_2OH} \quad (5)$$

$$CH_3CH_2CHTNH_2 \rightarrow \underset{97.1\%}{CH_3CH_2CHTOH} + \underset{1.3\%}{CH_3CHTCH_2OH} + \underset{1.6\%}{CH_3TCH_2CH_2OH} \quad (6)$$

More recently, Karabatsos et al. [78] studied the deamination of 1-aminopropane-1-d_2, -2 2-d_2, and -3,3-d_2 and obtained the results shown in Eqs. (7) and (8). These findings demonstrated clearly that in the deamination of isotopically labeled 1-aminopropanes, the resulting 1-propanol products showed a total of only 2 to 5% isotope-position rearrangement, with the isotope distributed between C-2 and C-3. Currently, the rearrangements of the labels in the 1-propanol products are believed to be due to the intervention of equilibrating protonated cyclopropane intermediates.

$$\underset{100\%\ d_2}{CH_3CH_2CD_2NH_2} \xrightarrow{40°} \underset{95.7\%}{C_2H_5\text{-}CD_2OH} + \underset{1.0\%}{C_2H_4D\text{-}CHDOH} + \underset{3.3\%}{C_2H_3D_2\text{-}CH_2OH} \quad (7a)$$

$$\underset{100\%\ d_2}{CH_3CD_2CH_2NH_2} \xrightarrow{40°} \quad 1.3\% \qquad 1.2\% \qquad 97.4\% \quad (7b)$$

$$\underset{100\%\ d_3}{CD_3CH_2CH_2NH_2} \longrightarrow \underset{1.0\%}{C_2H_4D\text{-}CD_2OH} + \underset{1.2\%}{C_2H_3D_2\text{-}CHDOH} + \underset{97.8\%}{C_2H_2D_3\text{-}CH_2OH} \quad (8)$$

The same type of isotope-position rearrangement of the *n*-propyl group has also been encountered in closely related solvolysis reactions. When Lee and Kruger [65] formolyzed 1-propyl-1-^{14}C tosylate in refluxing anhydrous formic acid, they found that the product had the following distribution of radioactivity:

$$CH_3CH_2\overset{*}{C}H_2OTs \longrightarrow \underset{0.68\%\ ^{14}C}{\overset{*}{C}H_3} - \underset{0.15\%\ ^{14}C}{\overset{*}{C}H_2} - \underset{99.1\%\ ^{14}C}{\overset{*}{C}H_2OCH(=O)}$$

The small amount (0.83%, overall) of isotope-position rearrangement was explained in terms of edge-protonated cyclopropane intermediates. However, when the solvolysis of isotopically labeled 1-propyl tosylates was carried out in 99% aqueous formic acid, the 1-propanol products obtained were shown to be exclusively isotope-position unrearranged [45,62].

In 1969, Winstein et al. [67] investigated the solvolysis of a number of primary tosylates, including n-PrOTs, in trifluoroacetic acid. Using NMR for analysis, they found that in the solvolysis of 1-PrOTs-1,1-d_2, no α signals were evident in the residual n-PrOTs or the nPrOCOCF$_3$ product. Accordingly, they concluded that this solvolysis is not appreciably accompanied by 1,3 hydride shifts or formation and reopening of cyclopropane.

In 1970, Lee and Chwang [86] carried out the trifluoroacetolysis of 1-propyl-1-^{14}C tosylate either with or without the presence of an equivalent amount of sodium trifluoracetate. They observed that the 1-propyl trifluoroacetate product had undergone isotope-position rearrangements from C-1 to C-2 and C-3 to the extent of about 2 and 17%, respectively, from the reactions performed with and without sodium trifluoroacetate. Moreover, they found that the rearranged ^{14}C label was almost equally distributed at C-2 and C-3. These authors proposed that the observed rearrangements could be explained by assuming that part of the overall reaction proceeded through equilibrating protonated cyclopropane intermediates [79]. In an addendum to their paper, Lee and Chwang [86] mentioned that their results are at variance with those of Winstein et al. [67], and that the discrepancy between them remains to be resolved.

With this background in the occurrence of isotope-position rearrangements of the n-propyl system in carbocation reactions, we shall now consider the possibility of such rearrangements under Friedel-Crafts conditions.

In 1963, Reutov and Shatkina [47,80,87] reported that upon treatment of 1-chloropropane-1-^{14}C with ZnCl$_2$-HCl, various extents of rearrangement to 1-chloropropane-3-^{14}C were found, depending on temperature and duration of reaction. For example, when the reaction was carried out at 50°C for 100 hr, 7.5% rearrangement was observed. Interestingly, no interconversion between 1- and 2-chloropropane was noted under the applied conditions. When the reaction was effected in the presence of aluminum chloride, however, rapid formation of isopropyl chloride was observed. The authors explained their results by suggesting that in the reaction with zinc chloride-hydrochloric acid a truly free propyl cation is not formed, and a 1,3 migration of the hydride ion occurs within a strongly polarized molecule.

$$CH_3CH_2{}^{14}CH_2Cl \xrightarrow[\text{very rapid}]{AlCl_3} CH_3\underset{\displaystyle Cl}{\underset{|}{C}}H{}^{14}CH_3$$

$$CH_3CH_2{}^{14}CH_2Cl \xrightarrow{ZnCl_2,HCl} {}^{14}CH_3CH_2CH_2Cl$$

With aluminum chloride, on the other hand, a free carbocation ion is formed which rapidly isomerizes to a more stable secondary cation by an irreversible 1,2 hydride shift. The 1,3-hydride-shift mechanism received more support when 1-bromopropane-3-d was reported to undergo conversion to 1-bromopropane-1-d upon treatment with ZnCl$_2$-HCl [88].

More recently, several attempts were made to verify this reported 1,3 hydride shift. In 1969, Lee and co-workers [48] studied the product of treatment of 1-chloropropane-1-t with ZnCl$_2$-HCl under reflux at 50 ± 2°C for 100 hr. The recovered 1-chloropropane was uncontaminated by 2-chloropropane as found by Reutov and Shatkina, but it showed no rearrangements of the t-label to either C-2 or C-3. Further confirmation to these results was gained when the treatment of 1-chloropropane-1-t with ZnCl$_2$-HCl was carried out with shaking in a sealed tube at 50 ± 2°C for 100 hr.

In 1970, Karabatsos et al. [49], reported on the question of protonated cyclopropanes in the reaction of 1-chloropropane with $ZnCl_2$-HCl. When 1-chloropropane-1,1-d_2 and -2,2-d_2 were treated with $ZnCl_2$-HCl at 50°C for 72 hr, the recovered products were shown to consist only of isotope-position unrearranged 1-chloropropanes; gas chromatographic analysis indicated less than 0.2% 2-chloropropane. These results, similar to those obtained by Lee et al. [48], with 1-chloropropane-1-t, failed to confirm the intervention of either protonated cyclopropanes or the 1,3-hydride shift reported by Reutov and Shatkina [47,80,87], and Susan [88] in the reaction of 1-chloropropane with $ZnCl_2$-HCl.

Previously, it was pointed out that Karabatsos et al. [45] studied the reaction of isotopically labeled 1-bromopropane in the presence of aluminum bromide. Accurate NMR and mass spectral analysis of the 1-bromopropane led to the isotopic distribution shown in Eqs. (9) to (11). In rationalizing

$$\underset{100\%\ d_2}{CH_3CH_2CD_2Br} \xrightarrow[\text{conversion}]{80\%} \underset{79.8\%}{C_2H_5\text{-}CD_2Br} + \underset{5.0\%}{C_2H_4D\text{-}CHDBr} + \underset{15.2\%}{C_2H_3D_2\text{-}CH_2Br} \qquad (9)$$

$$\underset{100\%\ d_2}{CH_3CD_2CH_2Br} \xrightarrow[\text{conversion}]{65\%} \quad 2.1\% \qquad 5.8\% \qquad 92.1\% \qquad (10)$$

$$\underset{100\%\ ^{13}C}{CH_3CH_2{}^{13}CH_2Br} \xrightarrow[\text{conversion}]{80\%} \underset{85.7\% \pm 0.2\%}{CH_3CH_2{}^{13}CH_2Br} + \underset{3.7 \pm 0.9\%}{CH_3{}^{13}CH_2CH_2Br} +$$

$$\underset{10.6 \pm 0.6\%}{^{13}CH_3CH_2CH_2Br} \qquad (11)$$

their results, the authors considered various possible mechanisms. However, after arguing against some of them, they concluded that the observed production of isotope-position-rearranged 1-bromopropanes could best be explained in terms of a protonated cyclopropane mechanism. In that mechanism, a single process involving equilibration between edge-protonated cyclopropane intermediates, as pictured in scheme 2, was held responsible for scrambling both the carbons and the hydrogens of the *n*-propyl system.

It was emphasized that for reactions proceeding by such a mechanism, various degrees of scrambling would be observed depending on the relative rates of edge-edge equilibration (k_1) and nucleophilic attack on the protonated cyclopropane (k_2). If $k_1 \gg k_2$, complete equilibration of all of the three carbons would occur; on the other hand, if $k_2 \gg k_1$, the mechanism would correspond to a nominal 1,3 hydride shift and only scrambling of C-1 and C-3 would take place. In the aqueous deamination of 1-aminopropane the k_1/k_2 ratio was estimated to be between 5 and 10 [78,89]. In the present reaction of 1-bromopropane with aluminum bromide, this ratio was suggested to be somewhat smaller than that of deamination, but still greater than 1.

Going back to Eq. 11, it is of importance to notice that, after the short reaction times (5 to 6 min) leading to these results, about one-third of the rearranged ^{13}C label had moved to C-2 and two-thirds to C-3. This particular finding served to rule out any significant contribution from the methyl-

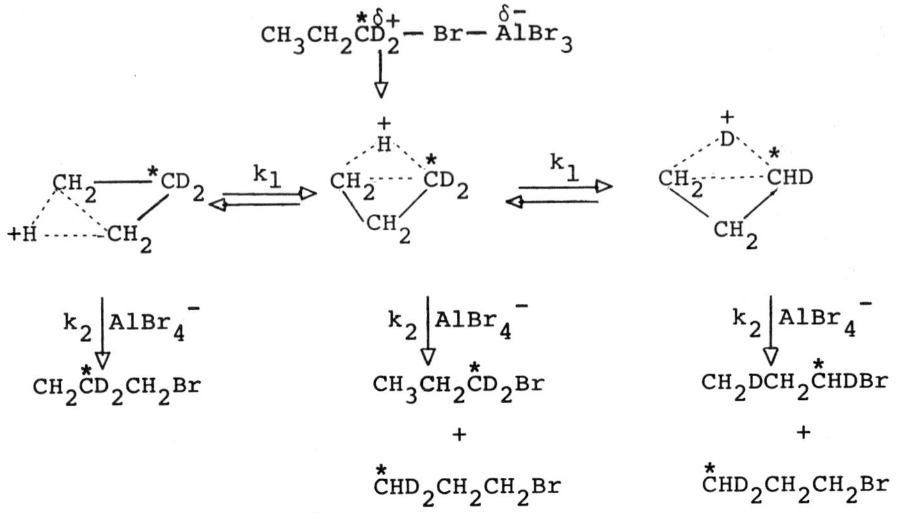

$$CH_3CH_2\overset{*\delta+}{C}D_2 - Br - \overset{\delta-}{A}lBr_3$$

$$\downarrow$$

$$\text{edge-protonated cyclopropanes: } CH_2\text{—}{}^{*}CD_2 \text{ (H}^+\text{ bridging } CH_2 \cdots CH_2) \underset{}{\overset{k_1}{\rightleftharpoons}} CH_2\cdots\overset{*}{C}D_2 \text{ (H}^+\text{ bridging)}, CH_2 \overset{k_1}{\rightleftharpoons} CH_2\cdots\overset{*}{C}HD \text{ (D}^+\text{ bridging)}, CH_2$$

$$k_2 \downarrow AlBr_4^-$$

$$CH_2\overset{*}{C}D_2CH_2Br \qquad CH_3CH_2\overset{*}{C}D_2Br + \overset{*}{C}HD_2CH_2CH_2Br \qquad CH_2DCH_2\overset{*}{C}HDBr + \overset{*}{C}HD_2CH_2CH_2Br$$

Scheme 2

bridged species 12 to the isotope-position rearrangement of recovered 1-bromopropane. Such a species, if formed prior to isomerizations, would lead to at least as much ^{13}C at C-2 as at C-3 of the propyl system. To account for the unequal distribution of activity between C-2 and C-3, it was assumed that "more product is formed from the initially formed edge protonated cyclopropane than from the other isotope-position isomers potentially in equilibrium with it" [89].

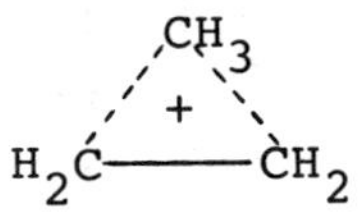

$$H_2C \overset{CH_3\;+}{\text{———}} CH_2$$

12

In the same investigation, Karabatsos et al. [45] carried out the treatment of isotopically labeled 1-bromopropanes with aluminum bromide for long reaction times. Under such conditions more extensive rearrangements were observed. For example, the 1-bromopropane-^{13}C recovered after treating 1-bromopropane-1-^{13}C for 180 min was found to contain 52% of the activity residing on carbons 2 and 3 (Eq. 12). The lack of statistical scrambling, even

$$CH_3CH_2{}^{13}CH_2Br \xrightarrow[AlBr_3]{180\ \text{min}} \underset{48\%}{CH_3CH_2{}^{13}CH_2Br} + \underset{52\%}{\overbrace{CH_3CH_2}^{13}CH_2Br} \tag{12}$$

after such prolonged reaction time, was attributed by the authors to deactivation of the catalyst by polymeric material formed during the reaction.

In connection with the results of Karabatsos et al. [45], it is noteworthy to mention that these authors excluded the bimolecular reactions shown in Eq. (13) as mechanistic paths responsible for the isotope-position-rearranged

```
                                     *C
 *                  *     *          |
 C— C=C + C-C-C  ⇌  C-C-C-C-C  ⇌  ⇌
              +         +     (several ~H and ~Me)           (13)
   *
   C
 * |
 C-C-C-C-C  ⇌  C-C-C + C-C=C
   +           +
               *   *
```

products, as no doubly labeled carbon-13 species were detected, even after reaction for 180 min.*

Much more recently, Lee and Woodcock [46] reported studies of the aluminum chloride-induced partial isomerization of 1-chloropropane-1-^{14}C to isotopically scrambled 1-chloropropane and 2-chloropropane at 0°C. The extents of isotopic scrambling from C-1 to C-2 and C-3 in the recovered 1-chloropropane were determined by degradation through conversions to propionic acid, acetic acid, and methylamine. The authors reported that when the isomerization of 1-chloropropane to 2-chloropropane was 90% complete, the recovered 1-chloropropane-1-^{14}C showed about 7 and 22% rearrangements of the label from C-1 to C-2 and C-3, respectively. As did Karabatsos et al. [45], they observed that the presence of traces of diethyl ether in the reaction mixture caused some deactivation of the $AlCl_3$ catalyst, resulting in lower extents of both primary-to-secondary rearrangement and isotopic scrambling. To account for the scrambling at C-2, these authors proposed the intervention of equilibrating protonated cyclopropane intermediates which would collapse to 1-chloropropane, with the ^{14}C label rearranging to both C-2 and C-3. Lee and Woodcock [46] also commented on the fact that, in these reactions, more of the rearranged activity was found at C-3 than at C-2. They pointed out that this finding could be reasonably explained in terms of two factors: (1) the involvement of edge-protonated cyclopropane intermediates with more of the isotopically rearranged 1-chloropropane produced via the first formed cyclopropane structure and (2) the reversible isomerizations between 1- and 2-chloropropane as depicted in Eq. (14). Apparently, a combination of both factors are responsible for locating more of the rearranged label at C-3 than C-2.

$$CH_3CH_2{}^{14}CH_2Cl \rightleftharpoons CH_3CHCl{}^{14}CH_3 \rightleftharpoons ClCh_2CH_2{}^{14}CH_3 \qquad (14)$$

In connection with the AlX_3-induced isomerizations of isotopically labeled *n*-propyl halides, it is noteworthy that in spite of all of the information available about the nature of the isomerized products, uncertainties still remain as to their mode of formation. This can be seen from the following quotations taken from a paper by Karabatsos et al. [45], in which they started their discussion by stating: "Accurate assessment of the role that protonated cyclopropane intermediates may play in the isomerization of 1-bromopropane to 2-bromopropane is rendered difficult by the fact that 2-bromopropane rearranges back to 1-bromopropane under the reaction conditions, and, therefore, part of the isotope-position rearranged 1-bromopropane must arise from this path." On the other hand, in closing their discussion the authors wrote: "Insofar as can be judged from the present results, acceptance or rejection of

---

*Such bimolecular reactions were shown by Karabatsos and his co-workers [98] to constitute a significant path in the isotope-position rearrangement of *t*-pentyl chloride with aluminum chloride.

the protonated cyclopropane path in the reaction of 1-bromopropane with aluminum bromide is a matter of personal taste."

In recent years the search for the intervention of protonated cyclopropanes in *n*-propyl to *n*-propyl-type isomerizations was also extended to other proplyating agents (e.g., cyclopropane and *n*-propyl alcohol). Baird and Aboderin [83] in a first paper published in 1963, reported that when cyclopropane was cycled through D_2SO_4 for varying lengths of time, it was found to undergo both solvolysis and acid-catalyzed deuterium exchange reactions, the rate of solvolysis being twice that of exchange. The recovered cyclopropane product was shown to be a mixture of cyclopropane and monodeuteriocyclopropane plus small amounts of dideuteriocyclopropane. The authors pointed out that their observations could be accommodated, among other possibilities, by a mechanism involving the reversible formation of a π-complex type of intermediate, $C\text{-}C_3H_6\text{-}D^+$, which could rearrange to $C\text{-}C_3H_5D\text{-}H^+$.

In a second, more detailed paper that appeared in 1964, Baird and Aboderin [84] reported that solvolysis of cyclopropane in 8.4 *M* H_2SO_4 led to the formation of 1-propanol (and its acid sulfate) as the only important product. When the reaction was conducted in D_2SO_4, deuterium was incorporated into all three positions of the 1-propanol carbon skeleton, resulting in an average deuterium distribution in the 1, 2, and 3 positions of 0.38, 0.17, and 0.46 deuterium atom, respectively. This scrambling was suggested to arise through addition of D^+ to cyclopropane to form intermediates 13 and 14. These, after partial equilibration with other possible isomeric forms, are trapped by DSO_4^- or D_2O, leading to a nonstatistical distribution of deuterium.

CH_2- - - -D (+), CH_2——CH_2

13

CH_2D (+), H_2C- - - -CH_2

14

In 1968, Deno and co-workers [91] repeated Baird and Aboderin's experiments and found that in D_2SO_4 of higher molarity (83 to 99% D_2SO_4), the deuterium scrambling in the isolated *n*-propyl sulfate approached the statistical 2:2:3 distribution of deuterium at C-1, C-2, and C-3, respectively. Statistical distribution of deuterium was also found by the authors in the *n*-propyl derivatives arising from the reaction of cyclopropane with other deuterated reagents, such as 20% CH_3COOD-80% D_2SO_4, CF_3COOD, and $C_6D_6 + D_2O + AlCl_3$. In all but the C_6D_6 experiment, the product was entirely the *n*-propyl product with no trace of isopropyl. Deno et al. considered their findings as providing further evidence for the existence of protonated cyclopropane as a chemical species. From their results they concluded that protonated cyclopropane (1) has an estimated stability between the isomeric 1-propyl and 2-propyl cations, (2) is curiously ineffective in abstracting hydride from branched alkanes, and (3) is unselective in alkylating toluene, indicating a high reactivity as an alkylating agent.

The involvement of protonated cyclopropane intermediates in the reactions of cyclopropane was also confirmed by Lee and co-workers [48,86,92,93], who found that the reaction of cyclopropane with tritiated sulfuric acid (H_2SO_4-t) [92], tritiated Lucas reagent ($ZnCl_2$-HCl-t) [93], and tritiated hydrochloric acid [48] gave *only* *n*-propyl-t products with no isopropyl isomers and that the *t* label was incorporated in all three carbon positions of the *n*-propyl chains. However, when the reaction with HCl-t was effected in the presence of 5% $AlCl_3$, the product 1-chloropropane-t showed not only contamination by 3 to 4% 2-chloropropane-t but also t distributions somewhat different from the same product obtained with no added aluminum chloride.

Other reactions of cyclopropane were also investigated and the formation of the products was rationalized in terms of protonated cyclopropane intermediates. These included the Lewis acid-catalyzed bromination of cyclopropane to yield a mixture of 1,1-, 1,2-, and 1,3-dibromopropanes, as well as some 1,1,2-tribromopropane [94], the $FeCl_3$-catalyzed addition of DCl to give a 1-chloropropane-D product with deuterium incorporated at all three carbon atoms [91], the reaction between bromocyclopropane and HBr, with or without added $AlCl_3$, to give a mixture of 1,1-, 1,2-, and 1,3-dibromopropane, and the $AlCl_3$-catalyzed acylation of cyclopropane to give a mixture of compounds 15-18 [95].

$CH_3COCH_2CH_2CH_2Cl$ (15)

$CH_3COCH(CH_3)-CH_2Cl$ (16)

$CH_3COCH(Cl)-CH_2CH_3$ (17)

$CH_3COC(CH_3)=CH_2$ (18)

The question of protonated cyclopropanes in the reaction of Lucas reagent ($ZnCl_2$-HCl) was also investigated to ascertain whether protonated cyclopropane intermediates would play any role in the conversion of 1-propanol to 1-chloropropane. This question was first examined by Lee et al. [93] using 1-propanol-1-t and ordinary Lucas reagent. From the reaction they obtained chiefly 1-chloropropane with 4 to 6% 2-chloropropane. Degradation of the 1-chloropropane product showed a total of about 1% isotope-position rearrangements with the rearranged label about equally distributed between positions 2 and 3. The latter finding was explained by assuming the intervention (1%) of a protonated cyclopropane intermediate in which complete equilibration had occurred prior to reaction with the nucleophile.

More recently, the same problem was approached by Karabatsos et al. [49] using deuterated 1-propanol and ordinary $ZnCl_2$-HCl. From the reactions of 1-propanol-1,1-d_2 and 1-propanol-2,2-d_2 with $ZnCl_2$-HCl, they obtained 1-chloropropane products contaminated with 11 and 5%, respectively, of the 2-chloro isomer. This difference in the chloropropane ratio was ascribed to a k_H/k_D isotope effect of the intramolecular 1,2 hydride shift involved in the formation of the secondary isomer. While the 1-chloropropane obtained from 1-propanol-2,2-d_2 was shown to be isotope-position unrearranged, consisting entirely of $C_2H_3D_2CH_2Cl$, that from 1-propanol-1,1-d_2 consisted of 98.3% $C_2H_5CD_2Cl$, 0.2% $C_2H_4DCHDCl$, and 1.5% $C_2H_3D_2CH_2Cl$. The latter finding was interpreted in terms of the intervention (about 2%) of a protonated cyclopropane path in which the rate of intramolecular rearrangements of the protonated cyclopropane was slower than the rate of reaction of the protonated cyclopropane with the chloride ion.

In connection with the reaction of 1-propanol with $ZnCl_2$-HCl, it is important to note that in all recent publications, mixtures consisting chiefly of the primary isomer with not more than 15% of the secondary chloride were reported [49,93,96].

(iii) Isopropyl to *n*-propyl rearrangements The interconversion between isopropyl and *n*-propyl halides was shown to occur under thermal conditions [53,97,99] and under neutron radiation [100]. Although the Lewis acid-catalyzed isomerization of *n*-propyl to isopropyl halide has been known since 1879 [37], it was only recently that the reverse process has been recognized.

In 1960, the rearrangement of isopropyl to *n*-propyl (1.3%) was observed, by the use of infrared spectrophotometry, when isopropyl bromide was subjected to heating with $AlBr_3$ and a trace of H_2O [53]. Four years later, this rearrangement (2 to 3%) was confirmed by Gerrard et al. [54], who applied GLPC for analysis. These authors also reported that a similar reaction between isopropyl chloride and aluminum chloride resulted in decomposition to hydrogen and polymer; formation of *n*-propyl chloride in trace amounts was suspected (GLPC), but was too small to be confirmed by infrared.

More recently, further confirmation of the conversion of isopropyl to *n*-propyl bromide was given by Karabatsos et al. [45], who found that the treatment of either *n*-propyl or isopropyl bromide with aluminum bromide at 0°C gave an equilibrium mixture consisting of 6% *n*-propyl bromide and 94% isopropyl bromide. They also pointed out that this equilibrium was attained within 10 to 12 min when approached from the side of *n*-propyl bromide and within 6 to 7 min when approached from the side of isopropyl bromide.

The occurrence of some reversible isomerizations of the propyl system was also demonstrated using the chloride combination. When isopropyl chloride was treated with anhydrous aluminum chloride at 0°C for 10, 20, or 30 min, GLPC analysis of the resulting material showed the presence of 1.5 to 2% *n*-propyl chloride [46].

In an attempt to determine the mechanism of isopropyl to *n*-propyl halide rearrangement, Karabatsos et al. [45] investigated the reaction of deuterium-labeled isopropyl bromide with $AlBr_3$. When these workers allowed 2-bromopropane-2-d to react with $AlBr_3$ at 0°C for 6 and 120 min, they recovered 74 and 52% yields, respectively, of bromopropane product mixture. These were shown to comprise the *n*-propyl isomer in 4 and 7% after 6 and 120 min, respectively. Separation (by preparative GLPC) and mass spectral analysis of the isomeric bromides obtained from the treatments above gave the results summarized below (Eqs. 15 and 16).

$$CH_3CD(Br)CH_3 \xrightarrow[6\ min]{AlBr_3,\ 0°} C\text{-}C\text{-}CBr + C\text{-}C(Br)\text{-}C \qquad (15)$$

$CH_3CD(Br)CH_3$	C-C-CBr	C-C(Br)-C
99.0% d_1	0.6% d_2	94.5% d_1
1.0% d_0	95.5% d_1	5.5% d_0
	3.9% d_0	

$$CH_3CD(Br)CH_3 \xrightarrow[120\ min]{AlBr_3,\ 0°} C\text{-}C\text{-}C\text{-}Br + C\text{-}C(Br)\text{-}C \qquad (16)$$

$CH_3CD(Br)CH_3$	C-C-C-Br	C-C(Br)-C
99.0% d_1	4.0% d_2	2.8% d_2
1.0% d_0	78.7% d_1	77.4% d_1
	17.3% d_0	19.8% d_0

When 2-bromopropane-1,1,1,3,3,3-d_6 (96.0% d_6 and 4.0% d_5) was similarly treated with $AlBr_3$ for 10 min at 0°C, bromopropane recovery was reported to be essentially quantitative, consisting of 3% 1-bromopropane and 97% 2-bromopropane. Mass spectral analysis of the 1-bromopropane produced in this reaction indicated that it was 94.0% d_6.

The following features of the results above were significant in regard to the rearrangement paths: (i) the fact that the 3% 1-bromopropane obtained from the reaction of 2-bromopropane-1,1,1,3,3,3-d_6 (96.0% d_6 and 4.0% d_5) was 94.1% d_6 suggested that the major rearrangement path was intramolecular in nature and: (2) the observation that the recovered 1- and 2-bromopropane products have not only lost deuterium but also acquired some doubly labeled molecules coupled with the formation of propane in these reactions implied the intervention of intermolecular hydride transfer. Relying on these considerations, the authors proposed that the rearrangement of 2-bromopropane to 1-bromopropane occurs by two paths: an intramolecular 1,2 hydride shift, which is the major path, and a minor intermolecular hydride transfer path. The latter path was believed to involve the sequence of reactions depicted in Eqs. 17 to 21. In seeking literature support for

$$CH_3CDBrCH_3 + AlBr_3 \rightleftharpoons CH_3\overset{+}{C}DCH_3 + AlBr_4^- \quad (17)$$

$$CH_3CDBrCH_3 + CH_3\overset{+}{C}DCH_3 \rightarrow CH_3CD_2CH_3 + CH_3\overset{+}{C}BrCH_3 \quad (18)$$

$$CH_3\overset{+}{C}BrCH_3 + CH_3CDBrCH_3 \rightleftharpoons CH_3CDBrCH_3 + CH_3\overset{+}{C}BrCH_3 \quad (19)$$

$$CH_3\overset{+}{C}BrCH_3 + CH_3CDBrCH_3 \rightleftharpoons CH_3CHBrCH_3 + CH_3\underset{\diagdown \overset{+}{Br} \diagup}{CD\text{—}CH_2} \quad (20)$$

$$CH_3\underset{\diagdown \overset{+}{Br} \diagup}{CD\text{—}CH_2} + CH_3CDBrCH_3 \rightarrow CH_3CD_2CH_2Br + CH_3\overset{+}{C}BrCH_3 \quad (21)$$

their proposed steps, Karabatsos et al. [45] referred to the hydride abstraction of the methine hydrogen of 2-propanol by the trityl cation [101] and the observation of propylene-bromonium ion by NMR [102].

c. *n-Butyl and Higher n-Alkyl Rearrangements*

In the course of investigating factors affecting the formation of isomeric and optically active alkyl halides by the decomposition of alkyl dihalogenoborinates and dialkyl halogenoboronates, Gerrard et al. [103-105] made the observation that Lewis acids ($AlCl_3$, $FeCl_3$) increased both the rate of decomposition of the chloro esters and the percentage rearrangement of the alkyl groups. The overall pattern of rearrangements found was of *n*-alkyl groups (*n*-propyl, *n*-butyl, *n*-pentyl, *n*-octyl) rearranging to *sec*-alkyl groups and of *sec*-alkyl groups rearranging among themselves: for example,

$$CH_3(CH_2)_6\overset{+}{C}H_2 \longrightarrow CH_3(CH_2)_5\overset{+}{C}H\text{-}CH_3 \rightleftharpoons$$
$$CH_3(CH_2)_4\overset{+}{C}H\,CH_2CH_3 \leftrightarrows CH_3(CH_2)_3\overset{+}{C}H\,(CH_2)_2\,CH_3.$$

These authors also indicated that *sec*-alkyl compounds usually did not give *n*-alkyl halides except that some *n*-propyl bromide was formed in certain isopropyl reactions.

n-Butyl chloride was shown by Gerrard et al. [54] to isomerize to the secondary isomer on heating with the corresponding aluminum halide. On the other hand, *sec*-butyl chloride [103] and bromide [54] were not isomerized to the *n*-butyl isomer by the use of the corresponding aluminum halide. The degree of isomerization of *n*- to *sec*-butyl chloride by aluminum chloride was found to be dependent on the proportion of reactants used. The treatment of *n*-butyl chloride at reflux temperature with 0.02 M concentration of $AlCl_3$ gave a 71% yield of mixture consisting of 91% *n*-butyl and 9% *sec*-butyl chloride [54]. With a 0.5 M concentration of aluminum chloride, formation of hydrocarbon by-products was high and from *n*-butyl chloride some tertiary halide was obtained in addition to the main product, *sec*-butyl chloride [103]. This dependence on the catalyst/substrate ratio was attributed to the deactivation of the catalyst caused by the polymer by-products produced [105].

The effect of other catalysts, such as zinc halides, boron halides, and concentrated sulfuric acid, on some butyl halides was also investigated. Gerrard et al. [103] reported that zinc chloride had no effect on *n*-butyl chloride even after reflux for 24 hr. Zinc bromide, however, caused rearrangement of *n*-butyl bromide at reflux temperature (11.5% in 8 hr), and finally caused complete decomposition to hydrogen bromide and polymer [54]. Zinc iodide was reported to have no effect on *n*-butyl iodide under reflux [54].

Boron trichloride was reported to be without effect on *n*-butyl chloride under reflux or in a sealed tube at 100°C (3 hr), but at 225°C (5 hr) it gave 2.5% *sec*-butyl and 1% *t*-butyl chloride [100].

Concentrated sulfuric acid did not isomerize either *n*-butyl chloride or *n*-butyl bromide to the *sec*-isomers when heated under reflux, but caused slow decomposition of *n*-butyl bromide to hydrogen bromide and polymer [54]. However, when a mixture of 87% *n*-butyl and 13% *sec*-butyl chloride was heated under reflux with concentrated sulfuric acid, the *sec* isomer gradually decomposed with the evolution of hydrogen chloride, leaving *n*-butyl chloride of over 99% purity. Accordingly, treatment by concentrated sulfuric acid was suggested as a means of removing impurities from the crude *n*-butyl chloride prepared by the $ZnCl_2$-HCl method [107].

The isomerization of 2- and 3-halopentanes was carried out by various investigators using a variety of acid catalysts. When Pines et al. [108] subjected mixtures of 6 to 13% 2- and 94 to 87% 3-bromopropane to the action of $HBr_{(g)}$, $HBr(g) + H_2O$, PBr_3, $HBr_{(aq)} + H_2SO_4$, and concentrated H_2SO_4 reagents, they obtained mixtures in which the percentage of the 2 isomer increased to 21 to 28%. These results were later confirmed by Cason and Correia [109], who reported that the treatment of either 2- or 3-bromopentane by concentrated sulfuric acid or by zinc chloride resulted in an equilibrated mixture containing 66% of the 2 isomer and 34% of the 3 isomer. Both phosphorus tribromide and concentrated hydrochloric acid led to similar results. In explaining their results the authors stated the following: "If the equilibration in both instances proceeds through the carbocation, as these data indicate, then the equilibrium concentrations must depend on the relative energies of the two isomers. As there are two 2-positions, the data indicate equal energy for the two secondary halides."

Baker [110] working with R. M. Roberts, investigated the equilibration of 2- and 3-chloropentane in the presence of aluminum chloride. In their investigation they performed reactions at -20, 0, and 25°C and analyzed the reaction mixtures after various time intervals, starting with 1 min. Using infrared spectrophotometry for analysis, they found that all reactions at all temperatures and after all times gave equilibrium mixtures containing approximately 65% 2-chloropentane and 35% 3-chloropentane. They also conducted one experiment in which hexamethylbenzene was added to a mixture of 3-chloropentane and aluminum chloride at -20°C for 1 min to see what effect it would have on the isomerization, and the result was none. The product ratio in this case was the same as in cases without added hexamethylbenzene. It was also pointed out that in the course of the isomerizations above, considerable polymerization occurred even in reactions carried out at -20°C.

The results of Roberts and Baker were explained in terms of a carbocation mechanism in which the cations produced by the action of aluminum chloride on the organic chlorides could either polymerize or undergo rapid hydride shifts, resulting in the 65:35 ratio of 2-/3-chloropentane. This ratio was suggested to result from a statistical distribution of the intermediate secondary cations, as the pentyl chain has two 2-carbon atoms and one 3-carbon atom, which would theoretically produce a 2:1 ratio, provided that the two isomeric halides have equal energy.

The aforementioned results of Roberts and Baker should be compared with the later results of Streitwieser et al. [111] on the treatment of 2-pentanol with boron trifuloride. According to these authors, when a solution of 2-pentanol in pentane or cyclohexane was saturated with boron trifluoride and was maintained at 0°C for 1 hr, the 2-pentanol was recovered unchanged. However, when the 2-pentanol in pentane or heptane solution was saturated with boron trifluoride and maintained at 0°C for 1 hr with 0.2 equivalent of hexamethylbenzene, a rapid reaction occurred which was accompanied by complete decomposition of the alcohol to a polymeric unsaturated oil. The formation of polymers in the latter reaction was explained in terms of a process involving dissociation of a π-complex intermediate, priorly formed by interaction between the aromatic and $ROH \cdot BF_3$, to a carbocation which could subsequently form elefin and polymerize. This is shown in Eq. (22).

$$ROH \cdot BF_3 + C_6(CH_3)_6 \longrightarrow [C_6(CH_3)_6 \cdot R^+] \longrightarrow R^+ + BF_3OH^- \qquad (22)$$

$$R^+ \rightleftharpoons \text{olefin} + H_2O \cdot BF_3; \quad \text{olefin} \longrightarrow \text{polymers}$$

The fact that in the reactions with 2- or 3-chloropentane and aluminum chloride the alkyl halide was isomerized without any arene present, whereas in the treatment of 2-pentanol with boron trifluoride reaction occurred only in the presence of added aromatic, suggests that the chloropentanes are probably freer to reach the statistical 65:35 ratio in an aluminum chloride-chloropentane complex.

The isomerizations of 2- and 3-chloropentanes by nitromethane-moderated aluminum chloride as well as of 2- and 3-chlorohexane by both moderated and nonmoderated aluminum chloride catalysts were more recently studied by R. M. Roberts et al. [112]. Using infrared spectrophotometry to analyze chloropentanes and gas chromatography to analyze chlorohexanes, they obtained the results summarized in Table 1. Examination of the data in the table shows clearly that the isomerization of 2- and 3-chloropentane by nitro-

Table 1. Isomerizations of 2- and 3-Chloropentanes and Chlorohexanes

Starting materials	Temperature, °C	Time, min	Chloroalkanes %2–	%3–
2-Chloropentane + aluminum chloride + nitromethane	25°	60	63	37
3-Chloropentane + aluminum chloride + nitromethane	25°	60	62	38
2-chlorohexane + aluminum chloride	−20°	1	53	47
		10	52	48
3-Chlorohexane + aluminum chloride	−20°	1	52	48
		10	52	48
3-Chlorohexane + aluminum chloride + nitromethane	25°	10	53	47
2-Chlorohexane + aluminum chloride + nitrobenzene	25°	10	55	45

methane-moderated aluminum chloride, as by aluminum chloride, gave isomer distributions very close to that of a statistical equilibrium value.

Predictably, if thermodynamic factors did not interfere, the isomerizations of 2- and 3-chlorohexane by moderated or nonmoderated aluminum chloride catalysts should give a 50:50 distribution since the alkyl group in this case has two carbon atoms to which chlorine may bond in the 2 position and another two to which chlorine may bond in the 3 position. As evident from the data in Table 1, the treatment of either chlorohexane with aluminum chloride at -20°C or with moderated aluminum chloride at 25°C resulted, quite rapidly, in the production of statistically determined equilibrium mixtures. The observed small deviations from the exact ratio of 50:50 in favor of the 2 isomer could perhaps be due to small differences in the relative thermodynamic stabilities of the two chlorohexane isomers, the 2 isomer being somewhat more stable.

A concentrated sulfuric acid-catalyzed isomerization of 2-chlorooctane was conducted by Gerrard and Hudson [104]. According to these authors, when optically active 2-chlorooctane was shaken with an equal volume of concentrated sulfuric acid at 20°C, loss of optical purity and rearrangement to 3-chlorooctane (3.5%) and 4-chlorooctane (0.5%) occurred, together with the evolution of hydrogen chloride. The same authors also reported that 2-bromooctane suffered no loss in optical rotatory power when shaken with concentrated hydrobromic acid or heated with it at 100°C for 5 hr in a sealed tube.

The fact that in the latter case, nonequal distribution of chlorine occurred between the 3 and 4 positions of the octyl chain, while in the lower pentyl and hexyl systems almost statistical equilibrations were rapidly achieved is in accordance with the findings of Karabatsos et al. [113], who, on the basis of deamination studies, concluded that intramolecular 1,2 hydride shifts are subject to strong conformational control and they occur faster in the smaller alkyl systems than in the larger ones.

2. Rearrangements in branched acyclic systems

a. *Isobutyl Rearrangements*

The simplest branched alkyl chloride, isobutyl chloride, was reported to isomerize to *t*-butyl chloride in the presence of aluminum chloride or boron halides [114]. It was also reported that anhydrous barium or thallium chlorides induced the isomerization of isobutyl chloride or isobutyl bromide to the corresponding *t*-butyl isomer upon heating in a sealed tube at 250°C [115].

The decomposition of isobutyl dihaloborinates and disobutyl haloboronates in the presence of $AlCl_3$ or $FeCl_3$ was found by Gerrard et al. [54] to produce not only *t*-butyl halide, by hydrogen migration, but also *sec*-butyl halide by methyl shift. Notably, products resulting from methyl shift were also obtained from related carbocation reactions, including aqueous acid deamination of isobutyl-1,1-d_2-amine [116], diazotization of deuterium-labeled [117-119] and unlabeled [117,120] isobutylamines, and thermal decomposition of the corresponding N-isobutyl-N-nitrosoacetamide [117,121]. Aside from isobutylene and *n*-butenes, methylcyclopropane was also one of the hydrocarbon products of the latter reactions. The formation of methylcyclopropane as well as of *n*-butenes was explained in terms of a mechanism involving an equilibrating protonated methylcyclopropane intermediate, 19, which could either be deprotonated to give methylcyclopropane or open to *sec*-butyl cation. Deprotonation of the latter gave 1- and 2-butenes [118].

$$(CH_3)_2CH\text{—}\overset{+}{C}H_2 \longrightarrow CH_3\overset{+}{C}HCH_2CH_3 \xrightarrow{-H^+} CH_2{=}CHCH_2CH_3 + CH_3CH{=}CHCH_3$$

$$\downarrow$$

$$\left[CH_3\text{-}CH\text{----}CH_2\text{(}CH_2\text{---}H\text{)}^{+} \rightleftarrows CH_3\text{—}CH\text{—}CH_2\text{(}H\cdots CH_2\text{)}^{+} \rightleftarrows CH_3\text{—}CH\text{---}CH_2\text{(}CH_2\text{)(}H\text{)}^{+} \right]$$

19

$$\downarrow$$

$$CH_3CH\text{(}CH_2\text{—}CH_2\text{)} \quad \text{(methylcyclopropane)}$$

b. *t-Butyl, t-Pentyl, and t-Heptyl Rearrangements*

In recent years, studies of rearrangements involving substances of identical carbon skeletons and nearly identical energies were made possible by the application of ^{14}C or ^{13}C as a tracer. For example, ^{14}C was used by J. D. Roberts et al. [122,123] to detect rearrangements of the types shown in Eqs. (23) and (24) which might occur when labeled *t*-butyl and *t*-pentyl chlorides

$$\text{C-}^{14}\text{C(C)(C)-X} \rightleftharpoons \text{C-C(}^{14}\text{C)(C)-Cl} \tag{23}$$

$$\text{C-}^{14}\text{C(C)(Cl)-C-C} \rightleftharpoons \text{C-C(}^{14}\text{C)(Cl)-C-C} \rightleftharpoons \text{C-C(C)(Cl)-}^{14}\text{C-C} \rightleftharpoons \text{C-C(C)(Cl)-C-}^{14}\text{C} \tag{24}$$

are subjected to the action of aluminum chloride. The results demonstrated that rearrangements such as those shown in Eqs. (23) and (24) did occur, but much more readily with *t*-pentyl chloride in the presence of $AlCl_3$. Thus treatment of 2-methyl-2-chloropropane-2-^{14}C with $AlCl_3$ at room temperature for 10 min resulted in only 1 to 3% rearrangement to 2-methyl-2-chloropropane-1-^{14}C, whereas treatment of ^{14}C-labeled 2-methyl-2-chlorobutane with $AlCl_3$ at 0°C for 1.5 min gave the results shown in Table 2. From Table 2 it can be seen that the experiment with 2-methyl-2-chlorobutene-1-^{14}C (Experiment 1) gave a product in which equilibrium was essentially established between C-1 and C-4 whereas that with 2-methyl-2-chlorobutane-2-^{14}C (Experiment 2) gave a partially rearranged product with about 70% of the statistically possible exchange between C-2 and C-3 and feeble activities at C-1 and C-4. These results, together with those of Experiment 3, demonstrated that little, if any, exchange occurred between the methyl groups and the carbons in the center of the chain during the isotopic equilibration between C-2 and C-3 or between C-1 and C-4. In seeking an explanation for their results, Roberts et al. [123] stated that the observed isotope-position rearrangements appeared to be most satisfactorily interpreted on the basis of the carbocation processes shown in Eqs. (25) to (27).

$$CH_2\text{-}^{14}\overset{+}{C}(CH_3)\text{-}CH_3 \xrightarrow{\sim H:^-} CH_3\text{-}^{14}C(CH_3)(H)\text{-}\overset{\oplus}{C}H_2 \xrightarrow{\sim CH_3:^-} CH_3\text{-}^{14}\overset{\oplus}{C}H\text{-}CH_2\text{-}CH_3 \xrightarrow{\sim H:^-}$$

$$CH_3\text{-}^{14}CH_2\text{-}\overset{\oplus}{C}H\text{-}CH_3 \xrightarrow{\sim CH_3:^-} {}^{14}\overset{\oplus}{C}H_2\text{-}C(CH_3)(H)\text{-}CH_3 \xrightarrow{\sim H:^-} {}^{14}CH_3\text{-}\overset{\oplus}{C}(CH_3)\text{-}CH_2 \tag{25}$$

$$CH_3\text{-}CH_3\text{-}^{14}\overset{\oplus}{C}(CH_3)\text{-}CH_3 \xrightarrow{\sim H:^-} CH_3\text{-}\overset{\oplus}{C}H\text{-}^{14}C(CH_3)(H)\text{-}CH_3 \xrightarrow{\sim CH_3:^-}$$

$$CH_3\text{-}C(CH_3)(H)\text{-}^{14}\overset{\oplus}{C}H\text{-}CH_2 \xrightarrow{\sim H:^-} CH_3\text{-}\overset{\oplus}{C}(CH_3)\text{-}^{14}CH_2\text{-}CH_3 \tag{26}$$

$$CH_3\text{-}CH_2\text{-}\overset{\oplus}{C}(^{14}CH_3)\text{-}CH_3 \xrightarrow{\sim CH_3:^-} \overset{\oplus}{C}H_2\text{-}C(^{14}CH_3)(CH_3)\text{-}CH_3 \xrightarrow{14\ \sim CH_3:^-} {}^{14}CH_3\text{-}CH_2\text{-}\overset{\oplus}{C}(CH_3)\text{-}CH_3 \tag{27}$$

In an attempt to determine the relative rates of reactions (26) and (27), the authors investigated the rearrangement of a mixture of C-1 and C-2 ^{14}C-

Table 2. Rearrangement of ^{14}C Labeled 2-Methyl-2-chlorobutane (*t*-Pentyl Chloride) with Water-Activated $AlCl_3$ at 0°

$$\begin{array}{c} \quad\; C \\ \quad\; | \\ C_1 - C_2 - C_3 - C_4 \\ \quad\; | \\ \quad\; Cl \end{array}$$

Experiment no.	Activity in starting chloride, %		Activity in resulting chloride, %			
	C_1	C_2	C_1	C_2	C_3	C_4
1	100	-	66.3	0.0	1.9	31.8
2	-	100	0.1	63.7	35.1	1.1
3	54.8	45.2	40.1	26.6	18.6	14.8

labeled *t*-pentyl chloride.* It was anticipated that if reaction (27) was the only path of isotopic equilibration, the isotopic scrambling of C-2 $\rightleftarrows$ C-3 of 2-methyl-2-chlorobutane-2-^{14}C with aluminum chloride would be expected to be twice as fast as the scrambling of C-1 $\rightleftarrows$ C-4 of 2-methyl-2-chlorobutane-1-^{14}C. It was demonstrated, however, that the rate of scrambling of C-2 $\rightleftarrows$ C-3 versus C1 $\rightleftarrows$ C-4 was 1.55 and this was interpreted as corresponding to 87% reaction by sequence (26) and 13% reaction by sequence (27).

In an effort to accommodate other findings to their mechanism, J. D. Roberts and co-workers [123] ascribed the fact that *t*-butyl chloride rearranged considerably less readily than *t*-pentyl chloride to the assumption that the simplest reaction sequence (Eq. 28) for *t*-butyl choride involves

$$CH_3\text{-}{}^*C(CH_3)_2\text{-}Cl \longrightarrow CH_3\text{-}{}^*\overset{+}{C}(CH_3)_2 \xrightarrow{\sim H:^-} \overset{+}{C}H_2\text{-}{}^*CH(CH_3)\text{-}CH_3 \xrightarrow{\sim CH_3:^-} CH_3\text{-}CH_2\text{-}{}^*\overset{+}{C}H\text{-}CH_3$$

$$\xrightarrow{\sim H:} CH_3\text{-}\overset{+}{C}H\text{-}{}^*CH_2\text{-}CH_3 \;(\text{written: } CH\ \text{-}\overset{+}{C}H\text{-}{}^*CH_2,\ CH_3) \xrightarrow{\sim CH_3:^-} CH_3\text{-}CH(CH_3)\text{-}{}^*\overset{+}{C}H_2 \xrightarrow{\sim H:^-} CH_3\text{-}\overset{+}{C}(CH_3)\text{-}{}^*CH_3 \longrightarrow CH_3\text{-}C(Cl)(CH_3)\text{-}{}^*CH_3 \qquad (28)$$

two primary and two secondary cations, while with *t*-pentyl chloride the simplest sequences have either two secondary (Eq. 26) or one primary (Eq. 27) cations as intermediates. In a similar manner, they rationalized the lack of

*It was pointed out that this mixture was, for all practical purposes, the equivalent of double-labeled material.

significant isotope exchange between the methyl groups and C-2 and C-3 of the *t*-pentyl halides by assuming that the simplest carbocation processes that could cause them would involve two primary and two secondary cationic intermediates (Eq. 29).

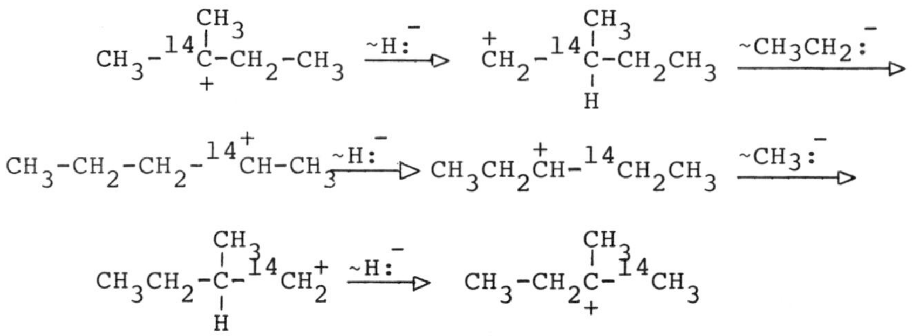

$$CH_3\text{-}{}^{14}\overset{+}{C}(CH_3)\text{-}CH_2\text{-}CH_3 \xrightarrow{\sim H:^-} \overset{+}{C}H_2\text{-}{}^{14}CH(CH_3)\text{-}CH_2CH_3 \xrightarrow{\sim CH_3CH_2:^-}$$

$$CH_3\text{-}CH_2\text{-}CH_2\text{-}{}^{14}\overset{+}{C}H\text{-}CH_3 \xrightarrow{\sim H:^-} CH_3CH_2\overset{+}{C}H\text{-}{}^{14}CH_2CH_3 \xrightarrow{\sim CH_3:^-}$$

$$CH_3CH_2\text{-}CH(CH_3)\text{-}{}^{14}CH_2^+ \xrightarrow{\sim H:^-} CH_3\text{-}CH_2\overset{+}{C}(CH_3)\text{-}{}^{14}CH_3 \qquad (29)$$

Although the mechanisms above explained the observed results, they were criticized by Karabatsos et al. [90,124,125] because the intramolecular rearrangements involved the formation of primary carbocations from the more stable secondary and tertiary ones. These authors referred to calculations [126] indicating that at least 22 kcal/mol of activation energy are needed for rearrangements of secondary to primary carbocations,* and 33 kcal/mol for rearrangements of tertiary to primary ones.

To avoid the involvement of primary carbocations in rationalizing these reactions, Karabatsos offered an alternative explanation based on bimolecular processes. Equations (30) and (32) were proposed to account for the isotopic scrambling observed in the *t*-pentyl system. Equation (30), as far as

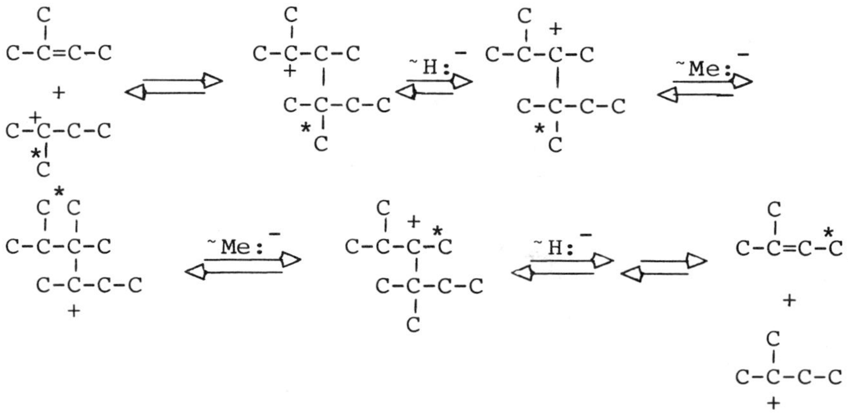

$$\text{C-C(C)=C-C} + \text{C-}\overset{+}{\text{C}}(\overset{*}{\text{C}})\text{-C-C} \rightleftharpoons \text{C-}\overset{+}{\text{C}}(\text{C})\text{-C(C-C(}\overset{*}{\text{C}})\text{-C)-C} \underset{}{\overset{\sim H:^-}{\rightleftharpoons}} \text{C-}\overset{+}{\text{C}}(\text{C})\text{-C(C-C(}\overset{*}{\text{C}})\text{-C)-C} \overset{\sim Me:^-}{\rightleftharpoons}$$

$$\text{C-C(C)-C(}\overset{*}{\text{C}}\text{)(C-}\overset{+}{\text{C}}\text{-C-C)-C} \overset{\sim Me:^-}{\rightleftharpoons} \text{C-C(C)-}\overset{+}{\text{C}}\text{-}\overset{*}{\text{C}}\text{(C-C(C)-C-C)} \overset{\sim H:^-}{\rightleftharpoons} \rightleftharpoons \text{C-C(C)=C-}\overset{*}{\text{C}} + \text{C-}\overset{+}{\text{C}}\text{(C)-C-C} \qquad (30)$$

$$\overset{*}{\text{C}}\text{-C(C)(Cl)-C-C} + \overset{*}{\text{C}}\text{-C(C)(Cl)-C-C} \overset{AlCl_3}{\rightleftharpoons} \overset{*}{\text{C}}\text{-C(C)(Cl)-C-}\overset{*}{\text{C}} + \text{C-C(C)(Cl)-C-C} \qquad (31)$$

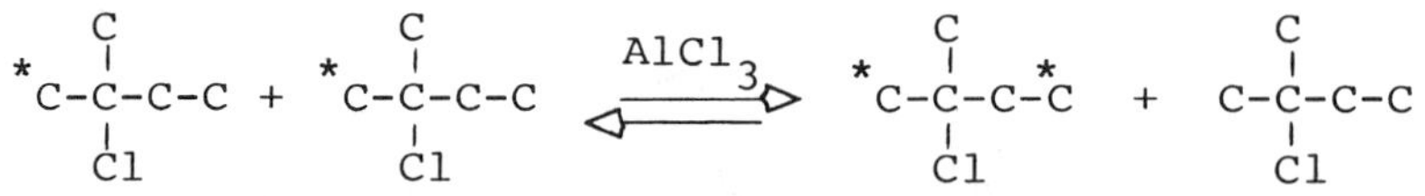

*More recently, a somewhat lower value of 15.3 kcal/mol was estimated [183].

```
                                                    C
   C                  C C                          +|
 * | +              * | |                          C-C-C
 C-C-C-C              C-C-C-C-C                     |
     |      ~Et:-         |+     ~i-Pr:-         C-C-C-C*     ~Me:-
   C-C-C-C  <====>     C-C        <====>         *|  |        ====>
   *|                  *|                          C  C
    C                   C
```

```
  C                                C
  |                              * |   *
C-C-C-C                          C-C=C-C
 +|  *        ~H:-
  C-C-C       <====>  <====>         +                     (32)
 *|  |
  C  C                               C
                                     |
                                   C-C-C-C
                                     +
```

```
  C              C          AlCl3           C              C
 *|             *|                         *| *            |
C-C-C-C   +   C-C-C-C      <=====>      C-C-C-C    +    C-C-C-C       (33)
  |              |                          |              |
  Cl             Cl                         Cl             Cl
```

its effect on isotopic equilibration is concerned, is identical with Eq. (27) in equilibrating only C-1 and C-4. Equation (32), unlike (30), leads not only to C-1, C-4 equilibration but also to C-2, C-3 equilibration; the net effect on isotope-position equilibration is the same as that of Eq. (26). It was pointed out that, according to these bimolecular paths, if both the cation and the olefin are labeled, the recovered *t*-pentyl chloride should contain dilabeled species. This can be seen from the net effects of Equations (30) and (32) as given by Eqs. (31) and (33).

Experimental support for the bimolecular mechanism was found by Karabatsos and Vane in the data for the treatment of ordinary [124] and ^{13}C-labeled [90,125] 2-chloro-2-methylbutane with aluminum chloride. Studying the reactions of 2-chloro-2-methylbutane-1-^{13}C and 2-chloro-2-methylbutane-2-^{13}C with aluminum chloride at 0°C for 5 min these authors found (by both mass spectral and proton NMR analysis) that the recovered *t*-pentyl chlorides contained, in addition to unlabeled and monolabeled molecules, dilabeled molecules. Furthermore, they found that this ^{13}C scrambling was accompanied by the formation of both disproportionation and hydride transfer products which included C_4, C_5, and C_6 alkanes as well as C_4, C_5, and C_6 alkyl chlorides. Based on these and other findings about the distribution of the label in the products, the authors reached the following conclusions: (1) Isotope-position rearrangements of the *t*-pentyl cations occurred via unimolecular reaction (Eq. 26) and via bimolecular reactions (Eqs. 30 and 32) involving C_{10} carbocations. (2) In the course of the reaction, the six methyl carbon atoms of the C_{10} carbocations became statistically distributed while the four nonmethyl carbon atoms did not. (3) The carbocations involved are classical tertiary and secondary ions. Protonated cyclopropanes did not intervene, as this would lead to equilibration between methyl and nonmethyl carbon atoms, which was not observed. (4) Under the mild conditions applied, only tertiary carbocations were formed.

The question of the intermediacy of primary carbocations and the relative contributions of unimolecular reactions and bimolecular reactions to carbocation rearrangements in liquid-phase reactions of alkyl halides, paraffins, olefines, and related compounds under ionizing conditions was generally considered in a detailed study by Karabatsos and Vane [124]. In this study they investigated the quantitative and qualitative effects of time, temperature, concentration of aluminum chloride, methylbutenes, hydrogen chloride, and water on the reaction of *t*-pentyl chloride with aluminum chloride. In all cases, the products obtained were shown to be mixtures of C_4 to C_6 alkanes and alkyl chlorides resulting from hydride transfer and disproportionation reactions. The qualitative composition of these products was demonstrated to be practically unaffected by reaction time, temperature, and aluminum chloride concentration. This fact was considered by the authors as supporting a suggestion that carbocations during their life span undergo many reactions of comparable rapidity. The authors also found that at higher temperatures and aluminum chloride concentrations the disproportionation of *t*-pentyl chloride increased, and this finding was considered to be in accord with the expected increase in the occurrence of bimolecular reactions.

In the light of their results, Karabatsos and Vane viewed the disproportionation of *t*-pentyl to *t*-butyl chloride and hexyl chlorides as occurring mainly via the two reaction paths of Eqs. (34) and (35) [127-130]. They also proposed that the simplest path leading to the conversion of *t*-pentyl chloride to 2-chloro-3-methylbutane could be visualized as in Eq. (36); the latter compound was always present in the product mixtures. In connection with

```
  C                    C               C              C
  |                    |               |+             |
CC=CC                CCCC            CCCC           C C
                      +|      ~H:⁻     |     ~Et:⁻  | |      Me:⁻
  +        <===>     CCCC    <===>   CCCC   <===>  CC-CC    <===>
 +                     |               |              |
CCCC                   C               C             CC+
 |                                                    |
 C                                                    C

                      C
                      |
            ~H:⁻    CC-CCC              C           C
                      + |      <===>    |           |
 C                     CCC            CC=CCC   +   CCC
 |+                    |                            +
CCCCC                  C
 |
CCC                   C
 |        ~Me:⁻      +|
 C                  CCCCC
                      |                 C           C
                     CCC       <===>    |           |
                      |               CC=CCC   +   CCC
                      C                             +
```

(34)

```
   C                    C             C
   |                    |             |
 C-CCC     ~Me:⁻      CC-C+   Me:⁻   CC-CC                    C        C
  +|       <===>       |  |   <===>  | ⤵|           <===>     |        |
  CCCC                 C CCCC         C |                   CC=CCC + CCC
   |                      |            CCCC                             +
   C                      C             +
                            ~Et:⁻
                                      C
                                      |
                                     CC-CCC                  C        C
                                      | ⤵|          <===>    |        |
                                      C |                  CC=CCC + CCC
                                       CCC                                +
                                        +
```

(35)

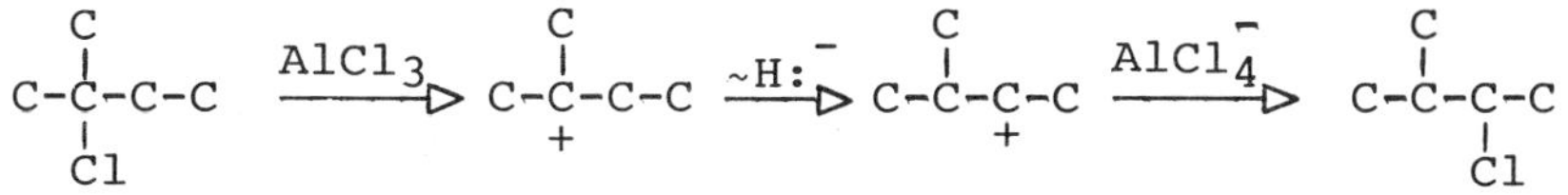

```
  C                    C              C      -          C
  |       AlCl3        |      ~H:-    |      AlCl4      |
C-C-C-C  -------> C-C-C-C  ------> C-C-C-C  -------> C-C-C-C
  |                  +                  +                 |
  Cl                                                      Cl
```

(36)

the work of Karabatsos and Vane, it is noteworthy that the reaction of *t*-pentyl chloride with aluminum chloride always stopped after about 5 min. Examination of various factors suggested that the most likely reason for this phenomenon was the deactivation of the catalyst by the polymers formed in the reaction.

The isomerization of *t*-pentyl bromide under Friedel-Crafts conditions was examined by Bartlett et al. [131]. In 1944, these workers reported that *t*-pentyl bromide on treatment with aluminum bromide always went into an equilibrium containing 15 to 20% of 2-bromo-3-methylbutane. Although they suggested that this reaction could be represented as an elimination followed by addition of hydrogen halide, in partial opposition to Markownikoff's rule, they did not rule out the possibility that this isomerization might have occurred by a reversible rearrangement in the carbocation. In reviewing this reaction, Nenitzescu favored the latter possibility, which, in fact, is analogous to the intramolecular hydride-shift mechanism postulated previously by Karabatsos and Vane [124] (Eq. 36) for the rearrangement of the corresponding chloride.

The extent of rearrangement in the carbocation-type reactions of 2-hydroxy- and 2-chloro-2,3,3-trimethylbutane-1-^{14}C with Lucas reagent ($ZnCl_2$-HCl) was found by J. D. Roberts and Yancey [132] to depend markedly on the degree of reversibility of the processes as determined by the reaction conditions. For example, when the alcohol was treated with Lucas reagent for 3 hr at room temperature, it gave a chloride in which the ^{14}C was distributed essentially equally among the five methyl groups (i.e., a 60:40 mixture of 20 and 21). Under comparable conditions, the corresponding chloride was shown to equilibrate completely with Lucas reagent containing radioactive chloride ion. In circumstances under which the reactions were less reversible, fewer rearrangements were observed. To account for the rearrangement and the chlorine exchange, the authors proposed the equilibria depicted in Eq. (37) involving solvated carbocations. The authors also concluded

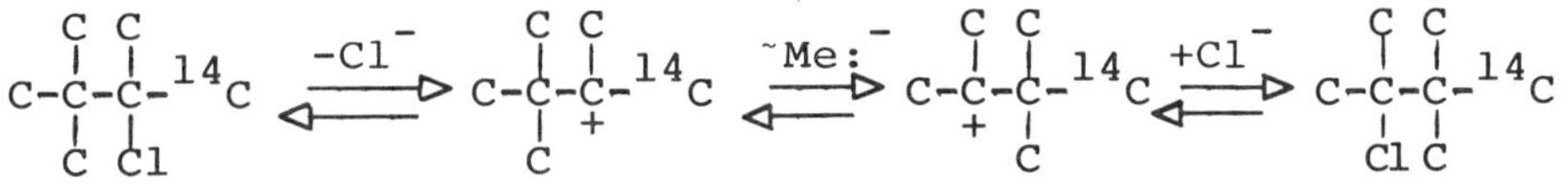

```
  C C                    C C                    C C                    C C
  | |      -Cl-          | |       ~Me:-        | |       +Cl-         | |
C-C-C-14C  <====>  C-C-C-14C   <====>   C-C-C-14C  <====>   C-C-C-14C
  | |                    | +                    + |                    | |
  C Cl                   C                        C                   Cl C

   20                                                                  21
```

(37)

that the nonclassical-cation intermediate 22 was not an important intermediate in irreversible carbocation-type reactions of pentamethylethyl derivatives in aqueous solutions, and that such an intermediate was unlikely to be as stable as the classical cations shown in Eq. (37).

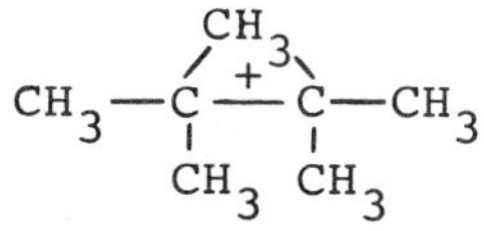

```
            CH3
           / + \
    CH3 -- C --- C -- CH3
           |     |
          CH3   CH3

            22
```

c. Neopentyl Rearrangements

The rearrangement of neopentyl compounds to *t*-pentyl compounds was recognized by numerous early workers, who noted that generation of the neopentyl cation in any irreversible reaction resulted in rapid and complete rearrangement to a *t*-pentyl cation [51,133,134].

In recent years, several studies on the conversion of the neopentyl group to the *t*-pentyl group were conducted using suitably labeled neopentyl derivatives. The chief aim of these studies was to elucidate the path of rearrangement and to assess the effect of mode of formation of the neopentyl cation on the relative importance of 1,2 shifts, 1,3 shifts, and formation of protonated cyclopropanes in this rearrangement. In 1960 it was shown that conversion of the neopentyl group to the *t*-pentyl group by the reacton of neopentyl-1-^{13}C alcohol with hydrogen bromide [135] or by deoxidation of neopentyl-1,1-d_2 alcohol [136] proceeded with no perceptible intervention of protonated cyclopropanes. In 1964, Karabatsos et al. [137] used NMR and mass spectroscopy to analyze the *t*-pentyl alcohol obtained from the deamination of neopentyl-1-^{13}C and neopentyl-1,1-d_2-amines, from the solvolysis of neopentyl-1-^{13}C and neopentyl-1,1-d_2 tosylates, and from the solvolysis of neopentyl-1-^{13}C iodide. They also analyzed (by NMR) the *t*-pentyl chloride obtained from the reaction of neopentyl-1-^{13}C alcohol with triphenyl phosphite and benzyl chloride. In all these cases, they found that the label originally present at C-1 of the neopentyl compounds always ended up at C-3 of the *t*-pentyl compounds. On the basis of these findings, the authors ruled out the intervention of 1,3-hydride shifts, protonated cyclopropanes, or hydrogen-bridged ions during the neopentyl rearrangement. The rearrangement to the *t*-pentyl cation was thus considered as proceeding by a 1,2-methyl shift. It was argued, however, that the methyl shift can proceed either by a synchronous C-X breaking and methyl migration, path (a) of Eq. (38), or by a two-step process involving the intermediacy of the neopentyl cation, path (b) of Eq. (38). Mechanisms intermediate between these two extremes could also be

```
   C              C                       C
   |             / \                      |
               / + \     -
C-C-CX   (a)   C-C—C--X    ——▷     C-C-C-C
   |     ——▷      |                       +
   C              C
                  ts                      △
   |                                      |
   |              C                       |
   |              | +  -       ~CH3:⁻     |
   |______(b)——▷ C-C-C X     ____________|     (38)
                  |
                  C
```

visualized. Based on results with optically active neopentyl-1-^{13}C or d systems [138], path (b) was suggested to intervene in the reaction of neopentyl iodide with silver nitrate, as in this case optically active neopentyl-1-d iodide led to inactive *t*-pentyl alcohol. On the other hand, path (a) was implicated in the deoxidation of neopentyl alcohol, because optically active (presumbably inverted) 2-methyl-1-butene-3-d resulted from optically active neopentyl-1-d alcohol. However, it was pointed out by Karabatsos et al. [137] that the results are equally compatible with path (b) of Eq. (380, provided that the methyl migration is faster than, or competes favorably with, rotation of the -CHD group with respect to X.

The neopentyl to *t*-pentyl rearrangement was also observed under solvolytic conditions [139-143]. In these reactions it was pointed out that the neopentyl solvolysis is always assisted by methyl participation.

On the basis of the data given above about the behavior of the neopentyl cation under various carbocation conditions, it is expected that generation of this cation under the related Friedel-Crafts conditions should also lead to rapid and complete rearrangement to *t*-pentyl cation. In fact, as we shall see later in the alkylation section, this is actually the case.

3. Rearrangements in dihalo acyclic systems

Very few reports concerning isomerizations of acyclic dihaloalkanes under the influence of metal halide catalysts have appeared. In 1963, Sommer [144] studied the rearrangement of dibromobutanes in CS_2 in the presence of $AlBr_3$ at various temperatures. On the basis of GLPC analysis, he concluded that all the dibromobutanes were produced in equilibrium composition when any isomer was treated with the catalyst. Starting with 1,3-dibromobutane at 25°C, an equilibrium mixture having the mol % composition shown in Eq. (39)

```
  Br                                                        Br
  |         AlBr3                                           |
C-C-C-C-Br  ------->  Br-C-C-C-C-Br  +  Br-C-C-C-C
            CS2,25°
             3 hr         1.5%               42.5%

                          Br                Br  Br
                          |                 |   |
                    Br-C-C-C-C       +   C-C---C-C
                          8.1%               19.1%

                         Br
                         |                Br
                       C-C-C-C       +      \C-C-C-C
                         |                  /
                         Br               Br
                         10%                 13.1%          (39)
```

was obtained. Moreover, the results obtained at various temperatures showed that the amount of 1,3-dibromobutane in the equilibrium mixtures decreased from 45.0 mol % to 37.2 mol % as the temperature was increased from 0°C to 35°C.

More recently, Billups and co-workers [145] published a study on the halogen migration in the reaction of dichloropropanes, dichlorobutanes, and dichloropentanes in the presence of $AlCl_3$. The results showed that treatment of 1,1-, 1,2-, or 1,3-dichloropropane with $AlCl_3$ at 25°C gave an equilibrium mixture of the four possible isomers. The composition of the equilibrium mixture from 1,3-dichloropropanes, for example, was that given in Eq. (40).

```
                                                       Cl
              AlCl3                                    |
Cl-C-C-C-Cl  ------->  Cl-C-C-C-Cl  +  Cl-C-C-C
             25°,4 hr
                          8.5%             90.0%

                        Cl                    Cl
                          \                   |
                     +     C-C-C       +    C-C-C              (40)
                          /                   |
                        Cl                    Cl
                          1.2%                0.3%
```

The migration of the chlorine to a primary carbon as in the conversion of 1,1- or 1,2-dichloropropane to 1,3-dichloropropane was explained on the basis of intermediate equilibrating protonated cyclopropanes.

As to dichlorobutanes, the isomerization results were in contradiction to those reported previously by Sommer, as these more recent workers were unable to obtain mixtures of all the possible isomers from any of the starting dichlorobutanes. When either 1,2- or 1,3-dichlorobutane was treated with $AlCl_3$ at 25°C, an equilibrium mixture of the two compounds was rapidly obtained which consisted of ca. 92% 1,3- and 8% 1,2-dichlorobutane. Under comparable conditions, 1,4-dichlorobutane was partially isomerized to a mixture of the 1,2- and 1,3-isomers, and both 2,3-dichlorobutane diastereomers underwent partial equilibration accompanied by 2,2-dichlorobutane formation. The results are formulated in scheme 3.

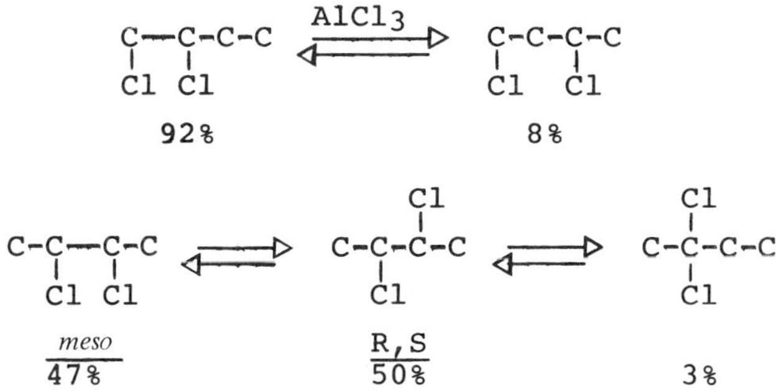

```
                 AlCl3
C—C-C-C       <=======>      C-C-C-C
|  |                         |   |
Cl Cl                        Cl  Cl

  92%                          8%

                     Cl               Cl
                     |                |
C-C—C-C   <====>  C-C-C-C   <====>  C-C-C-C
  |  |              |                 |
  Cl Cl             Cl                Cl

  meso             R,S
  47%              50%                3%
```

Scheme 3

The results from dichloropentanes were analogous to those from the dichlorobutanes. Thus both 2,3- and 2,4-dichloropentane diastereomers underwent rapid equilibration, whereas 1,2-dichloropentane formed predominantly 1,3- and 1,4-dichloropentane, besides small amounts of the four possible 2,3- and 2,4-dichloropentane diastereomers. The results are shown in Table 3.

These results were explained in terms of a mechanism involving intermediate chloronium ions. In fact, the existence of such ions has been conclusively demonstrated by Olah and his co-workers [146]. In terms of such a mechanism, the isomerization of 1,2- to 1,3- and 1,4-dichloropentanes can be en-

Table 3. Equilibration of Dichloropentanes at 25°

<table>
<tr><th></th><th colspan="7">% Composition of dichloropentanes</th></tr>
<tr><th></th><th>dl-
2,4-</th><th>meso-
2,4-</th><th>erythro-
2,3-</th><th>threo
2,3-</th><th>1,2-</th><th>1,3-</th><th>1,4-</th></tr>
<tr><td>2,4-Dichloropentane</td><td>64.3</td><td>26.1</td><td>5.0</td><td>4.6</td><td>—</td><td>—</td><td>—</td></tr>
<tr><td>2,3-Dichloropentane</td><td>64.2</td><td>25.9</td><td>4.4</td><td>5.5</td><td>—</td><td>—</td><td>—</td></tr>
<tr><td>1,2-Dichloropentane</td><td colspan="2">5.2</td><td colspan="2">0.3</td><td>6.5</td><td>23.9</td><td>64.1</td></tr>
</table>

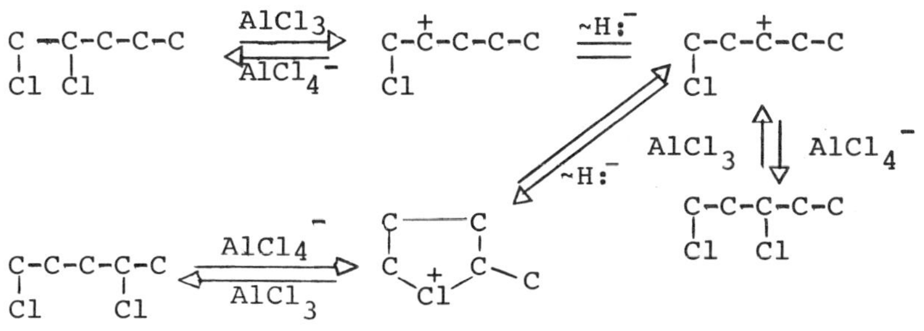

Scheme 4

visioned as shown in scheme 4. The predominance of 1,4-dichloropentane was attributed to the considerable stability of the five-membered-ring chloronium ion intermediate. The involvement of protonated cyclopropanes in this case was discarded on the ground that their presence should have caused formation of appreciable amounts of isomers with a branched carbon skeleton.

B. Rearrangements in Cyclic Systems

1. Cyclopropylcarbinyl and cyclobutyl rearrangements

In recent years considerable interest has been focused on the question of how best to formulate the intermediate or intermediates involved in the carbocation-type reactions of cyclopropylcarbinyl, cyclobutyl, and allylcarbinyl derivatives, which frequently result in mixtures of products of all three systems [147-150]. In 1951, J. D. Roberts and Mazur [147] investigated, among other things, the isomerization of cyclopropylcarbinyl, cyclobutyl, and allylcarbinyl alcohols and halides under the influence of Lucas reagent ($ZnCl_2$-HCl). They found that the treatment of a roughly equimolar mixture of cyclopropylcarbinyl and cyclobutyl chlorides or bromides with Lucas reagent (or its bromine analog) at or below room temperature resulted in complete isomerization to allylcarbinyl halides. Similarly, allylcarbinyl chloride was apparently the sole product of the reaction at room temperature between cyclopropylcarbinol or cyclobutanol with Lucas reagent. These authors also demonstrated that a 2:1 mixture of cyclopropylcarbinyl and cyclobutyl chlorides could be partially isomerized to a 1:1 mixture of allylcarbinyl and cyclobutyl chloride by using a shorter reaction time and less Lucas reagent than was employed for complete isomerization to allyl carbinyl chloride. Moreover, they showed that allylcarbinol was by far the least reactive alcohol toward Lucas reagent since no reaction occurred below or at room temperature, and on boiling, extensive decomposition resulted. These and other findings were explained by the authors in terms of a carbocation mechanism in which allylcarbinyl chloride is formed by a direct and essentially irreversible slow reaction between either or both of the cations and chloride ion or hydrogen chloride as shown in equation 41. Alternative mechanisms involving equilibration, between the various cations with irreversible reaction between the allylcarbinyl cation and chloride ion were also considered by the authors. These, however, were ruled out by the failure to observe some formation of crotyl and α-methylallyl chlorides, which would be expected in accordance with the modes of reaction of the allylcarbinyl cation observed in the allylcarbinylamine-nitrous acid reaction.

$$\text{cyclobutyl-Cl} \underset{\text{fast}}{\overset{ZnCl_2}{\rightleftharpoons}} \text{cyclobutyl}^{\oplus} \xrightarrow[\text{slow}]{\overset{\ominus}{Cl}(HCl)} CH_2{=}CH{-}CH_2{-}CH_2Cl$$
$$\text{cyclopropyl-}CH_2Cl \underset{\text{fast}}{\overset{ZnCl_2}{\rightleftharpoons}} \text{cyclopropyl-}\overset{+}{C}H_2 \quad (41)$$

In seeking direct evidence for the intramolecularity of the reaction above, the authors studied the extent of chlorine exchange between allylcarbinyl chloride and a mixture of cyclopropylcarbinyl and cyclobutyl chlorides with Lucas reagent containing ^{38}Cl. They found that under conditions where the mixture of cyclopropylcarbinyl and cyclobutyl chlorides was 77% isomerized to allylcarbinyl chloride, 107% of the theoretical chlorine exchange took place.* Under the same conditions, however, allylcarbinyl chloride showed but 13% exchange, demonstrating that allylcarbinyl chloride was not readily converted to the corresponding carbocation in the presence of Lucas reagent. On the basis of their findings, the authors concluded that the order of stability of the carbocations involved in these processes is probably

$$\text{cyclopropyl-}\overset{+}{C}H_2 \gtrsim \text{cyclobutyl}^{+} >> CH_2{=}CH{-}CH_2{-}\overset{+}{C}H_2$$

In 1968, Olah and Lin [151] investigated the aluminum chloride-catalyzed isomerization of cyclopropylcarbinyl and cyclobutyl chlorides. Because of the especially high reactivity of these chlorides toward aluminum chloride, it was found necessary to dilute the chlorides with an inert solvent or to complex the catalyst in order to conduct a kinetic study of the rearrangement of these halides. Consequently, the kinetic studies were conducted in a nitromethane-chlorobenzene mixed solvent system. Analysis of the kinetic data obtained under these conditions afforded the following conclusions:

1. The aluminum chloride-catalyzed isomerization of cyclopropylcarbinyl and cyclobutyl chlorides showed no tendency to form any equilibrium mixture, and the final product was always exclusively allylcarbinyl chloride.
2. Whereas cyclopropylcarbinyl chloride rearranged to cyclobutyl and allylcarbinyl chloride at -40 to -20°C, the rearrangement of cyclobutyl chloride to cyclopropylcarbinyl chloride was observed only at temperatures above 15°C.
3. No formation of cyclopropylcarbinyl or cyclobutyl chloride from allylcarbinyl chloride was observed under the reaction conditions, polymerization was the only reaction observed.
4. Rearrangement of cyclobutyl chloride to cyclopropylcarbinyl chloride was effected at 20 to 40°C in nitromethane solution but was also accompanied by substantial polymerization, resulting in the formation of brown polymers.

---

*The difference between 107% and 100% was attributed either to experimental error or to some addition of $H^{38}Cl$ from the Lucas reagent to the double bond.

5. Arrhenius plots showed activation energies of 18.7 and 20.4 kcal for the rearrangement of cyclopropylcarbinyl chloride to cyclobutyl chloride and to allylcarbinyl chloride, respectively, and 27.3 kcal for that of cyclobutyl chloride to allylcarbinyl chloride. On the other hand, an estimated value of 45 kcal of activation energy was derived for the rearrangement of cyclobutyl chloride to cyclopropylcarbinyl chloride.
6. The relative rate constants for rearrangement were $1:6 \times 10^{6}:4.5 \times 10^{-3}$ for cyclopropylcarbinyl-cyclobutyl-allylcarbinyl chlorides at 40°C.

As to the mechanisms of rearrangement, the authors pointed out that, contrary to expectation, the evidence obtained in their work indicated that at −40 to −20°C, cyclopropylcarbinyl chloride rearranges to allylcarbinyl chloride through an S_N2' mechanism, whereas cyclobutyl chloride rearranges to allylcarbinyl chloride through both S_N2' and S_N1' mechanisms, with the latter preferred in more polar media. The overall reaction pathways suggested for these isomerizations are illustrated in scheme 5.

In this mechanism the rearrangement of cyclopropylcarbinyl chloride was visualized as taking place by the attack of a molecule of the catalyst on the alkyl chloride-aluminum chloride complex. In the structure resulting from this attack, the cleaving of a C-C σ bond is compensated by the forming of a C-C σ bond and also by the partial release of ring strain.

It is noteworthy to mention that in the same investigation, Olah and Lin [151] also studied the rearrangement of their chlorides with zinc chloride as catalyst. As in the reactions with aluminum chloride, the results with zinc chloride showed a constant decrease of the cyclobutyl chloride concentration after it reached a maximum, as well as a constant increase of allylcarbinyl

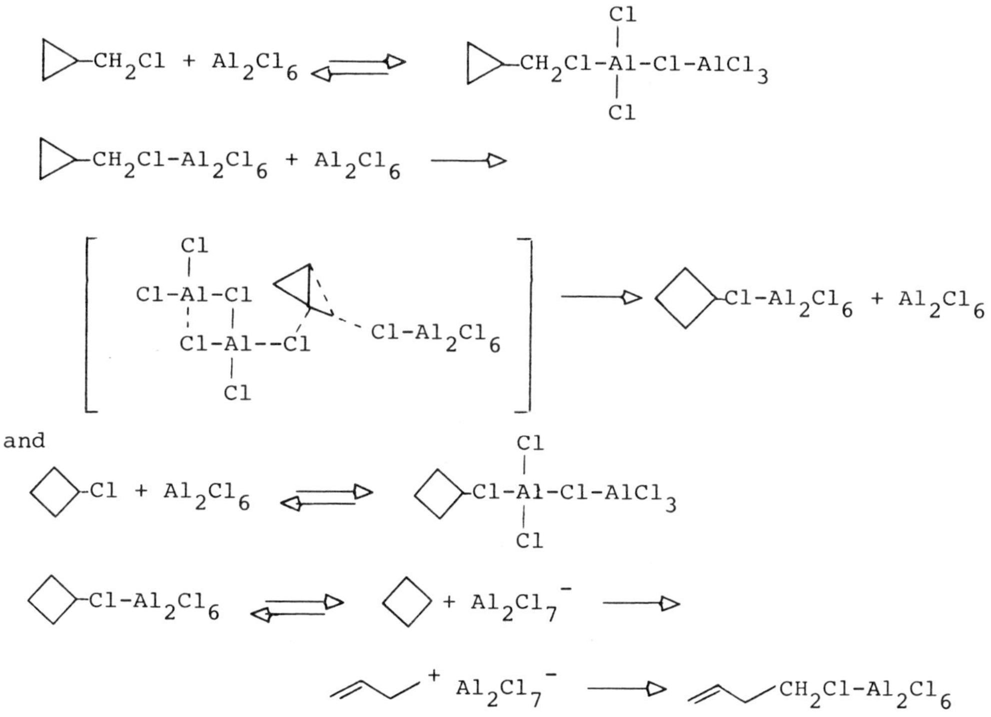

$$\triangleright\text{-}CH_2Cl + Al_2Cl_6 \rightleftharpoons \triangleright\text{-}CH_2Cl\text{-}AlCl_2\text{-}Cl\text{-}AlCl_3$$

$$\triangleright\text{-}CH_2Cl\text{-}Al_2Cl_6 + Al_2Cl_6 \longrightarrow$$

$$\left[Cl_3Al \cdots Cl\text{-}AlCl_2\text{-}Cl \cdots \triangleleft \cdots Cl\text{-}Al_2Cl_6 \right] \longrightarrow \diamond\text{-}Cl\text{-}Al_2Cl_6 + Al_2Cl_6$$

and

$$\diamond\text{-}Cl + Al_2Cl_6 \rightleftharpoons \diamond\text{-}Cl\text{-}AlCl_2\text{-}Cl\text{-}AlCl_3$$

$$\diamond\text{-}Cl\text{-}Al_2Cl_6 \rightleftharpoons \diamond^{+} + Al_2Cl_7^{-} \longrightarrow$$

$$CH_2{=}CH\text{-}CH_2\text{-}CH_2^{+}\ Al_2Cl_7^{-} \longrightarrow CH_2{=}CH\text{-}CH_2\text{-}CH_2Cl\text{-}Al_2Cl_6$$

Scheme 5

chloride concentration. This again implied that these reactions are consecutive ones in which cyclopropylcarbinyl chloride rearranges quickly to cyclobutyl chloride, which in turn rearranges slowly to allylcarbinyl chloride. It was pointed out that the mechanistic features are similar to those of the aluminum chloride-catalyzed rearrangement.

2. Cyclopentyl and higher cycloalkyl rearrangements

The reaction of cyclopentylcarbinyl, 3-methylcyclopentylcarbinyl, cyclohexylcarbinyl, and cyclohelptyl bromides with the Lucas reagent was examined [152,153]. Cyclopentylcarbinyl bromide (23) gave, after 16 hr at 50°C, 35.1% starting bromide, 0.7% cyclohexene, and 64% cyclohexyl chloride [152]. Similar ring expansion was also noted during the solvolysis of cyclopentylcarbinyl tosylate [154]. The observation that cyclohexyl bromide under

CH_2Br (23) $\xrightarrow[16\ hr,\ 50°]{ZnCl_2\text{-}HCl}$ CH_2Br (35.1%) + (0.7%) + Cl (64%)

these conditions exchanged only ca. 8% of its Br for Cl, coupled with the absence of cyclopentylcarbinyl chloride and cyclohexyl bromide, suggested that the Lucas reaction proceeded through simultaneous ring expansion and nucleophilic substitution at the carbocation center.

Treatment of 3-methylcyclopentylcarbinyl bromide (24) and cyclohexylcarbinyl bromide (25) with Lucas reagent under similar conditions gave the results summarized in Eqs. (42) and (43), respectively [152]. Cycloheptyl bromide and the Lucas reagent, at 55°C in 10 hr, underwent a rearrangement of the halide with ring contraction to the extent of 25.2%. Analysis of the

CH_2Br, CH_3 (24) $\xrightarrow[16\ hr,50°]{ZnCl_2\text{-}HCl}$ Cl, CH_3 (*trans*, 1.5%) + Cl, CH_3 (*cis*, 6.5%; *trans*, 7.9%) + Cl, CH_3 (*cis*, 6.7%; *trans*, 6.0%) + CH_3 (7.4%) (42)

CH_2Br (25) $\xrightarrow[16\ hr,50°]{ZnCl_2\text{-}HCl}$ Cl, CH_3 (*cis*, 0.6%; *trans*, 1.5%) + Cl, CH_3 (*cis*, 1.6%; *trans*, 0.8%) + Cl, CH_3 (*cis*, 0.5%; *trans*, 1.0%) + CH_3 (3.7%) (43)

reaction mixture by GLPC showed the distribution of products shown in Eq. (44) [153]. Although the mechanism of formation of secondary and ter-

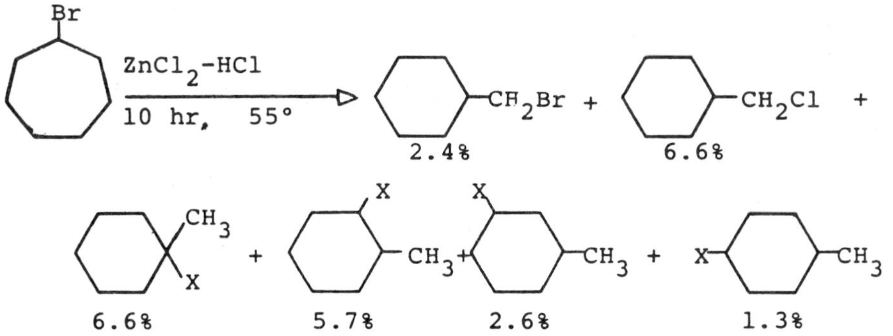

(44)

X = Cl and/or Br

tiary halomethylcyclohexanes was said to be uncertain, the primary halides were suggested to be formed as follows:

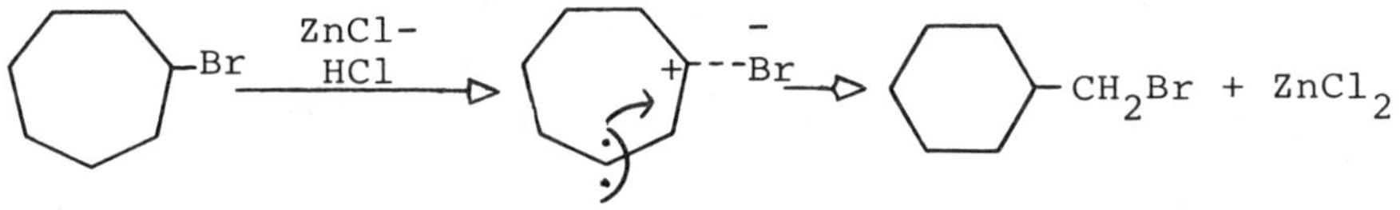

Since halogen exchange in the cyclohexylcarbinyl halides under the employed conditions was shown to be nil, it was suggested that the ring contraction process resulted to a considerable degree in the ion pair and was accompanied by return attack of the bromide ion. It is to be emphasized, however, that the ring contractions observed in Eq. (44) can be explained in terms of protonated cyclopropane intermediates (cf. Eq. 124 and Sec. III).

Dihalocyclohexanes have also been isomerized in the presence of metal halide catalysts. In 1957, Goering and Sims [155] observed the formation of *trans*-1,4-dibromocyclohexane in the course of HBr addition to 1-chlorocyclohexene (26) in the presence of $FeCl_3$. They attributed this to a series of reactions involving addition, halogen exchange, and isomerization (Eq. 45). Evidence for this proposition was found in a number of experiments, including the isomerization of *trans*-1,2-dibromocyclohexane to *trans*-1,4-dibromocyclohexane in the presence of $FeCl_3$-HBr.

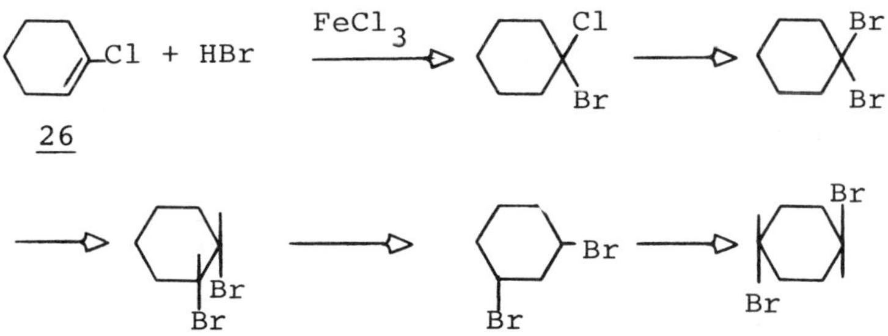

(45)

In 1959, Nenitzescu and Dinulescu [156] found that treatment of *trans*-1,2-dibromocyclohexane with $AlBr_3$ in CS_2 at 0°C gave 70 to 87% yield of an equilibrium mixture containing ca. 25% crystalline *trans*-1,4-dibromocyclohexane and 75% liquid products, mostly *trans*-1,3-dibromocyclohexane. This same equilibrium composition was also obtained from 1,1-, 1,3-, and *trans*-

1,4-dibromocyclohexane. The reaction was explained in terms of carbocation formation and hydride migration (Eq. 46).

Br, Br $\xrightarrow[CS_2, 0°]{AlBr_3}$ Br, + $\xrightarrow{\sim H:}$ Br, + $\underset{AlBr_3}{\overset{AlBr_4^-}{\rightleftarrows}}$ Br, Br

$\updownarrow$ $\sim H:^-$

Br, Br $\xrightarrow[AlBr_3]{AlBr_4^-}$ Br, +

(46)

More recently, Nozaki et al. [157] carried out the isomerization of *trans*-1,2-dichlorocyclohexane with $AlCl_3$ in CS_2 at ambient temperature. The dichlorocyclohexane fraction obtained in ca. 55% yield was found by GLPC and NMR to consist of *trans*-1,2-, *cis*-1,4-, *trans*-1,4-, *cis*-1,2-, and 1,3-dichlorocyclohexanes in a ratio of 1:0.99:1.13:0.48:0.25.

Rearrangement of larger polycyclic hydrocarbons related to bornane, norbornane, adamantane, protoadamantane, and diadamantane (congressane) structures were also effected by Lewis acid catalysts. These will be presented in connection with similar rearrangements induced by superacid catalysts near the end of this chapter.

II. REARRANGEMENTS OF ALKYLATING AGENTS AND RELATED COMPOUNDS IN THE PRESENCE OF SUPERACIDS

Needless to say, the field of carbocation chemistry has witnessed its greatest advances in the past two decades. This indeed can be traced to two major factors. First, the utilization of strong proton acids, or *superacids*, not only as catalysts for carbocation generation under mild conditions, but also as media for measurements leading to accurate information about the structure, rearrangement, and stability of carbocations. Second, the efficient application of instrumental techniques such as proton and carbon-13 nuclear magnetic resonance and infrared, Raman, ultraviolet, and ESCA spectroscopy as means of analysis in carbocation chemistry.

Accompanying this development of experimental techniques and findings there have also been parallel achievements in theory and understanding of carbocation chemistry. The continuous accumulation of results and interpretations has inspired the preparation of a number of reviews [151,158-161] and monographs [162,163] covering various aspects of the subject. In this short section we present only some of the recent results related to rearrangements of Friedel-Crafts alkylating agents under the catalytic influence of strong acid media under stable ion conditions. In discussing these rearrangements, we shall concentrate primarily on reactions induced by acids that are of current interest as "super" [164] or "magic" [159,165,166] acid systems. The initial systems included SbF_5, HSO_3F, SbF_5-HSO_3F, HF-SbF_3 and HF-SbF_5, either neat or in SO_2, SO_2ClF, and other relatively nonbasic solvents. However, as we advance through this section, we shall see that other superacid systems have been designed. Examples of these are $AlBr_3$-SO_2ClF, HF-TaF_5, HCl-$AlCl_3$, HBr-$AlBr_3$, perfluorinated sulfonic acid polymers (e.g., Nafion-H),* and organometallic superacids (e.g., complexes of $AlCl_3$ with polystyrenesulfonic acid).

For convenience, the material in this section is discussed under three headings: rearrangements in acycloalkyl systems, rearrangements in cycloalkyl and related cycloalkylcarbinyl systems, and mechanistic aspects.

A. Rearrangements of Acyclic Systems

1. Rearrangements of acyclic monohalides

The generation of carbocations from alkyl halides in the presence of superacids has been a subject of great concern in recent years. For example, in 1972, Olah and his co-workers [31] presented their findings on the nature of the complexes formed when methyl fluoride and labeled and nonlabeled ethyl fluorides were dissolved in the superacid media SbF_5, SbF_5-SO_2, and SbF_5-SO_2ClF. Using ^1H, ^{13}C, and ^{19}F NMR in combination with laser Raman spectroscopy for analysis, they concluded that in SbF_5-SO_2, CH_3F formed a tightly bound donor-acceptor complex, $CH_3F \rightarrow SbF_5$, which underwent rapid intramolecular fluorine exchange. Ethyl fluoride also formed a similar donor-acceptor complex, but showed a greater tendency to ionize in the medium than did methyl fluoride.

Besides fluorine exchange, there was also observed an additional exchange process with ethyl fluoride which resulted in rapid scrambling of hydrogen and carbon in the ethyl group. This rapid intramolecular scrambling of hydrogen and carbon of ethyl fluoride in excess SbF_5-SO_2, $FSO_3H(D)$-SbF_5-SO_2, and H(D)F-SbF_5-SO_2 was explained in terms of equilibrating 1,2 hydride shifts taking place through an intimate ion pair 27 (Eq. 47).

$$^*CH_3CH_2F + SbF_5 \rightleftarrows \underset{\mathbf{27}}{^*CH_3\overset{+}{C}H_2\,\overset{-}{SbF_6}} \rightleftarrows CH_3\overset{+}{C}H_2\,\overset{-}{SbF_6} \qquad (47)$$

The scrambling of both carbon and hydrogen, coupled with the observed fast intramolecular and slow intermolecular proton exchange, were said to be consistent only with the presence of an intermediate ethyl cation in the intimate ion pair. By analogy to other systems [167], Olah et al. [31] preferred the classical equilibrating system 28 over the bridged nonclassical system 29 to represent the ethyl cation $[CH_3CH_2]^+$.

$$\underbrace{^*CH_3CH_2^+ \rightleftarrows CH_3{}^*CH_2^+}_{\mathbf{28}}$$

$$\underbrace{\left[\overset{H}{CH_2\text{—}CH_2}\right]^+ \text{ or } \left[\overset{H}{CH_2\uparrow CH_2}\right]^+}_{\mathbf{29}}$$

---

*Nafion is a registered trademark for the perfluorinated resinsulfonic acid polymer produced by the DuPont Company.

At this point it is interesting to note that these results, which offered the first direct evidence for the intermediacy of the ethyl cation in solution, were later confirmed by other workers [167a]. Moreover, these results showed that $CH_3F \rightarrow SbF_5$ and $CH_3CH_2F \rightarrow SbF_5$ were so reactive as to O-alkylate solvent SO_2, forming the O-alkylated complexes $CH_3\text{-}O^+SO$ and $C_2H_5\text{-}O^+SOClF$, respectively. On the basis of this and similar observations, Olah and co-workers stated that the alkylating power of methyl and ethyl fluoroantimonates toward various types of bases, including the π-based alkenes, and arenes, was unmatched by that observed for any previously reported methylating or ethylating agents.

More recently, Olah and Donavan [32] reported in full a systematic investigation of the complexation and ionization of some C_1 to C_5 alkyl halides with SbF_5 and AsF_5 in SO_2, SO_2ClF, SO_2F_2, and CH_2F_2 as studied by ^1H, ^{13}C, and ^{19}F NMR spectroscopy. The results suggested that complexation (or ionization) was dependent on the stability of the species, the strength of the catalyst, and the nucleophilicity of the solvent. Thus in SO_2 (the most nucleophilic of the solvents studied), methyl, ethyl, and isopropyl halides formed O-alkylated complexes ($R\text{-}O^+{=}S{=}O$; $R{=}CH_3$, C_2H_5, $i\text{-}C_3H_7$), but *t*-butyl, *t*-pentyl, and cyclopentyl halides formed *t*-butyl, *t*-pentyl, and cyclopentyl cations, respectively. This fact, coupled with others, suggested the following order of stability for isopropyl, *t*-butyl, and cyclopentyl cations:

$$t\text{-}C_4H_9^+ > c\text{-}C_5H_{11}^+ > i\text{-}C_3H_7^+$$

The effect of the catalyst strength on the stability of the resulting donor-acceptor complexes was also evident from the results of the work. For example, the $CH_3F \rightarrow SbF_5$ complex was much stronger and more polar than the $CH_3F \rightarrow AsF_5$ complex. The stability and polarity of the complexes were directly related to the strength of the Lewis acid ($SbF_5 > AsF_5$). More interestingly, the results of this paper demonstrated several facts about the ethyl systems. For example, it was shown that there was no intermolecular hydrogen-deuterium exchange of the ethyl fluoride-SbF_5-SO_2 complex with DSO_3F-SbF_5 and DF-SbF_5 at −78°C, but when the temperature of the solution was raised, deuterium incorporation could be detected. Moreover, labeling experiments with CD_3CH_2F and 90% ^{13}C-enriched $^{13}CH_3CH_2F$ in SbF_5-SO_2 solution at -78°C showed an approximately statistical distribution of the label in the ethyl group of the complex. These results implied that both intermolecular and intramolecular exchanges were taking place but at different rates, the latter being much faster than the former at low temperatures. It was thus concluded that the exchanges took place through the O-ethylated sulfur dioxide (30). Since no intermolecular alkylation between 32 and 33 was observed until the

$$\underset{\underline{30}}{CH_3CH_2\text{-}\overset{+}{O}=S=O} \rightleftarrows \underset{\underline{31}}{[CH_3CH_2^+\cdots O)=S=O]} \underset{+SO_2}{\overset{-SO_2}{\rightleftarrows}}$$

$$\underset{\underline{32}}{[CH_3CH_2^+]} \underset{+H^+}{\overset{-H^+}{\rightleftarrows}} \underset{\underline{33}}{CH_2=CH_2}$$

temperature was raised, it was also concluded that the intramolecular exchange noted with CD_3CH_2F and $^{13}CH_3CH_2F$ cannot proceed through an entirely free ethyl cation. "If it did, one would expect approximately the same rates for both the intramolecular and intermolecular exchange." Since this was not the case, the intramolecular exchange was said to be best explained through the solvated cation 31.

The effect of superacids on other methyl and ethyl halides (chlorides, bromides, iodides) was similarly investigated. With these halides, the major reaction involved a self-condensation of alkyl halides to symmetrical dialkylhalonium ions $R\text{-}X^{+}\text{-}R$ (X = Cl, Br, I) [168-173]. This reaction was suggested to proceed by a mechanism analogous to that of the acid-catalyzed condensation of alcohols to ethers in which a primarily formed hydridohalonium ion intermediate ($R\text{-}X^{+}\text{-}H$) undergoes subsequent nucleophilic attack by excess alkyl halide [173]:

$$H^{+} + CH_3X \rightleftharpoons CH_3\text{—}\overset{+}{X}\text{—}H \;(\leftarrow :\ddot{X}\text{—}CH_3) \rightarrow CH_3\overset{+}{X}CH_3 + HX \qquad X = Cl,\ Br,\ I$$

Similar to methyl and ethyl fluorantimonates [31], methyl-, ethyl-, and other higher dialkylhalonium ions were found to be effective alkylating agents for a variety of n and π bases [168,170,172]:

$$R\text{-}\overset{+}{X}\text{-}R + ArH \longrightarrow Ar\text{-}R + R\text{-}X + H^{+}$$

$$R\text{-}\overset{+}{X}\text{-}R + (R')_n\text{-}Y \longrightarrow (R')_n\text{-}Y\text{-}R^{+} + R\text{-}X$$

(R = Me, Et; R' = alkyl or aryl; Y = $-\ddot{O}-$, $-\ddot{S}-$, $-\ddot{N}-$. n = 1,2,3)

More recently, the behavior of methyl, ethyl, propyl, and benzyl halides in the presence of $HF\text{-}TaF_5$, $HCl\text{-}AlCl_3$, and $HBr\text{-}AlBr_3$ strong acid systems was investigated by Siskin and co-workers [173a, 173b, 173c]. With alkyl chlorides the major products were those formed by direct reduction to the corresponding alkanes [173a,173b]. These reductions involved hydride abstraction from the dialkylhalonium ion by a second alkyl halide molecule, as shown in the following equation:

$$2RCH_2Cl \underset{}{\overset{HX}{\rightleftharpoons}} RCH_2\text{-}\overset{+}{Cl}\text{-}CH_2R\cdot X^{-} \xrightarrow{RCH_2Cl} RCH_2\text{-}H + RCHCl\text{-}\overset{+}{Cl}\text{-}CH_2R\cdot X^{-}$$

$X^{-} = TaF_3Cl^{-},\ AlCl_4^{-},\ AlBr_3Cl^{-}$

$R = Me, Et, Pr, PhCH_2$

The results of both Olah and Siskin demonstrated that: (1) in these strong acid systems, the two major reactions taking place after ionization are self-condensation and alkylation of added basic substrates; and (2) the ability to carry out alkylations using primary alkyl carbocations as reagents depends on the nature of the ion generated. With chlorides, where it is possible to form a halonium ion intermediate, alkylation is inhibited in favor of direct reduction to the corresponding alkane. With fluorides, which do not readily form halonium ions, other competing reactions predominate [173c].

However, in the presence of added alkanes, the alkylating ability of alkyl fluorides could be maximized. Thus when methyl fluoride reacted with HF-TaF_5 in the presence of ethane, the composition of the gas phase after 2 hr of reactions, excluding reactants, was methane (73.1%), propane (23.5%), *n*-butane (3.4%), and hydrogen (<0.1%). No alkylation reaction products were present in a similar reaction with methyl chloride and ethane [173a].

In other works Olah and his co-workers investigated the generation of carbocations from β-arylethyl chlorides in SbF_5 solutions. In 1967, Olah et al. [174] succeeded in showing that β-anisylethyl, β-mesitylethyl, and β-(pentamethylphenyl)ethyl chlorides underwent ionization in SbF_5-SO_2 solution at -60°C via strong aryl participation to form the corresponding bridged arylonium ions. In 1970, Olah and Porter [175] described the preparation of the parent ethylene phenonium ion and ethylene-*p*-toluonium ion through treatment of the corresponding chlorides with SbF_5 in SO_2ClF at -78°C.

X-C6H4-CH_2-CH_2Cl —(SbF_5 / SO_2ClF)→ [bridged phenonium ion, H_2C—CH_2] + X-C6H4-C^+(H)(CH_3)

X = H, CH_3

In both cases spectral and chemical evidence revealed the concurrent formation of substituted styryl cations. In 1971, Olah and Porter [176] elaborated on their previous work and reported new results relating to the nature of the cations generated in SbF_5-SO_2ClF solution from the β-(2,4-dimethylphenyl) ethyl chloride, 34, the 2,3-dimethyl-3-(*p*-X-phenyl)-2-butyl system, 35, the 3,3-dimethyl-2-(*p*-X-phenyl)-2-butyl system, 36, and the 3-aryl-2-butyl system, 37. In SbF_5-SO_2ClF at -78°C, β-(2,4-dimethylphenyl)ethyl

2,4-(CH_3)$_2$C6H3-CH_2-CH_2-Cl

34

X-C6H4-C(CH_3)$_2$-C(CH_3)$_2$-Cl,OH

35, X=H,Br,CF_3, CH_3,OCH_3

X-C6H4-C(CH_3)(Cl(OH))-C(CH_3)$_3$

36, X=H,Br,CF_3, CH_3,OCH_3

CH_3-CH(ArX)-CH(Cl,OH)-CH_3

37, *p*-X=H,Br,CF_3

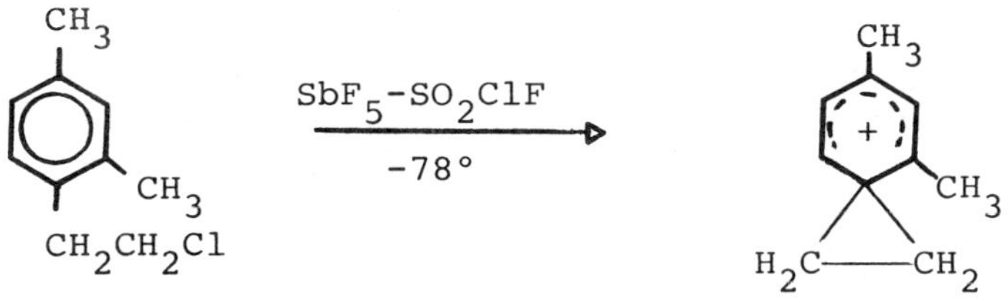

$\xrightarrow[-78°]{SbF_5-SO_2ClF}$

34 38

chloride (34) gave the unsymmetrical ethylenarenium ion, 38. Under similar conditions, ionization of either 35 or 36 gave rise to the set of equilibrating ions shown in scheme 6. In this case, the electronic nature of the *para* substituent controlled which of the ions 39-41 was the most stable; with X = OCH_3 or CH_3, ion 41 was found to be the most stable, but with X = H, Br, or CF_3, bridged ion 39 was the most stable.

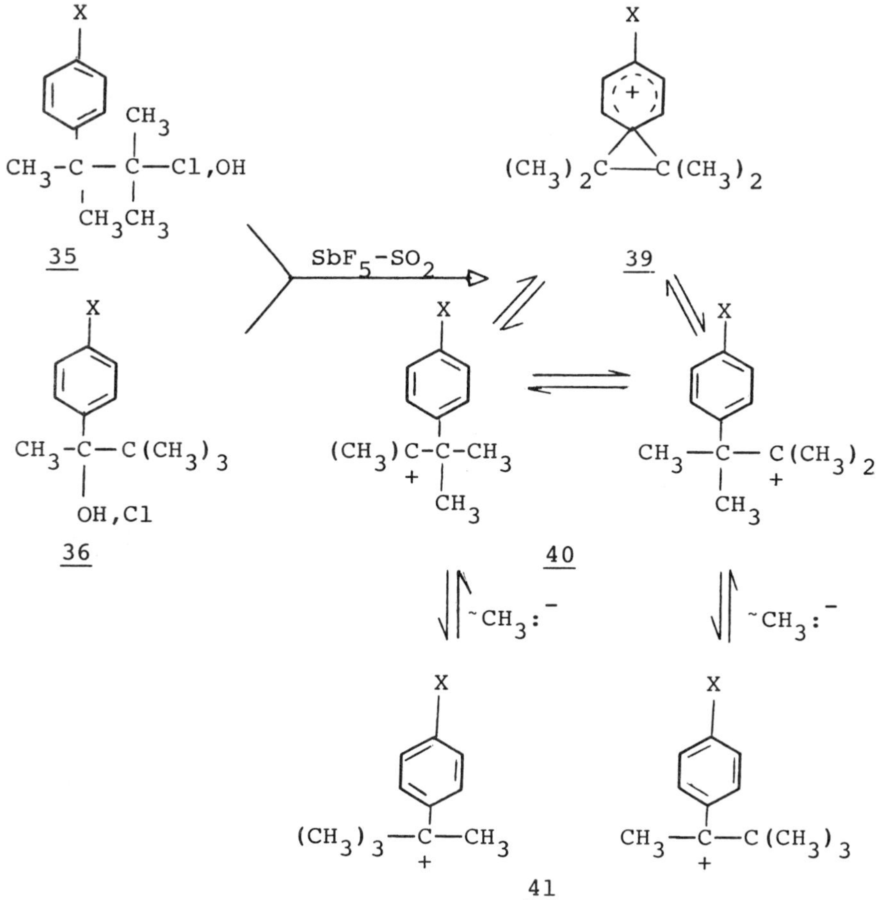

SbF_5-SO_2

35 36 39 40 41

$\sim CH_3:^-$

X = H, Br, CF_3, CH_3, OCH_3

Scheme 6

In contrast to the behavior of 34 and 36, ionization of the 3-aryl-2-butyl system 37 did not form aryl-bridged ions 42 but led to the formation of benzylic ions 43. This is because of the rapid rearrangement of the initially formed secondary ions to the more stable benzylic ions 43.

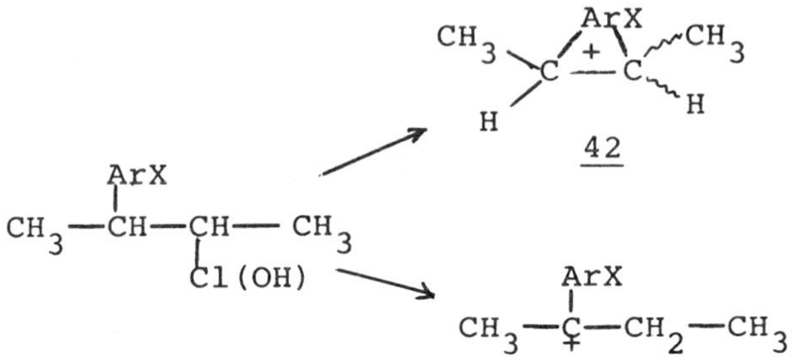

$$CH_3-CH(ArX)-CH(Cl(OH))-CH_3 \rightarrow \underset{42}{\text{aryl-bridged ion}} \quad ; \quad \rightarrow \underset{43}{CH_3-\overset{+}{C}(ArX)-CH_2-CH_3}$$

p-X = H,Br,CF_3

Apart from the preceding results, several other reports appeared describing the generation of carbocations from propyl and higher alkyl halides under a variety of conditions. For example, Olah et al. [177] described the ionization of propyl, butyl, and pentyl fluorides in the presence of SbF_5-SO_2 solution at 18°C. Under these conditions, it was shown by NMR analysis of both labeled and unlabeled alkyl fluorides that both n-propyl and isopropyl fluoride yielded only the secondary isopropyl cation (Eq. 48), all four isomeric butyl fluorides gave the t-butyl cation (Eq. 49), and all seven isomeric pentyl

$$C_3H_7F + SbF_5 \rightleftarrows CH_3\overset{+}{C}HCH_3SbF_6^- \quad (48)$$

$$C_4H_9F + SbF_5 \rightleftarrows CH_3\overset{+}{C}(CH_3)CH_3 \; SbF_6^- \quad (49)$$

$$C_5H_{11}F + SbF_5 \rightleftarrows CH_3\overset{+}{C}(CH_3)CH_2CH_3 \; SbF_6^- \quad (50)$$

fluorides gave the t-pentyl cation in the form of their hexafluorantimonate salts. This indicated that under the applied conditions, complete isomerization to the most thermodynamically stable carbocation took place.

The chemical reactivity of the prepared alkyl hexafluoroantimonate salts was also investigated in hydrolysis and in reactions with alcohols, H_2S, mercaptans, amines, acids, and aromatic hydrocarbons. Aside from by-products resulting from multiple side reactions, the main reaction paths for t-butyl hexafluoroantimonate are shown in scheme 7.

Behavior exactly analogous to that of Eqs. (44) to (46) was discovered later by Kramer et al. [178], but in a different catalyst system consisting of $AlBr_3$ in SO_2ClF. This new system, which was said to possess many of the properties of the SbF_5 systems, induced the generation of t-butyl and t-pentyl ions from butyl and pentyl halides, respectively, at -50°C. The C_7 and C_8 alkyl halides yielded mixtures of tertiary cations. The NMR spectra of solutions of C_4 to C_7 alkyl halides were the same as those found with SbF_5 in many solvents. Moreover, the ions in the $AlBr_3$-SO_2ClF system underwent extremely rapid intermolecular hydride transfer reactions with hydride donors, thus substantiating the existence of ionic intermediates.

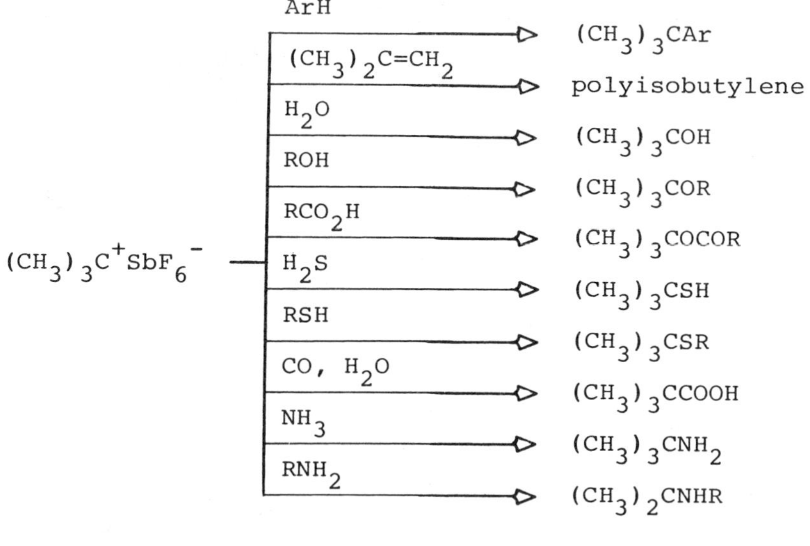

Scheme 7

In a subsequent paper, Kramer et al. [179] found four other new low-temperature acid systems which could also generate and stabilize higher concentrations of tertiary cations. These are $AlBr_3$-CH_2Cl_2, $AlBr_3$-CH_2Br_2, $GaCl_3$-CH_2Cl_2, and $GaCl_3$-1,2-$Cl_2C_2H_4$. In these systems, NMR data indicated that $AlBr_3$ formed $R^+Al_2Br_6X^-$ salts while $GaCl_3$ formed both $R^+Ga_2X_7^-$ and $R^+GaX_4^-$. According to a scale based on equilibrium measurements,* the ability of these aprotic media to stabilize cations was shown to increase in the order $AlBr_3$-CH_2Br_2 = $GaCl_3$-CH_2Cl_2 < $AlBr_3$-SO_2ClF < $AlBr_3$-CH_2Cl_2.

In the continued effort to prepare *sec*-alkyl cations under stable ion conditions, Saunders et al. [182-184] investigated the ionization of alkyl halides in SbF_5-SO_2ClF in a vacuum line at low temperatures. Under these conditions, isopropyl [183] and *sec*-butyl [184] chlorides gave isopropyl and *sec*-butyl cations, respectively, but like *t*-pentyl chloride [182], both 1- and 2-chloropentane gave only *t*-pentyl cation [184]. NMR studies of the resulting cations showed that the isopropyl cation was fairly stable below 50°C and exhibited interchange between its two types of protons over the range 0 to 40°C. This interchange, which was attributed to reversible rearrangments to *n*-propyl cation, was suggested to occur via protonated cyclopropane. On the other hand, *sec*-butyl cation was characterized by a rapid 2,3-hydride interchange even at -112°C and underwent rapid conversion to *t*-butyl cation above -40°C. Again, it was assumed that these processes proceed via protonated methylcyclopropanes, as suggested by Brouwer and Oelderik [142, 185] to explain the acid-induced isotopic interchange observed in *n*-butane [142,185].

The observation of *sec*-butyl cation by Saunders et al. [184] was very interesting. In fact, this was the first successful observation of this cation, as Olah et al. reported that its attempted preparation from 2-fluorobutane

---

*Notably, Kramer [180,181] developed a scale to rank strong acids via a selectivity parameter. Comparison of the data indicated that $AlBr_3$-HBr apparently provided the most stabilizing acid medium.

Table 4. Trapping Butyl Ions with Methylcyclopentane, −50°

Butyl chloride	Kinetically controlled products, %	
	n-Butane	Isobutane
1-Chlorobutane	>99.5	<0.5
2-Chlorobutane	99.4	0.6
Isobutyl chloride	18.6	81.4
t-Butyl chloride	0	100

with SbF_5 [177], *n*-butane with SbF_5-HSO_3H [186], or *sec*-butyl alcohol with SbF_5-HSO_3H [187] led directly to *t*-butyl cation.

In 1969, Kramer [188] reported a study in which he used methylcyclopentane to trap the cations generated by the action of SbF_5-HSO_3F on butyl, pentyl, and hexyl halides, both before and after equilibrium distribution of the cations was reached. His results indicated that tertiary cations predominated at equilibrium in these systems. Moreover, he also demonstrated that under kinetic control tertiary ions could be trapped with little or no rearrangement, whereas secondary and primary cations had greater tendencies to isomerize. These conclusions are supported by the data of Tables 4 to 6, which show the products trapped from isomeric butyl, pentyl, and hexyl halides after treatment with the catalyst under conditions favoring kinetic control.

The interesting production of *n*-butane from isobutyl chloride was explained in terms of the intervention of a protonated cyclopropane intermediate which could isomerize to a secondary butyl cation (Eq. 50a). It may be noted

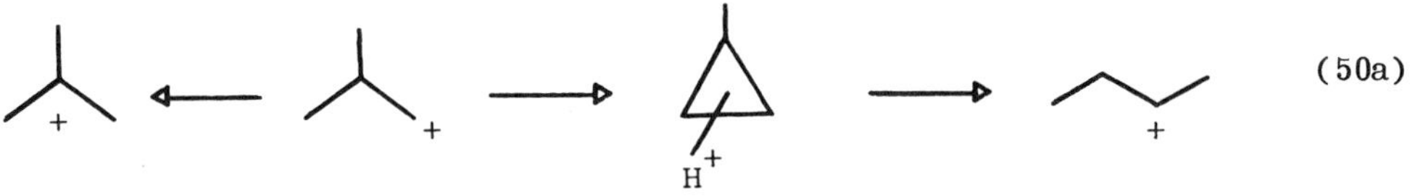

(50a)

Table 5. Trapping Pentyl Ions with Methylcyclopentane, −50°

Pentyl halide	Kinetically controlled product, %	
	n-Pentane	Isopentane
1-Chloropentane	16.7	83.3
2- and 3-Chloropentane	14.8	85.2
1-Chloro-2,2-dimethylpropane	0	100
2-Chloro-2-methylbutane	0	100
1-Bromo-2-methylbutane	0	100

Table 6. Trapping Hexyl Ions with Methylcyclopentane, −50°

Compound	Kinetically controlled products, %				
	n-C_6	2-Me-C_5	3-Me-C_5	2,3-Me_2C_4	2,2-Me_2C_4
1-Chlorohexane	3.4	66.3	30.3	Trace	0
2-Chloro-2-methylpentane	0	97.6	2.4	0	0
2-Chloro-2,3-dimethylbutane	0	0	0	100	0
3-Bromo-3-methylpentane	0	15.3	84.7	0	0
2-Chloro-4-methylpentane	0	88.1	10.5	1.4	0
1-Bromo-4-methylpentane	0	86.7	13.3	0	0
1-Bromo-2-methylpentane	0	80.7	19.3	0	0

here that the steps of Eq. (50a) are reversible, since dehydration of the four isomeric butyl alcohols by H_2SO_4 in H_3PO_4 at 164°C gave mixtures of isobutane and the isomeric butenes [189]. The formation of isobutane from n^- and *sec*-butyl alcohol during dehydration undoubtedly involved a protonated cyclopropane intermediate.

2. Rearrangements of acyclic dihalides

The ionization of dihaloalkanes under the influence of superacids has been studied extensively by Olah and his collaborators [102,146,190-197]. In most of the reactions investigated, assisted ionization of one of the halogens resulted in the formation of stable long-lived three- or five-membered ring halonium ions provided directly or via rearrangements leading to either ring contraction or ring expansion [102,146,190-193,198,199]. As these reactions are numerous and most of their interesting chemistry has already been reviewed [200,201], we present only a few illustrative examples in Eqs. (51) to (59) and in schemes 8 to 10.

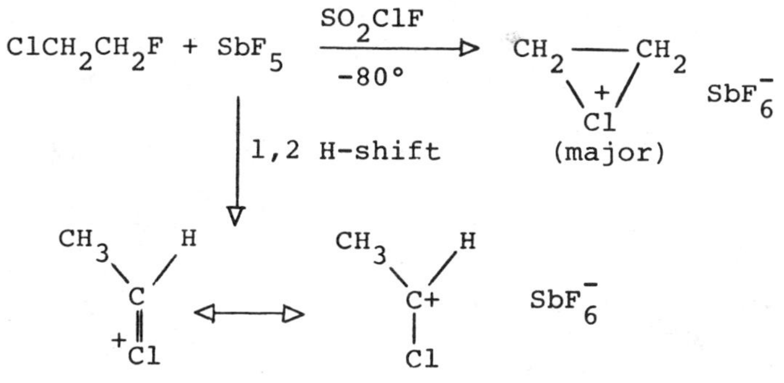

$$ClCH_2CH_2F + SbF_5 \xrightarrow[-80°]{SO_2ClF} \underset{\text{(major)}}{CH_2\text{—}CH_2\ (\backslash{+}/\ Cl)}\ SbF_6^-$$

↓ 1,2 H-shift

$$CH_3(H)C{=}Cl^+ \longleftrightarrow CH_3(H)C^+\text{—}Cl \quad SbF_6^-$$

Scheme 8 [196]

$$Cl(CH_2)_3Cl \text{ or } CH_3CHClCH_2Cl \text{ or } CH_3CH_2CHCl_2 \xrightarrow[-60°]{SbF_5-SO_2ClF} CH_3CH_2\overset{+}{C}HCl \rightleftharpoons CH_3\overset{+}{C}HCH_2Cl \rightleftharpoons CH_3-CH—CH \text{ (bridged } \overset{+}{Cl}\text{)}$$

(51) [146]

$$CH_2=C(X)-CH_2Cl \xrightarrow[-70°]{SbF_5-SO_2} CH_2=C—CH_2 \text{ (bridged } \overset{+}{X}\text{)}$$

(X = Cl, I)

(52) [194]

$$CH_3CH(X)CH_2CH_2I \xrightarrow[-60°]{SbF_5-SO_2} CH_3\overset{+}{C}HCH_2CH_2-I \xrightarrow{\sim H:^-} CH_3CH_2\overset{+}{C}HCH_2-I \longrightarrow CH_3CH_2-CH—CH \text{ (bridged } \overset{+}{I}\text{)}$$

(X-Cl, Br)

(53) [146]

$$CH_2(I)CH(F)CH_2CH_3 \xrightarrow[-78°]{SbF_5-SO_2} CH_2-CH-CH_2CH_3 \text{ (bridged } \overset{+}{I}\text{)}$$

(54) [192]

$$X-CH_2CH_2CH_2CH_2-X \xrightarrow[-60°]{SbF_5-SO_2} \text{(five-membered ring with } \overset{+}{X}\text{)}$$

(X = Cl,Br,I)

(55) [192,199]

$$(Cl)BrCH_2CH_2CH_2CH_2CH_2-Br(Cl) \xrightarrow{-60° [192,199]}$$

or

$$CH_3CH(Cl)CH_2CH_2CH_2Br \xrightarrow{-60° [192]}$$

or

$$CH_3CH(Br)CH_2CH(Br)CH_3 \xrightarrow{-40° [146]}$$

$$\xrightarrow{SbF_5-SO_2} \text{(2-methyl five-membered ring with } \overset{+}{Br}\text{)}$$

(56)

$$Cl(CH_2)_6Cl \xrightarrow{SbF_5-SO_2} \text{(2,2-dimethyl five-membered ring with } \overset{+}{Cl}\text{)}$$

(57) [199]

$$X-CH(X)CH_2CHCH_2CH_2CH_2X \xrightarrow[-60°]{SbF_5-SO_2} \text{(five-membered ring with } \overset{+}{X}\text{, substituted with } CH_2X\text{)}$$

(X = Cl,Br)

(58) [199]

$$(CH_3)_3C—CCl_2CH_3 \xrightarrow[-60°]{SbF_5-SO_2} (CH_3)_3C—\overset{+}{C}(Cl)CH_3 \xrightarrow{\sim CH_3:^-} (CH_3)_2C—C(CH_3)_2 \text{ (bridged } \overset{+}{Cl}\text{)}$$

(59) [190]

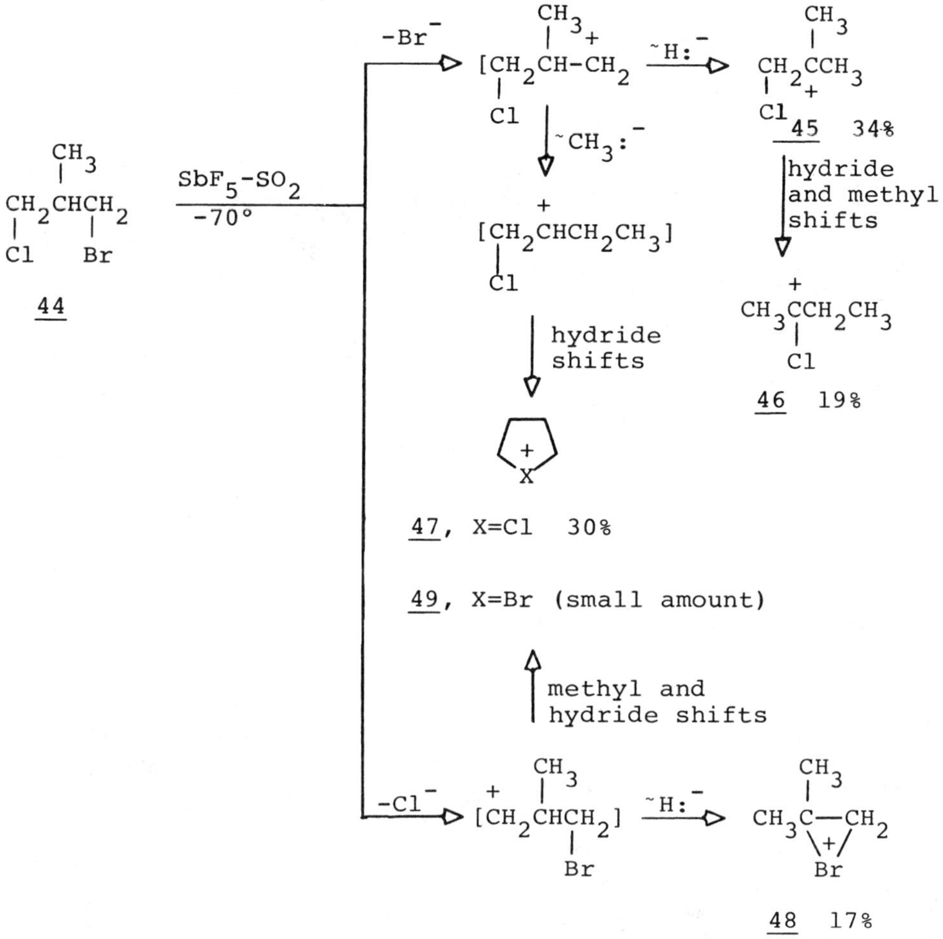

$$CH_2(Cl)CH(CH_3)CH_2Br \;(\underline{44}) \xrightarrow[-70°]{SbF_5-SO_2}$$

$\xrightarrow{-Br^-}$ $[CH_2(Cl)CH(CH_3)-\overset{+}{C}H_2$ $\xrightarrow{\sim H:^-}$ $CH_2(Cl)\overset{+}{C}(CH_3)CH_3$ — $\underline{45}$ 34%

$[CH_2(Cl)CH(CH_3)-\overset{+}{C}H_2$ $\xrightarrow{\sim CH_3:^-}$ $[CH_2(Cl)\overset{+}{C}HCH_2CH_3]$ $\xrightarrow{\text{hydride shifts}}$ cyclic X^+ (five-membered ring)

$\underline{45}$ $\xrightarrow{\text{hydride and methyl shifts}}$ $CH_3\overset{+}{C}(Cl)CH_2CH_3$ — $\underline{46}$ 19%

$\underline{47}$, X=Cl 30%

$\underline{49}$, X=Br (small amount)

$\xrightarrow{-Cl^-}$ $[\overset{+}{C}H_2CH(CH_3)CH_2Br]$ $\xrightarrow{\sim H:^-}$ $CH_3C(CH_3)\text{—}CH_2$ bridged by Br^+ — $\underline{48}$ 17%

$[\overset{+}{C}H_2CH(CH_3)CH_2Br]$ $\xrightarrow{\text{methyl and hydride shifts}}$ $\underline{47}$ / $\underline{49}$

Scheme 9 [146]

In 1972, Olah and co-workers [146] investigated the ionization behavior of 1-bromo-3-chloro-2-methylpropane (44) in SbF_5-SO_2 at -70°C. The proton magnetic resonance (PMR) spectrum of the resulting solution showed the presence of four major separate ions (45, 34%; 46, 19%; 47, 30% and 48, 17%) as well as a small amount of 49 (scheme 9). This was the first observation of bromine ionizing more readily than chlorine in a bromo chloride in SbF_5-SO_2 solution. As evident, the preference was nearly 5:1, corresponding to 83% of the products formed by bromide loss and 17% by chloride loss.

More recently, Henrichs and Peterson [202], investigated the ionization of 1,7-dichloropentane (49), 1,5-dichloro-5-methylhexane (50), and 1,4-dichloro-4-methylhexane (51) in SbF_5-SO_2 solution at -78°C. Both the ^1H and the ^{13}C spectra of the ionized solutions obtained from these chlorides indicated the presence of a mixture consisting principally of three cyclic seven-carbon chloronium ions [1,1-dimethylpentamethylenechloronium ion (52), 1-methyl-1-ethyltetramethylenechloronium ion (53), and 1,1,4-trimethyltetramethylenechloronium ion (54)] in equilibrium with their appropriate acyclic tertiary carbocations (scheme 10).

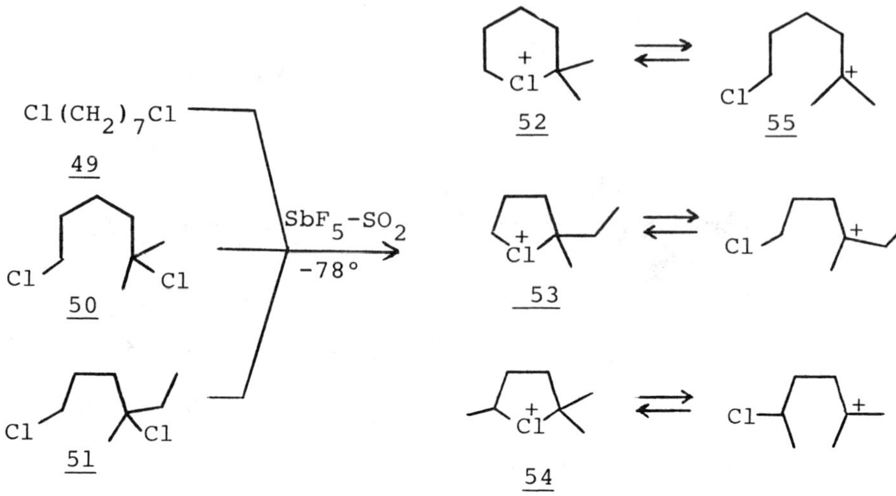

$Cl(CH_2)_7Cl$ 49, 50, 51 → SbF_5-SO_2, −78° → 52 ⇄ 55; 53 ⇄; 54 ⇄

Scheme 10

Notably, the results of Henrichs and Peterson showed that, in accordance with cyclic halonium ion stability order [201,203] the six-membered ring 1,1-dimethylpentamethylenechlorium ion (52) was predominantly opened to the 6-chloro-2-methyl-2-hexyl cation (55), whereas five-membered rings (53 and 54) were all present in the closed form. The observed rearrangement of the dimethylated six-membered ring 1,1-dimethylpentamethylenechloronium ion (52) to the trimethylated five-membered 1,1,4-trimethyltetramethylenechloronium ion (56) was suggested to proceed as follows:

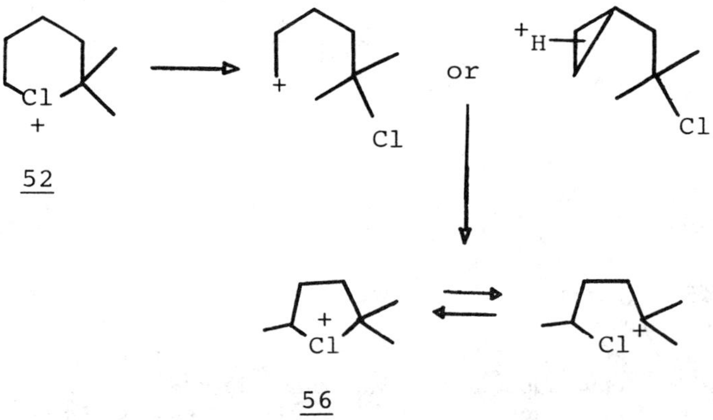

52 → or → 56

It is to be noted that, in all of the investigations of Olah and his coworkers, no fluoronium ions were encountered. Apparently, the only report of a fluoronium ion formation was by Peterson and Bopp [204], who visualized the fluoronium ion, 57, as an intermediate in the reaction of 5-fluoro-1-pentyne and trifluoroacetic acid (Eq. 60).

$\xrightarrow{H^+}$ 57 $\xrightarrow{CF_3COO^-}$ $OCOCF_3$ $\xrightarrow{CF_3COOH}$ CH_3COO, H_3C, F, $OCOCF_3$ (60)

The failure to obtain fluoronium ions can be illustrated by the ionization of 2,3-dihalo-2,3-dimethylbutanes, 58, in SbF_5-SO_2 solution, whereby all halonium ions, 59, with the exception of fluoronium ion were observed [190] (Eq. 61). With fluorine, only rapidly equilibrating open-chain classical fluorinated ions, 60, (Eq. 62), could be seen [190].

58 $\xrightarrow[60°]{SbF_5\text{-}SO_2}$ 59 (61)

58: X = Cl; Y = F,Cl
X = Br; Y = F,Br
X = I; Y = F

59: X = Cl
X = Br
X = I

$\xrightarrow[-90°]{SbF_5\text{-}SO_2}$ 60 (62)

It can be seen from the previous results of this section that the ionization of primary and secondary dihaloalkanes under the influence of superacids showed no evidence for dicarbocationic intermediates. This is probably due to the lack of stabilizing factors. In searching for stable dicarbocations, Olah and co-workers [198,205] studied the ionization of a series of tetramethyl- and tetraphenyl-substituted dihalides and/or diols under stable ion conditions. In these compunds the two functionalized carbons were separated by an ethylidene, *trans*-ethene, *trans*-1,2-dicyclopropyl, or 1,4-phenylene group. In most of the cases examined, both halide (or hydroxyl) groups were ionized to form dicarbocations, although proton elimination, ring opening, and other side reactions led only to monocarbocations in several instances. Some of the reported ionizations of dihalide precursors are shown in Eqs. (63) to (65). The ionization of 2,3-dichloro-2,3-dimethylbutane (61, Eq. 65) represents the only example in Olah's study where halonium ion formation successfully competed with dicarbocation formation.

$$(CH_3)_2C(Cl)CH_2C(Cl)(CH_3)_2 \xrightarrow[-78°]{SbF_5-SO_2ClF} [(CH_3)_2\overset{+}{C}CH_2\overset{+}{C}(CH_3)_2] \xrightarrow{-H^+} (CH_3)_2C\cdots CH \cdots C(CH_3)_2 \; (+) \tag{63}$$

$$(CH_3)_2C(Cl)(CH_2)_nC(Cl)(CH_3)_2 \xrightarrow[-78°]{SbF_5-SO_2ClF} (CH_3)_2\overset{+}{C}(CH_2)_n\overset{+}{C}(CH_3)_2 \tag{64}$$

$n = 2,3$

$$\underset{\underline{61}}{(CH_3)_2C(Cl)—C(Cl)(CH_3)_2} \xrightarrow[-78°]{SbF_5-SO_2ClF} (CH_3)_2C—C(CH_3)_2 \text{ (bridged by } \overset{+}{Cl}\text{)} \tag{65}$$

Examples of the related ionizations of diol precursors are shown in Eqs. (66) to (70).

$$CH_3—C(CH_3)(OH)—C_6H_4—C(CH_3)(OH)—CH_3 \xrightarrow[-78°]{SbF_5-SO_2ClF} CH_3—\overset{+}{C}(CH_3)—C_6H_4—\overset{+}{C}(CH_3)—CH_3 \tag{66}$$

$$Ph_2C(OH)—X—C(OH)Ph_2 \xrightarrow[\text{or } FSO_3H-SbF_5-SO_2ClF,\ -60°]{FSO_3H-SO_2ClF} Ph_2\overset{+}{C}—X—\overset{+}{C}Ph_2 \tag{67}$$

X = CH_2CH_2, *trans* -CH=CH, C≡C, *trans* -CH — CH (bridged by CH_2), $-C_6H_4-$

$(CH_3)_2C(OH)$-CH=CH-$C(OH)(CH_3)_2$ *cis* or *trans*

or

$(CH_3)_2C(OH)$-C≡C-$C(OH)(CH_3)_2$

$\xrightarrow{\text{variety of superacid conditions}}$ complex mixtures of ions (68)

$$\underset{\underline{62}}{\text{cyclopropane-1,2-bis}[C(CH_3)_2OH]} \xrightarrow[SO_2ClF,\ -78°]{FSO_3H} \text{cyclopropane-1,2-bis}[C(CH_3)_2\overset{+}{O}H_2] \qquad (69)$$

$$\underline{62} \xrightarrow[SO_2ClF,\ -78°]{FSO_3H\text{-}SbF_5\text{-}} \underset{\underline{63}}{CH_3C(CH_3)\overset{+}{=\!\!=}CH\text{-}C(H)\text{-}CH\overset{}{=\!\!=}C(CH_3)CH_3}$$

Notably, the ionization of diol 62 (Eq. 69) could not be induced in FSO_3H even by heating to 5°C. However, under the more forcing conditions of FSO_3H-SbF_5 or SbF_5 in SO_2ClF at -78°C, ionization to 2,6-dimethylhepta-2-dienyl cation (63) took place. The formation of cation 63 was attributed to the failure of the cyclopropyl ring to provide sufficient stabilization for the incipient dication, resulting in ring opening and simultaneous proton elimination. It can be seen from Eq. (70) that attempted generation of the dication from 2,3,4-tetramethylpentane-1,3-diol (64) resulted in a quantitive cleavage reaction affording a *t*-hexyl carbocation and protonated acetone.

$$\underset{\underline{64}}{(CH_3)_2C(OH)\text{—}C(CH_3)_2\text{—}C(OH)(CH_3)CH_2} \xrightarrow[SO_2\ -78°C]{FSO_3H\text{-}SbF_5} (CH_3)_2\overset{+}{C}\text{—}C(CH_3)_2\text{—}C(OH)(CH_3)_2 \qquad (70)$$

$$\longrightarrow (CH_3)_2C{=}C(CH_3)_2 + (CH_3)_2C{=}\overset{+}{O}H$$

$$(CH_3)_2C{=}C(CH_3)_2 \xrightarrow{H^+} (CH_3)_2\overset{+}{C}\text{—}CH(CH_3)_2$$

3. Rearrangements of acyclic alcohols, ethers, and related derivatives

In 1967, Olah et al. [187] studied the behavior of methyl, ethyl, *n*-propyl, isopropyl, *n*-butyl, isobutyl, *sec*-butyl, *t*-butyl, *n*-pentyl, isopentyl, neopentyl, *t*-pentyl, *n*-hexyl and neohexyl alcohols in HSO_3F-SbF_5-SO_2 at -60°C to +60°C. At -60°C, all normal and secondary aliphatic alcohols gave stable O-protonated species. However, observation of protonated *t*-butyl or *t*-pentyl alcohols was unsuccessful at this temperature due to a very fast cleavage to the corresponding tertiary carbocations.

Cleavage and rearrangement to the most stable corresponding carbocations also took place with protonated primary and secondary alcohols, but at different temperatures, depending on the structure of the alcohol. For example, protonated *n*-propyl, *n*-butyl, *n*-pentyl, and *n*-hexyl alcohols started to cleave at temperatures higher than 0°C to isopropyl-, *t*-butyl-, *t*-pentyl-, and a mixture of the three possible *t*-hexyl carbocations, respectively. Protonated isobutyl and isopentyl alcohols cleaved above -30°C to *t*-butyl and *t*-pentyl carbocations, respectively, whereas protonated neopentyl and neohexyl alcohols cleaved above -60°C to *t*-pentyl and isomeric *t*-hexyl carbocations, respectively. In general, the more the branching and the longer the chain, the lower the temperature at which cleavage started. The results above are represented diagrammatically in Eqs. (71) to (74).

$$C_3H_7OH \xrightarrow{HSO_3F\text{-}SbF_5\text{-}SO_2} C_3H_7OH_2^+ \longrightarrow CH_3\overset{+}{C}HCH_3 \tag{71}$$

$$C_4H_9OH \xrightarrow{HSO_3F\text{-}SbF_5\text{-}SO_2} C_4H_7OH_2^+ \longrightarrow CH_3\underset{\displaystyle CH_3}{\overset{+}{C}}CH_3 \tag{72}$$

$$C_5H_{11}OH \xrightarrow{HSO_3F\text{-}SbF_5\text{-}SO_2} C_5H_{11}OH_2^+ \longrightarrow CH_3\underset{\displaystyle CH_3}{\overset{+}{C}}CH_2CH_3 \tag{73}$$

$$C_6H_{13}OH \xrightarrow{HSO_3F\text{-}SbF_5\text{-}SO_2} C_6H_{13}OH_2^+ \longrightarrow CH_3\underset{\displaystyle CH_3}{\overset{+}{C}}CH_2CH_2CH_3$$

$$+\ CH_3\underset{\displaystyle CH_3}{CH}\text{—}\underset{\displaystyle CH_3}{\overset{+}{C}}CH_3 \ +\ (CH_3CH_2)_2\overset{+}{C}CH_3 \tag{74}$$

In a manner similar to the behavior of simple alkyl halides and alcohols, aliphatic ethers [206], haloalkyl ethers [207], mercaptans [208], and sulfides [208] were quantitatively protonated in $FSO_3H\text{-}SbF_5$ diluted with SO_2 at -60°C. Cleavage of the onium salts also yielded the corresponding most stable carbocations, but at varying temperatures which were generally higher than those required to cleave protonated alcohols. Examples of the reported rearrangements are shown in Eqs. (75) to (82).

$$CH_3CH_2\underset{\displaystyle CH_3}{CH}\text{-O-}CH_3 \xrightarrow[-60^\circ]{FSO_3H\text{-}SbF_5\text{-}SO_2} CH_3CH_2\underset{\displaystyle CH_3}{CH}\text{-}\overset{\displaystyle H}{\overset{+}{O}}\text{-}CH_3\ \ SbF_5SO_3^-$$

$$\xrightarrow{H^+,\ -30^\circ} CH_3OH_2^+ + [CH_3CH_2\overset{+}{C}H\text{-}CH_3] \longrightarrow (CH_3)_3C^+ \tag{75}$$

[206]

$$CH_3CH_2\underset{CH_3}{\underset{|}{C}}H\text{-}SH \xrightarrow[-60°]{FSO_3H\text{-}SbF_5\text{-}SO_2} CH_3CH_2\underset{CH_3}{C}H\text{-}SH_2^+ \quad SbF_5FSO_3^-$$

$$\xrightarrow{H^+,\ 0°} H_3S^+ + [CH_3CH_2\overset{+}{C}HCH_3] \longrightarrow (CH_3)_3C^+ \tag{76}$$

[208]

$$CH_3CH_2\underset{CH_3}{\underset{|}{C}}H\text{-}S\text{-}\underset{CH_3}{\underset{|}{C}}HCH_2CH_3 \xrightarrow[-60°]{FSO_3H\text{-}SbF_5\text{-}SO_2}$$

$$(CH_3CH_2\underset{CH_3}{\underset{|}{C}}H)_2\overset{+}{S}H \quad SbF_5FSO_3^- \xrightarrow{H^+,\ 70°}$$

$$H_3S^+ + 2[CH_3CH_2\overset{+}{C}HCH_3] \longrightarrow 2(CH_3)_3C^+ \tag{77}$$

[208]

$$CH_3\text{-}\underset{X}{\underset{|}{\overset{CH_3}{\overset{|}{C}}}}\text{—}\underset{OCH_3}{\underset{|}{\overset{CH_3}{\overset{|}{C}}}}\text{-}CH_3 \xrightarrow[\text{or } SbF_5\text{-}SO_2,\ -78°]{FSO_3H\text{-}SbF_5\text{-}SO_2} \begin{cases} (CH_3)_2C\text{—}C(CH_3)_2 \text{ bridged by } \overset{+}{X} \text{ (halonium ion)} + CH_3OH_2^+ \\ CH_3\text{-}\underset{CH_3}{\underset{|}{\overset{CH_3}{\overset{|}{C}}}}\text{—}\overset{\overset{+}{O}CH_3}{\overset{\|}{C}}\text{-}CH_3 \end{cases} \tag{78}$$

$(X = Cl, Br, I)$

[190]

$$CH_3\underset{CH_3O}{\overset{CH_3}{C}}\text{—}\underset{OCH_3}{\overset{CH_3}{C}}CH_3 \xrightarrow[\text{or } SbF_5\text{-}SO_2,\ -78°]{FSO_3H\text{-}SbF_5\text{-}SO_2} CH_3\underset{CH_3}{\overset{CH_3}{C}}\text{—}\underset{CH_3}{\overset{\overset{+}{O}CH_3}{\overset{\|}{C}}} \tag{79}$$

[190]

$$ClCH_2O(CH_2)_nCH_3 \xrightarrow[SO_2ClF,\ -78°C]{FSO_3H\text{-}SbF_5} ClCH_2\overset{+}{O}H(CH_2)_nCH_3 \xrightarrow[-60°]{-HCl} CH_2{=}\overset{+}{O}(CH_2)_nCH_3 \tag{80}$$

[207]

$$ClCH_2OCH_2CH_2X \xrightarrow[-78°]{FSO_3H-SbF_5-SO_2} ClCH_2O^+HCH_2CH_2X \xrightarrow[-HCl]{-60°} CH_2{=}O^+CH_2CH_2X \qquad (81)\ [207]$$

(X = Br, I)

$$BrCH_2CH_2CH_2CH_2OCH_3 \xrightarrow[SO_2]{SbF_5} CH_3-\overset{+}{O}\text{(tetrahydrofuranium ring)} \qquad (82)\ [207]$$

The behavior of allylic alcohols has also been investigated in superacid media. In 1975, Olah and Spear [209] investigated the action of superacids on a number of 1,2- and 1,1-disubstituted allyl alcohols. The resulting 1,3-disubstituted alkenyl cations (Eq. 83) were shown to adopt the *trans,trans* configuration and to exhibit strong charge delocalization between C_1 and C_3. On the other hand, the charge in the 1,1-disubstituted alkenyl cations (Eq. 84) was substantially higher at C_1, the tertiary carbon. The results also demonstrated that the relative charge delocalization afforded by the substituents is cPr $\gtrsim$ Ph >> CH_3.

$$R_1CH(OH)CH=CHR_2 \xrightarrow{FSO_3H} R_1\text{–}\overset{H}{C_1}\text{–}\overset{H}{C_2}\text{–}\overset{H}{C_3}\text{–}R_2\ (+) \qquad (83)$$

Alcohol	Cation
$R_1 = R_2 = CH_3$	$R_1 = R_2 = CH_3$
R_1 = cPr; $R_2 = CH_3$	$R_1 = CH_3$; R_2 = cPr
R_1 = Ph; $R_2 = CH_3$	$R_1 = CH_3$; R_2 = Ph
$R_1 = R_2$ = cPr	$R_1 = R_2$ = cPr
R_1 = cPr; R_2 = Ph	R_1 = cPr; R_2 = Ph
$R_1 = R_2$ = Ph	$R_1 = R_2$ = Ph

$$R_1R_2C(OH)CH=CH_2 \xrightarrow{FSO_3H} R_1R_2\overset{1}{C}\text{=-=}\overset{2}{C}H_2\text{–}\overset{3}{C}H_3H_3\ (+) \qquad (84)$$

Alcohol	Cation
$R_1 = R_2 = CH_3$	$R_1 = R_2 = CH_3$
$R_1 = CH_3$; R_2 = cPr	$R_1 = CH_3$; R_2 = cPr
$R_1 = CH_3$; R_2 = Ph	$R_1 = CH_3$; R_2 = Ph
$R_1 = R_2$ = cPr	$R_1 = R_2$ = cPr
R_1 = cPr; R_2 = Ph	R_1 = cPr; R_2 = Ph
$R_1 = R_2$ = Ph	$R_1 = R_2$ = Ph

Further work on the rearrangement of allylic alcohols in superacids under stable ion conditions was published by Deno and Lastomirsky in 1975 [209a] and by Olah and co-workers in 1977 [210] and 1978 [211,212]. The 2-methyl-3-hexen-2-yl cation (65) was formed on addition of either 2,3-dimethyl-4-penten-2-ol (66) or 5-methyl-4-hexen-3-ol (67) to HSO_3F at -78°C (Eq. 84a) [209a]. A novel feature of Eq. (84a) is the observed elongation of the five-carbon chain of 66 to the six-carbon chain of 65. Addition of 2,3-dimethyl-3-penten-2-ol (68) to HSO_3F at -40°C produced ion 68a, which on warming to 25°C formed a mixture of cyclopentenyl cations (Eq. 84b) [209a]; accordingly, ion 68a could not be an intermediate in the formation of carbocation 65 from alcohol 66.

66 $\xrightarrow[-78°]{HSO_3F}$ 65 $\xleftarrow[-78°]{HSO_3F}$ 67 (84a)

68 $\xrightarrow[-40°]{HSO_3F}$ 68a $\xrightarrow{25°}$ cyclopentenyl cation mixtures (84b)

Treatment of allyl alcohols 69 (Eq. 85) [210] with FSO_3H-SbF_5 in SO_2ClF resulted in allyl cations 70 which rearranged via homoallylic cations 71 to the isomeric cyclopropylcarbinyl cations 72. At higher temperatures, ring-opening reactions of 72 took place, leading to the more highly substituted

$CH_3C(R)(H)$—$C(H)(OH)$—$C(R)$=CH_2 (69) $\xrightarrow{FSO_3H\text{-}SbF_5}$ CH_3—$C(R)(H)$—$C(H)$⋯$C(R')$⋯CH_2 $^{+}$ (70) ⟶

$CH_3\overset{+}{C}(R)CH_2C(R')$=$CH_2$ (71) ⟶ cyclopropyl-$C(R')$—$\overset{+}{C}(R)(CH_3)$ (72) $\xrightarrow{\text{higher temperatures}}$

CH_3—$C(R)$⋯$C(H)$⋯$C(R')$—CH_3 $^{+}$ (72a)

R = CH_3, R' = H
R = R' = CH_3
R = R' = H

(85) [210]

allyl cations 72a. Based on these results, the general order of thermodynamic stability was derived as follows:

$$R-CH\cdots\overset{+}{C}H\cdots CH-R \;>\; \triangleright-\overset{+}{C}(R)H \;>\; R-CH\cdots\overset{+}{C}H\cdots CH_2$$

The rearrangement of allyl alcohols to aldehydes (or ketones) was studied in superacid solutions and in the gas phase over a solid perfluorinated resinsulfonic acid catalyst (Nafion-H) [212]. Both catalytic conditions caused the conversion of allyl alcohol into propionaldehyde, of crotyl alcohol into *n*-butyraldehyde, of 2-methallyl alcohol into isobutyraldehyde (Eqs. 86, 87, and 88, respectively). In FSO_3H-SbF_5-SO_2ClF solution the tertiary allylic 2,3,3-trimethylallyl alcohol was rearranged similarly to *t*-butyl methyl ketone (Eq. 89). The mechanism of these conversions was suggested to follow the steps shown in Eq. (90) using allyl alcohol as an example.

$$CH_2{=}C(CH_3)CH_2OH \xrightarrow[SO_2ClF,\ -90°C]{FSO_3H\text{-}SbF_5} (CH_3)_2CH\overset{\overset{+}{O}H}{\overset{\|}{C}}\text{-}H \qquad (86)$$

$$CH_3CH{=}CHCH_2OH \xrightarrow[SO_2ClF,\ -90°C]{FSO_3H\text{-}SbF_5} CH_3CH_2CH_2\overset{\overset{+}{O}H}{\overset{\|}{C}}\text{-}H \qquad (87)$$

$$CH_2{=}C(CH_3)CH(CH_3)OH \xrightarrow[SO_2ClF,\ -90°C]{FSO_3H\text{-}SbF_5} (CH_3)_2CH\,\overset{\overset{+}{O}H}{\overset{\|}{C}}\text{-}CH_3 \qquad (88)$$

$$CH_2{=}C(CH_3)C(CH_3)_2OH \xrightarrow[-90°C]{FSO_3H\text{-}SbF_5\text{-}SO_2ClF} (CH_3)_3C\overset{\overset{+}{O}H}{\overset{\|}{C}}CH_3 \qquad (89)$$

$$CH_2{=}CH\text{-}CHO \xrightarrow[-90°]{HSO_3F} CH_2{=}CH\text{-}CH_2\overset{+}{O}H_2 \underset{>-80°}{\overset{HSO_3F}{\rightleftharpoons}} CH_3\underset{}{\overset{SO_3F}{\overset{|}{C}}}HCH_2\overset{+}{O}H_2$$

$$\xrightarrow[HSO_3F]{-40°} [CH_3CH^+CH_2OH] \xrightarrow{\sim H:^-} CH_3CH_2\overset{\overset{+}{O}H}{\overset{\|}{C}}H \qquad (90)$$

In contrast to the behavior of alkyl-substituted allyl cations, which were shown to be stable in superacid solutions under stable ion conditions, aryl allyl cations underwent facile intramolecular cyclization to the corresponding indanyl cations [211]. For example, when 2-phenyl-3-penten-2-ol (74) was treated with magic acid solution at -120°C, it gave 2-phenyl-2-penten-4-yl cation (75). The latter rearranged first to cation 76 at -80°C and finally to the indanyl cation 77 at -70°C (Eq. 91). These rearrangements were shown (by deuterium labeling) to proceed via a series of intramolecular hydride shifts.

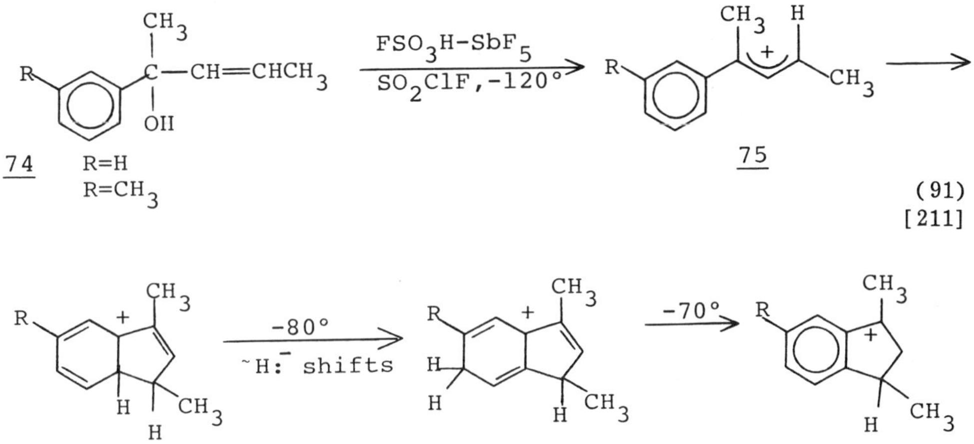

$FSO_3H\text{-}SbF_5$, SO_2ClF, -120°

74 R=H, R=CH_3 75

(91) [211]

-80° ~H:⁻ shifts; -70°

76 77

4. *Rearrangements of acyclic alkenes and alkanes*

Extensive rearrangements also accompanied the generation of carbocations from alkenes and alkanes in the presence of strong proton acid catalysts. The extent of rearrangement was dependent on catalyst and conditions. Thus migration of the double bond was effected mainly by treatment of linear alkenes with catalysts such as benzenesulfonic acid, phosphoric acid, phosphoric acid on silica gel, sulfuric acid, or perchloric acid/acetic anhydride [213-215]; e.g., treatment of 1-undecene with 70% perchloric acid and acetic anhydride at 100°C gave an isomerization mixture consisting of 4% 1-, 40% 2-, 28.1% 3-, 18.4% 4-, and 9.5% 5-undecene [216]. Isomerizations involving migration of alkene double bonds by the catalytic action f alumina have also been extensively studied [214,215,217-220]. In one report [221], it was indicated that the 1- into 2-butene isomerization on a silica-alumina catalyst at 66°C took place sterospecifically to yield the less stable but kinetically favored *cis* isomer. The results were explained by invoking the intermediacy of the nonclassical carbocations 78 and 79 with the stereospecificity of the reaction being the result of a restricted rotation of the terminal carbon atom in ion 79 to a *trans* position.

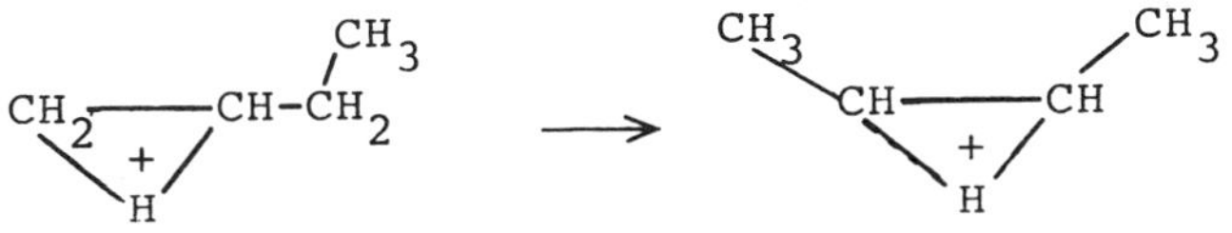

78 79

In another report [222], *trans*-2-butene was isomerized over deuterated *p*-toluenesulfonic acid at room temperature for the purpose of studying *cis-trans* isomerization and double-bond migration simultaneously. Analysis of the products by microwave spectrometry suggested that the *cis-trans* isomerization of 2-butene occurred via the *sec*-butyl cation, producing approximately equal amounts of the d_0 and 2-d_1 species of 2-butene, while the isomerization of 2- to 1-butene occurred by transfer of a proton from the acid to the =CH- group followed by another proton transfer from the methyl group on the opposite side to the acid catalyst. The latter mechanism is in accord with the observation that at the initial stages of the reaction, the only 1-butene produced by the double-bond migration from *trans*-2-butene was the 3-d_1 species. Nonconjugated dienes have been shown to isomerize under acidic conditions into conjugated dienes [223-225]: e.g., 1,5-hexadiene forms 2,4-hexadiene (Eq. 92 [225] and 1,7-octadiene formed 3,5-octadiene (Eq. 93) [225].

$$H_2C{=}CH{-}CH_2CH_2CH{=}CH_2 \xrightarrow[300°]{Al_2O_3} H_3C{-}CH{=}CH{-}CH{=}CH{-}CH_3 \quad (92)$$

$$H_2C{=}CH{-}(CH_2)_4{-}CH{=}CH_2 \xrightarrow[250°]{Cr_2O_3{-}Al_2O_3} H_5C_2{-}CH{=}CH{-}CH{=}CH{-}C_2H_5 \quad (93)$$

Under acid catalysis, branched alkenes have been shown to isomerize into the most stable alkene having the highest number of alkyl groups at the double bond. For example, in the presence of sulfuric acid, 2,3-dimethyl-1-butene (80) isomerized into 2,3-dimethyl-2-butene (81) (Eq. 94) [107].

$$\underset{\underline{80}}{CH_2{=}C(CH_3){-}CH(CH_3)CH_3} \xrightarrow{H_2SO_4} \underset{\underline{81}}{CH_3C(CH_3){=}C(CH_3){-}CH_3} \quad (94)$$

In many cases, however, bond migrations have been shown to be accompanied by competing processes, such as alkyl group migration, disproportionation, fragmentation, and polymerization [57,158,226-229]. For example, in the presence of sulfuric acid 1,1-di-*t*-butylethylene (82) completely rearranged into the more stable 1-methyl-1-triptylethylene (83), and the latter fragmented completely to tetramethylethylene (84) and *t*-butyl alcohol (scheme 11) [229].

Compared to the results from milder catalysts, much deeper rearrangements have been found to accompany carbocation generations from alkenes and alkanes in the presence of the superacid media of SbF_5, HSO_3F, SbF_5-HSO_3F, and HF-SbF_5.

In 1967, Olah and Lukas [186,230] reported observations relating to the generation of carbocations from alkanes via hydride ion abstraction with the strong acids HSO_3F-SbF_5 and HF-SbF_3 at temperatures between -125 and +25°C. NMR studies showed the following: (1) All the observed carbocations were tertiary, indicating that abstraction of a secondary or primary hydrogen atom resulted in immediate rearrangement via 1,2 hydride or 1,2 methyl shift

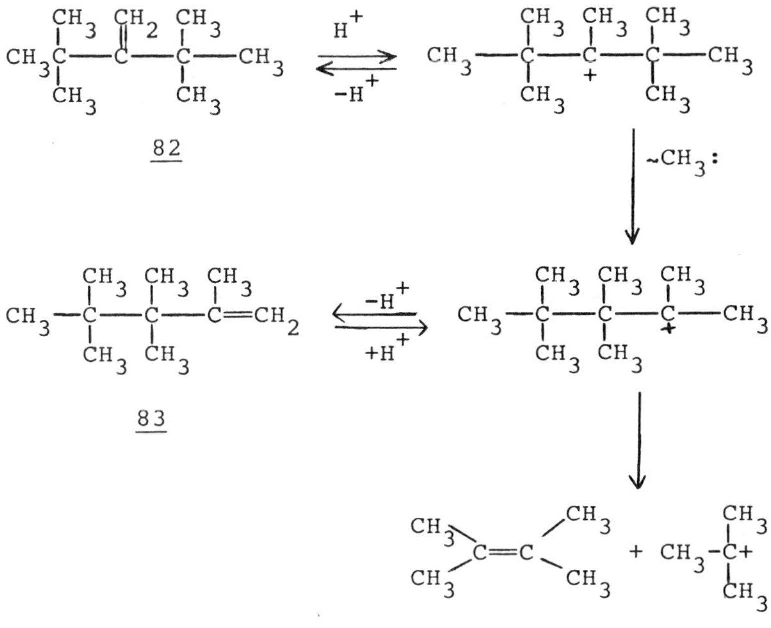

$$CH_3C(CH_3)_2\text{—}C(=CH_2)\text{—}C(CH_3)_2\text{—}CH_3 \underset{-H^+}{\overset{H^+}{\rightleftharpoons}} CH_3\text{—}C(CH_3)_2\text{—}\overset{+}{C}(CH_3)\text{—}C(CH_3)_2\text{—}CH_3$$

82

$\downarrow$ ~CH_3:

$$CH_3\text{—}C(CH_3)_2\text{—}C(CH_3)_2\text{—}C(CH_3)\text{=}CH_2 \underset{+H^+}{\overset{-H^+}{\rightleftharpoons}} CH_3\text{—}C(CH_3)_2\text{—}C(CH_3)_2\text{—}\overset{+}{C}(CH_3)\text{—}CH_3$$

83

$\downarrow$

$$(CH_3)_2C\text{=}C(CH_3)_2 + CH_3\text{-}\overset{+}{C}(CH_3)_2$$

84

Scheme 11

or cleavage to give tertiary carbocations, even at -100°C. (2) No evidence was found for reversible 1,2 hydride or 1,2 methyl shifts. (3) Ease of ionization increased in the normal order, primary < secondary < tertiary hydrogen atoms. Thus, whereas the secondary and tertiary systems underwent ionization at temperatures as low as -100°C, primary hydrogen atoms generally could not be ionized below -50°C. (4) On heating, all carbocations were converted ultimately, via isomerization, fragmentation, and dimerization, to the *t*-butyl cation, which was shown to be stable in the acid media up to 150°C. The results of some of the reactions examined are shown in Eqs. (95) to (97).

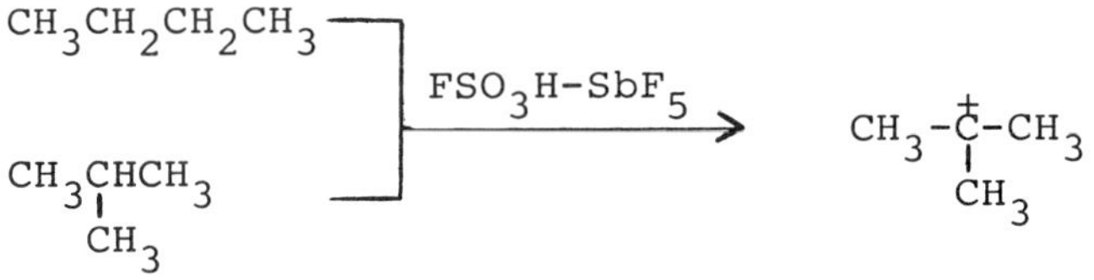

$$\left.\begin{array}{l} CH_3CH_2CH_2CH_3 \\ CH_3CH(CH_3)CH_3 \end{array}\right] \xrightarrow{FSO_3H\text{-}SbF_5} CH_3\text{-}\overset{+}{C}(CH_3)\text{-}CH_3 \quad (95)$$

$$\left.\begin{array}{l} CH_3CH_2CH_2CH_2CH_3 \\ CH_3CH(CH_3)CH_2CH_3 \end{array}\right] \xrightarrow[-30°]{FSO_3H\text{-}SbF_5} CH_3\overset{+}{C}(CH_3)\text{-}CH_2CH_3 \quad (96)$$

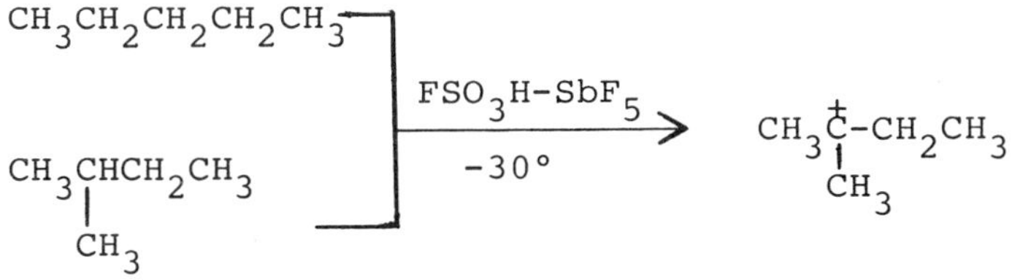

$$\left.\begin{array}{l} CH_3CH_2CH_2CH_2CH_2CH_3 \\ CH_3CH(CH_3)CH_2CH_2CH_3 \\ CH_3CH_2CH(CH_3)CH_2CH_3 \\ CH_3CH(CH_3)CH(CH_3)CH_3 \\ CH_3C(CH_3)_2CH_2CH_3 \end{array}\right\} \xrightarrow[\text{room temp}]{FSO_3H\text{-}SbF_5} \begin{array}{c} CH_3\text{-}CH(CH_3)\text{-}\overset{+}{C}(CH_3)\text{-}CH_3 \\ + \\ CH_3CH_2\overset{+}{C}(CH_3)CH_2CH_3 \\ + \\ CH_3\overset{+}{C}(CH_3)\text{-}CH_2CH_2CH_3 \end{array} \tag{97}$$

The ionization pattern of neopentane in FSO_3H-SbF_5-SO_2ClF was found to differ sharply with temperature, giving *t*-pentyl cation at -20°C and *t*-butyl cation and CH_4 at +25°C (Eq. 98).

$$(CH_3)_3C^+ + CH_4 \xleftarrow{25°} H_3C\text{-}C(CH_3)_2\text{-}CH_3 \xrightarrow{-20°} \left[CH_3\text{-}C(CH_3)_2\text{-}CH_2^+\right] \longrightarrow CH_3\overset{+}{C}(CH_3)CH_2CH_3 \tag{98}$$

The NMR data also excluded the nonclassical structures 85 and 86 for 2,3-dimethyl-2-butyl cation and 2,3,3-trimethyl-2-butyl cation, respectively [230].

$$(CH_3)_2C\overset{H^+}{\frown}C(CH_3)_2 \qquad (H_3C)_2C\overset{CH_3^+}{\frown}C(CH_3)_2$$

85 86

n-Hexane was also found to undergo both isomerization and cracking reactions upon treatment in a flow reactor at 385°K with an "organometallic superacid" polymer catalyst (made from $AlCl_3$ and beads of sulfonated poly (styrene-divinyl-benzene) [231]. The product consisted of varying proportions of isobutane, isopentane, *n*-pentane, and branched C_6 isomers, depending on time on stream.

In 1968, Brouwer [232,233] published the results of a detailed investigation on the structure, stabilities, and rearrangement reactions of C_4 to C_7 tertiary carbocations in HF-SbF_5. The ions were generated by the acid in various ways, including protonation of alkenes; acid-catalyzed ring opening of methylcyclopropane, 1,1- and 1,2-dimethylcyclopropane; solvolysis of tertiary alcohols, chlorides, bromides, and ethers; hydride abstraction from unbranched and branched alkanes; and dealkylation of *t*-butylbenzene. The results were essentially similar to those of Olah and Lukas [186,230] in that regardless of the kind of generation, all of the ions observed were tertiary. Thus C_4, C_5, C_6, and C_7 sources gave tertiary butyl, pentyl, hexyl, and heptyl cations, respectively. The NMR data again revealed that tertiary hexyl and heptyl cations underwent rapid rearrangements to equilibrium mixtures of the three isomeric hexyl cations (87-89) and the seven isomeric heptyl cations, 90-93, respectively, in ratios corresponding to their thermodynamic stabilities. Integration of peak areas showed that in the *t*-hexyl cation spectrum, the equilibrium composition of the cation mixture was 32% 87, 38% 88, and 30% 89. In the *t*-heptyl spectrum, the cations (94 + 95), 96, and (90 + 91 + 92 + 93) were shown to be present in an approximate ratio of 1:1:2.

$H_3C\text{–}\overset{+}{C}(CH_3)\text{–}CH_2\text{–}CH_2\text{–}CH_3$

87

$H_3C\text{–}CH_2\text{–}\overset{+}{C}(CH_3)\text{–}CH_2\text{–}CH_3$

88

$H_3C\text{–}\overset{+}{C}(CH_3)\text{–}CH(CH_3)\text{–}CH_3 \rightleftarrows H_3C\text{–}CH(CH_3)\text{–}\overset{+}{C}(CH_3)\text{–}CH_3$

89

$H_3C\text{–}\overset{+}{C}(CH_3)\text{–}CH_2\text{–}CH_2\text{–}CH_3$

90

$H_3C\text{–}CH_2\text{–}\overset{+}{C}(CH_3)\text{–}CH_2\text{–}CH_2\text{–}CH_3$

91

$H_3C\text{–}CH_2\text{–}\overset{+}{C}(CH_2\text{–}CH_3)\text{–}CH_2\text{–}CH_3$

92

$H_3C\text{–}\overset{+}{C}(CH_3)\text{–}CH_2\text{–}CH(CH_3)\text{–}CH_3$

93

$H_3C\text{–}\overset{+}{C}(CH_3)\text{–}CH(CH_3)\text{–}CH_2\text{–}CH_3 \rightleftharpoons H_3C\text{–}CH(CH_3)\text{–}\overset{+}{C}(CH_3)\text{–}CH_2\text{–}CH_3$

94 95

$H_3C\text{–}C(CH_3)_2\text{–}\overset{+}{C}(CH_3)\text{–}CH_3 \rightleftharpoons H_3C\text{–}\overset{+}{C}(CH_3)\text{–}C(CH_3)_2\text{–}CH_3$

96

In 1969, Brouwer [234] studied the rates of rearrangement of degenerate 2,3-dimethylbutyl ions (89) to degenerate 2-methyl-2-pentyl ions (87) and 3-methyl-3-pentyl ion (88) and of the much faster interconversion of the latter two ions (87 and 88) in a variety of solvents. The results showed that the rates of both reactions, supposed to proceed by different mechanisms [235,236], were but little affected by the nature of the solvents employed.

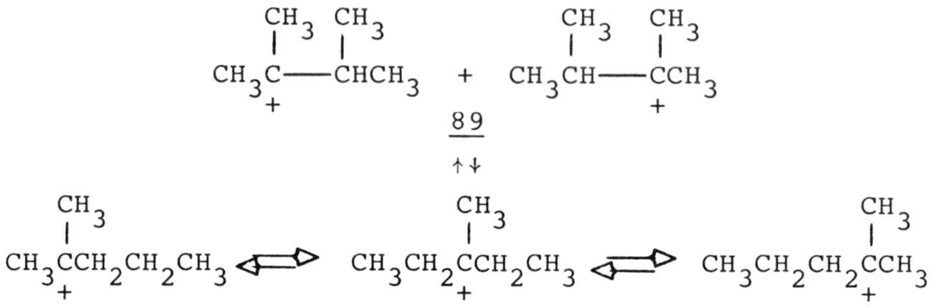

$$CH_3\overset{+}{C}(CH_3)\text{—}CH(CH_3)CH_3 \;+\; CH_3CH(CH_3)\text{—}\overset{+}{C}(CH_3)CH_3$$

89

↑↓

$$CH_3\overset{+}{C}(CH_3)CH_2CH_2CH_3 \rightleftharpoons CH_3CH_2\overset{+}{C}(CH_3)CH_2CH_3 \rightleftharpoons CH_3CH_2CH_2\overset{+}{C}(CH_3)CH_3$$

87 88 87

Again, in 1969, Brouwer and Van Doorn [237] deduced from spectroscopic observations that the degenerate rearrangement of ion 93 (prepared from the chloride in FSO_3H-SbF_5-SO_2ClF), occurred by the 1,3 hydride shift of Eq. (99) rather than by the double 1,2 hydride shift of Eq. (100). These authors further deduced that the proposed 1,3 hydride shift did not proceed via protonated cyclopropane ring intermediates such as 97 or 98.

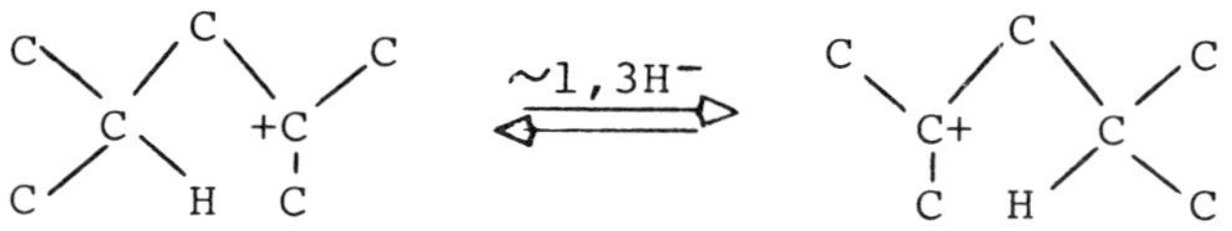

$$(C)_2CH\text{—}CH_2\text{—}\overset{+}{C}(C)_2 \;\xrightleftharpoons{\sim 1,3H^-}\; (C)_2\overset{+}{C}\text{—}CH_2\text{—}CH(C)_2 \qquad (99)$$

93 93

$$\underset{H\;\;H\;\;+}{C\text{-}\overset{C}{C}\text{-}\overset{H}{C}\text{-}\overset{C}{C}\text{-}C} \;\xrightleftharpoons{\sim 1,2H^-}\; \underset{H\;\;+\;\;H}{C\text{-}\overset{C}{C}\text{-}\overset{H}{C}\text{-}\overset{C}{C}\text{-}C} \;\xrightleftharpoons{\sim 1,2H^-}\; \underset{+\;\;H\;\;H}{C\text{-}\overset{C}{C}\text{-}\overset{H}{C}\text{-}\overset{C}{C}\text{-}C} \qquad (100)$$

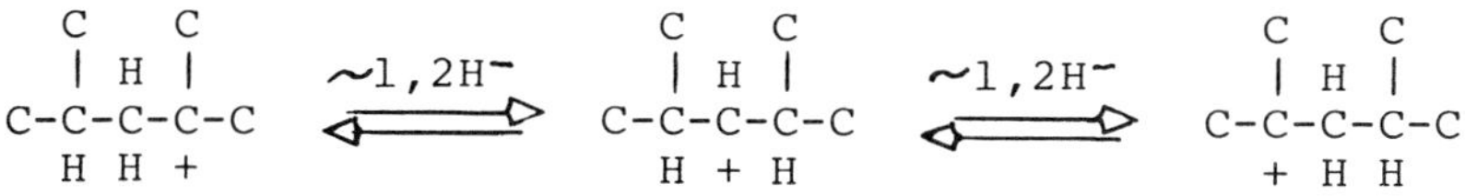

93 88 93

97: cyclopropane ring with CH_2 at the apex (H_2C^2), two C atoms each bearing H_3C / CH_3 groups, with the C---C edge bridged by H (+) (H, +, H_3C, H_3C, CH_3, CH_3)

98: cyclopropane ring with C at the apex bearing H (C^2), two C atoms each bearing H_3C / CH_3 groups (H_3C, H_3C, CH_3, CH_3)

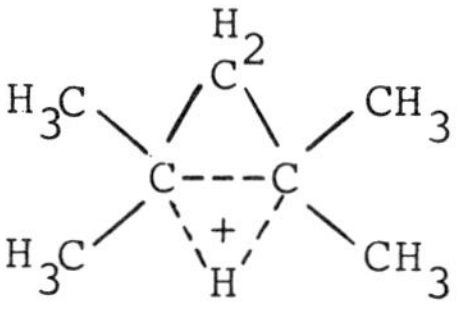

97 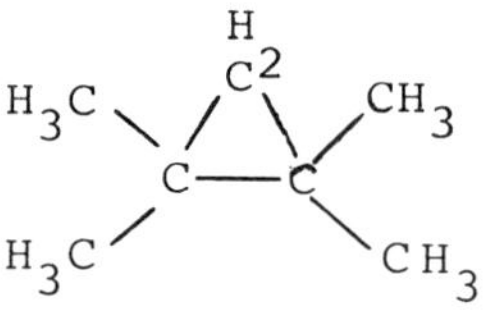 98

B. Rearrangements in Cyclic Systems

1. Rearrangements of monocyclic and related systems

a. Cyclopropyl Rearrangements

As mentioned previously, Baird and Aboderin [83] found that in deuteriosulfuric acid cyclopropane underwent both deuterium exchange and ring opening, and that the latter was twice as fast as exchange [Sec. IA.1.b(ii)]. Ring opening of cyclopropane derivatives was also observed in superacid media.

In 1968, Olah and Lukas [238] found that cyclopropane was protonated and cleaved in FSO_3H-SbF_5-SO_2ClF or HF-SbF_5-SO_2ClF solution at temperatures above -80°C to give acyclic products resulting from polymerization and fragmentation (Eq. 161). In the same acids, ethylcyclopropane underwent

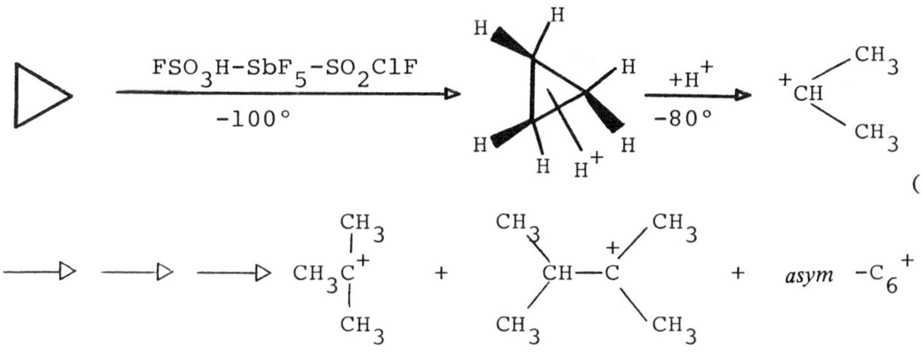

$$\text{cyclopropane} \xrightarrow[-100^\circ]{FSO_3H\text{-}SbF_5\text{-}SO_2ClF} \text{protonated cyclopropane } (H^+) \xrightarrow[-80^\circ]{+H^+} {}^{+}CH(CH_3)_2$$

$$\longrightarrow \longrightarrow \longrightarrow (CH_3)_3C^+ + (CH_3)_2CH\text{—}C^+(CH_3)_2 + asym\ \text{-}C_6^+ \qquad (101)$$

protolytic ring opening and rearrangement to *t*-pentyl cation under all conditions (Eq. 102). At -100°C, isopropylcyclopropane underwent hydride abstraction to 2-cyclopropyl-2-propyl cation concurrent with protonation and ring opening to *t*-hexyl cations (scheme 12).

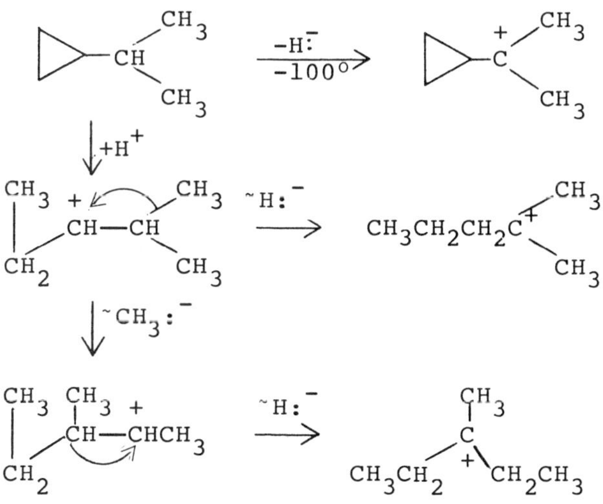

$$\text{c-}C_3H_5\text{—}CH(CH_3)_2 \xrightarrow[-100^\circ]{-H:^-} \text{c-}C_3H_5\text{—}C^+(CH_3)_2$$

$$\downarrow +H^+$$

$$CH_3\text{—}CH_2\text{—}CH^+\text{—}CH(CH_3)_2 \xrightarrow{\sim H:^-} CH_3CH_2CH_2C^+(CH_3)_2$$

$$\downarrow \sim CH_3:^-$$

$$CH_3\text{—}CH_2\text{—}CH(CH_3)\text{—}C^+HCH_3 \xrightarrow{\sim H:^-} CH_3CH_2\text{—}C^+(CH_3)\text{—}CH_2CH_3$$

Scheme 12

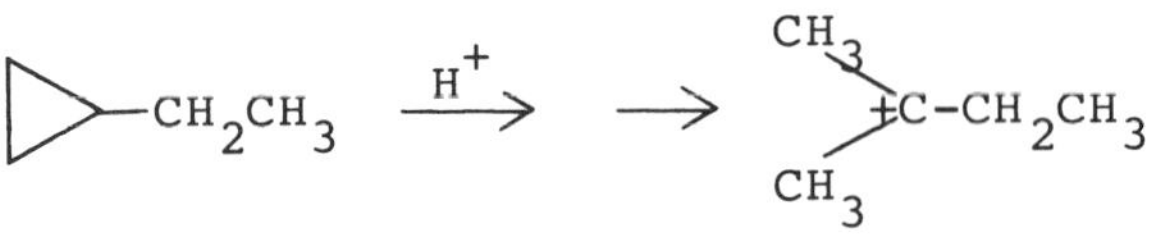

(102)

Depending on structure and reaction temperature, cyclopropyl halides in SbF_5-HSO_3F-SO_2 were found to give either the corresponding halonium ions through protonation, or the substituted allylic cations through ionization, with or without rearrangement (Eqs. 103-108) [239]. The accompanying rearrangements were rationalized in terms of normal carbocation processes.

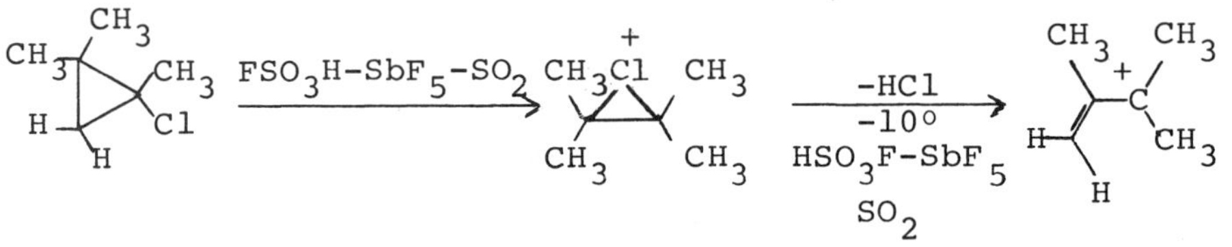

(103)

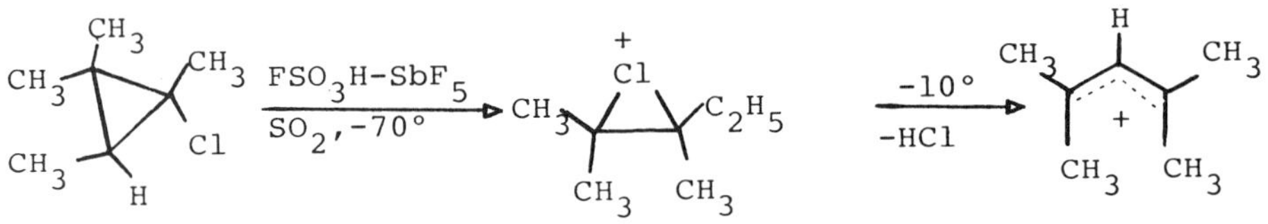

(104)

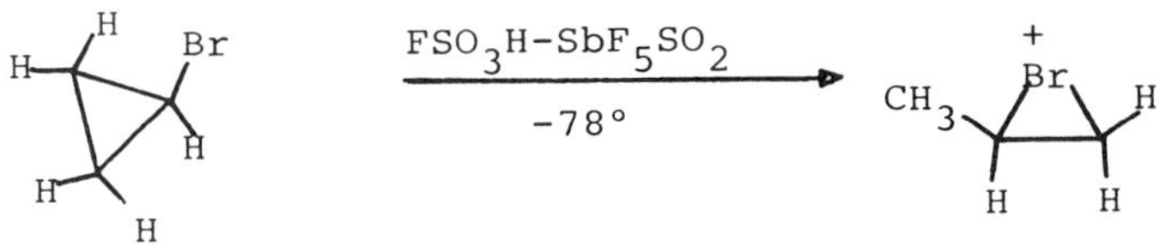

(105)

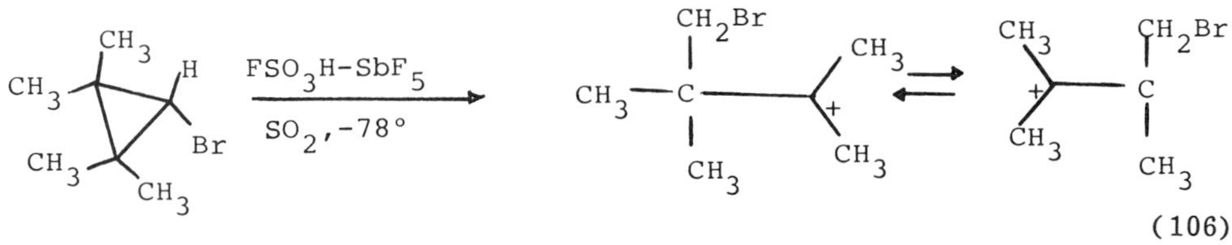

(106)

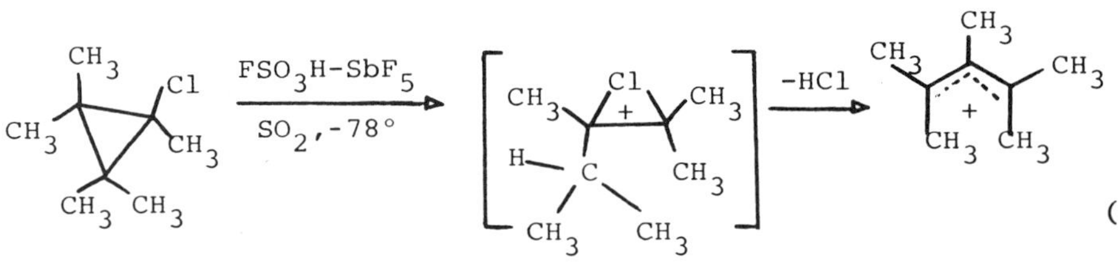

(107)

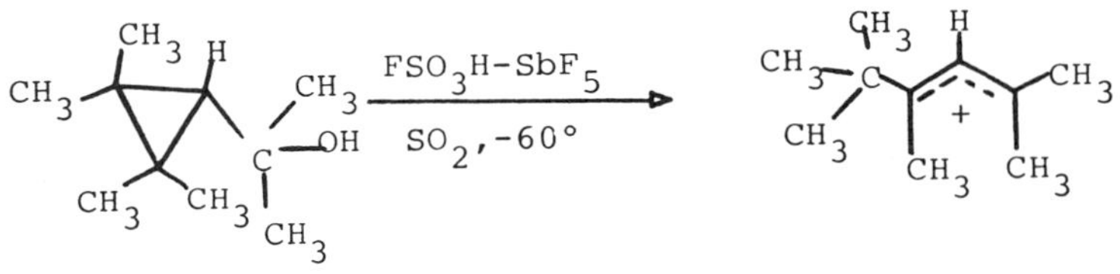

(108)

$$\text{99 (1,1-dihalo-2,2-dimethyl-3-}R_1\text{-3-}R_2\text{-cyclopropane)} \xrightarrow[-60^\circ]{SbF_5-SO_2} \text{100 } (H_3C-C(CH_3)\cdots C(X)\cdots C(R_2)-R_1)^{+} \quad (109)$$

(X = Cl or Br)

(R_1 and R_2 = H or CH_3)

Ionization of dihalocyclopropanes (99) in SbF_5-SO_2 at -60°C proceeded with ring opening to substituted 2-halomethylallyl cations (100) (Eq. 109) [194].

b. Cyclobutyl and Cyclopropylcarbinyl Rearrangements

The cyclobutyl (101) and cyclopropylcarbinyl (107) cations are very closely related, as they interconvert easily when either is generated. An illustration of this is the fact that attempted generation of cyclobutyl cation by the action of SbF_5-HSO_3F-SO_2ClF on cyclobutane at -100°C gave a solution whose NMR spectrum was assignable mainly to the cyclopropylcarbinyl cation. Moreover, quenching of this solution with methanol gave an 80% yield of a mixture consisting of 90% methyl cyclopropylcarbinyl ether and 10% methyl cyclobutyl ether (Eq. 110) [238].

$$\text{cyclobutane} \xrightarrow[-100^\circ]{SbF_5-HSO_3F-SO_2ClF} \text{cyclobutyl}^{+} \rightleftarrows \text{cyclopropyl-}\overset{+}{C}H_2 \xrightarrow{CH_3OH} \underset{10\%}{\text{cyclobutyl-}OCH_3} + \underset{90\%}{\text{cyclopropyl-}CH_2OCH_3} \quad (110)$$

In 1972, Olah et al. [240] published a detailed study on the generation and structure of cyclopropylcarbinyl and cyclobutyl cations. Their work showed that cyclopropylcarbinyl cation (102) could be generated in good purity either from cyclopropylcarbinol or from cyclobutanol in SbF_5-SO_2ClF. Ionization of the corresponding halides in SbF_5-SO_2 gave similar results. Again, quenching experiments produced product mixtures in which a 4:1 ratio of cyclobutanol to cyclopropylcarbinol was found. The results are summarized in scheme 13. In a manner similar to the behavior of cyclobutyl cation, attempted generation of methylcyclobutyl cation at -78 to -80°C by the methods shown in scheme 14 led only to the detection of 1-methylcyclopropylcarbinyl cation 103, which isomerized to 1-cyclopropylethyl cation 104 at -25°C [240]. Ion 103 was said to be observed by Saunders and Rosenfeld

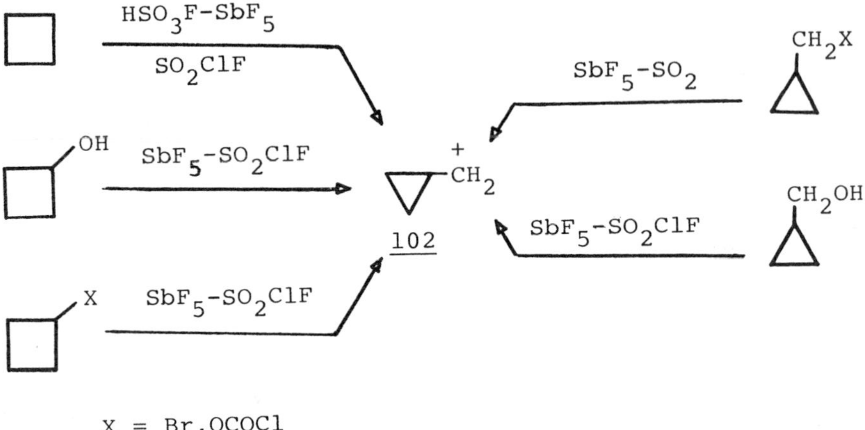

X = Br, OCOCl

Scheme 13

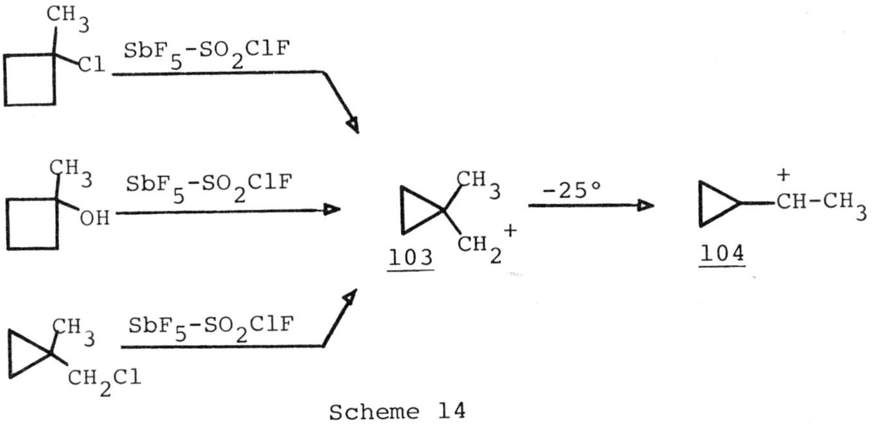

CH_3 Cl SbF_5-SO_2ClF

CH_3 OH SbF_5-SO_2ClF → 103 (CH_3, CH_2^+) $\xrightarrow{-25°}$ 104 ($\overset{+}{C}H-CH_3$)

CH_3 CH_2Cl SbF_5-SO_2ClF

Scheme 14

[24] in 1970, but they suggested its rapid rearrangement to the energetically less favorable methylcyclobutyl cation (104) (Eq. 111).

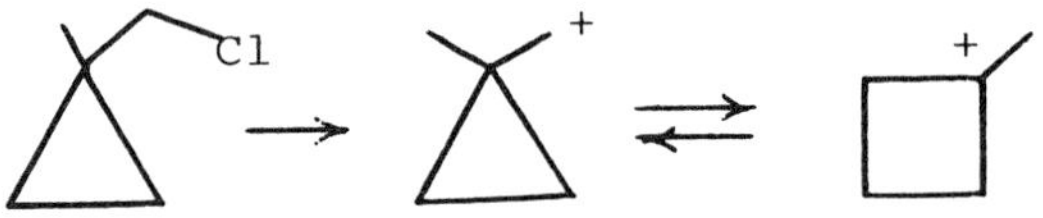

103 ⇌ 104 (111)

Olah et al. [240] also attempted the generation of the related secondary and tertiary 1-methylcyclopropylcarbinyl cations 105 and 106 but failed; in-

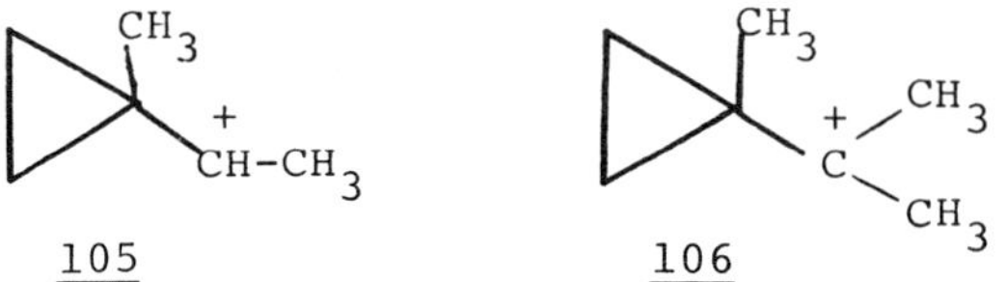

105 106

stead, the spectral data indicated formation of opened allylic ions (107, Eq. 112). To rationalize the relative instability of ions 105 and 106 and hence the

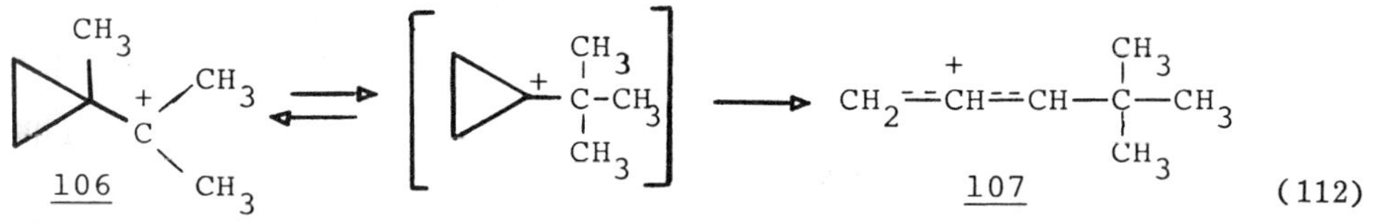

$$106 \rightleftharpoons [\ldots] \longrightarrow CH_2\text{=-=}CH\text{=-=}CH\text{—}C(CH_3)_2\text{—}CH_3 \quad 107 \qquad (112)$$

failure in their observation under the condition employed, it was suggested that the 1-methyl substituent causes sufficient steric interaction so as to render the bisected conformation of ions 105 and 106 somewhat less favorable than the in-plane conformation. More recently [240a], however, the α-1-dimethylcyclopropylcarbinyl cation 105 was observed when the corresponding α,1-dimethylcyclopropylcarbinol was treated with SbF_5 in SO_2ClF solution at -78°C. ^1H and ^{13}C NMR data suggested that ion 105 is a static carbocation at temperatures below -100°C, whereas at -70°C it undergoes equilibration through the 1,2-dimethylcyclobutyl cation. At -39°C, 105 irreversibly rearranges to 1,1,3-trimethylallyl cation:

(Scheme: α,1-dimethylcyclopropylcarbinol $\xrightarrow{SbF_5,\ SO_2ClF,\ -100°}$ 105 $\xrightleftharpoons{-70°}$ 1,2-dimethylcyclobutyl cations $\xrightarrow{-39°}$ 1,1,3-trimethylallyl cation)

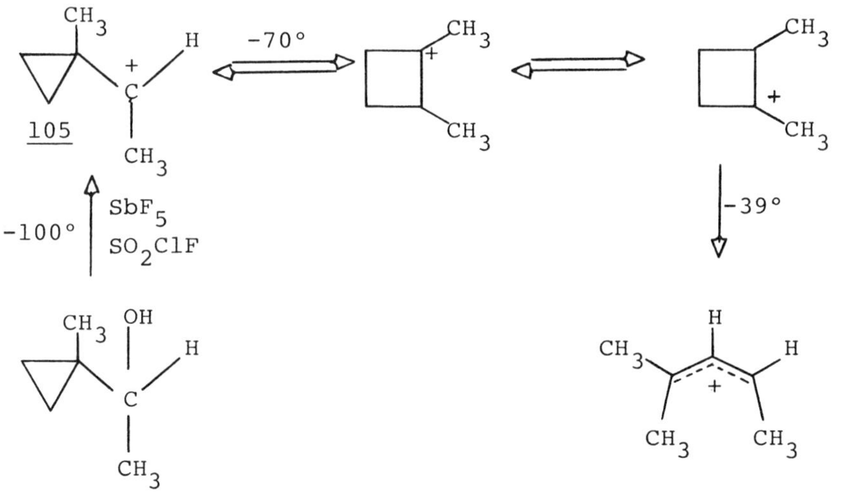

As the previous results showed that a methyl group was not capable of stabilizing a long-lived cyclobutyl cation, attempts were made to obtain 1-phenylcyclobutyl cation (109) by ionizing 1-phenylcyclobutylcarbinol (108) in SbF_5-SO_2 solution. These attempts were successful and led to the first classical static cyclobutyl cation [240] (Eq. 113). Other related phenyl-stabilized cyclobutyl cations were also prepared (see Eqs. 131 and 132) [242].

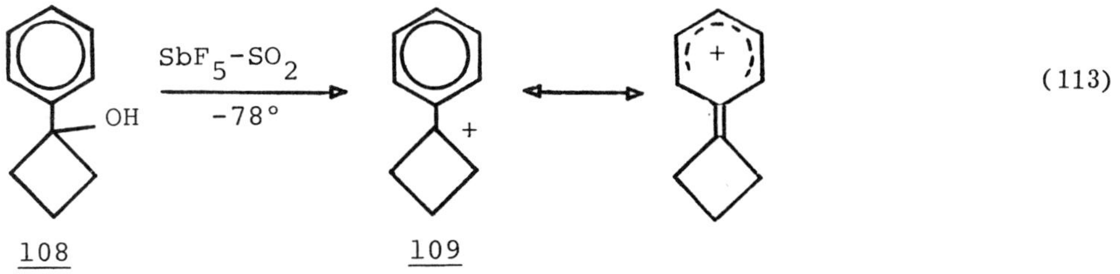

$$108 \xrightarrow[-78°]{SbF_5-SO_2} 109 \longleftrightarrow \qquad (113)$$

At this point, it seems useful to comment briefly on the structural nature of cyclopropylcarbinyl cations and also on the stabilizing effect of the cyclopropyl group on a carbocation center. Starting with the latter, it can be stated that numerous studies of the solvolysis of cyclopropylcarbinyl derivatives [243-247] and direct observation of cyclopropylcarbinyl cations under stable ion conditions [236,240,248-250] have demonstrated the remarkable ability of the cyclopropyl group to stabilize a carbocation center to which it is attached. Indeed, ^{13}C NMR studies of substituted cyclopropylcarbocations proved that the stabilizing effect of a cyclopropyl group on a carbenium ion center is comparable to the effect of a phenyl group [251]. Moreover, in the gas phase, the cyclopropyl group was shown to be slightly more effective than a phenyl group in stabilizing a tertiary carbocation center [252]. The stabilizing effect of cyclopropyl groups on carbenium ions can again be seen from the results summed up in schemes 13 and 14. These results indicated that both cyclobutyl and cyclopropylcarbinyl precursors yielded the obviously more stable cyclopropylcarbinyl cations (102 and 103) upon ionization in superacid media.

The structural nature of cyclopropylcarbinyl and cyclobutyl cations has been also a matter of considerable concern [240,241,248-250,253-256 and references cited therein]. Based on ^{1}H and ^{13}C NMR spectra under stable ion conditions, it has been suggested that the primary cyclopropylcarbinyl (102) and 1-methylcyclopropylcarbinyl (103) cations undergo a rapid equilibration among the three degenerate σ-delocalized nonclassical ions (110), in which there is little or no contribution by the cyclobutyl cation [240,241,249,250].

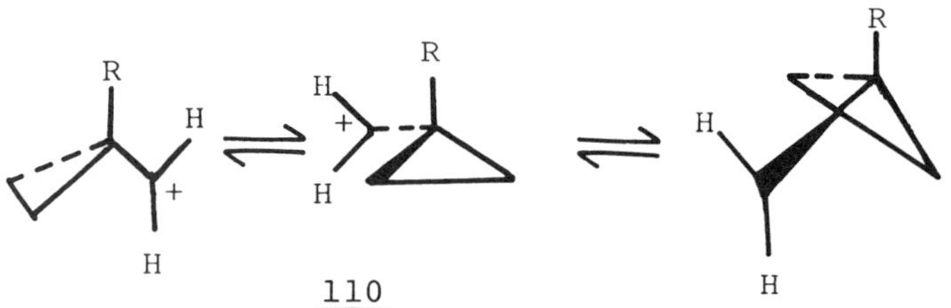

110

R = H, CH_3

In contrast, the corresponding secondary (104 and 105) and tertiary (106 and 111) cyclopropylcarbinyl cations, as well as the related tertiary 1-phenylcyclobutyl cation 109, were all proposed to be static classical carbocations with varying degrees of charge delocalization into the cyclopropane ring. Furthermore, these ions were shown to adopt the following bisected geometry [250]:

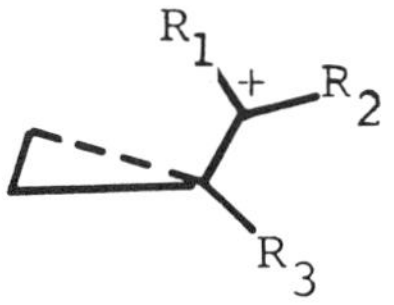

102: $R_1 = R_2 = R_3 = H$

103: $R_1 = R_2 = H$, $R_3 = CH_3$

104: $R_1 = R_3 = H$, $R_2 = CH_3$

105: $R_1 = H$, $R_2 = R_3 = CH_3$

106: $R_1 = R_2 = R_3 = CH_3$

111: $R_1 = R_2 = CH_3$, $R_3 = H$

To explore further the nature of cyclopropylcarbinyl cations, studies have been extended to cyclopropylcarbinyl cations with rigid carbon skeletons [236,242,249,256]. For that matter, the 8,9-dehydro-2-adamantyl cations (112) [236], the 2,4-dehydro-5-homoadamantyl cation (113) [257], the 3-nortricyclyl cation (114) [235], the 3-homotricyclyl cations (115) [256], and the 2-bicyclo[n.1.o]alkyl cations (116) [242] were generated from suitable precursors in superacid media under stable ion conditions and investigated by ^1H

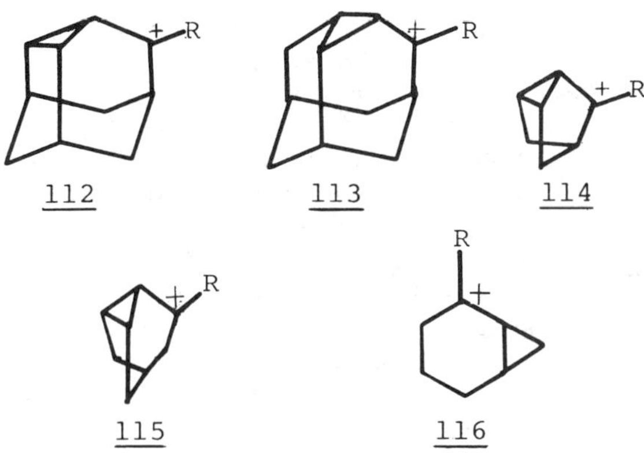

and ^{13}C NMR spectroscopy. The results from these studies demonstrated again that (1) both the secondary and the tertiary cations of structures 112 to 116 (R = H, CH_3, C_6H_5) were classical carbenium ions, and (2) the parent secondary cations of these systems (R = H) were mostly degenerate in nature, undergoing cyclopropylcarbinyl-cyclopropyl-carbinyl rearrangements (e.g., Eqs. 114 to 117), but the tertiary systems (R = CH_3 or C_6H_5) were mostly static, with charge delocalization into the cyclopropane ring.

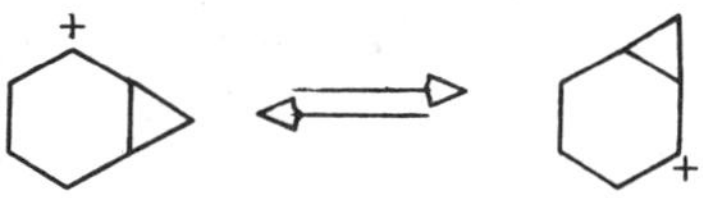

(114)

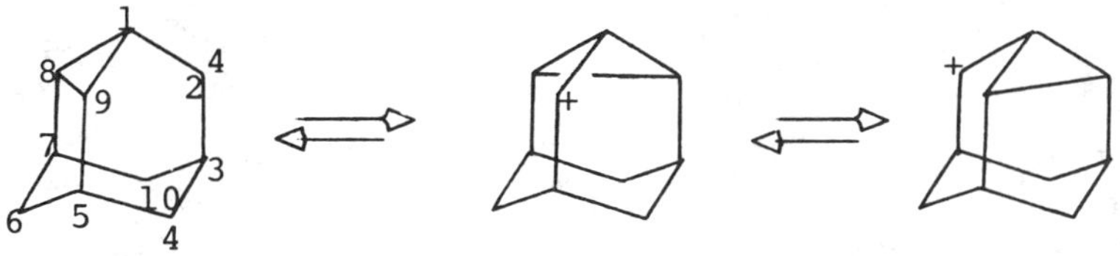

(115)

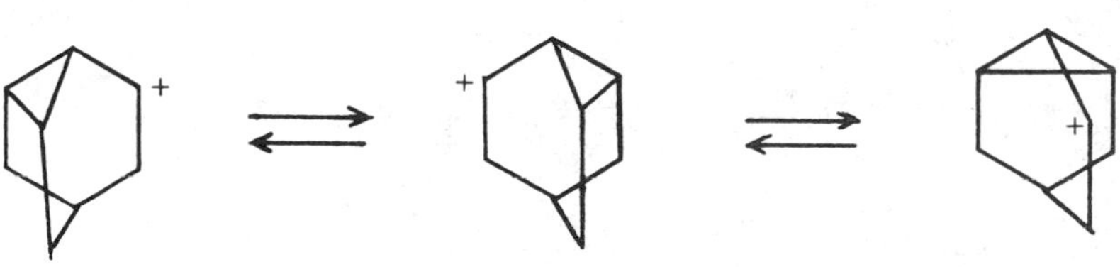

(116)

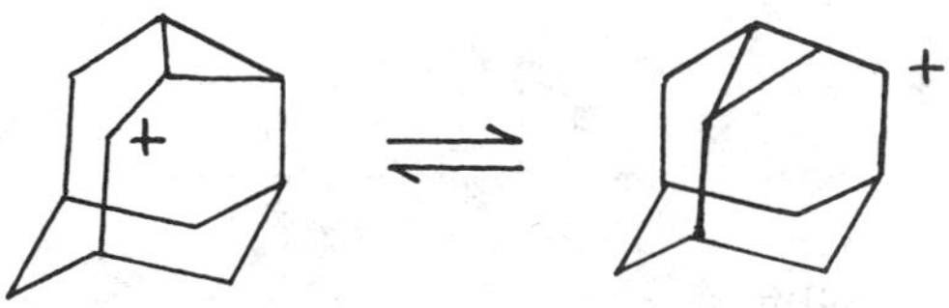

(117)

In concluding this brief discussion about the structure of cyclopropylcarbinyl cation, it is interesting to note that after all of the work that was subsequently carried out, the following early conclusion of Olah et al. [240] is still valid: "All of the cyclopropylcarbinyl cations show large delocalization by the cyclopropyl ring, but only with the primary system is the unique behavior observed which puts it in a class distinct from the secondary and tertiary ions."

c. *Cyclopentyl rearrangements*

The effect of superacids on cyclopentane, cyclopentene, and mono- and dihalocyclopentane derivates was investigated by Olah and his co-workers [167,197,238,258]. Cyclopentane produced cyclopentyl cation (117) by hydride abstraction in FSO_3H-SbF_5 or HF-SbF_5 solution diluted with SO_2ClF below -10°C. Above -10°C, cyclopentane was protonated and yielded, via cleavage and rearrangement, the *t*-pentyl cation [258]. Cyclopentyl cation (117) was also generated from cyclopentene by protonation in SO_2ClF solution [258]. Cyclopentyl chloride in SbF_5-SO_2(SO_2ClF) was first reported to give only cyclopentyl cation [167], but in a subsequent study [258], careful ionization of cyclopentyl chloride revealed the presence of another minor product, believed to be 1-chlorocyclopentenyl cation. Cyclopentyl bromide gave cyclopentyl cation (117) when ionized in SbF_5-SO_2ClF at -78°C, but in SbF_5-SO_2 it gave the cyclopentyl cation 118. As to the nature of the cyclopentyl cation 117, ^1H and ^{13}C NMR studies suggested that it is a rapidly equilibrating ion in which all carbons and hydrogens are magnetically equivalent [167,259]. The results described are formulated in scheme 15.

The effect of superacids on dihalocyclopentanes and hydroxycyclopentenes was similarly investigated [197,258]. Both 1,1-difluoro-(119) and 1,1-dichloro-(120) cyclopentanes gave the corresponding 1-halo-1-cyclopentyl cations (121 and 122, respectively) upon treatment with SbF_5-SO_2ClF at -78°C. However, warming the 1-fluoro-1-cyclopentyl cation 121 to above -30°C, or the 1-chloro-1-cyclopentyl cation (122) to above -60°C, resulted in collapse to the cyclopentyl cation 118. In contrast to the behavior of 119 and 120, when 1,1-dibromocyclopentane (123) was treated with SbF_5-SO_2ClF solution at -120°C, the corresponding 1-bromo-1-cyclopentyl cation (124) was not observed; only formation of the cyclopentyl cation 118 could be noted. Collapse to the cyclo-

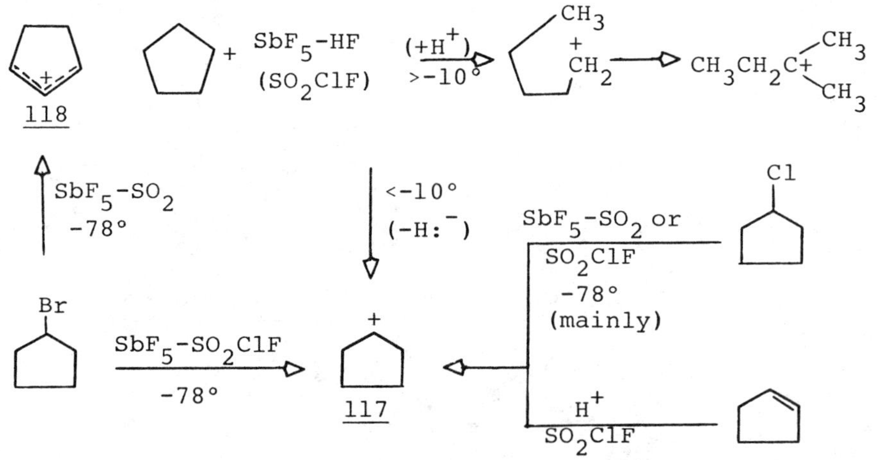

118

+ SbF_5-HF (SO_2ClF) → (+H^+) >-10° → $CH_3CH_2C^+(CH_3)_2$

SbF_5-SO_2, -78° ; <-10° (-H:$^-$) ; SbF_5-SO_2 or SO_2ClF, -78° (mainly) ; SbF_5-SO_2ClF, -78° ; H^+, SO_2ClF

117

Scheme 15

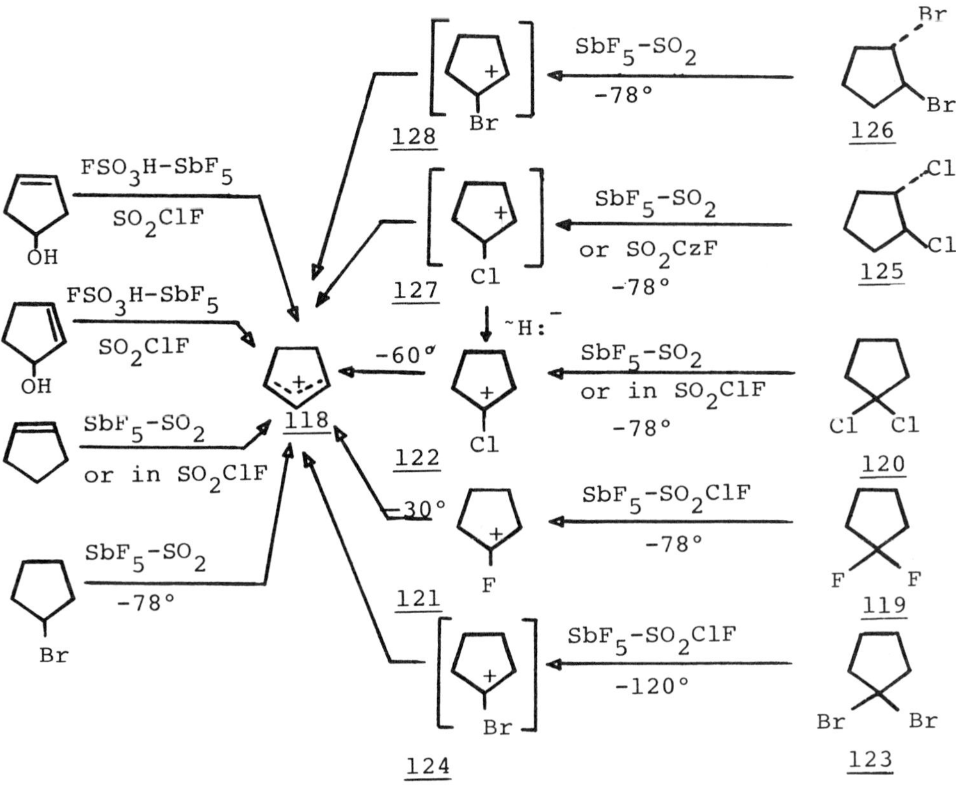

Scheme 16

pentyl cation (118) was again observed when the *trans*-1,2-dichloro (125)- and 1,2-dibromo (126)-cyclopentanes were ionized to the corresponding 2-chloro (127)- and 2-bromo (128)-1-cyclopentyl cations. The formation of cyclopentyl cation 118 from different cyclopentyl and cyclopentyl precursors is depicted in scheme 16.

The mechanism suggested for the formation of cyclopentenyl cation 118 from 1,1- and 1,2-dihalocyclopentanes is depicted in Eq. (118).

SbF_5-SO_2 / -78° ; $-H^+$; SbF_5 ; 118

X=Cl, Br

~H:⁻

SbF_5-SO_2 / -78°

X=F, Cl, Br

(118)

The fact that 1-fluoro-1-cyclopentyl cation (121) was unstable above -60°C, 1-chloro-1-cyclopentyl cation (122) was unstable above -30°C, and

1-bromo-1-cyclopentyl cation (124) could not be observed even at -120°C, suggested the following stability order for these cations:

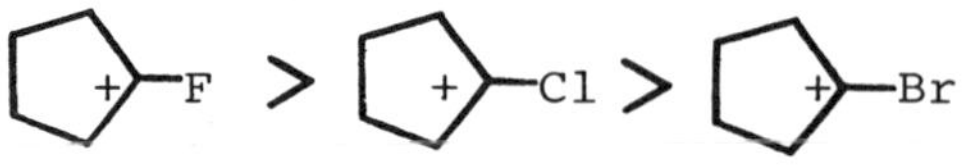

121 > 122 > 124

This stability order of ions 121, 122, and 124, and accordingly the different rates of their transformation into cyclopentyl cation 118, were explained in terms of the degree of halogen "back donation," which was suggested to be in the order F > Cl > Br [197].

d. Methylcyclopentyl and Cyclohexyl Rearrangements

These two systems are closely related to each other, as their ions are interconvertible [260]. In fact, the acid-catalyzed isomerization of methylcyclopentane and cyclohexane has long been recognized [261-263]. Ionization of both cyclohexane and methylcyclopentane in FSO_3H-SbF_5 gave the methylcyclopentyl cation, 129, formed directly from the latter and through rearrangement of the incipient cyclohexyl cation from the former [238]. The reaction was also accompanied by protonation, followed by ring opening and rearrangement to the stable 2,3-dimethyl-2-butyl cation (Eq. 119).

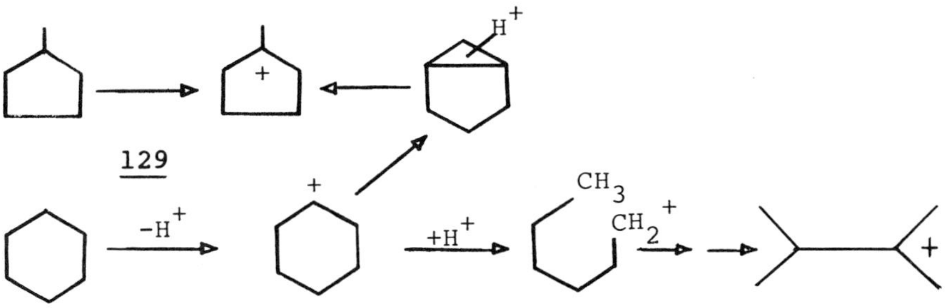

(119)

In 1967, Olah et al. [260] reported the rearrangements accompanying carbocation formation from 1-methylcyclopentyl, cyclopentylcarbinyl and cyclohexyl precursors in strong acid media at -60°C. A summary of their results is given in scheme 17.

As can be seen from this scheme, all precursors 130 to 138 led to the formation of the more thermodynamically stable 1-methylcyclopentyl cation (129). In fact, rearrangement of any initially formed cyclohexyl cation occurred too rapidly to be observed by NMR under the conditions employed.

More recently (1978), the ionization and cleavage behavior of several cycloalkanes, including cyclopentane, methylcyclopentane, cyclohexane, and methylcyclohexane, in the presence of HF-TaF_5 as catalyst and hydrogen as a hydride donor at 50°C was also investigated by Siskin [263a]. The results obtained suggested that, in general, the reactions involved rapid initial isomerization of the cycloalkanes, where possible, followed by a series of selective hydrogenolysis (ring cleavage and hydrogenation) reactions to form mixtures of isomeric alkanes. Thus cyclopentane underwent rapid hydrogenolysis via a secondary carbocation to form a mixture of *n*- and isopentane

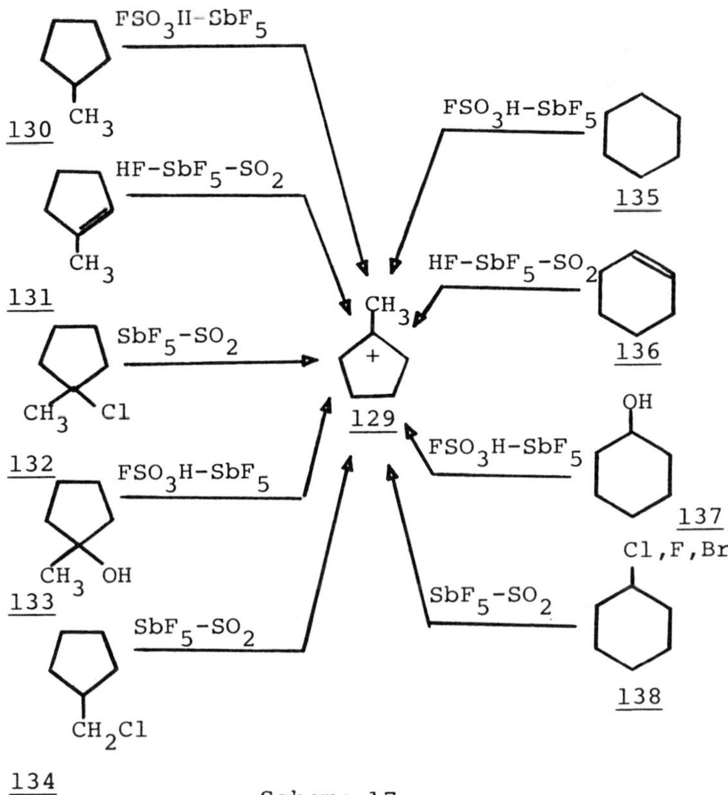

Scheme 17

(Eq. 119a). Similarly, methylcyclopentane underwent hydrogenolysis to an equilibrium mixture of t-hexyl cations, but at a slower rate than cyclopentane.

$$\xrightarrow[H_2]{H^+} \qquad (119a)$$

This difference in rate was attributed to the higher reactivity of the secondary carbocation in the case of cyclopentane. Cyclohexane underwent rapid isomerization to its equilibrium composition with methylcyclopentane, followed by hydrogenolysis and isomerization to form t-hexyl ions:

CH_3 $\xrightarrow{H^+}$ CH_3 H $H^{\delta+}$ $\longrightarrow$ [CH_3 + $\rightleftharpoons$ CH_3 H + H $\rightleftharpoons$ CH_3 H + H]

$\rightleftharpoons$ CH_3 + $\xrightarrow[(2)\,+H^+,\ (3)\,+H^-]{(1)\,+H^-}$ $\xrightarrow[H_2]{H^+}$ C_6H_{14}'s

The generation of *t*-hexyl carbocations in the foregoing reactions was confirmed by NMR analysis and also by trapping by benzene to give as final products 1,3,3-trimethylindanyl cation and hexanes [263b]:

$$3C_6H_{13}^+ + PhH \longrightarrow \text{(1,3,3-trimethylindanyl cation; } CH_3, CH_3, CH_3\text{)} + 2C_6H_{14} + 2H^+$$

The generation of 1-halo-1-cyclohexyl cations by ionization of 1,1-difluoro- or 1,1-dichlorocyclohexane in superacid media was also investigated [197]. The 1-fluorocyclohexyl cation (139) and the 1-chlorocyclohexyl cation (146) were observed below -45 and -90°C, respectively, in SbF_5-SO_2ClF, but at higher temperatures both ions were transformed to the methylcyclopentenyl

F(Cl), F(Cl); SbF_5-SO_2ClF; SO_2, -78° → 139 (F); -120° → 140 (Cl); -45°; -90° → 141 (CH_3)

(120)

cation 141 (Eq. 120). Similar behavior was observed when 2-halo-1-cyclohexyl cations were prepared from *trans*-1,2-dihalocyclohexanes in SbF_5-SO_2ClF [258] (Eq. 121). The formation of 1-methyl-2-cyclopentenyl cation 141 from other precursors under superacid conditions is summarized in scheme 18 [258].

Cl, Cl; SbF_5-SO_2 or in SO_2ClF, -60° → [Cl] → [H^+, Cl] → CH_3, Cl → CH_3, Cl → 141 (CH_3)

Br, Br; SbF_5-SO2, -40° → 141

(121)

The mechanisms proposed for the transformation of cyclohexenyl and bicyclohexyl derivatives into 1-methyl-2-cyclopentenyl cation 141 are worthy of mention. Based on the results of ionizaton in DF-SbF_5-SO_2ClF solution at -78°C, which showed no incorporated deuterium in final ion 141, the

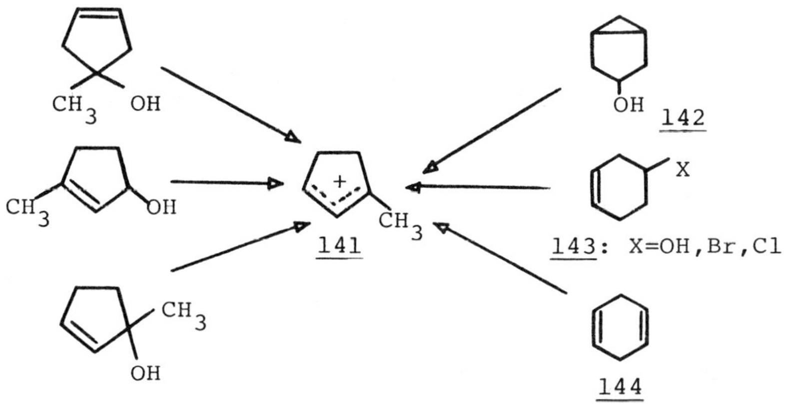

Scheme 18

mechanism in Eq. (122) was postulated for the ionization of 3-bicyclo[3.1.0]-hexanol, 142.

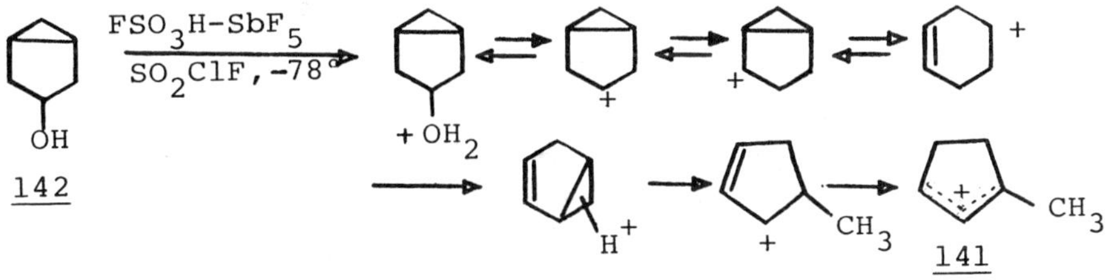

(122)

The mechanisms involved in the transformation of 3-hexenyl derivatives (143) and 1,4-cyclohexadiene 144 were suggested to be as shown in Eq. (123).

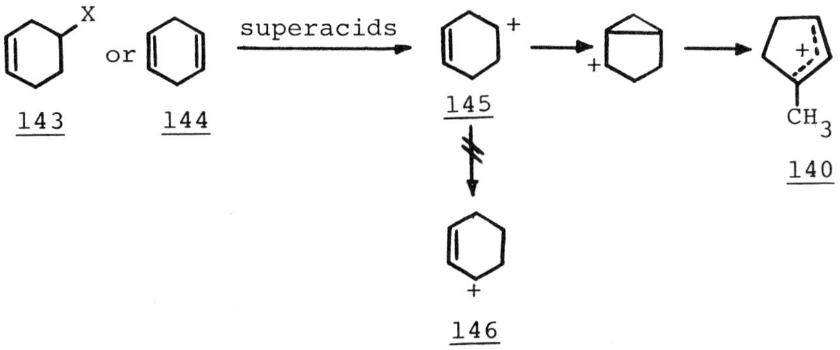

(123)

The fact that 2-cyclohexenyl cation 146 was completely absent, although it is known to be stable under the conditions employed, implied that ring contraction of the 3-cyclohexenyl cation (145) must have taken place prior to any intramolecular hydride shift to 146. However, in a more recent investigation, Farcasiu and Craine [264] concluded that, in contrast to Olah et al. [258], the rate-determining step for ring contraction occurred after the hydride shift.

e. Methylcyclohexyl, Cycloheptyl, and Higher Cycloalkyl Rearrangements

At 60°C in FSO_3H-SbF_5, cycloheptane was ionized to the cycloheptyl cation, which immediately rearranged to the thermodynamically more stable 1-methylcyclohexyl cation (147) [74]. As in other cases, (see Eq. 119), the ionization reaction was accompanied by ring opening and rearrangements leading to heptyl and butyl cations (Eq. 124). Similar behavior was also observed when cycloheptane was allowed to react in the presence of the strong acid HF-TaF_5

superacid media, $-H:^-$ → 147

superacid media, $+H^+$, -25° → t-$C_7H_{13}^+$ + t-$C_4H_9^+$ (124)

and hydrogen [263a]. Under the latter conditions, however, isobutane and propane were the final stable products resulting from heptane hydrocracking reactions.

Similar behavior was noted when 1-fluoro- and 1-chloro-1-cycloheptyl cations (148 and 149, respectively) were prepared from 1,1-difluoro- and 1,1-dichlorocycloheptane in SbF_5-SO_2ClF solution at low temperatures. At higher temperatures both halo cations rearranged into the methylcyclohexenyl cation (150) [197] (Eq. 125).

F(Cl), F(Cl) SbF_5-SO_2ClF → -78° → 148 (F) → -45° → 150 (CH_3); → -120° → 149 (Cl) → -100° → 150 (125)

The latter ring contraction as well as that of 1-halo [197] and 2-halocyclohexyl [258] cations to methylcyclopentenyl cation (141) were rationalized in terms of protonated cyclopropane intermediates generally formulated in scheme 19.

In 1970, Olah et al. [265] described the generation of 7- to 12-membered cycloalkyl cations by the action of superacids on the corresponding alcohols at -78°C. Under these conditions, the initially formed cycloheptyl, cyclooctyl, cyclononyl, cycloundecyl, and cyclododecyl cations underwent rapid rearrangement (ring contraction) to the correspondingly substituted 1-n-alkyl-1-cyclohexyl tertiary cations, 151. The latter cations, which were recently shown to have the twist-boat ground-state conformation 152 [266],

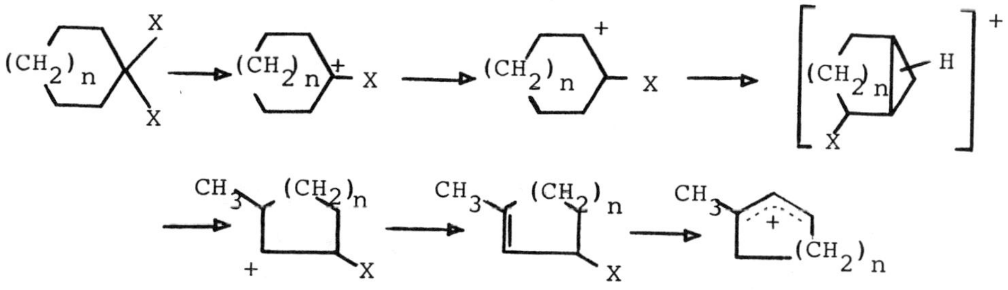

(X = F,Cl)

(n = 1,2)

Scheme 19

were also generated, in some cases, by similar treatment of isomeric pentanols and hexanols, as well as from other suitable cycloalkane precursors (scheme 20).*

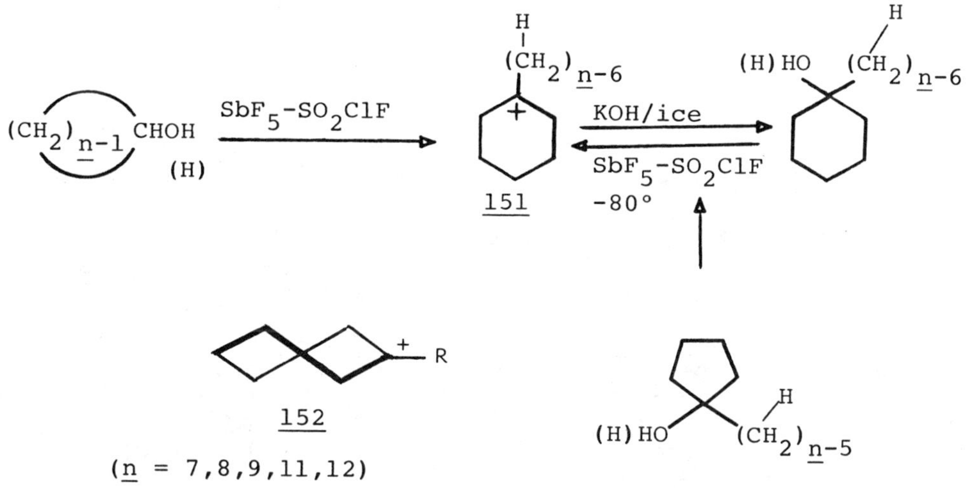

$(CH_2)_{n-1}$ CHOH (H) $\xrightarrow{SbF_5-SO_2ClF}$ 151 $\underset{SbF_5-SO_2ClF,\ -80°}{\overset{KOH/ice}{\rightleftharpoons}}$

152

(n = 7,8,9,11,12)

Scheme 20

The cyclodecyl cation behaved differently, however, as its attempted generation from cyclodecanol (153) was accompanied by effervescence of hydrogen, and NMR analysis of the resulting yellow solution gave no evidence of a cyclodecyl ion (154) or of a 1-*n*-butylcyclohexyl ion (151, n = 10). Instead, the NMR data indicated the formation of a rapidly equilibrating bridgehead decalyl ion, 155, through transannular bond formation. The latter ion was also obtained by dissolving the *cis*, *trans* isomers of decalin (156 and 157 and of 9-decalol (158 and 159) in the same acid systems [238] (scheme 21).

*In the case of methylcyclohexyl cation, the twist-boat conformation (152, R = CH_3) was shown by NMR analysis to be 500 cal/mol more stable than the chair conformation [266].

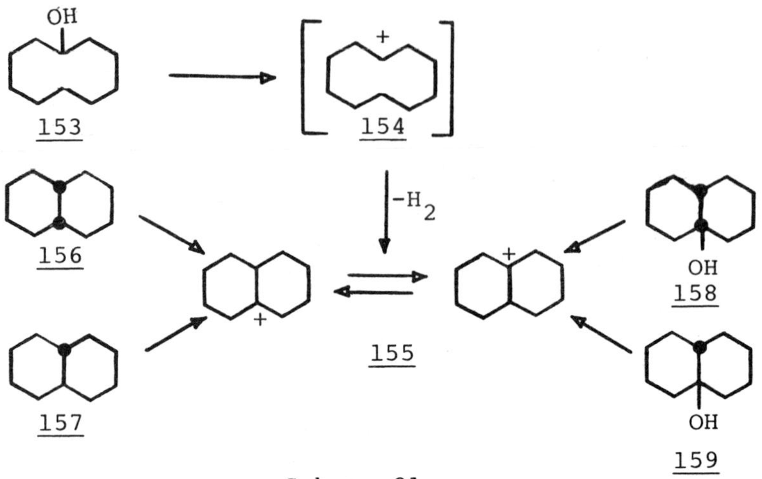

OH

153

154

$-H_2$

156

155

OH

158

157

OH

159

Scheme 21

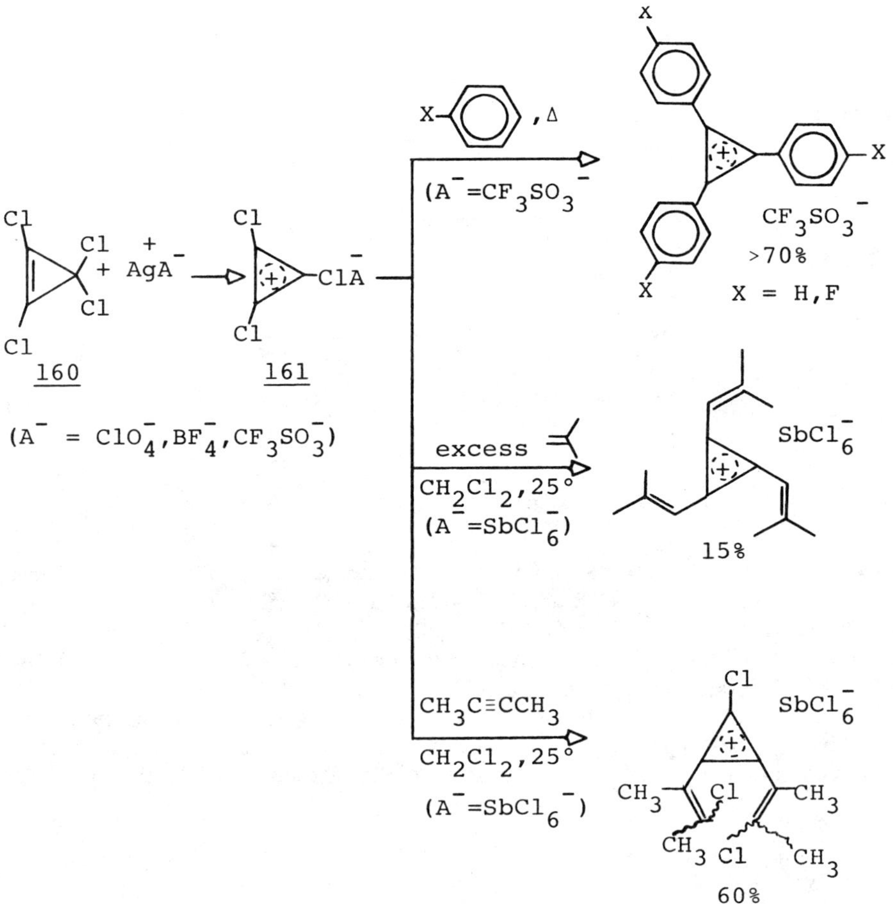

X, Δ

(A^- = $CF_3SO_3^-$)

X

X

X

$CF_3SO_3^-$

>70%

X = H,F

Cl Cl + AgA$^-$ Cl Cl

Cl ClA$^-$ Cl

160

161

(A^- = ClO_4^-, BF_4^-, $CF_3SO_3^-$)

excess

CH_2Cl_2, 25°

(A^- = $SbCl_6^-$)

$SbCl_6^-$

15%

$CH_3C{\equiv}CCH_3$

CH_2Cl_2, 25°

(A^- = $SbCl_6^-$)

Cl

$SbCl_6^-$

CH_3 Cl CH_3

CH_3 Cl CH_3

60%

Scheme 22

2. Rearrangements in monocycloalkenyl and related systems

Various groups have investigated the generation of cycloalkenyl carbocations either in concentrated sulfuric acid or in much stronger superacid solvents [158,267]. The simplest possible cycloalkenyl cation, the cyclopropenyl cation, has been reported in the form of various derivatives [268,269]. A recent example is the preparation of the trichlorocyclopropenium cation 161 from tetrachlorocyclopropene (160) and various silver salts (e.g., ClO_4^-, BF_4^-, $SbCl_6^-$, and $CF_3SO_3^-$) [268]. Interestingly, ion 161 was found to be a very reactive Friedel-Crafts alkylating agent in reactions with aromatics as well as with olefinic and acetylenic bonds [268] (scheme 22).

In the latter reaction with 2-butyne, the net result is the unprecedented insertion of an acetylene into the C-Cl bond of an aromatic carbocation. Although the stereochemistry of this process is unknown, NMR data suggested that the product consisted of a 4:1 mixture of two isomers.

The next-higher number of the series, the cyclobutenyl cation, was the subject of a recent study by Olah and co-workers [270]. By generating and examining the nature of a number of cyclobutenyl cations, these authors demonstrated that, depending on structure, the behavior of these cations can progress from classical allylic delocalization (i.e., 162a ↔ 162b), through species displaying increasing contributions from 1,3-π-orbital overlap (162c), and finally, the homoaromatic homocyclopropenium ions, (162d).

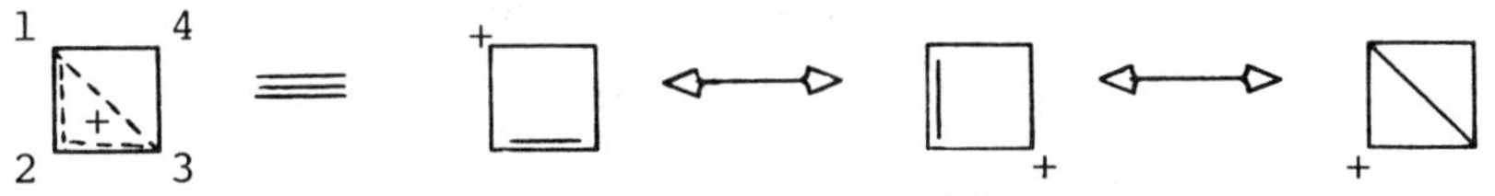

1 4 2 3 162d ≡ 162a ↔ 162b ↔ 162c

In 1972, Olah et al. [258] presented data on the generation and characterization of cyclopentenyl and cyclohexenyl cations and their methylated derivatives. These data revealed two significant points: (1) no NMR spectral characteristics corresponding to the bishomocyclopropenyl structure 163

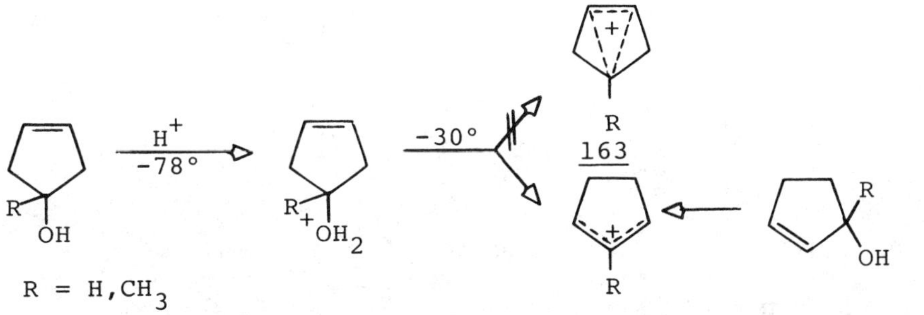

H+ −78° −30° R 163 R OH OH2 R R = H,CH3 (126)

(Eq. 126) were observed, indicating that no significant 1,3-orbital interaction was taking place in cyclopentenyl cations; and (2) the formation of the allylic cyclohexenyl cations from different precursors took place directly and without rearrangement when the leaving group was directly attached to the carbon atom α to the double bond. In contrast, when the leaving group was attached to the carbon atom at the homoallylic position, ring contraction or other rearrangement reactions occurred.

In another 1972 paper, Olah and Liang [271] investigated the generation and the behavior of the larger-ring cycloheptenenyl, cyclooctenyl, and cyclononenyl cations and some of their derivatives in superacid media at low temperature. In this study, the parent cycloheptenyl cation was successfully generated, whereas the parent cyclooctenyl and cyclononenyl cations were found to be unstable, although a series of their substituted derivatives were obtained. In agreement with others [267,272], Olah and Liang concluded that the stabilities of cycloalkenyl cations decrease as the ring size increases. A similar trend of stability has also been noted in simple cycloalkyl carbocations [265].

3. Rearrangements in saturated and unsaturated multicyclic systems

In recent years there has been growing interest in the generation of saturated and unsaturated multicyclic carbocations under conditions that allow the determination of the structure and isomeric composition of stable carbocations. Since most of the earlier work on these cations is reviewed elsewhere, e.g., [158,162,163,273-277], in this section we present only a few of the recent examples to illustrate some of the interesting aspects of long-lived multicyclic carbocations.

In the continuing effort to explore the effect of substituents on the degenerate cyclopropylcarbinyl cation rearrangements, a number of 2-bicyclo [n.l.o]alkan-2-ols (164 and 165) and bicyclo[3.2.0]heptan-6-ols (166) were

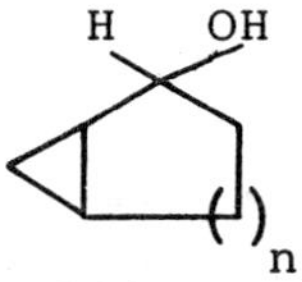

164 ($n = 1,2,3,4$)

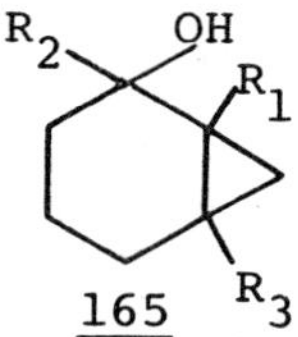

165 ($R_1=R_3=H$; $R_2=CH_3$) ($R_1=CH_3$; $R_2=R_3=H$) ($R_1=Ph$; $R_2=R_3=H$) ($R_1=R_2=H$; $R_3=CH_3$)

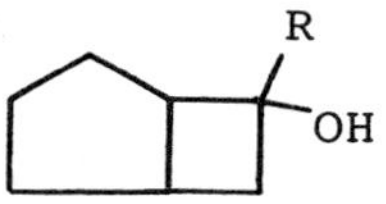

166 ($R=H,CH_3,Ph$)

ionized in superacidic solutions of SbF_5, FSO_3H, and SbF_5-FSO_3H in SO_2ClF at -140°C, and the resulting ions were spectroscopically investigated [242]. Some of the ionizations and accompanying rearrangements are summarized in Eqs. (127) to (133).

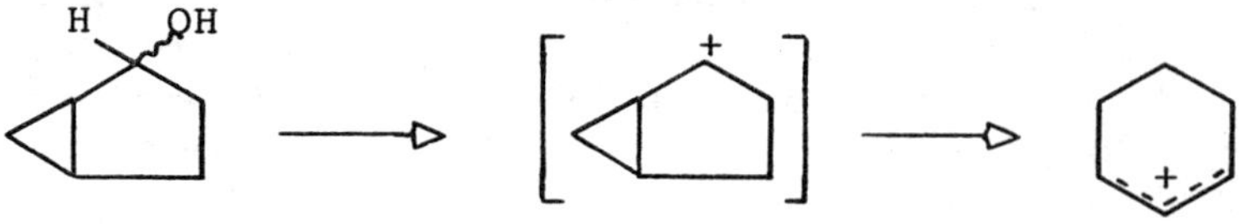

(127)

H OH

+

R

OH

(128)

>-50°

+

H OH

+

+

+

+

(129)

CH_3 OH

CH_3

+

H OH

CH_3

(130)

H OH

CH_3

-140°

+

CH_3

CH_3

+

>-80°

+

CH_3

(131)

CH_3

OH

CH_3

+

Ph

OH

-140°

Ph

+

(132)

H OH

+

Wagner-
Meerwein
shifts

+

OH

(133)

On the basis of these results, it was concluded that the barriers for the degenerate cyclopropylcarbinyl cation equilibration processes seemed to depend on both the ring size and the nature of the substituents at the C_1 or C_2 carbons in the bicyclic system [242].

In 1968, Olah and Lukas [238] studied, among other things, the behavior of a number of polycycloalkanes under superacid conditions (Eq. 134 to 137). Bicyclo[1.1.0]butane (167, Eq. 134) was found to polymerize in FSO_3H-SbF_5-SO_2ClF solution even at -125°C, and no conditions were found suitable to observe any well-identified products. Under similar conditions, bicyclo [3.1.0]hexane (168, Eq. 135) underwent rapid rearrangement to 1-methylcyclopentyl cation (129). Since hydride abstraction could not be detected in the latter rearrangement, it was explained in terms of protonation of the cyclopropyl ring followed by ring opening and rearrangement to 129 (Eq. 135).

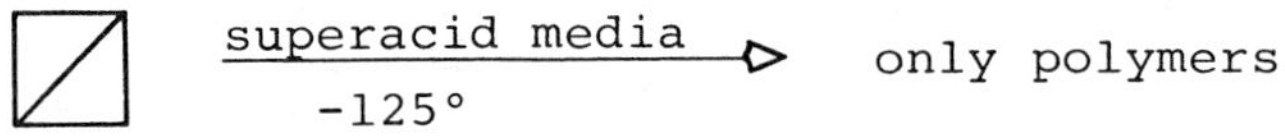

167 —superacid media, -125°→ only polymers (134)

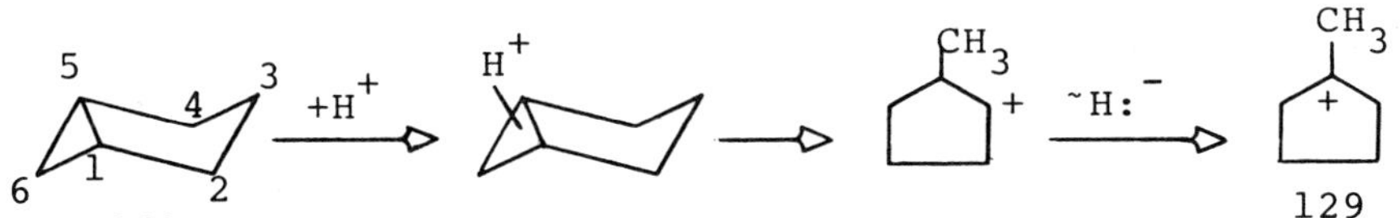

168 —$+H^+$→ … → CH_3 (+) —$-H:^-$→ 129 (135)

In contrast, in superacid media, both bicyclohexyl (169) and decalin (171) were deprotonated, leading to the rapidly equilibrating cations 170 and 172 (Eqs. 136 and 137), respectively.

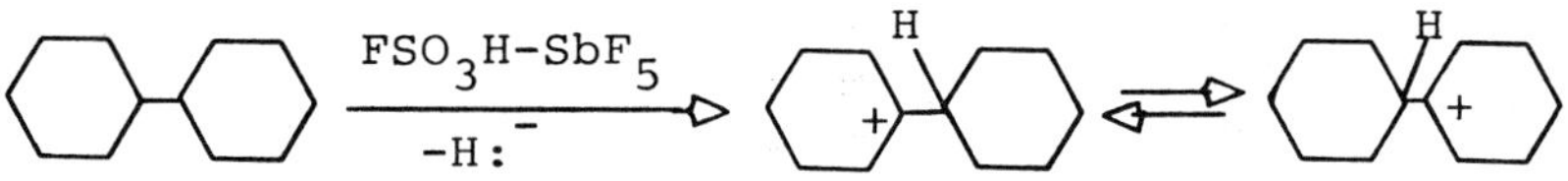

169 —FSO_3H-SbF_5, $-H:^-$→ 170 (136)

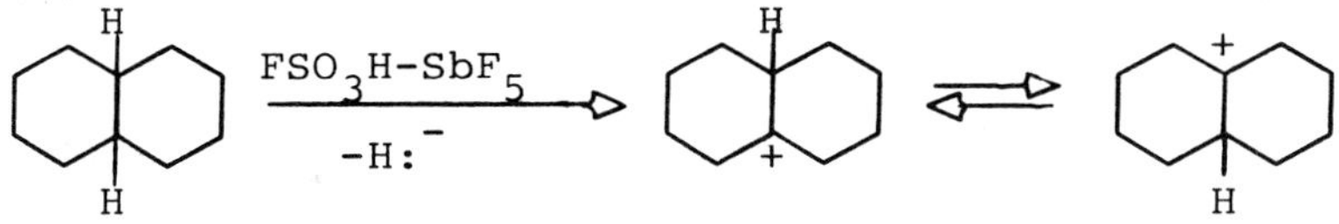

171 —FSO_3H-SbF_5, $-H:^-$→ 172 (137)

The behavior of decalin, bicyclopentyl, and bicyclohexyl in the presence of the strong acid HF-TaF_5 and hydrogen at 50°C was also investigated [263a]. Under these conditions, *cis*-decalin or a mixture of isomeric decalins underwent rapid isomerization to the *trans*-decalin isomer followed by a smooth hydrogenolysis (ring cleavage then hydrogenation) reaction to form methylcyclopentane and isobutane as the primary products (Eq. 137a). Further re-

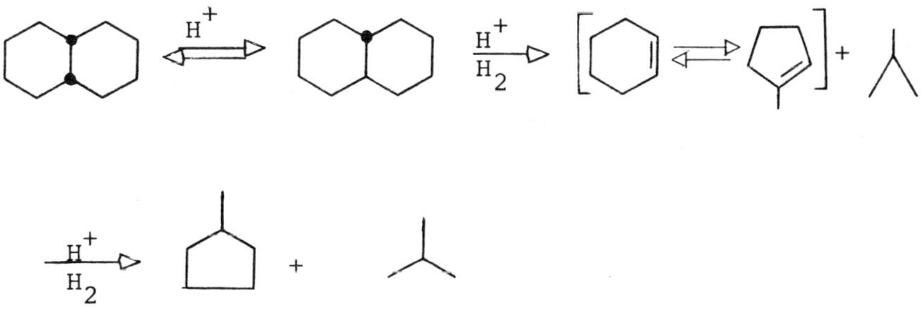

(137a)

action of methylcyclopentane, as described previously in Eq. 119a, leads to isomeric hexanes.

Under the same conditions, bicyclopentyl did not undergo hydrogenolysis, but isomerized rapidly to *cis*-decalin, then to *trans*-decalin (Eq. 137b); the

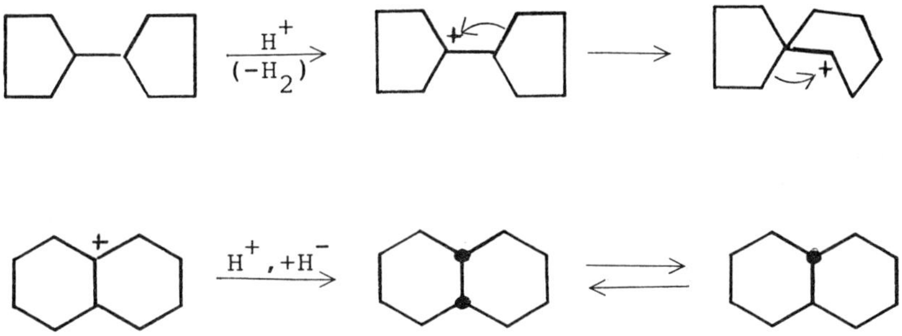

(137b)

latter subsequently underwent the transformations of Eq. (137a). Bicyclohexyl rearranged by an analogous mechanism directly to *cis*-dimethyldecalin(s), which then isomerized to the *trans* isomer(s). The latter can be isolated or allowed to react further to form dimethylcyclohexane(s) and isobutane as the final reaction products, as follows:

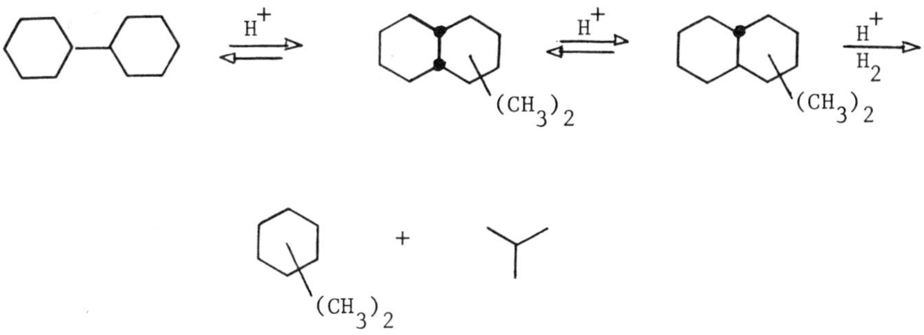

This behavior of decalin in the presence of $HF-TaF_5-H_2$ (Eq. 137a) is in contrast with that in the presence of other catalysts. In earlier accounts, decalin was reported to give dimethylbicyclo[3.3.0]octane with HI at 300°C or $AlBr_3$ at 100°C [277a] or bicyclo[3.2.1]octane with $AlCl_3$ at 130°C [277b]. Other products in the presence of $AlCl_3$ or $AlBr_3$ catalysts were cyclohexane, methylcyclohexane, and 1,3,5-trimethylcyclohexane [277b,277c].

The acid-catalyzed isomerization of *cis*-hydrindane (*cis*-bicyclo[4.3.0]nonane) was also examined. In the presence of $AlBr_3$, *cis*-hydrindane underwent isomerization into the *trans* isomer [277c]. Under the hydrogenolysis conditions of $HF-TaF_5-H_2$, similar isomerization was observed, but was followed by a rapid cleavage of the 4-9 bridge to cyclononane, then isomerization to the much more stable 1,2,4-trimethylcyclohexane isomer (263a).

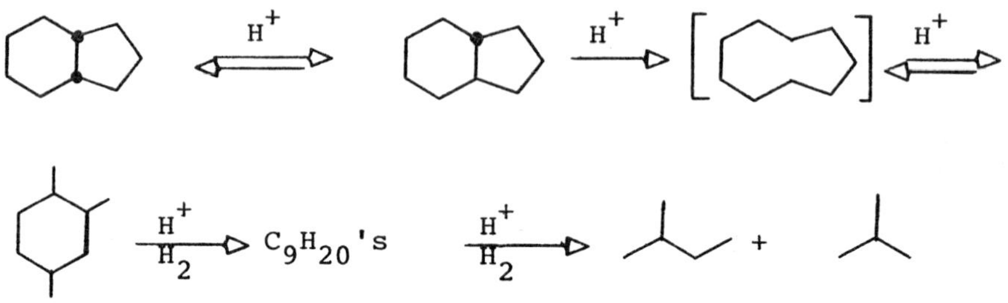

H^+ H^+ H^+

H^+/H_2 C_9H_{20}'s H^+/H_2 +

The isomerization of norbornane was investigated under different superacid conditions. When norbornane was dissolved in $HF-SbF_5$ solution at room temperature, an equilibrium mixture of methylcyclohexyl and 1,2- and 1,3-dimethylcyclopentyl cations were formed [277d]. On reduction with hydrogen, this mixture gave methylcyclohexane as the sole product. Similar behavior was again noted when norbornane was treated with $HF-TaF_5$ and hydrogen at 50°C; rapid and quantitative hydrogenolysis to methylcyclohexane took place [263a].

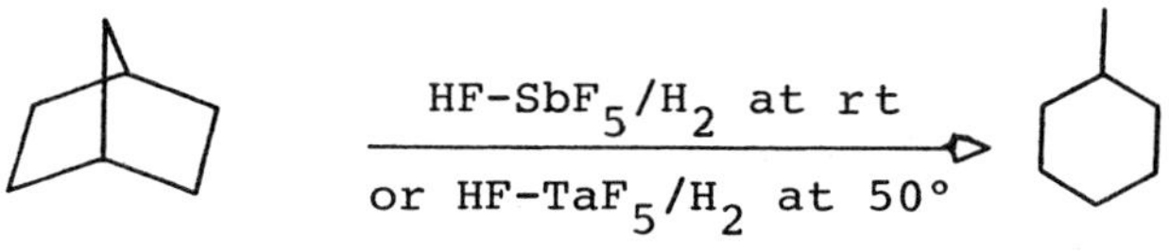

$HF-SbF_5/H_2$ at rt
or $HF-TaF_5/H_2$ at 50°

In 1973, Seybold et al. [278] generated the long-lived 2-bicyclo[2.1.1]hexyl cation 174 by the reaction of 2-chlorobicyclo[2.1.1]hexane 173 with antimony pentafluoride in sulfuryl chorofluoride at -125°C (Eq. 138). Based on NMR analysis of cations from labeled and nonlabeled starting material, they considered the resulting cation to be an equilibrating σ-bridged ion, in agreement with the solvolytic studies of earlier workers [279,280].

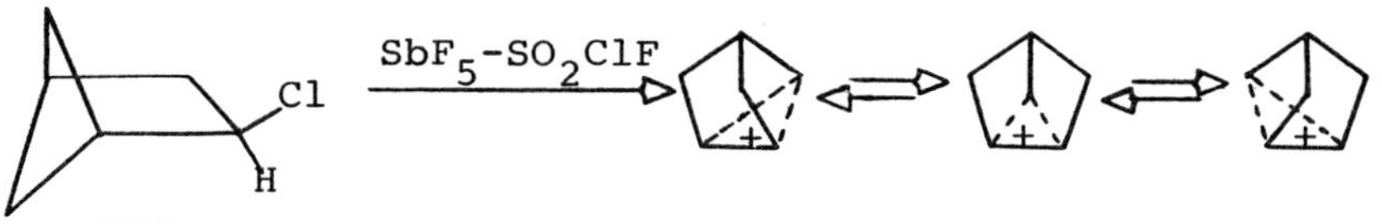

SbF_5-SO_2ClF (138)

Cl H

173 174

Prompted by the preceding findings, Olah et al. [281] reported the results of their proton and ^{13}C NMR spectroscopic studies of the parent cation, 174, as well as the 2-methyl (175), 2-phenyl (176), and 2-hydroxy 177 2-bicyclo[2.2.1]hexyl cations. The parent secondary ion 174 was again prepared from 173 in SbF_5-SO_2ClF at -130°C, and the tertiary ions 175 and 176 were prepared from the corresponding alcohols 178 and 179 in FSO_3H-SO_2ClF

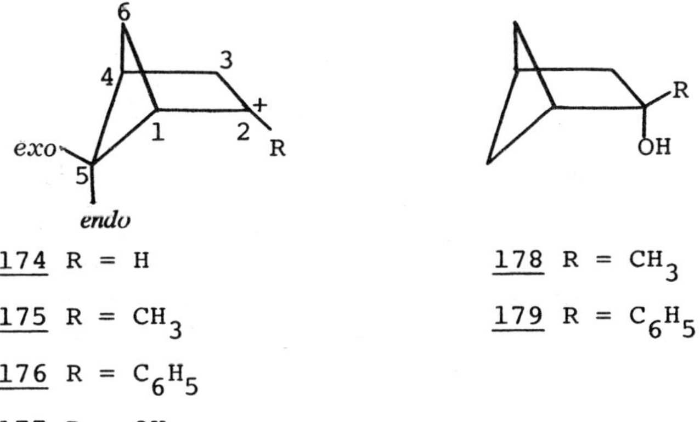

solution at -130 and -78°C, respectively. The results of Olah et al., while confirming those of Seybold et al. concerning the degenerate nature of the parent secondary 2-bicyclo[2.1.1]hexyl cation (174), indicated that the tertiary cations 175 and 176 are typical carbocations with limited σ delocalization.

Olah et al. [282] also prepared a series of bicyclo[3.1.0]hexenyl and benzobicyclo[3.1.0]hexenyl cations under stable ion conditions. The conditions employed to generate the parent ions (180 and 181) are presented in Eqs. (139) and (140).

OCH_3 —— FSO_3H-SbF_5-SO_2ClF, -78° ——> 180 + $CH_3OH_2^+$ (139)

or H OH —— FSO_3H-SO_2ClF, -78° ——> 181 H (140)

Based on ^{13}C NMR data, ions 180 and 181 were considered to be cyclopentenyl-like cations with charge delocalization into the fused cyclopropane ring. Notably, ion 181 was found to undergo slow ring opening to give the naphthalenium ion 182 upon warming to above -50°C (Eq. 141).

181 H —— FSO_3H-SO_2ClF, >-50° ——> 182 H H (141)

The structure and the stability of the norbornyl (bicyclo[2.2.1]heptyl), norbornenyl (bicyclo[2.2.1]heptenyl), and norbornadienyl (bicyclo[2.2.1] heptadienyl cations have been subjects of much interest and lengthy controversy [90,132,158,162,163,168,259,273-276,283-291]. The heart of the debate has been whether these cations exist as rapidly equilibrating classical ions or rather as σ-bridged nonclassical ions. Based on ^1H and ^{13}C NMR studies of parent and substituted norbornyl, norbornenyl, and norbornadienyl cations (prepared from halo or hydroxy precursors under stable ion conditions), Olah and his co-workers [259,288,290,292] concluded that the structures of the 2-norbornyl cation 183, the 7-norbornenyl cation 184 and the 7-norbornadienyl cation 135 are those of typical nonclassical carbocations ions. Examples illustrating the preparation of these ions are given in Eqs. (142)-(144).

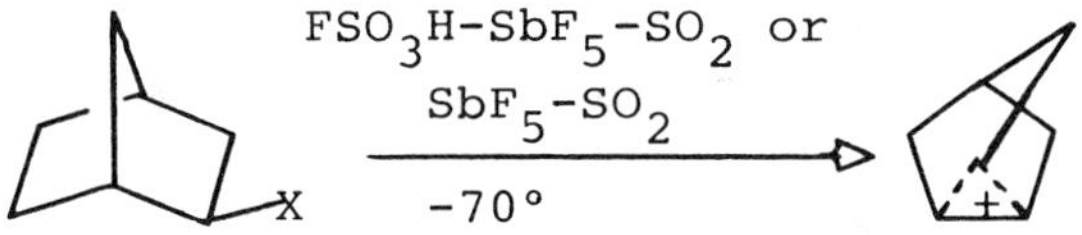

FSO_3H-SbF_5-SO_2 or SbF_5-SO_2, −70° → 183 (142) [168,259,288]

(X = F,Cl,Br,OH, etc.)

FSO_3H-SO_2ClF, −78° → 184 (143) [292]

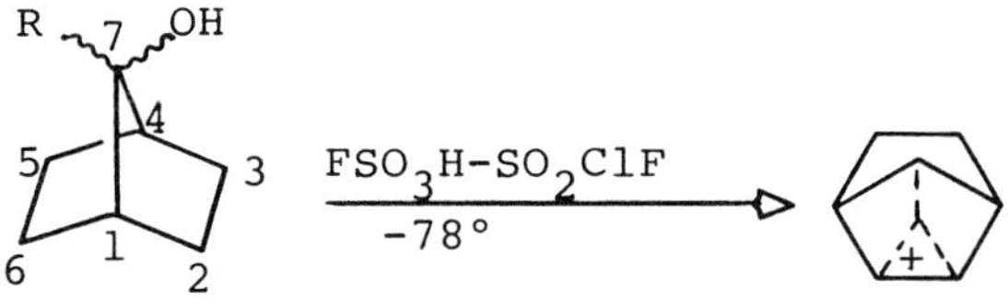

(R = H,$CH_3CH_2CH_3$,C_6H_5,OH)

FSO_3H-SbF_5-SO_2, −70° → 185 (144) [259]

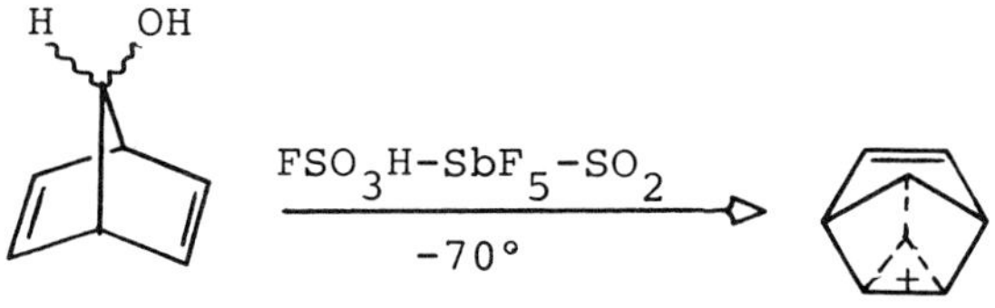

Support for the nonclassical nature of norbornyl cations was more recently gained by Solomon and Field [293] through stability measurements using the technique of time-resolved high-pressure mass spectrometry. These measurements indicated that the norbornyl cation has 10 kcal/mol more stability than that which would be expected for a secondary species. In contrast to the nonclassical nature of 2-norbornyl (183), 7-norbornenyl (184), and 7-norbornadienyl (185) cation, Olah and Liang [253] found that both secondary and tertiary 3-nortricyclyl cations (Eqs. 145 to 147) are classical cations with charge delocalization into the cyclopropyl ring.

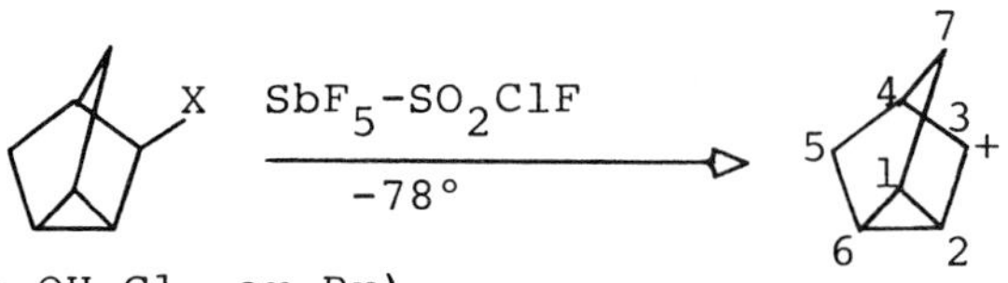

SbF_5-SO_2ClF, −78° (145)

(X = OH,Cl, or Br)

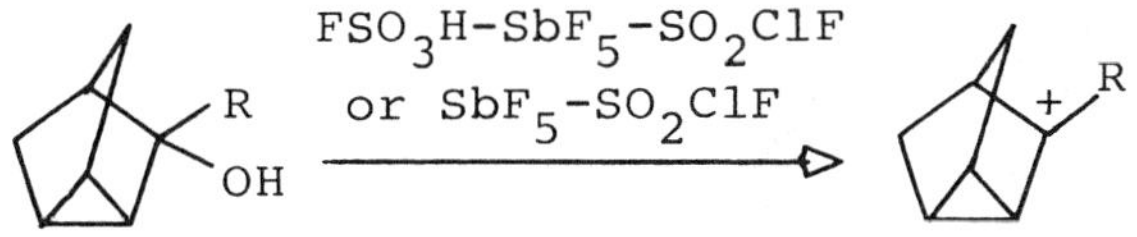

$$\text{(R-substituted tricyclic alcohol)} \xrightarrow[\text{or } SbF_5\text{-}SO_2ClF]{FSO_3H\text{-}SbF_5\text{-}SO_2ClF} \text{(cation)} \quad (146)$$

($R = CH_3, CH_2CH_3, C_6H_5, OH$)

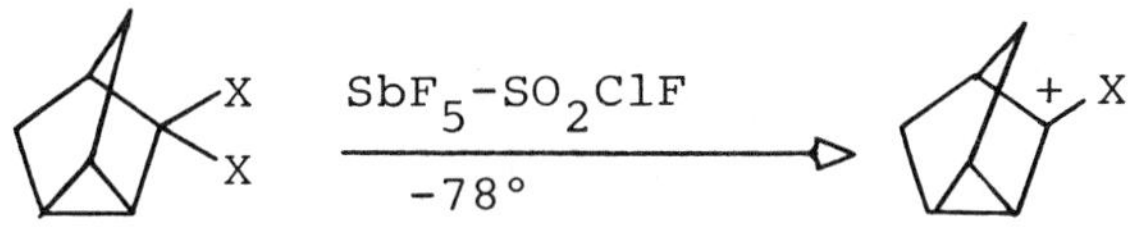

$$\text{(X,X-disubstituted tricyclic)} \xrightarrow[-78°]{SbF_5\text{-}SO_2ClF} \text{(cation, X)} \quad (147)$$

($X = Cl, F$)

The generation and behavior of the bicyclo[3.3.0]-1-octyl cation (186) in superacid media was reported in 1971 [294]. This ion was generated either in FSO_3H-SbF_5-SO_2ClF or in SbF_5-SO_2ClF at -78°C from a series of bicyclo [2.2.2] octyl (187), bicyclo [3.3.0] octyl (188), -bicyclo [3.2.1] octyl (189) 2-norbornanecarbinyl (190), and tricyclo[$3.3.3.0^{2,6}$]octyl (189), 2-norbornanecarbinyl (190), and tricyclo[$3.3.3.0^{2,6}$]octyl (191) derivatives. Examples of these precursors are shown in scheme 23.

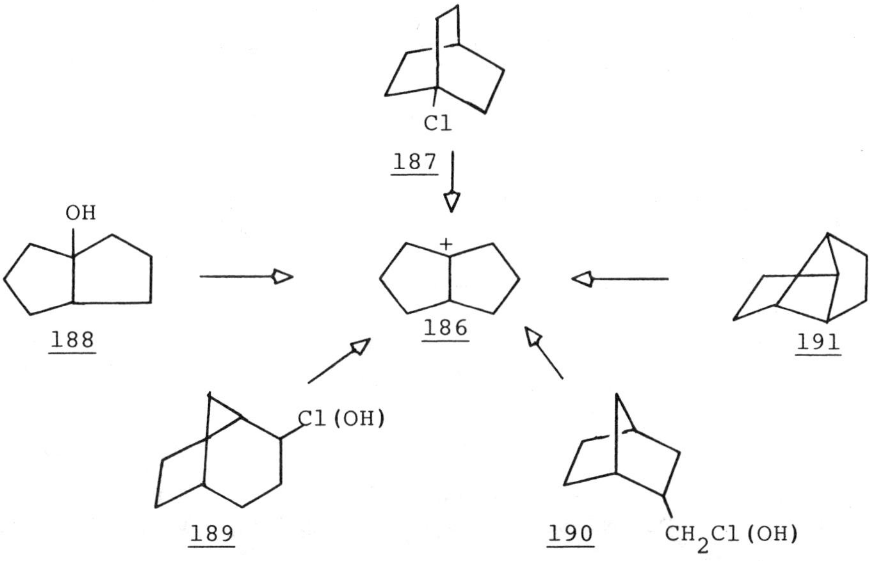

Cl

187

OH

188

186

191

Cl (OH)

189

190 CH_2Cl (OH)

Scheme 23

It might be mentioned here that in Olah's study [294], the bicyclo[3.3.0]-1-octyl cation (186) was found to be stable at temperatures below -30°C, but warming to 0°C resulted in an irreversible rearrangement to the 2-methylnorbornyl cation (192). The PMR spectrum of the latter cation was found to be identical with that generated directly from 2-methylnorbornyl derivatives (193) [295-296] (Eq. 148). The rearrangement of (186) to (192) in indeed a novel one, as both isomeric 2-methylnorbornanes (194) [297,298] and *cis*-bicyclo [3.3.0]octane (194a) [299] were found to isomerize to bicyclo[3.2.1] octane (195) in the presence of $AlCl_3$ at 75°C (Eq. 149); i.e., the direction of the rearrangement is exactly the reverse of that observed under the in-

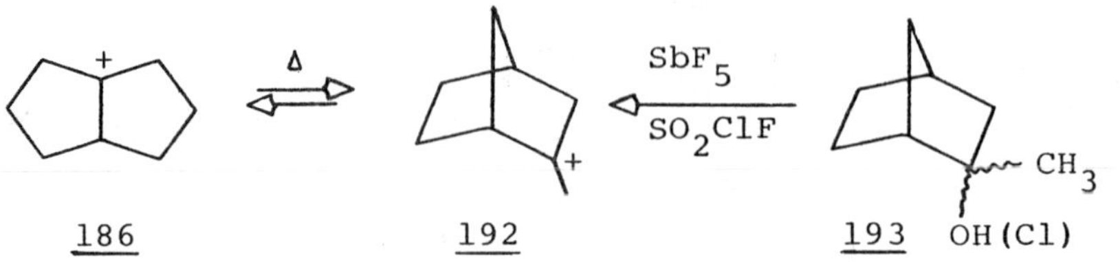

186 ⇌(Δ) 192 ←(SbF5 / SO2ClF) 193 OH(Cl) CH3 (148)

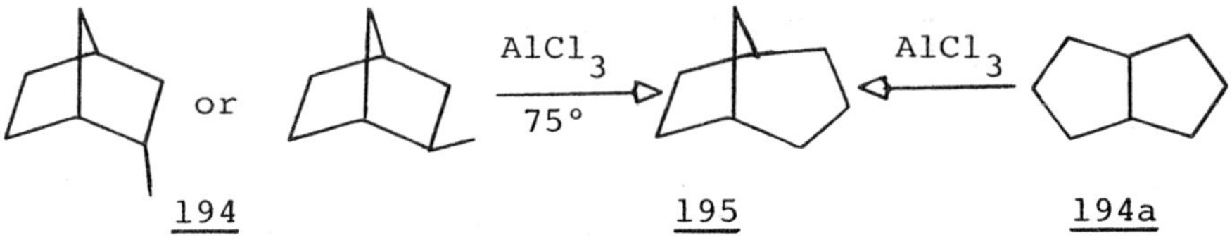

194 or → (AlCl3, 75°) 195 ←(AlCl3) 194a (149)

fluence of superacids. This novel rearrangement was explained in terms of the transformation shown in Eq. (150).

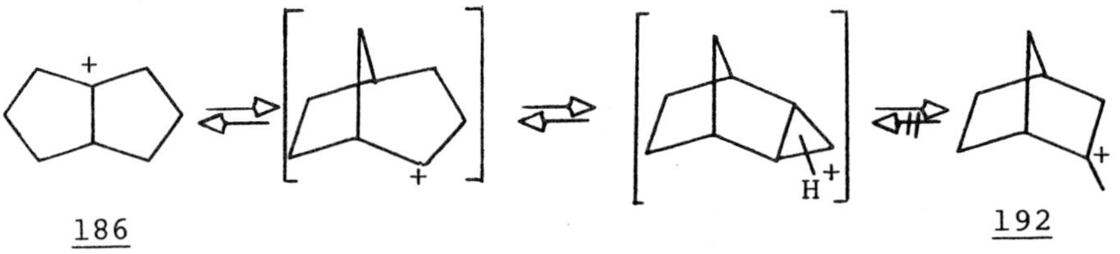

186 ⇌ [ ] ⇌ [ ] ⇸ 192 (150)

It might also be mentioned that the results of Olah and Liang [294], together with those of numerous others on the interconversions of bicyclooctanes [297-300] and their simple monofunctional derivatives by carbocation processes [301], have revealed the following stability order:

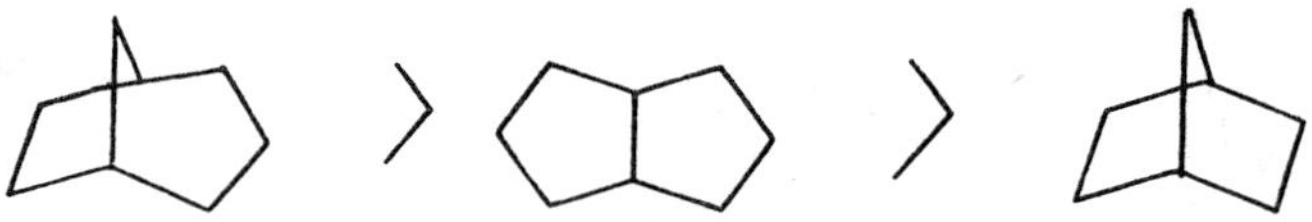

In accordance with this stability order, treatment of either 2-methylnorbornane (194) or 1-methylnorbornane (196) with $AlCl_3$ at 50°C gave the following distribution of products:

194 or 196 →(50°, Δ) trace 3% 55.8% 37.5% 3.6%

In an attempt directly to observe and investigate bicyclononyl and methylbicyclooctyl cations, Olah et al. [302] studied the ionization of precursors 197 to 204 in FSO_3H-SbF_5-SO_2ClF or SbF_5-$SO_2ClF(SO_2)$ at low temperature. The product obtained from all these precursors was suggested to be the 2-methyl-2-bicyclo[3.2.1]octyl cation 205. Later, this conclusion was criticized

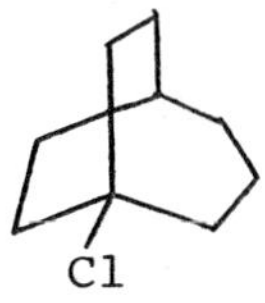

197 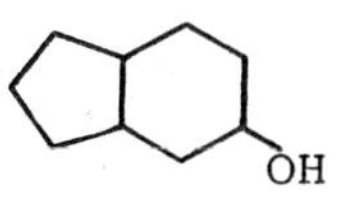 198 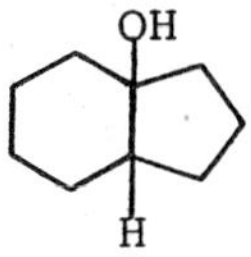 199 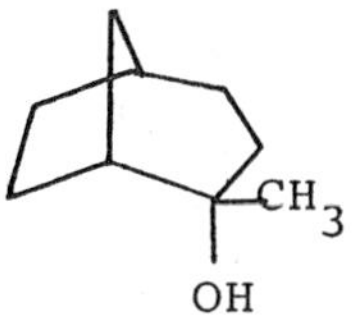 200 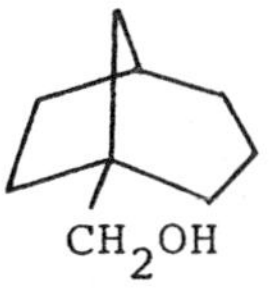 201

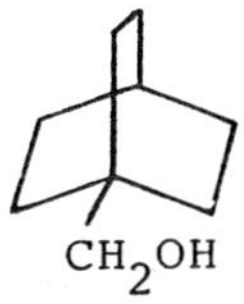

202 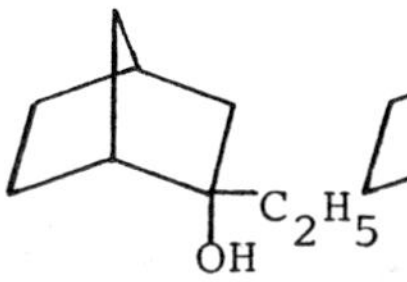 203 204 205 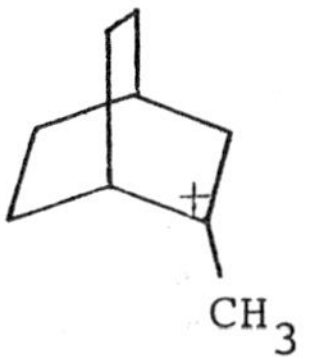 206

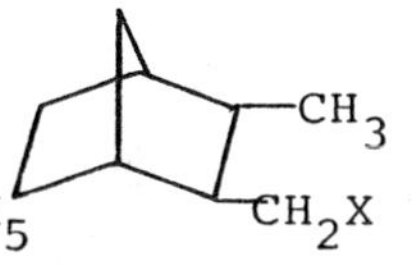

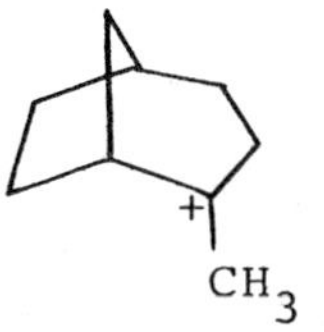

(X=OH,Cl)

by Kirchen and Sorensen [266], who suggested that the detectable product was not 205, but instead the 2-methyl-2-bicyclo[2.2.2]octyl cation 206. Moreover, it was indicated that successful preparation of carbocation 206 was achieved when the corresponding chloride was added to SbF_5-SO_2ClF at -135°C. At -103°C ion 206 rearranged rapidly into the bicyclo[2.2.2] system (205) by way of double Wagner-Meerwein shift (Eq. 151) [266]. The nearly complete

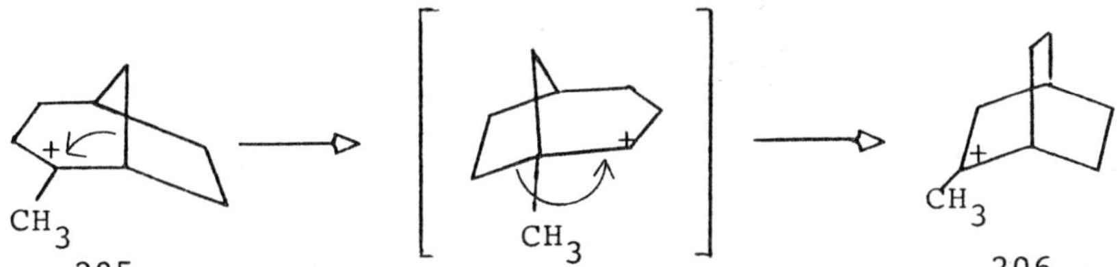

(151)

conversion of 205 into 206 at equilibrium is in direct contradiction to an earlier statement [303] claiming that transposition from the bicyclo[2.2.2]octyl cation represented an increase in thermodynamic stability because the change involved the conversion of a boat conformation of the cyclohexyl cation to a chair conformation.

Many rearrangements are known involving the interconversion of adamantane derivatives [279,304-309]. For example, 2-adamantanol (207) is converted in concentrated H_2SO_4 at 28°C to an equilibrium mixture containing over 98% 1-adamantanol (208) [310]. Similarly, 2-methyl-2-adamantanol (209) undergoes extensive isomerization in concentrated H_2SO_4 [311,312]. At room temperature or below, 2-methyl-1-adamantanol (210) and the *syn* and *anti* isomers

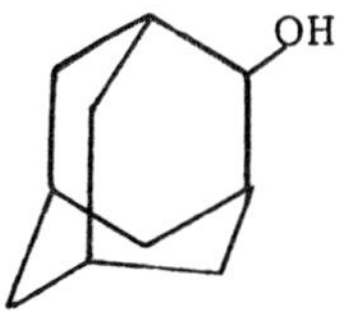

207 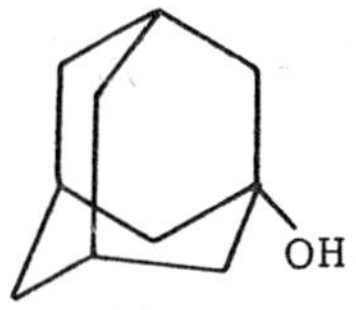 208 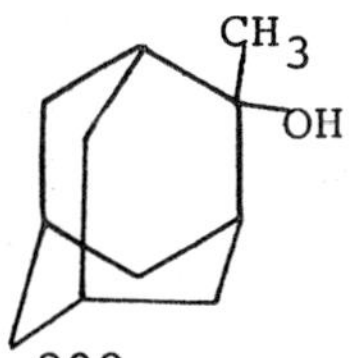 209 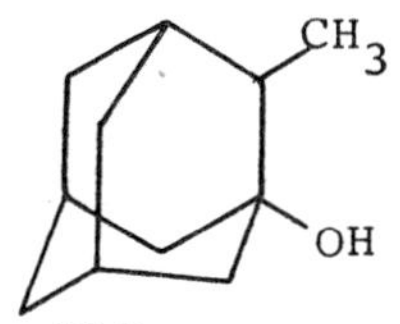 210

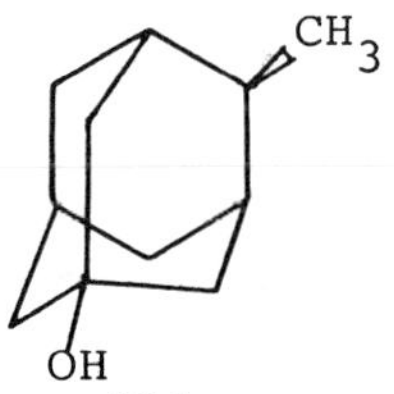

syn-211 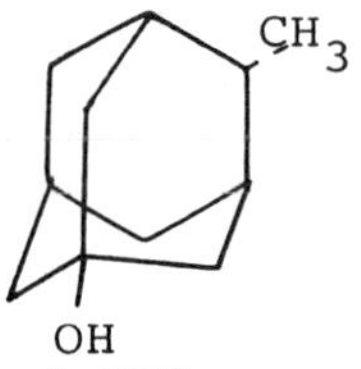 anti-211 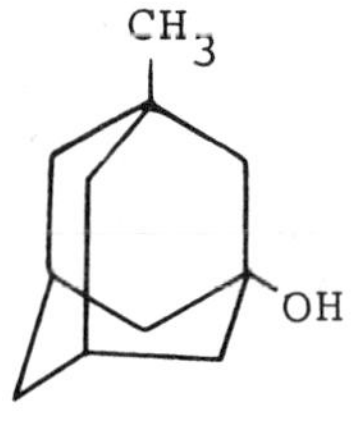 212 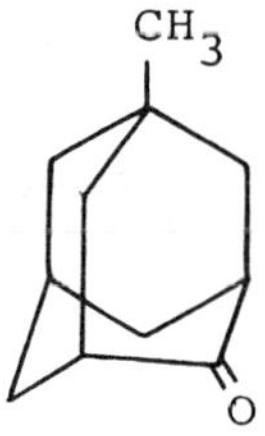 213

of 4-methyl-1-adamantanol (211) are the major products. At higher temperatures, however, 209 gives other products, such as 3-methyl-1-adamantanol (212) and 5-methyl-2-adamantanone (213), in which both the alkyl and the oxygen functions have shifted.

Adamantane rearrangements, apparently involving only 1,2 alkyl shifts, have also been reported [304,313,314]. The simplest example is the interconversion of 2-methyl-(214) and 1-methyladamantane (215) [304,314]. Equilibria established by aluminum halide catalysis strongly favor the 1-methyl-isomer (98% at 25°C). Since an intermolecular methyl-shift mechanism was

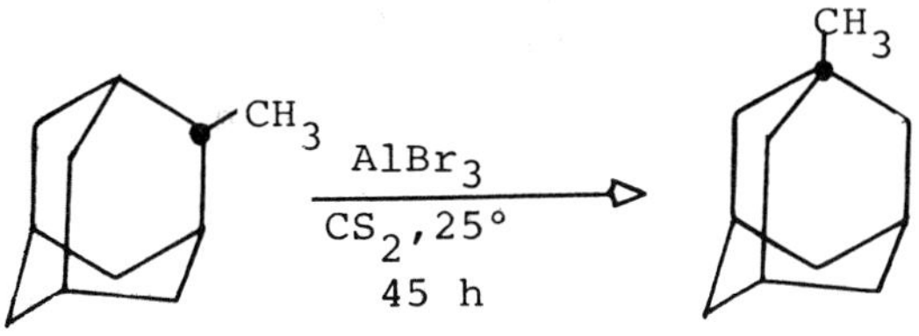

$$\mathbf{214} \xrightarrow[\text{CS}_2,\ 25°,\ 45\text{ h}]{\text{AlBr}_3} \mathbf{215}$$

ruled out on the basis of various chemical and stereochemical grounds, isotopic labeling techniques were employed to distinguish among three possible types of intramolecular processes [314]. The results demonstrated that the methyl group remains attached predominantly to the same carbon atom during the rearrangement of (214) to (215). The mechanism shown in scheme 24 was suggested by Majerski et al. [314] to account for the labeling patterns observed. In fact, this type of mechanism has been considered as a general

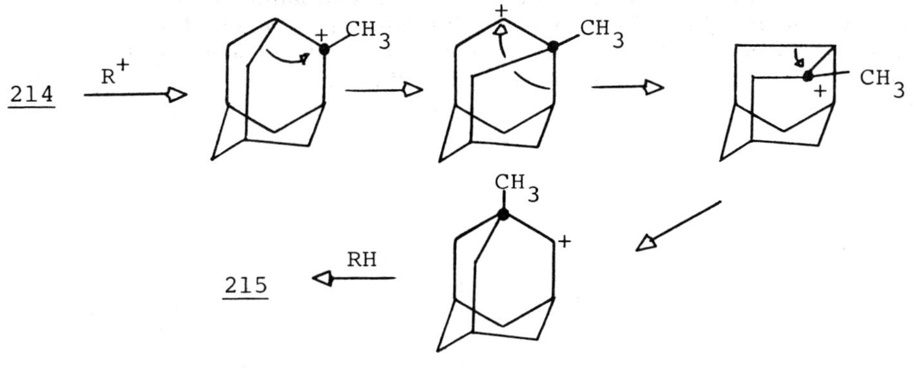

$$\mathbf{214} \xrightarrow{R^+} \cdots \xrightarrow{RH} \mathbf{215}$$

Scheme 24

one for many adamantane rearrangements, especially for those involving molecules not possessing favorable geometry for direct 1,2 shifts [304,315, 316].

Both 2,2-dichloro- and 2,2-dibromoadamantane (216, X = Cl, Br) were found to undergo rearrangement to the corresponding 1,3-dihaloadamantane (217, X = Cl, Br) upon treatment with $AlCl_3$ in CS_2 at room temperature [308,309]. The dibromide rearranged much more readily than did the dichloride. This was clear from the finding that 1,3-dibromoadamantane was obtained in 75% yield after 1 hr and 1,3-dichloroadamantane in 70% yield

$AlCl_3$ / CS_2, 25°

216 (X=Cl,Br) 217

after 3 days. It is of interest to point out here that contrary to this behavior in the presence of Lewis acid catalysts, the 2-halo-2-adamantyl cations did not show any rearrangement in antimony pentafluoride-based superacids [197,317].

Acid-catalyzed rearrangements in the homoadamantane system have been studied by Majerski and co-workers [280,318,319] utilizing ^{13}C-labeled molecules. Interestingly, "4-homoisotwistane" (tricyclo[5.3.1.$0^{3,8}$]undecane,

218

218, which was shown to be one of the most stable tricycloundecane isomers on the basis of molecular orbital calculations [320], was found to be a key intermediate in many other acid-catalyzed adamantane rearrangements [320-323].

In recent years Olah et al. [140,255,317,324] were able to generate and examine the rearrangements of a number of long-lived adamantyl [140,317], dehydroadamantyl [255], and dehydrohomoadamantyl [255] cations by the action of superacids on hydroxy compounds and other precursors. Whereas the tertiary 1-adamantyl and 2-alkyl-, 2-phenyl-, and 2-halo-2-adamantyl cations were stable enough to be observed directly by NMR, the secondary 2-adamantyl cation was too unstable to be observed directly.

In addition to the examples cited so far in this section, there have appeared numerous other recent reports on the rearrangements of more complex tri- and higher cyclic systems. For example, both *endo*- and *exo*- 5,6-trimethylenenorbornane (219) were found to interconvert and to rearrange to adamantane under the influence of $AlCl_3$ (Eq. 152) or $AlBr_3$ [325]. Adamantane was also obtained from the isomeric 4,5-*exo*-trimethylenenorbornane (220) but under

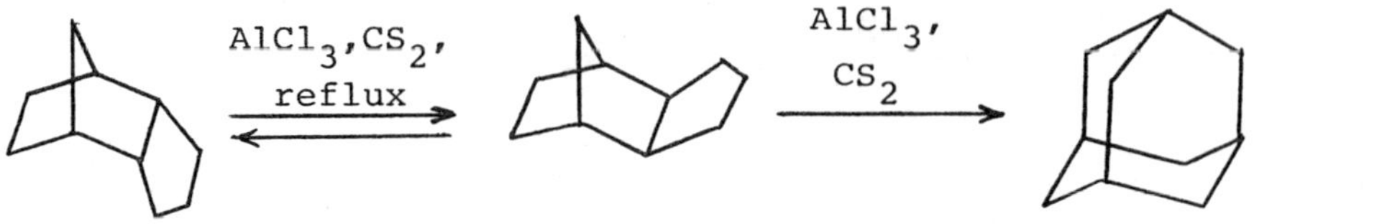

$AlCl_3, CS_2$, reflux; $AlCl_3$, CS_2

endo-219 ⇌ *exo*-219 → [+ other products]

(152)

milder conditions [326] (Eq. 153). Notably, the conditions to produce adamantane from *exo*-220 were much milder than those necessary from 219. Thus whereas 220 gave adamantane in minutes at 25°C, 219 required refluxing for considerably longer times [325-328].

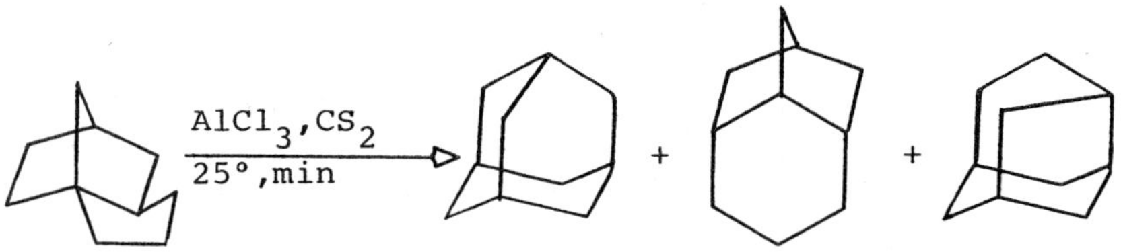

$AlCl_3, CS_2$, 25°, min

220 → + + (153)

The behavior of *exo*-219 was also investigated under superacid conditions. Olah and Lukas [132] found that 219 can undergo ionization and rearrangement in FSO_3H-SbF_5 solution to give the 1-adamantyl cation (221, Eq. 154). Addition of isopentane to the acid solution yielded adamantane.

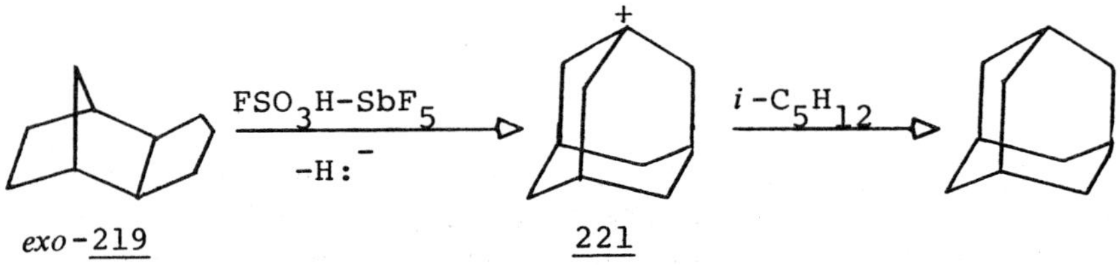

FSO_3H-SbF_5, $-H:^-$; i-C_5H_{12}

exo-219 → 221 → + t-$C_5H_{11}^+$

(154)

Takaishi et al. [329,330] found that 4-homoisotwistane (tricyclo[5.3.1.$0^{3,8}$]undecane, 222) can be produced by the isomerization of either 223 or 224 in CH_2Cl_2 at 0°C in the presence of acid catalysis such as CF_3SO_3H and $AlCl_3$. Compound 222 was also obtained in 33% yield upon treatment of the tricyclocarbinol 225 with 98% sulfuric acid in *n*-pentane for 5 min [330a]. Further isomerization of 222 in refluxing CH_2Cl_2 in the presence of either CF_3SO_3H or $AlCl_3$ gave 1-methyladamantane (226). The same workers examined other acid-catalyzed rearrangements of tricycloundecane derivatives [331,332]. The complex rearrangements and reactions were explained in terms of carbocation intermediates which undergo a variety of 1,2 methylene shifts.

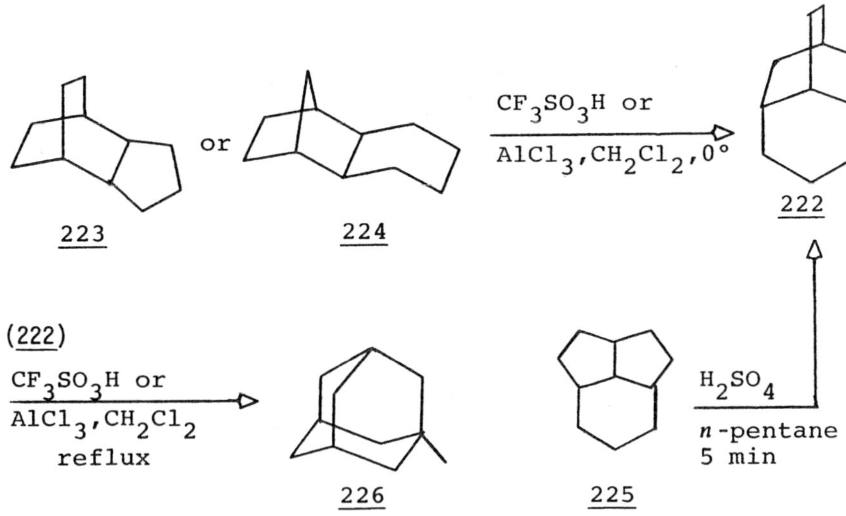

223 or 224 —(CF_3SO_3H or $AlCl_3, CH_2Cl_2, 0°$)→ 222

(222) —(CF_3SO_3H or $AlCl_3, CH_2Cl_2$ reflux)→ 226

225 —(H_2SO_4, n-pentane, 5 min)→ 222

Ordinary protoadamantane (tricyclo[4.3.1.$0^{3,8}$]decane) is known to undergo facile rearrangement to adamantane by treatment with $AlBr_3$ catalyst at room temperature [333-335]. Similar treatment of protoadamantane-4-^{13}C (227) with $AlBr_3$ in CS_2 at room temperature led to adamantane-1-^{13}C (228) in which the excess ^{13}C label was completely at the 1 position (Eq. 155)

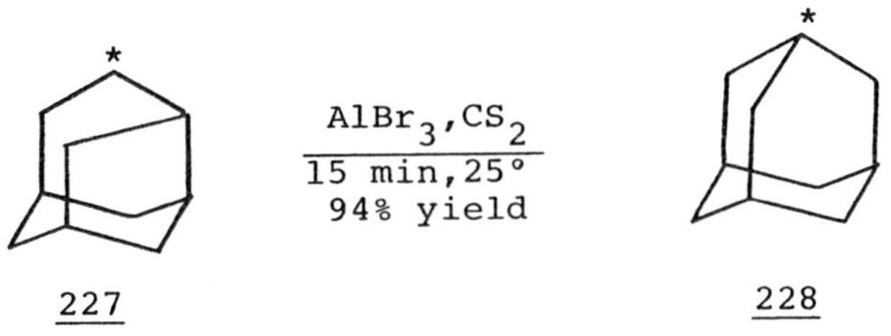

$$\underline{227} \xrightarrow[\substack{15\ \text{min}, 25°\\ 94\%\ \text{yield}}]{AlBr_3, CS_2} \underline{228} \tag{155}$$

[336,337]. This result helped to eliminate the possible intervention of a degenerate 4-protoadamantyl → 4-protoadamantyl type of rearrangement suggested to explain the solvolysis of 4-*endo*-protoadamantyl systems [335,338,339]. A degenerate rearrangement would, of course, result in complete scrambling of the label.

A convenient, high-yield preparation of diadamantane (pentacyclo [7.3.1.$1^{4,12}0^{2,7}0^{6,11}$]tetradecane, also named diamantane or congressane) (229), utilized the rearrangement of readily available isomeric norbornene dimers (e.g., 230-232) under Friedel-Crafts conditions. The significant stages in this interesting development which has now made diadamantane as readily available as adamantane are summarized in Eqs. (156) to (158). As can be seen from Eq. (158), the readily available norbornadiene dimer heptacyclo[8.4.0.$0^{2,12}0^{3,7}0^{4,9}0^{6,8}0^{11,13}$]tetradecane (simply called Binor-S, 233) proved to be most ideal for the preparation of diamantane via hydrogenation to an undetermined tetrahydro product, followed by rearrangement with $AlBr_3$ (Eq. 158) [342].

230 ($C_{14}H_{20}$) $\xrightarrow[\text{or AlBr}_3]{\text{AlCl}_3}$ $C_{14}H_{22}$ (major) + 229 ($C_{14}H_{20}$) only 1%

(156) [340, 341, 342]

231 (CH_3) $\xrightarrow[\sim 30\%]{\text{AlBr}_3\text{-sludge}}$ 229 $\xleftarrow[\sim 30\%]{\text{AlBr}_3\text{-sludge}}$ 232 (C_2H_5)

(157) [342]

233 ($C_{14}H_{16}$) $\xrightarrow[\text{HOAc, HCl; 90\%}]{2H_2, PtO_2}$ $C_{14}H_{20}$ $\xrightarrow[\sim 65\%]{\text{AlBr}_3, CS_2}$ 229

(158) [342]

The behavior of diadamantane (229) in FSO_3H-SbF_5 solution was examined [132]. The NMR spectrum of the resulting ion was indicative of ionization of one of the six tertiary belt hydrogens in the 1, 2, 6, 7, 11, or 12 position to give cation 234.

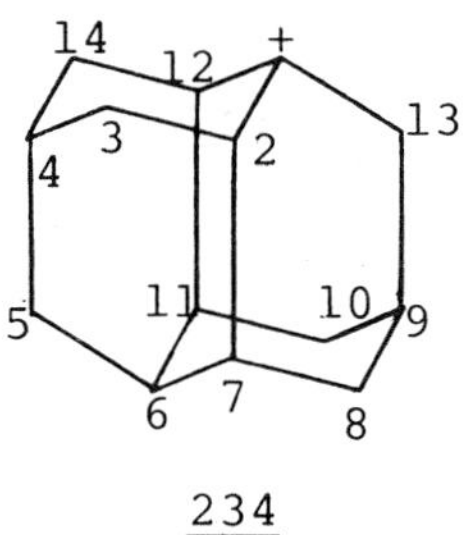

234

Further work in this area has been carried out by Stohrer and Hoffmann [343], Hart and Kuzuya [344-346], Paquette et al. [347], and Schleyer et al. [325,326,335,339,340,343,348-351, and references cited therein].

III. MECHANISTIC ASPECTS

In our previous discussion of the rearrangements of alkylating agents under conditions favoring formation of carbocations, we have encountered numerous examples of skeletal and isotope-position rearrangements. Skeletal isomerizations, including those of ring expansions and ring contractions, can conveniently be divided into two main types: (1) Rearrangements between species with the same number of branches; for example, the rearrangement of 2-methylpentane to 3-methylpentane or of 2,3-dimethylhexane to 2,4- and 2,5-dimethylhexane. For such rearrangements the term *nonbranching rearrangement* has been suggested [116]. (2) Rearrangements involving an increase or a decrease in the degree of chain branching, for example, the *n*-butane to isobutane or cyclohexane to methylcyclopentane rearrangements. These have been termed "branching rearrangements" [116].

In the course of such studies it has generally been found that reactions involving a change in branching are slower than those in which there is no change in branching [232,352-354]. Moreover, nonbranching rearrangements require less powerful catalysts, shorter reaction periods, and lower temperatures than do branching rearrangements. This is, in fact, a reflection of the differences in the activation energies required for the two types of rearrangements.

Throughout this chapter we have reported some of the mechanistic interpretations offered by various authors to account for their experimental findings. We should like to summarize next the major aspects of the mechanistic interpretations of carbocation rearrangements.

A. Nonbranching Rearrangement Mechanisms

These rearrangements proceed by simple 1,2 shifts of hydride and alkide moieties. Both secondary and tertiary carbocations can be involved. The rearrangements of 2-methylpentane [235a) [142] and 2-methylhexane [235b) [354] to 3-methylpentane (236a) and 3-methylhexane (236b), respectively (Eq. 159); of 1,2-dimethylcyclohexane (237) to 1,3-dimethylcyclohexane (238,

235 $\underset{RH}{\overset{R^+}{\rightleftharpoons}}$ $\underset{}{\overset{\sim H:^-}{\rightleftharpoons}}$ $\sim CH_3:^-$

236 $\underset{RH}{\overset{R^+}{\rightleftharpoons}}$ $\overset{\sim H:^-}{\rightleftharpoons}$

a, R=H b, R=CH_3 (159)

Eq. 160 [355]; and of methylcyclooctane (239) to propylcyclohexane (240, Eq. 161) [356] are illustrative.

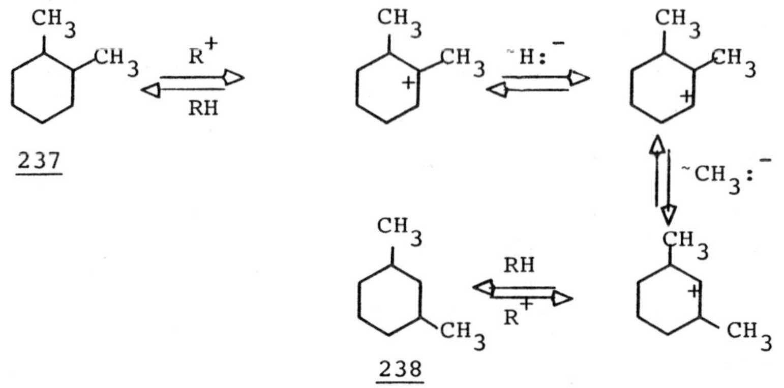

(160)

237

238

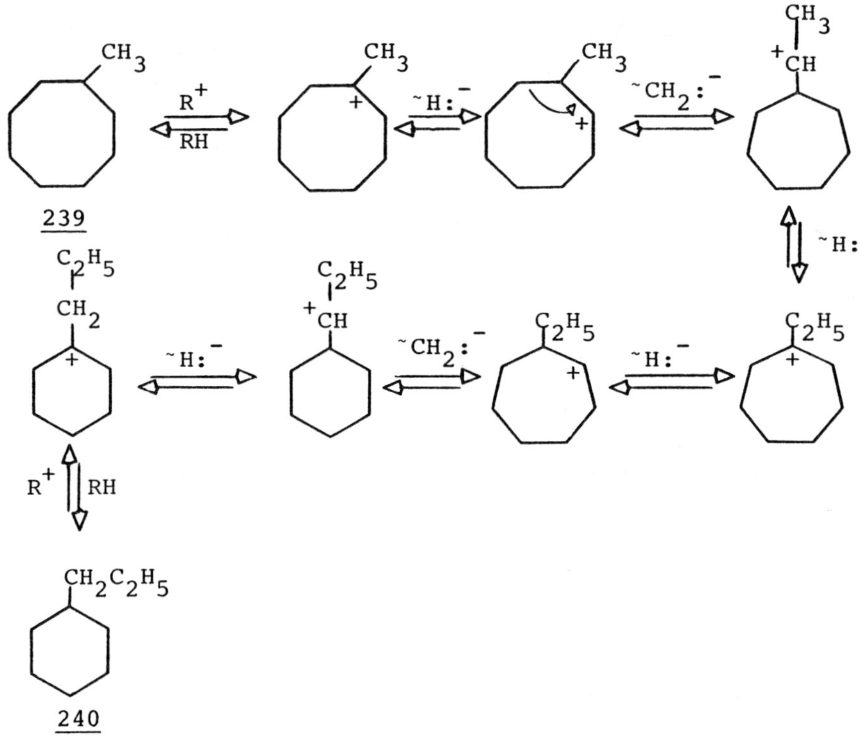

239

240

(161)

B. Branching Rearrangement Mechanisms

In the past, chemists have explained some of their results in terms of mechanisms involving primary carbocations. For example, the isomerization of *n*-butane to isobutane was suggested by Bloch et al. in 1946 [357] to proceed as follows:

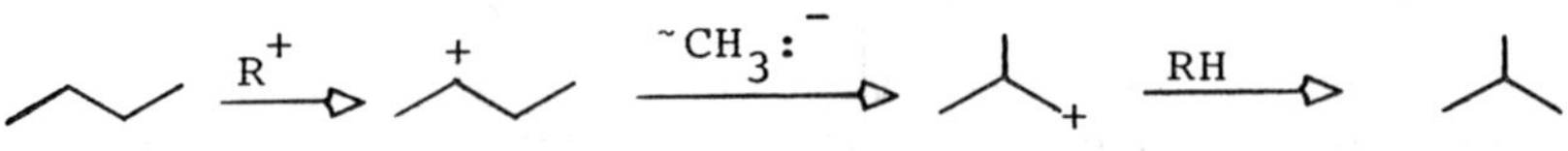

The isomerization of 2-methylpentane to 2,3-dimethylbutane was explained by Schneider and Kennedy in 1951 [358] to proceed as follows:

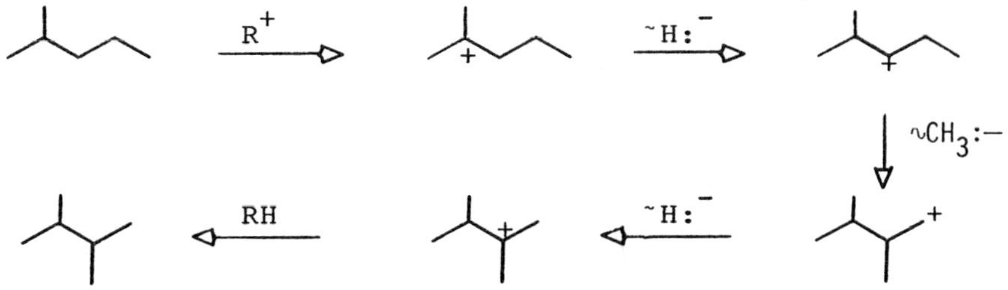

$$\xrightarrow{R^+} \quad \xrightarrow{\sim H:^-} \quad \downarrow \sim CH_3:^- \quad \xleftarrow{\sim H:^-} \quad \xleftarrow{RH}$$

Similarly, the isomerization of cyclohexane to 1-methylcyclopentane was suggested by Pines et al. in 1948 [359] to follow the path

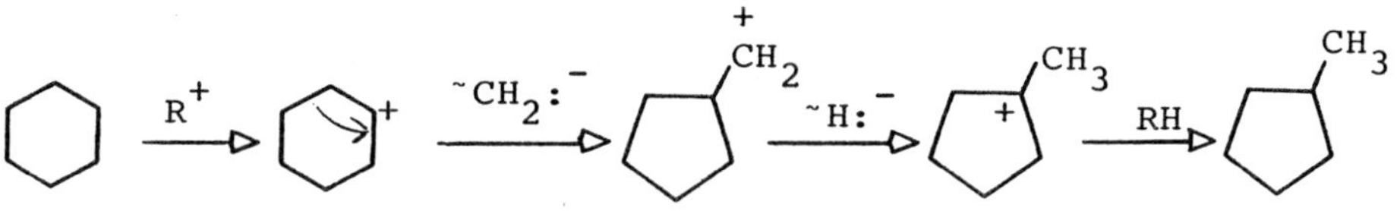

$$\xrightarrow{R^+} \quad \xrightarrow{\sim CH_2:^-} \quad \overset{+}{C}H_2 \quad \xrightarrow{\sim H:^-} \quad CH_3 \quad \xrightarrow{RH} \quad CH_3$$

Since going from a secondary to a primary carbocation would require about 33 kcal/mol in solution [126,360], the preceding mechanisms seem to be irreconcilable with modern theory. Accordingly, in recent years several alternative possibilities have been suggested to avoid the involvement of primary carbocations. These include nonclassical three- and four-membered cyclic intermediates [81,361-365], simultaneous hydrogen and methyl shifts [366], and bimolecular reaction pathways [113,115,125,367].

The bimolecular reaction pathway was first proposed by Schneider and Kennedy [367] and elaborated later by Karabatsos and co-workers [90,113, 125] (See Sec. I.A2b). A more recent proposal involves the concept of protonated cyclopropane intermediates [142,187,188,366]. This concept was apparently introduced by Roberts and Lee [361], who suggested a nonclassical "face-protonated" cyclopropane intermediate, 241,

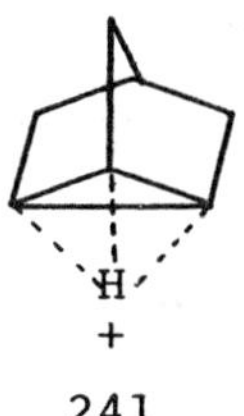

241

to account for the isotope-position rearrangements accompanying the solvolysis of 2,3-$^{14}C_2$-2-norbornyl brosylates. Protonated cyclopropane intermediates have been widely accepted and applied to rationalize the observation of various skeletal rearrangements, including ring expansions and contractions, and the scrambling of isotopic labels. For example, protonated cyclopropanes were proposed as intermediates in the isomerizations of isobutyl cation to *sec*-butyl cation with HSO_3F-SbF_5 (Eq. 167) [188], of *n*-butane-1-^{13}C to *n*-butane-2-^{13}C with HF-SbF_5 (Eq. 163) [233]; of *n*-hexane to 2-methylbutane and methylpentanes, respectively, with HF-SbF_5 (scheme 25) [142,161,368]; of 2,2,3-trimethylbutane to 2,4-dimethylpentane with H_2SO_4 (Eq. 164) [369]; and of cyclohexane to methylcyclopentane with $AlCl_3$ (Eq. 165) [355].

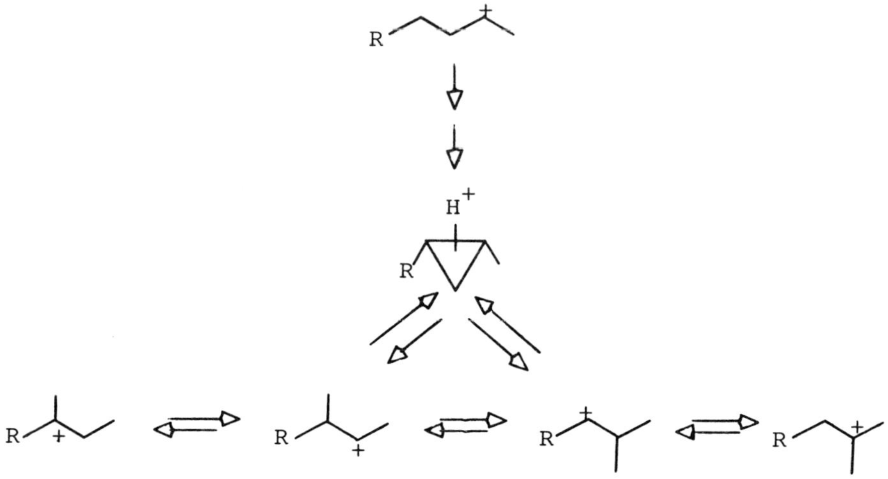

$R = CH_3, C_2H_5$

Scheme 25

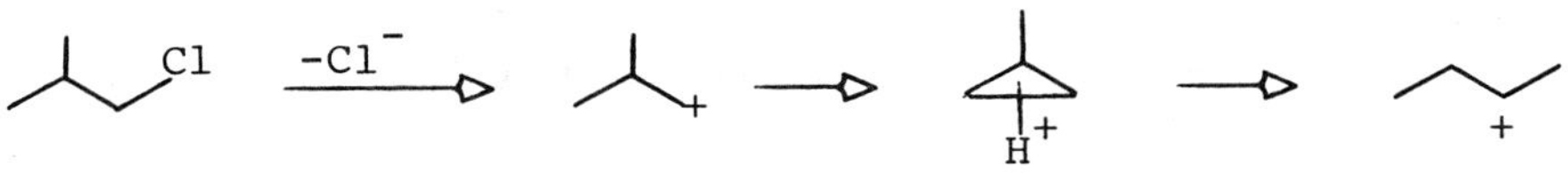

-Cl⁻

(162)

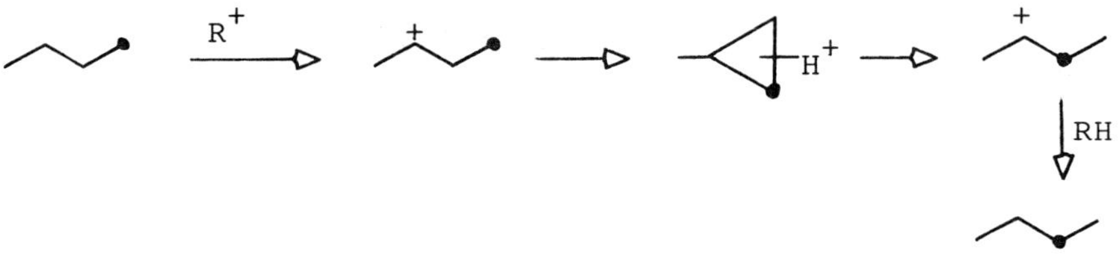

R^+

RH

(163)

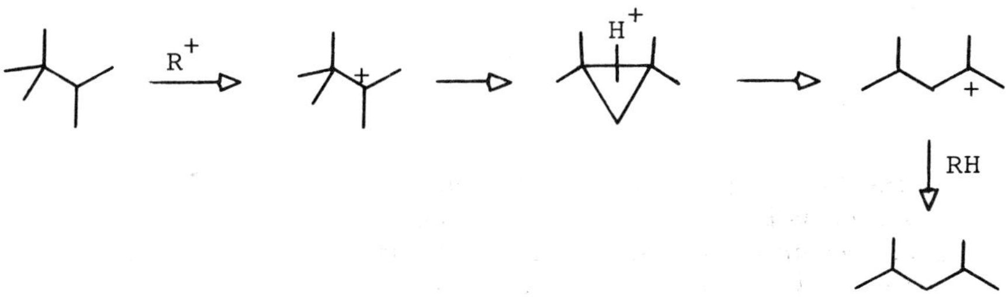

R^+

H^+

RH

(164)

R^+

H^+

RH

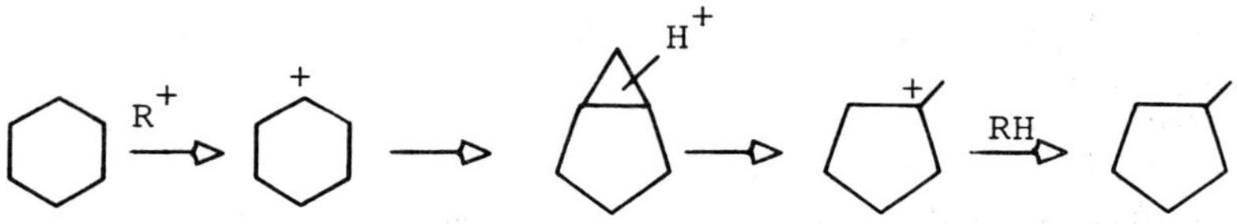

(165)

```
    C                            C              C               C
    |     AlBr3(H2O)             |              |               |
14C-C-C-C ----------->  C-C-C-14C  +  C-C-14C-C  +  C-14C-C-C
            25°
```

(166)

To explain the isomerization of n-pentane and n-hexane by HF-SbF_5 to 2-methylbutane and methylpentanes, respectively, whereas n-butane was not isomerized to isobutane under the same conditions, Brouwer and Oelderik [142] suggested that the isomerizations proceed via protonated cyclopropane intermediates. The cleavage of the analogous intermediate from n-butane to produce isobutane would involve an incipient primary carbocation, which is not true of the reactions of the higher homologs. In support of this theory Brouwer [233] demonstrated that isotopic rearrangement of 1-^{13}C-butane to 2-^{13}C-butane took place at about the same rate as the isomerization of n-pentane (Eq. 163).

On the basis of his findings, Brouwer concluded that alkylcarbocation rearrangements that involve a change in the degree of chain branching proceed via protonated cyclopropane intermediates, whereas those that do not involve a change in the degree of branching proceed via one or more 1,2 shifts of hydrogen or alkyl anions.

Nenitzescu [355] suggested that the isotope rearrangements observed by Roberts and Coraor [370] during the $AlBr_3$ treatment of 2-methylbutane-1-^{14}C (Eq. 166) could be explained in terms of protonated cyclopropane intermediates.

Nenitzescu did not elaborate or propose any specific steps, but the observed more rapid isomerization of the starting material to 2-methylbutane-4-^{14}C than to 2-methylbutane-2-(or 3)-^{14}C can be explained in terms of the following steps (Eq. 167):

```
    C                  C                    C
    |       R+         |        ~H:-        |       ~CH3:-
14C-C-C-C ----->  14C-C-C-C  ------->  14C-C-C-C  ---------->
                       +                    +

    C                   C                C
    |      ~H:-         |                |
14C-C-C-C  ----->  14C-C-C-C  ----->  C--C-C  ----->
    +                     +            14C  H+

      C                  C
      |       RH         |
14C-C-C-C  ----->  14C-C-C-C
      +
```

(167)

The observed equilibration of the label in the isopropyl cation generated from 2-chloropropane with a 50% ^{13}C enrichment at C_2 in SO_2ClF-SbF_5 solution at -60°C was attributed to the intermediacy of a protonated cyclopropane (Eq. 168) [167].

$$(CH_3)_2\overset{*}{C}H^+ \rightleftharpoons H_2C{-}CH_2{-}\overset{*}{C}H_2 \cdots H^+ \text{ (protonated cyclopropane)} \rightleftharpoons CH_3\overset{+}{C}H\overset{*}{C}H_3 \qquad (168)$$

The interchange of the methyl groups in the 3-methyl-3-pentyl cation was envisioned as occurring via a protonated cyclopropane intermediate (Eq. 169) [167].

$$(CH_3CH_2)_2\overset{+}{C}{-}\overset{*}{C}H_3 \rightleftharpoons CH_3CH{-}CH\overset{*}{C}H_3 \text{ with } CH(CH_3) \cdots H^+ \text{ (protonated cyclopropane)} \rightleftharpoons CH_3\overset{+}{C}(CH_2CH_3){-}CH_2\overset{*}{C}H_3 \qquad (169)$$

The fact that the reaction product of ring contraction of *cis*-1,2-dimethylcycloheptane in the presence of $AlBr_3$ consisted mainly of isopropylcyclohexane was rationalized in terms of an intermediate having a three-membered ring [356], but the formation of this product can also be explained by simple 1,2 shifts of methyl and methylene groups.

In connection with protonated cyclopropanes and their intermediacy in certain reactions, it might be mentioned that chemists have also searched for the possible intermediacy of the next-higher homolog, protonated cyclobutane. In 1973, protonated cyclobutane intermediates were suggested as being involved in the mass spectral fragmentation of butyl cation [371]. Recent theoretical calculations by Pakkanen and Whitten [372] predict that the protonation energy for cyclobutane is quite large, 126 kcal/mol, but is significantly lower than that required for edge protonation of cyclopropane, 176.9 kcal/mol, by similar theoretical calculations. In the experimental search for protonated cyclobutane, Lee and Ko [373] attempted the protonation of cyclobutane with H_2SO_4-*t* or CF_3COOH-*t*. When cyclobutane was shaken with excess CF_3SO_3H-*t* at 75°C for 3 days, the recovered cyclobutane was found to be tritiated. The similarity to the formation of monodeuterocyclopropane in this exchange reaction [cf. Sec. I.A.1.b(ii)] gave support to the probability of a role for protonated cyclobutane. However, since only a very small amount (ca. 1%) of ring cleavage was observed in the reaction, it was considered unlikely that processes via protonated cyclobutanes could be major pathways in thermochemical reactions. In line with this, the isotopic scrambling results observed during the hydrolysis of 1-butyl-1-^{14}C tosylate suggested the possibility of some involvement of protonated methylcyclopropane rather than protonated cyclobutane [374].

REFERENCES

1. Friedel, C., and J. M. Crafts, Bull Soc. Chim. Fr., (2),*27*, 482 (1877).
2. Friedel, C., and J. M. Crafts, Bull. Soc. Chim. Fr., (2), *27*, 530 (1877).

3. Friedel, C., and J. M. Crafts, Compt. Rend. Acad. Sci., *84*, 1450 (1877).
4. Gustavson, G. G., Chem. Ber., *11*, 1251 (1878).
5. Roberts, R. M., G. A. Ropp, and O. K. Neville, J. Am. Chem. Soc., *77*, 1764 (1955).
6. Sixma, F. L. J., and H. Hendriks, Recl. Trav. Chim. Pays-Bas, *75*, 169 (1956).
7. Lee, C. C., M. C. Hamblin, and H. James, Can. J. Chem., *36*, 1597 (1958).
8. Brown, H. C., and W. J. Wallace, J. Am. Chem. Soc., *75*, 6279 (1953).
9. Adema, E. H., and F. L. J. Sixma, Recl. Trav. Chim. Pays-Bas, *81*, 323 (1962).
10. Adema, E. H., H. Arnoldy, R. Visser, and F. L. J. Sixma, Recl. Trav. Chim. Pays-Bas, *79*, 1111 (1960).
11. Adema, E. H., and F. L. H. Sixma, Recl. Trav. Chim. Pays-Bas, *81*, 336 (1962).
12. Price, C. C., Chem. Rev., *29*, 37 (1941).
13. Price, C. C., Org. React., *3*, 1 (1946).
14. Hammett, L. P., *Physical Organic Chemistry*, McGraw Hill, New York, 1940, p. 309.
15. Williams, J. E., Jr., J. Buso, L. C. Allen, P. v. R. Schleyer, W. A. Latham, W. J. Hehre, and J. A. Pople, J. Am. Chem. Soc., *92*, 2141 (1970).
16. Pfeiffer, M. V., and J. G. Jwett, J. Am. Chem. Soc., *92*, 2144 (1970).
17. Clark, D. T., and D. M. Lilley, Chem. Commun., 549 (1970).
18. Sustmann, R., J. E. Williams, M. J. S. Dewar, L. C. Allen, and P. v. R. Schleyer, J. Am. Chem. Soc., *91*, 5350 (1969).
19. Bach, R. D., and H. F. Henneike, J. Am. Chem. Soc. *92*, 5589 (1970).
20. Brown, H. C., H. Pearsall, and L. P. Eddy, J. Am. Chem. Soc., *72*, 5347 (1950).
21. Walker, J. W., J. Chem. Soc., *85*, 1082 (1904).
22. Wohl, A., and E. Wertyporoch, Chem. Ber., *64*, 1357 (1931).
23. Wertyporoch, E., and T. Firla, Justus Liebigs Ann. Chem., *500*, 287 (1933).
24. Katsnelson, I. L., Mem. Int. Chem. Acad. Sci. Ukr. SSR, *4*, 393 (1938).
25. Jacober, W. J., and C. A. Kraus, J. Am. Chem. Soc., *71*, 2405, 2409 (1948).
26. Olah, G. A., S. Kuhn, and J. Olah, J. Chem. Soc., 2174 (1957).
27. Fairbrother, F., Trans. Faraday Soc., *37*, 763 (1941).
28. Fairbrother, F., J. Chem. Soc., 503 (1945).
29. Fairbrother, F., J. Chem. Soc., 503 (1937); 293 (1941).
30. Sixma, F. L. J., and H. Hendriks, Proc. Kl. Ned. Akad. Wet., *59B*, 61 (1956).
31. Olah, G. A., J. R. DeMember, R. H. Schlosberg, and Y. Halpern, J. Am. Chem. Soc., *94*, 156 (1972).
32. Olah, G. A., and D. J. Donovan, J. Am. Chem. Soc., *100*, 5163 (1978).
33. Sixma, F. L. J., H. H. Hendriks, and D. Holtzapffel, Recl. Trav. Chem. Pays-Bas, *74*, 127 (1955).
34. Brezhneva, N. E., S. Z. Roginskii, and A. I. Shilinskii, Russ. J. Phys. Chem. *9*, 752 (1937); Chem. Abstr., *31*, 8340 (1937).

35. Trotter, P. J., J. Org. Chem., *28*, 2093 (1963).
36. Brown, H. C., L. P. Eddy, and R. Wong, J. Am. Chem. Soc., *75*, 6275 (1953).
37. Kekule, A., and H. Schrotter, Bull. Soc. Chim. Fr., (2), *34*, 485 (1879).
38. Gustavson, G. G., J. Russ. Phys.-Chem. Soc., *14*, 354 (1882).
39. Gustavson, G. G., Chem. Ber., *16*, 958 (1883).
40. Gustavson, G. G., J. Prakt. Chem., (2), *34*, 161 (1886).
41. Nagai, W., J. Chem. Soc. Jpn, *61*, 864 (1940).
42. Gustavson, G. G., Chem. Ber., *13*, 157 (1880); J. Chem. Soc. Abstr., 370 (1880).
43. Thomas, C. A., *Anhydrous Aluminum Chloride in Organic Chemistry*, Reinhold, New York, 1941, p. 91.
44. Cronwell, T. I., and G. L. Jones, Jr., J. Am. Chem. Soc., *73*, 3506 (1951).
45. Karabatsos, G. J., J. I. Fry, and S. Meyerson, J. Am. Chem. Soc., *92*, 614 (1970).
46. Lee, C. C., and D. J. Woodcock, J. Am. Chem. Soc., *92*, 5992 (1970).
47. Reutov, O. A., and T. N. Shatkina, Izv. Akad. Nauk SSSR, Otd. Khim. Nauk, 195 (1963); Chem. Abstr., *58*, 10063c (1963).
48. Lee, C. C., B. Hahn, K. Wan, and D. J. Woodcock, J. Org. Chem., *34*, 3210 (1969).
49. Karabatsos, G. J., C. Zioudrou, and S. Meyerson, J. Am. Chem. Soc., *92*, 5996 (1970).
50. Topchiev, A. V., B. A. Krentsel, and L. N. Andreev, Dokl. Akad. Nauk SSSR, *92*, 781 (1953); Chem. Abstr., *49*, 3039 (1955).
51. Whitmore, F. C., E. L. Wittle, and A. H. Popkin, J. Am. Chem. Soc., *61*, 1586 (1939).
52. Mayo, F. R., and A. A. Dolmick, J. Am. Chem. Soc., *66*, 985 (1944).
53. Andreevakii, D. N., Dokl. Akad. Nauk SSSR, *135*, 312 (1960); Chem. Abstr., *55*, 10030c (1961).
54. Gerrard, W., H. R. Hudson, and W. S. Murphy, J. Chem. Soc., 2314 (1964).
55. McKenna, J. F., and F. J. Sowa, J. Am. Chem. Soc., *59*, 470 (1937).
56. McKenna, J. F., and F. J. Sowa, J. Am. Chem. Soc., *59*, 1204 (1937).
57. Whitmore, F. C., J. Am. Chem. Soc., *54*, 3274 (1932).
58. Nash, L. M., T. I. Taylor, and W. von E. Doering, J. Am. Chem. Soc., *71*, 1516 (1949).
59. Winstein, S., B. K. Morse, E. Grunwald, H. W. Jones, J. Corse, D. Trifan, and H. Marshall, J. Am. Chem. Soc., *74*, 1127 (1952).
60. Beers, M. J., Thesis, University of Amsterdam, 1958; cited in Ref. 61.
61. Douwes, H. S. A., and E. C. Kooyman, Recl. Trav. Chim. Pays-Bas, *83*, 276 (1964).
62. Karabatsos, G. J., and J. I. Fry, Tetrahedron Lett., *38*, 3735 (1967).
63. Lee, C. C., and D. J. Woodcock, Can. J. Chem., *48*, 858 (1970).
64. Coe, J. S., and V. Gold, J. Chem. Soc., 4940 (1960).
65. Lee, C. C., and E. J. Kruger, Can. J. Chem., *44*, 2343 (1966).
66. Myhre, P. C., and K. S. Brown, J. Am. Chem. Soc., *91*, 5639 (1969).
67. Reich, I. L., A. Diaz, and S. Winstein, J. Am. Chem. Soc., *91*, 5635 (1969).
68. Roberts, J. D., and M. Halmann, J. Am. Chem. Soc., *75*, 5759 (1953).

69. Karabatsos, G. J., and C. E. Orzech, J. Am. Chem. Soc., *84*, 2838 (1962).
70. Lee, C. C., J. E. Kruger, and E. W. C. Wong, J. Am. Chem. Soc., *87*, 3985 (1965).
71. Lee, C. C., and J. E. Kruger, J. Am. Chem. Soc., *87*, 3986 (1965).
72. Frigerio, N. A., M. J. Shaw, and D. J. Rausch, Argonne National Laboratory Biol. Med. Res. Div. Ann. Rep. ANL-7535, 1968, pp. 49-51.
73. Frigerio, N. A., private communications, cited in ref. 46.
74. Frigerio, N. A., and M. J. Shaw, 156th Nat. Meet. Am. Chem. Soc., Atlantic City, N.J., Sept. 1968, Abstr. ORGN 6.
75. Reutov, O. A., and T. N. Shatkina, Dokl. Akad. Nauk SSSR, *133*, 606 (1960).
76. (a) Reutov, O. A., and T. N. Shatkina, Tetrahedron, *18*, 237 (1962); (b) Aboderin, A. A., and R. L. Baird, J. Am. Chem. Soc., *86*, 2300 (1964).
77. Karabatsos, G. J., C. E. Orzech, Jr., and S. Meyerson, J. Am. Chem. Soc., *87*, 4394 (1965).
78. Karabatsos, G. J., C. E. Orzech, Jr., J. L. Fry, and S. Meyerson, J. Am. Chem. Soc., *92*, 606 (1970).
79. Mayer, V., and F. Forster, Chem. Ber., *9*, 535 (1876).
80. Reutov, O. A., Pure Appl. Chem., *7*, 203 (1963).
81. Skell, P. S., and I. Starer, J. Am. Chem. Soc., *82*, 2971 (1960).
82. Edwards, O. E., and M. Lesage, Can. J. Chem., *41*, 1562 (1963).
83. Baird, R. L., and A. A. Aboderin, Tetrahedron Lett., 235 (1963).
84. Baird, R. L., and A. A. Aboderin, J. Am. Chem. Soc., *86*, 252 (1964).
85. Lee, C. C., and J. E. Kruger, Tetrahedron, *23*, 2539 (1967).
86. Lee, C. C., and W. K. Chwang, Can. J. Chem., *48*, 1025 (1970).
87. Reutov, O. A., and T. N. Shatkina, Bull. Acad. Sci. USSR, Div. Chem. Sci., 180 (1963).
88. Susan, A. B., Acad. Repub. Pop. Rom. Stud. Cercet. Chim., *14*, 23 (1966); Chem. Abstr., *64*, 18484d (1966).
89. Fry, J. L., and G. J. Karabatsos, in *Carbonium Ions*, Vol. 2 (G. A. Olah and P. v. R. Schleyer, eds.), Wiley-Interscience, New York, 1970, Chap. 14.
90. Karabatsos, G. J., F. M. Vane, and S. Meyerson, J. Am. Chem. Soc., *83*, 4297 (1961).
91. Deno, N. C., D. LaVietes, J. Mockus, and P. C. Scholl, J. Amer. Chem. Soc., *90*, 6457 (1968).
92. Lee, C. C., and L. Gruber, J. Am. Chem. Soc., *90*, 3775 (1968).
93. Lee, C. C., K. Chwang, and K. Wan, J. Am. Chem. Soc., *90*, 3778 (1968).
94. Deno, N. C., and D. N. Lincoln, J. Am. Chem. Soc., *88*, 5357 (1966).
95. Hart, H., and R. H. Schlosberg, J. Am. Chem. Soc., *88*, 5030 (1966); *90*, 5189 (1968).
96. Roberts, R. M., and Dep Shiengthong, J. Am. Chem. Soc., *82*, 732 (1960).
97. Brouwer, L. G., and J. P. Wibaut, Recl. Trav. Chim. Pays-Bas, *53*, 1001 (1934); Chem. Abstr., *29*, 447 (1935).
98. Oberfell, G. G., and F. E. Frey, U. S. Patent 2,315,871(1943); Chem. Abstr., *37*, 5418 (1943).
99. Favorskii, A. E., Justus Liebigs Ann. Chem., *354*, 325 (1907).

100. McFadden, W. H., R. G. McIntosh, and W. E. Harris, J. Phys. Chem., *64*, 1076 (1960).
101. Bartlett, P. D., and J. D. McCollumn, J. Am. Chem. Soc., *78*, 1441 (1956).
102. Olah, G. A., J. M. Bollinger, and J. Brinich, J. Am. Chem. Soc., *90*, 2587 (1968).
103. Gerrard, W., H. R. Hudson, and W. S. Murphy, J. Chem. Soc., 1099 (1962).
104. Gerrard, W., and H. R. Hudson, J. Chem. Soc., 2310 (1964).
105. Gerrard, W., and H. R. Hudson, Chem. Rev., *65*, 697 (1965).
106. Gerrard, W., H. R. Hudson, and G. K. Sandhu, unpublished data, cited in Ref. 105.
107. Norris, J. F., Org. Synth., *5*, 27 (1925).
108. Pines, H., A. Rudin, and V. N. Ipatieff, J. Am. Chem. Soc., *74*, 4063 (1952).
109. Cason, J., and J. S. Correia, J. Org. Chem., *26*, 3645 (1961).
110. Baker, J. R., M. A. thesis, University of Texas at Austin, 1966.
111. Streitwieser, A., Jr., W. D. Schaeffer, and S. B. Andreades, J. Am. Chem. Soc., *81*, 1113 (1959).
112. Roberts, R. M., S. E. McGuire, and J. R. Baker, J. Org. Chem. *41*, 659 (1976).
113. Karabatsos, G. J., M. Anand, D. O. Rickter, and S. Meyerson, J. Am. Chem. Soc., *92*, 1254 (1970).
114. de Bataafsche, N. V., Petroleum Maalschappij, Br. Patent 535,435 (1941); Chem. Abstr., *36*, 1614 (1942).
115. Sabatier, P., and A. Maihle, C. R. Acad. Sci., *156*, 658 (1913).
116. Karabatsos, G. J., N. Hsi, and S. Meyerson, J. Am. Chem. Soc., *92*, 621 (1970).
117. Bayless, J. H., and L. Friedman, J. Am. Chem. Soc., *89*, 147 (1967).
118. Jurewics, A. T., and L. Friedman, J. Am. Chem. Soc., *89*, 149 (1967).
119. Bayless, J. H., A. T. Jurewics, and L. Friedman, J. Am. Chem. Soc., *90*, 4466 (1968).
120. Bayless, J. H., F. D. Mendicino, and L. Friedman, J. Am. Chem. Soc., *87*, 5790 (1965).
121. White, E. H., J. Am. Chem. Soc., *77*, 6011 (1955).
122. Roberts, J. D., R. E. McMahon, and J. S. Hine, J. Am. Chem. Soc., *71*, 1896 (1949).
123. Roberts, J. D., R. E. McMahon, and J. S. Hine, J. Am. Chem. Soc., *72*, 4237 (1950).
124. Karabatsos, G. J., and F. M. Vane, J. Am. Chem. Soc., *85*, 729 (1963).
125. Karabatsos, G. J., F. M. Vane, and S. Meyerson, J. Am. Chem. Soc., *85*, 733 (1963).
126. Evans, A. G., *The Reactions of Organic Halides in Solution*, The Manchester University Press, Manchester, England, 1946, p. 15.
127. Condon, F. E., Ph.D. dissertation, Harvard University, 1944.
128. Schneider, A., and R. M. Kennedy, J. Am. Chem. Soc., *73*, 5017 (1951).
129. Friedman, B. S., and F. L. Morritz, J. Am. Chem. Soc., *78*, 3430 (1956).
130. Bartlett, P. D., and G. J. Karabatsos, unpublished results, cited in Ref. 124.

131. Bartlett, P. D., F. E. Condon, and A. Schneider, J. Am. Chem. Soc., *66*, 1531 (1944).
132. Roberts, J. D., and J. A. Yancey, J. Am. Chem. Soc., *77*, 5558 (1955).
133. Whitmore, F. C., and H. S. Rothrock, J. Am. Chem. Soc., *54*, 3431 (1932).
134. Dostrovsky, I., and E. D. Hughes, J. Chem. Soc., 166 (1946).
135. Karabatsos, G. J., and J. D. Graham, J. Am. Chem. Soc., *82*, 5250 (1960).
136. Skell, P. S., I. Starer, and A. P. Krapcho, J. Am. Chem. Soc., *82*, 5257 (1960).
137. Karabatsos, G. J., C. E. Orzech, Jr., and S. Meyerson, J. Am. Chem. Soc., *86*, 1994 (1964).
138. Sanderson, W. A., and H. S. Mosher, J. Am. Chem. Soc., *83*, 5033 (1961).
139. Winstein, S., and H. Marshall, J. Am. Chem. Soc., *74*, 1120 (1952).
140. Olah, G. A., M. B. Comisarow, C. A. Cupas, and C. U. Pittman, Jr., J. Am. Chem. Soc., *87*, 2997 (1965).
141. Fraser, G. M., and H. M. R. Hoffman, Chem. Commun., 561 (1967).
142. Brouwer, D. M., and J. M. Oelderik, Recl. Trav. Chim. Pays-Bas, *87*, 721 (1968).
143. Diaz, A., I. L. Reich, and S. Winstein, J. Am. Chem. Soc., *91*, 5637 (1969).
144. Sommer, J. M., Bull. Soc. Chim. Fr., 1831 (1963); Chem. Abstr., *59*, 15156h (1963).
145. Billups, W. E., A. N. Kurtz, and M. L. Farmer, Tetrahedron, *26*, 1095 (1970).
146. Olah, G. A., J. M. Bollinger, Y. K. Mo, and J. M. Brinich, J. Am. Chem. Soc., *94*, 1164 (1972).
147. Roberts, J. D., and R. H. Mazur, J. Am. Chem. Soc., *73*, 2509 (1951).
148. Mazur, R. H., W. N. White, D. A. Semenow, C. C. Lee, M. S. Silver, and J. D. Roberts, J. Am. Chem. Soc., *81*, 4390 (1959).
149. Caserio, M. C., W. H. Graham, and J. D. Roberts, Tetrahedron, *11*, 171 (1960).
150. Hanack, M., and H. J. Schneider, Angew. Chem., Int. Ed. Engl., *6*, 666 (1967).
151. Olah, G. A., and C. Lin, J. Am. Chem. Soc., *90*, 6468 (1968).
152. Bundel, Yu. G., A. M. Yuldashev, N. M. Chistovalova, and O. A. Reutov, Dokl. Akad. Nauk SSSR, *186*, 99 (1969); Chem. Abstr., *71*, 60800m (1969).
153. Bundel, Yu. G., A. M. Yuldashev, and O. A. Reutov, Izv. Akad. Nauk SSSR, Ser. Khim., 1397 (1969); Chem. Abstr., *71*, 123666v (1969).
154. Reutov, O. A., T. N. Shatkina, E. V. Leont'eva, E. Lippmaa, and T. Pehk, Dokl. Akad. Nauk SSSR, *183*, 846 (1968); Chem. Abstr., *70*, 67349j (1969).
155. Goering, H. L., and L. L. Sims, J. Am. Chem. Soc., *79*, 6270 (1957).
156. Nenitzescu, C. D., and I. G. Dinulescu, Acad. Repub. Pop. Rom. Stud. Cercet. Chim., *7*, 7 (1959); Chem. Abstr., *53*, 21708 (1959).
157. Nozaki, H., M. Kawanishi, M. Okazaki, M. Yamae, Y. Nisikawa, T. Hisida, and K. Sisido, J. Org. Chem., *30*, 1303 (1965).

158. Deno, N. C., Prog. Phys. Org. Chem., *2*, 129-193 (1964).
159. Olah, G. A., Chem. Eng. News, *45*, 77 (1967); Science, *168*, 1298 (1970).
160. Olah, G. A., J. Am. Chem. Soc., *94*, 808 (1972); Chimia, *28*, 275 (1973); Angew. Chem., Int. Ed. Engl., *12*, 173 (1973).
161. Brouwer, D. M., and H. Hogeveen, Prog. Phys. Org. Chem., *9*, 179 (1972).
162. Bethell, D., and V. Gold, *Carbonium Ions, an Introduction*, Academic Press, New York, 1967.
163. Olah, G. A., and P. R. Schleyer, eds. *Carbonium Ions*, Wiley-Interscience, New York, Vol. 1 (1968), Vol. 2 (1970), Vol. 3 (1972), Vol. 4 (1973), and Vol. 5 (1976).
164. (a) Gillespie, R. J., Acc. Chem. Res., *1*, 202 (1968); (b) Gillespie, R. J., and T. E. Peel, Adv. Phys. Org. Chem., *9*, 1 (1971); (c) Gillespie, R. J., and T. E. Peel, J. Am. Chem. Soc., *93*, 5083 (1971); (d) Gillespie, R. J., and T. E. Peel, J. Am. Chem. Soc., *95*, 5173 (1973).
165. Arnett, E. M., and J. W. Larsen, Chem. Eng. News, *46*, 36 (Feb. 26, 1968).
166. Commeyras, A., and G. A. Olah, J. Am. Chem. Soc., *91*, 2929 (1969).
167. Olah, G. A., and A. M. White, J. Am. Chem. Soc., *91*, 5801 (1969).
167a. Siskin, M., J. Am. Chem. Soc., *98*, 5413 (1976).
168. Olah, G. A., and J. R. DeMember, J. Am. Chem. Soc., *91*, 2113 (1969).
169. Olah, G. A., and J. R. DeMember, J. Am. Chem. Soc., *92*, 718 (1970).
170. Olah, G. A., and J. R. DeMember, J. Am. Chem. Soc., *92*, 2562 (1970).
171. Olah, G. A. and J. J. Svoboda, Synthesis, 203 (1973).
172. Olah, G. A., J. R. DeMember, Y. K. Mo, J. J. Svoboda, P. Schilling, and J. A. Olah, J. Am. Chem. Soc., *96*, 884 (1974).
173. Olah, G. A., Y. Yamada, and R. J. Spear, J. Am. Chem. Soc., *97*, 680 (1975).
173a. Schlosberg, R. H., M. Siskin, W. P. Kocsi, and F. J. Parker, J. Am. Chem. Soc., *98*, 7723 (1976).
173b. Siskin, M., and R. H. Schlosberg, J. Am. Chem. Soc., *100*, 1842 (1978).
173c. Siskin, M., Tetrahedron Lett., 527 (1978).
174. Olah, G. A., M. B. Comisarow, E. Namanworth, and B. Ramsey, J. Am. Chem. Soc., *89*, 5259 (1967).
175. Olah, G. A., and R. D. Porter, J. Am. Chem. Soc., *92*, 7627 (1970).
176. Olah, G. A., and R. D. Porter, J. Am. Chem. Soc., *93*, 6877 (1971).
177. Olah, G. A., E. B. Baker, J. C. Evans, W. S. Tolgyesi, J. S. McIntyre, and I. J. Bastien, J. Am. Chem. Soc., *86*, 1360 (1964).
178. Kramer, G. M., J. Org. Chem., *44*, 2616 (1979).
179. Mirda, D., D. Rapp, and G. M. Kramer, J. Org. Chem. *44*, 2619 (1979).
180. Kramer, G. M., J. Org. Chem., *40*, 298 (1975).
181. Kramer, G. M., J. Org. Chem., *40*, 302 (1975).
182. Saunders, M., and E. L. Hagen, J. Am. Chem. Soc., *90*, 2436 (1968).

183. Saunders, M., and E. L. Hagen, J. Am. Chem. Soc., *90*, 6881 (1968).
184. Saunders, M., E. L. Hagen, and J. R. Rosenfeld, J. Am. Chem. Soc., *90*, 6882 (1968).
185. Brouwer, D. M., and J. M. Oelderik, 155th Nat. Meet. Am. Chem. Soc., San Francisco, Apr. 1968, Abstr. No. R58.
186. Olah, G. A., and J. Lukas, J. Am. Chem. Soc., *89*, 2227 (1967).
187. Olah, G. A., J. Sommer, and E. Namanworth, J. Am. Chem. Soc., *89*, 3576 (1967).
188. Kramer, G. M., J. Am. Chem. Soc., *91*, 4819 (1969).
189. Warkentin, J., and K. E. Hine, Can. J. Chem., *48*, 3545 (1970).
190. Olah, G. A., and J. M. Bollinger, J. Am. Chem. Soc., *89*, 4744 (1967).
191. Olah, G. A., and J. M. Bollinger, J. Am. Chem. Soc., *90*, 947 (1968).
192. Olah, G. A., and P. E. Peterson, J. Am. Chem. Soc., *90*, 4675 (1968).
193. Olah, G. A., and J. M. Bollinger, and J. Brinich, J. Am. Chem. Soc., *90*, 6988 (1968).
194. Bollinger, J. M., J. M. Brinich, and G. A. Olah, J. Am. Chem. Soc., *92*, 4025 (1970).
195. Olah, G. A., Y. K. Mo, E. G. Mo, E. G. Melby, and H. C. Lin, J. Org. Chem., *38*, 367 (1973).
196. Olah, G. A., D. A. Beal, and P. W. Westerman, J. Am. Chem. Soc., *95*, 3387 (1973).
197. Olah, G. A., G. Liang, and Y. K. Mo, J. Org. Chem., *39*, 2394 (1974).
198. Olah, G. A., J. L. Grant, R. J. Spear, J. M. Bollinger, A. Serianz, and G. Sipos, J. Am. Chem. Soc., *98*, 2501 (1976).
199. Peterson, P. E., P. R. Clifford, and F. J. Slama, J. Am. Chem. Soc., *92*, 2840 (1970).
200. Olah, G. A., and Y. K. Mo, Adv. Fluorine Chem., *7*, 69 (1973).
201. Olah, G. A., *Halonium Ions*, Wiley-Interscience, New York, 1975.
202. Henrichs, P. M., and P. E. Peterson, J. Org. Chem., *41*, 362 (1976).
203. (a) Peterson, P. E., B. R. Bonazza, and P. M. Henrichs, J. Am. Chem. Soc., *95*, 2222 (1973); (b) Larsen, J. W., and A. V. Metzner, J. Am. Chem. Soc., *94*, 1614 (1972); (c) Page, M. I., Chem. Soc. Rev., *2*, 295 (1973), and references cited therein.
204. Peterson, P. E., and R. J. Bopp, J. Am. Chem. Soc., *89*, 1248 (1967).
205. Bollinger, J. M., C. A. Cupas, K. J. Friday, M. L. Woolfe, and G. A. Olah, J. Am. Chem. Soc., *89*, 156 (1967).
206. Olah, G. A., and D. H. O'Brien, J. Am. Chem. Soc., *89*, 1725 (1967).
207. Olah, G. A., Yu Simon, G. Liang, G. D. Mateescu, M. R. Bruce, D. J. Donovan, and M. Arvanaghi, J. Org. Chem., *46*, 571 (1981).
208. Olah, G. A., D. H. O'Brien, and C. U. Pittman, Jr., J. Am. Chem. Soc., *89*, 2996 (1967).
209. Olah, G. A., and R. J. Spear, J. Am. Chem. Soc., *97*, 1539 (1975).
209a. Deno, N. C., and R. R. Lastomirsky, J. Org. Chem., *40*, 514 (1975).
210. Mayr, H., and G. A. Olah, J. Am. Chem. Soc., *99*, 510 (1977).
211. Olah, G. A., G. Asenio, and H. Mayr, J. Org. Chem., *43*, 1518 (1978).

212. Olah, G. A., D. Meidar, and G. Liang, J. Org. Chem., *43*, 3890 (1978).
213. Kirrmann, A., and E. Saito, Bull. Soc. Chim. Fr., 809 (1945).
214. Turkevich, J., and R. K. Smith, J. Chem. Phys., *16*, 466 (1948); Nature, *157*, 874 (1946).
215. Asinger, F., B. Fell, and G. Collin, Chem. Ber., *96*, 716 (1963).
216. Hubert, A. J., and H. Reimlinger, Synthesis, 405 (1970).
217. McCarthy, W. W., and J. Turkevich, J. Chem. Phys., *12*, 405, 461 (1944).
218. Kazanskii, B. A., I. V. Gastunskaya, and N. B. Dobroserdova, Dokl. Akad. Nauk SSSR, *130*, 82 (1960); Chem. Abstr., *54*, 10825h (1960).
219. Asinger, F., and B. Fell, Erdoel Kohle, Erdgas, Petrochem., *19*, 258 (1966).
220. Pisman, I. I., V. V. Kasyanov, I. I. Ninalalov, and M. A. Dalin, Khim. Prom., *44*, 98 (1968); Chem. Abstr., *69*, 66668e (1968).
221. Lucchesi, P. J., D. L. Baeder, and J. P. Longwell, J. Am. Chem. Soc., *81*, 3235 (1959).
222. Sakurai, Y., Y. Kaneda, S. Kondo, E. Hirota, T. Onishi, and K. Tamaru, Bull. Chem. Soc. Jpn., *41*, 1496 (1968); Chem. Abstr., *69*, 85962n (1968).
223. Henne, A. L., and A. Turk, J. Am. Chem. Soc., *64*, 826 (1942).
224. Plate, A. F., and M. I. Batuev, Dokl. Akad. Nauk SSSR, *59*, 1305 (1948); Chem. Abstr., *42*, 7234b (1948).
225. Levina, R. Y., and E. A. Victorova, Vestn. Mosk. Univ. 6, No. 2, Ser. Fiz.-Mat. Estestv. Nauk No. 1, 89 (1951); Chem. Abstr., *46*, 8605g (1952).
226. Butlerow, A., Justus Liebigs Ann. Chem., *180*, 245 (1876).
227. Whitmore, F. C., Ind. Eng. Chem., *26*, 94 (1934).
228. Deno, N. C., D. B. Boyd, J. D. Hodge, C. U. Pittman, Jr., and J. O. Turner, J. Am. Chem. Soc., *86*, 1745 (1964).
229. Shiner, V. J., Jr., and G. F. Meier, J. Org. Chem., *31*, 137 (1966).
230. Olah, J. A., and J. Lukas, J. Am. Chem. Soc., *89*, 4739 (1967).
231. Magnotta, V. L., B. C. Gates, and B. C. A. Schuit, J. Chem. Soc., Chem. Commun., 342 (1976).
232. Brouwer, D. M., Recl. Trav. Chim. Pays-Bas, *87*, 210 (1968).
233. Brouwer, D. M., Recl. Trav. Chim. Pays-Bas, *87*, 1435 (1968).
234. Brouwer, D. M., Recl. Trav. Chim. Pays-Bas, *88*, 9 (1969).
235. Olah, G. A., and G. Liang, J. Am. Chem. Soc., *95*, 3792 (1973); *97*, 1920 (1975).
236. Olah, G. A., G. Liang, K. A. Babiak, T. M. Ford, D. L. Goff, T. K. Morgan, Jr., and R. K. Murray, Jr., J. Am. Chem. Soc., *100*, 1494 (1978).
237. Brouwer, D. M., and J. A. Van Doora, Recl. Trav. Chim. Pays-Bas, *88*, 573 (1969).
238. Olah, G. A., and J. Lukas, J. Am. Chem. Soc., *90*, 933 (1968).
239. Olah, G. A., and J. M. Bollinger, J. Am. Chem. Soc., *90*, 6082 (1968).
240. Olah, G. A., C. I. Jeuell, D. P. Kelly, and R. D. Porter, J. Am. Chem. Soc., *92*, 2544 (1970); *94*, 146 (1972).

240a. Olah, G. A., D. J. Donovan, and K. S. Prakash, Tetrahedron Lett., 4779 (1978).
241. Saunders, M., and J. Rosenfeld, J. Am. Chem. Soc., *92*, 2548 (1970).
242. Olah, G. A., G. K. Surya Prakash, and T. N. Rawdah, J. Org. Chem., *45*, 965 (1980).
243. Roberts, J. D., and R. H. Mazur, J. Am. Chem. Soc., *73*, 2509 (1951).
244. Breslow, R., in *Molecular Rearrangements*, Part 1 (P. de Mayo, ed.), Wiley-Interscience, New York, 1963, Chap. 4.
245. Schleyer, P. V. R., and G. W. Van Dine, J. Am. Chem. Soc., *88*, 2321 (1966).
246. Hanack, M., and H. J. Schneider, Angew. Chem., Int. Ed. Engl., *6*, 666 (1967).
247. Wilberg, K. B., B. A. Hess, Jr., and A. J. Ashe III, in *Carbonium Ions*, Vol. 3, (G. A. Olah and P. v. R. Schleyer, eds.), Wiley-Interscience, New York, 1972, Chap. 26.
248. Richey, H. G., Jr., in *Carbonium Ions*, Vol. 3, (G. A. Olah and P. v. R. Schleyer, eds.), Wiley-Interscience, New York, 1972, Chap. 25. A review covering literature to early 1971.
249. Olah, G. A., and R. J. Spear, J. Am. Chem. Soc., *97*, 1539 (1975).
250. Olah, G. A., R. J. Spear, P. C. Hiberty, and W. J. Hehre, J. Am. Chem. Soc., *98*, 7470 (1976).
251. Olah, G. A., G. K. Surya Prakash, and G. Liang, J. Org. Chem. *42*, 2666 (1977).
252. Wolf, J. W., P. G. Hatch, R. W. Taft, and W. J. Hehre, J. Am. Chem. Soc., *97*, 2902 (1975).
253. Olah, G. A., and G. Liang, J. Am. Chem. Soc., *97*, 1920 (1975).
254. Kelly, D. P., and H. C. Brown, J. Am. Chem. Soc., *97*, 3897 (1975); *98*, 642 (1976).
255. Olah, G. A., G. Liang, K. A. Babiak, T. K. Morgan, Jr., and R. K. Murray, Jr., J. Am. Chem. Soc., *98*, 576 (1976).
256. Olah, G. A., and G. Liang, J. Am. Chem. Soc., *98*, 7026 (1976).
257. Olah, G. A., G. Liang, R. K. Murray, Jr., and K. A. Babiak, J. Am. Chem. Soc., *98*, 576 (1976).
258. Olah, G. A., G. Liang, and Y. K. Mo, J. Am. Chem. Soc., *94*, 3544 (1972).
259. Olah, G. A., and A. M. White, J. Am. Chem. Soc., *91*, 3954, 3956, 6883 (1969).
260. Olah, G. A., J. M. Bollinger, C. A. Cupas, and J. Lukas, J. Am. Chem. Soc., *89*, 2692 (1967).
261. Nenitzescu, C. D., and J. P. Cantuniari, Chem. Ber., *66*, 1097 (1933).
262. Glasebrook, A. L., and W. G. Lovell, J. Am. Chem. Soc., *61*, 1717 (1939).
263. Stevenson, D. P., and J. H. Morgan, J. Am. Chem. Soc., *70*, 2773 (1948).
263a. Siskin, M., J. Am. Chem. Soc., *100*, 1838 (1978).
263b. Farcasiu, D., M. Siskin, and R. P. Rhodes, J. Am. Chem. Soc., *101*, 7671 (1979).
264. Farcasiu, D., and L. Craine, J. Chem. Soc., Chem. Commun., 687 (1976).
265. Olah, G. A., D. P. Kelly, and R. G. Johanson, J. Am. Chem. Soc., *92*, 4137 (1970).

266. Kirchen, R. P., and T. S. Sorensen, J. Am. Chem. Soc., *100*, 1487 (1978).
267. Deno, N. C., in *Carbonium Ions*, Vol. 2, (G. A. Olah, and P. v. R. Schleyer, eds.), Wiley-Interscience, New York, 1970, Chap. 18.
268. Weiss, R., H. Kolbl, and C. Schlierf, J. Org. Chem., *41*, 2258 (1976), and references therein.
269. West, R., Acc. Chem. Res., *3*, 130 (1970).
270. Olah, G. A., J. S. Staral, R. J. Spear, and G. Liang, J. Am. Chem. Soc., *97*, 5489 (1975).
271. Olah, G. A., and G. Liang, J. Am. Chem. Soc., *94*, 6434 (1972).
272. Farnum, D. G., A. Mostashari, and A. A. Hagedorn, J. Org. Chem. *36*, 6398 (1971).
273. Brown, H. C., *Nonclassical Intermediates*, Org. React. Mech. Conf. Brookhaven, N.Y., Sept. 1962.
274. Bartlett, P. D., *Nonclassical Ions*, W. A. Benjamin, New York, 1965.
275. Winstein, S., Q. Rev. Chem. Soc., 141 (1969).
276. Brown, H. C., Acc. Chem. Res., *6*, 37 (1973), and references therein.
277. Deno, N. C., Isot. Org. Chem., *1*, 1 (1975).
277a. Zelinski, N. D., and M. B. Turowa-Pollack, Chem. Ber., *58*, 1292 (1925).
277b. Jones, R. L., and R. P. Linstead, J. Chem. Soc., 616 (1936).
277c. Zelinski, N. D., and M. B. Turowa-Pollack, Chem. Ber., *62*, 1658 (1929).
277d. Hogeveen, H., and C. J. Gaasbeek, Recl. Trav. Chim. Pays-Bas, *88*, 719 (1969).
278. Seybold, G. P. Vogel, M. Saunders, and K. B. Wilberg, J. Am. Chem. Soc., *95*, 2045 (1973).
279. McKervey, M. A., Chem. Soc. Rev., *3*, 479 (1974), for a review.
280. Majerski, Z., and K. Mlinaric, J. Chem. Soc., Chem. Commun., 1030 (1972).
281. Olah, G. A., G. Liang, and S. P. Jindal, J. Am. Chem. Soc., *98*, 2508 (1976).
282. Olah, G. A., G. Liang, and S. P. Jindal, J. Org. Chem., *40*, 3259 (1975).
283. Schleyer, P. v. R., W. E. Watts, R. C. Fort, Jr., M. D. Comisarow, and G. A. Olah, J. Am. Chem. Soc., *86*, 5679 (1964).
284. Diaz, A., M. Brookhart, and S. Winstein, J. Am. Chem. Soc., *88*, 3133, 3135 (1966).
285. Jensen, F. R., and B. H. Beck, Tetrahedron Lett., 4287 (1966).
286. Richey, H. G., Jr., and R. K. Lustgarten, J. Am. Chem. Soc., *88*, 3136 (1966).
287. Tanida, H., Acc. Chem. Res., *1*, 239 (1968).
288. Olah, G. A., A. M. White, J. R. DeMember, A. Commeyras, and C. Y. Lui, J. Am. Chem. Soc., *92*, 4627 (1970).
289. Olah, G. A., G. D. Mateescu, and J. L. Riemenschneider, J. Am. Chem. Soc., *94*, 2529 (1972).
290. Olah, G. A., G. Liang, G. D. Mateescu, and J. L. Riemenschneider, J. Am. Chem. Soc., *95*, 8698 (1973).
291. Brown, H. C., and J. H. Kawakami, J. Am. Chem. Soc., *97*, 5521 (1975).
292. Olah, G. A., and G. Liang, J. Am. Chem. Soc., *97*, 6803 (1975).
293. Solomon, J. J., and F. H. Field, J. Am. Chem. Soc., *98*, 1567 (1976).

294. Olah, G. A., and G. Liang, J. Am. Chem. Soc., *93*, 6873 (1971).
295. Olah, G. A., J. R. DeMember, C. Y. Lui, and A. M. White, J. Am. Chem. Soc., *91*, 3958 (1969).
296. Olah, G. A., J. R. DeMember, C. Y. Lui, and R. D. Porter, J. Am. Chem. Soc., *93*, 1442 (1971).
297. Turova-Polyurk, M. B., J. E. Sosnina, I. G. Gohetrina, and T. P. Yudkina, Zh. Obshch. Khim., *29*, 1078 (1959); Chem. Abstr., *54*, 1356 (1960).
298. Belikoona, N. A., H. A. Bibleva, and A. F. Plate, Zh. Org. Khim., *2*(11), 2031 (1966).
299. Barrett, J. W., and R. B. Linstead, J. Chem. Soc., 611 (1936).
300. Schleyer, P. v. R., K. R. Blanchard, and C. C. Woody, J. Am. Chem. Soc., *85*, 1358 (1963), and earlier references cited therein.
301. Georing, H. L., and M. F. Sloan, J. Am. Chem. Soc., *83*, 1397, 1992 (1961); in addition, see references cited in Refs. 161 and 204.
302. Olah, G. A., G. Liang, J. R. Wiseman, and J. A. Chong, J. Am. Chem. Soc., *94*, 4927 (1972).
303. Moriarty, R. M., and C. W. Jefford, *Organic Chemistry, A Problems Approach*, W. A. Benjamin, New York, 1975, p. 154.
304. Fort, R. C., Jr., and P. v. R. Schleyer, Chem. Rev., *64*, 277 (1964).
305. Lustgarten, R. K., M. Bookhart, and S. Winstein, J. Am. Chem. Soc., *89*, 6350 (1967).
306. Geluk, H. W., and J. L. M. A. Schlatmann, Chem. Commun., 426 (1967).
307. Leone, R. E., J. C. Barborak, and P. v. R. Schleyer in *Carbonium Ions*, Vol. 4 (G. A. Olah and P. v. R. Schleyer, eds.), Wiley Interscience, New York, 1972, p. 1931.
308. Cuddy, B. D., D. Grant, A. Karim, M. A. McKervey, and J. F. Rea, J. Chem. Soc., Perkin Trans. *1*, 2701 (1972).
309. Alford, J. R., B. D. Cuddy, D. Grant, and M. A. McKervey, J. Chem. Soc., Perkin Trans. *1*, 2707 (1972).
310. Geluk, H. W., and J. L. M. A. Schlatmann, Tetrahedron, *24*, 5361 (1968).
311. McKervey, M. A., J. R. Alford, J. F. McGarrity, and E. J. F. Rea, Tetrahedron Lett., 5165 (1968).
312. Geluk, H. W., and J. L. M. A. Schlatmann, Recl. Trav. Chim. Pays-Bas, *88*, 13 (1969).
313. Schneider, A., and R. W. Warren, Am. Chem. Soc. Div. Petrol. Chem., Prepr. *15*(2), B56 (1970); 159th Nat. Meet. Am. Chem. Soc., Houston, Feb. 1970, Abstr. PETR 54.
314. Majerski, Z., P. v. R. Schleyer, and A. P. Wolf, J. Am. Chem. Soc., *92*, 5731 (1970).
315. Balaban, A. T., D. Farcasiu, and R. Banica, Rev. Chim. (Bucharest), *11*, 1205 (1966).
316. Brouwer, D. M., and H. Hogeveen, Recl. Trav. Chim. Pays-Bas, *89*, 211 (1970).
317. Olah, G. A., G. Liang, and G. D. Mateescu, J. Org. Chem., *39*, 3750 (1974).
318. Mlinaric-Majerski, K., Z. Majerski, and E. Pretsch, J. Org. Chem., *40*, 3772 (1975).
319. Mlinaric-Majerski, K., Z. Majerski, and E. Pretsch, J. Org. Chem., *41*, 686 (1976).

320. Farcasiu, M., K. R. Blanchard, E. M. Engler, and P. v. R. Schleyer, Chem. Lett., 1189 (1973).
321. Schleyer, P. v. R., and R. D. Nicholas, Tetrahedron Lett., 305, (1961).
322. Blanchard, K. R., Ph.H. thesis, Princeton University, 1966.
323. Takaishi, N., Y. Inamoto, K. Tsuchihashi, K. Yashima, and K. Aigami, J. Org. Chem., *40,* 2929 (1975), and other references therein.
324. Olah, G. A., G. Liang, K. A. Babiak, and R. K. Murray, Jr., J. Am. Chem. Soc., *96,* 6794 (1974).
325. Schleyer, P. v. R., J. Am. Chem. Soc., *79,* 3292 (1957).
326. Engler, E. M., M. Farcasiu, A. Sevin, J. M. Cense, and P. v. R. Schleyer, J. Am. Chem. Soc., *95,* 5769 (1973).
327. Vorobeva, H. W., Jr., O. A. Arefev, V. I. Epishev, and A. A. Petrov, Neftekhimiya, *11,* 163 (1971).
328. Arefev, O. A., N. S. Vorobeva, and A. A. Petrov, Neftekhimiya, *11,* 32 (1971).
329. Takaishi, N., Y. Inamoto, and K. Tsuchihashi, Ger. Offen. 2,432, 219 (1975); Chem. Abstr., *83,* 58275y (1975).
330. Takaishi, N., Y. Inamoto, and K. Tsuchihashi, Ger. Offen. 2,432, 193 (1975); Chem. Abstr., *83,* 58276z (1975).
330a. Fujikura, Y., Y. Inamoto, and N. Takaishi, Jpn. Kokai Tokkyo Koho 79, 259 (1979); Chem. Abstr., *91,* 192915a (1979).
331. Takaishi, N., Y. Inatomo, and K. Aigami, J. Chem. Soc., Perkin Trans. 1, (9), 789 (1975); Chem. Abstr., *83,* 58237n (1975).
332. Takaishi, N., Y. Inamoto, K. Tsuchihashi, K. Aigami, and Y. Fujikura, J. Org. Chem., *41,* 771 (1976).
333. Whitlock, H. W., Jr., and M. W. Shiefken, J. Am. Chem. Soc., *90,* 4929 (1968).
334. Sinnott, M. L., H. J. Sotesund, and M. C. Whiting, Chem. Commun., 1000 (1969).
335. Lenoir, D., R. E. Hall, and P. v. R. Schleyer, J. Am. Chem. Soc., *96,* 2138 (1974).
336. Farcasiu, M., D. Farcasiu, J. Slutsky, and P. v. R. Schleyer, Tetrahedron Lett., 4062 (1974).
337. Farcasiu, M., E. W. Hagman, E. Wenkert, and P. v. R. Schleyer, 173rd Nat. Meet. Chem. Soc., New Orleans, Mar. 21-25, 1977, Abstr. 187.
338. Lenoir, D., P. Mison, E. Hyson, P. v. R. Schleyer, M. Saunders, P. Vogel, and L. A. Telkowski, J. Am. Chem. Soc., *96,* 2157 (1974).
339. Lenoir, D., D. J. Raber, and P. v. R. Schleyer, J. Am. Chem. Soc., *96,* 2149 (1974).
340. Cupas, C., P. v. R. Schleyer, and D. J. Trecher, J. Am. Chem. Soc., *87,* 917 (1965); cf. I. L. Karle and J. Karle, J. Am. Chem. Soc., *87,* 918 (1965).
341. Hala, S., J. Novak, and S. Landa, Sb. Vys. Sk. Chem.-Tekhnol. Praze, Technol. Paliv, *D-19,* 31 (1969).
342. Gund, T. M., V. Z. Williams, Jr., E. Osawa, and P. v. R. Schleyer, Tetrahedron Lett., 3877 (1970).
343. Stohrer, W. -D., and R. Hoffmann, J. Am. Chem. Soc., *94,* 1661 (1972).
344. Hart, H., and M. Kuzuya, J. Am. Chem. Soc., *97,* 2459 (1975).

345. Hart, H., and M. Kuzuya, J. Am. Chem. Soc., *98*, 1545 (1976).
346. Hart, H., and M. Kuzuya, J. Am. Chem. Soc., *98*, 1551 (1976).
347. Paquette, L. A., M. Oku, and W. B. Farnham, J. Org. Chem., *40*, 700 (1975).
348. Fort, R. C. Jr., and P. v. R. Schleyer, Chem. Rev., *64*, 277 (1964).
349. Schleyer, P. v. R., Angew. Chem., Int. Ed. Engl. *8*, 529 (1969).
350. Lenoir, D., and P. v. R. Schleyer, Chem. Commun., 941 (1970).
351. Schleyer, P. v. R., et al., J. Org. Chem., *45*, 2985 (1980).
352. Evering, B. L., and R. C. Waugh, Ind. Eng. Chem., *43*, 1820 (1951).
353. Kramer, G. M., and A. Schriesheim, J. Phys. Chem., *65*, 1283 (1961).
354. Kramer, G. M., J. Org. Chem., *34*, 2919 (1969).
355. Nenitzescu, C. D., in *Carbonium Ions*, Vol. 2, (G. A. Olah and P. v. R. Schleyer, eds.), Wiley-Interscience, New York, 1970, pp. 490-501.
356. Morozova, O. E., and A. A. Petrov, Petrol. Chem. USSR, *9*, 99 (1969).
357. Bloch, H. S., H. Pines, and L. Schmerling, J. Am. Chem. Soc., *68*, 153 (1946).
358. Schneider, A., and R. M. Kennedy, J. Am. Chem. Soc., *73*, 5013 (1951).
359. Pines, H., B. M. Abraham, and V. N. Ipatieff, J. Am. Chem. Soc., *70*, 1742 (1948).
360. Muller, N., and R. S. Mulliken, J. Am. Chem. Soc., *80*, 3489 (1958).
361. Roberts, J. D., and C. C. Lee, J. Am. Chem. Soc., *73*, 5009 (1951).
362. Roberts, J. D., C. C. Lee, and W. H. Saunders, Jr., J. Am. Chem. Soc., *76*, 4501 (1954).
363. Concon, F. E., in *Catalysts*, Vol. 6, (P. H. Emmett, ed.), Reinhold, New York, 1958, Chap. 2; see for a review.
364. Collins, C. J., Chem. Rev., 543 (1969).
365. Lee, C. C., Prog. Phys. Org. Chem., *7*, 129-187 (1970); see for a review.
366. McCaulay, D. A., Div. Petrol. Chem., Am. Chem. Soc. Meet. Boston, Apr. 1959; J. Am. Chem. Soc., *81*, 6437 (1959).
367. Schneider, A., and R. M. Kennedy, J. Am. Chem. Soc., *73*, 5024 (1951).
368. Brouwer, D. M., and J. A. Van Doorn, Recl. Trav. Chim Pays-Bas, *91*, 895 (1972).
369. Stevenson, D. P., C. D. Wagner, O. Beeck, and J. W. Otvos, J. Am. Chem. Soc., *74*, 3269 (1952).
370. Roberts, J. D., and G. R. Coraor, J. Am. Chem. Soc., *74*, 3586 (1952).
371. Liardon, R., and L. Gaumann, Helv. Chim. Acta, *54*, 1968 (1971).
372. Pakkanen, T., and J. L. Whitten, J. Am. Chem. Soc., *97*, 6337 (1975).
373. Lee, C. C., and E. C. F. Ko, Can. J. Chem., *54*, 1722 (1976).
374. Lee, C. C., A. J. Cessna, E. C. F. Ko, and S. Vassie, J. Am. Chem. Soc., *95*, 5688 (1973).

3

Alkylations of Arenes with Alkyl Halides

I. ALKYLATIONS WITH METHYL AND BENZYL HALIDES

A. The Kinetic Approach

The alkylation of arenes with methyl and benzyl halides is as old as the Friedel-Crafts reaction. In fact, Friedel and Crafts [1] were among the first to alkylate benzene with methyl and benzyl chlorides. They reported obtaining a mixture of toluene, xylenes, trimethylbenzenes, durene, and penta- and hexamethylbenzene from a reaction of benzene with methyl chloride and $AlCl_3$ at 80°C, and diphenylmethane from a reaction of benzene with benzyl chloride and $ZnCl_2$ at 80°C. Following this earliest report, there were several other accounts of methylation and benzylation. Those which appeared prior to 1946 were reviewed by Price [2]. Since that time reactions of methyl [3-10] and benzyl [11-25] halides under Friedel-Crafts conditions have been extensively utilized in studies related to mechanisms of Friedel-Crafts alkylations.

In 1953, Brown et al. [26,27] studied the addition complexes of metal halides with alkyl halides. In their first paper [26], they reported that gallium chloride was readily soluble in methyl chloride at -78.5°C and that a stable 1:1 addition complex, CH_3Cl-$GaCl_3$, existed in the solution phase. They also established the existence of similar 1:1 addition complexes between $GaCl_3$ and both CH_3Br and CH_3I. In their second paper [27], they reported that aluminum bromide dissolved in CH_3Br and C_2H_5Br and that molecular weight data supported the formation of stable 1:1 addition complexes, $AlBr_3$-CH_3Br and $AlBr_3$-C_2H_5Br. In contrast, $AlCl_3$ was not appreciably soluble in either CH_3Cl or C_2H_5Cl and showed no tendency to form simple 1:1 addition compounds of the type formed by $GaCl_3$ [26]. Thus, although it appears reasonable that the formation of an addition compound between an alkyl halide and aluminum bromide or gallium chloride constitutes the first step in Friedel-Crafts reactions promoted by these catalysts, the situation is not so simple for aluminum chloride catalyzed alkylations. Even in the case of the catalysts that give demonstrable 1:1 complexes with alkyl halides, the role these play in the mechanisms of alkylation has not been determined without a great deal of investigation. In 1954, Baddeley gave a summary of the state of knowledge of the various complexes formed by metal halides, alkyl halides, hydrogen halides, and arenes [28]. A more recent extensive treatment of this topic was presented in the Olah 1963 monograph [29], and as it is not our intention to give here a comprehensive survey of this large and complex topic, the reader is referred to this source. In this respect, we shall concentrate only on some of the interesting numerous investigations that have been carried out to determine the exact nature of the catalyst-alkyl halide addition compounds.

Among the many types of investigations that have been applied through the years to learn more about the nature of the intermediate complexes formed between Friedel-Crafts catalysts and reactants are the following: studies of (1) halogen exchange reactions; (2) the kinetics of Friedel-Crafts alkylations, especially in regard to reactivity and selectivity; and (3) the molecular rearrangements which accompany Friedel-Crafts alkylations.

Rearrangements accompanying Friedel-Crafts alkylations will be discussed extensively in connection with alkylations by ethyl and higher groups. We shall direct our discussion at this stage to studies of halogen exchange and kinetic experiments in which, again, *methyl* and *benzyl* halides played a major role.

In 1948, Fairbrother [30] estimated from measurements of the dielectric polarizability of solutions of $AlBr_3$ in ethyl bromide that ionization to the extent of 3 to 4% occurred. Jacober and Kraus [31] examined the conductivity of solutions of $AlBr_3$ in methyl bromide and reported conductivities much smaller than those noted by others for systems involving higher alkyl halides [32].

Besides these earlier studies, there have been a number of more recent reports on the subject. A report by Sixma and co-workers [33] in 1956 showed that the rate law for the bromine exchange reaction between ethyl bromide and aluminum bromide in CS_2 is

$$\text{rate} = R_3[AlBr_3]^2[C_2H_5Br]$$

In more recent reports, Choi and co-workers [7,34] found that radioactive bromine exchange between $GaBr_3$ and either methyl [7] or ethyl [34] bromide in either 1,2,4-trichorobenzene or nitrobenzene followed the rate law

$$\text{rate} = k_3[GaBr_3]^2[RBr]$$

They also observed that $GaBr_3$ exchanged bromine significantly faster with ethyl bromide than with methyl bromide. The following mechanism was then proposed for the bromine exchange reactions of $GaBr_3$ with alkyl bromides:

$$RBr + S\text{-}GaBr_3 \rightleftharpoons R^{\delta+}Br^{\delta-}\text{-}GaBr_3 + S$$

$$R^{\delta+}Br^{\delta-}\text{-}GaBr_3 + S\text{-}GaBr_3 \xrightarrow{\text{slow}} R^+Ga_2Br_7^- + S$$

$$S + R^+Ga_2Br_7^- \rightleftharpoons RBr + S\text{-}Ga_2Br_6$$

In this mechanism, the breaking of the C-Br bond in the alkyl bromide molecules was assumed to take an important part in determining the energy of the transition state of the exchange reaction. Accordingly, the observation that $GaBr_3$ exchanges bromine faster with ethyl bromide than with methyl bromide was attributed to the greater ability of an ethyl group to tolerate positive charge than methyl. In contrast, however, Polaczek and Halpern [35] noted from specific conductivity values that exchange between AlI_3, GaI_3, and InI_3 and alkyl iodides decreased in the order Me > Et > *n*-propyl.

Much more recently, kinetic studies have been made by Brown and co-workers of the chloride-36 exchange between radioactive $GaCl_3$ and both methyl [10] and ethyl [36] chlorides. With methyl chloride (in excess methyl chloride), the exchange reaction was stated to be second order in $GaCl_3$ and

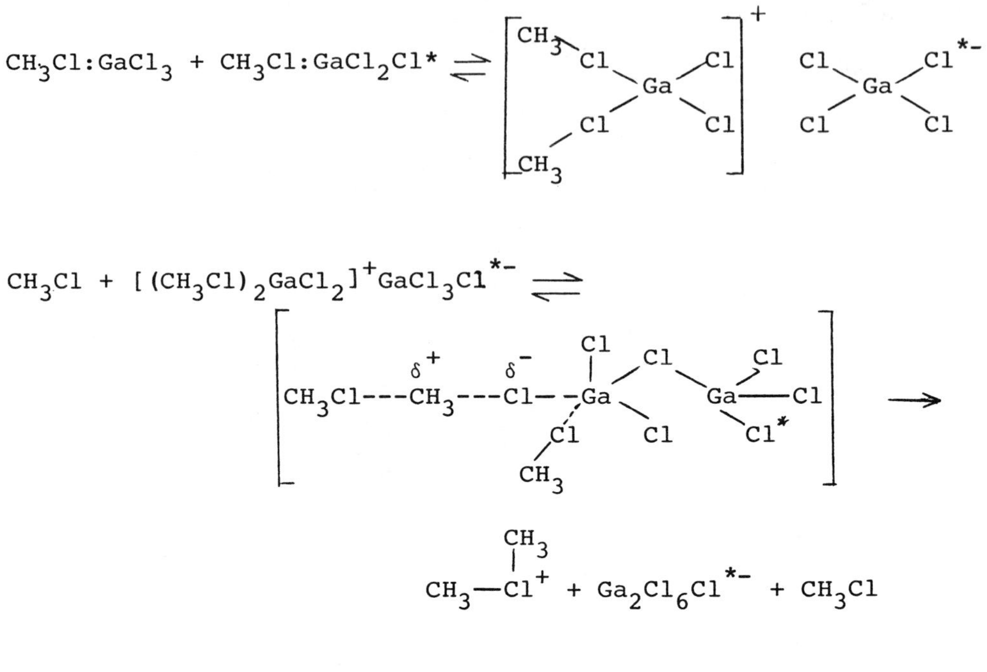

$$CH_3Cl{:}GaCl_3 + CH_3Cl{:}GaCl_2Cl^* \rightleftharpoons \left[\begin{matrix} CH_3 - Cl \\ CH_3 - Cl \end{matrix} > Ga < \begin{matrix} Cl \\ Cl \end{matrix} \right]^+ \; \begin{matrix} Cl \\ Cl \end{matrix} > Ga < \begin{matrix} Cl^{*-} \\ Cl \end{matrix}$$

$$CH_3Cl + [(CH_3Cl)_2GaCl_2]^+GaCl_3Cl^{*-} \rightleftharpoons$$

$$\left[CH_3Cl{-}{-}{-}\overset{\delta+}{CH_3}{-}{-}{-}\overset{\delta-}{Cl}{-}{-}Ga(Cl)(ClCH_3)(\mu\text{-}Cl)_2Ga(Cl)Cl^* \right] \longrightarrow$$

$$CH_3{-}\overset{CH_3}{\overset{|}{Cl^+}} + Ga_2Cl_6Cl^{*-} + CH_3Cl$$

Scheme 1

probably first order in CH_3Cl. Similar results were also noted with ethyl chloride. It was suggested that both reactions followed the kinetic expression

$$\text{rate} = k_3[RCl{:}\ GaCl_3]^2[RCl]$$

Consistent with the latter expression, a mechanism for halogen exchange was proposed in which a rate-determining attack by alkyl halide on an ion-pair dimer resulted in the formation of the dialkylchloronium ion as an unstable intermediate. This mechanism is shown in scheme 1 with CH_3Cl as an example. An alternative mechanism of exchange in which the reaction could involve nonionic intramolecular transfer of the alkyl group (Eq. 1) was disqualified on the

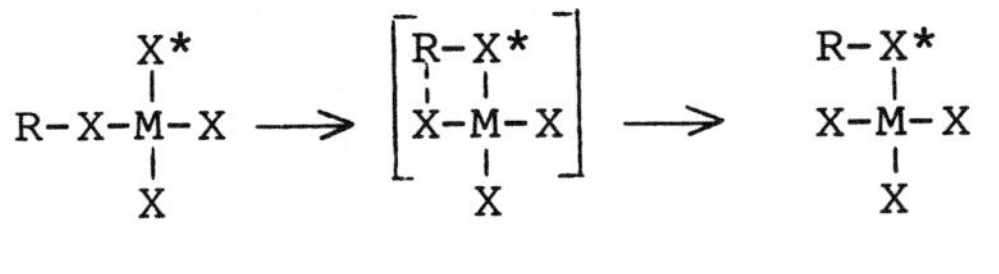

$$R{-}X{-}\underset{X}{\overset{X^*}{M}}{-}X \longrightarrow \left[\begin{matrix} R{-}X^* \\ \vdots \quad \vdots \\ X{-}M{-}X \\ \quad | \\ \quad X \end{matrix} \right] \longrightarrow \begin{matrix} R{-}X^* \\ \\ X{-}\underset{X}{M}{-}X \end{matrix} \qquad (1)$$

ground that comparison of the $GaCl_3$ catalyzed exchange and methylation reactions revealed that the rate of alkylation could exceed the rate of exchange. Had the center species of Eq. (1) been present as an unstable intermediate, and had the corresponding alkylation reaction involved an attack by the arene on this species, the rate of exchange would provide an upper limit to the rate of alkylation. In the light of their results, DeHaan et al. [10] stated that the earlier results of Sixma et al. [33] could best be represented by the expression

$$\text{rate} = k[C_2H_5Br{:}\ AlBr_3]^2[C_2H_5Br]$$

In recent years, the question of structure of alkyl fluoride-antimony fluoride mixtures in SO_2 and other solvents has received detailed studies in Olah's [37-40], Gillespie's [41], and Peterson's [42] groups.

According to Olah's group [38], both methyl and ethyl halides formed tightly bound donor-acceptor complexes ($CH_3F \rightarrow SbF_5$ and $C_2H_5F \rightarrow SbF_5$) which underwent rapid intramolecular fluorine exchange. The mechanisms suggested for intramolecular fluorine exchange are shown by Eqs. (2) and (3).

$$H_3C{-}F' \rightarrow SbF_5 \rightleftharpoons \left[H_3C \cdots \begin{matrix} F' \rightarrow \\ F'' \leftarrow \end{matrix} SbF_4 \right] \rightleftharpoons H_3C{-}F'' \rightarrow SbF_5 \qquad (2)$$

$$CH_3{-}F \rightarrow SbF_5 \rightleftharpoons \overset{+}{C}H_3\,SbF_6^- \qquad (3)$$

As stated, it should be kept in mind that Eqs. (2) and (3) represent limiting cases, and any degree of intermediate character between them should be possible. The slightly more deshielded PMR and CMR chemical shifts observed for C_2H_5F relative to CH_3F in SbF_5-SO_2 and related media were attributed to a greater ease of polarization and/or cleavage of the C-F bond of ethyl fluoride.

It can be inferred from the various reports on halogen exchange and their mechanistic interpretation that the carbon-halogen bond in the addition compounds formed between catalysts and alkyl halides is partially polarized, and that the degree of polarization increases with branching at the carbon atom. For halogen exchange, however, both ionic [10,36] and nonionic [38] mechanisms were invoked.

In their work, Olah and co-workers [38] investigated the reactivity of the CH_3F-SbF_5-SO_2 solution and found it to be an extremely powerful reagent, capable of methylating n, π, and σ donor bases. Examples of these methylations are shown in Eqs. (4) to (9).

$$CH_3F \rightarrow SbF_5 + CO \rightarrow CH_3\overset{+}{C}O\,SbF_6^- \xrightarrow{H_2O} CH_3\overset{O}{\overset{\|}{C}}{-}\overset{+}{O}H_2\,SbF_6^- \xrightarrow{-H^+SbF_6^-} CH_3COOH \qquad (4)$$

$$CH_3F \rightarrow SbF_5 + RX \xrightarrow[SO_2ClF]{SO_2 \text{ or}} R\overset{+}{X}CH_3 + SbF_6^- \qquad (5)$$

$$CH_3F \rightarrow SbF_5 + H_2C{=}C(CH_3)_2 \xrightarrow{SO_2} CH_3CH_2\overset{+}{C}(CH_3)_2\ SbF_6^- \quad (6)$$

$$C_6H_6 + CH_3F{\rightarrow}SbF_5 \longrightarrow C_6H_6CH_3^+SbF_6^- \xrightarrow[H_2O,0°]{HCO_3Na} C_6H_5CH_3 \quad (7)$$

$$CH_3C_6H_5 + CH_3F \rightarrow SbF_5 \xrightarrow[-78°,<1\ min]{SO_2ClF} CH_3C_6H_5CH_3^+SbF_6^- \xrightarrow[H_2O,0°]{HCO_3Na}$$

$$\text{o-xylene (54\%)} + \text{m-xylene (18\%)} + \text{p-xylene (28\%)} \quad (8)$$

$$FCH_3 + CH_3F \rightarrow SbF_5 \rightleftharpoons \left[FCH_2\text{---}\langle^{H}_{CH_3}\right]^+ \underset{}{\overset{-H^+}{\rightleftharpoons}} FC_2H_5 \xrightarrow{R\text{-}F \rightarrow SbF_5}$$

$$CH_3CH_2^+ \longrightarrow (CH_3)_3C^+,\ \text{etc.} \quad (9)$$

In addition to the examples above, the recent results of Olah's [40] and Peterson's [42] groups indicated that both SO_2 and SO_2ClF are O-methylated by the CH_3F-SbF_5 system, giving stable nonexchanging ions 1 and 2 in which the accompanying anion may be SbF_6^- or $Sb_2F_{11}^-$. Peterson et al. further proposed that 1 is the prinicipal carbon-containing species present in SO_2 solution of CH_3F-SbF_5.

$$CH_3\overset{+}{O}{=}S{=}O \cdot SbF_6^- \quad (\text{or } Sb_2F_{11}^-)$$

1

$$CH_3\overset{+}{O}{=}S(Cl)(F){=}O \cdot SbF_6^- \quad (\text{or } Sb_2F_{11}^-)$$

2

Interestingly, the alkylating power of both $CH_3F \rightarrow SbF_5$ and $C_2H_5F \rightarrow SbF_5$ surpassed the more conventional Friedel-Crafts systems, trialkyloxonium ions [43], dialkylhalonium ions [44], and alkylfluorosulfonates [45,46]. In fact, they truly behaved like methyl and ethyl fluoroantimonates and were considered to be the most reactive methylating and ethylating agents known.

Apart from other methods, kinetic studies of Friedel-Crafts reactions have provided very useful information regarding their nature. In the absence of these studies, confusing ideas about the alkylation mechanism prevailed and it was difficult for authors to choose between limiting extremes.

For example, the alkylation of an arene by an alkyl halide might proceed by a carbocation mechanism as suggested by Price [2], by a displacement mechanism as suggested by Schmerling [47], or by a displacement mechanism in the case of the primary halides with a shift to a carbocation mechanism for the more highly branched halides as suggested by Brown and co-workers [11,27]. Of course, detailed kinetic studies should offer some help in deciding among these alternatives. If the reaction proceeded by an ionization mechanism, the rate of alkylation should be determined solely by the rate of ionization of the alkyl halide and should be independent of the concentration or nucleophilic properties of the aromatic compound undergoing the alkylation. On the other hand, if a displacement reaction occurred, the rate should depend both on the concentration and the nucleophilicity of the aromatic present.

With these facts in mind, Brown and his co-workers directed part of their effort to careful studies of the kinetics of some Friedel-Crafts reactions. In their first report of research in this direction, Brown and Grayson [11] made the following statement to clarify the situation at that time (1953): "Unfortunately, in spite of the immense quantity of literature dealing with the Friedel-Crafts reaction, there is not now available a single kinetic study in which the order of the reaction has been established with regard to the aromatic component. The majority of the investigations involve heterogeneous systems or make use of unsatisfactory analytical methods."

In this first kinetic report, results from the reactions of some substituted benzyl halides with aromatic compounds in the presence of $AlCl_3$-$C_6H_5NO_2$ catalyst were given. The rate data revealed that the reaction was third order overall, first order in aromatic component, first order in $AlCl_3$, and first order in the benzyl halide. Brown and Grayson stated that a displacement mechanism involving a rate-determining nucleophilic attack by the aromatic component on a polar alkyl halide-aluminum chloride addition compound was consistent with these steps:

$$R\text{-}Cl + AlCl_3 \underset{k_{-1}}{\overset{k_1}{\rightleftharpoons}} RCl{:}AlCl_3$$

$$ArH + RCl{:}AlCl_3 \xrightarrow{k_2} \left[Ar\begin{smallmatrix} H \\ R \end{smallmatrix} \right]^+ AlCl_4^-$$

$$\left[Ar\begin{smallmatrix} H \\ R \end{smallmatrix} \right]^+ AlCl_4^- \rightleftharpoons ArR + HCl + AlCl_3$$

The transition state in the displacement step (3) was visualized as being a σ complex containing a partially formed C-C bond and a partially broken

$$\overset{\delta+}{C_6H_5}(H)\cdots R\text{---}\overset{\delta-}{Cl{:}AlCl_3}$$

3

C-Cl bond. Brown and Grayson extrapolated the interpretation of their results with benzyl halides to account for the *unrearranged* products of alkylation by primary alkyl halides such as *n*-propyl.

Several other reports on kinetics and relative reactivities followed from Brown's laboratory. A 1955 report [3] showed that both the toluene/benzene reactivity ratio and the *o, m, p*-isomer distribution in the methylation of toluene varied with the nature of the halogen in the methyl halide. The data on the relative rate of methylation of toluene and benzene at 0°C indicated a ratio of 3.8 for methyl bromide and of 4.8 for methyl iodide. The isomer distribution in methylation of toluene is shown by Eqs. (10) and (11). As

$$C_6H_5CH_3 + CH_3Br \xrightarrow[0°, 20\ min]{AlBr_3} o\text{-}C_6H_4(CH_3)_2\ (53.8\%) + m\text{-}C_6H_4(CH_3)_2\ (17.3\%) + p\text{-}C_6H_4(CH_3)_2\ (28.9\%) \quad (10)$$

$$C_6H_5CH_3 + CH_3I \xrightarrow[0°, 20\ min]{AlBr_3} o\text{-}C_6H_4(CH_3)_2\ (48.3\%) + m\text{-}C_6H_4(CH_3)_2\ (11.9\%) \quad p\text{-}C_6H_4(CH_3)_2\ (39.8\%) \quad (11)$$

such differences in the isomer distributions and the relative rate values were not compatible with a mechanism involving attack by a methyl cation, it was concluded that the methylation reaction involved a nucleophilic attack by the arene on a polarized methyl halide-aluminum bromide addition compound (Eq. 12).

$$CH_3X + Al_2Br_6 \rightleftharpoons CH_3X{:}Al_2Br_6 \xrightarrow{ArH} \left[Ar\begin{matrix}H\\CH_3\end{matrix}\right]^+ Al_2Br_6X^-$$

$$\rightleftharpoons ArCH_3 + Al_2Br_6 + HX \quad (12)$$

Brown continued his studies on alkylation by attempting to get good kinetics for the reactions of the series of methyl, ethyl, isopropyl, and *t*-butyl halides with aluminum bromide as catalyst [48], but difficulties were encountered with side reactions, such as isomerization and disproportionation. Using 1,2,4-trichlorobenzene as solvent allowed control of these side reactions, and the realtive rates of reaction of methyl and ethyl bromide with arenes were reported [49]. However, isopropyl bromide reacted too rapidly to allow accurate estimation of its rate. Brown and his co-workers next turned to gallium bromide as catalyst, and they obtained relative rates for the alkylation of toluene by methyl, ethyl, *n*-propyl, isopropyl, and *t*-butyl bromides of 1, 13.7, 15.9, 3 X 10^5, and 8 X 10^5, respectively [5]. The very rapid rates for isopropyl and *t*-butyl bromides were estimated from extent of reaction in a flow reactor. On the basis of the kinetic data, Smoot and Brown described the mechanism of reaction of the primary alkyl bromides as a displacement reaction in which the transition state "may have considerable carbonium ion character." A gradual change in mechanism was postulated in which "the carbonium ion character of the transition state increases with increasing branching of the alkyl group. With isopropyl and *t*-butyl bromides the carbonium ion character of the transition state dominates, and the data may best be interpreted by an ionization mechanism." In later work [50], Choi and Brown utilized a high-vacuum technique and obtained kinetic data which were considered more reliable than those from the flow system by measuring the vapor pressure of hydrogen bromide liberated in the reactions. They obtained a much lower rate for isopropyl bromide than that reported in the 1956 paper; the relative rates of alkylation of toluene for methyl, ethyl, and isopropyl bromides were found to be 1, 13, 7, and 20,000 respectively. The rate difference between ethyl and isopropyl bromides was still so large as to suggest a change from a displacement mechanism to a predominantly ionization mechanism, but the authors stated that such a change in mechanism should be accompanied by a marked change in entropy of activation, and this was not observed. The entropies of activation for methyl, ethyl, and isopropyl bromides were -20.0, -21.5, and -19.3, respectively. Choi and Brown thus concluded that there was no significant change in mechanism.

In 1963, Choi and Brown reported still another kinetic study in this area [51]. In the earlier studies, quite different kinetics were observed for the aluminum bromide-catalyzed reactions of alkyl halides with arenes in 1,2,4-trichlorobenzene as solvent [49] and for the gallium bromide-catalyzed alkylations in which excess arene served as solvent [4,50]. To establish whether the kinetic differences were a consequence of the different catalysts or the presence of 1,2,4-trichlorobenzene as a solvent, Choi and Brown carried out alkylations using gallium bromide catalyst in 1,2,4-trichlorobenzene solutions. The results indicated that the different kinetics arose from the use of the 1,2,4-trichlorobenzene as a solvent in one study and not in the other, rather than to a difference in the mode of operation of the two catalysts.

The kinetic results indicated that the reactions were third order overall, first order with respect to each reactant: the aromatic, the alkyl bromide, and the catalyst. The observed values of the third-order rate constants revealed that ethyl bromide alkylated toluene 6.5 times more rapidly than benzene at 18.9°C and that methyl bromide alkylated toluene 9.8 times more rapidly than benzene at 25°C. In comparing the results of this study with those of the previous one which was performed with $AlBr_3$ in excess arenes, the authors emphasized the following points: (1) the $GaBr_3$-catalyzed reactions were slower than the corresponding reactions under the influence of $AlBr_3$ indicating that $GaBr_3$ is a milder Friedel-Crafts catalyst than $AlBr_3$; (2) the methylation reactions were slower in both systems than the corresponding ethylation reactions; (3) the toluene/benzene reactivity ratio, k_T/k_B, was greater for methylation than for ethylation reactions under the same conditions; and (4) the coordinating ability of the solvent played an important role in the observed reaction kinetics.

On the basis of the last point, the authors suggested different mechanisms, one for alkylation in 1,2,4-trichlorobenzene and another for alkylation in excess arenes. These are given in schemes 2 and 3, respectively.

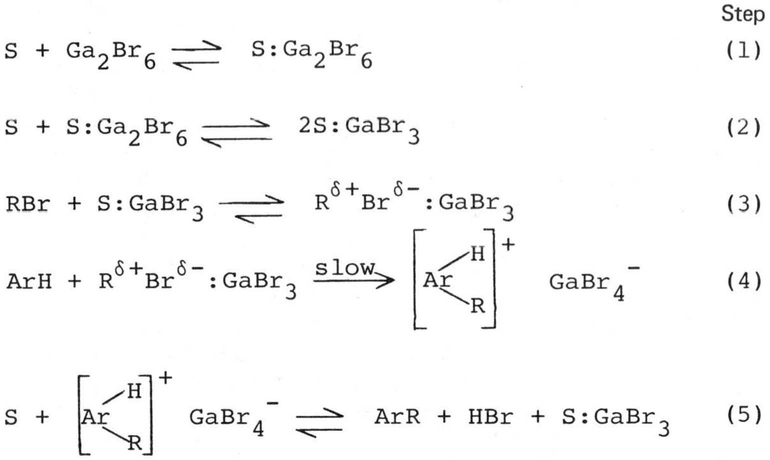

Step

$$S + Ga_2Br_6 \rightleftharpoons S{:}Ga_2Br_6 \qquad (1)$$

$$S + S{:}Ga_2Br_6 \rightleftharpoons 2S{:}GaBr_3 \qquad (2)$$

$$RBr + S{:}GaBr_3 \rightleftharpoons R^{\delta+}Br^{\delta-}{:}GaBr_3 \qquad (3)$$

$$ArH + R^{\delta+}Br^{\delta-}{:}GaBr_3 \xrightarrow{slow} \left[Ar\begin{smallmatrix}H\\R\end{smallmatrix}\right]^+ GaBr_4^- \qquad (4)$$

$$S + \left[Ar\begin{smallmatrix}H\\R\end{smallmatrix}\right]^+ GaBr_4^- \rightleftharpoons ArR + HBr + S{:}GaBr_3 \qquad (5)$$

(S = 1,2,4-trichlorobenzene)

Scheme 2

The above mechanism of scheme 2 led to the kinetic expression

$$\text{rate} = k_3[ArH][S{:}\ GaBr_3][RBr]$$

Step (4) was chosen as the rate-controlling step, although the observed slower rate of methylation compared to ethylation is in agreement with the ability of the alkyl group to tolerate a positive charge, which develops in step (3). This second mechanism (scheme 3) led to the kinetic expression

$$\text{rate} = k_2[ArH][RBr{:}\ GaBr_3]^2$$

Step (8) was postulated as the rate-controlling step—the formation of a σ complex—but the σ complex was assumed to be formed through the interaction

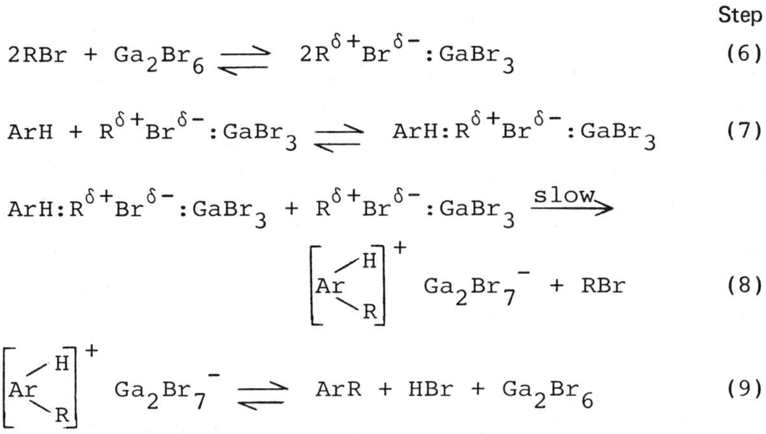

Step

$$2RBr + Ga_2Br_6 \rightleftharpoons 2R^{\delta+}Br^{\delta-}{:}GaBr_3 \qquad (6)$$

$$ArH + R^{\delta+}Br^{\delta-}{:}GaBr_3 \rightleftharpoons ArH{:}R^{\delta+}Br^{\delta-}{:}GaBr_3 \qquad (7)$$

$$ArH{:}R^{\delta+}Br^{\delta-}{:}GaBr_3 + R^{\delta+}Br^{\delta-}{:}GaBr_3 \xrightarrow{slow} \left[Ar\begin{smallmatrix}H\\R\end{smallmatrix}\right]^+ Ga_2Br_7^- + RBr \qquad (8)$$

$$\left[Ar\begin{smallmatrix}H\\R\end{smallmatrix}\right]^+ Ga_2Br_7^- \rightleftharpoons ArR + HBr + Ga_2Br_6 \qquad (9)$$

Scheme 3

of the alkyl bromide-gallium bromide addition compound with the π complex instead of with the arene molecules themselves. Thus the coordinating ability of the solvent 1,2,4-trichlorobenzene, was considered to play a dominant role in altering the kinetics from those observed in the hydrocarbon system alone.

In more recent studies, Brown and co-workers [8,9] investigated the kinetics of the $GaCl_3$-catalyzed methylation of deuterated and nondeuterated benzene [8] and of toluene and the xylenes [9] in excess methyl chloride at -35.6°C. The methylations of both benzene and hexadeuteriobenzene were first order in the aromatic and second order in gallium chloride. They also proceeded at essentially the same rate, suggesting that the rate-determining step did not involve the breakdown of the σ complex. The methylations of toluene and the xylenes were found to follow kinetics similar to those of the benzene reaction. For all these reactions, in excess methyl chloride, the rate expression was

$$\text{rate} = k_3[GaCl_3]^2[ArH]$$

One of two alternative mechanisms proposed for the $GaCl_3$-catalyzed methylation of benzene in methyl chloride at low temperature is shown in scheme 4. With the next-to-last step rate determining, the appropriate rate expression becomes

$$\text{rate} = k_3[CH_3\text{: }GaCl_3]^2[C_6H_6]$$

The squared rate expression term in $GaBr_3$ (in excess aromatic) [4,5] and in $GaCl_3$ (in excess CH_3Cl) [8,9] was suggested to result from the need for an additional pull of a second gallium halide molecule in order to transfer the alkyl group from the halogen atom to the aromatic.

Recently, the competitive alkylation concept was applied to the case of naphthalene in order to study both positional and substrate selectivities and to clarify the nature of the kinetically versus the thermodynamically controlled product composition. For that purpose, the $AlCl_3$-catalyzed competi-

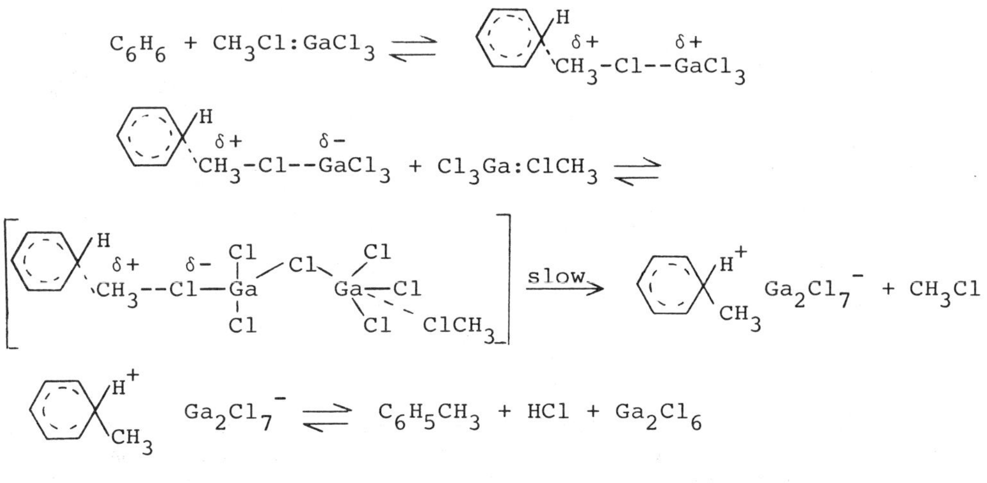

$$C_6H_6 + CH_3Cl{:}GaCl_3 \rightleftharpoons C_6H_6(H)\cdots\overset{\delta+}{CH_3}\text{-}Cl\text{--}\overset{\delta+}{GaCl_3}$$

$$C_6H_6(H)\cdots\overset{\delta+}{CH_3}\text{-}Cl\text{--}\overset{\delta-}{GaCl_3} + Cl_3Ga{:}ClCH_3 \rightleftharpoons$$

$$\left[C_6H_6(H)\cdots\overset{\delta+}{CH_3}\text{--}\overset{\delta-}{Cl}\text{—}GaCl_2\text{—}Cl\text{—}GaCl_2\cdots ClCH_3\right] \xrightarrow{\text{slow}} C_6H_6^{+}(H)(CH_3)\ Ga_2Cl_7^- + CH_3Cl$$

$$C_6H_6^{+}(H)(CH_3)\ Ga_2Cl_7^- \rightleftharpoons C_6H_5CH_3 + HCl + Ga_2Cl_6$$

Scheme 4

tive reaction of naphthalene and benzene with alkyl halides (MeI, EtI, *i*-PrCl, and *t*-BuCl) in benzene, nitromethane, and carbon disulfide solutions was carried out [51a]. The results showed that in both benzene and CS_2 solutions the substrate and positional selectivities, as expressed by the relative reaction rates of naphthalene and benzene (k_N/k_B) and by α-/β-alkylnaphthalene ratios, respectively, varied widely with reaction conditions, especially with time and temperature. In contrast, in nitromethane solution, k_N/k_B and α-/β-alkylnaphthalene ratios remained quite constant for methylation and ethylation, and showed relatively minor variations in the case of isopropylation; only in the case of *t*-butylation were major changes observed, even at 0°C. For methylation as an example, the data taken from 2 min to 6 hr showed a change in k_N/k_B ratio from 44 to 25 and a change in composition from 65% α- and 35% β- to 32% α- and 68% β-methylnaphthalene when $AlCl_3$ was used as catalyst in benzene solution. In nitromethane solution, however, a k_N/k_B value of 44 and an α-/β-methylnaphthalene percent ratio of 65: 33 stayed constant over a reaction period of 6 hr. The varying α-/β-alkylnaphthalene isomer ratios observed when benzene was the solvent were attributed to subsequent isomerization of initially formed α-alkylnaphthalenes to the more thermodynamically stable β isomers. Indeed, separate isomerization experiments confirmed this view, whereby both α- and β-methylnaphthene, for example, yielded an equilibrium mixture composed of 24.5% α and 75.5% β isomer upon treatment with $AlCl_3$ in CS_2 at reflux temperature.

Returning to the benzyl system, we pointed out before that in his first kinetic study of a Friedel-Crafts reaction, Brown [11] used substituted benzyl halides as the alkylating species. In a subsequent paper published in 1959, Brown and Bolto [13] studied the relative rates and isomer distribution in the gallium bromide-catalyzed benzylation of benzene and toluene at 25°C. At this temperature the reactions were exceedingly rapid for individual kinetic measurements, but competitive studies established that the relative rate of benzylation of toluene and benzene was 4.0. In the benzylation of toluene, the isomer distribution exhibited marked variations with reaction times even when these were quite short. Following are listed the observed isomer distributions after 0.01, 0.09, and 10 sec:

$$C_6H_5CH_3 + PhCH_2Cl \xrightarrow[25°]{GaBr_3} o\text{-}CH_3C_6H_4CH_2Ph + m\text{-}CH_3C_6H_4CH_2Ph + p\text{-}CH_3C_6H_4CH_2Ph$$

Time, sec	o-%	m-%	p-%
0.01	40.2	21.1	38.7
0.09	33.8	31.6	34.8
10.00	23.1	45.7	31.2

In recent years Friedel-Crafts benzylation was again tackled by Olah et al. [15,16,19] and others [20-25,52] in connection with continuing attempts to gain a better understanding of the mechanism of electrophilic aromatic substitution and to solve problems associated with anomalous positional and substrate

Table 1. Competitive $AlCl_3 \cdot CH_3NO_2$-Catalyzed Benzylation of Benzene and Substituted Benzenes with Benzyl Chloride in Nitromethane Solution at 25°C

Ar	Relative Rate	% Substituted Diphenylmethanes isomer distribution		
		ortho	*meta*	*para*
Benzene	1.00			
Toluene	3.20	43.5	4.5	52.0
Ethylbenzene	2.45	42.4	5.0	52.6
n-Propylbenzene	2.22	39.6	8.1	52.3
n-Butylbenzene	2.08	39.1	8.6	52.3
o-Xylene	4.25	Only 3,4-dimethyldiphenylmethane detectable		
m-Xylene	4.64	19.8% 2,6- and 80.2% 2,4-dimethyldiphenylmethane		
Fluorobenzene	0.46	14.7	0.2	85.1
Chlorobenzene	0.24	33.0	0.6	66.4
Bromobenzene	0.18	32.5	0.7	66.8
Iodobenzene	0.28	30.6	0.7	68.7

selectivities [21,25,53,54]. In 1962, Olah et al. investigated the $AlCl_3$-CH_3NO_2-catalyzed alkylation of benzene [15], *n*-alkylbenzenes [15], and halobenzenes [16] with benzyl chloride in nitromethane solution. The observed relative reactivities of the *n*-alkylbenzenes and halobenzenes compared to that of benzene, together with the isomer distributions of the monoalkylated products, are shown in Table 1. It is clear from the table that these benzylations gave rise to high positional but low substrate selectivity. This, together with the observations that the reactions were first order in aromatic hydrocarbons and that the benzylation of benzene showed a small secondary, direct isotope effect, suggested the following mechanism:

$$C_6H_5CH_2Cl + AlCl_3 \rightleftharpoons C_6H_5CH_2Cl{:}AlCl_3$$

$$ArH + C_6H_5CH_2Cl{:}AlCl_3 \rightleftharpoons [ArH\cdots CH_2{}^{+}C_6H_5]AlCl_4^-$$

$$[ArH\cdots \overset{+}{C}H_2C_6H_5]AlCl_4^- \rightleftharpoons [H^+\cdots ArCH_2C_6H_5]AlCl_4^- \rightleftharpoons$$

$$ArCH_2C_6H_5 + HCl + AlCl_3$$

In this mechanism, the rate-determining step was interpreted as the displacement step by the aromatic substrate on the benzyl chloride-Lewis acid catalyst system (as established previously by Brown) but leading to an oriented π-complex type rather than to a σ-complex-type activated state. The transition state in the displacement step was visualized as an oriented π complex containing a partially formed C-C bond and a partially broken C-Cl bond.

$$\begin{array}{c} C_6H_5 \\ | \\ ArH\cdots \overset{\delta+}{C}H_2\cdots \overset{\delta-}{Cl:AlCl_3} \end{array}$$

More recently (1972), Olah et al. [19] published a detailed report on Friedel-Crafts benzylation of benzene and toluene with benzyl and substituted benzyl halides in the presence of various Lewis acid catalysts. The results showed that the substrate (or intermolecular) selectivity (k_T/k_B) was very sensitive to the precise reaction conditions, such as the nature of the benzyl cationic species, while positional (or intramolecular) selectivity was less sensitive to conditions when unsubstituted benzyl halides were used. The $TiCl_4$-catalyzed competitive benzylation of toluene and benzene with 16 substituted benzyl chlorides, in which all parameters were kept constant and only the electrophilicity of the reagent was changed systematically by substitutents, revealed a close relationship between electrophilicity and selectivity of the reagents. As to the nature of the intermediates involved in the benzylation mechanism, the results were interpreted as an indication that the transition state of highest energy is not a *fixed one* but rather an easily variable one. In reaction with strongly eletrophilic reagents or with strongly basic aromatics, the transition state lies early on the reaction coordinate, resembling starting aromatic, thus having more of a π-complex nature. On the other hand, in reactions with relatively weak electrophiles, or in reactions with weakly basic aromatics, it lies late on the reaction coordinate, resembling the intermediate σ complex. This view was, however, recently criticized by Decoret et al. [21], who preferred to interpret the k_T/k_B and *o*/2*p* ratios in terms of the solvation energy influence on the reactivity of the intermediate benzyl cations. In fact, theoretical calculations showed that the electron affinities of the solvated cations are reversed compared to the order of the naked cation electron affinities.

In their work Olah et al. [19] also noted that for a given cation the k_T/k_B ratios decreased with increasing catalyst and solvent concentrations. Based on this relationship between intermolecular selectivity (k_T/k_B) values and catalyst activity, the following relative order of catalyst activity was indicated: $SbF_5 > SbF_5\text{-}CH_3NO_2 > AlCl_3\text{-}CH_3NO_2 > TaCl_5 > SbCl_5\text{-}CH_3NO_2 > AlBr_3\text{-}CH_3NO_2 > TiCl_4 > ZrI_4 > FeCl_3 > FeCl_3\text{-}CH_3NO_2 > ZrCl_4 > ZrBr_4 > SbCl_5 \gg SnCl_4 > AuCl$ [19].

In 1978, Nakane and co-workers [25] carried out the competitive benzylation of benzene and toluene with benzyl, chlorobenzyl, alkylbenzyl, and alkoxybenzyl chlorides catalyzed by boron trifluoride hydrate in hexane solutions and by boron fluoride in nitromethane solution. Their aim was to perform the benzylation reaction under conditions prone to inhibit the formation of the donor-acceptor complex, $RC_6H_4CH_2\overset{\delta+}{Cl}\cdot\overset{\delta-}{BF_3}$. The results obtained showed effects similar to those found earlier by Olah et al. [19]. For example, it was found that the isomer distribution was almost independent of the nature of the solvent, but the k_T/k_B value was very sensitive to the reaction conditions. Also, the formation of the *m* isomer amounted to only a small percentage in most of the reactions. These facts, which are illustrated by the

Table 2. Competitive Benzylation of Benzene and Toluene with Benzyl and Substituted Benzyl Chlorides at Room Temperature

R of $RC_6H_4CH_2Cl$	Reaction conditions	k_T/k_B	Isomer Distribution, %			
			o-	*m*-	*p*-	½*o*/*p*
o-Cl	A	0.5	46.5	16.1	37.4	0.62
	B	0.7	42.2	16.0	41.8	0.53
	C	2.8	38.2	15.8	46.0	0.42
H	A	0.9	43.6	6.1	50.3	0.43
	B	0.7	43.1	4.6	52.3	0.41
	C	3.7	40.5	6.5	53.0	0.38
p-CH_3	A	4.4	29.1	2.6	68.3	0.20
	B	4.4	24.4	2.1	73.4	0.17
	C					
p-CH_3O	A	34	27.4	1.3	71.3	0.19
	B	23	28.4	3.8	60.8	0.21
	C	106	29.8	1.4	68.8	0.22

A = Boron trifluoride hydrate catalyst in boron trifluoride solvent
B = Boron trifluoride hydrate catalyst in hexane solvent
C = Boron trifluoride catalyst in nitromethane solvent

selected data shown in Table 2, added to the reasonable assumption that in boron trifluoride hydrate there could exist no free BF_3 to form a donor-acceptor complex with the benzyl halide, led to the conclusion that the electrophile was the cation in the boron fluoride hydrate-catalyzed benzylation, whereas the electrophile was the polar complex in the boron fluoride-catalyzed benzylation. Therefore, the following mechanism was suggested for reactions catalyzed by boron trifluoride hydrate:

$$H_2O + BF_3 \rightleftharpoons H^+BF_3OH^-$$

$$H^+BF_3OH^- + C_6H_5CH_2Cl \longrightarrow C_6H_5CH_2{}^+BF_3OH^- + HCl$$

$$C_6H_5CH_2{}^+BF_3OH^- + ArH \longrightarrow C_6H_5CH_2Ar + H^+BF_3OH^-$$

It was emphasized that if the rupture of the C-Cl bond in the mechanism above determines the rate of the benzylation, the relative rate should always become 1 whether a substituted group of benzyl chloride is electron attractive or electron donative. However, the fact that the k_T/k_B values in the benzylation with *p*-alkyl- and *p*-alkoxybenzyl chlorides varied from 4 to 34 suggested that the rupture of the C-Cl bond is not involved in the rate-determining step.

Aside from benzylation studies in the presence of typical Friedel-Crafts catalysts, workers have also examined the catalytic activity and selectivity for benzylations over other catalysts, such as arenetricarbonylmolybdenum [$ArMo(CO)_3$] [55]; metal oxides [22]; calcined sulfate salts of Fe, Zn, Co, Mn, and Cu [2,24,56-58]; polyphosphoric acid [60]; and other Lewis acids [61]. Notably, the results with calcined sulfates suggested that ferrous sulfate has an unexpected effectiveness for the benzylation reaction. Its activity was more than 100 times as large as those of other catalysts, whereas the selectivity of all these catalysts was the same [57]. The results also showed that in benzylation of toluene with benzyl chloride the activation energies over $FeSO_4$ calcined at 400 and 600°C were quite small, 2.7 and 5.1 kcal/mol, respectively, and that the products were as follows [23]:

$$C_6H_5CH_3 + PhCH_2Cl \xrightarrow[53.5^\circ C]{FeSO_4} o\text{-}CH_3C_6H_4CH_2Ph\ (43\%) + m\text{-}CH_3C_6H_4CH_2Ph\ (6\%) + p\text{-}CH_3C_6H_4CH_2Ph\ (51\%)$$

The ratios of rates in toluene and benzene k_T/k_B were 6.29 and 6.60 for $FeSO_4$ and $Fe_2(SO_4)_3$, respectively. Similar product distribution was again noted when the benzylation of toluene was carried out over ferrous and ferric sulfates calcined at 700 to 900°C [59].

B. Synthetic Applications of Benzylation

Besides its application in mechanistic studies, the Friedel-Crafts benzylation reaction has also been used as a synthetic tool. For example, Montaudo et al. [62] prepared 54 new, variously substituted diphenylmethanes and dibenzylbenzenes through Friedel-Crafts benzylation, starting from appropriate aromatic hydrocarbon and chloromethyl derivatives. However, they pointed out that the choice of mold conditions (low temperature and reaction time, catalyst, solvent, excess of hydrocarbon and nitrogen stream) was crucial to avoid or minimize isomerization and disproportionation [15,17,19]. Variously substituted diarylalkanes (4) were also prepared by Schmerling [63] through the interaction of the proper aromatic with benzyl chloride in the presence of $CuCl_2$:

$$C_6H_5CHRCl + C_6H_4R^1R^2 \xrightarrow{CuCl_2} C_6H_5CHR\text{-}C_6H_3R^1R^2 \quad (\underline{4})$$

R = H, Me

R^1, R^2 = H, Me, OH

Friedel-Crafts benzylation was also used to prepare a number of 2-halobenzyl-4-nitrophenols (5) through the reaction of 4-nitrophenol with benzyl chloride and 2,4-disbustituted benzyl chlorides [64] (Eq. 13).

$$R^2\text{-}C_6H_3(R^1)\text{-}CH_2Cl + HO\text{-}C_6H_4\text{-}NO_2 \xrightarrow{ZnCl_2} R^2\text{-}C_6H_3(R^1)\text{-}CH_2\text{-}C_6H_3(OH)\text{-}NO_2 \;(\underline{5}) \quad (13)$$

$R^1 = H, Cl;\ R_2 = F, Cl, Br$

The reactions of phenols and their ethers with benzyl [60,65-70] and methylbenzyl [20] halides have been a subject of current concern. Examples of the results obtained are shown in Eqs. (14) to (19a). In the reactions of

$$C_6H_5OH + PhCH_2Cl \xrightarrow[\text{10-20 min, 80-100°, 70-80\% yield}]{Fe^{3+}} o\text{-}PhCH_2C_6H_4OH\ (50\text{-}70\%) + p\text{-}PhCH_2C_6H_4OH\ (30\text{-}50\%) \quad [67]\ (14)$$

$$C_6H_5OCH_3 + PhCH_2Cl \xrightarrow[\text{10-20 min, 80-100°, 70-80\% yield}]{Fe^{3+}} o\text{-}PhCH_2C_6H_4OCH_3\ (45\%) + p\text{-}PhCH_2C_6H_4OCH_3\ (55\%) \quad [67]\ (15)$$

$$C_6H_5OH + CH_3O\text{-}C_6H_4\text{-}CH_2Cl \xrightarrow[\text{or } FePO_4\cdot 2H_2O,\ 70\text{-}80°,\ \text{short time, 80-92\% yield}]{FeCl_3\cdot 12H_2O} o\text{-}RO\text{-}C_6H_4\text{-}CH_2\text{-}C_6H_4\text{-}OCH_3 + p\text{-}RO\text{-}C_6H_4\text{-}CH_2\text{-}C_6H_4\text{-}OCH_3 \quad [68]\ (16)$$

$$p\text{-}Cl\text{-}C_6H_4\text{-}OR + PhCH_2Cl \xrightarrow{\text{Fe, Zn, or Sn salts}} \text{2-}CH_2Ph\text{-4-}Cl\text{-}C_6H_3OR\ \text{(major)} + \text{2,6-}(CH_2Ph)_2\text{-4-}Cl\text{-}C_6H_2OR\ \text{(minor)} \quad [69]\ (17)$$

$(R = H, CH_3)$

Phenol + R-$C_6H_4CH_2Cl$ $\xrightarrow[\text{60-100°, 15-20 min, 53-80\% yield}]{FeCl_3 \cdot 12H_2O}$ o-($CH_2C_6H_4R$)phenol + p-($CH_2C_6H_4R$)phenol (18) [20]

R =		
o-CH_3	54%	46%
m-CH_3	60%	40%
p-CH_3	58%	42%

Anisole (OCH_3) + R-$C_6H_4CH_2Cl$ $\xrightarrow[\text{90-110°, 10-20 min, 80-83\% yield}]{FeCl_3 \cdot 12H_2O}$ o-($CH_2C_6H_4R$)anisole + p-($CH_2C_6H_4R$)anisole (19) [20]

R =		
o-CH_3	37%	63%
m-CH_3	44%	56%
p-CH_3	33%	67%

Phenol (OH) + $PhCH_2Cl$ $\xrightarrow[80°]{PPA}$ o-CH_2Ph phenol (39%) + p-CH_2Ph phenol (61%) (19a) [60]

phenol and anisole with methylbenzyl chloride (Eqs. 18 and 19) it was found that the dominant isomer was the *ortho* with phenol but the *para* with anisole. It was also found that the chlorides fall into the following sequence of activities: *p*-methylbenzyl chloride > *o*-methylbenzyl chloride > *m*-methylbenzyl chloride [20].

II. ALKYLATIONS WITH LINEAR PRIMARY ALKYL HALIDES

A. Ethyl Halides

Alkylations of arenes by ethyl halides under Friedel-Crafts conditions have been carried out by numerous early workers. The data appearing prior to 1946 have been compiled by Price [2]. It can be seen from his compilation that depending on the conditions applied, varying amounts of di- to higher ethylated arenes were usually produced, and in certain cases they constituted the main part of the reaction product. In general, the ratio of mono- to higher-alkylated products increased with increasing reaction temperature, reaction-time concentration of catalyst, and amount of ethyl halide.

Ethylations of substituted benzenes usually result in mixtures of isomeric products. The ratios of these isomers also vary with the reaction conditions. As in other alkylations [2,14,71-75] strong conditions and/or catalysts result in predominance of the *meta*-oriented isomers [76-78]. Some illustrative ethylations and the conditions producing them are shown in Eqs. (20) to (24).

$$\text{benzene} + C_2H_5Br \xrightarrow[0\text{-}25°,\ 24\ hr]{AlCl_3} \text{benzene with } C_2H_5,\ C_2H_5,\ C_2H_5 \qquad (20)\ [77]$$

$$\text{benzene with } CH_3 + C_2H_5Br \xrightarrow[-10°,\ 18\ hr]{AlCl_3} \text{benzene with } CH_3,\ C_2H_5,\ C_2H_5 \qquad (21)\ [77]$$

$$\text{benzene with } CH_3 + C_2H_5Br \xrightarrow[25°,\ 40\text{-}80\ min]{GaBr_3} \text{(} CH_3,\ C_2H_5\text{) } 38.3\% + \text{(} CH_3,\ C_2H_5\text{) } 21.1\% + \text{(} CH_3,\ C_2H_5\text{) } 40.6\% \qquad (22)\ [6]$$

$$\text{benzene with } CH_3,\ CH_3 + C_2H_5Cl \xrightarrow[6.5\ hr]{AlCl_3} \text{(} CH_3,\ C_2H_5,\ CH_3\text{)} + \text{(} CH_3,\ C_2H_5,\ C_2H_5,\ CH_3\text{)} \qquad (23)\ [79]$$

$$\text{benzene with } \overset{O}{\overset{\|}{C}}\text{-}CH_3 + C_2H_5Cl \xrightarrow[0°,\ 2\ min]{HF\text{-}SbF_5} \text{(} COCH_3,\ C_2H_5\text{) } 78\% + \text{(} C_2H_5,\ COCH_3,\ C_2H_5\text{) } 14\%$$

$$+ \text{(} COCH_3,\ C_2H_5,\ C_2H_5\text{) } 4\% + \text{(} C_2H_5,\ COCH_3,\ C_2H_5,\ C_2H_5\text{) } 4\% \qquad (24)\ [79a]$$

Similar effects were noted during the $AlCl_3$-catalyzed alkylation of napthalene with ethyl bromide in CS_2 and nitromethane solutions [51a]. Ethylation in CH_3NO_2 showed the least isomerization, giving 69% α and 31% β substitution after 6 hr of reaction. In CS_2, however, the isomeric α- and β-ethylnaphthalene ratios varied with reaction conditions from 51% α and 69% β after 0.5 min to 9.9% α and 90.1% β after 6 hr of reaction. As with methylation, these variations were again attributed to subsequent isomerization under the conditions employed.

In view of the well-known rearrangements of propyl and higher alkyl groups accompanying Friedel-Crafts alkylations and with the advent of isotopic tracer techniques, it became of interest to determine whether rearrangement might accompany alkylations with labeled ethyl halides. The earliest of several investigations in this direction was made in 1955 by R. M. Roberts et al. [80] through a study of the alkylation of benzene with ethyl-2-^{14}C chloride in the presence of $AlCl_3$. When the alkylation was carried out by adding a benzene solution of the radioactive chloride to a stirred aluminum chloride-benzene mixture followed by heating to reflux for 1.25 hr, the resulting ethylbenzene showed no isotope-position rearrangement in the ethyl side chain:

$$C_6H_6 + {}^{14}CH_3CH_2Cl \xrightarrow[\text{-HCl, reflux, 1.25 hr}]{AlCl_3} C_6H_5CH_2{}^{14}CH_3$$

$$C_6H_5CH_2{}^{14}CH_3 \xrightarrow{(O)} {}^{14}CO_2 + C_6H_5COOH$$

However, when the ethyl-2-^{14}C chloride was allowed to stand over aluminum chloride at room temperature for 1 hr before the addition of benzene, the ethylbenzene produced showed almost equal distribution of the ^{14}C between the 1- and 2-carbon positions of the ethyl side chain. The equilibration of ^{14}C in the latter case was attributed to prior equilibration of the ^{14}C label in the ethyl chloride during its contact with aluminum chloride. The results clearly demonstrated that isomerization of ethyl-2-^{14}C chloride by aluminum chloride took place under mild conditions in the absence of aromatic hydrocarbon. The possibility that rearrangement of the ethyl group might have taken place subsequent to reaction was excluded on the ground that treatment of ethyl-2-^{14}C-benzene with $AlCl_3$ at the boiling point for 2.5 hr resulted in no chain rearrangement. This fact was also substantiated by other results since most disproportionations and alkyl group migrations were shown to occur with little or no skeletal isomerization [81,82].

Two alternative mechanisms were considered to account for the lack of rearrangement observed in the reaction of benzene and ethyl-2-^{14}C chloride in the presence of aluminum chloride: (1) a bimolecular displacement mechanism involving no free ethyl carbocations, and (2) a carbocation mechanism in which the intermediate ion reacted with the aromatic ring so rapidly as to exclude internal hydrogen shifts. The information available did not allow a decision as to whether one of these explanations is exclusively correct and, if so, which one. In considering a possible carbocation process for the reactions, the authors suggested that their results might be interpreted as demonstrating the relative ease of primary-to-primary carbocation conversion in the absence and presence of aromatic compounds. In that respect they referred to previous work by J. D. Roberts et al. [83] and by Pines et al. [83a] in which similar inhibitions of isomerization by aluminum chloride were attributed to the presence of aromatic hydrocarbon. More recently, R. M. Roberts and co-workers [84,85] found several cases of inhibition by aromatic hydrocarbons of the aluminum chloride-catalyzed isomerization of alkyl halides.

Extension of the ^{14}C-labeled ethyl chloride alkylation studies was carried out by R. M. Roberts and Panayides [86] with the aim of learning more about the relationship of isomerizaton and alkylation. Their results suggested that the isomerization reaction could be competitive, to a small degree, with the alkylation process at lower reaction temperatures. Thus the aluminum chloride-catalyzed alkylation of benzene with ethyl-1-^{14}C chloride at room temperature (using the same amounts as in the previous experiments) afforded ethylbenzene and diethylbenzene in which the distribution of ^{14}C corresponded to 4% isotopic rearrangement accompanying the alkylation. The use of 10 times as much aluminum chloride in a similar alkylation yielded a product whose ^{14}C distribution amounted to about 8% isotopic rearrangement.

In 1958, Lee et al. [87] reported the aluminum chloride-catalyzed alkylation of benzene with ethyl-2-^{14}C iodide in excess benzene overnight at reflux temperature. The results obtained were substantially in agreement with those reported by Roberts et al. [83] for the alkylation of benzene with ethyl-2-^{14}C chloride, i.e., no rearrangement accompanying alkylation; however, the aluminum chloride-induced isomerization was much faster for ethyl chloride than for ethyl iodide. This fact is evident from the observation that after 1 hr of treatment with aluminum chloride at room temperature, the percentage isomerization was 83.6% in the case of ethyl-2-^{14}C chloride and 1.7% in the case of ethyl-2-^{14}C iodide. The higher electronegativity of chlorine relative to iodine was assumed to result in a greater tendency for ethyl cation formation from ethyl chloride than from ethyl iodide under the influence of aluminum chloride.

In the same investigation, Lee and co-workers applied an isotope dilution technique to obtain kinetic data on the alkylation process. The data obtained suggested that in the Friedel-Crafts reaction between benzene and ethyl-2-^{14}C iodide at 40°C, the rate of ethylbenzene formation is faster than the rate of a possible rearrangement of the ethyl iodide caused by aluminum chloride. The authors, however, were not able to reach any conclusion about the actual mechanism of alkylation since the results obtained could be rationalized equally well by either of the two explanations described by Roberts et al. [80].

In 1969, Nakane et al. [88] studied the aluminum bromide-catalyzed ethylation of benzene with ethyl-2-^{14}C iodide in *n*-hexane solution, in which the proportions were *n*-hexane, benzene, ethyl-2-^{14}C iodide, and aluminum bromide in a mole ratio of 100:10:1:7; the reaction was carried out at 25°C for 30 sec, giving only ca.5% reaction. Oxidation of the ethylbenzene produced in the experiment with nitric acid gave *p*-nitrobenzoic acid in which 20% of the starting radioactivity of ethyl-2-^{14}C iodide was found. In contrasting their alkylation results in nonpolar *n*-hexane with those of Roberts et al. [80, 86] and Lee et al. [87] in excess benzene, the present workers suggested that ethyl cation is an electrophile in ethylations carried out in nonpolar organic solvents and that an internal hydride shift occurred partly in the ethyl group before the ethyl cation combined with the benzene nucleus. Accordingly, they proposed that the Friedel-Crafts ethylation of arenes by ethyl halides proceeds by one of two mechanisms. In basic organic solvents or in excess arenes, the ionization of ethyl halide is minimized because of the competing donor effect of the solvent on the catalyst, and as a result ethyl halide forms a nonionized complex with metallic halide, $\overset{\delta+}{C_2H_5X}\cdot\overset{\delta-}{MX_3}$, and reacts by S_N2 displacement. On the other hand, in nonpolar organic solvents ethyl halide can be ionized directly to the ethyl cation as rupture of the C-X bond by the metal halide catalyst is probably complete before reaction with the arene.

In 1975, DeHaan et al. [54] described a study in which they determined the kinetics of the $AlBr_3$-catalyzed reaction of ethyl bromide with benzene or toluene in *n*-hexane, using a vacuum-line procedure to optimize the exclusion of water. With both arenes, separation into two layers was noted to occur within 30 min of reaction. Kinetically, the rates for both reactions showed a second-order dependence on $AlBr_3$. Calculation of the relative reaction rates of toluene and benzene (k_T/k_B) gave a value of 3.7 at the early stages, which dropped to 0.53 at long times. Sampling of the initially homogeneous reaction mixture showed that the relative amount of *meta*-ethylated toluene product remained constant at about 20%, then fell to 15% during the appearance of turbidity, then to 6% during the early existence of the second layer, and finally rose to the thermodynamic value of 40% at long times (> 120 min), which is a sure sign of isomerization. A continued increase in diethylated products and an appearance of significant amounts of benzene (3 to 5%) during the two-phase period was also observed which, taken together, indicated a concomitant disproportionation reaction. These observations led to the interesting suggestion that side reactions, including disproportionation and isomerization, become important with the onset of the formation of the second layer. Furthermore, the low k_T/k_B value, together with the relatively high percentage of *meta* isomer found at the early stages of the reaction (i.e., prior to the formation of the isomerizing second phase), supported the earlier proposition [19,88] that Friedel-Crafts ethylations in nonpolar media involve the highly reactive ethyl carbocation.

Apart from these studies involving aluminum halides, studies relating to the mechanism of Friedel-Crafts ethylation have also been conducted in the presence of other metal halide catalysts. The gallium halide-catalyzed alkylation of arenes with ethyl halides was thoroughly explored by Brown et al. [4,5,6,10,89] and by others [34]. Most of their results were presented earlier in connection with the methylation reactions to which the ethylation reactions are closely related. However, there remains one report on ethylation which will be presented here.

The boron trifluoride-catalyzed alkylation of arenes with ethyl halides has been studied in connection with the growing interest in this catalyst [88, 90-101]. In 1958, Olah and Kuhn [91] studied the reactions of toluene, *m*-xylene, and mesitylene with methyl, ethyl, *n*-propyl, and isopropyl halides in the presence of BF_3 or $AgBF_4$ catalyst. The monoalkylated products were obtained with a yield above 90%, besides small amounts of di- and higher alkylated products. Based on the results of their study, Olah and Kuhn envisioned the mechanism of these reactions as occurring in two steps: one involving formation of a 1:1 addition complex between the alkyl halide and the catalyst and another involving attack by the 1:1 addition complex on the aromatic substrate to give a 1:1:1 alkyl benzenonium tetrafluoroborate σ complex such as 6. The latter complex, when allowed to warm up above its decomposition point (-80°C) gave the alkylated product in addition to BF_3 and HF. The steps of the suggested mechanism are exemplified by the ethylation of toluene as shown in scheme 5.

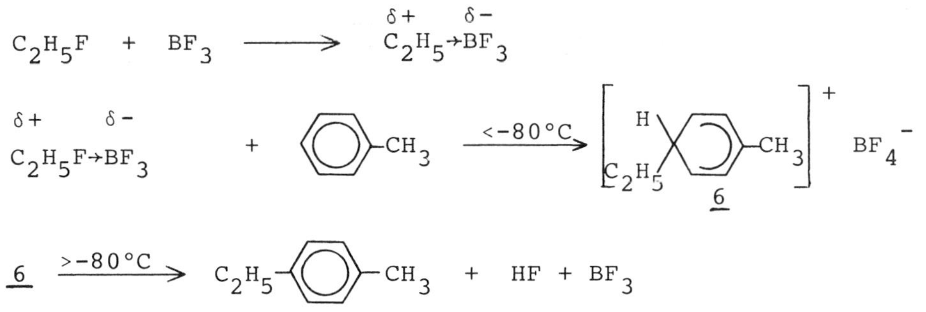

$$C_2H_5F + BF_3 \longrightarrow \overset{\delta+}{C_2H_5}\rightarrow \overset{\delta-}{BF_3}$$

$$\overset{\delta+}{C_2H_5F}\rightarrow \overset{\delta-}{BF_3} + C_6H_5\text{-}CH_3 \xrightarrow{<-80°C} \left[\text{H}(C_2H_5)C_6H_5\text{-}CH_3\right]^+ BF_4^- \quad \underline{6}$$

$$\underline{6} \xrightarrow{>-80°C} C_2H_5\text{-}C_6H_4\text{-}CH_3 + HF + BF_3$$

Scheme 5

Nakane et al. extended their studies with the aim of gaining a better understanding of the ethylation mechanism in particular and the alkylation mechanism in general [88,92-96,99-102]. These studies led to interesting conclusions regarding alkylation kinetics, the nature and composition of alkylation intermediates, and the effect of solvents on alkylation pathways. A summary of these conclusions is as follows:

1. The study of solvent and catalyst effects on the extent of rearrangements accompanying the ethylation of benzene with labeled ethyl halides [88,100-102] suggested that the Friedel-Crafts ethylation of arenes with ethyl halides proceeds by one of two mechanisms, depending on the nature of solvent and catalyst. In basic organic solvents such as nitromethane or excess aromatic, in which the ionization of ethyl halide is not favored because of the competing donor effect of the solvent, ethyl halide forms a nonionized complex with the catalyst and reacts by S_N2 displacement. On the other hand, in nonpolar organic solvents such as hexane or in the presence of a strong proton-acid catalyst, ethyl halide can be ionized to the ethyl cation through complete rupture of the C-X bond before reaction. Thus it was found that in ethylations of benzene with ethyl-2-^{14}C fluoride and BF_3 in hexane [100] or with ethyl-1-^{13}C fluoride and ethyl-1,1-d_2 fluoride in boron trifluoride hydrate medium, where donor-acceptor complex formation is supressed [102], almost a 50% isotope position rearrangement in the side-chain ethyl group was found, whereas in the boron trifluoride-catalyzed ethylation in nitromethane [100] or in excess benzene [102] no rearrangement was observed. These two types of mechanisms can be illustrated by those suggested to explain the results of ethylating benzene with labeled ethyl fluoride using boron trifluoride or its hydrate as catalyst (schemes 6 and 7, respectively).

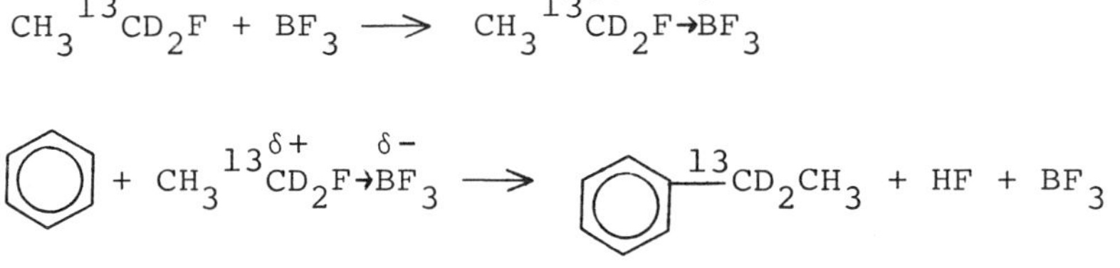

$$CH_3{}^{13}CD_2F + BF_3 \longrightarrow CH_3{}^{13}\overset{\delta+}{CD_2F}\rightarrow \overset{\delta-}{BF_3}$$

$$C_6H_6 + CH_3{}^{13}\overset{\delta+}{CD_2F}\rightarrow \overset{\delta-}{BF_3} \longrightarrow C_6H_5\text{-}{}^{13}CD_2CH_3 + HF + BF_3$$

Scheme 6: Mechanism of ethylation with BF_3 catalyst

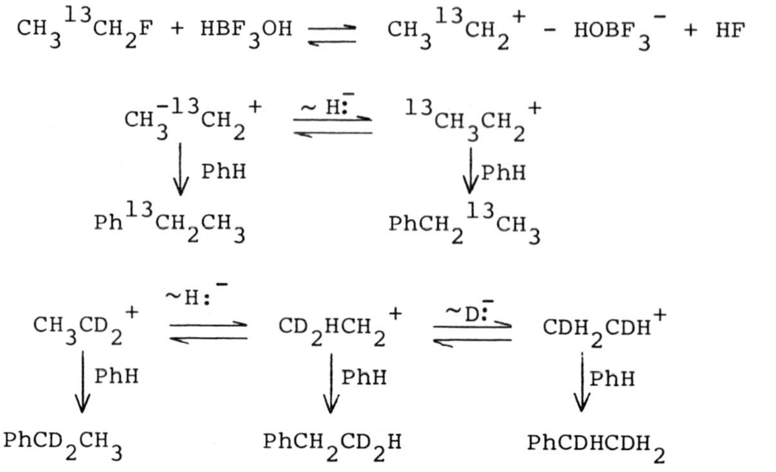

$$CH_3{}^{13}CH_2F + HBF_3OH \rightleftharpoons CH_3{}^{13}CH_2{}^{+} - HOBF_3{}^{-} + HF$$

$$CH_3{}^{-13}CH_2{}^{+} \overset{\sim H:^-}{\rightleftharpoons} {}^{13}CH_3CH_2{}^{+}$$

$$\downarrow PhH \qquad \downarrow PhH$$

$$Ph^{13}CH_2CH_3 \qquad PhCH_2{}^{13}CH_3$$

$$CH_3CD_2{}^{+} \overset{\sim H:^-}{\rightleftharpoons} CD_2HCH_2{}^{+} \overset{\sim D:^-}{\rightleftharpoons} CDH_2CDH^{+}$$

$$\downarrow PhH \qquad \downarrow PhH \qquad \downarrow PhH$$

$$PhCD_2CH_3 \qquad PhCH_2CD_2H \qquad PhCDHCDH_2$$

Scheme 7: Mechanism of ethylation with $BF_3 \cdot H_2O$ catalyst

2. In all Friedel-Crafts ethylations, the electrophile, whether an incipient ethyl carbocation (7a) or a nonionized complex (7b), attacks the aromatic ring to form an unstable termolecular oriented π complex (8a or 8b) and then a σ complex (8c), the formation of which determines both substrate and positional selectivities (scheme 8).

3. In the BF_3-catalyzed ethylation of toluene, low-temperature spectral analysis [93,96,99] indicated that the BF_3-EtF-$CH_3C_6H_5$ 1: 1: 1 addition complex is not the ethylbenzenonium tetrafluoroborate σ complex (6, scheme 5) suggested earlier by Olah and Kuhn [91], but rather a termolecular, oriented

$$C_6H_6 + \left\{ \begin{array}{c} C_2H_5{}^{+}MX_4{}^{-} \ (\underline{7a}) \\ \text{or} \\ \overset{\delta+}{C_2H_5}\overset{\delta-}{X}\cdot MX_3 \ (\underline{7b}) \end{array} \right\} \rightleftharpoons \left\{ \begin{array}{c} C_6H_6 \cdots C_2H_5{}^{+}MX_4{}^{-} \ (\underline{8a}) \\ \text{or} \\ C_6H_6 \cdots \overset{\delta+}{C_2H_5}\overset{\delta-}{X}\cdot MX_3 \ (\underline{8b}) \end{array} \right\} \xrightarrow{\text{rate determining}}$$

$$C_6H_6{}^{+}(H)C_2H_5MX_4{}^{-} \ (\underline{8c}) \longrightarrow C_6H_5C_2H_5 + HX + MX_3$$

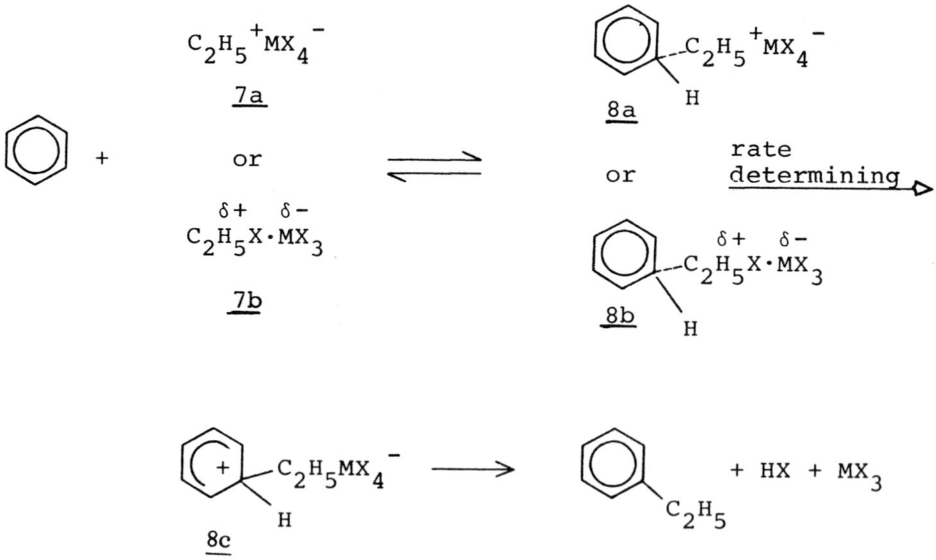

Scheme 8

π complex in which a nonionized ethyl fluoride-boron fluoride complex is oriented near an *ortho* or *para* position in the ring:

$$\text{CH}_3\text{C}_6\text{H}_4(\text{H})\cdots\overset{\delta+}{\text{C}_2\text{H}_5\text{F}}\cdot\overset{\delta-}{\text{BF}_3}$$

4. The kinetics of competitive and noncompetitive alkylations revealed first-order dependence on the catalyst and showed anomalous substrate selectivity for ethylation [88,102] and isopropylation [101] in nonpolar solvents. In *n*-hexane, for example, the relative ratio of toluene to benzene (k_T/k_B) in BF_3-catalyzed ethylation and isopropylation were 0.61 and 0.70, respectively. Another anomalous ratio of 0.80 was obtained for a competitive ethylation in boron trifluoride hydrate. As to positional selectivity, the isomer distributions of ethyltoluene formed during the ethylation of toluene were not statistical in any of the solvents employed. For example, the distributions were 49.2% *o*-, 27.1% *m*-, and 23.7% *p*-ethyltoluene in *n*-hexane; 46.2% *o*-, 28.7% *m*-, and 25.1% *p*-ethyltoluene in excess toluene; and 48.6% *o*-, 23.7% *m*-, and 27.7% *p*-ethyltoluene in chloroform. In accounting for these data [88], it was stated that "a powerful electrophile like ethyl carbonium ion can distinguish slightly the difference between each site in benzene and the *ortho* or *para* site in toluene, but readily the difference between each site in benzene and the *meta* site in toluene."

The ethylation of benzene with ethyl fluoride was also catalyzed by SbF_5 at low temperature. As pointed out before, Olah and his co-workers [38] were able to prepare the $CH_3F \rightarrow SbF_5$ and $CH_3CH_2F \rightarrow SbF_5$ complexes by interaction of CH_3F and C_2H_5F with SbF_5 in SO_2 or SO_2ClF solvent. The results showed that the ethyl complex was more polarized than the methyl complex. However, like the $CH_3F \rightarrow SbF_5$-SO_2 solution, the $C_2H_5F \rightarrow SbF_5$-$SO_2$ solution proved to be an extremely powerful alkylating mixture capable of ethylating n-donor, π-donor, and σ-donor bases. Alkylation of benzene and toluene by CH_3CH_2F and SbF_5-SO_2ClF at -78°C resulted in irreversible formation of intermediate arenium fluoroantimonate complexes, which upon quenching gave > 70% yields of the monoethylates. Diethylated products were also produced in yields varying from 1 to 10% depending on the concentration of reagents. The ethylation of toluene at -78°C gave a monoethyltoluene mixture whose composition was as follows:

$$\text{C}_6\text{H}_5\text{CH}_3 + \text{C}_2\text{H}_5\text{F-SbF}_5 \xrightarrow[-78°]{\text{SO}_2\text{ClF}} [\text{CH}_3\text{C}_6\text{H}_4\text{C}_2\text{H}_5]^+\text{SbF}_6^- \longrightarrow$$

$$\underset{39\%}{o\text{-CH}_3\text{C}_6\text{H}_4\text{C}_2\text{H}_5} + \underset{19\%}{m\text{-CH}_3\text{C}_6\text{H}_4\text{C}_2\text{H}_5} + \underset{41\%}{p\text{-CH}_3\text{C}_6\text{H}_4\text{C}_2\text{H}_5}$$

B. Arylethyl Halides

The alkylation of benzene with phenylethyl halides and alcohols was studied by early workers [103-105]. For example, treatment of benzene with 1-phenylethyl bromide and Zn gave an unspecified yield of 1,1-diphenylethane [103]. On the other hand, treatment of benzene with 2-phenylethyl chloride and $AlCl_3$ gave an 85% yield of 1,2-diphenylethane [105].

In recent years Friedel-Crafts alkylations with 2-phenylethyl halides were studied to provide an opportunity for determining the relative contributions of S_N1 and S_N2 pathways and the extent of phenyl participation to form a symmetrical intermediate ion. Accordingly, studies of reaction kinetics, of isotope-position rearrangement, and of the effect of substituents on rates of reaction and rearrangement, were all carried out using labeled and unlabeled 2-phenylethyl derivatives [106,107].

In 1957, Lee et al. [106] found that the alkylation of anisole with either 2-phenylethyl-1-^{14}C chloride or 2-phenylethanol-1-^{14}C in the presence of $AlCl_3$ gave a reaction product with nearly equal distribution of the label in the two alkyl carbons. In the absence of supporting data, the authors were unable to decide between several alternative explanations such as simultaneous operation of S_N1 and S_N2 pathways, rearrangement prior to or subsequent to alkylation, or involvement of a common intermediate for both rearrangement and alkylation.

With hopes of clarifying the situation, McMahon and Bunce [107] studied a number of Friedel-Crafts alkylations with substituted 2-phenylethyl-1-^{14}C chlorides. The reaction of toluene with 2-phenylethyl-1-^{14}C chloride and aluminum chloride catalyst at 0°C for 10 min gave 2-p-tolylethylbenzene in which 52.0 to 53.3% of the carbon-14 was contained in the 2 position. The 2-phenylethyl choride recovered from contact with aluminum chloride in toluene at -5°C for 10 min showed no isotope-position rearrangement. In a similar manner, the alkylation of naphthalene and of phenol with 2-phenylethyl-1-^{14}C chloride gave 48.7% and 52.7% of the ^{14}C in the 1 position.

In rationalizing these results, the authors considered two alternative reaction pathways: one involving a classical carbocation (path A) and another involving a nonclassical phenonium ion (path B):

$$C_6H_5CH_2{}^{14}CH_2^+ \rightleftharpoons C_6H_5{}^{14}CH_2\overset{+}{C}H_2 \qquad (A)$$

$$C_6H_5CH_2{}^{14}CH_2Cl \longrightarrow \text{phenonium ion } (H_2C\text{---}CH_2 \text{ bridged, } ^{14}C) \qquad (B)$$

Path A, in which a simple carbocation (or ion pair) equilibrates with another in which a phenyl group has shifted, could be applied only if the reaction of the carbocation with toluene was very much faster than its recombination with chloride. Also, in such a case, the loss of chloride from 2-phenylethyl-1-^{14}C chloride would probably be rate controlling, a step that would be sub-

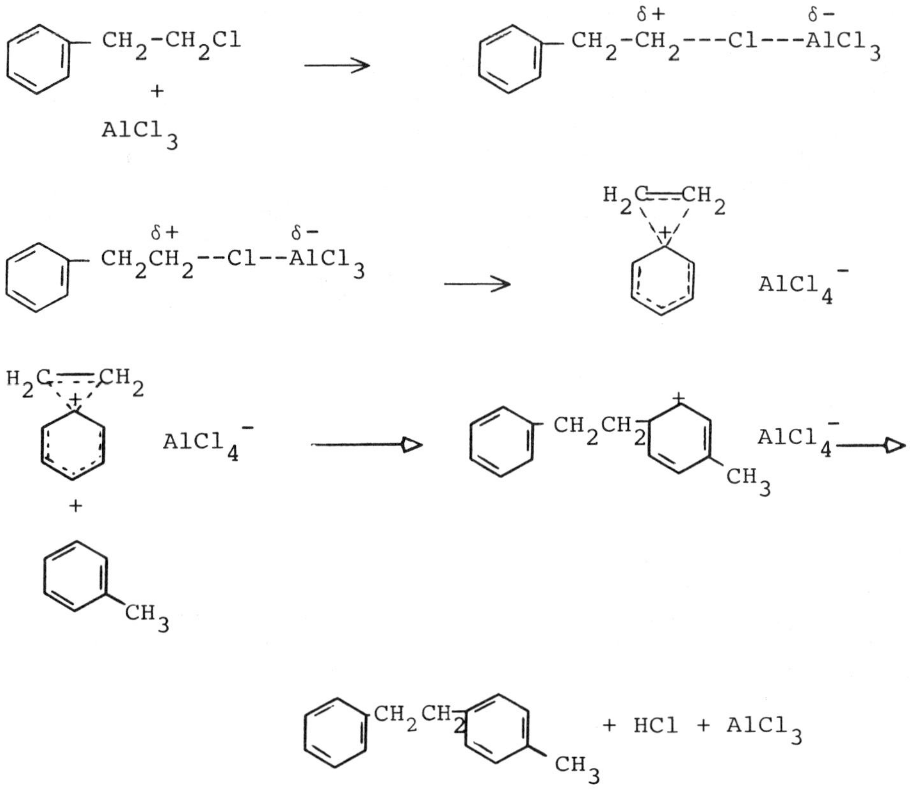

$$C_6H_5-CH_2-CH_2Cl + AlCl_3 \longrightarrow C_6H_5-\overset{\delta+}{CH_2-CH_2}---Cl---\overset{\delta-}{AlCl_3}$$

$$C_6H_5-\overset{\delta+}{CH_2CH_2}--Cl--\overset{\delta-}{AlCl_3} \longrightarrow \text{(phenonium ion, } H_2C\text{=}CH_2 \text{ bridged, } +) \quad AlCl_4^-$$

$$\text{(phenonium ion)}\ AlCl_4^- + C_6H_5CH_3 \longrightarrow C_6H_5CH_2CH_2\text{-}(C_6H_5^+\text{-}CH_3)\ AlCl_4^- \longrightarrow$$

$$C_6H_5CH_2CH_2C_6H_4CH_3 + HCl + AlCl_3$$

Scheme 9

ject to an intermolecular isotope effect. Accordingly, path A was disqualified on the ground that no intermolecular isotope effect was found. Instead, there was observed an intramolecular isotope effect, the magnitude of which indicated discrimination in a bond-breaking step involving nucleophilic attack by toluene on a symmetrical phenonium ion intermediate. The mechanism of scheme 9 was thus suggested for the reaction of 2-phenylethyl chloride with toluene and aluminum chloride: In this mechanism, it was proposed that the second step must be irreversible, to account for the recovery of 2-phenylethyl-1-^{14}C chloride totally unrearranged after contact with aluminum chloride in toluene. For the same reason, it was postulated that the stability of the phenonium ion intermediate increased its ability to react preferentially with toluene by discriminating against the chloride ion, whose nucleophilic reactivity was weakened by being involved in the $AlCl_4^-$ complex.

The mechanism of scheme 9 received direct support from the observation that the reaction of toluene with 2-(4-nitrophenyl)ethyl-1-^{14}C chloride gave only 8% isotope position rearrangement, whereas alkylation of naphthalene and of phenol with 2-phenylethyl-1-^{14}C chloride gave 48.7 and 52.7%, respectively, of the label in the α position. As pointed out, the smaller rearrangement observed during toluene alkylation with 2-(4-nitrophenyl)ethyl-1-^{14}C chloride indicated a considerably smaller contribution from a symmetrical arylonium intermediate whose stability would be decreased by the electron-withdrawing group. In such a case, most probably, a displacement reaction competes importantly with the formation of the symmetrical intermediate. Indirect support for the mechanism above was also gained when alternative mechanisms, such as one involving reversible hydride transfer and alkylation-dealkylation processes [108-111] and another involving intramolecular migration of both phenyl and *p*-tolyl in protonated 4-methylbibenzyl [112,113], were disqualified on various grounds. Further indirect support of the same mechanism came in recent years from the direct observation of ethylene phenonium, *p*-toluonium, *p*-anisonium, and other ethylene arylonium ions in superacid media under stable ion conditions [114,115].

In 1972, Schwartz et al. [116], carried out an interesting study on the comparative Friedel-Crafts reactions of 2-phenylethyl chloride (9) and benzocyclobutene (10) with benzene and toluene in the presence of $AlCl_3$. Their

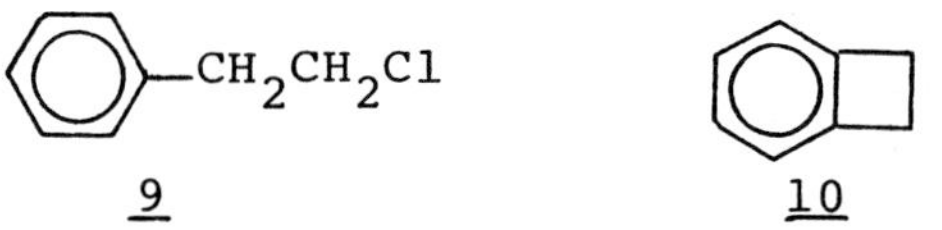

$C_6H_5-CH_2CH_2Cl$ (9)   benzocyclobutene (10)

goal was to test the possible involvement of a common intermediate in these reactions, which might clear up some of the uncertainty regarding the mode of ring opening of benzocyclobutene [117,118]. In the literature, it had earlier been suggested that this ring opening occurred either via a multicentered transition state (path A, scheme 10) [117] or via electrophilic attack at a bridgehead carbon (path B, scheme 10) [118] with the resultant formation of *ortho*-substituted 2-phenylethyl derivatives.

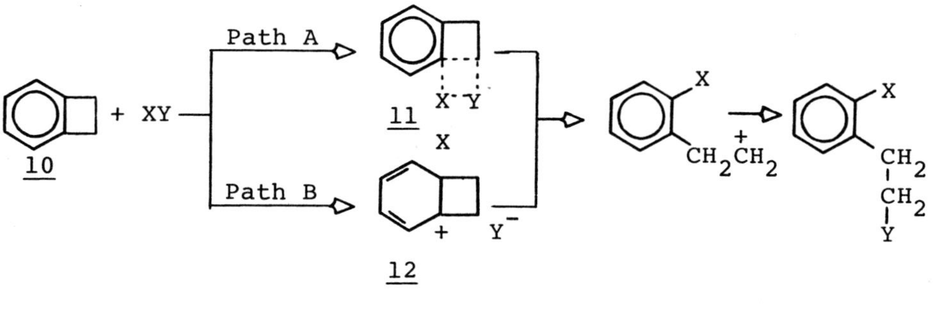

10 + XY — Path A → 11 (X···Y four-centered) / Path B → 12 (X, +, Y^-) → o-X-C_6H_4-$CH_2\overset{+}{C}H_2$ → o-X-C_6H_4-CH_2-CH_2-Y

Scheme 10

Schwartz et al. found that the reaction of either 2-phenylethyl choride (9) or benzocyclobutene (10) with benzene in the presence of $AlCl_3$ gave a quantitative yield of bibenzyl (13) (Eq. 25). Similar alkylations of toluene with

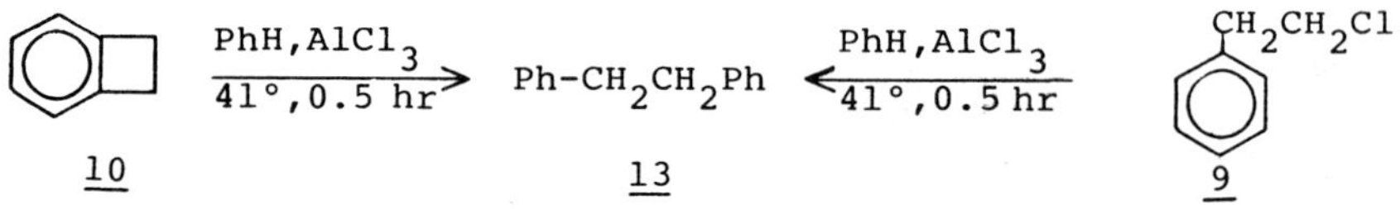

$$\underset{\mathbf{10}}{\text{benzocyclobutene}} \xrightarrow[41°, 0.5\ hr]{PhH, AlCl_3} \underset{\mathbf{13}}{Ph\text{-}CH_2CH_2Ph} \xleftarrow[41°, 0.5\ hr]{PhH, AlCl_3} \underset{\mathbf{9}}{C_6H_5CH_2CH_2Cl} \qquad (25)$$

either 9 or 10 gave almost identical mixtures of the three possible *ortho*, *meta*, and *para* 1-phenyl-2-tolylethane isomers (14-16, respectively) (Eq. 26).

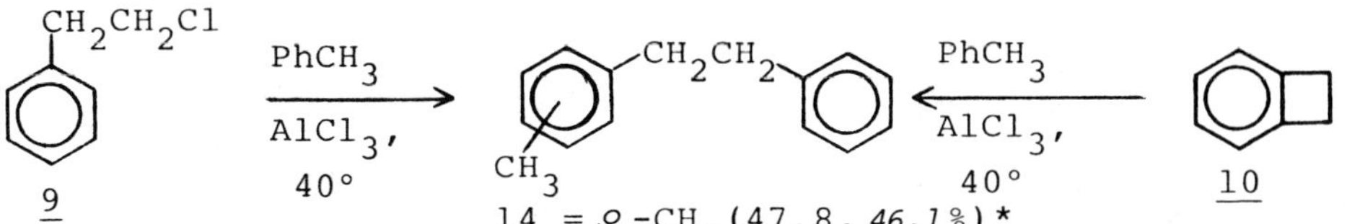

$$9 \;(\text{PhCH}_2\text{CH}_2\text{Cl}) \xrightarrow[\text{AlCl}_3,\ 40°]{\text{PhCH}_3} \text{CH}_3\text{C}_6\text{H}_4\text{-CH}_2\text{CH}_2\text{-C}_6\text{H}_5 \xleftarrow[\text{AlCl}_3,\ 40°]{\text{PhCH}_3} 10 \tag{26}$$

14 = *o*-CH_3 (47.8, *46.1*%)*

15 = *m*-CH_3 (18.2, *18.8*%)

16 = *p*-CH_3 (34.0, *35.1*%)

*Figures in italic are those from 10

These findings, while suggesting that the reactions of 9 and 10 proceeded through a common intermediate such as 11 or 12 (scheme 10), gave no definite answer to the question of which of these may actually be involved. The answer was, however, found in the alkylation of benzene with 2-(*m*-tolyl)ethyl chloride (17) and 2-(*p*-tolyl)ethyl chloride (18) in the presence of $AlCl_3$

$$m\text{-CH}_3\text{C}_6\text{H}_4\text{-CH}_2\text{CH}_2\text{-Cl} \;(17) + \text{PhH} \xrightarrow{\text{AlCl}_3} m\text{-CH}_3\text{C}_6\text{H}_4\text{-CH}_2\text{CH}_2\text{-Ph} \;(15) \tag{27}$$

$$p\text{-CH}_3\text{C}_6\text{H}_4\text{-CH}_2\text{CH}_2\text{-Cl} \;(18) + \text{PhH} \xrightarrow{\text{AlCl}_3} p\text{-CH}_3\text{-C}_6\text{H}_4\text{-CH}_2\text{CH}_2\text{-Ph} \;(16) \tag{28}$$

(Eqs. 27 and 28). The fact that 17 gave only 1-phenyl-2-*m*-tolylethane (15, Eq. 27) and 18 gave only 1-phenyl-2-*p*-tolylethane (16, Eq. 28) disqualified

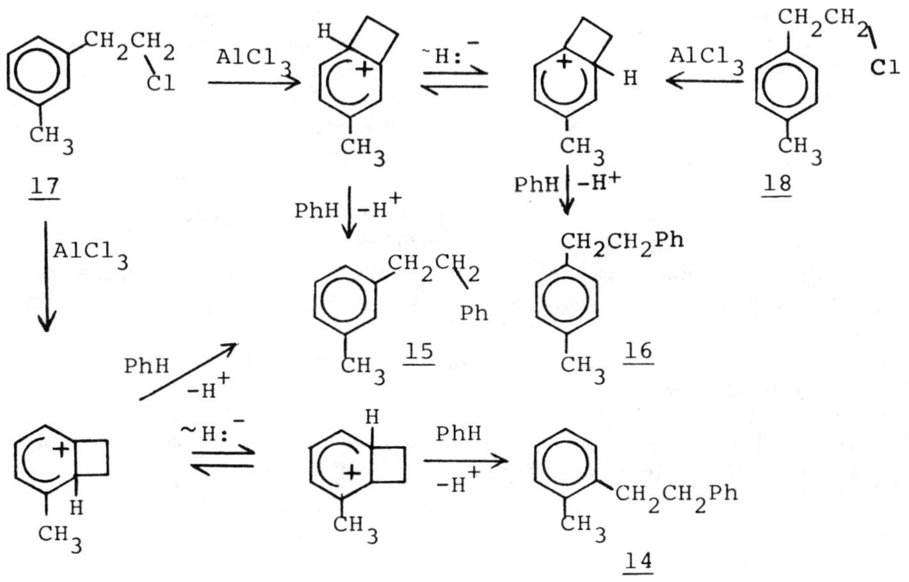

Scheme 11

the possible intermediacy of the benzenonium ion 12 (scheme 10). This conclusion was based on the assumption that had the methyl homolog of 12 been an intermediate in these reactions, rapid 1,2 hydride shifts of the type shown in scheme 11 would have been likely to occur and both 17 and 18 could yield mixtures of the isomeric 1-phenyl-2-tolylethanes 14-16.

The results described above, in addition to the isolation of 2-phenylethyl chloride during alkylations with benzocyclobutene, the earlier labeling results of McMahon and Bunce [107], the authors' findings of complete equilibration between CH_2 and CD_2 groups during the reaction of 1,1-dideuterio-2-*p*-tolyethyl chloride (19) with benzene and $AlCl_3$ (Eq. 29), and various reason-

$$p\text{-}CH_3C_6H_4CH_2CD_2Cl \xrightarrow[AlCl_3,\,40°]{PhH} p\text{-}CH_3C_6H_4CH_2CD_2Ph \;(\underline{20}\text{-}1,1\text{-}d_2\;47\%) + p\text{-}CH_3C_6H_4CD_2CH_2Ph \;(\underline{20}\text{-}2,2\text{-}d_2\;53\%) \quad (29)$$

(19)

able stereochemical arguments presented in the paper, led to the following conclusions: (1) in the presence of $AlCl_3$, benzocyclobutane was directly converted through path A of scheme 10 to 2-phenylethyl chloride, prior to its reaction with the arene; and (2) the alkylation of arenes by 2-arylethyl halide proceeds through a symmetrical phenonium ion or its equivalent.

It is of interest to point out that Schwartz et al. [116] were able to follow the rearrangement of deuterium label in the starting chloride (18-1,1-d_2) and the product (16-1, 1-d_2 and 20-2,2-d_2) during an alkylation experiment carried out at 7°C. These results showed that the extent of the partial rearrangement of the recovered starting chloride increased with time during reaction, whereas the rearrangement of the product was found to be complete at all stages of the reaction. This significant result, which was also noted by others with 2-phenylethyl-1, 1-d_2 and 2-(*p*-chlorophenyl)ethyl-1,1-d_2 chlorides [119], not only strengthened the concept of involvement of a symmetrical carbocation but also helped to eliminate any possibility of a mechanism that would predict equal scrambling of product and reactant throughout the reaction. However, one should not forget that this observed partial scrambling of the deuterium in the starting chlorides during alkylation differs from the previous results of McMahon and Bunce [107], who did not find any isotope-position rearrangement in recovered 2-phenylethyl-1-^{14}C chloride. To account for this difference, Schwartz et al. suggested that in some cases the intermediate phenonium ion, or its equivalent, might revert in part to starting material [116].

Before closing this section, it is worthwhile to mention that optically active 1-phenylethyl chloride was used by Hart et al. [120,121] to C-alkylate phenols in connection with the exploration of the stereochemistry of aromatic alkylations. Due to the reactivity of both reactants, nuclear alkylation occurred spontaneously with evolution of HCl just on mixing at room temperature without any added catalyst. Whereas benzyl chloride alkylated phenol only one-twentieth as fast as *t*-butyl chloride [122], 1-phenylethyl chloride reacted more readily than *t*-butyl chloride. Examination of the optical properties of the products showed that they were optically active. Whereas *ortho* alkylation (of p-cresol and p-chlorophenol) proceeded with partial net retention of configuration, *para* alkylation (of 2,6-xylenol) proceeded with partial net inversion. The results with 2,6-xylenol were accounted for in terms of the commonly accepted ionic alkylation mechanism, with a slight predominance of inversion resulting from shielding [123]. On the other hand, a concerted nucleophilic displacement mechanism analogous to S_Ni was suggested to account for the observed retention in the *ortho* alkylation of p-cresol and p-chlorophenol. In this cyclic mechanism the *ortho* carbon of phenol acts as a nucleophilic site, while the hydroxyl group functions as the acid catalyst in removing the chlorine from the alkyl halide. The whole mechanism is thus a concerted one, resulting finally in retention of configuration at the asymmetric carbon:

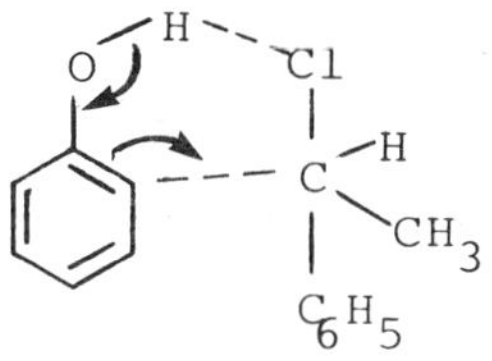

C. n-Propyl Halides

For a long time there was widespread dispute about the occurrence and extent of rearrangement in the aluminum halide-catalyzed alkylation of aromatics by n-propyl halides. This dispute, which resulted in such confusion that it has been perpetuated in textbooks [124-126] as well as in advanced books on Friedel-Crafts [127] and carbocation chemistry [128], can be dated back to the early days of Friedel-Crafts chemistry. Only one year after the discovery of Friedel-Crafts reactions, Gustavson [129] reported that the alkylation of benzene with either n-propyl or isopropyl bromide yielded the same product, isopropylbenzene (cumene). This was shortly supported by Silva [130] using the two isomeric propyl chlorides. Following these two reports, several other reports appeared in which it was claimed that the reaction of benzene with n-propyl halide gave chiefly n-propylbenzene with $AlCl_3$ at 0°C and lower temperatures [131,132], but chiefly isopropylbenzene with $AlCl_3$ [131, 133] at higher temperatures. In contrast, *sym*-triisopropylbenzene was claimed to result from alkylation of benzene with $AlCl_3$ at -10°C [134].

Confusion also existed in regard to alkylations of other aromatics with n-propyl halides and aluminum halides. Thus, exclusive production of rearranged isopropyl derivatives was reported from the reactions of ethylbenzene [133], m-xylene [135], tetralin [136], naphthalene [137,138], and methyl 2-furoate [139] with n-propyl halide and aluminum chloride. In contrast, nonrearranged hexa-n-propylbenzene was reported to result from the reaction of di-n-propylbenzene with n-propyl chloride and $AlCl_3$ [140]. However, a mixture of isomers was also suspected in the latter case [141].

The validity of some of the reports cited above remained unchallenged for almost four decades. It was in 1940 that Ipatieff et al. [141a] studied the effect of temperature on product composition in the aluminum chloride-catalyzed alkylation of benzene with n-propyl chloride. Using the formation of acetamido derivatives to identify and determine the yield of products, they concluded that the ratio of n-propyl- to isopropylbenzene in the monopropylbenzene fraction varied from 40:60 at 35°C to 60:40 at -6°C. This presented the first case in which the formation of a mixture of the propylbenzenes was confirmed. Reports of the production of various mixtures subsequently followed. For example, between 1940 and 1959 it was reported that the alkylation of benzene with *n*-propyl halide using catalysts such as Al(Hg)*x* [142], $AlCl_2HSO_4$ [143] or Al shavings [144] gave propylbenzene mixtures consisting mainly of the isopropylbenzene isomer. In 1956, Nightingale and Shackelford [145] reported that the alkylation of *m*-xylene with *n*-propyl chloride and aluminum chloride at 20°C yielded a monoalkylation mixture consisting of 1,3-dimethyl-5-*n*-propylbenzene (21), 1,3-dimethyl-5-isopropylbenzene (22), and a small amount of unspecified 1,3-dimethyl-4-propylbenzene (23) (Eq. 30). The first two isomers were produced in a ratio of 1.8:1. The predominance of the 5-rather than the 4-propyl-*m*-xylene isomers was attributed to subsequent isomerization of the initially formed 4-propyl isomers under the influence of the catalyst.

$$m\text{-}CH_3C_6H_4CH_3 + CH_3CH_2CH_2Cl \xrightarrow[\text{overnight, }20^\circ]{AlCl_3} \underset{\mathbf{21}}{3,5\text{-}(CH_3)_2C_6H_3CH_2CH_2CH_3} + \underset{\mathbf{22}}{3,5\text{-}(CH_3)_2C_6H_3CH(CH_3)_2} + \underset{\mathbf{23}}{2,4\text{-}(CH_3)_2C_6H_3C_3H_7} \quad (30)$$

In connection with their effort to explore the alkylation mechanism, Smoot and Brown [4,5] investigated the $GaBr_3$-catalyzed alkylation of benzene and toluene with *n*-propyl bromide. In the reaction of *n*-propyl bromide with benzene at 25°C there was formed a mixture of 28% *n*- and 72% isopropylbenzene (Eq. 31). The product from a similar reaction with toluene was shown

$$C_6H_6 + CH_3CH_2CH_2Br \xrightarrow[25^\circ]{GaBr_3} \underset{72\%}{n\text{-}C_3H_7C_6H_5} + \underset{28\%}{i\text{-}C_3H_7C_6H_5} \quad (31)$$

by infrared to be a mixture of six isomers. Brown predicted on the basis of kinetic analysis that the proportions of *n*-propylarenes in the alkylation mixtures from *p*-xylene, *m*-xylene, mesitylene, and pentamethylbenzene should be 53.6, 74.2, 90.7, and 95.9%, respectively. (Interestingly, the prediction for mesitylene has been quoted as experimental fact [146].)

It should be mentioned at this point that although mixtures were recognized in the more recent reports, the compositions of these mixtures were determined by analytical techniques which were qualitative at best and certainly liable to failure in detecting minor components.

The first application of modern techniques to the study of rearrangements accompanying Friedel-Crafts alkylation of benzene by *n*-propyl and butyl halides was carried out by R. M. Roberts and Shiengthong [147] in 1960. Using both infrared spectrophotometry and gas chromatography for analysis, they determined the products from the alkylation of benzene with *n*-propyl chloride and aluminum chloride over a temperature range extending from -18 to +80°C. In contrast to earlier claims, the results showed that the composition of the reaction product was essentially independent of the reaction temperature. The ratio of *n*-propylbenzene to isopropylbenzene varied only from 34:66 at -18°C to 30:70 at 80°C. It is interesting to note that the proportion of *n*-propylbenzene to isopropylbenzene at 35°C found by Roberts and Shiengthong was 33:67, not very different from the 40:60 ratio reported earlier by Ipatieff et al. [144a] at the same temperature. However, there was wide disagreement about the proportions of the two isomers formed at -6°C; whereas Roberts and Shiengthong found a ratio of *n*-propyl- to isopropylbenzene of 34:66, the earlier groups reported a ratio of 60:40. This is too wide a discrepancy to ascribe to the limited precision of the analytical method of the earlier workers, and the best explanation seems to lie in a mixup in the identification of the acetamido derivatives used by Ipatieff et al. to indicate the proportion of propylbenzene isomers.

In 1964, R. M. Roberts and Shiengthong [148] conducted a similar study on the Friedel-Crafts alkylations of *p*-xylene and mesitylene with propyl halides in the presence of varying amounts of aluminum halides and at different temperatures. Unlike the alkylations of benzene, the composition of the alkylates from these two arenes showed notable dependence on both reaction conditions and steric interactions. This can be seen from the data below Eq. (32), where yields and product compositions for the reactions of *p*-xylene (0.50 mol) with *n*-propyl chloride (0.25 mol) and $AlCl_3$ (0.025 mol) at -17, 10, 30, and 50°C are summarized.

p-xylene + *n*-propyl chloride (Cl) $\xrightarrow[\text{2 hr}]{AlCl_3}$ 2-*n*-propyl-*p*-xylene + 2-isopropyl-*p*-xylene + 5-isopropyl-*m*-xylene (32)

Temp.	Yield	2-*n*-propyl-*p*-xylene	2-isopropyl-*p*-xylene	5-isopropyl-*m*-xylene
-17°	(19% yield)	73%	27%	--
10°	(51% yield)	62%	38%	Tr
30°	(55% yield)	54%	41%	5%
50°	(68% yield)	53%	31%	16%

As is evident from Eq. (32), the reaction of *n*-propyl chloride and aluminum chloride at temperatures from -17 to +50°C gave 2-*n*-propyl-*p*-xylene as the major product. The highest proportion of *n*-propyl isomer (73%) was ob tained at the lowest temperature, as expected. Besides the other expected isomer, 2-isopropyl-*p*-xylene, an isomer resulting from methyl reorientation, 5-isopropyl-*m*-xylene, was also produced in the alkylations at 30 and 50°C. Further experiments showed that longer reaction times and larger amounts of catalysts had little effect on the proportion of isomers produced (at 10°C). Experiments also showed that *n*-propyl bromide, with either aluminum chloride or bromide catalyst, gave less rearrangement (22 to 25% at 10 to 14°C). *n*-Propyl chloride with aluminum bromide catalyst gave less rearrangement than with aluminum chloride; halide exchange undoubtedly occurred (see Chap. 2).

The interesting finding that the proportion of *n*-propyl-*m*-xylene at 30°C (54%) was exactly that predicted by Smoot and Brown (53.6%) [4,5] was considered by Roberts and Shiengthong to be probably only a coincidence. In that respect they stated: "Besides the fact that the prediction was for reaction of *n*-propyl bromide instead of the chloride and for gallium bromide instead of aluminum chloride catalyst, one must consider that the basis of the prediction was the electronic effect of the methyl groups, with no allowance for steric effects."

The effect of reaction conditions and steric interactions on product composition were even more pronounced in the case of mesitylene. Thus, in the reaction of mesitylene (0.50 mol) with *n*-propyl chloride (0.25 mol) and $AlCl_3$ (0.025 mol), the major isomer produced at low temperatures (0 to 10°C) was *n*-propylmesitylene (24), in less than 10% yield. At 30°C, the major product was still *n*-propylmesitylene, but appreciable amounts of 5-isopropylpseudocumene (25) and traces of 5-*n*-propylpseudocumene (26) and 5-isopropylhemimellitene (27) were formed. At the same temperature with double and ten times the amount of catalyst, the production of 5-isopropylpseudocumene (25)

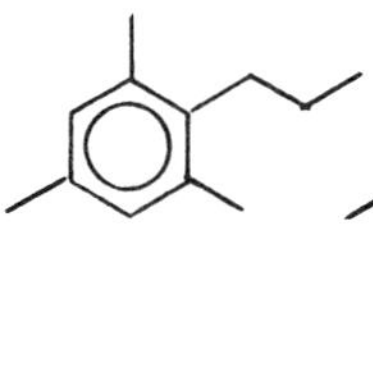

24 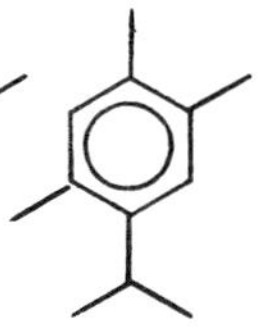 25 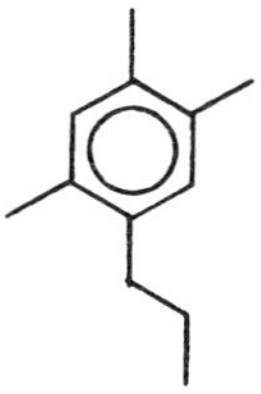 26 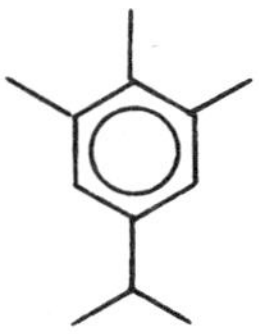 27 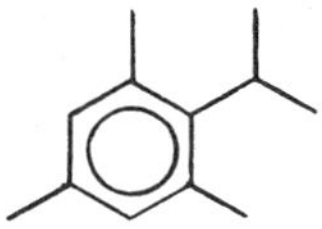 28 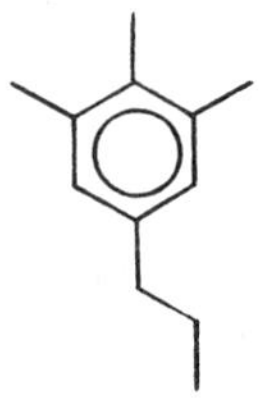 29

and 5-isopropylhemimellitene (27) was increased. At 50 and 70°C, with 2-hr reaction times, the major product was 5-isopropylpseudocumene (25). Under the more strenuous conditions of 70°C with a 4-hr reaction time, the major product was 5-isopropylhemimellitene (27). Based on reorientation studies, this observed predominance of 25 and 27 at higher reaction temperatures was attributed to methyl reorientation. Notably, in none of the eight experiments carried out by Roberts and Shiengthong was more than a trace amount of isopropylmesitylene (28) or of 5-*n*-propylhemimellitene (29) produced. The yields at 70°C were lower than at 30 and 50°C, due to the instability of the isopropylarenes toward dealkylation.

In the same paper, Roberts and Shiengthong also conducted some competitive alkylation, transalkylation, and reorientation experiments. When *n*-propyl chloride was allowed to react with equimolar amounts of benzene and mesitylene in the presence of $AlCl_3$ at 35°C the product contained isopropylbenzene (27% yield, based on *n*-propyl chloride) and *n*-propylmesitylene (24, 26%) with only about 1% isopropylmesitylene (28). When the mole ratio of benzene to mesitylene was 3:1, the product contained 1% isopropylbenzene and 8% *n*-propylmesitylene (24). A similar competitive reaction in which isopropyl chloride reacted with equimolar amounts of benzene and mesitylene at 30°C gave isopropylbenzene (39%), diisopropylbenzene (16%), and triisopropylbenzene (1%), with only traces of isopropylmesitylene.

The transalkylation experiments showed that transalkylations between *n*-propylmesitylene and benzene and between *n*-propylbenzene and mesitylene were insignificant under the conditions employed in the alkylations. Also, transalkylations between isopropylmesitylene and benzene did take place, but to a limited extent.

The results of the reorientation experiments were also significant. They showed that, under the conditions of alkylation, *p*-xylene and mesitylene gave little reorientation, *n*-propylmesitylene produced no reorientation or rearrangement, and reaction of isopropylmesitylene resulted in considerable methyl reorientation. Accordingly, it was concluded that the methyl-reoriented, isopropylated products observed in the alkylations did not result from reorientation of *p*-xylene and mesitylene to *m*-xylene and pseudocume, followed by isopropylation.

In rationalizing the results above, Roberts and Shiengthong emphasized the facts that the reorientations accompanying the alkylations and the products from competitive and noncompetitive alkylations point to steric effects which play an important role in determining both the rates and the product distribution of the alkylations. In fact, steric interactions were held partly or fully responsible for (1) the production of *n*-propyl-*p*-xylene and *n*-propylmesitylene as major products from the reactions of *p*-xylene and mesitylene with *n*-propyl halides at low temperatures; (2) the methyl reorientations leading to the production of 5-isopropyl-*m*-xylene from *p*-xylene and of 5-isopropylpseudocumene (25) and 5-isopropylhemimellitene (27) from mesitylene; (3) the very limited reorientation of *n*-propyl-*p*-xylene and *n*-propylmesitylene in the alkylations; and (4) the anomalous faster rate of reaction of *n*-propyl chloride with benzene than with the more nucleophilic mesitylene. Besides steric reasons, the latter anomaly was also attributed to the deactivation of the $AlCl_3$ catalyst by more extensive acid-base complex formation with the stronger Lewis base mesitylene than with benzene.

The established production of nonrearranged *n*-propyl derivatives besides the rearranged isopropyl derivatives in alkylations of arenes with *n*-propyl halides and metal halide catalysts required an explanation. On the basis of his kinetic studies, Brown [4,5] suggested that in such alkylations two concurrent reactions are possible for the initially formed *n*-propyl halide-metal

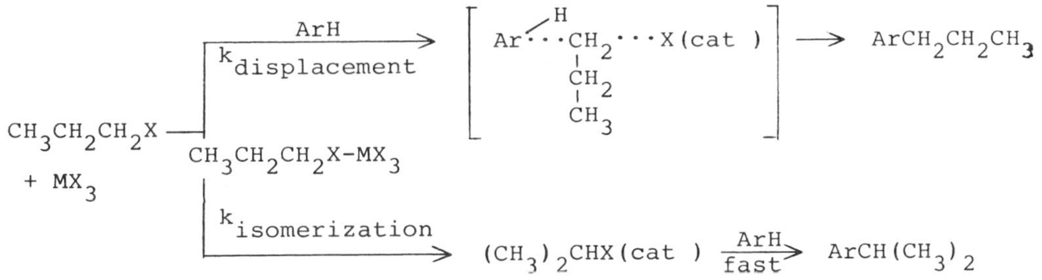

$$CH_3CH_2CH_2X + MX_3 \longrightarrow CH_3CH_2CH_2X\text{-}MX_3 \xrightarrow[k_{displacement}]{ArH} \left[Ar\overset{H}{\cdots} CH_2(CH_2CH_3) \cdots X(cat) \right] \longrightarrow ArCH_2CH_2CH_3$$

$$CH_3CH_2CH_2X\text{-}MX_3 \xrightarrow{k_{isomerization}} (CH_3)_2CHX(cat) \xrightarrow[fast]{ArH} ArCH(CH_3)_2$$

Scheme 12

halide complex (scheme 12): (1) a displacement-type reaction to give unrearranged product in which the rate-determining step involves nucleophilic attack by the aromatic moiety upon the polarized complex, and (2) rearrangement of the *n*-propyl complex to an isopropyl complex followed by extremely rapid alkylation to give isopropyl derivatives. The experimental rate constant for the total reaction will be the sum of the rate constants for the displacement and isomerization:

$$k_{total} = k_{displ} + k_{isomer}$$

Support for this view was given in terms of observations such as the increase in displacement rate with increasing basisity of the aromatic component; the similarity between the energies of activation for the reaction of benzene with methyl, ethyl, and *n*-propyl halides (12.5, 12.4, and 12.9 kcal/mol at 25°C, respectively); the low values for these activation energies relative to those anticipated for an ionization process in nonpolar media; the decrease in rate in going from ethyl to *n*-propyl halide (presumably as a consequence of steric considerations); and the pronounced difference between the relative rates of alkylations with methyl, ethyl, and *n*-propyl halides on the one hand and with isopropyl and *t*-butyl halides on the other hand.

An alternative explanation for the formation of primary derivatives during alkylations of arenes with *n*-propyl and *n*-butyl halides was later suggested by Roberts and Shiengthong [147], who favored a mechanism analogous to that proposed by Streitwieser et al. [149] for alkylations with alcohols and boron trifluoride (Chap. 4).

In an effort to test the possible involvement of protonated cyclopropane intermediates in alkylations with *n*-propyl halide, Lee and Woodcock [150] studied the $AlCl_3$-catalyzed reaction of 1,1-d_2-1-chloropropane with benzene in excess benzene or in 1,2,4-trichlorobenzene solvent at low temperatures. The ratios of isopropylbenzene to *n*-propylbenzene in the products were about 55:45 in benzene and 85:15 in trichlorobenzene, regardless of the length of reaction time. Examination of the two isomers by NMR and mass spectroscopy (MS) indicated only very minor amounts (5%) of isotope-position scrambling in the propyl side chain (Eq. 33), suggesting no major involvement of proton-

$$C_6H_6 + CH_3CH_2CD_2Cl \xrightarrow{AlCl_3} C_6H_5CD_2CH_2CH_3 + CH_3\underset{\underset{C_6H_5}{|}}{C}HCD_2H \qquad (33)$$

ated cyclopropane intermediates in the reaction. The major result was thus rationalized on the basis of the alkylation mechanism proposed previously by Brown and co-workers [4,5]. The small amount (5%) of accompanying isotope-position rearrangement of the D label in the side chain was attributed to intermolecular transfer of both hydride and deuteride ions in a manner similar to that proposed by Roberts and co-workers [111] (scheme 13).

The observation that more isopropylbenzene was formed when 1,2,4-trichlorobenzene instead of excess benzene was used as solvent was attributed to a decreased rate of alkylation of benzene in the inert solvent, thus giving more opportunity for *n*-propyl chloride to isomerize prior to alkylation. Since isomerization of the primary halide prior to alkylation could not be found in similar cases [82], the observation can better be explained as suggested by Nakane et al. [88], in terms of favorable reaction by an ionic path in the inert solvent and by a larger contribution from bimolecular displacement in the excess aromatic solvent.

The alkylation of aromatics with *n*-propyl chloride over Nafion-H was investigated recently by Olah and Meider [151]. In contrast to alkylations by metal

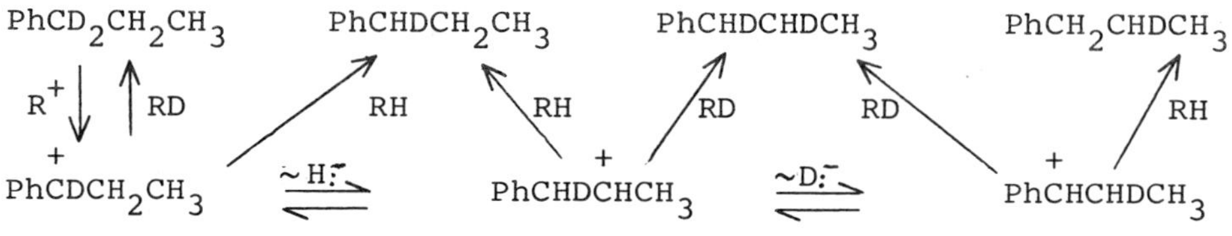

$PhCD_2CH_2CH_3$ $\quad$ $PhCHDCH_2CH_3$ $\quad$ $PhCHDCHDCH_3$ $\quad$ $PhCH_2CHDCH_3$

(R^+ ↓, ↑ RD; RH; RH; RD; RD; RH)

$$Ph\overset{+}{C}DCH_2CH_3 \underset{}{\overset{\sim H:^-}{\rightleftharpoons}} PhCHD\overset{+}{C}HCH_3 \overset{\sim D:^-}{\rightleftharpoons} Ph\overset{+}{C}HCHDCH_3$$

Scheme 13

halides which yield mixtures of *n* and *iso* alkylates, the only products obtained in the alkylation of toluene by *n*-propyl chloride were cymenes (isopropyltoluenes); no *n*-propyltoluenes were detected. Although the initial formation of Cl-protonated *n*-propyl chloride (28, Eq. 34) was assumed as the

$$CH_3CH_2CH_2\text{-}Cl \underset{}{\overset{H^+}{\rightleftharpoons}} \underset{\underline{28}}{CH_3CH_2CH_2\text{-}\overset{+}{Cl}\text{-}H} \xrightarrow{-HX} \longrightarrow CH_3\overset{+}{C}HCH_3 \quad (34)$$

first step in the reaction, the sole formation of cymenes ruled out the possibility of alkylation by this intermediate via an S_N2-type process. Instead, rapid rearrangement to the isopropyl cation was proposed.

D. *n*-Butyl Halides

As in the *n*-propyl case, the old reports concerning the occurrence and extent of rearrangements of *n*-butyl groups during alkylation by *n*-butyl halides are in many cases misleading and some of them are clearly in error. In some instances, these alkylations were described as proceeding without rearrangement, while in others partial or complete rearrangement was reported. For example, the alkylation of benzene by *n*-butyl halides and aluminum chloride was reported to give *sec*-butylbenzene [152], a mixture of *n*- and *sec*-butylbenzene [153,154], *n*-butylbenzene [155], and *t*-butylbenzene [156]. The same reactants were also reported to give *sec*-butylbenzene in the presence of Al-$HgCl_2$ [153] or $SbCl_5$ [154], a mixture of *n*- and *sec*-butylbenzene in the presence of Al-$HgCl_2$ [153] or $Al(Hg)_x$ [142], and an isomer believed to be isobutylbenzene in the presence of Al shavings at 70°C [144].

Again, the reaction of *n*-butyl chloride and aluminum chloride was claimed to give only rearranged *sec*-butyl derivatives with toluene [157] and *m*-xylene [158], only nonrearranged *n*-butyl derivatives with phenol [159], anisole [159], and ethyl α-naphthoate [139], only rearranged *t*-butyl derivatives with 2-furfural [160] and methyl 2-furoate [139], and a mixture of nonrearranged, 1,3-dimethyl-5-*n*-butylbenzene and rearranged 1,3-dimethyl-5-*sec*-butylbenzene with *m*-xylene [160a]. In fact, it was stated that "the alkylation products of methyl 2-furoate were somewhat unusual in the sense that the alkyl group appeared to undergo rearrangement so that, for example, each of the four butyl chlorides (as well as 'butylene') gave some of the same product: methyl 5-*t*-butyl-2-furoate" (Eq. 35) [139].

$$\text{(2-furyl)}COOCH_3 + C_4H_9Cl \longrightarrow (CH_3)_3C\text{-(2,5-furandiyl)-}COOCH_3 \quad (35)$$

The first reliable literature data concerning the aluminum chloride-catalyzed alkylation of benzene by *n*-butyl chloride is found in the work of Roberts and Shiengthong [147], who applied modern instrumental techniques for analysis of the butylbenzenes produced. Using benzene, *n*-butyl chloride, and aluminum chloride in a molar ratio of 5:1:0.09, they investigated the effect of temperature and reaction time on the composition of the product. The results showed that at 0°C for 2.5 hr, benzene and *n*-butyl chloride gave a mixture of *n*- and *sec*-butylbenzene in a ratio of ca. 1:2. At 80°C for 2.5 hr, a third isomer, isobutylbenzene, was also produced and the product consisted of isobutylbenzene, *n*-butylbenzene, and *sec*-butylbenzene in a percent ratio of ca. 17:23:60 (Eq. 36). Extension of the reaction time to 4 hr resulted

$$PhH + C\text{-}C\text{-}C\text{-}C(Cl) \xrightarrow{AlCl_3} \begin{cases} \xrightarrow[2.5\ hr]{0°} C\text{-}C\text{-}C\text{-}C(Ph)\ 34\% + C\text{-}C(Ph)\text{-}C\text{-}C\ 66\% \\ \xrightarrow[2.5\ hr]{80°} C\text{-}C\text{-}C\text{-}C(Ph)\ 22\% + C\text{-}C(Ph)\text{-}C\text{-}C\ 62\% + Ph\text{-}C\text{-}C(C)(C)\ 16\% \end{cases} \quad (36)$$

in no significant change in the latter ratio, suggesting that it represents a pseudoequilibrium value for the three isomers. The production of isobutylbenzene was explained in terms of initial formation of *sec*-butylbenzene, followed by its rearrangement at the higher reaction temperature. The nature of this rearrangement is discussed in Chap. 8 in connection with alkylbenzene rearrangements. As to the alkylation mechanism, Roberts and Shiengthong assumed it to be the same as for alkylation by *n*-propyl halides. The discrepancies in the earlier work may be attributed to unreliable analytical methods and, in some instances, to impure starting halides; e.g., in the preparation of halides from alcohols, rearrangements to isomeric alkyl halides may occur.

E. *n*-Pentyl Halides

The alkylation of benzene by *n*-pentyl halide and aluminum halide is of particular historical interest since it represents one of the first alkylation examples studied by Friedel and Crafts [161]. The reaction was described, among others, in a paper by Friedel and Crafts, "On a new general method of synthesis of hydrocarbons, ketones, etc.", which directed the first attention to the synthetic potentiality of their discovery. In that paper, Friedel and Crafts described the reaction of *n*-pentyl chloride with benzene and aluminum chloride and, being unaware of accompanying rearrangements, they reported the product as pure *n*-pentylbenzene. They visualized the formation of the product as taking place simply by the combination of the hydrocarbon radical from the chloride with the aromatic less hydrogen. This was represented empirically by the equation

$$C_6H_6 + CH_3(CH_2)_4Cl \xrightarrow{AlCl_3} C_6H_5(CH_2)_4CH_3 + HCl$$

This earliest historical report was later followed by a number of contradictory ones. Gilman and co-workers [139,160,162] claimed extensive rearrangement with formation of *t*-alkylaromatics in the $AlCl_3$-catalyzed reactions of substituted furans with isomeric butyl, *n*-pentyl, and *n*-hexyl halides.

Other workers [167,168] suspected the formation of mixtures of unspecified isomeric pentylbenzenes from the reactions of benzene with *n*-pentyl chloride and aluminum chloride. Although the suspicion of mixtures is acceptable, the claimed transformation of the *n*-alkyl group into a tertiary one during alkylation is unlikely. Such rearrangements are currently known to occur under certain specific Friedel-Crafts conditions [163,164], but not in alkylations of this type [82,84,85,147,165,166]. It now seems very probable that these apparently anomalous results may be attributed to the authors' incorrect assignments of structure to their products.

Another alkylation of benzene by *n*-pentyl chloride and aluminum chloride was also described by Thomas [168] to illustrate the factors that should be observed in controlling a Friedel-Crafts alkylation. In describing the product from his experiment, he stated that owing to isomerization by aluminum chloride, a mixture of isomeric pentylbenzenes was obtained. However, he did not indicate what isomers were present.

Almost a century after the first alkylation of benzene by *n*-pentyl chloride was reported by Friedel and Crafts [161], Roberts and Baker [169] made a careful study of the reaction in which they examined both temperature and time effects on the composition of the alkylation products. This was done by analyzing samples removed and quenched after various time intervals from reactions carried out at 0 and 25°C.

In line with the behavior of the simpler homologs, *n*-propyl and *n*-butyl halides, *n*-pentyl chloride was found to react with benzene and aluminum chloride producing mixtures of the corresponding primary and linear secondary alkylates, namely, 1-, 2-, and 3-phenylpentanes (Eq. 37). The relative

$$\text{PhH} + \underset{\text{Cl}}{\text{C-C-C-C-C}} \xrightarrow[155\ \text{min}]{AlCl_3} \underset{\text{Ph}}{\text{C-C-C-C-C}} + \underset{\text{Ph}}{\text{C-C-C-C-C}} + \underset{\text{Ph}}{\text{C-C-C-C-C}} \tag{37}$$

	1-phenyl	2-phenyl	3-phenyl
0° :	30	52	18
25° :	26	55	19

amount of the 1 isomer in the product showed greater dependence on reaction temperature than on contact time. Thus samples taken after 1, 3, 5, 25, 45, and 155 min were found to contain 30% 1-phenylpentane at 0°C and 26% at 25°C, with the rest of the monoalkylate being a mixture of 2- and 3-phenylpentane in a ratio of about 3:1. The latter ratio was approximately equal to the thermodynamic equilibrium value obtained by the same workers from the rearrangements of pure 2- and 3-phenylpentane induced by aluminum chloride under conditions comparable to those of the alkylation.

The results from the alkylation of benzene by *n*-pentyl chloride and aluminum chloride were explained by assuming the intermediacy of a primary penty cation or a polarized catalyst complex, which rapidly rearranges to the 2- and 3-pentyl cations by consecutive 1,2 hydride shifts (scheme 14).

In the light of such a mechanism, statistical distribution of the charge on the secondary carbon would result in a product containing the 2 and 3 isomers in a 2:1 ratio if no other rearrangements were operating during the alkylation.

The fact that thermodynamic rather than statistical ratio of the 2- and 3-phenylpentane isomers was observed in reaction of 1-chloropentane with benzene and aluminum chloride at 0 or 25°C was attributed to the subsequent equilibration of the two isomers under the reaction conditions. As pointed

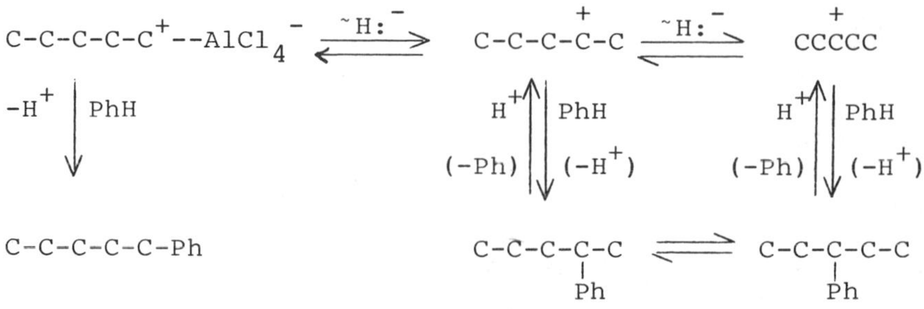

```
                          ~H:-                   +      ~H:-         +
C-C-C-C-C+--AlCl4-     <=======>     C-C-C-C-C      <=======>     CCCCC

-H+ | PhH                      H+ ^| PhH                  H+ ^| PhH
    |                      (-Ph)  |v (-H+)           (-Ph)  |v (-H+)
    v

C-C-C-C-C-Ph                   C-C-C-C-C       <=======>   C-C-C-C-C
                                     |                         |
                                     Ph                        Ph
```

Scheme 14

out before, the workers demonstrated the occurrence of such equilibration under comparable isomerization conditions.

In explaining the findings that a higher proportion of 1-phenylpentane was formed at 0°C than at room temperature, while a 3:1 ratio of 2-:3-phenylpentane was obtained at both temperatures, Roberts and Baker suggested that the isomerization of the primary carbocation to the secondary one was more temperature dependent than the isomerization of 2- to 3-pentyl cation.

F. *n*-Hexyl Halides

Apparently, the first use of *n*-hexyl halide as an alkylating agent in Friedel-Crafts chemistry was that reported by Gilman and Calloway in 1933 [139]. As mentioned before, these authors claimed that in the presence of aluminum chloride, *n*-hexyl bromide gave a branched-chain 5-hexyl derivative with methyl 2-furoate [139] and a 4-*t*-butyl derivative with ethyl 5-bromo-2-furoate [160,162].

In later years, Meals and Gilman [165,166] reported the alkylation of benzene with *n*-hexyl bromide and aluminum chloride. The product obtained was found to be a mixture of 1-, 2-, and 3-phenylhexane. However, the authors were unable to determine the relative amounts of these isomers because of difficulties encountered in the separation of the mixed derivatives. To account for the formation of the 2- and 3-phenylhexane isomers, the authors suggested the isomerization of the *n*-hexyl bromide prior to alkylation via consecutive elimination-addition reactions of hydrogen bromide. This mechanism was later disqualified by the discovery that the aluminum chloride-catalyzed isomerization of *n*-propyl halide to isopropyl halide did not involve the intermediate formation of propene [170].

Much more recently, the alkylation of benzene by both *n*-hexyl chloride and bromide in the presence of aluminum chloride was reinvestigated by Roberts and Khalaf [85], who studied the effect of temperature and time on the product composition. The results of these alkylations showed that under all conditions and within reasonable approximation the 1 isomer comprised one-third of the product, with the rest being a mixture of the 2 and 3 isomers in the ratio 2:1, respectively. The data also revealed that, in line with *n*-pentyl alkylations [169], the amount of the primary isomer in the mixture was to some extent dependent on the reaction temperature and to a lesser extent on the type of halogen in the alkylating agents. This amount increased with decreasing reaction temperature and was higher for bromine than for chlorine.

G. *n*-Heptyl and Higher *n*-Alkyl Halides

In 1935, Gilman and Burtner [160] reported that the reaction of *n*-octadecyl bromide ($C_{18}H_{37}Br$) with ethyl 5-bromo-2-furoate and aluminum chloride gave a 46% yield of ethyl 4-*t*-butyl-5-bromo-2-furoate, as well as 32% of isobutane and 7% of *n*-butane (Eq. 38).

$$\text{Br-}C_4H_2O\text{-}COOC_2H_5 \xrightarrow{AlCl_3} \underset{46\%}{\text{Br-}(CH_3)_3C\text{-}C_4HO\text{-}COOC_2H_5} + \underset{32\%}{CH_3\overset{CH_3}{\overset{|}{C}}HCH_3} + \underset{7\%}{CH_3CH_2CH_2CH_3} \quad (38)$$

One year later, Seidel and Engelfried [171] described the preparation of octadecylbenzene from reaction of *n*-octadecyl bromide with benzene and aluminum chloride. Although they reported the preparation of the sulfonamide derivative of their product, they did not try to identify its structure.

In 1939, Gilman and Turck [162] reexamined the reaction above and concluded that the major alkylation product obtained was *n*-octadecylbenzene. They established this by comparing the Friedel-Crafts product with the product of the Wurtz-Fittig reaction between *n*-octadecyl iodide and iodobenzene, and with the product obtained by the reduction of stearophenone. The identity of the octadecylbenzene obtained by these three different procedures was authenticated by the mixture melting points of their sulfonamides. These results were shortly criticized by Gilman and Meals [166], who (in 1942) discovered that the sulfonamide procedure, as applied by Gilman and Turck, was satisfactory only for the detection of 1-phenylalkanes. Gilman and Meals found that while the sodium alkylbenzenesulfonates derived from *n*-dodecyl- and higher *n*-alkylbenzenes were practically insoluble in ether, those derived from the respective *sec*-alkylbenzenes were, in general, freely soluble in ether and could be lost in the extraction process. On the basis of these findings, Gilman questioned the validity of his former conclusions about the formation of only pure 1-phenyloctadecane from *n*-octadecyl bromide, and suggested that all eight secondary octadecylbenzenes could have been present, but were lost during the preparation of the derivative.

In their 1942 paper, Gilman and Meals reported the aluminum chloride-catalyzed alkylation of benzene by *n*-hexyl, *n*-heptyl, *n*-dodecyl, *n*-tetradecyl, *n*-hexadecyl, and *n*-octadecyl bromides. Although no satisfactory way was available to identify the isomers or to estimate their proportions produced in these reactions, the authors believed that in all of these cases the products contained not only the normal primary alkylbenzenes, but also all of the possible linear secondary isomers.

In 1962, Sharman [82] carried out a thorough investigation of the aluminum halide-catalyzed alkylation of arenes by *n*-alkyl halides using *n*-octyl and *n*-dodecyl halides as representative examples. In this investigation the author studied the effects on product composition of reaction variables such as mole ratios of reactants, reaction temperature, contact time, and type of arene and solvent. Analysis was by means of infrared (IR) spectroscopy and GLPC. The results obtained led to the following significant conclusions regarding alkylations by *n*-alkyl halides.

1. In general, the alkylations were fast and, in excess arene at 6°C or above, many of them were complete in less than 20 min.

2. The normal and all of the possible linear secondary isomers were produced, but no evidence for skeletal rearrangement of the alkyl group was obtained (Eqs. 39 and 40).

$$PhH + n\text{-}C_8H_{17}Br \xrightarrow[\text{80 min, petroleum ether}]{AlBr_3, -15°} \underset{46.5\%}{Ph(CH_2)_7CH_3} + \underset{25.8\%\ (48.3\%)^*}{CH_3CH(Ph)(CH_2)_5CH_3} +$$

$$\underset{15.2\%\ (28.4\%)^*}{CH_3CH_2CH(Ph)(CH_2)_4CH_3} + \underset{12.5\%\ (23.3\%)^*}{CH_3(CH_2)_2CH(Ph)(CH_2)_3CH_3} \quad (39)$$

$$PhH + n\text{-}C_{12}H_{25}Br \xrightarrow[\text{90 min, petroleum ether}]{AlBr_3, -15°} \underset{47.8\%}{Ph(CH_2)_{11}CH_3} + \quad (40)$$

$$\underset{16.3\%\ (31.2\%)^*}{CH_3CH(Ph)(CH_2)_9CH_3} + \underset{9.4\%\ (18\%)^*}{CH_3CH_2CH(Ph)(CH_2)_8CH_3} +$$

(4- + 5- + 6-phenyldodecane)

26.5% (50.8%)*

3. No rearranged alkyl bromide was found in samples removed from reactions that had not reached 100% conversion. This excluded the possibility that rearrangement during alkylation was due to rearrangement of the halide prior to alkylation.

4. The product distributions obtained in such alkylations are not necessarily those intrinsic to the alkylation reaction itself. This is because the alkylation process is usually accompanied by side reactions involving isomerization, disproportionation, dealkylation, and reorientation. The selective destruction of the secondary alkylarenes by these side reactions accounts for the sharp increase with time at 35°C in the concentration of the primary alkylates in the alkylation mixtures. On the other hand, the facile interconversion between the secondary isomers in the presence of aluminum halide catalysts accounts for the observation that, in all alkylations of benzene with *n*-dodecyl halide and aluminum halide, the distribution of the secondary dodecylbenzene isomers did not vary from the known equilibrium composition (32% 2-phenyl-, 20% 3-phenyl-, and 48% 4- + 5- + 6-phenylbenzene). In fact, experiments under simulated alkylation conditions proved the stability of the 1-phenyldodecane isomer and the rapid involvement of the secondary isomers in isomerization, dealkylation, reorientation, and disproportionation reactions [82,172,173]. An unexpected side reaction observed [82] was the formation of branched alkanes corresponding in carbon content to the original side chain. This is surprising in view of the fact that the author found no evidence for skeletal rearrangements of the side chains of the alkylbenzenes.

---

*Percent distribution of secondary alkylates normalized to 100.

This result is discussed in Chap. 8 in connection with other dealkylation reactions.

5. For a given arene, the primary/secondary product ratio intrinsic to alkylation alone was found to increase with decreasing temperature. For example, in alkylations of benzene with dodecyl bromide and $AlBr_3$, a change in temperature from 35°C to -34°C caused an increase in the percentage of primary alkylbenzene from about 31% to approximately 50%. Moreover, a plot of percentage primary isomer formed against reaction temperature showed a rather good straight line, indicating an increase in primary alkylbenzene content of approximately 3% per 10°C reduction in temperature.

6. Increase in the nucleophilic character of the arene resulted in a dramatic rise in the amount of nonrearranged product obtained. This is evident from the fact that alkylations by *n*-dodecyl bromide and aluminum bromide gave *n*-alkylaromatics that comprised up to 41% of thc product with benzene but up to 88% with mesitylene. Interestingly, the latter value is very close to that predicted by Smoot and Brown [5] for alkylations with *n*-propyl bromide and gallium bromide at 25°C.

7. Both product composition and reaction rate were affected, to varying degrees, by type and ratio of reactants. For example, the primary/secondary product ratio was found to be slightly smaller in alkyl chloride-aluminum chloride than in alkyl bromide-aluminum bromide alkylations. Also, a minor reduction in the proportion of the primary alkylate and a dramatic reduction in the rate were caused by dilution of the arene with petroleum ether. Moreover, variation in the mole ratio of aluminum bromide to alkyl bromide in the range applied (0.09 to 0.25:1.0) slightly altered the reaction rate, but had a negligible effect on product composition.

To account for the preceding conclusions, Sharman suggested an alkylation mechanism (scheme 15) in which an initially formed π-complex intermediate is responsible for the production of both normal and all secondary alkylbenzenes.

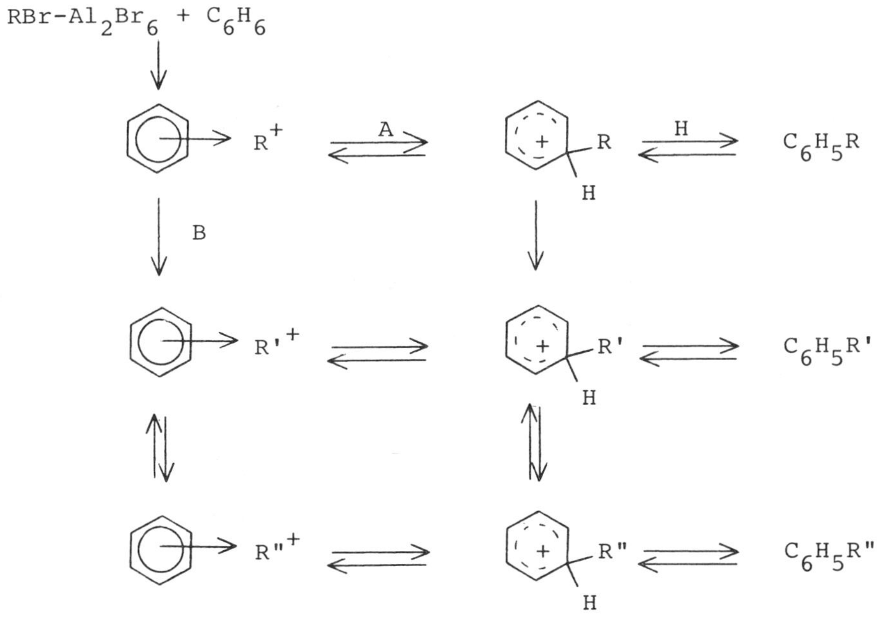

$RBr\text{-}Al_2Br_6 + C_6H_6$

$\downarrow$

π-complex ($C_6H_6 \rightarrow R^+$) $\overset{A}{\rightleftharpoons}$ σ-complex ($C_6H_6R^+$, with H) $\overset{H}{\rightleftharpoons}$ C_6H_5R

$\downarrow$ B $\qquad\qquad\qquad \downarrow$

π-complex ($C_6H_6 \rightarrow R'^+$) $\rightleftharpoons$ σ-complex ($C_6H_6R'^+$, with H) $\rightleftharpoons$ C_6H_5R'

$\updownarrow \qquad\qquad\qquad \updownarrow$

π-complex ($C_6H_6 \rightarrow R''^+$) $\rightleftharpoons$ σ-complex ($C_6H_6R''^+$, with H) $\rightleftharpoons$ C_6H_5R''

Scheme 15

According to scheme 15, the formation of both nonrearranged primary and rearranged secondary products reflects a competition between paths A and B. It was indicated that both pathways may be very fast but must be similar in rate and that increase in the nucleophilicity of the aromatic undergoing alkylation should increase the rate of A but have little effect on B. The latter suggestion, which follows from data compiled by Brown and Nelson [174] on the relative stability of σ complexes as a function of the nucleophilic character of the aromatic, was offered to explain the observed reduction in rearrangement accompanying increase in the nucleophilic character of the alkylated arene. One interesting feature of scheme 15 is that "it provides for rearrangement of the side chain (by path B) with involvement of the aromatic in a π-complex and does not require that collapse to a secondary bromide occur or that the intermediates be the same as those encountered when alkylations are carried out with secondary bromide."

In line with other alkylations by *n*-alkyl halides, the alkylation of arenes with *n*-heptyl halides (halogen = F, Cl, Br, I) and $AlCl_3$, $AlBr_3$, or AlI_3 catalyst was shown by Asinger et al. [174a] to give mixtures of the four possible linear arylheptanes, including the 1 isomer. In these mixtures, the average distribution ratio of 2- 3- 4-phenylheptane was 55:33:12. The results also demonstrated that (1) a more easily alkylated arene such as cumene gave higher proportions of the 1 isomer than benzene, and (2) the amount of the 1 isomer was smallest with *n*-heptyl fluoride and largest with *n*-heptyl iodide. These findings are consistent with Brown's predictions [4,5,74] and also with the experimental results of these and other authors [85].

In concluding this discussion of Friedel-Crafts alkylations with normal straight-chain halides, the following points can be emphasized: (1) The product should contain all of the possible straight-chain alkylbenzene isomers unless the procedure is adapted for the selective destruction of the secondary alkylbenzene isomers. (2) The primary/secondary product ratio is sensitive to various factors, such as reaction temperature, nucleophilicity of the aromatic, ratios of reactants, solvent, time of reaction, and even the type of halogen in the alkyl halide and metal halide catalyst. (3) As in all alkylations, side reactions such as isomerization, dealkylation, disproportionation, reorientation, and fragmentation play a significant role in shaping the final picture of the product mixture. The effect of these additional reactions is always dependent on the reaction conditions applied. In general, these side reactions increase with increasing reaction severity.

III. ALKYLATIONS WITH LINEAR SECONDARY ALKYL HALIDES

A. Isopropyl Halides

The Friedel-Crafts alkylation of arenes by isopropyl halides was reported by both early [2] and recent workers to yield exclusively unrearranged isopropyl derivatives. Consequently, the term *isopropylation* was frequently applied for this reaction. All of these workers have also recognized that as in other alkylations of alkylbenzenes with alkyl halides, alkenes, and other alkylating agents [2,71,72,73], considerable amounts of *meta*-oriented isopropyl isomers are formed [6,48,50,101,130,131,151,160,175-182]. Prior to 1940, for example, isopropyl halide and aluminum halide were reported to give *m*-cymene with toluene [130,175-177], 5-isopropyl-*m*-xylene with *m*-xylene [135], *m*-isopropylbenzaldehyde with benzaldehyde [160], *m*- and *p*-ethylisopropylbenzene with ethylbenzene [179], and *m*- and *p*-propylisopropylbenzene with propylbenzene [131]. However, considering the tedious and unreliable methods used for the analysis of isomeric alkylaromatics in early work, one should not take these old results for granted unless they have been verified by modern analysis.

During the last three decades, both competitive and noncompetitive Friedel-Crafts isopropylations of aromatics have been applied exhaustively to investigate the question of positional and substrate selectivity in electrophilic aromatic substitution. Most of the accumulated results have been reviewed repeatedly [101,174,180,183-185] and their implications on the alkylation mechanism in particular, and the mechanism of electrophilic aromatic substitution in general, have been treated thoroughly [53,174,183-186]. Accordingly, we shall only emphasize the most important experimental results and the significant conclusions that have been reached from them.

The leading work of Brown and his collaborators has demonstrated that compared to methylation and ethylation, the aluminum halide-catalyzed isopropylation of benzene and toluene with isopropyl halide is considerably faster [6,48,50] and is invariably accompanied by both disproportionation and reorientation reactions [73,180,187-190].

For example, the alkylation of benzene with isopropyl bromide and aluminum bromide was essentially complete within 0.005 sec, and in the alkylation of toluene the isomer distribution was 14.5% *p*- and 85.5% *m*-isopropyltoluene, with no *ortho* isomer [48]. In view of the large yields of the *ortho* isomer obtained in isopropylations with nonisomerizing catalysts, [6,73,188], the absence of the *ortho* isomer was attributed to concurrent isomerization to the more stable *meta* isomer [6,13,48].

The kinetic data of Brown et al. also furnished significant information about the comparative behavior of the isopropyl group and other groups. For example, the toluene/benzene reactivity ratios were shown to be 2.95 for methylation, 2.4 for ethylation, and 1.65 for isopropylation [48]. Moreover, the order of activities was shown to be methylation < ethylation < isopropylation < *t*-butylation [50]. In Brown's opinion, the facts suggested that the aromatic must be involved in the rate-determining step [48] and that the nature of the transition state changes markedly with increasing branching of the alkyl halide [6].

Since this work by Brown and co-workers was published, several other papers have appeared in which isopropylation of aromatics has been described and the isomer-distribution data have been correlated with the mechanism of Friedel-Crafts alkylation [35,37-40]. In the following, we shall describe some of these papers, stressing first the experimental facts and later on theories offered to accommodate them.

In 1961, Allen and Yats [180] studied the monoalkylbenzene isomer distributions obtained by Friedel-Crafts methylation, ethylation, and isopropylation of toluene using a variety of catalysts and alkyl derivatives. From kinetic evaluation of their results and of the previously reported results, these authors concluded that in spite of the great variety of catalysts and alkylating agents used for the propylation of toluene, all of the isopropyltoluene isomers could have been obtained by alkylation to produce a mixture containing 36.5% *p*-, 21.5% *m*-, and 42.0% *o*-isopropyltoluene followed by isomerization of this mixture. This hypothesis was later criticized by Olah et al. [181] on the ground that the *o*-isomer value (42%) seemed to be too low and the *m*-isomer value (21.5%) seemed too high for a typical electrophilic aromatic substitution of toluene in which steric hindrance is not of major importance.

In 1964, Olah et al. investigated the competitive and noncompetitive Friedel-Crafts isopropylation of methylbenzenes [181] and halobenzenes [182]

Table 3. $AlCl_3/CH_3NO_2$ Catalyzed Isopropylation of Benzene and Halobenzenes with Isopropyl Bromide in Nitromethane Solution at 25°C

Aromatic	$k_{Ar}:k_{benzene}$	Isomer distribution, %		
		ortho	*meta*	*para*
Benzene	1.00			
Fluoro-	0.23	41.8	1.9	56.3
Chloro-	0.10	49.8	7.9	42.3
Bromo-	0.08	51.4	11.3	37.3

with isopropyl bromide and propylene in the presence of different catalysts and solvents. They found the following:

1. Competitive isopropylation of toluene in homogeneous organic solutions (nitromethane, tetramethylenesulfone, sulfur dioxide, carbon disulfide) showed low substrate ($k_{toluene}/k_{benzene} = \sim 2$) but higher positional selectivity. Depending on type of catalyst, the isomer distribution in the isopropylation of toluene was 44 to 60% *ortho*, 25 to 40% *para*, while the amount of the *meta* isomer in general was 14 to 18%. The isopropylation of monohalobenzenes showed similar effects. The observed reactivities of the halobenzenes relative to benzene, together with the isomer distributions of the monoalkylated products obtained from the $AlCl_3$-CH_3NO_2-catalyzed isopropylation of benzene and halobenzenes with isopropyl bromide, are summarized in Table 3. Results similar to those of Table 3 were also obtained when the isopropylation was carried out with $FeCl_3$-CH_3NO_2 and isopropyl bromide or with $AlCl_3$-CH_3NO_2 and propene.

2. The isomer distributions were seriously influenced by concurrent isomerization, as well as by steric factors involving heterogeneous systems with bulky catalyst-reagent complexes. These conclusions were deduced from the following observations:

(a) A typical heterogeneous system, $AgSbF_6$-*i*-C_3H_7Br, gave the isomer distribution 59.6% *o*-, 14.7% *m*-, and 25.7% *p*-isopropyltoluene, an *ortho*/*para* ratio higher than any of the ratios obtained in alkylations by the same system carried out under homogeneous reaction conditions. The explanation proposed was that in homogeneous solution the solvation of the alkylating agent (*i*-C_3H_7Br-$AgSbF_6$) was more complete, and owing to increased steric requirements, gave a somewhat low *ortho*/*para* isomer ratio.

(b) The $AlCl_3$-CH_3NO_2-catalyzed alkylations using isopropyl bromide in nitromethane solutions showed considerably increased steric requirements over similar isopropylations with propylene. Whereas up to 15.8% of 1,2-dimethyl-3-isopropylbenzene was formed from *o*-xylene and propylene, none of this isomer was produced from isopropyl bromide.

(c) In the $AlCl_3$-CH_3NO_2-catalyzed alkylations with isopropyl bromide in nitromethane, the relative reactivity of mesitylene was shown to be less than one-tenth of that observed in similar reaction with propylene.

3. The observed relative reactivities of halobenzenes in the case of isopropylations showed good agreement with relative stabilities of complexes of halobenzenes with π acids and also with reactivities in nitration with $NO_2^+BF_4^-$ [191] and benzylations with benzyl chlorides [16].

4. A small secondary kinetic isotope effect was observed in the isopropylation of benzene-d_6.

In 1964, R. M. Roberts and Shiengthong [148] studied the Friedel-Crafts alkylations of *p*-xylene and mesitylene with propyl halides and aluminum halide. With *p*-xylene (0.50 mol) the reaction of isopropyl chloride (0.25 mol) and aluminum chloride (0.025 mol) at 0 and 10°C gave only 2-isopropyl-*p*-xylene. However, with a tripled amount of catalyst, even at 0°C some methyl reorientation occurred with the formation of 5-isopropyl-*m*-xylene (ca. 5%). At higher temperatures this reorientation became pronounced, so that at 70°C the major product was 5-isopropyl-*m*-xylene (52%) (Eq. 41).

+ Br, $AlCl_3$, 2 hr, (0.025 mol); 10°, 60%; 70°, 58%; +

(41)

The isopropylation of mesitylene (0.5 mol) with isopropyl chloride (0.25) and $AlCl_3$ (0.025) was similarly investigated. Reorientations were found to be more prevalent. Even at 0°C almost equal amounts of isopropylmesitylene and 5-isopropylpseudocumene were produced; with a doubled amount of catalyst the latter isomer predominated by more than a 2:1 ratio. At 30°C and above, 5-isopropylhemimellitene was produced in significant amounts, actually predominating in some experiments. Isopropylmesitylene was the least abundant isomer in all alkylations conducted at temperatures above 0°C. Some of the results obtained are shown in Eq. (42). As in alkylations with *n*-propyl

+ Br, $AlCl_3$, 2 hr, (0.025 mol); + ; +

(42)

(0.50 mole)	(0.25 mole)	Temp., °C	Yield, %	Product Distribution		
		0°	27	57%	43%	-
		30°	67	19%	48%	33%
		50°	58	17%	52%	31%
		70°	39	19%	39%	42%

halides, the reorientations accompanying the isopropylations of *p*-xylene and mesitylene pointed to steric effects which play an important part in determining the product distribution of the alkylations. This indication was confirmed by the competitive alkylation of benzene and mesitylene with isopropyl chloride, whereby the products were mono-, di-, and triisopropylbenzene, with only traces of isopropylmesitylene being produced.

Steric effects similar to those encountered in the isopropylation of methylbenzenes have been manifested in the isopropylation of other aromatics, including cresols and naphthalene. The reaction of *o*- and *p*-cresols with isopropyl chloride and $AlCl_3$ at 15 to 25°C was reported by Dewar and Puttman [192] to give the products shown in Eqs. (43) and (44), respectively. In

OH + Br $\xrightarrow[15-25°]{AlCl_3}$ OH + OH + OH

+ OH + OH + OH

29 (not found)

(43)

OH + Br $\xrightarrow[15-25°]{AlCl_3}$ OH 20.5% + OH 4.6% + OH 30 17.4%

+ OH trace + OH 31, trace + OH 32 (not found)

(44)

these reactions, steric repulsions were held to be fully or partially responsible for the following observations: (1) the lack of formation of 3,4-diisopropyl-*o*-cresol (29, Eq. 43) from *o*-cresol, (2) the lack of formation of 2,3-diisopropyl-*p*-cresol (32, Eq. 44) from *p*-cresol, and (3) the facile demethylation of 2,5-diisopropyl-*p*-cresol (31) to produce considerable amounts of 2,5-diisopropylphenol (30, Eq. 44).

The alkylation of naphthalene with isopropyl and other alkyl halides was carried out by Roberts and Zee-Cheng in 1963 [193] and by Olah and Olah in 1976 [51a]. Both groups of workers found that the $AlCl_3$-catalyzed alkylation of naphthalene with isopropyl halide in CS_2 solution gave isopropylnaphthalene mixtures in which the β-isomer constituted over 90% (Eq. 45). The

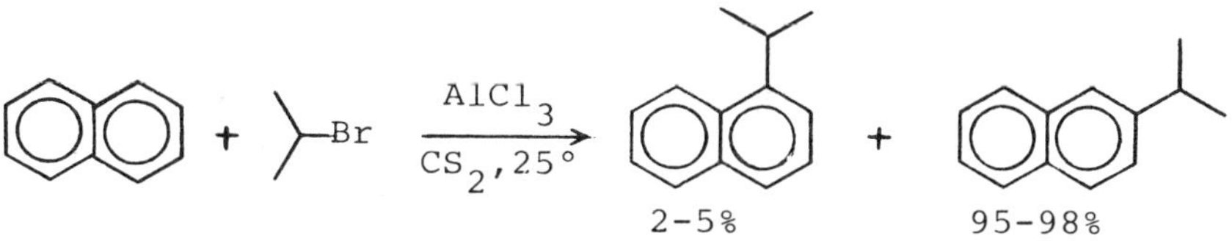

$$\text{naphthalene} + (CH_3)_2CHBr \xrightarrow[CS_2,\,25°]{AlCl_3} \alpha\text{-isopropylnaphthalene}\ (2\text{-}5\%) + \beta\text{-isopropylnaphthalene}\ (95\text{-}98\%) \quad (45)$$

lowest β-isomer value (90.4%) was obtained after 1 min at 25°C [51a]. These investigators also noted that variations in reaction temperature (from 0 to 80°C) [193] and/or reaction time (from 1 min to 2 hr) resulted in slight changes in product composition until a constant α/β equilibrium ratio was attained. This α-/β-isopropylnaphthalene equilibrium ratio, which was attained by the $AlCl_3$-catalyzed isomerization of both α and β isomers [193,51a], was estimated as 5:95 by IR spectroscopy [193] and as 1.5:98.5 by GLPC [51a]. In contrast to reactions in CS_2 solutions, reactions in nitromethane solutions at 25°C gave much higher α/β ratios, ranging from 82.5:17.5 after 5 min to 70:30 after 1 hr [51a]. This was attributed to slower reorientation of the initially formed α-isopropylnaphthalene in the weak $AlCl_3$-CH_3NO_2 catalyst medium [51a].

In their paper, Olah and Olah [51a] also reported the $AlCl_3$-catalyzed competitive alkylation of naphthalene and benzene in the presence of various solvents. Whereas the substrate (k_N/k_B) and positional (α/β) selectivities varied widely with reaction time and temperature in carbon disulfide and benzene solutions, they showed smaller variations in nitromethane solution. In the latter solvent, the k_N/k_B ratio for isopropylation at 25°C varied from 4.7 after 5 min to 7.5 after 1 hr. This change in value with time could be attributed to disproportionation (intermolecular alkyl transfer from benzene to naphthalene) under the alkylation conditions. The finding that both the α- and β-isopropylnaphthalene isomers were produced under all conditions applied should be contrasted with earlier work of Haworth et al. [141], who reported obtaining only β-isopropylnaphthalene from an alkylation of naphthalene with isopropyl bromide and $AlCl_3$.

In addition to accompanying reorientations and transalkylations, isopropylation reactions have also been found to be complicated by hydride transfer reactions. Such reactions were held responsible for the formation of large amounts of the hexamethylindene (33) and smaller amounts of the hexamethylindan (34) in the alkylation of mesitylene with isopropyl bromide and $AlCl_3$

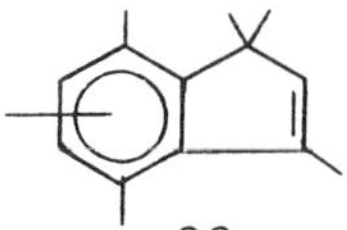

33

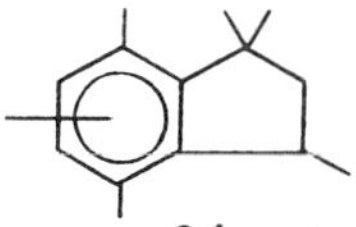

34

[194]. Hydride transfer reactions were also invoked to explain the production of considerable amounts of olefinic, alcoholic, and condensed products, together with the usual Friedel-Crafts products from the alkylation of acetophenone with *n*-propyl or isopropyl chloride in the presence of HF-SbF_5 catalyst [79a].

The isopropylation of furfural and 2-acetylfuran with isopropyl chloride and aluminum chloride was reported [195]. The results obtained are shown in Eqs. (46) and (47), respectively.

CHO (1 mol) + Cl (1 mol) $\xrightarrow[CS_2]{AlCl_3 \ (1.2\ mol)}$ CHO 40% + CHO 31% + CHO 26% (46)

CH_3 + Cl $\xrightarrow[CS_2]{AlCl_3}$ 19% + 47% + 34% (47)

The alkylation of arenes by isopropyl halide has also been accomplished using boron halide catalysts. In 1943, Hennion and Kurtz [196] observed that benzene and toluene could be alkylated with alkyl halides in the presence of boron fluoride activated by cocatalysts such as water, alcohol, or sulfuric acid. Unaware of the presence of other isomers, they reported the product from toluene and isopropyl bromide to be pure *p*-isopropyltoluene. Interestingly, Hennion and Kurtz explained their results in terms of what they called "a positive fragment mechanism," illustrated as follows:

$$R\text{-}X + BF_3 \cdot H_2O \longrightarrow R^+ \ [BF_3OH]^- + HCl$$

$$R^+ + ArH \longrightarrow ArR + H^+$$

In 1957, Olah et al. [12] found that methyl, ethyl, propyl, isopropyl, *t*-butyl, or cyclohexyl fluoride in the presence of boron fluoride gave good yields of monoalkylated products. In alkylating toluene by isopropyl fluoride and boron trifluoride, these workers obtained an 84% yield of isomeric isopropyltoluenes. In a following paper, Olah and Kuhn [91] reported over 90% yield of isomeric isopropylated methylbenzenes by allowing the intermediate toluene-, *m*-xylene-, or mesitylene-isopropyltetrahaloborate complex to warm up to the decomposition temperature.

In another development, Olah and Kuhn [197] found that arenes could also be alkylated with mono- and dihaloalkanes in the presence of BCl_3, BBr_3, and BI_3 as the catalyst. The results showed that the order of reactivity of the boron trihalide catalysts is $BI_3 > BBr_3 > BCl_3 > BF_3$, whereas the reactivity of the carbon-halogen bonds is C-F > C-Cl > C-Br > C-I.

With the aim of explaining the mechanism of BF_3-catalyzed Friedel-Crafts alkylation, Nakane et al. [101] investigated the competitive isopropylations of benzene and toluene with isopropyl fluoride in various solvents at different temperatures. The results showed that (1) the isopropylations were fast and were accelerated by a trace of water, (2) the relative rates of toluene to benzene were always smaller than 1 (ca. 0.70) in nonpolar solvents such as hexane and cyclohexane but greater than 1 (2.18) in the polar solvent nitromethane, and (3) the isopropylations of benzene-d_6 and toluene-d_6 at 25°C experienced a small secondary inverse kinetic isotope effect. These findings, which are consistent with more recent ones obtained for the BF_3-catalyzed benzylation [25] and ethylation [102], led to the following suggestions:

1. The alkylation catalyst is a hydrate of BF_3.
2. The electrophile is an isopropyl carbocation or its equivalent.
3. The transition state in the rate-determining step in the substitution of toluene closely resembles the termolecular-oriented π-complex as suggested by Olah et al. [181].
4. With toluene, the oriented π-complex transition state would be quinoid in structure and composed of the separate transition states corresponding to the *ortho*, *meta*, and *para* positions; the *meta* position would represent the highest energy barrier because the *ortho* and *para* counterparts would be stabilized by resonance.
5. The difference in rates between benzene and toluene is due primarily to the difference in the entropies of activation.
6. The electrophilic character of electrophiles increases as follows: $\overset{\delta+}{C_2H_5F}\cdot\overset{\delta-}{BF_3} < \overset{\delta+}{i\text{-}C_3H_7F}\cdot\overset{\delta-}{BF_3} < i\text{-}C_3H_7^+ < C_2H_5^+$.
7. The low substrate but high positional selectivity is a characteristic of aromatic substitutions involving strong electrophiles.

The mechanism in scheme 16 was suggested to account for isopropylations catalyzed by water-activated BF_3 in nonpolar solvents.

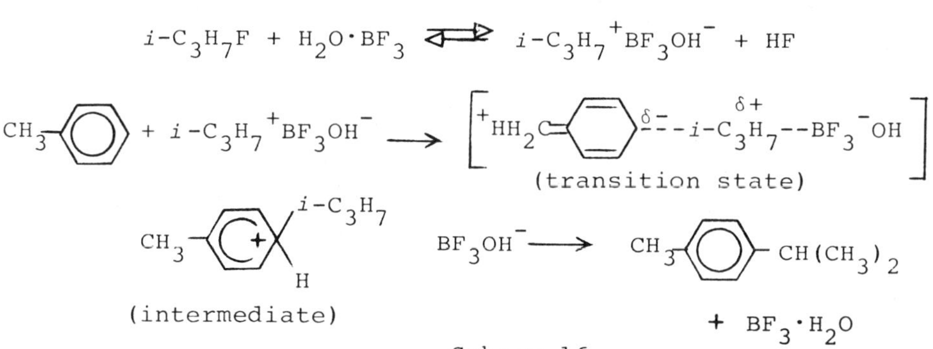

$$i\text{-}C_3H_7F + H_2O\cdot BF_3 \rightleftharpoons i\text{-}C_3H_7^+\,BF_3OH^- + HF$$

$$CH_3C_6H_5 + i\text{-}C_3H_7^+\,BF_3OH^- \longrightarrow \left[\overset{+}{H}H_2C{=}C_6H_5\overset{\delta-}{\cdots}\overset{\delta+}{i\text{-}C_3H_7}\cdots BF_3^-OH\right]$$

(transition state)

$$CH_3C_6H_5^+(H)(i\text{-}C_3H_7) \quad BF_3OH^- \longrightarrow CH_3C_6H_4CH(CH_3)_2 + BF_3\cdot H_2O$$

(intermediate)

Scheme 16

Apart from alkylations under conventional conditions, the alkylation of arenes with isopropyl and other alkyl halides was conducted in the gas phase over a perfluorinated resinsulfonic acid catalyst (Nafion-H) [151]. Compared to other conditions, these reactions gave lesser amounts of di- and trialkylation products, which were comprised mainly of the 1,3 or 1,3,5 isomers, respectively, indicating predominant thermodynamic control. In the isopropylation of benzene with isopropyl chloride, for example, with a molar ratio of benzene/ isopropyl chloride of 5:2, only 10 to 12% of the alkylated products at 150 to 196°C were diisopropylbenzenes.

Notably, the results with Nafion-H catalyst [151] deomonstrated that the alkylation rate is controlled by the formation of the reactive intermediate, suggested to be a protonated alkyl halide (Eq. 48), rather than by σ-complex

$$R'\text{-}\overset{R''}{\underset{H}{C}}\text{-}X \underset{}{\overset{H^+}{\rightleftharpoons}} R'\text{-}\overset{R''\,+}{\underset{H}{C}}\text{-}X\text{-}H \overset{-HX}{\rightleftharpoons} R'\text{-}\overset{R''}{\underset{H}{C}}+ \tag{48}$$

formation between the electrophile and the aromatic compound. This is because various substrates showed only small differences in reactivity. As with other catalysts, however, the results with Nafion-H showed that the order found of alkylation ability of alkyl halides is RF > RCl > RBr and secondary > primary; tertiary reactions are very susceptible to elimination and polymerization under the reaction conditions.

With this review of isopropylation coming to an end and with the results of Brown's [6, 48, 50], Olah's [181], Roberts' [148], and Nakane's [101] groups in mind, let us sum up their views on the isopropylation mechanism. The kinetic results of Brown and co-workers [48,50] revealed two significant points. First, the kinetic order of the isopropylation of arenes with isopropyl bromide and gallium bromide followed pseudo-zero-order kinetics, identical with the order observed for the corresponding reactions of methyl and ethyl bromides. Second, the rate of isopropylation was much faster than the rates of methylation and ethylation under similar conditions. At 25°C, the relative rates of reaction of methyl, ethyl, and isopropyl bromides with toluenes were 1.00, 13.7, and 20,000, respectively. This large increase in the rate of isopropyl bromide suggested the possibility of a change to a predominantly ionization mechanism. These facts combined with others led to the conclusion that the isopropylation mechanism is best described in terms of nucleophilic attack by the aromatic on a "strongly polarized" alkyl bromide-aluminum bromide addition complex [48]. As to intermediates, Brown et al. [48,49] first believed that a high-energy localized π complex is involved as a common intermediate in alkylation and in the accompanying isomerization and disproportionation reactions [48,49]. Later, however, Stock and Brown [53] expressed another view, concluding that the transition state for these reactions generally resembles a σ complex and that "a more general mechanism provides for the involvement of π-complexes which are usually low-energy intermediates." Their reason was that a high-reactivity but low-selectivity Friedel-Crafts alkylation reaction involving a σ-complex type of transition state sufficiently well explained the reaction in accordance with Brown's selectivity relationships [14,53]. As expressed by Stock and Brown [53], this change in view about the π- and σ-complex involvement was largely due to the repeated isolation and characterization of benzenonium ion intermediates by many investigators [53,183-185].

In subsequent years, the aforementioned view of Brown and his group prevoked numerous criticisms and inspired a great deal of discussion [18, 101,181,185,198,199]. Major criticisms came from Olah [185], who emphasized that the existence of σ complexes as stable intermediates "does not necessarily prove that the transition states are always closely related to them".

Being aware of this, and of all the essential findings presented earlier in this part, Olah suggested that the transition states of electrophilic aromatic substitutions are not rigidly fixed, always resembling the σ intermediates as later believed by Brown's group, but frequently represent a much earlier state on the reaction coordinate resembling starting aromatics [198]. Accordingly, to account for a low substrate, but high positional selectivity in aromatic substitutions, Olah [184] stated that "activated states corresponding both to π- and σ-complex nature play an important role, the first being, however, dominant in the rate determining step, whereas the latter effects the isomer distribution". Generally, the greater the reactivity of an electrophilic reagent, the smaller its selectivity [48,49,53,101,148,181,185] and the closer the activated state is to an oriented π complex in the rate-determining step [181].

Although the concept of an oriented or localized π complex represents a considerable modification of Dewar's original proposal [200,201], it is currently favored by many chemists, including Melander [202], Brown et al. [48,49], Olah et al. [181,203], Roberts and Shiengthong [148], Zollinger et al. [204] and Nakane et al. [93,94,96,101]. According to this oriented π-complex concept, there exists a greater probability for the electrophile to be localized in the vicinity of the higher electron density positions in the ring.

In shaping their views on the isopropylation mechanism, Olah et al. [181] pointed out that a typical aromatic substitution reaction involving a strong electrophile such as the incipient carbocations [181,182,185,205], halonium ions [206], and nitronium ion [207] can be represented by the potential energy diagram shown in the accompanying figure. In this diagram, T_1 represents the transition state corresponding to a π complex and T_2 corresponds to the transition state of a σ-complex nature.

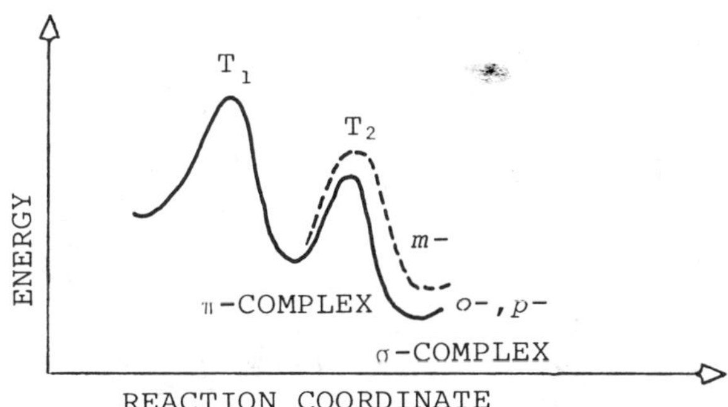

In line with the arguments above, Olah et al. [181] suggested the mechanism of scheme 17 for isopropylation and related alkylation reactions. In this mechanism it was emphasized that the driving force for the alkyl shift to the *m*-position is of thermodynamic rather than kinetic nature.

A mechanism essentially similar to that of Olah's was adopted by Roberts and Shiengthong [148] to explain the results of isopropylation of mesitylene with isopropyl chloride and $AlCl_3$ and also by Nakane et al. [101] (scheme 16) to explain the results of isopropylation of benzene and toluene with isopropyl fluoride and BF_3 in nonpolar solvents.

$$CH_3C_6H_5 + R^+ \rightleftarrows CH_3C_6H_5R^+$$

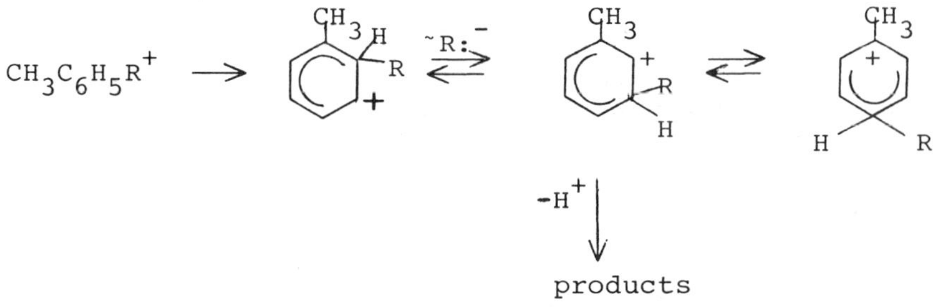

$$CH_3C_6H_5R^+ \rightarrow$$

$\sim R:^-$

$-H^+$

products

Scheme 17

In relation to isopropylation by isopropyl halides, the alkylation of benzene with optically active R-(-)-2-chloro-1-phenylpropane (35) and the isomeric R-(+)-1-chloro-2-phenylpropane (36) was recently reported [208]. In the presence of $FeCl_3$ or $AlCl_3$ below room temperature, both halides alkylated benzene to give similar mixtures of (-)-1,2-diphenylpropane (37), 1,1-diphenylpropane (38), and polymer. Since the reactions with both chlorides proceeded with almost complete retention of configuration, it was suggested that the reactions take place through an intermediate of the phenonium ion type (39) as shown in scheme 18. Alkylation of *p*-xylene with 35 in the pre-

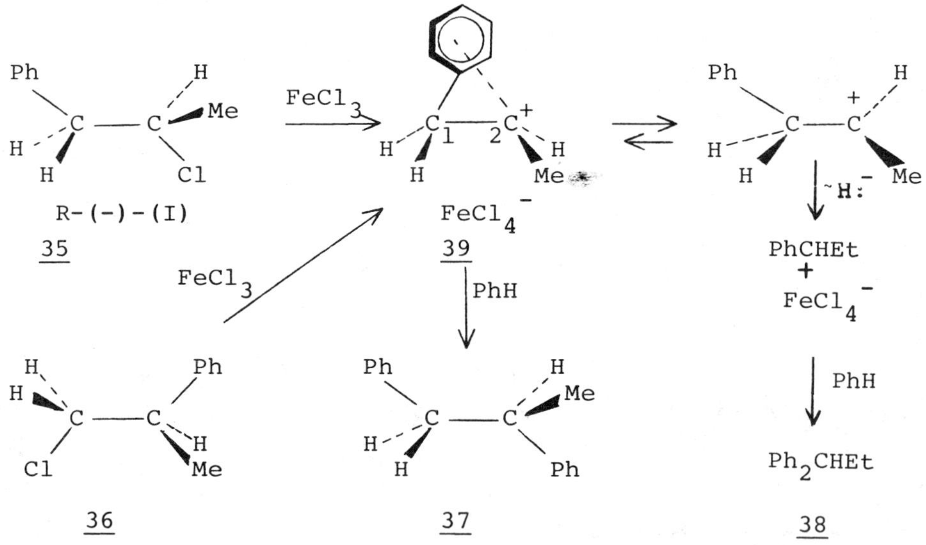

R-(-)-(I)
35

$FeCl_3$

$FeCl_4^-$
39

PhH

$\sim H:^-$

PhCHEt + $FeCl_4^-$

PhH

Ph_2CHEt

36

37

38

Scheme 18

sence of $FeCl_3$ at room temperature gave a mixture of 1-phenyl-2-(2,5-dimethylphenyl)propane (40), 1-phenyl-1-(2,5-dimethylphenyl)propane (41), and polymer (Eq. 49). The fact that none of the 1-(2,5-dimethylphenyl)-2-phenylpropane (42) expected from attack at C-1 in intermediate 39 was obtained suggested that the C-2 atom has such a strong carbocation character that each electrophile attacks it exclusively.

$$\text{p-xylene} + \underset{\underline{35}}{\text{Ph-CH}_2\text{CH(Cl)CH}_3} \xrightarrow[\text{room temp}]{\text{FeCl}_3} \underset{\underline{40}}{\text{PhCH}_2\text{CH(Ar)CH}_3} + \underset{\underline{41}}{\text{ArCH(Ph)CH}_2\text{CH}_3} + \text{polymer} \quad \left(\underset{\underline{42}}{\text{ArCH(Ph)CHCH}_3}\right) \qquad (49)$$

B. sec-Butyl Halides

The early reports on alkylations of arenes with *sec*-butyl halides are few and contradictory. In 1900, one of the first alkylations by a *sec*-butyl halide was carried out by Estreicher [153], who reported that the reaction of benzene with *sec*-butyl chloride and aluminum amalgam at 0°C produced *sec*-butylbenzene. Diuguid [142], in 1941, reported that the same reactants at 25°C gave *t*-butylbenzene as the principal product. A similar claim of extensive rearrangement was also made somewhat earlier by Gilman and Calloway [139], who reported that the alkylation of methyl 2-furoate by *sec*-butyl chloride and aluminum chloride in CS_2 at 0°C gave a very poor yield of methyl 5-*t*-butyl-2-furoate. The results from these early experiments with *sec*-butyl halides must be considered as of questionable reliability, owing to the primitive analytical techniques available in these times.

The newer work on the alkylation of arenes with *sec*-butyl halides has revealed that (1) as in the reactions with isopropyl halides, the alkylations are fast and accompanied by reorientation, disproportionation, polyalkylation, and hydride transfer reactions; and (2) as a consequence of the fact that *sec*-butyl halides have one more carbon than isopropyl, there exists the possibility that these alkylations may also be accompanied by two new types of rearrangement reactions, a degenerate one resulting in equilibration of the two secondary carbons (which can be detected only by the use of isotopes) and a skeletal one resulting in the branching of the side chain. Since the reactions in (1) have been discussed in detail in connection with isopropylation, we intend here to concentrate on the accompanying rearrangements of the side chain.

Utilizing modern analytical techniques, Roberts and Shiengthong [147] conducted a reinvestigation of the alkylation of benzene with *sec*-butyl chloride in the presence of $AlCl_3$ or aluminum amalgam catalyst. In this investigation the effects of reaction variables such as time, temperature, and catalyst ratio on product composition were extensively explored. With benzene, *sec*-butyl chloride and aluminum chloride in a mole ratio of 2.5:0.5:0.045, the results showed that (1) reaction at 0°C for 6 hr or at 30°C for 2.5 hr gave pure *sec*-butylbenzene; (2) reaction at 80°C for 1 to 2.5 hr gave a product composed of a 2:1 mixture of isobutylbenzene and *sec*-butylbenzene (Eq. 50); doubling the amount of $AlCl_3$ did not alter the composition of the

$$\text{PhH} + \text{CH}_3\text{CH(Cl)CH}_2\text{CH}_3 \xrightarrow{\text{AlCl}_3} \begin{cases} \xrightarrow[\text{70\% yield}]{\text{0°, 6 hrs or 30°, 2.5 hr}} \text{CH}_3\text{CH(Ph)CH}_2\text{CH}_3 \\ \xrightarrow[\text{68\% yield}]{\text{80°, 1 or 2.5 hr}} \underset{35\%}{\text{CH}_3\text{CH(Ph)CH}_2\text{CH}_3} + \underset{65\%}{\text{PhCHCH(CH}_3\text{)CH}_3} \end{cases} \qquad (50)$$

butylbenzene mixture but did decrease the yield; and (3) reaction using aluminum amalgam as catalyst at 25°C, under conditions duplicating those of Diuguid [142], resulted in 78% yield of a product consisting of 92% *sec*-butylbenzene and 8% isobutylbenzene, with no trace of *t*-butylbenzene, which was the major product reported by Diuguid.

Both side-chain rearrangement and hydride transfer products were also observed, but at lower temperature, during the alkylation of acetophenone with *n*-, *sec*-, and isobutyl chloride in superacid media [79a]. The results are summarized below Eq. (51). The formation of a mixture of *sec*- and iso-

$$\text{PhCOCH}_3 + \text{R-Cl} \xrightarrow[-20°,\,5\,\text{min}]{\text{HF-SbF}_5} (\text{products}) + (\text{products}) + (\text{products}) + (\text{products}) + \text{Bu-C}_6\text{H}_4\text{COCH}_3 \quad (51)$$

RCl	Yield,%	Product Distribution,%				
n-BuCl	58	15	3	55	7	20
s-BuCl	58	17	some	57	7	19
i-BuCl	5	21	9	67	3	-

butylarenes in the reactions above was attributed to isomerization of the initially formed *sec*-butylarene to isobutylarene by a process apparently not related to alkylation, which is discussed in Chap. 8.

Apart from studies of accompanying side-chain rearrangements, the alkylation of benzene with *sec*-butyl chloride was also used to compare the catalytic activity of various Lewis acids in alkylation reactions [209]. Using the percent conversion as a measure of catalyst activity, the following relative order of catalyst activity has been established: $AlBr_3$, $AlCl_3$, $MoCl_5$, SbF_5 > $TaCl_5$ > $NbCl_5$ > $FeCl_3$ > $FeCl_3{:}CH_3NO_2$(10:1 mol) ≈ $AlCl_3{:}CH_3NO_2$ (1.3:1 mol) > $ZrCl_4$ > $AlCl_3{:}CH_3NO_2$ (10:1 mol) > $AlCl_3{:}PhNO_2$ (10:1 mol) > $TiCl_4$ ≫ WCl_6. According to this method, metal halides such as Cu_2Cl_2, $ZnCl_2$, $CdCl_2$, $HgCl_2$, $SnCl_4$, $BiCl_3$, VCl_3, $NiCl_2$, $CoCl_2$, and $PdCl_2$ were classified as inactive catalysts.

The alkylation of phenol [210], anisole [210], and cresols [192] with *n* and *sec*-butyl halide and aluminum chloride was investigated by Dewar and Puttnam in connection with the study of the mechanism of acid-catalyzed rearrangements of alkyl aryl ethers. These alkylations were carried out at ambient temperature with the aromatic, alkyl halide, and aluminum chloride used in equimolar amounts. The results from phenol, anisole, *o*-cresol, and *p*-cresol, as determined by IR spectroscopic analysis, are summarized in Eqs. (52) to (55), respectively. It is to be noted that product compositions are expressed in mole percent yields, that values between parentheses are those from *n*-butyl alkylations, and that owing to incomplete conversion, the recovered starting aromatics account for the rest of the yield. Some of the important characteristics of the results presented in these equations should be emphasized: (1) the alkylation of phenol by *n*- and *sec*-butyl chloride gave product mixtures of almost identical composition; (2) in the case of *p*-cresol (Eq. 55) some demethylation has taken place; and (3) in contrast to

OH + Cl or (Cl) $\xrightarrow[\text{36 or (60) hr}]{AlCl_3}$ OH OH OH OH (52)

8.7% (8.7%) 2.0% (1.9%) 42.3% (43.4%) 13.3% (13.2%)

OCH_3 + Cl or (Cl) $\xrightarrow{AlCl_3}$ OH OCH_3 OH OCH_3 (53)

4.9% (3.3%) 18.5% (58.7%) 40.4% (12.4%) 13.6% (1.8%)

OH CH_3 + Cl $\xrightarrow{AlCl_3}$ OH CH_3 + OH CH_3 + OH CH_3 + OH CH_3 + OH CH_3 (54)

OH CH_3 + Cl $\xrightarrow{AlCl_3}$ OH CH_3 + OH CH_3 (55)

23.7% 6.0%

+ OH + OH CH_3 + OH CH_3

17.3% trace trace

benzene [147], no n-alkylaromatics were detected in alkylations of phenol or anisole with n-butyl chloride. Whereas the first two points are predictable and can be accounted for in terms of accompanying reorientations and disproportionations, the third point is rather unexpected.

In their work, Dewar and Puttnam [210] also examined the alkylation of phenol with *sec*-butyl bromide and $AlBr_3$; after 24 hr at 5 to 10°C, 95% of the phenol was recovered unchanged, together with a small amount (4%) of mixed butylphenols. This demonstrated the slowness of bromides relative to chlorides in Friedel-Crafts alkylations.

Friedel-Crafts alkylations by *sec*-butyl halides were also extended to the naphthalene series. In 1941, Diuguid [142] claimed pure α-*sec*-butylnaphthalene to be the product from an alkylation of naphthalene with *sec*-butyl chloride in the presence of amalgamated aluminum catalyst in chlorobenzene solvent at room temperature for 18 hr. Duplication of this reaction by Roberts and Zee-Cheng in 1963 [193] showed the product to be a mixture of 4% α- and 96% β- *sec*-butylnaphthalene.

The latter authors also studied the alkylation of naphthalene (12 mol) by *sec*-butyl chloride (6 mol) and aluminum chloride (1 mol) in CS_2 under reaction conditions ranging from treatment for 30 min at 0°C to 2 hr at 80°C. Reaction at 0°C for 30 min gave in 50% yield a mixture of α-*sec*-butylnaphthalene and β-*sec*-butylnaphthalene in a percent ratio of 60:40. Reaction at 0°C for 2 or more hours gave 58% yield with an α/β isomer ratio of 4:96 (Eq. 56). This

Cl

(56)

Reaction Conditions	Distribution, %	
Al(Hg),25°,18 hrs	4	96
$AlCl_3$,0°,0.5 hr	60	40
$AlCl_3$,0°,2 hrs	4	96

ratio, which was repeatedly obtained in alkylations carried out at 25°C or higher, was shown by the authors to represent the thermodynamically controlled distribution value for these two isomers. These results suggested that in the aluminum chloride-catalyzed alkylation of naphthalene by *sec*-butyl halide, the reaction conditions could be adapted to yield mainly the kinetically controlled α isomer or the thermodynamically controlled β isomer. Interestingly, the results of the alkylations of naphthalene by *sec*-butyl chloride and aluminum chloride indicated that no detectable amounts of isobutylnaphthalene were found under any of the conditions applied. This is to be contrasted with the corresponding alkylations of benzene, in which isobutylbenzene became the major final product in alkylations carried out at 80°C.

Several studies of the stereochemistry of alkylation have been made in which optically active *sec*-butyl derivatives have been chosen, because they are the simplest alkylating agents with the required chirality. In 1940, Price and Lund reported that alkylation of benzene by optically active *sec*-butyl alcohol at -5 to +25°C with BF_3 catalyst gave *sec*-butylbenzene with a low but authentic optical rotation [211(a)]. With $AlCl_3$ catalyst, completely racemic *sec*-butylbenzene was obtained. Two years later, Burwell and Archer found that reactions of this alcohol catalyzed by HF, H_2SO_4, and H_3PO_4 also gave products with small (0.3 to 0.7%) but definite maintenance of configuration

[211(b)]. More recently, Spanninger and von Rosenberg reported that with $AlBr_3$ catalyst at 0 to 5°C, alkylation occurred with 27% inversion [211(c)]. Most recently, Suga and co-workers studied the $AlCl_3$-catalyzed alkylation of benzene with optically active *sec*-butyl chloride [211(d)]. At -30°C with a reaction time of 50 sec, an optical yield of 24% (inversion) was obtained. With longer reaction periods the optical yield decreased to 14% (100 sec) and 0% (20 min). The decrease in the optical activity of the product with time was attritubted to racemization of both the starting material and the product. Suga has demonstrated that greater sterospecificity can occur in alkylations with difunctional alkylating agents such as chloro alcohols, acids, and esters (see Chap. 5).

C. Pentyl and Higher Secondary Halides

Apparently, the earliest alkylation of an arene with a linear *sec*-pentyl halide was that described by Hennion and Kurtz [196] in 1943. Unaware of accompanying rearrangements, these authors claimed that the reaction of toluene with 3-bromopentane in the presence of BF_3 catalyst and alcohol, water, or water-sulfuric acid as cocatalyst yielded a single monoalkylation product, 3-*p*-tolylpentane. This claim has always been questionable on the ground that similar alkylations of arenes with other secondary pentyl derivatives such as 2-methoxypentane [212] and 2- and 3-pentanols [149,163,213,214] were shown to give mixtures of 2- and 3-aryl pentane. The real challenge, however, came almost 39 years later when Roberts et al. [215] published their study on the alkylations and competing rearrangements in the aluminum chloride-catalyzed reactions of secondary alkyl chlorides with arenes. In this study an extensive investigation of the various factors affecting the alkylation of benzene and *p*-xylene with 2- and 3-chloropentane and 2- and 3-chlorohexane in the presence of $AlCl_3$ or the moderated $AlCl_3$-CH_3NO_2 catalyst was carried out. The most important findings of this investigation may be condensed to the following points:

1. In contradiction to the earlier claim of Hennion and Kurtz [196], rearrangements were prevalent, and regardless of the halide, arene, catalyst, temperature, and/or solvent used, the products were invariably mixtures of the corresponding 2- and 3-arylalkane isomers. None of the primary 1-arylalkane isomer could be found.
2. The alkylations were accompanied by simultaneous isomerization of the chloroalkanes and rearrangement of the product arylalkanes. The isomerization mainly determines the product distribution of 2- and 3-arylalkanes from 2- and 3-chloroalkane and $AlCl_3$ at -20°C or $AlCl_3$-CH_3NO_2 at 25°C.
3. In reactions catalyzed by $AlCl_3$ at 25°C, rearrangement of the arylalkanes to a constant thermodynamic equilibrium composition is the dominant product-determining factor. The equilibrium distributions produced from alkylations at 25°C are shown in Table 4.
4. The 2-/3-arylalkanes ratios in the initially formed alkylates showed dependence not only on the conditions, but also on the nucleophilicity of the arene and the position of the chlorine on the starting chloropentane. The following two examples are illustrative. First, alkylations with 2-chloropentane and $AlCl_3$ at -20°C gave a 2-/3-arylpentane ratio of 73:27 from benzene, but a 67:33 ratio from *p*-xylene. Second, alkylations at -20°C, in which the $AlCl_3$ catalyst was added to an arene-alkyl chloride mixture, gave quite different results from 2- and 3-chloroalkane; benzene at -20°C gave a 64:36 distribution of 2-/3-phenylhexane starting with 2-chlorohexane, but a 57:43 distribution

Table 4. Equilibrium Compositions of Products from $AlCl_3$-Catalyzed Alkylations of Benzene and *p*-Xylene with 2- and 3- Chloropentane and 2- and 3-Chlorohexane at 25°C

	2-Arylalkane	3-Arylalkane
Benzene + 2- or 3-Chloropentane	73	27
p-Xylene + 2- or 3-Chloropentane	75	25
Benzene + 2- or 3-Chlorohexane	63	37
p-Xylene + 2- or 3-Chlorohexane	66	34

starting with 3-chlorohexane. This and other similar observations were attributed to differences in rates of alkylation by 2- and 3-chloroalkanes (cf. point 5).

5. Rough kinetic analysis of alkylations of benzene and *p*-xylene by 2- and 3-chlorohexane indicated that (a) the rates of isomerization and alkylation of the chloroalkanes are of the same order of magnitude, and (b) the rate of alkylation of 2-chlorohexane is about two to three times the rate of alkylation by 3-chlorohexane.

6. The effect of moderating the $AlCl_3$ catalyst by addition of CH_3NO_2 is apparently to exclude rearrangement of the arylhexane products and to reduce the rates of alkylation by the chlorohexanes more than it reduces their rate of isomerization. This is illustrated by the fact that with $AlCl_3$-CH_3NO_2 catalyst, in contrast to alkylations of benzene with $AlCl_3$ alone at -20°C, the products were 2- and 3-phenylpentane in a 74:26 ratio and 2- and 3-phenylhexane in a 66 ± 1:34 ± 1 ratio, regardless of whether 2- or 3-chloroalkane was used and of whether or not the chloride was subjected to equilibration prior to reaction with the arene.

7. Both the starting 2- and 3-chloroalkanes and the resulting 2- and 3-arylalkanes involved in the study have all been shown to undergo facile equilibration in the presence of $AlCl_3$. The equilibrium compositions for the arylalkanes are similar to those presented in Table 4. For the chlorides, both 2- and 3-chloropentane yielded a 64:36 equilibrium distribution of 2-/3-chloropentane and both 2- and 3-chlorohexane yielded a 52:48 ratio of 2-/3-chlorohexane.

The alkylation of benzene by 2- and 4-bromooctane in the presence of $AlCl_3$ at -15°C was carefully studied by Sharman in 1962 [82]. In essence, the results obtained paralleled those of Roberts et al. [215] for alkylations with secondary pentyl and hexyl chlorides, especially with regard to accompanying rearrangement and noticeable dependence of initial product composition on the position of bromine in the chain. For example, after 7 min, 2-bromooctane gave a 50.3:29.0:20.7 distribution of 2-/3-/4-phenyloctane, but 4-bromooctane gave a 44.1:29.4:26.5 distribution of the same isomers. After 30 min of reaction, however, the mixtures from both chlorides reach a constant 2-/3-/4-phenyloctane equilibrium ratio of 48:29:23. Comparing the initial ratios with the equilibrium ratio reveals that the product from 2-bromooctane was higher in 2-phenyloctane, but lower in 4-phenyloctane, than the equilibrium composition. The exact opposite is found with 4-bromooctane.

These results are again probably due to differences in rates of alkylations with the two bromides, and to competition between rearrangement and alkylation.

The alkylation of benzene by decyl, dodecyl, and hexadecyl chlorides was similarly investigated by Mazonski and Hopfinger [216] in the presence of aluminum chloride catalyst. They found that an optimal yield of alkylbenzene was obtained by using a molar ratio of $AlCl_3$ to alkyl chloride of 0.1:1 and by carrying out the reaction at 70°C for 2.5 hr. They also found that the yield was independent of both the length of the alkyl chain and the position of the halogen on the alkyl chain, and that excess benzene promoted the formation of the monoalkylated benzene. As to side reactions, they pointed out that the most important one was hydrogenation of the alkyl chlorides to alkanes. Such alkane formation was also reported by other workers [82,217, 218].

In addition to the reports noted above, several others on the alkylation of arenes with linear secondary derivatives have appeared in which the emphasis was on industrial applicability [219-223]. Nevertheless, they confirmed the generality of rearrangements in these reactions and the multiplicity of conditions under which they can be achieved.

In conclusion, it is to be emphasized that alkylations of arenes with linear secondary halides lead invariably to product mixtures in which all the possible secondary alkylarenes are present. The distribution of these isomers is somewhat dependent on conditions and on the position of halogen in the alkyl chain, but with a strong catalyst like $AlCl_3$ at room temperature or higher, a constant equilibrium distribution of the isomers is reached in a few minutes.

IV. ALKYLATIONS WITH BRANCHED ALKYL HALIDES

A. Butyl Halides

As in the case of isopropylations, there is unanimous agreement among workers that Friedel-Crafts alkylations by *t*-butyl halides proceed without alkyl-group rearrangement to yield mono- and higher-*t*-butylated derivatives. Because of this unanimity, the Friedel-Crafts alkylations of aromatics by *t*-butyl derivatives have been termed *t*-butylation reactions. In alkylations by isobutyl halides, however, most reactions were described as proceeding with accompanying rearrangement to give *t*-butyl derivatives. A few early cases, however, were reported otherwise. For example, the alkylation of benzene with isobutyl chloride and $AlCl_3$ was said to yield nonrearranged isobutylbenzene [224,225]. These claims were disproved by other investigators [132,148,152,226]. According to the most recent work of Roberts and Shiengthong [148], the $AlCl_3$-catalyzed alkylation of benzene with isobutyl chloride at temperatures ranging from -18 to 80°C gave only *t*-butylbenzene, with no evidence of even traces of other isomers indicated by infrared analysis.

The early work on alkylations of arenes by *t*-butyl and isobutyl halides has been compiled and reviewed by others [2,168,185,225(a)]. This work was mainly directed toward evaluation of the synthetic potential of these reactions. More recently, studies have focused on the kinetic aspects, as well as the positional and substrate selectivities of the *t*-butylation reaction [6,12,48,51a, 73,79a,90,151,180,181,182,187-190,197,205,227,228]. The results obtained, although similar in many respects to those obtained from isopropylation, illustrate clearly the pronounced role of steric interactions in determining the course of *t*-butylations. In fact, in the case of *t*-butylation, steric hindrance prevents the formation of *ortho* isomers.* Together, the size and the stability of the *t*-butyl cation govern both positional and substrate selectivities in *t*-butylation reactions.

The kinetic work of Brown and Junk [48] demonstrated that *t*-butylation is extraordinarily fast (essentially complete within 0.005 sec) and is usually accompanied by considerable disproportionation (transalkylation) and reorientation. With toluene, the relative rates of the gallium bromide-catalyzed reaction of methyl, ethyl, *n*-propyl, isopropy, and *t*-butyl bromides were reported to be 1.00, 13.7, 15.9, 3×10^5, and 8×10^5, respectively [5]. (As mentioned earlier, the rate value for isopropyl bromide was later found to be considerably lower [50]). By running the alkylation reactions with a contact time of 0.005 sec it was possible to eliminate the disproportionation reaction in the ethylation and isopropylation of arenes, but not in the *t*-butylation reaction [48]. Whereas disproportionation accounts for the large amounts of polyalkylated products, the reorientations explain the high proportions of *meta* isomers. In the *t*-butylation of toluene with *t*-butyl bromide and $AlBr_3$, the product obtained after a short contact time (0.005 sec) contained 25% of poly-*t*-butylated derivatives, and the distribution of isomers in the monoalkylate amounted to 69.5% *m*- and 30.5% *p*-*t*-butyltoluene, with no *ortho* isomer present. Also, the *t*-butylation of *t*-butylbenzene [75] or *p*-di-*t*-butylbenzene [229] with *t*-butyl chloride and aluminum chloride at temperatures below 5°C resulted in the formation of 1,3,5-tri-*t*-butylbenzene. The process involved concurrent alkylation and reorientation reactions (Eq. 57) (see Chap. 7).

$$\text{PhC(CH}_3)_3 \xrightarrow[\text{AlCl}_3,\ -10^\circ]{(CH_3)_3CCl} p\text{-C}_6\text{H}_4[\text{C(CH}_3)_3]_2 \xrightarrow{\text{AlCl}_3} m\text{-C}_6\text{H}_4[\text{C(CH}_3)_3]_2 \xrightarrow[\text{AlCl}_3]{(CH_3)_3CCl} 1,3,5\text{-C}_6\text{H}_3[\text{C(CH}_3)_3]_3 \qquad (57)$$

In general, it was concluded that in the alkylation of toluene, the particular catalysts and alkylating agents used influence the extent of isomerization [180,205,230]. Under isomerizing conditions, with equilibration practically complete, the isomer distribution of *t*-butyltoluenes was 35.6% *p* and 64.6 to 68% *m* isomer, with no *o*-*t*-butyltoluene formed [6,205,227]. This isomer distribution was established recently by GLPC through isomerization of *o*-, *m*-, and *p*-*t*-butyltoluene with water-promoted $AlCl_3$ or $AlCl_3$-CH_3NO_2 catalyst [231]. The equilibrium mixture starting from any of the isomers contained about 63 to 64% *m*- and 36 to 37% *p*-*t*-butyltoluene.

The isomer distribution under nonisomerizing conditions was also determined experimentally by some [205,227] and calculated from kinetic data by

---

**t*-Butyl-*p*-xylene has been prepared, however, by a *trans*alkylation reaction, as described in Chap. 7.

others [180]. From a series of alkylations of toluene with a variety of *t*-butylating agents and catalysts, Schlatter and Clark [227] obtained limiting distributions of 67% *m*- and 33% *p*-, and 7% *m*- and 93% *p*-*t*-butyltoluene, with several cases at each extreme. Their experiments showed that the former distribution corresponds to the thermodynamically controlled equilibrium value. The recurrent observation of the latter distribution strongly suggested that this too is a true limiting initial value corresponding to direct *meta*-*t*-butylation. To examine further the validity of their suggestion, Schlatter and Clark reinvestigated several *t*-butylations which had been reported to give only the *m* [232-234] or the *p* isomer [235-238] and found the products in all cases to be mixtures of *meta* and *para* isomers falling in the previously defined range. The same authors also studied the *t*-pentylations of toluene and *t*-butylations of ethylbenzene, but the results obtained were inadequate to establish the limiting compositons of the alkylates in these systems.

In a paper published in 1961, Allen and Yats [180] reviewed the then current status of simultaneous alkylation and isomerization of alkylbenzenes with various alkylating agents. Elaborating on Schlatter's results, they stated that all *t*-butylations of toluene could have been obtained by alkylations producing isomeric mixtures containing 7% *m*- and 93% *p*-*t*-butyltoluene and subsequent or concurrent isomerization to the equilibrium composition of 67% *m*- and 33% *p*-*t*-butyltoluene. The fact that the first distribution was produced by mild ("nonisomerizing") catalysts whereas the second was produced by strong ("isomerizing") catalysts was considered as supporing evidence for the authors' conclusion.

Further support for Schlatter's and Allen's findings was provided later by the work of Olah et al. [205], who studied the *t*-butylation of benzene and methylbenzenes in nitromethane solution, using various Lewis and protonic acid catalysts. The results of this work demonstrated the following: (1) In confirmation of Schlatter's and Allen's findings, the isomer distribution of *t*-butyltoluenes formed under nonisomerizing conditions was 5.7 to 7% *m* and 94.3 to 93% *p* isomer, (2) Competitive *t*-butylation of toluene and benzene under nonisomerizing conditions showed intermediate substrate selectivity (k_B/k_T = 14 to 16). (3) No *t*-butylation *ortho* to the methyl group was observed. (4) Larger amounts of *m*-*t*-butylbenzene isomer (46.3 to 64.8%) were produced when $AlCl_3$, $FeCl_3$, or $AgBF_4$ was used as catalyst. It was pointed out that the conjugate acids of these catalysts (formed by proton elimination from the arenes) must be sufficiently strong acids even in nitromethane solution to effect isomerization.

The *t*-butylation of monohalobenzenes with *t*-butyl and isobutyl halides was also investigated [2,182,239]. In the presence of strong catalysts, these reactions are complicated by accompanying disproportionation and reorientation of both the halogen atom [118,240-242] and the alkyl group (see Chap. 7). With respect to halogens, the tendency for such *inter*- and *intra*molecular transfer increases in the order I > Br > Cl > F [241,242]. As with monoalkylbenzenes, no substitution *ortho* to a halogen could be observed except for the smallest halogen, fluorine. This can be seen from Table 5, in which Olah et al.'s [182] data on the observed reactivities of the halobenzenes relative to benzene, together with the isomer distribution of the monoalkylated products, are summarized. As can be seen from the table, with the exception of bromobenzene, both catalysts resulted in similar distributions of products. This is probably a result of the nonisomerizing conditions under which the reactions were conducted.

Table 5. $AlCl_3 \cdot CH_3NO_2$ and $(SnCl_4)^a$ Catalyzed *t*-Butylation of Benzene and Halobenzenes with *t*-Butyl Bromide in Nitromethane at 25°C

Aromatic	$k_{AR}:k_{benzene}$	Isomer distribution, %		
		ortho	*meta*	*para*
Ph-H	1.00 (1.0)			
Ph-F	0.16 (0.12)	3.6 (3.6)	0.1 (0.1)	96.3 (96.3)
Ph-Cl	0.03 (0.07)		5.5 (5.0)	94.5 (95.0)
Ph-Br	0.02 (0.02)		3.0	97.0 (100)

[a]Values between parentheses are those from $SnCl_4$.

So far in our discussion we have been concerned mainly with alkylations of benzene or monosubstituted benzenes with isobutyl and *t*-butyl halides. In the following few paragraphs, we consider the alkylation of higher methylated benzenes with these two butyl derivatives.

The early interest in the preparation of *t*-butylxylene arose because of its practical importance in the perfume industry in the manufacture of xylene musk. Excellent yields of 5-*t*-butyl-*m*-xylene were reported from the alkylation of *m*-xylene with *t*-butyl chloride and $AlCl_3$ below 50°C [243] or with isobutylene and $AlCl_3$ in the presence of a little isobutyl chloride [244].

Besides interest in synthesis, there has been growing concern about how steric factors may retard or divert the course of the *t*-alkylation of hindered molecules. Friedman et al. [81] demonstrated that whereas *p*-xylene could be alkylated readily with straight-chain alkenes, alcohols, and alkyl halides, it always resisted alkylation by the corresponding branched-chain isomers. In alkylations by the latter, considerable steric hindrance must be overcome to accomplish introduction of a tertiary group *ortho* to a methyl group. This steric effect, which was suggested to be felt mainly during collapse of the intermediate σ complex to product [245,246], is powerful enough to enable separation of *o*- and *m*-xylene from *p*-xylene through *t*-butylation [247-249]. The retardation of *t*-butylation by steric factors has been expected to cause other reactions to occur. This has been verified by Friedman [81] and Schmerling [250] and their collaborators. They showed that the principal products from the reaction of *p*-xylene with *t*-butyl chloride and $AlCl_3$ were isobutane, toluene, *m*- and *p*-*t*-butyltoluene, 3,5-di-*t*-butyltoluene, 2-(*p*-methylbenzyl)-*p*-xylene (43), and di-*p*-xylylmethane (44). A combination of hydride transfer, alkyl transfer, and alkylation processes were proposed to account for the results (scheme 19).

The alkylation of *p*-xylene with *t*-butyl chloride was also studied in the presence of other catalysts. Whereas zirconium chloride [250] and $AlCl_3$-CH_3NO_2 [81] gave product mixtures very similar to those obtained with $AlCl_3$, $FeCl_3$ gave only a nonspecified butylxylene even at 78 to 86°C. The results with $AlCl_3$-CH_3NO_2 demonstrated the ability of even this mild catalyst to effect both methyl and hydride transfers in reactions with xylene. The failure of $FeCl_3$ to exert similar effects is probably due to a tighter complex formation with the *t*-butyl chloride. Steric hindrance resulted in similar processes

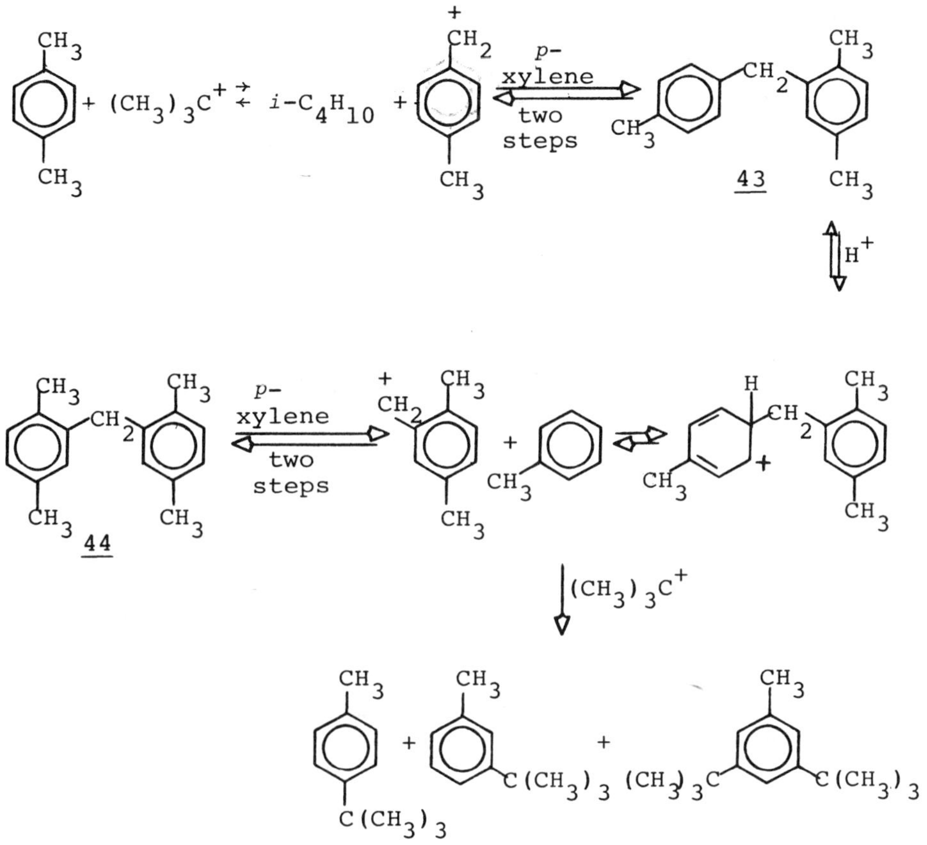

$$\text{p-xylene} + (CH_3)_3C^+ \rightleftarrows i\text{-}C_4H_{10} + \text{p-methylbenzyl cation} \xrightarrow[\text{two steps}]{p\text{-xylene}} \underline{43} \overset{H^+}{\rightleftarrows} \underline{44} \xrightarrow[\text{two steps}]{p\text{-xylene}} \ldots \xrightarrow{(CH_3)_3C^+} \ldots$$

Scheme 19

during the attempted alkylation of *p*-chlorotoluene with *t*-butyl chloride and $AlCl_3$ at ambient temperature [250]. The products, together with their reported yields, are shown in Eq. (58).

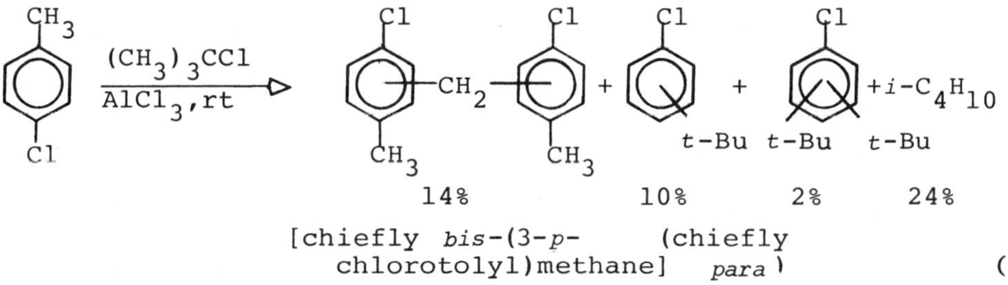

$$\text{p-chlorotoluene} \xrightarrow[AlCl_3,\,rt]{(CH_3)_3CCl} \text{bis(chlorotolyl)methane} + \text{t-Bu-chlorobenzene} + \text{di-t-Bu-chlorobenzene} + i\text{-}C_4H_{10} \quad (58)$$

14% [chiefly *bis*-(3-*p*-chlorotolyl)methane]; 10% (chiefly *para*); 2%; 24%

To explore further the abnormalities associated with sterically hindered *t*-butylations, Roberts and McGuire [246] conducted a study of the comparative alkylation of *p*-xylene with *sec*-, *i*-, and *t*-butyl chlorides. Their object was to test whether steric resistance to *t*-butylation might enforce primary or secondary butylation, as visualized in scheme 20. To ensure the elimina-

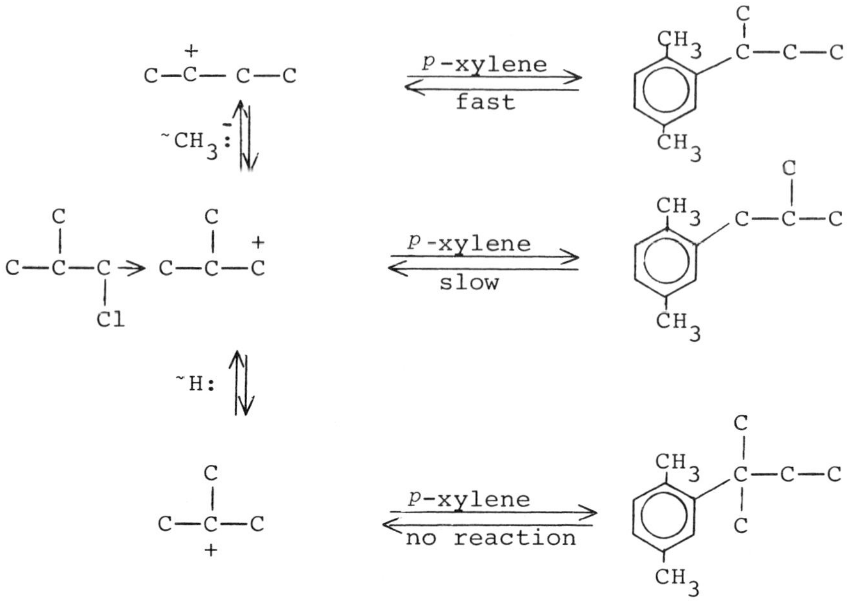

Scheme 20

tion of any accompanying methyl reorientation, the reactions were carried out in hexane solution at 0°C with small amounts of $AlCl_3$ catalyst. The products obtained, together with their present yields, are summarized in Eq. (59).

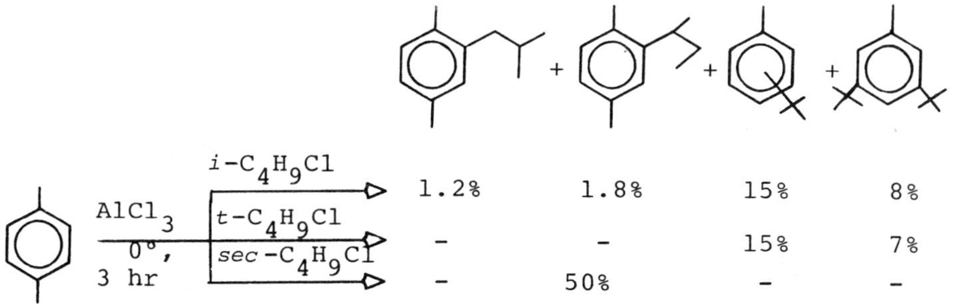

$AlCl_3$, 0°, 3 hr

i-C_4H_9Cl	1.2%	1.8%	15%	8%
t-C_4H_9Cl	-	-	15%	7%
sec-C_4H_9Cl	-	50%	-	-

(59)

Examination of the results in Eq. (59) reveals that the alkylations of p-xylene with isobutyl and t-butyl chloride gave very similar results, with one interesting exception. Alkylation with isobutyl chloride gave a small but appreciable yield of a mixture of isobutyl- and *sec*-butyl-p-xylene, whereas alkylation with t-butyl chloride gave neither of these products in detectable amounts. At this point, it should be recalled that neither isobutyl- nor *sec*-butylbenzene were found as products of the alkylation of benzene by either isobutyl or t-butyl chloride [147]. These results confirmed the expectation that the steric hindrance encountered by the bulky t-butyl cation would give the isobutyl and *sec*-butyl cations (which have much smaller steric requirements) an opportunity to alkylate the p-xylene ring. Another interesting observation was that the butyl-p-xylene mixtures isolated from alkylations

Table 6. Competitive $SnCl_4$-Catalyzed *t*-Butylation of Benzene and Methylbenzenes with *t*-Butyl Bromide in Nitromethane Solution at 25°C

Aromatic	$k_{Ar}:k_{PhH}$	*t*-Butylalkylbenzenes, %
Benzene	1.00	*t*-Butylbenzene (1.00)
Toluene	16.60	*m*- (6.4) and *p*-*t*-Butyltoluene (93.6)
o-Xylene	44.30	4-*t*-Butyl-*o*-xylene (100)
m-Xylene	2.54	5-*t*-Butyl-*m*-xylene (100)
p-Xylene	No *t*-butylation observed in competition with benzene	
Mesitylene		
1,2,4-Trimethylbenzene		
1,2,3-Trimethylbenzene	170.00	1,2,3-Trimethyl-5-*t*-butylbenzene (100)

with isobutyl chloride were found by NMR analysis to be composed of a nearly 2:1 ratio of *sec*-butyl- to isobutyl-*p*-xylene. This apparently represents direct competitive alkylation by the secondary and primary alkylating agents, since interconversion of *sec*-butyl and isobutyl-*p*-xylene did not occur under the conditions of alkylation. The latter fact was borne out by the finding that alkylation by *sec*-butyl chloride gave only *sec*-butyl-*p*-xylene.

The effect of methyl substitutents on the relative reactivity of mono-, di-, and trimethylbenzene in *t*-butylation reactions was studied by Olah et al. [205] under nonisomerizing conditions. The results obtained using *t*-butyl chloride and $SnCl_4$ are shown in table 6.

The isomer distributions were characterized by the fact that no *ortho* substitution relative to a methyl group was found. Ratewise, the *t*-butylation of *p*-xylene was only one-twentieth of that of the *t*-butylation of *o*-xylene. This is in good agreement with the observed 5 to 6% *m*-*t*-butyltoluene isomer formed in the *t*-butylation of toluene, thus giving a *meta/para* ratio of about 1:20 and providing strong evidence for direct, kinetically controlled alkylation in the position *meta* to the methyl groups.

As to *p*-xylene, mesitylene, and pseudocumene (1,2,4-trimethylbenzene), they were not *t*-butylated in competitive alkylation with benzene in nitromethane solution, since no positions other than those *ortho* to a methyl group are available.

Both *t*-butyl and isobutyl halides were used to alkylate phenols under Friedel-Crafts conditions. In the reaction of resorcinol with *t*-butyl chloride and ferric chloride, the initial product was the *t*-butyl ether, which was easily converted to *di*-*t*-butyl resorcinol [251]. When aluminum chloride was used as catalyst, however, *di*-*t*-butylresorcinol was obtained directly without isolation of the intermediate ether.

The alkylation of phenol with *n*-butyl-, isobutyl-, and isopentyl chlorides was carried out in the presence of equimolar amounts of aluminum chloride [159]. The best yields were obtained in the absence of solvents and by heating the reactants at 110°C for 4 to 6 hr. Under these conditions, both isomerization of the alkyl group and formation of alkyl phenyl ethers were observed.

Table 7. Some Early Literature Data on the *t*-Butylation of Naphthalene with *t*-Butyl and Isobutyl Halides

		Reaction Conditions					
Entry	Alkylating agent	Catalyst	Temp., °C	Time, hr	Yield, %	*t*-Butyl naphthalenes	Ref.
1	*t*-C_4H_9Cl	$AlCl_3$	50-60°	2	30	1- and 2-*t*-Butyl-	[254]
2	*t*-C_4H_9Cl[a]	HF	80°	-	76	*mono*-*t*-Butyl- (46%)+di-*t*-Butyl (30%)	[236]
3	*t*-C_4H_9Cl	$ZnCl_2$	95-105°	6	33	2-*t*-Butyl	[255]
4	*t*-C_4H_9Cl	$AlCl_3$	0°	6	45	2-*t*-Butyl	[256
5	*i*-C_4H_9Br	$ZnCl_2$	100°	24	24	2-*t*-Butyl	[255

[a]Carbon tetrachloride was used as solvent.

The *t*-butylation of phenols having free *o* and *p* positions relative to the hydroxyl group have been shown to give mixtures in which one or both of the two free positions were *t*-butylated [252]. The use of an inert solvent, in which the reaction mixture can be kept fluid at temperatures below 50°C, was reported to improve the yield of the most desired 4-*t*-alkyl phenolic derivative [253]. We shall see in Chap. 4 that, in general, the alkylation of phenols can best be achieved by the use of olefins and alcohols as alkylating agents, with a variety of acidic catalysts.

The *t*-butylation reaction has also been extended to the naphthalene series. Both *t*-butyl and isobutyl halides were used as alkylating agents. Some of the data reported between 1936 and 1943 are presented in Table 7.

More recently, Roberts and Zee-Cheng [193] conducted a thorough investigation of the $AlCl_3$-catalyzed alkylation of naphthalene by both *t*-butyl and isobutyl chloride. They conducted the alkylations in carbon disulfide or petroleum ether solvent under conditions ranging from 6 min at 0°C to 4 hr at 80°C. From an examination of their results, the following conclusions can be drawn:

1. Under all conditions employed, both *t*-butyl and isobutyl chlorides reacted with naphthalene to give exclusively 2-*t*-butylnaphthalene with no detectable amounts of the 1 isomer.
2. Maximum yields were obtained after 30 min at 0°C with *t*-butyl chloride (81%) and after 2 hr at 0°C with isobutyl chloride (59%). Longer reaction times and/or higher temperatures lowered the yield due to destruction of the product by dealkylation and disproportionation to naphthalene and di-*t*-butylated naphthalene. After 4 hr at 80°C, the yield dropped to 10%.
3. Lower yields of 2-*t*-butylnaphthalene were always obtained from isobutyl than from *t*-butyl chloride under identical conditions. This suggested a difference in the reaciton of the catalyst with isobutyl chloride and with *t*-butyl chloride to form the alkylating agent.
4. The exclusive formation of 2-*t*-butylnaphthalene observed was ascribed to steric retardation to attack in the 1-position. The bulky *t*-butyl complex, formed directly from *t*-butyl chloride or by rearrangement of isobutyl chloride, is impeded in its attack on the 1-position by the hydrogen in the 8-position.

Still more recently, Olah and Olah [51a] studied the $SnCl_4$-catalyzed competitive *t*-butylation of naphthalene and benzene with *t*-butyl bromide in nitromethane solution at 0 and 25°C. As with isopropylation, the substrate and positional selectivities showed wide variations with reaction conditions, primarily with reaciton time and temperature. For example, at 0°C, the 1-/2-*t*-butylnaphthalene percent ratio changed from 95:5 after 0.5 min to 7:93 after 60 min. Over the same period, the relative rates of alkylation of naphthalene and benzene (k_N/k_B) changed from 21.6 to 14. Reactions carried out at 25°C showed a change in the 1/2 isomer ratio from 13.5:85.5 after 0.25 min to 2.3:97.7 after 2 min. These data clearly show that under predominantly kinetic conditions 1-*t*-butylnaphthalene may be the major product. Moreover, the data also illustrate the ease with which such kinetically controlled alkylations can be affected by thermodynamically controlled isomerizations. This has been verified through a study of the behavior of pure 1- and 2-*t*-butylnaphthalene under the influence of $AlCl_3$ or CS_2. Whereas the 1 isomer was transformed completely to the 2 isomer in 15 min at 0°C, the 2 isomer showed no sign of change to the 1 isomer; only dealkylation and disproportionation was observed.

Other naphthalene derivatives were also *t*-butylated. The *t*-butylation of 2-naphthol and its methyl ether with both *t*-butyl and isobutyl derivatives has been the subject of several reports [228,257-261]. However, as yet there is uncertainty in the literature as to the constitution of some of the reaction products. To illustrate, the *t*-butylation of 2-naphthol and 1-chloro-2-naphthol with *t*-butyl chloride and $AlCl_3$ was reported by Buu-Hoi et al. [260] to give 6-*t*-butyl-2-naphthol and 6-*t*-butyl-1-chloro-2-naphthol, respectively, as monoalkylates. A dialkylated product from the former reaction was assigned the structure 1,6-di-*t*-butyl-2-naphthol. This assignment, which has been accepted [262] and even supported [263] by later workers, was more recently proved to be incorrect. Instead, Brady et al. [228] assigned to it the structure 3,6,-di-*t*-butyl-2-naphthol (45, Eq. 60).

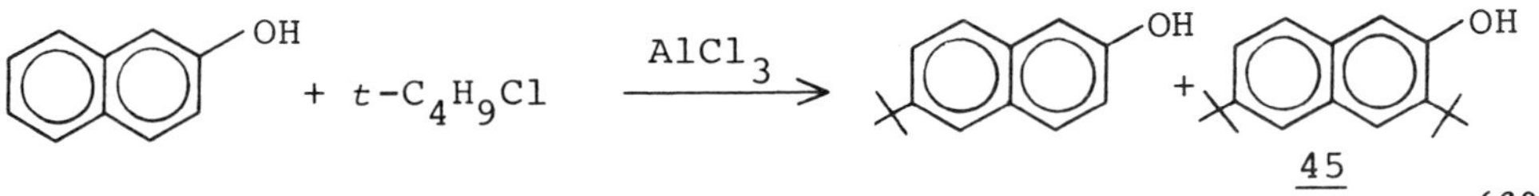

OH + t-C_4H_9Cl $\xrightarrow{AlCl_3}$ OH + OH

45

(60)

Brady et al. also reinvestigated the *t*-butylation of 2-methoxynaphthalene and 1-bromo-2-methoxynaphthalene with *t*-butyl chloride and $AlCl_3$. In contrast to the earlier claim [261] that the first substrate gave 1-*t*-butyl-2-methoxynaphthalene and the second gave 4-*t*-butyl-1-bromo-2-methoxylnaphthalene, the results in Eqs. (61) and (62) were obtained.

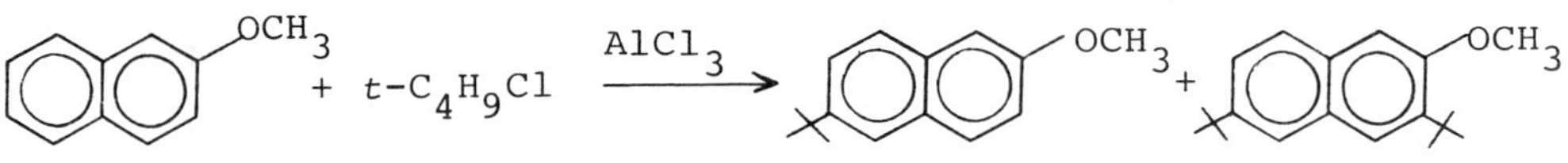

OCH_3 + t-C_4H_9Cl $\xrightarrow{AlCl_3}$ OCH_3 + OCH_3

(61)

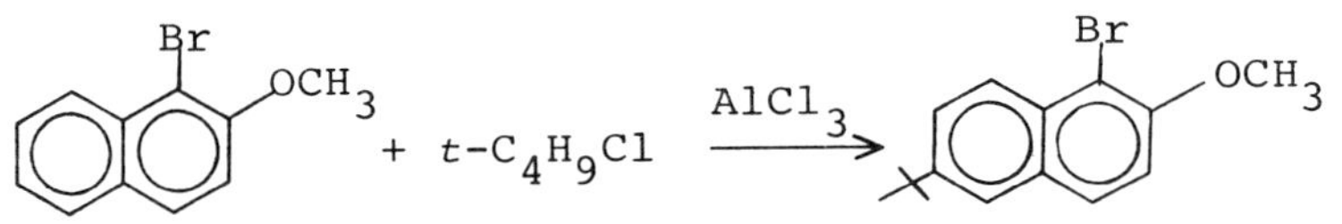

Br OCH_3 + t-C_4H_9Cl $\xrightarrow{AlCl_3}$ Br OCH_3 (62)

Based on their results, Brady et al. concluded that in Friedel-Crafts alkylations of 2-naphthol with small alkyl groups, electronic factors are more important than steric ones, but that in *t*-butylation the reverse is true. On electronic grounds the order of reactivity of the positions in 2-naphthol is $1 > 6 > 3 > 8$, but on steric grounds the order is $6 > 3 > 8 > 1$.

Besides the alkylations of the aromatics discussed previously, the *t*-butylation reacton has been effectively applied to many other aromatic and aliphatic substrates under varying conditions. The following recent examples are illustrative:

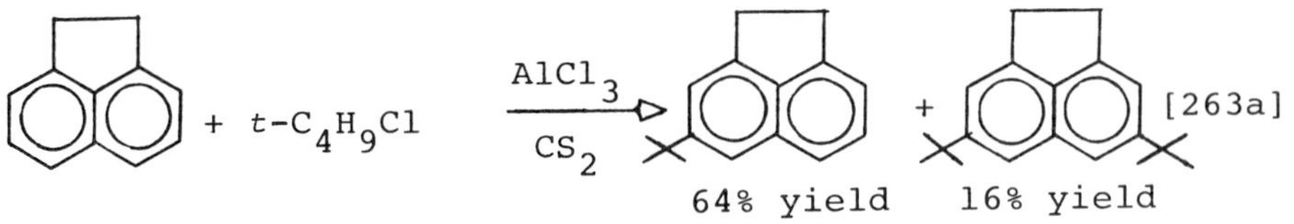

+ t-C_4H_9Cl $\xrightarrow[CS_2]{AlCl_3}$ + [263a]

64% yield 16% yield

$$\text{benzofuran} + t\text{-}C_4H_9Cl \xrightarrow[\text{Hexane, 60-100°}]{ZnCl_2} \text{2-}t\text{-butylbenzofuran} + \text{3-}t\text{-butylbenzofuran}$$

1 : 2 [264]

$$\text{indole (2-R)} + R'X \xrightarrow[ZnCl_2]{ZnCl_2\text{-dipyridine}} \text{3-R'-indole (2-R)}$$ [264a]

R = H, CH_3

R' = *n*-Bu, *sec*-Bu, *i*-Bu, *t*-Bu, PhCMeEt

X = Cl, Br

$$\text{acetophenone} + t\text{-}C_4H_9Cl \xrightarrow[0°,\ 60\ \text{min}]{HF\text{-}SbF_5} \text{3-}t\text{-butylacetophenone}$$ [79a]

3%

$$CH_2(CN)_2 + t\text{-}C_4H_9Cl \xrightarrow[3°,\ 16\ \text{hr}]{AlCl_3, CH_3NO_2} (CH_3)_3CCH(CN)_2 + HBr$$

60% [265]

$$\text{1-(trimethylsiloxy)cyclohexene (OSiMe}_3\text{)} + t\text{-}C_4H_9Cl \xrightarrow[-23°,\ 3\ \text{hr}]{TiCl_4, CH_2Cl_2} \xrightarrow{H_2O} \text{2-}t\text{-butylcyclohexanone}$$ [266]

48%

$$\text{Ph-C(OSiMe}_3\text{)=CH}_2 + t\text{-}C_4H_9Cl \xrightarrow[-23°,\ 3\ \text{hr}]{TiCl_4, CH_2Cl_2} \xrightarrow{H_2O} \text{PhCOCH}_2C(CH_3)_3$$

43%

[266]

$$\text{cyclohexylidene-CH-OSiMe}_3 + t\text{-}C_4H_9Cl \xrightarrow[-23°,\ 3\ \text{hr}]{TiCl_4, CH_2Cl_2} \xrightarrow{H_2O} \text{1-}t\text{-butylcyclohexane-1-CHO}$$

40%

[266]

In the foregoing syntheses of α-*t*-butyl derivatives of aldehydes and ketones by the Friedel-Crafts alkylation of the corresponding enol ethers, the efficacy of the catalyst followed the order $TiCl_4$ > $ZnCl_2$ > $SnCl_4$ > $AlCl_3$. With $TiCl_4$ as the catalyst, a substrate/catalyst ratio of 1:1 gave the best yield [266].

The *t*-butylation reaction has also been conducted by Schmerling [267] in the presence of an isoparaffin. The product formed by the alkylation of benzene with *t*-butyl chloride and $AlCl_3$ in the presence of isoparaffins such as isopentane, 3-methylhexane, methylcyclopentane, and methylcyclohexane consisted not only of *t*-butylbenzene but also of an alkylbenzene in which the alkyl group is derived from the isoparaffin. The results obtained showed great dependence on the structural nature of both the halide and the isoparaffin. For example, the gradual addition of *t*-butyl chloride to a stirred mixture of benzene, isopentane, and $AlCl_3$ at room temperature yielded chiefly pentylbenzenes during the early part of the reaction, but as the reaction proceeded, the relative amount of *t*-butylbenzene increased (Eq. 63). When methylcyclo-

$$+ \ t\text{-}C_4H_9Cl + i\text{-}C_5H_{12} \xrightarrow[rt]{AlCl_3} \quad + \quad + \quad + i\text{-}C_4H_{10} \tag{63}$$

hexane was used instead of isopentane, the entire alkylation product was (methylcyclohexyl)benzene, formed via hydrogen transfer (Eq. 64). Sub-

$$+ \ t\text{-}C_4H_9Cl \ + \quad \xrightarrow[rt]{AlCl_3} \quad + \tag{64}$$

16% 64%

stitution of isopropyl chloride for *t*-butyl chloride in the reaction with isopentane, benzene, and $AlCl_3$ gave isopropylbenzene as the chief product (Eq. 65).

$$+ \ i\text{-}C_3H_7Cl + i\text{-}C_5H_{12} \xrightarrow[rt]{AlCl_3} \quad + \quad (C_5H_{11}) \tag{65}$$

54% 4%

The rationale for the formation of the products from the reaction of benzene with *t*-butyl chloride, isopentane, and $AlCl_3$ is shown in scheme 21.

After this much discussion of the *t*-butylation reaction, let us now summarize the main views about its mechanism. This is desirable because *t*-butylation is an ideal example of alkylation by a *t*-alkyl halide. According to Brown et al. [6,48,49], the *t*-butylation (like isopropylation) proceeds by an S_N1-type ionization mechanism in which a localized π complex appears to be involved as a high-energy intermediate. In view of the Olah et al. [185,205], the mechanism of *t*-butylation is similar to that of isopropylation in that it involves two transition states: "one (T_1) of an oriented π-complex nature which is involved in the substrate rate-determining step, and a subsequent second one (T_2) corresponding to a σ-complex which is involved in determin-

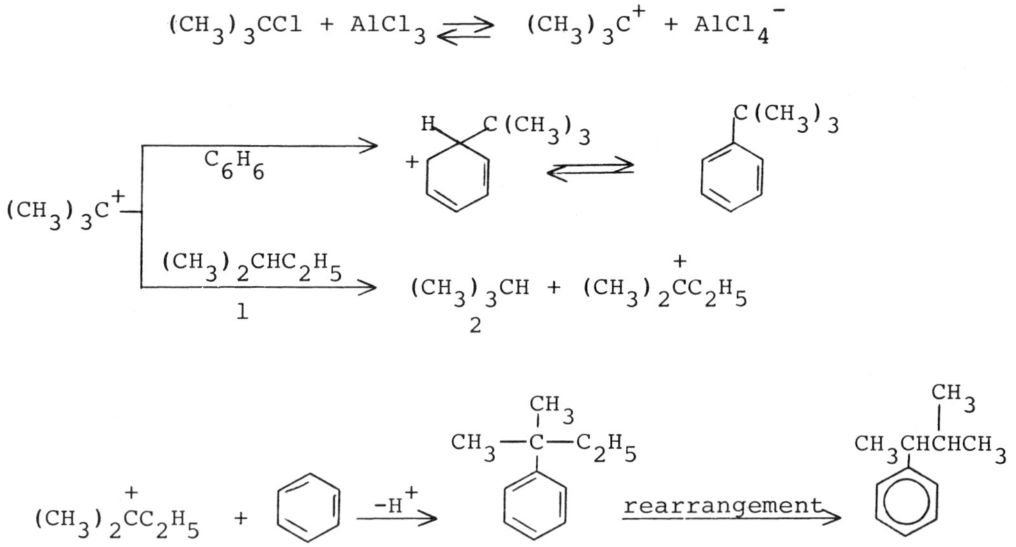

$$(CH_3)_3CCl + AlCl_3 \rightleftharpoons (CH_3)_3C^+ + AlCl_4^-$$

$$(CH_3)_3C^+ \xrightarrow{C_6H_6} \text{[}\sigma\text{ complex: H, }C(CH_3)_3\text{ on cyclohexadienyl cation]} \rightleftharpoons C_6H_5C(CH_3)_3$$

$$(CH_3)_3C^+ + \underset{1}{(CH_3)_2CHC_2H_5} \longrightarrow \underset{2}{(CH_3)_3CH} + (CH_3)_2\overset{+}{C}C_2H_5$$

$$(CH_3)_2\overset{+}{C}C_2H_5 + C_6H_6 \xrightarrow{-H^+} C_6H_5C(CH_3)_2C_2H_5 \xrightarrow{\text{rearrangement}} C_6H_5CH(CH_3)CH(CH_3)CH_3$$

Scheme 21

ing the positional selectivity." The fact that *t*-butylation of benzene and toluene showed higher substrate selectivity than isopropylation was attributed to a weaker electrophilic character of the *t*-butyl cation (or its precursor) compared to the isopropyl cation (or its precursor). With the *t*-butyl cation the electron deficiency of the carbocation center is decreased by hyperconjugation with one more methyl group [205].

For reactions catalyzed by boron trifluoride [12,196,197], Nakane and Natsubori [95] presented another view, especially in regard to the nature of the alkylating species. This view was based on the observation that Friedel-Crafts *t*-butylation resulting from decomposition of the alkylbenzene-*t*-butyl chloride-boron fluoride-oriented 1:1:1 π complexes at low temperatures was different from conventional Friedel-Crafts *t*-butylation at room temperature [6,205,268]. Rather, it resembled Friedel-Crafts acetylation [269] and benzoylation [270] or noncatalyzed chlorination [271] or bromination [272] in showing high substrate and positional selectivities. Thus in the decomposition process, only *p*-*t*-butyltoluene was formed from the corresponding toluene complex. Neither *m*- nor *o*-*t*-butyltoluene was formed, nor were di- or higher *t*-butylated products obtained. Similarly, 1,2-dimethyl-4-*t*-butylbenzene was the sole product obtained from *o*-xylene. These and other results led to the conclusion that the electrophile under such conditions is neither a *t*-butyl cation nor an incipient *t*-butyl cation but a polar donor-acceptor complex, $(CH_3)_3\overset{\delta+}{C}Cl{-}\overset{\delta-}{B}F_3$, whose electrophilic character is very weak. Once formed, the latter complex attacks the arene, giving an oriented π complex whose transformation into the σ complex constitutes the rate-determining step. Preferential attack at the *para* position will thus result in a σ complex which is not only conjugatively stabilized, but is also free from steric hindrance. The mechanism outlined in scheme 22 was accordingly depicted.

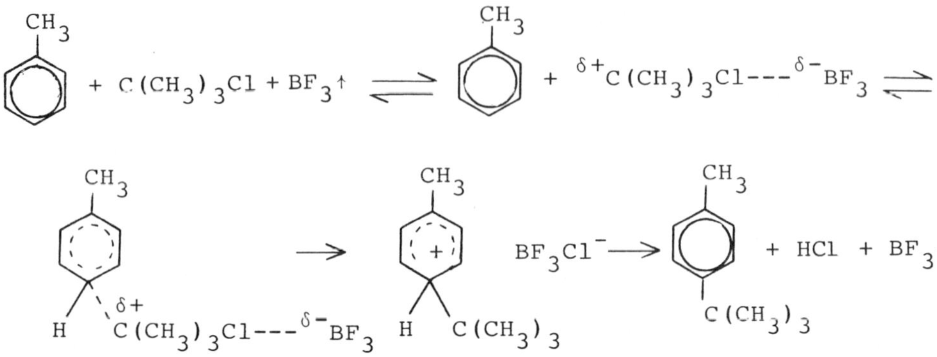

$$C_6H_5CH_3 + C(CH_3)_3Cl + BF_3\uparrow \rightleftharpoons C_6H_5CH_3 + {}^{\delta+}C(CH_3)_3Cl\text{---}{}^{\delta-}BF_3 \rightleftharpoons$$

$$[CH_3C_6H_5\cdots{}^{\delta+}C(CH_3)_3Cl\text{---}{}^{\delta-}BF_3] \rightarrow [CH_3C_6H_5(H)C(CH_3)_3]^+\ BF_3Cl^- \rightarrow CH_3C_6H_4C(CH_3)_3 + HCl + BF_3\uparrow$$

Scheme 22

In concluding this part, the following points can be emphasized:

1. Alkylations with *t*-butyl halides invariably and exclusively lead to *t*-butylated products.
2. Alkylations with isobutyl halides behave similarly; only in cases where steric factors interfere (e.g., *p*-xylene) could minor amounts of isobutyl and *sec*-butyl derivatives be produced.
3. With the exception of the smallest halogen, fluorine, no *t*-butylation *ortho* to any other atom or group could be observed.
4. In *t*-butylations of monosubstituted benzenes, mixtures of *meta* and *para* *t*-butylated isomers are usually produced; only with BF_3 at very low temperatures could *para* substitution be selectively enforced. This behavior was verified even in gas-phase *t*-butylations in which the *t*-butyl cation was produced by γ radiolysis [274].
5. No single mechanism can be offered to account for all *t*-butylations; depending on conditions, mechanisms range from one in which the attacking electrophile is strong and is essentially a free cation, and the formation of the π complex is rate determining, to one in which the attacking electrophile is weak and has the form of a polar donor-acceptor complex, and the formation of the σ complex is rate determining.

B. Pentyl Halides

The history of the Friedel-Crafts reaction of arenes with branched-chain pentyl halides is littered with misinterpretations. For example, examination of the earlier data compiled by Price [2] reveals that there had been widespread belief among early investigators that alkylations of arenes with *t*-pentyl and isopentyl halides in the presence of $AlCl_3$ afforded only *t*-pentyl derivatives. Also, the alkylation of benzene with neopentyl chloride and $AlCl_3$ at 0°C was reported by Pines et al. [275] to give exclusively 2-methyl-3-phenylbutane. The earliest exception to this unanimity came in a report published by Konowalow and Jegerow in 1898 [273]. Using nitration for product identification, these authors concluded that the product from the $AlCl_3$-catalyzed alkylation of benzene with isopentyl chloride consisted of a mixture of isopentylbenzene, *t*-pentylbenzene, and 2-methyl-3-phenylbutane. This prevalent misinterpretation of results by early workers was undoubtedly due to the lack of our present knowledge of carbocation behavior and the limited identification facilities available at their time.

With the advances in both theory and technique made in the 1950s and later, workers have become aware of the possible involvement of side-chain rearrangements during and after alkylations with branched-chain pentyl halides [85, 246,276-279]. Moreover, they have recognized that both the occurrence and the extent of accompanying rearrangements are highly dependent not only on the specificity of catalyst action and the severity of the employed conditions [85,246,278], but also on steric interactions [81,250].

With this background, let us now examine the reliable literature data on the alkylations of arenes with the five isomeric branched-chain pentyl halides (48-52) in order to gain more insight into their nature. In doing this, we

```
  C          C                 C             C           C
  |          |                 |             |           |
C-C-C-C    C-C-C-C-Br    Cl-C-C-C-C       C-C-C-C     C-C-C-Cl
  |                                            |         |
  Cl                                           Cl        C

  48         49                50            51          52
```

shall start with alkylations of the simplest arene, benzene, where orientation (position selectivity) and steric factors are absent. The results of Boord and co-workers [276] in 1952, Schmerling and West [277] in 1954, Friedman and Morritz [278] in 1956, and Roberts and co-workers [85,246] in 1970 on the alkylations of benzene with isomeric pentyl halides under various conditions are compiled in Table 8. Taken together, the results of Table 8 support the assumption that the initial alkylation product from benzene and all five isomeric branched pentyl halides (48-52) is *t*-pentylbenzene (53). When mild catalysts and/or conditions are employed, the initial product is stable and remains to be the sole or major product. When stronger catalysts and/or conditions are employed, *t*-pentylbenzene undergoes successive isomerization to 2-methyl-3-phenylbutane (54), and then, much more slowly, to neopentylbenzene (55) (Eq. 66).

```
                                  C                       C C
                                  |                       | |
C5H11X + C6H6    catalyst  →  Ph-C-C-C      slow  →   Ph-C-C-C
                   -HX            |
                                  C
   48-52                          53                      54

                                               C
                                               |
                             slower  →    Ph-C-C-C                (66)
                                               |
                                               C

                                               55
```

Combining their views with those of others [277,278] Roberts et al. [85, 246; see Chap. 8] suggested scheme 23 to rationalize the alkylations of benzene with these branched-chain pentyl halides. According to scheme 23, treatment of the halides 49-52 gives the corresponding carbocations 56-59

Table 8. Alkylations of Benzene with Branched-Chain Pentyl Halides

Entry no.	Reactants, mol: Benzene	Reactants, mol: Pentyl halide	Catalyst: Kind	Catalyst: Mol	Temp, °C	Time, hr	Yield, %[a]	Pentylbenzene isomers, %[b]: *t*-Pentylbenzene	Pentylbenzene isomers, %[b]: 2-Methyl-3-phenylbutane	Pentylbenzene isomers, %[b]: Neopentylbenzene	Ref.
					With *t*-Pentyl Chloride (48)						
1	100	25	$AlCl_3$	0.74	25–30	—[c]	52	Isomers[d]		0	276
2	2	1	$AlCl_3$/	0.037	25–30	—[c]	18	100	0	0	276
			CH_3NO_2	0.29							
3	1.35	0.27	$FeCl_3$	0.07	10–25	—[c]	60	100	0	0	276
4	1.35	0.27	$FeCl_3$/	0.07							
			CH_3NO_2	0.29	10–25	—[c]	33	100	0	0	276
5	5.18	1.13	$AlCl_3$	0.15	1	3	43	15	85	0	277
6	1.55	0.35	$ZrCl_4$	0.013	80	8	71	85	15	0	277
7	1.5	0.43	BF_3	—[e]	33	2	28	100	0	0	277
8	5	1	$AlCl_3$	0.025	24	2.5	71	100	0	0	278
9	5	1	$AlCl_3$	0.20	0	0.51	57	35	65	0	278
					With Isopentyl Bromide (49)						
10	5.15	1	$AlCl_3$	0.70	1	3	7	15	85	0	277

					With 1-Chloro-2-methylbutane (50)						
11	5	1	$AlCl_3$	0.1	25	0.25		20	80	0	246
						4	65	17	83	0	
12	5	1	$AlCl_3$	0.1	0–5	0.25		56	44	0	246
						4	69	18	82	0	
13	5	1	$AlCl_3$/	0.1	25	0.25		100	0	0	246
			CH_3NO_2	1		23	42	96	4	0	
					With 2-Chloro-3-methylbutane (51)						
14	5	1	$AlCl_3$	0.1	0–5	0.25		43	57	0	246
						4		18	82	0	
15	5	1	$AlCl_3$/	0.1	25	0.25		100	0	0	246
			CH_3NO_2	1		24		98	2	0	
					With Neopentyl Chloride (52)						
16	4	3	$AlCl_3$	0.8	0–25	0.25		20	76	4	85
						2		11	50	39	
						24	40	5	16	79	
17	5	1	$AlCl_3$/	0.1	25	5		100	0	0	85
			CH_3NO_2	1		24	35	98	2	0	

[a]Combined yield of pentylbenzene isomers based on starting pentyl halide.
[b]Relative percentages of pentylbenzene isomers: determined by infrared, permanganate oxidation, selective adsorption and physical property studies in entries 1-4; by infrared in entries 5-10; and by glpc in entries 11-17.
[c]Until the evolution of hydrogen chloride gas ceased.
[d]Incorrectly claimed to contain 20% *t*-pentylbenzene, with the remainder being a mixture chiefly of 1-phenyl-2-methylbutane and 2-methyl-3-phenylbutane.
[e]Undetermined amount.

```
     C                            C                      C                      C
     |                            |    +      ~H:-       |  +                   | .
C-C-C-C-Br      -Br-  ->     C-C-C-C        ------>  C-C-C-C      <----     C-C-C-C
                                                                                |
                                                                                Cl
     49                           50                     58                     51
                                                         ^|
                                                         || ~H:-
                                                         |v
       C                    +  C                         C                      C
       |                       |          ~H:-           |         -Cl-         |
Cl-C-C-C-C     -Cl-  ->     C-C-C-C       ------>   C-C-C-C     <------     C-C-C-C
                                                         +                      |
                                                                                Cl
       50                      57                        60                     48

    C                          C                    ~CH3:-  (59 -> 60)
    |                          |  +                            ^|   
C-C-C-Cl      -Cl-  ->     C-C-C                        -PhH   ||   +PhH
    |                          |                        +H+    ||   -H+
    C                          C                               |v
    52                         59

       C          AlCl3        C C        AlCl3            C
       |                       | |                         |
Ph - C-C-C     <------     C-C-C-C       <====>      Ph-C-C-C
       |                     |                             |
       C                     Ph                            C

       55                    54                            53
```

Scheme 23

(or their equivalent complexes with the catalyst), and these isomerize by 1,2-hydride or 1,2 methyl shifts to the tertiary carbocation 60. The latter is also produced directly from *t*-pentyl chloride 48. Regardless of which of the five pentyl halides is used as alkylating agent, and of whether $AlCl_3$ or milder catalyst is used, 60 rapidly alkylates benzene to produce *t*-pentylbenzene (53). Under suitable nonisomerizing conditions, 53 will remain to be the sole or the major product. With reference to Table 8, suitable non-isomerizing conditions include a very small amount of $AlCl_3$ (entry 8) [278] or a milder catalyst such as $FeCl_3$ (entry 3) [276], $FeCl_3$-CH_3NO_2 (entry 4) [276], $AlCl_3$-CH_3NO_2 (entries 2, 13, 15, and 17) [85,246,276], $ZrCl_4$ (entry 6) [277], BF_3 (entry 7) [277], or $AgBF_4$ [279].

On the other hand, in the presence of a strong catalyst such as $AlCl_3$ or $AlBr_3$, the rapid initial formation of *t*-pentylbenzene (53) is followed by a slower isomerization of 53 to 2-methyl-3-phenylbutane (54) until a limiting equilibrium composition (18% 53, 82% 54) is attained. Further isomerization of 54 to 55 also took place when larger amounts of $AlCl_3$ were used (entry 16). This behavior represents another example of kinetic versus thermodynamic control. The experiments in which 53 is found to be the sole or the major product exhibit kinetic control, whereas those in which 54 is the major product exhibit thermodynamic control. The carbocation processes leading to the sequence 53 → 54 → 55 are discussed in Chap. 8.

In view of the now well-understood results on the alkylations of benzene with branched pentyl halides, one should view the few contradictory results with great doubt. In fact, claims such as the formation of isopentylbenzene from benzene alkylations with isopentyl bromide and $AlCl_3$ [273], of only 2-methyl-3-phenylbutane from benzene alkylation with neopentyl chloride and

$AlCl_3$ [275], or of 2-methyl-1-phenylbutane from benzene alkylation with *t*-pentyl chloride and $AlCl_3$ [276] should be considered as totally erroneous.

Besides benzene, alkylations with branched pentyl halides were also extended to other benzene derivatives. In these cases, steric interactions manifested themselves by hindering alkylations at *ortho* positions [277] or by causing unusual rearrangements in the alkylating agent [81,246,250]. For example, the alkylation of toluene with *t*-pentyl chloride in the presence of HF, $FeCl_3$, or $AlCl_3$-CH_3NO_2 catalyst at 0 to 4°C gave varying mixtures of *m*- and *p*-*t*-pentyltoluene with no substitution at the *ortho* position [277]. The results are summarized below Eq. (67). These results were considered

$$C_6H_5CH_3 + t\text{-}C_5H_{11}Cl \xrightarrow{\text{catalyst}} m\text{-}CH_3C_6H_4C(CH_3)_2CH_2CH_3 + p\text{-}CH_3C_6H_4C(CH_3)_2CH_2CH_3 \quad (67)$$

(mol)	(mol)	catalyst (mol)	time, hr	yield, %	composition, %	
4.00	2.0	HF (7.6)	5.0	94	20	80
0.24	5.7	$AlCl_3$ (0.6)	4.5	86	27	73
		$PhNO_2$ (1.2)				
2.50	2.0	$FeCl_3$ (0.4)	7.0	78	72	28

to be inadequate to establish limiting values for the kinetic versus thermodynamic compositions of the alkylates as in the case of *t*-butylation. However, for this system the kinetic limit was suggested to be lower than 20% *meta* and higher than 80% *para*.

The alkylation of *p*-xylene by *t*-pentyl chloride reported by Friedman et al. [81] in 1958 and by Schmerling et al. [250] in 1959 gave interesting results. These workers found that the alkylation of *p*-xylene with *t*-pentyl chloride and $AlCl_3$ [81], $AlCl_3$-CH_3NO_2 [250], or $FeCl_3$ [81], or even with methylbutenes and various catalysts [81], gave the rearranged compound, 2-methyl-3-*p*-xylylbutane (61) as the sole monoalkylation product (Eq. 68).

$$p\text{-}CH_3C_6H_4CH_3 + CH_3C(CH_3)(Cl)CH_2CH_3 \xrightarrow[\text{or } AlCl_3\cdot CH_3NO_2]{AlCl_3} \underset{\mathbf{61}}{2,5\text{-}(CH_3)_2C_6H_3CH(CH_3)CH(CH_3)CH_3} + \underset{\mathbf{44}}{[2,5\text{-}(CH_3)_2C_6H_3]_2CH_2} \quad (68)$$

In addition, a small amount of di-*p*-xylylmethane (44) was also produced in reactions catalyzed by $AlCl_3$ or $AlCl_3$-CH_3NO_2 (Eq. 68), suggesting that, as in *t*-butylation, steric hindrance to tertiary pentylation results in the occurrence of hydrogen transfer and disproportionation reactions similar to those of scheme 19.

The production of 2-methyl-3-*p*-xylylbutane (61) in the reactions above was rationalized on steric grounds. When the *t*-pentyl cation is prevented from reacting with *p*-xylene by steric effects, the *sec*-pentyl cation in equilibrium with it, although in low concentraiton, becomes the predominant if not the only alkylating species (Eq. 69) [28,81,146].

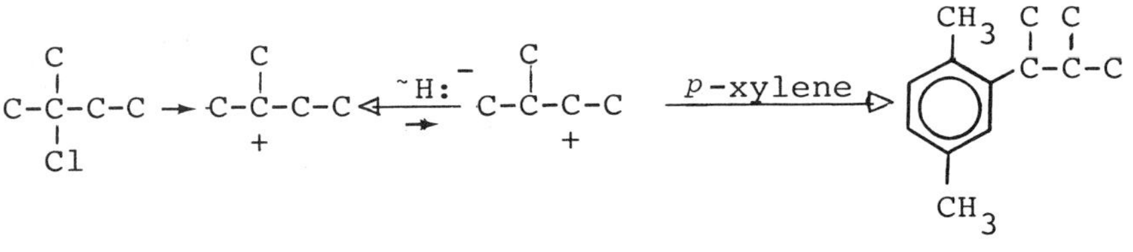

$$\mathrm{C{-}C(C)(Cl){-}C{-}C \rightarrow C{-}\overset{+}{C}(C){-}C{-}C \underset{\rightarrow}{\overset{\sim H:^-}{\leftarrow}} C{-}C(C){-}\overset{+}{C}{-}C \xrightarrow{p\text{-xylene}} \underset{\mathbf{61}}{2,5\text{-}(CH_3)_2C_6H_3{-}CH(C){-}CH(C){-}C}} \quad (69)$$

The alternative rationalization of the formation of the secondary alkylate, 61, in terms of initial production of the tertiary alkylate 2-methyl-2-*p*-xylylbutane followed by subsequent isomerization to 61 by the catalyst was also considered but was disqualified on various grounds [81,277,280,281].

The alkylation of *p*-xylene with the 1-chloro-2-methylbutane and $AlCl_3$ was studied by Roberts and McGuire in 1970 [246]. The results obtained (Table 9) closely paralleled those reported for alkylation with *t*-pentyl chloride [81,250]. The only monoalkylation product from the primary chloride was the same as from the tertiary chloride, 2-methyl-3-*p*-xylylbutane (61). The 3-methyl-2-butyl cation (58), which is produced by isomerization from the primary chloride (50) (Eqs. 70 and 71), has a much smaller steric re-

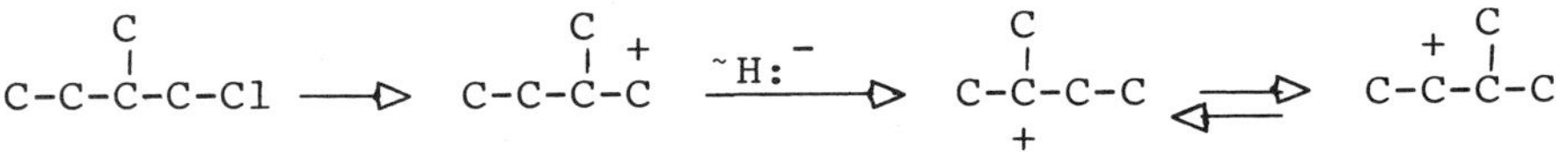

$$\underset{\mathbf{50}}{\mathrm{C{-}C{-}C(C){-}C{-}Cl}} \rightarrow \underset{\mathbf{57}}{\mathrm{C{-}C{-}\overset{+}{C}(C){-}C}} \xrightarrow{\sim H:^-} \underset{\mathbf{60}}{\mathrm{C{-}C(C){-}\overset{+}{C}{-}C}} \rightleftarrows \underset{\mathbf{58}}{\mathrm{C{-}\overset{+}{C}{-}C(C){-}C}} \quad (70)$$

$$\underset{\mathbf{58}}{\mathrm{C{-}\overset{+}{C}{-}C(C){-}C}} + p\text{-xylene} \longrightarrow \underset{\mathbf{61}}{\mathrm{2,5\text{-}(CH_3)_2C_6H_3{-}CH(C){-}CH(C){-}C}} \quad (71)$$

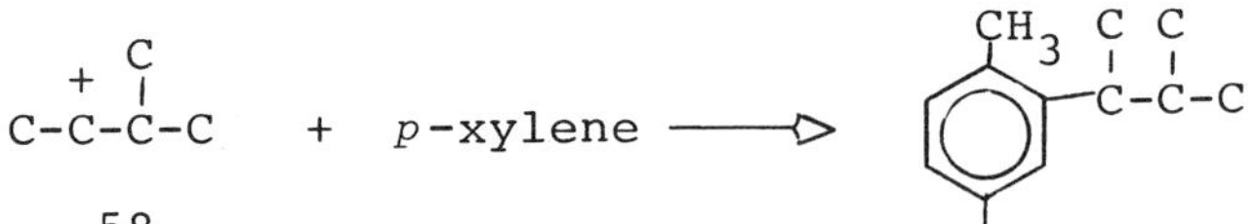

quirement than the *t*-pentyl cation and alkylates *p*-xylene readily. As in alkylations with *t*-butyl and *t*-pentyl chlorides, a considerable quantity of 2-(*p*-methylbenzyl)-*p*-xylene (43) and di-*p*-xylylmethane (44) was produced.

Examination of the pentyltoluene fraction produced in these reactions (Table 9) disclosed the interesting fact that it consisted not only of *meta* (62) and *para* (63) *t*-pentyltoluene, as previously described [81,250], but also of *meta* (64) and *para* (65) 2-methyl-3-tolylbutane (Eq. 71). Moreover, the

Table 9. Alkylation of *p*-Xylene with 1-Chloro-2-methylbutane[a] at 0°C

	Yield of products, %		
Time, hr	3-Methyl-2-*p*-xylylbutane	*t*-Pentyltoluene (*meta*/*para*)	2-Methyl-3-tolylbutane (*meta*/*para*)
0.25	80		12 (64:36)
1.0	66		25 (59:41)
3.0	40	6 (60: 40)	47 (51:49)
6.0	31		53

[a]Mol. ratio of *p*-xylene/pentyl chloride/$AlCl_3$ = 4:1:0.1.

p-xylene + C-C-C(C)-C-Cl —($AlCl_3$, 3 hr)→ 61 (40%) + 62 (2.6%) + 63 (2.4%) + 64 (24%) + 65 (23%) (72)

amounts of 2-methyl-3-tolylbutanes (47%) and *t*-pentyltoluenes (6%) were in a ratio of 88:12, which is similar to the 82:18 equilibrium distribution of 2-methyl-3-phenylbutane/*t*-pentylbenzene observed during the alkylation of benzene by the same pentyl halide.

Steric factors also played a role in the BF_3-catalyzed alkylation of *p*-cresol with *t*-pentyl chloride or 2-methyl-2-butene [282]. When the alkylation was carried out with a low concentration of BF_3 (1 to 25 mol %) it gave a high yield of the expected 2-*t*-pentyl-4-methylphenol (66, Eq. 73). With a higher

p-cresol + C-C(C)(Cl)-C-C —(25 mol % RF_3, <50°, 4 hr)→ 66 (73) concentration of BF_3 (50 mol %) present, the alkylation products included less than 20% of the expected 2-*t*-pentyl-4-methylphenol (66). The remainder of the phenolic material was composed of a complex mixture separated by chromatography into four general classes of compounds: 2-*t*-alkyl-4-methylphenols (67), 2-*sec*-alkyl-4-alkylphenols (68), *p*-*t*-alkylphenols (69) and 2,2'-bis-methylenediphenols (70). The enhanced formation of by-products 67-70 at

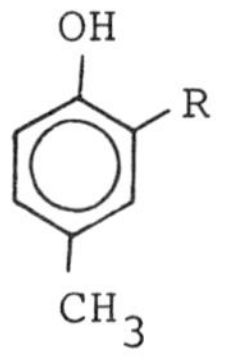

67:	68:	69:	70:
R=*t*-butyl	R_1= *sec*-alkyl	R=*t*-butyl or *t*-pentyl	$R_1=R_2=CH_3$ or $R_1=CH_3, R_2=H$

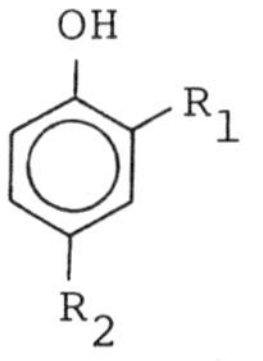

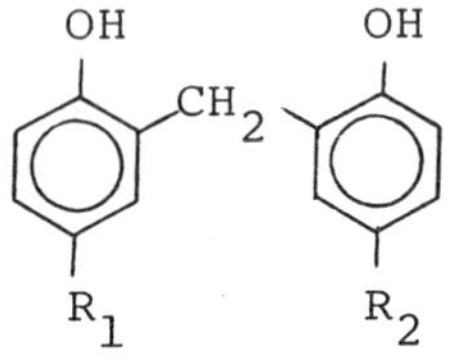

high catalyst concentrations was attributed to steric factors. At low BF_3 concentrations, normal alkylation resulted in *ortho* substitution by reaction of the *t*-pentyl cation with an uncomplexed *p*-cresol molecule. At high BF_3 concentrations, however, BF_3-complexed *p*-cresol predominates. Since an $-OBF_3$ group apparently cannot exist in a position *ortho* to a tertiary alkyl group because of steric opposition, the alkylation process is retarded. The result is the enhancement of hydride transfer and other side reactions similar to those observed previously during the attempted *t*-pentylation of *p*-xylene (see scheme 19).

C. Hexyl and Higher Alkyl Halides

Examination of the compilation of Price [2] reveals that most of the early work on the Friedel-Crafts alkylation of aromatics by hexyl and higher alkyl halides was performed by Schreiner [283] in 1910 and by Halse [284] in 1914. These two workers, being unaware of complicating factors, thought that their alkylations simply involved direct displacement of the chlorine by the phenyl group, resulting in the formation of unrearranged *t*-alkylates. Moreover, the two authors made their starting chlorides by treating the corresponding alcohols with dry hydrogen chloride, a procedure that has been established to involve carbocation rearrangements and probably gave them impure chlorides. Because of these and other factors, the data reported by these early investigators should be regarded as being highly questionable.

In 1954, Schmerling and West [277] investigated the alkylation of benzene with both 2-chloro-2,3-dimethylbutane (71) and 1-chloro-3,3-dimethylbutane (72) with a variety of catalysts and under different conditions. The results showed that alkylation of benzene with the *t*-hexyl chloride, 70, in the presence of aluminum chloride at 1°C or zirconium chloride at 85°C gave a 47 to 56% yield of hexylbenzenes composed of 10 to 15% 2,3-dimethyl-2-phenylbutane (73) and 85 to 90% 2,2-dimethyl-3-phenylbutane (74) (Eq. 74). On

```
                C C                                    C C              C C
                | |                                    | |              | |
        C-C-C-C       strong                     /\    C-C-C          /\ C-C-C
          |         ------------->              (O)     |      +     (O)    |
          Cl        |  catalysts                 \/     C             \/    C
                    |                            73 (minor)          74 (major)
          71        |
  (O) +             -
          or        |                                                       (74)
          C         |
          |         |  weak
     C-C-C-C-Cl     ------------->              73 (sole or major)
          |            catalysts
          C
          72
```

the other hand, when the reaction was carried out in the presence of $FeCl_3$ or $AlCl_3$-CH_3NO_2, the product was practically pure 2,3-dimethyl-2-phenylbutane (73).

Similarly, the alkylation of benzene with the isomeric primary halide 1-chloro-3,3-dimethylbutane (72) in the presence of $AlCl_3$ at 2°C or 42°C, $AlBr_3$ at 2°C or $ZrCl_4$ at 85°C, gave 49 to 56% yields of mixtures composed of 0 to 15% (73) and 85 to 100% (74). A 7 to 15% yield of hexanes, predominantly 2,3-dimethylbutane, was observed in reactions catalyzed by $AlCl_3$. In reactions catalyzed by $FeCl_3$ or $AlCl_3$-CH_3NO_2 at 83 to 84°C, 72 gave only 6% and 35% yields, respectively, of hexylbenzenes composed of 90 to 94% (73) and 6 to 10% (74). These low yields with the primary chloride should be contrasted with the 71% and 62% yields of pure 73 obtained in similar alkylations of benzene with the isomeric tertiary chloride 71 in the presence of the same $FeCl_3$ and $AlCl_3$-CH_3NO_2 catalysts.

As to mechanism, Schmerling and West [277] favored the ionic mechanism presented in scheme 24 to account for their results. According to this mechanism, the *t*-carbocation 75 produced directly from chloride 71 and by double rearrangement from chloride 72 alkylates benzene rapidly to give the tertiary alkylate 2,3-dimethyl-2-phenylbutane (73). In the presence of strong catalysts, the latter subsequently rearranges to yield the 2,2-dimethyl-3-phenylbutane (74). The fact that mild catalysts and/or conditions gave chiefly 73 by suppressing its transformation to 74, lent logical support for the proposed mechanism.

```
  C C                    C C                     C                C                    C
  | |      AlCl3         | |      ~CH3:-         | +    ~H:-      |   +     AlCl3      |
C-C-C-C  --------->    C-C-C-C  <---------   C-C-C-C  <-----  C-C-C-C  <---------  C-C-C-C-Cl
  |        -Cl           +                       |                |         -Cl-       |
  Cl                                             C                C                    C
  71                     75                                                            72

 C C                     C C                 C C                       C                     C
 | |     PhH,-H+         | |       R+        | |        ~CH3:-       + |        RH           |
C-C-C-C <=========>   C-C-C-C   <=====>   C-C-C-C   ------------>  C-C-C-C  <======>     C-C-C-C
 +       H+,-PhH         |         RH        | +                     | |        R+         | |
                         Ph                  Ph                      Ph C                  Ph C
 75                      73                                                                74
```

Scheme 24

Further experimental support for the mechanism of scheme 24 was found in the observation that *t*-hexylbenzene 73 was easily isomerized to the isomeric secondary alkyl benzene, 74 by treatment with the strong catalyst aluminum chloride. The rearrangement of this *t*-hexylbenzene and the analogous rearrangement of *t*-pentylbenzene described in the same paper [277] were the first examples of the "alkylbenzene rearrangements," which are discussed in detail in Chap. 8.

Some early alkylations of arenes with higher branched-chain alkyl halides were also compiled [2]. A few other examples have appeared as patents in recent years. In one of these [285], 3-chloro-3-ethylpentane alkylated benzene in the presence of $FeCl_3$ and C_4H_{10} to give 95.6% *t*-alkyl- and 4.4% *sec*-alkylbenzene (Eq. 75). In another patent [286] benzene and naphthalene

$$PhH + (Et)_3CCl \xrightarrow[30\ min,\ 13\text{-}20°]{FeCl_3, C_4H_{10}} \underset{95.6\%}{Ph\text{-}C(Et)_3} + \underset{4.4\%}{sec\text{-alkylate}} \qquad (75)$$

were alkylated with 7-chloro-7-methylpentadecane (76, X = Cl) at 3°C in the presence of $FeBr_3$-HCl catalyst. Benzene gave a product containing 9.8% monoalkylbenzene and 33.5% dialkylbenzenes (Eq. 76). The monoalkylbenzene

$$C_6H_6 + \underset{\mathbf{76},\ X=Cl,Br}{CH_3(CH_2)_5C(CH_3)(X)(CH_2)_7CH_3} \text{ or } \underset{\mathbf{77}}{CH_3(CH_2)_5C(=CH_2)(CH_2)_7CH_3} \xrightarrow[HCl,\ 3°]{FeBr_3} \underset{\mathbf{78}}{CH_3(CH_2)_5C(CH_3)(C_6H_5)(CH_2)_7CH_3} + \underset{\mathbf{79}}{[CH_3(CH_2)_5C(CH_3)(CH_2)_7CH_3]_2C_6H_4} \qquad (76)$$

was shown to contain 89% (1-methyl-1-hexyloctyl)benzene (78) and the dialkylbenzenes contained 26.5% di(1-methyl-1-hexyloctyl)benzene (79). Similar products were also obtained when 7-bromo-7-methylpentadecane (76, X = Br) and 2-hexyl-1-decene (77) were used instead of the chloride.

V. ALKYLATIONS WITH CYCLOALKYL AND RELATED HALIDES

Relative to those of cycloalkanols and cycloalkenes, there have been few applications of cycloalkyl halides as cycloalkating agents. In view of this and of the fact that most of the interesting mechanistic investigations have involved cycloalkanols and cycloalkenes, it will be satisfactory at this point to present some illustrative alkylations of arenes with cycloalkyl halides, deferring most of the mechanistic details to Chap. 4. Starting with the simple cyclopentylation and cyclohexylation reactions, we have selected the most recent examples, whose conditions and results are summarized in Eqs. (77) to (97).

Benzene + cyclohexyl chloride —($AlCl_3$, 5-25% yield)→ cyclohexylbenzene (50-60% yield) + dicyclohexylbenzenes [287,288] (77)

Benzene + cyclohexyl fluoride —(BF_3)→ cyclohexylbenzene, 80% yield [197] (78)

Toluene + cyclohexyl chloride —($AlCl_3$)→ o-cyclohexyltoluene + p-cyclohexyltoluene [289] (79)

Toluene + cyclohexyl chloride —(HF)→ p-cyclohexyltoluene, 8% yield [290] (80)

Toluene + cyclohexyl fluoride —(BX_3 (X=F,Cl, Br or I), -20 to -30°)→ cyclohexyltoluene, 81-90% yield [197] (81)

Toluene + cyclohexyl fluoride —(HF or BF_3, 0°)→ p-cyclohexyltoluene, 76% yield [290,291] (82)

m-Xylene + cyclohexyl bromide —($FeCl_3$, 20-50°, 3 hr)→ 1-cyclohexyl-3,5-dimethylbenzene, 75% yield [292] (83)

1,2,4-Trimethylbenzene + cyclohexyl chloride —($AlCl_3$)→ cyclohexyltrimethylbenzene, 91.3% yield [293] (84)

Cl Cl

(chlorobenzene) + (chlorocyclohexane)

$AlCl_3$, 25° → Cl–(cyclohexylbenzene), 70% yield [288] **(85)**

$AgBF_4$, 25° → Cl–(cyclohexylbenzene) + (Cl)(cyclohexylbenzene)

44% yield 33% yield [294] **(86)**

Br Cl

(bromobenzene) + (chlorocyclohexane) $\xrightarrow[25°, 12\ hr]{AlCl_3}$ Br(cyclohexylbenzene) + Br(benzene)Br **(87)**

65% yield (*o*-, *m*- and *p*- isomers) [288,295 296]

OH Cl

(phenol) + (chlorocyclohexane) $\xrightarrow{ZnCl_2}$ HO(cycloheptylbenzene) 20% yield [297] **(88)**

OH

(o-cresol) CH_3 + R-Cl $\xrightarrow[120\text{-}130°,\ 20\ min]{FeCl_3}$ OH, CH_3, R (major) + OH, R, CH_3 (minor) [298] **(89)**

OH

(m-cresol) CH_3 + R-Cl $\xrightarrow[140\text{-}150°,\ 20\ min]{FeCl_3}$ OH, CH_3, R + OH, R, CH_3 + OH, R, CH_3 [298,299] **(90)**

R = cyclopentyl or cyclohexyl

OR Cl

(alkoxybenzene) + (chlorocyclohexane) $\xrightarrow[25°]{AlCl_3}$ OR(cycloheptyl) + RO(cyclohexylbenzene) [300] **(91)**

R = CH_3 at 25°	65.6%	34.4%
R = C_2H_5 at 25°	28.3%	71.7%
R = C_2H_5 at 50°	37.0%	63.0%

OR

(4-chloro-OR-benzene) Cl + R'-Cl $\xrightarrow[\text{Zn salts}]{\text{Trace amounts of Fe, Sn or}}$ OR, R', Cl + OR, R', R', Cl [69] **(92)**

R = H, CH_3; R' = cyclopentyl, cyclohexyl, benzyl

OC_2H_5 + Cl → ($AlCl_3$) C_2H_5O ... N [301] **(93)**

Cl + N→O → ($AlCl_3$ or $FeCl_3$) N + N O [302] **(94)**

+RCl → ($AlCl_3$ (1 mol) heptane 50-60°) R ... 60% + R 38% + 2- and 4-isomers 2%

→ ($FeCl_3$ 100-135°) R (70%) + R (15%)

(R = cyclopentyl or cyclohexyl) + R + R 15% [303] **(95)**

+ Cl →

$AlCl_3$, (1 mol), heptane 40°, 5 hr → 27% yield

$AlCl_3$ (1 mol) heptane 40°, 0.5 hr → 50% of total + +

$AlCl_3$ (0.3 mol) heptane 40°, 5 min →

[304] **(96)**

$FeCl_3$, 115-120°, 15 min → 12% + 88%, 30% yield

$FeCl_3$, 115-120°, 5 min →

$AlCl_3$, heptane, 40-50°, 5 hr → 25% yield

(97) [305]

The alkylation of benzene with the two isomers 1-chloro-1-methylcyclohexane (80) and 1-chloro-3-methylcyclohexane (81) in the presence of $AlCl_3$ was investigated by Sidorova [306]. At 40°C both chlorides gave mixtures of 1-methyl-4-phenylcyclohexane (82) and the 1,3 isomer (83), together with a little dialkylated product (Eq. 98); none of the tertiary alkylate, 1-methyl-1-phenylcyclohexane (84), was found under the temperature employed.

$AlCl_3$, 40°, 50-60% yield

80 or 81 → 82 + 83 + 84 + dialkylates (98)

The fact that none of the expected tertiary alkylate, 1-methyl-1-phenylcyclohexane (84), could be found in the reaction above suggested that its initial formation was followed by complete conversion to the isomeric secondary alkylates that were isolated. This suggestion was supported by two experimental observations: (1) the isolation of some 20% pure 1-methyl-1-phenylcyclohexane (84) after repeated mild alkylation of benzene with 1-chloro-1-methylcyclohexane (80) at 20°C for 15 min, and (2) the irreversible isomerization of 1-methyl-1-phenylcyclohexane (84) to a mixture of 1,3- and 1,4-

84 —$AlCl_3$→ 82 + 83 (99)

methylphenylcyclohexane isomers upon treatment with $AlCl_3$ in the presence of a trace of H_2O overnight at room temperature or for 3 hr at 50 to 85°C (Eq. 99).

The isomerization of cyclooctyl chloride and its alkylation of benzene in the presence of $AlCl_3$ was investigated by Akhmedov et al. [307]. Isomerization of the compound at 0 to 40°C gave 12.4 to 60.2% ethylchlorocyclohexane, 2.9 to 34.8% bicyclo[4.2.0] octane, 2.5 to 18.7% ethylcyclohexane, 2.3 to 11.1% dimethylcyclohexane, and 6 to 9.7% methylcyclohexane. Alkylation of benzene with cyclooctyl chloride over the same temperature range gave 28.5 to 88.7% phenylcyclooctane and ethylphenycyclohexanes (Eq. 100).

Cl; $AlCl_3$; Ph; C_2H_5; Ph (100)

Some alkylations with multicycloalkyl halides were also reported. An interesting case of Friedel-Crafts alkylation with a symmetrical geminal dihalobicyclic system was reported by Sandler [308], who investigated the reaction of 6,6-dichlorobicyclo-[3.1.0]hexane (85) with benzene in the presence of $AlCl_3$. Far from the author's own expectation, the product of this reaction was found to be mainly a mixture of mono- and diphenylcyclohexane (Eq.

Cl, Cl; 85; $AlCl_3$, 25-30°, 24 hr; Ph; Ph, Ph; + tars; 14.6%; 16.1% (101)

101). The results were justifiably attributed to a reductive type of Friedel-Crafts reaction with accompanying isomerization (Eq. 102). A similar reduc-

Cl, Cl; 85; $AlCl_3$; Cl, Cl; $AlCl_3$; +, Cl, $AlCl_4^-$; +, Cl, $AlCl_4^-$ (102)

tive isomerization was also observed when the related 7.7-dichloronorcarane (86) was treated with benzene and aluminum chloride at 7 to 8°C (Eq. 103)

Cl, Cl; 86; $AlCl_3$, 7-8°; CH_2; 35% of total (103)

[309]. The formation of 1,4-diphenylcyclohexane in the reaction of benzene with 85 was attributed to a reaction of phenylcyclohexane with benzene as suggested earlier [310].

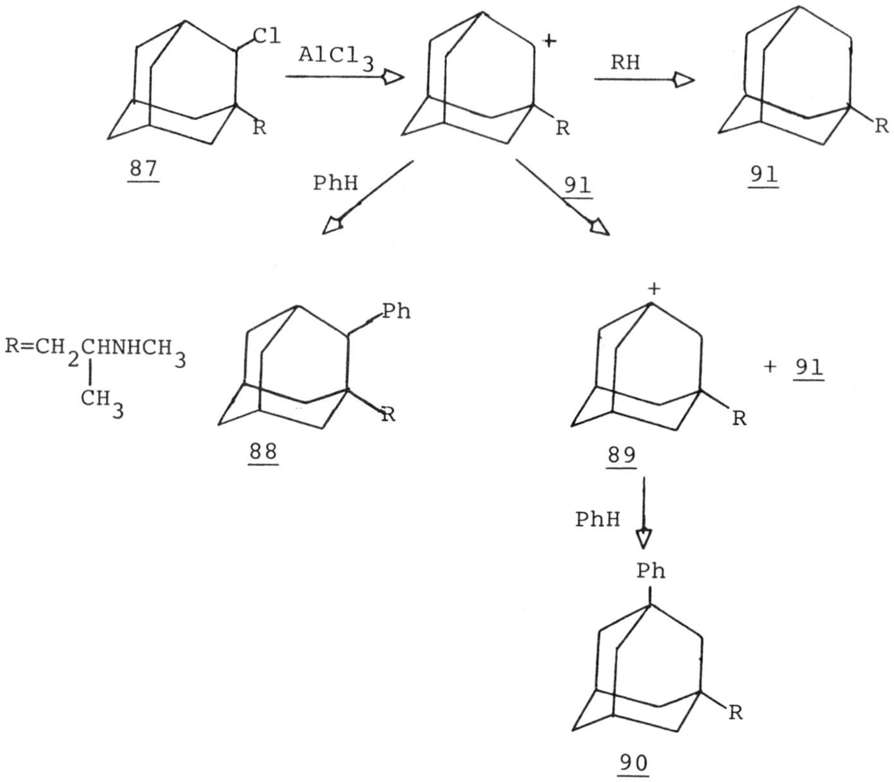

Scheme 25

Another interesting alkylation of benzene with 1-(2-chloroadamant-1-yl)-2-methylaminopropane (87) and $AlCl_3$ was reported by Chakrabarti et al. [311] (scheme 25). When the reaction was carried out with 1% w/v solution of 87 in benzene under reflux for 2 to 3 hr, direct displacement of chlorine with the phenyl group occurred, giving almost a quantitative yield of a 1:1 mixture of the two diastereomers of 1-(2-phenyladamant-1-yl)-2-methylaminopropane (88). On the other hand, when the reaction was conducted in 20% w/v solution of 87 in benzene, it led to a mixture of products consisting of 20% 88, 50% 1-(3-phenyladamant-1-yl)-2-methylaminopropane (90), 15% dehalogenated hydride transfer product 91, and 15% minor constituents and polymeric material. The formation of the bridgehead isomer 90 was rationalized by a process involving intermolecular hydride shifts generating a tertiary adamantyl cation 89. The presence of a considerable amount of the disproportionation product 91, coupled with the fact that the rearrangement was depressed in dilute solutions, gave experimental support to the suggested intermolecular reaction pathway.

Examination of the previous cycloalkylation reactions reveals that they show many of the characteristics of acycloalkylations. They could be complicated by side reactions such as reorientation, disproportionation, polyalkylation, hydride transfer, and skeletal rearrangement, including ring contraction [307] and ring expansion [312]. As is normally expected, the occurrence and the extent of these side reactions are greatly dependent on the nature of the reactants and the severity of the employed conditions. Thc morc severe the conditions are, the more the side reactions will interfere. This can be seen from the effect of reaction variables on the yields and the compositions of the products form cyclopentylation and the cyclohexylation of acenaphthene. As can be seen from Eq. (96), the alkylation of acenaphthene with cyclopentyl chloride in the presence of an equimolar amount of $AlCl_3$ in heptane at 40°C for 5 hr gave 27% 4-cyclopentylacenaphthene. Reduction of reaction duration to 0.5 hr gave a mixture of 3-, 4-, and 5-cyclopentylacenaphthene with the 4 isomer amounting to but 50%, while reduction of catalyst ratio to 0.3 mol and of duration to 5 min gave only the 3 isomer [304]. Again, Eq. (97) shows that heating acenaphthene with cyclohexyl chloride in the presence of $FeCl_3$ for 15 min at 115 to 120°C gave 30% 3- and 5-cyclohexylacenaphthenes in a 1:9 ratio, whereas heating thc rcaction mixture for a period of 0.5 hr gave only the 5-isomer. Moreover, when the cyclohexylation was conducted in heptane solvent in the presence of $AlCl_3$ catalyst and the alkylation mixture kept for 5 hr at 40 to 50°C, the product was only 4-cyclohexylacenaphthene [305].

Another sound example of the effect of conditions on the course of cycloalkylation can be found in the reported reaction of benzene with cyclohexyl bromide under various conditions [312]. Under mild conditions, the cycloalkylation proceeded without isomerization. When more drastic conditions were used, in addition to cycloalkylation, ring contraction to benzylcyclopentane (92) and tolylcyclopentane (93), as well as fragmentation to phenylcyclopentane (94) also occurred (Eq. 104). Moreover, the results showed that $AlBr_3$ had a larger isomerization influence than did $AlCl_3$.

Br; $AlCl_3$ or $AlBr_3$; CH_2; 92; CH_3; 93; 94

(104)

Besides cycloalkylations with halogenated cycloalkanes, some alkylations of arenes with halogenated heterocyclic derivatives were also reported. For example, the alkylation of toluene with the halocyclic ether epichlorohydrin (95) in the presence of aluminum chloride was first indicated by Furakawa and Oda [313] in 1957 to give pure 1-chloro-3-*p*-tolyl-2-propanol (*para*, 96) in 23% yield. A reinvestigation of the same reaction by Sadykh-Zade et al. [314] in 1970 showed that the alkylation product was a mixture of *ortho*, *meta*, and *para* isomers in proportions of 60 to 65:20 to 25:15 to 20, respectively, (Eq. 105). In a similar manner, the alkylation of benzene, ethylbenzene,

$$C_6H_5CH_3 + \underset{\textbf{95}}{CH_2\text{-}CH\text{-}CH_2Cl\ (\text{epoxide})} \xrightarrow[\text{6 hr, 52\% yield}]{AlCl_3,\ 0\text{-}25^\circ} \underset{\textbf{96}}{CH_3C_6H_4CH_2CH(OH)CH_2Cl} \quad (105)$$

(*o*-, *m*- and *p*-isomers)

isopropylbenzene, xylenes, mesitylene, anisole, or ethoxylbenzene with epichlorohydrin gave mixtures of isomers [315].

In a detailed investigation of the synthetic utility of 4-halotetrahydropyrans, Stapp and Drake [316] explored the behavior of 4-chlorotetrahydropyran (97) and 4-chloro-3-methyltetrahydropyran (99) in Friedel-Crafts reactions. Using aluminum chloride as the catalyst, these authors found that 97 smoothly alkylated benzene and toluene to give, respectively, 97% yield of 4-phenyltetrahydropyran (98, R = H) and 76% yield of 4-tolyltetrahydropyran (98), R = CH_3) as a mixture of *ortho*, *meta*, and *para* isomers (Eq. 106).

$$RC_6H_5 + \underset{\textbf{97}}{\text{4-chlorotetrahydropyran}} \xrightarrow[\text{room temp, 1 hr}]{AlCl_3} \underset{\textbf{98}}{RC_6H_4\text{-(4-tetrahydropyranyl)}} \quad (106)$$

R = H, CH_3

A similar alkylation of benzene with *cis-trans*-4-chloro-3-methyltetrahydropyran (99) under conditions chosen to give about 50% conversion gave the rearranged product 3-methyl-3-phenyltetrahydropyran (100, Eq. 107). The recovered 99 was shown to have the same *cis/trans* ratio as before reaction.

$$\underset{\textbf{99}}{\text{4-chloro-3-methyltetrahydropyran}} + C_6H_6 \xrightarrow[50\text{-}60^\circ,\ 30\text{ min}]{AlCl_3} \underset{\textbf{100}\text{, 77\% yield}}{\text{3-methyl-3-phenyltetrahydropyran}} \quad (107)$$

In comparing the alkylation behavior of chlorotetrahydropyrans (97 and 99) with that of epichlorohydrin (95), it is interesting to note that the former involved replacement of the chlorine atom while the latter involved cleavage of the heterocyclic three-membered ring. This difference is clearly the result of the greater thermodynamic stability of the essentially strain-free six-membered tetrahydropyran ring system over the highly strained three-membered cyclic structure. In line with this we shall see in Chap. 4 that related three-, four-, and even five-membered heterocycles have similarly been found to alkylate aromatics through ring cleavage under the influence of Friedel-Crafts catalysts. Another related example of an alkylation with a heterocycle in which the halogen is attached to an aromatic moiety has been described [317]. The results are shown in Eq. (108).

$$\text{2-chloroquinoline} + \text{resorcinol} \xrightarrow[\text{PhNO}_2]{\text{AlCl}_3} \text{2-(2,4-dihydroxyphenyl)quinoline} \tag{108}$$

SUMMARY

We shall summarize some of the important facts about the nature of alkylation reactions. These facts are applicable not only to alkylations with alkyl halides, but also to alkylations with other alkyl derivatives. A consideration of these facts will be helpful in shaping the general picture of a given alkylation reaction.

1. Most of the Friedel-Crafts results published prior to the advent of modern instrumentation are controversial and full of contradictions. Thus one should have great doubt of their validity unless they are substantiated by modern analysis. Correct theories could not be formed until consistent data were available.

2. Friedel-Crafts alkylations are frequently complicated by the involvement of side reactions such as reorientation, disproportionation, dealkylation, fragmentation, hydride transfer, and skeletal rearrangements in the side chain. The occurrence and the extent of such complicating reactions are determined mostly by the severity of the employed catalyst and/or experimental conditions. Because of these side reactions, it is difficult in most alkylations to predict accurately either the structure of the entering alkyl group or its orientation in the nucleus.

3. Skeletal rearrangements in the side chain can occur at a stage prior to, during, or subsequent to the alkylation step. Those occurring in the alkylating agent prior to alkylation or in the alkylation product after its formation can be eliminated partly or completely by the proper choice of catalyst and conditions. Rearrangements intrinsic to the alkylating intermediate carbocation are unavoidable. For example, in the following reactions one cannot prevent the rearrangement in the first steps, but can satisfactorily avoid the rearrangements of the second and third steps.

$$C_6H_6 + \text{C-C(C)-C-Cl} \xrightarrow{(1)} C_6H_5\text{-C(C)(C)-C}$$

$$C_6H_6 + \text{C-C-C(Cl)-C} \xrightarrow{(1)} C_6H_5\text{-C(C-C-C)} \underset{}{\overset{(2)}{\rightleftharpoons}} C_6H_5\text{-C(C)(C)-C}$$

```
(O) + C-C-C-C-C  --(1)-->  C-C-C-C-C  +  C-C-C-C-C  --(2)-->
            |                      |            |
            Cl                     Ph           Ph

            C                    C
            |                    |
          C-C-C-C  --(3)-->    C-C-C
          |                    |  |
          Ph                   Ph C

                                   C                    C C
                                   |                    | |
                                 C-C-C-C                C-C-C
         C                         |                    |
         |                                            
(O) +  C-C-C-C  --(1)-->         (O)   <==(2)==>       (O)
           |
           Cl
```

4. Steric opposition to normal alkylation may cause unusual rearrangements to occur. A good example is the alkylation of *p*-xylene with *t*-butyl and *t*-pentyl halides, in which secondary alkylation is favored over tertiary alkylation.

5. The rate of a given alkylation reaction is a function of many factors, including the electronic and structural nature of the reactants, the severity of the conditions, and the basicity of the solvent, if any. In this regard, the following generalizations may be helpful.

(a). For a given alkyl group the order of halogen reactivity is F > Cl > Br > I.

(b) For a given halogen the order of alkyl group reactivity is tertiary, allyl, benzyl > secondary > primary > methyl. With toluene and $GaBr_3$, for example, the relative rates of reaction of methyl, ethyl, *n*-propyl, isopropyl, and *t*-butyl bromides are 1.00, 13.7, 15.9, 2×10^4 and 8×10^5, respectively. With the less basic benzene, the rates are not so different [4,6,48].

(c) Based on product yields in the alkylation of benzene with *sec*-butyl chloride [208], the following relative order of catalyst activity has recently been established: $AlBr_3$, $AlCl_3$, $MoCl_5$, SbF_5 > $TaCl_5$ > $NbCl_5$ > $FeCl_3$ > $FeCl_3$-CH_3NO_2 (10 mol)-$AlCl_3$-CH_3NO_2 (1.3 mol) > $ZrCl_4$ > $AlCl_3$-CH_3NO_2 (10 mol) > $AlCl_3$-$PhNO_2$ (10 mol) > $TiCl_4$ ≫ WCl_6.

(d) The rates of alkylations also show dependence on steric hindrance and nucleophilicity of the arene. Due to one or both factors, the following general sequence of declining activity of the benzene ring results [181, 182, 188, 318]: 1,2,3- and 1,2,4-trimethyl > dimethyl > Me > Et > Me_2CH > $(Et)_2CH$ > Me_3C > H > I > Br > Cl > F > $COCH_3$ > COOR > NO_2 > CN. Also, although all six of the hydrogen atoms in benzene can be substituted by methyl, ethyl, or *n*-propyl groups [319-321] only four can be replaced by isopropyl groups [320,322] and only three can be replaced by *meta*-oriented *t*-butyl groups [75,323].

6. Whereas alkylations with simple primary alkyl halides require the presence of a strong catalyst such as AlX_3 or GaX_3, those with "assisted" primary alkyl halides (e.g., benzyl, allyl, or neophyl halides) or with secondary and tertiary alkyl halides can be initiated by milder catalysts, such as $FeCl_3$, $AlCl_3$-CH_3NO_2, and BX_3.

7. The tendency for polysubstitution may be minimized by the use of a higher aromatic/alkyl halide ratio, by the use of a mutual solvent for the hydrocarbon and catalyst layers, by high-speed stirring, by operating in

the vapor phase, or by operating at a temperature sufficiently high that $AlCl_3$ is soluble in the hydrocarbon layer [151,185,324].

8. Alkylations by linear primary halides yield mixtures in which the aryl group will be found attached to various carbon atoms of the chain; e.g., from *n*-propyl halide both normal and iso isomer will be obtained, and from *n*-pentyl or *n*-hexyl halide a mixture of 1-, 2-, and 3-arylalkanes will be formed. In such alkylations, the ratio of the primary to secondary alkylate has been shown to be dependent on reaction conditions as well as the type of halogen present in the alkylating agent.

9. Alkylations by linear secondary halides yield mixtures of all possible secondary alkylates; e.g., from *n*-pentyl halides a mixture of 2- and 3-arylpentane is produced. Moreover, there is a tendency among the lower members of this class to produce the various secondary isomers in ratios reflecting statistical distribution under kinetically controlled conditions and thermodynamic distribution under thermodynamically controlled conditions. The latter conditions include the use of strong catalysts and/or high temperatures. In general, it has been noted that deviation from statistical distribution increases by increasing the length of the alkylating group.

10. Branched-chain alkyl halides yield tertiary alkylates under nonisomerizing conditions and mixtures of secondary and tertiary alkylates under isomerizing conditions; e.g., alkylations of benzene by branched pentyl halides give *t* pentylbenzene in the presence of nonisomerizing catalysts and mixtures of this and 2-methyl-3-phenylbutane in the presence of isomerizing catalysts. However, when severe steric inhibition to *t*-alkylation exists, as in the alkylation of *p*-xylene with *t*-pentyl halides, only secondary alkylation occurs. In alkylations of substituted aromatics, the general rules governing orientation in electrophilic substitution reactions are followed in reactions carried out under kinetically controlled conditions, but change to the more stable *m*-oriented isomers prevails in thermodynamically controlled reactions. However, no tertiary alkylation occurs *ortho* to another group on an aromatic nucleus.

11. In the alkylation of naphthalene, variations in the catalytic conditions make a considerable difference in the ratio of the 1- and 2-alkylnaphthalenes formed. Whereas the formation of 1-alkylnaphthalenes is favored by mild conditions, including the use of weak catalysts, increasing amounts of the strong catalyst aluminum chloride and longer reaction times favor rearrangement to 2-alkylnaphthalenes.

12. No one mechanism can account for all alkylations [325]. Depending on conditions and solvent medium, the mechanism for the same alkylation may vary from one in which the alkylating agent is the free carbocation to another in which the alkylating agent is a donor-acceptor complex. This will lead to variations in transitions states and hence in product compositions [93,95, 101,181,182].

13. Because of the activity of the attacking electrophiles, most Friedel-Crafts alkylations are characterized by a low substrate but high positional selectivity [185].

REFERENCES

1. Friedel, C., and J. M. Crafts, Ann. Chim., *1*, 449 (1884).
2. Price, C. C., *Organic Reactions*, Vol. 3, Wiley, New York, 1946, Chap. 1.

3. Brown, H. C., and H. Junk, J. Am. Chem. Soc., *77*, 5584 (1955).
4. Smoot, C. R., and H. C. Brown, J. Am. Chem. Soc., *78*, 6245 (1956).
5. Smoot, C. R., and H. C. Brown, J. Am. Chem. Soc., *78*, 6249 (1956).
6. Brown, H. C., and C. R. Smoot, J. Am. Chem. Soc., *78*, 6255 (1956).
7. Choi, S. U., and J. E. Willard, J. Am. Chem. Soc., *87*, 3072 (1965).
8. DeHaan, E. P., and H. C. Brown, J. Am. Chem. Soc., *91*, 4844 (1969).
9. DeHaan, E. P., H. C. Brown, and J. C. Hill, J. Am. Chem. Soc., *91*, 4850 (1969).
10. DeHaan, E. P., H., C. Brown, D. C. Conway, and M. G. Gibby, J. Am. Chem. Soc., *91*, 4854 (1969).
11. Brown, H. C., and M. Grayson, J. Am. Chem. Soc., *75*, 6285 (1953).
12. Olah, G. A., S. J. Kuhn, and J. A. Olah, J. Chem. Soc., 2174 (1957).
13. Brown, H. C., and B. A. Bolto, J. Am. Chem. Soc., *81*, 3320 (1959).
14. Stock, L. M., and H. C. Brown, J. Am. Chem. Soc., *81*, 3323 (1959).
15. Olah, G. A., S. J. Kuhn, and S. Flood, J. Am. Chem. Soc., *84*, 1688 (1962).
16. Olah, G. A., S. J. Kuhn, and S. H. Flood, J. Am. Chem. Soc., *84*, 1695 (1962).
17. Olah, G. A., and J. A. Olah, J. Org. Chem., *32*, 1612 (1967).
18. Olah, G. A., M. Tashiro, and S. Kobayashi, J. Am. Chem. Soc., *92*, 6369 (1970).
19. Olah, G. A., S. Kobayashi, and M. Tashiro, J. Am. Chem. Soc., *94*, 7448 (1972).
20. Tadzhimkhamedov, Kh. S., A. R. Abdurasuleva, K. N. Akhmedov, E. R. Riskibaev, and N. U. Yusupov, Zh. Org. Khim., *11*, 1665 (1975); J. Org. Chem. USSR, *11*, 1655 (1975).
21. Decoret, C., J. Royer, and O. Chalvet, Tetrahedron, *31*, 973 (1975).
22. Arata, K., N. Azumi, and H. Sawamura, Bull. Chem. Soc. Jpn., *48*, 2944 (1975).
23. Arata, K., K. Sato, and I. Toyoshima, J. Catal., *42*, 222 (1976).
24. Arata, K., Y. Katsumasa, and I. Toyoshima, J. Catal., *44*, 385 (1976).
25. Takematsu, A., K. Sugita, and R. Nakane, Bull. Chem. Soc. Jpn., *51*, 2082 (1978).
26. Brown, H. C., and W. J. Wallace, J. Am. Chem. Soc., *75*, 6268 (1953).
27. Brown, H. C., L. P. Eddy, and R. Wong, J. Am. Chem. Soc., *75*, 6275 (1953).
28. Baddeley, G., Q. Rev. (Lond), *8*, 355 (1954).
29. Olah, G. A., and M. W. Meyer, in *Friedel-Crafts and Related Reactions*, Vol. 1, G. A. Olah, ed., Wiley-Interscience, New York, 1963, pp. 623-765.
30. Fairbrother, F., Trans. Faraday Soc., *37*, 763 (1941); J. Chem. Soc., 503 (1948).
31. Jacober, W. J., and C. A. Kraus, J. Am. Chem. Soc., *71*, 2405, 2409 (1949).
32. Dove, M. F. A., and D. B. Sowerky, in *Halogen Chemistry*, Vol. 1, (V. Gutmann, ed.), Academic Press, New York, 1967, Chap. 2, See references therein.
33. Sixma, F. L. J., H. Hendriks, and D. Holtzapffel, Recl. Trav. Chim. Pays-Bas, *75*, 127 (1956).
34. Kwun, O. C., and S. U. Choi, J. Phys. Chem., *72*, 3148 (1968).
35. Polaczek, A., and A. Halpern, Nukleonika, *8*, 667 (1963); Nature, *199*, 1286 (1963); Chem. Abstr. *59*, 14828g (1963).
36. DeHaan, F. P., M. G. Gibby, and D. R. Aebersold, J. Am. Chem. Soc., *91*, 4860 (1969).

37. Olah, G. A., J. R. DeMember, and R. H. Schlosberg, J. Am. Chem. Soc., *91*, 2112 (1969).
38. Olah, G. A., J. R. DeMember, R. H. Schlosberg, and Y. Halpern, J. Amer. Chem. Soc., *94*, 156 (1972).
39. Olah, G. A., P. Schilling, J. M. Bollinger and J. Niahimura, J. Am. Chem. Soc., *96*, 2221 (1974).
40. Olah, G. A., D. J. Donovan, and H. C. Lin, J. Am. Chem. Soc., *98*, 2661 (1976).
41. Bacon, J., and R. J. Gillespie, J. Am. Chem. Soc., *93*, 6914 (1971).
42. Peterson, P. E., R. Brockington, and D. W. Vidrine, J. Am. Chem. Soc., *98*, 2660 (1976).
43. Meerwein, H., in Houben-Weyl, *Methoden der organischen Chemie*, 2nd ed., Vol. VI/3,6 G. Thieme, Stuttgart, 1965, pp. 325-366. See references therein.
44. Olah, G. A., and J. R. DeMember, J. Am. Chem. Soc., *91*, 2113 (1969); *92*, 718, 2562 (1970).
45. Ahmed, M. G., R. W. Alder, G. H. James, M. L. Sinnott, and M. C. Whiting, Chem. Commun., 1533 (1968).
46. Ahmed, M. G., and R. W. Alder, Chem. Commun., 1389 (1969).
47. Schmerling. L., Ind. Eng. Chem., *45*, 1462 (1953).
48. Brown, H. C., and H. Junk, J. Am. Chem. Soc., *78*, 2182 (1956).
49. Junk, H., C. R. Smoot, and H. C. Brown, J. Am. Chem. Soc., *78*, 2185 (1956).
50. Choi, S. U., and H. C. Brown, J. Am. Chem. Soc., *81*, 3315 (1959).
51. Choi, S. U., and H. C. Brown, J. Am. Chem. Soc., *85*, 2596 (1963).
51a. Olah, G. A., and J. A. Olah, J. Am. Chem. Soc., *98*, 1843 (1976).
52. Tsuge, O., and M. Tashiro, Bull. Chem. Soc. Jpn., *40*, 125 (1967).
53. Stock. L. M., and H. C. Brown, Adv. Phys. Org. Chem., *1*, 35 (1963), and other references therein.
54. Carter, B. J., W. D. Corey, and F. P. DeHaan, J. Am. Chem. Soc., *97*, 4783 (1975).
55. White, J. F., and M. F. Farona, J. Organomet. Chem., *63*, 329 (1973).
56. Tanabe, K. T., Yamagatsu, and T. Takeshita, J. Res. Inst. Catal., Hokkaido Univ., *12*, 230 (1965); Chem. Abstr., *63*, 8145f (1965).
57. Arata, K., T. Takeshita, and K. Tanabe, Shokubai, *8*, 226 (1966).
58. Takeshita, T., K. Arata, T. Sano, and K. Tanabe, Kogyo Kagaku Zasshi, *69*, 916 (1966); Chem. Abstr., *66*, 45989h (1967).
59. Arata, K., and I. Toyoshima, Chem. Lett., 929 (1974).
60. Murohashi, S., Nagaoka Koto Semmon Gakko Kenkyu Kiyo, *9*, 103, 107 (1973); Chem. Abstr., *80*, 132917r, 132946z (1974).
61. Warshawsky, A., R. Kalir, and A. Patchornik, J. Org. Chem. *43*, 3151 (1978).
62. Montaudo, G., P. Funocchiaro, S. Caccamese, and F. Bottino, J. Chem. Eng. Data, *16*, 249 (1971).
63. Schmerling, L., U. S. Patent 3,679,760 (1972); Chem. Abstr., *77*, 101092b (1972).
64. Institut de Recherches Chimiques et Biologiques Appliquees, Fr. Demande 2,113,806 (1972); Chem. Abstr., *78*, 58032p (1973).
65. Abdurasuleva, A. R., and K. N. Akhmedov, Zh. Org. Khim., *8*, 134 (1972); Chem. Abstr., *76*, 112816w (1972).
66. Abdurasuleva, A. R., K. N. Akhmedov, A. Yusupov, and Kh. S. Tadzhimukhamedov, USSR Patent 327,150 (1972); Chem. Abstr., *76*, 140201k (1972).

67. Abdurasuleva, A. R., K. N. Akhmedov, and Kh. S. Tadzhimukhamedov, Uzb. Khim. Zh., No. 2, 54 (1972); Chem. Abstr., *77*, 74942v (1972).
68. Abdurasuleva, A. R., K. N. Akhmedov, Kh. S. Tadzhimukhamedov, and B. Kh. Pulatov, Uzb. Khim. Zh., *16*, 62 (1972); Chem. Abstr. *77*, 100999r (1972).
69. Abdurasuleva, A. R., K. N. Akhmedov, and M. K. Tureva, Zh. Org. Khim., *9*, 132 (1973); Chem. Abstr., *78*, 110954s (1973).
70. Abdurasuleva, A. R., K. N. Akhmedov, Kh. S. Tadzhimukhamedov, M. N. Khodzhaeva, and M. N. Nurmiraeva, Tr. Tashk. Gos. Univ. Vopr. Khim., *419*, 144 (1972); Chem. Abstr., *79*, 183105 (1973).
71. Simons, J. H., and H. Hart, J. Am. Chem. Soc., *69*, 979 (1947).
72. Serijan, K. T., H. F. Hipsher, and L. C. Gibbons, J. Am. Chem. Soc., *71*, 873 (1949).
73. Condon, F. E., J. Am. Chem. Soc., *71*, 3544 (1949).
74. Brown, H. C., and K. L. Nelson, J. Am. Chem. Soc., *75*, 6292 (1953).
75. Myhre, P. C., T. Rieger, and J. T. Stone, J. Org. Chem. *31*, 3425 (1966).
76. Stahl, J., Ber., *23*, 988 (1890).
77. Norris, J. F., and D. Rubinstein, J. Am. Chem. Soc., *61*, 1163 (1939).
78. Snyder, H. R., R. R. Adams, and A. V. McIntosh, Jr., J. Am. Chem. Soc., *63*, 3280 (1941).
79. Babin, E. P., V. I. Lozovoi, N. A. Goryunova, and N. I. Danilova, Ref. Zh., Khim, Abstr. No. 10zh240 (1971); Chem. Abstr., *78*, 29335j (1973).
79a. Yonida, N., T. Fukuhara, Y. Takahashi, and A. Suzuki, Chem. Lett., 1003 (1979); Chem. Abstr., *91*, 192604y (1979).
80. Roberts, R. M., G. A. Ropp, and O. K. Neville, J. Am. Chem. Soc., *77*, 1764 (1955).
81. Friedman, B. S., F. L. Morritz, C. J. Morrissey, and R. Koncos, J. Am. Chem. Soc., *80*, 5867 (1958).
82. Sharman, S. H., J. Am. Chem. Soc., *84*, 2945 (1962).
83. Roberts, J. D., R. E. McMahon, and J. S. Hine, J. Am. Chem. Soc., *72*, 4237 (1950).
83a. Pines, H., E. Aristoff, and V. N. Ipatieff, J. Am. Chem. Soc., *71*, 749 (1949).
84. McGuire, S. E., Ph. D. dissertation, University of Texas at Austin, 1967.
85. Roberts, R. M., and A. A. Khalaf, unpublished results.
86. Roberts, R. M., and S. G. Panayides, J. Org. Chem., *23*, 1080 (1958).
87. Lee, C. C., M. C. Hamblin, and N. James, Can. J. Chem., *36*, 1957 (1958).
88. Nakane, R., O. Kurihara, and A. Natsubori, J. Am. Chem. Soc., *91*, 4528 (1969).
89. Brown, H. C., and A. R. Neyens, J. Am. Chem. Soc., *84*, 1233 (1962).
90. Olah, G. A., A. E. Pavlath, and J. A. Olah, J. Am. Chem. Soc., *80*, 6540 (1958).
91. Olah, G. A., and S. J. Kuhn, J. Am. Chem. Soc., *80*, 6541 (1958).
92. Nakane, R., O. Kurihara, and A. Natsubori, J. Phys. Chem., *68*, 2876 (1964).
93. Nakane, R., A. Natsubori, and O. Kurihara, J. Am. Chem. Soc., *87*, 3597 (1965).
94. Nakane, R., and T. Oyama, J. Phys. Chem., *70*, 1146 (1966).
95. Nakane, R., and A. Natsubori, J. Am. Chem. Soc., *88*, 3011 (1966).

96. Nakane, R., T. Oyama, and A. Natsubori, J. Org. Chem., *33,* 275 (1968).
97. Ransley, D. L., J. Org. Chem., *33,* 1517 (1968).
98. Ransley, D. L., J. Org. Chem., *34,* 2618 (1969).
99. Oyama, T., and R. Nakane, J. Org. Chem., *34,* 949 (1969).
100. Natsubori, A., and R. Nakane, J. Org. Chem., *35,* 3372 (1970).
101. Nakane, R., O. Kurihara, and A. Takematsu, J. Org. Chem., *36,* 2753 (1971).
102. Oyama, T., T. Hamano, K. Nagumo, and R. Nakane, Bull. Chem. Soc. Jpn., *51,* 1441 (1978).
103. Radziszewski, Br., Chem. Ber., *7,* 141 (1874).
104. Huston, R. C., and T. E. Friedmann, J. Am. Chem. Soc., *40,* 785 (1918).
105. Ninetzescu, C. D., D. A. Isacescu, and C. N. Ionescu, Justus Liebigs Ann. Chem., *491,* 210 (1931).
106. Lee, C. C., A. G. Forman, and A. Rosenthal, Can. J. Chem., *35,* 220 (1957).
107. McMahon, M. A., and S. C. Bunce, J. Org. Chem., *29,* 1515 (1964).
108. Streitwieser, A., Jr., and L. Reif, J. Am. Chem. Soc., *82,* 5003 (1960).
109. Streitwieser, A., Jr. and W. J. Downs, J. Org. Chem., *27,* 625 (1962).
110. Unsersen, E., and A. P. Wolf, J. Org. Chem., *27,* 1509 (1962).
111. Roberts, R. M., A. A. Khalaf, and R. N. Greene, J. Am. Chem. Soc., *86,* 2846 (1964).
112. Roberts, R. M., and S. G. Brandenberger, J. Am. Chem. Soc., *79,* 5484 (1957).
113. Roberts, R. M., Y. W. Han, C. H. Schmid, and D. A. Davis, J. Am. Chem. Soc., *81,* 640 (1959).
114. Olah, G. A., M. B. Comisarow, E. Namanworth, and B. Ramsey, J. Am. Chem. Soc., *89,* 5259 (1967).
115. Olah, G. A., and R. D. Porter, J. Am. Chem. Soc., *92,* 7627(1970).
116. Schwartz, L. H., J. Landies, S. B. Lazarus, and S. H. Stoldt, J. Org. Chem., *37,* 1979 (1972).
117. Lloyd, J. B. F., and P. A. Ongley, Tetrahedron, *21,* 245 (1965).
118. Horner, L., P. V. Subramaniam, and E. Eiben, Justus Liebigs Ann. Chem. *714,* 91 (1968).
119. Ichii, M., T. Sugiyama, and S. Oka, Bull. Inst. Chem. Res., Kyoto Univ., *50,* 404 (1972); Chem. Abstr., *78,* 97221u (1973).
120. Hart, H., and H. S. Eleuterio, J. Am. Chem. Soc., *76,* 516 (1954).
121. Hart, H., W. L. Spliethoff, and H. S. Eleuterio, J. Am. Chem. Soc., *76,* 4547 (1954).
122. Hart, H., and J. H. Simons, J. Am. Chem. Soc., *71,* 345 (1949).
123. Price, C. C., *Mechanisms of Reactions at Carbon-Carbon Double Bonds,* Interscience, New York, 1946, p. 51.
124. Fuson, R. C., *Advanced Organic Chemistry,* Wiley, New York, 1950, p. 324.
125. Royals, E. E., *Advanced Organic Chemistry,* Prentice-Hall, New York, 1954, p. 472.
126. Cram, D. J., and G. S. Hammond, *Organic Chemistry,* McGraw-Hill, New York, 1959, p. 375.
127. Olah, G. A., in *Friedel-Crafts and Related Reactions,* Vol. 1, (G. A. Olah, ed.), Wiley-Interscience, New York, 1963, Chap. 2, p. 68.

128. Bethell, D., and V. Gold, *Carbonium Ions, an Introduction*, Academic Press, New York, 1967, p. 181.
129. Gustavson, G., Bull. Soc. Chim. Fr., *30*, 22 (1878); J. Russ. Phys.-Chem. Soc., *10*, 269 (1878); Chem. Ber., *11*, 1251 (1878).
130. Silva, R. D., Bull. Soc. Chim. Fr., (2), *43*, 317 (1885); J. Chem. Soc. Abstr., *48*, 1054 (1885).
131. Heise, R., Ber., *24*, 768 (1891).
132. Konowalow, M. I., J. Russ. Phys.-Chem., *27*, 456 (1895); Bull. Soc. Chim. Fr., *16*, 864 (1896).
133. Becke, P., Ber., *23*, 3191 (1890).
134. Gustavson, G., C. R. Acad. Sci., *140*, 940 (1905).
135. Nightingale, D., and B. Carton, Jr., J. Am. Chem. Soc., *62*, 280 (1940).
136. Boedtker, E., and O. Rambech, Bull. Soc. Chim. Fr., (4), *35*, 631 (1924).
137. Roux, M. L., Bull. Soc. Chim. Fr., (2), *41*, 379 (1884).
138. Roux, M. L., Ann. Chim. Phys., *12*, 289 (1887).
139. Gilman, H., and N. O. Calloway, J. Am. Chem. Soc., *55*, 4197 (1933).
140. Wertyporoch, E., and T. Firla, Justus Liebigs Ann. Chem., *500*, 287 (1933).
141. Haworth, R. D., B. M. Letsky, and C. R. Mavin, J. Chem. Soc., 1784 (1932).
141a. Ipatieff, V. N., H. Pines, and L. Schmerling, J. Org. Chem. *5*, 253 (1940).
142. Diuguid, L. T., J. Am. Chem. Soc., *63*, 3527 (1941).
143. Topchiev, A. V., B. A. Krentsel, and L. N. Andreev, Dokl. Akad. Nauk SSSR, *92*, 781 (1953); Chem. Abstr., *49*, 3039 (1955).
144. Turova-Pollak, M. B., and M. A. Maslova, Zh. Obshch. Khim., *26*, 2185 (1956); Chem. Abstr., *51*, 4972c (1957).
145. Nightingale, D. V., and J. M. Shackelford, J. Am. Chem. Soc., *78*, 1225 (1956).
146. Gould, E. S., *Mechanism and Structure in Organic Chemistry*, Henry Holt, New York, 1959, p. 450.
147. Roberts, R. M., and D. Shiengthong, J. Am. Chem. Soc., *82*, 732 (1960).
148. Roberts, R. M., and D. Shiengthong, J. Am. Chem. Soc., *86*, 2851 (1964).
149. Streitwieser, A., Jr., W. D. Schaeffer, and S. Andreades, J. Am. Chem. Soc., *81*, 1113 (1959).
150. Lee, C. C., and D. J. Woodcock, Can. J. Chem., *48*, 858 (1970).
151. Olah, G. A., and D. Meider, Nouv. J. Chim., *3*, 269 (1979).
152. Schramm, J., Monatsh. Chem., *9*, 613 (1888); J. Chem. Soc. Abstr., *127*, 127 (1889).
153. Estreicher, T., Ber., *33*, 436 (1900); J. Chem. Soc. Abstr., *78*, 213 (1900).
154. Petrova, A. M., Zhr. Obshch. Khim., *24*, 491 (1954); Chem. Abstr., *49*, 6150i; (1955).
155. Firla, T., Rocz. Chem., *14*, 87 (1934); Chem. Abstr., *28*, 6426 (1934).
156. Callaway, N., O., J. Am. Chem. Soc., *59*, 1474 (1937).
157. Shoesmith, J. B., and J. F. McGechen, J. Chem. Soc., 2231 (1930).

158. Nightingale, D., and L. I. Smith, J. Am. Chem. Soc., *61*, 101 (1939).
159. Tsukervanik, I. P., and V. D. Tambovtseva, Bull. Univ. Asie Cent., *22*, 221 (1938); Chem. Abstr., *34*, 4729 (1940).
160. Gilman, H., and R. R. Burtner, J. Am. Chem. Soc., *57*, 909 (1935).
160a. Nightingale, D. V., and J. M. Shackelford, J. Am. Chem. Soc., *76*, 5767 (1954).
161. Friedel, C., and J. M. Crafts, C. R. Acad. Sci., *84*, 1392 (1877); J. Chem. Soc. Abstr., 725 (1877); C. R. Acad. Sci., *85*, 74 (1877).
162. Gilman, H., and J. A. V. Turck, Jr., J. Am. Chem. Soc., *61*, 473 (1939).
163. Streitwieser, A., Jr., D. P. Stevenson, and W. D. Schaffer, J. Am. Chem. Soc., *81*, 1110 (1959).
164. Olah, G. A., and J. Lukas, J. Am. Chem. Soc., *89*, 2227 (1967).
165. Meals, R. N., and H. Gilman, Proc. Iowa Acad. Sci., *48*, 250 (1941).
166. Gilman, H., and R. N. Meals, J. Org. Chem., *8*, 126 (1942).
167. Nightingale, D., and O. G. Shanholtzer, J. Org. Chem., *7*, 6 (1942).
168. Thomas, C. A., *Anhydrous Aluminum Chloride in Organic Chemistry*, Reinhold, New York, 1941, pp. 78-79, 174-178.
169. Baker, J. R., M. A. thesis, University of Texas at Austin, 1966.
170. Nash, L. M., T. I. Taylor, and W. E. Doering, J. Am. Chem. Soc., *71*, 1516 (1949).
171. Seidel, F., and O. Engelfried, Ber., *69*, 2567 (1936); Br. Chem. Abstr., B, 116 (1937).
172. Olson, A. C., Ind. Eng. Chem., *52*, 833 (1960).
173. Sharman, S. H., J. Am. Chem. Soc., *84*, 2951 (1962), and references therein.
174. Brown, H. C., and K. L. Nelson, in *The Chemistry of Petroleum Hydrocarbons*, Vol. 3, (B. T. Brooks, S. S. Kulz, Jr., C. E. Boord, and L. Schmerling, eds.), Reinhold, New York, 1955, pp. 465-578.
174a. Asinger, F., B. Fell, H. Verbeck, and J. Fernandez-Bustillo, Erdoel Kohle, Erdgas, Petrochem., *20*, 852 (1967); Chem. Abstr., *68*, 48718w (1968).
175. Silva, R. D., Bull. Soc. Chim. Fr., (2), *29*, 193 (1878).
176. Kelbe, W., Justus Liebigs Ann. Chem., *210*, 25 (1881).
177. von Auwers, K., and H. Kollings, Ber., *55*(B), 3872 (1922); J. Chem. Soc. Abstr., *124* (I), 99 (1923).
178. (a) Roberts, R. M., Chem. Eng. News, *43*, 96 (1965); (b) Francis, A. W., Chem. Rev., *43*, 257 (1948). See references therein.
179. von der Becke, P., Ber., *23*, 3191 (1890); J. Chem. Soc. Abstr., *60* (I), 183 (1891).
180. Allen, R. H., and L. D. Yats, J. Am. Chem. Soc., *83*, 2799 (1961).
181. Olah, G. A., S. H. Flood, S. J. Kuhn, M. E. Moffatt, and N. A. Overchuck, J. Am. Chem. Soc., *86*, 1046 (1964).
182. Olah, G. A., S. H. Flood, and M. E. Moffatt, J. Am. Chem. Soc., *86*, 1065 (1964).
183. Ref. 127, Vol. 1, Chaps. 8 and 11.
184. Olah, G. A., in *Organic Reaction Mechanisms*, Spec. Publ. *19*, The Chemical Society, London, 1965.
185. Olah, G. A., *Friedel-Crafts Chemistry*, Wiley, New York, 1973.

186. For other reviews see: (a) Breslow, R., *Organic Reaction Mechanisms*, W. A. Bemjamin, New York, 1969; (b) Stock, L., *Aromatic Substitution Reactions*, Prentice-Hall, Englewood Cliffs, N. J., 1968; (c) Norman, R. O. C., and R. Taylor, *Electrophilic Substitution in Benzenoid Compounds*, American Elsevier, New York, 1965; (d) De La Mare, P. B. D., and J. Ridd, *Aromatic Substitutions—Nitration and Halogenation*, Academic Press, New York, 1959; (e) Berliner, E., Prog. Phys. Org. Chem., *2*, 253 (1964); (f) Ingold, C. K., *Structure and Mechanism in Organic Chemistry*, 2nd ed., Cornell University Press, Ithaca, N. Y., 1969, pp. 264-417; (g) March, J., *Advanced Organic Chemistry, Reaction Mechanisms, and Structure*, McGraw-Hill, New York, 1968, pp. 376-441.
187. Brown, H. C., and J. D. Brady, J. Am. Chem. Soc., *74*, 3570 (1952).
188. Condon, F. E., J. Am. Chem. Soc., *70*, 2265 (1948).
189. Brown, H. C., and H. Junk, J. Am. Chem. Soc., *77*, 5579 (1955).
190. Brown, H. C., and C. R. Smoot, J. Am. Chem. Soc., *78*, 2176 (1956).
191. Olah, G. A., S. J. Kuhn, and S. H. Flood, J. Am. Chem. Soc., *83*, 4851 (1961).
192. Dewar, M. J. S., and N. A. Puttnam, J. Chem. Soc., 959 (1960).
193. Roberts, R. M., and K. Y. Zee-Cheng, abstracted from the Ph.D. Dissertation of the latter, University of Texas at Austin, 1963.
194. Gambacorta, A., R. Nicoletta, and A. Oratore, Chem. Ind. (Milan), *54*, 997 (1972); Chem. Abstr., *78*, 124328t (1973).
195. Valenta, M., and I. Koubek, Collect. Czech. Chem. Commun., *41*, 78 (1976); Chem. Abstr., *85*, 5432 (1976).
196. Hennion, G. F., and R. A. Kurtz, J. Am. Chem. Soc., *65*, 1001 (1943).
197. Olah, G. A., and S. J. Kuhn, J. Org. Chem., *29*, 2317 (1964).
198. Olah, G. A., and S. Kobayashi, J. Am. Chem. Soc., *93*, 6964 (1971).
199. Nakane, R., Yuki-Gosei Kagaku Kyokai Shi, *29*, 1020 (1971); Chem. Abstr., *76*, 98564d (1972).
200. Dewar, M. J. S., Annu. Rept. Prog. Chem., *53*, 132 (1956).
201. Dewar, M. J. S., *Electronic Theory of Organic Chemistry*, Oxford University Press, New York, 1949, p. 432.
202. Melander, L., Acta Chem. Scand., *3*, 95 (1949); Nature, *163*, 599 (1949); Ark. Kemi, *2*, 213 (1950).
203. Olah, G. A., S. J. Kuhn, and S. H. Flood, J. Am. Chem. Soc., *83*, 4571 (1961).
204. Christen, M., W. Koch, W. Simon, and H. Zollinger, Helv. Chim. Acta, *45*, 2077 (1962).
205. Olah, G. A., S. H. Flood, and M. E. Moffatt, J. Am. Chem. Soc., *86*, 1060 (1964).
206. (a) Olah, G. A., S. J. Kuhn, S. H. Flood, and B. A. Hardie, J. Am. Chem. Soc., *86*, 1039, 1044 (1964); (b) Olah, G. A., S. J. Kuhn, and B. A. Hardie, J. Am. Chem. Soc., *86*, 1055 (1964).
207. (a) Olah, G. A., S. J. Kuhn, and S. H. Flood, J. Am. Chem. Soc., *83*, 4571 (1961); (b) Olah, G. A., and S. J. Kuhn, J. Am. Chem. Soc., *86*, 1067 (1964); (c) Olah, G. A., P. E. Alexander, S. C. Narang, and J. A. Olah, J. Org. Chem., *46*, 3533 (1981).
208. Masuda, S., T. Nakajima, and S. Suga, J. Chem. Soc., Chem. Commun., 954 (1974).

209. Segi, M., T. Nakajima, and S. Suga, Bull. Chem. Soc. Jpn., *53,* 1465 (1980).
210. Dewar, M. J. S., and N. A. Puttnam, J. Chem. Soc., 4080 (1959).
211. (a) Price, C. C., and M. Lund, J. Am. Chem. Soc., *62,* 3105 (1940); (b) Burwell, R. L., and S. Archer, J. Am. Chem. Soc., *64,* 1032 (1942); (c) Spanninger, P. A., and J. L. von Rosenberg, J. Am. Chem. Soc., *94,* 1973 (1972); (d) Suga, S., M. Segi, K. Kitano, S. Masuda, and T. Nakajima, Bull. Chem. Soc. Jpn., *54,* 3611 (1981).
212. Burwell, R. L., Jr., L. M. Elkin, and A. D. Shields, J. Am. Chem. Soc., *74,* 4570 (1952).
213. Tsukervanik, I. P., and Z. N. Nazarova, Zh. Obshch. Khim., *7,* 623 (1937); Chem. Abstr., *31,* 5778 (1937).
214. Pines, H., D. W. Huntsman, and V. N. Ipatieff, J. Am. Chem. Soc., *73,* 4483 (1951).
215. Roberts, R. M., S. E. McGuire, and J. R. Baker, J. Org. Chem., *41,* 659 (1976).
216. Mazonski, T., and A. Hopfinger, Przem. Chem., *40,* 453 (1961); Chem. Abstr., *62,* 3957f (1965).
217. Alul, H. R., J. Org. Chem., *33,* 1522 (1968).
218. Kondelik, P., J. Pasek, and M. Ranny, Tenside Deterg., *16,* 20 (1979); Chem. Abstr., *90,* 139335g (1979).
219. McGuire, S. E., and G. E. Nicks, J. Am. Oil Chem. Soc., 48, 872 (1971).
220. Yoneda, N., Sekiyu Gakkai Shi, *15,* 894 (1972); Chem. Abstr., *78,* 74369j (1972).
221. Sakic, A., M. Orlov, S. Mijovic, and Z. Binenfeld, Hem. Ind., *29,* 343 (1975); Chem. Abstr., *83,* 163725t (1975).
222. Aliev, S. M., N. K. Rzaev, and V. N. Nagiev, Azerb. Neft. Khoz., 37 (1976); Chem. Abstr., *87,* 184101j (1977).
223. Kondelik, P., J. Pasek, and M. Kovar, Sb. Vys. Sk. Chem.-Technol. Praze, Org. Chem. Technol., *C25,* 65 (1978); Chem. Abstr., *91,* 192922a (1979).
224. Gossin, E., Bull. Soc. Chim., Fr., (2), *41,* 446 (1884); J. Chem. Soc. Abstr., 1312 (1884).
225. (a) Boedtker, E., Bull. Soc. Chim. Fr., (3), *31,* 965 (1904); J. Chem. Soc. Abstr., *86,* 801 (1904); (b) Gilman, H., N. O. Calloway, and R. R. Burtner, J. Am. Chem. Soc., *57,* 906 (1935).
226. Senkowski, M., Ber., *23,* 2412 (1890).
227. Schlatter, M. J., and R. D. Clark, J. Am. Chem. Soc., *75,* 361 (1953).
228. Brady, P. A., J. Carnduff, and D. G. Leppard, Tetrahedron Lett., 4183 (1972).
229. Bartlett, P. D., M. Roha, and M. Stiles, J. Am. Chem. Soc., *76,* 2349 (1954).
230. Allen, R. H., J. Am. Chem. Soc., *82,* 4856 (1960).
231. (a) Ono, A., Nippon Kagaku Zasshi, *91,* 1005 (1970); Chem. Abstr., *74,* 76101g (1971); (b) Olah, G. A., M. W. Meyer, and N. A. Overchuck, J. Org. Chem., *29,* 2310 (1964).
232. Kelbe, W., and A. Bauer, Ber., *16,* 2559 (1883).
233. Bauer, A., Ber., *24,* 2832 (1891).
234. Buu-Hoi, N. G., and P. Cagniant, Bull. Soc. Chim. Fr., *9,* 887 (1942).
235. von Auwers, K., Ber., *49,* 2403 (1916).

236. Simons, J. H., and S. Archer, J. Am. Chem. Soc., *60,* 2953 (1938).
237. Sprauer, J. W., and H. Simons, J. Am. Chem. Soc., *64,* 648 (1942).
238. Simons, J. H., and H. Hart, J. Am. Chem. Soc., *66,* 1309 (1944).
239. Boedtker, E., Bull. Soc. Chim. Fr., (3), *35,* 825 (1906); J. Chem. Soc. Abstr., *90,* (I), 942 (1906).
240. Brown, H. C., and C. W. McGray, Jr., J. Am. Chem. Soc., *77,* 2309 (1955).
241. Olah, G. A., and M. W. Meyer, J. Org. Chem., *27,* 3464 (1962).
242. Olah, G. A., W. S. Tolgyesi, and E. A. Dear, J. Org. Chem., *27,* 3441, 3449, 3455 (1962).
243. Gerhardt, O., Reichstoffind, *5,* 67 (1930); Chem. Abstr., *24,* 4897 (1930).
244. Wirth, W. V., U.S. Patent 2,023,566 (1935); Chem. Abstr., *30,* 738 (1936).
245. Nelson, K. L., and H. C. Brown, in *The Chemistry of Petroleum Hydrocarbons* (B. J. Brooks, S. S. Kurtz, C. E. Boord, and L. Schmerling, eds.), Reinhold, New York, 1955, p. 501.
246. Roberts, R. M., and S. E. McGuire, J. Org. Chem., *35,* 102 (1970).
247. Schlatter, M. J., J. Am. Chem. Soc., *76,* 4952 (1954).
248. Schneider, A., U.S. Patent 2,648,713 (1953); Chem. Abstr., *48,* 8258 (1954).
249. Carson, B. B., W. J. Heintzelman, R. C. Odioso, H. E. Tiefenthal, and F. J. Pavlik, Ind. Eng. Chem., *48,* 1180 (1956).
250. Schmerling, L., J. P. Luvisi, and R. W. Welch, J. Am. Chem. Soc., *81,* 2718 (1959).
251. Gurewitsch, A., Ber., *32,* 2424 (1899).
252. Perkins, R. P., A. J. Dietzler, and J. T. Lundquist, U.S. Patent 1,972,599 (1934); Chem. Abstr., *28,* 6532 (1934).
253. Putnam, M. E., E. C. Britton, and R. P. Perkins, U.S. Patent 2,039,344 (1936); Chem. Abstr., *30,* 4176 (1936).
254. Fieser, L. F., and C. C. Price, J. Am. Chem. Soc., *58,* 1838 (1936).
255. Bromby, N. G., A. T. Peters, and F. M. Power, J. Chem. Soc., 144 (1943).
256. Whitmore, F. C., and W. H. James, J. Am. Chem. Soc., *65,* 2088 (1943).
257. Cahen, P. M., Bull. Soc. Chim. Fr., (3), *19,* 1007 (1898).
258. Tschitschibabin, A. E., Bull. Soc. Chim. Fr., (5), *2,* 497 (1935).
259. Contractor, R. B., A. T. Peters, and F. M. Rowe, J. Chem. Soc., 1993 (1949).
260. Buu-Hoi, N. P., H. LeBihan, F. Binon, and P. Ryet, J. Org. Chem., *15,* 1060 (1950).
261. Ferris, R. T., and D. Hamer, J. Chem. Soc., 1409 (1960).
262. Rieker, A., N. Zeller, K. Schurr, and E. Muller, Justus Liebigs Ann. Chem., *697,* 1 (1966).
263. Razuvaev, G. A., M. L. Khidekel, and V. B. Berlina, Dokl. Akad. Nauk SSSR, *145,* 1071 (1962).
263a. Nurster, H. E., and A. T. Peters, J. Chem. Soc., 729 (1950).
264. Karakhanov, E. A., G. V. Drovyannikova, and E. A. Viktorova, Khim. Geterotsikl. Soedin., *7,* 156 (1971); Chem. Abstr., *75,* 35677b (1971).
264a. Budylin, V. A., M. S. Ermolenko, and A. M. Kost, Khim. Geterotsikl. Soedin, 921 (1978); Chem. Abstr., *89,* 146706a (1978).

265. Boldt, P., H. Militzen, W. Thielecke, and L. Schultz, Justus Liebigs Ann. Chem., *718,* 101 (1968).
266. Chan, T. H., I. Paterson, and J. Pinsonnault, Tetrahedron Lett., 4183 (1977).
267. Schmerling, L., J. Am. Chem. Soc., *97,* 6134 (1975).
268. Olah, G. A., and N. A. Overchuck, J. Am. Chem. Soc., *87,* 5786 (1965).
269. (a) Olah, G. A., S. J. Kuhn, S. H. Flood, and B. A. Hardie, J. Am. Chem. Soc., *86,* 2203 (1964); (b) Brown, H. C., G. Marino, and L. M. Stock, *81,* 3310 (1959); Brown, H. C., and G. Marino, J. Am. Chem. Soc., *81,* 5611 (1959).
270. (a) Brown, H. C., and H. L. Young, J. Org. Chem., *22,* 719, 724 (1957); (b) Brown, H. C., and G. Marino, J. Am. Chem. Soc., *81,* 3308 (1959).
271. Brown, H. C., and L. M. Stock, J. Am. Chem. Soc., *79,* 5175 (1957).
272. Brown, H. C., and L. M. Stock, J. Am. Chem. Soc., *79,* 1421 (1957).
273. Konowalow, M. I., and J. Jegerow, J. Russ. Phys.-Chem. Soc., *30,* 1031 (1898); Chem. Zentralbl., *1,* 776 (1899).
274. (a) Cacace, F., and P. Giacomello, J. Am. Chem. Soc., *95,* 5851 (1973); (b) Cacace, F., R. Cipollini, P. Giacomello, and E. Possagno, Gazz. Chim. Ital., *104,* 977 (1974); Chem. Abstr., *84,* 58135y (1976).
275. Pines, H., L. Schmerling, and V. N.Ipatieff, J. Am. Chem. Soc., *62,* 2901 (1940).
276. Inatome, M., K. W. Greenlee, J. M. Defer, and C. E. Boord, J. Am. Chem. Soc., *74,* 292 (1952).
277. Schmerling, L., and J. P. West, J. Am. Chem. Soc., *76,* 1917 (1954).
278. Friedman, B. S., and F. L. Morritz, J. Am. Chem. Soc., *78,* 2000 (1956).
279. Olah, G. A., unpublished results, reported by G. A. Olah, and H. W. Quinn, in *Friedel-Crafts and Related Reactions,* Vol. 4, G. A. Olah, ed.), Wiley-Interscience, New York, 1965, p. 278.
280. Friedman, B. S., and F. L. Morriz, J. Am. Chem. Soc., *78,* 3430 (1956).
281. Schmerling, L., R. W. Welch, and J. P. Luvisi, J. Am. Chem. Soc., *79,* 2636 (1957).
282. Malchick, S. P., and R. B. Hannan, J. Am. Chem. Soc., *81,* 2119 (1959).
283. Schreiner, E., J. Prakt. Chem., *82,* 294 (1910); Chem. Abstr., *5,* 282 (1911).
284. Halse, O. M., J. Prakt. Chem., *89,* 451 (1914); Chem. Abstr., *8,* 2702 (1914).
285. Boggs, J. K., U.S. Patent 3,739,040 (1973); Chem. Abstr., *79,* 65979b (1973).
286. Hammann, W. C., and C. F. Hobbs, U.S. Patent 3,657,370 (1972); Chem. Abstr., *77,* 34122w (1972).
287. Kursanow, N., Justus Liebigs Ann. Chem., *318,* 311 (1901).
288. Mayes, H. A., and E. E. Turner, J. Chem. Soc., *500* (1929).
289. Kursanow, N., J. Russ. Phys.-Chem. Soc., *38,* 1304 (1907).
290. Simons, J. H., and G. C. Bassler, J. Am. Chem. Soc., *63,* 880 (1941).

291. Burwell, R. L., Jr., and S. Archer, J. Am. Chem. Soc., *64*, 1032 (1942).
292. Battegay, M., and M. Kappeler, Bull. Soc. Chim. Fr., (4), *35*, 992 (1924).
293. Pashaev, T. A., F. A. Mamedov, and F. A. Iseeva, Azerb. Khim. Zh. 95 (1970); Chem. Abstr., *75*, 19811 (1971).
294. Beak, P., R. J. Trancik, J. B. Mooberry, and P. Y. Johnson, J. Am. Chem. Soc., *88*, 4288 (1966).
295. Brown, J. H., and C. S. Marvel, J. Am. Chem. Soc., *59*, 1248 (1937).
296. Marvel, C. S., and C. M. Himel, J. Am. Chem. Soc., *62*, 1550 (1940).
297. Bartlett, J. F., and C. E. Garland, J. Am. Chem. Soc., *49*, 2098 (1927).
298. Akhmedov, K. N., and A. R. Abdurasuleva, Zh. Org. Khim., *3*, 896 (1967); J. Org. Chem. USSR, *3*, 861 (1967).
299. Abdurasuleva, A. R., and K. N. Akhmedov, Uzb. Khim. Zh., No. 5, 31 (1964); Chem. Abstr., *60*, 7933 (1964).
300. Shakhgel'diev, M. A., R. M. Shamkhalov, and E. K. Mamedov, Kratk. Tezisy-Vses. Soveshch. Probl. Mekh. Geterolitichesk. Reakts, 78 (1974); Chem. Abstr., *85*, 62749g (1976).
301. Saidova, F. M., O. V. Mikhailova, and Ch. Sh. Kadyrov, Deposited Doc., VINITI, 1994 (1975); Chem. Abstr., *87*, 102133j (1977).
302. Saidova, F. M., Zh. A. Amanturdyeva, and Ch. Sh. Kadyrov, Deposited Doc., VINITI, 2212 (1975); Chem. Abstr., *87*, 134968b (1977).
303. Saidova, F. M., and N. G. Sidorova, Zh. Obshch. Khim., *34*, 1601 (1964); Chem. Abstr., *61*, 5580g (1964).
304. Sidorova, N. G., and F. M. Saidova, Zh. Obshch. Khim., *34*, 38 (1964); Chem. Abstr., *60*, 10615a (1964).
305. Sidorova, N. G., and F. M. Saidova, Zh. Obshch. Khim., *33*, 2213 (1963); Chem. Abstr., *59*, 13899 (1963).
306. Sidorova, N. G., Zh. Obshch. Khim., *32*, 2642 (1962); Chem. Abstr., *58*, 7845h (1963).
307. Akhmedov, V. M., F. R. Alieva, and M. A. Mardanov, Zh. Org. Khim., *9*, 1653 (1973); Chem. Abstr., *79*, 125968j (1973).
308. Sandler, S. R., Chem. Ind., 565 (1970).
309. Nefedov, O. M., E. S. Agavelyan, and O. S. Chizhov., Izv. Akad. Nauk SSSR, Ser. Khim., 2084 (1969); Chem. Abstr., *72*, 12887 (1970).
310. Carson, B. B., and V. N. Ipatieff, J. Am. Chem. Soc., *60*, 747 (1938).
311. Chakrabarti, J. K., M. R. J. Jolley, and A. Todd, Tetrahedron Lett., 391 (1974).
312. Mirzaeva, A. K., N. G. Sidorova, and A. K. Spiropulu, Nauchn. Tr., Tashk. Gos. Univ., No. 462, 41 (1974); Chem. Abstr., *84*, 30547k (1976).
313. Furakawa and Olda, Ref. Zh., Khim., 173 (1957).
314. Sadykh-Zade, S. I., S. B. Kurbanov, and R. I. Mustafaev, Zh. Org. Khim., *6*, 989 (1970); J. Org. Chem. USSR, *6*, 993 (1970).
315. Sadykh-Zade, S. I., S. B. Kurbanov, and R. I. Mustafaev, Dokl. Akad. Nauk Az. SSR, *26*, 22 (1970); Chem. Abstr., *74*, 53194a (1971).
316. Stapp, P. R., and C. A. Drake, J. Org. Chem., *36*, 522 (1971).

317. Ref. 127, Vol. 2, Part 1, p. 433.
318. Kharasch, M. S., and A. L. Flenner, J. Am. Chem. Soc., *54,* 674 (1932).
319. Cullinane, N. M., S. J. Chard, and C. W. C. Dawkins, Org. Synth., Coll. Vol. 4, 520 (1963).
320. Koch, H., and H. Steinbrink, Brennst.-Chem., *19,* 277 (1938); Chem. Abstr., *33,* 150 (1939).
321. Krespan, C. G., J. Org. Chem., *44,* 4924 (1979).
322. Brown, H. C., and W. S. Higley, U.S. Patent 2,767,230 (1956); Chem. Abstr., *51,* 8785b (1957).
323. Bass, T. M. A., H. Van Bekkum, M. A. Hoefnagel, and B. M. Wepster, Recl. Trav. Chim. Pays-Bas, *88,* 1110 (1969); Chem. Abstr., *71,* 123738v (1969).
324. Francis, A. W., Chem. Rev., *43,* 257 (1948).
325. For reviews on the alkylation mechanisms see: (a) Ref. 127, Vol. 2, Parts 1 and 2; (b) Refs. 183-186 and 199; (c) Brown, H. C., H. W. Pearsall, L. P. Eddy, W. J. Wallace, M. Grayson, and K. L. Nelson, Ind. Eng. Chem., *45,* 1462 (1953); (d) Nenitzescu, C. D., Rev. Chim. Acad. Repub. Pop. Roum. *7,* 349 (1962); Chem. Abstr., *59,* 6278 (1963); (e) Nenitzescu, C. D., Rev. Chim. (Bucharest), *9,* 5 (1964); Chem. Abstr., *61,* 11861a (1964).

4

Alkylations of Arenes with Alcohols, Ethers, Esters, and Alkenes

I. INTRODUCTION

The alkylation of arenes by alcohols, ethers, esters, and alkenes has been the subject of considerable interest to Friedel-Crafts chemists [1-15, for reviews]. Although the alkylating ability of these derivatives was recognized by early workers in the field, the isomerizations accompanying their use escaped notice in many cases and were discovered much later. On the other hand, the early workers were misled into the belief that alkylations with some primary derivatives and mild catalysts such as sulfuric acid, boron trifluoride, and zinc chloride took place without rearrangements.

The aim of this section is to review the literature briefly with special emphasis on recent work that led not only to correction of the older literature but also to better understanding of the mechanisms of these alkylations.

II. ALKYLATIONS WITH LINEAR PRIMARY ALCOHOLS, ETHERS, AND ESTERS

A. In the Presence of Strong Lewis Acid Catalysts

Since alkylations in which strong Lewis acids such as $AlCl_3$ or SbF_5 are used as catalyst give significantly different results from those in which milder catalysts are used, it will be convenient to discuss the different catalyst systems separately.

Apparently, the first attempt to alkylate arenes by primary alcohols in the presence of aluminum chloride was carried out by Huston and Sager in 1926 [16]. Using alcohol, benzene, and aluminum chloride in a ratio of 1:1:0.5, these workers failed to accomplish any alkylation by methyl, ethyl, *n*-propyl, *n*-butyl, isobutyl, isopentyl, phenylethyl, and phenylpropyl alcohols at a reaction temperature of 25 to 30°C. The successful alkylation of arenes by benzyl [17,18] and allyl [16] alcohols led these workers to conclude that only certain reactive primary alcohols could be utilized as alkylating agents.

Almost 10 years later, Tsukervanik and co-workers succeeded in alkylating phenol [19], benzene [20], toluene [20], naphthalene [22], and haloarenes [21] with several C_1 to C_5 primary and secondary alcohols in the presence of $AlCl_3$. In spite of their failure to recognize accompanying side-chain rearrangements, the work of these authors established two important facts about alkylations with primary alcohols. First, compared to alkylations with alkyl halides or secondary alcohols, alkylations with primary alcohols require larger proportions of catalyst (ca. 2 mol for 1 mol of alcohol) and longer heating at higher temperatures (10 hr at 110 to 140°C). Second, as in other alkylating systems, the relative ease of alkylations with alcohols follows the order benzyl, allyl, tertiary > secondary > primary > methyl. Thus even under conditions effecting alkylations with primary alcohols, methyl alcohol failed to alkylate phenol [19] or benzene [20]. Toluene [20], however, gave products of varying degrees of methylation, but these were attributed to disproportionation rather than to alkylation.

Since this work of Tsukervanik's group was reported, several other papers describing $AlCl_3$-catalyzed alkylations by primary alcohols, esters, and ethers have appeared. In 1938, Bowden [23] described the $AlCl_3$-catalyzed alkylation of benzene with a number of esters, including *n*-propyl and *n*-butyl. According to his report, the product from the former was nonrearranged *n*-propylbenzene [66%) and from the latter was the rearranged *sec*-butylbenzene (41 to 55%). In 1938-1939, Norris and co-workers obtained toluene, ethylbenzene [24], mesitylene [25], and *sym*-triethylbenzene by reactions involving methyl alcohol, ethyl alcohol, or ethyl ether and aluminum chloride. In accordance with Tsukervanik's report, they found that a higher temperature was required for alkylating with methyl than with ethyl alcohol. In 1940, Ipatieff et al. [26] reported that the $AlCl_3$-catalyzed alkylation of benzene with *n*-propyl alcohol (at 110 to 120°C for 10 hr) gave solely nonrearranged *n*-propylbenzene. These authors emphasized the view that any mechanism advanced for the interpretation of the catalytic action of aluminum chloride in the alkylation of an aromatic hydrocarbon must explain the fact that the alkylation is not accompanied by isomerization. In 1942, the alkylation of benzene with *n*-butyl ether and $AlCl_3$ was reported to give pure *sec*-butylbenzene (16% yield) [27].

Since 1948, workers have become aware of accompanying side-chain rearrangements during alkylations of arenes with propyl and higher primary alcohols and $AlCl_3$. In that year, Tsukervanik and Poletaev [28] alkylated benzene and toluene with *n*-butyl, isobutyl, and isopentyl alcohols in the presence of $AlCl_3$ and they stated that in all cases the products contained isomerized side chains. In the 1950s several reports of accompanying rearrangements were given, but differences regarding their extent were substantial. For example, the $AlCl_3$-catalyzed alkylation of benzene was reported to give a mixture of 85% *n*-butyl- and 15% *sec*-butylbenzene with *n*-butyl alcohol [29] and a mixture of 73.7% isopropylbenzene, 16.5% *n*-propylbenzene, and 9.8% ethylbenzene with *n*-propyl ether [30]. In the latter report, Searles pointed out that ethylbenzene resulted from fragmentation. Moreover, in sharp contrast to the earlier statement of Ipatieff et al. [26], he stated that "rearrangement is as extensive in the aluminum chlorde catalyzed alkylation of benzene with *n*-propyl ether as with *n*-propyl chloride."

More recently (1963), Tsukervanik and Khakimov [31-33] conducted a systematic investigation of the alkylation of benzene with primary alcohols in the presence of aluminum chloride at 80°C. Their results with *n*-propyl [31], *n*-butyl [31], *n*-pentyl [32], and *n*-hexyl [32] alcohols are summarized in Eqs. (1) to (4).

$$PhH + CH_3CH_2CH_2OH \xrightarrow[80°]{AlCl_3} \underset{33\%}{Ph\text{-}CH_2CH_2CH_3} + \underset{67\%}{\underset{\displaystyle Ph}{CH_3\underset{|}{C}HCH_3}} \quad (1)$$

$$PhH + CH_3(CH_2)_2CH_2OH \xrightarrow[80^\circ]{AlCl_3} \underset{50\text{–}55\%}{CH_3\underset{|}{\underset{Ph}{C}}HCH_2CH_3} + \underset{45\text{–}50\%}{Ph\text{–}CH_2\overset{CH_3}{\overset{|}{C}}H\text{–}CH_3} \quad (2)$$

$$PhH + CH_3(CH_2)_3CH_2OH \xrightarrow{AlCl_3} \underset{10\%}{Ph\text{–}CH_2(CH_2)_2CH_3} + \underset{60\%}{CH_3\underset{|}{\underset{Ph}{C}}HCH_2CH_2CH_3}$$

$$+ \underset{30\%}{CH_3CH_2\underset{|}{\underset{Ph}{C}}HCH_2CH_3} \quad (3)$$

$$PhH + CH_3(CH_2)_4CH_2OH \xrightarrow[80^\circ]{AlCl_3} \underset{50\%}{CH_3\underset{|}{\underset{Ph}{C}}HCH_2CH_2CH_2CH_3} + \underset{50\%}{CH_3CH_2\underset{|}{\underset{Ph}{C}}HCH_2CH_2OH_3} \quad (4)$$

Similar to alkylations with n-alkyl halides (see Chap. 3), higher n-heptyl, n-octyl, and n-hexadecyl alcohols [33-35] were reported to give complex product mixtures consisting of all the possible secondary alkylbenzene isomers with up to 14% of the corresponding primary isomer. In the light of the results with other alcohols, one should wonder why no primary isomers were reported from alkylations with either n-butyl or n-hexyl alcohols. However, these authors did not even raise the question [31,32].

A year later this question was partly resolved by Nield [36,37] who, in contrast to the results of Eq. (2), obtained mixtures of primary and secondary alkylates from the $AlCl_3$-catalyzed alkylations of benzene with n-propyl and n-butyl alcohols or ethers (Eqs. 5 and 6). These results of Nield were in accord with those of Roberts and Shiengthong [38] for alkylations with n-propyl and n-butyl chlorides.

$$PhH + \begin{matrix} CH_3CH_2CH_2OH \\ \text{or} \\ (CH_3CH_2CH_2)_2O \end{matrix} \xrightarrow[80^\circ,\ 5\text{–}6.5\ hr]{AlCl_3} PhCH_2CH_2CH_2 + Ph\overset{CH_3}{\overset{|}{C}}HCH_3 \quad (5)$$

$$PhH + \begin{matrix} CH_3CH_2CH_2CH_2OH \\ \text{or} \\ (CH_3CH_2CH_2CH_2)_2O \end{matrix} \xrightarrow[80^\circ,\ 6.5\ hr]{AlCl_3} PhCH_2CH_2CH_2CH_3$$

$$+ Ph\overset{CH_3}{\overset{|}{C}}HCH_2CH_3 + PhCH\overset{CH_3}{\overset{|}{C}}HCH_3 \quad (6)$$

A compilation of the results of the $AlCl_3$-catalyzed alkylations of benzene with *n*-propyl and *n*-butyl alcohols, ethers, and esters reported in the period 1940-1964 is given in Table 1. Examination of these results reveals a number of discrepancies which are hard to ignore. Whether these differences were due to inaccurate analyses or to variations in reaction conditions called for resolution. This inspired Roberts and co-workers in 1969 to reexamine the most glaring conflicting reports [39]. These workers believed that selective destruction of isopropylbenzene by dealkylation and other reactions occurring under the severe conditions of the reaction employed by Ipatieff and Nield provided a reasonable explanation for the observed low proportion of isopropylbenzene in the product. They suggested that the initial major product was probably isopropylbenzene, as expected, but prolonged heating in the presence of the catalyst decomposed isopropylbenzene by dealkylation more rapidly than *n*-propylbenzene. As a result, the relative amount of the latter isomer became increasingly larger until it was the only isomer detectable by the poor identification technique used in the early work. As experimental support for this view, they found that reaction at 25°C for 24 to 60 hr gave isopropylbenzene as the major product (74 to 77%). On the other hand, as the amount of isopropylbenzene decreased during heating at 80°C for 7 hr, that of *n*-propylbenzene increased until the latter became the major alkylation product. Significant amounts of methyl-, ethyl-, and butylbenzenes as well as propane and isobutane were also produced. These were attributed to accompanying fragmentation and dealkylation reactions [40,41].

The problems concerning side-chain and positional rearrangements during alkylations with esters and haloesters were recently resolved. In 1974, Olah and Nishimura [42] undertook a detailed study of the $AlCl_3$- or SbF_5-catalyzed alkylation of benzene and toluene with alkyl esters and haloesters in various solvents. The ester function included chlorosulfite, arenesulfinate, chloro- and fluorosulfate, triflate, pentafluorobenzenesulfonate, and trifluoroacetate, whereas the alkyl group varied from C_1 to C_4. As in solvolysis, the leaving ability of the ester group decreased in the order $CF_3SO_3^-$, $FSO_3^- > ClSO_3^-$ $> CF_3COO^-$. Both positional and substrate selectivity ratios showed dependence on solvent and, to a lesser extent, on temperature, but were insignificantly affected by the nature of the leaving group. In excess arene, methylene chloride, and carbon disulfide solutions, the $AlCl_3$-catalyzed reactions showed low substrate selectivity and high *meta*-isomer ratios due to accompanying reorientation and disproportionation. In nitromethane, side reactions were suppressed, resulting in higher substrate (k_T/k_B) but lower positional *meta*-isomer ratios. Alkylations with *n*-propyl, *n*-butyl, and isobutyl chlorosulfite or chlorosulfate using aluminum chloride catalyst showed effects of substantial isomerization, similar to those observed in alkyl halide reactions. With chlorosulfite, for example, these reactions gave isopropylation, *sec*-butylation, and *t*-butylation accompanied by *n*-propylation, *n*-butylation, and isobutylation, respectively (Eqs. 7 to 9).

$$\text{ArH} + \text{C-C-C-OSOCl} \xrightarrow[\text{or } SbF_5]{AlCl_3/CH_3NO_2} \text{Ar-C-C-C} + \text{Ar-}\overset{\text{C}}{\overset{|}{\text{C}}}\text{-C} + SO_2 + HCl \tag{7}$$

$$\text{ArH} + \text{C-C-C-C-OSOCl} \xrightarrow[\text{or } SbF_5]{AlCl_3/CH_3NO_2} \text{Ar-C-C-C-C} + \text{Ar-}\overset{\text{C}}{\overset{|}{\text{C}}}\text{-C-C} + SO_2 + HCl \tag{8}$$

Table 1. Alkylation of Benzene with *n*-Propyl and *n*-Butyl Alcohols, Ethers and Esters in the Presence of $AlCl_3$ (1940-1964)

Reactants (mol)[a]	Temp., °C	Time, hr	Yield, %	Product composition, %	Ref.
Benzene (120g), *n*-propyl alcohol (20g), $AlCl_3$ (87g)	110	10	26	*n*-Propylbenzene	26
Benzene (large excess), *n*-propyl alcohol (1), $AlCl_3$ (2.5)	80	6.5	38	*n*-Propylbenzene (47), isopropylbenzene (53)	36,37
Benzene (750 ml), *n*-propyl ether (60g), $AlCl_3$ (154g)	25° Reflux	15 6	9.8	*n*-Propylbenzene (17), isopropylbenzene (83) + small amounts of ethylbenzene	30
Benzene (20), *n*-propyl alcohol (1) $AlCl_3$ (1.3) with or without added water (3%)	80	4-6	70	*n*-Propylbenzene (67),	31
Benzene (large excess), *n*-propyl formate (1), $AlCl_3$ (0.8)	25-60	4	36	*n*-Propyl- and isopropylbenzene	36,37

Benzene (large excess), *n*-butyl alcohol (1), $AlCl_3$ (2)	-	-	16	*sec*-Butylbenzene	27
Benzene (3.38), *n*-butyl alcohol (0.33), $AlCl_3$ (0.42) in a continuous reactor	75	3	60	Mixture of isomers	29
Benzene (20), *n*-butyl alcohol (1), $AlCl_3$ (1.3)	80	4-6	73	*sec*-Butylbenzene (50-55), isobutylbenzene (45-50)	31
Benzene (-), *n*-butyl ether (-), $AlCl_3$ (-)	-	-	16	*sec*-Butylbenzene	27
Benzene (large excess), *n*-butyl ether (1), $AlCl_3$ (2.3)	50	5.5	11	Mixture of *n*-butyl-, *sec*-butyl- and isobutylbenzene	36

[a]Unless otherwise specified.

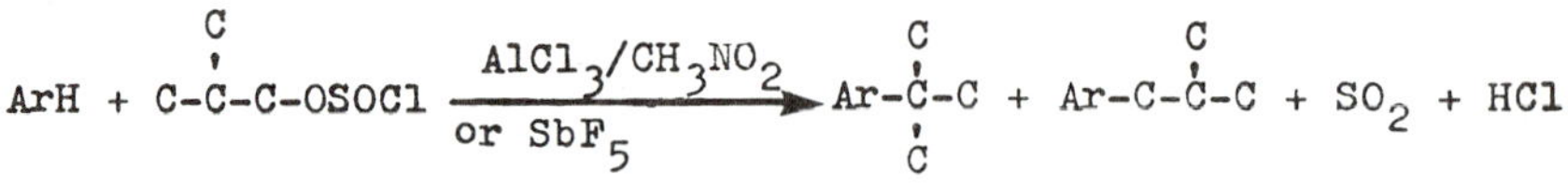

$$\text{ArH} + \text{C-}\overset{\text{C}}{\overset{|}{\text{C}}}\text{-C-OSOCl} \xrightarrow[\text{or SbF}_5]{\text{AlCl}_3/\text{CH}_3\text{NO}_2} \text{Ar-}\underset{\text{C}}{\underset{|}{\overset{\text{C}}{\overset{|}{\text{C}}}}}\text{-C} + \text{Ar-C-}\overset{\text{C}}{\overset{|}{\text{C}}}\text{-C} + \text{SO}_2 + \text{HCl} \qquad (9)$$

Ar = Phenyl or tolyl (*o*-, *m*-, and *p*)

The formation of isobutyl derivatives in the alkylation with isobutyl chlorosulfate is very significant (Eq. 9). To our knowledge, it represents one of the very few substantiated cases [43] of direct primary alkylation of benzene with an isobutyl derivative under Friedel-Crafts conditions. Alkylations with all other isobutyl derivatives under almost all employed catalysts and conditions have resulted in complete rearrangement to *t*-butylated arenes.* The results were rationalized in terms of a mechanism involving competition between an S_N2-type path giving nonrearranged products and an S_N1-type path with rearrangement in the alkyl cation [42].

In accordance with this mechanism, the alkylation with carboxylic acid esters was suggested to involve initial attack by the catalyst on the carbonyl oxygen followed by subsequent rearrangement to the ether oxygen. Cleavage of the R-O bond, either in S_N1 fashion with rearrangement or in S_N2 fashion without rearrangement, gives the alkylated arenes (scheme 1). In scheme 1, the competing acyl-oxygen cleavage would be much decreased or absent if R is a tertiary group and/or R's a poor nucleophilic leaving group.

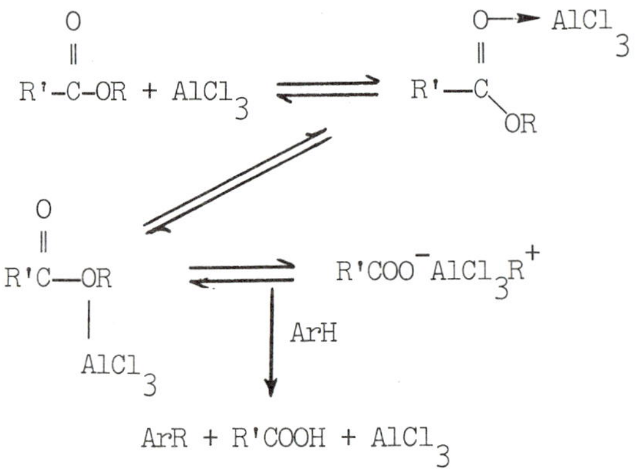

$$\text{R'-}\overset{\text{O}}{\overset{\|}{\text{C}}}\text{-OR} + \text{AlCl}_3 \rightleftharpoons \text{R'-}\overset{\text{O}\rightarrow\text{AlCl}_3}{\overset{\|}{\text{C}}}\text{-OR}$$

$$\rightleftharpoons \text{R'}\overset{\text{O}}{\overset{\|}{\text{C}}}\text{-}\underset{\text{AlCl}_3}{\underset{|}{\text{OR}}} \rightleftharpoons \text{R'COO}^-\text{AlCl}_3\text{R}^+ \xrightarrow{\text{ArH}} \text{ArR} + \text{R'COOH} + \text{AlCl}_3$$

Scheme 1

Two possible mechanisms were also suggested for Friedel-Crafts alkylations with sulfonyl chlorides [77]. The first (A) is an S_N1-type mechanism involving initial ionization of the sulfonyl chloride by the catalyst. The second (B) is an S_N2-type mechanism involving displacement of the alkyl group from the polarized sulfonyl chloride-catalyst complex. Although the available experimental data did not allow a differentiation, mechanism A was slightly preferred.

---

*A small amount of isobutyl- (and *sec-butyl-*)*p*-xylene was obtained from the reaction of isobutyl chloride with *p*-xylene [44]. In this case steric hindrance blocks the attachment of a bulky t-butyl group *ortho* to the methyl groups.

B. In the Presence of Mild Catalysts

Besides the strong SbF_5 and $AlCl_3$ catalysts, various other milder catalysts have been used for the alkylations of aromatics with benzyl, methyl, ethyl, and higher *n*-alkyl alcohols, ethers, and esters. These included weak acidic metal halides (Lewis acids), such as $AlCl_3$-CH_3NO_2, BF_3, $FeCl_3$, $TiCl_4$, $SnCl_4$, or $ZnCl_2$; ordinary protonic acids (Brönsted acids) such as BF_3, H_3PO_4, HF, H_2SO_4, H_3PO_4, or polyphosphoric acid (PPA); solid superacids, such as the perfluorinated sulfonic acid Nafion-H; and inorganic acidic oxides, such as phosphorus pentoxide or alumina, natural and synthetic aluminosilicates, and zeolites. Since these mild catalysts show similar characteristics by leading solely or mostly to rearranged secondary alkylates, we chose to treat them in this separate section.

In line with the well-established order of reactivity, one would expect to have the following sequence of alkylating ability of the *n*-alkyl derivatives: methyl < ethyl < *n*-propyl and higher *n*-alkyl < benzyl. In practice, this was found to be the case, as conditions shown to be suitable for alkylations with other alcohols did not induce any appreciable alkylations with either methyl or ethyl alcohol [45,45a]. Moreover, when conditions suitable for alkylation were used, ethyl derivatives gave higher alkylation yields than the corresponding methyl derivatives, indicating an easier reaction [45a-48]. For example, a competitive alkylation of toluene with an equimolar mixture of methyl and ethyl alcohols in the gas phase at 185°C over a perfluorinated resinsulfonic acid (Nafion-H) catalyst gave four times more ethyltoluenes than xylenes [46]. Similar effects were observed during the alkylation of benzene, toluene, and chlorobenzene with dimethyl and diethyl sulfates in the presence of HSO_3Cl catalyst [47] and of toluene with methyl and ethyl chloroformate in the gas phase over Nafion-H catalyst [48]. Again, in the BF_3-catalyzed alkylation of biphenyl with alcohols under pressure at 165 to 170°C, the conversions were 1% with methyl, 8% with ethyl, and 72% with *n*-propyl alcohol [45a].

In contrast to methyl and ethyl alcohols, ethers, or esters [23,24,49-68], similar benzyl derivatives proved to be very active in alkylations, even when milder catalysts and conditions were employed [17,69-80]. Whereas the alkylation of arenes with methyl alcohol [24,49-55,65-67] or ether [56-58] required heating at 70 to 95°C for several hours even with the strong catalyst $AlCl_3$, alkylations with benzyl alcohol were accomplished at 25 to 30°C with $AlCl_3$ [69,71,75] or P_2O_5 [69], at 40°C with H_2SO_4 [70], at reflux temperature with $H_4SiMo_{12}O_{40}$ [78,79], and at 100°C with aluminosilicate [80]. Even milder catalysts such as $AgClO_4$-$SnCl_4$ [74], $AgClO_4$-Sn [73], $TiCl_4$ [76], or $AlCl_3$-CH_3NO_2 [77] induced the benzylation of arenes with benzyl alcohol [73,74,77], benzyl ethers [74], or esters [77,77a] at room temperature (Eqs. 10 to 13).

$$R\text{-}C_6H_5 + C_6H_5CH_2OH \xrightarrow[25^\circ]{AgClO_4\text{-}SnCl_4} R\text{-}C_6H_4CH_2C_6H_5 \quad (88\text{-}90\%) \tag{10}$$

R = H, CH_3, OCH_3, Cl (*o*, *m*, and *p*)

$$\text{1,3,5-trimethylbenzene} + C_6H_5CH_2OH \xrightarrow{AgClO_4-SnCl_4} \text{benzylmesitylene (96\%)} \quad (11)$$

$$PhCH_2OCH_2Ph + 2\ PhH \xrightarrow[\text{benzene, } 25°]{AgClO_4-1/4SnCl_4} 2\ PhCH_2Ph \quad (12)$$

$$PhCH_2OCH_2CH(CH_3)_2 + PhH \xrightarrow[\text{benzene, } 25°]{AgClO_4-1/4SnCl_4} PhCH_2Ph + (CH_3)_2CHCH_2OH \quad (13a)$$

$$C_6H_6 + RC_6H_4CH_2SO_2Cl \xrightarrow[25°/1\ hr]{AlCl_3/CH_3NO_2} C_6H_5\text{—}CH_2\text{—}C_6H_4\text{—}R\ (+SO_2) \quad (13b)\ [77]$$

	o	*m*	*p*
R=H	34.4	3.3	62.4
R=CH_3	30.3	3.1	66.7

$$PhH + PhCH_2\text{-}OSO_2Ph \xrightarrow{PhSO_3H} PhCH_2Ph + PhSO_3H \quad (13c)\ [77a]$$

The mechanism of the last reaction (Eq. 13c) was studied thoroughly by Nenitzescu et al. in 1955 [77a]. They found that the reaction was first order in the alkylating ester and second order in the catalyst and that the relative rates of reaction of benzyl benzenesulfonate with benzene, toluene, and *m*-xylene were 1:2.6:3.7. Based on this and on a study of the alkylation of benzene with *p*-methyl-, *p*-chloro-, and *m*-dinitrobenzyl benzenesulfonates, they concluded that the arene participates in the rate-determining step and that the essential factor determining the rate of alkylation of an arene is the breaking of a bond, not the formation of a bond.

Methyl, benzyl, ethyl, and higher *n*-alkyl alcohols, ethers, and esters were frequently used not only to alkylate benzene and substituted benzenes, but also to alkylate higher ring systems in the presence of various protonic acids and metal oxides. For example, methyl alcohol was used to alkylate napthalene [59], α- and β-dimethylnaphthalenes [60], biphenyl [66], fluorene [67], cyclopentaphenanthrene [67], and indene [67]. The latter three compounds were also alkylated by ethyl, benzyl, isopropyl, and 1-phenylethyl alcohols [67]. Higher alcohols and ethers were also used to alkylate biphenyl [45a], naphthalene [82-87], acenaphthene [88], and thiaindenes [89].

The alkylations of naphthalene and biphenyl were mostly carried out by Romadane et al. in the presence of BF_3 catalyst [68,81-84,87]. Based on their work in the 1950s, these authors offered the following generalizations:

1. The products were mixtures of mono- and dialkylates.
2. In the case of naphthalene, the major components were the 1 isomer among the monoalkylates and the 1,4 isomer among the dialkylates.
3. In the case of biphenyl, the products were mixtures of 4-alkyl- and 4,4-dialkylbiphenyl.
4. Alkylations with normal alcohols involved direct condensation to nonrearranged normal alkylates [83(a)], but those with isoalkyl alcohols proceeded through intermediate alkenes and led to rearranged secondary or tertiary alkylates [83(b),84].
5. Products with only linear alkyl groups were obtained from alkylations with normal alcohols.

The latter two of these conclusions are highly questionable. We shall see later in this section that various workers demonstrated that BF_3-catalyzed alkylations with normal alcohols were accompanied by extensive rearrangements to tertiary alkylates. Moreover, the claimed exclusive production of nonrearranged 1-*n*-alkylnaphthalenes in alkylations with normal alcohols was disproved by Zee-Cheng in 1963 [85] and by Romadane with Dokuchaeva in 1974 [87]. The alkylation of naphthalene by *n*-propyl or *n*-butyl alcohol in the presence of 85% H_2SO_4 at 85°C gave, respectively, a 1% yield of isomeric isopropylnaphthalenes and a 9% yield of a mixture of 61% 1- and 32% 2-*sec*-butylnaphthalene [85]. Also, a reinvestigation of the BF_3-catalyzed reactions of naphthalene with *n*-alkanols showed the products to be mainly 2-*sec*-alkylnaphthalenes [87].

Besides traditional catalysts in liquid-phase alkylations, the application of solid supported catalysts in heterogeneous gas-phase alkylations with C_1 to C_3 alcohols, ethers, and esters has gained importance in recent years [46, 48,90-93]. A striking example of this application is the use of highly acidic perfluorinated resin sulfonic acid catalysts, such as Nafion-H, for the methylation [48,90,91], ethylation [48,92,93], and propylation [48,92] of arenes. Among the advantages possible with such insoluble polymeric catalysts are the use of truly catalytic rather than molar amounts of catalyst, the use of relatively mild conditions, the ease of product workup, the formation of clean reaction products in good yields, and the recycling of the catalyst on a continuous or batchwise basis [75,92,93,94].

Nafion-H readily catalyzed both the rearrangement of anisoles and the methylation of phenols with methyl alcohol [90]. Both O- and C-methylated products were obtained. The results showed that methylation gave the *ortho* and *para* isomers in the same amount, indicating a preference of the *para* position by a factor of 2. *Meta* methylated products comprised only 1% of the product. As to mechanism, the alkylation was shown to proceed via fast initial O-methylation of the phenol followed by an intermolecular rearrangement of the aryl methyl ether to methylphenol [90]. A similar view was expressed by other workers [95], who suggested the following reaction sequence for alkylations of phenol with C_1 to C_3 alcohols in the presence of activated alumina at 200 to 450°C:

$$\text{C}_6\text{H}_5\text{OH} \xrightarrow{\text{ROH}} \text{C}_6\text{H}_5\text{OR} \longrightarrow o\text{-RC}_6\text{H}_4\text{OH} \longrightarrow \text{RC}_6\text{H}_4\text{OH}$$

Benzene, toluene, and the three isomeric xylenes were alkylated also with methyl alcohol over Nafion-H [91] in the gas phase. The reactions were clean and showed low substrate selectivity, indicating that a highly energetic methylating species was involved in the alkylation process.

Dimethyl ether was also investigated as a methylating agent with Nafion-H catalyst [91] but was found not only to be an inferior methylating agent relative to methyl alcohol, but also to deactivate the catalyst by esterifying its sulfonic acid groups. These results, however, helped to rule out the possible intermediacy of the methyl ether in methylations involving methyl alcohol as alkylating agent. A contrasting view was given earlier (1963) by Romadane [96] for the alkylation of arenes with molecular complexes of the type $2ROH \cdot BF_3$. From the dependence of reaction rate on temperature this author estimated two maxima (at 100 to 130°C and at 160 to 180°C), the former corresponding to ether formation and the latter giving the carbocations and alkenes.

Very recently, Olah et al [48] investigated the Nafion-H-catalyzed alkylation of toluene and phenol with dimethyl and diethyl oxalates, as well as with methyl, ethyl, and propyl chloroformates, in both liquid- and gas-phase reactions. The reactions proceeded smoothly, giving good yields of clean alkylation products with no acylation side products. The alkylating ability of dimethyl oxalate was comparable to that of methyl alcohol [90,91], but that of diethyl oxalate was much higher than that of ethyl alcohol [91,92] or ethyl chloride [97]. Whereas ethyl alcohol gave only 3 to 5% yield in its ethylation of benzene [92] and ethyl chloride gave only 5.2% yield in its ethylation of toluene [97], diethyl oxalate showed 70% conversion in its ethylation of toluene.

Alkylations with chloroformates revealed some important advantages. First, they proved superior to alkyl halides and alcohols as alkylating agents for batch reactions because they allow sufficiently high boiling points to allow reactions under atmospheric pressure with high reactivity. Second, due to their higher reactivity compared to alcohols, higher yields of alkylates usually resulted. A 59% conversion of methyl chloroformate was observed in the alkylation of toluene at 200°C, compared to about 10% conversion using methyl alcohol [91]. Third, their by-products, CO_2 and HCL, are volatile (Eq. 14).

$$\text{ArH} + \text{R-OCO-Cl} \xrightarrow{\text{Nafion-H}} \text{ArR} + CO_2\uparrow + \text{HCl}\uparrow \qquad (14)$$

As to product distributions, alkylations of toluene and phenol with both the oxalates and the chloroformates gave mixtures of the three possible *ortho*, *meta*, and *para* isomers. Under the conditions employed, intramolecular reorientation resulted in increased *meta* substitution. For example, in the alkylation of toluene with ethyl chloroformate, the isomer distribution at 197°C was very similar to the $AlCl_3$-catalyzed equilibrium isomer distribution of ethyltoluene in the liquid phase (*o-/m-/p*-ethyltoluene, 9:64:26) [98].

Besides accompanying reorientation, the alkylation of toluene with *n*-propyl chloroformate showed another interesting characteristic. Instead of all the product being exclusively rearranged isopropyltoluenes, as expected on the basis of earlier studies of the Nafion-H-catalyzed alkylation of toluene with *n*-propyl alcohol [91] or chloride [97], about 3% of the monoalkylate was *n*-propyltoluenes, with the balance being cymenes (isopropyltoluenes). The explanation was that the major reaction path proceeded through the intermediacy of the isopropyl cation, but some proceeded by S_N2 type attack of toluene on the carbon of a protonated *n*-propyl chloroformate.

We have mentioned that rearranged products were reported by some workers, whereas others reported unrearranged products from alkylations

Table 2. Alkylation of Benzene with Normal Alcohols, Ethers and Esters in the Presence of Mild Catalysts (Prior to Early 1940s)

Benzene (mol)	Alkylating agents, (mol)	Catalyst (mol)	Temp., °C	Time, hr	Product (yield %)	Ref.
-	*n*-Propyl alcohol (-)	80% H_2SO_4 (-)	65	-	Cumene (45%), *p*-diisopropylbenzene, 1,2,4-triisopropylbenzene	104,105
1	*n*-Propyl alcohol (1)	BF_3 (1)	60	9	Cumene (20%), *p*-diisopropylbenzene (20%)	101
2	*n*-Propyl alcohol (0.5)	BF_3-P_2O_5 (0.5)	80	3	Cumene (60%), *p*-diisopropylbenzene (13%)	106
1	*n*-Propyl formate (1)	BF_3 (0.8)	-	-	Cumene (30%), *p*-diisopropylbenzene (30%)	107
1	*n*-Propyl sulfate (1)	BF_3 (0.1)	-	-	Cumene (40%), *p*-diisopropylbenzene (25%)	107
-	*n*-Butyl alcohol (-)	80% H_2SO_4 (-)	70	-	*sec*-Butylbenzene, *p*-di-*sec*-butylbenzene	105
1	*n*-Butyl alcohol (1)	BF_3 (1)	60	9	*sec*-Butylbenzene (35%), *p*-di-*sec*-butylbenzene (25%)	101
2	*n*-Butyl alcohol (0.5)	BF_3-P_2O_5 (0.5)	80	-	*sec*-Butylbenzene (75%), *p*-di-*sec*-butylbenzene (5-10%)	106
1	*n*-Butyl formate (1)	BF_3 (1)	80	5	*sec*-Butylbenzene (30%), *p*-di-*sec*-butylbenzene (30%)	107

Table 2. (continued)

Benzene (mol)	Alkylating agents, (mol)	Catalyst (mol)	Temp., °C	Time, hr	Product (yield %)	Ref.
-	*n*-Butyl acetate (-)	HF (-)	80	-	*sec*-Butylbenzene (60%)	108
1	*n*-Butyl phosphate (1)	BF_3 (1)	-	-	*sec*-Butylbenzene (8%), *p*-di-*sec*-butylbenzene (20%)	107
0.5	*n*-Pentyl alcohol (0.3)	80% H_2SO_4 (7)	70	6	2- and 3-Phenylpentane in *ca.* 3:2 ratio (60%)	104
4	*n*-Pentyl alcohol (0.5)	BF_3-P_2O_5 (0.5)	80	-	*sec*-Pentylbenzene (85%)	106
2	*n*-Pentyl ether (1)	BF_3 (-)	150	3	*sec*-Pentylbenzene (20%)	109
3.3	*n*-Propyl benzyl ether (1)	BF_3 (1)	80-90	2	Cumene (11%), diphenylmethane (46%)	110
8	*n*-Octyl alcohol (1)	BF_3-P_2O_5(1)	80	-	*sec*-Octylbenzene (79%)	111
4	*n*-Dodecyl alcohol (0.5)	BF_3-P_2O_5 (0.5)	80	-	*sec*-Dodecylbenzene (33%)	106
2	*n*-Dodecyl alcohol (0.5)	BF_3-$C_6H_5SO_3H$ (0.5)	80	-	*sec*-Dodecylbenzene (45%)	106

with normal alcohols, ethers, or esters. We shall now examine the question of rearrangement or lack of it in more detail.

For many years, conflicting claims were made regarding alkyl-group rearrangements in alkylations of aromatics with *n*-propyl and higher *n*-alkyl derivatives in the presence of mild Friedel-Crafts catalysts. The conflicts were most evident among workers prior to the 1940s. While some of them believed that these alkylations proceeded without rearrangement to yield normal alkylates, other assumed complete rearrangement to yield secondary alkylates. For example, it was claimed that the $ZnCl_2$-catalyzed alkylations produced *p*-*n*-propylaniline from aniline and *n*-propyl alcohol [99], *p*-*n*-octylaniline from aniline and *n*-octyl alcohol [100], *n*-octyl-*o*-toluidine from *o*-toluidine and *n*-octyl alcohol [102]. Also, the Al_2O_3-catalyzed alkylation of phenol with *n*-propyl alcohol was claimed to give *n*-propylphenol derivatives [103].

In contrast to the lack of rearrangement reported in the aforementioned examples, many reports appeared prior to the early 1940s in which similar alkylations were described as proceeding with complete primary-to secondary rearrangement. Examples of these are given in Table 2 and in Eq. (15).

Phenol + *n*-propyl alcohol $\xrightarrow[115\text{-}160°]{BF_3}$ 2-isopropylphenol (32%) + 4-isopropylphenol (16%) + isopropyl 2,4-diisopropylphenyl ether (13%) (15) [112]

Since 1940, however, there has been almost unanimous agreement among Friedel-Crafts chemists concerning the formation of rearranged secondary alkylates as sole or major products from the alkylation of aromatics with *n*-alkyl alcohols, ethers, and esters when mild catalysts were used [35,42,48, 85,87,92,106,110,111,113-125a]. Accordingly, alkylations with *n*-propyl and *n*-butyl derivatives have been expected to yield isopropyl- and *sec*-butylarenes, respectively. Some illustrative examples are given in Eqs. (16) to (29).

PhH + C-C-C-OH →

Conditions	Ph-C(C)-C	Ph-C-C-C	Ref.
84% H_2SO_4	99.65%	0.35 %	(16) [115]
80% H_2SO_4, 80°, 6 hr	100%	--	[123,125]
Nafion-H, in gas phase, 175°, 17% conversion	100%	--	[92]
BF_3, 60°, 24 hr, 55% yield	100%	--	[113]

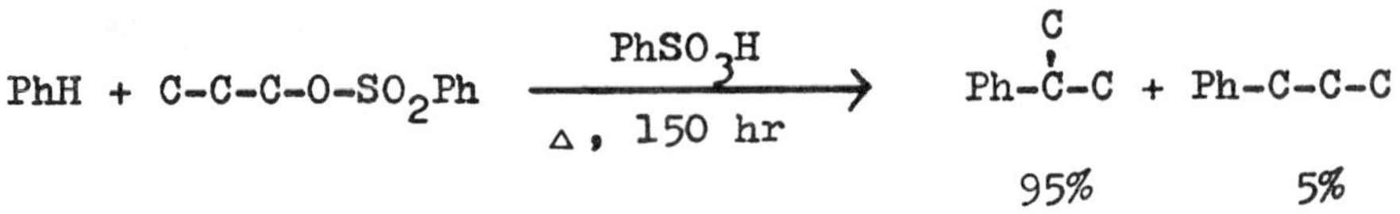

PhH + C-C-C-O-SO_2Ph $\xrightarrow[\Delta,\ 150\ hr]{PhSO_3H}$ Ph-C(C)-C + Ph-C-C-C

95% 5%

(16a)
[125a]

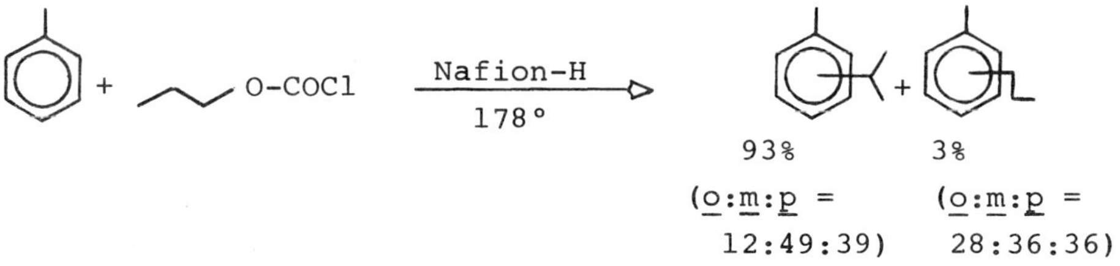

Nafion-H

178°

93% (o:m:p = 12:49:39)

3% (o:m:p = 28:36:36)

(17)
[48]

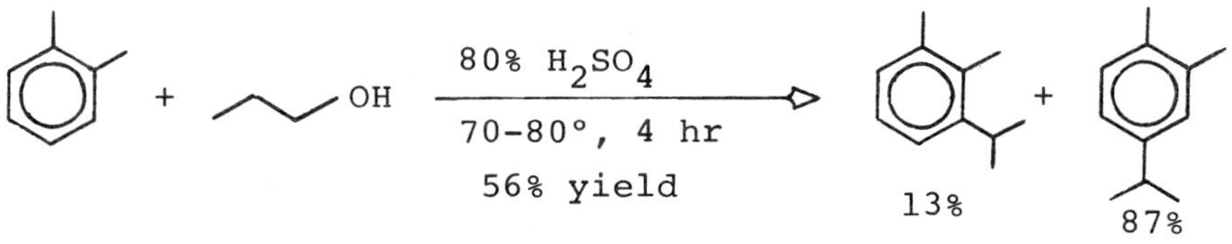

+ OH

80% H_2SO_4

70-80°, 4 hr

56% yield

13% 87%

(18)
[116]

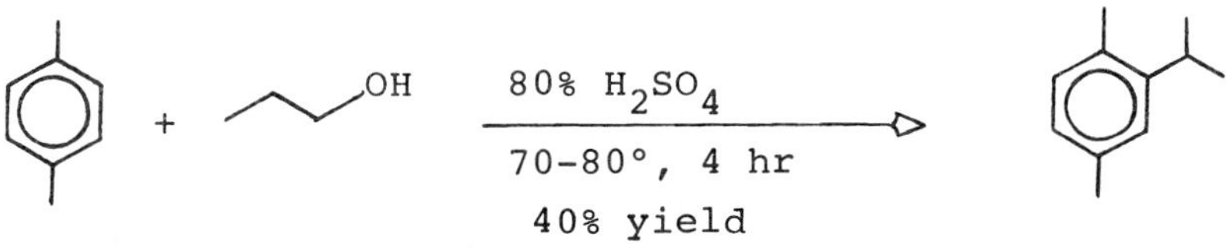

+ OH

80% H_2SO_4

70-80°, 4 hr

40% yield

(19)
[116]

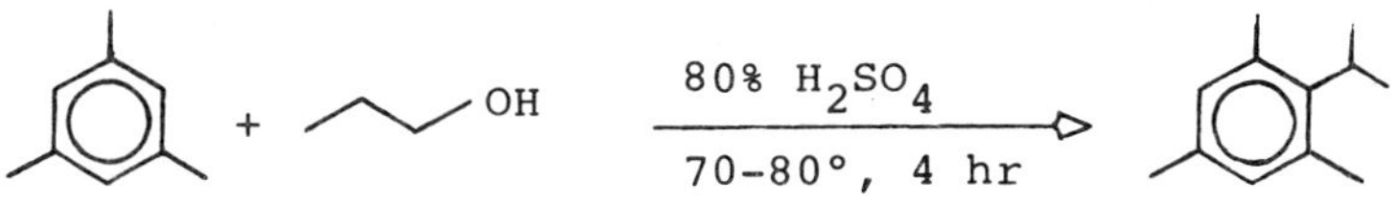

+ OH

80% H_2SO_4

70-80°, 4 hr

(20)
[116]

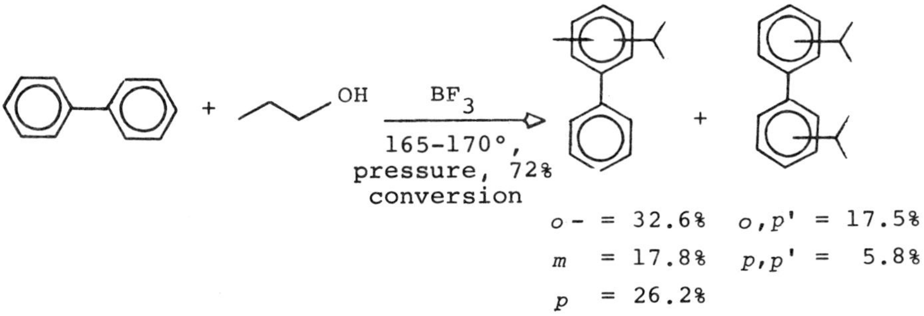

+ OH

BF_3

165-170°, pressure, 72% conversion

(21)
[41a]

o- = 32.6% *o,p'* = 17.5%

m = 17.8% *p,p'* = 5.8%

p = 26.2%

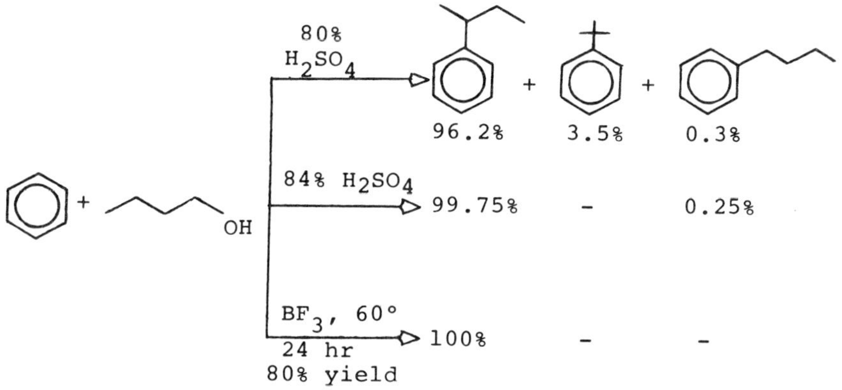

80% H_2SO_4 — 96.2% + 3.5% + 0.3% [123,125]

84% H_2SO_4 — 99.75% - 0.25%

OH (22) [115]

BF_3, 60° 24 hr 80% yield — 100% - - [113]

$AlCl_3/CH_3NO_2$ at 25°, 80% H_2SO_4 at 80° or polyphosphoric acid at 80°

OH

(*o*-; *m*-; *p*- = 40:20:40) traces

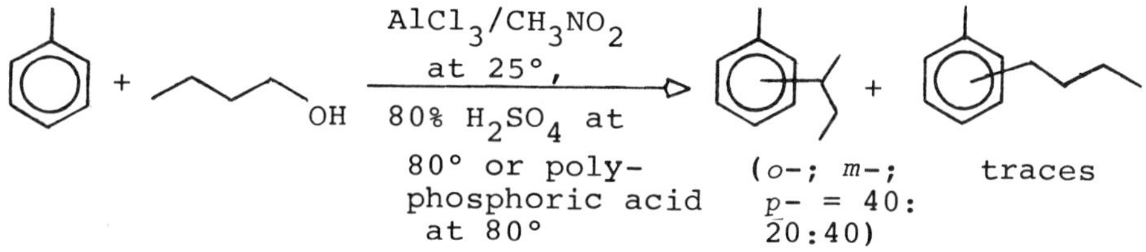

(23) [119]

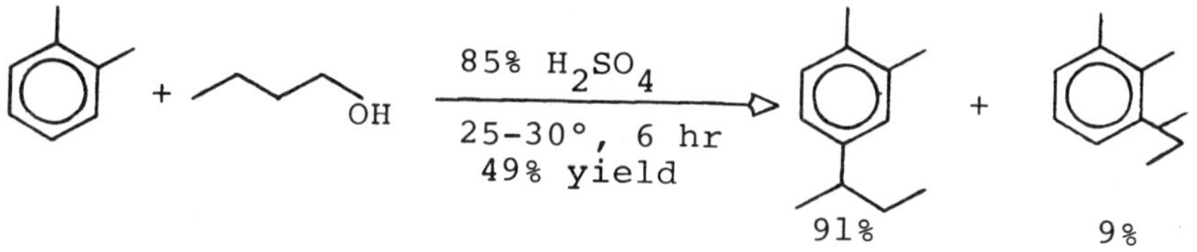

OH 85% H_2SO_4 25-30°, 6 hr 49% yield

91% 9%

(24) [116]

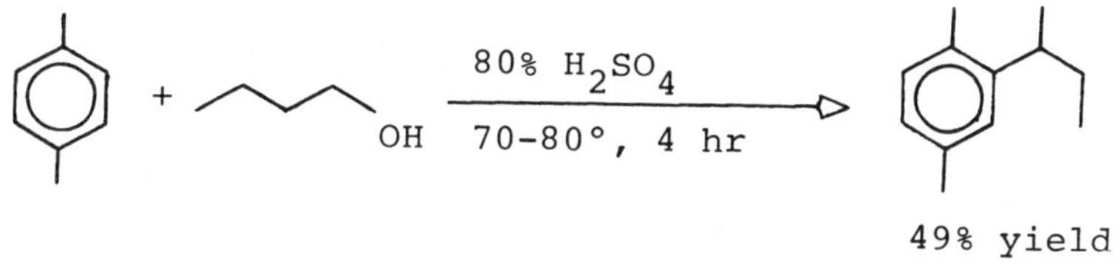

OH 80% H_2SO_4 70-80°, 4 hr

49% yield

(25) [116]

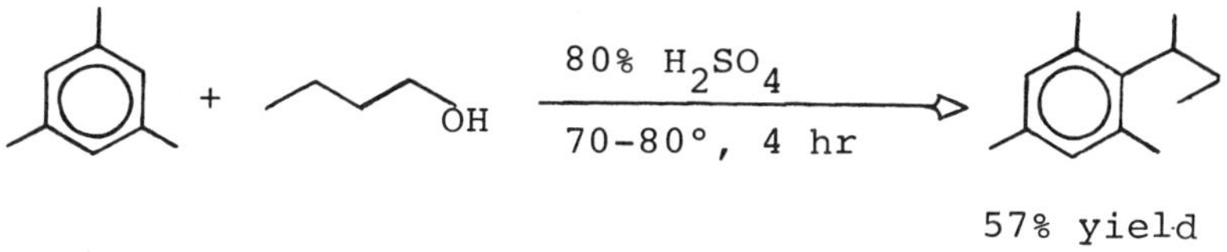

OH 80% H_2SO_4 70-80°, 4 hr

57% yield

(26) [116]

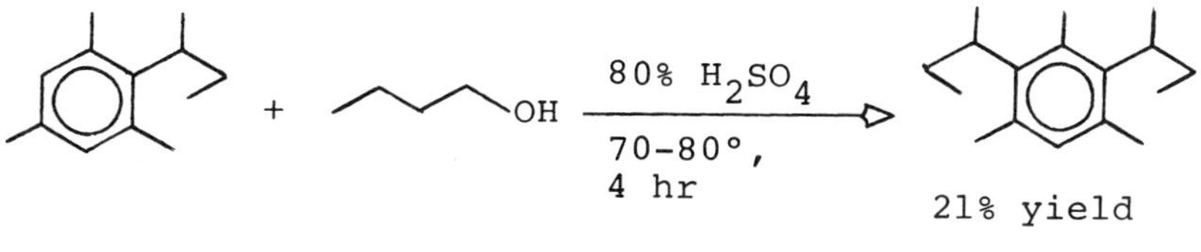

OH 80% H_2SO_4 70-80°, 4 hr

21% yield

(27) [116]

$$\text{Phenol} + n\text{-BuOH} \xrightarrow[15\ \text{min},\ 150°]{BF_3\ \text{in}\ CCl_4} p\text{-sec-butylphenol}\ (67\%) \quad (28)\ [120]$$

$$\text{Biphenyl} + n\text{-BuOH} \xrightarrow[165\text{-}170°,\ 44\%\ \text{conversion}]{BF_3} \text{sec-butylbiphenyl} + \text{di-sec-butylbiphenyl} \quad (29)\ [45a]$$

o = 17.0% o,p'=36.1%

m = 20.0% p,p'=13.0%

p = 24.5%

In a similar manner, alkylations with n-pentyl and higher n-alkyl derivatives have been expected to yield the corresponding secondary alkylarenes. However, in such alkylations, there remained a controversy about whether the secondary alkylates obtained were composed of one [109,111,126-128] or two [118] specific secondary isomers as claimed by some, or of all possible secondary isomers as claimed by others [26,32,35,113,114,121].

To illustrate, the alkylation of benzene with n-pentyl ether and BF_3 [109], with n-pentyl, n-octyl, and n-dodecyl alcohols and BF_3 [106], and with n-heptyl, n-octyl, n-nonyl, and n-hexadecyl alcohols and H_2SO_4 [33] was reported to yield the respective 2-phenylalkanes. Similarly, it was reported that the alkylation of o-cresol, p-cresol, o-chlorophenol, and p-chlorophenol with n-capryl, n-cetyl, and n-lauryl alcohols in the presence of zinc chloride gave alkylated products in which the aryl groups were exclusively located at C-2 of the alkyl side chains [127]. Moreover, the alkylation of phenol, cresols, and alkyl phenyl ethers with normal hexyl, heptyl, and octyl alcohols yielded 2- and 3-(p-hydroxyphenyl)alkanes; other isomers were said to be absent [118]. Although the chemists whose work is reported above recognized the rearrangements of alkyl groups from primary to secondary, they failed to recognize the concurrent secondary-to-secondary rearrangements which lead to the formation of all possible secondary alkylates.

Expectedly, these ambiguous results led to misleading explanations. At one stage, it was suggested that the pure 2-arylalkanes were produced by an elimination-addition mechanism, as in Eqs. (30) and (31) [101,112,129].

$$CH_3(CH_2)_3CH_2OR \longrightarrow CH_3(CH_2)_2CH{=}CH_2 + ROH \quad (30)$$

$$CH_3(CH_2)_2CH{=}CH_2 + C_6H_6 \longrightarrow CH_3(CH_2)_2\underset{\substack{|\\ Ph}}{C}HCH_3 \quad (31)$$

However, this mechanism was opposed by Price et al. [130-132] (1938-1940), who demonstrated that cyclohexanol with BF_3 gave no cyclohexene under conditions which in the presence of naphthalene resulted in rapid alkylation [131]. Instead, Price and Ciskowski [131] suggested a heterolytic cleavage of the alcohol-catalyst complex to a carbocation, followed by electrophilic attack on the aromatic nucleus. This view, which seemed consistent with facts such as the high reactivity of secondary relative to primary alcohols, the extensive rearrangements often observed, and the noted racemization during the BF_3-catalyzed alkylation of benzene by optically active 2-butanol [132] was shortly adopted by others. In 1941, Hennion and co-workers [110,133] suggested the following mechanism for the BF_3-catalyzed alkylations of arenes with normal alcohols:

$$n\text{-ROH} + BF_3 \longrightarrow sec\text{-}R^+ + (HO \rightarrow BF_3)^-$$

$$C_6H_6 + sec\text{-}R^+ \longrightarrow [C_6H_6R]^+ \longrightarrow C_6H_5R + H$$

In spite of the adoption of what they chose to call the "positive fragment theory," the latter authors did not consider possible secondary-to-secondary rearrangements, and the only products reported were 2-arylalkanes.

Apparently, the earliest substantiated discovery of the occurrence of secondary-to-secondary rearrangements during alkylations with n-alkyl derivatives should be credited to Ipatieff et al. [26]. In 1940, these authors reported that the products of alkylation of benzene with either n-pentyl alcohol and BF_3 or 1-pentane and H_2SO_4 comprised mixtures of the two possible secondary isomers, 2- and 3-phenylpentane. To account for these results, they suggested that the reaction proceeded through formation of 2-pentyl hydrogen sulfate, which underwent reversible isomerization to the 3-pentyl isomer followed by alkylation to yield mixture of 2- and 3-phenylpentane (Eq. 32).

$$\text{C–C–C–C–C} \xrightarrow{H_2SO_4} \text{C–C(OSO}_3\text{H)–C–C–C} \xrightarrow{\text{PhH}} \text{C–C(Ph)–C–C–C}$$
$$\text{C–C(OSO}_3\text{H)–C–C–C} \rightleftharpoons \text{C–C–C(OSO}_3\text{H)–C–C} \xrightarrow{\text{PhH}} \text{C–C–C(Ph)–C–C} \qquad (32)$$

This reported formation of isomeric secondary alkylates was later supported by other workers. In 1959, Streitwieser et al. [113,114] investigated the BF_3-catalyzed alkylation of benzene with a number of alcohols, including ethyl, n-propyl, n-butyl, and n-pentyl alcohols. In agreement with previous workers [101,134] they found that n-propyl and n-butyl alcohols gave good yields of isopropyl- and *sec*-butylbenzenes, respectively. Ethyl alcohol did not react under similar conditions [101,106]. On the other hand, n-pentyl alcohol gave a mixture of monoalkylated products which was shown by infrared and mass spectral analyses to have the composition shown in Eq. (33).

```
                         BF3
PhH + C-C-C-C-C-OH  ----------->  C-C-C-C-C    +    C-C-C-C-C
                         60°        |                    |
                                    Ph                   Ph

                                  53-54%               20%

                                    C                 C                    (33)
                                    |                 |
                              +   C-C-C-C    +     C-C-C-C
                                    |              |
                                    Ph             Ph

                                  22-25%           3%
```

More recently, Khalaf and Roberts [121] repeated the reaction with *n*-pentyl alcohol. Using both GLPC and NMR analyses they found the products to be similar to those of Eq. (33), consisting of 59% 2-phenylpentane, 21% 3-phenylpentane, and 20% *t*-pentylbenzene; no 2-methyl-3-phenylbutane could be detected, however. Notably, the results of both groups of workers showed that the 2- and 3-phenylpentane isomers were produced in a ratio corresponding to their established equilibrium composition (2-/3 isomer, 71:29) [135].

The reported rearrangements of linear secondary alkyl groups to tertiary ones are of uncertain validity. Rearrangements observed in 1939 by Nightingale and Smith [135a], who identified 1,3-dimethyl-5-*t*-butylbenzene among the products of reaction of 1,3-dimethyl-4-*sec*-butylbenzene with $AlCl_3$ at steam-bath temperature, were later shown to be incorrect [135b]. However, similar extensive rearrangements were noted during the alkylation of toluene with *n*-butyl or *sec*-butyl alcohol and 80% H_2SO_4 or polyphosphoric acid at 80°C [119]; the alkylation of phenol with *sec*-butyl alcohol and BF_3 at 130°C [136]; and the alkylation of naphthalene with *n*-butyl alcohol and BF_3 under pressure at 165°C [137]. In the latter two cases, the product from phenol was said to consist mainly of *p*-*t*-butylphenol and that from naphthalene to contain ca. 3% 2-*t*-butylnaphthalene. Moreover, the product of alkylation of benzene with *n*-pentyl alcohol and 80% H_2SO_4 at 80°C was reported to contain 9% of *t*-pentylbenzene (Eq. 34) [117].

$$PhH + n\text{-}C_5H_{11}OH \xrightarrow[80°,\ 63\%\ \text{yield}]{80\%\ H_2SO_4} \text{2-phenylpentane } (63\%) + \text{3-phenylpentane } (18\%) + t\text{-pentylbenzene } (9\%) \quad (34)\ [117]$$

These observed isomerizations of linear primary or secondary alkyl groups to tertiary ones have been explained by some workers in terms of cationic transformations that involved hydride and methyl shifts and primary carbocation formation, as outlined in Eq. (35) [122,124,135a,138]. Other workers

```
                                                                  C              C
      +   ~H:⁻        +      ~H:⁻       +       ~CH3:⁻           |  +  ~H:⁻      |
RCCC    --------> RCCC    <=======>  RCCC    ------------>  RCC   --------->  RCC     (35)
                                                                                 +
                     (R = Me or Et)
```

[113,114], however, attributed such transformations to nonspecified processes which they termed "deep-seated rearrangements."

Since the involvement of such intramolecular rearrangements of secondary to primary carbocations have been criticized on thermodynamic grounds [139], we believe that better explanations of authenticated rearrangements could be given in terms of either protonated cyclopropanes [140] or bimolecular reactions [139], as discussed in Chap. 2.

In the 1970s Lipovich et al. [122-125] applied ^{14}C tracer techniques to gain more insight into the nature of the side-chain rearrangements accompanying alkylations with linear acyclic alcohols and alkenes. The ^{14}C-labeled compounds 1-7 were used to alkylate benzene in the presence of $AlCl_3$, BF_3, or H_2SO_4 catalyst.

```
C-C-14C-OH      C-14C-C-OH      14C-C-14C      C-C=14C
                                    |
                                    OH

    1               2               3             4

C-C-C-14C-OH    C-14C-C-C       C-C-C=14C
                      |
                      OH

    5               6               7
```

The product compositions showed remarkable dependence on the structure of the alkylating agent and the nature of the catalyst employed. Regardless of catalyst type, alkylations with secondary alcohols or alkenes gave only secondary alkylates. On the other hand, alkylations with normal alcohols gave mixtures of secondary and normal alkylbenzene with BF_3 or $AlCl_3$ catalyst, and only secondary alkylbenzene with H_2SO_4 catalyst. These results, which are similar to previous ones, suggested a rapid conversion of the primary carbocation to the secondary one prior to alkylation via a 1,2-hydride shift. The rate of such conversions by H_2SO_4, however, apparently exceeded those by $AlCl_3$ and BF_3, judging by the considerable amount (ca. 20%) of *n*-propylbenzene from the *n*-propyl alcohol alkylation with the latter catalysts. Radioassay of the products showed that the alkylations were accompanied by some scrambling of the label in the alkyl groups. The extent of scrambling depended on the structure of the alkylating species, the type of catalyst, and the reaction conditions. In H_2SO_4-catalyzed alkylation of benzene with the ^{14}C-labeled propyl alcohols (1-3) and propene (4) [123,125],

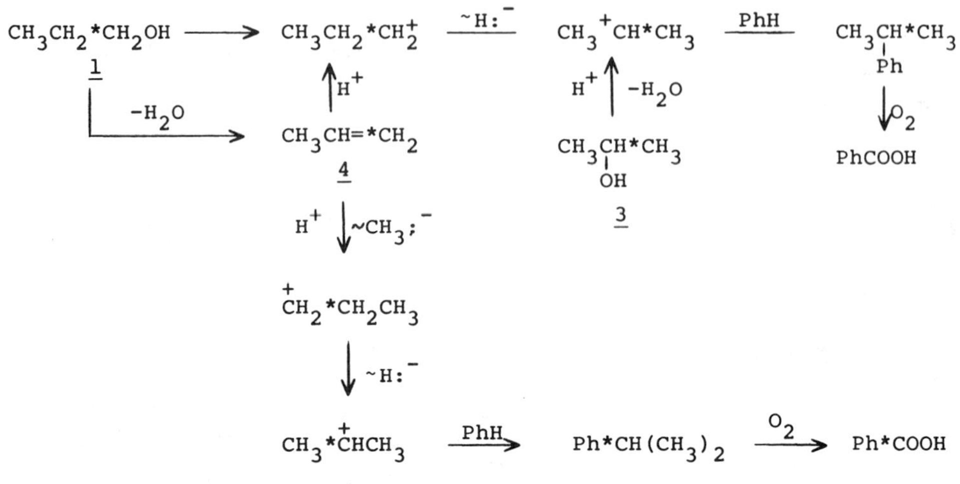

$$\underset{\underline{1}}{CH_3CH_2{*}CH_2OH} \longrightarrow CH_3CH_2{*}CH_2^+ \xrightarrow{\sim H:^-} CH_3\overset{+}{C}H{*}CH_3 \xrightarrow{PhH} CH_3CH(Ph){*}CH_3 \xrightarrow{O_2} PhCOOH$$

$$\underset{\underline{1}}{CH_3CH_2{*}CH_2OH} \xrightarrow{-H_2O} \underset{\underline{4}}{CH_3CH{=}{*}CH_2} \xrightarrow{H^+} CH_3CH_2{*}CH_2^+$$

$$\underset{\underline{3}}{CH_3CH(OH){*}CH_3} \xrightarrow{H^+,\ -H_2O} CH_3\overset{+}{C}H{*}CH_3$$

$$\underset{\underline{4}}{CH_3CH{=}{*}CH_2} \xrightarrow{H^+,\ \sim CH_3:^-} \overset{+}{C}H_2{*}CH_2CH_3 \xrightarrow{\sim H:^-} CH_3{*}\overset{+}{C}HCH_3 \xrightarrow{PhH} Ph{*}CH(CH_3)_2 \xrightarrow{O_2} Ph{*}COOH$$

Scheme 2

the observed scrambling of the label between the α and β-carbons of the propyl side chain in the products ranged from 2.3 to 9.1%. This scrambling was explained in terms of mechanisms that involved primary carbocation intermediates (scheme 2). Since no isotopic rearrangements could be induced in labeled isopropylbenzene by treatment with H_2SO_4 under the alkylation conditions, the rearrangements must have occurred in the alkylating moiety prior to the alkylation step. However, the involvement of primary carbocation intermediates in the mechanism above makes it hard to accept. Instead, we believe that the observed isotopic rearrangements can better be explained in terms of bimolecular mechanism of the type proposed by Karabatsos et al. [139].

The alkylation of benzene with $Et^{14}CH_2OH$ (1) and $AlCl_3$ or BF_3 gave mixtures of labeled *n*-propyl and isopropylbenzene (mainly) in which some scrambling of the label between the α and the β-positions had taken place [125]. The authors again explained the results in terms of mechanisms similar to those of scheme 2. In the presence of these two Lewis acid catalysts, however, the results can better be accounted for, at least in part, by the diphenylalkane mechanism of Roberts et al. [41,141,142] and/or the bimolecular mechanism of Farcasiu [143]. Details of these mechanisms are given in Chap. 8.

The alkylation of benzene with $CH_3CH_2CH_2{}^{14}CH_2OH$ (5) and 80% H_2SO_4 at 80°C gave a labeled mixture composed of *sec*-butylbenzene as the main product with small quantities of *t*-butylbenzene (3.5%) and *n*-butylbenzene (0.3%) [122,124]. Radioassay of the *sec*-butylbenzene obtained showed that the radioactive label was distributed to the extent of 93% at C-1, 6% at C-2, and 1% at C-3 of the secondary butyl group. Under similar conditions benzene and $CH_3CH_2CH = {}^{14}CH_2$ (7) gave 14.8% ^{14}C rearrangement, indicating that butene is not a major intermediate in the reaction with *n*-butyl alcohol. On the other hand, under similar conditions at 80°C or at 10 to 12°C in the presence of 98% H_2SO_4, $CH_3{}^{14}CH_2CH(OH)CH_3$ (6) yielded 90% *sec*-butylbenzene, in which 39% of label was scrambled. These results were explained in terms of carbocation processes involving methyl, ethyl, and hydride shifts with primary carbocations as intermediates.

In summing up their views, Lipovich et al. emphasized the following points: (1) Alkylations with alcohol proceeded via initial protonation and loss of H_2O, followed by methyl, ethyl or hydride 1,2 shifts within the carbocation intermediates. (2) The rate of conversion of primary carbocations to secondary ones by H_2SO_4 via 1,2 hydride shift exceeded that by $AlCl_3$ or BF_3. (3) In H_2SO_4-catalyzed alkylations, alkenes are major intermediates from *sec*-alcohols, but not normal alcohols. It is to be emphasized, however, that Lipovich et al. explained their results in terms of mechanisms that involved primary carbocations as intermediates. This subjects the explanations to heavy criticism that can be avoided by suggesting alternative mechanisms that involve protonated cyclopropanes and/or bimolecular reaction pathways (see Chaps. 2 and 8).

In conclusion, for alkylations with *n*-alkyl derivatives in the presence of mild catalysts, rearrangements are rather extensive, leading conclusively to secondary alkylates in which the aryl group becomes attached to all possible secondary carbons in the chain. The literature reports claiming the production of pure primary alkylates [62,83(a),99,101,102], pure secondary alkylates [109,111,126,127,128], or of some, but not all, possible secondary alkylates [118] should be considered as highly questionable.

III. ALKYLATIONS OF ARENES WITH LINEAR ALKENES AND LINEAR SECONDARY ALCOHOLS, ETHERS, AND ESTERS IN THE PRESENCE OF VARIOUS CATALYSTS

A. Ethylene, Propylene, and Isopropyl Alcohol, Ethers, and Esters

The alkylation of arenes with ethylene and propylene comprises two of the most important industrial applications of Friedel-Crafts reactions. The products from benzene, ethylbenzene, and isopropylbenzene (cumene) are two of the most important petrochemical raw materials [144,145]. In 1979 the production of ethyl- and isopropylbenzene in the United States alone amounted to about 7 and 4 billion pounds, respectively. This explains why ethylation and isopropylation reactions have always been subjects of lively concern.

The huge volume of literature data on alkylations with alkenes, including ethylene and propylene, has been frequently reviewed by other authors [1-15]. The most comprehensive survey through 1961 is that given by Patinkin and Friedman [146] in Olah's treatise. In this survey, the essential information regarding reactants, catalysts, conditions, and products has been abstracted and tabulated.

Since 1961, a vast number of investigations have been reported in which ethylene and propylene, as well as other ethyl and propyl derivatives, were used to alkylate arenes [61,116,119,147-244]. Between 1968 and 1974 only, Refs. [157] to [244] could be traced.

However, as can be seen from the cited references, most of these investigations were conducted by Russian workers and were primarily aimed at evaluating conditions for optimal production in the presence of various catalyst systems. These included protonic acids [122-125,178,189,208], metal halides [189,193,194,202,242,244], P_2O_5-H_2O [244], boron trifluoride complexes [175,176,187,199,200,201,204,206,210,211,224,226,228,229,232,233], metal oxides [62-64,155,161,197,203], organoaluminum complex catalysts [219], ion exchange resins [182,251], zeolites [182], and a wide variety of synthetic mixtures [163,190]. Besides these conventional catalysts, there is now growing interest in the application of solid superacids such as Nafion-H to

Table 3.

Alkylating agent	Solvent	Catalyst type	Temp, °C	Yield, %	Composition, % 1-	2-	Ref.
$CH_3CH = CH_2$	—	$AlCl_3$	100	—	4	96	254
$CH_3CH = CH_2$	—	BF_3Et_2O	5-10	69	70	30	198
$CH_3CH = CH_2$	Heptane	HF	74-82	91.2	97	3	218
$CH_3CH(OH)CH_3$	Petroleum ether	$AlCl_3$	80	19	46	54	85

catalyze the alkylation of arenes with alkenes in heterogeneous vapor-phase reactions [92,93].

Besides industrial applications, both ethylation with ethylene and isopropylation with propylene or isopropyl alcohol have found laboratory applications in synthesis and in kinetic studies of positional and substrate selectivities. In connection with the latter studies, ethylene was used to alkylate toluene [245] and propylene was used to alkylate methylbenzenes [245-247] and monohalobenzenes [248] in the presence of various catalysts and solvents. The results of these studies demonstrated that alkylations with the alkenes showed effects similar to those noted before in alkylations with ethyl or isopropyl halide (see Chap. 3). In line with expectations, alkylations of toluene with propylene gave more of the *m* isomer than with ethylene, probably as a result of faster reorientation of the isopropyl group [245]. Moreover, in the $AlCl_3$-CH_3NO_2-catalyzed isopropylation of *m*-xylene in nitromethane solution, propylene showed considerably decreased steric factors over similar alkylations with isopropyl bromide [246].

The question concerning positional selectivity during the alkylation of napthalene and other polynuclear aromatics with ethylene and propylene or isopropyl alcohol was also addressed. Recently, the alkylation of naphthalene with ethylene and BF_3-Et_2O or P_2O_5 at 185°C was reported to give 81% ethylnaphthalenes in which the selectivity for the 1-isomer was 79% [239].

In the years 1939-1961, propylene was reported to alkylate naphthalene to give mono- and higher isopropylated derivatives, but without specified orientation [249,250]. The work of more recent investigators showed that in ethylation or propylation, mixtures of 1 and 2 isomers usually result whose composition ratios depend largely on the type of catalyst and the severity of conditions. Nonisomerizing catalysts such as BF_3-H_3PO_4 [198], HF [218], BF_3-Et_2O [239], or P_2O_5 [239] give higher 1/2 isomer ratios than the isomerizing $AlCl_3$ catalyst. The results obtained under various conditions are summarized in Table 3.

Table 4. Product Distributions in Alkylations of Biphenyl with C_1-C_4 Alkenes and Alkanols

				Product distribution, %[b]					
				Monoalkylates			Dialkylates		
Entry No.	Alkylating agent	Catalyst[a]	Conversion, %	*o*	*m*	*p*	*o,p'*		*p,p'*
1	MeOH[c]	BF_3	1	—	—	—	—		—
2	EtOH[c]	BF_3	8	32.5	30.2	33.3		3.8	
3	C_2H_4	$BF_3 \cdot H_3PO_4$	1	—	—	—	—		—
4	C_3H_6	H_2SO_4	100	22.0	11.8	16.2	35.8		14.2
5	C_3H_6	$BF_3 \cdot H_3PO_4$	100	21.6	12.0	16.6	36.1		13.7
6	*i*-PrOH[c]	BF_3	69	33.2	18.1	26.8	16.8		5.2
7	*i*-PrOH[c]	$BF_3 \cdot H_3PO_4$	17	36.8	22.1	28.3	9.5		3.7
8	*n*-PrOH[c]	BF_3	72	32.6	17.8	26.2	17.5		5.8
9	*n*-BuOH	BF_3	67	17.0	10.0	24.5	36.1		13.0
10	*s*-BuOH	BF_3	44	17.9	10.2	25.8	34.0		9.0
11	*cis*-C_4H_8	$BF_3 \cdot H_3PO_4$	83	12.8	6.1	3.0[d] 22.2 13.8[d]	35.6		9.5
12	*t*-BuOH	BF_3	12	0	6.7	95.3	0		0
13	*i*-C_4H_8	$BF_3 \cdot H_3PO_4$	26	0	6.6	95.5	0		0

[a]Concentration of all catalysts except BF_3, 3% by weight of biphenyl.
[b]Combined weight per cent.
[c]Run in pressure apparatus.
[d]*tert*-Butyl-biphenyl.
Source: Ref. 45a.

Table 5. Alkylations of Arenes with Isopropyl and Isobutyl Alcohols and Polyphosphoric Acid Catalyst

	Yield of isopropyl derivatives[a], %			Yield of *tert*-butyl derivatives[a], %	
Aromatic hydrocarbon	mono-	di-	tri-	mono-	di-
Benzene	59.7	16.8	2.0	44.0	15.6
Toluene	51.0	10.6	0.7	73.2	2.5
Ethylbenzene	48.2	8.6	—	32.6	-
n-Propylbenzene	43.0	3.4	-	28.2	-
Isopropylbenzene	42.2	14.0	-	32.3	-
n-Butylbenzene	32.9	-	-	35.8	-
sec-Butylbenzene	34.6	-	-	33.1	-
t-Butylbenzene	33.4	-	-	36.0	-
o-Xylene	46.8	8.2	-	39.0	-
m-Xylene	51.3	17.1	-	6.7	-
p-Xylene	38.4	12.4	-	4.2	-
Mesitylene	41.7	2.8	-	-	-
Pseudocumene	74.9	1.1	-	6.8	-
p-Cymene	29.1	-	-	1.2	-

Mixture of cymenes	48.9	8.1	-	18.1	-
1-Ethyl-4-n-propylbenzene	15.7	-	-	-	-
1-Ethyl-3-propylbenzene	12.1	-	-	-	-
Fluorobenzene	65.0	10.2	2.0	35.8	4.3
Chlorobenzene	31.7	-	-	16.4	-
Bromobenzene	16.2	-	-	10.5	-
Iodobenzene	9.8	-	-	5.5	-
o-Fluorotoluene	34.0	2.5	-	41.0	-
p-Fluorotoluene	34.5	2.9	-	8.2	-
o-Chlorotoluene	31.8	2.3	-	25.1	-
m-Chlorotoluene	53.8	3.5	-	6.1	-
p-Chlorotoluene	29.9	1.9	-	5.3	-
p-Bromotoluene	12.6	-	-	1.6	-
o-Dichlorobenzene	3.8	-	-	6.2	-
m-Dichlorobenzene	5.9	-	-	2.1	-
p-Dichlorobenzene	3.5	-	-	-	-
p-Dibromobenzene	1.8	-	-	-	-

[a]Based on the amount of starting arene.
Source: Refs. 262 and 263.

Table 6. Isopropylation of Toluene with Esters and Haloesters

Alkylating agent	Solvent	Temp, °C	Time, hr	Catalyst	Isomer Distribution, %			Ref.
					o	*m*	*p*	
i-$PrOSO_2Cl$	CH_3NO_2	30	0.25	$AlCl_3$-CH_3NO	42.3	20.3	37.4	42
i-$PrSO_2Ph$	CH_3NO_2	30	0.25	$AlCl_3$-CH_3NO_2	45.2	21.5	33.2	42
i-$PrSO_2Ph$	CS_2	30	0.25	$AlCl_3$	0.0	82.3	17.7	42
i-$PrOSO_2F$	SO_2	−78	0.25	$AlCl_3$	43.3	23.4	33.3	42
i-$PrOSO_2$-*p*-Tol	CH_3NO_2	30	0.25	$AlCl_3CH_3NO_2$	46.4	21.8	31.8	42
i-$PrOSO_2$-*p*-Tol	CS_2	30	0.25	$AlCl_3$	0.0	81.6	18.4	42
i-PrOCOCl[a]	—	192	—	Nafion-H	4.0	65.0	31.0	48
i-PrOCOCl[b]	—	110	12	Nafion-H	42.0	21.0	37.0	48
i-$PrSO_2$-*p*-Tol	Benzene	60	1	$AlCl_3$	2.2	65.1	32.5	77
i-$PrSO_2Cl$	CH_3NO_2	25	4	$AlCl_3$-CH_3NO_2	46.7	22.6	30.7	77

[a]Gas-phase alkylation.
[b]Liquid-phase alkylation.

These data should be contrasted with earlier ones in which composition was either unspecified [249,250] or contradictory [22]. An example of the latter is the reported production of only the 1-isopropyl isomer from a reaction of naphthalene with isopropyl alcohol and $AlCl_3$ [22].

The isopropylation reactions of other polynuclear aromatics, such as phenanthrene [256,257], fluorene [62], indene [62], cyclopentaphenanthrene [62], and biphenyl [45a,225], were also reported. These gave mixtures of isomeric mono and higher isopropylation products whose composition depended not only on catalyst type and reaction conditions, but also on the steric bulk and reactivity of the alkylating species [45a,258]. In 1969, Priddy [45a] conducted a thorough study of the alkylation of biphenyl using alkylating agent-catalyst combinations known to produce the kinetically controlled product distributions without concurrent isomerization to thermodynamic equilibrium compositions. For comparison, the results obtained for methylation (entry 1), ethylation (entries 2 and 3), isopropylation (entries 3 to 8), *sec*-butylation (entries 9 to 11), and *t*-butylation (entries 12 and 13) are summarized in Table 4.

The results of Table 4 revealed that the positional selectivity in the alkylation of biphenyl under nonisomerizing conditions was significantly affected by both the reactivity and the size of the attacking electrophile. Comparison of the *o/m/p* ratios obtained for ethylation, isopropylation, *sec*-butylation, and *t*-butylation shows an obvious decrease in *ortho* selectivity due to increasing bulk and decreasing reactivity of the respective cations. The results are also in contrast with those reported earlier by Akhmedov et al. in 1962 [259], Zavgorodnii and Sidel'nikova in 1958 [258], Romadane and Berga in 1958 [260], and Romadane in 1957 [261], in which much higher proportions of *para* isomers were claimed.

The synthetic utility of propylene and isopropyl alcohol is well recognized and growing. Besides the previously discussed and compiled data [6,7,146], several reports of other utilizations have been published in recent years. For example, in 1968, Kozlov and Klein [262,263] studied the alkylation of a number of arenes with isopropyl and isobutyl alcohols in the presence of polyphosphoric acid catalyst. The results of these reactions are given in Table 5. Based on infrared analysis, it was concluded that the monoisopropylation products of Table 5 were mixtures of *ortho* and *para* isomers, mixed with small amounts of the *meta* isomers in the case of halobenzenes.

The predicted *ortho/para* ratios as well as the reported yields pointed to some steric opposition to the entering isopropyl group. This is reflected in the observation that the relative amounts of the *o*-isoemrs and of the total yields decreased with increase in the size of the original substitutent in the ring. This opposition is, of course, much milder than that experienced in reactions with isobutyl or *t*-butyl derivatives. It is evident from Tables 5 and 6 as well as from the related isopropylation results shown schematically in Eqs. (36) to (47) that no complete hindrance to *ortho* alkylation or even to alkylation in between two *meta*-oriented methyl groups was encountered. For example, the isopropylation of toluene with isopropyl alcohol [92], with isopropyl esters and haloesters (Table 6) [42,48], with propylene (Eq. 36) [246], or with isopropyl ether (Eq. 37) [175] gave mixtures in which the *ortho*, *meta*, and *para* isomers were found. Similar isopropylations gave the 3- and 4-isopropyl isomers with *o*-xylene (Eq. 38) [116]; the 2-, 4-, and 5-isopropyl isomers with *m*-xylene (Eq. 39) [246]; 2-isopropyl-*p*-xylene with *p*-xylene (Eq. 40) [116]; 2-isopropylmesitylene with mesitylene (Eq. 41) [116]; and 3-, 5-, and 6-isopropylpseudocumene with cumene (Eq. 42) [227]. Equations (43) to (47) showed similar effects.

Various catalysts, 25°

44-47% 15-18% 36-41%

(36) [246]

$80\%\ H_2SO_4$, 1.5 hr, 70°

50% 20% 5%

o- = 39% 2,4- = 45.4% 2,4,5- = 61.2%

m- = 27.1% 2,5- = 36.0% 2,3,6- = 38.2%

p- = 33.6% 2,6- = 18.6%

(37) [175]

OH

$80\%\ H_2SO_4$, 4 hr, 70-80°, 58% yield

13-16% 84-87%

(38) [116]

$AlCl_3/CH_3NO_2$ in CH_3NO_2, 25°

15.8% 74.0% 10.2%

(39) [246]

OH

$80\%\ H_2SO_4$, 4 hr, 70-80°

55% yield

(40) [116]

OH

$80\%\ H_2SO_4$, 4 hr, 70-80°

43% yield

(41) [116]

+ or $R-C_6H_4SO_3CH(CH_3)_2$ — $R-C_6H_4SO_3H$, 120° → 4% + 70% + 26%

(R = H, CH_3)

(42) [227]

Br + → $BF_3 \cdot H_3PO_4$ → Br 40% + Br 60%

(43) [222]

OH + OH → 33% H_3PO_4 at 90°, 3 hr or pyrophosphoric acid, 0.5 hr → OH (70% yield)

(44) [61]

OH + → P_2O_5/H_2O or $AlCl_3$ → OH + OH + OH ( )2

2,5- and 3,5- diisopropyl isomers

(45) [224]

OCH_3 Br + → 92% H_2SO_4, 50° → OCH_3 Br 20.5%

(46) [192a]

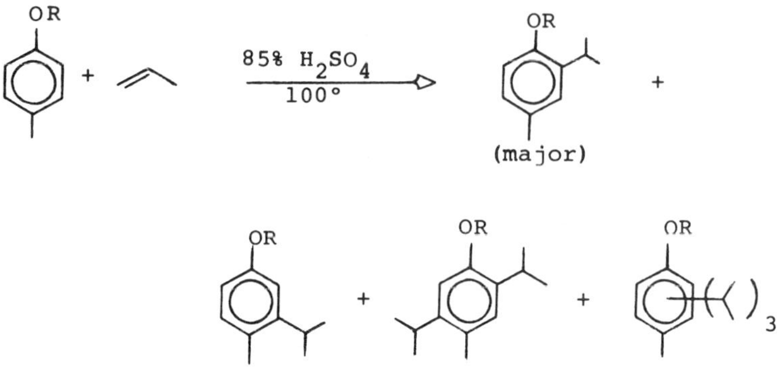

OR + ⟶ (85% H_2SO_4, 100°) OR (major) +

(47) [192b,c]

OR + OR + OR ( )$_3$

(R = Me, Et, Bu)

In the latter two reactions (Eq. 46 and 47), product distributions were determined by both electronic and steric factors. In both cases, the alkoxyl group showed predominant orientation effect over either bromine or methyl. Moreover, in the last reaction (Eq. 47), the overall conversion and product yields decreased in the order R = Me > Et > Bu.

In a few cases, the isopropylation reaction showed abnormal orientation. For example, far from expectation, the isopropylation of benzo[b]thiophene with several alkylating agents using various methods and catalysts gave mixtures of the 2- and 3-isopropylated derivatives 8 and 9, respectively, in which the 2-isomer predominated (Eq. 48) [150]. No reasonable explanation is yet available for this unusual behavior.

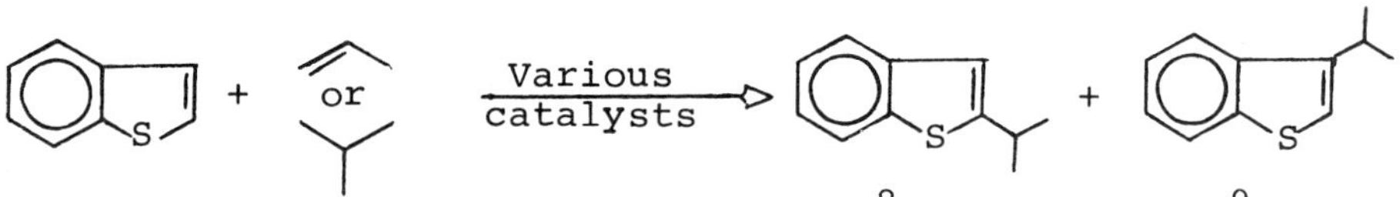

S + or OH,Cl ⟶ (Various catalysts) S (8, 78-92%) + S (9, 8-22%) (48)

The isopropylation reaction was also successful in cases involving deactivated ring systems. For example, Lespagnol et al. [45] were able to alkylate the benzene ring of benzoxazolinones (10) by reaction with isopropyl, isobutyl, and isopentyl alcohols in the presence of H_2SO_4 at 90 to 100°C. With isopropyl alcohol, a mixture of the mono- and diisopropylated derivatives was obtained (Eq. 49).

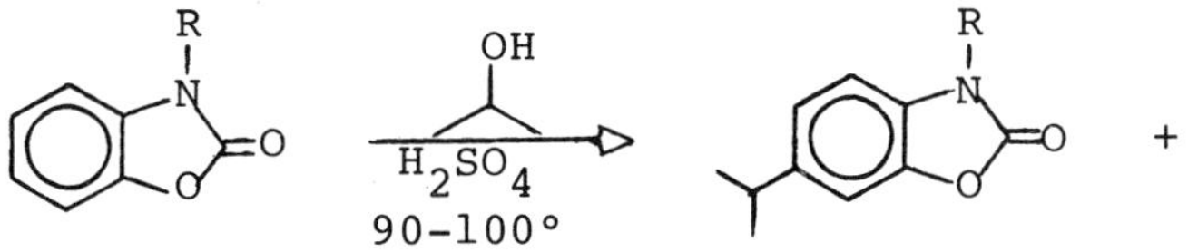

10 ⟶ (OH, H_2SO_4, 90-100°) R = H,CH_3COPh +

(49)

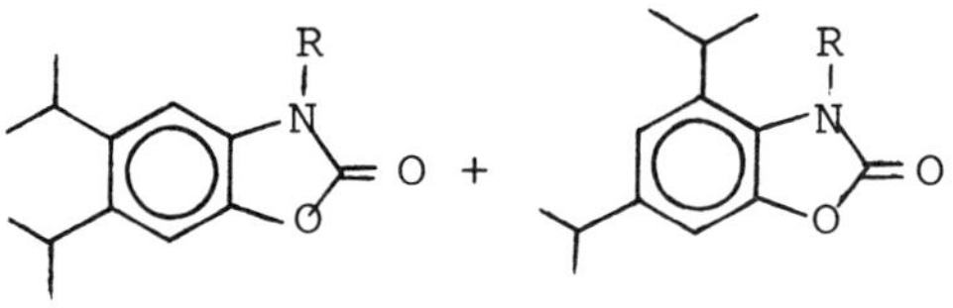

R = H,CH_3,COPh + R = H,COPh

Other highly deactivated aromatics such as *o*-nitrophenol [152(b)], nitrobenzene [178], benzene-sulfonic acid, the polyfluoroarenes 11 and 12, *o*-nitroanisole [249], methyl-2-pyrrolecarboxylate [252], and furfural [253] were also isopropylated. The conditions and results of some of these reactions are given in Eqs. (50) to (54).

NO_2 + (diisopropyl ether) —94% H_2SO_4→ (NO_2, isopropyl) + unidentified nitrocumene isomers (50)

SO_3H + (isopropyl)OH —91% H_2SO_4, 150 hr, 25°→ (SO_3H, ortho-isopropyl) 3.5% + (SO_3H, meta-isopropyl) 88.5% + (SO_3H, para-isopropyl) 8% + (SO_3H, polyisopropyl) 23% (n=2,3) (51)

OH, F, F, F, F (11) —OH or OH, H_2SO_4 or HF; Cl or Cl, $AlCl_3$, 15-22°, 20 hr→ OH, F, F, F, F, isopropyl 40-82% (52)

F, F, F, F, F (12) + Cl(isopropyl) —$AlCl_3$, reflux, 15 hr→ F, F, F, F, F, isopropyl 40% (53)

OCH_3, NO_2 + (isopropyl)OH —HF→ OCH_3, NO_2, isopropyl 85% (54)

The mechanism of alkylation of arenes with alkenes, in general, and propylene, in particular, has received considerable attention. Besides standard kinetic methods [211], it was also explored by the application of labeling techniques using both ^{14}C [123,125] and deuterium [210,243]. For the case of propylene containing the radioactive isotope ^{14}C in a fixed position, Lipovich et al. [123,125] showed that under the conditions of alkylation with sulfuric acid catalyst, a skeletal rearrangement took place. This was explained in terms of a mechanism involving intramolecular hydride and methyl transfers (see scheme 2).

In the alkylation of benzene with propylene in the presence of BF_3-D_2O complex catalyst, Hasegawa et al. [210] observed that the main product in the early stages of the reaction was $CH_3CH(Ph)CH_2D$. They also observed that the BF_3-D_2O complex catalyst was gradually changed into BF_3-HOD and BF_3-H_2O. Based on these results, the mechanism in scheme 3 was proposed.

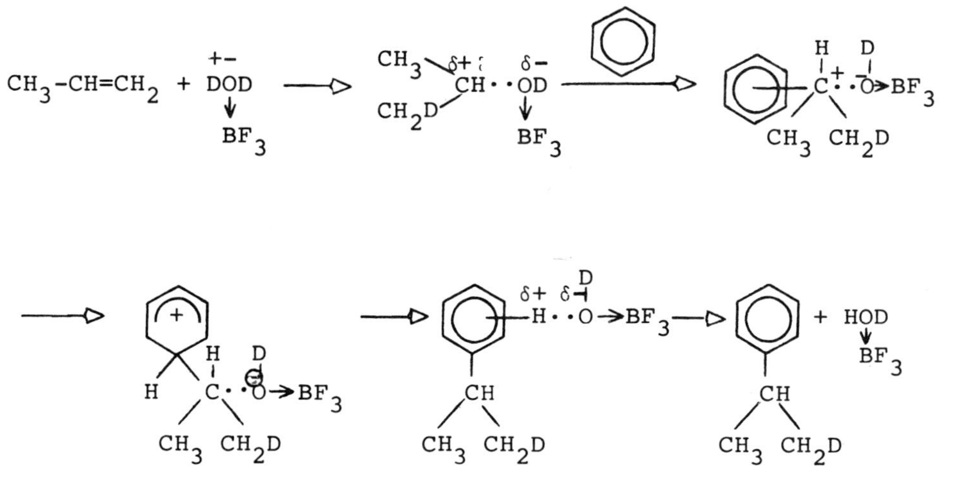

CH_3-CH=CH_2 + DOD ($\rightarrow BF_3$) ⟶ CH_3(CH_2D)$\overset{\delta+}{CH}\cdots\overset{\delta-}{O}D$ ($\rightarrow BF_3$) ⟶ (benzene) ⟶ C_6H_6–$\overset{+}{C}$(H)(CH_3)(CH_2D)$\cdots\bar{O}$(D)$\rightarrow BF_3$

⟶ (arenium ion, H)–C(H)(CH_3)(CH_2D)$\cdots\overset{\ominus}{O}(D)\rightarrow BF_3$ ⟶ C_6H_5[CH(CH_3)(CH_2D)]$\cdots\overset{\delta+}{H}\cdots\overset{\delta-}{O}(D)\rightarrow BF_3$ ⟶ C_6H_5CH(CH_3)(CH_2D) + HOD ($\rightarrow BF_3$)

Scheme 3

To explore further the mechanism of alkylation with alkenes, Polubentseva et al. [243] studied the $AlCl_3$-catalyzed reactions of ethylene, propylene, and 1-butene with benzene-d_6. The results showed that at 10 to 25°C the deuterium of the benzene ring passed mainly into the β position of the side chain of the alkylbenzene obtained. The mechanism in scheme 4 was proposed to account for the results.

Comparing the mechanisms of schemes 3 and 4 reveals that they have much in common. Both mechanisms assume that the initial step is attack by the hydrated form of the catalyst* [264-267], $H^+[BF_3OH]^-$ or $H^+[AlCl_3OH]^-$, on the alkene with the formation of an intermediate alkene-catalyst complex ($R^{\delta+}\cdots Cat^{\delta-}$). In this complex, the degree of polarization could vary according to the type of the alkene, the nature of the catalyst, the basicity of the solvent, if any, and the reaction conditions. The two proposed ex-

*In the case of $AlCl_3$, this form is produced by interaction of the catalyst with the traces of moisture usually available.

$$AlCl_3 + H_2O \rightleftharpoons H_2O \cdot AlCl_3 \rightleftharpoons H^+[AlCl_3OH]^-$$

$$RCH{=}CH_2 + H^+[AlCl_3OH]^- \rightleftharpoons [RCHCH_3][AlCl_3OH]^-$$

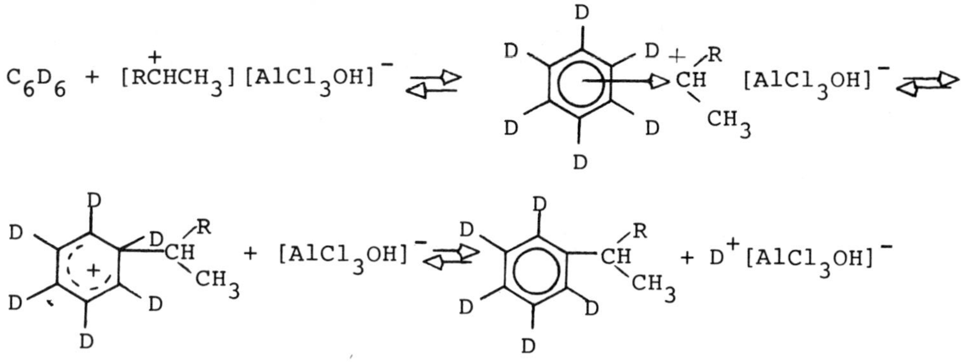

$$C_6D_6 + [R\overset{+}{C}HCH_3][AlCl_3OH]^- \rightleftharpoons C_6D_6 \rightarrow \overset{+}{C}H(R)CH_3\ [AlCl_3OH]^- \rightleftharpoons$$

$$[C_6D_6CH(R)CH_3]^+ + [AlCl_3OH]^- \rightleftharpoons C_6D_5CH(R)CH_3 + D^+[AlCl_3OH]^-$$

Scheme 4

tremes of $R^{\delta+} \cdots Cat^{\delta-}$ are (carbocation + acid anion) and alkyl ester (R-Cat) [211]. With $AlCl_3$ and BF_3-H_2O catalysts, it is probably the former extreme [268]. The next step would be the attack of the catalyst complex on the aromatic ring to form a π complex, then a σ complex. The latter, on reacting with any proton acceptor such as the anions $[AlCl_3OH]^-$ or $[BF_3OH]^-$ forms the product with the regeneration of the hydrated catalyst complex. In the alkylation of benzene-d_6, the deuterium split out from the σ complex in the subsequent stages attacks the initial alkene, forming a new deuterium-containing alkyl cation,

$$RCH{=}CH_2 + D^+[AlCl_3OH]^- \rightleftharpoons [R\overset{+}{C}HCH_2D]\ [AlCl_3OH]^-$$

The alkylation of benzene-d_6 with ethylene or propylene and $AlCl_3$ was also investigated at an elevated temperature of 75°C [243]. With propylene, the deuterium was redistributed between the α and β carbons of the isopropyl side chain. This was suggested to take place by 1,2-hydride (deuteride) and methyl transfers in the intermediate isopropyl cation *prior* to the alkylation step as shown in scheme 5. Support for this suggestion came from the

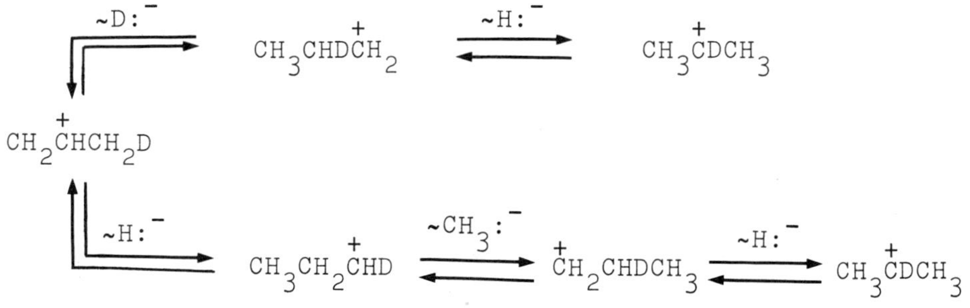

$$CH_3CHD\overset{+}{C}H_2 \underset{}{\overset{\sim H:^-}{\rightleftharpoons}} CH_3\overset{+}{C}DCH_3$$

$$\overset{+}{C}H_2CHCH_2D \xrightarrow{\sim D:^-} CH_3CHD\overset{+}{C}H_2$$

$$\overset{+}{C}H_2CHCH_2D \xrightarrow{\sim H:^-} CH_3CH_2\overset{+}{C}HD \overset{\sim CH_3:^-}{\rightleftharpoons} \overset{+}{C}H_2CHDCH_3 \overset{\sim H:^-}{\rightleftharpoons} CH_3\overset{+}{C}DCH_3$$

Scheme 5

fact that no transfer of deuterium took place when ethylbenzene and isopropylbenzene containing the deuterium in fixed positions were treated with the catalyst under simulated alkylation conditions. However, this mechanism has the same drawback as that of scheme 2, in that primary carbocations are involved. We believe that a better explanation can be made in terms of a bimolecular process analogous to that proposed by Karabatsos et al. [139], which does not require primary carbocation intermediates. (See Chap. 2 for Karabatsos' mechanism applied to the ^{14}C scrambling in *t*-pentyl chloride.)

In contrast to alkylation with propylene, alkylation with ethylene at 75°C resulted in only negligible transfer of deuterium to the α position of the ethylbenzene produced. It was suggested that the ethyl cations produced by protonation reacted rapidly with the benzene ring before any appreciable deuterium scrambling could occur.

Deuterium labeling was also used to investigate the mechanism of alkylation of arenes with isopropyl alcohol. In 1959, Streitwieser et al. [114] found that the reaction of 2-propanol-1-d with benzene and BF_3 at 0°C gave 2-phenylpropane-1-d, which showed no observable loss or rearrangement of deuterium. This, added to the fact that no propene could be found in the exit gases, suggested that alkylations with at least some secondary alcohols at 0°C do not involve significant reaction via alkene intermediates.

In concluding this discussion of ethylation and isopropylation of arenes with ethylene and propylene as well as with ethyl and isopropyl alcohols, ethers, and esters, the following points can be emphasized:

1. Under all conditions and in the presence of all types of catalysts, alkylations of arenes by propene and isopropyl derivatives resulted exclusively in the formation of isopropylarenes. Accordingly, these reactions have been called the "isopropylation" reactions.
2. Both ethylation and isopropylation reactions gave alkylation mixtures consisting of varying proportions of mono and higher alkylates.
3. As usual, the ratio of mono- to higher alkylates, as well as the orientation of the groups in the aromatic nucleus, are dependent on various factors, which include the type and ratio of reactants, and the duration and temperature of the reaction.

B. Linear Butenes and *sec*-Butyl Alcohols, Ethers, and Esters

Prior to the 1950s, there had been a widespread belief among workers that, regardless of catalyst type or reaction condition, the alkylation of arenes with linear butenes or *sec*-butyl derivatives invariably yielded *sec*-butylarenes [1,4-6,8,12,269]. This belief was first challenged in 1950 by Pines et al. [270], who investigated the alkylation of benzene with 2-butene and isobutylene in the presence of catalysts such as silicophosphoric acid, alumina-zinc chloride, and silica-alumina at 200 to 300°C. Their main object was to determine whether or not such strenuous reactions were accompanied by skeletal isomerization of the alkylating group. Their results (see Table 7) demonstrated that rearrangements could occur, but both their occurrence and extent showed great dependence on type of catalyst and severity of conditions. For example, silicophosphoric acid catalyzed the reaction between benzene and 2-butene to give *sec*-butylbenzene mixed with ca. 12 to 21% of the rearrangement product *t*-butylbenzene (entries 1 to 3). In the same reaction, zinc chloride-alumina catalyst caused only 2% of rearrangement to *t*-butylbenzene (entry 4), while silica-alumina catalyst produced exclusively *sec*-butylbenzene (entry 5). On the other hand, the reaction of benzene with isobutylene and silicophosphoric acid at 250°C gave a 31% yield of a mixture of 87% *t*-butylbenzene and 13% *sec*-butylbenzene (entry 6). In this reaction, the degree of polymerization was more pronounced than in similar experiments carried out with 2-butene.

Since this pioneering report by Pines et al. several other reports of accompanying side-chain rearrangements during alkylations with linear butenes and *sec*-butyl alcohol and ether have followed. Some of these rearrangements and the conditions inducing them are also included in Table 7.

Comparison of the literature data summarized in Table 7 reveals the interesting fact that we are dealing with two different types of accompanying side-chain rearrangements. An extensive one leading to the interconversion *sec*-butyl $\rightleftarrows$ *t*-butyl and a less extensive one leading to the interconversion *sec*-butyl $\rightleftarrows$ isobutyl. These two types of rearrangements require different catalysts and conditions and seem to be quite independent of each other. Thus reactions leading to the interconversion *sec*-butyl $\rightleftarrows$ *t*-butyl did not produce any rearrangement to isobutyl and those leading to the *sec*-butyl $\rightleftarrows$ isobutyl interconversion did not produce any *t*-butyl side chains. It can also be gathered from the available data that the *sec*-butyl $\rightleftarrows$ isobutyl rearrangements require the presence of a hydride abstracting catalyst such as $AlCl_3$ or SbF_5 and somewhat elevated temperatures (entries 13, 14, and 16). Being comparable to the rearrangements observed by Roberts and Shiengthong [38] during the alkylation of benzene with butyl chlorides at 80°C, the *sec*-butyl $\rightleftarrows$ isobutyl rearrangements can be explained similarly. The *sec*-butylbenzene produced initially in the reaction undergoes subsequent transformation into the rearrangement product isobutylbenzene (see Chap. 8). Support for this can be found in entries 9, 10, and 12, whereby only *sec*-butylbenzene was found under the milder conditions employed. On the other hand, the more extensive *sec*-butyl $\rightleftarrows$ *t*-butyl interconversion can be explained by branching of the *sec*-butyl group prior to alkylation either via protonated cyclopropane intermediates or via bimolecular reactions (Chap. 2). The latter route seems more likely in this case, since varying amounts of butene dimers were usually found among the product of alkylation of benzene with 2-butene [270,272]. However, the finding that *sec*-butyl- and *t*-butylbenzene underwent partial interconversion to each other in the presence of silicophosphoric acid or zinc chloride-alumina under simulated alkylation conditions suggests that the observed rearrangements could have occurred, at least in part, by subsequent isomerization of the initially formed butylbenzenes.

Rearrangement of the *sec*-butyl to *t*-butyl group also accompanied the alkylation of biphenyl [45a,258] and naphthalene [137] with *sec*-butyl alcohol and linear butenes in the presence of BF_3 or BF_3-H_3PO_4 catalyst. In 1958, Zavgorodnii and Sidel'nikova [258] found that the alkylation of biphenyl with 2-butene in the presence of BF_3-H_3PO_4 catalyst at 50 to 100°C gave mixtures of *p*-*sec*-butyl- and *p*-*t*-butylbiphenyl, whose relative proportions depended on reaction temperature and other reaction variables. More recently (1969), the latter finding was confirmed by Priddy [45a], who also found that secondary to tertiary butyl rearrangement accompanied the BF_3-catalyzed alkylation of biphenyl with *sec*-butyl alcohol at 165 to 170°C, but to a lesser extent (Eq. 55, see Table 4).

Table 7. Skeletal Rearrangements Accompanying Alkylations of Benzene with Linear Butenes and *sec*-Butyl Derivatives in the Presence of Various Lewis or Protonic Acid Catalysts

Entry no.	Benzene, mol	Alkylating agent, (mol)	Catalyst		Temp., °C	Time, hr	Total yield, %[b]	Products				Ref.
								Monobutylbenzenes			Other products, % composition	
			Type	mol[a]				*Sec*	*Tert*	*Iso-*		
1[c]	1.70	2-Butene (0.66)	Silico-phosphoric acid	45-55g	200	1	62	88	12	—	Dimer (20), dibutylbenzenes (8)	270
2[c]	1.25	2-Butene (0.50)	Silico-phosphoric acid	45-55g	255	1	75	87	13	—	Dimer (9), dibutylbenzenes (8)	270
3[c]	2.50	2-Butene (1.00)	Silico-phosphoric acid	45-55g	300	1	67	79	21	—	Dimer (20), dibutylbenzenes (3)	270
4[c]	2.50	2-Butene (1.00)	$ZnCl_2$-alumina	45-55g	200	1	43	98	2	—	Dimer (7), dibutylbenzenes (32)	270
5[c]	2.50	2-Butene (1.00)	Silica-alumina	45-55g	250	1	56	100	0	—	Dimer (9), dibutylbenzenes (18)	270
6[c]	2.50	Isobutylene (1.00)	Silicophosphoric acid	45-55g	250	1	31	13	87	—	Dimer (34), dibutylbenzenes (2)	270
7	1	*n*-Butenes (0.46)	$BF_3 \cdot H_3PO_4$ H_2SO_4(95/5) (5-6 vol %)	—	50-60	-	56	30	70	—	Polymer (10-12)	271

8	2.50	2-Butene (1)	Al_2O_3/3.6%-8.2%	360-370 F	1	31-35	95-96	5-6	—	Dimer (20-25) dibutylbenzenes (18-28)	272
9	1	1-Butene (0.37)	$AlCl_3$ (0.03)	10	4	28.4	100		—	—	243
10	1	1-Butene (0.37)	$AlCl_3$ (0.03)	25	3.5	38.5	100	—	—	—	243
11	1	1-Butene (0.37)	$AlCl_3$ (0.03)	50	2.5	29.4	Mixture of *sec*- and isobutyl-benzene			—	243
12	Large excess	*sec*-BuOH(1)	$AlCl_3$ (1.4)	15	—	77.5	100	—	—	—	36
13	Large excess	*sec*-BuOH(1)	$AlCl_3$ (1.5)	80	1.25	30	30	—	70	—	37
14	Large excess	Di-*sec*-butyl ether (1)	$AlCl_3$ (2.3)	75	6	33	35	—	65	—	36, 37
15	0.06	*sec*-BuOH (0.03)	80% H_2SO_4 (12.5 ml)	80	—	60-65	88	12	—	—	117
16[d]	2	Butane (20)	HF-SbE_5 (1)	25	24	2.5-	36	—	64	Dibutylbenzenes (0.5)	273

[a]Unless otherwise specified.
[b]Yields were based on alkylating agent.
[c]These experiments were carried out under a pressure of 40 atmospheres at an hourly liquid space velocity (ml of hydrocarbons/ml of catalyst/hour) of 1.0.
[d]Reaction carried out under pressure.

Biphenyl + cis-2-butene $(BF_3 \cdot H_3PO_4)$ or sec-butyl alcohol $(BF_3)^*$ ⟶ sec-butylbiphenyl + 4-tert-butylbiphenyl + di-sec-butylbiphenyl (55)

sec-butylbiphenyl	tert-butylbiphenyl	di-sec-butylbiphenyl
o:12.8 (17.9)	--	*o,p'*:35.6 (34.0)
m:6.1 (10.2)	--	*p,p'*:9.5 (9.0)
p:22.2 (25.8)	13.8 (3.0)	*m,p'*: --

We have seen from the preceding discussion that alkylations with linear butenes and *sec*-butyl derivatives were complicated by side-chain rearrangements when strong catalysts and/or strenuous conditions were employed. However, when mild catalysts and/or conditions were used, these reactions yielded exclusively *sec*-butylated products. This fact, which has been affirmed repeatedly by both early and recent workers [6,21 for reviews], is illustrated by the following examples:

PhH + sec-BuOH $\xrightarrow{\text{Catalyst}^\dagger}$ sec-BuPh + di-*sec*-butylbenzene

PhH + sec-BuOSO$_2$Ph $\xrightarrow[\text{reflux, 7 hr}]{PhSO_3H}$ sec-BuPh (48%) [276]

PhH + 2-butene $\xrightarrow[\text{200°, 20-30 atm}]{\text{CaY and rare earth zeolites}}$ sec-BuPh (70%) [277]

---

*Values in parenthes are those from *sec*-butyl alcohol.

†80% H_2SO_4 at 50°C [274] or 70°C [105]; $AlCl_3$ at 0 to 25°C [132] or at 30°C [132,275]; BF_3 at 20°C [274] or at 25°C [101,106,132]; BF_3-P_2O_5 [106]; H_3PO_4 at 70°C [274]; HF at 16°C [274].

+ OH → [119]

o-:*m*-:*p*-

$AlCl_3/CH_3NO_2$ at 25°
80% H_2SO_4 at 80° } 40:20:40
PPA at 80°

$AlCl_3$ at 25°
$AlCl_3/CH_3NO_2$ at 80° } 2-5:70-60:28-35

Br + → ($BF_3 \cdot H_3PO_4$; 80°, 8 hr; 32% yield) Br (36%) + Br (58%) [222]

+ unidentified (6%)

+ OH → (80% H_2SO_4; 70-80°, 4 hr; 57% yield) 91% + 9% [116]

+ OH → (85% H_2SO_4; 7 hr, rt; 42% yield) + + [278]

+ OH → (80% H_2SO_4; 70-80°, 4 hr) [116]

42% yield

+ OH → (80% H_2SO_4; 70-80°, 4 hr; 27% yield) [116]

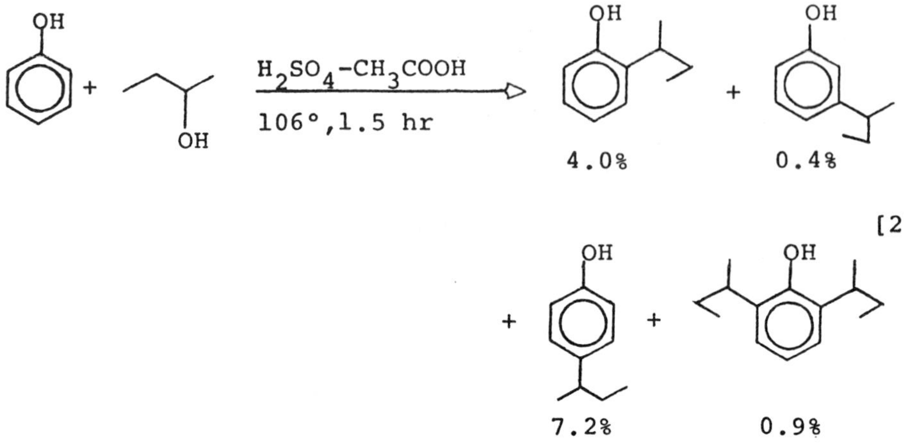

OH + OH $\xrightarrow[106°, 1.5\ hr]{H_2SO_4-CH_3COOH}$ OH (4.0%) + OH (0.4%) + OH (7.2%) + OH (0.9%) [279]

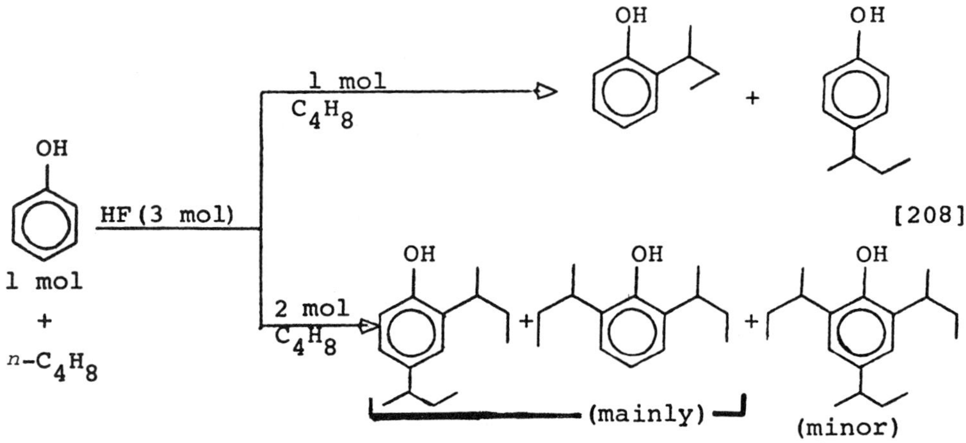

OH (1 mol) + n-C_4H_8 — HF (3 mol) → 1 mol C_4H_8 → OH + OH [208]

→ 2 mol C_4H_8 → OH + OH + OH (mainly) (minor)

The alkylation of naphthalene with *sec*-butyl alcohol and linear butenes was also reported to proceed without skeletal rearrangement of the *sec*-butyl group [22,85,198]. Thus, in contrast to the reaction with benzene, Zee-Cheng [85] found that the $AlCl_3$-catalyzed alkylation of naphthalene with *sec*-butyl alcohol under varying temperatures, catalyst ratio, and in different solvents gave monoalkylates consisting of varying proportions of 1- and 2-*sec*-butylnaphthalene with no detectable amounts of isobutylnaphthalene. The results obtained are partly summarized in Table 8. Notably, the results of entry 1 of Table 8 disqualified Tsukervanik's earlier claim that the product from a similar reaction was pure 1-*sec*-butylnaphthalene [22].

Naphthalene was also exclusively *sec*-butylated with *sec*-butyl alcohol in the presence of BF_3 [137] and with linear butenes in the presence of BF_3-H_3PO_4 [198]. The latter nonisomerizing catalyst was chosen by Friedman and Nelson [198] to determine the kinetically controlled distribution isomers during the alkylation of naphthalene with propene, isomeric butenes, and diisobutylene. The results with linear butenes are summarized in Eq. (56).

$$\text{Naphthalene} + \text{1-butene or trans-2-butene or cis-2-butene} \xrightarrow[\text{5-10°, 6 hr; 75-90\% yield}]{BF_3 \cdot H_3PO_4} \text{1-sec-butylnaphthalene (70-74\%)} + \text{2-sec-butylnaphthalene (26-30\%)} \quad (56)$$

Two important characteristics of Eq. (56) should be emphasized. First, the similarity of the product distribution of the three isomeric butenes, and second, the dominance of the 1 isomer. The first characteristic indicated that the position and the sterochemistry of the double bond in the alkylating species was immaterial. This excluded the possibility that the attacking electrophile was an olefin-catalyst complex [246]. If such a complex was operable, one would anticipate substantial dependence upon both steric factors and degree of ionization in the complex. Accordingly, the electrophile in these alkylations was suggested to be a common one, the *sec*-butyl cation, produced by protonation of the alkene shown in Eq. (57).

$$n\text{-}C_4H_8 + H_3PO_4 + BF_3 \rightleftharpoons CH_3CH_2\overset{+}{C}HCH_3(H_2PO_4BF_3)^- \quad (57)$$

On the other hand, the enhanced alkylation at the 1 position was attributed to the low temperature of the reaction and the lower transition-state energy for the formation of the corresponding naphthalenonium ion complex in the kinetically controlled mechanism [281,282].

So far, we have seen that *sec*-butylations of arenes with linear butenes and other *sec*-butyl derivatives can proceed with or without accompanying skeletal branching of the butyl group, depending on type of catalyst and

Table 8. Monoalkylation of Naphthalene with *sec*-Butyl Alcohol

$$\text{Naphthalene} + \text{2-butanol} \xrightarrow{AlCl_3} \text{1-sec-butylnaphthalene} + \text{2-sec-butylnaphthalene}$$

Entry no.	Reactant ratio ArH:ROH:$AlCl_3$	Solvent type	Temp, °C	Time, hr	Yield, %	Composition, % 1-*sec*-Bu	Composition, % 2-*sec*-Bu
1	0.41:0.5:0.55	Pet. ether	80	5	10	4	96
2	0.5:0.5:0.33	Pet. ether + PhH	80	4	25	34	66
3	0.5:0.5:0.33	PhH	80	4	26	31	69

Source: Ref. 85.

severity of conditions. However, another kind of rearrangement of *sec*-butyl group might be expected on the basis of the behavior of higher linear alkylating agents. This rearrangement, which could be common to all *sec*-butylation reactions, would result from intramolecular 1,2 hydride shifts in the intermediate *sec*-butyl cation before its attack on the aromatic ring,

$$\mathrm{C{-}\overset{+}{C}{-}C{-}C} \xrightleftharpoons{\sim H:^-} \mathrm{C{-}C{-}\overset{+}{C}{-}C}$$

In the *sec*-butyl case, an investigation of this degenerate type of rearrangement can only be made by the application of isotopically labeled molecules. For this purpose the H_2SO_4-catalyzed alkylation of benzene with 1-butanol-1-^{14}C (13), 2-butanol-3-^{14}C (14), and 1-butene-1-^{14}C (14) was investigated by Lipovich et al. [124].

$$\underset{\mathbf{13}}{\mathrm{C{-}C{-}C{-}{}^{14}C{-}OH}} \qquad \underset{\mathbf{14}}{\mathrm{C{-}{}^{14}C{-}\underset{OH}{\underset{|}{C}}{-}C}} \qquad \underset{\mathbf{15}}{\mathrm{C{-}C{-}C{=}{}^{14}C}}$$

The results with 1-butanol-1-^{14}C (13) and 1-butene-1-^{14}C (15) were considered previously and their implications on the alkylation mechanism were emphasized. The alkylation of benzene with 2-butanol-3-^{14}C (14) was carried out under two sets of conditions, at 80°C with 80% H_2SO_4 and at 10°C with 98% H_2SO_4. The *sec*-butylbenzene, produced in 80% and 17% yields from these alkylations, respectively, was shown to consist of 2-phenyl-3-^{14}C-butane (16) and 2-phenyl-2-^{14}C-butane (17, Eq. 58). Radiochemical analysis showed

$$\underset{\mathbf{14}}{\mathrm{C{-}{}^{14}C{-}\underset{OH}{\underset{|}{C}}{-}C}} + \mathrm{PhH} \xrightarrow{H_2SO_4} \underset{\mathbf{16}\ (40\%)}{\mathrm{C{-}{}^{14}C{-}\underset{Ph}{\underset{|}{C}}{-}C}} + \underset{\mathbf{17}\ (60\%)}{\mathrm{C{-}\underset{Ph}{\underset{|}{{}^{14}C}}{-}C{-}C}} \tag{58}$$

that exchange of *sec*-carbon atoms of the butyl group in *sec*-butylbenzene proceeded to the extent of about 40% and did not depend on temperature. These results, while confirming the occurrence of degenerate secondary-secondary rearrangement, also suggested that the rate of alkylation was somewhat faster than that of 1,2-hydride transfer.

Further confirmation of the occurrence of degenerate rearrangements in the *sec*-butyl case came from the alkylation of benzene-d_6 with 1-butene [243] and of the rearrangement of 2-phenyl-2-^{14}C-butane [283] under the influence of aluminum chloride. In the alkylation of benzene-d_6 with 1-butene and $AlCl_3$, the resulting *sec*-butylbenzene had the deuterium scrambled between C-1 and C-4 in the *sec*-butyl side chain. This indicated that the intermediate *sec*-butyl cation, formed by protonation (deuteration), had time to undergo a 1,2 hydride shift, at least in part, prior to alkylation. The C-1/C-4, deuterium content ratios at 10 and 25°C were 60:40 and 52:48, respectively, suggesting that the rate of alkylation exceeded the rate of hydride transfer at the lower temperature. The carbocation transformations leading to the observed scrambling of deuterium in the *sec*-butyl group were suggested to proceed as follows:

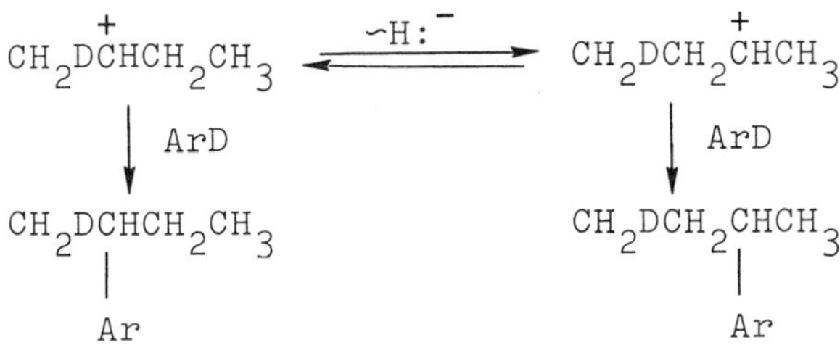

In a related study on the mechanism of alkylation of arenes by secondary alcohols using deuterium-labeling techniques, Streitwieser et al. [113] found that the BF_3-catalyzed alkylation of benzene with 2-butanol-d at 0°C gave 2-phenylbutane with no detectable C-D stretching band in the infrared. This finding demonstrated that, under such conditions, alkylation with secondary alcohols do not involve significant reaction via elimination to alkenes. If the alkylation proceeded significantly by way of carbocations generated via butene, partially deuterated product would be expected because of the following reactions:

$$C_4H_9OD.BF_3 \longrightarrow C_4H_8 + HOD.BF_3$$

$$\downarrow \qquad\qquad\qquad \downarrow$$

$$C_4H_9^+ + BF_3OD^- \qquad\qquad C_4H_8D^+ + BF_3OH^-$$

In concluding this section on the alkylation of arenes with *sec*-butyl derivatives and linear butenes the following points may be emphasized:

1. Depending on type of catalyst and severity of conditions, the *sec*-butylation of arenes may proceed with or without accompanying skeletal rearrangement.
2. Due to their carbocationic nature, the *sec*-butylation reactions are invariably accompanied by degenerate rearrangement between the secondary carbons of the *sec*-butyl group.
3. In alkylations with *sec*-butyl alcohol, the possible mechanism of elimination to an alkene followed by addition of arenes across the double bond has been disproved for some conditions on two essential grounds:
   (a) The reaction of 2-butanol-d with benzene and BF_3 at 0°C to give 2-phenylbutane with no incorporated deuterium [114]
   (b) The reaction of optically active 2-butanol and BF_3 at 25°C with benzene to yield 2-phenylbutane which is about 99.5% racemized [274,275]

The elimination-addition mechanism was also shown to be unlikely in similar alkylations with other secondary alcohols. For example, Price and Ciskowski [131] demonstrated that cyclohexanol with BF_3 gave no cyclohexene under conditions which in the presence of naphthalene resulted in rapid alkylation.

The following general mechanism was suggested for the alkylation of benzene with linear secondary alcohols in the presence of BF_3 at 0°C [114]:

$$ROH\cdot BF_3 \xrightarrow[\text{slow step}]{PhH} \text{[PhH]}\cdots R^+ \underset{\text{rearrangements}}{\overset{sec.-sec.}{\rightleftharpoons}} \text{[PhH]}\cdots R'$$

$$\text{PhR} \xleftarrow{-H^+} [\text{C}_6\text{H}_6\text{R}]^+ \qquad [\text{C}_6\text{H}_6\text{R}']^+ \longrightarrow \text{PhR}'$$

C. C_5 and Higher Linear Alkenes and Linear Secondary Alcohols, Ethers, and Esters

In the foregoing section we have just seen that *sec*-butylations of arenes are always accompanied by a degenerate type of rearrangement resulting from rapid 1,2 hydride shifts between the two secondary carbons of the *sec*-butyl chain. An observation of such rearrangement in the butyl case required the use of labeled molecules. Turning now to alkylations with linear pentyl and higher systems, we can see that the above-mentioned rearrangement can easily be detected in ordinary (nonlabeled) molecules since they result in the formation of different isomeric carbocations which, upon reaction with the arene, yield isomeric arylalkanes. For example, in the pentyl system the processes represented diagrammatically in Eq. (59), when they occur, lead

$$\begin{array}{ccccc} C\text{-}C\text{-}C\text{-}\overset{+}{C}\text{-}C & \underset{}{\overset{\sim H:^-}{\rightleftharpoons}} & C\text{-}C\text{-}\overset{+}{C}\text{-}C\text{-}C & \overset{\sim H:^-}{\rightleftharpoons} & C\text{-}\overset{+}{C}\text{-}C\text{-}C\text{-}C \\ {}^{-ArH,\ H^+}\uparrow\downarrow{}^{ArH} & & {}^{-ArH,\ H^+}\uparrow\downarrow{}^{ArH,\ -H^+} & & {}^{-ArH,\ H^+}\uparrow\downarrow{}^{ArH,\ -H^+} \\ C\text{-}C\text{-}C\text{-}\underset{Ar}{\underset{|}{C}}\text{-}C & \rightleftharpoons & C\text{-}C\text{-}\underset{Ar}{\underset{|}{C}}\text{-}C\text{-}C & \rightleftharpoons & C\text{-}\underset{Ar}{\underset{|}{C}}\text{-}C\text{-}C\text{-}C \end{array} \qquad (59)$$

to the formation of a mixture of isomeric 2- and 3-arylalkanes. If equilibration of the secondary pentyl cations is complete prior to the alkylation step, and if subsequent isomerization of the resulting arylpentanes does not occur, the formation of 2- and 3-arylalkane in a statistical ratio of 2:1 would be anticipated. We shall see in the forthcoming discussion that these considerations went almost completely unnoticed by early workers and they are still being overlooked by some recent ones. Illustation of this can be found in the literature data of Table 9, which reveals the following distinct features. Prior to 1950, [Table 9, part (a)] most workers were apparently unaware of the involvement of the secondary-to-secondary rearrangements within the alkylating moiety.

Table 9. Alkylations of Arenes with C_5 and Higher Linear Alkenes and Secondary Alcohols, Ethers, or Esters

Arene, mol	Alkylating agent, mol	Catalyst, mol	Temp, °C	Time, hr	Yield, %	Composition	Ref.
		A. Alkylations Reported Prior to 1950					
Benzene (2.5)	2-Pentanol (1)	$AlCl_3$ (0.5)	30	24	25	2-Phenylpentane	275 (1936)
Phenol (—)	2-Pentanol (1)	$AlCl_3$ (—)	100	2	58	2- and 3-(*p*-Hydroxyphenyl)-pentane	19 (1937)
1,2,3,4-Tetrahydronaphthalene (9.9 g)	2-Pentanol (6.6g)	$ZnCl_2$ (20g) sealed tube	170-200	12	77	2-*sec*-Pentyl-1,2,3,4-tetrahydronaphthalene	302 (1949)
Benzene (0.4)	1-Pentene (0.3)	H_2SO_4 (96%) (0.6)	5	1.2	65	2- and 3-Phenylpentanes in 3:2 ratio	26, (1940)
Benzene (—)	2-Pentene (—)	HF (—)	0	—	47	Probably a mixture of 2- and 3-phenylpentanes	284 (1938)
Benzene (2.8)	2-Pentene (2.8)	BF_3-CH_3OH (42g)	32-38	50	68	2-Phenylpentane	304 (1947)
Benzene (3)	2-Pentene (2)	BF_3-H_3PO_4 (-)	30	–	80	2-Phenylpentane	305 (1946)
Benzene (240 pts)	2-Pentene (30 pts)	HF (20 pts)	0	3	35 pts	2-Phenylpentane	306 (1947)
Acenaphthene (1)	3-Hexanol (1.17)	$ZnCl_2$ (-)	180	-	32	2-Acenaphthylhexane	307 (1939)

Table 9. (continued)

Arene, mol	Alkylating agent, mol	Catalyst, mol	Temp, °C	Time, hr	Yield, %	Composition	Ref.
Benzene (1)	1-Hexene (1)	H_2SO_4 (0.1)	25	1	50	2-Phenylhexane	308 (1893)
Benzene (1)	3-Hexene (0.05)	H_2SO_4 (0.1)	25	-	50	3-Phenylhexane	307 (1939)
Benzene (1)	3-Hexene (0.66)	$H_3BO_2F_2$ (5g)	reflux 40	0.65 3-4	24	3-Phenylhexane	307 (1939)
Benzene (1)	3-Hexene (0.66)	HF (3.3)	10	24	59	3-Phenylhexane	307 (1939)
Benzene (1)	3-Hexene (3)	HF (10)	10	24	13 41	3-Phenylhexane 1,4-di-(1-ethyl-butyl)benzene	307 (1939)
Chlorobenzene (1)	3-Hexene (0.66)	HF (66g)	10	24	25	2-(*p*-Chlorophenyl) hexane	307 (1939)
Toluene (1)	3-Hexene (0.66)	HF (66g)	10	24	63	3-(*p*-Tolyl)hexane	307 (1939)
m-Xylene (1)	3-Hexene (0.75)	HF (66g)	10	24	80	Hexylxylenes	307 (1939)
m-Xylene (1)	3-Hexyl ether (0.18)	HF (66g)	10	24	61	3-(*m*-Xylyl)hexane	307 (1939)
Naphthalene (1)	3-Hexene (1)	HF (66g)	10	24	30	3-(Naphthyl)hexane,	307 (1939)

					28	+Poly-*sec*-hexyl-naphthalene	
Toluene (1.5)	2-Octanol (0.33)	HF (3.2)	rt	2	42	2-(*p*-Octyl)toluene	309 (1941)
Benzene (390g)	1-Octene (280g)	H_2SO_4 (100 ml)	5-15	0.5	81	2-Phenyloctane, dioctylbenzene	154 (1949)
					8		303 (1936)
Toluene (1.5)	1-Octene (0.33)	HF (3.2)	rt	2	73	2-(*p*-Octyl)-toluene	309 (1941)
Benzene (-)	1-Decene (-)	H_2SO_4	-	-	88	2-Phenyldecane	309 (1949)
Toluene (50 pts)	1-Decene (10 pts)	HF (15 pts)	0-10	18	-	2-(*p*-Methylphenyl)-decane	310 (1949)
m-Xylene (50 pts)	1-Decene (10 pts)	HF (15 pts)	0-10	18	-	2-(2,4-Dimethyl-phenyl)decane	310 (1949)
Benzene (40 pts)	6-Dodecene (17 pts)	HF (21 pts)	0-5	18	-	6-Phenyldodecane	311 (1949)
		B. Alkylations Reported in 1950s					
Benzene (1.25)	1-Pentanol (0.164)	BF_3 (until saturation)	60	24	85	2-Phenylpentane (53), 3-phenylpentane (20), *t*-pentylbenzene (25) and other products (3%)	113 (1959)

Table 9. (continued)

Arene, mol	Alkylating agent, mol	Catalyst, mol	Temp, °C	Time, hr	Yield, %	Composition	Ref.
Benzene (1)	2-Pentanol (-)	$AlCl_3$ (-)	-	-	40	2- and 3-Phenylpentanes in 60:40% ratio	126 (1951)
Benzene (1)	3-Pentanol (0.114 ml)	HF (1)	-	-	78	2- and 3-Phenylpentanes in 36:44% ratio	126 (1951)
Benzene (1)	3-Pentanol (0.2)	$AlCl_3$ (0.2)	<35 rt	1 overnight	80	2- and 3-Phenylpentanes in 76:24% ratio	126 (1951)
Benzene (1.25)	2-Pentanol (0.113)	BF_3 (until saturation)	0	1	67	2-Phenylpentane (66), 3-phenylpentane (23), *t*-pentylbenzene (9) and other products (2)	113 (1959)
Benzene (1.25)	2-Pentanol (0.113)	BF_3 (until saturation)	60	1	58	2-Phenylpentane (55), 3-phenylpentane (21), *t*-pentylbenzene (13), and other products (11)	113 (1959)
Benzene (1.25)	3-Pentanol (0.113)	BF_3 (until saturation)	0	1	62	2-Phenylpentane (64), 3-phenylpentane (23), *t*-pentylbenzene (10), and other products (3)	113 (1959)

Benzene (0.326)	2-Methoxypentane (0.0477)	BF_3 (0.0474), H_2O (2 drops)	rt	several days	-	2- and 3-Phenylpentanes in 75:25% ratio	312 (1952)
Benzene (1)	1-Heptene (0.2-0.25)	92-98% H_2SO_4 (40-50g)	3-4	0.75-1.16	72-79	2-, 3-, and 4-Phenylheptanes in 1:1:1 ratio	288 (1959)
Benzene (117.1g)	1-Octene (42.1g)	96% H_2SO_4 (77g)	3-4	45-100 min	91.9	2-Phenyloctane	285 (1954)
Benzene (117.1g)	1-Decene (63.1g)	96% H_2SO_4 (77g)	3-4	45-100 min	90.2	2-Phenyldecane	285 (1954)
Benzene (117g)	1-Dodecene (63.1g)	96% H_2SO_4 (77g)	3-4	45-100 min	80.5	2-Phenyldodecane	285 (1954)
Benzene (6)	1-Dodecene (1)	$AlCl_3$ (2)-CCl_4	60	-	72	Mainly 2- and 3-phenyldodecanes beside small amounts of the 4- and 5-isomers	287 (1955)
Benzene (202.2g)	1-Tetradecene (84.6g)	96% H_2SO_4 (200g)	3-4	45-100 min	70.4	2-Phenyltetradecane	285 (1954)
Benzene (202.2g)	1-Hexadecene (56.1g)	96% H_2SO_4 (103g)	3-4	45-100 min	75.1	2-Phenylhexadecane	285 (1954)
Benzene (219.0g)	1-Octadecene (88.6g)	96% H_2SO_4 (200g)	3-4	45-100 min	70.4	2-Phenyloctadecane	285 (1954)

Table 9. (continued)

Arene, mol	Alkylating agent, mol	Catalyst, mol	Temp, °C	Time, hr	Yield, %	Composition	Ref.
		C. Alkylations Reported Since 1960					
Benzene (0.06)	1-Pentanol (0.03)	80% H_2SO_4 (12.5 ml)	80	-	60-65	2-Phenylpentane (63), 3-phenylpentane (18), and 2-methyl-2-phenylbutane (9)	117 (1963)
Benzene (0.06)	2-Pentanol (0.03)	80% H_2SO_4 (12.5 ml)	80	-	60-65	2-Phenylpentane (59), 3-phenylpentane (35.5) and 2-methyl-2-phenylbutane (5.5)	117 (1963)
Benzene (0.06)	3-Pentanol (0.03)	80% H_2SO_4 (12.5 ml)	80	-	60-65	2-Phenylpentane (62), 3-phenylpentane (31.5) and 2-methyl-3-phenylbutane (6.5)	117 (1963)
Benzene (-)	1- or 2-Pentene (-)	H_2SO_4 (-)	5-10	1	60-65	2-Phenylpentane	128 (1963)
Benzene (1)	*n*-Pentenes (0.20-0.25)	HSO_3Cl (3.0)	70	-	-	2- and 3-Phenylpentanes	313 (1979)
Ethylbenzene (5)	1- or 2-Pentene (1)	BF_3-H_3PO_4 (0.3)	20-40	-	70	*o*-, *m*- and *p*-*sec*-Pentylethylbenzene in 39:1:60% ratio[a,b]	314 (1970)
n-Propylbenzene (5)	1-Pentene (1) or 2-Pentene (1)	BF_3-H_3PO_4 (0.3)	20-40	-	63	*o*-, *m*-, and *p*-*sec*-Pentylpropylbenzene in a 25:2:73% ratio	314 (1970)

Isopropylbenzene (-)	1-Pentene (-)	BF_3-H_3PO_4	20-80	-	68	*o*-, *m*-, and *p*-*iso*-Pr-C_6H_4CHMePr in 10:22:68% ratio	286 (1971)
n-Butylbenzene (5)	1-Pentene (1) or 2-Pentene (1)	BF_3-H_3PO_4 (0.3)	20-40	-	54	*o*-, *m*- and *p*-*sec*-Pentylbutylbenzene in 23:2:75% ratio[a,b]	314 (1970)
t-Butylbenzene (-)	1-Pentene (-)	BF_3-H_3PO_4 (-)	20-80	-	59	*o*-, *m*- and *p*-*t*-Bu C_6H_4CHMePr in 6.8:14.2:55.9% ratio +0.1, 1.9 and 12.7% yield of *o*-, *m*- and *p*-di-*t*-butylbenzene	286 (1971)
Benzene (10)	1-Hexene (1)	HF (10)	16	-	-	2- and 3-Phenylhexane in 37:63% ratio	290 (1960)
Benzene (3.9)	1-Hexene (0.4) in PhHCl	HF (4)	0-5	1	90	2- and 3-Phenylhexanes in 57.8:42.7% ratio	209 (1972)
Benzene (3.9)	1-Hexene (0.4) in PhH (1)	HF (4)-BF_3 (5% based on HF)	0-5	1	62	2- and 3-Phenylhexanes in 63.5:36.5% ratio	298 (1972)
Benzene (3.9)	1-Hexene (0.4) in PhH (1)	HF (4)	155	1	90	2- and 3-Phenylhexanes in 62.4:37.4% ratio	298 (1972)
Benzene (3.9)	*trans*-3-Hexene (0.4) in PhH (1)	HF (4)	0-5	1	90	2- and 3-Phenylhexanes in 57.0:43.0% ratio	298 (1972)
Benzene (3.9)	*trans*-3-Hexene (0.4) in PhH (1)	HF (4)-BF_3 (5% based on HF)	0-5	1	62	2- and 3-Phenylhexanes in 61.0:39.0% ratio	298 (1972)

Table 9. (continued)

Arene, mol	Alkylating agent, mol	Catalyst, mol	Temp, °C	Time, hr	Yield, %	Composition	Ref.
Benzene (3.9)	*trans*-3-Hexene (0.4)	HF (4)	155	1	62	2- and 3-Phenyl hexanes in 59.8:40.2% ratio	298 (1972)
p-Chloroanisole (3)	1-Hexene (1)	BF_3-H_3PO_4	60	1 2/3	53	2-*sec*-Hexyl-4-chloroanisole	315 (1969)
o-Chloroanisole (3)	1-Hexene (1)	BF_3-H_3PO_4 (0.4)	80	1.5-2.0	83	4-*sec*-Hexyl-2-chloroanisole	315 (1969)
Benzene (78g)	1-Heptene	100% H_2SO_4 (50g)	Cooling	1	-	2-, 3- and 4-Phenyl heptane in 46.5-37.5:16.0% ratio	292 (1965)
Benzene (2g)	1-Heptene (1) (24g)	$AlCl_3$ (15 wt% based on the alkene) in $PhNO_2$	37	2	57.5	2-, 3- and/or 4-Heptylbenzene in 1.05:1.0 ratio	300 (1969)
Benzene (2)	3-Heptene	$AlCl_3$ (15 wt% based on the alkene) in $PhNO_2$	37	2	55	2-, 3- and/or 4-Heptylbenzenes in 1.09:1 ratio	300 (1969)
o-Chloroanisole (3)	1-Heptene (1)	BF_3-H_3PO_4 (0.4)	90	1.5-2.0	88	4-*sec*-Heptyl-2-chloroanisole	315 (1969)
p-Chloroanisole (3)	1-Heptene (1)	BF_3-H_3PO_4 (0.4)	90	1 2/3	71	2-*sec*-Heptyl-4-chloroanisole	315 (1969)
Benzene (10)	1-Octene (1)	HF (10)	16	-		2-, 3- and 4-Phenyl octanes in 33:32:35% ratio	290 (1960)

Benzene (8)	1-Octene (1)	$AlCl_3$ (0.0085)	20	5 min	100% conversion	2-, 3- and 4-Phenyl octanes in a 52.8:29.7:17.5% ratio	296 (1969)
Benzene (8)	4-Octene (1)	$AlCl_3$ (0.0085)	20	60 min	40% conversion	2-, 3- and 4-Phenyl octanes in 33.4:32.5:34.1% ratio	296 (1969)
Benzene (8)	1-Octene (1)	$EtAlCl_2$ (5% based on the alkene)	20	10 min	100% conversion	2-, 3- and 4-Phenyl octanes in a 64.4:24:11.4% ratio	296 (1969)
Benzene (2)	1-Octene (1)	$AlCl_3$ (15% wt based on the alkene) in $PhNO_2$	37	2	65.6	2-, 3- and 4-Phenyl octanes in 1.8:1.2:1 ratio	300 (1969)
Benzene (2)	2-Octene (1)	$AlCl_3$ (15 wt% based on the alkene) in $PhNO_2$	37	2	67.2	2-, 3- and 4-Phenyl octanes in 1.6:1.3:1 ratio	300 (1969)
Benzene (2)	3-Octene	$AlCl_3$ (15% wt based on the alkene) in $PhNO_2$	37	2	62.4	2-, 3- and 4-Phenyl octanes in 1.6:1.2:1 ratio	300 (1969)
Benzene (3.9)	1-Octene (0.4) in PhH (1)	HF (4)	0-5	1	90	2-, 3- and 4-Phenyl octanes in 32.1:32.4:35.5% ratio	298 (1972)
Benzene (3.9)	1-Octene (0.4) in PhH (1)	HF (4)-BF_3 (5% based on HF)	0-5	1	70	2-, 3- and 4-Phenyl-octanes in 47.3:28.3:24.4% ratio	298 (1972)

Table 9. (continued)

Arene, mol	Alkylating agent, mol	Catalyst, mol	Temp, °C	Time, hr	Yield, %	Composition	Ref.
Benzene (3.9)	4-Octene (0.4) in PhH (1)	HF (4)	0-5	1	90	2-, 3- and 4-Phenyl octanes in 28.5:32.4:39.1% ratio	298 (1972)
Benzene (3.9)	4-Octene (0.4) in PhH (1)	HF (4)-BF_3 (5% based on HF)	0-5	1	70	2-, 3- and 4-Phenyl octanes in 40.0:29.2:30.8% ratio	298 (1972)
Benzene (10)	1-Decene (1)	HF (10)	16	-	-	2-, 3-, 4- and 5-Phenyldecanes in 23:22:17:38% ratio	290 (1960)
Benzene (3.9)	1-Decene (0.4) in PhH (1)	HF (4)	0-5	1	90	2-, 3-, 4- and 5-Phenyldecanes in 22.6:21.8:23.4:32.2% ratio	298 (1972)
Benzene (3.9)	1-Decene (0.4) in PhH (1)	HF (4)-BF_3 (5% based on HF)	0-5	1	77	2-, 3-, 4- and 5-Phenyldecanes in 41.7:22.1:16.4:19.8% ratio	298 (1972)
Benzene (3.9)	5-Decene (0.4) in PhH (1)	HF (4)	0-5	1	90	2-, 3-, 4- and 5-Phenyldecanes in 16.2:19.3:25.5:39.0% ratio	298 (1972)
Benzene (3.9)	5-Decene (0.4) in PhH (1)	HF (4)-BF_3 (5% based on HF)	0-5	1	77	2-, 3-, 4- and 5-Phenyldecanes in 29.1:20.7:20.3:29.9 % ratio	298 (1972)

m-Xylene (10)	1-Decene (1)	HF(10-20)	16±3	-	-	2-, 3-, 4- and 5-(2,4-dimethyl-phenyl) decane in a wt % ratio of 20:24:20:29 + 8 wt % of two unidentified products	290 (1960)
Benzene (10)	1-Docedene (2)	$AlCl_3$ (0.1)-H_2O (0.1g)	3-	0.25	68	2-, 3-, 4-, 5- and 6-Phenyldodecanes in 32:22:16:15:15 % ratio	290 (1960)
Benzene (10)	1-Dodecene (1)	HF (10)	16	-	-	2-, 3- and 4-Phenyl dodecanes in 20:17:16 % ratio	290 (1960)
Benzene (1.5)	1-Dodecene (0.15)	98% H_2SO_4 (5 ml)	0-10	2	72	Phenyldodecane in 0:41:20:13:13:13 % ratio + 9% yield didodecylbenzene	290 (1960)
Benzene (10)	1-Dodecene (1)	HF (5)	16	-	92	2-, 3-, 4-, 5- and 6-Phenyldodecanes in 20:17:16:23:23% ratio	290 (1960)
					5	+ didocylbenzene	
Benzene (3.6)	1-Dodecene (0.6)	$AlCl_3$ (0.037)-HCl	35-37	0.7	-	2-, 3-, 4-, 5- and 6-Phenyldodecanes in 32:19:16:17:16% ratio	291 (1961)

Table 9. (continued)

Arene, mol	Alkylating agent, mol	Catalyst, mol	Temp, °C	Time, hr	Yield, %	Composition	Ref.
Benzene (3.0)	1-Dodecene (0.3)	$AlCl_3$[a] (0.037)-HCl	35-37	45	83	2-, 3-, 4-, 5- and 6-Phenyldodecanes in 31.8:20.8:17.2:30.2 % ratio	295 (1968)
Benzene (3.0)	*trans*-6-Dodecene (0.15)	$AlCl_3$ (0.02) + 20 wt % of red oil from previous reaction	0-5	30	75	2-, 3-, 4-, 5- and 6-Phenyldodecanes in 57.0:21.5:9.0:12.5 % ratio	295 (1968)
Benzene (9)	1-Dodecene (1)	Silicotungstic acid (20 wt % on silica gel)	300F 500 psi	-	84-90	Similar to distributions obtained with $AlCl_3$	316 (1971)
Benzene (3.9)	1-Dodecene (0.4) in PhH (1)	HF (4)	0-5	1	90	2-, 3-, 4-, 5- and 6-Phenyldodecanes in 18.5:15.5:18.3:47.7 % ratio	298 (1972)
Benzene (3.9)	1-Dodecene (0.4) in PhH (1)	HF (4)-BF_3 (5% based on HF)	0-5	1	77	2-, 3-, 4-, 5- and 6-Phenyldodecanes in 41.0:18.4:13.8:26.8 % ratio	298 (1972)
Benzene (3.9)	1-Dodecene (0.4) in PhH (1)	HF (4)-BF_3 (5% based on HF)	35-40	1	77	2-, 3-, 4-, 5- and 6-Phenyldodecanes in 33.5:17.8:16.2:32.5 % ratio	298 (1972)

Benzene (3.9)	6-Dodecene (0.4) in PhH (1)	HF	0-5	1	90	2-, 3-, 4-, 5- and 6-Phenyldodecanes in 10.6:12.7:17.9:58.6 % ratio	298 (1972)
Benzene (3.9)	6-Dodecene (0.4) in PhH (1)	HF (4)-BF_3 (5% based on HF)	0-5	1	77	2-, 3-, 4-, 5- and 6-Phenyldodecanes in 24.7:15.8:15.9:43.6 % ratio	298 (1972)
Toluene (10)	1-Dodecene (1)	HF (10-20)	16±3	-	-	2-, 3-, 4-, 5- and 6-(4-Methylphenyl)-dodecane in a wt % ratio of 13.3:15:17:44; + 8 wt % of what was believed to be dialkylbenzenes	290 (1960)
Benzene (50)	1-Dodecene (5)	HF (50)	16	-	92	2-, 3-, 4-, 5- and 6-Phenyldodecanes in 20:17:16:23:24 wt % ratio + 5% yield didodecylbenzene	290 (1960)
Benzene (10)	1-Dodecene (2)	$AlCl_3$ (0.1) + 0.1g H_2O	16	-	68	2-, 3-, 4-, 5- and 6-Phenyldodecanes in 32:22:16:15:15 wt % ratio + 15% yield didodecylbenzene	290 (1960)
Benzene (1.5)	1-Dodecene (0.15)	98% H_2SO_4 (5 ml)	16	-	72	2-, 3-, 4-, 5- and 6-Phenyldodecanes in 41:20:13:13:13 wt % ratio	290 (1960)

[a]The corresponding dialkylates, 2,4-di(*sec*-pentyl)C_6H_3R, were also produced in 10-16% yields.
[b]All the secondary pentyl derivatives were presumably mixtures of 2- and 3-arylpentanes.

Accordingly, they believed that the alkylations of aromatic with linear pentyl and higher derivatives yielded only one nonrearranged arylalkane. The latter was assumed to result from direct Markovinkov-type addition of the arene on the alkene double bond or direct displacement of the functional group (OH, OR) in alcohols and ethers, respectively.

There were a few exceptions, however, in which mixtures of secondary alkylates were either recognized [19] or suspected [284]. In one of these, Tsukervanik and Nazarova [19] reported that the condensation of phenol with 2-pentanol in the presence of $AlCl_3$ resulted chiefly in a mixture of *o*- and *p*-(2-pentyl)phenol and a little *o*- and *p*-(3-pentyl)phenol. The latter was thought to result from the isomerization of the alcohol or of its alkylated phenyl ether by the action of the catalyst prior to the alkylation step.

In another exception, Ipatieff et al. [26] reported that the H_2SO_4-catalyzed alkylation of benzene with 1-pentene gave a mixture of 2- and 3-phenylpentanes. Using the formation of acetamino derivatives for product identification, these authors concluded that the mixture comprised 55 to 60% of the 2-phenyl isomer and 40 to 45% of the 3-phenyl isomer. This conclusion, in fact, provided the first clearly substantiated evidence for this type of linear secondary-to-secondary rearrangement. Ipatieff et al. attributed the observed rearrangement to the isomerization of intermediate esters,

$$[R\text{-}CH_2CH(OSO_3H)CH_3 \rightleftharpoons RCH(OSO_3H)\ CH_2CH_3]$$

rather than of intermediate carbocations,

$$(R\text{-}CH_2\overset{+}{C}HCH_3 \rightleftharpoons R\overset{+}{C}HCH_2CH_3)$$

From 1950 onward, most workers became fully aware of the secondary-to-secondary rearrangements accompanying alkylations with linear systems. This can be seen from part (b) of Table 9, where the products were reported as mixtures in which all of the possible secondary alkylates were shown to be present. Nevertheless, a few exceptions also existed in which some workers totally overlooked the reality of accompanying rearrangements. For example, it was claimed that (1) the H_2SO_4-catalyzed alkylation of benzene with 1- or 2-pentene gave only 2-phenylpentane [128]; (2) the H_2SO_4-catalyzed alkylation of benzene with 1-octene, 1-decene, 1-dodecene, 1-tetradecene, 1-hexadecene, and 1-octadecene gave the corresponding pure 2-pheylalkanes [285]; and (3) the alkylation of isopropyl- or *t*-butylbenzene with 1-pentene and BF_3-H_3PO_4 gave, as monoalkylates, mixtures of the corresponding *ortho*, *meta*, and *para* 2-arylpentanes (Eq. 60) [286].

$$C_6H_5R + C{=}CCCC \xrightarrow[20\text{-}80°]{BF_3\cdot H_3PO_4} R\text{-}C_6H_4\text{-}C(C)CCC \quad (60)$$

o, *m*, and *p*

R = *i*-Pr, *t*-Bu

Since rearrangements leading to all possible linear secondary alkylates have been recognized, workers directed their effort to investigate the effect of various factors on their extent. Because of poor analysis, there had been considerable disagreements for some time as to the distributions of isomers in the resulting mixtures. For example, it was stated that the ratio of 2/3-phenylpentane in the products from the $AlCl_3$-catalyzed alkylation of benzene was 1.5 with 2-pentanol and 3.2 with 3-pentanol [126]. When BF_3 was used as catalyst, it was stated [113] that "within experimental error 2-pentanol and 3-pentanol gave identical mixtures of products."

Also, it was indicated in 1955 that the $AlCl_3$-catalyzed alkylation of benzene with 1-dodecene gave a mixture consisting chiefly of 2- and 3-phenyldodecanes, besides small amounts of 4- and 5-phenyldodecanes [287]. However, in 1959 it was reported that the alkylation of benzene with 1-heptane in the presence of H_2SO_4 [288] or with 1-dodecane in the presence of various technical catalysts [289] resulted in the formation of mixtures containing approximately equal amounts of the theoretically possible secondary phenylalkanes.

With the aid of modern instrumentation, workers have become able to determine the exact distribution of isomers. Moreover, with the growing understanding in theory, they have also become able to account for any existing deviations from statistical ratios. In fact, the exhaustive work that has been carried out by workers in the last two decades has disclosed many interesting aspects of alkylation by linear alkenes and secondary derivatives. With reference to Table 9, and to the discussion presented by various authors [35, 288,290-299], some of these aspects will be described in the following outline:

1. Alkylations of arenes with linear alkenes and linear secondary alcohols or ethers afford all of the theoretically possible isomeric *sec*-arylalkanes.

2. In the presence of strong catalysts such as $AlBr_3$, $AlCl_3$, or HF-BF_3 at room temperature or higher, the alkylations are thermodynamically controlled and the distribution of the products is the same regardless of the position of the double bond in the starting alkene. For example, in the presence of $AlCl_3$ at 35 to 37°C, both 1-dodecene and *trans*-6-dodecene alkylated benzene to give nearly identical distributions, including about 32% 2-, 20% 3-, 17% 4-, and 31% 5- + 6-phenyldodecane [295]. The alkylation of benzene with 1-heptene in the presence of $AlCl_3$ or $AlBr_3$ under varying conditions and catalyst ratios gave product mixtures in which 2-, 3-, and 4-phenylpentane were always present in a ratio of 55:53:12, respectively [297].

3. In contrast, the alkylations become kinetically controlled when carried out in the presence of weak catalyst such as HF, H_2SO_4, $AlCl_3$-CH_3NO_2, $AlCl_2$-HSO_4, and $EtAlCl_2$, or even in the presence of $AlCl_3$ at low temperatures [295].

4. Under kinetically controlled conditions, the distribution of various isomers shows varying degrees of dependence on factors such as chain length, location of the double bond in the chain, nucleophilicity of arene, type of catalyst and solvent, ratio of reactants, temperature, and homogeneity of the reaction medium. The following comparisons illustrate some of these effects:

(a) The alkylation of benzene with 1-dodecene under nonisomerizing kinetically controlled conditions afforded 2-, 3-, 4-, and 5- +6-phenyldecanes in a percent ratio of 20:17:16:47 with HF catalyst [290] and in a ratio of 57.0:21.5:8.0:12.5 with $AlCl_2$-HSO_4 catalyst [295].

(b) The HF-catalyzed alkylation of benzene at 0 to 5°C gave 2-, 3-, 4-, and 5- + 6-phenyldodecane in a percent ratio of 18.5:15.5:18.3:47.7 with 1-dodecene and in a ratio of 10.6:12.7:17.9:58.6 with trans-6-dodecene [298]. Other workers claimed different results, however. For example, Asinger et al. [35] reported that the alkylation of benzene with 1-heptene, *cis*- and *trans*-2-heptene, *cis*- and *trans*-3-heptene, or a mixture of heptenes in the presence of 92 to 100% H_2SO_4 or anhydrous HF gave nearly the same isomer distributions of 2-, 3-, and 4-phenylheptane. Thus, according to the latter authors, the location and stereochemistry of the double bond is immaterial.

(c) In general, reactions of 1-alkenes are faster and produce more of the 2-isomer than do internal alkenes [293,296,300]. Thus, under similar conditions (using 0.0085 mol $AlCl_3$ per mol of alkene at 20°C), the alkylation of benzene with 1-octene was complete after 5 min, but that with 4-octene was only 40% complete after 60 min. Moreover, the percent ratio of 2-/3-/4-phenyloctane in the product was 52.8:29.7:17.5 in the case of 1-octene and 33.4:32.5:34.1 in the case of 4-octene [296]. By substituting the milder catalyst $EtAlCl_2$ for $AlCl_3$, the content of 2-phenyloctane in the alkylate (after complete conversion in 10 min) was raised to 64.6% and that of 3- and 4-phenyloctanes was lowered to 24 and 11.4%, respectively [296].

(d) In the HF-catalyzed alkylations with 1-decene, the percent ratio of 2-/3-/4-/5-aryldecanes was 23:22:17:38 with benzene and 21.5:26:21.5:31 with *m*-xylene [290].

(e) At 0°C, the alkylation of benzene with 1-dodecene in the presence of a separate HF phase invariably resulted in greater amounts of the internal isomers than at 55°C [293].

(f) In alkylation of benzene with either 1-dodecene or *trans*-6-dodecene at 55°C, higher ratios of the 2-phenylalkane isomer were produced in the absence than in the presence of a separate liquid phase [293].

(g) Alkylations carried out in *n*-hexane and a separate HF phase gave the same isomer distributions from 1-dodecene and *trans*-6-dodecene, indicating that under such conditions the position of the double bond is immaterial. Dilution with *n*-hexane apparently slowed down the alkylation step sufficiently to allow the intermediate ions to isomerize to the equilibrium distribution [293].

(h) In general, the amount of the 2-isomer produced from 1-alkenes increases with increasing chain length [290,294]. This can best be illustrated by the useful data of Table 10, which show the analysis of the products of HF-catalyzed alkylation of benzene with different linear 1-alkenes.

5. Under nonisomerizing conditions, alkylations seem to be faster than isomerizations. In the alkylation of benzene with a long-chain alkene such as 1-dodecene, the alkylation step is too fast to permit the intermediate secondary carbocation to attain statistical equilibrium before attacking the arene. It was only when the chain was shortened to six carbons that such statistical carbocation distribution was reached. Thus both 1-hexene and *trans*-3-hexene afforded similar isomer distributions in the HF-catalyzed alkylation of benzene even at 0°C [298].

$$\text{C-C-C-C-}\overset{+}{\text{C}}\text{-C} \rightleftharpoons \text{C-C-}\overset{+}{\text{C}}\text{-C-C-C}$$

6. Besides secondary-to-secondary rearrangements, the alkylations of arenes with linear alkenes and related derivatives are sometimes accompanied by disproportionation, transalkylation, and hydride transfer reactions. Such

Table 10[a]. Products of Alkylation of Benzene with Linear 1-Alkenes in the Presence of HF at 16°

	Phenylalkane isomers, %[b]			
Alkylating alkene	2-	3-	4-	Combined 5-, 6-, 7-, 8-, 9-
1-Hexene	37	63	—	—
1-Octene	33	32	35	—
1-Decene	23	22	17	38
1-Hendecene	20	19	16	45
1-Dodecene	20	17	16	47
1-Tridecene	17	15	12	56
1-Tetradecene	18	14	14	54
1-Pentadecene	18	14	14	54
1-Hexadecene	24	14	10	52
1-Heptadecene	23	14	9	54
1-Octadecene	31	17	9	44

[a]Mole ratio of benzene:alkene:HF = 1:10:10
[b]Determined by a combination of glpc, ir, and ms analysis
Source: Ref. 290.

reactions lead to by-products such as linear alkanes [295], branched alkanes [295], cyclic tetralins [301], and polyalkylated arenes [290]. For example:

(a) In the $AlCl_3$/HCl-catalyzed alkylation of benzene with dodecene at 35°C, a 6% yield of alkane was obtained of which only 5.5% was normal and 94.5% was branched dodecane [295].

(b) In the alkylation of benzene with 1-dodecene in the presence of various catalysts didodecylbenzene was obtained in 5%, 9%, and 21% yields with HF, 98% H_2SO_4, and $AlCl_3$-H_2O, respectively.

(c) Under certain conditions branching of the side chain in the alkylate might also be observed. For example, 2-methyl-2-phenylundecane (2%) was identified among the products of HF-catalyzed alkylation of benzene with 1-dodecene in *n*-hexane at 55°C [293].

7. The isomer distributions of the products determined under kinetic control suggested that the internal secondary carbocations are slightly more stable than those near the end of the chain. For example, the alkylbenzenes from benzene alkylation with *trans*-6-dodecene and HF showed the amount of internal isomers (i.e., 5- and 6-phenyldodecanes) to be nearly twice as great as the 2-phenyldodecane [293].

8. The generally accepted mechanism for alkylations with alkenes involves interaction of the alkene with the catalyst to form a carbocation, the corresponding ion pair, or a polarized complex. This undergoes rapid secondary-to-secondary rearrangement in varying degrees followed by attack on the arene nucleus in a rate-determining step to form the product [146].

Further to establish the mechanism, especially in regard to the stage or stages at which isomerization occurs, Olson [290] studied the effect of catalysts on both alkylating agents and products, as well as the change in product composition with time. He found that neither HF nor H_2SO_4 had any effect on pure phenylalkanes, indicating that the isomerization occurs prior to final attachment of the phenyl group to the alkyl chain with these catalysts. He also found that the $AlCl_3$-catalyzed alkylation of benzene with 1-dodecene gave the same mixture of isomeric phenyldecanes obtained by treating either 2- or 6-phenyldodecane with $AlCl_3$ in benzene under alkylation conditions. Samples of the product mixture first contained 40% 2-phenyldodecane, but the concentration changed with time to an equilibrium value of 32% with compensating increases in the amounts of the 4, 5, and 6 isomers. Thus it was concluded that the manner in which the isomeric phenylalkanes were formed in H_2SO4 and HF catalysis was different from that in $AlCl_3$ catalysis. In the case of H_2SO4 and HF, the isomerization was assumed to occur by a repeated addition and elimination of a molecule of acid, giving a mixture of all the possible secondary ions, which in turn alkylate benzene to yield the phenylalkanes

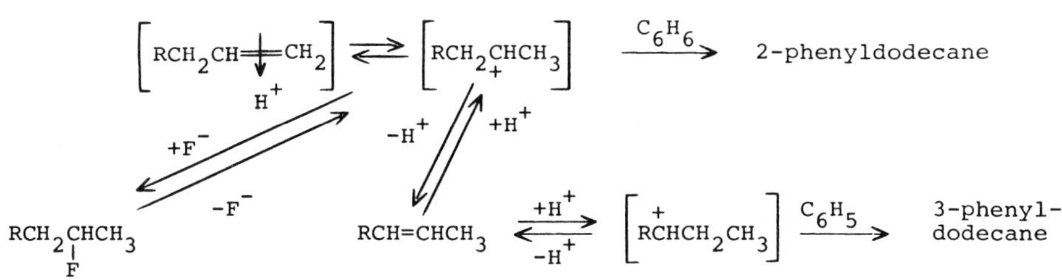

Scheme 6

(scheme 6). For the aluminum chloride-catalyzed reaction, Olson concluded that the isomerization may occur at one or both of the following stages:

1. Prior to alkylation within a π complex between catalyst, alkene, and aromatic.
2. Subsequent to alkylation through reverse alkylation-dealkylation steps and isomerization in the catalyst-alkene-aromatic complex, and/or reverse abstraction of hydride ion from the arylalkane followed by a series of hydride shifts and phenyl migrations

These conclusions and similar ones reached by Swisher et al. in 1961 [291], Alul and McEwan in 1967 [293], and Alul in 1968 [294,295] led to an expanded mechanism for the $AlCl_3$-catalyzed alkylation of arenes with linear alkenes illustrated in scheme 7 by the alkylation of benzene with 1-dodecene.

As can be seen from scheme 7, rearrangement can be achieved along two paths, at the π-complex stage via <u>18</u>, or by hydride abstraction via ion <u>19</u> or its equivalent, phenonium ion <u>20</u>. Credence for the latter path is found in the isolation of dodecanes from the reaction of benzene with dodecenes in the presence of $AlCl_3$ [295]. On the other hand, credence for the former path is found in the fact that such alkylations afford all of the possible secondary alkylates regardless of the type of arene, catalyst, or conditions employed. Thus, in addition to all of the examples cited before, when phenol was alkylated with 1-nonene under the influence of a variety of acid catalysts

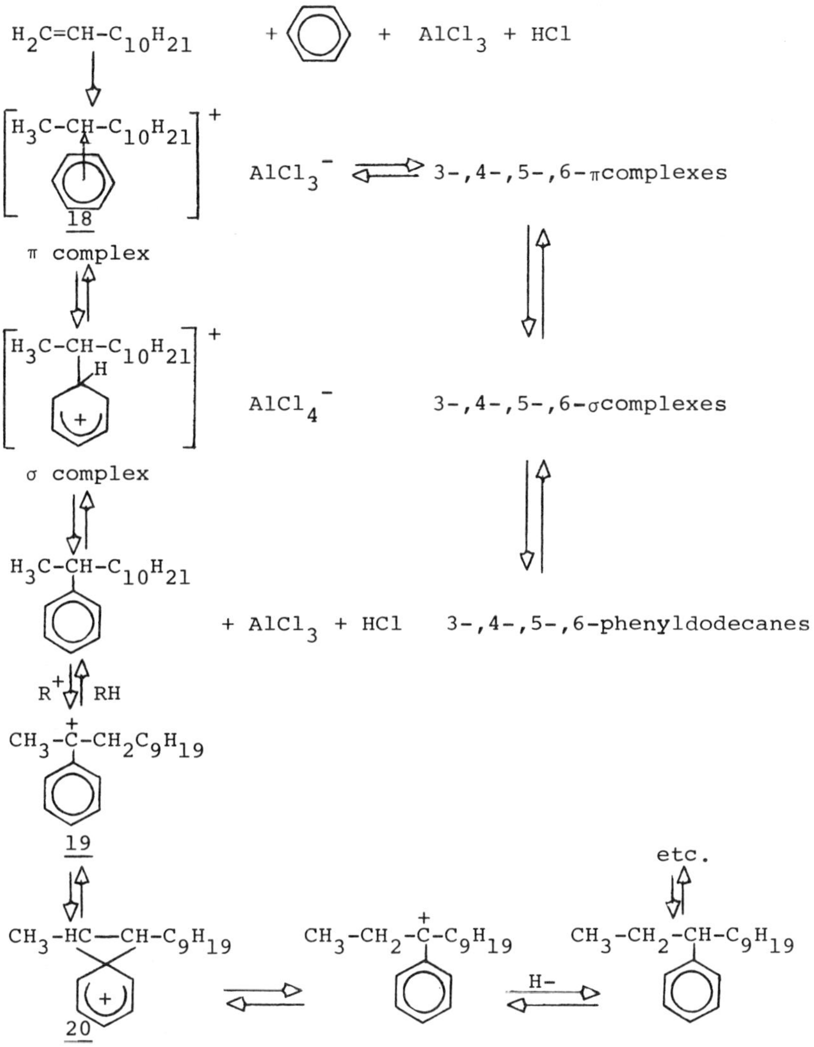

$H_2C{=}CH{-}C_{10}H_{21}$ + (benzene) + $AlCl_3$ + HCl

$[H_3C{-}CH{-}C_{10}H_{21}]^+$ 18 π complex $AlCl_3^-$ ⇌ 3-,4-,5-,6-πcomplexes

$[H_3C{-}CH{-}C_{10}H_{21}]^+$ σ complex $AlCl_4^-$ 3-,4-,5-,6-σcomplexes

$H_3C{-}CH{-}C_{10}H_{21}$ + $AlCl_3$ + HCl 3-,4-,5-,6-phenyldodecanes

R^+ ⇌ RH

$CH_3{-}\overset{+}{C}{-}CH_2C_9H_{19}$ 19

$CH_3{-}HC{-}CH{-}C_9H_{19}$ 20 ⇌ $CH_3{-}CH_2{-}\overset{+}{C}{-}C_9H_{19}$ ⇌ (H-) $CH_3{-}CH_2{-}CH{-}C_9H_{19}$ ⇌ etc.

Scheme 7

analysis confirmed the presence of seven out of the eight expected *ortho* and *para* secondary alkylates [317].

IV. ALKYLATIONS OF ARENES WITH BRANCHED ALKENES, ALCOHOLS, ETHERS, AND ESTERS

A. Isobutylene and *t*-Butyl Alcohol, Ethers, and Esters

Examination of the literature results on the alkylation of aromatics with isobutylene and *t*-butyl alcohol, ethers, and esters in the presence of most Friedel-Crafts catalysts indicates the exclusive formation of *t*-butylated derivatives [36, 37, 45a, 77, 92, 116, 119, 192(c), 198, 200, 219, 318-322]. This in fact justifies the use of the term *t*-butylation for these reactions. To our knowledge, however, a few exceptions existed in which *sec*- and/or isobutylated isomers were claimed to form as minor or sole products in such reactions. For example, it was reported that the alkylation of benzene with isobutylene in the presence of silicophosphoric acid at 250°C gave a monobutylbenzene mixture consisting of 87% *t*-butylbenzene and 13% *sec*-butylbenzene [270]. It was also claimed that the alkylation of benzene, isopropylbenzene, and isobutylbenzene with isobutylene in the presence of BF_3-H_3PO_4 or $AlCl_2$-H_2PO_4 gave good yields of isobutylbenzene, *p*-isobutylcumene, and *p*-diisobutylbenzene, respectively [333]. Whereas the former claim seems justifiable on the basis of the strenuousness of catalyst and conditions, the latter one seems rather unlikely. In fact, examination of the physical properties reported for the products from the latter reaction reveals that they are closer to those of *t*-butyl than to isobutyl derivatives.

In practice, many classes of aromatics have been successfully *t*-butylated with isobutylene and *t*-butyl alcohol, ethers, and esters. These include aromatic hydrocarbons [36, 37, 45a, 77, 92, 116, 119, 192(c), 198, 200, 219, 318-333], haloarenes [248,263], phenols [334-345a], phenol ethers [345a-352], aromatic amines [353], and heteroaromatics [354,355]. The yields, however, depended on various factors, of which size, electronic nature, and position of ring substituents were most determining [324]. For example, Carpenter et al. [323] found that the presence of a *meta*-trifluoromethyl group greatly deactivated phenol and anisole and prevented their alkylation with isobutylene in the presence of H_2SO_4, BF_3, or HF catalyst.

The same authors also found that the introduction of one nitro group into the ring of 1,3-dimethoxybenzene deactivated the ring to the extent that alkylation with isobutylene was impossible in the presence of boron trifluoride hydrate or a 20 molar ratio of hydrogen fluoride catalyst.

Steric effects are also important in *t*-butylation of substituted benzenes. It has been confirmed by many workers that the introduction of a *t*-butyl group *ortho* to a methyl group is either impossible or negligible [116, 119, 200, 262, 324-330]. For example, depending on the particular catalysts and alkylating agents used, the *t*-butylation of toluene was reported to give *t*-butyltoluene mixtures ranging in composition from a 67% *meta*/33% *para* thermodynamic equilibrium ratio to a 7% *meta*/93% *para* limiting kinetic ratio with no *ortho* isomer [119, 120, 324, 325, 327, 328]. As concluded by Allen and Yats [320], all *t*-butylations of toluene could have been obtained by alkylations producing isomeric mixtures containing 7% *m*- and 93% *p*-*t*-butyltoluene with subsequent or concurrent isomerizations of the mixture.

The *t*-butylation of higher methylbenzenes [116, 262, 324, 326, 327, 329] and cymenes [330] showed similar steric effects. Thus the *t*-butylation of *o*- and *m*-xylene gave the following results:

+ OH

PPA, 100° → [262]

80% H_2SO_4, 25-30° → [116]

98-99% 1-2%

$$\text{toluene} + \text{isobutylene}, AlCl_3 \ [324] \text{ or } t\text{-BuOH}, H_2SO_4 \ [326] \xrightarrow{20°} \text{3,5-di-}t\text{-butyltoluene}$$

On the other hand, *p*-xylene, mesitylene, and pseudocumene were not *t*-butylated in competitive *t*-butylations with benzene in nitrobenzene solution, because no positions other than those *ortho* to methyl group are available [327]. Even in the absence of competition with benzene, only a minor amount of substitution *ortho* to the methyl was reported to take place with either *p*-xylene [116,329] or *p*-cymene [330].

In 1970 a detailed investigation of the acid-catalyzed reaction of cymenes with isobutylene and related alkenes was reported by Boone et al. [331]. In conformity with an earlier observation by Ipatieff et al. [332], they found that the reaction of cymenes with isobutylene gave complex mixtures of products resulting from alkylation and/or cyclialkylation. *o*-Cymene gave only alkylation products, whereas *p*-cymene gave predominatnly cyclialkylation products with only one case of alkylation. *m*-Cymene, however, occupied an intermediate position, giving both alkylation and cyclialkylation products. The complex array of products found in the reactions of *o*-, *m*-, and *p*-cymenes with isobutylene were accounted for in terms of carbocation intermediates.

Both steric and electronic factors also play determining roles during the *t*-butylations of other substituted benzenes. The observed reactivities of the halobenzenes relative to that of benzene, together with the isomer distributions of the mono-*t*-butylated products obtained during the $AlCl_3$-CH_3NO_2-catalyzed *t*-butylation of benzene and halobenzenes with isobutylene in nitrobenzene at 25°C, are summarized in Table 11. As is the case with the *t*-butylation of methylbenzenes with the exception of the smallest halogen, fluorine, no ortho substitution occurred.

The *t*-butylation of phenol through alkylation with isobutylene in the presence of various acid catalysts [335,336] or through transalkylation with *t*-butylbenzene in the presence of BF_3 or $AlCl_3$-CH_3NO_2 catalyst [337] gave the sterically more favorable *p*-*t*-butylphenol as major [335] or sole [336,337] product. The alkylation of catechol with isobutylene and H_2SO_4 gave 4-*t*-

Table 11. $AlCl_3/CH_3NO_2$-Catalyzed *t*-Butylation of Benzene and Halobenzenes with Isobutylene in CH_3NO_2 at 25°

		Isomer distribution, %		
Aromatic	$k_{AR}:k_{PhH}$	*ortho*	*meta*	*para*
Benzene	1.00			
Fluorobenzene	0.19	1.8	0.1	98.1
Chlorobenzene	0.06	—	5.5	94.5
Bromobenzene	0.03	—	-	100.0

butylcatechol of high purity [338]. Studies on cresols, xylenols, and anisoles demonstrated the following:

1. *t*-Butyl groups are not inserted between two methyl groups, or between a methyl and a hydroxyl group, placed *meta* to each other [339]. Thus 3,5-dimethylphenol did not react with isobutylene [339,340].
2. If the choice is given, the *t*-butyl group inserts *ortho* to hydroxyl rather than *ortho* to a methyl [340-345a] or a methoxyl [346-348] group. The following examples are illustrative:

OH + isobutylene —($AlCl_3/HCl$ [341], H_2SO_4 [342,343] or $POCl_3$ [344])→ OH (mainly)

OH + isobutylene —(KU-2 cation exchange, 120-150°, 24% yield)→ OH (96%) + OH (3%) [345]

OH + isobutylene —(H_2SO_4)→ (OH)* 100% [340]

OH + R-olefin —(H_2SO_4)→ OH, R [345a]

($R = CH_3$ or C_2H_5)

OH, OCH_3 + isobutylene —(H_2SO_4)→ OH, OCH_3 [346,347,348]

*Stevens and Friedman [345a], however, reported the formation of the 4-*t*-butyl isomer from a similar alkylation.

3. In the *t*-butylation of alkyl- and haloanisoles, the *t*-butyl group inserts *ortho* to the methoxyl rather than *ortho* to the alkyl or halo substituent [323,349-352]. The following three examples are illustrative:

OCH_3 ... + isobutylene $\xrightarrow{H_2SO_4}$ ... [323,350]

(R = Me, Et, *i*-Pr)

R = C_2H_5; $AlCl_3$ [349,351]
R = CH_3; $FeCl_3$ [352]

$\xrightarrow[50°]{52\%\ H_2SO_4}$ [192c]

R = H (20.5% yield)
R = CH_3 (31% yield)

The *t*-butylation of polynuclear aromatics was also expected. Two such examples are the reported alkylations of biphenyl and naphthalene with isobutylene and isobutyl alcohol.

According to Priddy [45a], the alkylation of biphenyl with isobutyl alcohol and BF_3 or with isobutylene and BF_3-H_3PO_4 gave similar product mixtures, consisting mainly of the *p*-*t*-butyl isomer (Eq. 61). The conversion was better with the alkene (26%) than with the alcohol (12%).

OH, BF_3 or , $BF_3 \cdot H_3PO_4$ (61)

95% 5%

Examination of the literature data concerning *t*-butylation of naphthalene reveals considerable discrepancy. For example, the alkylation of naphthalene with isobutyl alcohol and $AlCl_3$ [269] or BF_3 [334] and with *t*-butyl alcohol and $AlCl_3$ [22] was reported to give mixtures of 1- and 2-*t*-butylnaphthalene as monoalkylation products. However, the alkylation of naphthalene with *t*-butyl alcohol and BF_3 [130] or with isobutylene and $AlCl_3$ [319] was reported to yield only 2-*t*-butylnaphthalene. This confusion was clarified by more recent workers who found that 2-*t*-butylnaphthalene was the only product of monoalkylation of naphthalene with both isobutylene and diisobutylene in the presence of BF_3-H_3PO_4 [198] or with *t*-butyl alcohol in the presence of H_2SO_4 or $AlCl_3$ [324]. The results with isobutylene and diisobutylene are summarized in Eq. (62).

C=C(C)-C or C-C(C)(C)-C=C(C)=C; BF_3-H_3PO_4, 5-10°

70-73.4%

11.6-15.6% 8-8.8%

+ minor products
4.9-7%

(62)

The minor products in Eq. (62) were similar from both isobutylene and diisobutylene and included 2.0 to 4.4% of 2-(1,1-dimethylpropyl)-, 0.6 to 0.7% of 2-(1,1-dimethylbutyl)-, 1.1 to 0.9% of 2-(1,1,2-trimethylpropyl)-, and 1.2-1.0% of 2-(1,1,3-trimethylbutyl)naphthalene. The formation of these minor products was said to occur through a polymerization-depolymerization-alkylation mechanism [356]. On the other hand, the lack of formation of 1-*t*-butylnaphthalene under such nonisomerizing conditions was attributed to steric effects. It was noted that small alkyl groups branched to a maximum extent at the α-carbon atom cause considerable steric hindrance, and in alkylations with such groups the 2-isomer becomes the exclusive kinetic product [198,324].

Few studies have been devoted to establishing the behavior of heterocyclic aromatics toward *t*-butylation. Both furan [354] and thiophene [354a,354b] were *t*-butylated with isobutylene or *t*-butyl alcohol under suitable conditions to give the corresponding 2-*t*-butyl derivatives as monoalkylates:

Catalyst

X = O,S

However, thianaphthene was reported to react with isobutylene in the presence of acid catalysts to give good yields of the pure 3-*t*-butyl isomer, with no mention of any 2-*t*-butyl isomer (Eq. 63) [355].

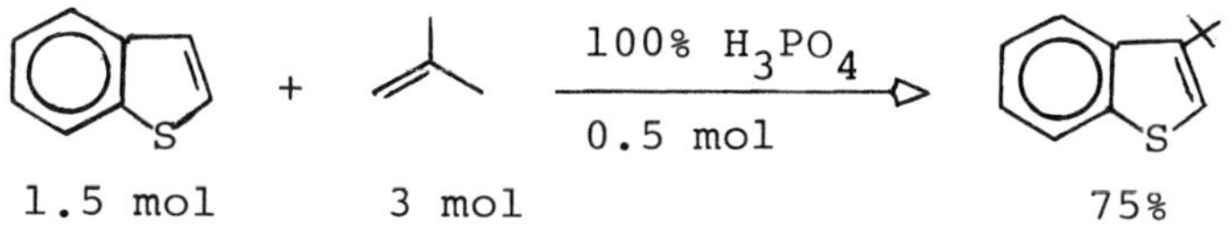

1.5 mol + 3 mol —(100% H_3PO_4, 0.5 mol)→ 75% (63)

1. Utilization of *t*-butylation for separation and synthetic purposes

The ease with which both introduction (by alkylation) and removal (by dealkylation) of the *t*-butyl group can be achieved has been utilized advantageously in both separation and synthetic problems. A thorough study of this utilization was carried out by Schlatter [357,358], who described the *t*-butyl group as "an unusual removable blocking group" for three reasons:

1. It can readily be introduced and removed from aromatic nuclei without disturbing other alkyl substituents.
2. It blocks not only the position it occupies, but the two adjacent *ortho* positions as well.
3. As an electron releasing group, it enhances further substitution in the *meta* and particularly in the *para* positions. Schlatter used these advantages to synthesize several 1,2-dialkylbenzenes and 1,2,3-trialkylbenzenes. Applications of the method to the synthesis of 1,2,3-trimethylbenzene and 1,3-dimethyl-2-*n*-alkylbenzenes are illustrated in schemes 8 and 9, respectively.

The activating effect of the *t*-butyl group toward further electrophilic substitution is illustrated by the HF-catalyzed ethylation of 4-*t*-butyl-1,3-dimethylbenzene, *m*-xylene, and toluene. Under conditions at which ready reaction occurred with the butyl derivative, *m*-xylene reacted slowly and toluene and benzene did not react at all. In further applications, Schlatter [359], Corson et al. [360] and Schneider [361] separated *p*-dialkylbenzenes

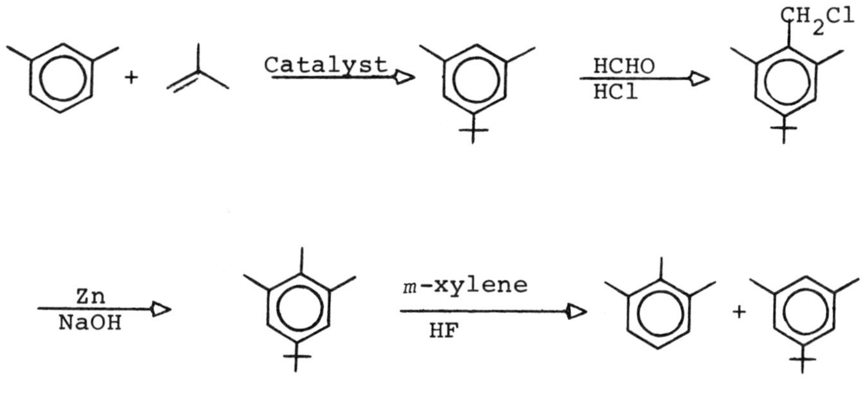

+ —Catalyst→ —(HCHO, HCl)→ CH_2Cl

—(Zn, NaOH)→ —(*m*-xylene, HF)→ +

Scheme 8

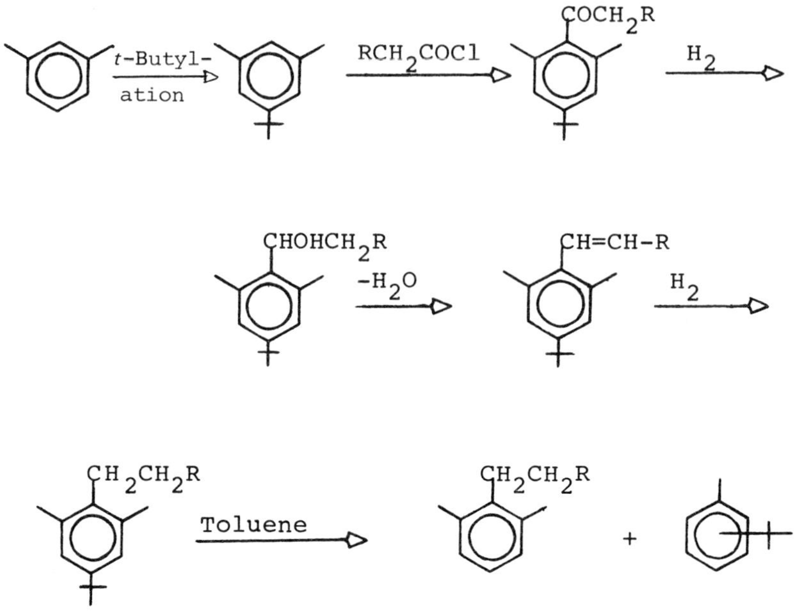

Scheme 9

from their isomers by selective alkylation of the *ortho* and *meta* isomers with isobutylene. De Pierre et al. [362] claimed the separation of mesitylene from other C_9 arenes by treating mesitylene concentrate with isobutylene in the presence of $AlCl_3$. Schlatter [363] separated hemimellitine from its isomers by selective *t*-butylation followed by debutylation in the presence of hydrogen fluoride catalyst. Similarly, Stevens [339] separated mixtures of *meta* and *para* cresols, xylenols, and ethylphenols by processes involving alkylation with isobutylene, separation of the *t*-butylated phenols by fractional distillation, and debutylation of the isolated derivatives. Also, Roberts and Rose [364] successfully separated 2,4-dimethylphenol from its 2,5-isomer by treating the mixture with diisobutylene in the presence of sulfuric acid. Distillation isolated the unchanged 2,5-isomer from octylated 2,4-dimethylphenol.

B. Isobutyl Alcohol and Isobutyl Ether

Until fairly recently, the Friedel-Crafts alkylations of aromatics by isobutyl alcohol and ether had been reported by almost all workers to yield *t*-butyl derivatives, regardless of the strength of catalyst and the severity of reaction conditions. The earliest report of an exception to the *t*-butylation rule was the claim made by Goldschmidt in 1882 [102] and by Senkowski in 1891 [365] that the $ZnCl_2$-catalyzed alkylation of benzene by isobutyl alcohol at 260 to 270°C gave a mixture of iso- and *t*-butylbenzene.

Much more recently (1964), Nield [36,37] proved by GLPC and IR analysis that the monoalkylation product resulting from the $AlCl_3$-catalyzed alkylation of benzene by both isobutyl alcohol and diisobutyl ether at 50°C consisted not only of *t*-butylbenzene (38 to 41%) but also of *sec*- and isobutylbenzene

Table 12. Alkylations of Benzene (50 ml) with Butyl Chlorides and Alcohols (0.1 mol) and Aluminum Chloride (0.15 mol)

	Temp, °C	Products, yield, %				
		i-C_4H_{10}	*i*-PrPh	*t*-BuPh	*i*/*s*-BuPh	PePh
i-BuCl	80	64	2.0	5.0	2.0[a]	48g
i-BuCl	25	24	2.0	9.0	5.0	4
i-BuOH	80	70	0.3	0.3	1.0	0
i-BuOH	25	0	Trace	0.2	0.5	0
t-BuCl	25	20	Trace	34	1.0	0
t-BuCl	25	20	1.0	14.0	3.0	3

[a]60% *i*-BuPh and 40% *s*-BuPh

(56 to 61%). Similar results were also noted for isobutyl chloride under the employed conditions.

Still more recently (1968), Romadane [43] reported that the BF_3-catalyzed alkylation of benzene and toluene with isobutyl alcohol at high temperature (210 to 215°C) and pressure (100 to 110°C atm) resulted solely or mainly in isobutylation. However, at lower temperatures (170 to 180°C) and pressures (24 to 40 atm) *t*-butylation took place. In a related development, Romadane et al. [137] reported that the alkylation of naphthalene with isobutyl alcohol and BF_3 produced a mixture of 1- and 2- *sec*-butylnaphthalene.

To explain the formation of *sec*- and isobutylbenzene from reagents containing the isobutyl group under vigorous conditions, Nield [36,37] suggested that drastic conditions favor primary over tertiary alkylation and hence" the isobutyl alkylating moiety alkylates before it has had the opportunity to give a *t*-butyl cation." Nield excluded the possibility that *sec*- and isobutylbenzene might have resulted from the rearrangement of the side chain of an initially formed *t*-butylbenzene, or of a *t*-butyl cation on grounds that these isomers were not found with the *t*-butylbenzene produced from reagents containing the *t*-butyl group.

The question of Friedel-Crafts alkylation with isobutyl and other related primary alcohols under drastic conditions was reconsidered by Roberts et al. [366] in 1969. Alkylations of benzene with isobutyl alcohol and isobutyl chloride using massive amounts of $AlCl_3$ (1.5 mol of $AlCl_3$ per mole of RX) gave low yields of butylbenzene mixtures composed mainly of *t*-butylbenzene mixed with minor amounts of isobutyl- and *sec*-butylbenzene (Table 12). Alkylations of benzene with *t*-butyl chloride or treatment of *t*-butylbenzene under similar conditions also yielded iso- and *sec*-butylbenzene.

To avoid the ad hoc theory [36,37] that drastic experimental conditions reverse the usual order of isomerization [primary → secondary, primary (iso) → tertiary], Roberts and co-workers proposed that the observed rearrangements can be rationalized by the mechanisms outlined in scheme 10.

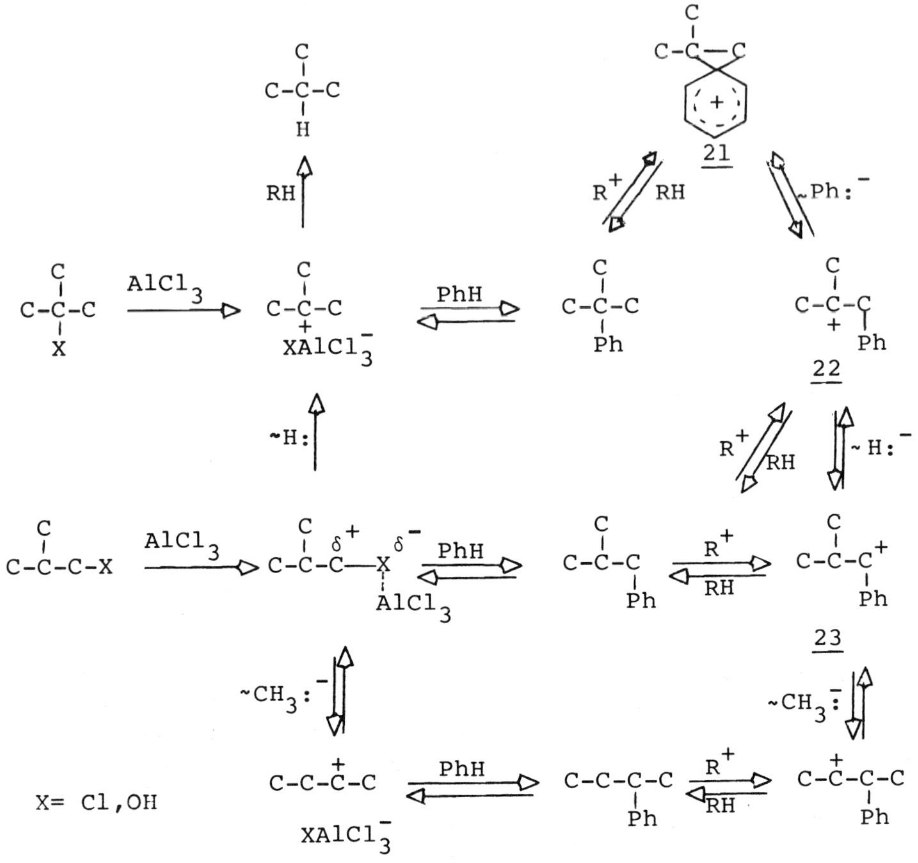

X= Cl,OH

Scheme 10

Roberts et al. proposed that some isobutylbenzene was produced by direct alkylation by isobutyl alcohol or isobutyl chloride complex, with a higher ratio of direct alkylation occurring with the alcohol than with the chloride. It was suggested that the formation of isobutyl- and/or secondary butylbenzene in the alkylation with *t*-butyl chloride and in treatments of *t*-butylbenzene with $AlCl_3$ proceeds via intermediates 21, 22, and possibly 23, none of which is a primary carbocation. As proposed, the formation of the phenonium ion 21 involves participation of neighboring phenyl in the hydride abstraction process.

To gain more insight into the alkylation behavior of isobutyl-like cycloalkyl carbinols, Roberts et al. [39] reinvestigated the alkylation of benzene with the related cyclobutyl-, cyclopentyl-, and cyclohexylcarbinols in the presence of $AlCl_3$. It was reported by Huston and Goodemoot [367] that these carbinols alkylated benzene without rearrangement to give the corresponding benzylcycloalkanes. The results of the reinvestigation with these cycloalkylcarbinols showed that these reactions in fact took place with accompanying rearrangements to yield mainly the products shown in Eqs. (64) to (66).

$$\text{cyclobutyl-}CH_2\text{-OH} \xrightarrow[AlCl_3]{PhH} \text{cyclopentyl-Ph} \quad \underline{24} \qquad (64)$$

$$\text{cyclopentyl-}CH_2\text{-OH} \xrightarrow[AlCl_3]{PhH} \text{cyclohexyl-Ph} \quad \underline{25} \qquad (65)$$

$$\text{cyclohexyl-}CH_2\text{-OH} \xrightarrow[AlCl_3]{PhH} \text{1-}CH_3\text{-1-Ph-cyclohexane} \ \underline{26} + CH_3\text{-cyclohexyl-Ph} \ \underline{27} \qquad (66)$$

Besides the major products noted above, the presence of only very minor amounts of the unrearranged products 28-30 from the reactions of Eqs. (64) to (66), respectively, was observed. Also, minor amounts of isomeric methylphenylcyclopentanes (31) were detected in alkylations of benzene with cyclopentylcarbinol (Eq. 65).

$$\text{cyclobutyl-}CH_2\text{-Ph} \ \underline{28} \qquad \text{cyclopentyl-}CH_2\text{-Ph} \ \underline{29} \qquad \text{cyclohexyl-}CH_2\text{-Ph} \ \underline{30} \qquad CH_3\text{-cyclopentyl-Ph} \ \underline{31}$$

To account for the corrected results of Eqs. (64) to (66), Roberts and coworkers suggested that highly polarized $AlCl_3$-carbinol complexes were formed which behaved as carbocation precursors. The rearrangements outlined in Eqs. (67) to (67) then occurred, leading to the rearranged products 24, 25, and 26 + 27, respectively.

$$\text{cyclobutyl-}\overset{\delta+}{CH_2}\text{--}\overset{\delta-}{O}\text{-} \xrightarrow{\sim CH_2:^-} \text{cyclopentyl}^+ \xrightarrow{PhH} \text{cyclopentyl-Ph} \quad \underline{24} \qquad (67)$$

$$\text{cyclopentyl-}\overset{\delta+}{CH_2}\text{--}\overset{\delta-}{O}\text{-} \xrightarrow{\sim CH_2:^-} \text{cyclohexyl}^+ \xrightarrow{PhH} \text{cyclohexyl-Ph} \quad \underline{25} \qquad (68)$$

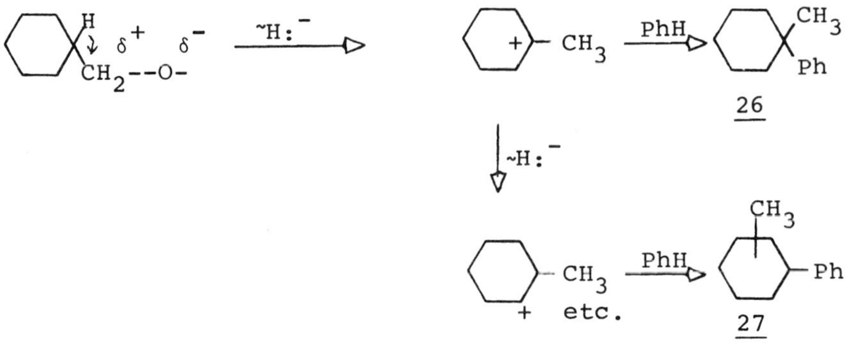

H δ^+ δ^- CH_2--O- —~H:⁻→ (+)–CH_3 —PhH→ CH_3, Ph **26**

↓ ~H:⁻ (69)

(+)–CH_3 etc. —PhH→ CH_3, Ph **27**

In alkylations with cyclobutyl- and cyclopentylcarbinols, the shift of a ring methylene group was said to be favored over a hydrogen shift. The release of angle and torsional strains in going from a four-membered to a five-membered ring and from a five-membered to a six-membered ring was said to be favored over the formation of a tertiary carbocation by a hydrogen shift. In the case of cyclohexylcarbinol, however, a hydrogen shift was said to be favored over ring expansion to a seven-membered ring.

The detection of minor amounts of isomeric methylphenylcyclopentanes (31) in alkylations of benzene with cyclopentylcarbinol was explained in terms of competition between a hydrogen shift and a ring-enlarging methylene shift as shown in scheme 11.

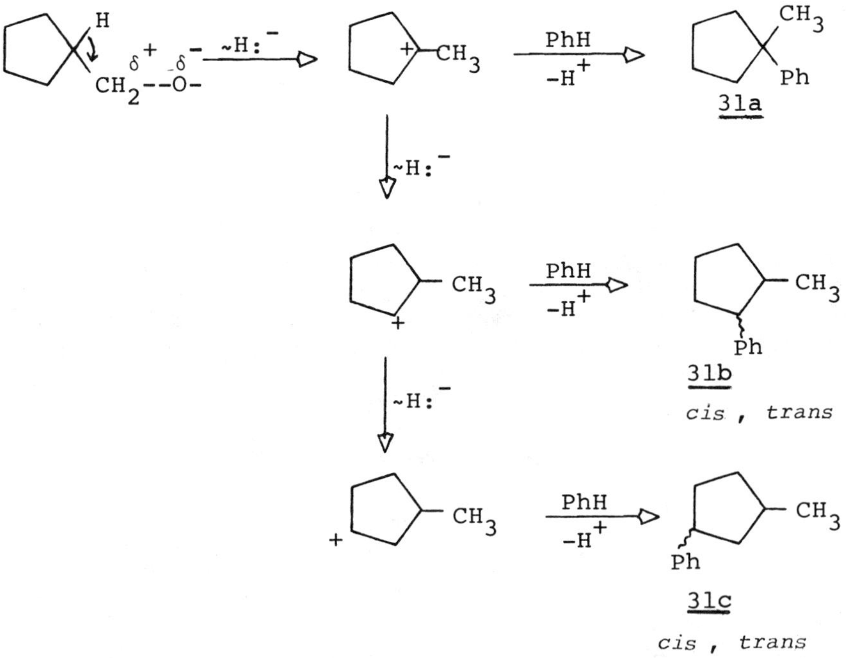

H δ^+ δ^- CH_2--O- —~H:⁻→ (+)–CH_3 —PhH, $-H^+$→ CH_3, Ph **31a**

↓ ~H:⁻

CH_3, + —PhH, $-H^+$→ CH_3, Ph **31b** *cis*, *trans*

↓ ~H:⁻

+, CH_3 —PhH, $-H^+$→ CH_3, Ph **31c** *cis*, *trans*

Scheme 11

In recent years, Kozlov et al. [177,262,263,368-371] have studied alkylations of aromatics with various primary and secondary alcohols in the presence of polyphosphoric acid. They found that alkylations of monoalkylbenzenes [262] and halobenzenes [263] with isobutyl alcohol gave *t*-butyl derivatives, the IR spectra of which showed considerable preponderance of the *p*-*t*-butyl isomers (Table 5). In some cases, however, there was a small amount of the *meta* (3 to 5%) and traces of the *ortho* isomers. As anticipated, Kozlov and co-workers found that in alkylations with isobutyl alcohol steric factors were important. While the yield of 4-*t*-butyl-*o*-xylene was 39%, virtually no reaction occurred with *m*- and *p*-xylenes [262]. Again, under similar conditions, isobutyl alcohol and polyphosphoric acid alkylated fluoro- and chlorotoluenes and *p*-difluorobenzene but failed to alkylate *p*-bromotoluene, *p*-dichlorobenzene, or *p*-dibromobenzene [263].

In continuation, Koslov and Klein [371] studied the alkylation of biphenyl with isobutyl alcohol in the presence of polyphosphoric acid under various reaction conditions. The yield of *t*-butylbiphenyl was found to increase considerably with increase of the biphenyl/alcohol molar ratio and also with the use of emulsifiers (e.g., cetylbenzyldimethylammonium chloride) to homogenize the reaction mass. As expected on a steric basis, the *t*-butylbiphenyl obtained consisted mainly of the *para* isomer with small amounts of the other isomers.

C. Branched Pentenes and Pentyl Alcohols, Ethers, and Esters

The alkylation of aromatics with branched C_5 compounds has been a matter of considerable concern for a number of Friedel-Crafts researchers. Alkylations by all of the five isomeric alcohols (32-36) and the three isomeric isopentenes (37-39) have been investigated. The results of these alkylations were extensively reviewed and compiled by earlier authors [6,12]. A re-compilation of some of these data according to the type of alkylating agent is depicted in Table 13.

```
  C                 C                 C
  |                 |                 |
C-C-C-C-OH        C-C-C-C           C-C-C-C
                      |               |
                      OH              OH

 (32)              (33)              (34)

       C             C               C
       |             |               |
HO-C-C-C-C         C-C-C-OH        C-C-C=C
                     |
                     C

 (35)              (36)              (37)

        C                 C
        |                 |
      C-C=C-C           C=C-C-C

        38                39
```

Table 13. Alkylations of Aromatics with Branched C_5 Alkenes, Alcohols, Ethers, and Esters

Entry no.	Aromatic mol	Alkylating agent, mol	Catalyst mol	Temp, °C	Time, hr	Product yield and composition, %	Ref.
			A. Isopentyl Alcohol, Ether, and Ester				
1	Benzene (0.8)	Isopentyl alcohol (0.8)	80% H_2SO_4 (6)	65	5	*t*-Pentylbenzene (36%)	26
2	Benzene (2)	Isopentyl ether (1)	BF_3 (-)	150	3	*t*-Pentylbenzene (10%)	109
3	Chlorobenzene (1)	Isopentyl alcohol (0.5)	$AlCl_3$ (0.6)	80-90	2-3	*p*-*t*-Pentylchlorobenzene (35%)	21
4	Tetralin (13.29)	Isopentyl alcohol (8.89)	H_3PO_4 (53 ml, d, 1.88)	100-130	5-9	6-*t*-Pentyltetralin (45-54%)	302
5	Naphthalene (-)	Isopentyl alcohol (-)	$AlCl_3$ (-)	—	-	β-*t*-Pentylnaphthalene (62%)	269
6	Naphthalene (-)	Isopentyl alcohol (-)	BF_3 (-)	165-170	2.5-3	α-and β-*t*-Pentylnaphthalenes (36-40%), 1,4-di-*t*-pentylnaphthalene (-)	83a
7	Phenol (-)	Isopentyl alcohol (-)	$ZnCl_2$ (-)	180	1	*p*-*t*-Pentylphenol (40%)	373,374
8	Phenol (0.2)	Isopentyl chloroformate (0.2)	$FeCl_3$ (-)	25-80	-	*p*-*t*-Pentylphenol	375
9	Aniline (-)	Isopentyl alcohol (-)	$ZnCl_2$ (-)	280	-	*p*-Isopentylaniline (40%)	365 374b

10	Aniline (-)	Isopentyl alcohol (-)	P_2O_5 (-)	250	-	*p*-Isopentylaniline (40%)	365 376 377
			B. 3-Methyl-2-butanol				
11	Benzene (2-5)	3-Methyl-2-butanol (1.0)	$AlCl_3$ (0.5)	30	24	2-Methyl-3-phenylbutane (25%)	275
			C. *t*-Pentyl Alcohol				
12	Benzene (7)	*t*-Pentyl alcohol (1)	HF (-)	-	-	*t*-Pentylbenzene (40%), di-*t*-pentylbenzene (50%)	378
13	Benzene (0.55)	*t*-Pentyl alcohol	$TiCl_4$ (0.1-0.2)	-	1-48	t-Pentylbenzene (41-74), *p*-di-*t*-pentylbenzene (11-33)	372
14	Benzene (4)	*t*-Pentyl alcohol (1)	$AlCl_3$ (0.5)	25-50	-	No *t*-pentylbenzene	379
15	Benzene (2)	*t*-Pentyl alcohol	H_2SO_4 (80%) (5)	60-65	3.5	t-Pentylbenzene (5)	379
16	Chlorobenzene (1)	*t*-Pentyl alcohol (0.5)	$AlCl_3$ (0.2)	80-90	2.3	*p*-and *m*-*t*-Pentylchlorobenzenes (50%)	21
17	Toluene (-)	t-Pentyl alcohol (-)	$TiCl_4$ (0.1-0.3)	18-40	3-18	*p*-*t*-Pentyltoluene (57-70%)	372
18	Tetralin	*t*-Pentyl alcohol (17g)	$AlCl_3$ (13.5g)	30	3	Tetralin (9g), 6-*t*-pentyltetralin (3.5g)	302
19	Tetralin	*t*-Pentyl alcohol	$ZnCl_2$ (27g) in sealed tube	200	12	Tetralin (2.3g), 6-*t*-pentyltetralin (82%)	302

Table 13. (continued)

Entry no.	Aromatic, mol	Alkylating agent, mol	Catalyst, mol	Temp, °C	Time, hr	Product yield and composition, %	Ref.
20	Tetralin (132g)	t-Pentyl alcohol (8.8g)	H_3PO_4 (53 ml, d, 1.88)	100-300	5-9	6-*t*-Pentyltetralin (45-54%)	302
21	Naphthalene (0.2)	*t*-Pentyl alcohol (0.25)	$AlCl_3$ (0.12) in ligroin	90	2	α-and β-*t*-Pentyl-naphthalenes (34%), di-*t*-pentylnaphth-alene (20%)	22
22	1-Chloronaphth-alene (0.8)	*t*-Pentyl alcohol (0.5)	$AlCl_3$ (0.2)	80-90	2-3	α-*t*-Pentyl-1-chloro-naphthalene (60%)	21
23	Phenol (1)	*t*-Pentyl alcohol (1)	$ZnCl_2$ (2)	180	-	*p*-*t*-Pentylphenol (65%)	380
24	Phenol (-)	*t*-Pentyl alcohol (-)	$AlCl_3$ (0.125) in petroleum ether	25-30	3-4	*p*-*t*-Pentylphenol (45-60%)	275
25	Aniline (0.2)	*t*-Pentyl alcohol (0.1)	$ZnCl_2$ (-)	270	9	*p*-*t*-Pentylaniline (-)	377
			D. Neopentyl Alcohol				
26	Benzene (0.25)	Neopentyl alcohol (0.25)	80% H_2SO_4 (3)	65	6	*t*-Pentylbenzene (30%)	381
27	Benzene (1.0)	Neopentyl alcohol (0.25)	$AlCl_3$ (0.33)	80	8	Neopentylbenzene (9%)	381

E. 3-Methyl-1-butene

28	Butene (2)	3-Methyl-1-butene (1)	88.7% BF_3-H_3PO_4 + 11.3% BF_3-H_2O (15)	0,30,80	1.33	*t*-Pentylbenzene (30%, 42%, 57%), higher (47%, 37%, 25%)	382
29	Benzene (5)	3-Methyl-1-butene (3)	96% H_2SO_4 (1.8)	5	2	*t*-Pentylbenzene (20%), di-*t*-pentylbenzene (56%)	104
30	Benzene (0.5)	3-Methyl-1-butene (0.25)	$AlCl_3$-HCl (0.08)	5	1.7	2-Methyl-3-phenylbutane (12%)	26
31	Benzene (5)	3-Methyl-1-(1)	$AlCl_3$ (0.2)-HCl	0 21	0.68 0.60	Pentylbenzenes (38%, 42%)[a] (55:45, 11:87)[b]	383 26
32	Benzene (5)	3-Methyl-1-butene (1)	HF (5)	35 100	0.65 0.6	Pentylbenzenes (27%, 32%)[a], (100:1), (70:30)[b], *t*-butylbenzene (0%, 2%)	383
33	Benzene (4)	3-Methyl-1-butene (1)	$NiCl_2$-Al_2O_3	300 40 atm	0.5	t-Pentylbenzene (58%)	384
34	*p*-Xylene (5)	3-Methyl-1-butene (1)	$AlCl_3$ (0.2)-CH_3NO_2 (100g), *t*-pentyl chloride (5 ml)	25	0.58	2-*p*-Xylyl-3-methylbutane (7%), *t*-Butyl-*p*-xylene (8%), much polymer	329
35	*p*-Xylene (2)	3-Methyl-1-butene (1)	HF (5.5)	23-35	1.43	2-*p*-Xylyl-3-methylbutane (34%)	329
36	*m*-Cresol (-)	3-Methyl-1-butene (-)	$ZnCl_2$-Al_2O_3	150 3-4 atm	6	3-Methyl-6-isopentylphenol	385
37	Aniline (55g)	3-Methyl-1-butene (20g)	Act. Kaolin (25g)	250-260 25 atm	13.5	*p*-*t*-Pentylaniline (8.8g)	386

Table 13. (continued)

Entry no.	Aromatic, mol	Alkylating agent, mol	Catalyst, mol	Temp, °C	Time, hr	Product yield and composition, %	Ref.
38	Phenol (949)	3-Methyl-1-butene (70g)	BF_3 (1.8-26%)	100	0.5	*p-t*-Pentylphenol (95-96%), liquid products (2-3%)	387
			F. 2-Methyl-2-butene				
39	Benzene (10 pts)	—	$AlCl_3$ (-)	—	-	*t*-Pentylbenzene (55%)	388
40	Benzene (5)	2-Methyl-2-butene (1)	$AlCl_3$ (0.2)-HCl	21	0.56	Pentylbenzenes (42%, 54%, 39%)[a], (55:45, 75:25, 100:0)[b], *t*-butylbenzene (5%, 1%, 1%)	389
41	Benzene (5)	2-Methyl-2-butene (1)	$AlBr_3$ (27g)-HBr	25 25	1 sec 0.42	Pentylbenzenes (22%, 21%)[a], (90:10, 70:30)[c], *t*-butylbenzene (17%, 13%)	389
42	Benzene (0.55)	2-Methyl-2-butene (-)	$TiCl_4$ (0.1)	80	6	*t*-Pentylbenzene (40%)	390
43	Benzene (5)	2-Methyl-2-butene (1)	BF_3 (1)	28 119	0.65 0.5	Pentylbenzenes (10%, 11%)[a], (100:0, 80:20)[b], *t*-butylbenzene (6%, 6%)	383
44	Benzene (5)	2-Methyl-2-butene (1)	BF_3 (105g)-H_2O (36g)	24	0.5	*t*-Pentylbenzene (54%)	393

45	Benzene (5)	2-Methyl-2-butene (1)	H_2SO_4 (185g)	0	1.11	*t*-Pentylbenzene (4%), *t*-butylbenzene (2%)	383
46	Benzene (5)	2-Methyl-2-butene (3)	96% H_2SO_4 (1.8)	5	2	*t*-Pentylbenzene (18%), di-*t*-pentylbenzene (50%)	104
47	Benzene (5)	2-Methyl-2-butene (-)	HF (-)	0	-	*t*-Pentylbenzene (21%) *p*-di-*t*-pentylbenzene (60%)	284
48	Benzene (5)	2-Methyl-2-butene (1)	HF (100g)	100	0.56	Pentylbenzenes (32%), (70:30)[b], *t*-butylbenzene (2%)	389
49	Chlorobenzene (15)	2-Methyl-2-butene (1)	$AlCl_3$ (0.2)-HCl	24	1	Pentylchlorobenzene (8%), *t*-butylchlorobenzene (12%)	389
50	Toluene (-)	2-Methyl-2-butene (-)	$TiCl_4$ (0.1)	20	3	*p*-*t*-Pentyltoluene (33%)	390
51	*p*-Xylene (1.7)	2-Methyl-2-butene (1)	HF (5.7)	24-28	1.58	2-*p*-Xylyl-3-methylbutane (40%)	329
52	*p*-Cymene (2)	2-Methyl-2-butene (1)	HF (8.4)	0-7	0.5	*t*-Pentyl-*p*-cymene (25%), 1,3,3,6-tetramethyl-1-*p*-tolylindan (46%)	332
53	Phenol (-)	2-Methyl-2-butene (-)	H_2SO_4 (-)	-	-	*t*-Pentylphenol (-)	391
54	Phenol (2)	2-Methyl-2-butene (-)	$FeCl_3$-HCl	90	2-4	4-*t*-Pentylphenol (63%)	341
55	Phenol (1)	2-Methyl-2-butene (1)	BF_3 (1.6%)	100-120	0.5	4-*t*-Pentylphenol (95-96%)	387

Table 13. (continued)

Entry no.	Aromatic, mol	Alkylating agent, mol	Catalyst, mol	Temp, °C	Time, hr	Product yield and composition, %	Ref.
56	2-Cyclohexylphenol (500 pts)	2-Methyl-2-butene (-)	H_2SO_4 (25 pts)	70	-	2,6-Di-*t*-pentyl-2-cyclohexylphenol	392
57	*p*-Cresol (1)	2-Methyl-2-butene (1)	BF_3 (1-23 mol%) or BF_3 (50 mol%) in isopentane	rt	2-4	2-*t*-Pentyl-4-methylphenol (high yield)	393
58	*p*-Cresol (1)	2-Methyl-2-butene (1)	BF_3 (50 mol%)	rt	2-4	2-*t*-Pentyl-4-methylphenol (20%), 4-*t*-C_4- and C_5-phenols; 2-*sec*-alkyl-4-methylphenols, and 2,2'-bismethylenediphenols, isobutane, isopentane, and methylpentane	393

59	Anisole (0.2)	2-Methyl-2-butene (-)	$TiCl_4$ (0.2)	0-60	1	*p*-*t*-Pentylanisole (61-67%)	390
60	Anisole (0.3)	2-Methyl-2-butene (0.4)	KU-1 cation exchange resin (12.8g)	95-100	6-8	*p*-*t*-Pentylanisole (14%)	394
61	Phenetole (0.3)	2-Methyl-2-butene (0.4)	KU-1 cation exchange resin (12.8g)	95-100	6-8	*p*-*t*-Pentylphenetole (20%)	394
62	Thiophene (840g)	2-Methyl-2-butene (350g)	70% H_2SO_4 (300g)	40-50	3	*t*-Pentylthiophene (57%)	395, 396
			G. 2-Methyl-1-butene				
63	*p*-Xylene (2)	2-Methyl-1-butene (1)	$AlCl_3$-HCl (13g)	0	0.43	2-*p*-Xylyl-3-methylbutane (41%), isopentane (0.19 mol), some di-*p*-xylylmethane	329

[a]Yields at different temperatures.
[b]Ratio of 2-phenyl-2-methylbutane to 2-phenyl-3-methylbutane.
[c]Ratio of 2-phenyl-2-methylbutane to 2-phenyl-3-methylbutane at different contact times.

Careful examination of the data of Table 13 reveals the following obvious features. First, similar to alkylations with branched C_5 halides, the present alkylations show complications by side reactions such as reorientation, disproportionation, fragmentation, polymerization, hydride transfer, and skeletal rearrangement of the alkylating group. Second, in line with other *t*-alkylations, steric factors are important, favoring *p*-substitution in alkylations of monosubstituted benzenes and secondary over tertiary substitution in alkylations of *p*-disubstituted benzenes. Third, besides apparent contradiction in some cases, other cases seemed questionable in view of current knowledge of Friedel-Crafts alkylations and rearrangements.

Whereas some reactions were said to occur with no postalkylation rearrangement to yield only *t*-pentyl derivatives, others were said to occur with partial or complete rearrangement to 3-methyl-2-butyl or neopentyl derivatives. Sound examples can be found in the following reports: the formation of only neopentylbenzene from the reaction of benzene with neopentyl alcohol and $AlCl_3$ [381]; the formation of only 2-methyl-3-phenylbutane from the reaction of benzene with 3-methyl-2-butanol and $AlCl_3$ [381] or with 3-methyl-1-butene and $AlCl_3/HCl$ [26]; the formation of only *t*-pentyl derivatives from the $AlCl_3$-catalyzed alkylations of benzene with pentylene [397], 2-methyl-2-butene [388] or 3-methyl-2-butanol [275], toluene with pentylene [398], and chlorobenzene with isopentyl or *t*-pentyl alcohol [21]; and even the formation of *p*-isopentylaniline from the reaction of aniline with isopentyl alcohol in the presence of $ZnCl_2$ [365,374(b)] or P_2O_5 [365,376,377].

In the following, we shall elaborate on the features above, placing special emphasis on benzene pentylations, which have received the most attention in the past three decades.

One of the leading studies on the effect of reaction variables on the extent of rearrangements accompanying the alkylation of benzene with isopentenes was conducted by Friedman and Morritz in 1956 [383]. Their results showed that the extent of rearrangement of *t*-pentylbenzene into 2-methyl-3-phenylbutane increased with an increase in operating temperature, reaction time, and catalyst ratio. The type of alkene used was also reported to affect the ratio. For example, the aluminum chloride-catalyzed alkylation of benzene with 3-methyl-1-butene produced pure *t*-pentylbenzene at -40°C and pure 2-methyl-3-phenylbutane at 21°C. A mixture of both pentylbenzene isomers was produced in benzene alkylations with 2-methyl-2-butene at 21°C. Similarly, the hydrogen fluoride-catalyzed alkylation of benzene with 3-methyl-1-butene gave pure *t*-pentylbenzene at 35°C, whereas with 2-methyl-2-butene it gave a mixture of *t*-pentylbenzene (70 to 75%) and 2-methyl-3-phenylbutane (30 to 25%) at either 0 or 100°C.

This dependence of product composition on the type of starting alkene is inconsistent with the currently accepted view that the *t*-pentyl cation is a common intermediate leading to similar products regardless of which of the isopentene isomers is used (Eq. 70).

$$\mathrm{C{=}\underset{|}{\overset{C}{C}}{-}C{-}C} \quad \underset{-H^+}{\overset{+H^+}{\rightleftharpoons}} \quad \mathrm{C{-}\overset{C}{\underset{+}{C}}{-}C{-}C} \quad \underset{+H^+}{\overset{-H^+}{\rightleftharpoons}} \quad \mathrm{C{-}\overset{C}{C}{=}C{-}C} \tag{70}$$

A duplication of the reaction of the aluminum chloride-catalyzed alkylation of benzene with 3-methyl-1-butene by the reviewers [121] gave a complex mixture containing not only 3-methyl-2-phenylbutane, as originally reported, but also *t*-pentylbenzene, neopentylbenzene, *t*-butylbenzene, and hexylbenzenes (Eq. 71). The foramtion of the latter two compounds was also noted

```
            C                        C          C C           C
            |     AlCl3/HCl          |          | |           |
PhH  +  C-C-C=C  ----------->  Ph-C-C-C  +  Ph-C-C-C  +  Ph-C-C-C
                  21,0.6 hr          |                        |
                                     C          48%           C
                                  36%                      1%

                                  + Ph C6H13                     (71)
                                    5%
```

by other workers [113,383,389,398a] and was accounted for by Friedman and Morritz [389] in terms of the dimerization-rearrangement-cleavage-alkylation mechanism shown in scheme 12.

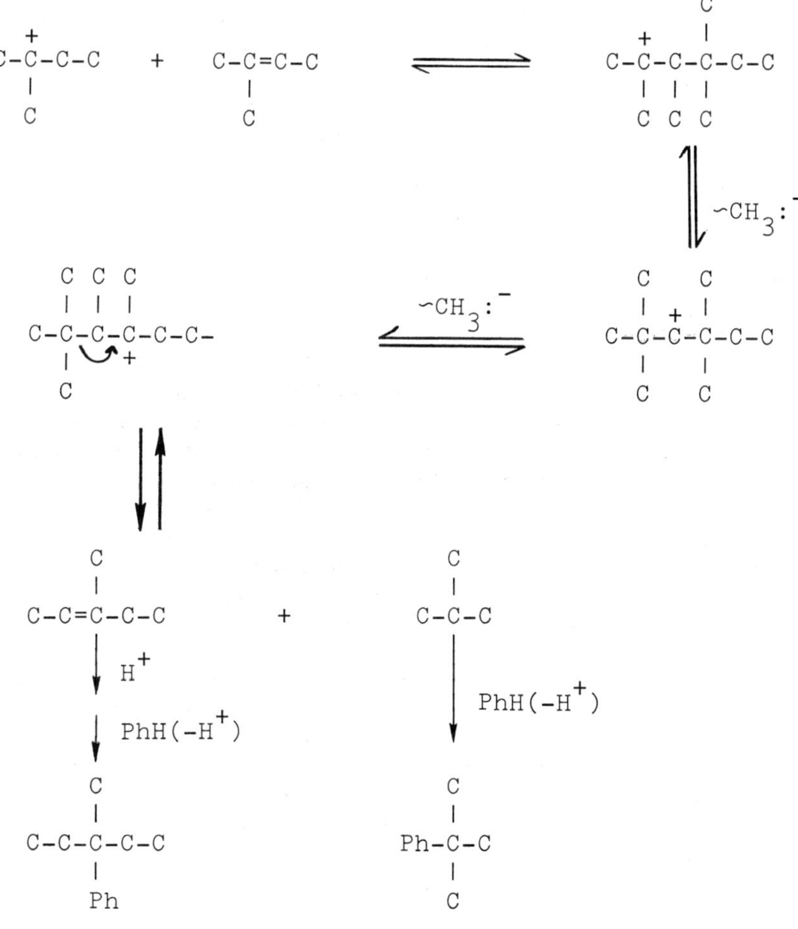

Scheme 12

It is of interest to note that whereas the formation of *t*-butyl derivatives seems to be rather general in alkylations of aromatics by isopentenes, they probably do not form in alkylations with *t*-pentyl halides [44,389]. This difference was attributed [389] to the following reasons: (1) the formation of alkene from *t*-pentyl cation (or its equivalent) and the subsequent dimerization are both slow compared to the alkylation step; and (2) the alkylation may proceed via an S_N2 concerted nucleophilic displacement rather than through an S_E attack involving carbocations. In view of current knowledge, the second reason appears unrealistic.

Experimental support for the first assumption, however, was found when alkylation of chlorobenzene and benzene with 2-methyl-2-butene in the presence of aluminum chloride under similar conditions gave 2.5 times more *t*-butylchlorobenzene than *t*-butylbenzene. The less reactive chlorobenzene would yield more *t*-butyl derivative since its rate of alkylation would be retarded while the competing dimerization-cleavage reactions would proceed unimpeded. When toluene, a more reactive aromatic, was similarly alkylated with 2-methyl-2-butene, the yield of *t*-butyl derivatives was not materially less than that from benzene, but the yield of pentyltoluene was shown to be considerably larger (46 versus 11%).

As with alkenes, the early literature data about the alkylation of benzene with branched alcohols are highly controversial. The controversy seemed to be most pronounced in connection with reactions catalyzed by aluminum chloride catalyst. For example, benzene and 3-methyl-2-butanol were reported by Huston et al. to give 2-methyl-3-phenylbutane with $AlCl_3$ at room temperature [275], but *t*-pentylbenzene with $AlCl_3$-HCl at ice temperature [129(a)]. It seemed unlikely, however, that the differences in reaction conditions indicated above would have such a large effect on the product composition. So a thorough investigation of the alkylation of benzene with 3-methyl-2-butanol was recently carried out by Khalaf, Roberts, and coworkers [121,399]. In contrast to the report of Huston and Hsieh [275], the new data demonstrated that the product from the room-temperature reaction consisted not only of 2-methyl-3-phenylbutane (mainly), but also of *t*-pentylbenzene [399]. Besides, the new data demonstrated that the aluminum chloride-catalyzed alkylation of benzene with 3-methyl-2-butanol at ice temperature or room temperature, with or without the presence of hydrogen chloride and independent of solvent resulted initially in the formation of *t*-pentylbenzene. Subsequent side-chain isomerizations by the strong catalyst, aluminum chloride, would account for the formation of varying proportions of 2-methyl-3-phenylbutane and, in some cases, even of neopentylbenzene. Thus, in the presence of nonisomerizing catalysts such as $AlCl_3$-CH_3NO_2 or BF_3, no such isomerizations were observed and pure *t*-pentylbenzene was obtained in good yield [121].

The extent of isomerization in the alkylation above was found to be much more dependent on the proportion of aluminum chloride than on either temperature or the presence of the cocatalyst, hydrogen chloride [399]. This is evident from the fact that, after 15 hr at 0°C, when a 1:0.5 mole ratio of alcohol to aluminum chloride was used, *t*-pentylbenzene was obtained as the major end product (83%), but when the mole ratio was 1:0.75, rearranged 2-methyl-3-phenylbutane became the major end product (78%).

Confusion also existed with respect to the $AlCl_3$-catalyzed alkylation of benzene with *t*-pentyl alcohol. This reaction was reported by Huston and Hsieh [275] to yield 40% *t*-pentylbenzene and 1.8% 2-methylbutane. A similar alkylation by Inatome et al. [379] was reported to produce no *t*-pentylbenzene. A recent duplication of this reaction revealed the sensitivity of product composition/catalyst ratio [121]. Thus at 25°C for 4 hr with an alcohol/catalyst ratio of 1.0:0.5, the monoalkylation fraction consisted chiefly of *t*-pentylbenzene (66%) mixed with smaller amounts of 2-methyl-3-phenylbutane (34%) and traces of neopentylbenzene. With an alcohol/catalyst ratio of 1:1.5, however, 2-methyl-3-phenylbutane became the main product (57%) mixed with *t*-pentylbenzene (29%), neopentylbenzene (12%), and *t*-butylbenzene (2%) (Eq. 72).

```
          C                                 C                 C
          |      AlCl3(0.15 mol )           |                 |
PhH + C-C-C-C  ------------------>    C-C-C-C     +     C-C-C-C
          |      25°, 4 hr                    |               |
          OH                                  Ph              Ph

                                        57%              29%
                                                                        (72)
                                      C                C
                                      |                |
                         +   Ph-C-C-C        +     C-C-C
                                      |                |
                                      C                Ph

                                     12%              2%
```

The literature data about the alkylation of benzene with neopentyl alcohol deserve special comment. This alkylation was reported to yield pure *t*-pentylbenzene when catalyzed by either sulfuric acid [381] or boron trifluoride [113], but pure neopentylbenzene when the reaction was catalyzed by aluminum chloride at 80°C (Eq. 73) [381].

```
                        80% H2SO4                  C
                                                   |
                    ----------------------->  Ph-C-C-C
                   |    or BF3 at 60°              |
          C        |                               C
          |        |
PhH + C-C-C-OH ----|                                              (73)
          |        |
          C        |    AlCl3                         C
                   |                                  |
                    ----------------------->  Ph-C-C-C
                        Reflux, 8 hr                  |
                                                      C
```

While the experimental results are reasonable and consistant with new findings [121], the explanations invoked to rationalize them were incorrect. So in an attempt to account for the production of neopentylbenzene from the $AlCl_3$ alkylation, Schmerling and West (1954) [400] suggested a bimolecular displacement reaction between benzene and the alcohol $AlCl_3$ complex <u>40</u> (scheme 13).

To account for the rearrangement observed with the same catalyst when neopentyl chloride was the alkylating agent, they suggested a mechanism in which a neopentyl cation intermediate is formed, thus allowing a methyl shift to occur. This mechanism is not entirely satisfactory, of course, since it does not explain the fact that little or none of the primary alkylbenzene is formed in this and in other reported similar alkylations of benzene by other primary alcohols [401].

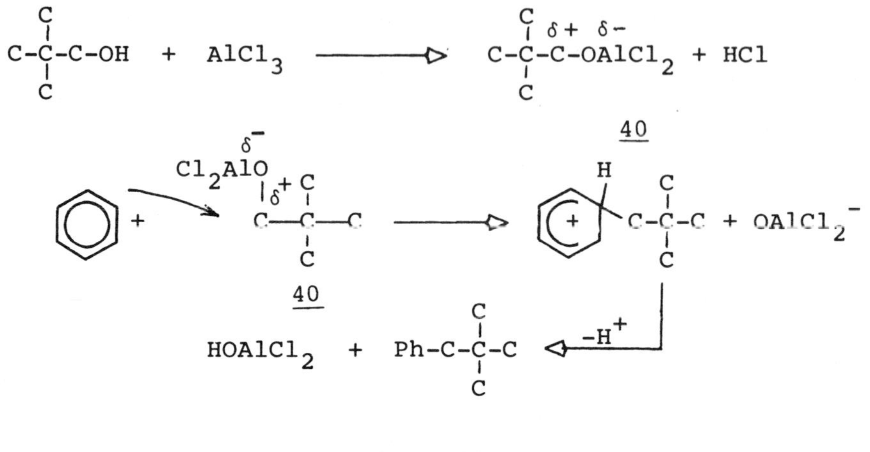

$$\mathrm{C\text{-}\underset{C}{\overset{C}{C}}\text{-}C\text{-}OH + AlCl_3 \longrightarrow C\text{-}\underset{C}{\overset{C}{C}}\text{-}\overset{\delta+}{C}\text{-}\overset{\delta-}{O}AlCl_2 + HCl}$$

40

$$\mathrm{C_6H_6 + \overset{\delta-}{Cl_2AlO}\text{-}\overset{\delta+}{C}\text{-}\underset{C}{\overset{C}{C}}\text{-}C\ (\mathbf{40}) \longrightarrow [C_6H_6(H)\text{-}C\text{-}\underset{C}{\overset{C}{C}}\text{-}C]^{+} + OAlCl_2^{-}}$$

$$\xrightarrow{-H^{+}} \mathrm{HOAlCl_2 + Ph\text{-}C\text{-}\underset{C}{\overset{C}{C}}\text{-}C}$$

Scheme 13

More recently, Nenitzescu et al. [402,403] and Roberts and Han [404] showed that the isolation of the so-called "unrearranged product" in the alkylation of benzene with neopentyl alcohol and aluminum chloride is actually the end result of multiple rearrangements. The first rearrangement is one of the alkylating agent before attachment to the benzene ring, yielding initially *t*-pentylbenzene. This hydrocarbon is subsequently isomerized by aluminum chloride, first to 2-methyl-3-phenylbutane and then to neopentylbenzene (Chap. 8). All three catalysts isomerize the alkylating agent,but only aluminum chloride is capable of isomerizing *t*-pentylbenzene Eq. (74).

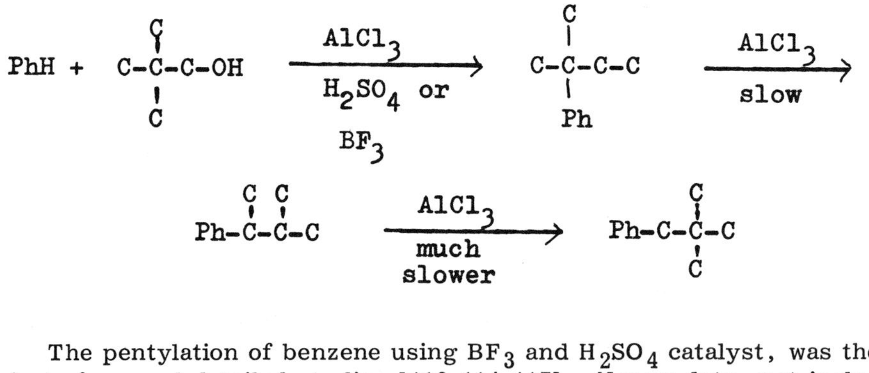

$$\mathrm{PhH + C\text{-}\underset{C}{\overset{C}{C}}\text{-}C\text{-}OH \xrightarrow[H_2SO_4\ or\ BF_3]{AlCl_3} C\text{-}\underset{Ph}{\overset{C}{C}}\text{-}C\text{-}C \xrightarrow[slow]{AlCl_3}}$$

$$\mathrm{Ph\text{-}\overset{C}{C}\text{-}\overset{C}{C}\text{-}C \xrightarrow[much\ slower]{AlCl_3} Ph\text{-}C\text{-}\underset{C}{\overset{C}{C}}\text{-}C} \tag{74}$$

The pentylation of benzene using BF_3 and H_2SO_4 catalyst, was the subject of several detailed studies [113,114,117]. Newer data, not included in Table 13, are given in Table 14. As can be seen from Table 14, the products from BF_3 reactions were more complex than from H_2SO_4 reactions. Except for 2-methyl-1-butanol (35), the latter catalyst produced only *t*-pentylbenzene from alkylations with the other isomeric alcohols.

For the BF_3-catalyzed alkylation of benzene with 3-methyl-2-butanol, Streitwieser et al. [113] obtained two sets of results (from MS and IR analysis), which differed significantly in product composition. In a recent reinvestigation of this reaction, we found the product to consist of 52% *t*-pentylbenzene, 3% 2-methyl-3-phenylbutane, 33% *t*-butylbenzene, and 12% hexylbenzenes (Eq. 75) [121]. The formation of the latter two products is not un-

Table 14. Reaction Products from the Alkylation of Benzene with Branched Pentyl Alcohols in the Presence of BF_3[a] and H_2SO_4[b] Catalysts

Alcohol	Catalyst	Temp, °C	Time, hr	Yield, %	Product Composition, %: 2-Phenylpentane	3-Phenylpentane	*t*-Pentylbenzene	Other products	Ref.
C C-C-C-C-OH	BF_3[c]	60	12	68	14[e] (9)	2(6)	81(83)	3	113[d]
32	80% H_2SO_4	80	—	60-5	0	0	100	-	117
C OH C-C-C-C	BF_3	0	1	42	10[e] (0)	1(0)	88(72)	28[f](21)[g]	113[d]
33	80% H_2SO_4	80	-	60-5	0	0	100	-	117
C C-C-C-C OH	BF_3	0	1	-	No alkylated benzenes			-	113[d]
34	80% H_2SO_4	80	-	60-5	0	0	100	-	117

Table 14. (Continued)

Alcohol	Catalyst	Temp, °C	Time, hr	Product Composition, %					Ref.
				Yield, %	2-Phenyl-pentane	3-Phenyl-pentane	*t*-Pentyl-benzene	Other products	
C-C(C)(OH)-C-C **35**	BF_3[c]	60	9	68	20[e] (15)	4(5)	76(78)	2	113[d]
	80% H_2SO_4	80	-	60-5	25	12	63	-	117
C-C(C)(C)-C-OH **36**	BF_3[c]	60	12	61	⩽4[e] (~2)	<1(~1)	⩾96(97)	0	113[d]
	80% H_2SO_4	80	-	60-5	0	0	47	-	117

[a]Reactants were 0.113 mol alcohol, 100 g benzene and BF_3 passed in until saturation was achieved.
[b]Reactants were 0.03 mol alcohol, 0.06 mol benzene and 12.5 ml 80% H_2SO_4.
[c]Activated by a few drops of water.
[d]First figure from ms, figure from ir in parenthesis.
[e]Includes 2-methyl-3-phenylbutane, if any.
[f]Contains 25% butylbenzenes.
[g]*t*-Butylbenzene.

$$PhH + C\text{-}\overset{C}{C}\text{-}\overset{OH}{C}\text{-}C \xrightarrow[0°,\,1\,hr]{BF_3} \underset{52\%}{C\text{-}\underset{Ph}{\overset{C}{C}}\text{-}C\text{-}C} + \underset{3\%}{Ph\text{-}\overset{C}{C}\text{-}\overset{C}{C}\text{-}C} + \underset{33\%}{Ph\text{-}\underset{C}{\overset{C}{C}}\text{-}C} + \underset{12\%}{PhC_6H_{13}} \quad (75)$$

expected and finds adequate explanation in the bimolecular mechanism shown in scheme 12. In contrast to the results of Streitwieser, et al. [113], no 2- or 3-phenylpentane was found in the product. In view of this recent result with 3-methyl-2-butanol, the presence of the secondary pentylbenzenes 2- and 3-phenylpentanes among the products from the BF_3-catalyzed alkylation of benzene with 3-methyl-1-butanol (32), 2-methyl-1-butanol (35), and neopentyl alcohol (36) seems questionable.

We pointed out before that steric factors play determining roles in tertiary alkylations. In alkylations of monosubstituted benzenes with branched C_5 compounds, there are no serious inhibitions since *para* substitution is not hindered. Thus *t*-pentyl alcohol gave *p*-*t*-pentylanisole with anisole and $TiCl_4$ at 25°C [372], *p*-*t*-pentylaniline with aniline and $ZnCl_2$ at 270°C [377], *p*-*t*-pentyltoluene with toluene and $TiCl_4$ at 180°C [372], and *p*-*t*-pentylphenol with phenol and $ZnCl_2$ at 180°C [380]. Similarly, isopentyl alcohol gave *p*-*t*-pentyltoluene with toluene and BF_3 at 170 to 180°C [43] and *p*-*t*-pentylphenol with phenol and $ZnCl_2$ at 180°C [382].

Similarly, 2-methyl-2-butene gave *p*-*t*-pentyltoluene with toluene and $TiCl_4$ at 20°C [390], *p*-*t*-pentylanisole with anisole and $TiCl_4$ at 0 to 60°C [390], or KU-1 cation exchange resin at 95 to 100°C [394] and *p*-*t*-pentylphenetole with phenetole and KU-1 cation exchange resin at 95 to 100°C [394].

The previous discussion of *t*-butylation indicated the difficulty of inserting a *t*-butyl group in a position that was *ortho* to an alkyl group. The insertion of the bulkier *t*-pentyl group should even be more difficult. Thus, while *para* alkyltoluenes such as *p*-xylene [262,329] and *p*-cymene [262, 330,331] were *t*-butylated with low yields (1 to 4%), attempts to accomplish *t*-pentylation always failed [329]. Instead, other reactions, including *sec*-pentylation by the less stable 3-methyl-2-butyl cation, probably in equilibrium with its more stable *t*-pentyl isomer, were favored. For example, the reaction of *p*-xylene with either 2-methyl-2-butene or 3-methyl-1-butene in the presence of $AlCl_3$ or HF catalyst gave a product consisting chiefly of the *sec*-alkylate 2-*p*-xylyl-3-methylbutane (41, 34 to 41%) mixed in some cases with varying amounts of by-products such as di-*p*-xylylmethane (42), *t*-butyl-*p*-xylene (43), and isopentane (Eq. 76) [329]. When the reaction above was

+ or $\xrightarrow[\text{or HF}]{AlCl_3}$ 41 42 43 (76)

carried out in the presence of H_2SO_4, BF_3, $FeCl_3$, or $AlCl_3$-CH_3NO_2 catalysts, the product consisted of polymers mixed with smaller amounts of the by-product, *t*-butyl-*p*-xylene [329]. Moreover, attempted *t*-pentylation of *p*-xylene with *t*-pentyl alcohol in the presence of H_2SO_4 at 0 to 70°C gave considerable amounts of polymers mixed with about 11% of unidentified C_{14} and higher products [329].

The *t*-pentylation of *p*-cymene with 2-methyl-2-butene was also attempted in the presence of HF at 0 to 7°C [332]. Expectedly, no *t*-pentylation occurred, but the product included 1,3,3,6-tetramethyl-1-*p*-tolylindan (44) and probably the secondary alkylate 2-*p*-cymyl-3-methylbutane (45)

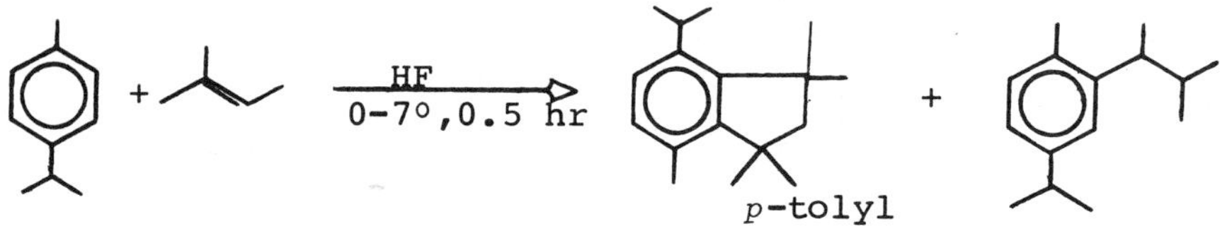

HF, 0-7°, 0.5 hr (77)

p-tolyl

44 46%   45 25%

$+ \ i\text{-}C_5H_{12} + C_{10}H_{22}$

(Eq. 77). The formation of 44 was attributed to hydride transfer reactions [331,405], which are discussed at length in Chap. 6.

Whereas attempted *t*-pentylation *ortho* to an alkyl group in *p*-xylene and *p*-cymene was met with failure, *t*-pentylation *ortho* to a hydroxyl group in *p*-cresol was achieved [393]. In 1959, Malchick and Hannan [393] found that alkylation of *p*-cresol with 2-methyl-2-butene in the presence of 1 to 25 mole % BF_3 as catalyst gave a high yield of 2-*t*-pentyl-4-methylphenol (46).

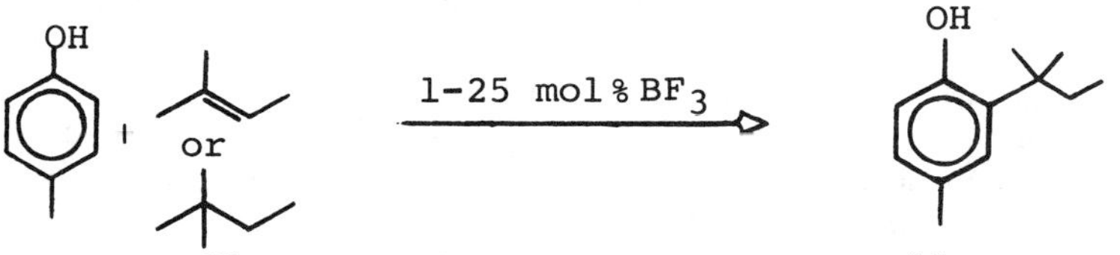

OH   or   Cl   1-25 mol % BF_3 OH

46

With 50 mol % BF_3, however, the reaction resulted in a complex mixture of phenolic products containing less than 20% of the *t*-pentyl-*p*-cresol (46). The structures and modes of formation of the rest of the phenolic components are discussed in connection with similar alkylations by branched C_5 halides (Chap. 3).

In conclusion, a comparison of these results with those of the alkylations of aromatics with the corresponding branched C_5 halides (Chap. 3) allows one to point out certain similarities.

1. In the absence of steric factors, alkylations with branched pentyl alcohols and pentenes give *t*-pentyl derivatives as the initial products. When mild (nonisomerizing) catalysts are used, the tertiary structure remains intact, but when strong catalysts are used rearrangements of the type

$$\text{Ar-C(C)}_2\text{-C-C} \xrightarrow{\text{slow}} \text{C-C(C)-C(Ar)-C} \xrightarrow{\text{slower}} \text{Ar-C-C(C)}_2\text{-C}$$

take place resulting in mixtures of isomers.

Thus *t*-pentylbenzene was the only monoalkylate formed in the alkylation of benzene with *t*-pentyl alcohol and $AlCl_3$-CH_3NO_2 [121], $TiCl_4$ [372], H_2SO_4 [121,374,379] or HF [378]; with isopentyl alcohol and H_2SO_4 [26] or BF_3 [113]; with neopentyl alcohol and H_2SO_4 [381] or BF_3 [113]; with isopentyl ether and BF_3 [109]; as well as with isopentenes and HF [383], H_2SO_4 [121, 104], BF_3-H_2O [383], $TiCl_4$ [390], or $ZnCl_2$-Al_2O_3 [406] (scheme 14). On the other hand, mixtures of *t*-pentylbenzene and 2-methyl-3-phenylbutane were formed in benzene alkylations with *t*-pentyl alcohol and $AlCl_3$ [121,275]; with 3-methyl-2-butanol and $AlCl_3$ [121]; and with isopentenes and $AlCl_3$ [121], $AlCl_3$-HCl [383,389], or $AlBr_3$-HBr [389]. In some cases where severe conditions are used, neopentylbenzene could be the end product [381].

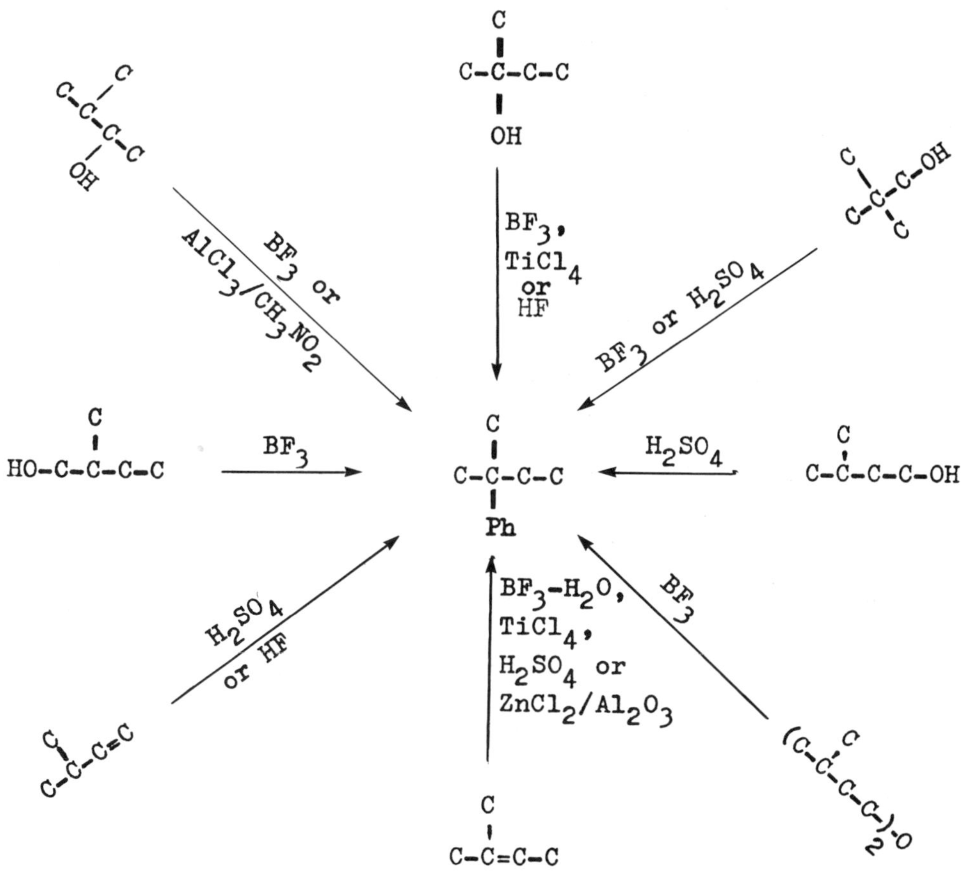

Scheme 14

2. In view of substantiated results and current knowledge, some of the early results seems questionable and should be taken with considerable doubt. Examples of these are found in the reported $AlCl_3$-catalyzed alkylations of benzene with 3-methyl-2-butanol [275], 2-methyl-2-butene [388], or pentylene [397] to give pure *t*-pentylbenzene, and of aniline with isopentyl alcohol in the presence of $ZnCl_2$ [365,374(b)] or P_2O_5 [365,736,377] to give *p*-isopentylaniline. Similar anomalous cases where pre- or postalkylation rearrangements escaped notice are listed elsewhere [6,12] and can easily be identified.

3. Steric factors favor *p* substitutions in reactions with monosubstituted benzenes and secondary over tertiary alkylations in reactions with *p*-dialkylated benzenes [332,383].

4. Depending on catalyst and conditions, the alkylations are subject to the usual complicating side reactions. These include hydride transfer, reorientation, fragmentation, dealkylation, and rearrangement of the alkylating group prior to and/or subsequent to the alkylation step. Further proof of the latter reaction was offered by Duganova et al. [407], who used ^{14}C labeling to study isomerization of alkyl groups during benzene alkylations with branched pentyl and hexyl alcohols. The results demonstrated that isomerization via alkyl shifts took place during H_2SO_4- catalyzed alkylation of benzene with $Me_2CH^{14}CH(OH)Me$, $Me_2{}^{14}CHCH(OH)Me$, $EtMe_2{}^{14}C\ OH$, $Me^{14}CH\ Me_2OH$, 2-methyl- and 2-ethylcyclohexanol-2-^{14}C, 1-methylcyclohexanol-1-^{14}C, and 1-methylcyclohexene-1-^{14}C. The results also showed that isomerization was more extensive in the cyclic system than in the linear ones and that the composition of the products was dependent on reaction temperatures.

D. Branched C_6 and Higher Alcohols and Alkenes

Before 1950, Huston et al. [111,129(a),275,408-411] conducted a number of studies on the aluminum chloride-catalyzed condensation of benzene and phenol with branched secondary and tertiary hexyl, heptyl, and octyl alcohols. Based on product identification by the conventional methods then available, Huston et al. [111,129(a),408] offered the following conclusions:

1. Both tertiary alcohols and secondary alcohols in which the hydroxyl is on a carbon adjacent to one carrying tertiary hydrogen give tertiary alkylarenes. This is illustrated by the results of Eqs. (78) to (81).

$$PhH + C\text{-}\underset{OH}{\overset{R}{C}}\text{-}R \xrightarrow[25-35°]{AlCl_3} C\text{-}\underset{Ph}{\overset{C}{C}}\text{-}R + C\text{-}\overset{C}{C}\text{-}R \qquad (78)\ [411]$$

27-33% 2-9%

$R = n\text{-}C_3H_7,\ n\text{-}C_4H_9,\ n\text{-}C_5H_{11}$

$$PhH + C\text{-}C\text{-}\underset{OH}{\overset{C}{C}}\text{-}R \xrightarrow[35°]{AlCl_3} C\text{-}C\text{-}\underset{Ph}{\overset{C}{C}}\text{-}R + C\text{-}C\text{-}\overset{C}{C}\text{-}R \qquad (79)\ [411]$$

24-30% 2-10%

$R = n\text{-}C_3H_7,\ n\text{-}C_4H_9$

```
          C C                                C C
          | |          AlCl3                 | |
PhH  +  C-C-C-C   ------------------->     C-C-C-C                 (80)
          |                                  |                     [275]
          OH                                 Ph

          C                                    C
          |            AlCl3                   |
PhH  +  C-C-C-C-C  ------------------->   C-C-C-C-C                (81)
          |                                    |                   [411]
          OH                                   Ph
```

2. Linear alcohols and alcohols having a branched methyl group remote from the carbinol carbon give mixtures of products assumed to form via elimination to alkene followed by addition of arene at either of the two ends of the double bond. Accordingly, both 5-methyl-2-hexanol (<u>47</u>) and 5-methyl-3-hexanol (<u>48</u>) were reported to yield similar mixtures of 2-methyl-4-phenylpentane (<u>49</u>) and 2-methyl-5-phenylpentane (<u>50</u>) (Eq. 82).

```
  C
  |
C-C-C-C-C-C
        |
        OH                C                    C
                          |                    |
   47        PhH       C-C-C-C-C-C    +     C-C-C-C-C-C             (82)
   or     -------->         |                       |
             AlCl3          Ph                      Ph
  C                        49                  50
  |
C-C-C-C-C-C
      |
      OH

   48
```

Similar behavior was predicted for 4,4-dimethyl-3-pentanol (<u>51</u>, Eq. 83) [129(a)].

```
          C        AlCl3        C                  C
          |                     |                  |
PhH  +  C-C-C-C-C  ------->   C-C-C-C-C   +      C-C-C-C-C          (83)
          | |                   | |                |   |
          C OH                  C Ph               C   Ph

          51                    52                 53
```

Examination of these early conclusions in terms of current results and ideas subjects them to heavy criticisms. Formulating them, the authors were not only unaware of carbocation behavior and the isomerizing influence of $AlCl_3$ both on the alkylating agents and the final alkylates (see Chaps. 2 and 8), but they also overlooked the simple fact that, in some cases, water elimination could possibly lead to two isomeric intermediate alkenes. These criticisms, added to the fact that the old mechanism of elimination to an alkene followed by addition of arene across the double bond was shown to be unlikely [114,131], suggests that the early results of Huston's and of others [380,412-415] should be questioned.

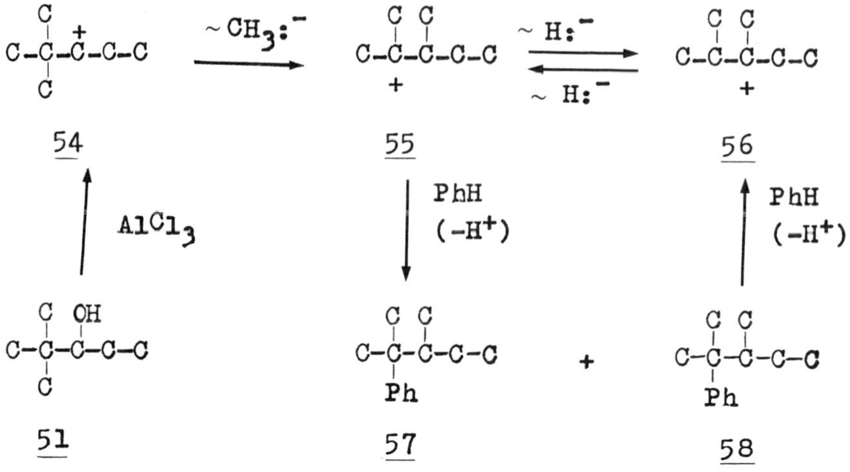

Scheme 15

As an example, let us examine the case of the $AlCl_3$-catalyzed alkylation of benzene with the neopentyl-type alcohol, 2,2-dimethyl-3-pentanol (51, Eq. 83). The reported formation of 2,2-dimethyl-3-phenylpentane (52) and 2,2-dimethyl-4-phenylpentane (53) from this reaction could hardly be explained. In recent terms, one can assume with assurance that such a reaction should involve the carbocation transformations outlined in scheme 15.

The originally formed neopentyl-type secondary carbocation 54 would undoubtedly rearrange rapidly and completely to the more stable tertiary carbocation 55 by a 1,2 methyl shift [416; see also Chap. 2]. The latter in turn equilibrates with another tertiary carbocation 56 by a 1,2 hydride shift. Alkylation of benzene by 55 and 56 would lead to the two isomeric tertiary heptylbenzenes 57 and 58, respectively. In the presence of the isomerizing $AlCl_3$ catalyst, further rearrangements of arenes 57 and 58 can still occur.

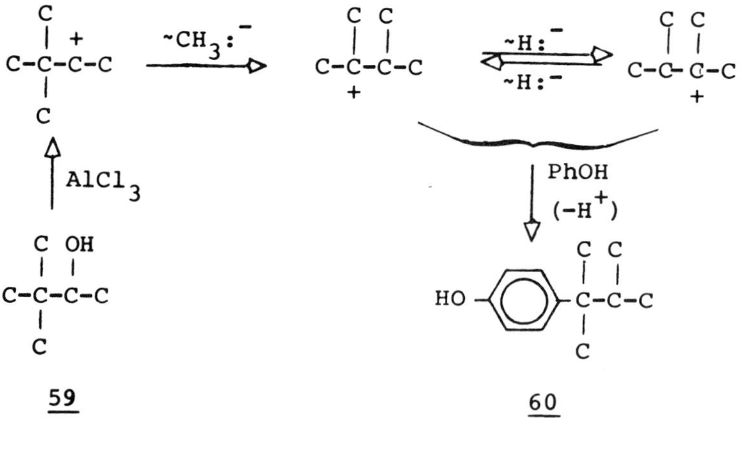

Scheme 16

In fact, these predictions were verified by Huston et al. [410], who found that the product from the $AlCl_3$-catalyzed alkylation of phenol with 3,3-dimethyl-3-butanol (59) consisted entirely of 2,3-dimethyl-2-(*p*-hydroxylphenyl)-butane (60). The formation of the latter was explained by Schriesheim [138] as shown in scheme 16. In this case, the originally formed tertiary cation will rapidly equilibrate with another degenerate cation, leading to the same alkylation product.

In the course of their investigations of alkylations with large alcohols, Huston et al. [111,129(a),408] noted that accumulation of alkyl groups on the carbon atom adjacent to the carbinol carbon caused marked reductions in the yields of alkylates produced in the presence of $AlCl_3$. The same structural feature also favored side reactions, resulting in the formation of unsaturated compounds, tertiary alkyl halides, and *t*-butylbenzenes. The isolation of the latter in a number of cases led Huston and Awuapara [409] to believe that the low yields obtained in alkylations by hindered alcohols were the result of fragmentation of the carbon chain rather than of a depressive steric influence on the condensing capability of the alcohols. This belief was supported by the finding that when an active aromatic substrate such as phenol was used instead of benzene, side reactions were minimized and good yields of alkylates were obtained. This was attributed to the faster rate of alkylation than of fragmentation.

Fragmentations during alkylations with large alcohols were also noted by other workers [71,127]. For example, the $ZnCl_2$-catalyzed reaction of phenol with both 2,4,4-trimethyl-2-pentanol (61) [127] and the 4-*t*-octylcyclohexanol (62) [71] gave *p*-*t*-butylphenol as the entire alkylate. The results were later explained in terms of the fragmentation-alkylation steps, outlined in schemes 17 and 18 [138].

Following Huston's early work on Friedel-Crafts alkylations with large branched alcohols, there was a period of relative inactivity in this field, and no serious studies were made until 1967. Since that year a number of alkylations with large branched alcohols and alkenes have been carried out to evaluate the effect of reaction variables on the extent of accompanying isomerizations. As in other cases, the extent of isomerizations showed varying degrees of dependence on experimental conditions, such as choice of catalyst, solvent, temperature, and duration of the reaction.

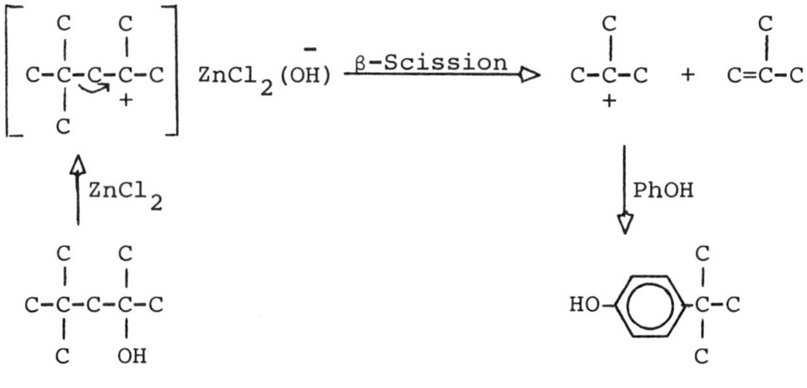

```
 ┌  C   C  ┐                                  C          C
    |   |                 -                   |          |
 | C-C-C-C-C |  ZnCl2(OH)  ──β-Scission──▷  C-C-C   +   C=C-C
    |   +                                       +
 └  C       ┘
                                               |
    △ ZnCl2                                    | PhOH
    |                                          ▽
    C   C                                          C
    |   |                                          |
  C-C-C-C-C                            HO-(C6H4)-C-C
    |   |                                          |
    C   OH                                         C

    61
```

Scheme 17

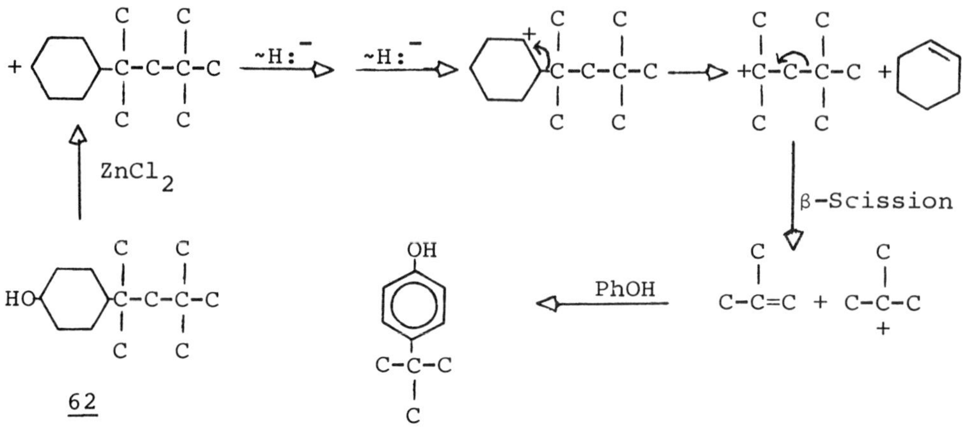

Scheme 18

Generally, the course of the rearrangement was quite profoundly influenced by catalyst type. With sulfuric acid catalyst, there was a considerable preference for formation of tertiary alkylates even when the tertiary carbon was far removed from the double bond. Geiseler and Braun [417] reported that the product from the H_2SO_4-catalyzed alkylation of benzene with 2-methyl-1-heptene (63) consisted chiefly of 2-methyl-2-phenylheptane (64) mixed with minor amounts of other isomers (Eq. 84).

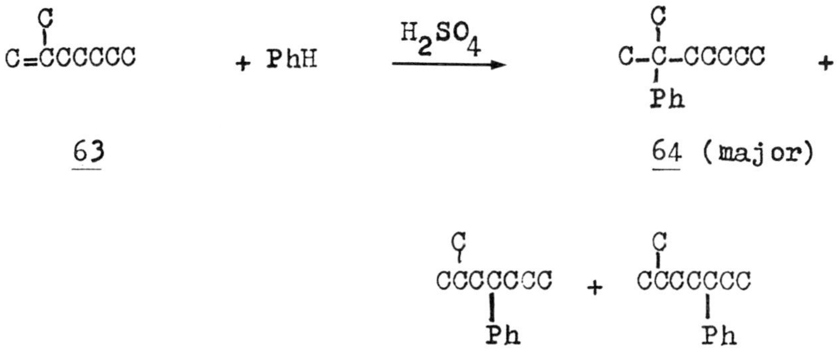

63 64 (major) (minor) (trace) (84)

Under similar conditions, Nooi et al. [418] obtained solely or predominantly tertiary alkybenzenes from the alkylation of benzene with 2-methyl-1-octene (65), 2-methyl-2-octene (66), 2-butyl-1-octene (67), 3,7-dimethyl-1-octene (68), 10-methyl-1-undecene (69), and 3,3-dimethyl-1-undecene (70). With a

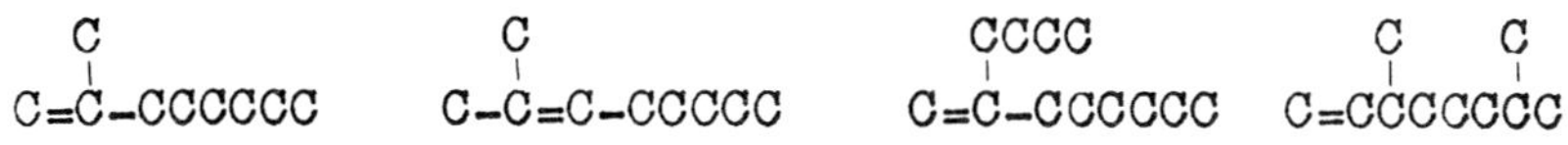

65 66 67 68

```
        C                          C
        |                          |
C=CCCCCCCCCC               C=C-CCCCCCCCC
                                   |
                                   C
       69                          70
```

mixture of the two isomeric methyloctenes 65 and 66, benzene gave 2-methyl-2-phenyloctane (71) as the only monoalkylate (Eq. 85).

```
 C
 |                                   C
C=CCCCCCC              H2SO4          |
    65       + PhH  ----------->   CCCCCCCC          (85)
                       5-10°          |
 C   +                               Ph
 |
CC=CCCCCC                          71 (51%)
    66
```

Nooi et al. [418] also examined the alkylation of benzene with the four methylated octenes 65-68 and the mono- and dimethylated undecenes 69 and 70, respectively, in the presence of aluminum chloride. Compared to H_2SO_4, $AlCl_3$ resulted not only in less preference for the formation of tertiary alkylbenzenes, but also in low yield of products. The yields were particularly poor in alkylations involving tertiary carbocation intermediates in the early stages of the reactions. Examples are found in alkylations with 2-methyl-1-octene (65), 2-methyl-2-octene (66), and 2-butyl-1-octene (67). In such cases, the low reactivity of the bulky tertiary carbocations coupled with the instability of the tertiary alkylates to $AlCl_3$ were suggested as possible causes for the reduction in yields.

The H_2SO_4-catalyzed alkylation of benzene with 3,3-dimethyl-1-undecene (70) is of particular interest because it bears some similarity to Huston's alkylation of benzene with 2,2-dimethyl-3-pentanol (51; see Eq. 83 and scheme 15). With 70, the monoalkylbenzene fraction consisted of many tridecylbenzene isomers, none of which was the direct (nonrearranged) product $CH_3CH(Ph)C(CH_3)_2(CH_2)_7CH_3$, however. The established absence of the latter product strengthens our earlier criticism of Huston's claim [109] that 2,2-dimethyl-3-phenylpentane (52) was produced by direct alkyaltìon of benzene with 2,2-dimethyl-3-pentanol (51).

Several other cases of alkylations with large branched alcohols and alkenes have also appeared in the more recent literature. Some of these were conducted in the presence of the moderated $AlCl_3$-$PhNO_2$ catalyst [419,420]. In contrast to $AlCl_3$, the moderated catalyst resulted in high yields of pure tertiary alkylates. This is illustrated by the alkylation of o-xylene with 2,4,4-trimethyl-1-pentene (Eq. 86) [420].

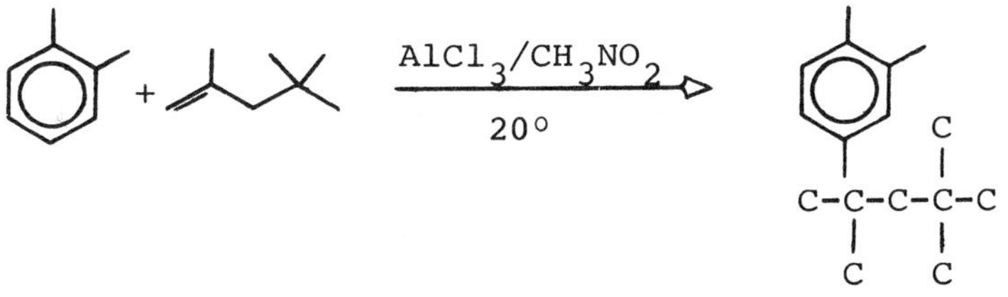

```
                    AlCl3/CH3NO2         (o-xylyl)
o-xylene + 2,4,4-trimethyl-1-pentene  ---------------->     |          C        (86)
                        20°                                 |          |
                                                        C-C-C-C-C
                                                          |     |
                                                          C     C
```

Both phenol [421] and cresols [422] were recently alkylated with large branched alcohols in the presence various acidic catalysts. Examples of these are shown in Eqs. (87) to (89).

$$PhOH + CH_3(CH_2)_n\underset{OH}{\overset{CH_3}{C}}C_2H_5 \xrightarrow{\text{Acid catalyst}} p\text{-}HOC_6H_4\text{-}\underset{CH_3}{C}(C_2H_5)(CH_2)_nCH_3 \quad (75\text{-}95\%) \qquad (87)\ [421]$$

$n = 4,5$

$$PhOH + CH_3(CH_2)_4\underset{OH}{\overset{CH_3}{C}}CH(CH_3)_2 \xrightarrow{\text{Acid catalyst}} p\text{-}HOC_6H_4\text{-}\underset{CH_3}{C}(CH(CH_3)_2)(CH_2)_4CH_3 \qquad (88)\ [421]$$

$$CH_3C_6H_4OH + (CH_3)_2C{=}C(CH_3)_2 \xrightarrow[\text{or } BF_3/AcOH]{ZnCl_2/HCl} CH_3(HO)C_6H_3\text{-}C(CH_3)_2CH(CH_3)_2 \qquad (89)\ [422]$$

In 1972, Alul and McEwan [299] investigated the effect of catalyst, solvent, and temperature on the extent of isomerizations accompanying the alkylation of benzene with 8-methyl-1-nonene. The goal of their research was to find out how far down the chain the positive charge can reach before the attack on benzene takes place. Table 15 lists the amounts of various secondary and tertiary isomers obtained.

It can be seen from the results of Table 15 that all alkylations of benzene with 8-methyl-1-none afforded various amounts of all secondary and tertiary alkylbenzenes. However, the relative amounts of these various isomers were highly dependent on reaction conditions. Thus, under conditions of minimum isomerization, as with $AlCl_2$-HSO_4, the positive charge barely reached the tertiary carbon atom of the chain. On the other hand, under conditions of maximum isomerization, as with large amounts of HF, the positive charge wound up almost exclusively on the tertiary atom. Other alkylation catalysts afforded results that lie between these two extremes. This difference in the rate of

Table 15. Products of Reaction of Benzene with 8-Methyl-1-nonene in the Presence of Various Catalysts

$$(C^1\text{-}\overset{\overset{\displaystyle C}{|}}{C^2}\text{-}C^3\text{-}C^4\text{-}C^5\text{-}C^6\text{-}C^7\text{-}C^8\text{-}C^9)\ Ph$$

Catalyst	Temp, °C	Yield, %	2-Phenyl	3-Phenyl	4-Phenyl	5-Phenyl	6-Phenyl	7-Phenyl	8-Phenyl
HF	0-5	88	56.5	0.8	2.9	8.9	6.5	8.9	15.5
HF + *n*-hexane	0-5	86	71.0	0.5	2.1	6.9	4.6	5.8	9.1
HF[a]	55	86	30.7	0.6	1.9	8.1	8.0	15.7	35.0
$AlCl_2 \cdot HSO_4$	35-40	75	9.7	0.1	0.8	4.3	5.1	18.7	61.3
$AlCl_3 \cdot CH_3NO_2$	35-40	75	24.8	0.5	1.8	8.6	8.3	16.7	39.3
$AlCl_3 \cdot HCl$	35-40	30	3.2	1.2	8.6	11.7	12.7	21.2	41.4
$AlCl_3 \cdot HCl$[b]	0-5	80	22.9	0.5	1.6	8.0	8.1	17.1	41.8

[a]Reaction was carried out in the absence of a liquid catalyst phase to minimize isomerization under these conditions.
[b]Recycled aluminum chloride was used in this experiment to prevent product isomerization or dealkylation.
Source: Ref. 299.

isomerization was attributed to the effect of the negative ion of the ion pair on the mobility of the positive ion across the chain.

An interesting feature of the isomer distributions reported in Table 15 is the sharp drop in the amount of the 3- and 4-isomers compared to other secondary alkylbenzenes. This was attributed to a slower rate of alkylation by the corresponding cations as a result of the steric effect of the adjacent isopropyl group.

V. ALKYLATION OF ARENES WITH CYCLOALKENES AND CYCLOALKANOLS

As pointed out before, the term *cycloalkylation* has long been used to refer to Friedel-Crafts alkylations of arenes with cycloalkyl derivatives to produce cycloalkylarenes. In Chap. 3 we reviewed cycloalkylations with cycloalkyl halides. Like acyclic alkylations, however, the cycloalkylating agent can also be a cyclic alcohol, alkene, ester, or even a cycloalkane under special promoting conditions [423-428]. At this point, we shall consider cycloalkylations with cycloalkenes and cycloalkanols.

The voluminous cycloalkylations compiled by earlier authors [6,12], together with the more recent ones, reveal considerable similarities to other types of alkylations. On the one hand, cycloalkylations with cycloalkenes and cycloalkanols were performed in the presence of a wide variety of acidic catalysts. For example, the cyclohexylation of benzene with cyclohexene was reported to take place in the presence of catalysts such as $AlCl_3$ [429-432], $AlCl_3$-H_2O [433], $AlBr_3$ [434], $ZnCl_2$ [435], $TiCl_4$-H_2O [436], BF_3-HCl [437], H_2SO_4 [432,438-441], HF [284,306], HCl [442], Al-C_2H_5Br [443], Al-C_4H_9Cl [444], BF_3-H_3PO_4 [222,445], and P_2O_5 [446]. On the other hand, depending on catalyst and conditions, cycloalkylations were in many cases complicated by accompanying reactions such as polyalkylation [101, 430,432,433,439-441,444,447-453], reorientation [439,449-452], fragmentation [138,454], and skeletal isomerization of the cycloalkylating group [425,433, 454-458]. These similarities and others relating to orientation during cycloalkylation will become clear from the following discussion in which the emphasis will be placed on recent literature. For convenience, this discussion will be presented under three separate headings: cycloalkylations with unbranched groups, cycloalkylations with branched groups, and mechanistic aspects of cycloalkylations.

A. Cycloalkylations with Unbranched Cycloalkenes and Cycloalkanols

The cycloalkylations of arenes with unbranched cycloalkenes and cycloalkanols were reported by most workers to proceed without accompanying skeletal rearrangement of the cycloalkyl group. For example, in the presence of $AlCl_3$ catalyst, cyclopentene gave as monoalkylates: cyclopentylbenzene with benzene [449], *m*- and *p*-cyclopentyltoluene with toluene [449], 1,2-dimethyl-4-cyclopentylbenzene with *o*-xylene [450], 1,3-dimethyl-5-cyclopentylbenzene with *m*-xylene [450], 1,4-dimethyl-2-cyclopentylbenzene with *p*-xylene [450], reoriented 1,2,4-trimethyl-5-cyclopentylbenzene with mesitylene [451], 1,2-dimethyl-4-*t*-butyl-6-cyclopentylbenzene with 1,2-dimethyl-4-*t*-butylbenzene [459], and *p*-cyclopentylbiphenyl with biphenyl [449]. Moreover, cyclopentene gave 1,3,5-trimethyl-2-cyclopentylbenzene with mesitylene in the presence of H_2SO_4 [451], *p*-chloro- and *p*-bromocyclopentylbenzenes with chloro- and bromobenzenes, respectively, in the presence of H_2SO_4 [460], a mixture of 2- and 3-cyclopentyl-1-methyl-4-bromobenzenes with *p*-bromotoluene in the presence of BF_3-H_3PO_4 [222], and cyclopentylbenzene, cyclopentyltoluenes, and cyclopentylethylbenzenes, with benzene, toluene, and ethylbenzene, respectively, in the presence of BF_3-H_3PO_4 [445]. Also, in the presence of BF_3-PPA, benzene gave an 88.2% yield of cyclopentylbenzene [460a].

As in cyclopentylation, most of the cyclohexylations of arenes with either cyclohexene or cyclohexanol in the presence of acid catalysts proceeded without rearrangement to yield the corresponding cyclohexyl derivatives. This is illustrated by the reported cyclohexylations of benzene [453,461,462], toluene [452,462], xylenes [324,457,463,464], *p*-cymene [464], pseudocumene [465], mesitylene [451,452], bromotoluenes [222,466], phenol [467-472], cresols [345,467,468,473,474], chlorophenols [475], anisole [476-478]], *o*-nitroanisole [476], biphenyl [439,479], naphthalene [130,324,480], and acenaphthene [481]. Examples of these cyclohexylations are depicted in Table 16.

In contrast to the numerous cyclohexylations without rearrangement compiled here (Table 16) and elsewhere [6,12], a few exceptions existed in which the cyclohexylations were reported to occur with minor skeletal rearrangements. For example, the cycloalkylation of benzene with cyclohexanol in the presence of H_2SO_4 was accompanied by 1 to 2% isomerization to 1-methyl-1-phenylcyclopentane [454]. In the presence of $AlCl_3$, the same cycloalkylation proceeded without isomerization when a five- to six-fold excess of benzene was used, but with less benzene the reaction proceeded not only with rearrangement to both 1-methyl-1-phenylpentane and benzylcyclopentane but also with fragmentation to phenylcyclopentane. Rearrangement to 1-methyl-1-arylpentanes also accompanied the cyclohexylations of benzene and toluene with cyclohexane in the presence of $AlCl_3$ as catalyst and isopentane as promotor [425] as well as in the reaction of hydroquinone and its methyl ethers with cyclohexanol in the presence of H_3PO_4 at 145°C [458].

A few reports have appeared about cycloalkylations with C_7 [17,455,482] and higher cycloalkenes and cycloalkanols. In 1940, Sidorova and Tsukervanik [482] reported that cycloheptanol on reaction with benzene and $AlCl_3$ yielded nonrearranged cycloheptylbenzene. In 1945, Pines et al. [455] reexamined this reaction and extended the study to include cycloheptene as alkylating agent and $AlCl_3$-HCl, H_2SO_4, and HF as catalysts. The results illustrated the dependence of products on reaction variables with strenuous conditions favoring isomerization to methylphenylcyclohexanes. Thus, in contrast to the earlier report, the product form the $AlCl_3$-catalyzed reaction contained only 10% of cycloheptylbenzene, the remainder being "probably isomeric methylphenylcyclohexanes of which the 1,4-isomer amounted to 20%." The reaction with 80% H_2SO_4 at 80°C was also accompanied by partial isomerization. The monoalkylated benzene fraction consisted of 40% cycloheptylbenzene and 60% methylphenylcyclohexanes (Eq. 90).

$$\text{cycloheptanol (OH)} + \text{PhH} \xrightarrow[\text{or } 80\%\ H_2SO_4,\ 80^\circ]{AlCl_3,\ \text{reflux}} \underset{\text{minor}}{\text{cycloheptyl-Ph}} + \underset{\text{major}}{\text{Ph-cyclohexane-}CH_3} \tag{90}$$

The $AlCl_3$-HCl catalyst was most isomerizing, giving only rearranged methylphenylcyclohexanes from benzene and cyclopentene even at 5 to 8°C.

Table 16. Alkylations of Arenes with Cyclohexene and Cyclohexanol

Aromatic substrate, mol[a]	Alkylating agent, mol[a]	Catalyst, mol[a]	Temp, °C	Time, hr	Product, % yield	Ref.
Benzene	Cyclohexene	BF_3-H_3PO_4	30	—	Cyclohexylbenzene (major), dicyclohexylbenzenes (minor)	461
Benzene (3)	Cyclohexanol (1)	96% H_2SO_4	5	3	Cyclohexylbenzene (90%), *p*-dicyclohexylbenzene	453, 101
Benzene (4)	Cyclohexanol (-)	$AlCl_3$ (1%)	80	2	Cyclohexylbenzene (94.5%), minor amounts of 1,3-, 1,4-di- and 1,3,5-tri-cyclohexylbenzenes	453
Benzene (200 ml)	Cyclohexanol (40g)	$AlCl_3$ (80g) (0.5 mol)	Water-bath		Cyclohexylbenzene (62%), mixture of di- and tri-cyclohexylbenzenes (35%),	462
Toluene	Cyclohexene or cyclohexanol	H_2SO_4	-	-	*o*- and *p*-Cyclohexyltoluenes (80%), dicyclohexylbenzene	452
Toluene (-)	Cyclohexanol (0.2)	$AlCl_3$ (0.13)	-	-	Cyclohexyltoluenes (56.4%), 2,4- and 3,5-dicyclohexyltoluenes (14%)	452
Toluene (4)	Cyclohexanol (1)	$AlCl_3$	80	2	*m*- and *p*-Cyclohexyltoluenes, 3,5-dicyclohexyltoluene	462
o-Xylene	Cyclohexene	?	25-50	-	1,2-Dimethyl-3-cyclohexylbenzene (30.7%), 1,2-dimethyl-4-cyclohexylbenzene (69.3%)	457

m-Xylene	Cyclohexene	?	25-50	-	1,3-Dimethyl-4-cyclohexylbenzene (86.1%), other isomers (13.9%)	457
m-Xylene	Cyclohexanol	95% H_2SO_4	40	1	1,3-Dimethyl-4-cyclohexylbenzene	324
p-Cymene	Cyclohexanol	H_2SO_4	0-5	3	1-Methyl-2-cyclohexyl-4-isopropylbenzene, dicyclohexyl-*p*-cymenes	464
Pseudocumene	Cyclohexene	H_2SO_4	-	-	Monocyclohexylpseudocumene (65.6%)	465
Mesitylene (-)	Cyclohexene (-)	$AlCl_3$ (-)	-	-	1,2,4-Trimethyl-5-cyclohexylbenzene	451
Mesitylene	Cyclohexene	H_2SO_4	0	-	1,3,5-Trimethyl-2-cyclohexylbenzene, 1,3,5-trimethyl-2,4-dicyclohexylbenzene	451
Mesitylene (-)	Cyclohexanol (0.2)	$AlCl_3$ (0.13)	-	-	Mixed cyclohexylmesitylene and 5-cyclohexylpseudocumene (28.4%), trimethyldicyclohexylbenzenes (10%)	452
Mesitylene	Cyclohexanol	H_2SO_4	-	-	Mixed monocyclohexyl derivatives (4.2%), dicyclohexyl derivatives (13.5%)	452
p-Bromotoluene (85.6g)	Cyclohexene (8.2g)	H_3PO_4 (6.8g)-BF_3(5.4g)	25	4	A mixture of 2- and 3-cyclohexylbromotoluene in a ratio of 42:58 (23%)	222
Phenol (2)	Cyclohexene (1)	PPA (200g)	120-140	5	*o*- and *p*-Cyclohexylphenol (80-85% yield)	467

Table 16. (continued)

Aromatic, substrate, mol[a]	Alkylating agent, mol[a]	Catalyst, mol[a]	Temp, °C	Time, hr	Product, % yield	Ref.
o-Cresol	Cyclohexene	KU-2 cation exchange (30% by wt of cresol)	120-150	1-6	2-Methyl-4-cyclohexylphenol and 2-methyl-6-cyclohexylphenol in a ratio of 52:47% (60-96%)	345
o-Cresol (3-4 mol)	Cyclohexene (1 mol)	PPA (37 wt%)	-	-	Monocyclohexyl-*o*-cresols (75%)	473
o-Chlorophenol	Cyclohexene	$ZnCl_2$	-	-	2-Chloro-4-cyclohexylphenol	475
p-Chlorophenol	Cyclohexene	$ZnCl_2$	-	-	2-Cyclohexyl-4-chlorophenol	475
Anisole (4-6)	Cyclohexene (1)	PPA (33 wt%)	100	1.5	A mixture of *o*- and *p*-cyclohexylanisole in a ratio of 51:49 (88.6%)	476
Anisole (0.50 mol)	Cyclohexene (0.10)	$AlCl_3$ (0.64)	rt	3	*o*-Cyclohexylanisole (66%), *p*-Cyclohexylanisole (34%)	477

o-Nitroanisole	Cyclohexanol	HF	10-20	18	2-Nitro-4-cyclohexylanisole	476
Biphenyl (-)	Cyclohexene (-)	H_2SO_4 (-)	50	-	*p,p'*-Dicyclohexylbiphenyl	439
Biphenyl (-)	Cyclohexene (-)	$AlCl_3$ (-)	75	3	Isomeric cyclohexylbiphenyls, higher alkylates	479
Naphthalene (-)	Cyclohexene (-)	H_2SO_4 (-)	40	1	1-Cyclohexylnaphthalene	324
Naphthalene (-)	Cyclohexene (-)	BF_3	25	24	2-Cyclohexylnaphthalene	130
Naphthalene (-)	Cyclohexanol (-)	96% H_2SO_4	5-10	2.5-5	1- and 2-Cyclohexylnaphthalene (75-88% yield)	480
Acenaphthene (-)	Cyclohexanol (-)	BF_3	30	1	Mixture of 3- and 5-cyclohexylacenaphthene	481
Acenaphthene (-)	Cyclohexanol (-)	$AlCl_3$ in heptane	30-70	12	5-Cyclohexylacenaphthene	481

[a]Unless otherwise specified

On the other hand, both HF and 96% H_2SO_4 catalysts gave only phenylcycloheptane when the reaction temperature was lowered to 0 to 5°C (Eq. 91) [455].

Cycloheptene + PhH — ($AlCl_3$/HCl, 5–8°) → Ph-substituted methylcyclohexane (CH_3); (HF or 96% H_2SO_4) → phenylcycloheptane (91)

The production of methylcyclohexylbenzenes in the reactions above can be accounted for in terms of prior isomerization of the cycloheptyl cation to a methylcyclohexyl cation through protonated cyclopropane intermediates as suggested for similar transformations [483; see also Chap. 2].

In related studies, phenol was alkylated with cyclooctene in the presence of BF_3-HOAc catalyst at 60°C to give primarily 2-cyclooctylphenol mixed with some of the 4-isomer [484]. Benzene was similarly cycloalkylated with cyclododecene and cyclopentadecene in the presence of ethylaluminum chloride ($EtAlCl_2$) catalyst [485]. The products consisted of mixtures of mono-, di-, and higher cycloalkylated benzenes in a ratio of 76:19:1 from the former and 66:25:7 from the latter. The dicycloalkylates from both reactions were shown to be mixtures of two isomers. Cyclododecene was also used to alkylate benzene and some of its derivatives in the presence of H_2SO_4 catalyst [486]. In reactions conducted at 10 to 20°C for 2 hr, yields ranging from 54.85 to 89.40% of the following monocycloalkylates were obtained: cyclododecylbenzene from benzene, *m*- and *p*-cyclododecyltoluene from toluene, 4-cyclododecyl-1,2-dimethylbenzene from *o*-xylene, 5-cyclododecyl-1,3-dimethylbenzene from *m*-xylene, 2-cyclododecyl-1,4-dimethylbenzene from *p*-xylene, *p*-cyclododecylethylbenzene from ethylbenzene, and *p*-cyclododecylisopropylbenzene from isopropylbenzene.

B. Cycloalkylations with Branched Cycloalkenes and Cycloalkanols

As in alkylations with branched acyclic alcohols and alkenes, the alkylation of aromatics with branched cyclic derivatives leads initially to tertiary cycloalkylates. These result from attack by tertiary carbocations generated either directly or via rearrangement from other carbocations prior to the alkylation step. With weak (nonisomerizing) catalysts, the initial tertiary cycloalkylates survive to be the end products, but with strong catalysts the products are usually mixtures of isomeric components resulting from isomerization and/or reorientation. In view of this, the early results of cycloalkylations with branched derivatives must be assessed with caution.

The following examples of cycloalkylations with branched derivatives illustrate the dependence of product composition on both catalyst and conditions:

1. Examples of Cycloalkylations with Branched Cyclopentenes and Cyclopentanols

PhH + [3-butylcyclopentene] —$AlCl_3$, 40°→ [1-phenyl-3-butylcyclopentane] + [1-phenyl-2-butylcyclopentane] [448]

ArH + [1-methylcyclopentene] and [3-methylcyclopentene] —H_2SO_4, 10-20°→ [1-aryl-1-methylcyclopentane] [487]

PhH + [1-methylcyclopentanol] or [3-methylcyclopentanol] —H_2SO_4, 5-25°, 2 hr→ [1-phenyl-1-methylcyclopentane] + [1,4-bis(1-methylcyclopentyl)benzene] [487a]

PhOH + [1-methylcycloalkene, $(CH_2)_n$], n = 1,2

—KU-2, 75-12°, 4 hr→ HO–C$_6$H$_4$–[1-methylcycloalkyl, $(CH_2)_n$] 65-85% [488(a)]

—Al(OPh)$_3$, 240-290°, 4-6 hr→ 2-(cycloalkyl, $(CH_2)_n$)phenol (OH) 46-74% [488(b)]

[C$_6$H$_5$OR] + [1-(R^1-phenyl)cyclopentene] —$H_3PO_4 \cdot BF_3 \cdot OEt_2$ or $H_3PO_4 \cdot BF_3$→ [1-(R^1-phenyl)-1-(4-OR-phenyl)cyclopentane] [489]

R = H,Me,Et

R^1 = H, *p*-Me, *p*-MeO, *m*-MeO

2. Examples of Cycloalkylations with Branched Cyclohexenes and Cyclohexanols

PhH —, H_2SO_4, 10-13°, 1.5 hr [490]; HF, 5° [490a] → Ph + C_6H_4 (2)

PhH + OH — H_2SO_4 → Ph [491]; $AlCl_3$ → Ph + Ph [492]

PhH + — HF → Ph [493]

+ — H_2SO_4 → + Hydrogen transfer products [494]

+ — HF → + [495]

+ + Other products

OH + CH_3 (1-,3- or 4-methyl isomer) $\xrightarrow{H_2SO_4}$ [496]

OH + $\xrightarrow[120\text{-}150^\circ]{KU\text{-}2}$ 96.8% + 3.2% [345]

OH + R^1, OH, R $\xrightarrow[\text{or KU-2, 100-150}^\circ]{H_3PO_4, 80\text{-}100^\circ}$ OH, R^2 [221]

R = H, R^1 = Me or R^1 = H, R = Me R^2 = Me

R = H, R^1 = Et or R^1 = H, R = Et R^2 = Et

OH, R $\xrightarrow[AlCl_3]{PhH, H_2SO_4 \text{ or}}$ Ph, R + Ph, R + R, R + R-R + Ph-R [497]

R = 1-,2-,3- or 4-cyclohexyl   major products   minor products

3. Examples of Cycloalkylation with Terpenes and Related Derivatives

An early case of cycloalkylation with terpene derivatives is the reported alkylation of benzene with dihydrolimonene (72), menthol (73), or dihydroterpineol (74) in the presence of HF to give similar products consisting of 1-methyl-1-phenyl-4-isopropylcyclohexane (75) mixed with small amounts of *p*-di(1-methyl-4-isopropylcyclohexyl)benzene (76) (scheme 19) [493].

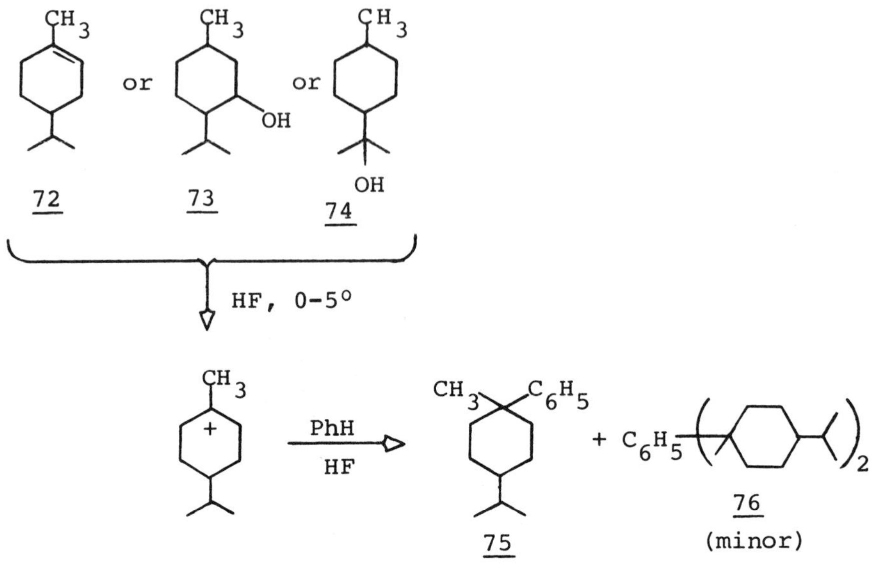

72 or 73 or 74

HF, 0-5°

PhH / HF → 75 + 76 (minor)

Scheme 19

In these alkylations, hydrogen shifts occurred so that the reaction always took place at the tertiary carbon atom holding the methyl group. This prefered alkylation at the methyl site is believed to be of steric origin. When the difference in size between the two groups was minimized, alkylation took place at both of the tertiary ring centers. Thus the alkylation of benzene with 1-methyl-4-ethylcyclohexanol (77), 1-ethyl-4-methylcyclohexanol (78), or 1-ethyl-4-methylcyclohexene (79) gave mixtures of the two possible tertiary alkylates 80 and 81 (Eq. 92) [493].

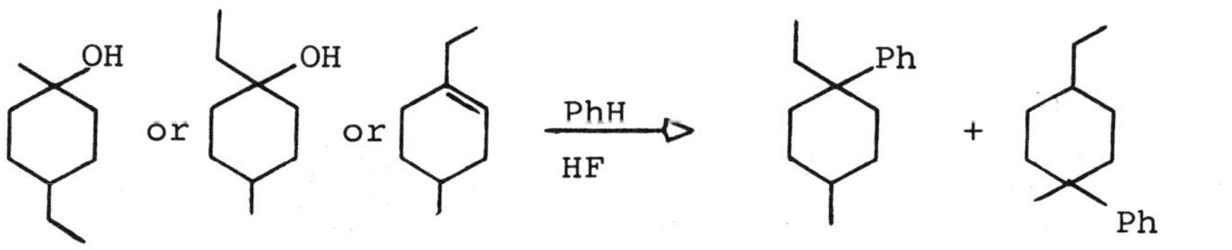

77 or 78 or 79 —PhH / HF→ 80 40% + 81 60% (92)

In contrast, mixtures of optically active secondary and tertiary alkylates were reported to result from the cycloalkylations of benzene, *m*-xylene, and *p*-xylene with menthol (73) in the presence of H_2SO_4 or $AlCl_3$ under a variety of conditions [498].

In recent years, there has been growing interest in the synthesis of terpenophenols through cycloalkylation of phenols with terpenes [499-515]. This interest arose because of the usefulness of these compounds as aromatic principles and antioxidants [513]. As anticipated, these alkylations were complicated by accompanying Wagner-Meerwein shifts and led to complex mixtures of nonrearranged and rearranged terpenophenol isomers. To illustrate, we shall consider the alkylation of phenol with both α-pinene [514] and α-fen-

OH
OH
HO
83
84
OH
85
HO
86
attack at C1
attack at C1
attack at C1
attack at C3
82

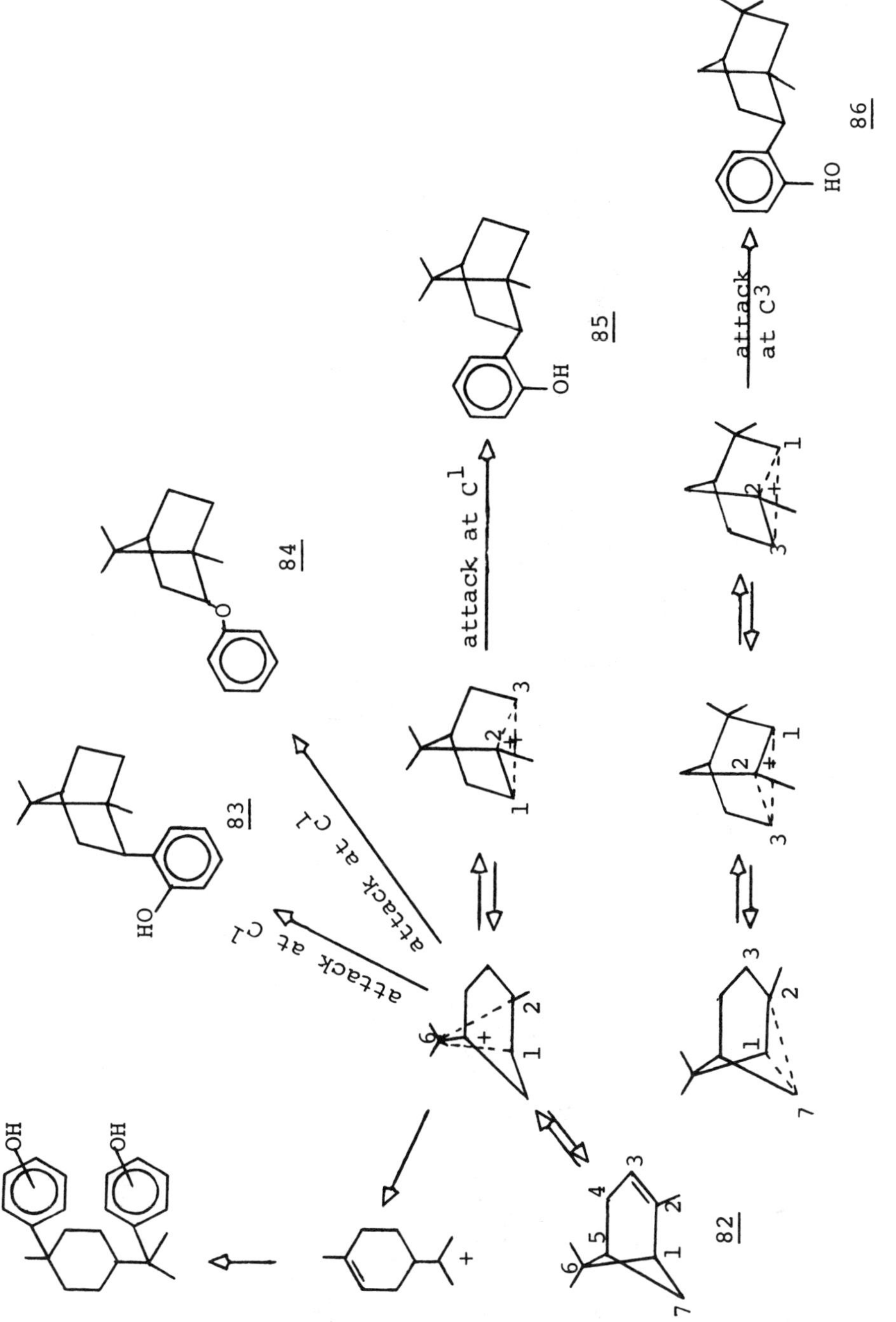

Scheme 20

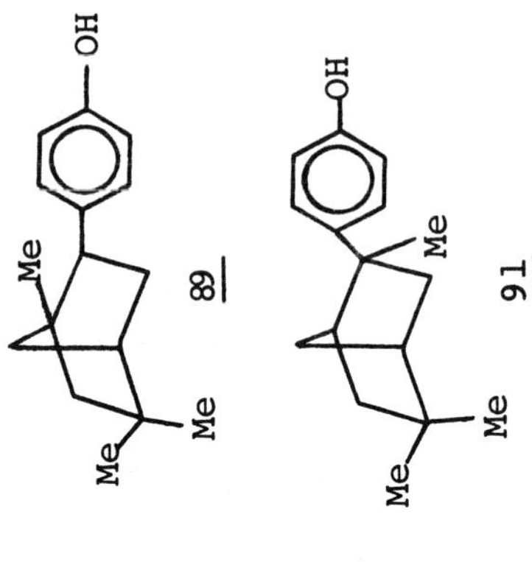

(93)

chene [515] in the presence of aluminum phenolate $Al(OPh)_3$ catalyst at various temperatures.

With α-pinene (82, scheme 20), the monoalkylation product consisted of 23% of *o*-bornylphenol (83), 62% of bornyl phenyl ether (84), 7% of isobornylphenol (85), 4% of *o*-isofenchylphenol (86), and 4% of an *o*-terpenophenol of unknown structure. At 100°C, the product consisted of 13% of *o*-bornylphenol (83), 22% of the ether (84), 12% of *o*-isobornylphenol (85), 3% of *o*-isofenchylphenol (86), 1% of *p*-terpenophenols, 40% of unknown phenols, and 7% of two unknown phenyl terpenyl ethers. These products and the carbocation transformations suggested for their formation are described in scheme 20.

The reaction with α-fenchene (87, Eq. 93) gave mixtures consisting of compounds 88-91 in ratios again depending on the reaction temperature used. Thus, whereas 2-*exo*-(*o*-hydroxyphenyl)-2,5,5-trimethylbicyclo[2.2.1]heptane (90) constituted the main part of the product at 50 to 130°C, the *para* isomer (89) accounted for 64% of the product at 180°C.

The formation of compounds 88-91 from the reaction of phenol with α-fenchene was accounted for in terms of carbocation transformations similar to those of scheme 20. Moreover, the observed preference for *ortho* and O-alkylations in these $Al(OPh)_3$-catalyzed reactions was attributed to steric proximity of the cationic centers of the bicyclic system to the carbon atoms in the *ortho* position to the oxygen and to the oxygen atoms themselves in the ion pair formed from the terpene and the acid $HAl(OPh)_4$ [514].

Other related alkylations with multicyclic systems include the reactions of halobenzenes with cyanonorbornene (92) and $AlCl_3$ to give *o*-, *m*- and *p*-(cyanonorbornyl)halobenzenes (93, Eq. 94) [516]; of benzene, toluene, and chlorobenzene with 1-azidoadamantane (94) and $AlCl_3$ to give the corresponding 3-aryl-4-azahomoadamantanes (95, Eq. 95) [517]; and of phenol and cresols with 1,2-dihydronaphthalene and H_2SO_4 or P_2O_5 to give tetrahydronaphthyl derivatives of phenol or cresols (96, Eq. 96) [518]. The reaction of

PhX + 92 —($AlCl_3$, 50-60°)→ 93 (94)

(X = Cl, Br, I)

ArH + 94 —($AlCl_3$, $-N_2$)→ 95 (95)

Ar = C_6H_5
= $C_6H_4CH_3$ (*o*-:*p*- 1:1)
= C_6H_4Cl (*o*-:*p*- ca. 1:1.86)

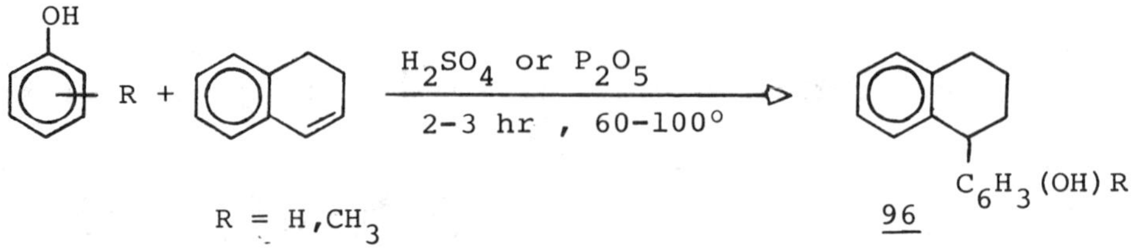

$$\text{HOC}_6\text{H}_4\text{R} + \text{1,2-dihydronaphthalene} \xrightarrow[\text{2-3 hr, 60-100°}]{H_2SO_4 \text{ or } P_2O_5} \text{1-}(C_6H_3(OH)R)\text{-tetralin } (\underline{96}) \quad (96)$$

R = H, CH_3

phenol with ethylidenenorbornene (97) and dicyclopentadiene (98) in the presence of KU-2 ion exchange showed that 97 reacted under milder conditions (80 to 85°C, 4 hr) than 98 (110 to 115°C, 5 hr). The products of the reaction were those of C- and O-cycloalkenylation together with chroman.

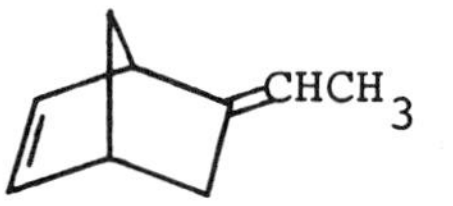

=CHCH$_3$

97

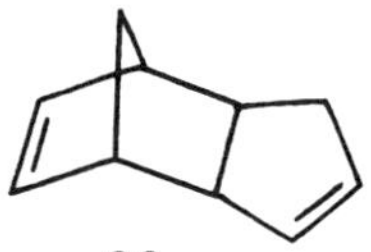

98

As to reactivity and selectivity during the alkylation of phenols with cyclic derivatives, the results of various studies demonstrated that the hydroxyl group not only causes lesser hindrance to *ortho* substitution but also has a much higher orienting effect than does the methyl or methyoxyl group [468, 519]. An illustrative example is the alkenylation of resorcinol and *p*-hydroquinone methyl ethers with 1,3-cyclopentadiene and 1,3-cyclohexadiene in the presence of phosphoric acid, where the predominating C-alkylenation isomers were those in which the cycloalkenyl groups were at the *ortho* position to the hydroxyl (Eq. 97) [519(a)]. No O-alkylenation occurred in the case of *p*-hydroquinone methyl ether.

(97)

Cyclic diene $(CH_2)_n$, n = 1,2

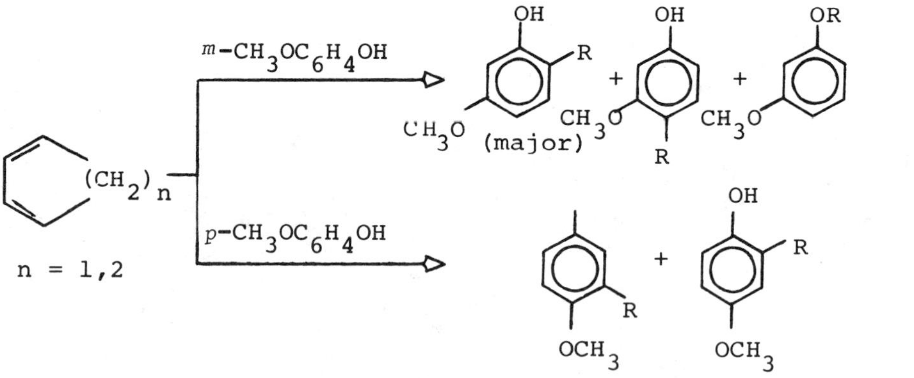

m-$CH_3OC_6H_4OH$ → OH, R, CH_3O (major) + OH, CH_3O, R + OR, CH_3O

p-$CH_3OC_6H_4OH$ → R, OCH_3 + OH, R, OCH_3

C. Mechanistic Aspects of Cycloalkylations

From previous discussions we noted that, depending on type of catalyst and structure of the alkylating group, some cycloalkylations are accompanied by varying degrees of rearrangements. Some of these rearrangements are observable, as they lead to isomeric products whose detection is possible by ordinary identification techniques [425,433,454-458]. Other rearrangements are still probable, but they are degenerate in nature and their detection requires the appliction of tracer techniques. To explore such rearrangements and also to clarify other mechanistic aspects of cycloalkylation reactions, Lipovich [407,520], Sidorova [492], and Roberts [521] and their co-workers conducted a number of reactions employing both radioactive and optically active cycloalkylating groups. Their aim was to determine whether the isomerizations observed in some cycloalkylations [455,456,522] occur before or after the cycloalkylation step. The goal was to find out whether the alkylations take place by an S_N2-type mechanism, involving direct attack of the aromatic on the hydroxyl carbon in an alcohol-catalyst complex, or by an S_N1-type reaction, involving an intermediate carbocation which undergoes rearrangements through the usual 1,2 hydride and/or alkyl shifts.

Since the probable occurrence of 1,2 hydride shifts during alkylations with ordinary unbranched cycloalkanols will take place without visible change in the hydrocarbon skeleton, Lipovich et al. [520(b),520(c)] studied the cycloalkylation of benzene with labeled cyclohexanol-1-^{14}C. With the strong $AlCl_3$ catalyst at 100°C, the resulting cyclohexylbenzene showed a statistical distribution of the tag in the cyclohexane ring [520(b)]. It was pointed out, however, that such a distribution could be the result of one or more of secondary processes, such as an alternative contraction-expansion of the cyclohexane ring, a migration of phenyl and the formation of diphenylcyclohexane as an intermediate product, and subsequent cleavage of one of the phenyl groups. In fact, it was established that intermolecular migration of the hexyl group took place when the $AlCl_3$ treatment of phenylcyclohexane-1-^{14}C in benzene resulted in a statistical distribution of the label in the cyclohexane ring [520(a)].

To eliminate some of the foregoing possible causes for label scrambling it was expedient to use a catalyst which would not cause the alkylation product to undergo substantial isomerization because of secondary carbocation processes. the catalyst of choice was 96% H_2SO_4 since it caused no label transfer from the methyl to the ring in methylcyclohexane-7-^{14}C nor inter- or intramolecular migration of the cyclohexyl group in the presence of ^{14}C-labeled benzene [520(c)]. When benzene was alkylated with cyclohexanol-1-^{14}C in the presence of 96% H_2SO_4, statistical distribution of the label in the cyclohexane ring occurred (Eq. 98).

$$C_6H_6 + HO\text{-}C_6H_{11}^{*} \xrightarrow[\text{or } 96\%\ H_2SO_4]{AlCl_3} C_6H_5\text{-}C_6H_{11}^{*} \qquad (98)$$

On the basis of these results, it was concluded that the observed uniform distribution of the label in the cyclohexane ring was caused by consecutive 1,2-hydride shifts that occurred at a carbocation stage prior to the alkylation step [520(c)].

The mechanism of cycloalkylation of arenes with branched cycloalkanols was also studied by Lipovich et al. [491]. It was known that the $AlCl_3$-catalyzed alkylation of benzene with isomeric methylcyclohexanols gave methylphenylcyclohexanes with composition depending on the conditions employed [490(a),522(b)]. Although it was evident from these results that isomerizations the tertiary 1-methyl-1-phenylcyclohexane was formed, regardless of which one of the isomeric methylcyclohexanols or methylcyclohexenes was used [490(a),522(b)]. Although it was evident from these results that isomerizations to give the relatively more stable *t*-methylcyclohexyl carbocation must have

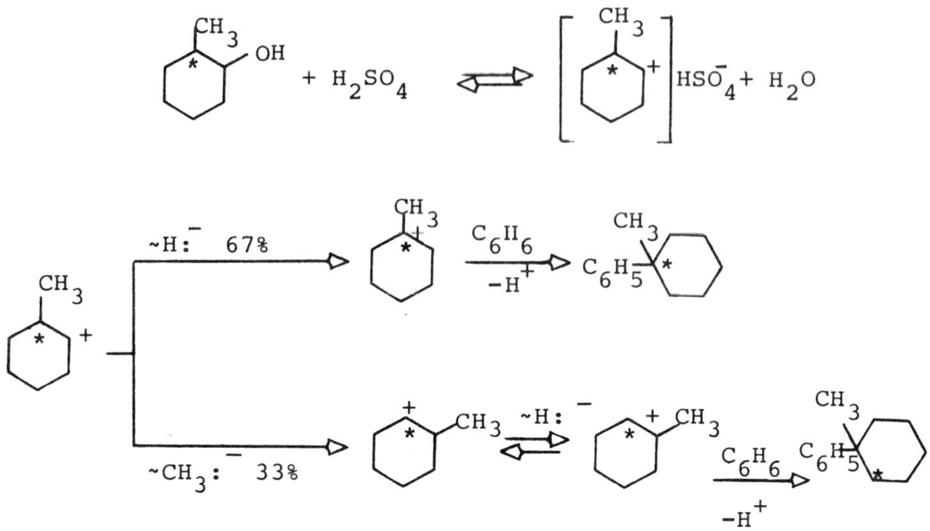

Scheme 21

taken place, it was not known whether hydride shifts, methyl shifts, or possibly both types of shifts were involved.

The answer to the latter question was found by Lipovich et al. [491] in the results of alkylation of benzene with 2-methylcyclohexanol-2-^{14}C and 1-methylcyclohexene-1-^{14}C in the presence of 96% H_2SO_4 at 5°C. The 1-methyl-1-phenylcyclohexane formed in these reactions contained 67 and 68%, respectively, of the original amount of radioactivity on the first carbons, suggesting that both hydride and methyl shifts occurred prior to product formation. The mechanism of cycloalkylation with the alcohol was explained in terms of scheme 21.

The fact that exactly the same distribution of radioactivity was observed when benzene was alkylated with 1-methylcyclohexene-1-^{14}C was rather unexpected, as it contradicts the belief that the more stable tertiary ion should be the one to form from 1-methylcyclohexene, resulting only in alkylation to give 1 methyl 1 phenylcyclohexane 1 ^{14}C. This anomalous behavior was assumed to be the result of a prior equilibrium between the initially formed tertiary cation with a secondary one (Eq. 99).

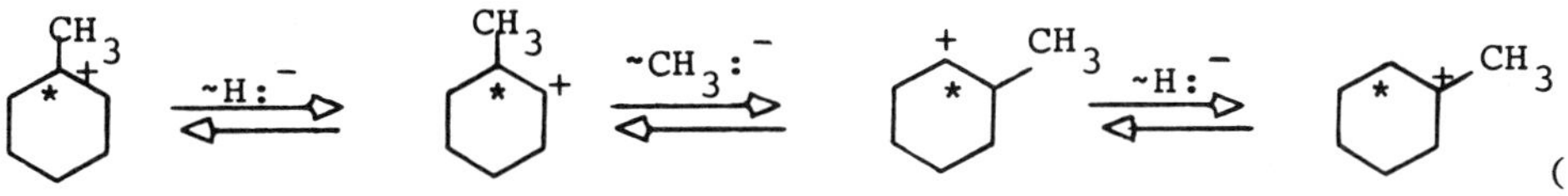

(99)

The alkylation of benzene with (+)-*trans*-methylcyclohexanol was investigated by Sidorova and Bokova [492] in the presence of $AlCl_3$ at 20 to 40°C. The product obtained was a mixture of the secondary alkylates 1-methyl-2-phenylcyclohexane and 1-methyl-3-phenylcyclohexane in which considerable optical activity was retained. The absence of the tertiary cycloalkylate, 1-methyl-1-phenylcyclohexane, together with the retention of considerable optical activity in the product, led to the statement that "alkylation with *sec*-methylcyclohexanols in the presence of $AlCl_3$ does not proceed via a free carbocation" [492]. These results, which are either exceptional or incorrect, are in accord neither with the reported production of racemic *sec*-butylbenzene from a similar $AlCl_3$-catalyzed alkylation of benzene with *d-sec*-butyl alcohol [132,492] nor with the noted generation of carbocations (or their precursors) under the influence of $AlCl_3$ [366,491,520(b),520(c)]. It is believed that the observed production of secondary alkylates in the alkylation above can best be explained in terms of subsequent rearrangement of the initially formed 1-methyl-1-phenylcyclohexane under the influence of $AlCl_3$ [366] (Eq. 100).

CH3 Ph ⇌ (R+ / RH) H3C [arenium σ-complex] ⇌ CH3, + Ph ⇌ (RH / R+) CH3 Ph → etc. (100)

VI. ALKYLATIONS OF ARENES WITH HETEROCYCLIC DERIVATIVES

A. Cyclic Ethers

1. Epoxides

It is generally believed that Friedel-Crafts alkylations of aromatics with epoxides produce β-hydroxyarylalkanes which, depending on catalyst type and reaction temperature, can react further with aromatics to produce 1,2-diarylalkanes (Eq. 101) [523].

$$\text{R-C(-O-)C-} \xrightarrow{\text{ArH}} \text{R-C(Ar)-C-OH} \xrightarrow{\text{ArH}} \text{R-C(Ar)-C-Ar} \quad (101)$$

Ethylene oxide, the simplest epoxyalkane, was reported to alkylate benzene to give a good yield of 2-phenylethanol (99) with $AlCl_3$ catalyst at low reaction temperatures [524-526(a)]. At a high operating temperature [525] or with BF_3 as the catalyst [527,528], partial conversion into dibenzyl was also observed. A similar behavior was also observed for the $AlCl_3$-catalyzed alkylation of other arenes, including toluene [525(b),529] *p*-xylene [526(b)], biphenyl [526(c)], naphthalene, and tetrahydronaphthalene [526(d)].

$$C_6H_6 + CH_2\text{—}CH_2 \text{ (epoxide, O)} \begin{cases} \xrightarrow[\text{Low temp}]{AlCl_3} PhCH_2CH_2OH \ (\underline{99}) \\ \xrightarrow[\text{or } AlCl_3\text{, high temp}]{BF_3} \underline{99} + PhCH_2CH_2Ph \end{cases}$$

Higher 1,2-epoxyalkanes have been repeatedly shown by recent workers [530-536] to alkylate arenes in the presence of Friedel-Crafts catalysts to give 2-aryl-1-alkanols by fission of the secondary C-O bond in the epoxide. For example, it was established by Milstein in 1968 [533] that the reaction of 1,2-epoxypropane with benzene and aluminum chloride under strictly anhydrous conditions gave rise to 2-phenyl-1-propanol (64% yield); the isomeric 1-phenyl-2-propanol was obtained only in traces (Rq. 102). Earlier, however,

$$PhH + CH_3\text{-}\overset{O}{CH\text{-}CH_2} \xrightarrow[20°,\ 1\ hr]{AlCl_3} CH_3\underset{Ph}{CH}\text{-}CH_2OH \qquad (102)$$

the same reaction was claimed by different workers to produce a mixture of 1,2-diphenylpropane (46% yield) and 1-phenyl-2-propanol (8% yield) [537]; only 2-phenyl-1-propanol [538]; and only 1-phenyl-2-propanol [539].

A similar reaction of 1,2-epoxybutane with benzene and $AlCl_3$ was found by Nakamoto et al. [535] to give a product mixture consisting of the phenylalkanols 2-phenyl-1-butanol (100) and 3-phenyl-1-butanol (101) together with the chlorobutanols 102-104 (Eq. 103).

$$CH_3CH_2\underset{O}{CH-CH_2} \xrightarrow[AlCl_3]{C_6H_6} \underset{\mathbf{100}}{CH_3CH_2\underset{Ph}{C}HCH_2OH} + \underset{\mathbf{101}}{CH_3\underset{Ph}{C}HCH_2CH_2OH} + \underset{\mathbf{102}}{CH_3CH_2\underset{Cl}{C}HCH_2OH} + \underset{\mathbf{103}}{CH_3\underset{Cl}{C}HCH_2CH_2OH} + \underset{\mathbf{104}}{CH_3CH_2\underset{OH}{C}HCH_2Cl} \qquad (103)$$

The results of various runs carried out in carbon disulfide at -20 to +30°C demonstrated that, in this mixture, the total yield of the phenylbutanols 100 and 101 and the relative amount of 101 to 100 increased with an increase in temperature, while the yield of the chlorobutanols 102-104 decreased with a rise of temperature, being only 2% at 30°C [540]. The presence of 3-phenyl-1-butanol and 3-chloro-1-butanol as products of isomerization was not unexpected, knowing that carbocation rearrangments do accompany many Friedel-Crafts alkylations. A similar isomerization was also reported by Hata et al. [536] in the reaction of 1,2-epoxyoctane with benzene in the presence of a Lewis acid catalyst. However, although both rearranged and nonrearranged products were obtained in alkylations with 1,2-epoxyalkanes, their modes of formation have been a matter of dispute among recent workers.

Hata et al. [536] suggested that the formation of isomeric monophenylated 1-octanols during the alkylation of benzene with 1,2-epoxyoctane and aluminum chloride proceeded through initial carbocation formation followed by a series of 1,2-hydride and phenyl shifts. In line with this, Milstein [533] rationalized the formation of only 2-phenyl-1-propanol in the reaction of 1,2-epoxypropane with benzene and aluminum chloride in terms of an opening of the aluminum chloride-coordinated propylene oxide complex to give the most stable ion prior to attack on benzene as formulated in Eq. (104).

$$CH_3\text{-}\underset{\underset{\ddot{O}\cdot AlCl_3}{\diagdown\ \diagup}}{CH\text{-}CH_2} \xrightarrow{-Cl^-} CH_3\overset{+}{C}HCH_2OAlCl_2 \xrightarrow[(2)\ H_2O]{(1)\ C_6H_6,\ AlCl_3} CH_3\underset{\underset{C_6H_5}{|}}{C}HCH_2OH \qquad (104)$$

Inconsistently, however, a study by Nakajima and co-workers [532,534] of the reaction of (+)-propylene oxide with benzene and aluminum chloride revealed that it proceeded with at least 95% inversion of configuration at the secondary carbon atom in the epoxide. Accordingly, it was stated by these authors that a mechanism involving a free carbocation should be ruled out in the reactions of epoxides in the presence of Lewis acids.

To gain more support for the preceding statement, Suga et al. [540] investigated the stereochemistry of the reaction of R-(+)-1,2-epoxybutane (105) with benzene in the presence of $AlCl_3$ or $SnCl_4$ catalyst in CS_2 solvent at -10 to +30°C (scheme 22). In this case, the resulting mixture consisted of optically active 2-phenyl-1-butanol (100), 3-phenyl-1-butanol (101), 2-chloro-1-butanol (102), 3-chloro-1-butanol (103), and 1-chloro-2-butanol (104) (cf. Eq. 103). On the one hand, the formation of R-(-)-(100) and S-(-)-(102) in almost 100% and 55 to 56% optical yields, respectively, suggested that in analogy with reaction of (+)-propylene oxide, the fission of the secondary C-O bond in the R-(+)-1,2-epoxybutane proceeded with an inversion of configuration. The observed lower yield of S-(-)-(102) was interpreted as involving an exchange of a chlorine atom. On the other hand, the production of optically active 100 and 103 invoked the involvement of an asymmetric induction at the carbon adjacent to the terminal methyl group of the epoxide. Accordingly, pathways (a) and (b) of scheme 22 were suggested to compete during the conversion of R-(+)-1,2-epoxybutane (105) to optically active 102. In these pathways asymmetric induction was rationalized by the involvement of the aluminum chloride salts of 102 and 103, in which new chiral centers are formed at the carbon adjacent to the terminal methyl group in the epoxide. In the overall reaction, the final optical yield of 101 should be determined by the difference in the reaction rates of pathways (a) and (b).

To rationalize the formation of S-(+)-(103) from R-(+)- 1,2-epoxybutane (105) in terms of asymmetric induction, the authors stated that "the epoxide reacts initially with $AlCl_3$ to give a complex in which the secondary C-O bond is released and either of the hydrogens on C_3 migrates to C_2 from the back

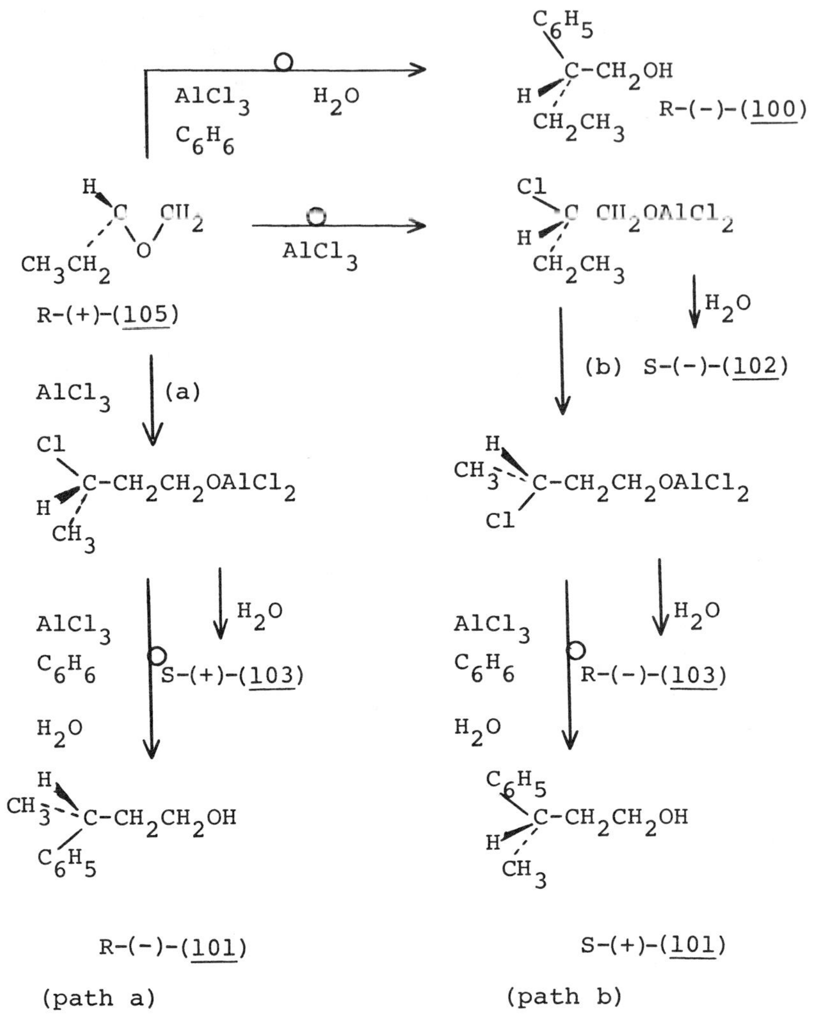

Scheme 22

of the C_2-O bond to form a hydrogen bridge as in 106. Thus the rotation about the C_2-C_3 bond would be restricted, and the chlorine atom of the complexing aluminum chloride would then attack C_3 intramolecularly from the back of the migrating hydrogen."

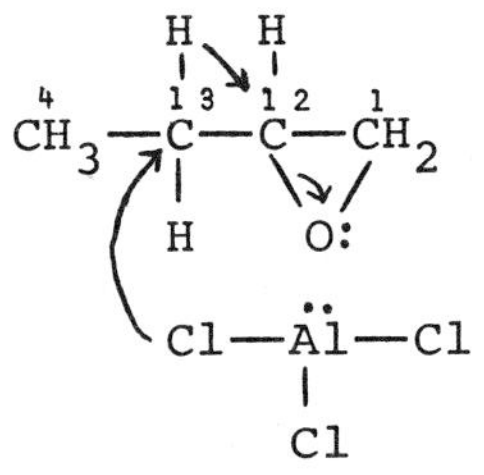

106

In previous discussion we noted that 1,2-diarylalkanes are the only diarylated products reported in some alkylations of arenes with epoxides. In none of the cases mentioned so far was a 1,1-diarylalkane reported even as a minor by-product. However, during a recent study of solvent effects on the ring-opening reaction of epoxides with aluminum chloride, Inoue et al. [541] made the interesting observation that the reaction of propylene oxide with anisole yielded a complex mixture including 1,1-dianisylpropanes, $CH_3CH_2CH(An)_2$, without any contamination of the 1,2-isomers, $CH_3CH(An)CH_2An$. Without solvent at 0 to 20°C, the resulting mixture contained at least eight components: 2-chloro-1-propanol (107, 24%), 1-chloro-2-propanol (108, 8%), 2-(*o*- and *p*-anisyl)-1-propanols (109, 17%), 1-phenoxy-2-propanol (110, 4%), and three isomers (*o,o*-, *o*-,*p*-, and *p,p*-) of 1,1-dianisylpropanes (111, 24%) (Eq. 105).

$$CH_3CH\overset{O}{—}CH_2 + PhOCH_3 \xrightarrow[0-2°]{AlCl_3} \underset{\underline{107}\ 24\%}{CH_3\underset{Cl}{C}HCH_2OH} + \underset{\underline{108}\ 8\%}{CH_3\underset{OH}{C}HCH_2Cl} \qquad (105)$$

$$+ \underset{\underline{109}\ 17\%}{CH_3\underset{An}{C}HCH_2OH} + \underset{\underline{110}\ 4\%}{CH_3\underset{OH}{C}HCH_2OPh}$$

$$+ \underset{\underline{111}\ 24\%}{CH_3CH_2CH(An)_2}$$

(An = *o*, *m*, or *p*-anisyl group)

In a subsequent publication concerned primarily with the mode of formation of isomeric 1,1-dianisylpropanes, Inoue et al. [542] studied the $AlCl_3$-catalyzed reaction of anisole with propylene oxide in various kinds of solvents. With reference to Table 17, the most characteristic features of the results can be summarized in the following points:

1. Except in nitromethane, reactions in all other solvents gave mixtures including both the normal Friedel-Crafts product 2-anisyl-1-propanol (109, Eq. 105) and 1,1-dianisylpropane (111). In nitromethane, only the abnormal product 111 could be found (entry 4).
2. The yield of 111 was solvent dependent and increased with increasing solvent basicity in the following order: hexane < chloroform < 1,2-dichloroethane < anisole < nitromethane.
3. 1,1-Dianisylpropanes (111) did not result from subsequent reaction of monoanisylpropanols with anisole. When a mixture of *o*- and *p*-2-anisyl-1-propanol (109) was allowed to react with excess anisole and $AlCl_3$ at 0 to 2°C, the predominant product was 1,2-dianisylpropane (112) (80%), and 111 was only a minor one (20%) (Eq. 106).

Table 17. Effect of Solvents on the Reaction of Propylene Oxide with Anisole[a,b]

Entry	Solvent	Yield of 2-anisyl-1-propanol[c] (109), %	Yield of 1,1-dianisylpropane (111), %	Isomer Distribution in 111		
				o,o-	*o*-,*p*-	*p*-,*p*
1	1,2-Dichloroethane	5.8	13.4	1.2	24.1	74.5
2	Chloroform	3.2	6.4	1.1	22.7	76.2
3	*n*-Hexane	1.3	0.6	1.8	13.2	85.0
4	CH_3NO_2	0.0	25.8	0.7	12.3	87.0

[a]Reactants were 40 ml of solvent, 20 m mol of epoxide, 40 m mol of anisole and 20 m mol of $AlCl_3$. Reaction temperature and time were 0-2°C and 3.5 hr.

[b]Yields of the products which did not include anisyl moiety were not determined. The other products were oligomers of epoxide and high-boiling tarry matter.

[c]Mixture of the *o*- and *p*-isomers.

$$\underset{\displaystyle An}{CH_3\overset{|}{C}HCH_2OH} + PhOCH_3 \xrightarrow{AlCl_3} \underset{80\%}{\underline{111}} + \underset{\underline{112}\ 20\ \%}{\underset{\displaystyle An}{CH_3\overset{|}{C}HCH_2An}}\ . \qquad (106)$$

The findings above added to the known facts that epoxides form aldehydes under the influence of mild acid catalysts such as magnesium bromide, zinc chloride, and boron trifluoride etherate [543], and that aldehydes condense with aromatics in the presence of Friedel-Crafts catalysts to give 1,1-diarylated products [544] led to the formulation of the mechanism shown in scheme 23 to account for the results.

According to scheme 23 the formation of the normal Friedel-Crafts product 109 is considered to be of the S_N2 type, as described before, while the formation of 111 involves isomerization from epoxide to aldehyde followed by condensation with the arene. To test the generality of the latter reaction, the authors [542] extended their study to include nine other expoxides. Interestingly enough, parallel behavior was again noticed, as can be seen from Table 18.

The results with other alkene oxides fit well with the mechanism above. Thus the reactions of benzene with 2,3-butylene, isobutylene [544a], and α-methylstyrene [544b] oxides gave alcohol fractions resulting from ring openings leading to the relatively more stable secondary or tertiary carbocations.

AlCl3:Solvent
(weak Lewis acid)
AlCl3
(strong Lewis acid)
AlCl3
Sv
ArH
Tarry matter
ArH (reactive
substrate)
high temp
R-CH2-C-H
R-CH2-CH
R-CH-CH2-Ar
R-CH-CH2
Ar OAlCl2

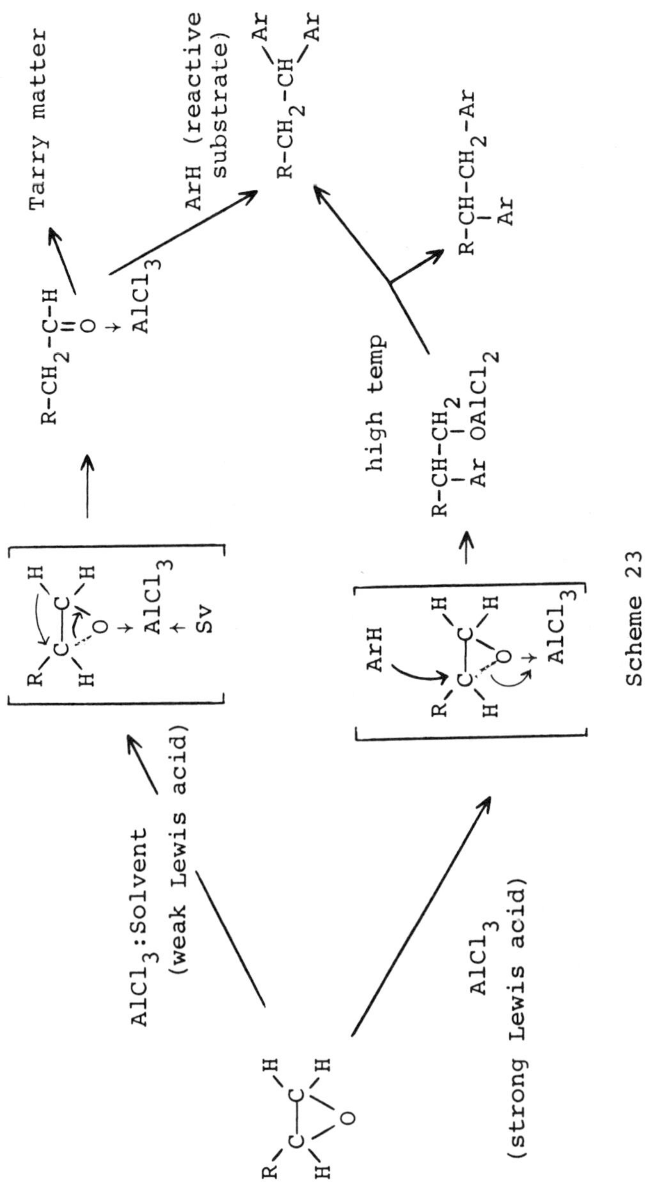

Scheme 23

Table 18 Products of the Reaction of Various Epoxides with Anisole (AnH) by Aluminum Chloride in Nitromethane[a]

Epoxide	Product	Yield, %	Isomer distribution in bis(methoxyphenyl) derivatives[b]		
			o,o-	*o,p*-	*p,p*-
MeCH—CH$_2$ (epoxide, O bridging)	EtCH(An)$_2$	28	1.4	24.8	73.8
EtCH—CH$_2$ (epoxide, O bridging)	PrCH(An)$_2$	33	1.7	24.8	73.5
n-BuCH—CH$_2$ (epoxide, O bridging)	$CH_3(CH_2)_4CH(An)_2$	32	2.5	23.3	74.2
$ClCH_2CH_2$—CH_2 (oxetane, O bridging)	None				
PhCH—CH$_2$ (epoxide, O bridging)	$PhCH_2CH(An)_2$	46	2.6	25.0	72.4
Ph(Me)C—CH$_2$ (epoxide, O bridging)	Ph(Me)CHCH(An)$_2$	76	0.7	22.2	77.1

$Et_2C\text{—}CH_2$ (—O—)	$Et_2CHCH(An)_2$	40	2.3	18.2	79.5
PhCH—CHPh (—O—)	$Ph_2CHCH(An)_2$	56	2.2	24.7	73.1
	$PhCH_2CH(Ph)(An)$	11			
	$(Ph)(An)C{=}CHAn$	10			
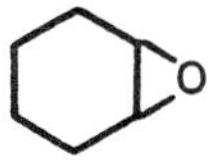 O	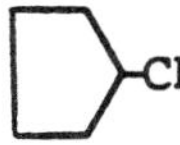 $CH(An)_2$	17	0.7	23.0	76.3
	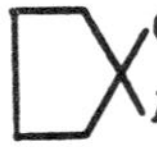 CH_2An, An	8			
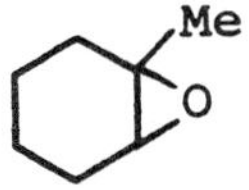 Me, O	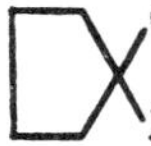 $CH(An)_2$, Me	5	0.9	22.6	76.5

[a]Reactions were conducted with anisole (100 mmol) and $AlCl_3$ (40 mmol) in nitromethane (40 ml), and the epoxide (20 mmol) in nitromethane (20 ml) at 0 to 2°C, the mixtures were homogeneous. Isolated yields calculated on the basis of the epoxides used.

[b]Calculated from GLPC.

Source: M. Inoue, T. Sugita, and R. Ichikawa, *Bull. Chem. Soc. Jpn.*, *51*, 174 (1978).

2. Four-membered and higher cyclic ethers

The Friedel-Crafts reaction of trimethylene oxide with arenes was reported long ago by Thiemer [538], but no information was given regarding the reaction conditions, the hydrocarbons used, or the structure of the products obtained except that they were primary alcohols. In 1954, Searles [526] showed that trimethylene oxide condensed with benzene and mesitylene in the presence of aluminum chloride to form 3-phenyl-1-propanol and 3-mesityl-1-propanol, respectively, in 50 to 70% yields (Eq. 107). The use of BF_3 as

$$\begin{matrix} CH_2-CH_2 \\ | \quad\quad | \\ CH_2-O \end{matrix} + ArH \xrightarrow[\text{0.5 hr at 10° and 10 hr at rt}]{AlCl_3} ArCH_2CH_2CH_2OH \quad (107)$$

(Ar = phenyl or mesityl)

catalyst was unsuccessful, even at 0 to 10°C, as it brought about polymerization of the ether instead of alkylation. Searles also found that the condensation of trimethylene oxide with benzene proceeded at a considerably slower rate than that of ethylene oxide, as shown by an experiment in which an equimolar mixture of both oxides was allowed to react simultaneously with benzene. When the reaction was quenched after a relatively short time (15 min), 2-phenylethanol was the only product detectable. This fact is consistent with the relative stability of the two ethers.

In an extension of their earlier experiments with optically active 1,2-epoxybutane [540], Suga and co-workers examined the reaction of benzene with (+)-2-methyloxetane, catalyzed by $AlCl_3$, $SnCl_4$, and $TiCl_4$ [540a]. The alkylation product was 3-phenyl-1-butanol, obtained in 62% yield at 0°C with $AlCl_3$, with 20% inversion of configuration at the chiral reaction center. 4-Chloro-2-butanol and 3-chloro-1-butanol were obtained in optically active form as by-products resulting from attack of the chlorine atom in the Lewis acid catalysts.

Several $AlCl_3$-catalyzed alkylations of arenes with tetrahydrofurans appeared in the literature. These include the alkylations of benzene, toluene, *m*-xylene, and phenol with tetrahydrofuran [545]; of mesitylene with tetrahydrofuran and 2-methyltetrahydrofuran [546]; of benzene with 2-methyl- and 2,2-dimethyltetrahydrofuran [547]; and of benzene with 2-methyltetrahydrofuran [548]. Some of the results obtained are illustrated in Eqs. (108) to (112).

PhH (31.2g) + tetrahydrofuran (14.4g) —$AlCl_3$ (54g), up to 105°, 10 hr→ n-propylbenzene (1.2g) + tetralin (1.8g) + 4-phenyl-1-butanol (5.5g) + octahydroanthracene + octahydrophenanthrene (6.2g) [545] (108)

PhOH + [tetrahydrofuran] —$AlCl_3$ (40.5g), up to 80°, 10 hr→ HO-C6H4-(CH2)3-OH + HO-tetralin (109) [545]

(56.4g) (14.4g) in cyclohexane (3.5g) (5.4g)

[mesitylene] + [2-methyltetrahydrofuran](CH3) —$AlCl_3$ (54g), 80°, 20 hr→ 10% (5%)*

(240g) (14.2g)

(110) [546]

+ 11% (5%)* + 10% (15%)* + 3% (0%)*

PhH + [2,2-dimethyltetrahydrofuran] —$AlCl_3$ (1.2 mol), 20°, 6 hr→ C-C(C)(Cl)CCC-OH + Ph-C(C)(C)CCC-OH

(20 mol) (1 mol) 15% 54%

(111) [547]

+ [1,1-dimethyltetralin] 10%

PhH + [tetrahydropyran] —excess $AlCl_3$→ 20°: C-C(Ph)-C-C-C-OH 63.5%; 80°: [1-methyltetralin] 66%

(112) [548]

*Values in parentheses are those from 2-methyltetrahydrofuran.

The question concerning relative reactivity and mechanism in alkylations with cyclic ethers was addressed by various workers. The ease with which cyclic ethers alkylate arenes showed considerable dependence on both ring size [526] and state of the oxygen-bonded carbons [547]. The relative reactivity decreased with ring size in the order

[oxirane] > [oxetane] > [tetrahydrofuran] > [tetrahydropyran]

It also decreased with the state of carbon in the order

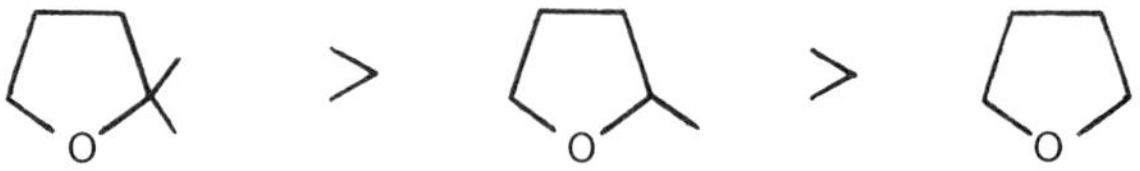

[2,2-dimethyltetrahydrofuran] > [2-methyltetrahydrofuran] > [tetrahydrofuran]

In fact, using the same amount of catalyst, 2-methyltetrahydrofuran condensed with benzene at 20°C to the same extent as tetrahydrofuran at 80°C [547].

In an effort to explore the mechanism, Brauman and co-workers [549,550] conducted two interesting cases of stereospecific Friedel-Crafts alkylations with optically active five-membered heterocycles. In the first case [549], it was proved that alkylation of benzene with optically active γ-valerolactone in the presence of aluminum chloride proceeded with nearly 50% net inversion of configuration (Eq. 113).

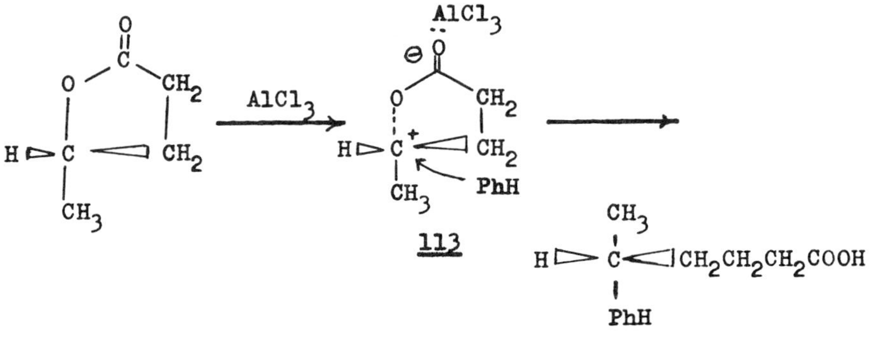

$$\gamma\text{-valerolactone} \xrightarrow{AlCl_3} \underline{113}\ (AlCl_3\text{-complexed carbocation, attacked by PhH}) \longrightarrow H{-}C(CH_3)(PhH){-}CH_2CH_2CH_2COOH \qquad (113)$$

In the second case [550] it was shown that the alkylation of benzene by S-(+)-2-methyltetrahydrofuran (114) gave R-(-)-4-phenyl-1-pentanol (115). The reaction occurred with at least 35% inversion of configuration (Eq. 114).

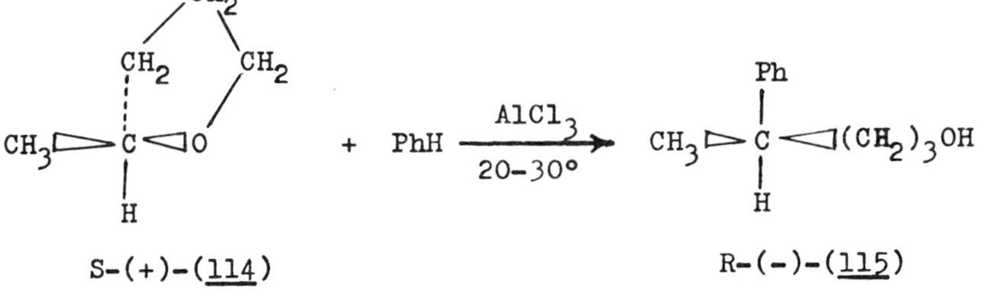

$$\text{S-(+)-(}\underline{114}\text{)} + PhH \xrightarrow[20-30°]{AlCl_3} CH_3{-}C(Ph)(H){-}(CH_2)_3OH\ \ \text{R-(-)-(}\underline{115}\text{)} \qquad (114)$$

The results of both cases have been taken to indicate that the reactions involved an S_N2 attack by the arene on the ion pairs of the type shown by structure 113 (Eq. 113) and that much of the stereospecificity was due to the cyclic nature and the enforced proximity of the leaving group. Consistent with such a mechanism are the observations that (1) the stereochemisty of the Friedel-Crafts alkylations with cyclic ethers is in marked contrast to the essentially complete racemization noted with acyclic ethers such as optically active *sec*-butyl methyl ether [312], and (2) the alkylations of benzene with S-(+)-3-chlorobutanoic acid and S-(+)-3-chloro-1-butanol proceed in stereospecific manners (43.2 and 20% inversion, respectively) [550a]. It was suggested that these compounds react with $AlCl_3$ to form quasi-ring intermediates, for example 116, followed by nucleophilic attack of benzene on this intermediate.

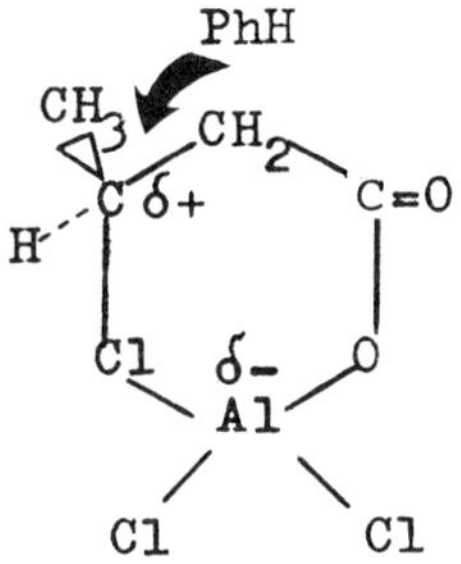

PhH
CH_3 CH_2 C=O
H C δ+
Cl δ- O
Al
Cl Cl

116

Based on their results with optically active methyltetrahydrofuran and γ-valerolactone, as well as on other reported examples of stereospecific Friedel-Crafts alkylations [532] with cyclic ethers, Brauman and Solladie-Cavallo suggested that stereospecificity might be general for Friedel-Crafts reactions of cyclic compounds.

3. Aziridines

The Friedel-Crafts alkylation of arenes with the three-membered nitrogen heterocycles, aziridines, was investigated by Braz [551] in 1952 and by Milstein [552] in 1968. In line with the behavior of epoxides, both workers established that β-arylethylamines are produced in the reaction of benzene and its derivatives with aziridines (Eqs. 115 and 116). In addition, Milstein

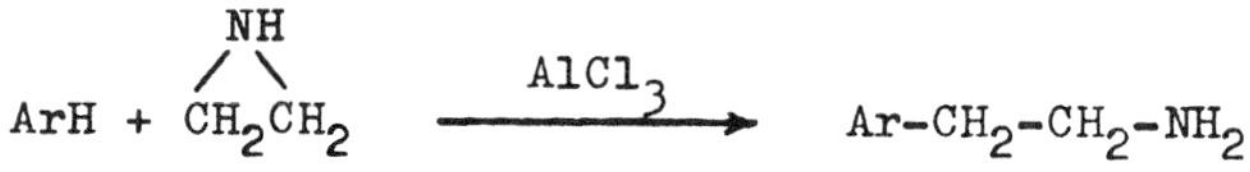

$$ArH + \underset{\text{(NH bridge)}}{CH_2CH_2} \xrightarrow{AlCl_3} Ar\text{-}CH_2\text{-}CH_2\text{-}NH_2 \qquad (115)$$

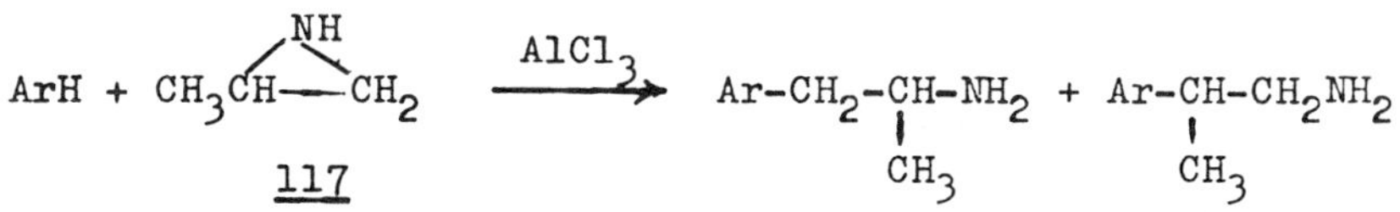

$$ArH + \underset{\mathbf{117}}{CH_3CH\text{—}CH_2 \text{ (NH bridge)}} \xrightarrow{AlCl_3} \underset{\mathbf{118}}{Ar\text{-}CH_2\text{-}CH(CH_3)\text{-}NH_2} + \underset{\mathbf{119}}{Ar\text{-}CH(CH_3)\text{-}CH_2NH_2} \qquad (116)$$

Table 19 Alkylation of Arenes with 2-Methylaziridine in the Presence of $AlCl_3$

Arene	Time, hr	Temp, °C	118/119 ratio	Total yield, %
Benzene	24	Reflux	22:78	15.7
Benzene	22	170	11:89	31.0
p-Xylene	16	Reflux	27:73	19.0
Mesitylene	4.5	Reflux	43:57	43.0

Source: Ref. 552.

showed that the α-methyl-β-phenethylamine (118)/β-methyl-β-phenethylamine (119) ratio increases with increasing nucleophilicity of the aromatic hydrocarbon in the reaction of 2-methylaziridine (117) with benzene, *p*-xylene, and mesitylene, whereas increasing the temperature has the opposite effect. He also showed that in the reactions of chlorobenzene and toluene with aziridine, the nature of the substituent has little effect on the *ortho/para* ratio. Milstein's results are summarized in Table 19 and in Eqs. (117a) and (117b).

$$C_6H_5CH_3 + \underset{\displaystyle NH}{CH_2-CH_2} \xrightarrow[\text{5 hr, reflux; 55\% yield}]{AlCl_3} p\text{-}CH_3C_6H_4CH_2CH_2NH_2\ (50\%) + o\text{-}CH_3C_6H_4CH_2CH_2NH_2\ (50\%) \quad (117a)$$

$$C_6H_5Cl + \underset{\displaystyle NH}{CH_2-CH_2} \xrightarrow[\text{16 hr, reflux; 69\% yield}]{AlCl_3} p\text{-}ClC_6H_4CH_2CH_2NH_2\ (45\%) + o\text{-}ClC_6H_4CH_2CH_2NH_2\ (55\%) \quad (117b)$$

To rationalize the formation of the two isomers α-methyl-β-arylethylamine and β-methyl-β-arylethylamine in alkylations with 2-methylaziridine, Milstein suggested the occurrence of two competing reaction paths, (a) and (b) (Eq. 118). In path (a), the aluminum chloride-coordinated aziridine undergoes ring fission of the secondary C-N bond to give the more stable carbocation, followed by attack on the arene resulting in β-methyl-β-phenethylamine. In path (b), the aromatic component functions as a nucleophile in which π electrons attack the aluminum chloride-coordinated aziridine ring from the less-hindered side, resulting in α-methyl-β-phenylethylamine.

$$CH_3\text{-}CH_2\text{-}\underset{\substack{\diagdown\ \diagup \\ NH \\ \cdot\cdot \\ AlCl_3}}{CH_2} \longrightarrow \begin{cases} \xrightarrow[-Cl^-]{(a)} CH_3\overset{+}{C}HCH_2NHAlCl_2 \xrightarrow[(2)\,H_2O]{(1)\,PhH} CH_3\underset{Ph}{C}HCH_2NH_2 \\ \xrightarrow[PhH,-HCl]{(b)} CH_3\underset{NHAlCl_2}{C}HCH_2Ph \xrightarrow{H_2O} CH_3\underset{NH_2}{C}HCH_2Ph \end{cases} \tag{118}$$

The reaction of 1-benzenesulfonyl-2-(bromomethyl)aziridine (120) with benzene in the presence of $AlCl_3$ was studied thoroughly by Gensler and co-workers [553-555]. In line with reactions of other aziridines, the product of this reaction consisted mainly of 3,3-diphenyl-1-benzenesulfonamidopropane (121) mixed with smaller amounts of the 2,3-diphenyl isomer 122 (Eq. 119).

$$\underset{\mathbf{120}}{Br\text{-}CH_2\overset{2}{C}H\text{—}\overset{3}{C}H_2 \ (\overset{1}{N}\text{-}SO_2Ph)} \xrightarrow[AlCl_3,\ reflux]{PhH} \underset{\mathbf{121}\ (major)}{(Ph)_2\overset{3}{C}H\overset{2}{C}H_2\overset{1}{C}H_2NHSO_2Ph} + \underset{\mathbf{122}\ (minor)}{PhCH_2\underset{Ph}{C}HCH_2NHSO_2Ph} \tag{119}$$

To gain some insight into the mechanism of this conversion, Gensler and co-workers [554,555] consequently conducted two tracer experiments and tested a number of compounds as candidate intermediates. One carbon-14 labeling experiment showed that the 3-carbon atom of the labeled aziridine 123 emerged as the 1-carbon of the product, as in 124 (Eq. 120) [554].

$$\underset{\mathbf{123}}{BrCH_2\overset{2}{C}H\text{—}\overset{*}{C}H_2{}^{3} \ (\overset{1}{N}\text{-}SO_2Ph)} \xrightarrow[AlCl_3]{PhH} \underset{\mathbf{124}}{(Ph)_2CHCH_2\overset{*}{C}H_2NHSO_2Ph} \tag{120}$$

Another tracer experiment with the aziridine 2-carbon atom labeled as in 125 gave a product in which the label was located at the 2 carbon, as in 126 (Eq. 121) [555].

*Position of labeled carbon atom.

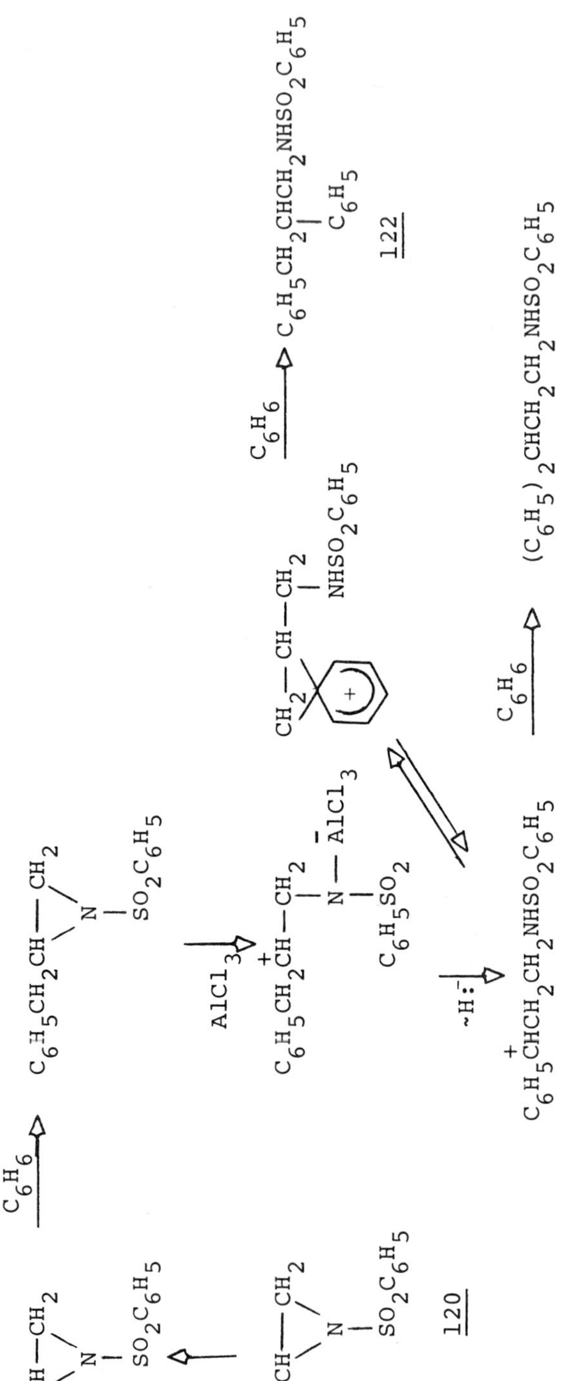

$BrCH_2CH(-N(SO_2C_6H_5)-)CH_2$ (120) $\rightarrow$ $^{+}CH_2CH(-N(SO_2C_6H_5)-)CH_2$ $\xrightarrow{C_6H_6}$ $C_6H_5CH_2CH(-N(SO_2C_6H_5)-)CH_2$

$\xrightarrow{AlCl_3}$ $C_6H_5CH_2\overset{+}{C}HCH_2N(SO_2C_6H_5)\overset{-}{A}lCl_3$

$\xrightarrow{\sim H:^-}$ $C_6H_5\overset{+}{C}HCH_2CH_2NHSO_2C_6H_5$ $\xrightarrow{C_6H_6}$ $(C_6H_5)_2CHCH_2CH_2NHSO_2C_6H_5$ (121)

$\rightleftharpoons$ spiro-cation CH_2—CH—$CH_2NHSO_2C_6H_5$ (with $C_6H_5^+$ ring) $\xrightarrow{C_6H_6}$ $C_6H_5CH_2CH(C_6H_5)CH_2NHSO_2C_6H_5$ (122)

Scheme 24

$$\underset{\mathbf{125}}{\overset{*}{\mathrm{BrCH_2CH}}\!-\!\mathrm{CH_2}\ (\text{ring via N}-\mathrm{SO_2Ph})} \xrightarrow[\mathrm{AlCl_3}]{\mathrm{PhH}} \underset{\mathbf{126}}{(\mathrm{Ph})_2\mathrm{CH_2}\overset{*}{\mathrm{C}}\mathrm{HCH_2NHSO_2Ph}} \tag{121}$$

To satisfy the demands of these tracer experiments and also to accommodate both the results of testing the possible intermediates and the formation of the 2,3-diphenyl isomer (122, Eq. 119), Gensler and Dheer [555] introduced the mechanism outlined in scheme 24. Notably, this mechanism disqualifies an earlier one which failed to account for all the available facts about the reaction [553,554].

B. Lactones

The use of lactones as Friedel-Crafts alkylating agents was first described in 1904 by Eijkmanm [556], who reported obtianing a nonspecified yield of γ-phenylvaleric acid from a reaction of γ-valerolactone with benzene in the presence of $AlCl_3$. Since then the alkylation of arenes with lactones under suitable conditions has provided a convenient route for the synthesis of arylalkanoic acids. However, depending on various factors, such as arene nature, catalyst concentration, as well as temperature and duration of the reaction, the resulting arylalkanoic acids can undergo partial or complete cyclization to the corresponding cyclic ketones. This behavior, which was first disclosed by Arnold et al. in 1947 [557], resolves most of the apparent contradictions in the reported $AlCl_3$-catalyzed alkylations of benzene to give: a mixture of β-phenylpropionic acid (62%) and phenyl vinyl ketone (15%) [558], 1-indanone (80%) [559] or no alkylation products [560] with β-propiolactone; γ-phenylbutyric acid [561] and/or 1-tetralone [557,562,563] with γ-butyrolactone; γ-phenylvaleric acid [556,557,560] and/or 4-methyl-1-tetralone [557, 561,562] with γ-valerolactone; only δ-phenylvaleric acid with δ-valerolactone [560]; and only 4,4-dimethyl-1-tetralone with δ,δ-dimethylbutyrolactone (70%) [557]. In fact, these reports can all be accommodated nicely in the following general equation, in which the closure step is enhanced by longer duration [561], higher temperature [561], and excess catalyst [557,562,563a].

$$\mathrm{C_6H_6} + \mathrm{R_2C(CH_2)_nC(=O)O}\ (\text{lactone}) \xrightarrow[\text{mild conditions}]{\mathrm{AlCl_3}} \mathrm{C_6H_5CR_2(CH_2)_nCOOH}$$

$$\xrightarrow[\text{stronger conditions}]{\mathrm{AlCl_3}} \text{cyclic ketone: } \mathrm{C_6H_4}\text{-fused ring } \mathrm{CR_2(CH_2)_nC{=}O}$$

(R = H,alkyl)

*Position of labeled carbon atom.

As is the case with benzene, alkylations of other arenes with lactones showed similar dependence on conditions. Whereas pure acids or ketones were reported in some cases, varying mixtures of both were reported in other cases. For example, in the presence of excess $AlCl_3$ catalyst, only cyclic ketones were obtained from the alkylations of chloro-, *m*-dichloro-, *p*-dichloro-, and *p*-difluorobenzene with γ-butyrolactone [564]. In another report, however, a mixture of isomeric acids and cyclic ketone was obtained from the alkylation of chlorobenzene with γ-butyrolactone [565] (see Eqs. 122 to 124).

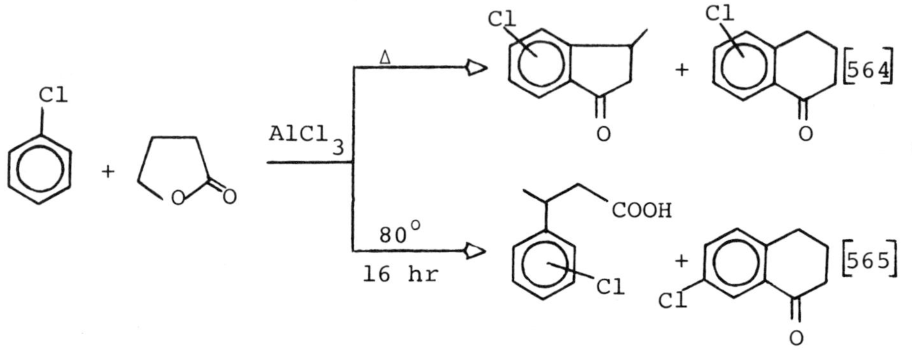

$AlCl_3$; Δ [564]; 80°, 16 hr [565] (122)

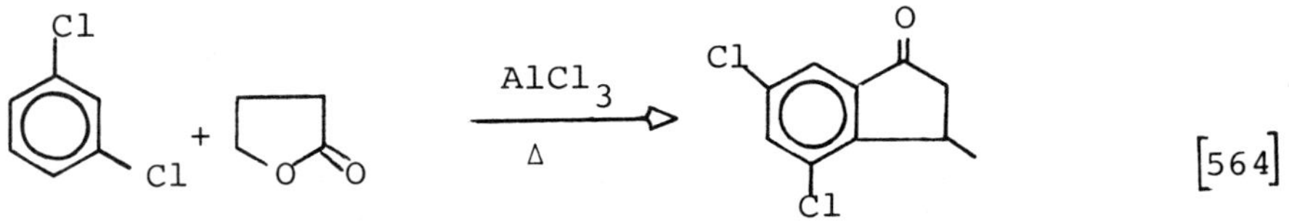

$AlCl_3$, Δ [564] (123)

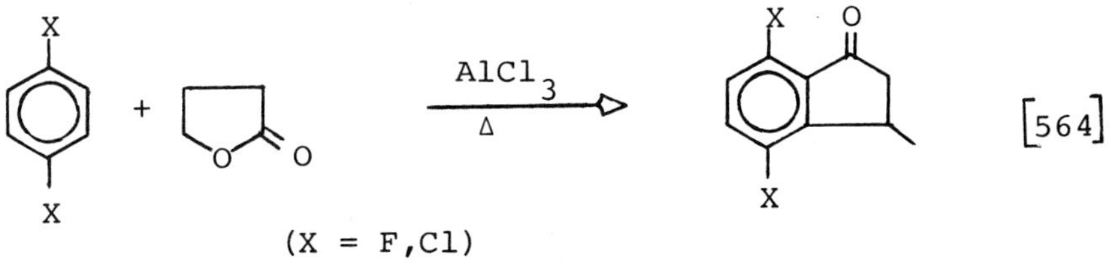

$AlCl_3$, Δ [564] (124)

(X = F,Cl)

In some cases, the alkylations of arenes with γ-lactones to produce γ-arylbutyric acids were not only complicated by subsequent cyclization to ketones but also by the abnormal addition of the arene to the lactone. For example, the $AlCl_3$-catalyzed alkylations of benzene, chlorobenzene [565], phenol, anisole [566], N-phenylbutyrolactam, and 2-quinolinone [567] with γ-butyrolactone gave β- rather than γ-arylbutyric acids (scheme 25).

Moreover, the alkylation of *p*-hydroquinone with γ-valerolactone or with the homologous γ-ethylbutyrolactone under strong conditions gave, not the expected tetralones, but the corresponding isomeric indanones (Eq. 125) [567a].

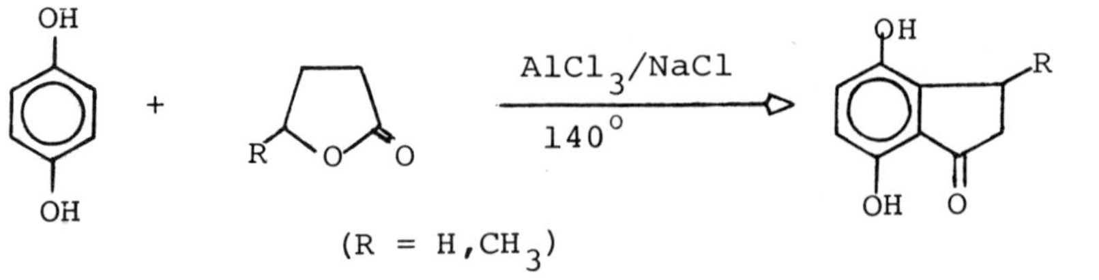

$AlCl_3$/NaCl, 140° (125)

(R = H, CH_3)

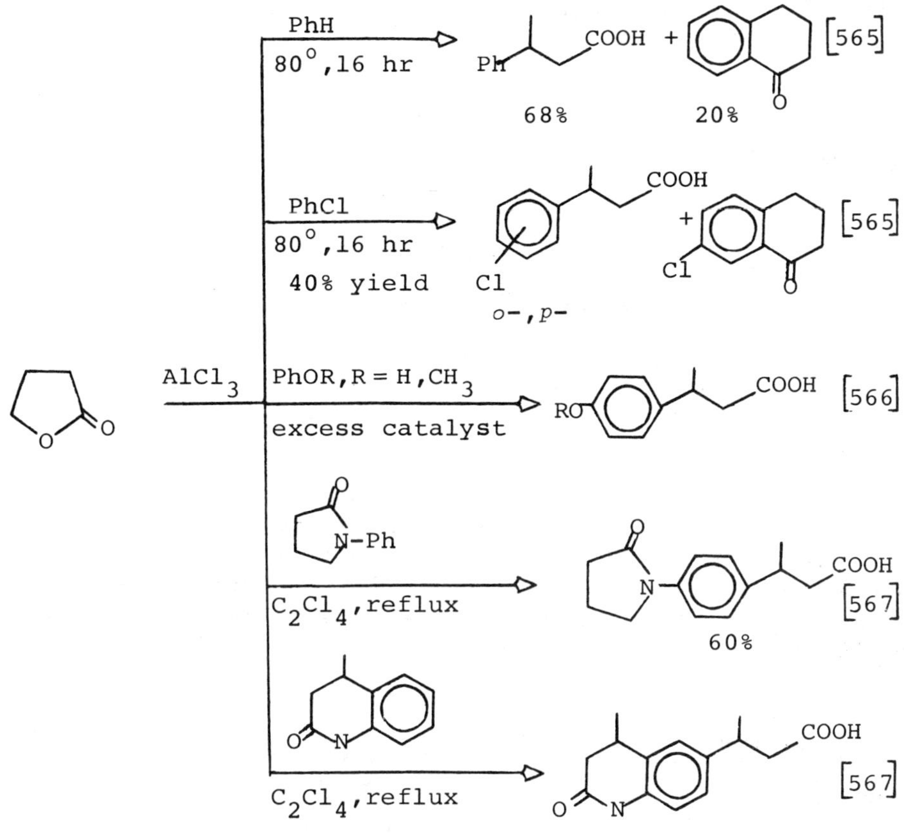

Scheme 25

The following explanation was offered to account for the result of Eq. 125.

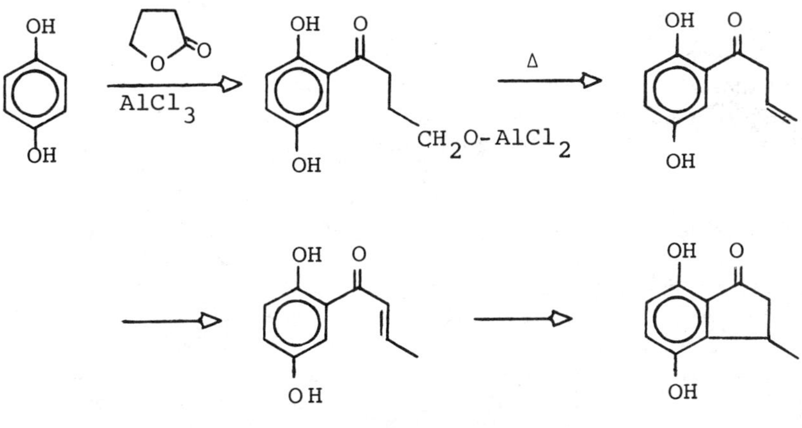

The reactions of lactones with substituted benzenes and polynuclear arenes were still further complicated by other accompanying reactions such

Table 20 Reactions of Halobenzenes with γ-Valerolactone and $AlCl_3$ at 20 to 22 and 98 to 100°C [568]

		Product composition, %						
		127				129		
Type of halogen	Overall yield (%)	o-	m-	p	128	o-	m-	p-
Cl	35-36	45-55	12-19	27-34	—	—	—	—
Br	12-47	26-31	8-16	42-53	6-20		Traces	
I	28-53	3-6	23-44	13-26	14-31	3-5	0.3-1.8	6-15

as disproportionation, reorientation, hydride transfer, and dealkylation. For example, the reactions of halobenzenes with γ-valerolactone and $AlCl_3$ gave products consisting mainly of isomeric γ-(halophenyl)valeric acids (127) formed by direct alkylation, together with smaller amounts of γ-phenylvaleric acid (128) and isomeric γ-(chlorophenyl)valeric acids (129) formed by halogen exchange [568]. The amounts of 127-129 varied considerably with both tem-

$$\text{γ-valerolactone} \xrightarrow[AlCl_3]{C_6H_5X} XC_6H_4\text{-C(C)CCCOOH} \; (\underline{127}) + C_6H_5\text{-C(C)CCCOH} \; (\underline{128}) + ClC_6H_4\text{C(C)CCCOOH} \; (\underline{129})$$

(X = Cl,Br,I)

perature and nature of halogen. This is evident from Table 20 which summarizes the results of alkylations carried out at 20 to 22 and 98 to 100°C. Competitive reactions of halobenzenes and benzene with γ-valerolactone and $AlCl_3$-CH_3NO_2 at 98 to 100°C in 1 hr gave relative rate constants for PhCl and PhBr of 0.119 and 0.0834, respectively.

As in the reactions of halobenzenes, alkylations of alkylbenzenes with lactones revealed considerable dependence of both reactivity and orientation on the nature of the alkyl group. Based on an extensive study of the $AlCl_3$-catalyzed alkylations of alkylbenzenes with γ-butyrolactone under various conditions, Kadyrov et al. [569] noted that (1) the reactivity of monoalkylbenzenes decreased in the order MePh > EtPh > *n*-PrPh > *i*-PrPh > *n*-BuPh > *t*-BuPh, (2) the formation of the *o*-isomer decreased in the same order, and (3) the formation of the *m*-isomer and dealkylation decreased in the reverse order. Thus, whereas no *o*-isomers could be found in the alkylation mixtures from cumene, *t*-butylbenzene, or biphenyl, the mixtures from toluene and ethylbenzene contained varying amounts of the corresponding *o*, *m*, and *p* isomers. At 100°C, for example, the mixture from toluene contained *o*-, *m*-, and *p*-tolybutyric acids in percent ratios of 51:36:9 and 13:81:6 after 1.5 and 4 hr, respectively.

In addition to the alkylations above, several other alkylations of substituted benzenes and polynuclear arenes with γ-butyro-, γ-valero-, and γ-undecalactones appeared in the literature. These include the $AlCl_3$-catalyzed reactions of γ-butyrolactone with phenol and anisole [570,571]; of γ-valerolactone with toluene [563a] and anisole [571]; of γ-butyro- [572], γ-valero- [561], and γ-undecalactones [573] with xylenes; of γ-butyrolactone with mesitylene [574]; of γ-butyro- and/or γ-valerolactone with naphthalene, 1-methylnaphthalene, and biphenyl [569,575-577]; and of γ-butyro- and γ-valerolactones with benzimidazole [578]. The courses followed in some of these alkylations are represented in the following equations.

$$\text{anisole} + \gamma\text{-butyrolactone} \xrightarrow{AlCl_3} o\text{-}MeOC_6H_4(CH_2)_3COOH \;(11\%)$$

$$+\; p\text{-}MeOC_6H_4(CH_2)_3COOH \;(24\%) \;+\; PhO(CH_2)_3COOH \;(4.5\%) \qquad [570]$$

$$+\; p\text{-}MeOC_6H_4CHMeCH_2COOH \;(60.5\%)$$

Toluene (excess) + γ-valerolactone (0.4 mol), 50°, 1 hr:

- $AlCl_3$ (0.4 mol) → 4-methylphenyl–C(C)CCCOOH, **130**, 63%
- $AlCl_3$ (0.8 mol) → **130** (45%) + methyl-substituted 4-methyl-1-tetralone (20%)

[563a]

$$\text{anisole} + \gamma\text{-valerolactone} \xrightarrow[\text{41\% yield}]{AlCl_3,\ 90\text{-}100^\circ,\ 1.8\ hr} o\text{-}MeOC_6H_4CH(CH_3)CH_2CH_2COOH + p\text{-}MeOC_6H_4CH(CH_3)CH_2CH_2COOH \qquad [571]$$

$AlCl_3$, 2 hr at 5°
15 min at 95-100°

COOH

78% 1.5%

[572]

$AlCl_3$

COOH

[561]

$R^1 = R^4 = CH_3$, $R^2 = R^3 = H$; 47% yield

$R^2 = R^4 = CH_3$, $R^1 = R^3 = H$; 84% yield

$R^2 = R^3 = CH_3$, $R^1 = R^4 = H$; 53% yield

$AlCl_3$ in PhCl
130°, 32 hr
60% yield

COOH

50%

[575,576]

COOH

50%

$AlCl_3$
98°, 4 hr

COOH

[576]

23% yield

$AlCl_3$
98°, 15 min
41% yield

COOH

COOH

[576]

50% 50%

+ $\xrightarrow[\text{98°, 3 hr; 70% yield}]{AlCl_3 \text{ in PhCl}}$ HOOC 80% [576]

(Note: The methyl group was reported to be in the 3-position, but was probably in the 4-position)

+ HOOC 20%

(0.1 m) + $\xrightarrow[\text{CH}_2\text{Cl}_4\text{; 1.5 hr at 0°, 2 hr at 100-20°}]{AlCl_3 \text{ (0.3 m)}}$ COOH [578]

$R^1 = R^2 = R^3 = H$; 35% yield

$R^1 = R^2 = H$; $R^3 = CH_3$; 20% yield

$R^1 = R^2 = CH_3$; $R^3 = H$; 20% yield

Besides the previous alkylations with monocyclic lactones, several alkylations with bi- and higher cyclic γ-lactones have been carried out in the course of investigations into the full synthesis of steroidal molecules. Examples of these include the alkylations of benzene, naphthalene [579,580], *p*-xylene, tetralin, and 1-methylnaphthalene [581] with the lactone of *trans*-2-hydroxycyclohexaneacetic acid (131, Eq. 126); of benzene toluene [582], *p*-xylene, and tetralin [583] with the lactone of 4-methyl-2-hydroxycyclohexylacetic

PhH (70 ml) + 131 (0.2 mol) $\xrightarrow[\text{45°; 87% yield}]{AlCl_3 \text{ (0.22 mol)}}$ Ph, COOH 15% + Ph, COOH 15% + Ph, COOH 70% (126) [580]

ArH + 132 —$AlCl_3$→ (major) + (minor) (127)

ArH = PhH Ar = 2-Ph (35%), 3-Ph (65%) [582]

ArH = $PhCH_3$ Ar = 2-*o*-tolyl, 2- and 3-*p*-tolyl [582]

ArH = *p*-$C_6H_4(CH_3)_2$ Ar = 2- and 3-*p*-xylyl [583]

ArH = Tetralin Ar - 2- and 3-(6-tetrahydronaphthyl) [583]

(excess) + 133 (0.25 mol) —$AlCl_3$ (0.25 mol), rt→ (128) [584]

$R^1 = R^2 = R^3 = H$; $R^1 = Cl$, $R^2 = R^3 = H$;

$R^1 = R^2 = CH_3$, $R^3 = H$; $R^1 = R^2 = R^3 = CH_3$

acid (132, Eq. 127); of benzene, chlorobenzene, *m*-xylene, and mesitylene with 1-methyl-1,3-cyclohexanecarbolactone (133, Eq. 128) [584] and of benzene with the lactones of *cis*-1-hydroxy-2-indanylacetic acid (134, Eq. 129) [585], *trans*-2-(*m*-methoxyphenyl)-2-hydroxycyclohexyl)acetic acid (135, Eq. 130) [586], and *trans*-6-methoxycarbonyl-6-methyl-*trans*-2-(*m*-methoxybenzyl)-1-hydroxycyclohexyl)acetic acid (137, Eq. 131) [587].

134 —PhH, $AlCl_3$→ Ph COOH [585] (129)

135 $\xrightarrow[\text{AlCl}_3/\text{HCl}]{\text{PhH}}$ 136 [586] (130)

137 $\xrightarrow{\text{PhH, AlCl}_3}$ 138 [587] (131)

Comparing the last two reactions, it is to be noted that whereas lactone 135 underwent intramolecular alkylation with cyclization to form the half-ester of marrianolic acid 136, lactone 137 alkylated benzene intermolecularly to give 138. The formation of the latter is explicable in terms of rearrangement of the initially formed 3° carbocation to a more stable 2° benzylic carbocation [79] *prior* to the alkylation step.

In relation to bicyclic lactones, the phthalide derivatives 139 were recently used to alkylate aniline and indoles to give the respective products 140 (Eq. 132) [588].

139 + ArH $\xrightarrow{\text{Catalyst}}$ 140 (132)

R = NMe_2, MeO, Me; R^1 = H, OMe

R, R^1 as in 139; Ar = *p*-$Me_2NC_6H_4$, 2,4-$Me(Et_2N)C_6H_3$, 2-methyl-3-indolyl or 2-phenyl-3-indolyl

A few alkylations of arenes with unsaturated as well as halo, hydroxy, and keto functionalized γ-lactones were cited in the literature. Early, in 1937, Beyer [589] studied the alkylation of benzene [589] and toluene [590] with δ-chloro-γ-valerolactone in the presence of $AlCl_3$. The products from benzene and toluene are shown in the following equation:

$$\text{5-(chloromethyl)-}\gamma\text{-butyrolactone} \xrightarrow[\text{2 hr, 80-90}^\circ]{AlCl_3}$$

PhH → PhC(Ph)CCCCOOH (80%) + PhCCCCCOOH (15-20%) + 9,10-bis(CCCCOOH)anthracene (7-10%)

$PhCH_3$ → p-$CH_3C_6H_4CC(C_6H_4CH_3\text{-}p)CCCOOH$ (75%) + 4-($CH_2C_6H_4CH_3$-p)-7-methyl-1-tetralone + 2,6- and 2,7-dimethylenthracene (smaller amounts)

In 1940, Boese [591] found that diketene (141) interacted with benzene and $AlCl_3$, in a manner similar to ketene [592-596], to form benzoylacetone (Eq. 133).

$$PhH + \underset{\mathbf{141}}{CH_2{=}C{-}O{-}C({=}O){-}CH_2} \xrightarrow{AlCl_3} Ph{-}\overset{O}{\overset{\|}{C}}{-}CH_2{-}\overset{O}{\overset{\|}{C}}{-}CH_3 \qquad (133)$$

Recently (1978-1981), Canevet and co-workers [597-599] applied 5-chloro- and 5-hydroxyfuranones of general formula 142 (Eq. 134) to alkylate methyl and methoxybenzenes using H_2SO_4 and $AlCl_3$ catalysts. The products obtained depended on the catalyst used. With H_2SO_4, 5-hydroxyfuranones 142 (X = Cl, OH) simply alkylated the arene to give the corresponding 5-arylfuranones 144 as end products. With $AlCl_3$ catalyst, however, both 5-chloro- and 5-hydroxyfuranones 142 (X= Cl, OH) formed initially the corresponding 5-arylfuranones 144, but some of these subsequently isomerized into 1H-indenecarboxylic acids (145) by ring opening, allylic-type rearrangement, and intramolecular alkylation.

$$142 + 143 \xrightarrow[\text{or } AlCl_3 \ (X=OH,Cl)]{H_2SO_4 \ (X=OH)} 144 \xrightarrow[\Delta]{AlCl_3} 145 \quad (134)$$

X = Cl,OH R^4 = H,OMe

R^1 = H,Me,Ph

R^2 = H,Me R^1,R^2,R^3,R^4 as in 142 and 143

R^3 = H,Me

$R^2 + R^3 = (CH_2)_4$

The mechanism of the transformation of 142-145 as applied to furanone 142 ($R^1 = R^2 = R^3$ = Me) as well as some specified alkylations with furanones is shown in the following equations:

HO + PhOMe $\xrightarrow[\text{18 hr, reflux}]{AlCl_3, CH_2Cl_2}$ (not isolated; OMe) $\xrightarrow{AlCl_3}$ $COO\bar{A}lCl_3$ (OMe) [597]

$\longrightarrow$ MeO, $COO\bar{A}lCl_3$ $\longrightarrow$ MeO, COOH

H_2SO_4, 15 hr, rt [597]

PhOMe, H_2SO_4; $AlCl_3$, CH_2Cl_2, 15 hr, reflux [599]

In reactions analogous to those of the furanones 142 (X = OH), the 3-hydroxyphthalides 146 (R = H, CH_3, Ph) alkylated anisole and *m*-dimethoxybenzene in the presence of H_2SO_4 to give the corresponding 3-arylphthalides 147 (Eq. 135) [599].

H_2SO_4 (135)

146 147

R = H, CH_3, Ph R^1 = H, OMe

R = R^1 = H; 56% yield

R = CH_3, R^1 = H; 22% yield

R = CH_3, R^1 = OMe; 15% yield

R = Ph, R^1 = H; 49% yield

R = Ph, R^1 = OCH_3; 80% yield

In cases where H was hydrogen, this reaction was found to continue with phthalide ring opening and another intermolecular alkylation to give the triarylmethanes 148 (Eq. 136).

$$\mathbf{147} + \text{arene} \xrightarrow[\text{or } H_2SO_4]{AlCl_3} \mathbf{148a\text{-}d} \qquad (136)$$

a: $R^1=R^3=H$; $R^2=R^4=OCH_3$; 88%   c: $R^1=R^2=H$; $R^3=R^4=OCH_3$; 31%

b: $R^1=R^2=R^3=H$; $R^4=OCH_3$; 12%   d: $R^1=R^2=R^3=R^4=OCH_3$; 80%

In recent years, Broquet et al. [600-605] studied the alkylations of arenes with a number of keto and hydroxy γ-lactones. These included the condensation of benzene [600,601,605], toluene [601], *o*-xylene [602], and fluorobenzene [602] with 2-hydroxy-3,3-dimethyl-1,4-butanolide (149) and of benzene with 2-*oxo*-3,3-dimethyl-1,4-butanolide (150) [603]. The results of some of these reactions are represented in Eqs. (137) to (139), and in scheme 26.

$$\mathbf{149} \xrightarrow[\text{63\% yield}]{AlCl_3,\ 90^\circ,\ 5\ hr} \text{COOH product (51\%)} + \text{OH product (49\%)} \quad [602] \qquad (137)$$

$$\text{tetralin} + \mathbf{149} \xrightarrow{AlCl_3,\ 90^\circ, 5\ hr} \text{COOH, OH product (65\%)} + \text{OH product (35\%)} \quad [602] \qquad (138)$$

$$PhH + \mathbf{150} \xrightarrow[\text{reflux, 4.5 hr}]{AlCl_3} \text{Ph, Ph lactone (60\%)} + \text{HO Ph COOH product (10–15\%)} \quad [603] \qquad (139)$$

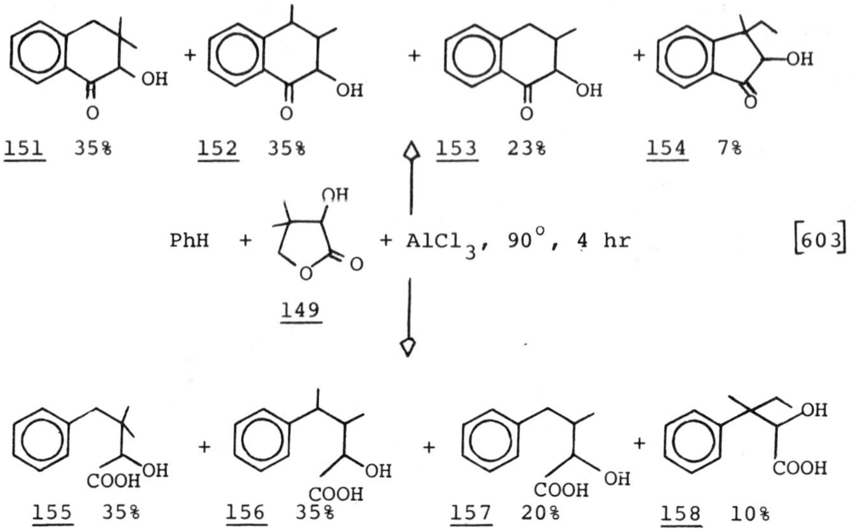

151 35% + 152 35% + 153 23% + 154 7%

PhH + 149 + $AlCl_3$, 90°, 4 hr [603] (140)

155 35% + 156 35% + 157 20% + 158 10%

Scheme 26 [605]

In concluding this discussion of Friedel-Crafts alkylations with lactones, the following points will be emphasized:

1. Like other alkylations, these reactions are sometimes complicated by side processes such as reorientation, disproportionation, hydride transfer, polyalkylation, dealkylation, rearrangement, and abnormal addition of the arene to the lactone.

2. Depending on catalyst ratio and other reaction variables, the end product may be the corresponding arylalkanoic acid and/or the cyclic ketone resulting from it by intramolecular closure.

3. The generally accepted mechanism for alkylations with lactones calls for ionization of the lactone-coordinated catalyst complex to an ion pair of type

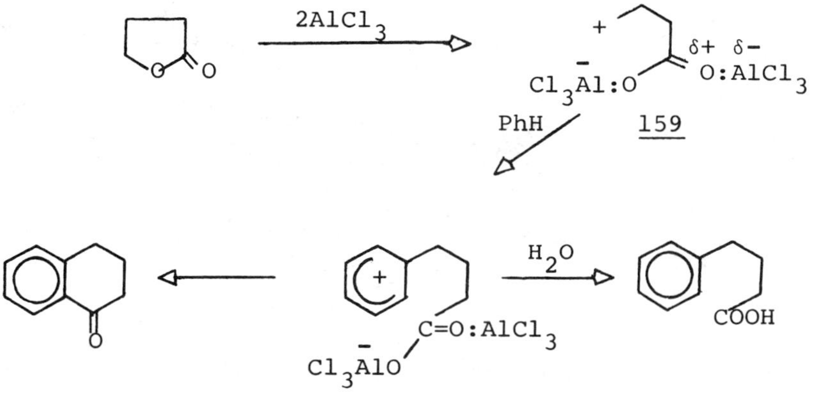

$2AlCl_3$; $Cl_3\bar{A}l{:}O$, $\delta+$ $\delta-$ $O{:}AlCl_3$ 159; PhH; H_2O; $C{=}O{:}AlCl_3$, $Cl_3\bar{A}lO$; COOH

Scheme 27

159 followed by nucleophilic attack by the arene on this ion pair with complete dissociation to the product [549,605-607]. This is illustrated in scheme 27 by the reaction of γ-butyrolactone with benzene and $AlCl_3$.

4. In some cases, rearrangements involving 1,2 hydride and/or alkide shifts take place within the ion pair prior to attack by the arene. This explains the formation of rearranged products during some reactions. For example, the formation of the products of reaction of 2-hydroxy-3,3-dimethyl-1,4-butanolide (149, scheme 26) was accounted for as follows [605]:

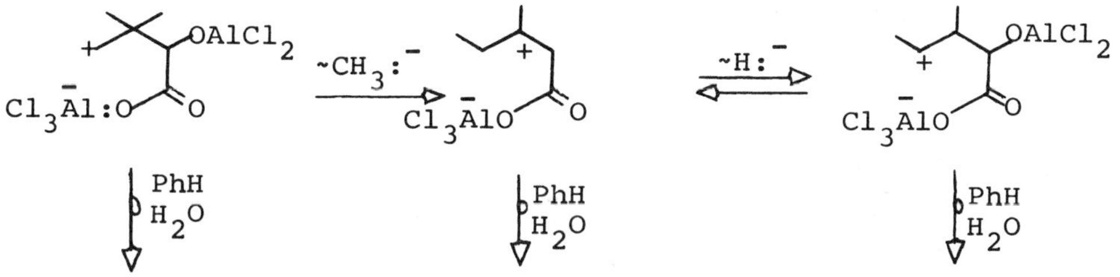

$Cl_3\bar{A}l:O$ … $OAlCl_2$ —$\sim CH_3:^-$→ $Cl_3\bar{A}lO$ … ⇌$\sim H:^-$ $Cl_3\bar{A}lO$ … $OAlCl_2$

PhH, H_2O ↓ (each)

151 + 155   152 + 156   154 + 158

Products 153 and 157 of the same scheme are, probably, the result of fragmentation under the conditions employed.

Similarly, the formation of isomeric cyclohexaneacetic acids during the alkylations of arenes with the lactones of *trans*-2-hydroxycyclohexaneacetic acid (131, Eq. 126) and 4-methyl-2-hydroxycyclohexaneacetic acid (132, Eq. 127) was explained in terms of carbocation rearrangments occurring in the ion pair prior to the alkylation step:

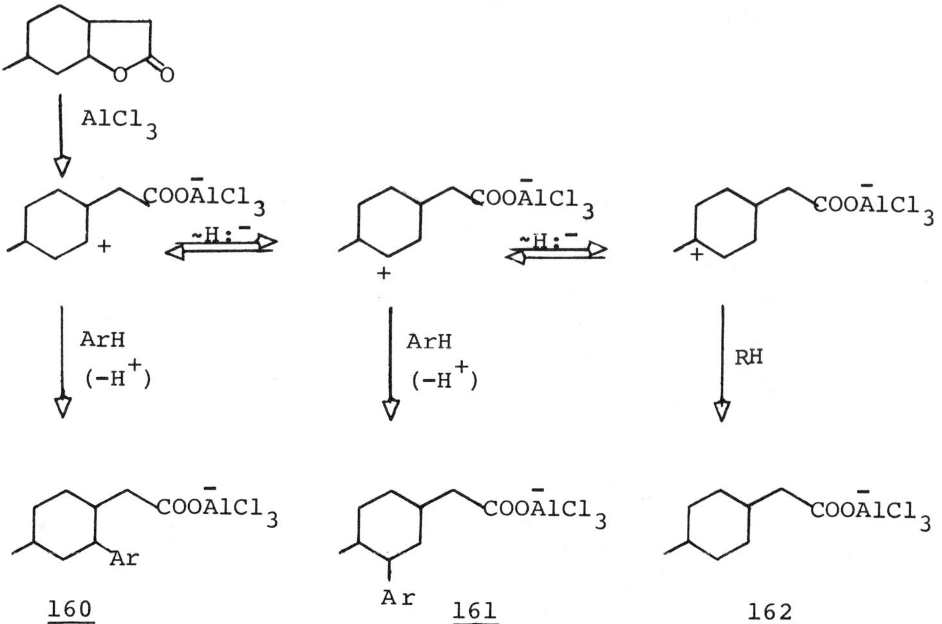

↓ $AlCl_3$

$CO\bar{O}AlCl_3$ ⇌$\sim H:^-$ $CO\bar{O}AlCl_3$ ⇌$\sim H:^-$ $CO\bar{O}AlCl_3$

ArH $(-H^+)$ ↓   ArH $(-H^+)$ ↓   RH ↓

$CO\bar{O}AlCl_3$ (Ar) 160 $CO\bar{O}AlCl_3$ (Ar) 161 $CO\bar{O}AlCl_3$ 162

However, the latter rearrangements can alternatively occur via hydride abstraction after the initial formation of 160 as follows:

160 $\xrightarrow[RH]{R^+ (-H^+)}$ [cation, $COOAlCl_3$, Ar] $\xrightarrow{\sim Ar:^-}$ [cation, $COOAlCl_3$, Ar]

H^+ (-ArH)
Two 1,2-hydride shifts

R^+ / RH

[cation, $CO\bar{O}AlCl_3$] $\xrightarrow{RH}$ 162

161

Also, the formation of the hydride transfer product 162 can take place by subsequent protonation and dealkyaltion of 160 or 161.

5. Sultones, the sulfur analogs of lactones, have also been shown to alkylate arenes in the presence of $AlCl_3$ to give arylalkanesulfonates (Eq. 141). With toluene and anisole, mixtures of isomers were obtained [608].

$$ArH + \text{[sultone: R, O, } SO_2\text{]} \xrightarrow[(2)\ Na_2CO_3]{(1)\ AlCl_3} ArCH(R)CH_2CH_2CH_2SO_3Na \qquad (141)$$

$$Ar = C_6H_5,\ CH_3C_6H_4,\ CH_3OC_6H_4$$

REFERENCES

1. Calloway, N. O., Chem. Rev., *17*, 327 (1935).
2. Ellis, C., *The Chemistry of Petroleum Derivatives*, Vol. 1, Chemical Catalogue Company, New York, 1934; Vol. 2, Reinhold Company, New York, 1937.
3. Nightingale, D. V., Chem. Rev., *25*, 329 (1939).
4. Thomas, C. A., *Anhydrous Aluminum Chloride in Organic Chemistry*, Reinhold, New York, 1941.
5. Price, C. C., Chem. Rev., *29*, 44, 64 (1941).
6. Price, C. C., Org. React., *3*, 58 (1946).
7. Francis, A. W., and E. E. Reid, Ind. Eng. Chem., *38*, 1194 (1946).
8. Francis, A. W., Chem. Rev., *43*, 257 (1948).
9. McAllister, S. H., in *The Chemistry of Petroleum Hydrocarbons*, Vol. 3 (B. T. Brooks, S. S. Kulz, Jr., C. E. Boord, and L. Schmerling, eds.), Reinhold, New York, 1955, Chap. 57.
10. Dalin, M. A., P. I. Markosov, R. I. Shenderova, and G. V. Prokof'eva, *Alkylation of Benzene with Olefins*, Gopudarst. Nauch. Tech. Izd. Khim. Lit., Moscow, 1957; Chem. Abstr., *52*, 15563h (1958).

11. Topchiev, A. V., S. V. Zavgorodnii, and Ya. M. Paushkin, *Boron Fluoride and Its Compounds as Catalysts in Organic Chemistry*, Pergamon Press, New York, 1959; (b) Topchiev, A. V., S. V. Zavgorodnii, and V. G. Kryuchkova, *Alkylation with Olefins*, Elsevier, 1964.
12. For comprehensive reviews, see *Friedel-Crafts and Related Reactions*, Vol. 2, Parts 1 and 2 (G. A. Olah, ed.), Wiley-Interscience, New York, 1963.
13. Tarama, K. Yuki Gosei Kagaku Kyokai Shi, *27*, 1 (1970); Chem. Abstr., *70*, 77194u (1970).
14. Abdurasuleva, A. R., and A. D. Grebenyuk, Nauchn. Tr., Tashk. Gos. Univ., No. 399, 5 (1979); Chem. Abstr., *78*, 29333g (1973).
15. Ohmori, H., Yuki Gosei Kagaku Kyokai Shi, *36*, 512 (1978); Chem. Abstr., *89*, 165304c (1978).
16. Huston, R. C., and D. D. Sager, J. Am. Chem. Soc., *48*, 1955 (1926).
17. Nef, J. U., Justus Liebigs Ann. Chem., *298*, 254 (1897).
18. Huston, R. C., and T. E. Friedemann, J. Am. Chem. Soc., *38*, 2527 (1916).
19. Tsukervanik, I. P., and Z. N. Nazarova, J. Gen. Chem. USSR, *7*, 623 (1937); Chem. Abstr., *31*, 57784 (1937).
20. Tsukervanik, I. P., and G. Vikhrova, J. Gen. Chem. USSR, *7*, 632 (1937); Chem. Abstr., *31*, 57795 (1937).
21. Tsukervanik, I. P., J. Gen. Chem. USSR, *8*, 1512 (1938); Chem. Abstr., *33*, 45873 (1939).
22. Tsukervanik, I. P., and I. Terenteva, J. Gen. Chem. USSR, *7*, 63 (1937); Chem. Abstr., *31*, 5780 (1937).
23. Bowden, E., J. Am. Chem. Soc., *60*, 645 (1938).
24. Norris, J. F., and B. M. Sturgis, J. Am. Chem. Soc., *61*, 1413 (1939).
25. Norris, J. F., and J. N. Ingraham, J. Am. Chem. Soc., *60*, 1421 (1938).
26. Ipatieff, V. N., H. Pines, and L. Schmerling, J. Org. Chem., *5*, 253 (1940).
27. Prindle, R. E., Ph. D. dissertation, Washington University, St. Louis, Mo., 1942.
28. Tsukervanik, I. P., and A. V. Poletaev, Zh. Obshch. Khim. (J. Gen. Chem.), *17*, 2240 (1948); Chem. Abstr., *43*, 164b (1949).
29. Tsukervanik, I. P., and Kh. Taveeva, J. Gen. Chem. USSR, *22*, 1019 (1952); Chem. Abstr., *47*, 8025i (1953).
30. Searles, S., J. Am. Chem. Soc., *76*, 2313 (1954).
31. Tsukervanik, I. P., and G. Kh. Khakimov, Zh. Obshch. Khim., *32*, 1296 (1962); Chem. Abstr., *58*, 3333c (1963).
32. Khakimov, G. Kh., and I. P. Tsukervanik, Zh. Obshch. Khim. *33*, 493 (1963); Chem. Abstr., *59*, 2673h (1963).
33. Khakimov, G. Kh., and I. P. Tsukervanik, Uzb. Khim. Zh., *7*, 76 (1963); Chem. Abstr., *59*, 5041g (1963).
34. Asinger, F., B. Fell, H. Verbeck, and J. Fernandez-Bustillo, Erdoel Kohle, Erdgas, Petrochem., *20*, 852 (1967); Chem. Abstr., *68*, 48718w (1968).
35. Asinger, F., B. Fell, H. Verbeck, and J. Fernandez-Bustillo, Erdoel Kohle, Erdgas, Petrochem., *20*, 786 (1967); Chem. Abstr., *68*, 86904m (1968).
36. Nield, P. G., J. Chem. Soc., 2278 (1964).

37. Nield, P. G., J. Chem. Soc. C, 712 (1966).
38. Roberts, R. M., and D. Shiengthong, J. Am. Chem. Soc., *82*, 732 (1960).
39. Roberts, R. M., Y. -T. Lin, and G. P. Anderson, Jr., Tetrahedron, *25*, 4173 (1969).
40. Roberts, R. M., E. K. Baylis, and G. J. Fonken, J. Am. Chem. Soc., *85*, 3454 (1963).
41. Roberts, R. M., A. A. Khalaf, and R. N. Greene, J. Am. Chem. Soc., *86*, 2846 (1964).
42. Olah, G. A., and J. Nishimura, J. Am. Chem. Soc., *96*, 2214 (1974).
43. Romadane, I., Lav. PSR Zinat. Akad. Vestis, Kim. Ser., 338 (1967); Chem. Abstr., *68*, 39203x (1968).
44. Roberts, R. M., and S. E. McGuire, J. Org. Chem., *35*, 102 (1970).
45. Lespagnol, C., D. Bar, A. Marcincal-Lefebvre, P. Marcincal, L. Masse, and N. Garot, Bull. Soc. Chim. Fr. (2), 552 (1971); Chem. Abstr., *74*, 125520d (1971).
45a. Priddy, D. B., Ind. Eng. Chem., Prod. Res. Dev., *8*, 239 (1969).
46. Kaspi, J. K., and G. A. Olah, J. Org. Chem., *43*, 3142 (1978).
47. Kiersznicki, T., J. Mzyk, and W. Pawlus, Zesz. Nauk. Politech. Slask. Chem., 135 (1978); Chem. Abstr., *91*, 210534g (1979).
48. Olah, G. A., D. Meidar, R. Malhotra, J. A. Olah, and S. C. Narang, J. Catal., *61*, 96 (1980).
49. Kunckell, F., and G. Ulex, J. Prakt. Chem., *87*, 228 (1913).
50. Ipatieff, V. N., N. Orloff, and G. Rasuvaeff, Bull. Soc. Chim. Fr., *37*, 1576 (1925).
51. Hey, D. H., and E. R. B. Jackson, J. Chem. Soc., 1783 (1936).
52. Norris, J. F., and D. Rubinstein, J. Am. Chem. Soc., *61*, 1163 (1939).
53. Kursanov, D. N., and R. R. Zel'vin, J. Gen. Chem. USSR, *9*, 2173 (1939); Chem. Abstr., *34*, 4062 (1940).
54. Norris, J. F., and P. Arthur, Jr., J. Am. Chem. Soc., *62*, 874 (1940).
55. Ebelberg, J., and A. Lowy, J. Am. Chem. Soc., *63*, 101 (1941).
56. Given, D. H., and D. L. Hammick, J. Chem. Soc., 1774 (1941).
57. Given, D. H., and D. L. Hammick, J. Chem. Soc., 428 (1947).
58. Armstrong, G. P., D. H. grove, D. L. Hammick, and H. W. Thompson, J. Chem. Soc., 1700 (1948).
59. Cullinane, N., and S. J. Chard, J. Chem. Soc., 804 (1948).
60. Galich, P. M., V. S. Gutirya, and O. D. Konoval'chikov, Dopov. Akad. Nauk Ukr., RSR, Ser. B., *30*, 930 (1968); Chem. Abstr., *70*, 47144k (1969).
61. Bertholon, G., and R. Perrin, C. R. Acad. Sci., Ser. C., *268*, 1413 (1969); Chem. Abstr., *71*, 21786s (1969).
62. Ogata, Y., K. Sakanishi, and H. Hosoi, Kogyo Kagaku Zasshi, *72*, 1102 (1969); Chem. Abstr., *71*, 80842e (1969).
63. Ismailov, R. G., G. M. Mamedaliev, Yu. I. Tokarev, and T. I. Skameikina, Dokl. Akad. Nauk Az. SSR, *25*, 17 (1969); Chem. Abstr., *73*, 25024f (1970).
64. Ismailov, M. V., and S. M. Aliev, Azerb. Khim. Zh., 32 (1969); Chem. Abstr., *73*, 66157v (1970).
65. Sosnovsky, G., and M. W. Shende, Z. Naturforsch. B, *27*, 1339 (1972); Chem. Abstr., *78*, 97213t (1973).

66. Mekhtiev, S. D., A. A. Kudinov, N. M. Indyukov, and R. I. Zeinalov, Azerb. Khim. Zh., (1), 8 (1972); Chem. Abstr., *78*, 71576p (1973).
67. Douris, J., and A. Mathieu, Bull. Soc. Chim. Fr., (2, Pt. 2), 707 (1973); Chem. Abstr., *78*, 124334s (1973).
68. Romadane, I., Zh. Obshch. Khim., *29*, 102 (1959); Chem. Abstr., *53*, 21821 (1959).
69. Zonew, J., J. Russ. Phys. Chem. Soc., *48*, 550 (9116); Chem. Zentrabl., *1*, 1497 (1923).
70. Meyer, H., and K. Bernhauer, Monatsh. Chem., *53/54*, 743 (1929); Chem. Abstr., *24*, 346 (1930).
71. McGreal, M. E., and J. B. Niederl, J. Am. Chem. Soc., *57*, 2625 (1935).
72. Nenitzescu, C. D., Rev. Chim. (Bucharust), *9*, 5 (1964).
73. Mukaiyama, T., M. Kuwahara, T. Izawa, and M. Ueki, Chem. Lett., 287 (1972).
74. Mukaiyama, T., M. Ueki, T. Izawa, and M. Kuwahara, Chem. Lett., (6), 443 (1972).
75. Warshawsky, A., R. Kalir, and A. Patchornik, J. Org. Chem., *43*, 3151 (1978).
76. Lysenko, Yu. A., and L. I. Khokhlova, Ukr. Khim. Zh., *40*, 799 (1974); Chem. Abstr., *82*, 3872q (1975).
77. Olah, G. A., J. Nishimura, and Y. Yamada, J. Org. Chem., *39*, 2430 (1974).
77a. Nenitzescu, D., S. Tzitzeica, and V. Ioan, Bull. Chim. Soc. Fr, 1272, (1955); Chem. Abstr., *50*, 2259h, 2260a (1956).
78. Nomiya, K., T. Ueno, and M. Miwa, Bull. Chem. Soc. Jpn., *53*, 827 (1980).
79. Nomiya, K., Y. Sugaya, S. Sasa, and M. Miwa, Bull. Chem. Soc., Jpn., *53*, 2089 (1980).
80. Samsonova, I. N., and V. M. Eshakova, Nizkotemp. Katal., Leningr. Gos. Univ., 115 (1962); Chem. Abstr., *59*, 2673e (1963).
81. Romadane, I. A., N. I. Shuken, and Yu. P. Egorov, Izv. Akad. Nauk SSSR, Otd. Khim. Nauk, 648 (1957).
82. Romadane, I. A., Zh. Obshch. Khim., *27*, 1833 (1957); Chem. Abstr., *52*, 4579 (1958); Zh. Obshch. Khim., *27*, 1939 (1957); Chem. Abstr., *52*, 5356 (1958).
83. (a) Romadane, I., Zh. Obshch. Khim., *27*, 1833 (1957); J. Gen. Chem. USSR, *27*, 1898 (1957); Chem. Abstr., *53*, 5206e (1959); Chem. Abstr., *52*, 4579f (1958); (b) Romadane, I., Zh. Obshch. Khim., *27*, 1939 (1957); J. Gen. Chem. USSR, *27*, 2000 (1957); Chem. Abstr., *53*, 5206f (1959); Chem. Abstr., *52*, 5356a (1958).
84. Romadane, I. A., Latv. Valsts Univ. Kim. Fak., Zinat. Raksti, *14*, 49 (1957); Chem. Abstr., *54*, 2456s (1960).
85. Zee-Cheng, K. Y., Ph. D. dissertation, University of Texas at Austin, 1963.
86. Chumakov, S. Ya., A. G. Klein, and M. Sh. Pulina, Ref. Zh., Khim.; Chem. Abstr., *77*, 75047u (1972).
87. Romadane, I., and V. M. Dokuchaeva, USSR Patent 433,118 (1974); Chem. Abstr., *81*, 91250t (1974).
88. Levchenko, A. I., and N. P. Komlev, Ref. Zh., Khim., Abstr. 4Zh276 (1973); Chem. Abstr., *79*, 78451y (1973).
89. Nasyrov, I. M., I. Y. Numanov, and R. Usmanov, Dokl. Akad. Nauk Tadzh. SSR, *17*, 35 (1974); Chem. Abstr. *81*, 37447z (1974).

90. Kaspi, J. K., and G. A. Olah, J. Org. Chem., *43*, 3142 (1978).
91. Kaspi, J. K., D. D. Montgomery, and G. A. Olah, J. Org. Chem., *43*, 3147 (1978).
92. Olah, G. A., J. Kaspi, and J. Bukala, J. Org. Chem., *42*, 4187 (1977).
93. Krespan, C. G., J. Org. Chem., *44*, 4924 (1979).
94. Olah, G. A., R. Malhotra, S. C. Narang, and J. A. Olah, Synthesis, 672 (1978).
95. Ogata, Y., K. Sakanishi, and H. Hosi, Kogyo Kagaku Zasshi, *72*(5), 1102 (1969); Chem. Abstr., *71*, 80842e (1969).
96. Romadane, I., Latv. PSR Zinat. Akad. Vestis, Kim. ser., 279 (1961); Chem. Abstr., *58*, 2384d (1963).
97. Olah, G. A., and D. Meider, Nouv. J. Chim., *3*, 269 (1979).
98. Allen, R. H., L. D. Yats, and D. S. Erley, J. Am. Chem. Soc., *82*, 4853 (1960).
99. Louis, E., Chem. Ber., *16*, 105 (1883).
100. Beran, A., Chem. Ber., *18*, 182 (1885).
101. McKenna, J. F., and F. J. Sowa, J. Am. Chem. Soc., *59*, 470 (1937).
102. Goldschmidt, H., Chem. Ber., *15*, 1067 (1882).
103. Ipatieff, V. N., N. Orlov, and A. Petrov, Chem. Ber., *60*, 1006 (1927).
104. Ipatieff. V. N., H. Pines, and L. Schmerling, J. Am. Chem. Soc., *60*, 353 (1938).
105. Meyer, H., and K. Bernhauer, Monatsh. Chem., *53/54*, 721 (1929); Chem. Abstr., *24*, 346 (1930).
106. Toussaint, N. F., and G. F. Hennion, J. Am. Chem. Soc., *62*, 1145 (1940).
107. McKenna, J. F., and F. J. Sowa, J. Am. Chem. Soc., *59*, 1204 (1937).
108. Simons, J. H., S. Archer, and D. I. Randall, J. Am. Chem. Soc., *61*, 1821 (1939).
109. O'Connor, M. J., and F. J. Sowa, J. Am. Chem. Soc., *60*, 125 (1938).
110. Monacelli, W. J., and G. F. Hennion, J. Am. Chem. Soc., *63*, 1722 (1941).
111. Huston, R. G., R. L. Guile, I. T. Sculati, and W. N. Wasson, J. Org. Chem., *6*, 252 (1941).
112. Sowa, F. J., G. F. Hennion, and J. A. Nieuwland, J. Am. Chem. Soc., *57*, 709 (1935).
113. Streitwieser, A., D. P. Stevenson, and W. D. Schaeffer, J. Am. Chem. Soc., *81*, 1110 (1959).
114. Streitwieser, A., Jr., W. D. Schaeffer, and S. Andreades, J. Am. Chem. Soc., *81*, 1113 (1959).
115. Ioffe, B. V., and B. V. Stolyarov, Zh. Obshch. Khim., *32*, 3452 (1962); Chem. Abstr., *58*, 8938b (1963).
116. Ioffe, B. V., and Tsan-Hsi Yang, Zh. Obshch. Khim., *33*, 2196 (1963); Chem. Abstr., *59*, 13844h (1963).
117. Ioffe, B. V., and B. V. Stolyarov, Dokl. Akad. Nauk SSSR, *161*, 1339 (1965); Chem. Abstr., *63*, 8230 (1965).
118. Romadane, I., Uch. Zap. Rizh. Politekh. Inst., *16*, 155 (1965); Chem. Abstr., *68*, 59203a (1968).
119. Ioffe, B. V., R. Lehmann, and V. B. Stolyarov, Neftekhimiya, *9*, 386 (1969); Chem. Abstr., *71*, 90477d (1969).

120. Kozlova, L. N., and I. Romadane, Neftekhimiya, *10*, 845 (1970); Chem. Abstr., *74*, 64007t (1971).
121. Khalaf, A. A., and R. M. Roberts, Rev. Chim. (Bucharest), submitted for publication.
122. Lipovich, V. G., M. F. Polubenteseva, V. V. Duganova, K. F. Kosygina, and M. P. Ivanova, Pererab. Tverd. Topl., No. 2, 301 (1970); Chem. Abstr., *77*, 139123a (1970).
123. Duganova, V. V., M. F. Polubentseva, and V. G. Lipovich, Izv. Nauch.-Issled. Inst. Nefte-Uglekhim. Sin. Irkutsk. Univ., *12*, 11 (1970); Chem. Abstr., *75*, 117800a (1971).
124. Lipovich, V. G., M. F. Polubentseva, V. V. Duganova, and G. N. Pugalkina, Zh. Org. Khim., *8*, 1458 (1972); Chem. Abstr., *77*, 139075m (1972).
125. Lipovich, V. G., M. F. Polubentseva, V. V. Duganova, V. A. Vykhovanets, and L. G. Elshina, Zh. Org. Khim., *8*, 317 (1972); Chem. Abstr., *76*, 152794u (1972).
125a. Nenitzescu, C. D., V. Ioan, and L. Teodorescu, Chem. Ber., *90*, 585 (1957); Chem. Abstr., *51*, 15434b (1957).
126. Pines, H., W. D. Hunstman, and V. N. Ipatieff, J. Am. Chem. Soc., *73*, 4483 (1951).
127. Degering, E. F., H. J. Gryting, and P. A. Tetrault, J. Am. Chem. Soc., *74*, 3599 (1952).
128. Musaev, M. R., I. G. Ismailzade, and F. A. Mamedov, Azerb. Khim. Zh., No. 4, 33 (1962); Chem. Abstr., *58*, 3333f (1963).
129. (a) Huston, R. C., and I. A. Kaye, J. Am. Chem. Soc., *64*, 1576 (1942); (b) McKenna, J. F., and F. J. Sowa, J. Am. Chem. Soc., *59*, 1204 (1937); (c) O'Connor, M. J. and F. J. Sowa, J. Am. Chem. Soc., *60*, 125 (1938); (d) Sowa, F. J., J. Am. Chem. Soc., *60*, 654 (1938); (e) Francis, A. W., Chem. Rev., *24*, 111 (1948).
130. Pirce, C. C., and M. Meister, J. Am. Chem. Soc., *61*, 1595 (1939).
131. Price, C. C., and J. M. Ciskowski, J. Am. Chem. Soc., *60*, 2499 (1938).
132. Price, C. C., and M. Lund, J. Am. Chem. Soc., *62*, 3105 (1940).
133. Welsh, C. E., and G. F. Hennion, J. Am. Chem. Soc., *63*, 2603 (1941).
134. McKenna, J. F., and F. J. Sowa, J. Am. Chem. Soc., *60*, 124 (1938).
135. Roberts, R. M., S. E. McGuire, and J. R. Baker, J. Org. Chem., *41*, 659 (1978).
135a. Nightingale, D. V., and L. I. Smith, J. Am. Chem. Soc., *61*, 101 (1939).
135b. Nightingale, D. V., and J. M. Shackelford, J. Am. Chem. Soc., *76*, 5767 (1954).
136. Koslova, L. M., and I. Romadane, Latv. PSR Zinat. Akad. Vestis, Kim. Ser., No. 1, 103 (1969); Chem. Abstr., *71*, 3078h (1969).
137. Romadane, I., V. M. Dokuchaeva, L. M. Koslova, and V. A. Pestunovich, Latv. PSR Zinat. Akad. Vestis, Kim. Ser. No. 6, 725 (1971); Chem. Abstr., *76*, 99392h (1972).
138. Schriesheim, A., in Ref. 12, pp. 502, 511, 512.
139. (a) Karabatsos, G. J., F. M. Vane, and S. Meyerson, J. Am. Chem. Soc., *83*, 4297 (1961); (b) Karabatsos, J. G., and F. M. Vane, J. Am. Chem. Soc., *85*, 799 (1963); (c) Karabatsos, J. G., F. M. Vane, and S. Meyerson, J. Am. Chem. Soc., *85*, 733 (1963).

140. G. M. Kramer, J. Org. Chem., *34*, 2919 (1969), and references therein.
141. Roberts, R. M., and T. L. Gibson, J. Am. Chem. Soc., *93*, 7340 (1971).
142. Roberts, R. M., T. L. Gibson, and M. B. Abdel-Baset, J. Org. Chem., *42*, 3018 (1977).
143. Farcasiu, D., Rev. Chim. (Bucharest), *10*, 457 (1965).
144. Schriesheim, A., and I. Kirshenbaum, Chemtech, 310 (1978).
145. Hatch, L. F., and S. Matar, *From Hydrocarbons to Petrochemicals*, Gulf Publishing Co., Houston, 1981.
146. Patinkin, S. H., and B. S. Friedman, in Ref. 12, Chap. 14, pp. 1-288.
147. Akhmadov, Sh. T., G. Yu. Gadzhiev, and R. A. Akhmedova, Uch. Zap. Azerb. Gos. Univ. ser. Fiz.-Mat. Khim. Nauk, 73 (1961); Chem. Abstr., *59*, 6278f (1963).
148. Radzevenchuk, I. F., Zh. Prikl. Khim., *35*, 2538 (1962); Chem. Abstr., *59*, 2673c (1963), Zh. Org. Khim., *1*, 1017 (1965); Chem. Abstr., *63*, 11407d (1965).
149. Askerov, A. K., T. P. Kamysheva, S. L. Sadykh-Zade, I. G. Ismailzade, and I. M. Mamedov, Azerb. Khim. Zh., 31 (1964); Chem. Abstr., *63*, 6888g (1965).
150. Bedell, S. F., E. C. Spaeth, and J. M. Bobbitt, J. Org. Chem., *27*, 2026 (1962).
151. Davies, K. M., and W. J. Hickinbottom, J. Chem. Soc., 2295 (1965); Chem. Abstr., *62*, 14531 (1965).
152. (a) Babakhanov, R. A. and E. E. Gaidarova, Azerb. Khim. Zh., *67*, (1964); Chem. Abstr., *63*, 2912d (1965); (b) Burmistrov, S. I., V. N. Zaitsev, and Zh. G. Dzyra, Zh. Organ. Khim., *1*, 1000 (1965); Chem. Abstr., *63*, 11407b (1965); (c) Plyusnin, V. G., and Maksimov, Zh. Prikl. Khim, *38*, 1191 (1965); Chem. Abstr., *63*, 5544c (1965); Ogino, Y., and T. Kawakami, Kogyo. Kagaku Zasshi, *68*, 45 (1965); Chem. Abstr., *63*, 14656c (1965).
153. Venuto, P. V., C. A. Hamilton, P. S. Landis, and J. J. Wise, J. Catal., *4*, 81 (1966).
154. Emerson, W. S., V. E. Lucas and R. A. Heimsch, J. Am. Chem. Soc., *71*, 1742 (1949).
155. Ismailov, R. G., S. M. Aliev, G. M. Mamedaliev, N. I. Guseinov, and Sh. S. Vezirov, Azerb. Neft. Khoz., *46*, 34 (1967); Chem. Abstr., *67*, 43481h (1967).
156. Bolton, A. P., M. A. Lanawela, and P. E. Pickert, J. Org. Chem., *33*, 1513 (1968).
157. Panchenkov, G. M., I. M. Kolesnikov, E. A. Morozov, and V. M. Prusenko, Tr. Mosk. Inst. Neftekhim. Gasov. Prom., No. 69, 5 (1967); Chem. Abstr., *68*, 86900g (1968).
158. Tret'yakova, V. A., Tr. Mosk. Inst. Neftekhim. Gazov. Prom., No. 69, 75 (1967); Chem. Abstr., *68*, 86903k (1968).
159. Takaya, H., K. Ogawa, and H. Uchida, Kogyo Kagaku Zasshi, *70* (12), 2280 (1967); Chem. Abstr., *68*, 104317d (1968).
160. Arase, A., and A. Suzuki, Aromatikkusu, *20*(3), 129 (1968); Chem. Abstr., *69*, 106022y (1968).
161. Lobanova, N. S., and M. A. Popov, Zh. Prikl. Khim., *41*, 366 (1968); Chem. Abstr., *69*, 10142m (1968).

162. Gakh, I. G., E. P. Babin, and L. G. Gakh, Metody Poluch. Khim. Reakt. Prep., No. 15, 58 (1967); Chem. Abstr., *69*, 18721y (1968).

163. Ismailov, R. G., S. M. Aliev, and Sh. S. Vezirov, Dokl. Akad. Nauk Azb. SSSR, *23*(8), 19 (1967); Chem. Abstr., *69*, 18723x (1968).

164. Aliev, S. M., R. G. Ismailov, N. I. Guseinov, G. M. Mamedaliev, and R. I. Guseinov, Dokl. Akad. Nauk Az. SSSR, *23*(8), 11 (1967); Chem. Abstr., *69*, 18725z (1968).

165. Ismailov, R. G., S. M. Aliev, N. I. Guseinov, G. M. Mamedaliev, R. I. Guseinov, and N. Yu. Ibragimov, Azerb. Khim. Zh., No. 3, 71 (1967); Chem. Abstr., *69*, 18722w (1968).

166. Terpugova, M. P., N. A. Svetlichnaya, R. Z. Sagdeev, and I. L. Kotylarevskii, Izv. Akad. Nauk SSSR, Ser. Khim., No. 12, 2757 (1967); Chem. Abstr., *69*, 26874n (1968).

167. Beginina, M. S., E. P. Babin, V. P. Marshtupa, and N. L. Zotova, Zh. Prikl. Khim., *41*, 370 (1968); Chem. Abstr., *69*, 35109u (1968).

168. Chekhuta, V. G., and S. I. Burmistrov, Probl. Poluch. Poluprod. Prom. Org. Sin. Akad. Nauk SSSR, Otd. Obshch. Tekh. Khim., 138 (1967); Chem. Abstr., *69*, 67011x (1968).

169. Allakhverdieva, D. T., A. H. Gonzalez Morales, B. V. Romanovskii, and K. V. Topchieva, Izv. Vyssh. Uchebn. Zaved., Neft Gaz., *10*, 63 (1967); Chem. Abstr., *69*, 76159d (1968).

170. Askerov, A. K., T. P. Kamysheva, R. A. Eivazova, S. I. Sadykh-Zade, and A. G. Ismailov, Azerb. Neft. Khoz., *47*, 38 (1968); Chem. Abstr., *69*, 76761u (1968).

171. Pis'man, I. I., N. A. Smirnova, A. A. Bakhshi-Zade, and M. A. Dalin, Dokl. Akad. Nauk SSSR, *179*(5), 1117 (1968); Chem. Abstr., *69*, 76763w (1968).

172. Marshtupa, V. P., E. P. Babin, L. I. Maryshkina, and N. S. Morozov, Metody Poluch. Khim. React. Prep., No. 15, 127 (1967); Chem. Abstr., *68*, 77863y (1968).

173. Bodre, R. J., U.S. Patent 3,381,050 (1968); Chem. Abstr., *69*, 96157e (1968).

174. Prisyazhnyuk, K. I., and M. N. Alekseeva, Khim. Prom., Ukr., No. 3, 56 (1968); Chem. Abstr., *69*, 106021x (1968).

175. Chekula, V. G., and E. P. Badin, Ukr. Khim. Zh., *33*, 1275 (1967); Chem. Abstr., *68*, 114178y (1968).

176. (a) Yoneda, N., K. Aomura, and H. Ohtsuka, Bull. Jpn. Petrol. Inst., *9*, 26 (1967); Chem. Abstr., *67*, 43084f (1967); (b) Yoneda, N., K. Aomura, and H. Ohtsuka, Hokkaido Daigaku Kogakubu Kenkyu Hokoku, 149 (1966); Chem. Abstr., *68*, 39202w (1968).

177. (a) Kozlov, N. S., and A. G. Klein, Uch. Zap. Permsk. Gos. Pedagog. Inst., No. 2, 63 (1967); Chem. Abstr., *70*, 3366e (1969); (b) Uch. Zap. Permsk. Gos. Pedagog. Inst., No. 2, 105 (1967); Chem. Abstr., *70*, 3368g (1969).

178. Chekhuta, V. G., Zh. Org. Khim., *5*, 789 (1969); Chem. Abstr., *71*, 21775n (1969).

179. Feizkhanov, F. A., R. Sh. Fattakhova, and Yu. V. Churkin, Khim. Tekhnol., Topl. Masel, *14*(5), 9 (1969); Chem. Abstr., *71*, 38049b (1969).

180. Vodchaeva, N. N., V. U. Lobkina, and A. A. Bakhshi-Zade, Khim. Prom., *45*(5), 337 (1969); Chem. Abstr., *71*, 60882g (1969).

181. Morita, Y., M. Takayasu, and H. Matsumoto, Kogyo Kagaku Zasshi, *71*(9), 1492 (1968); Chem. Abstr., *70*, 46968v (1969).
182. (a) Romanovskii, B. V., D. T. Allakhverdieva, and K. V. Topchieva, Kinet. Katal., *9*(6), 1384 (1968); Chem. Abstr., *70*, 76973k (1969); (b) Kolesnikov, I. M., G. M. Panchenkov, and V. A. Tret'yakova, Zh. Fiz. Khim., *45*, 1707 (1971); Chem. Abstr., *75*, 98021d (1971); (c) Morita, Y., N. Sakata, R. Kuwayama, and M. Kagaya, Sekiyu Gakkai, *15*, 197 (9172); Chem. Abstr., *77*, 19265b (1972).
183. Kozorezov, Yu. I., and T. S. Novozhilova, Neftekhimiya, *8*(6), 858 (1968); Chem. Abstr., *70*, 76984g (1969).
184. Volkov, R. N., and G. I. Popova, Neftekhimiya, *8*(6), 864 (1968); Chem. Abstr., *70*, 76985r (1969).
185. Butina, I. V., N. I. Plotkina, N. A. Shevchenko, and N. Y. Gein, Tr. Inst. Khim., Akad. Nauk SSSR, Ural. Fil., *16*, 9 (1968); Chem. Abstr., *70*, 93010m (1969).
186. Plyusnin, W. G., N. I. Plotkina, and N. V. Gein, Tr. Inst. Khim., Akad. Nauk SSSR, Ural Fil. *16*, 3 (1968); Chem. Abstr., *70*, 96279s (1969).
187. Kolesnikov, I. M., and I. G. Mirgaleev, Neftepererab. Neftekhim. (Moscow), No. 2, 22 (1969); Chem. Abstr., *70*, 106071j (1969).
188. Pliev, T. N., Yu. T. Gordash, V. N. Poletova, A. K. Sopkina, and A. E. Bondar, Khim. Tekhnol. Topl. Masel, *14*(3), 19 (1969); Chem. Abstr., *70*, 106110w (1969).
189. Gramenitskaya, V. N., E. P. Kaplan, and G. I. Nikishin, Neftekhimiya, *9*(1), 17 (1969); Chem. Abstr., *70*, 114721m (1969).
190. Strohmeyer, M., K. H. Hiller, and K. Witte, Angew. Chem., Int. Ed. Engl., *8*, 281 (1969); Chem. Abstr., *70*, 114863j (1969).
191. Berg. A. , H. J. Jakobson, and S. R. Johansen, Acta Chem. Scand., *23*(2), 567 (1969); Chem. Abstr., *71*, 21519g (1969).
192. (a) Ismailov, R. G., R. S. Alimardanov, G. O. Eminov, and Sh. S. Vezirov, Azerb. Neft. Khoz., *48*, 36 (1969); Chem. Abstr., *72*, 3151x (1970); (b) Ismailov, R. G., R. S. Alimardanov, and E. Kh. Mustafaev, Dokl. Akad. Nauk Az. SSR, *26*, 30 (1970); Chem. Abstr., *74*, 125045e (1971); (c) Ismailov, R. G., R. S. Alimardanov, and E. Kh. Mustafaev, Dokl. Akad. Nauk Az. SSR, *27*, 27 (1971); Chem. Abstr., *75*, 118059c (1971).
193. Yoneda, N., E. Hasegawa, K. Aomura, and H. Ohtsuka, Bull. Jpn. Petrol. Inst., *11*, 54 (1969); Chem. Abstr., *71*, 90481a (1969).
194. Hasegawa, E., N. Yoneda, T. Yotsuyanagi, and K. Aomura, Hokkaido Daigaku Koyakubu Kenkyu Hokoku, 233 (1969); Chem. Abstr., *71*, 95357z (1969).
195. Allakhverdieva, D. T., B. V. Romanovskii, and K. V. Topchieva, Neftekhimiya, *9*, 373 (1969); Chem. Abstr., *71*, 100994m (1969).
196. R. G. Ismailov, S. M. Aliev, G. M. Mamedaliev, and E. I. Maister, Dokl. Akad. Nauk Az. SSR, *25*, 17 (1969); Chem. Abstr., *71*, 112539e (1969).
197. Kozorezov, Yu. I., and T. S. Novozhilova, Neftekhimiya, *9*(6), 837 (1969).
198. Friedman, H. M., and A. L. Nelson, J. Org. Chem., *34*, 3211 (1969).
199. Mirgaleev, I. G., and I. M. Kolesnikov, Tr., Mosk. Inst. Neftekhim. Gazov. Prom., No. 86, 168 (1969); Chem. Abstr., *71*, 103768b (1969).
200. Hasegawa, E., N. Yoneda, T. Yotsayanagi, K. Aomura, and H. Ohtsuka, Hokkaido Daigaku Kogakubu Kenyu Hokoku, 105 (1970); Chem. Abstr., *72*, 55720x (1970).

201. Kozorezov, Yu. I., and A. P. Rusakov, Neftepererab. Neftekhim., *60,* (1970); Chem. Abstr., *72,* 66151p (1970).
202. Kolpakchi, A. A., and E. P. Babin, Zh. Fiz. Khim., *44*(1), 69 (1970).
203. Kozorezov, Yu. I., and A. P. Rusakova, Neftepererab. Neftekhim., (1), 42B (1970).
204. Yoneda, N., and H. Ohtsuka, Kogyo Kagaku Zasshi, *72,* 1743 (1969); Chem. Abstr., *72,* 21414q (1970).
205. Alimardanov, R. S., G. O. Eminov, and R. B. Agabekov, Azerb. Khim. Zh., (4), 22 (1970); Chem. Abstr., *74,* 19842w (1971).
206. Hasegawa, E., N. Yoneda, K. Aomura, and H. Ohtsuka, Kogyo Kagaku Zasshi, *73,* 1124 (1970); Chem. Abstr., *74,* 25437p (1971).
207. Kozorezov, Yu. I., A. P. Rusakov, and A. N. Kuleshova, Khim. Prom., *46,* 815 (1970); Chem. Abstr., *74,* 42024u (1971).
208. Ismailov, R. G., R. S. Alimardanov, Yu. A. Ustynyuk, and E. Kh. Mustafaev, Dokl. Akad. Nauk Az. SSR, *26,* 39 (1970); Chem. Abstr., *74,* 53211d (1971).
209. Feizkhanov, F. A., R. Sh. Fattakhova, Yu. V. Churkin, M. F. Khabibbulin, and M. N. Khannanova, Tr. Nauch.-Issled. Inst. Neftekhim. Prooizved., No. 2, 27 (1970); Chem. Abstr., *74,* 76098m (1971).
210. Hasegawa, E., N. Eishi, K. Aomura, and H. Ohtsuka, Kogyo Kagaku Zasshi, *74,* 903 (1971); Chem. Abstr., *75,* 48074c (1971).
211. N. Yoneda, E. Hasegawa, and H. Ohtsuka, Kogyo Kagaku Zasshi, *74,* 908 (1971); Chem. Abstr., *75,* 48075d (1971).
212. Tsypin, A. N., et al., Inf. Soobshch. Gos. Nauch.-Issled. Proektn., Inst. Azotn. Prom. Prod. Org. Sin., No. 2 (Pt. 2), 54 (1969); Chem. Abstr., *75,* 63272j (1971).
213. Romanenko, V. I., A. E. Pinsker, and L. F. Polezhaeva, Inf. Soobshch. Gos. Nauch.-Issled. Proektn. Inst. Azotn. Prom. Prod. Org. Sin., No. 2 (Pt. 2), 78 (1969); Chem. Abstr., *75,* 63273k (1971).
214. Bakin, E. P., V. P. Marshtupa, and Z. S. Borodina, Tr. Vses. Nauch-Issled. Inst. Khim. Reakt. Osobo Chist. Khim. Veshchestv., No. 31, 94 (1969); Ref. Zh., Khim., Abstr. 12B1650 (1970); Chem. Abstr., *75,* 129066d (1971).
215. Gakh, I. G., E. Baben, V. P. Marshtupa, and L. G. Gakh, Tr. Vses. Nauch.-Issled. Inst. Khim. Reakt. Osobo Chist. Khim. Veshchestv, No. 31, 85 (1969); Ref. Zh., Khim., Abstr. 12zh300 (1970); Chem. Abstr., *76,* 13439u (1972).
216. Ermishina, A. M., and L. D. Gluzman, Khim. Tverd. Topl., 1 No. 6, 114 (1971); Chem. Abstr., *76,* 45986m (1972).
217. Khaziakhmetov, F. G., and N. M. Levedeva, Sb. Aspir. Rab. Kazan. Khim.-Tekhnol. Inst., Khim. Nauki, 3 (1970); Chem. Abstr., *76,* 72128q (1972).
218. Vesely, J. A., U.S. Patent 3,637,884 (1972); Chem. Abstr., *76,* 72309z (1972).
219. Popov, A. F., N. N. Korneev, and G. S. Solov'eva, Khim. Prom. (Moscow), *48,* 90 (1972); Chem. Abstr., *76,* 126489d (1972).
220. Bineeva, N. T., et al., Khim. Prom. (Moscow), *48,* 428 (1972); Chem. Abstr., *77,* 78984k (1972).
221. Abdurasuleva, A. R., and N. Ismailov, Tr. Ferg. Politekh. Inst., No. 2, 68 (1969); Chem. Abstr., *77,* 101005a (1972).
222. Zavgorodnii, S. V., G. K. Bereznitskii, and D. A. Pisanenko, Zh. Org. Khim., *8,* 1484 (1972); Chem. Abstr., *77,* 113919q (1972).

223. Morita, Y., M. Kagaya, and T. Kimura, Sekiyu Gakkai Shi, *15,* 1027 (1972); Chem. Abstr., *78,* 71579s (1973).
224. Yoneda, N., Sekiyu Gakkai Shi, *15,* 894 (1972); Chem. Abstr., *78,* 74369j (1973).
225. Plotkina, N. I., N. V. Gein, N. A. Shevehenko, V. G. Plyusnin, and M. I. Kachalkova, Neftekhimiya, *12,* 806 (1972); Chem. Abstr., *78,* 82923n (1973).
226. (a) Yoneda, N., M. Somai, K. Aomura, and H. Ohtsuka, Hokkaido Daigaku Kogakubu Kenkyu Hokoku, 109 (1972); Chem. Abstr., *78,* 164623s (1972); (b) Yoneda, N., A. Chiba, and H. Ohtsuka, Hokkaido Daigaku Kogakubu Kenkyu, 137 (1972); Chem. Abstr., *78,* 164624t (1973).
227. Horie, T., and Y. Katsuyama, Sekiyu, Gakkai Shi., *16,* 117 (1973); Chem. Abstr., *79,* 55634w (1973).
228. Yoneda, N., T. Musha, K. Aomura, and H. Ohtsuka, Hokkaido Daigaku Kogakubu Kenkyu Hokoku, 119 (1972); Chem. Abstr., *78,* 164627w (1973).
229. Yoneda, N., T. Musha, K. Aomura, and H. Ohtsuka, Hokkaido Daigaku Kogakubu Kenkyu Hokoku, 127 (1972); Chem. Abstr., *79,* 10266m (1973).
230. Il'in, V. F., and Yu. N. Usov, Neftekhimiya, *13,* 514 (1973); Chem. Abstr., *79,* 125970d (1973).
231. Kozorezov, Yu. I., T. S. Novozhilova, and Yu. V., Ryabtseva, Khim. Prom., *49,* 629 (1973); Chem. abstr., *79,* 125976k (1973).
232. Yoneda, N., E. Hasegawa, H. Yoshida, K. Aomura, and H. Ohtsuka, Mem. Fac. Eng., Hokkaido Univ., *13,* 227 (1973); Chem. Abstr., *79,* 146068f (1973).
233. Yoneda, N., M. Somai, K. Aomura, and H. Ohtsuka, Hokkaido Daigaku Kogakubu Kenkyu Hokoku, 109 (1972); Chem. Abstr., *79,* 164623s (1973).
234. Plaksunova, S. L., L. A. Efendieva, V. V. Lobkina, and A. M. Bakhshizade, Khim. Prom., No. 5, 340 (1974); Chem. Abstr., *81,* 49347e (1974).
235. Morita, M., S. Sato, S. Mihara, and T. Yamada, Nippon Kagaku Kaishi, No. 12, 2332 (1973); Chem. Abstr., *80,* 70452z (1974).
236. Kozorezov, Yu. I., and L. G. Molchanova, Khim. Tekhnol. Topl. Masel, *18,* 5 (1973); Chem. Abstr., *80,* 70456e (1974).
237. Nel'kenbaum, Yu. Ya, and Yu. A. Sangalov, Khim. Vysokomol. Soedin. Neftekhim., 112 (1973); Chem. Abstr., *80,* 14560q (1974).
238. Canfield, R. C., T. E. Neta, and R. F. Wiesenborn, U.S. Patent 3,813,451 (1974); Chem. Abstr., *81,* 25320q (1974).
239. Morimoto, T., and I. Isas, Jpn. Kokai 74-36,659 (1974); Chem. Abstr., *81,* 49470q (1974).
240. Isakov, Ya. I. Kh. M. Minachev, Ya. T. Eidus, A. L. Lapidus, and V. P. Kalinin, USSR Patent 417,405 (1972); Chem. Abstr., *81,* 145723z (1974).
241. Ellis, A. F., E. F. Harper, and R. C. Williamson, U.S. Patent 3,803,255 (1974); Chem. Abstr., *81,* 145726c (1974).
242. (a) Applegath, F., L. E. Du Pree, Jr., A. C. MacFarlane, and J. D. Robinson, U.S. Patent 3,848,012 (1974); Chem. Abstr., *82,* 97803t (1975); (b) MacFarlane, A. C., Oil Gas J., *74,* 99 (1976); Chem. Abstr., *85,* 46111v (1976).

243. Polubentseva, M. F., V. I. Nikolaeva, G. A. Kalabin, V. I. Glukhikh, B. A. Bazhenov, and V. G. Lipovich, Zh. Org. Khim., *11*, 773 (1975).
244. Bassus, J., G. Bertholon, C. Decoret, and R. Perrin, Bull. Soc. Chim. Fr., *12*, 3031 (1974); Chem. Abstr., *83*, 8699u (1975).
245. Askerov, A. K., T. P. Kamysheva, S. I. Sadykh-Zade, I. G. Ismailzade, and I. M. Mamedov, Azerb. Khim. Zh., No. 5, 31 (1964); Chem. Abstr., *63*, 6888g (1965).
246. Olah, G. A., S. H. Flood, S. J. Kuhn, M. E. Moffatt, and N. A. Overchuck, J. Am. Chem. Soc., *86*, 1046 (1964).
247. Topchiev, A. V., R. N. Volkov, and S. V. Zavgorodnii, Dokl. Akad. Nauk SSSR, *134*, 1101 (1960), Chem. Abstr., *55*, 6432g (1961).
248. Olah, G. A., S. H. Flood, and M. E. Moffatt, J. Am. Chem. Soc., *86*, 1065 (1964).
249. Calcott, W. S., J. M. Tinker, and V. Weinmayr, J. Am. Chem. Soc., *61*, 1010 (1939).
250. (a) Andreev, D. N., A. P. Meshoheryakov, and A. D. Petrov, J. Appl. Chem. USSR, *19*, 705 (1946); (b) Topchiev, A. V., Ya. M. Paushkin, and M. V. Kurashev, Proc. Acad. Sci. USSR, Sect. Chem., *108*, 219 (1956); (c) Gilyarovskaya, L. A., and S. L. Lyubimova, Tr. Mosk. Inst. Tonkoi Khim. Tekhnol, No. 8, 21 (1958); (d) Topchiev, A. V., M. V. Kurashev, and I. F. Gavrilenko, Dokl. Akad. Nauk SSSR, *139*, 124 (1961); Chem. Abstr., *56*, 2391c (1962).
251. Olah, G. A., *Friedel-Crafts Chemistry*, New York, 1973, pp. 356-358.
252. Anderson, H. J., and L. C. Hopkins, Can. J. Chem., *42*, 1279 (1964).
253. Belen'kii, L. I., G. P. Gromova, and Ya. L. Gol'dfarb, Khim. Geterotsikl. Soedin., 591 (1972); Chem. Abstr., *77*, 151767w (1972).
254. Enos, H. I, Jr., Ger. Patent 936089 (1955); Chem. Abstr., *53*, 6189b (1959).
255. Kyska, K., Chem. Prum., *21*, 264 (1971), Chem. Abstr., *75*, 75887b (1971).
256. D'yachenko, N. L., et al., Khim. Tverd. Topl., No. 3, 141 (1973); Chem. Abstr., *79*, 66077t (1973).
257. Zil'berbrandt, A. M., and A. A. Tyazhelova, Tr. Inst. Goryuch. Iskop. Moscow, *28*, 9 (1972); Chem. Abstr., *79*, 79556g (1973).
258. Zavgorodnii, S. V., and V. I. Sidel'nikova, Dokl. Akad. Nauk, SSSR *118*, 9 (1958); Chem. Abstr., *52*, 11784f (1958).
259. Akhmedov, Sh. T., G. Yu. Gadzhiev, and M. A. Salimov, Uch. Zap. Azerb. Gos. Univ., ser. Fiz.-Mat. Khim. Nauk, 66 (1962).
260. Romadane, I., and S. Berga, Obshch. Khim., *28*, 413 (1958).
261. Romadane, I., Latv. Valsts Univ. Khim. Fak. Zinat. Raksti, *14*, 49 (1957).
262. Koslov, N. S., and A. G. Klein, Zh. Org. Khim., *4*, 1407 (1968) [J. Org. Chem., *4*, 1353 (1968)]; Chem. Abstr., *69*, 86481s (1968).
263. Kozlov, N. S., and A. G. Klein, Zh. Org. Chem., *4*, 1409 (1968) [J. Org. Chem., *4*, 1357 (1968)]; Chem. Abstr., *69*, 85888t (1968).
264. Schmerling, L., Ind. Eng. Chem., *45*, 1447 (1953).
265. Nenitzescu, C. D., M. Avram, and E. Sliam, Bull. Soc. Chem. Fr., 1266 (1955).
266. Sergeev, P. G., Khim. Prom, *144*, 208 (1956).
267. Lebedev, N. N., Zh. Obshch. Khim., *34*, 1782 (1964).

268. (a) Nakane, R., O. Kurihara, and A. Takematsu, J. Org. Chem., *36,* 2753 (1971); (b) Nakane, R., O. Kurihara, and A. Natsubori, J. Am. Chem. Soc., *91,* 4528 (1969).
269. Pavelkina, A. E., J. Appl. Chem. USSR, *12,* 1422 (1939); Chem. Abstr., *34,* 3485 (1940).
270. Pines, H., J. D. LaZerte, and V. N. Ipatieff, J. Am. Chem. Soc., *72,* 2850 (1950).
271. Baev, N. P., A. V. Topchiev, Y. M. Paushkin, and M. V. Kurashev, Khim. Pererab. Neft. Uglevodorodov, Tr. Vses. Soveshch. Po. Kompleksn. Khim. Pererab. Neft. Gaz., *422* (1956); Chem. Abstr., *51,* 17782d (1957).
272. Pines, H., and V. N. Ipatieff, U. S. Patent 2,584,103 (1952).
273. Olah, G. A., P. Schilling, J. S. Staral, Y. Halpern, and J. A. Olah, J. Am. Chem. Soc., *97,* 6807 (1975).
274. Burwell, R. L., and S. Archer, J. Am. Chem. Soc., *64,* 1032 (1942).
275. Huston, R. C., and T. J. Hsieh, J. Am. Chem. Soc., *58,* 439 (1936).
276. Nenitzescu, C. D., V. Ioan, and L. Teodorescu, Chem. Ber., *90,* 585 (1957); Chem. Abstr., 15434b (1957).
277. Mortikov, E. S., et al., Zh. Prikl. Khim. (Leningrad), *50,* 1867 (1977); Chem. Abstr., *87,* 184105p (1977).
278. Nightingale, D. V., and J. M. Shackelford, J. Am. Chem. Soc., *76,* 5767 (1954).
279. Dewar, M. J. S., and N. A. Puttnam, J. Chem. Soc., 4080, 4090 (1959).
280. Lysenko, A. P., G. I. Yakunina, V. G. Plyusnin, and M. I. Zelentsova, Neftekhimiya, *8*(1), 42 (1968); Chem. Abstr., *69,* 58894k (1968).
281. Dewar, M. J. S., T. Mole, and E. W. T. Warford, J. Chem. Soc., 3581 (1956).
282. Brown, H. C., J. Chem. Educ., 36, 424 (1959).
283. Roberts, r. M., G. P. Anderson, Jr., and N. L. Doss, J. Org. Chem., *33,* 4259 (1968).
284. Simons, J. H., and S. Archer, J. Am. Chem. Soc., *60,* 2952 (1938).
285. Lenneman, W. L., R. D. Hites, and V. I. Komarewsky, J. Org. Chem., *19,* 463 (1954).
286. Ovchinnikov, V. G., S. V. Zavgorodnii, and S. I. Bass, Ukr. Khim. Zh., *37,* 465 (1971); Chem. Abstr., *75,* 48567r (1971).
287. Asahara, T., and Y. Takagi, J. Chem. Soc. Jpn., Ind. Chem. Sect., *58,* 147 (1955); Chem. Abstr., *49,* 13922b (1955).
288. Asinger, F., G. Geiseler, and W. Beetz, Chem. Ber., *92,* 755 (1959); Chem. Abstr., *53,* 14975e (1959).
289. Tjepkema, J. J., B. Pavlis, and H. W. Huijser, Proc. 5th World Petrol. Congr., New York, 1959, Sect. IV, Paper 21.
290. Olson, A. C., Ind. Eng. Chem., *62,* 833 (1960).
291. Swisher, R. D., F. Koelble, and S. K. Liu, J. Org. Chem. *26,* 4066 (1961).
292. Geiseler, G., P. Herrmann, and G. Kurzel, Chem. Ber., *98,* 1695 (1965); Chem. Abstr., *63,* 6888f (1965).
293. Alul, H. R., and G. J. McEwan, J. Org. Chem., *32,* 3365 (1967).
294. Alul, H. R., Ind. Eng. Chem., Prod. Res. Dev., 7,7 (1968).
295. Alul, H. R., J. Org. Chem., *33,* 1522 (1968).

296. Lesment, T., S. Faingold, and T. Liiv, Neftekhimiya, *9*(3) 379 (1969); Chem. Abstr., *71*, 90474e (1969).
297. Asinger, F., B. Fell, H. Verbeck, and J. Fernandez-Bustillo, Erdoel. Kohle, Erdgas, Petrochem., *20*, 852 (1967); Chem. Abstr., *68*, 48718w (1968).
298. Alul, H. R., and G. J. McEwan, J. Org. Chem., *37*, 3323 (1972).
299. Alul, H. R., and G. J. McEwan, J. Org. Chem., *37*, 4157 (1972).
300. Stepanova, G. G., and S. I. Faingd'd Esti NSV Tead. Akad. Toim. Keem. Geol., *18*, 69 (1969); Chem. Abstr., *70*, 105611y (1969).
301. Flanagan, P. W., M. C. Hamming, and F. M. Evans, J. Am. Oil. Chem. Soc., *44*, 30 (1967).
302. Tatarenko, A. N., and I. P. Tsukervanik, Zh. Obshch. Khim. (J. Gen. Chem.), *18*, 106 (1948); Chem. Abstr., *43*, 1369g (1949).
303. Ipatieff, V. N., B. B. Corson, and H. Pines, J. Am. Chem. Soc., *58*, 919 (1936).
304. Schulze, W. A., and W. M. Axe, U.S. Patent 2,245,839 (Aug. 19, 1947).
305. Axe, W. N., U.S. Patent 2,412,595 (Dec. 17, 1946).
306. Simons, J. H., U.S. Patent 2,423,470 (1947); Chem. Abstr., *42*, 3777 (1948).
307. Spiegler, L., and J. M. Tinker, J. Am. Chem. Soc., *61*, 1002 (1939).
308. Brochet, A., C. R. Acad. Sci., *117*, 115, 235 (1893).
309. Simons, J. H., and G. C. Bassler, J. Am. Chem. Soc., *63*, 880 (1941).
310. Hunt, M., U.S. Patent 2,467,131 (1949), Chem. Abstr., *43*, 6232c (1949).
311. Hunt, M., V. Weinmayr, and A. V. Willett, Jr., U.S. Patent 2,467,132 (1949), Chem. Abstr., *43*, 6232e (1949).
312. Burwell, R. L., Jr., L. M. Elkin, and A. D. Shields, J. Am. Chem. Soc., *74*, 4570 (1952).
313. Kiersznicki, T., J. Majewski, and J. Mzyk, Zesz. Nauk. Politech. Slask. Chem., 43 (1979); Chem. Abstr., *91*, 192921z (1979).
314. Ovchinnikov, B. G., and S. V. Zavgorodnii, Neftekhimiya, *9*, 516 (1969); Chem. Abstr., *72*, 42983q (1970).
315. Kryuchkova, V. E., and S. V. Zavgorodnii, Izv. Vyssh. Uchebn. Zaved., Khim. Khim. Tekhnol., *11*(12), 1353 (1968); Chem. Abstr., *71*, 21789v (1969).
316. (a) Sebulsky, R. T., and A. M. Henke, Am. Chem. Soc., Div. Petrol. Chem., Prepr., *14*, A115 (1969); Chem. Abstr., *74*, 111674v (1971); (b) Sebulsky, R. T., and A. M. Henke, Ind. Eng. Chem. Process Des. Dev., *10*, 272 (1971); Chem. Abstr., *74*, 113606y (1971).
317. Anderson, R. G., and S. H. Sharman, Symp. New Olefin Chem. Am. Chem. Soc., Houston, Feb. 22-27, 1970 Abstr. E27.
318. For reviews, see Refs. 1,3-6,8 and 10-12.
319. Whitmore, F. C., and W. H. James, J. Am. Chem. Soc., *65*, 2088 (1943).
320. Allen, R. H., and L. D. Yats, J. Am. Chem. Soc., *83*, 2799 (1961).
321. Ismailov, R. G., R. S. Alimardanov, S. A. Aliev, Sh. T. Akhmedov, and M. R. Bairamov, Azer. Khim. Zh., *17*, (1967); Chem. Abstr., *69*, 58878h (1968).
322. Mortikov, E. S., et al., Zh. Prikl. Khim. (Leningrad), *50*, 1867 (1977); Chem. Abstr., *87*, 184105p (1977).

323. Carpenter, M. S., W. M. Easter, and T. F. Wood, J. Org. Chem., *16*, 586 (1951).
324. Voronenkov, V. V., V. I. Yulina, and Ye. A. Lazurin, Neftekhimiya, *11*, 857 (1971); Petrol. Chem. USSR, *11*, 242 (1971); Chem. Abstr., *76*, 152849r (1972).
325. Bondarenko, A. V., and M. I. Farberov, Zh. Prikl. Khim., *35*, 1584 (1962).
326. Rodionov, V. M., V. N. Belov, and S. A. Kore, Zh. Obshch. Khim., *23*, 1802 (1953).
327. Olah, G. A., S. H. Flood, and M. E. Moffatt, J. Am. Chem. Soc., *86*, 1060 (1964).
328. Schlatter, M. J., and R. D. Clark, J. Am. Chem. Soc., *75*, 361 (1953).
329. Friedman, B. S., F. L. Morritz, C. J. Morrisey, and R. Koncos, J. Am. Chem. Soc., *80*, 5867 (1958).
330. Schlatter, M. J., Am. Chem. Soc., Div. Petrol. Chem., Prepr. Symp., *1*(2), (Chemicals from Petrol), 77 (1956).
331. Boone, D. E., E. J. Eisenbraun, P. W. K. Flanagan, and R. D. Grigsby, Am. Chem. Soc., Div. Petrol. Chem. Prepr., *15*(1), A77 (1970); Chem. Abstr., *75*, 109649t (1971).
332. Ipatieff, V. N., H. Pines, and R. C. Olberg, J. Am. Chem. Soc., *70*, 2123 (1948).
333. Shvetsova, L. S., Tr. Voronezh. Gos. Univ., *60*(3), 23 (1957); Chem. Abstr., *55*, 27151i (1961).
334. Bethel, D., and V. Gold, J. Chem. Soc., 1905 (1958).
335. Yoneda, N., A. Naka, H. Ohtsuka, and K. Aomura, Nippon Kagaku Kaishi, 2385 (1972); Chem. Abstr., *78*, 71003z (1973).
336. Belov, P. S., V. I. Ponomarenko, and B. L. Irkhin, USSR Patent 434,077 (1974); Chem. Abstr., *81*, 77669e (1974).
337. Yoneda, N., M. Yamaguchi, and H. Ohtsuka, Nippon Kagaku Kaishi, 331 (1973); Chem. Abstr., *78*, 124187w (1973).
338. Matsukuma, A., I. Takagishi, and K. Yoshida, Jpn. Kakai 74-127, 932 (1974); Chem. Abstr., *82*, 155778b (1975).
339. Stevens, D. R., Ind. Eng. Chem., *35*, 655 (1943).
340. Parc, G., Rev. Inst. Fr. Petrol. Ann. Combust. Liq., *15*, 680 (1960); Chem. Abstr., *55*, 4402f (1961).
341. Bryner, F., U.S. Patent 2,726,270 (1955); Chem. Abstr., *50*, *12110e (1956)*.
342. Stevens, D. R., J. Org. Chem., *20*, 1232 (1955).
343. Parc, G., Rev. Inst. Fr. Petrol., Ann. Combust. Liq., *15*, 567 (1960).
344. Bryner, F., U.S. Patent 2,655,547 (1953); Br. Patent 725,873; Chem. Abstr., *49*, 3251g (1955).
345. Starkov, S. P., S. K. Starkova, L. A. Zhidkova, M. N. Volkotrub, and G. S. Leonova, Izv. Vyssh. Uchebn. Zaved. Khim. Teknol., *15*, 1186 (1972); Chem. Abstr., *77*, 164172y (1972).
345a. Stevens, D. R., and Friedman, B. S., unpublished results, cited in Ref. 12, pp. 82 and 92.
346. Rosenwald, R. H., U.S. Patent 2,470,902 (May 24, 1949); Chem. Abstr., *43*, 6235 (1949).
347. Cook, C. D., R. G. Inskeep, A. S. Rosenberg, and E. C. Curtis, J. Am. Chem. Soc., *77*, 1672 (1955).
348. Rosenwald, R. H., U.S. Patent 2,908,718 (Oct. 13, 1959).
349. Lur'e, S. I., J. Gen. Chem. USSR, *16*, 145 (1946).

350. Carpenter, M. S., and W. M. Easter, U.S. Patents 2,476,815 (1949); 2,450,877 (1948).
351. Wood, T. F., U.S. Patent 2,493,797 (1950); Chem. Abstr., *44,* 5548d (1950).
352. Moryashchev, A. K., I. I. Sidorov, E. Kh., Frenklakh, USSR Patent 346,299 (1972); Chem. Abstr., *77,* 139608n (1972).
353. Merkushev, E. B., and V. S. Raida, Zh. Org. Khim., *10,* 405 (1974); Chem. Abstr., *80,* 120419s (1974).
354. Hillers, S., and Z. Berzina, Latv. PSR Zinat. Akad., Vestis Kim. Ser., No. 1, 103 (1962); Chem. Abstr., *59,* 1564 (1963).
354a. Kutz, W. M., and B. B. Corson, J. Am. Chem. Soc., *68,* 1477 (1946).
354b. Caesar, P. D., J. Am. Chem. Soc., *70,* 3623 (1948).
355. Conary, R. E., and R. F. McCleary, U.S. Patent 2,652,405 (1953); Chem. Abstr., *48,* 12178 (1954); Corson, B. B., H. E. Tiefenthal, G. H. Atwood, W. J. Heintzelman, and W. L. Reilly, J. Org. Chem., *21,* 584 (1956).
356. Hofmann, J. E., and A. Schriesheim, J. Am. Chem. Soc., *84,* 953, 957 (1962).
357. Schlatter, M. J., Symp. Petrochemicals Postwar Years, Am. Chem. Soc., Div. Petrol. Chem., Prepr., Chicago, Sept. 9, 1953, p. 79.
358. Schlatter, M. J., J. Am. Chem. Soc., *76,* 4952 (1954).
359. Schlatter, M. J., U.S. Patent 2,801,271 (July 30, 1957).
360. Corson, B. B., W. J. Heintzelman, R. C. Odioso, H. E. Tiefenthal, and F. J. Pavlik, Ind. Eng. Chem., *48,* 1180 (1956).
361. Schneider, A., U.S. Patent 2,648,713 (1953).
362. De Pierre, W. G., Jr., W. R. Edwards, and H. G. Boynton, U.S. Patent 3,052,741 (1962).
363. Schlatter, M. J., U.S. Patent 2,816,940 (1957).
364. Roberts, E., and M. P. Rose, Br. Patent 645,446 (1950).
365. Senkowski, M., Chem. Ber., *24,* 2974 (1891).
366. Roberts, R. M., O. P. Anderson, Jr., and S. E. McGuire, Tetrahedron, *25,* 4523 (1969).
367. Huston, R. C., and K. Goodemoot, J. Am. Chem. Soc., *56,* 2432 (1934).
368. Kozlov, N. S., and A. G. Klein, Uch. Zap. Permsk., Gos. Pedagog. Inst., No. 2, 69 (1967); Ref. Zh., Khim., Abstr. 7B864 (1968); Chem. Abstr., *69,* 106023z (1968).
369. Kozlov, N. S., and A. G. Klein, Zh. Phys. Khim., *41,* 1092 (1968).
370. Kozlov, N. S., and A. G. Klein, Zh. Prikl. Khim. (Leningrad), *41,* 1360 (1968).
371. Kozlov, N. S., and A. G. Klein, Zh. Prikl. Khim. (Leningrad), *42,* 2863 (1969).
372. Cullinane, N. M., and D. M. Leyshan, J. Chem. Soc., 2952 (1957).
373. Gurewitsch, A., Chem. Ber., *32,* 2424 (1899).
374. (a) Liebmann, A., Chem. Ber., *14,* 1842 (1881); (b) Liebmann, A., Chem. Ber., *15,* 150 (1882); (c) Sears, C. A., Jr., J. Org. Chem., *13,* 120 (1948).
375. Meissel, N., Chem. Ber., *32,* 2419 (1899).
376. Merz, V., and W. Weith, Chem. Ber., *14,* 2343 (1881).
377. Anschutz, R., and H. Beckerhoff, Justus Liebigs Ann. Chem., *327,* 218 (1903).
378. Simons, J. H., and S. Archer, J. Am. Chem. Soc., *62,* 1623 (1940).

379. Inatome, M., K. W. Greenlee, J. M. Derfer, and C. E. Boord, J. Am. Chem. Soc., *74,* 292 (1952).
380. Fischer, B., and B. Grutzner, Chem. Ber., *26,* 1646 (1893).
381. Pines, H., L. Schmerling, and V. N. Ipatieff, J. Am. Chem. Soc., *62,* 2901 (1940).
382. Kurashev, M. A., A. V. Topchiev, and Ya. M. Paushkin, Proc. Acad. Sci. USSR, Chem. Sect., *107,* 203 (1956).
383. Friedman, B. S., and F. L. Morritz, J. Am. Chem. Soc., *78,* 2000 (1956).
384. Shuikin, N. I., N. A. Pozdnyak, and Y. P. Egorov, Izv. Akad. Nauk SSSR, Otd. Khim. Nauk, *10,* 1239 (1958).
385. Shuikin, N. I., E. A. Viktorova, and V. P. Litvinov, Vestn. Mosk. Univ. Ser. II, Khim., *12,* No. 5, 121 (1957); Chem. Abstr., *53,* 1211a (1959).
386. Vorozhtsov, N. N., and I. I. Ioffe, Zh. Obshch. Khim., *21,* 1659 (1951).
387. Topchiev, A. V., Ya. M. Paushkin, and M. V. Kurashev, Dokl. Akad. Nauk SSSR, *130,* 559 (1960); Chem. Abstr., *54,* 10921d (1960).
388. Tilicheev, M. D., and K. S. Kuruindin, Neft. Khoz., *19,* 586 (1930).
389. Friedman, B. S., and F. L. Morritz, J. Am. Chem. Soc., *78,* 3430 (1956).
390. Cullinane, N. M., and D. M. Leyshon, J. Chem. Soc., 2942 (1954).
391. Monsanto Chemical Company, Br. Patent 452,335 (1937); Chem. Abstr., *31,* 485 (1937).
392. Stevens, D. R., and W. A. Gruse, U.S. Patent 2,248,827 (1941); Chem. Abstr., *35,* 71768 (1941).
393. Malchick, S. P., and R. B. Hannan, J. Am. Chem. Soc., *81,* 2119 (1959).
394. Viktorova, E. A., N. A. Shuikin, G. S. Korosteleva, and N. G. Baranova, Izv. Akad. Nauk SSSR, Otd. Khim. Nauk, 1518 (1961); Chem. Abstr., *56,* 369h (1962).
395. Caeser, P. D., and P. G. Waldo, U.S. Patent 2,482,084 (1949); Chem. Abstr., *44,* 2030g (1950).
396. Texaco Development Co., Br. Patent, 625,173 (1949); Chem. Abstr., *44,* 2566g (1950).
397. Essner, J. C., Bull. Soc. Chim. Fr., *36,* 212 (1881).
398. Essner, J. C., Bull. Soc. Chim. Fr., *42,* 213 (1884).
398a. Polubentseva, M. F., M. F. Duganova, and V. G. Lipovich, Pererab. Tverd. Topl., No. 2, 292 (1970); Ref. Zh., Khim., Abstr. 7N166 (1971); Chem. Abstr., *76,* 153250a (1972).
399. (a) Khalaf, A. A., and R. M. Roberts, Rev. Chim. (Bucharest), submitted for publication; (b) Low, Chow-Eng, Ph. D. dissertation, University of Texas at Austin, 1970.
400. Schmerling, L., and J. P. West, J. Am. Chem. Soc., *76,* 1917 (1954).
401. Pines, H., W. D. Huntsman, and V. N. Ipatieff, J. Am. Chem. Soc., *73,* 4343 (1951).
402. Nenitzescu, C. D., Rev. Chim. (Bucharest), 7, 349 (1962).
403. Nenitzescu, C. D., I. Necsoiu, A. Glatz, and M. Zalman, Chem. Ber., *92,* 10 (1959).
404. Roberts, R. M., and Y. W. Han, Tetrahedron Lett. 5, (1959); J. Am. Chem. Soc., *85,* 1168 (1963).
405. Weber, S. H., D. B. Spoelstra, and E. H. Polak, Recl. Trav. Chim. Pays-Bas, *74,* 1179 (1955).

406. Shuikin, N. I., N. A. Pozdnyak, and Y. P. Egorov, Izv. Akad. Nauk SSSR, Otd. Khim. Nauk, 1988 (1959).
407. Duganova, V. V., M. F. Polubentseva, K. F. Kosygina, V. V. Chenets, and V. G. Lipovich, Izv. Nauch.-Issled. Inst. Nefte-Uglekhim. Sin. Irkutsk. Univ., *11*, 10 (1969); Chem. Abstr., *77*, 164156w (1972).
408. Huston, R. C., W. B. Fox, and M. N. Binder, J. Org. Chem., *3*, 251 (1938).
409. Huston, R. C., and J. Awuapara, J. Org. Chem., *9*, 401 (1944).
410. Huston, R. C., R. L. Guile, D. L. Bailey, R. S. Curtis, and T. E. Esterdahl, J. Am. Chem. Soc., *67*, 899 (1945).
411. Huston, R. C., and R. Smith, J. Org. Chem., *15*, 1074 (1950).
412. Cramer, C., Chem. Ber., *31*, 2813 (1848).
413. Bistrzycki, I., and W. Fattau, Chem. Ber., *28*, 989 (1895); *30*, 124 (1897); Bistrzycki, I., and P. Simons, Chem. Ber., *31*, 2812 (1898).
414. Simons, P., Chem. Ber., *31*, 2821 (1898).
415. Smith, R. A., and S. Natelson, J. Am. Chem. Soc., *53*, 3476 (1931).
416. Brouwer, D. M., and H. Hogeveen, in *Progress in Physical Organic Chemistry*, A. Streitweiser, Jr., and R. W. Taft, eds., Wiley-Interscience, New York, 1972, p. 206.
417. Geiseler, G., and D. Braun, Chem. Ber., *100*, 3231 (1967).
418. Nooi, J. R., J. J. Muller, M. C. Testa, and S. Willemse, Recl. Trav. Chim. Pays-Bas, *88*, 398 (1969); Chem. Abstr., *70*, 114716p (1969).
419. Voore, H., Esti NSV Tead. Acad. Tiom. Keem., Geol., *19*, 218 (1970); Chem. Abstr., *74*, 3275g (1971).
420. Davtyan, E. G., S. D. Mekhtiev, Sh. S. Shchegol, M. M., Lyushin, V. F. Mamedova, and V. B. Lapin, Azerb. Neft. Khoz. No. 8, 39 (1971), Chem. Abstr., *76*, 24796j (1972).
421. Navruzov, Kh., A. B. Kuchukarov, S. A. Sarankina, F. K. Kurbanov, and Kh. Z. Babaeva, Tr. Tashkt. Politekh. Inst., *74*, 23-24 (1971); Chem. Abstr., *80*, 108097c (1974).
422. Kheifits, L. A., A. S. Podberezina, and Yu. D. Kaznacheev, USSR Patent 427,921 (1974); Chem. Abstr., *81*, 91133g (1974).
423. Miethchen, R., and C. F. Kroeger, Z. Chem., *15*, 135 (1975); Chem. Abstr., *83*, 42418m (1975).
424. Shakhgel'diev, M. A., R. M. Shamkhalov, and E. K. Mamedov, Kratk. Tezisy-Vses. Soveshch. Probl. Mekh. Geterolitichesk. Reakt., *78* (1974); Chem. Abstr., *85*, 62749q (1976).
425. Miethchen, R., A. Gaertner, and C. F. Kroeger, Z. Chem., *17*, 443 (1977); Chem. Abstr., *88*, 104795w (1978).
426. Ndandji, C., M. Desbois, R. Gallo, and J. Metzger, C. R. Acad. Sci., Ser. C *285*, 591 (1977); Chem. Abstr., *88*, 135859z (1978).
427. Rhone-Poulenc Industries S. A., Ger. Offen. 2,740,495 (1978); Chem. Abstr., *89*, 75303x (1978).
428. Ndandji, C., and M. Desbois, Ger. Offen. 2,740,525 (1978); Chem. Abstr., *89*, 42738q (1978).
429. Berry, T. M., and E. E. Reid, J. Am. Chem. Soc., *49*, 3142 (1927).
430. Bodroux, D., Ann. Chim. (Paris), (10); *11*, 511 (1929).
431. Nametkin, S. S., and J. S. Pokrovskaya, J. Gen. Chem. USSR, 7, 69 (1937).
432. (a) Corson, B. B., and V. N. Ipatieff, J. Am. Chem. Soc., *59*, 645 (1937); (b) Truffault, R., C. R. Acad. Sci., *202*, 1286 (1936).
433. Tudor, E., and M. Iovu, Rev. Chim. (Bucharest), *20*, 813 (1975); Chem. Abstr., *84*, 98913g (1976).

434. Tilecheev, M. D., J. Appl. Chem. USSR, *12*, 735 (1939).
435. Institut Francais du Petrole, des Carburants et Lubrifiants, Br. Patent 870,772 (1961).
436. Pajeau, R., C. R. Acad. Sci., *252*, 3060 (1961).
437. Hoffman, F., and C. Wulff, Br. Patent 307,802 (1928); U.S. Patent 1,898,627 (1933).
438. Corson, B. B., and V. N. Ipatieff, Organic Synthesis, Vol. 19, Wiley, New York, 1939 p. 36.
439. Buu-Hoi, Ng. Ph., and P. Cagniant, C. R. Acad. Sci., *216*, 381 (1943).
440. Buu-Hoi, Ng. Ph., P. Cagniant, and C. Mentzer, Bull. Soc. Chim. Fr., (5), *11*, 127 (1944).
441. Mekhtiev, S. D., and T. A. Pashaev, Azerb. Khim. Zh., No. 2, *39* (1959).
442. Simons, J. H., and H. Hart, J. Am. Chem. Soc., *66*, 1309 (1944).
443. Nicolescu, I. V., M. Iovu, and G. I. Nikishin, Izv. Akad. Nauk SSSR, Otd. Khim. Nauk, *94* (1960).
444. Sidorova, N. G., I. P. Tsukervanik, and E. Pak, Zh. Obshch. Khim. *24*, 94 (1954).
445. Zavgorodnii, S. V., V. G. Ovchinnikov, G. K. Ovchinnikova, S. A. Mramomova, and I. S. Bass, Neftekhimiya, *12*, 200 (1972); Chem. Abstr., *77*, 61386s (1972).
446. Truffault, R., C. R. Acad. Sci., *202*, 1286 (1936).
447. Ipatieff, V. N., and H. Pines, J. Am. Chem. Soc., *61*, 3374 (1939).
448. Colomb, Mlle., Ann. Min. Carb. Doc. Fr., *135*, 545 (1946).
449. Cagniant, P., A. Deluzarche, and G. Chatelus, C. R. Acad. Sci., *224*, 1064 (1947).
450. Monteils, Y., Bull. Soc. Chim. Fr., 747 (1953).
451. Pokrovskaya, E. S., and N. A. Shimanko, Dokl. Nauk SSSR, *123*, 109 (1958); Chem. Abstr., *53*, 6146b (1959).
452. Sidorova, N. G., and I. I. Kartseva, Zh. Obshch. Khim., *32*, 2785e (1962); Chem. Abstr., *58*, 8937 (1963).
453. Zalygin, L. L., G. N. Koshel, and M. I. Farberov, Uch. Zap. Yaroslav. Tekhnol. Inst., *22*, 62 (1972); Chem. Abstr., *80*, 59583x (1974).
454. Mirzaeva, A. K., N. G. Sidorova, and A. F. Pankratova, Nauchn. Tr. Tashk. Gos. Univ. (USSR), *462*, 35 (1974); Chem. Abstr., *84*, 30546j (1976).
455. Pines, H., A. E. Edeleanu, and V. N. Ipatieff, J. Am. Chem. Soc., *67*, 2193 (1945).
456. Sidorova, N. G., and I. P. Tsukervanik, J. Gen. Chem. USSR, 1543 (1957); Chem. Abstr., *52*, 3747g (1958).
457. Shakhgel'diev, M. A., R. M. Shamkhalov, and E. A. Kyazimov, Dokl. Akad. Nauk Azb. SSR, *30*, 28 (1974); Chem. Abstr., *81*, 49346d (1974).
458. Yusupov, A., A. R. Abdurasuleva, and D. Ziyaev, Deposited Doc. (USSR), VINITI 323-76 (1976); Chem. Abstr., *88*, 89259d (1978).
459. Shimanko, N. A., and E. S. Pokrovskaya, Dokl. Akad. Nauk SSSR, *129*, 1313 (1959).
460. Monteils, Y., Bull. Soc. Chim. Fr., 637 (1951).
460a. Orchinnikova, G. K., and D. A. Pisanenko, Izb. Vyssh. Uchebn. Zaved., Khim. Khim. Tekhnol., *23*, 392 (1980); Chem. Abstr., *93*, 132164m (1980).

461. Paushkin, Ya. M., E. A. Laryutina, and V. K. Pel'ttser, Vestsi Akad. Navuk B. SSR, Ser. Khim. Navuk, No. 6, 97 (1971); Chem. Abstr., *76*, 72130j (1972).
462. Tsukervanik, I. P., and N. G. Sidorova, J. Gen. Chem. USSR, *7*, 641 (1937); Chem. Abstr., *31*, 5780g (1937).
463. (a) Mekhtiev, S. D., and T. A. Pashev, Azerb. Khim. Zh., *2*, 39 (1959); (b) Nicolescu, I. V., M. Iovu, and G. I. Nikishin, Izv. Akad. Nauk SSSR, Otd. Khim. Nauk, *94* (1960).
464. Bordoux, D., Ann. Chim. (Paris), *11*, 511 (1929).
465. Pashaev, T. A., F. A. Mamedov, and F. A. Iseeva, Azerb. Khim. Zh., No. 4, 95 (1970); Chem. Abstr., *75*, 19811k (1971).
466. Pisanenko, D. A., and G. K. Bereznitskii, Vestn. Kiev. Politekh. Inst., Khim. Mashinostr. Tekhnol., *6*, (1979); Chem. Abstr., *93*, 26006b (1980).
467. Kozlov, N. S., A. G. Klein, and Yu. A. Galishevskii, Uch. Zap. Permsk. Gos. Pedagog. Instr., *85*, 3 (1970); Chem. Abstr., *77*, 139509f (1972).
468. Koslov, N. S., A. G. Klein, and Yu. A. Galishevskii, 4th Khim. Khim. Tekhnol. Obl. Nauchno. Tekh. Konf. (Mater), 1973, *2*, 61 (1973); Chem. Abstr., *82*, 85951b (1975).
469. Koslov, N. S., A. G. Klein, and Yu. A. Galishevskii, Neftekhimiya, *14*, 853 (1974); Chem. Abstr., *82*, 170256y (1975).
470. Kozlov, N. S., A. G. Klein, and Yu. A. Galishevskii, V sb., XI Mendeleevsk. Sezd po Obshch. Prikl. Khim. Ref. Dokl. Soobshch., 339 (1974); Chem. Abstr., *84*, 179802x (1975).
471. Koslova, N. S., A. G. Klein,Yu. A. Galishevskii, and D. M. Zimakova, Izb. Vyssh. Uchebn. Zaved., Khim. Khim. Tekhnol., *19*, 704 (1976); Chem. Abstr., *85*, 77792x (1976).
472. Drabkin, A. E., A. S. Fedotov, and N. F. Fedotova, Zh. Prikl. Khim. (Leningrad), *53*, 888 (1980); Chem. Abstr., *93*, 167782p (1980).
473. Koslov, N. S., A. G. Klein, and Yu. A. Galishevskii, Neftekhimiya, *15*, 699 (1975); Chem. Abstr., *84*, 30584v (1976).
474. Egidis, F. M., S. N. Lutskaya, I. V. Kokhavova, and M. N. Volkotrub, Zh. Prikl. Khim. (Leningrad), *41*, 404 (1968); Chem. Abstr., *69*, 58895m (1968).
475. Iovu, M., M. Manolescu, and N. Miclovici, Bull. Soc. Chim. Fr., *11*, 4004 (1970); Chem. Abstr., *74*, 53199f (1971).
476. Kozlov, N. S., A. G. Klein, and Yu. A. Galishevskii, Vestsi Akad. Navuk B. SSR, Ser. Khim. Navuk, No. 2, 97 (1973); Chem. Abstr., *78*, 159113w (1973).
477. Kretchmer, R. A., and M. B. McCloskey, J. Org. Chem., *37*, 1989 (1972).
478. Ismailov, N., and A. P. Abdurasuleva, Izv. Vyssh. Uchebn. Zaved. Khim. Khim. Tekhnol., *19*, 1282 (1976); Chem. Abstr., *85*, 159568t (1976).
479. Bertolini, N., B. Calcagno, and M. Chirga, U.S. Patent 3,801,665 (1974); Chem. Abstr., *81*, 145727d (1974).
480. Akhmedova, R. A., and Sh. A. Romazanova, Uch. Zap.-Minist. Vyssh. Sredn., Spets. Orbaz. Az. SSR, Ser. Khim. Nauk, *39* (1975); Chem. Abstr., *85*, 123652b (1976).
481. Sidorova, N. G., and F. M. Saidova, Zh. Obshch. Khim., *33*, 2213 (1936); Chem. Abstr., *59*, 13899e (1963).

482. Sidorova, N. G., and I. R. Tsukervanik, J. Gen. Chem. USSR, *10*, 2073 (1940).
483. Nenitzescu, C. D., in *Carbonium Ions*, Vol. 2, (G. A. Olah and P. v. R. Schleyer, eds.), Wiley-Interscience, New York, 1970, p. 501.
484. Jones, W. O., Br. Patents 743,153 (1956); 861,792 (1961).
485. Walousky, R., N. Maoz, and Z. Nir, Synthesis No. 12, 656 (1970).
486. Mardanov, M. A., V. M. Akhmedov, and F. R. Mlieva, Zh. Org. Khim., *7*, 953 (1971).
487. Pashaev, T. A., Ya. G. Abdullaev, and F. A. Iseeva, Uch. Zap. Azerb. Gos. Med. Instr., *30*, 202 (1969); Chem. Abstr., *76*, 126495c (1972).
487a. Sidorova, N. G., Zh. Obshch. Khim., *32*, 2649 (1962); Chem. Abstr., *58*, 7847c (1963).
488. (a) Rasulov, Ch. K., Sh. G. Sadykhov, I. I. Sidorchuk, T. N. Mashkova, and L. B. Zeinalova, Azerb. Khim. Zh., *45* (1977); Chem. Abstr., *89*, 59732f (1978); (b) Sadykhov, Sh. G., I. I. Sidorchuk, Ch. K. Rasulov, and Yu. K. Dzhafavov, Azerb. Khim Zh., 35 (1978); Chem. Abstr., *90*, 6027u (1979).
489. Zavgorodnii, S. V., and Ya. B. Kozlikovskii, Zh. Org. Khim., *8*, 799 (1972); J. Org. Chem. USSR, *8*, 809 (1972); Chem. Abstr., *77*, 19289n (1972).
490. Linsk, J., J. Am. Chem. Soc., *72*, 4257 (1950).
490a.. Ipatieff, V. N., E. E. Meisinger, and H. Pines, J. Am. Chem. Soc., *72*, 2772 (1950).
491. Lipovich, V. G., V. V. Chenets, A. I. Rudenkov, and I. V. Kalechits, Zh. Org. Khim., *4*, 1779 (1967); J. Org. Chem. USSR, *4*, 1717 (1967); Chem. Abstr., *70*, 28487h (1969).
492. Sidorova, N. G., and A. I. Bokova, Zh. Org. Khim., *1*, 2176 (1965); J. Org. Chem. USSR, *1*, 2218 (1965); Chem. Abstr., *64*, 11105c (1966).
493. Ipatieff, V. N., H. R. Appell, and H. Pines, J. Am. Chem. Soc., *72*, 4260 (1950).
494. Pines, H., D. R. Strehlau, and V. N. Ipatieff, J. Am. Chem. Soc., *71*, 3534 (1949).
495. Pines, H., D. R. Strehlau, and V. N. Ipatieff, J. Am. Chem. Soc., *72*, 5521 (1950).
496. Schrauth, W., and K. Quasebarth, Chem. Ber., *57*, 854 (1924).
497. (a) Sidorova, N. G., and V. Yu. Telly, Zh. Obshch. Khim., *31*, 2149 (1961); Chem. Abstr., *56*, 2353i (1962). (b) Sidorova, N. G., and N. I. Frakman, Zh. Obshch. Khim., *31*, 2155 (1961); Chem. Abstr., *56*, 2354i (1962); (c) Sidorova, N. G., and A. I. Ivanova, Zh. Obshch. Khim., *32*, 2790 (9162); Chem. Abstr., *58*, 8937f (1963).
498. Sidorova, N. G., and I. Sh. Shamakhmudova, Uzb. Khim. Zh., *6*, 57 (1962); Chem. Abstr., *58*, 2341b (1963).
499. Kheifits, L. A., G. I. Moldovanskaya, E. V. Broun, and V. N. Belov, Zh. Obshch. Khim., *30*, 1716 (1960).
500. Kheifits, L. A., L. M. Shulov, E. V. Broun, and V. N. Belov, Zh. Obshch. Khim., *31*, 672 (1961).
501. Kheifits, L. A., L. M. Shulov, and V. N. Belov, Zh. Obshch. Khim., *32*, 1474 (1962).
502. Kheifits, L. A., and A. E. Gol'dovskii, Zh. Obshch. Khim., *33*, 2048 (1963).

503. Kheifits, L. A., L. M. Shulov, A. V. Kokhmanskii, T. F. Gavrilova, and V. N. Belov, Zh. Obshch. Khim., *33*, 2051 (1963).
504. Kheifits, L. A., L. M. Shulov, A. V. Kokhmanskii, and V. N. Belov, Zh. Obshch. Khim., *33*, 2412 (1963).
505. Kheifits, L. A., L. M. Shulov, and V. N. Belov, Zh. Obshch. Khim., *33*, 2748 (1963).
506. Moldovanskaya, G. I., L. A. Kheifits, A. V. Kokhmanskii, and V. N. Belov, Zh. Obshch. Khim., *33*, 3392 (1963).
507. Kheifits, L. A., G. I. Moldovanskaya, and L. M. Shulov, Zh. Org. Khim., *1*, 1057 (1965).
508. Kheifits, L. A., and L. M. Shulov, Zh. Org. Khim., *1*, 1063 (1965).
509. Aul'chenko, I. S. and L. A. Kheifits, Zh. Org. Khim., *2*, 2055 (1966).
510. Zakharova, N. A., K. E. Kruglyakova, G. N. Bodganov, L. A. Kheifits, and N. M. Emanuel', Zh. Obshch. Khim., *37*, 801 (1967).
511. Kheifits, L. A., I. S. Aul'chenko, and G. M. Shchegoleva, Zh. Org. Khim., *2*, 2059 (1966).
512. Shulov, L. M., T. F. Gavrilova, and L. A. Kheifits, Zh. Org. Khim., *3*, 1819 (1967).
513. Shulov, L. M., T. F. Gavrilova, and L. A. Kheifits, Zh. Org. Khim., *5*, 77 (1969); J. Org. Chem. USSR, *5*, 75 (1969); Chem. Abstr., *70*, 87986t (1969); see references therein.
514. Moskvichev, V. I., and L. A. Kheifits, Zh. Org. Khim., *9*, 2256 (1973); J. Org. Chem. USSR, *9*, 2273 (1973); Chem. Abstr., *80*, 83267t (1974).
515. Gavrilova, T. F., I. S. Kul'chenko, L. A. Kheifits, N. D. Antonova, and O. A. Subbotin, Zh. Org. Khim., *9*, 2260 (1973); J. Org. Chem. USSR, *9*, 2277 (1973); Chem. Abstr., *80*, 83268u (1974) and references therein.
516. Babakhanov, R. A., N. A. Dzhafarova, and I. G. Mursakulov, Azerb. Khim. Zh., *47*, (1971); Chem. Abstr., *76*, 140008c (1972).
517. Margosian, D., D. Sparks, and P. Kovacic, J. Chem. Soc., Chem. Commun., 275 (1980).
518. Vasil'eva, L. L., Neft Gas, 128 (1974); Chem. Abstr., *84*, 30579x (1976).
519. (a) Piasanenko, D. A., and S. A. Nesterenko, Zh. Org. Khim., *15*, 1911 (1979); (b) Piasanenko, D. A., and S. A. Nesterenko, Izv. Vyssh. Uchebn. Zaved., Khim. Khim. Tekhnol., *21*, 1095 (1978); Chem. Abstr., *90*, 6028v (1979).
520. (a) Lipovich, V. G., V. V. Chenets, and A. G. Sakhabutdinov, Zh. Org. Khim., *1*, 974 (1965); J. Org. Chem. USSR, *1*, 981 (1965); Chem. Abstr., *63*, 6887h (1965); (b) Lipovich, V. G., and V. V. Chenets, Zh. Org. Khim., *1*, 2151 (1965); J. Org. Chem. USSR, *1*, 2192 (1965); (c) Lipovich, V. G., and V. M. Kazantseva, Zh. Org. Khim., *2*, 2035 (1965); J. Org. Chem. USSR, *2*, 1997 (1965).
521. Roberts, R. M., Intra-Sci. Chem. Rep., *6*, 89 (1972); Lin, Y. -T. Ph.D. dissertation, University of Texas at Austin, 1966.
522. (a) Sidorova, N. G., Zh. Obshch. Khim., *24*, 255 (1954); (b) Sidorova, N. G., Zh. Org. Khim., *32*, 2642 (1962); (c) Sidorova, N. G., Zh. Obshch. Khim., *21*, 869 (1951).
523. (a) Rosowsky, A., *Heterocyclic Compounds with Three and Four-*

Membered Rings, Part I, (A. Weissberger, ed.), Wiley-Interscience, New York, 1964, p. 432; (b) Schriesheim, A., in ref. 12, p. 543.

524. Schaarschmidt, A., L. Herman, and B. Szemzo, Chem. Ber., *58*, 1914 (1925).
525. (a) I. G. Farbenindustrie, Ger. Patent, 523,436 (1931); (b) Kipstein, K. H., Can. Patent 340,555 (1934); Chem. Abstr., *28*, 40678 (1934).
526. (a) Colonge, J., and P. Rochas, Bull. Soc. Chim. Fr., 818 (1948); Chem. Abstr., *43*, 2182d (1949); (b) Bull. Soc. Chim. Fr., 822 (1948); Chem. Abstr., *43*, 2182g (1949); (c) Bull. Soc. Chim. Fr., 825 (1948); Chem. Abstr., *43*, 2182i (1949); (d) Bull. Soc. Chim. Fr., 827 (1948); Chem. Abstr., *43*, 2182i (1949).
527. I. G. Farbenindustrie, Br. Patent 354,992 (1930); Chem. Abstr., *26*, 5574 (1932); Fr. Patent 716,604 (1931); Chem. Abstr., *26*, 2198 (1932).
528. Houston, R. C., R. L. Guile, J. J. Sculatis, and W. N. Wasson, J. Org. Chem., *6*, 252 (1941).
529. (a) Shorygina, N. V., Zh. Obshch. Khim., *21*, 1273 (1951); Chem. Abstr., *46*, 2000i (1952); (b) Iyers, Ya., and Kh. Urbel', Zh. Org. Khim., *15*, 790 (1979).
530. Suga, S., T. Nakajima, A. Takagi, and T. Takahashi, Kogyo Kagaku Zasshi, *70*, 487 (1967); Chem. Abstr., *68*, 59198c (1968).
531. Hata, S., H. Matsuda, and S. Matsuda, Kogyo Kagaku Zasshi, *70*, 2291 (1967).
532. Nakajima, T., S. Suga, T. Sugita, and K. Ichikawa, Bull. Chem. Soc. Jpn, *40*, 2980 (1967).
533. Milstein, N., J. Heterocycl. Chem., *5*, 337 (1968).
534. Nakajima, T., S. Suga, T. Sugita, and K. Ichikawa, Tetrahedron, *25*, 807 (1969).
535. Nakamoto, Y., T. Nakajima, and S. Suga, Kogyo Kagaku Zasshi, *72*, 2594 (1969).
536. Hata, S., K. Ono, H. Matsuda, and S. Matsuda, Kogyo Kagaku Zasshi, *73*, 1680 (1970).
537. Smith, R. A., and S. L. Natelson, J. Am. Chem. Soc., *53*, 3476 (1931).
538. Theimer, E. T., U.S. Patent 2,125,968 (1938), and Abstract, Div. Org. Chem., 99th Meet. Am. Chem. Soc., Cincinnati, Ohio, Apr. 1940, p. 42.
539. Likhterov, V. R., and V. S. Etlis, Zh. Obshch. Khim., *27*, 2903 (1957).
540. Nakajima, T., Y. Nakamoto, and S. Suga, Bull. Chem. Soc. Jpn, *48*, 960 (1975).

540a. Segi, M., M. Takebe, S. Masuda, T. Nakajima, and S. Suga, Bull. Chem. Soc. Jpn., *55*, 167 (1982).

541. Inoue, M., T. Sugita, Y. Kiso, and K. Ichikawa, Bull. Chem. Soc. Jpn., *49*, 1063 (1976).
542. Inoue, M., T. Sugita, and K. Ichikawa, Bull. Chem. Soc., Jpn, *51*, 174 (1978).
543. Parker, R. E., and N. S. Isaacs, Chem. Rev., *59*, 768 (1959).
544. Hofmann, J. E., and A. Schriesheim, in Ref. 12, Chap. 19.

544a. Somerville, W. T., and P. E. Spoerri, J. Am. Chem. Soc., *72*, 2185 (1950).

544b. Somerville, W. T., and P. E. Spoerri, J. Am. Chem. Soc., *73*, 697 (1951).

545. Tsukervanik, I. P., and Ch. Sh. Kadyrov, Uzb. Khim. Zh., 45 (1960); Chem. Abstr., *55*, 13394e (1961).
546. Kadyrov, Ch. Sh., and I. P. Tsukervanik, Dokl. Akad. Nauk Uzb. SSR, 33 (1960); Chem. Abstr., *56*, 5855c (1962).
547. Kadyrov, Ch. Sh., and I. P. Tsukervanik, Dokl. Akad. Nauk Uzb. SSR, 27 (1960); Chem. Abstr., *58*, 2420d (1963).
548. Tsukervanik, I. P., and Ch. Sh. Kadyrov, Dokl. Akad. Nauk Uzb. SSR, 31 (1958); Chem. Abstr., *53*, 16083i (1959).
549. Brauman, J. I., and A. J. Pandell, J. Am. Chem. Soc., *89*, 5421 (1967).
550. Brauman, J. I., and A. Solladie-Cavallo, Chem. Commun., 1124 (1968).
550a. Nakajima, T., S. Masuda, S. Nakashima, T. Kondo, Y. Nakamoto, and S. Suga, Bull. Chem. Soc. Jpn., *52*, 2377 (1979).
551. Braz, G. I., Dokl. Akad. Nauk SSSR, *87*, 589 (1952); *87*, 747 (1952).
552. Milstein, N., J. Heterocycl. Chem., *5*, 339 (1968).
553. Gensler, W. J., and J. C. Rockett, J. Am. Chem. Soc., *77*, 3262 (1955).
554. Gensler, W. J., and W. R. Koehler, J. Org. Chem., *27*, 2754 (1962).
555. Gensler, W. J., and S. K. Dheer, J. Org. Chem., *46*, 4051 (1981).
556. Eijkman, J. F., Chem. Weekbl. *1*, 421 (1904); *2*, 229 (1905); *4*, 191, 727 (1907).
557. Arnold, R. T., J. S. Buckley, and J. Richter, J. Am. Chem. Soc., *69*, 2322 (1947).
558. Shaver, F. W., U.S. Patent 2,587,540 (1952); Chem. Abstr., *46*, 9603c (1952).
559. Rinehart, K. L., and D. H. Gustafson, J. Org. Chem., *26*, 1836 (1960).
560. Christian, R. V., J. Am. Chem. Soc., *74*, 1591 (1952).
561. Mosby, W. L., J. Am. Chem. Soc., *74*, 2564 (1952).
562. Truce, W. E., and C. E. Olson, J. Am. Chem. Soc., *74*, 4721 (1952).
563. Olson, C. E., and A. R. Bader, Org. Synth., *35*, 95 (1955).
563a. Phillips, D. D., J. Am. Chem. Soc., *77*, 3658 (1955).
564. Kerr, C. A., and I. D. Rae, Aust. J. Chem., *31*, 341 (1978); Chem. Abstr., *88*, 152280w (1978).
565. Kadyrov, Ch. Sh., Uzb. Khim. Zh., No. 8, 52 (1964); Chem. Abstr., *61*, 4254a (1964).
566. Kadyrov, Ch. Sh., and V. A. Barashkin, Zh. Org. Khim., *4*, 1302 (1968); Chem. Abstr., *69*, 67058t (1968).
567. Kadyrov, Ch. Sh., Kh. A. Suerbaev, and A. A. Azizov, Khim. Geterotsikl. Soedin., 77 (1973); Chem. Abstr., *78*, 111087p (1973).
567a. Bruce, D. B., A. J. S. Sorrie, and R. H. Thomson, J. Chem. Soc., 2403 (1953).
568. Laipanov, D. Z., and Ch. Sh. Kadyrov, Uzb. Khim. Zh., *14*, 44 (1970); Chem. Abstr., *74*, 53231k (1971).
569. Kadyrov, Ch. Sh., M. Kholmatov, D. Z. Laipanov, and A. A. Azizov, Biol. Aktiv. Soedin, 120 (1968); Chem. Abstr., *71*, 112567n (1969).
570. Barashkin, V. A., and Ch. Sh. Kadyrov, Zh. Org. Khim., *11*, 2341 (1975); Chem. Abstr., *84*, 43553j (1976).
571. Kretchmer, R. A., and M. B. McCloskey, J. Org. Chem., *37*, 1989 (1972).

572. Kadyrov, Ch. Sh., and D. Z. Laipanov, Zh. Org. Khim., *2*, 1272 (1966); Chem. Abstr., *66*, 46175b (1967).
573. Kadyrov, Ch. Sh., and S. V. Levushkind, Uzb. Khim. Zh., *12*, 40 (1968); Chem. Abstr., *70*, 37393b (1969).
574. Kadyrov, Ch. Sh., D. Z. Laipanov, and S. V. Levushkina, Dokl. Akad. Nauk Uzb. SSR, *27*, 32 (1970); Chem. Abstr., *74*, 31623n (1971).
575. Kadyrov, Ch. Sh., Uzb. Khim. Zh., No. 8, 89 (1964); Chem. Abstr., *62*, 6442g (1965).
576. Kholmatov, M., and Ch. Sh. Kadyrov, Tr. Ferg. Politekh. Inst., No. 2, 78 (1969); Chem. Abstr., *78*, 71741p (1973).
577. Kadyrov, Ch. Sh., A. V. Gordeeva, M. Khalmatov, and A. R. Akhmadzhanov, Tr. Taskh. Fer. Politekh. in-tov, 158 (1974); Chem. Abstr., *83*, 9598x (1975).
578. Kadyrov, Ch. Sh., and M. N. Kosyakovskaya, Uzb. Khim. Zh., *10*, 29 (1966); Chem. Abstr., *65*, 13690a (1966).
579. Phillips, D. D., Chem. Inds., 54 (1956).
580. Phillips, D. D., and D. N. Chatterjee, J. Am. Chem. Soc., *80*, 1360 (1956).
581. Phillips, D. D., and D. N. Chatterjee, J. Am. Chem. Soc., *80*, 1911 (1956).
582. Chatterjee, D. N., and S. P. Bhattacharjee, Tetrahedron, *27*, 4153 (1971); Chem. Abstr., *75*, 117715b (1971).
583. Chatterjee, D. N., and S. P. Bhattacharjee, Indian J. Chem., *12*, 958 (1974); Chem. Abstr., *87*, 97794r (1975).
584. Ismailov, A. G., M. A. Rustamov, and A. A. Akhmedov, Zh. Org. Khim., *7*, 1427 (1977).
585. Chatterjee, D. N., and M. Sarkar, Curr. Sci., *48*, 347 (1979); Chem. Abstr., *91*, 56680t (1979).
586. Andryushina, V. A., E. V. Popova, O. S. Anisimova, and G. S. Grinenko, Zh. Org. Khim., *10*, 222 (1974); Chem. Abstr., *80*, 121186a (1974).
587. Andryushina, V. A., E. V. Popova, O. S. Anisimova, and G. S. Grinenko, Zh. Org. Khim., *12*, 897 (1976); Chem. Abstr., *85*, 77725c (1976).
588. Kondo, M., K. Yasui, M. Miake, T. Shiraishi, and H. Iwasaki, Nippon Nogei Kagaku Kaishi, 276 (1978); Chem. Abstr., *88*, 192736f (1978).
589. Beyer, H., Ber., *70B*, 1011 (1937); Chem. Abstr., *31*, 4968 (1937).
590. Beyer, H., Ber., *70B*, 1482 (1937); Chem. Abstr., *31*, 66397 (1937).
591. Boese, A. B., Jr., Ind. Eng. Chem., *32*, 16 (1940).
592. Hurd, C. D., J. Am. Chem. Soc., *47*, 2777 (1925).
593. Ploeg, W., Recl. Trav. Chim Pays-Bas, *45*, 342 (1926).
594. Packendroff, K., N. D. Zelinsky, and L. Leder-Packendorff, Chem. Ber., *66*, 1069 (1933).
595. Spring, F. S., and T. Vickerstaff, J. Chem., Soc., 1873 (1935).
596. Williams, J. W., and J. M. Osborn, J. Am. Chem. Soc., *61*, 3438 (1939).
597. Canevet, J. C., and Y. Graff, Tetrahedron, *34*, 1935 (1978).
598. Canevet, J. C., and Y. Graff, Helv. Chim. Acta, *62*, 1613 (1979); Chem. Abstr., *91*, 157371g (1979).
599. Canevet, J. C., and O. Guilloton, Bull. Soc. Chim. Fr., 96 (1981).
600. Broquet, C., and R. Quelet, C. R. Acad. Sci., *252*, 3066 (1961); Chem. Abstr., *55*, 24660i (1961).

601. Broquet, C., and R. Quelet, Bull. Soc. Chim. Fr., 1882 (1962); Chem. Abstr., *58*, 12480a (1963).
602. Broquet, C., and J. Bedin, C. R. Acad. Sci., *261*(5) (Group 8), 1335 (1965); Chem. Abstr., *63*, 16251h (1965).
603. Broquet, C., and J. Bedin, C. R. Acad. Sci, Ser. C, *262*(26), 1891 (1966); Chem. Abstr., *65*, 13535g (1966).
604. Broquet, C., and J. Bedin, Bull. Soc. Chim. Fr., 1909 (1967); Chem. Abstr., *67*, 82010g (1967).
605. Bedin, J., and C. Broquet, Bull. Soc. Chim. Fr., 4340 (1972); Chem. Abstr., *78*, 71575n (1973).
606. Waples, D. W., Diss. Abstr. Int. B, *32*(8), 4505 (1972).
607. Waples, D. W., and J. I. Brauman, Chem. Commun., 1075 (1971).
608. Truce, W. E., and E. P. Hoeger, J. Am. Chem. Soc., *76*, 5357 (1954).

5

Alkylations with Di- and Polyfunctional Alkylating Agents

I. ALKYLATIONS WITH DI- AND POLYHALOALKANES

A. Nonselective Alkylations with Linear Dihaloalkanes

1. Dihalomethanes

As early as 1884, Friedel and Crafts investigated the action of dichloromethane on both benzene and toluene in the presence of aluminum chloride [1]. They reported that the final products from benzene were diphenylmethane, anthracene, and toluene and those from toluene were ditolylmethane and dimethylanthracene, together with *m*-, *p*-, and "probably" *o*-xylene. Equations (1) to (3) were offered to account for the results from benzene.

$$2\,C_6H_6 + CH_2Cl_2 \xrightarrow{AlCl_3} C_6H_5{-}CH_2{-}C_6H_5 + 2HCl \qquad (1)$$

$$C_6H_5{-}CH_2{-}C_6H_5 + 2CH_2Cl_2 \xrightarrow{AlCl_3} C_{14}H_{10} \text{ (anthracene)} + 3HCl \qquad (2)$$

$$C_6H_6 + CH_3Cl \xrightarrow{AlCl_3} C_6H_5CH_3 + HCl \qquad (3)$$

A few years later, the aluminum chloride-catalyzed reaction of toluene with dichloromethane was carefully reinvestigated by Lavaux [2-5], who found the products to be more complex than originally claimed by Friedel and Crafts. Thus, in addition to benzene and isomeric xylenes, Lavaux [2,3] also characterized three isomeric dimethylanthracenes, 2-methylanthracene, di-*m*-tolylmethane, and di-*p*-tolylmethane, Finding that pure di-*p*-tolylmethane gave a mixture of isomeric dimethylanthracenes upon treatment with dichloromethane and $AlCl_3$, Lavaux concluded that the formation of these isomers was the result of methyl migrations [4]. The production of benzene was explained in terms of toluene demethylation, and that of 2-methylanthracene in terms of reaction of benzene and toluene with dichloromethane.

The reactions of higher methylated benzenes with dihalomethanes was also studied. In 1887, Friedel and Crafts described the action of dichloromethane on *m*-xylene and pseudocumene in the presence of aluminum chloride [6]. Crystalline tetramethylanthracene (mp 162 to 163°C), and some liquid products were formed from *m*-xylene. Pseudocumene was shown to yield durene, a small amount of tetramethylanthracene (identical with that obtained from *m*-

xylene), a hexamethylanthracene (mp 220°C), and a hepta-or octamethylanthracene (mp 290°C). Being unaware of possible methyl reorientation reactions, Friedel and Crafts believed that the tetramethylanthracene obtained from *m*-xylene should have either structure 1 or 2 and the hexamethylanthracene from pseudocumene should have either structure 3 or 4.

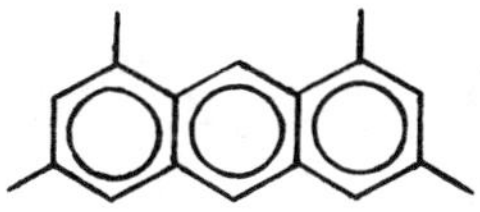

1

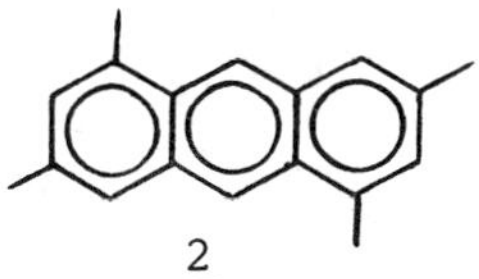

2

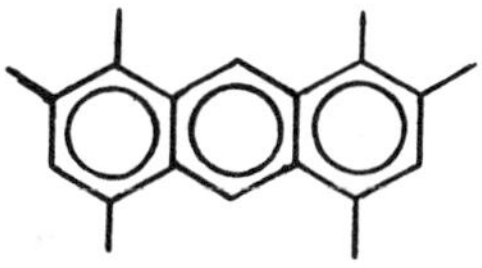

3

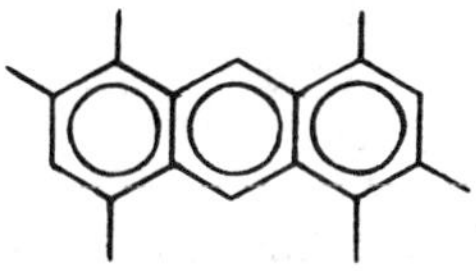

4

In 1914, Wenzel and co-workers reported the aluminum chloride-catalyzed reactions of mesitylene and pseudocumene with dihaloalkanes [7,8]. They found that dimesitylmethane and 1,2-dimesitylethane could be readily prepared by the condensation of methylene and ethylene dibromides, respectively, with mesitylene in the presence of $AlCl_3$ [7]. Under similar conditions, a reaction between mesitylene and trimethylene dibromide failed to produce any dimesitylpropane [8], and reaction between pseudocumene and methylene dichloride gave only a small yield of dipseudocymylmethane [7]. Further, Wenzel and Kugel pointed out that dimesitylmethane could best be prepared by the interaction of mesitylene and chloromethyl acetate in the presence of $AlCl_3$ [7].

Dichloromethane was also reported to condense with biphenyl in the presence of $AlCl_3$ to give fluorene [9]. The products from naphthalene were reported to be a mixture of α- and β-methylnaphthalene, ethylnaphthalenes, and β,β-dinaphthyl [10], or only β,β-dinaphthyl [11].

2. Dihaloethanes

a. 1,2-Dihaloethanes

From the aluminum chloride-catalyzed reaction of benzene with 1,2-dichloroethane, 1,2-diphenylethane was obtained as the main product [12]. However, small amounts of ethylbenzene and triphenylethane were also formed [13]. With 1,2-dibromoethane and toluene in the presence of $AlCl_3$ a mixture of isomeric ditolylethanes was produced [14,15]. Similarly, mesitylene and 1,2-dibromoethane were reported to give 1,2-dimesitylethane [7].

The reaction of naphthalene with both 1,1- and 1,2-dihaloethane was studied by various investigators. From the reaction of naphthalene with 1,2-dichloroethane and $AlCl_3$, Silva reported isolating α- and β-methylnaphthalene and ethylnaphthalenes, in addition to β,β-dinaphthyl [13]. In addition to these products, Roux obtained a product which he considered to be dinaphthylnaphthalene, $C_{10}H_6(C_{10}H_7)_2$ [16]. Later, this product was shown by Lespieau [17] and by Homer [11] to be picene (5). The production of picene instead of 1,2-dinaphthylethane was attributed by Homer to a secondary reaction brought about either by a dehydrogenating effect of the aluminum chloride or by the action of heat on the dinaphthylethane first formed, as shown in Eq. (4).

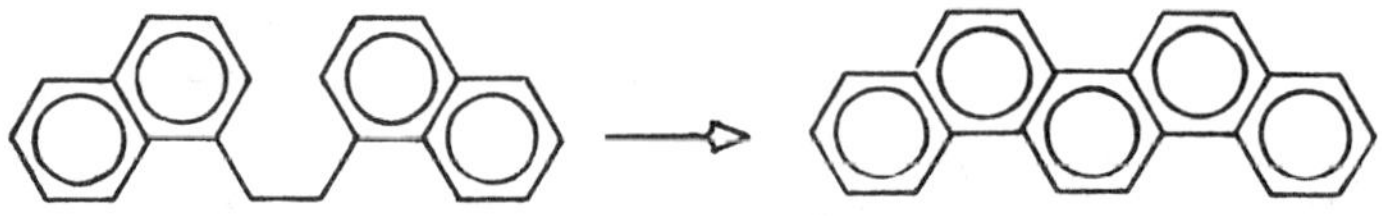

(4)

5

The observation that elastic, plastic materials were produced during the reaction of ethylene dihalides with arenes under Friedel-Crafts conditions initiated numerous studies on the influence the conditions have on the course of the reaction and on the structure of the resulting polymeric compounds [15,18-24]. Generally, it has been shown that the formation of polymers is favored by an excess of dihalide and by extended reaction time [21], whereas an excess of arene and shorter reaction time favors production of 1,2-diarylethane [15,19,20,24].

Attempts were also made to determine the structure of the resulting polymers. The fact that the yellow elastic polymer obtained by treating 1,2-dichloroethane with chlorobenzene in the presence of $AlCl_3$ gave 4-chloroisophthalic acid upon oxidation by $KMnO_4$ led Sisido to propose structure 6 for

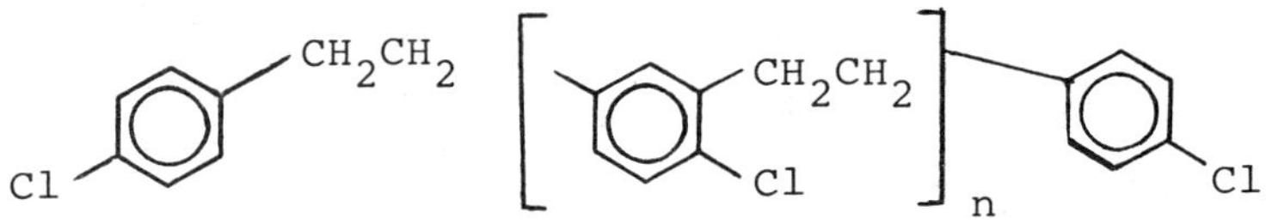

CH_2CH_2 CH_2CH_2 Cl Cl n Cl

6

the polymer [21]. Sisido's assignment was also supported by the finding that considerable amounts of 4,4'-dichlorodibenzyl (7) could be isolated after a short reaction time.

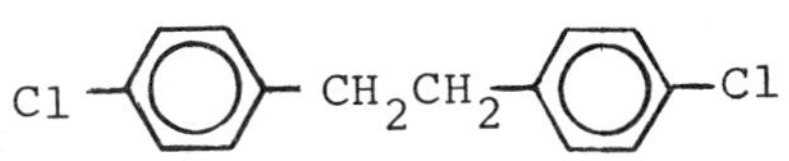

Cl CH_2CH_2 Cl

7

It is noteworthy that in all of these Friedel-Crafts reactions with 1,2-dihaloalkanes, 1,1-diarylethanes were never found to be among the products. In 1950, however, Dolgov and Larin studied the reaction of benzene with 1,2-dichloroethane in the presence of aluminum metal; in addition to the chief product, 1,2-diphenylethane, they detected small amounts of the isomeric 1,1-diphenylethane [25]. The lack of formation, or the formation in only small amounts, of the isomeric 1,1-diarylethanes in such reactions is believed by the present authors to be due to primary reaction of the dihalide to give an arylhaloethane, and subsequent reaction of the latter via a phenonium-type intermediate, resulting in the formation of the 1,2 isomer as shown in Eq. (5).

$$X\text{-}CH_2CH_2\text{-}X + ArH \xrightarrow[-HX]{cat} ArCH_2CH_2X \longrightarrow$$

$$\underset{Ar}{\overset{+}{CH_2\text{---}CH_2}} \xrightarrow{ArH} ArCH_2CH_2Ar \quad (5)$$

This explanation is supported by two experimental facts: (1) the reaction of α-phenylethyl chloride with arenes in the presence of $AlCl_3$ gave no detectable amounts of 1,1-diphenylethane [26,27], and (2) the reaction of ^{14}C-labeled α-phenylethyl chloride with arenes in the presence of $AlCl_3$ gave β-arylethylbenzenes in which complete equilibration of the label between C_1 and C_2 had occurred [28,29].

It is also interesting that none of the investigators was able to detect the presence of either acenaphthene (8) or its dehydrogenation product acenaphthylene (9), which would be expected to form by the steps shown in Eq. (6).

$$\text{naphthalene} + Cl\text{-}CH_2CH_2Cl \xrightarrow{AlCl_3} \text{1-}(Cl\text{-}CH_2\text{-}CH_2)\text{naphthalene} \xrightarrow{-HCl} \underset{8}{\text{acenaphthene}} \xrightarrow{-H_2} \underset{9}{\text{acenaphthylene}} \quad (6)$$

An explanation for the lack of formation of 8 and 9 may lie in the difficulty encountered in the closure of the primary chloride to give a five-membered ring [30].

A number of 1,1-diaryl-2-haloethanes were prepared in 40 to 73% yield by the reaction of 1,2-dihalo-2-phenylethane with arenes in the presence of $TiCl_4$ catalyst at room temperature [31] (Eq. 7).

$$C_6H_5\text{-}CH(X)\text{-}CH_2\text{-}X + ArH \xrightarrow[4\text{ hr-}2\text{ days}]{TiCl_4} C_6H_5\text{-}CH(Ar)\text{-}CH_2\text{-}X \quad (7)$$

X = Cl or Br

Ar = Ph, p-MeC_6H_4, p-$Me_2CHC_6H_4$, 3,4-, 2,4- and 2,5-$Me_2C_6H_3$, p-$PhCH_2C_6H_4$, p-PhC_6H_4, p-$PhOC_6H_4$, p-cyclohexylphenyl and α-$C_{10}H_7$

Dialkylations of xylenes and mesitylene were also effected under similar conditions (Eq. 8).

$$C_6H_5\text{-}\underset{X}{CH}\text{-}CH_2\text{-}X + 2\ ArH \longrightarrow C_6H_5\text{-}\underset{Ar}{CH}CH_2Ar \qquad (8)$$

X = Cl or Br

Ar = 3,4-, 2,4- and 2,5-$Me_2C_6H_3$ and 2,4,6-$Me_3C_6H_2$

b. 1,1-Dihaloethanes

The reaction of benzene with 1,1-dichloroethane was reported to give chiefly 1,1-diphenylethane [32-37]. However, depending on the ratio of reactants and the reaction conditions, products such as 9,10-dihydro-9,10-dimethylanthracene [34-36] and ethylbenzene [35-37] have also been identified. The production of the former product has been rationalized in terms of Eq. (9).

$$C_6H_6 + 2\ CH_3CHCl_2 + C_6H_6 \xrightarrow{AlCl_3} \text{9,10-dihydro-9,10-dimethylanthracene} \qquad (9)$$

Similarly, the aluminum chloride-catalyzed reaction of 1,1-dichloroethane with toluene gave 1,1-di-*p*-tolylethane, *p*-ethyltoluene, and tetramethyldihydroanthracene. With *m*-xylene, 1,1-dixylylethane and ethyl-*m*-xylene were obtained, but no anthracene derivatives were found [35]. The tetramethylanthracene obtained from toluene was assigned structure 10 on the assumption that it was formed by subsequent reaction of 1,1-*p*-ditolylethane with another mole of 1,1-dichloroethane (Eq. 10) [38].

$$H_3C\text{-}C_6H_4\text{-}CH_2\text{-}C_6H_4\text{-}CH_3 + Cl_2CH\text{-}CH_3 \longrightarrow \text{2,6,9,10-tetramethylanthracene} \qquad (10)$$

10

The $AlCl_3$-catalyzed reaction of naphthalene with 1,1-dichloroethane was reported by Bodroux [10] to result in the formation of α- and β-methylnaphthalenes and ethylnaphthalenes and β,β-dinaphthyl. The same reaction was later reinvestigated by Homer [11], who reported obtaining different products by varying the reaction conditions and the pressure under which the products were distilled. Thus, on distillation under atmospheric pressure, the reaction gave methylnaphthalenes and β,β-dinaphthyl, as reported by Bodroux. However, on allowing the initial stage of the reaction to proceed at ordinary temperature and by distillation of the products under reduced pressure, dinaphthylmethane and what was probably a small amount of 1,2-di(β-naphthyl)-ethane were obtained. Practically no methylnaphthalenes or β,β-dinaphthyl were found, however.

The formation of β,β-dinaphthylmethane rather than of 1,1-di(β-naphthyl)-ethane in the reaction above was attributed by Homer [11] to subsequent conversion of the latter compound into the former and methane "by the hydrogen evolved by the action of the aluminum chloride" (Eq. 11):

$$CH_3CH(C_{10}H_7)_2 + 2H \longrightarrow CH_4 + CH_2(C_{10}H_7)_2 \quad (11)$$

The author cited Bodroux's observation that methane and hydrogen chloride were evolved during the earlier stages of the reaction as experimental support for his view.

3. Dihalopropanes

a. *1,2-Dihalopropanes*

In a very early report (1879), the aluminum chloride-catalyzed reaction of benzene with 1,2-dichloropropane was said to yield only 1,2-diphenylpropane [33,39]. Much more recently (1966), the same reaction was reinvestigated by Ransley, who followed the progress of the reaction by analyzing samples taken from the alkylation mixture after various reaction times [40]. Contrary to earlier findings, this author reported the product to contain not only 1,2-diphenylpropane, but also small amounts of 1,1-diphenylpropane and isopropylbenzene. An appreciable amount of 1,1-diphenylpropane was produced even after only 7 min of reaction time.

In discussing his results, Ransley suggested that the first step in Friedel-Crafts alkylations with 1,2-dichloropropane is the initial ionization of the secondary chloride, followed by reaction with benzene to give 1-chloro-2-phenylpropane. Simple assisted ionization of the latter primary chloride with subsequent alkylation on benzene would give the 1,2-diphenylpropane. Ransley also suggested that rearrangement may accompany ionization as in the solvolysis of 2-phenylpropyl brosylate [41]. The resulting secondary ion may react directly to give the 1,2-diphenyl isomer or, after a favorable hydride shift, to give the 1,1-diphenyl isomer.

b. *2,2-Dihalopropanes*

The reaction of 2,2-dichloropropane with arenes has also been investigated. In early work with $AlCl_3$ catalyst this was reported to give 2,2-diphenylpropane with benzene [42] and 2,2-di-*m*-xylylpropane with *m*-xylene [43]. More recently, the reaction of benzene with 2,2-dichloropropane in the presence of $AlCl_3$ was reinvestigated by Greene [44]. Using modern analytical tools, these workers found the products to be a mixture of 2,2-diphenylpropane (40.2%), isopropylbenzene (0.36%), α-methylstyrene (17.1%), and 1,3,3-trimethyl-1-phenylindan (1.3%). The later indan derivative was undoubtedly formed by the acid-catalyzed dimerization of α-methylstyrene.

c. *1,3-Dihalopropanes*

In 1901, 1,3-dibromopropane was reported to interact with benzene in the presence of $AlCl_3$, resulting in the formation of a mixture of 1,3-diphenylpropane and *n*-propylbenzene [45]. A similar reaction with mesitylene failed to produce the corresponding 1,3-dimesitylpropane [7]. The products of the aluminum chloride-catalyzed reaction of benzene with 1-bromo-3-chloropropane were shown to vary with temperature. At 6 to 12°C 1-bromo-3-phenylpropane was formed in 60% yield, but at a higher temperature 1,3-diphenylpropane was the product obtained [46].

In 1968, Ransley examined the aluminum chloride-catalyzed alkylation of benzene with a series of 1,3-dichloroalkanes, including 1,3-dichloropropane [47]. The reactions were carried out with 5 mol % $AlCl_3$ at 0 to 5°C in excess benzene for 2 hr. Under these conditions, 1,3-dichloropropane gave a mixture consisting of 1-chloro-3-phenylpropane (19.8%), 1,3-diphenylpropane (51.8%), 1,2-diphenylpropane (22.3%), and 1,1-diphenylpropane (4.2%). The formation of this mixture was explained in terms of competitions between direct alkylations and the usual rearrangement involving carbocation intermediates. It was shown that 1-chloro-3-phenylpropane alkylated benzene in the presence of $AlCl_3$, giving a mixture of 1,1-, 1,2-, and 1,3-diphenylpropane in the ratio 4.3:23.5:72.2.

4. Dihalobutanes

a. *1,1-Dihalobutanes*

The behavior of 1,1-dihalobutane is similar to that of 1,1-dihaloethane and 1,1-dihalopropane. Thus the reaction of benzene with 1,1-dichlorobutane in the presence of $AlCl_3$ gave a product consisting principally of 1,1-diphenylbutane mixed with small amounts of *n*-butylbenzene [37]. The latter product is probably formed by a hydride exchange reaction; i.e., the cation resulting from 1-chloro-1-phenylbutane would be a good hydride-abstracting agent. It will be noted that monoarylalkanes have been reported as products from reactions of various dihaloalkanes.

b. *1,2-Dihalobutanes*

The reaction of 1,2-dichlorobutane with benzene in the presence of $AlCl_3$ at 4 to 6°C was reported by Schmerling et al. [37] to result in a 37% yield of 1-chloro-2-phenylbutane and only an 8% yield of a diphenylbutane mixture, consisting of about 85% 1,2-diphenylbutane and 15% *meso*-2,3-diphenylbutane. More recently (1966), the same reaction was reinvestigated by Ransley [40], who found that the addition of 6.6 mol % of $AlCl_3$ to a 1 *M* solution of 1,2-dichlorobutane in benzene gave a product consisting of 80.2% 1-chloro-3-phenylbutane (11) and 14.3% of diphenylbutanes (12-15) after 3 hr at room temperature.

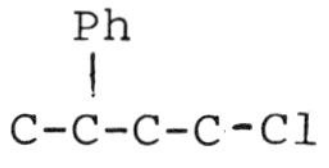

```
Ph                                  Ph
|                  /Ph              |
C-C-C-C-Cl   C-C-C-C               C-C-C-CPh
                   \Ph
 11                 12                13

        Ph            Ph Ph
        |             |  |
    C-C-C-C         C-C--C-C
        |
        Ph
        14               15
```

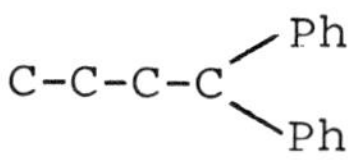

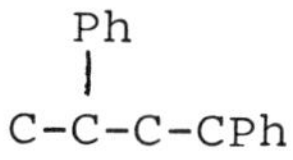

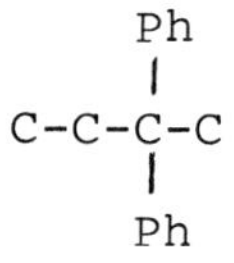

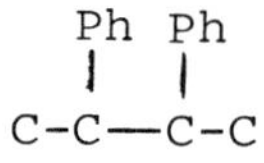

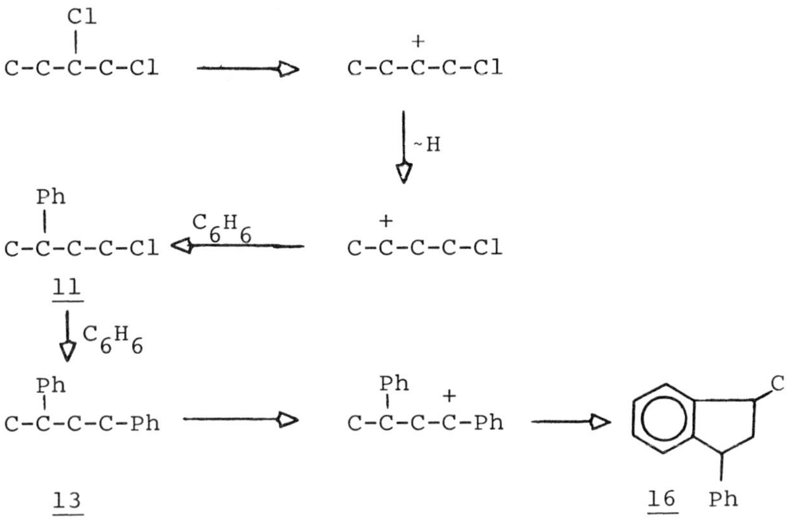

```
      Cl
      |                        +
C-C-C-C-Cl  ------>  C-C-C-C-Cl
                          |
                          | ~H
                          v
  Ph
  |       C6H6           +
C-C-C-C-Cl <------  C-C-C-C-Cl
  11
  | C6H6
  v
  Ph                 Ph  +
  |                  |
C-C-C-C-Ph ------>  C-C-C-C-Ph ------>  [indan, C, Ph]

  13                                     16  Ph
```

Scheme 1

When the reaction was carried out under more vigorous conditions (50% $AlCl_3$ at 40 to 80°C for 3 hr) the products were shown to be a mixture of isobutylbenzene (19.4%), *sec*-butylbenzene (19.4%), diphenylbutanes (7.3%), 1-phenyl-3-methylindan (16, 13.0%), and higher-boiling materials (36.9%). Ransley proposed the mechanism shown in scheme 1 to explain some of his products. In addition, the author emphasized the following points:

1. Under mild conditions, the initially formed 1-chloro-3-phenylbutane (11) would not be expected to react readily, as it may undergo only the unfavored Ar_1-4 or Ar_2-5 participation in the ionization of the primary chloride [48]:

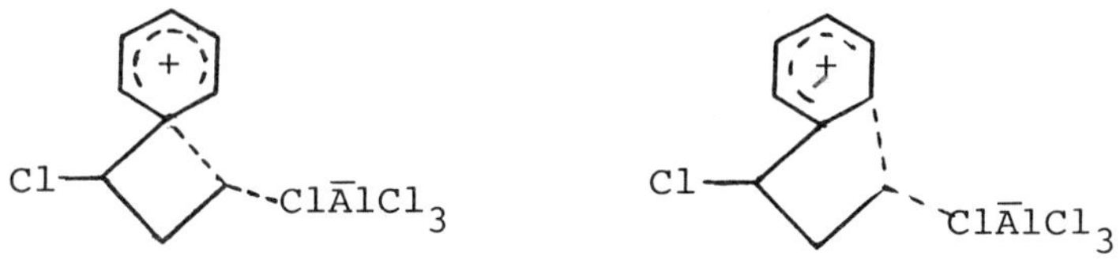

```
Cl–[ring]--ClĀlCl3        Cl–[ring]--ClĀlCl3
    Ar1-4                      Ar2-5
```

2. In view of the results of Roberts et al. [49], the formation of isobutylbenzene under the more vigorous conditions could be explained in terms of subsequent isomerization of initially formed *sec*-butylbenzene (see Chap. 8).
3. Speculation as to the mechanism of formation of the various diphenylbutanes was said to be complicated by the known fact that isomerization of diphenylbutanes occurs under quite mild conditions with $AlCl_3$ [50-53].
4. The absence of 1-methylindan from the reaction products ruled against its intermediacy in the formation of 1-methyl-3-phenylindan (by hydride abstraction followed by alkylation of benzene).

Hence, it was concluded that hydride abstraction from 1,3-diphenylbutane followed by cyclialkylation to give 16, as shown in scheme 1, is a more likely mechanism.

c. *1,3-Dihalobutanes*

The aluminum chloride-catalyzed alkylation of benzene by 1,3-dichlorobutane was reported by Sisido and Nozaki in 1947 [54] to produce 2,3-diphenylbutane and *sec*-butylbenzene, but no 1,3-diphenylbutane. However, in Ransley's hands (1966), this reaction gave 1-chloro-3-phenylbutane as the major product [40]. Thus after 5 hr of interaction between benzene and 1,3-dichlorobutane in the presence of 33% $AlCl_3$, 1-chloro-3-phenylbutane constituted 70% of the product. However, Ransley made no attempt to identify the other products.

d. *2,3-Dihalobutanes*

The product obtained from the reaction of 2,3-dibromobutane with excess benzene in the presence of $AlCl_3$ was shown to consist mainly of 2,3-diphenylbutane mixed with small amounts of butylbenzene [37].

The reaction was found not only to be fast but also to give a high yield of 2,3-diphenylbutane. This, of course, is understandable in terms of both the secondary nature of the halogen atoms and the neighboring group assistance expected in the ionization steps.

e. *1,4-Dihalobutanes*

In 1943, Yura and Oda reported that tetralin was the main product of condensation of 1,4-dichlorobutane with benzene by the Friedel-Crafts reaction [18]. In 1957, Tsukervanik and Bugrova alkylated benzene with 1-bromo-4-chlorobutane and obtained reaction mixtures that contained tetralin, octahydroanthracene, octahydrophenanthrene, and dodecahydrotriphenylene [55]. The proportions of these components in the product mixture were shown to depend on the ratio of reactants used; a lower concentration of benzene gave more decahydrotriphenylene.

Between 1957 and 1964, Shadmanov published a series of papers on the $AlCl_3$-catalyzed alkylation of arenes with 1,4-dichlorobutane that included alkylations of benzene [56], toluene [57], *o*-xylene [58], *m*-xylene [58], *p*-xylene [59], mesitylene [60], and tetralin [61]. The products identified from each of these reactions are represented in Eqs. (12) to (18).

$$C_6H_6 + Cl(CH_2)_4Cl \xrightarrow[80^\circ]{AlCl_3} \underline{17} + \underline{18} + \underline{19} + \underline{20} + \underline{21} \qquad (12)$$

$$\text{toluene} + Cl(CH_2)_4Cl \xrightarrow[0-20^\circ]{AlCl_3} \underline{22} + \underline{23} + \underline{24} \quad (13)$$

$$\text{m-xylene} + Cl(CH_2)_4Cl \xrightarrow[75-80^\circ]{AlCl_3} \underline{25} + \underline{26} + \underline{27} \quad (14)$$

$$\text{p-xylene} + Cl(CH_2)_4Cl \xrightarrow[75-80^\circ]{AlCl_3} \underline{25} \quad (15)$$

$$\text{o-xylene} + Cl(CH_2)_4Cl \xrightarrow[75-80^\circ]{AlCl_3} \underline{25} \quad (16)$$

CH_3 / H_3C / CH_3 + $Cl(CH_2)_4Cl$ $\xrightarrow[18-20^\circ]{AlCl_3}$ CH_3 / H_3C / CH_3 28 +

H_3C / CH_3 / CH_3 29 + H_3C / CH_3 / $(CH_2)_4$ / CH_3 / CH_3 / CH_3 / CH_3 30 (17)

+ $Cl(CH_2)Cl$ $\xrightarrow[20-22^\circ]{AlCl_3}$ 18 + (18)

19 + 31

Shadmanov pointed out that the distribution of the products in the alkylation mixtures from the reactions above was dependent not only on the ratio of reactants but also on the duration and temperature of the reaction. He also examined the effect of $AlCl_3$ on the isomeric octahydrophenanthrenes 19 and 31 and found that the former rearranged to a mixture of 18 and 31, but the latter gave only 18 as shown in Eqs. (19) and (20) [61].

19 (2.6g) $\xrightarrow[\text{in benzene (50 ml), } 18-20^\circ]{AlCl_3 \text{ (1.8g)}}$ 18 (0.6g) + 31 (1.25g) (19)

31 (3.4g) $\xrightarrow[\text{in benzene (50 ml), } 18-20^\circ]{AlCl_3 \text{ (2.4g)}}$ 18 (1.12g) [+ 31 (1.73g)] (20)

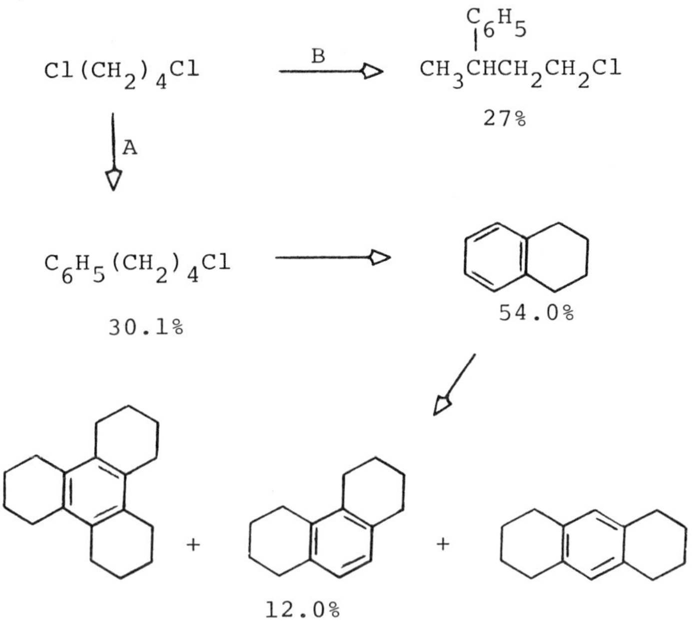

$Cl(CH_2)_4Cl \xrightarrow{B} CH_3CH(C_6H_5)CH_2CH_2Cl$ 27%

A ↓

$C_6H_5(CH_2)_4Cl$ → 54.0%

30.1%

12.0%

Scheme 2

These isomerization experiments suggested that 31 was most probably produced by the acid-catalyzed rearrangement of 19. Although it is easy to rationalize the isomerization of 19 to 18, it is not clear how the transformations 19 → 31 and 31 → 18 occurred. More recently (1968), the reaction of 1,4-dichlorobutane with benzene in the presence of $AlCl_3$ was reinvestigated by Ransley [47]. The reaction was conducted with 5 mol % $AlCl_3$ at 0 to 5°C in excess benzene for 2 hr, resulting in the product distribution shown in scheme 2.

Comparison of scheme 2 with equation (12) demonstrates the agreement between Ransley's [47] and Shadmanov's [56] findings concerning the type of products obtained. The formation of triphenylene observed by Shadmanov can be attributed to his more strenuous conditions which lead to the aromatization of octahydrotriphenylene.

It is interesting to note that after Ransley examined the aluminum chloride-catalyzed alkylation of benzene with a series of α,ω-dichloroalkanes, he concluded that the reaction with 1,4-dichlorobutane was exceptional because (1) it did not fit into the general pattern of a gradual rate change in the α,ω-dichloroalkane series, (2) in contrast to the behavior of both lower and higher homologs, the reaction rate was faster than that of the corresponding monochloroalkane, and (3) it gave less rearrangement than would normally be anticipated.

The facts above were attributed to two factors: First, the ionization of the first chloro group of 1,4-dichlorobutane occurs with anchimeric assistance of the remaining chloro group, thereby generating a five-membered chloronium ion. Second, after reaction with benzene, the favored phenyl participation in the ionization of the remaining chloride assists in maximizing the amount of product formed via the unrearranged route. These two factors have been pictured as shown in scheme 3.

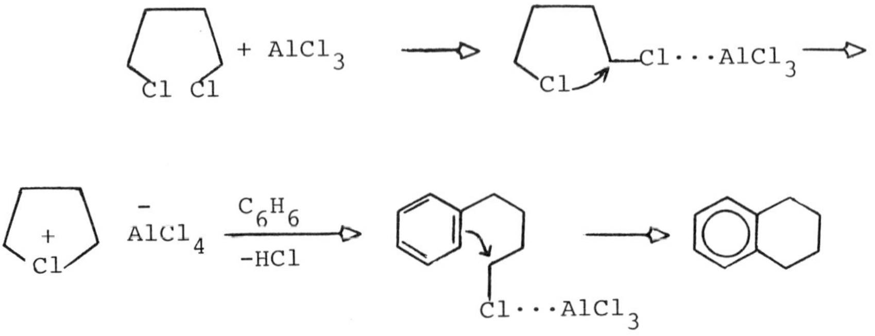

Scheme 3

The reaction of 1,4-dihalobutanes with benzene and $AlCl_3$ was also investigated by Flanagan and Shadan [62]. Using benzene and 1,4-dichlorobutane in a mole ratio of 10:1 and taking $AlCl_3$ in amount of 8 mol %, based on chloroalkane, they obtained surprisingly high yields (ca. 90%) of dodecahydrotriphenylene (31a).

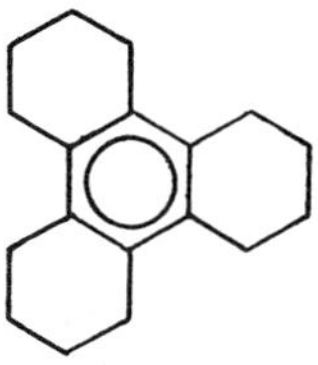

31a

5. C_5 and higher linear dihaloalkanes

In the early Friedel-Crafts literature there were a few scattered reports of the reaction of arenes with linear dihaloalkanes. For example, the $AlCl_3$-catalyzed reaction of benzene with 1,1-dichloroheptane was reported by Krafft in 1886 [63] and by Auger in 1887 [64] to yield 1,1-diphenylheptane. Krafft also found that when the reaction was conducted in the presence of a larger proportion of $AlCl_3$, heptylbenzene was the product obtained. The alkylation of benzene with both 3,4-dichlorohexane [65] and 1,2-dibromohexadecane [66] was investigated by Sisido and Nozaki, who reported the results shown in Eqs. (21) and (22). In explaining the results above, the authors suggested

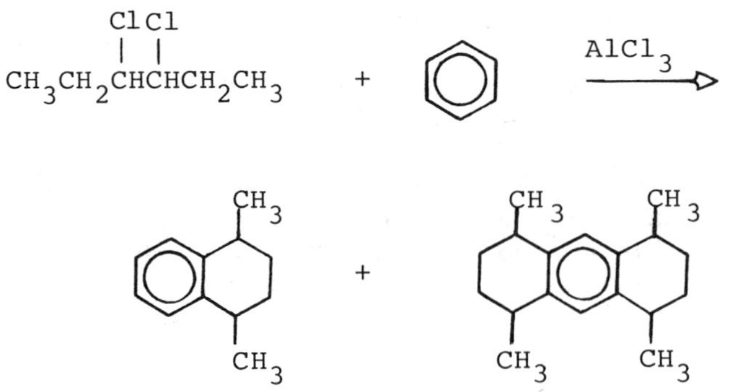

$$CH_3CH_2CHClCHClCH_2CH_3 + C_6H_6 \xrightarrow{AlCl_3} \qquad (21)$$

$$C_{14}H_{29}CH(Br)-CH_2Br + C_6H_6 \xrightarrow[0-5^\circ]{AlCl_3}$$

$C_{12}H_{25}$-substituted tetralin + bis($C_{12}H_{25}$)-octahydroanthracene (or isomeric bis($C_{12}H_{25}$)-octahydroanthracene) (22)

prior isomerization of the alkylating agents to 2,5-dichlorohexane and 1,4-dibromohexadecane, respectively, followed by cyclialkylation to yield the products. It is believed by the present authors that the mechanism more probably involves rearrangement of the carbocations produced by initial ionization at the secondary chloride positions.

In addition to the foregoing early findings a number of interesting papers have appeared more recently in which the alkylation of arenes by dihaloalkanes was carefully investigated. The results of these more recent studies are important not only because of the better techniques employed in product analysis, but also because of the critical discussion of the mechanistic aspects of these reactions. A summary of the most important recent results will comprise the rest of this section.

In 1963, Shadmanov investigated the aluminum chloride-catalyzed reaction of benzene with 1,4-dichloropentane under a variety of conditions [56]. The product in all cases was shown to be 1-methyltetralin.

In 1965, Sisido and co-workers studied the reaction of benzene with 1,5-dibromopentane and 1,6-dibromohexane in the presence of $AlCl_3$ [67]. The reaction conditions, including time, temperature, mole ratios of reactants, and the percent yields of identified products for each of the two reactions are shown in Eqs. (23) and (24).

$$C_6H_6 \text{ (10 mol)} + Br(CH_2)_5Br \text{ (1 mol)} + AlCl_3 \text{ (0.5 mol)} \xrightarrow[\text{15 hr}]{\text{room temp,}}$$

$Ph(CH_2)_5Ph$ (mainly, plus other isomers) 19% + $Ph(CH_2)_4CH_3$ 19% + 1-methyltetralin 25% (23)

$$C_6H_6 + Br(CH_2)_6Br + AlCl_3 \xrightarrow[(2)\ 80\text{-}83^\circ]{(1)\ \text{room temp}}$$

(10 mol) (1 mol) (0.125 mol) 15 hr

$Ph(CH_2)_6Ph$ + Isomeric diphenylhexane(s) + $Ph(CH_2)_5CH_3$

(1) 13%, (2) 8% (1) 3.8%, (2) 8% (1) 11%, (2) 14%

+ [1,4-dimethyltetralin] + [1-ethyltetralin] (24)

(1) 14%, (2) 6.2% (1) 26%, (2) 29%

The increase noted in the yield of *n*-hexylbenzene at the higher temperature may reasonably be explained in terms of dephenylation [52,53] of diphenylhexane and/or ring opening of 1-ethyltetralin [30,68,69] under the more vigorous conditions. Similar reactions were observed under milder conditions [30,52,53,68,69]. Again, in line with the old concept [65], Nozaki et al. [67] explained their results in terms of migration of halogen atoms rather than of carbocation rearrangements prior to reaction with benzene.

In 1966, Ransley started a systematic investigation of the alkylation of benzene with different series of structurally related dihaloalkanes. This author's first article on the subject [40] was concerned with the alkylation of benzene with several 1,2-dichloroalkanes. These included 1,2-dichloropropane, 1,2-dichlorobutane, 1,2-dichlorohexane, and 1,2-dichloro-4-methylpentane. The catalysts were either $AlCl_3$ or liquid HF. Since the results with the first two members of the series were presented earlier in this section and the results with the last member will be discussed in a later section, at this point we shall concentrate only on the results with 1,2-dichlorohexane.

The alkylation of benzene with 1,2-dichlorohexane was performed both in the presence of HF at 80°C and of $AlCl_3$ at 25°C. With HF as catalyst, the reaction gave a low conversion (5%) but an almost quantitative yield of 1-chloro-5-phenylhexane (32). With $AlCl_3$ as catalyst, on the other hand, products 32-35 together with other diphenylhexanes and higher-boiling materials were obtained.

C-C(Ph)-C-C-C-C-Cl — 32 27.5%

[1-ethyltetralin] — 33 19.0%

[1,4-dimethyltetralin] — 34 25.1%

C-C(Ph)-C-C-C-C-Ph — 35 8.2%

The product distribution is given in weight percent, the balance being other diphenylhexanes or higher-boiling materials. The reaction was followed as a function of time by GLPC, and it was observed that 33 was formed early in the reaction and did not increase greatly with time.

```
     Cl
     |                                  +
C-C-C-C-C-C-Cl      ------->      C-C-C-C-C-C-Cl
                                        | ~H:⁻
                                        v
 +                   ~H:⁻                +
C-C-C-C-C-C-Cl      ------->      C-C-C-C-C-C-Cl
 | C6H6
 v                   ~H:⁻
 Ph                       \
 |                         v      +
C-C-C-C-C-C-Cl                    C-C-C-C-C-C-Cl
 |                                   | C6H6
 v                                   v
      C-C                        Ph
                                 |
                                 C-C-C-C-C-C-Cl

     33                               32
                         C        /        \
                                 v          v Ph
                                              |
                                              C-C-C-C-C-C-Ph
                         C
                     34                       35
```

Scheme 4

There was a gradual buildup of 34 at the expense of 32, which decreased with time after reaching a maximum of 53%.

In discussing his findings, Ransley stated that two 1-chloro-x-phenylhexanes, namely the 4 and 5 isomers, were produced in the initial stages of the reaction. He proposed that the former isomer reacts rapidly with phenyl-assisted ionization of the primary chloride giving 33, and the latter isomer reacts more slowly with concurrent 1,2-dihydride shift to yield 34 and with hydride exchange to yield 35. The overall mechanism as visualized by Ransley is shown in scheme 4.

By correlating the results from 1,2-dichlorohexane and those from other 1,2-dihalides, Ransley reached the following conclusions:

1. The products obtained are consistent with a mechanism in which the secondary chloride reacts initially to give preferentially the 1-chlorophenylalkane in which the phenyl group is in the penultimate position. This behavior was rationalized in terms of carbonium ion stability. It was stated that "the most stable carbonium ion in the systems studied is on the penultimate carbon, largely owing to the decrease in the inductive effect with distance [70], and possibly to the greater hyperconjugative stabilization of methyl or methylene."

PRODUCTS FROM THE ALKYLATION OF BENZENE WITH α,ω-DICHLOROALKANES

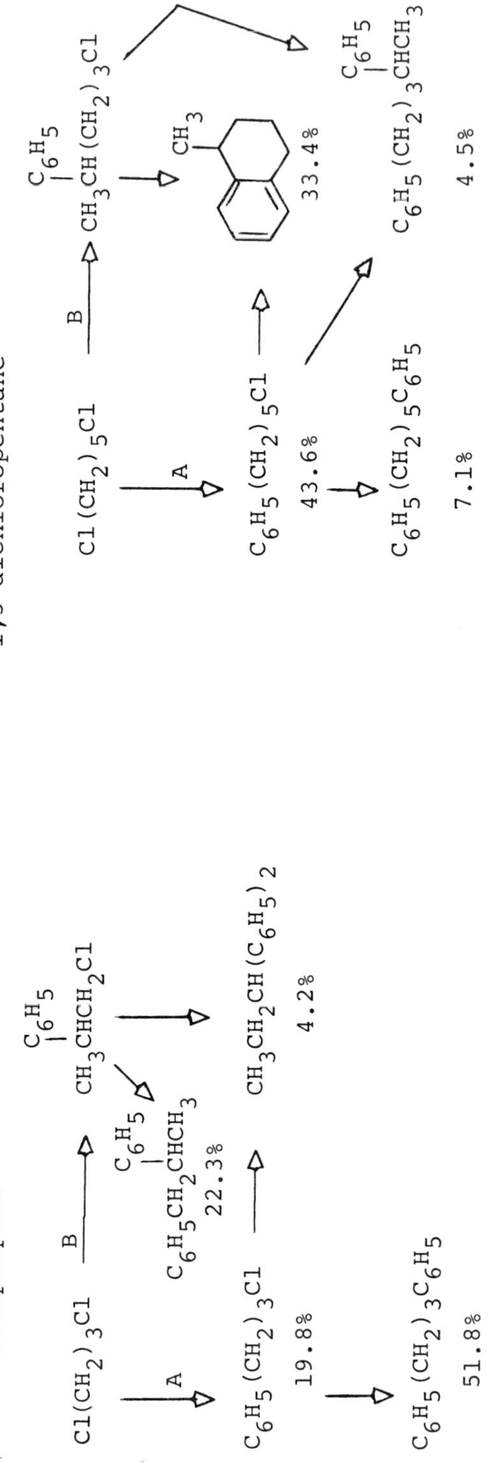

Scheme 5

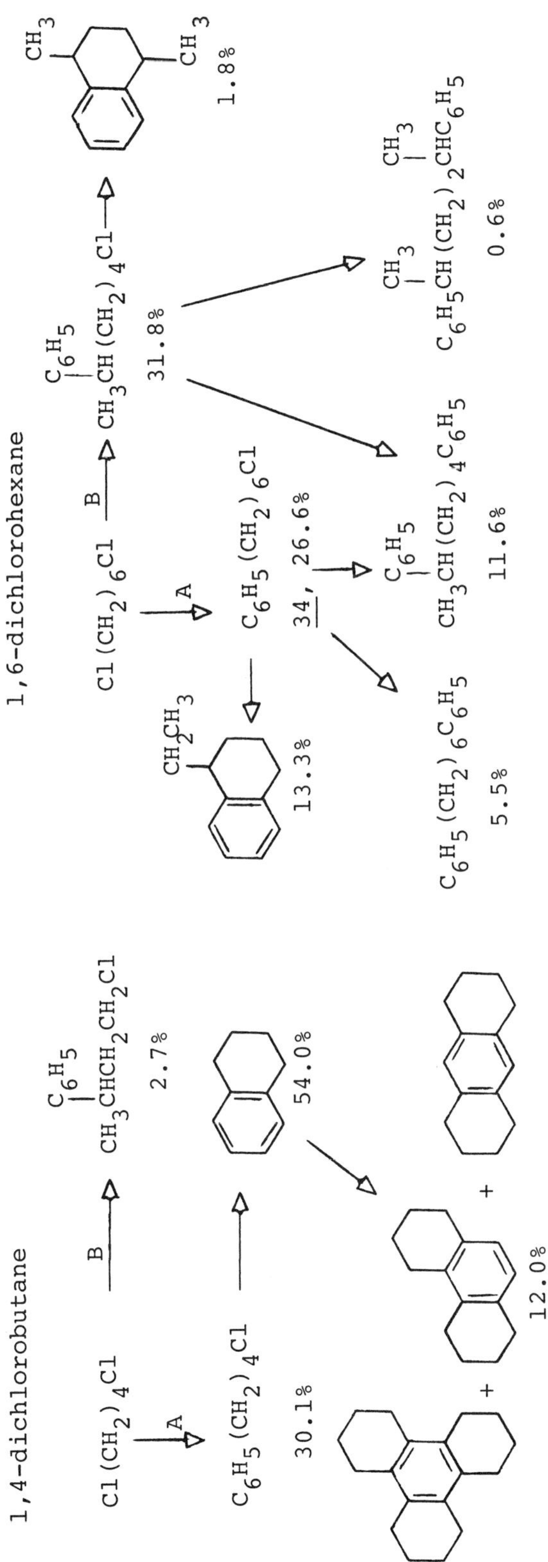

Scheme 5 (continued)

Table 1. Rearrangement in the Alkylation of Benzene with Mono- and Dichloroalkanes

Reactant	Products (wt%)		
	Path A unrearranged	Path B rearranged	Ambiguous
1-Chloropropane	44.4	55.6	
1,3-Dichloropropane	90.6	7.5	1.9
1-Chlorobutane	36.0	60.6	
1,4-Dichlorobutane	96.1	2.7	1.2
1-Chloropentane	40.5	59.5	
1,5-Dichloropentane	66.8	23.4	9.8
1-Chlorohexane	32.5	67.5	
1,6-Dichlorohexane	47.4	49.8	2.8

2. Subsequent products are determined largely by the influence of the phenyl group in assisting ionization of the primary chloride. This, in turn, is dependent on the relative positions of the phenyl and chloro groups. In fact, it has been shown that the order of anchimeric assistance in these studies is perfectly consistent with that found in the solvolysis studies of Heck and Winstein [48]. This order is as follows: Ar_1-3 $\gg$ Ar_1-4 $\ll$ Ar_1-5 $\gg$ Ar_1-6 and Ar_2-5 $\ll$ Ar_2-6 $\gg$ Ar_2-7, where Ar_x-Y signifies aryl (Ar) participation, the aromatic ring position (x) involved, and the size (Y) of the ring formed in the transition state.

In a second paper published in 1968, Ransley described the aluminum chloride-catalyzed alkylation of benzene with a series of α,ω-dichloroalkanes, including 1,3-dichloropropane, 1,4-dichlorobutane, 1,5-dichloropentane, and 1,6-dichlorohexane [47]. The reactions were carried out with 5 mol % $AlCl_3$ at 0 to 5°C in excess benzene for 2 hr. Product distributions and rate characteristics were examined and compared with those for similar reactions of the corresponding 1-chloroalkanes. Summaries of Ransley's results are given in scheme 5 and in Table 1. Scheme 5 shows the product distribution for each reaction: paths B and A signify reactions with and without accompanying rearrangement, respectively. In Table 1, the percentages of rearranged and unrearranged products are compared for the mono- and dichloroalkanes.

Ransley also used competitive rate studies to estimate the relative first-order rate constants for the reactions. The first-order rate constant for 1,3-dichloropropane ($k_{1,3}$) was set at unity. The relative rates were then $k_{1,3} = 1$, $k_{1,4} = 108$, $k_{1,5} = 24$, $k_{1,6} = 45$, and $k_{mono} = 50$.

From these data, Ransley made two generalizations: First, with the exception of 1,4-dichlorobutane, the rate of reaction of dichloroalkanes increases as the distance between the two chloro groups is increased. Second, α,ω-dichloroalkanes undergo initial reaction with less rearrangement than the corresponding monochloroalkanes, but this difference in behavior becomes smaller as the distance between the chloro groups increases.

These generalizations were rationalized in terms of a destabilizing electron-withdrawing influence of the reference primary chlorine on the ionization of the other chloride. This influence decreases as the distance between the two chlorines increases. Thus a rearrangement of an α,ω-dichloroalkane is energetically less favorable than a rearrangement of a corresponding monochloride because of the necessity of transferring a partial positive charge to a carbon atom closer to the second chlorine atom.

Alkylation with methoxychloroalkanes was investigated by Matsuda [71]. The reaction of 1-methoxy-2-, 1-methoxy-5-, and 1-methoxy-7-chlorooctane with benzene in the presence of $AlCl_3$ gave 1-methoxy-*n*-phenyloctanes (n = 3 to 7) with various isomer compositions. The electron-withdrawing inductive effect of the methoxy group was seen in retarded alkylation at the carbon atoms near the ether function.

In 1969, Gelin et al. published a detailed account of the products obtained from the reaction of benzene with 2,3- and 2,4-dichloropentane in the presence of $AlCl_3$ [72,73]. Using GLPC and NMR, the authors were able to verify the presence of compounds 36-50 in the alkylation mixtures obtained from the reaction of 2,4-dichloropentane.

$CH_3\text{-}CH(Ph)\text{-}CH_2\text{-}CH_2\text{-}CH_3$ **36**

$CH_3CH_2CH(Ph)CH_2CH_3$ **37**

$CH_3CH(Ph)\text{-}CH_2CH_3$ **38**

cis and *trans* **39**

$CH_3\text{-}CH(Ph)\text{-}CH_2\text{-}CH(Ph)\text{-}CH_3$ **40**

$CH_3\text{-}CH(Ph)\text{-}CH(Ph)\text{-}CH_2\text{-}CH_3$ **41**

$Ph_2CH\text{-}CH(CH_3)\text{-}CH_2\text{-}CH_3$ **42**

$Ph_2C(CH_3)H\text{-}CH_2\text{-}CH_2\text{-}CH_3$ **43**

$PhCH_2\text{-}C(CH_3)(Ph)\text{-}CH_2\text{-}CH_3$ **44**

$Ph\text{-}CH_2\text{-}CH(CH_3)\text{-}CH(Ph)\text{-}CH_3$ **45**

46

Ph **47**

Ph **48**

Ph **49**

Ph **50**

Two distillation fractions were collected, bp 76 to 100°C (15 mm) and 110 to 200°C (15 mm). In the lower-boiling fraction, 36 and 37 predominated and their yields increased with increasing amounts of $AlCl_3$; 40 and 41 were the major higher-boiling products. Reaction of 2,3-dichloropentane with benzene gave the same products, but in slightly different proportions.

In an extension of this work, Gelin and co-workers investigated the reactions of 2,4-dichloro- and 2,4-dibromohexane with benzene in the presence of $AlCl_3$ [74]. The major product from both dihalides was 1,4-dimethyltetralin (51), accompanied by smaller amounts of 2,5- and 2,4-diphenylhexane (54 and 55), 1,4-dimethyl-6-(1-ethylbutyl)tetralin (56), and the three tricyclic products 52, 53, and 57. The fact that ca. 90% of the products consisted of 1,4-dimethyltetralin and compounds derived from it by further reaction was cited by Gelin and co-workers as being in accord with the observations of

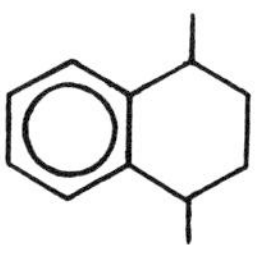

51 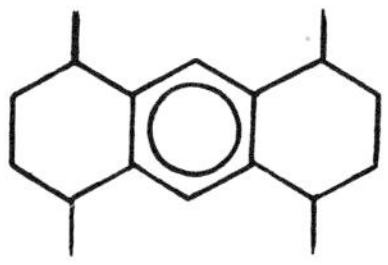 52 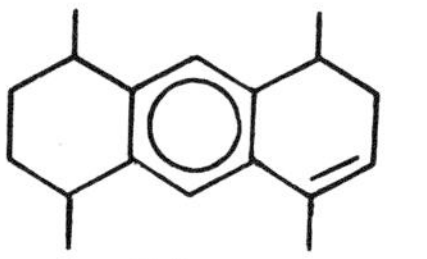 53 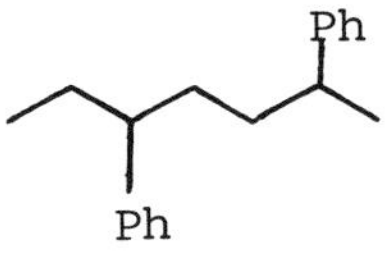 Ph Ph 54

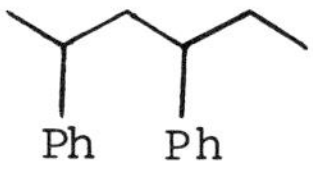

Ph Ph 55 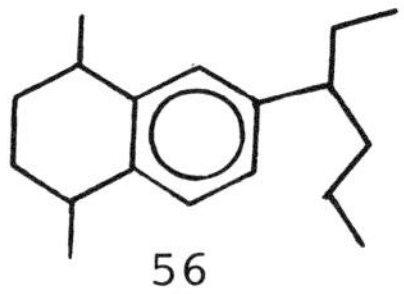 56 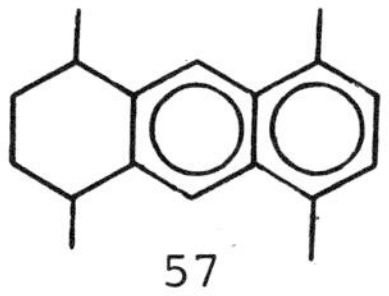 57

Roberts and Khalaf regarding the case of formation of tetralins by cyclialkylation compared to indans [30] (see Chap. 6).

The reaction of two other dichlorohexanes in Friedel-Crafts alkylations had been studied earlier. Sisido and Nozaki reported that 3,4- and 2,5-dichlorohexane reacted with benzene in the presence of $AlCl_3$ to give 1,4-dimethyltetralin (51) mainly, and small amounts of the tricyclic compounds 52 and/or 58 [65,75].

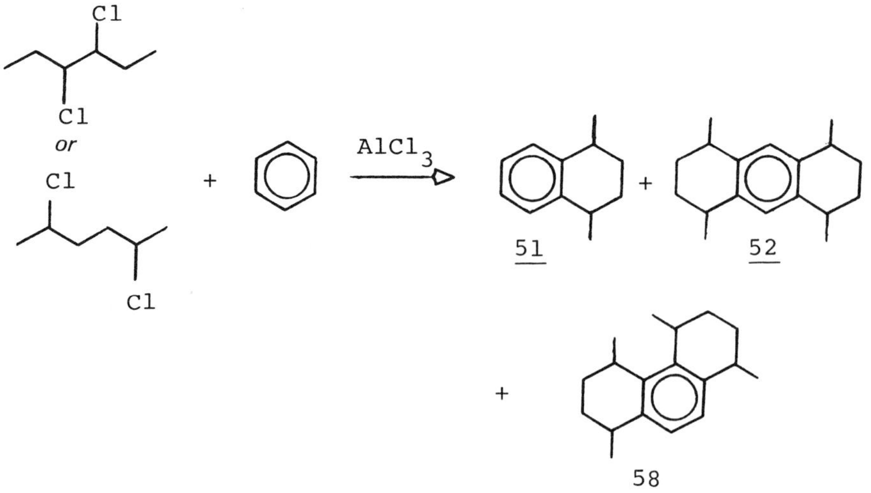

Cl Cl *or* Cl Cl + $\xrightarrow{AlCl_3}$ 51 + 52 + 58

Flanagan and Shadan investigated the aluminum chloride-catalyzed alkylation of benzene with α,ω-dichloroalkanes in the molecular weight range of dichloropropane to dichlorodecane [62]. In accordance with other studies, the reaction products comprised mixtures of tetralins, mono- and dialkyltetralins, monophenyl- and diphenylalkanes, and in some cases phenylchloroalkanes. The yields of the latter three products were larger from lower-molecular-weight dihalides (dihalopropanes and dihalobutanes) than from higher-molecular-weight dihalides. However, these authors emphasized the fact that in no case could they detect the presence of indans produced from dichloro- or dibromoalkanes.

Recently, the alkylation of benzene with dichloroundecane (structure unspecified) in the presence of $AlCl_3$ was reported to give a product mixture containing mainly dialkyltetralins, tetraalkylhydrindacenes, and diphenylalkanes [76]. The composition of the reaction mixture was reported to depend on the dichloroundecane/benzene ratio.

6. Stereospecific alkylations with dichloroalkanes

A recent publication described the Friedel-Crafts alkylation of benzene with optically active 1,2-, 1,3-, 1,4-, and 1,5- dihaloalkanes in the presence of $AlCl_3$ to give the corresponding X-phenyl-1-haloalkanes, with stereospecificity ranging from 15 to 60% [77]. In the case of the 1,3- and 1,5-dihaloalkanes, the major result was alkylation with inversion of configuration at the secondary halogen positions. However, (S)-1,2-dichloropropane (59) gave (S)-1,2-diphenylpropane (61) with retention, which was explained as occurring via two inversions, the first by attack of benzene on the secondary position, followed by formation of a phenonium ion (60). The second inversion was proposed to occur by attack of the second benzene molecule on the phenonium ion (Eq. 25). (S)-1-Bromo-2-chloropropane (62) gave (R)-1,2-diphenylpropane (64). The mechanism proposed included initial formation of a propylene bromonium ion (63); two other sequential inversions gave an overall result of one inversion (Eq. 26). The stereochemical course of the reaction of 1,4-

Cl-CH$_2$-C(H)(Me)(Cl) (59) —PhH, −Cl$^-$→ Cl-CH$_2$-C(Ph)(Me)(H) → CH$_2$—C(Me)(H) bridged by $C_6H_5^+$ (60) —PhH→ PhCH$_2$-C(H)(Me)(Ph) (61)   (25)

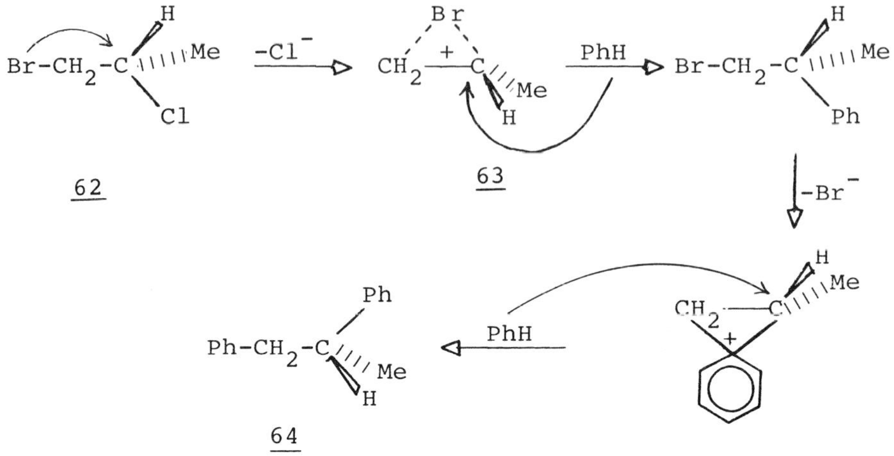

(26)

dichloropentane to produce 1-chloro-4-phenylpentane varied with the reaction temperature, giving retention at 0°C and inversion at -20°C. The retention was explained in terms of the intermediate formation of a five-membered halonium ion (65, Eq. 27). Earlier stereospecific Friedel-Crafts reactions reported

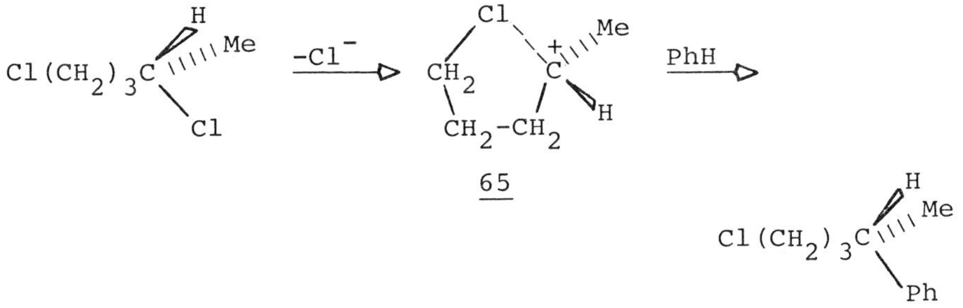

(27)

by Suga and co-workers are described in Sec. IV and in Chaps. 3 and 4.

B. Alkylations with Branched Dihaloalkanes

1. Branched dihalobutanes

The alkylation of benzene with isobutylene dihalides under Freidel-Crafts conditions has been carried out by various investigators [25,78-83]. Until recently the diphenylated fraction from this reaction was believed to consist of the nonrearranged alkylation product, 1,2-diphenyl-2-methylpropane, mixed with a small quantity of the rearranged alkylate, *meso*-2,3-diphenylbutane [25,78,79,81-83]. The presence of the latter isomer in the product mixture was discovered owing to the fortuitous circumstances of its high melting point and its low solubility in the medium.

In 1966, the product mixture from the reaction of benzene with isobutylene dibromide was carefully reinvestigated by Khalaf and Roberts [80]. Using modern techniques for identification, these authors found the product mixture from the aluminum chloride-catalyzed reaction to consist of four isomers: 1,2-diphenyl-2-methylpropane (75), 1,1-diphenyl-2-methylpropane (76), and R,S- and *meso*-2,3-diphenylbutane (81) (scheme 6). The first two isomers,

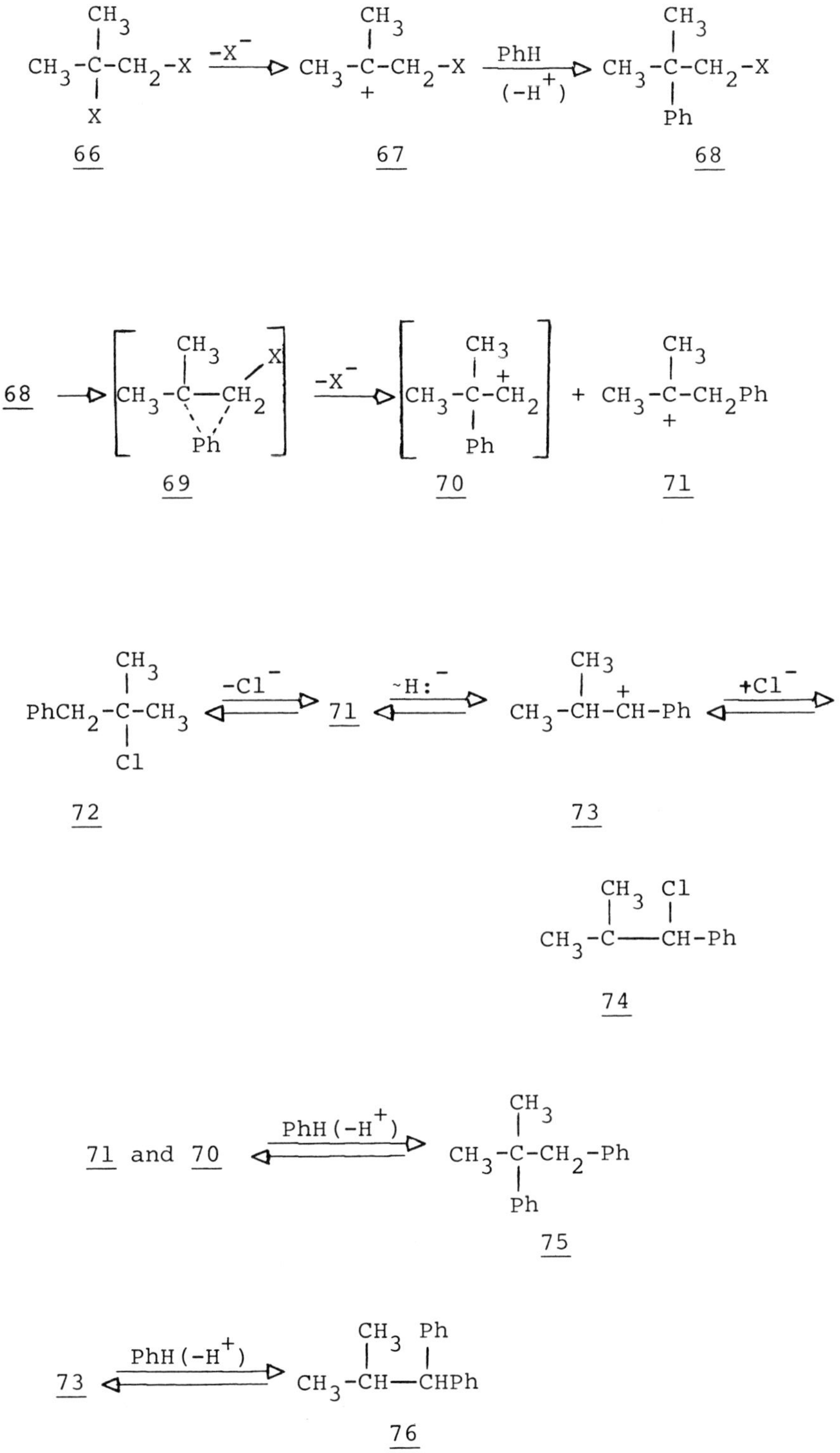

$$CH_3-\underset{X}{\overset{CH_3}{C}}-CH_2-X \xrightarrow{-X^-} CH_3-\underset{+}{\overset{CH_3}{C}}-CH_2-X \xrightarrow[(-H^+)]{PhH} CH_3-\underset{Ph}{\overset{CH_3}{C}}-CH_2-X \quad (28)$$

$$\underline{66} \qquad \underline{67} \qquad \underline{68}$$

$$\underline{68} \longrightarrow \left[CH_3-\overset{CH_3}{C}\text{—}CH_2(X)\ \text{(bridged by Ph)}\right] \xrightarrow{-X^-} \left[CH_3-\underset{Ph}{\overset{CH_3}{C}}-\overset{+}{C}H_2\right] + CH_3-\underset{+}{\overset{CH_3}{C}}-CH_2Ph \quad (29)$$

$$\underline{69} \qquad \underline{70} \qquad \underline{71}$$

$$PhCH_2-\underset{Cl}{\overset{CH_3}{C}}-CH_3 \underset{}{\overset{-Cl^-}{\rightleftharpoons}} \underline{71} \overset{\sim H:^-}{\rightleftharpoons} CH_3-\overset{CH_3}{C}H-\overset{+}{C}H-Ph \overset{+Cl^-}{\rightleftharpoons} \quad (30)$$

$$\underline{72} \qquad \underline{73}$$

$$CH_3-\overset{CH_3}{C}\text{—}\overset{Cl}{C}H-Ph$$

$$\underline{74}$$

$$\underline{71} \text{ and } \underline{70} \overset{PhH(-H^+)}{\rightleftharpoons} CH_3-\underset{Ph}{\overset{CH_3}{C}}-CH_2-Ph \quad (31)$$

$$\underline{75}$$

$$\underline{73} \overset{PhH(-H^+)}{\rightleftharpoons} CH_3-\overset{CH_3}{C}H\text{—}\overset{Ph}{C}HPh \quad (32)$$

$$\underline{76}$$

Scheme 6

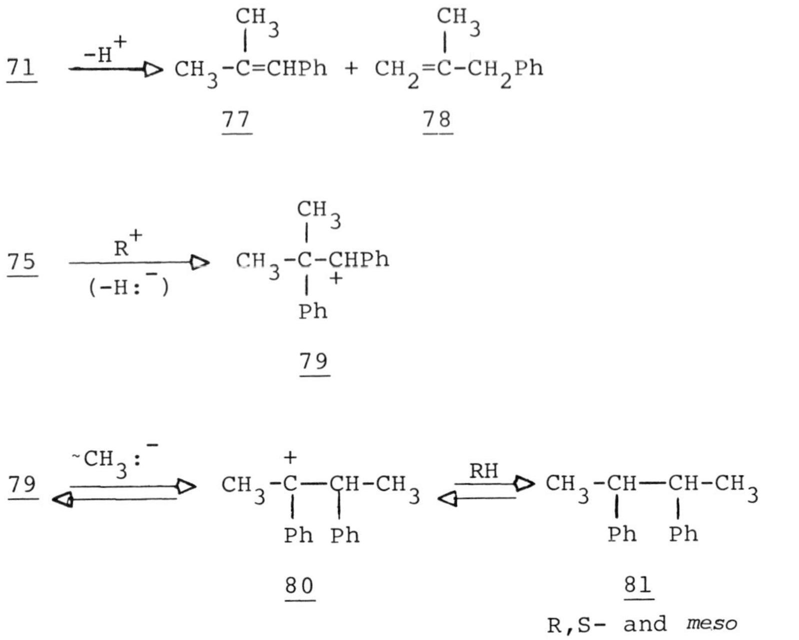

$$\underline{71} \xrightarrow{-H^+} CH_3-\overset{\overset{CH_3}{|}}{C}=CHPh + CH_2=\overset{\overset{CH_3}{|}}{C}-CH_2Ph \quad (33)$$

77 78

$$\underline{75} \xrightarrow[(-H:^-)]{R^+} CH_3-\underset{\underset{Ph}{|}}{\overset{\overset{CH_3}{|}}{C}}-\overset{+}{C}HPh \quad (34)$$

79

$$\underline{79} \underset{}{\overset{\sim CH_3:^-}{\rightleftharpoons}} CH_3-\underset{\underset{Ph}{|}}{\overset{+}{C}}-\underset{\underset{Ph}{|}}{C}H-CH_3 \overset{RH}{\rightleftharpoons} CH_3-\underset{\underset{Ph}{|}}{C}H-\underset{\underset{Ph}{|}}{C}H-CH_3 \quad (35)$$

80 81

R,S- and *meso*

Scheme 6 (continued)

75 and 76, were always produced in an apparent equilibrium ratio of approximately 2:1.

The reaction of isobutylene dibromide with benzene in the presence of milder catalysts such as ferric chloride and aluminum chloride-nitromethane was also investigated. With these catalysts, an equilibrium mixture of only 75 and 76 was produced. Both forms of 81 were absent from the product mixtures. This suggested to the authors that the formation of 81 is subsequent to alkylation, and requires a catalyst strong enough to abstract a hydride ion. This was substantiated by the finding that when pure samples of both 75 and 76 were treated with $AlCl_3$, they were easily equilibrated, and both forms of 81 were produced. When $FeCl_3$ or $AlCl_3$-CH_3NO_2 catalysts were used to treat 75 and 76, the respective diphenylalkanes were recovered unchanged.

Khalaf and Roberts [80,84] also investigated the reaction of the related chlorophenylbutanes 68 (X = Cl), 72 and 74 (scheme 6), with benzene in the presence of $AlCl_3$, $FeCl_3$, or $AlCl_3$-CH_3NO_2 catalysts. The results with these monochlorides were similar to those with isobutylene dibromide under comparable catalytic conditions. These new findings disproved an earlier report by Somerville and Spoerri [78] that chlorides 68 (X = Cl) and 72 alkylate benzene in the presence of $AlCl_3$ to give only the diphenylbutane isomer 75 and *meso* 81.

The similar products obtained in the alkylation of benzene with isobutylene dibromide (68, X = Br), neophyl chloride (68, X = Cl), 1-chloro-2-methyl-1-phenylpropane (74), and 2-chloro-2-methyl-1-phenylpropane (72) using Friedel-Crafts catalysts may be rationalized in terms of the mechanisms outlined in scheme 6 [80].

Commenting on these mechanisms the authors emphasized the following points:

1. In the case of isobutylene dibromide the first step is the formation of the tertiary carbocation (67, X = Br), which in turn alkylates benzene to give neophyl bromide (68, X = Br). This is, indeed, reasonable because of the comparative inactivity of the halogen atom attached to the primary carbon atom [40,47,85,86].
2. Participation of the phenyl group appears to be essential to the removal of the primary halogen atom from the neophyl halide (Eq. 29), since alkylation of benzene with neophyl chloride was readily achieved in the presence of the weak catalysts $FeCl_3$ and $AlCl_3$-CH_3NO_2. When the latter catalyst was used with 1,3-dichloro-3-methylbutane, reaction of the primary chlorine atom was negligible and the chief product was 1-chloro-3-methyl-3-methylbutane [85].
3. The production of an equilibrium mixture of 75 and 76 from the reaction of benzene with 66, 68, 72, and 74, even when the weak nonisomerizing catalysts were used, suggests that an equilibration of the tertiary ion (71) with the secondary (benzylic) ion (73) by a 1,2 hydride shift (Eq. 30) occurs rapidly prior to the formation of 75 and 76 (Eqs. 31 and 32).
4. The rearrangements of 75 and 76 to R,S- and *meso*-81 require the strong catalyst $AlCl_3$ for hydride abstraction steps such as that indicated in Eq. (34). A 1,2 methyl shift may then occur, followed by hydride exchange (Eq. 35).
5. The intermediacy of alkenes [probably 77 and 78 (Eq. 33)] among the reaction products was indicated by the observation of an alkene absorption band in the IR spectra of the crude mixtures and by the ability of these mixtures to decolorize $KMnO_4$ and Br_2 solutions.

2. Branched C_5 and higher dihaloalkanes

a. *Secondary and Tertiary Dihaloalkanes*

As described previously, the alkylation of alkenes with dihaloalkanes is sometimes accompanied by ring closure to give cyclic derivatives. However, the occurrence of the cyclialkylation process is always dependent on the structural features of the alkylating agent. These, of course, include the length of the chain and the stability of the intermediate cations. The results of alkylation with straight-chain dihalides demonstrated clearly that both 1,4-diprimary or 1,4-disecondary carbocations can react with arenes to yield cyclialkylation products. These results also indicated that the 1,4 position between the two halogens in straight-chain C_5 and higher dihaloalkanes is not an essential requirement for closure to a six-membered ring, since the 1,4 relationship can easily be attained through rearrangement of either the dihalides or the intermediate carbocations.

Observations similar to these have also been noted in Friedel-Crafts alkylations of aromatics by branched dihalides. This similarity will become apparent from examination of the following results.

In 1940, Bruson and Kroeger [85] investigated the Friedel-Crafts-catalyzed condensation of various aromatic compounds with 2,5-dichloro-2,5-dimethylhexane (82) and a number of related derivatives, such as 2,5-dihydroxy-2,5-dimethylhexane (83), 2,5-dimethyl-1,5-hexadiene (84), and 2,2,5,5-tetramethyltetrahydrofuran (85). They found that phenol condensed with compounds 82-85 in the presence of $AlCl_3$ to give the same crystalline compound in each case, the tetramethyltetrahydro-β-naphthol (86) and a small amount of a crystalline nonphenolic compound believed to be the chroman derivative (87).

$(CH_3)_2CCl\,CH_2CH_2CCl(CH_3)_2$ (82)

$(CH_3)_2C(OH)CH_2CH_2C(OH)(CH_3)_2$ (83)

$CH_2{=}C(CH_3)CH_2CH_2C(CH_3){=}CH_2$ (84)

$(CH_3)_2CCH_2CH_2C(CH_3)_2$ with O bridging (85)

82–85 $\xrightarrow[AlCl_3]{PhOH}$ 86 + 87

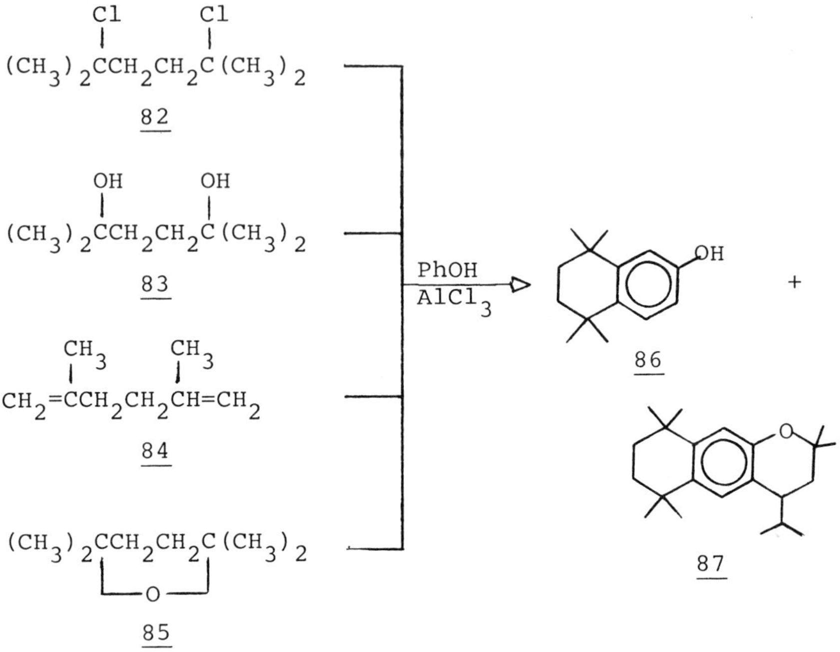

In accordance with known alkylation reaction of phenol with tertiary derivatives to yield *para* tertiary alkylphenols, it was suggested that compounds 82 to 84 first alkylated phenol in the *para* position to form monoalkyl derivatives. These intermediates then rapidly undergo cyclialkylation to form the β-naphthol derivative (86), which, in turn, may react with another molecule of the alkylating agent to form the chroman derivative (87). In the case of compound 85, the furan ring must open up before cyclidehydration can occur.

The same authors also studied the condensation of 2,5-dichloro-2,5-dimethylhexane (82) with other aromatic compounds, including aromatic ethers, substituted phenols, thiophenols, thiophene, *o*-chlorotoluene, and a number of aromatic hydrocarbons. They indicated that most aromatic ethers and substituted phenols underwent condensation with 82 in the presence of $AlCl_3$ to produce the corresponding analogs of 86.

The condensation of 2,5-dichloro-2,5-dimethylhexane (82) with benzene in the presence of $AlCl_3$ was reported to yield a liquid monocyclialkylation product (88) and a crystalline dicyclialkylation product 89 [85]. Bruson

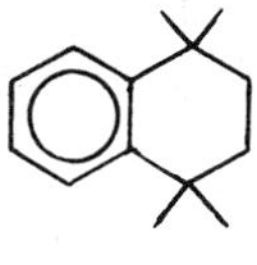

88 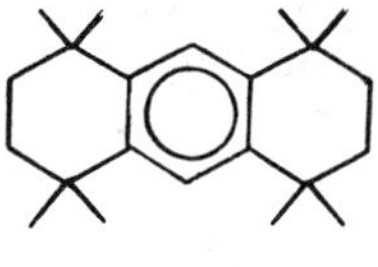 89

and Kroeger pointed out that nearly quantitative yields of the dicyclialkylation product 89 were obtained when only catalytic quantities of $AlCl_3$ were used, but larger amounts of catalysts favored the formation of the monocyclialkylation compuond (88). Toluene, however, gave only a liquid monocyclialkylation product, which was believed to be 90 (Eq. 36):

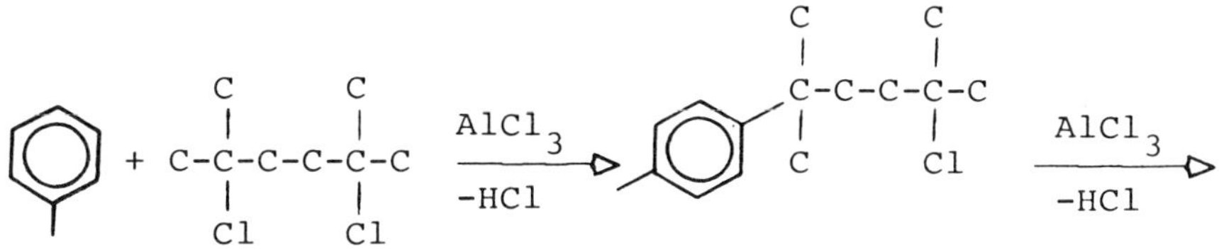

(36)

82

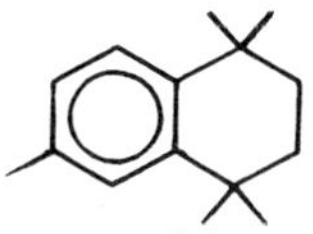

90

Similarly, the reaction of 2,5-dichloro-2,5-dimethylhexane (82) with *o*-chlorotoluene was reported to give 91, with tetralin to give 92, with indan to give 93, and with naphthalene to give 94 [85].

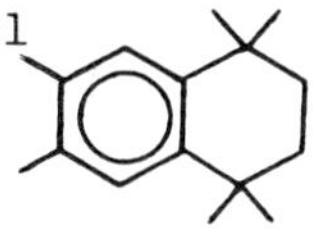

91 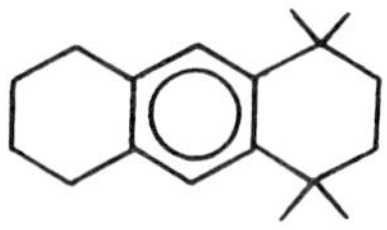 92 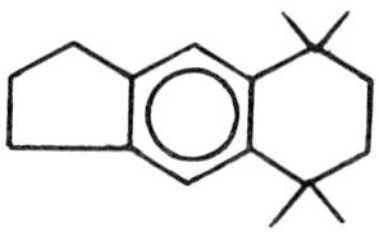 93

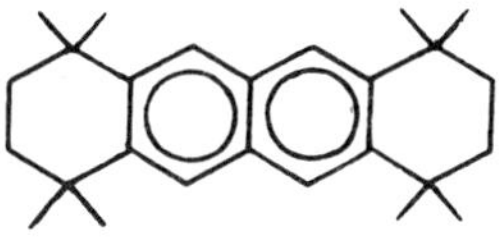

94

In 1947, Sisido and Nozaki investigated the aluminum chloride-catalyzed alkylation of benzene with 2,3-dichloro-2,3-dimethylbutane (95) and 3,4-dichloro-3,4-dimethylhexane (96) [54]. On the basis of dehydrogenation and/or oxidation experiments, the product from 95 was said to consist mainly of 1,1,2-trimethylindan (97) mixed with small amounts of hexylbenzene (Eq. 37). The product from 96 was considered to be 1,2,3,4-tetramethyltetralin 98 (Eq. 38).

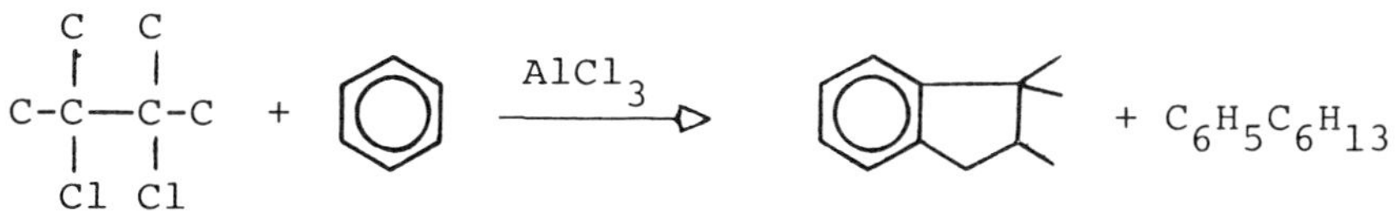

$$95 + C_6H_6 \xrightarrow{AlCl_3} 97 + C_6H_5C_6H_{13} \quad (37)$$

95 97

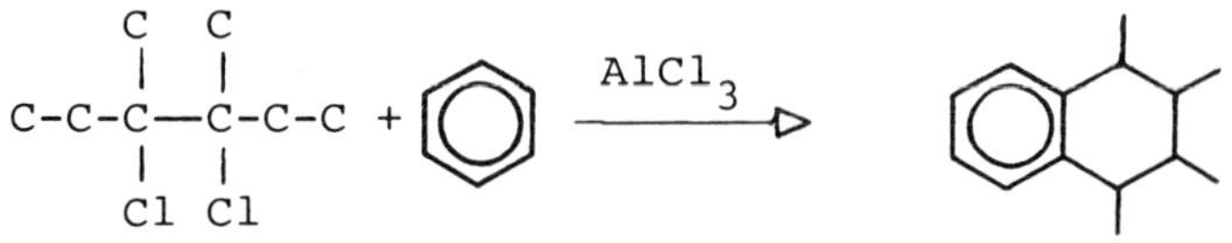

```
    C  C
    |  |
C-C-C--C-C-C  +  (benzene)   --AlCl3-->   98          (38)
    |  |
    Cl Cl

    96
```

In 1962, Shadmanov reinvestigated the condensation of 2,5-di-2,5-dichloro-dimethylhexane (82) with benzene in the presence of $AlCl_3$ [87]. After heating at 76 to 80°C for 24 hr, the product obtained was 1,1,4,4-tetramethyltetralin (88) but reaction at 20-22° for 72 hr gave 1,1,4,4,5,5,8,8-octamethyloctahydroanthracene (89).

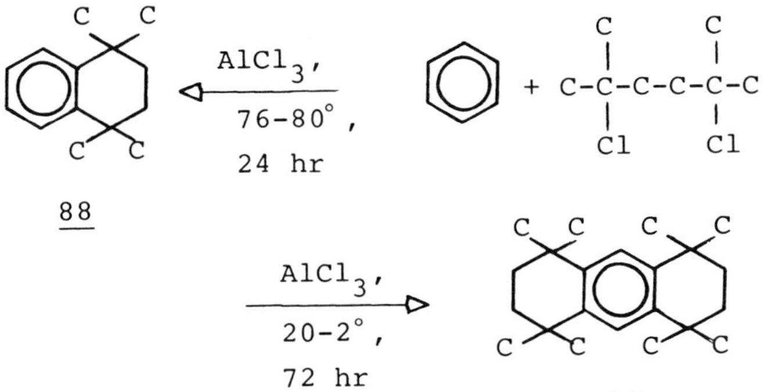

```
                                          C       C
                                          |       |
 88  <--AlCl3, 76-80°, 24 hr--  (benzene) + C-C-C-C-C-C
                                          |       |
                                          Cl      Cl

     --AlCl3, 20-2°, 72 hr-->  89
```

The results of Shadmanov were thus in accord with those obtained earlier by Bruson and Kroeger [85].

In an extension of their work with linear dichloropentanes [72,73] (Sec. I.B), Gelin and co-workers studied the Friedel-Crafts reactions with benzene of a number of secondary and tertiary halogen (Cl and Br) derivatives of 2- and 3-methylpentane (99, 100) and the phenylalkyl halides expected as intermediates from the initial reaction of 99 and 100 with benzene, compounds 101-104 [74,88].

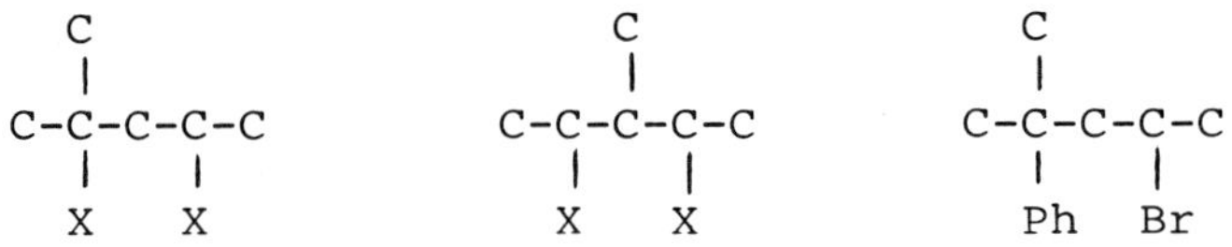

```
C                 C                 C
|                 |                 |
C-C-C-C-C       C-C-C-C-C       C-C-C-C-C
|   |             |   |             |   |
X   X             X   X             Ph  Br

99               100               101

     X = Cl,Br

Cl C              C                 C
|  |              |                 |
C-C--C-C-C      C-C--C-C-C      C-C-C-C-C
|                 |  |              |   |
Ph                Ph Cl             Ph  Br

102              103               104
```

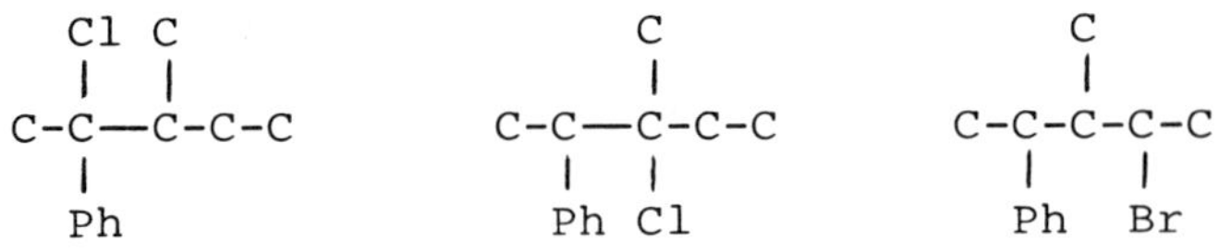

The products were reported to be mainly 105-108, produced in almost the same proportions from 99 and 100. Very similarly, mixtures of products were

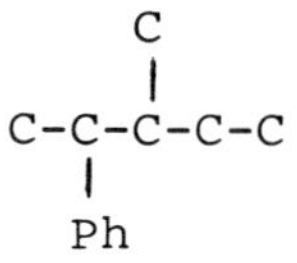

105

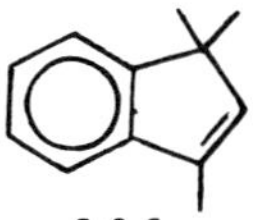

106

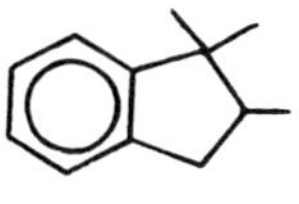

97

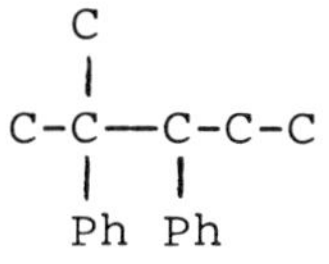

107

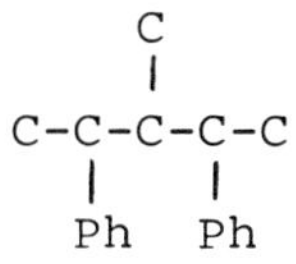

108

also obtained from reactions of 101-104 with benzene and $AlCl_3$. Gelin et al. comment that 97 was a minor component of their product mixtures whereas Sisido and Nozaki [54] reported it as a principal product of the reaction of 2,3-dichloro-2,3-dimethylbutane (95) with benzene. To explain the formation of their products, Gelin et al. proposed mechanisms involving carbocation intermediates which are isomerized by hydride and methyl shifts and are converted to products by cyclialkylations and hydride transfers. The mechanism for the formation of 97 was depicted as involving as an intermediate the primary carbocation 110 formed by a 1,2 hydride shift from the tertiary carbocation 109:

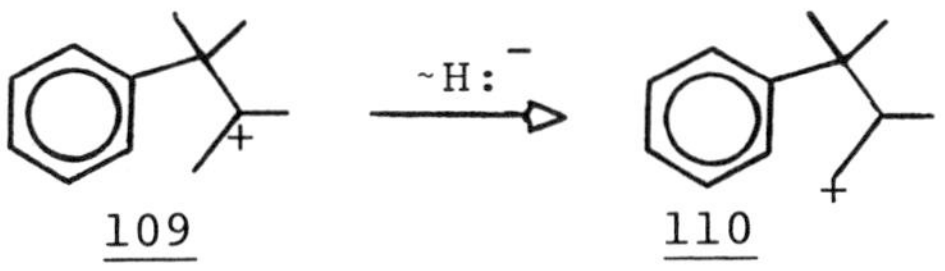

~H:⁻

109

110

An interesting case of Friedel-Crafts alkylation with a symmetrical dihalide was the preparation of di-α-arylcumylbenzenes (112, 114) by the reaction of the di-tertiary-aralkyl chloride 111 or its *m* isomer (113) with excess molar amounts of the appropriate benzene derivatives in the presence of aluminum

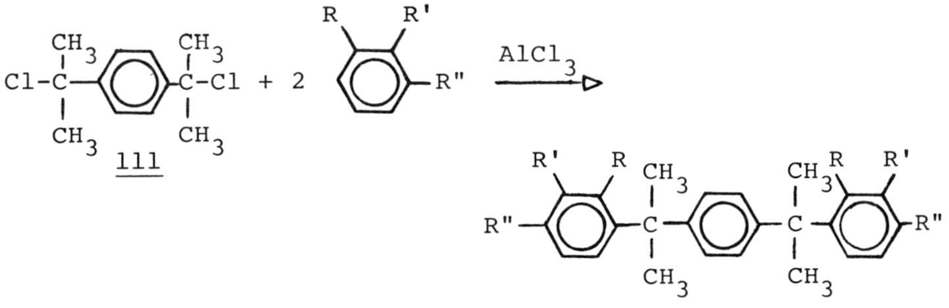

$AlCl_3$

111

112

113 + 2 (R"-substituted benzene) $\xrightarrow{AlCl_3}$ 114

R,R' = H or Me; R" = H,Me or Et

chloride [89]. The resulting alkylates (112, 114) are useful as starting materials for the preparation of monomers used in the syntheses of polymers.

b. Primary (1,1-)Dihaloalkanes

In 1955, Schmerling et al. investigated the alkylation of benzene with 1,1-dichloro-3,3-dimethylbutane (115) under a variety of conditions [90]. In the presence of $AlCl_3$ at 0°C, the reaction yielded 1,1-diphenyl-3,3-dimethylbutane (117, 26 to 28%) and a primary hexylbenzene, 1-phenyl-3,3-dimethylbutane (118, 19 to 20%). The yield of the latter was increased markedly (to 60%) at the expense of the former (yield, 6%) by carrying out the reaction in the presence of methylcyclopentane, which served as a hydride donor and was converted to (methylcyclopentyl)benzene (120, 48%).

When the reaction of benzene with 115 was carried out in the presence of zirconium chloride at 27°C, or of aluminum chloride monomethanolate at 84°C, the yield of monophenylhexane (chiefly 118) was only 5 to 6% while that of 117 was 20%. It was also noted that less hydrogen transfer occurred in the presence of these milder catalysts.

To account for the unexpected formation of the nonrearrangement products 1,1-diphenyl 3,3-dimethylbutane (117) and 1-phenyl-3,3-dimethylbutane (118) and for the role of methylcyclopentane in the reaction above, Schmerling and co-workers proposed the processes outlined in Eqs. (39) to (43).

```
                  C                              C
                  |     AlCl3                    |
PhH + Cl-C-C-C-C  ----------->  Ph-C-C-C-C + HCl              (39)
         |   |                     |   |
         Cl  C                     Cl  C
      115

          C                          C
          |                          |
Ph-C-C-C-C + AlCl3  <====>  Ph-C-C-C-C   AlCl4-               (40)
   |   |                       +   |
   Cl  C                           C
                               116
```

$$\underline{116} + PhH \rightleftharpoons Ph\text{-}C(\text{C}_6H_5^+)\text{-}C\text{-}C(C)_2\text{-}C \quad AlCl_4^- \rightleftharpoons Ph\text{-}CH(Ph)\text{-}C\text{-}C(C)_2\text{-}C + AlCl_3 + HCl \quad (\underline{117}) \tag{41}$$

$$\underline{116} + \text{methylcyclopentane} \rightleftharpoons Ph\text{-}C\text{-}C\text{-}C(C)_2\text{-}C\ (\underline{118}) + \text{methylcyclopentyl cation}\ AlCl_4^-\ (\underline{119}) \tag{42}$$

$$\underline{119} + C_6H_6 \longrightarrow \text{1-methyl-1-}C_6H_5\text{-cyclopentane}\ (\underline{120}) + HCl + AlCl_3 \tag{43}$$

It is noteworthy that the $AlCl_3$-catalyzed reaction of benzene with 1,1-dichloro-3,3-dimethylbutane (115) in the presence of a good hydride donor offers a simple means for obtaining the neohexylbenzene, 118. As suggested by Schmerling et al. [90], other primary alkylbenzenes containing a *gem*-dialkyl group can be obtained in a similar manner starting with the suitable 1,1-dichloro-3,3-dialkylalkane.

In 1957, Schmerling and co-workers studied the aluminum chloride-catalyzed reaction of benzene with 1,1-dichloro-3-methylbutane 121 and 1,1-dibromo-2-cyclohexylethane (123) in the presence and absence of added hydride donor [91]. In the absence of an added hydride donor, the product from 121 was shown to consist chiefly of pentylbenzenes (27% yield) mixed with a little 1,1-diphenyl-3-methylbutane (122). The pentylbenzene fraction was shown

$$C\text{-}C(C)\text{-}C\text{-}CCl_2\ (\underline{121}) + C_6H_6 + AlCl_3 \longrightarrow$$

$$Ph\text{-}C\text{-}C\text{-}C(C)\text{-}C\ (80\%) + \text{other pentylbenzenes or dimethylindan}\ (20\%)\ [27\%\ \text{yield}] + Ph_2C\text{-}C\text{-}C(C)\text{-}C\ (\underline{122})\ (\text{trace})$$

by infrared analysis to consist of about 80% of isopentylbenzene and 20% of 1,1-dimethylindan and/or 2-methyl-3-phenylbutane and *t*-pentylbenzene.

In the presence of methylcyclohexane, a higher yield (49%) of pentylbenzenes (approximately 85 to 90% isopentylbenzene and 10 to 15% of other pentylbenzenes or dimethylindan) was obtained, together with a 42% yield of (methylcyclohexyl)benzene.

It is evident from the results above that hydrogen transfer occurred with 1,1-dichloro-3-methylbutane (121), even in the absence of a hydride donor. This was reasonably attributed to the hydrogen attached to the tertiary carbon atom in the dichloride and/or the isopentylbenzene, which can take part in the hydrogen exchange reaction in much the same way as the analogous tertiary hydrogen in methylcyclohexane.

Further evidence that the presence of a tertiary hydrogen atom in a dihalide results in comparatively high yields of monoalkylbenzene, even in the absence of an added hydride donor, was obtained when Schmerling et al. carried out the reaction of 1,1-dibromo-2-cyclohexylethane (123) with benzene in the presence of $AlCl_3$. When no hydride donor was used, 1-cyclohexyl-2-phenylethane (124) was obtained in 37% yield. When isopentane was added, the yield of 124 was 40%. Interestingly, there was a relatively low yield of pentylbenzenes produced in the latter case. This indicates that the hydrogen

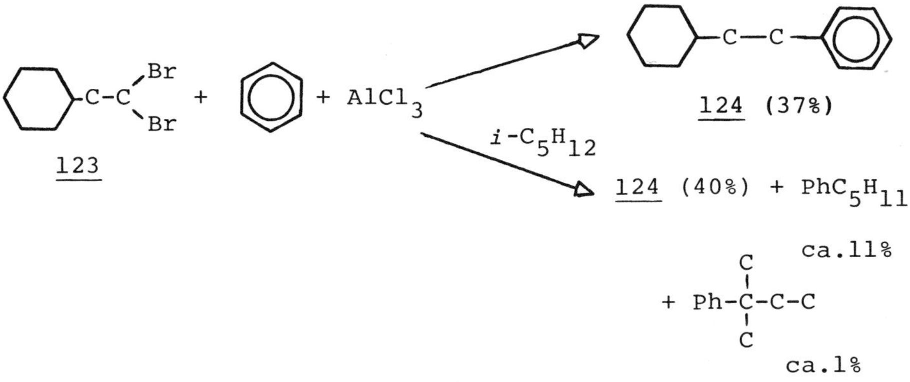

C₆H₁₁-C-CBr₂ (123) + C₆H₆ + $AlCl_3$ → C₆H₁₁-C-C-C₆H₅ 124 (37%)

with i-C_5H_{12} → 124 (40%) + PhC_5H_{11} ca.11% + Ph-C(C)(C)-C-C ca.1%

atom attached to the tertiary carbon in the dibromoethylcyclohexane is more readily abstracted than is that in isopentane.

In 1959, Schmerling et al. [92] found that hydrogen transfer involving benzylic hydrogen was the principal reaction even in the absence of saturated hydrocarbons when *p*-xylene was alkylated with 1,1-dichloroalkanes in the presence of $AlCl_3$. For example, the reaction of *p*-xylene with 1,1-dichloro-3,3-dimethylbutane (125) yielded (3,3-dimethylbutyl)-*p*-xylene (129), di-(*p*-xylyl)methane (132), and a hydrocarbon, $C_{14}H_{20}$, believed to be a pentamethylindan (133).

Schmerling et al. rationalized the production of 132 in terms of the intermediate formation of 130 via Eqs. (44) to (48).

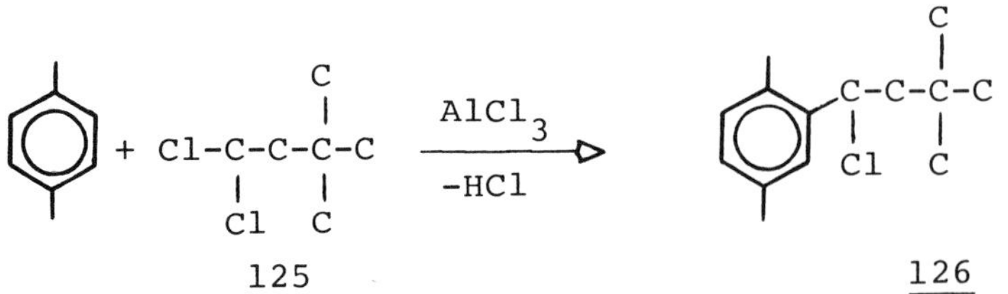

p-xylene + Cl-C(Cl)-C-C(C)(C)-C (125) —$AlCl_3$, −HCl→ *p*-xylyl-C(Cl)-C-C(C)(C)-C (126) (44)

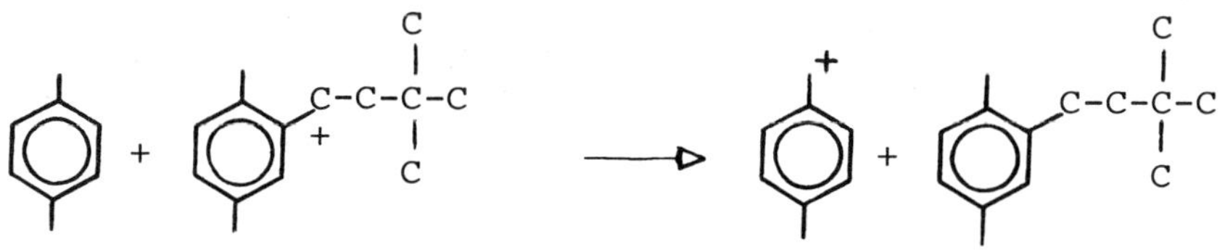

127 128 129 (45)

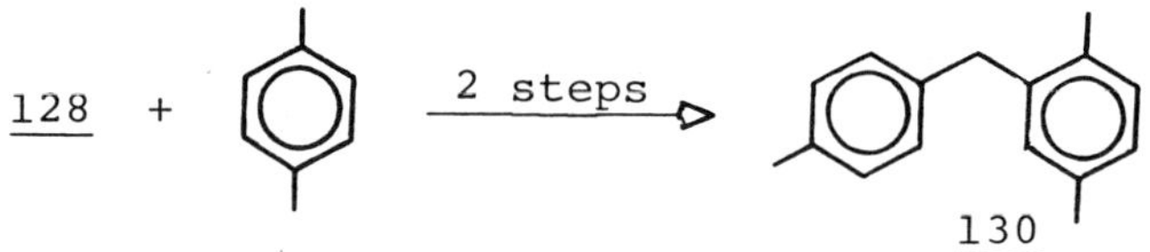

128 + 2 steps 130 (46)

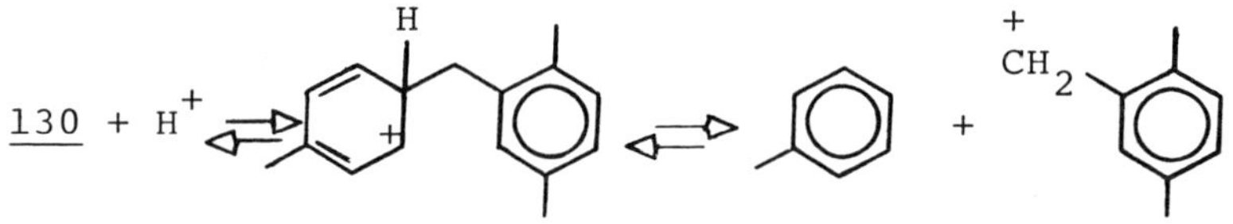

130 + H^+ ⇄ ⇄ + $\overset{+}{CH_2}$ 131 (47)

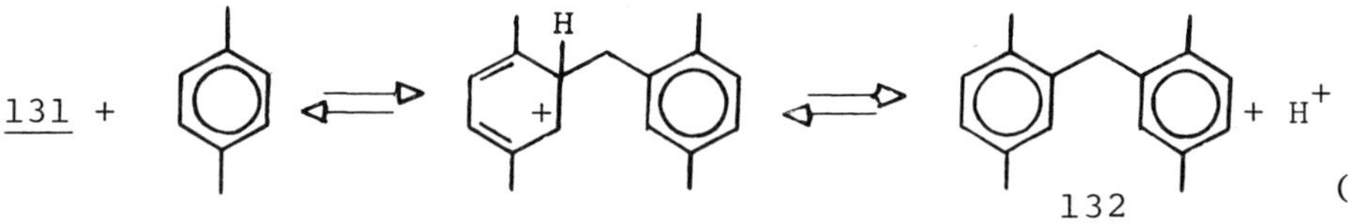

131 + ⇄ ⇄ 132 + H^+ (48)

The formation of 133 was suggested to occur according to scheme 7. The carbocation 127 could be obtained either by the ionization of the intermediate monochloride 126, produced by the reaction of the xylene with the dichloride (Eq. 44), or by abstraction of a hydride ion from the hexyl-*p*-xylene 129.

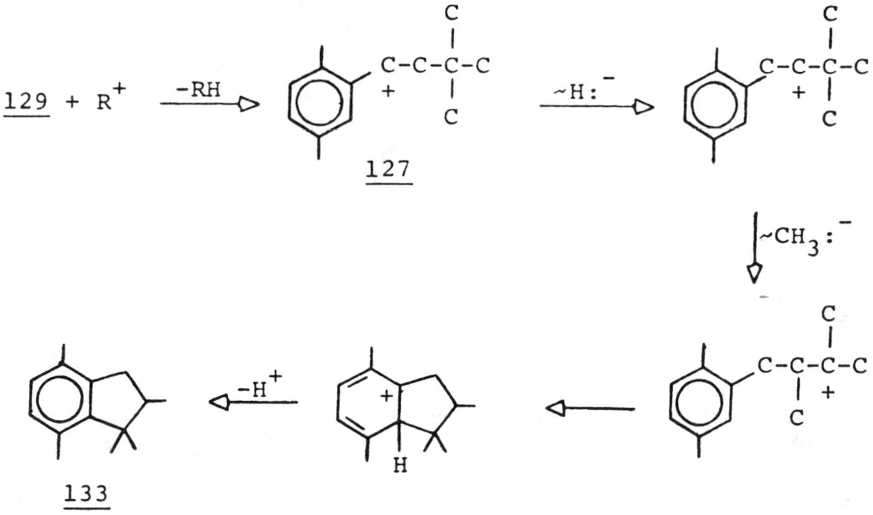

129 + R^+ —(−RH)→ 127 —(~H:⁻)→ —(~CH_3:⁻)→ → —($-H^+$)→ 133

Scheme 7

Schmerling et al. also investigated the reaction of *p*-xylene with 1,1-dichloro-3-methylbutane (125) in the presence of $AlCl_3$ and added methylcyclohexane. Under these conditions the product consisted chiefly of the hexyl-*p*-xylene (129) and (methylcyclohexyl)-*p*-xylene, together with only a minor amount of dixylylmethane (132). This indicated that the saturated hydrocarbon containing a tertiary hydrogen atom (methylcyclohexane) furnished the hydrogen necessary for the formation of the hexylxylene (129) more readily than did the *p*-xylene, indicating that a tertiary alkyl cation is more stable than a primary benzyl cation.

Other 1,1-dichloroalkanes were shown to react similarly with *p*-xylene. For example, the reactions with 1,1-dichloroethane and methylene chloride gave the product distributions shown in Eqs. (49) and (50) [92]. The

CH_3CHCl_2 + [p-xylene] $\xrightarrow[22-26^\circ\ (3-4\ hr)]{AlCl_3}$ 22% + 5% + 14% (49)

CH_2Cl_2 + [p-xylene] $\xrightarrow[28-31^\circ\ (0.9\ hr),\ 31-63^\circ\ (1.9\ hr)]{AlCl_3}$ 20% + 10% + 8% + 15% (50)

formation of toluene in the reaction with methylene chloride was again explained in terms of transalkylation as shown before in Eq. (47).

C. Alkylation with Polyhaloalkanes

1. Polyhalomethanes

As early as 1877, Friedel and Crafts reported that the alkylation of benzene with chloroform and carbon tetrachloride in the presence of $AlCl_3$ proceeded very easily in the cold to give triphenylmethane and tetraphenylmethane, respectively [93]. Reinvestigation of the reaction with chloroform by Schwarz [94], Emil and Otto Fischer [95], Friedel and Crafts [96], and Böeseken [97] showed that, in addition to triphenylmethane, diphenylmethane, tetraphenylethylene, and chlorotriphenylmethane were also products of the reaction.

The reaction between benzene and carbon tetrachloride in the presence of $AlCl_3$ was similarly reexamined. The results obtained were found to vary not only with reaction temperature and ratio of reactants, but also with the decomposition and distillation procedures. Although products such as dichlorophenylmethane [98-101], chlorotriphenylmethane [98], and triphenylmethane [94, 102-104] were always obtained, no tetraphenylmethane could be found in the mixtures. This disproved the claim first made by Friedel and Crafts [93] that tetraphenylmethane was produced in this reaction.

It is interesting to note that under certain conditions, it was possible to direct the C_6H_6-CCl_4-$AlCl_3$ reaction in such a way as to produce an 80 to 90% yield of dichlorodiphenylmethane [99,100] or a 68 to 84% yield of triphenylmethane [105-107]. Moreover, an 80 to 90% yield of benzophenone was obtained by first treating benzene with CCl_4 and $AlCl_3$ in the cold and then hydrolyzing the resulting dichlorodiphenylmethane by distillation with steam [108]. A similar reaction of carbon tetrachloride with 1,2,4- and 1,3,5-trichlorobenzene [109] and *m*-xylene [99] resulted in the formation of the corresponding benzophenone derivatives 134-136. Chlorobenzene and carbon tetrachloride gave a mixture consisting of almost equal amounts of 2,4'- and 4,4'-dichlorobenzophenone [110].

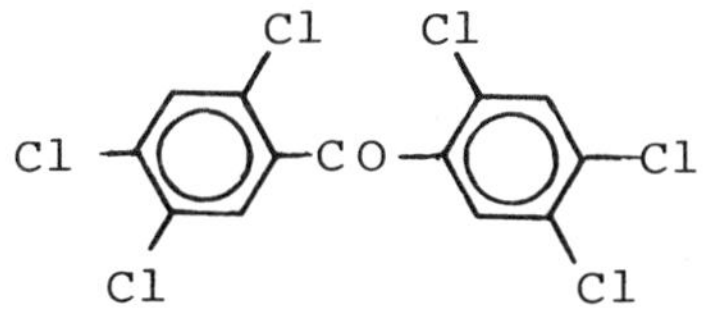

134

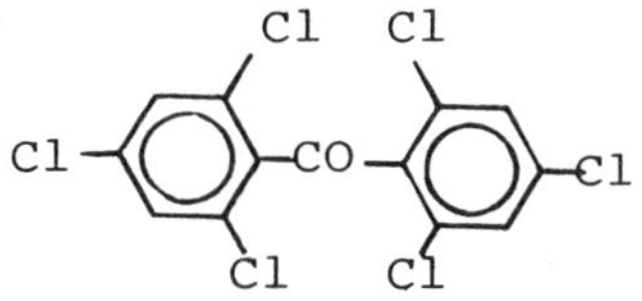

135

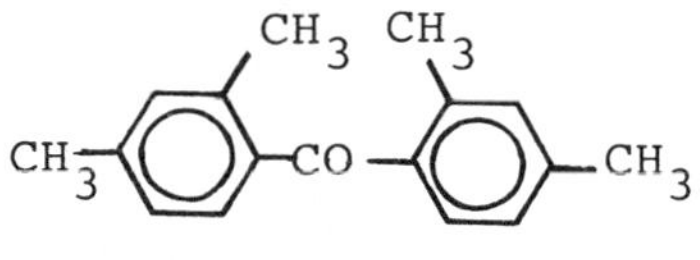

136

The aluminum chloride-catalyzed reactions of chloroform with toluene [111,112], halobenzenes [109], and naphthalene [94,113,114] and of carbon tetrachloride with toluene [99,115-117] and *m*-xylene [99] were also explored by early investigators. In almost all cases, the resulting products were shown to be analogous to those obtained with benzene.

From the reaction between toluene and $CHCl_3$, Lavaux [112] identified three isomeric dimethylanthracenes, as well as a β-methylanthracene. These were said to result from the reaction of the ditolylchloromethane first formed with another molecule of chloroform.

In addition to these early studies on the alkylation of aromatics with chloroform and carbon tetrachloride, more recent ones have also been reported. For example, in 1951, Dolgov et al. [118] investigated the effect of catalyst concentration on the relative yields of arylmethanes in the reaction of arenes with chloroform. These authors found that benzene with chloroform and $AlCl_3$ gave diphenylmethane, the yield of which increased with catalyst concentration. They also found that the addition of cuprous chloride favored the yield of diphenylmethane, and the addition of 40% of the salt gave up to 40% yield of pure product. In all cases the yield of triphenylmethane did not exceed 4%.

In 1957, Schmerling et al. [91] investigated the effect of added hydrogen donor on the aluminum chloride-catalyzed reaction of benzene with carbon tetrachloride. In the presence of isopentane, hydrogen transfer occurred, resulting in the formation of diphenylmethane in 24% yield and pentylbenzenes (65 to 70% 2-methyl-3-phenylbutane and 30 to 35% *t*-pentylbenzene) in 30% yield, based on a theoretical yield of 2 mol of pentylbenzene per mole of carbon tetrachloride (Eq. 51).

$$CCl_4 + 4\ C_6H_6 + 2\ C_5H_{12} \longrightarrow CH_2(C_6H_5)_2 + 2\ C_6H_5C_5H_{11} + 4\ HCl \quad (51)$$

In the absence of isopentane, the principal product of the reaction of CCl_4 with benzene was either chlorotriphenylmethane or triphenylmethane, depending on the reaction conditions.

In 1967, Beckert and Lowe [119] investigated the Friedel-Crafts alkylation of pentafluorobenzene with mono-, di-, and trichloromethane. Although their attempts at alkylation by standard Friedel-Crafts procedures were unsuccessful, the authors were able to obtain some products by carrying out the reactions in excess arene at reflux temperatures. Still better results were obtained when the experiments were carried out in a steel bomb at 150°C using an excess of pentafluorobenzene as solvent; methylene chloride and chloroform reacted readily with pentafluorobenzene in the presence of $AlCl_3$. The products were bis(pentafluorophenyl)methane (137) and tris(pentafluorophenyl)methane (138) in 77% and 92% yields, respectively.

$$CH_2Cl_2 + 2\ C_6F_5H \xrightarrow{AlCl_3} (C_6F_5)_2CH_2 \quad (52)$$

137

$$CHCl_3 + 3\ C_6F_5H \xrightarrow{AlCl_3} (C_6F_5)_3CH \quad (53)$$

138

A reaction that employed methyl chloride instead of methylene chloride as the alkylating agent produced 8% of pentafluorotoluene. The low yield was ascribed to the fact that methyl halides are the least reactive ones in the series of primary halides, and halogenated aromatics are alkylated only with difficulty [120].

Beckert and Lowe also examined the catalytic influence of elemental copper and cupric chloride on the extent of reaction [119]. The results of these experiments demonstrated that these cocatalysts had a negligible effect on the product yields.

One of the interesting aspects of the results of Beckert and Lowe was the fact that they always found unchanged pentafluorobenzene and the starting haloalkane in the reaction mixture. All attempts to isolate pentafluorophenylchloromethane, pentafluorophenyldichloromethane or bis(pentafluorophenyl)-chloromethane were unsuccessful. This suggested that the latter intermediates react more rapidly with pentafluorobenzene than the starting haloalkanes, methylene chloride, or chloroform. This view was strengthened by the finding that the Friedel-Crafts reaction of pentafluorobenzyl chloride with pentafluorobenzene produced 137 in 45% yield. A similar reaction between pentafluorobenzyl chloride and benzene gave pentafluorophenyl(phenyl)methane in 64% yield.

More recently the alkylation of pentafluorobenzene by trifluoromethane in the presence of SbF_5 was shown to yield 5% $(C_6F_5)_2CHF$, 1% $(C_6F_5)_3CH$, 56% $(C_6F_5)_2CHOH$, and 36% $[(C_6F_5)_2CH]_2O$ [121]. Also, the alkylation of ethylbenzene by carbon tetrachloride in the presence of $AlCl_3$ was studied under different conditions [122]. At 0 to 20°C with a CCl_4/PhEt ratio of 1:10 and 5% $AlCl_3$, the reaction primarily afforded benzene and p-$Et_2C_6H_4$. At 70°C with a CCl_4-PhEt ratio of 1:1-3 and 10% $AlCl_3$, condensation products predominated, including p-EtC_6H_4Ph, p-$EtC_6H_4CH_2Ph$, di- and triethylbiphenyl, and di- and triethyldiphenylmethane.

An interesting case of alkylation with carbon tetrachloride is the reported reaction with 1,2,3,4,5-pentachlorocyclopentadiene (139) to give octachloro-5-methylcyclopentadiene (140) [123].

$$\underset{\textbf{139}}{C_5Cl_5H} + CCl_4 \xrightarrow[75^\circ,\ 21\ hr]{AlCl_3} \underset{\textbf{140}}{C_5Cl_5(CCl_3)}$$

It was stated that "the mechanism of this unusual reaction is not known, since both 139 and CCl_4 are capable of interacting with this Lewis acid." One possibility is the initial ionization of 139 to produce the cyclopentadienyl anion, which can then behave as a good nucleophile in reaction with a polarized complex of CCl_4 and $AlCl_3$. The whole mechanism can be visualized as follows:

$$C_5Cl_5H \longrightarrow \underline{141}\ (C_5Cl_5^-) + H^+$$

141

$$CCl_4 + AlCl_3 \longrightarrow \overset{\delta+}{Cl_3CCl}:\overset{\delta-}{AlCl_3}$$

$$\underline{141} + \overset{\delta+}{Cl_3CCl}:\overset{\delta-}{AlCl_3} \longrightarrow C_5Cl_5(CCl_3) + AlCl_4^-$$

From the results summarized above it can be seen that in all reactions with tetrahalomethanes, triarylmethanes represent the highest arylation step reached. It can also be seen that in these reactions chlorotriarylmethanes were isolated.

Hine discussed the lack of formation of tetraarylmethanes in alkylations with tetrahalomethanes [124]. He argued that this cannot be due to the impossibility of the existence of the product, since tetraphenylmethane, for example, is a quite stable compound [125], nor can it be due to steric hindrance, since triphenylmethyl chloride and phenol are known to react with each other, yielding triphenyl (*p*-hydroxyphenyl)methane, even in the absence of added Friedel-Crafts catalysts [126].

The question of alkylating benzene with triphenylmethyl chloride was again considered by Fonken in 1963 [127]. After treatment of tetraphenylmethane with $AlCl_3$ and HCl in benzene solution, hydrolysis of the reaction mixture gave an almost quantitative yield of triphenylcarbinol, thus indicating facile dealkylation of tetraphenylmethane. With this evidence that tetraphenylmethane is unstable with respect to reactants under the alkylating conditions, Fonken postulated that the triphenylmethyl cation actually may alkylate benzene, but significant quantities of tetraphenylmethane are not observed because facile dealkylation occurs to produce an equilibrium mixture in which the stable triphenylmethyl cation and benzene will be favored. As a simple test of his hypothesis, Fonken studied the $AlCl_3$-catalyzed alky-

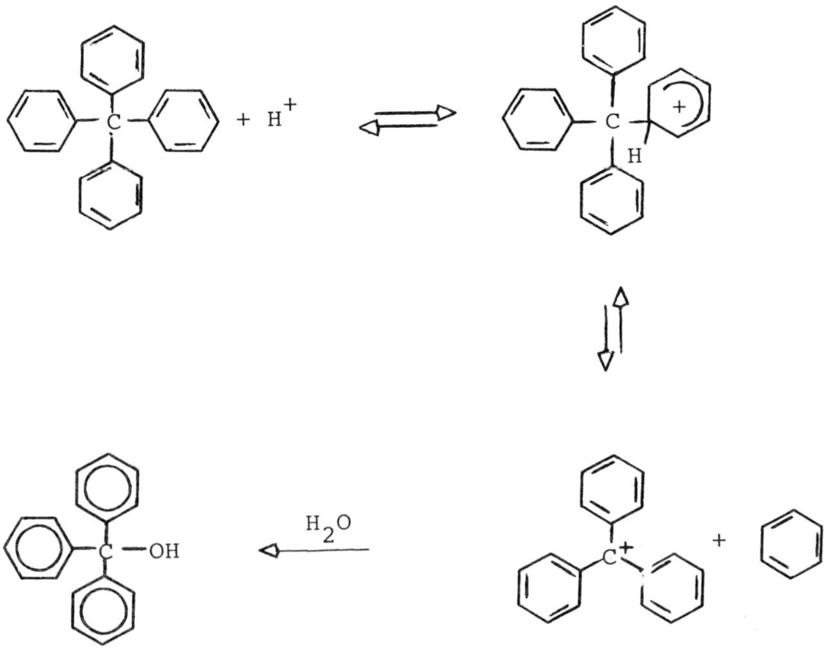

(54a)

(54b)

(54c)

lation of benzene with the "methyl-labeled" diphenyl-*p*-tolylmethyl chloride. Reversible alkylation and dealkylation should generate a mixture of benzene, toluene, triphenylmethyl cation, and the cation derived from the starting chloride (Eqs. 55a to 55c).

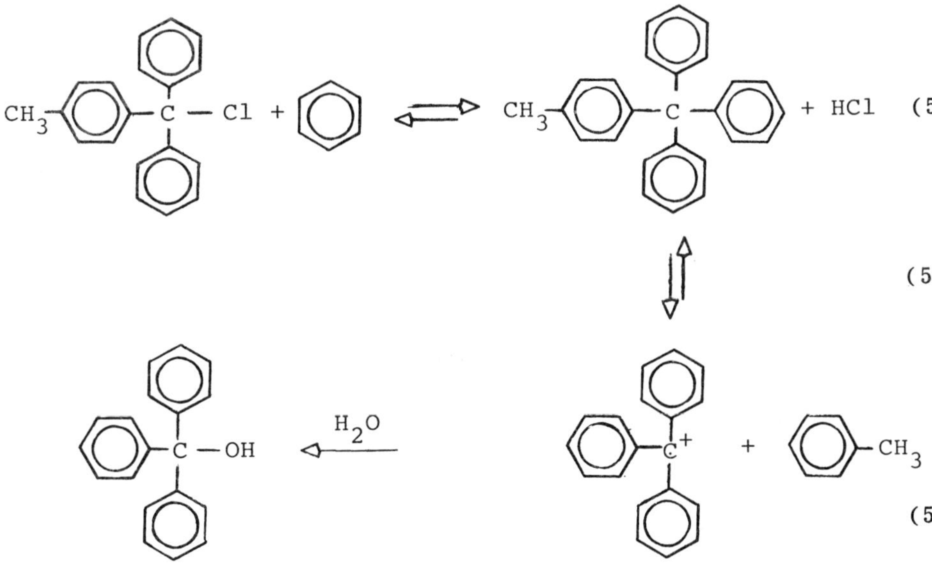

(55a)

(55b)

(55c)

The fact that hydrolysis of the reaction mixture afforded triphenylcarbinol, diphenyltolylcarbinol, and a liquid fraction containing mainly benzene and some toluene was considered supporting evidence for Fonken's view.

2. Polyhaloethanes and higher polyhaloalkanes

Schmerling et al. [91] reported that 1,1,2-trichloroethane reacts with benzene in the presence of $AlCl_3$ and methylcyclohexane to yield bibenzyl and methylcyclohexylbenzene. Bibenzyl was also shown to be a major product of the reaction of 1,1,2-trichloro- or tribromoethane and benzene in the absence of a hydrogen donor [128-131]. Other products included diphenylmethane, anthracene, and probably, dihydroanthracene [128].

The aluminum chloride-catalyzed reaction of benzene with 1,1,2,2-tetrachlorethane was reported to give a mixture of diphenylmethane and anthracene. A similar reaction with hexachloroethane gave a mixture of diphenylmethane, triphenylmethane, anthracene, and *p*-xylene [132].

The formation of anthracene derivatives was also reported from the aluminum chloride-catalyzed reaction of 1,1,2,2-tetrachloro- or tetrabromoethane with toluene [133,134], the three xylenes [133], and naphthalene [135], and from the reaction of pentachloro- or hexachloroethane with benzene [136]. No homolog of anthracene could be obtained from ethylbenzene and *sym*-tetrabromoethane in the presence of $AlCl_3$ [133]. Concerning the structure of the anthracenes produced, Lavaux [134] reported that 1,6- and 2,7-dimethylanthracene were the isomers formed from toluene and 1,1,2,2-tetrabromoethane. Homer [135] reported that 1,2,7,8-dibenzoanthracene was the isomer formed from the latter tetrabromide and naphthalene.

The formation of anthracene derivatives in the reactions of polyhaloethanes and arenes may be explained in terms of a concurrent alkylation and isomerization process as represented by Eq. (56):

$$C_6H_6 + X_2CH{-}CHX_2 + C_6H_6 \xrightarrow{AlCl_3} (\text{dihydroanthracene}) \longrightarrow \text{anthracene} \quad (56)$$

In 1950, Dolgov and Larin investigated the alkylation of benzene with some polyhalides in the presence of Al metal as the catalyst [137]. Some of the results they reported are given in Eqs. (57) to (60).

$$Cl_2CHCHCl_2 + C_6H_6 \xrightarrow{Al} \begin{cases} \xrightarrow{20^\circ} \text{anthracene} \\ \xrightarrow{70\text{-}75^\circ} PhCH_2CH_2Ph \end{cases} \quad (57)$$

$$BrCH_2CH(Br)CH_2Br + C_6H_6 \xrightarrow[20\ hr]{Al} Ph_2C_3H_6\ (10\%)\ \text{but no}\ Ph_3C_3H_5 \quad (58)$$

$$BrCH_2\underset{\displaystyle Br}{\overset{\displaystyle CH_3}{\overset{|}{\underset{|}{C}}}}CH_2Cl + C_6H_6 \xrightarrow{Al} Ph\text{-}\overset{\displaystyle CH_3}{\overset{|}{C}}H\text{—}\overset{\displaystyle CH_3}{\overset{|}{C}}H\text{-}Ph \qquad (59)$$

$$BrCH_2\text{-}\underset{\displaystyle Br}{\overset{\displaystyle CH_3}{\overset{|}{\underset{|}{C}}}}\text{-}CH_2Br + C_6H_6 \xrightarrow[20\ hr]{Al} \underset{(meso\text{- and R,S})}{Ph\text{-}\overset{\displaystyle CH_3}{\overset{|}{C}}H\text{—}\overset{\displaystyle CH_3}{\overset{|}{C}}HPh} + CH_3\underset{\displaystyle Ph}{\underset{|}{C}}HCH_2CH_3 \qquad (60)$$

It should be noted that the diphenylbutanes reported are the same ones obtained from alkylations with dibromobutanes. It is quite likely that some of the other diphenylbutanes identified by Khalaf and Roberts [80] went undetected (see Sec. I.B.1.) In 1958, Schmerling et al. reported the reaction shown in Eq. (61) [138].

$$\underset{Br}{\underset{|}{CH_2}}\text{-}\underset{Br}{\underset{|}{CH}}\text{-}\underset{Br}{\underset{|}{CH}}\text{-}\underset{Br}{\underset{|}{CH_2}} + C_6H_6 \xrightarrow{AlCl_3} \text{tetralin} + \text{2-phenyltetralin} + Ph\text{-}Ph \qquad (61)$$

D. Selective Alkylations with Di- and Polyhaloalkanes; Haloalkylation

1. Early examples

So far in our discussion of Friedel-Crafts alkylations with di- and polyhalides, we have been concentrating on cases in which the alkylating agents were simple (i.e., containing the same halogen) and/or of a symmetrical nature. It has been repeatedly shown that alkylations with such systems do not stop at the first formed haloalkylated product. Instead, higher condensation products have usually been produced, mainly because the initially formed haloarylalkanes are often more reactive than the starting halogen compounds. This is due to the introduction of an aromatic ring which often enhances the reactivity of the remaining halogens by anchimeric assistance.

In Friedel-Crafts alkylation by unsymmetrical or mixed halides, one of the halogen atoms usually has a greater reactivity, and hence selective reaction with aromatic substrates can take place. This may result in the formation of comparatively less reactive haloalkylated products, making it possible to isolate these initial haloalkylated products in good yields.

Very few examples of haloalkylations with di- and polyhalides were reported in the early literature. Apparently, it was Böeseken in 1903 who first observed the presence of the haloalkylation product, chlorotriphenylmethane, among the products of reaction of chloroform with benzene in the presence of $AlCl_3$ [97]. Subsequently, Gomberg and Jickling found that dichlorodiphenylmethane was the major product (90%) formed when benzene was allowed to react with carbon tetrachloride and $AlCl_3$ under ordinary conditions [100].

Since these early examples of haloalkylation, several other cases have been investigated in more detail. These newer cases of selective alkylations with di- and polyhaloalkanes can conveniently be divided into three main categories on the basis of the nature of the starting di- or polyhalides. These main categories include alkylations with (1) branched unsymmetrical simple di- and polyhalides, (2) straight-chain unsymmetrical simple di- and polyhalides, and (3) mixed di- and polyhalides. The following discussion will cover these three categories.

2. Selective alkylations with branched unsymmetrical di- and polyhalides

In 1956, Schmerling et al. investigated the Friedel-Crafts reactions of benzene with unsymmetrical dihaloalkanes in which one halogen atom is attached to a tertiary and the other halogen to a primary carbon atom [81]. They found that the alkylation of benzene with 1,3-dichloro-3-methylbutane (142) in the presence of $AlCl_3$ at 20 to 4°C resulted in the formation of 1-chloro-3-methyl-3-phenylbutane (144) and 2-methyl-2,3-diphenylbutane (148) in equal yields (28 to 29%). The former product was suggested to result from the condensation of benzene at the tertiary carbon atom via a tertiary carbocation (143) and the latter from 144 via a secondary carbocation (146) as shown in scheme 8.

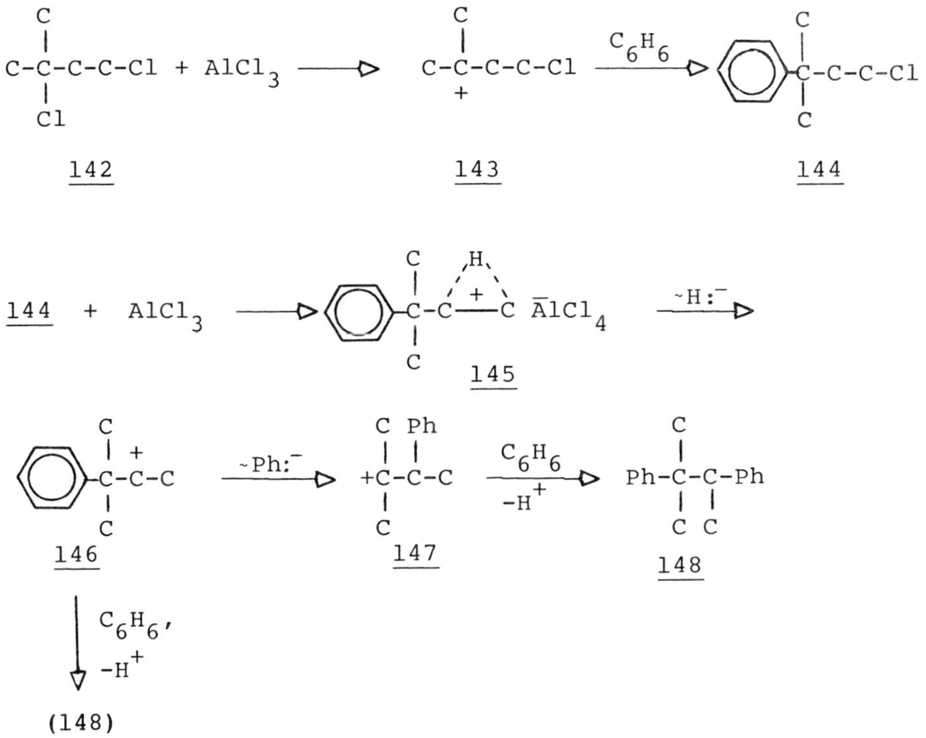

C-C(C)(Cl)-C-C-Cl (142) + $AlCl_3$ → C-C⁺(C)-C-C-Cl (143) —C_6H_6→ Ph-C(C)(C)-C-C-Cl (144)

144 + $AlCl_3$ → Ph-C(C)(C)-C(H⁺-bridged)C $\bar{A}lCl_4$ (145) —~H:⁻→

Ph-C⁺(C)(C)-C-C (146) —~Ph:⁻→ +C(C)(C)-C(Ph)-C (147) —C_6H_6, $-H^+$→ Ph-C(C)-C(C)(C)-Ph (148)

146 —C_6H_6, $-H^+$→ (148)

Scheme 8

The reaction of 1,3-dichloro-3-methylbutane (142) with benzene was also studied under other conditions. When benzene was treated with 142 in the presence of $AlCl_3$ and isobutane or methylcyclopentane, the mixture obtained consisted mainly of the intermediate alkylation product 144 and *t*-butylbenzene or methylcyclopentylbenzene, respectively. Other hydride transfer products, such as 2-methyl-3-phenylbutane, isopentyl chloride, and isopentane, were also observed, but in lesser amounts.

Reaction of the primary chlorine atom in the initially formed chloropentylbenzene (144) was almost completely suppressed when Schmerling et al. carried out the reaction in the presence of the weak catalysts $AlCl_3$-CH_3NO_2 or $FeCl_3$ at 20 to 27°C. With the former catalyst 144 was obtained in 55% yield. In this case the principal by-product (22%) was crystalline *p*-bis(3-chloro-1,1-dimethylpropyl)benzene (149).

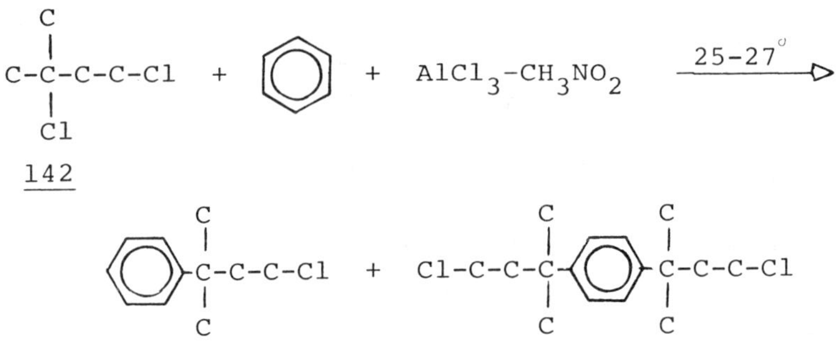

```
    C
    |
C-C-C-C-Cl   +   C6H6   +   AlCl3-CH3NO2   --25-27°-->
    |
    Cl
   142

        C                       C            C
        |                       |            |
C6H5-C-C-C-Cl    +    Cl-C-C-C-C6H4-C-C-C-Cl
        |                       |            |
        C                       C            C

   144  55%                     149  22%
```

The same authors found that no 1-chloro-2-methyl-2-phenylpropane (analogous to 144) was produced by the reaction of benzene with 1,2-dichloro-2-methylpropane (66, X = Cl, Sec. I.B.1.). This difference in behavior was rationalized in terms of the great reactivity of 1-chloro-2-methyl-2-phenylpropane (neophyl chloride, 68, X = Cl) relative to 144. The presence of a phenyl group on the carbon atom adjacent to the carbon atom holding the chlorine was said to aid in the elimination of the chloride ion, as shown in Eq. (62).

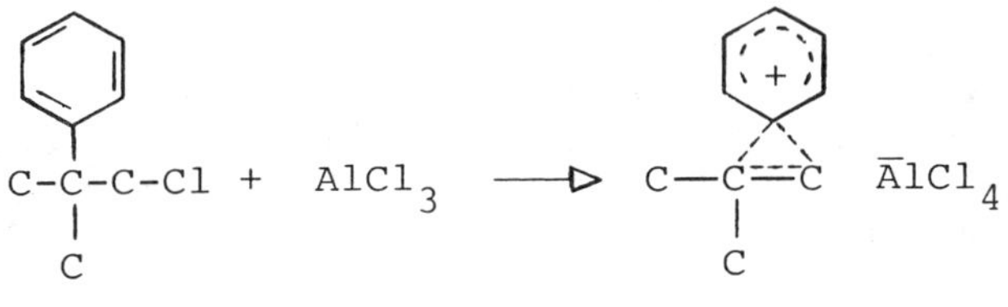

```
   C6H5                       C6H5(+)
    |                          / \
C-C-C-Cl  +  AlCl3  ---->  C—C---C   AlCl4(-)      (62)
  |                          |
  C                          C
```

In 1965, Bugrova and Tsukervanik investigated the possibility of haloalkylating benzene with the previously undescribed 1-chloro-3-bromo-3-methylbutane (150), 1-chloro-4-bromo-4-methylpentane (152), 3-chloromethyl-3-bromopentane (154), and 1-chloro-3-bromo-3-ethylpentane (156) [82]. To avoid the accumulation of diphenylalkanes, they carried out the experiments for only 3 to 4 min at 18 to 22°C, with the smallest possible amount of $AlCl_3$. The results of these experiments are shown in Eqs. (63) to (66).

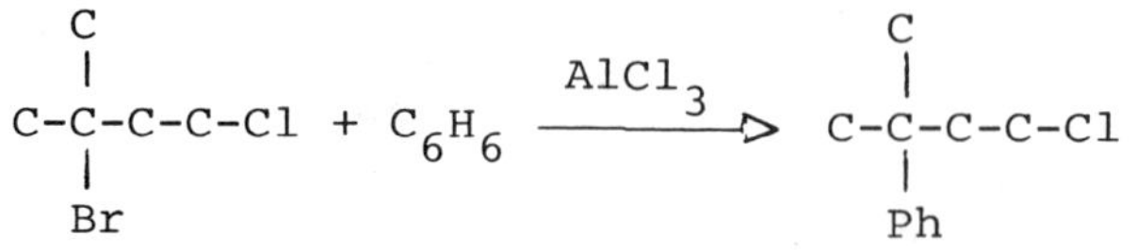

```
      C                                C
      |                AlCl3           |
C-C-C-C-Cl  +  C6H6  ---------->  C-C-C-C-Cl                    (63)
      |                                |
      Br                               Ph

    150                            151   40%
```

```
      C                                  C
      |                 AlCl3            |
C-C-C-C-C-Cl  +  C6H6  ---------->  C-C-C-C-C-Cl  +

      |                                  |
      Br                                 Ph

    152                              153   37%          10%
                                                                (64)
```

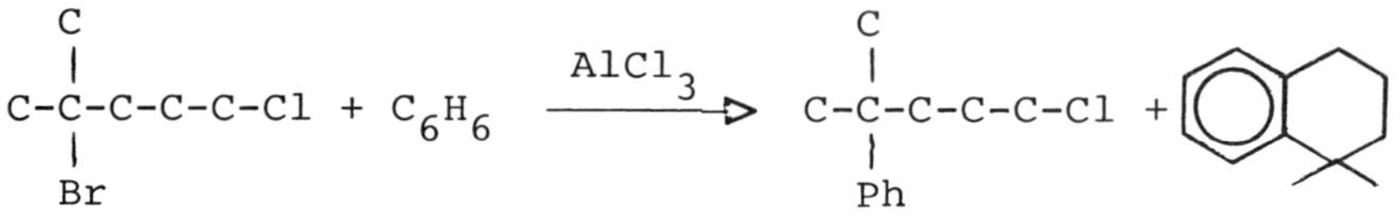

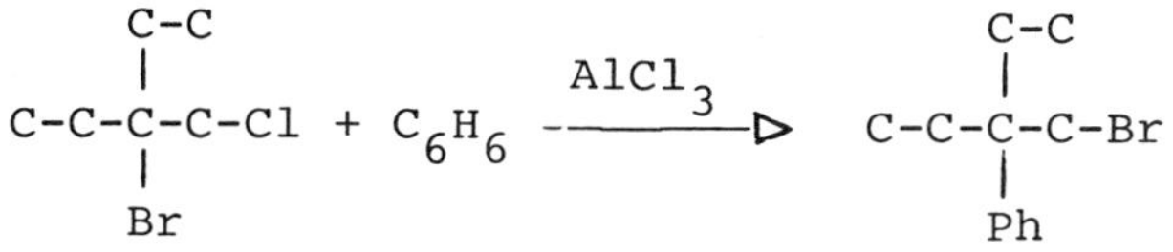

```
     C-C                                 C-C
      |                 AlCl3             |
C-C-C-C-Cl  +  C6H6  ---------->  C-C-C-C-Br                    (65)
      |                                   |
      Br                                  Ph

    154                              155   35%
```

```
     C-C                                  C-C
      |                 AlCl3              |
C-C-C-C-C-Cl  +  C6H6  ---------->  C-C-C-C-C-Ph                (66)
      |                                    |
      Br                                   Br

    156                               157   18%
```

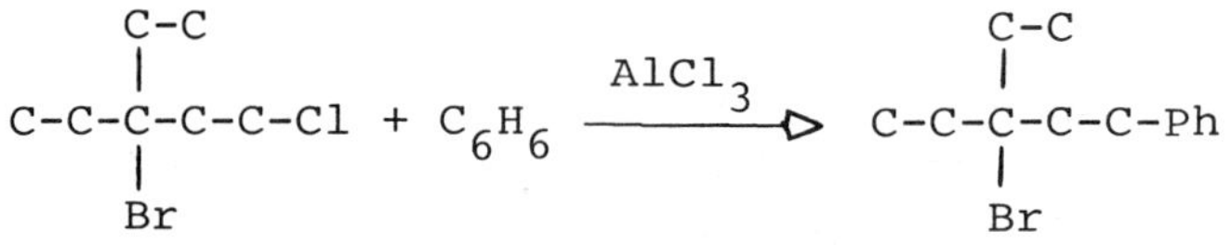

Examination of these equations indicates that the reaction with dihalides 150 and 152 proceeded by selective replacement of the tertiary bromide atom, in good accord with the previous findings of Schmerling et al. [81]. However, the results with dihalides 154 and 156 were rather different, and were attributed to favorable attack at the primary carbon as a result of the screening of the tertiary bromine atom by two ethyl groups in these systems. The production of a bromohexylbenzene containing primary bromine (155) instead of the expected tertiary bromide in the reaction of 154 was said to be the result of subsequent isomerization of an initially formed tertiary bromide. Provided that the experimental facts are valid, the present authors believe that a better explanation for the result from dihalide 154 can be offered in terms of the involvement of the bridged bromonium ion intermediate 158 which is attacked preferentially at one site by benzene to give the product reported (155):

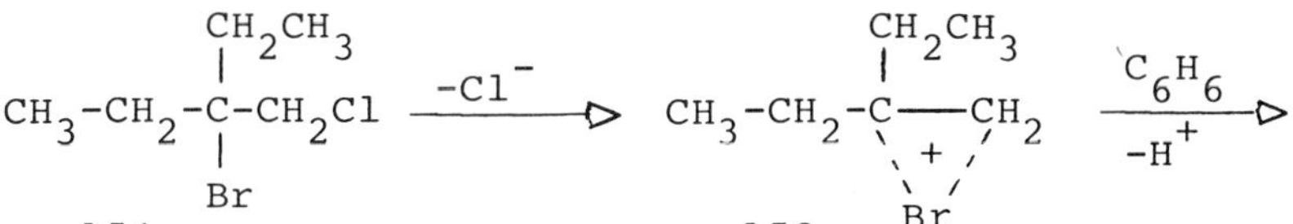

$$CH_3\text{-}CH_2\text{-}\underset{\underset{Br}{|}}{\overset{\overset{CH_2CH_3}{|}}{C}}\text{-}CH_2Cl \xrightarrow{-Cl^-} CH_3\text{-}CH_2\text{-}\overset{\overset{CH_2CH_3}{|}}{C}\text{—}CH_2 \;(\text{bridged } Br^+) \xrightarrow[-H^+]{C_6H_6}$$

154 158

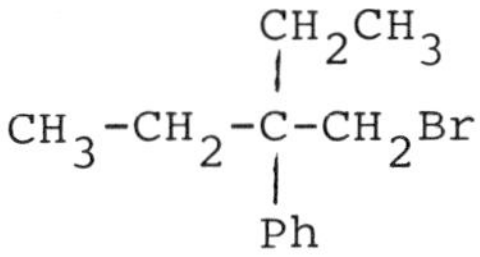

$$CH_3\text{-}CH_2\text{-}\underset{\underset{Ph}{|}}{\overset{\overset{CH_2CH_3}{|}}{C}}\text{-}CH_2Br$$

155

In the same paper, Bugrova and Tsukervanik reported an attempt to haloalkylate benzene with 1-chloro-2-bromoethane (159) and 1-chloro-2-bromo-2-methylpropane (160) but in each case the reaction proceeded by replacement of both halogen atoms, resulting in no haloalkylbenzenes. The products were said to be a mixture of bibenzyl and *p*-bis(β-phenylethyl)benzene from 159 and "only 2-methyl-1,2-diphenylpropane" (75) from 160. Apparently, the authors were completely unaware of the presence of other diphenylbutane isomers in the latter case (see Sec. I.B.1.).

In discussing their results, the authors stated that the latter two reactions proceeded in stages, but the intermediate haloalkylbenzenes rapidly reacted with a second molecule of benzene, possible through a readily formed bridged ion.

In 1966, Ransley investigated, among other things (see Sec. I.A.5), the reaction of benzene with 1,2-dichloro-4-methylpentane (161) in the presence of either $AlCl_3$ or liquid HF catalysts [40]. After 6 hr at 90°C with liquid HF as catalyst, 161 yielded 1-chloro-4-methyl-4-phenylpentane (162) as the major product, (ca. 80%). Reaction of 161 and benzene with $AlCl_3$ catalyst for only 3 hr allowed identification of 162 and the additional compounds 163-167 in the amounts shown.

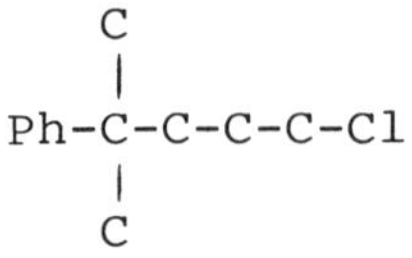

$$Ph\text{-}\underset{\underset{C}{|}}{\overset{\overset{C}{|}}{C}}\text{-}C\text{-}C\text{-}C\text{-}Cl$$

162

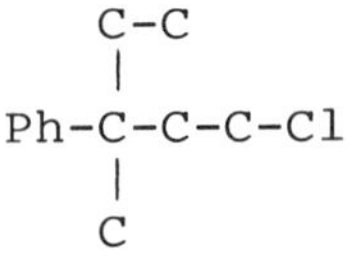

$$Ph\text{-}\underset{\underset{C}{|}}{\overset{\overset{C\text{-}C}{|}}{C}}\text{-}C\text{-}C\text{-}Cl$$

163 9.8%

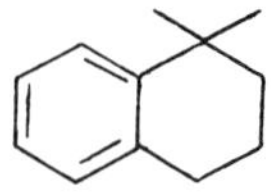

164 64.3%

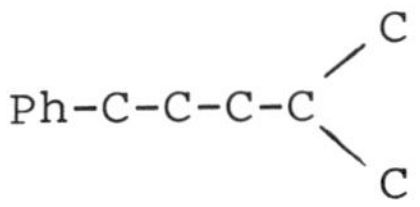

$$Ph\text{-}C\text{-}C\text{-}C\text{-}C\langle^{C}_{C}$$

165 4.9%

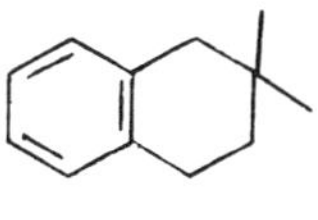

166 6.6%

Ph

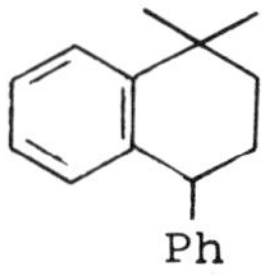

167 7.7%

The formation of 163-165 and 167 was explained in terms of the carbocation processes depicted in scheme 9.

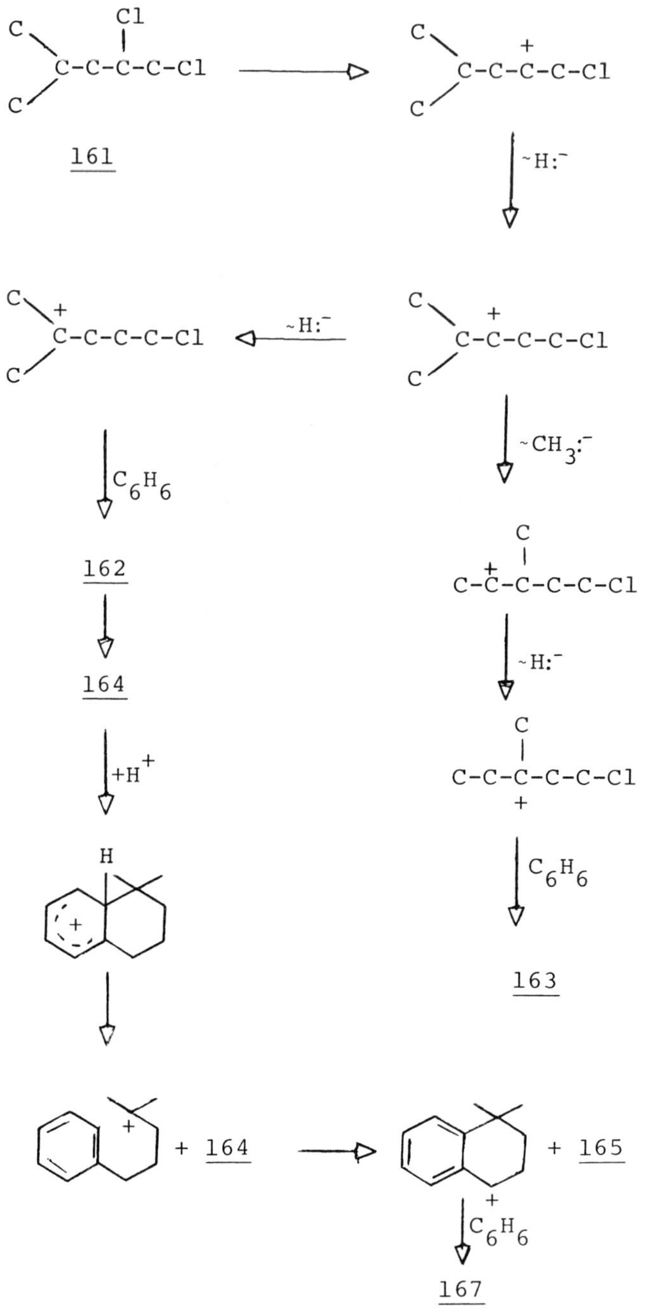

```
C     Cl                      C      +
 \    |                        \
  C-C-C-C-Cl   ------>          C-C-C-C-Cl
 /                             /
C                             C
    161                          | ~H:⁻
                                 v
C  +                          C      +
 \                             \
  C-C-C-C-Cl  <--- ~H:⁻ ---     C-C-C-C-Cl
 /                             /
C                             C
    | C6H6                       | ~CH3:⁻
    v                            v
   162                               C
    |                            +   |
    v                            C-C-C-C-C-Cl
   164                               | ~H:⁻
    | +H+                            v
    v                                C
                                     |
    H                            C-C-C-C-C-Cl
                                     +
                                     | C6H6
    |                                v
    v                               163

    + 164   ------>    + 165
                  | C6H6
                  v
                 167
```

Scheme 9

It was demonstrated that the addition of $AlCl_3$ in the absence of benzene caused the conversion of 162-164 and 164-165.

While no explanation was offered by Ransley for the formation of 2,2-dimethyltetralin (166), it was probably produced by rearrangement of 164 by a series of hydride and methyl shifts as shown in scheme 10.

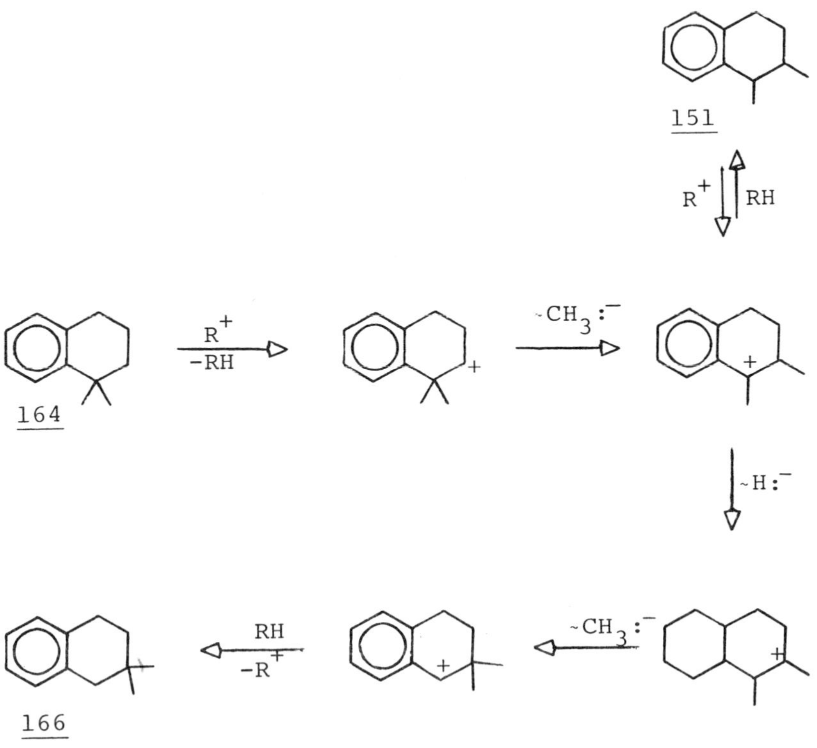

Scheme 10

More recently, both 1,1-dimethyltetralin (164) and 1,2-dimethyltetralin (168) were found to rearrange to 2,2-dimethyltetralin (166) under the influence of $AlCl_3$ [26].

3. Selective alkylations with linear unsymmetrical halides

Haloalkylation reactions have also been carried out with linear dihalides in which one of the halogen atoms is attached to a primary and the other to a secondary carbon atom. For example, a chlorophenylbutane was reported to be the major product formed from the aluminum chloride-catalyzed reaction of 1,2-dichlorobutane and benzene at 4 to 6°C [91] and at 25°C [40]. The chlorophenylbutane produced was claimed by Schmerling et al. [91] to be 1-chloro-3-phenylbutane. The latter compound was also found to be the major product of reaction of 1,3-dichlorobutane with benzene catalyzed by $AlCl_3$ [40].

The hydrogen fluoride-catalyzed reaction of 1,2-dichlorohexane gave a low conversion, but an almost quantitative yield of 1-chloro-5-phenylhexane; conversion was only 5% after 5 hr at 80°C. When a solution of 1,2-dichlorohexane in benzene was treated with 8.8 mol % $AlCl_3$ for 6 hr, it gave a complex mixture containing 27.5% 1-chloro-5-phenylhexane [40].

In 1968, Ransley examined the alkylation of benzene with 1,X-dichlorooctanes using a hydrogen fluoride-boron trifluoride catalyst (a system specific for alkylation by secondary halides) [47]. The alkylations were carried out by adding 2 mol of liquid HF to a cold solution of 0.2 mol of 1,X-dichloro-octane in 4 mol of benzene, followed by bubbling BF_3 through the well-stirred mixture at about 4 ml/min at 0°C. Under these conditions, Ransley found the following results:

1. The 1,1-, 1,2-, and 1,8-dichlorooctanes did not react.
2. In the case of other isomers, no reaction occurred for the first 30 to 35 min, but the reaction then proceeded smoothly, although the reaction rates were different for each isomer.
3. The rates of reaction for the various isomers were in the order 1,7 > 1,6 > 1,5 > 1,4 > 1,3, indicating that the rate of reaction increases as the distance between the two chlorine atoms is increased.
4. The reaction of each dichlorooctane isomer was shown to be first order in dichloride in the presence of excess benzene and hydrogen fluoride.
5. The products of reaction were 7-, 6-, 5-, and 4-phenyl-1-chlorooctanes in a ratio 53:29:14:4, which was independent of reaction time and degree of dichloride conversion. This ratio indicated that the products from 1,X-dichlorooctanes showed a strong tendency for phenyl attachment at secondary positions most distant from the unreactive primary chloride. In contrast, the author found that monochloroalkanes form products with random phenyl attachment to the secondary carbons.

Earlier, Ransley had examined the aluminum chloride-catalyzed alkylation of benzene with a series of α,ω-dichloroalkanes [40]. The reactions were carried out with 4 mol % $AlCl_3$ at 0 to 5°C in excess benzene for 2 hr (Sec. I.A.5). Under these conditions 1,3-dichloropropane, 1,4-dichlorobutane, 1,5-dichloropentane, and 1,6-dichlorohexane gave complex mixtures containing 19.8% 1-chloro-3-phenylpropane; 30.1% 1-chloro-4-phenylbutane and 2.7% 1-chloro-3-phenylbutane; 43.6% 1-chloro-5-phenylpentane and 1.6% 1-chloro-4-phenylpentane; and 26.6% 1-chloro-6-phenylhexane and 31.8% 1-chloro-5-phenylhexane, respectively. As in the 1,X-dichlorooctane series, the rates of reaction of the α,ω-dichloroalkanes increased as the distance between the two chlorine atoms was increased, but were slower than the rates of the corresponding monochloroalkanes. As indicated earlier, the exception to this generalization was the reaction of 1,4-dichlorobutane.

The above observations were interpreted by Ransley [40] in terms of a long-range electron-withdrawing effect of the reference primary chlorine atom on the reaction of the secondary chlorine atom, with the influence decreasing as the distance between the two chlorine atoms is increased. It may be of interest to mention that an analogous explanation was given by Peterson et al. [139] to account for the gradual change in reaction rate of the addition of CF_3COOH to $Cl(CH_2)_nCH = CH_2$ as n is increased, and by Stevenson and Williamson [140] to rationalize the observed gradual change in pK in a series of cyanoamines with the functional groups separated by as many as five methylene groups.

In 1969, Ransley extended his studies by reporting the HF-BF_3-catalyzed reaction of benzene with three other series of polyhaloalkane isomers [86]. These were the 1-bromo-X-chlorooctane, 1,1,X-trichlorooctane, and 1,1,1,X-tetrachlorooctane series. The alkylation procedure was identical with that used in previously described work with 1,X-dichlorooctanes. Likewise, in each series the substituents at C-1 were shown to be unreactive under the conditions used. The 1-bromo-1-chloro-, 1-bromo-2-chloro-, 1-bromo-8-chloro-, 1,1,1-trichloro-, 1,1,2-trichloro-, 1,1,3-trichloro-, and 1,1,8-trichlorooctanes were recovered unchanged after 6 hr, as shown by GLPC analysis using an internal standard.

In accord with his previous findings, Ransley reported the following: First, the reactions were first order in each alkylating agent, with the rate increasing as the distance between the secondary chlorine atom and the unreactive electron withdrawing group increased. Second, the product proportions were dependent on the substituents at C-1. This suggested that during the reaction there was a common carbocation intermediate within each series which underwent rapid isomerization by a series of 1,2-hydride shifts, leading to alkylation at the various secondary positions. There was also a greater tendency for phenyl attachment at positions remote from the unreactive electron-withdrawing group.

To summarize, Ransley's results showed that the 1-chloro-1-bromo, 1,1-dichloro, and 1,1,1-trichloro substituents decreased the rate of reaction of a secondary group, relative to the unsubstituted case, even when separated by seven methylene units, since in each series the 7-chloro isomer was found to react more slowly than the unsubstituted monochlorooctanes. The influence of these primary substituents was shown, by an empirical treatment [47,139], to decrease by a constant attenuation factor of 0.59 per methylene unit.

4. Selective alkylations with mixed di- and polyhalides

An early example of selective Friedel-Crafts alkylation with a mixed dihalide is the reported aluminum chloride-catalyzed alkylation of benzene with 1-bromo-3-chloropropane. This reaction was carried out in 1940 by Tsukervanik and Yatsirmirskii [46], who reported obtaining a 60% yield of 1-bromo-3-phenylpropane at 6 to 12°C, but only diphenylpropane at higher temperatures.

The observed difference between the reactivity of alkyl fluorides and the other alkyl halides [141-143] inspired Olah and Kuhn to carry out Friedel-Crafts haloalkylations with dihaloalkanes in which one of the halogen atoms was fluorine [144]. These authors found that fluorochloro-, fluorobromo-, and fluoroiodoalkanes are effective chloro-, bromo-, and iodoalkylating agents in Friedel-Crafts alkylations of benzene and alkylbenzenes in the presence of boron halide catalysts.

$$ArH + F(CH_2)_nX \xrightarrow{BX_3} Ar(CH_2)_nX$$

$$n > 1;\ X = Cl, Br, I$$

Fluorohaloalkanes used in the boron trifluoride-catalyzed haloalkylations included 1-chloro-2-fluoroethane, 1-bromo-2-fluoroethane, 1-iodo-2-fluoroethane, 1-bromo-3-fluoropropane, 1-bromo-2-fluoropropane, 1-chloro-4-fluorobutane, 1-bromo-4-fluorobutane, and 1-bromo-3-fluorobutane. The haloalkylated aromatics included benzene, toluene, *m*-xylene, and mesitylene.

On the basis of their studies Olah and Kuhn [144] concluded that in boron trihalide-catalyzed haloalkylations of arenes, the order of reactivity of the boron trihalide catalysts is $BI_3 > BBr_3 > BCl_3 > BF_3$, whereas the reactivity of the carbon-halogen bonds in the investigated dihalides is C-F > C-Cl > C-Br > C-I.

In rationalizing the reactivity sequences above, the authors noted that the relative order of catalyst activity is in agreement with the relative Lewis acid strengths of the boron halides as determined by measurement of the heats of formation of their complexes [145-147]. The substantially higher reactivity of the C-F bond over that of the C-Cl, C-Br, and C-I bonds in these reactions was attributed by the authors to three factors: (1) the high polarity of the C-F bond to be cleaved, (2) the high bond energy of the B-F bond to be formed by interaction with the Lewis halide catalyst, and (3) the small steric hindrance involved.

No halomethylation was effected when halofluoromethanes were used. This was explained in terms of the deactivating effect exerted by more than one halogen atom attached to the same carbon, which renders the C-F bond unreactive.

In the course of their study Olah and Kuhn made the interesting observation that when BF_3 was used as the catalyst in reactions of straight-chain fluorohaloalkanes, almost complete isomerization along the alkyl chain occurred, whereas in the case of the other boron halide catalysts, the isomerized product amounted to only 5 to 15%.

$$C_6H_6 + ClCH_2CH_2CH_2F \xrightarrow{BF_3} C_6H_5CH(CH_3)CH_2Cl \ (\underline{169}) \quad (67a)$$

$$C_6H_6 + ClCH_2CH_2CH_2F \xrightarrow{BCl_3 \text{ or } BBr_3} \underset{90\%}{C_6H_5CH_2CH_2CH_2Cl} + \underset{10\%}{\underline{169}} \quad (67b)$$

$$C_6H_6 + Cl(CH_2)_4F \xrightarrow{BF_3} C_6H_5CH(CH_3)CH_2CH_2Cl \ (\underline{170}) \quad (68a)$$

$$C_6H_6 + Cl(CH_2)_4F \xrightarrow{BCl_3 \text{ or } BBr_3} \underset{70\text{-}85\%}{C_6H_5CH_2CH_2CH_2CH_2Cl} + \underset{15\text{-}30\%}{\underline{170}} \quad (68b)$$

The isomerizations observed in reactions (67a) and (68a) were said to be due to the cocatalytic effect of HF formed in the reaction, providing the strong conjugate acid $HF + BF_3$. As indicated, no such conjugate acid is known with hydrogen halides and the other boron trihalides.

The difference in the reactivity of fluorine and bromine in Friedel-Crafts alkylations was used by Müller and Weyerstahl [148] to provide insight into the mechanism of the indene synthesis of Buddrus [149] and Skattebol and Boulette [150]. In this synthesis geminal dichloro- or dibromocyclopropanes with quaternary ring carbons react with arenes in the presence of $AlCl_3$ to give indenes. Müller and Weyerstahl used 1-bromo-1-fluorotetramethylcyclopropane in reaction with benzene, toluene, and anisole of low temperatures in the presence of $AlCl_3$. With benzene, toluene, and anisole at -5, -30, and -8°C, respectively, they obtained the bromoindanes 171 (R = H, CH_3) and 172. At higher temperatures (20°C for benzene, -5°C for toluene, and -8°C

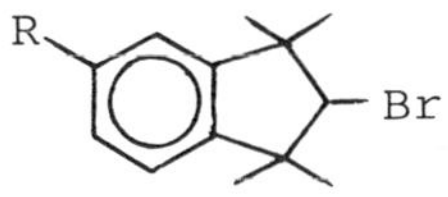

R Br

171

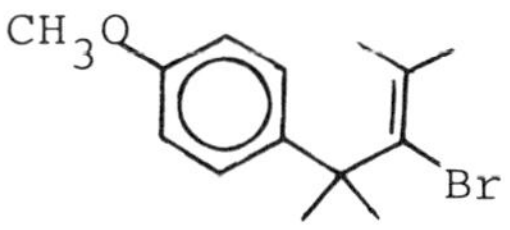

CH_3O Br

172

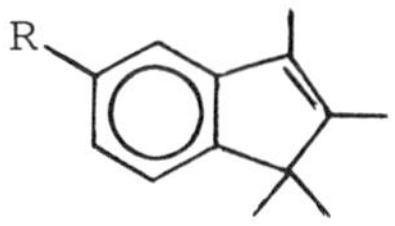

R

173

for anisole) the corresponding indenes (173) were obtained from benzene and toluene, whereas anisole gave 171 and 173, R = OCH_3. These results were taken to support the mechanism of scheme 11 for the Buddrus-Skattebol indene synthesis.

In 1974, Brovko et al. studied the alkylation of pentafluorobenzene by 1,1,2-trichloro-1,2,2-trifluoroethane (Freon-113, 174) in the presence of

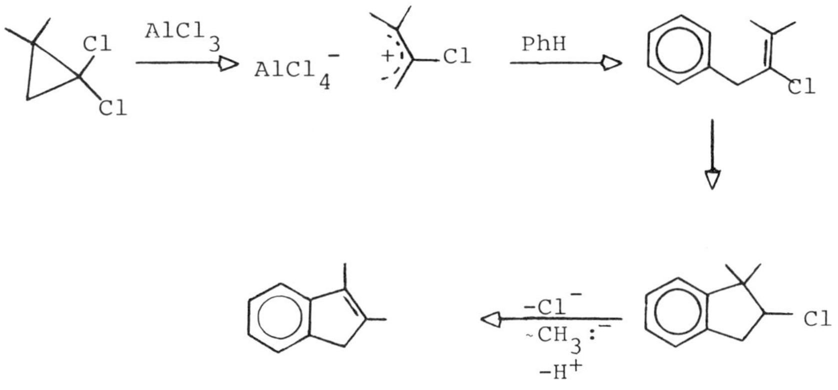

Cl Cl $\xrightarrow{AlCl_3}$ $AlCl_4^-$ Cl $\xrightarrow{PhH}$ Cl

Cl $\xrightarrow[-CH_3:^-]{-Cl^-}$ $-H^+$

Scheme 11

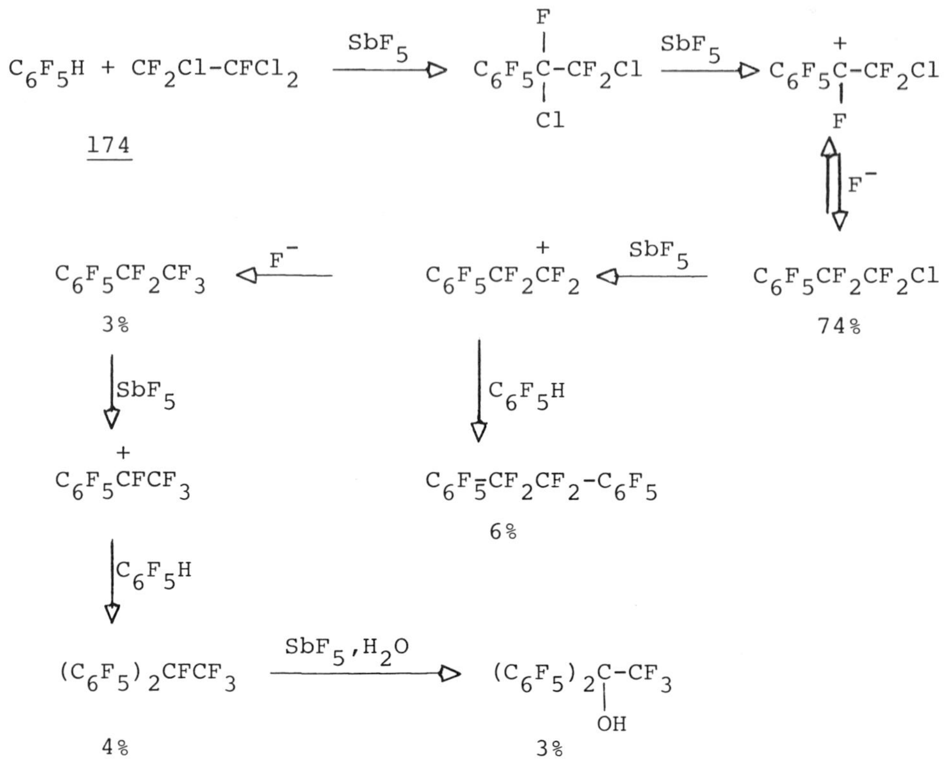

$$C_6F_5H + CF_2Cl\text{-}CFCl_2 \xrightarrow{SbF_5} C_6F_5C(F)(Cl)\text{-}CF_2Cl \xrightarrow{SbF_5} C_6F_5\overset{+}{C}(F)\text{-}CF_2Cl$$

$\underline{174}$

$$C_6F_5\overset{+}{C}(F)\text{-}CF_2Cl \overset{F^-}{\rightleftarrows} C_6F_5CF_2CF_2Cl \; (74\%)$$

$$C_6F_5CF_2CF_2Cl \xrightarrow{SbF_5} C_6F_5CF_2\overset{+}{C}F_2 \xrightarrow{F^-} C_6F_5CF_2CF_3 \; (3\%)$$

$$C_6F_5CF_2\overset{+}{C}F_2 \xrightarrow{C_6F_5H} C_6F_5CF_2CF_2\text{-}C_6F_5 \; (6\%)$$

$$C_6F_5CF_2CF_3 \xrightarrow{SbF_5} C_6F_5\overset{+}{C}FCF_3 \xrightarrow{C_6F_5H} (C_6F_5)_2CFCF_3 \; (4\%)$$

$$(C_6F_5)_2CFCF_3 \xrightarrow{SbF_5, H_2O} (C_6F_5)_2C(OH)\text{-}CF_3 \; (3\%)$$

Scheme 12

SbF_5 catalyst at -10 to -5°C [151]. The main products of this reaction, their amounts, and suggested modes of formation are given in scheme 12.

In formulating scheme 12, it was assumed on the basis of the chemical properties of the chlorofluoroalkanes [152] that the predominant process is the splitting out of a chlorine atom in the dichlorofluoromethyl group with the formation of $CF_2Cl\text{-}\overset{+}{C}FCl$. Support for this hypothesis was found in the formation of 1,2-dichlorotetrafluoroethane (175) by the reaction of the initial trichlorotrifluoroethane (174) with SbF_5 under the reaction conditions.

$$\underset{\underline{174}}{CF_2ClCFCl_2} \xrightarrow[\text{room temp, 2 hr}]{SbF_5} \underset{\underline{175}}{CF_2ClCF_2Cl}$$

An interesting fact about the alkylation under consideration was the presence of ca. 3% chloropentafluorobenzene in the reaction product. Its formation was attributed to the chlorination of the starting pentafluorobenzene by the antimony chloride fluorides formed in the reaction through halogen exchange. This hypothesis was confirmed by the finding that pentafluorobenzene was smoothly chlorinated by the action of HCl or NaCl in the presence of SbF_5.

$$C_6F_5H \xrightarrow[SbF_5]{HCl \text{ or } NaCl} C_6F_5Cl$$

Another interesting case is the reported alkylation of benzene with hexachlorofulvene (176) in the presence of $AlCl_3$ [153]. In this reaction, the number and position of the chlorines replaced could be controlled to some degree by choice of both solvent and temperature. At room temperature, 2-phenylpentachlorofulvene (177) was the only product isolated when a 1:1 molar ratio of 176 to benzene was allowed to react with $AlCl_3$. Lowering the temperature to 5°C resulted in a mixture of about equal amounts of 177 and 1-phenylpentachlorofulvene (178).

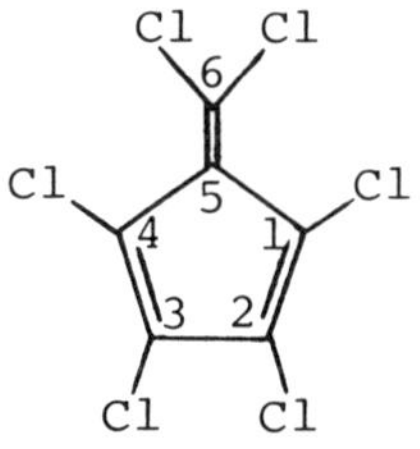

176 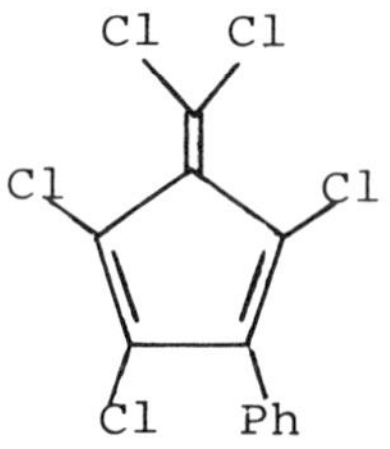 177 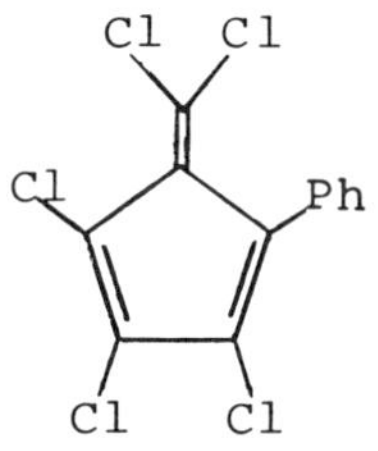 178

Although the alkylating agent was not a di- or polyhalide, a report of Tsukervanik and co-workers may appropriately be mentioned here since it illustrates selective alkylation by a bifunctional compound. A number of δ-arylbutyl alkyl ethers were prepared through selective alkylation of arenes with alkyl γ-chloro-*n*-butyl ethers in the presence of $AlCl_3$ [154].

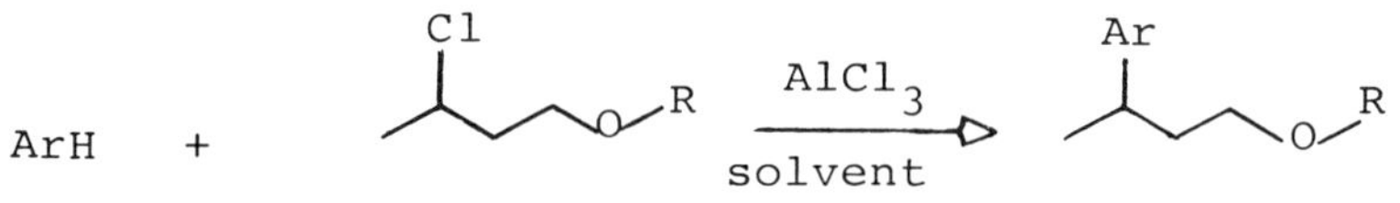

Ar = phenyl or alkylated phenyl

R = alkyl

II. ALKYLATIONS WITH UNSATURATED ALKYL HALIDES, ALCOHOLS, AND ETHERS

A. Vinylic Reagents

1. Vinyl monohalides and vinyl ethers

Depending on reaction conditions and type of catalyst, the Friedel-Crafts alkylation of arenes with vinyl halides may take place at one or more of the available reactive sites. Examination of the following examples will illustrate the effect of reaction variables and structure of the alkylating species on the course of these reactions.

Böeseken and Bastet [155] found that vinyl chloride and benzene reacted in the presence of a catalyst made from aluminum chloride and mercuric chloride, and produced chiefly 1,1-diphenylethane and a small quantity of what they believed to be 9,10-dimethyldihydroanthracene. Other investigators employed $AlCl_3$ with vinyl chloride or bromide and reported essentially the same results [156-159]. However, Davidson and Lowy [158] were also able to isolate some ethylbenzene and to verify the presence of considerable amounts of anthracene-type resin in the products of reaction of vinyl chloride with benzene and $AlCl_3$. Hanriot and Guilbert [160] obtained products believed to be β-bromoethylbenzene and *p*-di(β-bromoethyl)benzene in a similar reaction using vinyl bromide.

The repeated claim that 9,10-dihydro-9,10-dimethylanthracene was one of the products of the metal halide-catalyzed reaction of vinyl halide with benzene was examined by Sisido and co-workers first in 1942 [161] and subsequently in 1948 [162]. As a result of independent synthesis, these authors showed beyond doubt that the product isolated in the above reaction was not 9,10-dihydro-9,10-dimethylanthracene but rather the aromatized counterpart, 9,10-dimethylanthracene. Furthermore, they demonstrated that the dihydro derivative could easily be aromatized upon treatment with $AlCl_3$.

The formation of 9,10-dimethylanthracene can be rationalized as shown in Eqs. (69a) to (69c).

$$CH_2{=}CHX \; + \; C_6H_6 \xrightarrow{AlCl_3} C_6H_5{-}CHX{-}CH_3 \qquad (69a)$$

$$2\; C_6H_5{-}CH(X){-}CH_3 \xrightarrow{AlCl_3} \text{9,10-dihydro-9,10-dimethylanthracene} + 2HX \qquad (69b)$$

$$\text{9,10-dihydro-9,10-dimethylanthracene} + C_6H_5{-}CHX{-}CH_3 \xrightarrow{AlCl_3} \text{9,10-dimethylanthracene} + C_6H_5CH_2CH_3 + HX \qquad (69c)$$

In 1961, Tsukervanik and Yuldashev investigated the reaction of vinyl chloride with toluene and anisole in the presence of $AlCl_3$, using various ratios of reactants [163]. A summary of their findings is given in Eqs. (70) and (71).

Toluene + $CH_2{=}CHCl$ + $AlCl_3$ ⟶ 179 + 180 + 181 + 182 (70)

179: p-$CH_3C_6H_4CH_2CH_3$; 180: $(p\text{-}CH_3C_6H_4)_2CHCH_3$; 181: $(p\text{-}CH_3C_6H_4)_2C{=}CH_2$; 182: CH_3, CH_3, CH_3, CH_3-substituted anthracene

Toluene + $CH_2{=}CHCl$ + $AlCl_3$ $\xrightarrow[20^\circ]{\text{overnight,}}$ 183 + 184 (71)

5 : 1 : 1

183: p-$CH_3OC_6H_4CH_2CH_3$; 184: $(p\text{-}CH_3OC_6H_4)_2CHCH_3$

In an extension of this work, Yuldashev and Tsukervanik tested the alkylation of benzene with 1-chloro-1-butene, 2-chloro-1-butene, 1-bromo-2-methyl-1-propene, α-bromostyrene, and β-bromostyrene [164]. They found that with BF_3-H_3PO_4 as catalyst the vinyl halides substituted in the β position with either alkyl or aryl groups easily gave haloalkylbenzenes, and with $AlCl_3$ as catalyst they gave diarylalkanes, e.g., Eqs. (72) and (73).

$$\text{Ph-CH=CHBr} + C_6H_6 \xrightarrow{BF_3\text{-}H_3PO_4} Ph_2CH\text{-}CH_2Br \qquad (72)$$

$$Ph_2CH\text{-}CH_2Br + C_6H_6 \xrightarrow{AlCl_3} Ph_2CH\text{-}CH_2Ph \qquad (73)$$

Substitution of a vinyl halide in the α position with an alkyl group, and especially an aryl group, was said to diminish the reactivity of the double bond.

Thus attempts to alkylate benzene with α-bromostyrene in the presence of BF_3-H_3PO_4 or $AlCl_3$ and with 2-chloro-1-butene in the presence of BF_3-H_3PO_4 failed. Reaction of the latter compound with benzene in the presence of $AlCl_3$ at 44°C, however, gave a mixture of products as shown in Eq. (74):

$$CH_3CH_2C(Cl){=}CH_2 + C_6H_6 \xrightarrow{AlCl_3} CH_3\text{-}CH_2\text{-}C(Ph)_2\text{-}CH_3 + CH_3\text{-}CH(Ph)\text{-}CH(Ph)\text{-}CH_3\ (meso) + CH_3CH_2CH(Ph)\text{-}CH_3 \quad (74)$$

Although Yuldashev and Tsukervanik reported that α-bromostyrene (185) did not react with benzene in the presence of either $AlCl_3$ or BF_3-H_3PO_4, Roberts and Abdel-Baset found that 185 condensed rapidly with toluene in the presence of Al_2Br_6 to give mainly 1-phenyl-1-*p*-tolylethene (186) and small amounts of acetophenone and phenylacetylene [165]. Analysis of the

$$\underset{\mathbf{185}}{CH_2{=}CBr(C_6H_5)} + C_6H_5CH_3 \xrightarrow{Al_2Br_6} \underset{\mathbf{186}}{CH_2{=}C(C_6H_5)(C_6H_4\text{-}p\text{-}CH_3)} + CH{\equiv}C\text{-}C_6H_5 + CH_3\text{-}C(=O)\text{-}C_6H_5$$

reaction mixture by GLPC after various short reaction periods showed that although 186 was produced rapidly, it also underwent polymerization rapidly, which may explain why Yuldashev and Tsukervanik failed to report this product. The type of products from vinyl halides and arenes that Yuldashev and Tsukervanik did observe with $AlCl_3$ catalyst were those resulting from reaction at both the double bond and the halide site, and in the case of such reaction of 185, this would lead to the introduction of threee aromatic rings on one carbon atom, which would encounter some steric inhibition.

The reactions of 2-bromopropene with benzene and toluene were also investigated by Roberts and Abdel-Baset. The major products with Al_2Br_6 as catalyst were those expected [164], which resulted from reaction of both functional groups (188a, 188b, scheme 13). The minor products were 189a and 189b, 190a and 190b, and 191a and 191b. When $AlCl_3$ was used as catalyst for a reaction of 2-bromopropene with toluene, a small amount of 187b was identified among the reaction products. The spirobiindanes 191a and 191b had been reported before, but not starting with 2-bromopropene and benzene

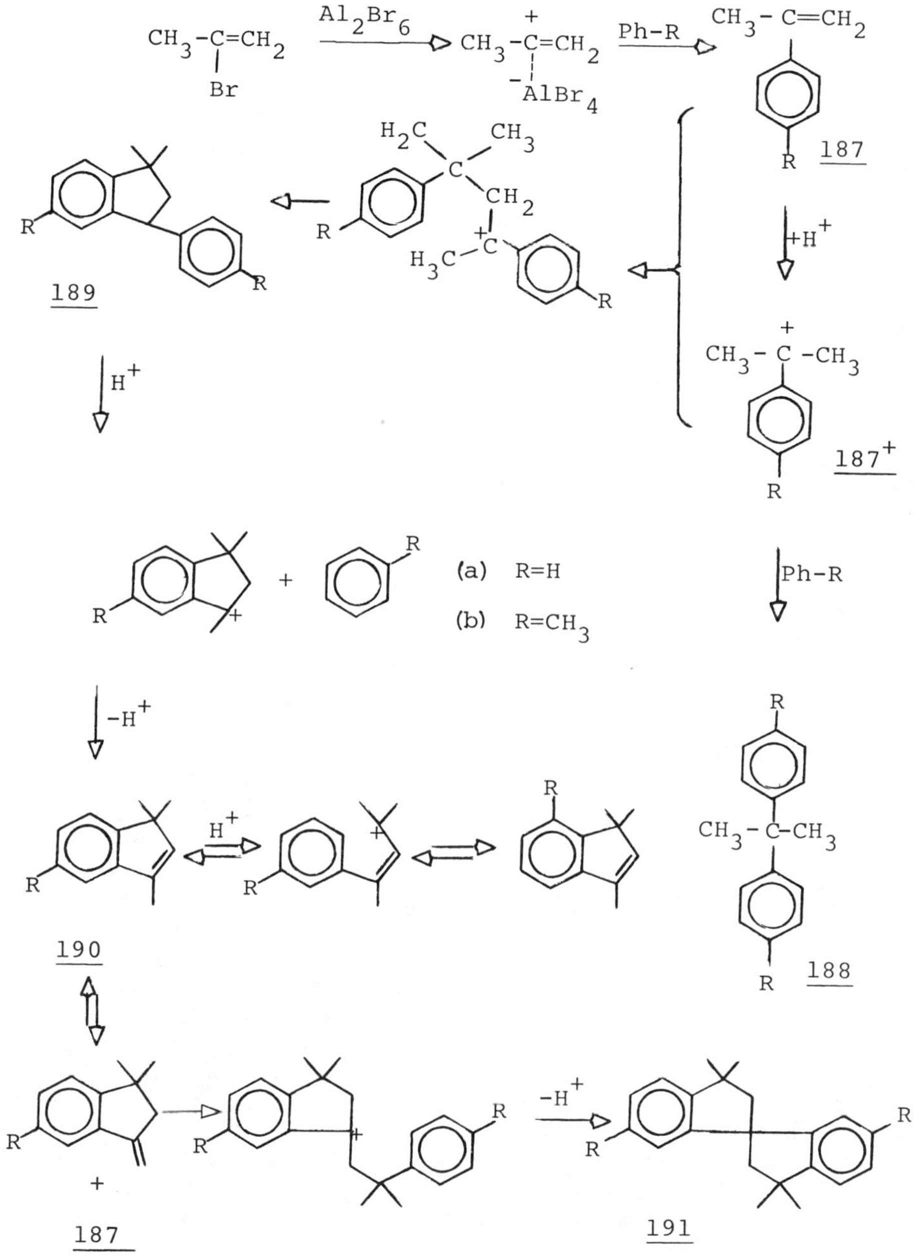

Scheme 13

and toluene. Compound 191a was first reported in 1929 [166], an incorrect structure was assigned in 1954 [167], and the correct structures for both 191a and 191b were proven in 1962 [168].

The reaction of benzene with 1-chloro- or 1-bromo-2-methyl-1-propene (192) was found to give the same product as that from methallyl halide, 2-halo-1,1-dimethylethylbenzene (193), when H_2SO_4 was used as catalyst [169,170]. When this reaction was catalyzed by $AlCl_3$ in small amounts, it was

$$CH_3-\overset{CH_3}{\overset{|}{C}}=CHBr(Cl) \ (\underline{192}) + C_6H_6 \xrightarrow{H_2SO_4} C_6H_5-\underset{CH_3}{\overset{CH_3}{C}}-CH_2Br(Cl) \ (\underline{193})$$

claimed to give 2-methyl-1,2-diphenylpropane (194) and some *t*-butylbenzene. With larger amounts of catalyst, an additional product, *meso*-2,3-diphenylbutane was observed.

$$C_6H_6 + CH_3-\overset{CH_3}{\overset{|}{C}}=CHBr + AlCl_3 \xrightarrow[30^\circ]{4\ hr} C_6H_5-\underset{CH_3}{\overset{CH_3}{C}}-CH_3 +$$

(a)	(100 ml)	(13.5g)	(1.67g)	(0.6g)
(b)	(100 ml)	(13.5g)	(3.35g)	(1.2g)

$$C_6H_5-\underset{CH_3}{\overset{CH_3}{C}}HCH_2-C_6H_5 \ (\underline{194}) + CH_3\underset{Ph}{CH}-\underset{Ph}{CH}CH_3$$

(a)	(14.0g)	(none)
(b)	(8.8 g)	(5.6g)

In view of their results from experiments with methallyl chloride, neophyl chloride, and isobutylene dibromide [50,51,80], the present authors believe that the product mixture from the aluminum chloride-catalyzed reaction probably also contained both 2-methyl-1,1-diphenylpropane and R,S-2,3-diphenylbutane (see Sec. I.B.1).

In 1958, Schmerling et al. investigated, among other things, the reaction of benzene with the vinylic chlorides 2-chloropropene, 1-chlorocyclohexene, and 1-chloro-2,4,4-trimethyl-1-pentene [171]. They found that neither 2-chlororpropene nor 1-chlorocyclohexene could be condensed with benzene in the presence of 96% H_2SO_4 at 0°C, results that are similar to those of Yuldashev and Tsukervanik with BF_3-H_3PO_4 catalyst [164]. These findings were presumed by the authors to be due to a deactivating effect of the chlorine substituent which decreases the reactivity of the double bond toward formation of the intermediate cation. The reaction of 1-chloro-2,4,4-trimethyl-1-pentene with benzene in the presence of H_2SO_4 at ice temperature resulted in a small yield (10%) of 1-chloro-2-phenyl-2-methylpropane and some *t*-butylbenzene. This was attributed to "depolymerization" of the chlorooctene to isobutylene and methallyl chloride, followed by reaction with benzene as shown in Eqs. (75) to (77).

$$CH_3-\underset{CH_3}{\overset{CH_3}{C}}-CH_2-\overset{CH_3}{\overset{|}{C}}=CH-Cl \longrightarrow CH_3-\overset{CH_3}{\overset{|}{C}}=CH_2 + CH_2=\overset{CH_3}{\overset{|}{C}}-CH_2Cl \qquad (75)$$

$$CH_2{=}C(CH_3){-}CH_3 + C_6H_6 \longrightarrow C_6H_5{-}C(CH_3)_2{-}CH_3 \qquad (76)$$

$$CH_2{=}C(CH_3){-}CH_2Cl + C_6H_6 \longrightarrow C_6H_5{-}C(CH_3)_2{-}CH_2Cl \qquad (77)$$

a. β-Chlorovinyl Ketones

Belyaev et al. prepared a number of aryl styryl ketones through the Friedel-Crafts alkylation of aromatic compounds by β-chlorovinyl ketones in the presence of $AlCl_3$, $AlCl_3$-CH_3NO_2, or $SnCl_4$ as catalyst [172-179]. These reactions included the ketovinylation of phenolic ethers [179], thiophene [176], 2-alkoxynaphthalene [172,173], methylbenzenes [174,178], and carbazoles [177]. Some typical results are shown in Eqs. (78) and (79).

$$R{-}C_6H_4{-}COCH{=}CH{-}Cl + \text{thiophene} \xrightarrow[-10^\circ,\ 1\ hr]{SnCl_4} R{-}C_6H_4{-}COCH{=}CH{-}(2\text{-thienyl}) \qquad (78)\ [176]$$

R = *p*-NO_2, *o*-Cl, *m*-Cl, *p*-Cl, *p*-Br

$$RCOCH{=}CH{-}Cl + ArH \xrightarrow[\text{room temp, } 3\ hr]{AlCl_3-CH_3NO_2} RCOCH{=}CH{-}Ar \qquad (79)\ [178]$$

R = Et or Pr and Ar = *o*-, *m*- or *p*-xylyl

2. Vinyl di- and polyhalides

a. 1,1-Dihaloethylenes

As early as 1879, Demole reported that 1,1-diphenylethylene was one of the major products of condensation of benzene with 1,1-dibromoethylene in the presence of a high ratio of $AlCl_3$ [180]. In 1948, a similar result was obtained by Korshak and Samplavskaya using 1,1-dichloroethylene [181].

b. 1,2-Dihaloethylenes

Demole reported that the condensation of 1,2-dibromoethylene with benzene in the presence of $AlCl_3$ gave a mixture of bibenzyl, anthracene, and bromobenzene [180]. Later Böeseken and Bastet found that the reaction of benzene with a mixture of *cis*- and *trans*-dichloroethylene in the presence of an aluminum-mercuric chloride catalyst produced bibenzyl, 1,1,2-triphenylethane and 1,1,2,2-tetraphenylethylene (Eq. 80) [155]:

$$\text{Cl-CH=CH-Cl} + \text{C}_6\text{H}_6 \xrightarrow{\text{Al-HgCl}_2} \text{C}_6\text{H}_5\text{-CH}_2\text{-CH}_2\text{-C}_6\text{H}_5 + \text{C}_6\text{H}_5\text{-CH}_2\text{CH}(\text{C}_6\text{H}_5)_2 + (\text{C}_6\text{H}_5)_2\text{CH-CH}(\text{C}_6\text{H}_5)_2 \quad (80)$$

More recently, the aluminum chloride-catalyzed condensation of benzene with *trans*-1,2-dichloroethylene was shown by Schmerling et al. to be accompanied by hydrogen transfer both in the presence and absence of added alkanes [171]. Thus the reaction of benzene with *trans*-1,2-diehloroethylene at about 40°C resulted in a 3% yield of bibenzyl. When the same reaction was carried out in the presence of added isopentane, bibenzyl was isolated in 12% yeild together with a 10% yield of a mixture of 80 to 85% 2-methyl-3-phenylbutane and 15 to 20% *t*-pentylbenzene. The formation of bibenzyl was presumed by the authors to involve the condensation of the benzene across the double bond of dichloroethylene to form 1,2-dichloro-1-phenylethane, followed by replacement of the benzylic chlorine by hydrogen and of the other chlorine by a phenyl group (Eqs. 81 to 83).

$$\text{Cl-CH=CH-Cl} + \text{PhH} \longrightarrow \underset{\underline{194}}{\text{Cl-CH(Ph)-CH}_2\text{-Cl}} \quad (81)$$

$$\text{Cl-CH(Ph)-CH}_2\text{Cl} \xrightarrow{-\text{Cl}^-} \text{Ph}\overset{+}{\text{C}}\text{HCH}_2\text{Cl} \underset{\text{R}^+}{\overset{\text{RH}}{\rightleftharpoons}} \underset{\underline{195}}{\text{PhCH}_2\text{CH}_2\text{Cl}} \quad (82)$$

$$\text{PhCH}_2\text{CH}_2\text{Cl} + \text{PhH} \xrightarrow{-\text{HCl}} \underset{\underline{196}}{\text{PhCH}_2\text{CH}_2\text{Ph}} \quad (83)$$

The source of hydrogen in the absence of added hydrocarbon was presumed to be triphenylethane formed by direct reaction of 194 with benzene, or polymeric products found in the catalyst layer.

c. *Trihaloethylenes*

Schmerling et al. also investigated the treatment of benzene with trichloroethylene in the presence of $AlCl_3$ at 40 to 45°C; a product was obtained from which triphenylethane (a hydrogen transfer product) was isolated in about 2% yield and 1,1,2,2-tetraphenylethane in up to 4% yield [171]. When the same reaction was carried out in the presence of isopentane, there was obtained a 22% yield of bibenzyl and 20% yield of pentylbenzenes, the latter yield being based on a theoretical yield of 2.0 mol of pentylbenzene per mole of trichloroethylene charged. The mechanism of the reaction was again presumed to involve the addition of trichloroethylene to benzene to yield 1,2,2-trichloro-1-phenylethane, followed by nucleophilic displacement of one of the primary chlorines by phenyl, and subsequent replacement of both benzylic chlorines by hydrogens. Evidence that the intermediate trichloroethylbenzene was the 1,2,2-trichloroethyl-1-phenylethane isomer rather than 1,1,2-trichloro-1-phenylethane was deduced from the fact that the tetraphenylethane produced was the symmetrical 1,1,2,2-isomer and not 1,1,1,2-tertraphenylethane.

More recently, Guseinov et al. investigated the condensation of benzene [182], toluene [183], and alkylbenzenes [184] with trichloroethylene in the presence of various proportions of $AlCl_3$ at -10 to +70°C. The reactions proceeded with saturation of the double bond and substitution of chlorines by aryl groups or hydrogen atoms. Using benzene and the trichloroethylene in a 3:1 molar ratio, the highest yield (82.4%) of tetraphenylethane was obtained at 10°C and the highest yields of diphenylethane and triphenylethane (17.8 and 27.3%, respectively) were obtained at 70°C when the yield of tetraphenylethane dropped to 32.9%. With toluene, the products were shown to contain varying amounts of ditolyl, *sym*-di-*p*-tolylethane, *sym*-di-*m*-tolylethane, ditolylbenzene, tritolylethane, tetratolylethane, and tars. The amounts of ditolylbenzene and ditolyl were shown to vary but little under the different conditions used. Ethylbenzene gave products similar in type to those obtained with toluene.

d. *Tetrahaloethylenes*

It has been reported that tetrachloroethylene was essentially unaffected when treated with aluminum [185] or aluminum-mercuric chloride [155] catalysts and benzene. When $AlCl_3$ was used as catalyst, however, a considerable quantity of anthracene was obtained [186].

B. Alkylation of Aromatics with Allylic Reagents

1. Allyl halides

a. *Allyl Monohalides*

As with vinyl halides, protonic acid catalysts favor reaction at the double bond of allyl halides, whereas strong metal halide catalysts favor reaction at the allylic carbon atom. It is thus convenient to present the discussion under two headings, one for protonic acids and another for metal halides.

(i) Protonic acid-catalyzed alkylations with haloallylic compounds. Before more recent work [187-191], the alkylation of arenes with allyl halides in the presence of sulfuric acid had been considered to proceed exclusively according to Markovinkov's rule. Thus, under the catalytic effect of sulfuric acid, benzene was reported to condense with allyl chloride [192,193], bromide [192,194-197], or iodide [197] to give the corresponding 1-halo-2-phenylpropane, and with methallyl chloride [169,179,198-201] or bromide [169] to give the corresponding 1-halo-2-methyl-2-phenylpropane (neophyl halide). The latter was also produced with HF catalyst [202].

Like benzene, other arenes were reported to condense with allyl chloride [191,192,203], bromide [191,203], or iodide [191,204] and with methallyl chloride [26,269,191,205] or bromide [169,191] in the presence of sulfuric acid catalyst. These reactions were shown to proceed in much the same manner as with benzene, producing predominantly the *para*-substituted alkylbenzene derivatives. Only minor amounts of *meta*- and *ortho*-substituted products have been detected [26].

Equations (84) to (86) give some illustrative examples of arene alkylations with allylic reagents. In most cases, small amounts of dialkylated products were produced.

$$CH_2{=}CH{-}CH_2Cl \xrightarrow{H_2SO_4} \begin{cases} \xrightarrow{C_6H_6} C_6H_5CH(CH_3)CH_2Cl\ (97\%) + C_6H_5CH_2CH_2CH_2Cl\ (3\%) \\ \xrightarrow{C_6H_5CH_3} p\text{-}CH_3C_6H_4CH(CH_3)CH_2Cl \end{cases} \quad (84)\ [203]$$

$$CH_2{=}C(CH_3){-}CH_2Cl \xrightarrow{H_2SO_4} \begin{cases} \xrightarrow{C_6H_6} C_6H_5C(CH_3)_2{-}CH_2Cl \\ \xrightarrow{C_6H_5CH_3} p\text{-}CH_3C_6H_4C(CH_3)_2{-}CH_2Cl\ (92\text{-}93\%) + o\text{- and } m\text{-isomers } (7\text{-}8\%) \end{cases} \quad (85)\ [26]$$

$$\text{biphenyl} + CH_2{=}CHCH_2I \xrightarrow[30^\circ,\ 1.5\ hr]{H_2SO_4} \text{2-}(CH_3CHCH_2I)\text{biphenyl}\ (64.3\%) + \text{3-}(CH_3CHCH_2I)\text{biphenyl}\ (6.7\%) + \text{4-}(CH_3CHCH_2I)\text{biphenyl}\ (28.1\%) \quad (86)\ [204]$$

Like arenes, thiophene was reported to react with methallyl chloride to give the 2-substituted thiophene in 50% yield with boron fluoride etherate or ethanesulfonic acid catalyst [206] and in only 6% yield with sulfuric acid catalyst [207]. Allyl chloride was also reported to alkylate phenol and cresols after prolonged contact (6 months) with H_2SO_4 at room temperature [208]. In these cases, however, the products were claimed to be the corresponding isopropenyl derivatives.

It is quite possible that the production of isopropenyl derivatives in these reactions may have involved dehydrochlorination of initially formed 1-chloro-2-arylpropanes. Experimental support for this view was provided when 197 and 199 were shown to dehydrobrominate to 198 and 200 upon treatment with H_2SO_4 [188,189].

$$\underset{\mathbf{197}}{PhCH(CH_3)CH_2Br} \xrightarrow{H_2SO_4} \underset{\mathbf{198}}{Ph\text{-}C(CH_3){=}CH_2}$$

$$\underset{\mathbf{199}}{PhCH_2CH(Br)CH_3} \xrightarrow{H_2SO_4} \underset{\mathbf{200}}{Ph\text{-}CH{=}CHCH_3}$$

Protonic acid-catalyzed alkylations have also been carried out with pentenyl halides. The reaction of 4-chloro-2-pentene with benzene in the presence of various acid catalysts was shown by Vdovtsova and Khar'yanov [209] to give 2-chloro-3-phenylpentane (201) and 2,3-diphenylpentane (202). In 18 experiments with various reaction times, temperatures, and proportions of reactants and catalysts (H_2SO_4, H_3PO_4, BF_3-H_3PO_4, and $AlCl_3$) these authors obtained maximum yields of 17 and 12%.

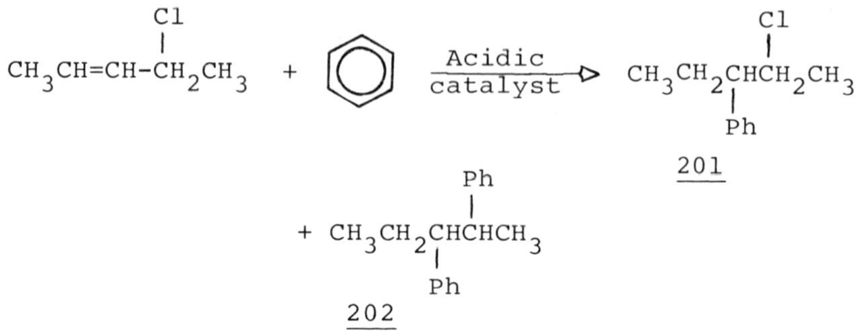

$$CH_3CH{=}CH{-}\underset{}{\overset{Cl}{CH}}CH_2CH_3 + C_6H_6 \xrightarrow[\text{catalyst}]{\text{Acidic}} CH_3CH_2\underset{Ph}{\overset{Cl}{C}}HCH_2CH_3 \;(\underline{201}) + CH_3CH_2\underset{Ph}{C}H\overset{Ph}{C}HCH_3 \;(\underline{202})$$

In subsequent paper, Vdovtsova and co-workers examined protonic acid-catalyzed reactions of 1-chloro-3-methyl-2-butene, and 4-chloro-2-pentene with a number of phenols and their ethers. These included the condensation of 1-chloro-3-methyl-2-butene with sodium phenoxide [210] and guaiacol [211], of 4-chloro-2-pentene with anisole [212], *o*-bromophenol, and *o*-bromoanisole [213], catechol, and veratrol [214], guaiacol [215], *o*-chlorophenol, *o*-chloroanisole [216], *m*-chlorophenol, and *m*-chloroanisole [217].

An example of some of their results is given for the reaction of guaiacol in Eq. (87), and the reaction pathways suggested to account for the products

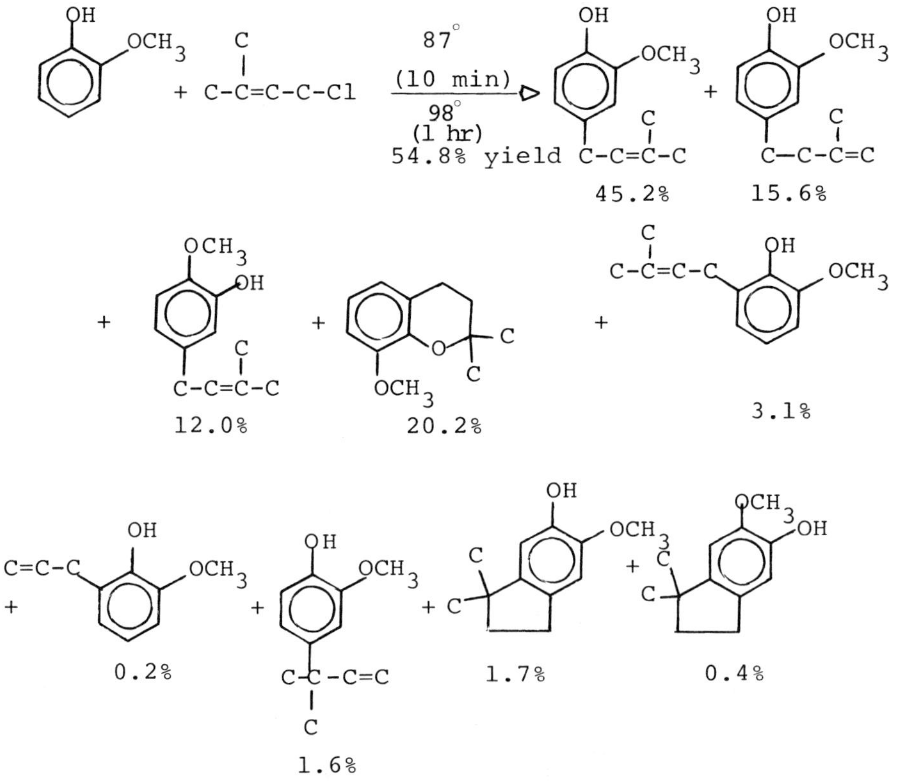

Guaiacol + C-C(C)=C-C-Cl —[87° (10 min), 98° (1 hr); 54.8% yield]→ 4-(C-C=C(C)-C)-guaiacol 45.2% + 4-(C-C-C(C)=C)-guaiacol 15.6% + 5-(C-C=C(C)-C)-guaiacol (2-methoxy-5-substituted phenol) 12.0% + 8-methoxy-2,2-dimethylchroman 20.2% + 6-(C-C(C)=C-C)-guaiacol 3.1% + 6-(C=C-C)-guaiacol 0.2% + 4-(C-C(C)-C=C)-guaiacol 1.6% + 1,1-dimethylindan (OH, OCH$_3$) 1.7% + 1,1-dimethylindan (OCH$_3$, OH) 0.4% (87)

obtained from anisole and 4-chloro-2-pentene in the presence of H_2SO_4 are shown in scheme 14.

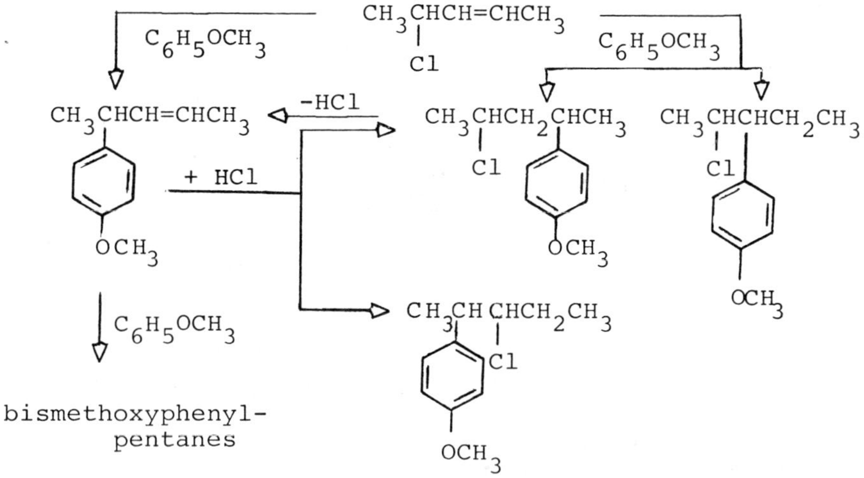

$CH_3CH(Cl)CH=CHCH_3$

$\xrightarrow{C_6H_5OCH_3}$ $CH_3CH(C_6H_4OCH_3)CH=CHCH_3$; $\xrightarrow{C_6H_5OCH_3}$ $CH_3CH(Cl)CH_2CH(C_6H_4OCH_3)CH_3$ + $CH_3CH(Cl)CH(C_6H_4OCH_3)CH_2CH_3$

$CH_3CH(C_6H_4OCH_3)CH=CHCH_3$ $\xrightarrow{+HCl}$ $CH_3CH(Cl)CH_2CH(C_6H_4OCH_3)CH_3$ + $CH_3CH(C_6H_4OCH_3)CH(Cl)CH_2CH_3$; $\xleftarrow{-HCl}$

$CH_3CH(C_6H_4OCH_3)CH=CHCH_3$ $\xrightarrow{C_6H_5OCH_3}$ bismethoxyphenyl-pentanes

Scheme 14

$CH_3-C(CH_3)(Cl)-CH=CH_2$ (203) or $CH_3-C(CH_3)=CH-CH_2Cl$ (204)

$\xrightarrow{H_3PO_4}$ $CH_3-\overset{+}{C(CH_3)\text{=}CH\text{=}CH_2}$

$\xrightarrow{PhOCH_3}$

$CH_3-C(CH_3)(C_6H_4OCH_3)-CH=CH_2$ (209)

$CH_3C(CH_3)=CH-CH_2-C_6H_4-OCH_3$ (207) $\xrightarrow{H_3PO_4}$ 208 (6-methoxy-1,1-dimethylindan; CH_3O, CH_3, CH_3)

$CH_3-C(CH_3)(Cl)-CH_2CH_2-C_6H_4-OCH_3$ (205) $\xrightarrow{PhOMe,\ H_3PO_4}$ $CH_3C(CH_3)(C_6H_4OCH_3)-CH_2CH_2-C_6H_4-OCH_3$ (206)

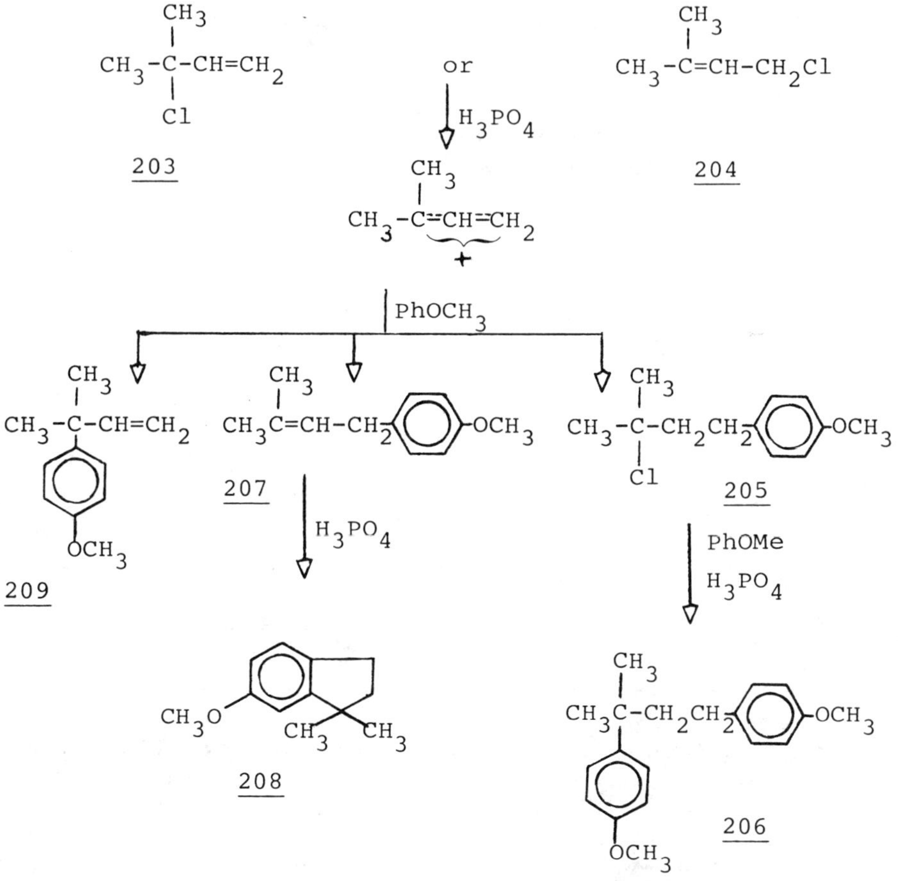

Scheme 15

It should be noted that in all of the reactions other higher products were also produced but usually in small amounts, and that the proportions of products were dependent on the alkylation conditions. For example, in the alkylation of anisole with 4-chloro-2-pentene, the composition of the alkylation product was found to depend mainly on the amount of catalyst used [212,218]. In the presence of 0.03 to 0.3 molecular proportion of H_2SO_4, H_3PO_4, or BF_3-H_3PO_4 the reaction went entirely at the expense of the more active chlorine in the direction of 1-methyl-2-butenylation, and secondary processes were practically absent. On the other hand, chlorine-containing products (chloropentylanisoles) resulting from reaction at the double bond were obtained predominantly with an excess of the acid catalyst (1 to 3 mol per mole of reactant) [212,219].

In addition to the aforementioned catalyst effects, temperature effects were also shown to be significant, especially when H_3PO_4 was used as the catalyst. Thus, at low (<0°C) and at high (96 to 98°C) temperatures alkenylation predominated even with an excess of catalyst. The optimum temperature for chloroalkylation was in the range 5 to 40°C [212].

In 1969, Vdovtsova and Yanichkin investigated the alkylation of anisole with the isoprene hydrochlorides 176 and 177 in the presence of anhydrous phosphoric acid and $FeCl_3$ [220]. Regardless of whether the alkylating agent was 203 or 204, similar reaction mixtures were obtained in which compound 205 and its alkylation product 206 and compound 207 and its cyclization product 208 were the main constituents. Compound 209 was also obtained, but in smaller yield. A summary of the reactions involved is given in scheme 15.

Until recently, the mechanism of reaction of arenes with allyl halides in the presence of protonic acid catalysts was visualized simply as involving the addition of a proton to the double bond of the allyl halide followed by alkylation [221]:

$$CH_2{=}C(R){-}CH_2X \xrightarrow{H^+} CH_3{-}\overset{+}{C}(R){-}CH_2X$$

$$CH_3\overset{+}{C}(R){-}CH_2X + C_6H_6 \longrightarrow CH_3{-}C(R)(C_6H_5){-}CH_2X + H^+$$

This mechanism was based on repeated claims that only Markovnikov-type products could be isolated from these reactions. On the basis of studies by Bodrikov et al. however, it was established that the sulfuric acid-catalyzed condensation of benzene with allyl halides [188,189,191] and substituted allyl halides [187,191] takes place both in accordance with and opposite to the Markovnikov rule. For example, the reaction of methallyl chloride with benzene gave a mixture of neophyl chloride (210) and the isomeric chloroisobutylbenzene (211) in a ratio 4:1. Besides, the presence of small amounts of the dehydrochlorination products 212 and 213 was also confirmed.

```
   C                                                    C              C
   |                  H2SO4                             |              |
C=C-C-Cl  +  (benzene) ---------->  Ph-C-C-Cl  +  Ph-C-C-C
                     56% yield                          |              |
                                                        C              Cl
                                                       210            211

                                        C                 C
                                        |                 |
                              +  Ph-C=C-C     +    Ph-C-C=C

                                    212                213
```

In a 1972 paper on this subject, Bodrikov and co-workers examined the products of benzene alkylation with nine substituted allyl halides in the presence of H_2SO_4 as catalyst and in CH_3NO_2 as solvent [191]. From the following allylic halides, the relative percentages of anti-Markovnikov isomers were as shown:

$CH_2 = CH\text{-}CH_2Cl$	(3%)	$CH_2 = CMeCH_2Br$	(22%)
$CH_2 = CH\text{-}CH_2Br$	(16%)	$MeCH = CH\text{-}Br$	(13%)
$CH_2 = CH\text{-}CH_2I$	(20%)	$Me_2C = CHCl$	(6%)
$CH_2 = CMeCH_2Cl$	(6%)	$Me_2C = CHBr$	(20%)

These results demonstrated that the extent of violation of the Markovnikov rule under these conditions was dependent on both the type of halogen and the degree of branching at the double bond. The use of nitromethane as solvent was also shown to enhance abnormal addition [189,190]. Thus the 16% ratio found for alkylation with allyl bromide in nitromethane dropped to only 3% in excess benzene. The marked increase in yield of anomalous product in nitromethane relative to benzene was attributed to the involvement of the ion pair 214 in benzene solution and of the bromonium ion 215 in nitromethane solution [188]. They proposed that in the latter the highly dissociated nitromethane molecules displace the equilibrium (Eq. 88) to the right and that the high-energy "open" cation (215) is thus stabilized by conversion to the cyclic form 216.

$$\underset{\mathbf{214}}{CH_3\overset{+}{C}HCH_2Br,\ HSO_4^-} \rightleftharpoons \underset{\mathbf{215}}{CH_3\overset{+}{C}HCH_2Br} + HSO_4^- \tag{88}$$

$$\mathbf{215} \longrightarrow \underset{\mathbf{216}}{CH_3\overset{+}{C}H\text{—}CH_2 \text{ (bridged by Br)}} \tag{89}$$

(ii) Alkylation of aromatics with allyl halides in the presence of metal and metal halide catalysts. In 1933, Nenitzescu and Isacescu reported the effect of various catalysts and reaction conditions on the composition of the products of reaction of benzene with allyl chloride [222]. A summary of their findings is shown in Eqs. (90) to (93).

$$CH_2{=}CHCH_2Cl + C_6H_6 \xrightarrow[\text{reflux}]{ZnCl_2} C_6H_5CH_2CHClCH_3\ (\underline{217}) + C_6H_5CH_2CH_2CH_3\ (\underline{218}) \quad (90)$$

$$\xrightarrow[-10\ \text{to}\ -20^\circ]{FeCl_3} C_6H_5CH_2CH{=}CH_2\ (\underline{219}) + C_6H_5CH_2CH(C_6H_5)CH_3\ (\underline{220}) + C_6H_5CH_2CHClCH_3\ (\underline{217}) \quad (91)$$

$$\xrightarrow[45^\circ\ \text{or}\ -14^\circ]{AlCl_3\ (\text{Dry})} C_6H_5CH_2\text{-}CH(C_6H_5)\text{-}CH_3\ (\underline{220}) \quad (92)$$

$$\xrightarrow[50^\circ]{AlCl_3\ (H_2O)} C_6H_5CH_2\text{-}CH(C_6H_5)\text{-}CH_3\ (\underline{220}) + C_6H_5CH_2CH_2CH_3\ (\underline{218}) + \text{9,10-}(CH_2CH_3)_2\text{-anthracene}\ (\underline{221}) \quad (93)$$

Similar results were observed by other investigators using $ZnCl_2$ [223], $AlCl_3$ [33, 224-227], Al [185], Zn [228, 229], or $AlCl_3$-CH_3NO_2 [230] catalysts. 2-Halo-1-phenylpropanes were reported to be the only products obtained with $AlCl_3$-CH_3NO_2 [230] or with $FeCl_3$ [231].

The formation of compounds 217-221 from the reaction of allyl halide with benzene was explained by Nenitzescu and Isacescu in terms of the processes shown in Eqs. (94) to (98).

$$C_6H_6 + CH_2{=}CHCH_2Cl \longrightarrow C_6H_5CH_2CH{=}CH_2\ (\underline{219}) + HCl \quad (94)$$

$$C_6H_5-CH_2CH=CH_2 + HCl \longrightarrow C_6H_5-CH_2CHCl-CH_3 \quad (\underline{217}) \qquad (95)$$

$$C_6H_5-CH_2CHCl-CH_3 + C_6H_6 \longrightarrow C_6H_5-CH_2CH(C_6H_5)-CH_3 \quad (\underline{220}) + HCl \qquad (96)$$

$$2\ C_6H_5-CH_2CHCl-CH_3 \longrightarrow \text{9,10-di}(CH_2CH_3)\text{-9,10-dihydroanthracene} + 2HCl \qquad (97)$$

$$\text{9,10-di}(CH_2CH_3)\text{-9,10-dihydroanthracene} + C_6H_5-CH_2-CHCl-CH_3 \longrightarrow \text{9,10-di}(CH_2CH_3)\text{anthracene} \quad (\underline{221}) + C_6H_5CH_2CH_2CH_3 \quad (\underline{218}) + HCl \qquad (98)$$

The reaction of toluene with allyl chloride or bromide has been found to proceed in much the same manner as with benzene [226,228,332]. More recently benzene, toluene, ethylbenzene, cumene, and *o*-, *m*-, and *p*-xylene were found to react with allyl halides in the presence of $FeCl_3$ or $ZnCl_2$-Al_2O_3 at temperatures from 25 to 80°C to give mainly isomers of the corresponding 1-aryl-2-chloropropane [223,231,233].

In contrast, the reaction of xylenes with allyl chloride in the presence of $FeCl_3 \cdot 2.5\ H_2O$ at 30 to 40°C was reported to proceed in an anti-Markovnikov manner to give isomeric 1-xylyl-3-chloropropanes (Eq. 99) [234]. The reactivity of the xylenes was in the order $m > p > o$. Reactions leading mainly to the displacement of the halogen atom were reported to occur in the presence of catalysts such as Cu [235], $CuCl_2$ [236], or $ZnCl_2$ [237].

$$\text{o-xylene} + CH_2{=}CH{-}CH_2Cl \xrightarrow[30-40^\circ]{FeCl_3\cdot 2.5\ H_2O} \underset{47\%}{\text{1,2-}(CH_3)_2\text{-3-}(CH_2CH_2CH_2Cl)C_6H_3} + \underset{53\%}{\text{1,2-}(CH_3)_2\text{-4-}(CH_2CH_2CH_2Cl)C_6H_3} \tag{99}$$

The reaction of arenes with C_4 and higher allylic halides has also been explored. In the presence of $AlCl_3$ or Al catalyst, benzene was reported to give, with crotyl chloride [239], mixtures of 1,2-diphenylbutane and a small amount of *sec*-butylbenzene; with methallyl chloride [185] or bromide [238] the products were 2-methyl-1,2-diphenylpropane, *meso*-2,3-diphenylbutane, and *t*-butylbenzene. More recently, however, the product from methallyl chloride, benzene, and $AlCl_3$ was shown to contain 2-methyl-1,1-diphenylpropane and R,S-2,3-diphenylbutane in addition to the previously observed compounds (Eq. 100) [26].

$$C{=}C(C){-}C{-}X + PhH \xrightarrow{AlCl_3} Ph{-}C(C)_2{-}C{-}Ph + (Ph)_2C{-}C(C)_2 + \underset{R,S\ \text{and}\ meso}{C{-}C(Ph){-}C(Ph){-}C} + Ph{-}C(C)_2{-}C + C{-}C(Ph){-}C{-}C + Ph{-}C{-}C(C){-}C \tag{100}$$

Vdovtsova and co-workers published a number of studies on the condensation of C_4 and C_5 allyl chlorides with phenols and their ethers in the presence of metal halide catalysts. The primary object of these studies was to investigate the reactivity of the chlorine atom in allylic chlorides of different structures and to determine the possibility of a selective course of the process. In one of the studies, Vdovtsova and Yanichkin [239] found that the reaction of 1-chloro-2-butene and 3-chloro-1-butene with anisole in the presence of small amounts of $FeCl_3$ proceeded only by displacement of the chlorine atom, resulting in similar mixtures of *o*- and *p*-*trans*-2-butylanisoles (222a-222b),

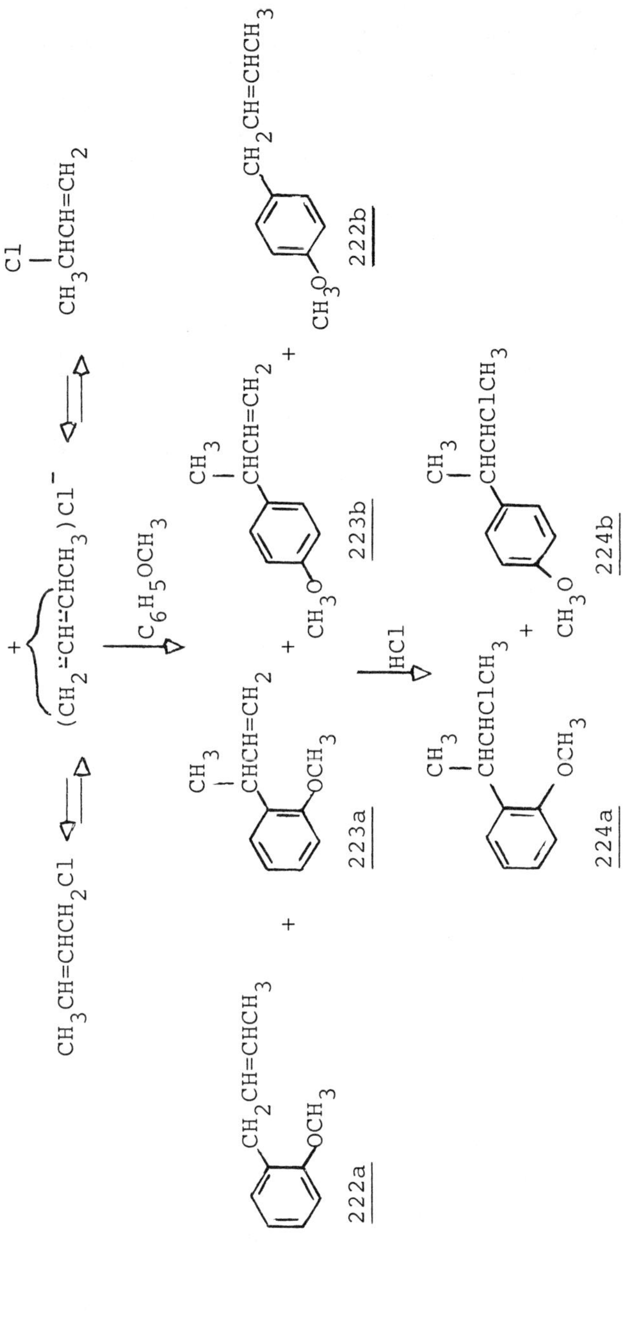

$$CH_3CH{=}CHCH_2Cl \rightleftharpoons (CH_2{\cdots}CH{\cdots}CHCH_3)^+ Cl^- \rightleftharpoons CH_3CH(Cl)CH{=}CH_2$$

$$\xrightarrow{C_6H_5OCH_3}$$

222a (o-$CH_3OC_6H_4CH_2CH{=}CHCH_3$) + 223a ($o$-$CH_3OC_6H_4CH(CH_3)CH{=}CH_2$) + 223b ($p$-$CH_3OC_6H_4CH(CH_3)CH{=}CH_2$) + 222b ($p$-$CH_3OC_6H_4CH_2CH{=}CHCH_3$)

$$\xrightarrow{HCl}$$

224a (o-$CH_3OC_6H_4CH(CH_3)CHClCH_3$) + 224b ($p$-$CH_3OC_6H_4CH(CH_3)CHClCH_3$)

Scheme 16

as well as small amounts of *o*- and *p*-2-chloro-1-methylpropylanisoles (224a, 224b) formed by hydrochlorination of the major products. These results demonstrated clearly that the isomeric composition of the alkenylation products was independent of the identity of the starting allylic chloride. In rationalizing their findings, Vdovtsova and Yanichkin suggested that the process proceeded by a unimolecular mechanism in which an allylic ion pair formed by interaction of the chloride with the catalyst attacks the aromatic nucleus preferentially at the primary carbon atom, giving an *ortho*/*para* ratio in accordance with the steric requirements of the attacking species. The course of the reactions was outlined as shown in scheme 16.

In other studies, Vdovtsova et al. investigated the reaction of 4-chloro-2-pentene (piperylene hydrochloride) with a number of phenols and halophenols and their ethers in the presence of metal and metal halide catalysts [202,216-218,240,241]. In view of the fact that the secondary α,γ-dimethylallyl cation is readily formed and possesses high reactivity, and the double bond of 4-chloro-2-pentene is relatively inactive, these authors succeeded in directing their reactions toward alkenylation and also in designing conditions for a stepwise and selective process. For example, the use of small amounts of metal or metal halide catalysts in the reaction of 4-chloro-2-pentene with anisole [218,240,241], phenetole [218], or phenol [218] enabled them to carry out the reaction in a stepwise and selective manner as a process of α,γ-dimethylallylation. Under such conditions, up to 90% yields of alkenylation products were obtained which consisted of mixtures of the *para* and *ortho* isomers with a 73 to 93% predominance of the former and a small amount of the *meta* isomer (Eq. 101).

$$CH_3CH{=}CH{-}CH(CH_3){-}Cl + C_6H_5OR \xrightarrow[<90^\circ]{\text{Metal or metal halide,}} p\text{-}RO{-}C_6H_4{-}CH(CH_3)CH{=}CH{-}CH_3 + o\text{-}RO{-}C_6H_4{-}CH(CH_3)CH{=}CH{-}CH_3 + \underset{\text{(very small amounts)}}{m\text{-}RO{-}C_6H_4{-}CH(CH_3)CH{=}CHCH_3} \quad (101)$$

$R = H, CH_3, C_2H_5$

The ultimate products of the reactions, diarylpentanes, were formed when the amount of catalyst and contact time were increased and the temperature was raised to 110 to 140°C (Eq. 102):

$$RO{-}C_6H_4{-}CH(CH_3)CH{=}CHCH_3 \xrightarrow{C_6H_5OR} RO{-}C_6H_4{-}CH(CH_3)CH(C_6H_4{-}OR)CH_2CH_3 \quad (102)$$

On the basis of their examination of various metal and metal halide catalysts, Vdovtsova et al. arrived at the following conclusions: (1) The reactions with metal chlorides took place more readily than with the metals themselves, which was taken as a confirmation of the ionic nature of the process. (2) In all cases, the reaction took place in the direction of α,γ-dimethylallylation, with $FeCl_3$ being the most active catalyst of those tested under similar conditions. (3) The nature of the catalyst had little influence on the isomeric composition of the alkenylation products, which consisted predominantly of the *para* isomers.

In accordance with these results, 4-chloro-2-pentene condensed with *o*-bromophenol and *o*-bromoanisole [213], *o*-chlorophenol, *o*-chloroanisole [216], *m*-chloroanisole [217], and guaiacol [202] in the presence of catalytic amounts of $FeCl_3$ by displacement of chlorine to give mixtures consisting mainly of isomers with the α,γ-dimethylallyl group on the *para* and *ortho* positions with respect to the hydroxyl (or alkoxyl) group (Eq. 102a).

$$CH_3CH{=}CHCH(CH_3){-}Cl + C_6H_4(OR)X \xrightarrow{FeCl_3} 4\text{-}(CH_3CHCH{=}CHCH_3)C_6H_3(OR)X + 6\text{-}(CH_3CH(CH_3)CH{=}CH)C_6H_3(OR)X \quad (102a)$$

X = Cl or Br

R = H or CH_3

The reaction of naphthalene and methylnaphthalenes with 4-chloro-2-pentene in the presence of a catalyst such as Fe, Cu, $FeCl_3$, $AlCl_3$, $SnCl_4$, or $ZnCl_2$ resulted in α,γ-dimethylallylation of the aromatic hydrocarbons (Eqs. 103 to 105) [242].

$$C_{10}H_8 + CH_3CH(Cl){-}CH{=}CH{-}CH_3 \xrightarrow{\text{catalyst}} 1\text{-}(CH_3{-}CH{-}CH{=}CH{-}CH_3)C_{10}H_7 \quad (103)$$

$$1\text{-}CH_3C_{10}H_7 + CH_3{-}CH(Cl){-}CH{=}CH{-}CH_3 \xrightarrow{\text{catalyst}} 1\text{-}CH_3\text{-}4\text{-}(CH_3{-}CH{-}CH{=}CH{-}CH_3)C_{10}H_6 + 1\text{-}CH_3\text{-}6\text{-}(CH_3CH(CH{=}CH{-}CH_3))C_{10}H_6 \quad (104)$$

$$\text{2-methylnaphthalene} + CH_3\text{-}\underset{Cl}{CH}\text{-}CH\text{=}CH\text{-}CH_3 \xrightarrow{\text{catalyst}} \text{1-}(CH_3\text{-}CHCH\text{=}CH\text{-}CH_3)\text{-2-methylnaphthalene} + \text{6-}(CH_3CH(CH\text{=}CH\text{-}CH_3))\text{-2-methylnaphthalene} \quad (105)$$

Chlorinated alkenylbenzenes were obtained by alkylation with compounds such as $Cl_2C = CH\text{-}\underset{R}{CH}CH_2Cl$ (R = Me, hexyl, or octyl) or $ClCH = CH\text{-}\underset{Me}{CH}CH_2Cl$ in the presence of $AlCl_3$ or $ZnCl_2$ catalyst [243]. Benzene, toluene, and *m*-xylene gave a mixture of the isomeric chlorinated alkenylbenzenes 225 and 226 where R' = H, 3-Me, or 3,5-Me_2.

$$Cl_2C\text{=}CHCH(CH_2R)\text{-}C_6H_4R' \quad \textbf{225} \qquad Cl_2C\text{=}CHCH_2CHR\text{-}C_6H_4R' \quad \textbf{226}$$

The condensation of halo derivatives of 1,3-dienes with arenes in the presence of $AlCl_3$ was also investigated (Eqs. 106 and 107) [244].

$$ArH + CH_2\text{=}\underset{Cl}{C}\text{-}CH\text{=}CH_2 \xrightarrow{AlCl_3} ArCH_2CH\text{=}\underset{Cl}{C}\text{-}CH_3 \quad (106)$$

$$ArH + CH_2\text{=}\underset{Cl}{C}\text{—}\underset{Cl}{C}\text{=}CH_2 \xrightarrow{AlCl_3} ArCH_2\underset{Cl}{C}\text{=}\underset{Cl}{C}\text{-}CH_3 \quad (107)$$

b. *Alkylation of Aromatics with Allyl Di- and Polyhalides*

Comparatively few examples of Friedel-Crafts alkylations with allyl di- and polyhalides have been described. In 1888, Willgerodt and Genieser reported that the aluminum chloride-catalyzed reaction of 1,1,1-trichloro-2-methyl-2-propanol with benzene, toluene, and *p*-xylene gave small amounts of compounds in which all of the chlorine atoms and the hydroxyl group of the trichlorobutanol were replaced [245]. In 1956, Kundiger and Pledger reinvestigated the same reaction with toluene and with chlorobenzene, bromobenzene, and anisole, but did not find the expected replacement products [246]. Instead, they obtained, as major products, mixtures of the corresponding

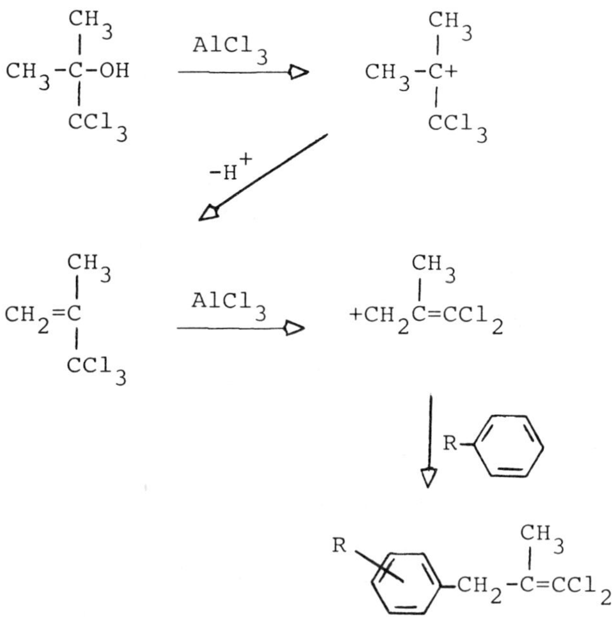

$$CH_3-C(CH_3)(CCl_3)-OH \xrightarrow{AlCl_3} CH_3-C^+(CH_3)(CCl_3) \xrightarrow{-H^+} CH_2=C(CH_3)(CCl_3) \xrightarrow{AlCl_3} {}^+CH_2C(CH_3)=CCl_2 \xrightarrow{R-C_6H_5} R-C_6H_4-CH_2-C(CH_3)=CCl_2$$

227a, R = Cl
b, Br
c, CH_3
d, CH_3O

Scheme 17

ortho- and *para*-3,3-dichloro-2-methylallylbenzenes (227) whose formations were explained in terms of the steps outlined in scheme 17.

Experimental support for this mechanism above was described in a second paper [247]. Kundiger and Pledger extended their work to include the reaction of 1,1,3-trichloro-2-methyl-1-propene and 3,3,3-trichloro-2-methyl-1-propene with phenols in the presence of $AlCl_3$ or $FeCl_3$ catalysts.

In 1948, Sisido and Nozaki reported that the Friedel-Crafts reactions of 1,4-dibromo-2-butene with benzene yielded tetralin and 2-phenyl-1,2,3,4-tetrahydronaphthalene (227, Eq. 108), along with a considerable amount of tarry matter [248].

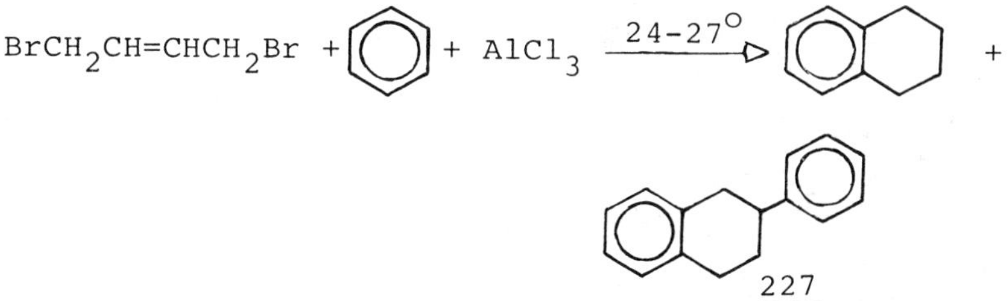

$$BrCH_2CH=CHCH_2Br + C_6H_6 + AlCl_3 \xrightarrow{24-27^\circ} \text{tetralin} + \text{2-phenyltetralin } (\mathbf{227}) \quad (108)$$

Neither naphthalene nor dihydronaphthalene was produced. Phenyl-substituted butanes which might be expected as a result of the ordinary Friedel-Crafts reaction were also not detected. Sisido and Nozaki pointed out that their results could be explained in terms of a mechanism similar to that proposed by Nenitzescu and Isacescu [222] for the reaction of benzene with allyl chloride (see Eqs. 94 to 98). However, on the basis of the observed behavior of both 1,4- and 1,2-dihydronaphthalene under the alkylation conditions, they suggested that the double bond in these molecules may act directly as a hydrogen acceptor. In 1972, Mironov et al. studied the silicoalkylation of adamantane by a number of alkenylsilanes, including two silicon analogs of allyl polyhalides, $CH_2 = CHSiCl_3$ (228) and $CH_2 = CHSiMeCl_2$ (229) [249,250]. As halosilanes do not easily produce silicocations upon interaction with Lewis acid catalysts [251-254], the reaction of 228 and 229 with adamantane was found to involve addition to the double bond rather than substitution of the halogen (Eq. 109).

$$\text{adamantane} + CH_2{=}CH{-}Si(R)Cl_2 \xrightarrow{AlCl_3} \text{adamantyl}{-}CH_2CH_2SCl_2(R) \tag{109}$$

R = Cl (228)

R = CH_3 (229)

Vinylic-allylic halides. Compounds containing both allylic and vinylic halogens have also been investigated. In such cases, it has been found that reaction at the more active allylic carbon is the primary reaction and that under controlled conditions selective replacement of the allyl halogen can be achieved. These facts will become clear from examination of Eqs. (110) to (114).

$$CH_3{-}C(Cl){=}CH{-}CH_2Cl + C_6H_6 \xrightarrow{Al} CH_3C(Cl){=}CH{-}CH_2{-}C_6H_5 \quad (\text{ca.}50\%) \tag{110}$$

[255]

$$CH_3{-}C(Cl){=}CH{-}CH_2Cl + C_6H_5{-}CH(CH_3)_2 \xrightarrow{AlCl_3} CH_3{-}C(Cl){=}CH{-}CH_2{-}C_6H_4{-}CH(CH_3)_2 \quad (\text{ca. }50\%) \tag{111}$$

[255]

$$Cl-CH=CH-CH_2Cl + C_6H_6 \xrightarrow[\text{or (b) } AlCl_3]{\text{(a) } AlCl_3-CH_3NO_2,\ 0^\circ\ (47\%)} C_6H_5-CH_2CH=CH-Cl \quad (112)$$

[(a)171, (b)256]

$$Cl-CH=C(CH_3)-CH_2Cl + C_6H_6 \xrightarrow[\text{or (b) } H_2SO_4\ (6\%)]{\text{(a) } AlCl_3-CH_3NO_2\ (51\%)} C_6H_5-C(CH_3)(CH_2Cl)-CH_2Cl \quad (113)$$

[(a)171, (b)169]

$$Cl-CH=C(CH_3)-CH_2Cl + C_6H_6 \xrightarrow{AlCl_3,\ 0^\circ} C_6H_5-CH_2-C(CH_3)=CHCl\ (23\%) \quad (114)$$

[171]

Mochida, Takeshita, and co-workers compared the reactivity of two allylic chlorides (230 and 231) and two allylic-vinylic chlorides (232 and 233) in alkylations of benzene at 0°C using tungsten hexachloride as catalyst [257]. The primary products were allyl benzenes (237), which, however, suffered some

230: $(CH_3)HC=CH(CH_2Cl)$ — CH_3 and H on one carbon; H and CH_2Cl on the other (CH₃ and CH₂Cl trans)

231: $CH_2=CH-CH_2-Cl$

232: $Cl(H)C=C(H)CH_2Cl$ — Cl and CH_2Cl trans

233: $Cl(H)C=C(H)CH_2Cl$ — Cl and CH_2Cl cis

consecutive hydrochlorinations. The order of reactivity was 230 > 231 > 232 > 233. No products corresponding to 236 were obtained from 230, 232, or 233, indicating that a carbocation (234) was not an intermediate, since it would be expected to rearrange to 235, leading to production of some 236 (scheme 18).

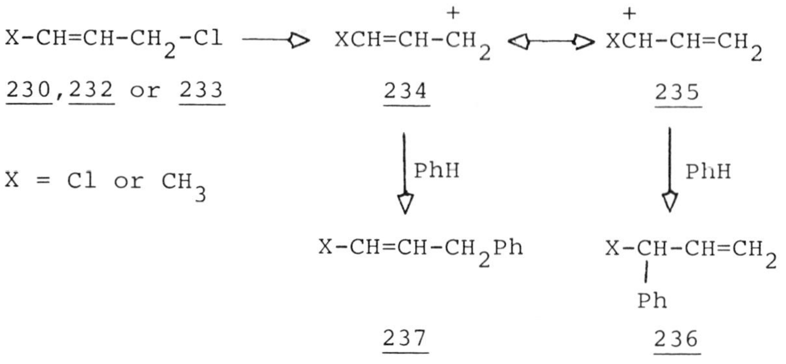

$$X{-}CH{=}CH{-}CH_2{-}Cl \longrightarrow X\overset{}{C}H{=}CH{-}\overset{+}{C}H_2 \longleftrightarrow X\overset{+}{C}H{-}CH{=}CH_2$$

230, 232 or 233 — 234 — 235

X = Cl or CH_3

234 $\xrightarrow{PhH}$ $X{-}CH{=}CH{-}CH_2Ph$ (237)

235 $\xrightarrow{PhH}$ $X{-}CH(Ph){-}CH{=}CH_2$ (236)

Scheme 18

The order of reactivity was explained in terms of the inductive effects of CH_3 and Cl in polarized-complex intermediates. The lowest reactivity of the *cis*-1,3-dichloropropene (233) were ascribed to a deactivating cyclic interaction as shown in 237.

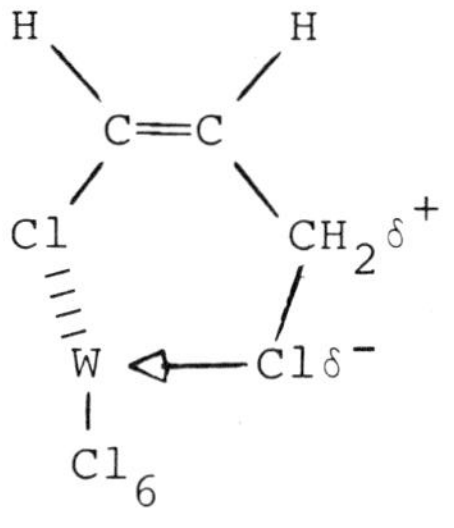

H, H on C=C; Cl on one carbon coordinated to W; $CH_2^{\delta+}$ on the other carbon, bonded to $Cl^{\delta-}$ $\rightarrow$ W; W bearing Cl_6

237

2. Alkylation of aromatics with allyl alcohols and ethers

The reactions of arenes with allyl alcohols and ethers are similar to those with allyl halides. Here, again, protonic acid catalysts favor attack at the double bond while metal halide catalysts favor attack at the allylic carbon atom. These similarities will become apparent from examination of the data of scheme 19. The reaction of benzene and allyl alcohol was reported to give 1,2-diphenylpropane with H_2SO_4 [258] or HF [259], a mixture of 1,2-diphenylpropane and allylbenzene with $AlCl_3$ [260], BF_3 [261] or HF [258], a good yield of 2-chloro-1-phenylpropane with $AlCl_3$-$C_3H_7NO_2$ [230], and a mixture of 1,1- and 1,2-diphenylpropane with $AlCl_3$ [262].

The data at the bottom of scheme 19 are from the work of Ackermann and Heesing [262], who found that the proportions of the four products were strongly dependent on the reaction conditions. 2-Phenyl-1-propanol was demonstrated to be the initial product of the reaction and was shown to be converted readily into the two diphenylpropane isomers by further reaction with benzene, $AlCl_3$, and HCl. Experiments with deuterium-labeled molecules showed that the alkylation was not reversible under the mild conditions used.

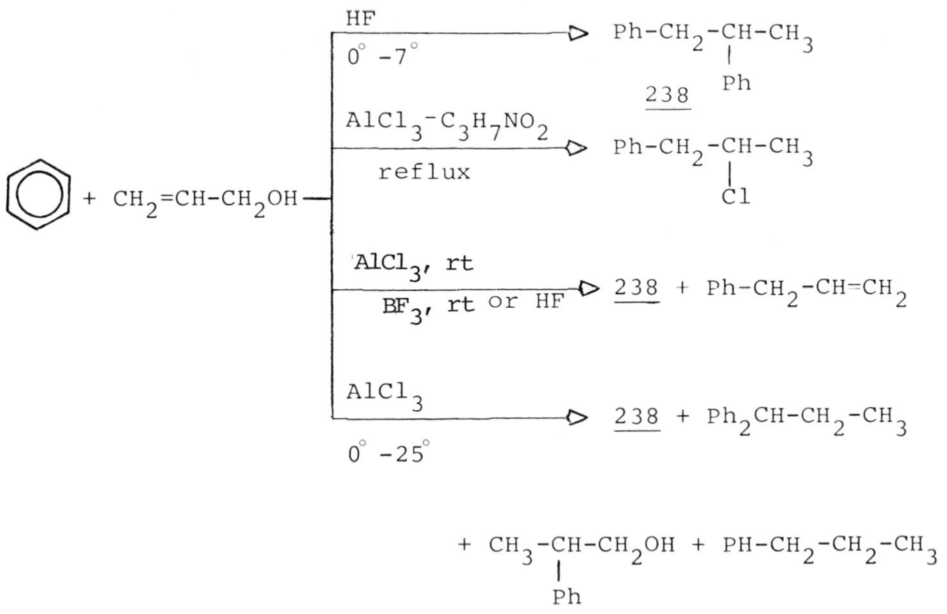

$$\text{C}_6\text{H}_6 + CH_2{=}CH{-}CH_2OH \begin{cases} \xrightarrow[0^\circ - 7^\circ]{HF} Ph{-}CH_2{-}CH(Ph){-}CH_3 \quad \underline{238} \\ \xrightarrow[\text{reflux}]{AlCl_3{-}C_3H_7NO_2} Ph{-}CH_2{-}CH(Cl){-}CH_3 \\ \xrightarrow[BF_3,\ \text{rt or HF}]{AlCl_3,\ \text{rt}} \underline{238} + Ph{-}CH_2{-}CH{=}CH_2 \\ \xrightarrow[0^\circ - 25^\circ]{AlCl_3} \underline{238} + Ph_2CH{-}CH_2{-}CH_3 \end{cases}$$

$$+ CH_3{-}CH(Ph){-}CH_2OH + PH{-}CH_2{-}CH_2{-}CH_3$$

Scheme 19

In 1968, the $ZnCl_2$-catalyzed alkylation of *m*-xylene, cumene, and chlorobenzene with allyl alcohol was reported to give mixtures of mono- and dialkylated products [263]. With *m*-xylene, for example, the products shown in Eq. (115) were observed.

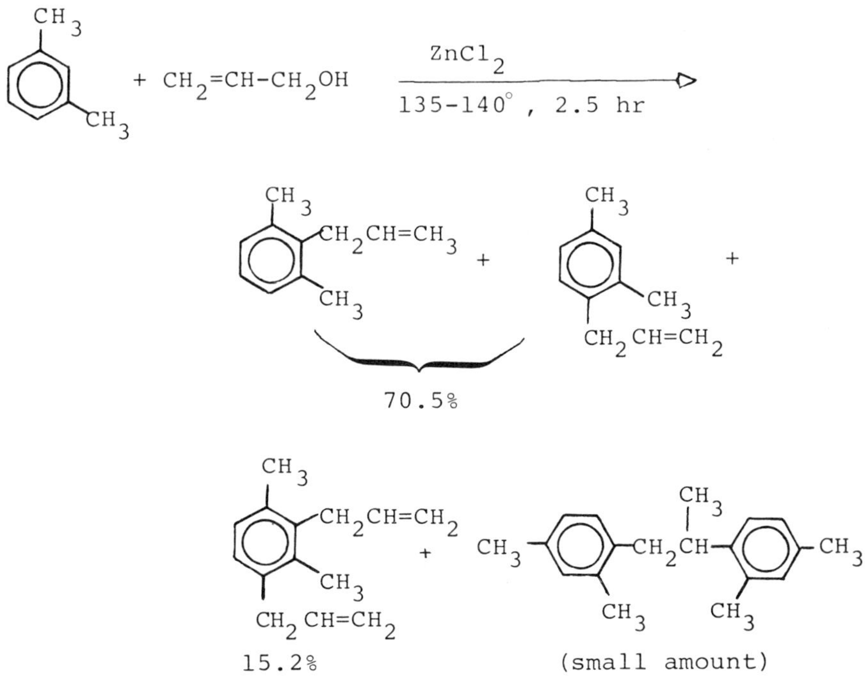

$$m\text{-}(CH_3)_2C_6H_4 + CH_2{=}CH{-}CH_2OH \xrightarrow[135\text{-}140^\circ,\ 2.5\ \text{hr}]{ZnCl_2}$$

$$2,6\text{-}(CH_3)_2C_6H_3CH_2CH{=}CH_3 + 2,4\text{-}(CH_3)_2C_6H_3CH_2CH{=}CH_2 \quad (70.5\%) + \tag{115}$$

$$(CH_3)_2C_6H_2(CH_2CH{=}CH_2)_2 \ (15.2\%) + CH_3C_6H_3(CH_3){-}CH_2CH(CH_3){-}C_6H_3(CH_3)CH_3 \ \text{(small amount)}$$

Similar reactions of biphenyl, diphenylmethane, diphenylethane, and cyclohexylbenzene with allyl alcohol and $ZnCl_2$ gave similar mixtures of the *ortho*- and *para*-monoalkylates [263].

The alkylation of naphthalene with allyl alcohol was reported to give monoallylnaphthalenes in up to 68% yield with $ZnCl_2$, but in only 15 to 20% yield with $FeCl_3$ or $SnCl_4 \cdot 6H_2O$ [264]. In all cases diallylated naphthalenes were obtained in varying amounts depending on the ratio of reactants.

The reaction of 9-alkylfluorenes with allyl alcohol and $ZnCl_2$ gave the corresponding 3-allylfluorenes in 85 to 95% yield (Eq. 116) [265]. The structures of the products were confirmed by infrared as well as by oxidation to 3-carboxy-9-fluorenone.

$$\text{9-R-fluorene} + CH_2{=}CH{-}CH_2OH \xrightarrow{ZnCl_2} \text{3-}(CH_2CH{=}CH_2)\text{-9-R-fluorene} \quad (85\text{-}95\%) \tag{116}$$

$R = CH_3, C_2H_5, C_3H_7$ and C_4H_9

The condensation of phenols and their ethers with various allylic alcohols and ethers have also been investigated. The results were similar to those found for arenes. In some cases, depending on reaction conditions, the resulting *ortho*-allylphenols may undergo intramolecular ring closure to form coumarans.

In 1931, Neiderl et al. [266] reported that the sulfuric acid-catalyzed reactions of phenol and *m*-cresol with allyl alcohol, ether, or acetate in the cold resulted in the production of *o*-isopropenylphenol and 3-methyl-6-isopropenylphenol, respectively (Eqs. 117 and 118).

$$C_6H_5OH + CH_2{=}CH{-}CH_2OR \xrightarrow{H_2SO_4} o\text{-}(CH_2{=}C(CH_3))C_6H_4OH \tag{117}$$

$$m\text{-}CH_3C_6H_4OH + CH_2{=}CH{-}CH_2OR \xrightarrow{H_2SO_4} \text{3-methyl-6-}(CH_2{=}C(CH_3))\text{phenol} \tag{118}$$

$R = H, CH_2{-}CH{=}CH_2$ or $COCH_3$

Two years later, Niederl and Storch reported further studies on the Friedel-Crafts alkylation of phenols with allyl alcohols in which they demonstrated that certain conditions favored cyclization to coumarans [267]. Thus, at elevated temperatures and in the presence of acetic acid as solvent, phenols gave methyl- and benzylcoumarans upon reaction with allyl and benzyl alcohols, respectively.

Allylation with or without accompanying allylic rearrangement has been observed in Friedel-Crafts alkylations of phenols with allylic alcohols. The condensation of phenol with cinnamyl alcohol in the presence of KU-2 cation exchange resin proceeded with no rearrangement to give the mixture shown by Eq. (119) in 75% total yield [268]. In this mixture compounds <u>239</u> and <u>240</u>

OH + $PhCH=CHCH_2OH$ —(5 wt % KU-2)→ OH, $CH_2CH=CHPh$ (<u>239</u>)

OH, $CH_2CH=CHPh$ (<u>240</u>) + OH, $CH_2CH=CHPh$, $CH_2CH=CHPh$ + O, Ph (119)

comprised 64 to 72% of the product mixture. In contrast to this, completely or partially rearranged products were observed from the reactions of phenylvinylcarbinol with *m*-cresol in the presence of polyphosphoric acid (Eq. 120)

OH, CH_3 + OH, $PhCHCH=CH_2$ —(PPA)→ OH, CH_3, $CH_2CH=CH-Ph$ (120)

[269], of 3-hydroxy-1-pentene [270], and 3-hydroxy-1-hexene [271] with phenol, *o*-cresol, and *m*-cresol in the presence of phosphoric acid, and of 4-hydroxy-2-octene with phenol and *p*-chlorophenol in the presence of boron trifluoride etherate (Eq. 121) [272].

OH, Cl + OH, $CH_3CH=CH-CH-(CH_2)_3CH_3$ —(BF_3 etherate; rt, overnight)→ (121)

OH, CH_3, $CH-CH=CH-(CH_2)_3CH_3$, Cl + OH, $CH=CHCH_3$, CH, $(CH_2)_3CH_3$, Cl

The alkylation of phenols with allylic alcohols in aqueous acids has been performed in studies connected with the biogenesis of the naturally occurring *ortho*-γ,γ-dimethallylphenols, cinnamylphenols, and neoflavanoids. In one study, Cardillo et al. reported the alkylation of 5-alkylresorcinols with monoterpenoid allylic alcohols in the presence of 5% citric acid solution or *p*-toluenesulfonic acid at room temperature [273]. Analysis of the products by various techniques showed that they consisted of nonrearranged mixtures of mono- and dialkylated resorcinols together with the cyclic ethers resulting from intramolecular ring closure of the monoalkylates under the influence of acid. These may be illustrated by the results of alkylation of orcinol (242) with piperitol (241, Eq. 122).

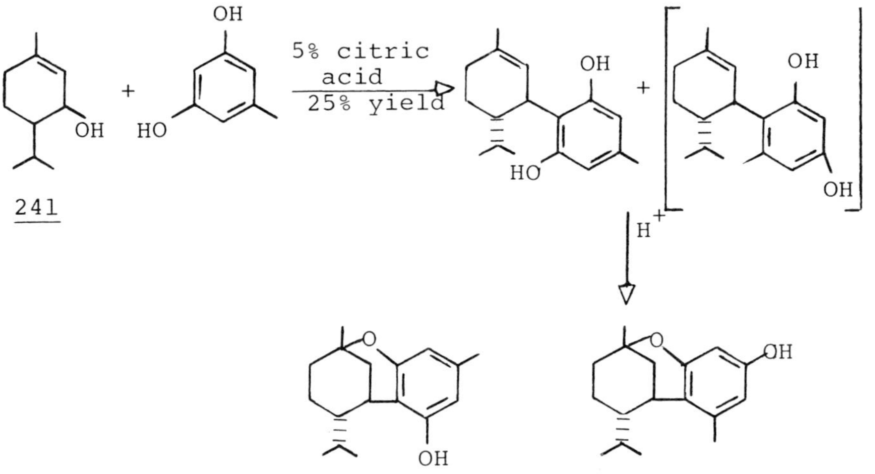

241 + orcinol $\xrightarrow[\text{25\% yield}]{\text{5\% citric acid}}$ (products) $\xrightarrow{H^+}$ (cyclic ethers) (122)

As in alkylations with allyl alcohols, the alkylations of phenols with tertiary and secondary dienic alcohols proceed with accompanying allylic type rearrangements to yield *para*-substituted phenols, corresponding not to the initial tertiary or secondary alcohols but to the primary alcohols isomeric with them [274-278]. For example, the reaction of guaiacol (243) with 5-hydroxy-6-methyl-1,3-heptadiene (244) and crystalline phosphoric acid gave 6-methyl-1-(3-methoxy-4-hydroxyphenyl)-2,4-heptadiene (245, Eq. 123) [278].

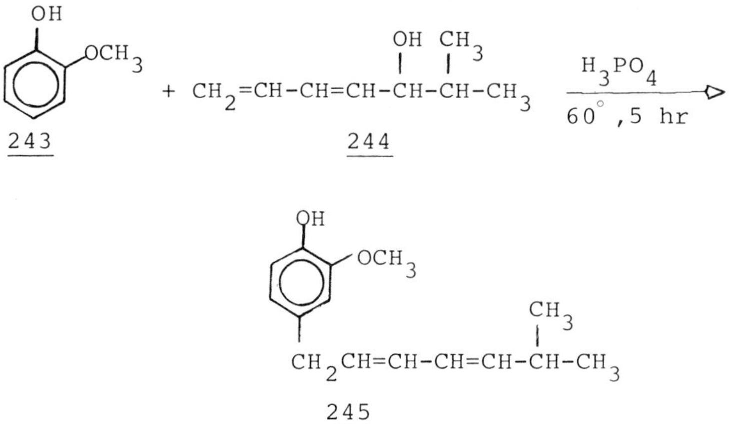

243 + $CH_2{=}CH{-}CH{=}CH{-}CH(OH){-}CH(CH_3){-}CH_3$ (244) $\xrightarrow[60^\circ, 5\ hr]{H_3PO_4}$ (123)

245: 4-hydroxy-3-methoxyphenyl–$CH_2CH{=}CH{-}CH{=}CH{-}CH(CH_3){-}CH_3$

The reaction of phenol and of *o*- and *m*-cresols with 5-hydroxy-1,3-hexadiene in the presence of phosphoric acid gave analogous products [277].

The observation of rearranged products during Friedel-Crafts alkylations with allylic alcohols was attributed by some workers to initial isomerization of the starting alcohol (e.g., $PhCH(OH)CH=CH_2 \rightarrow PhCH=CH-CH_2OH$) [269]. A more plausible explanation, however, can be given in terms of the initial formation of a resonating allyl (or dienyl) carbocation intermediate, followed by aromatic attack on the least sterically hindered positive carbon (schemes 20 and 21).

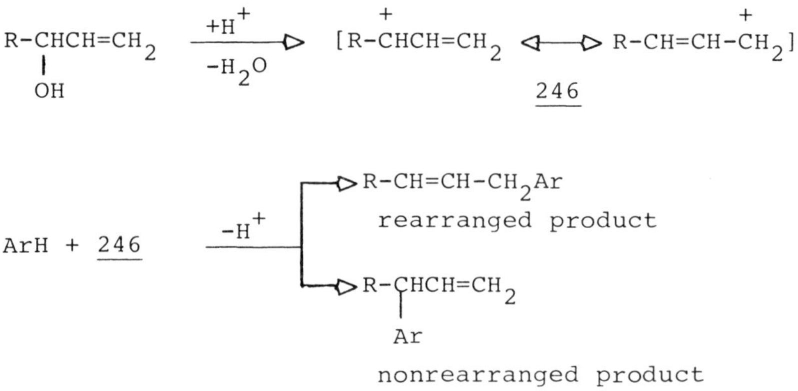

$$R-CH(OH)CH=CH_2 \xrightarrow[-H_2O]{+H^+} [R-\overset{+}{C}HCH=CH_2 \longleftrightarrow R-CH=CH-\overset{+}{C}H_2] \quad \underline{246}$$

$$ArH + \underline{246} \xrightarrow{-H^+} \begin{cases} R-CH=CH-CH_2Ar & \text{rearranged product} \\ R-CH(Ar)CH=CH_2 & \text{nonrearranged product} \end{cases}$$

Scheme 20

$$ArH + \underline{247} \xrightarrow{-H^+} \begin{cases} R-CH=CH-CH=CHCH_2Ar & \text{rearranged product} \\ R-CH(Ar)-CH=CH-CH=CH_2 & \text{nonrearranged product} \end{cases}$$

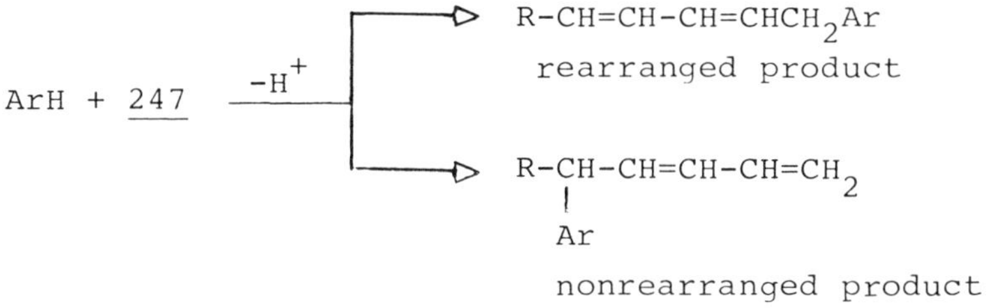

Scheme 21

Compared to alkylations with allyl alcohols, very few cases of alkylations with allyl ethers are known, mostly involving phenols, and these were found to proceed mainly by addition to the double bond (Eq. 124) [279]. However, displacement of the alkoxy group has been observed (Eq. 125) [266].

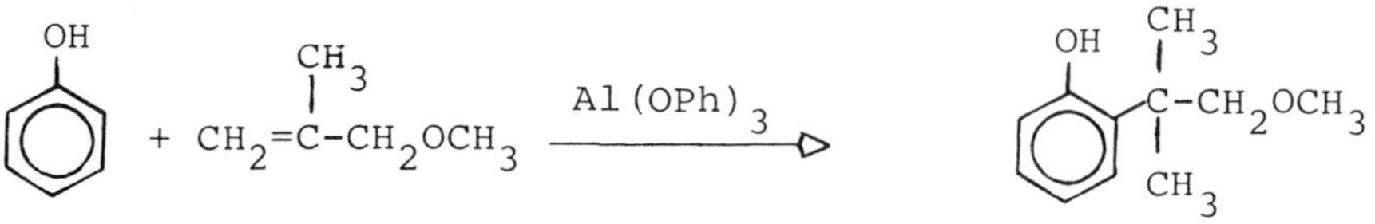

$$C_6H_5OH + CH_2=C(CH_3)-CH_2OCH_3 \xrightarrow{Al(OPh)_3} o\text{-}HOC_6H_4-C(CH_3)_2-CH_2OCH_3 \quad (124)$$

$$C_6H_5OH + CH_2{=}CH{-}CH_2OC_2H_5 \xrightarrow{H_2SO_4} o\text{-}HOC_6H_4{-}CH(CH_3){-}CH{=}CH_2 \quad [235] \qquad (125)$$

3. Alkylations with long-chain unsaturated alcohols and chloro alcohols

A study was made of the reactions of 2-, 4-, and 5-chlorooctan-1-ols, 2-chlorohexan-1-ol, and 4- and 7-octen-1-ols with benzene in the presence of $AlCl_3$ [280]. Isomeric x-phenylalkan-1-ols were obtained from both unsaturated alcohols and chloroalcohols, but the isomer distribution differed considerably depending on the starting compounds. 2-Chloroalkan-1-ol yielded x-phenylalkan-1-ols, in which x was between 2 and the penultimate carbon number, but neither 2- nor 3-phenyloctan-1-ol was detected from the reactions of a mixture of 4- and 5-chlorooctan-1-ol or reactions of 4- and 7-octen-1-ol. Both the cation isomerization and phenyl migration toward the 2- or 3-position appeared to be unfavorable. The experimental data suggested that the terminal OH group had little effect on reaction of the Cl or double bond at positions remote to it.

III. ALKLYATIONS OF AROMATICS WITH DIENES

In the past four decades a large number of investigations were conducted in which the alkylation of aromatic compounds with dienes was the subject of concern. The aromatics employed were mostly arenes, phenols, and phenol ethers. The following is a summary of some representative examples, chosen to illustrate the effect of reaction variables on product composition.

A. Alkylation of Arenes with Dienes

The early work on the alkylation of arenes by dienes through 1962 has been reviewed previously [221].

An investigation by Mamedaliev et al. of the alkenylation of benzene and toluene with butadiene in the presence of 95% H_2SO_4 indicated that the arene molecule was added to the butadiene molecule in the 1,4 positions and, in the case of toluene, that 1,2 addition also took place [281]. The butyenyltoluenes obtained consisted of 70% *ortho* and 30% *para* isomers.

Dienes higher than butadiene were also used to alkylate arenes. For example, the acid-catalyzed condensation of 2 mol of isoprene and 1 mole of benzene was reported to yield hydrindacene products accompanied by 1,1-dimethylindan in lower yield [282,283]. Guseinov et al. determined the best conditions for preparing isopentyl derivatives by the reaction of isoprene with toluene, xylenes, and trimethylbenzenes in the presence of H_2SO_4 [284].

4-Vinyl-1-cyclohexene was reported to give 1,2,3,4,4a,9,10,10a-octahydrophenanthrene with benzene and isopropyloctahydrophenanthrene with cumene in the presence of Friedel-Crafts catalysts [285]. Also, piperylene (1,3-pentadiene) was reported to alkylate toluene to give 1-(p-tolyl)-2-pentene (<u>246</u>) in the presence of BF_3-H_2O [286], BF_3-H_3PO_4 [287], or BF_3-CH_3CO_2H [288] catalysts (Eq. 126).

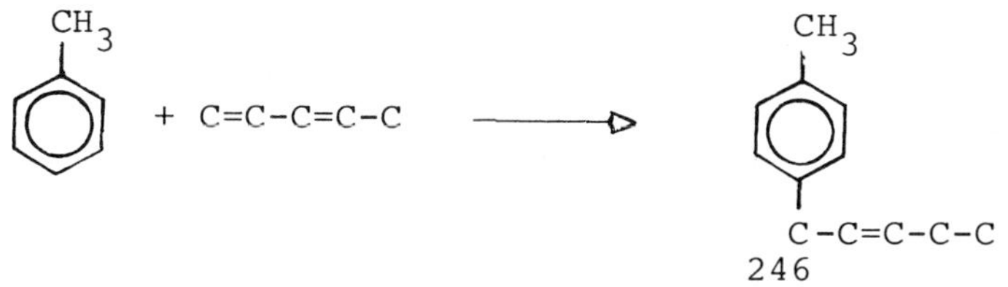

CH_3 + C=C-C=C-C → CH_3 / C-C=C-C-C (126)

246

In 1963, Wood and Angiolini described the sulfuric acid-catalyzed reactions between 1,3-dienes and arenes [283]. They noted that cyclialkylations to indans readily occurred during the condensation of isoprene or 2,3-dimethyl-1,3-butadiene with arenes having at least two adjacent empty positions. Thus benzene and isoprene produced a low yield of 1,1-dimethylindan, plus a considerable amount of what was believed to be 1,1,7,7-tetramethyl-5-hydrindacene (247).

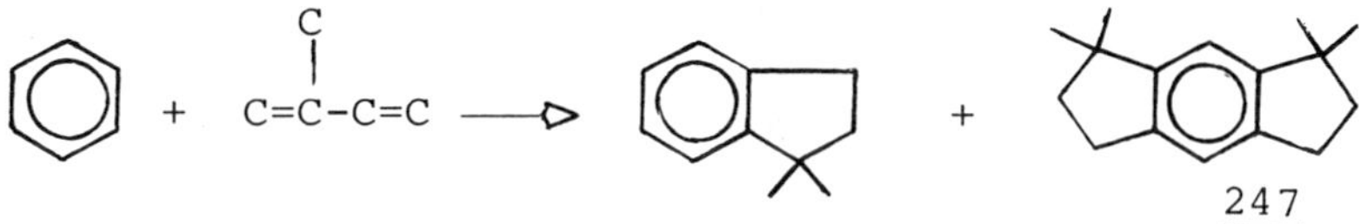

+ C=C(C)-C=C → (minor) + 247 (major)

From monoalkylbenzenes and isoprene there were obtained 6-alkyl-1,1-dimethylindans, with cumene and *t*-butylbenzene giving the best yields (53 and 61%) of 6-isopropyl and 6-*t*-butyl-1,1-dimethylindans, respectively. The condensation of xylenes and other di- and trialkylbenzenes with isoprene and with 2,3-dimethyl-1,3-butadiene was described as proceeding "very well" to yield substituted indans.

Although no definite decision was made on the exact structure of the products of reaction of *p*-cymene with isoprene and 2,3-dimethyl-1,3-butadiene, Wood and Angiolini favored structures 248 and 249 on the basis of mechanistic considerations, and 250 was assumed to be the product from pseudocumene and isoprene.

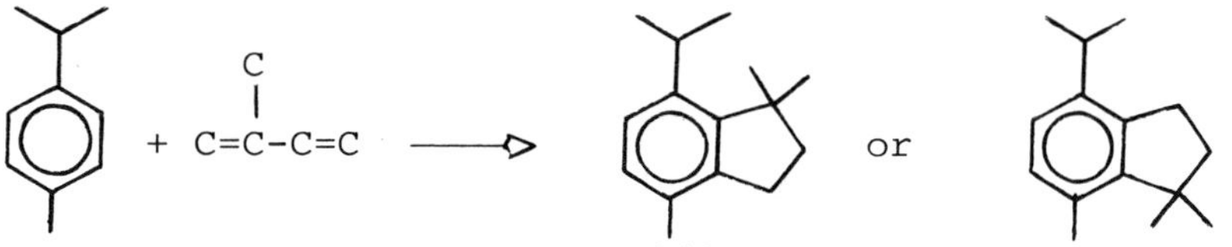

+ C=C(C)-C=C → 248 or (127a)

+ C=C(C)-C(C)=C → 249 or (127b)

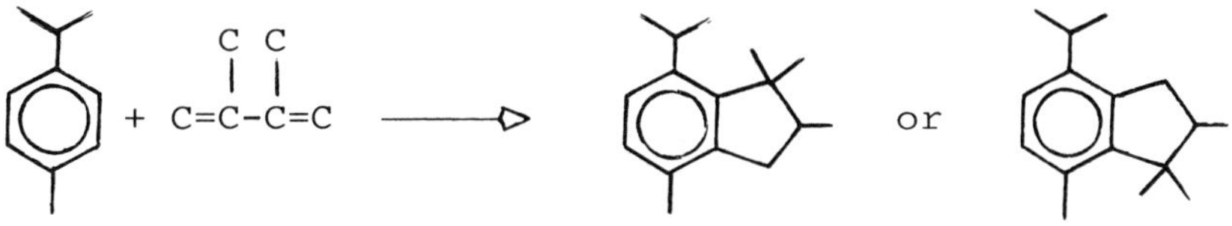

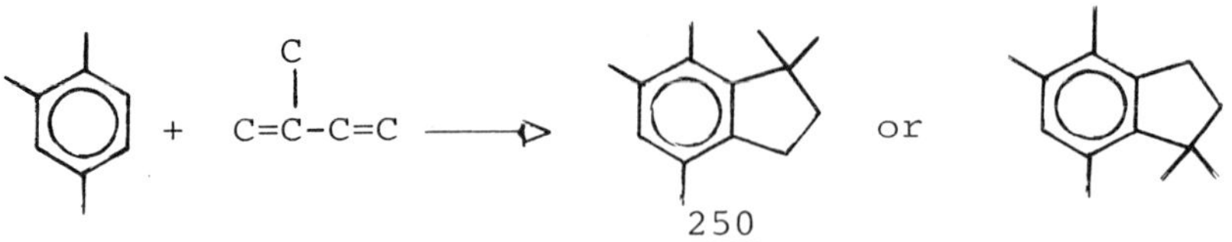

C=C(C)-C=C → 250 or

250

70% yield

(127c)

In 1966, good evidence for the structure 250 as that of the product from pseudocumene and isoprene was provided by Eisenbraun and co-workers [289,290] by carrying out the transformations shown in scheme 22.

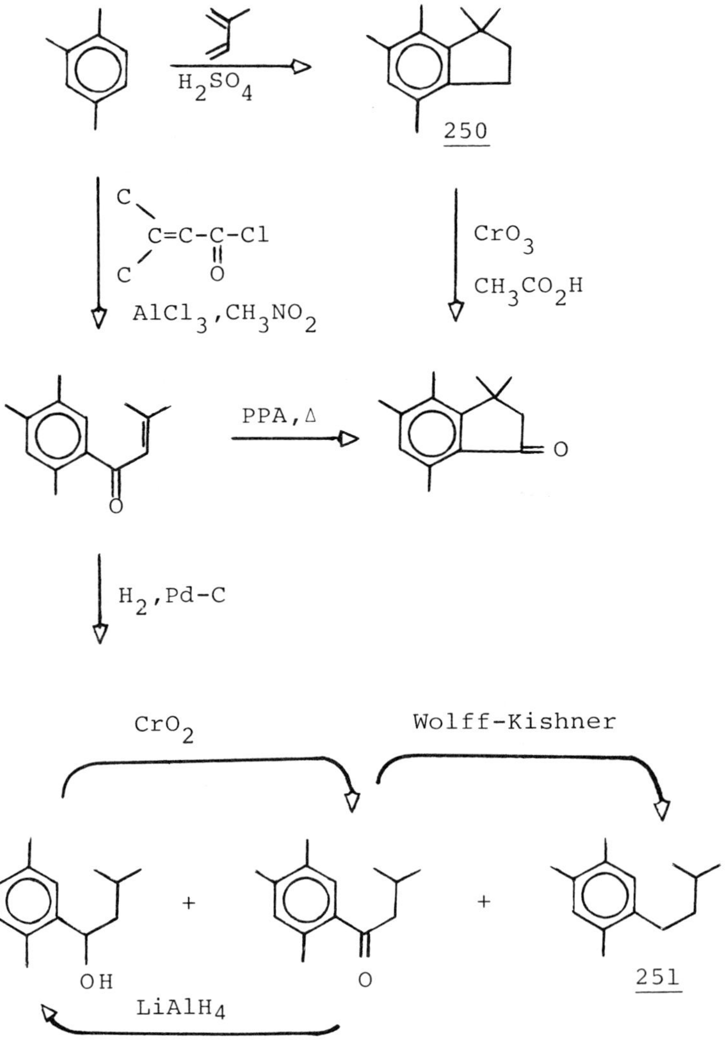

H_2SO_4

250

$(C)_2C=C\text{-}C(=O)\text{-}Cl$, $AlCl_3, CH_3NO_2$

CrO_3, CH_3CO_2H

PPA, Δ

H_2, Pd-C

CrO_2

Wolff-Kishner

OH + O + 251

$LiAlH_4$

Scheme 22

The structure of the key reaction product 251 was rigorously established by careful analysis of its infrared, mass, and NMR spectra.

In 1968, Eisenbraun et al. published a detailed account of the cyclialkylation of a number of mono-, di-, and trisubstituted benzenes with isoprene in the presence of protonic acids [291]. In agreement with other investigators, they noted that isoprene and arenes having unoccupied vicinal positions combine in the presence of H_2SO_4 to give indans. They also noted that the indan formation was complicated by competing reactions such as polycyclialkylation to form hydrindacenes and other higher alkylation derivatives, sulfonation of the aromatic ring, polymerization and reduction by hydride transfer of isoprene and other intermediate alkenes. Furthermore, the authors found that optimum conditions for monocyclialkylation required the use of excess arene, the rapid addition of isoprene, and the use of sulfuric acid concentrations ranging from 96% at below -15°C to 75% at 30 to 40°C.

The products of alkylation of *o*-, *m*-, and *p*-xylene with isoprene in the presence of H_2SO_4 and the mechanisms proposed by Eisenbraun et al. to account for their formation are presented in schemes 23 and 24. By reference to these schemes, one can see that the products consisted essentially of mixtures of mono- and dicyclialkylated products.

The reaction with *p*-xylene gave 59% of 251 and 15% of an isomeric mixture of the hydrindacenes 252 and 253 (scheme 23).

From *o*-xylene, the product consisted of a mixture of three tetramethylindans, 254-256, the three hexamethylhydrindacenes, 257-259, and an additional unidentified compound thought possibly to be an alkenylation or alkylation product of 256 at positions 4 or 7 (scheme 24).

m-Xylene readily combined with isoprene to give a 68% yield of 1,1,4,6-tetramethylindan 260, and smaller amounts of 253 and 258. The latter two compounds were also produced from *o*- and *p*-xylene, indicating that methyl reorientation may precede or accompany cyclialkylation. That 260 was the likely precursor to 253 and 258 was established by treating 260 with additional isoprene in the presence of H_2SO_4 to obtain a mixture of the two hexamethylhydrindacenes.

H^+ H^+ $-H^+$ H_2SO_4

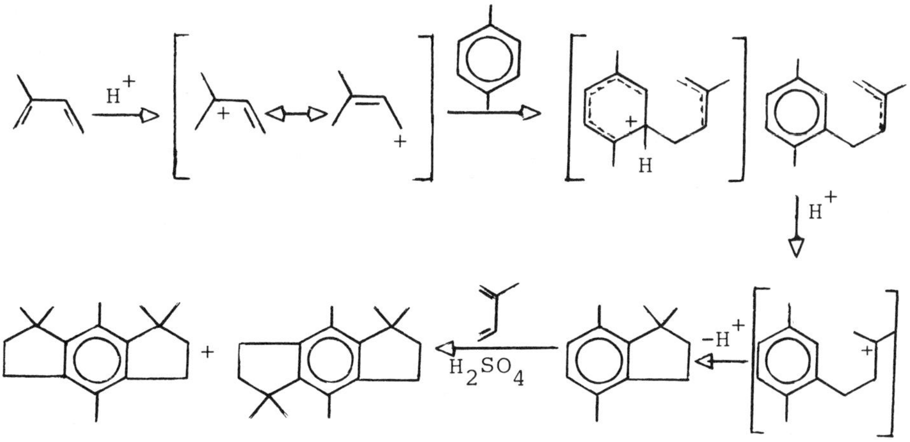

253   252   251

Scheme 23

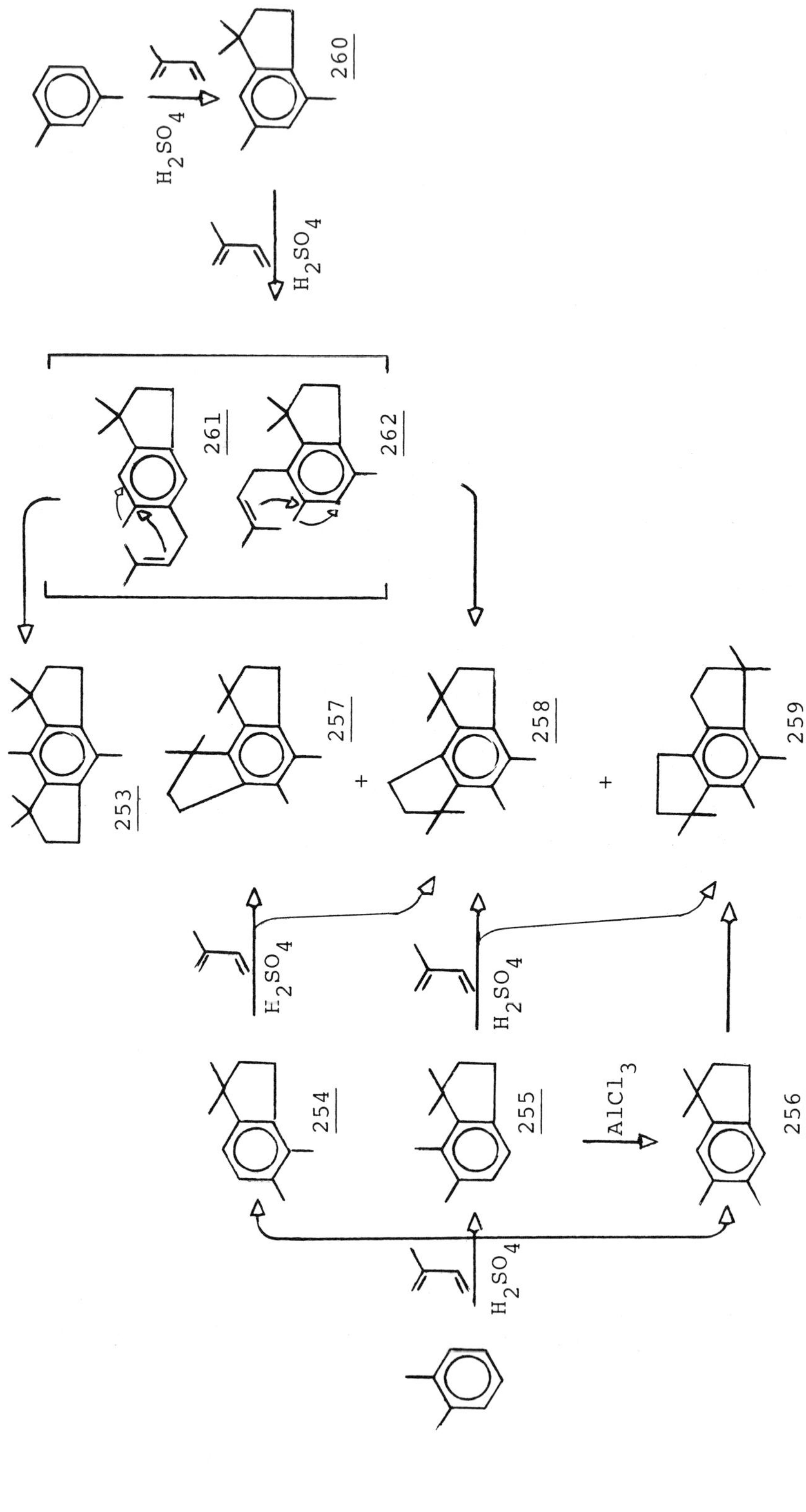

Scheme 24

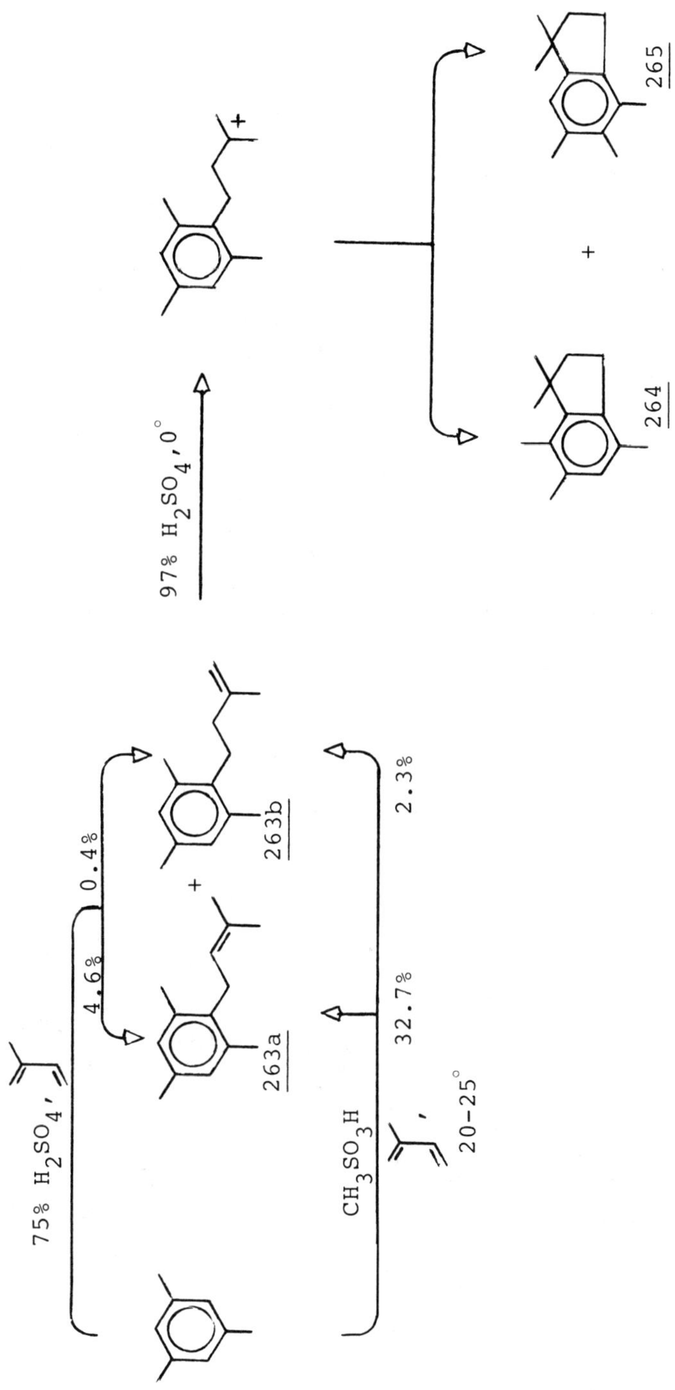

75% H_2SO_4,

4.6%

0.4%

263a

263b

CH_3SO_3H,

20-25°

32.7%

2.3%

97% H_2SO_4, 0°

264

265

Scheme 25

The reaction between mesitylene and isoprene was shown by Eisenbraun et al. to be critically dependent on the acidic strength of the catalyst used [291]. When mesitylene was condensed with isoprene in the presence of methanesulfonic acid, a 35% yield of the products 263a and 263b in the ratio of 15:1 resulted (scheme 25). A mixture of this approximate composition was also obtained in about 5% yield when sulfuric acid (93, 85, or 75%) was substituted for methanesulfonic acid. However, the two compounds 263a and 263b were cyclized to the pentamethylindans 264 and 265 (1:2 ratio) in 62% yield by contact with 97% H_2SO_4 at 25°C for 3 hr. It should be noted that the formation of 264 and 265 involves methyl reorientations preceding or accompanying the cyclialkylations, as was observed in the reactions of the xylenes with isoprene. These methyl reorientations are reminiscent of those observed by Roberts and Shiengthong in the alkylation of mesitylene with propyl chlorides [292] (see Sec. 3.III.A).

Eisenbraun and co-workers also cyclialkylated a number of monoalkylbenzenes with isoprene in the presence of 93% H_2SO_4. The structures and ratios of the cyclialkylation products obtained from these reactions are shown in Table 2.

It is of interest to note that examination of the products above as well as those previously obtained by xylenes and pseudocumenes indicates that they resulted from initial attack at positions *ortho* or *para* to alkyl groups, forming only those isomers ordinarily predicted for electrophilic substitution reactions of alkylbenzenes.

Shadan and Flanagan investigated the reactions of a series of α,ω-dienes with benzene in the presence of $AlCl_3$ [293]. 1,3-Butadiene, 1,4-pentadiene, 1,5-hexadiene, 1,7-octadiene, 1,8-nonadiene, 1,9-decadiene, and 1,11-dodecadiene in 20:1 benzene/diene molar ratio gave mono- and disubtituted alkanes, mono- and disubstituted tetralins, and 1,1-diphenylalkanes. No condensed polynuclear compounds were observed.

Table 2. Cyclialkylation Products[a] from Alkylbenzenes

		Products	
Starting mono-alkylbenzene	R	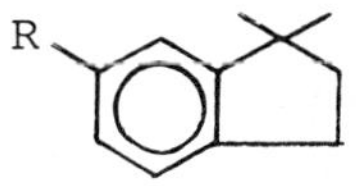 % [b]	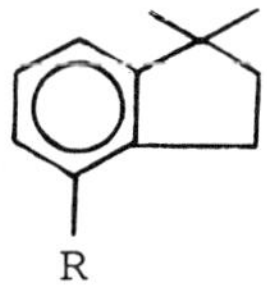 % [b]
Toluene	CH_3	68-71	20-32
Ethylbenzene	CH_2CH_3	70-83	17-21
Isopropylbenzene	$CH(CH_3)_2$	88-92	8-12
t-Butylbenzene	$C(CH_3)_3$ [c]	98	2

[a]Absolute yields of indans ranged from 55 to 70%.
[b]Relative percent yields obtained from ratios of gas chromatography peaks.
[c]An 8% yield of 1,4-di-*t*-butylbenzene was observed.

Table 3. Products Isolated from the Phenol-Isoprene Reaction

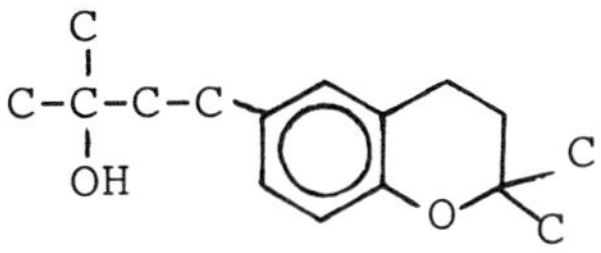

266 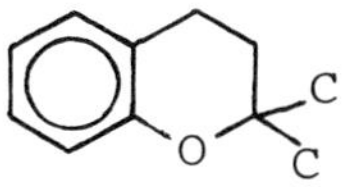 267 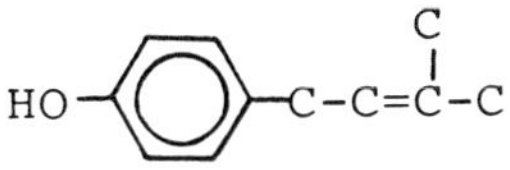 268

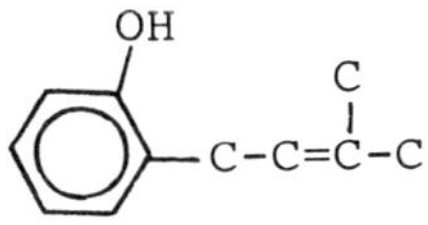

269 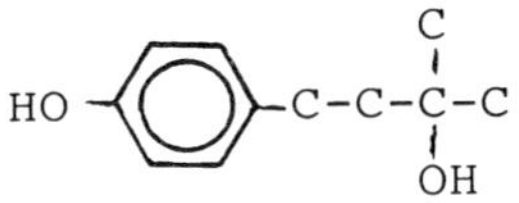 270 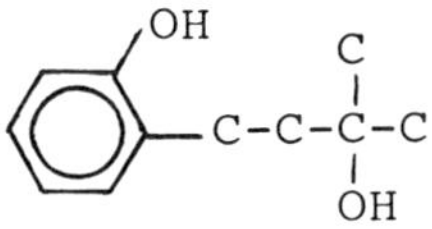 271

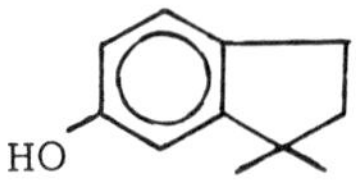

272 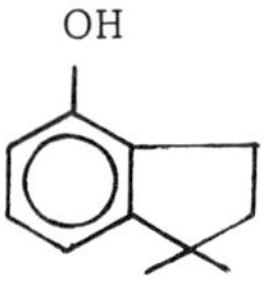 273 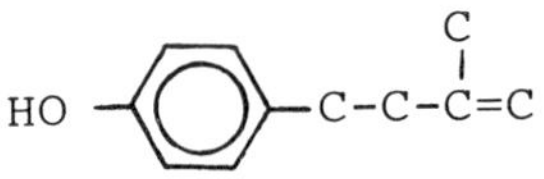 274

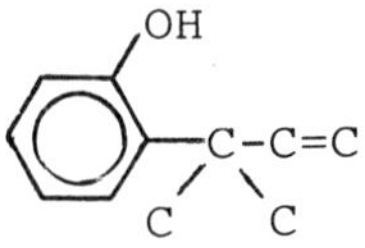

275 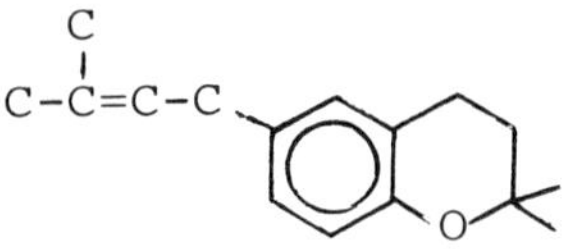 276

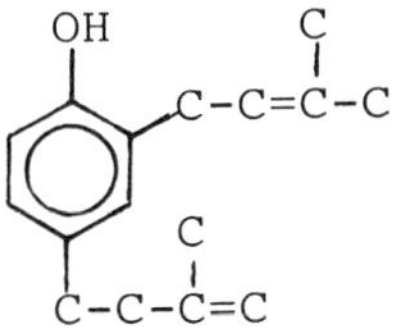

277 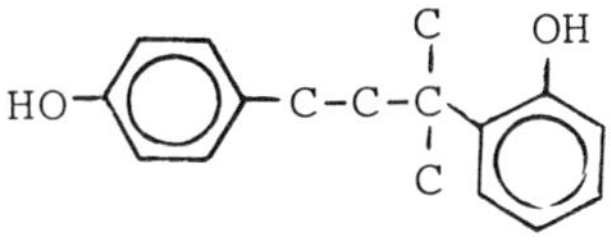 278

B. Alkylation of Phenols and Phenol Ethers with Dienes

As in the case of arenes, the Friedel-Crafts alkylation of phenols or phenol ethers with dienes results in mixtures consisting of mono- and higher alkyl derivatives as well as diarylalkanes and indans. Besides these products, the reaction with phenols has also been found to yield chromans. Because of the close structural relationship between chromans and both tocopherol and vitamin K_1, numerous studies on the condensation of phenols and hydroquinones with dienes have been made. A summary of the early work up until

1962 is given in the Olah monograph [221]; examples of the more recent work in this area will be described here.

1. Reactions of branched dienes with phenols and phenol ethers.

Using modern methods for product analysis, chemists in the 1960s were able to demonstrate that the products obtained from these reactions are more complex than previously indicated. For example, according to Bader and Bean [294], the alkylation of phenol with isoprene in the presence of 71% H_3PO_4 gave compounds 266-271 (Table 3). According to Vdovtsova, however [295], the same reaction gave compounds 268-275, along with four isomers of 3-methyl-1,2-bis(methoxyphenyl)butane. Again, the same alkylation was reported to give compounds 266, 268, 269, and 274 with the cation-exchange resin KU-2 catalyst [296], and compounds 267, 268, and 276-278 with aluminum phenoxide catalyst [297]. The product distribution in the presence of the latter catalyst varied considerably with reaction conditions. The formation of compounds 267-269 and 276-278 during the aluminum phenoxide-catalyzed alkylation of phenol was explained in terms of the series of transformations of scheme 26.

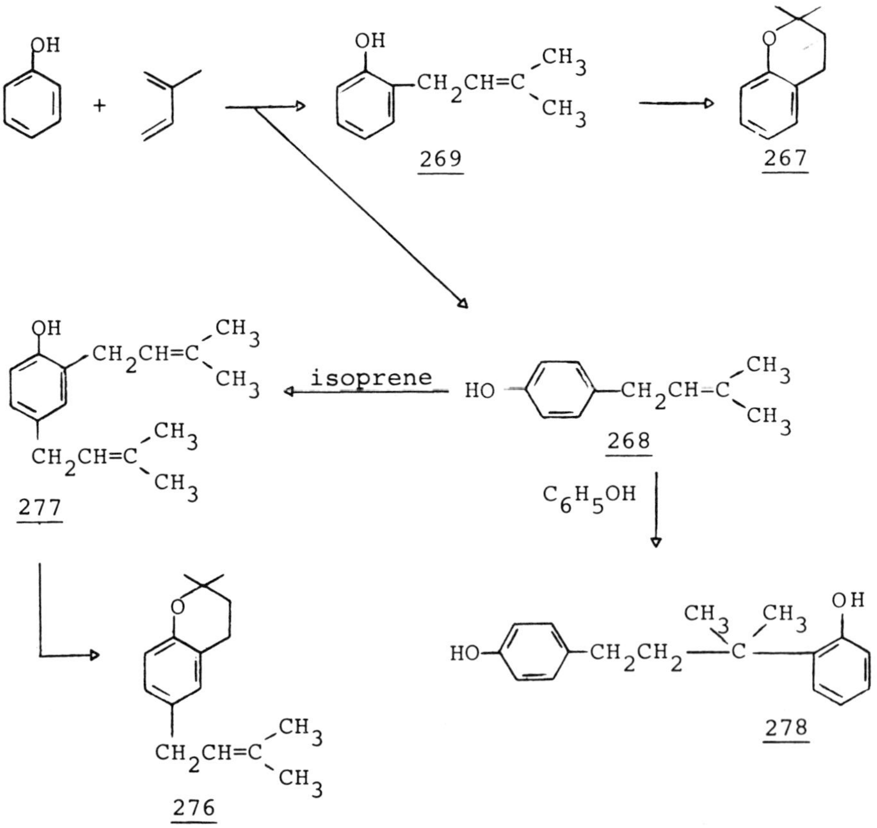

OH
+
OH
$CH_2CH=C(CH_3)_2$
269
267
OH
$CH_2CH=C(CH_3)_2$
$CH_2CH=C(CH_3)_2$
277
isoprene
HO
$CH_2CH=C(CH_3)_2$
268
C_6H_5OH
$CH_2CH=C(CH_3)_2$
276
HO
CH_2CH_2
CH_3 CH_3
C
OH
278

Scheme 26

Possible mechanisms for the various steps of scheme 26 were discussed and supported by additional experimental results.

In his investigations on the alkylation of phenols and their ethers with 1,3-dienes and their hydrochlorides, Vdovtsova studied the condensation of isoprene with anisole [210], phenetole [298], and guaiacol [211] in the presence of phosphoric acid as catalyst. Using a combination of spectroscopic and chromatographic methods of analysis, this author found the reaction products to be complex mixtures consisting essentially of 1,1-, 1,4-, and 3,4-addition products, as well as cyclization products, in proportions depending on the reaction conditions employed. Examination of these results leads to the following general observations:

1. As usual, product composition varied considerably with temperature, ratio of reactants, and concentration of catalyst. In fact, by choice of conditions, it was possible to direct the reaction toward the formation of alkenyl derivatives in 70 to 80% yield.
2. No selectivity was observed with isoprene in the course of addition, but the direction of addition could be correlated with the distribution of the electron density in the isoprene molecule. Thus, in all cases and under all conditions, 1,4 addition predominated (60 to 90%) and 3,4 and 1,2 additions always occurred to a smaller extent.
3. The data from anisole and phenetole showed that in the alkenylation process mixtures of *ortho* and *para* isomers of 1,4 and 3,4 adducts were formed in which the amounts of total *ortho* and total *para* isomers were about equal.
4. As expected on the basis of the polarity of the substituting groups, the alkenylation of phenol and of guaiacol showed preferential *ortho* and *para* orientation in the nucleus with respect to the phenolic hydroxyl group. The content of the *para* isomers with respect to the methoxyl group in the reaction mixtures did not exceed 15 to 30%.
5. In general, the reactions proceeded stepwise, with formation of alkenylation (monoaddition) products in the first stage and of diaddition, cyclization, and polycondensation products in subsequent stages.

In addition to indan derivatives, which were formed in all cases by intramolecular alkylation on the aromatic ring, the reaction with guaiacol gave also chromans and coumarans resulting from attack on the hydroxyl group.

Diagrammatic representations of the chemical transformations involved in the production of the various compounds isolated form the anisole-isoprene, phenetole-isoprene, and guaiacol-isoprene reactions are given by Vdovtsova in comprehensive schemes which may be consulted in his papers.

In 1967, Mishiev et al. [299] found that the reaction of methyl, ethyl, propyl, isopropyl, *n*-butyl, isobutyl, *n*-pentyl, and isopentyl phenyl ethers with isoprene in the presence of sulfuric acid catalyst gave the corresponding *ortho*- and *para*-isopentyl derivatives in yields ranging from 16% with the *n*-pentyl ether to 75.9% with the methyl ether. The reaction can generally be represented by Eq. (128).

$$C_6H_5OR + CH_2{=}C(CH_3){-}CH{=}CH_2 \xrightarrow{H_2SO_4} o\text{-}RO{-}C_6H_4{-}C{-}C{=}C(C){-}C + p\text{-}RO{-}C_6H_4{-}C{-}C{=}C(C){-}C \qquad (128)$$

Later, in 1970, Srebrodol'skaya et al. [300] conducted a detailed study of the alkenylation of methyl, ethyl, propyl, and butyl phenyl ethers with 2,3-dimethyl-1,3-butadiene in the presence of H_3PO_4, BF_3-H_3PO_4, and BF_3-$O(C_2H_5)_2$ as catalysts. Using a combination of spectral, chromatographic, and chemical methods for product analysis, these authors made the following findings:

1. The alkenylation occurred mainly by 1,4-addition of the ether to the diene to form 1-(*o*- and *p*-alkoxyphenyl)-2,3-dimethyl-2-butenes, with predominance of the *para* isomer in accordance with Eq. (129).

OR-C6H5 + C=C(C)-C(C)=C ⟶ p-(C-C(C)=C(C)-C)C6H4OR (major) + o-(C-C(C)=C(C)-C)C6H4OR (minor) (129)

R = CH_3, C_2H_5, n-C_3H_7, n-C_4H_9

2. Partial 1,2-addition also occurred with formation of 4-(*o*- and *p*-alkoxyphenyl)2,3-dimethyl-1-butenes, as shown in equation 130, but the relative content of these products did not exceed 2%.

OR-C6H5 + C=C(C)-C(C)=C ⟶ o-(C-C(C)-C(C)=C)C6H4OR + p-(C-C(C)-C(C)=C)C6H4OR (130)

3. Variation of temperature in the range 20 to 30°C, of reaction time from 3 to 6 hr, and of amounts of catalyst in the range 0.05 to 1 mol per mole of diene had a small effect on the isomeric composition of the alkenylation product. Under these conditions, the content of *ortho* isomer in the product ranged between 19 and 29%.
4. The most active alkenylation catalyst was orthophosphoric acid, in the presence of which alkenylated methyl, ethyl, propyl, and butyl phenyl ethers were obtained in yields of 84, 78, 69, and 72%, respectively.
5. The best alkenylation conditions were a molar proportion of alkyl phenyl ether, diene, and catalyst of 10:1.1, a temperature of 20°C, and a reaction time of 6 hr.

2. Alkylation of phenols and their ethers with straight-chain and cyclic dienes

a. 1,3-Butadiene

As in the case of branched dienes, the main products of alkenylation of phenol and its ethers with straight-chain dienes have been shown in almost all cases to be those resulting from 1,4-addition of the aromatic to the diene. In 1942, a patent by Schaad [301] claimed that *p*-crotylphenol was the major constitutent of the monoalkenylphenolic fraction obtained in 36% yield from the high-temperature reaction of phenol with 1,3-butadiene in the presence of solid phosphoric acid. In 1951, Proell described the reaction of 1,3-butadiene with phenol and alkanesulfonic acid as yielding largely a mixture of butenylphenols, with some chromans [302]. In 1957, Bader reported that the reaction of phenol with 1,3-butadiene in the presence of a number of Friedel-Crafts catalysts yielded products consisting chiefly of *ortho*- and *para*-2-butenylphenols, mixed with lesser amounts of higher phenols and ethers [303]. This author also indicated that the reaction temperature and the catalyst composition directed the orientation. Thus, at 15 to 25°C, *para* substitution predominated with H_3PO_4-BF_3 or RSO_3H catalyst, but the *ortho* isomer predominated with aqueous H_2SO_4 catalyst. This observed dependence of the product composition on the anion associated with the carbocation led Bader to conclude that his alkenylation reactions did not involve a simple electrophilic attack on phenol by the free resonance-stabilized carbocation derived from the diene.

More recently, 1,3-butadiene was used to alkylate monohalobenzenes [303a]. Alkenylation under optimal conditions (50°C, halobenzenes-butadiene-95% H_2SO_4 = 6:1:0.15) gave a 26.3% yield of butenylchlorobenzene, including *o*, *p*, and *m* isomers in the ratio 52:43:5%. Bromobenzene gave a 20% yield of butenylbromobenzenes, including *o*, *p*, and *m* isomers, in the ratio 60:34:6%. The yield of monoalkenyl halobenzenes decreased in the order PhBr < C_6H_6 < $PhCh_3$ < $PhOCH_3$.

For α-ethoxynaphthalene, the best conditions for alkenylation with 1,3-butadiene to obtain a 41.6% yield of mono(*trans*-β-butenyl) derivative were a 4:1:0.12 molar ratio of α-ethoxynaphthalene-butadiene-H_2SO_4, 50.5°C, and an addition rate of 1,3-butadiene of 5.6 liters/hr [304].

b. 1,3-Pentadiene (Piperylene)

In 1939, Smith and co-workers studied the reaction of trimethylhydroquinone (279) with 1,3-pentadiene and tentatively assigned structure 280 to the product obtained under the conditions shown in Eq. (131) [305].

HO, CH_3, CH_3, OH, CH_3 (279) + C-C=C-C=C —[$ZnCl_2$, H_2SO_4, reflux, 1 hr]→ HO, CH_3, CH_3, O, CH_2CH_3, CH_3 (280) (131)

In 1961, Vdovtsova and Romanikhin proposed that the product was actually 281 because, as we shall see later, similar reactions between anisole or phenol and 1,3-pentadiene gave 4-(*p*-methoxylphenyl)-2-pentene (282) and 4-(*p*-hydroxyphenyl)-2-pentene (283), respectively, as the major products [306].

281: HO, CH_3, CH_3, O, CH_3, CH_3, CH_3

RO–C_6H_4–C(C)-C=C-C

282 R = CH_3

283 R = H

In the course of an extensive systematic exploration of the alkylation of aromatic compounds with 1,3-dienic hydrocarbons and their hydrochlorides, Vdovtsova and collaborators investigated the alkylation of phenol [306], anisole [307-311], phenetole [312-316], cresols and their methyl ethers, *o*-bromophenol and *o*-bromoanisole, *o*-chlorophenol and *o*-chloroanisole [317], *m*-chlorophenol and *m*-chloroanisole [318], guaiacol [214], catechol, and veratrole [215] with 1,3-pentadiene, using a variety of catalysts under different reaction conditions. Using IR, GLPC, and thin-layer chromatography for product analysis, these authors identified the main reaction products and provided detailed schemes suggested pathways for their formation. They also determined the proportions of the products in the resulting mixtures and, in tabulated forms, they listed numerous data on changes in product composition with variations in reaction conditions.

Examination of Vdovtsova's original papers leads to the following observations:

1. The yield and composition of the products in these reactions were highly dependent on variables such as type of catalyst, ratio of reactants, polarity of solvent, temperature, and duration of reaction. For example, in the reaction of anisole with 1,3-pentadiene, using the diene, PhOMe, and 100% H_3PO_4 in a molar ratio of 1:4:0.25 gave the best yield of pentenylanisoles [307,308,311], whereas using a diene/PhOMe/BF_3 molar ratio of 1:8:0.4 gave the best yield of dianisylpentane [309, 310]. Also, in alkylating phenetole with peperylene, the reaction temperature had a major effect on the yield, but a minor effect on the composition of the alkylation mixture. Thus a rise in temperature from 15°C to 38°C caused an increase in the yield of pentenylphenetole from 54% to 80%, whereas the content of the *ortho* isomer rose only from 9.9% to 11.6% [315].

2. The reactions proceeded as a *trans*-α,α-dimethylallylation in which the addition of aromatic to 1,3-pentadiene took place selectively at the 1,4 positions in accordance with the character of the distribution of electron density in a diene molecule, i.e., via the relatively more stable secondary allyl carbocation 284 (scheme 27). As to alkylations with 4-chloro-2-pentene, the reaction appeared as a selective replacement of the halogen atom.

3. Although the process proceeded mainly as an alkenylation, it was complicated by the secondary reactions of O and C cyclizations of the pentenyl

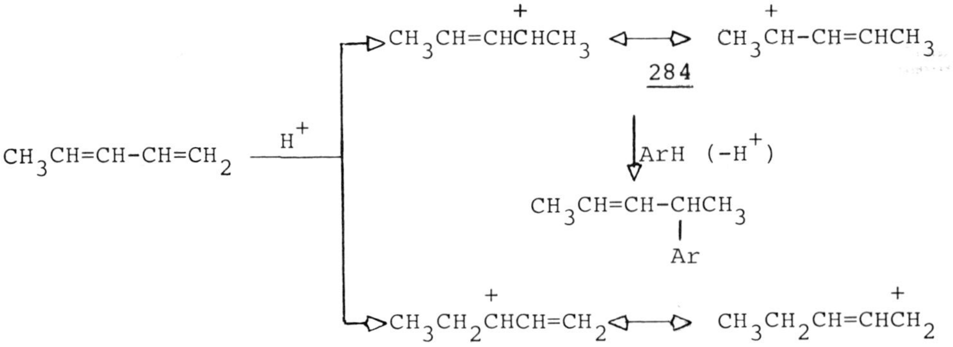

$$CH_3CH=CH-CH=CH_2 \xrightarrow{H^+} \begin{cases} CH_3CH=CH\overset{+}{C}HCH_3 \longleftrightarrow CH_3\overset{+}{C}H-CH=CHCH_3 \quad (\mathbf{284}) \xrightarrow{ArH\ (-H^+)} CH_3CH=CH-CH(Ar)CH_3 \\ CH_3CH_2\overset{+}{C}HCH=CH_2 \longleftrightarrow CH_3CH_2CH=CH\overset{+}{C}H_2 \end{cases}$$

Scheme 27

derivatives, which occurred readily, leading to substituted indans, chromans, and coumarans. The ratio of these cyclic derivatives in the products was shown to be dependent on the polarity of the medium, the nucleophilicity of the aromatic, the nature of the alkylating agent, and the severity of the reaction conditions. Generally, cyclization was favored by severe conditions and by the utilization of 4-chloro-2-pentene as the alkylating species.

4. Orientation of alkenylation was in agreement with the known electronic and steric effects of the substituents. Thus the major monocondensation products obtained from the reaction of 1,3-pentadiene with the monosubstituted phenol, anisole, and phenetole were the *p*-alkenyl derivatives 4-(*p*-hydroxyphenyl)- (285a), 4-(*p*-methoxyphenyl)- (285b), and 4-(*p*-ethoxyphenyl)-2-pentene (285c). The corresponding *ortho* isomers 286a-286c were produced in minor amounts.

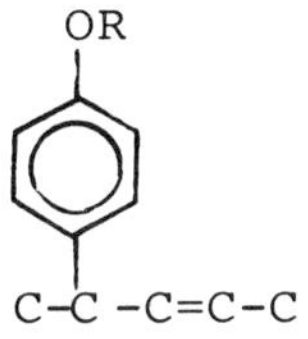

OR

C-C-C=C-C

285

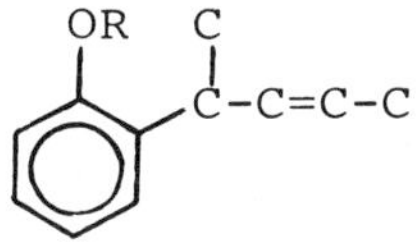

OR C

C-C=C-C

286

R = H (a)

R = CH_3 (b)

R = C_2H_5 (c)

In disubstituted aromatics, when the substituents did not orient to the same positions and there were free *ortho* and *para* positions on the ring (e.g., *o*- and *p*-cresols and their ethers, guaiacol, catechol, or veratrole), a predominant *para*-orienting effect of the hydroxyl (or methoxyl) group was observed. In most cases the *para*-alkenyl derivative comprised over 90% of the total monoalkenylation mixtures. On the other hand, when the substituents oriented to the same position, as in *m*-cresol or *m*-methylanisole, the amount of *ortho* isomers increased noticeably.

5. The yield of *ortho* isomer among the products of the alkenylation of the phenol ethers was far less than for the phenols themselves.

6. Free phenols were generally more reactive than their ethers. For example, while the alkenylation of catechol took place under mild conditions in the presence of the weak protonic acid catalyst H_3PO_4, giving good yields of alkenylation products, that of veratrole required either severe conditions or the use of more active catalysts, such as $AlCl_3$ or $AlCl_2$-H_2PO_4.

c. *Cyclic Dienes*

The reactions of cyclopentadiene and dicyclopentadiene with phenol using phosphoric acid as catalyst were investigated by Bader in 1953 [319], and the reaction of the latter diene and phenol with sulfuric acid catalyst by Bruson and Riener in 1946 [320]. Cyclopentadiene gave *o*- and *p*-2-cyclopentenylphenol; dicyclopentadiene gave some *o*-2-cyclopentenylphenol and isomeric solid and liquid products thought to be 287 and 288.

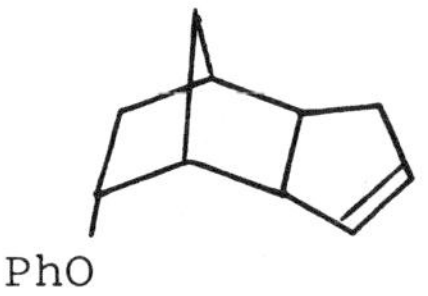

PhO

287

PhO

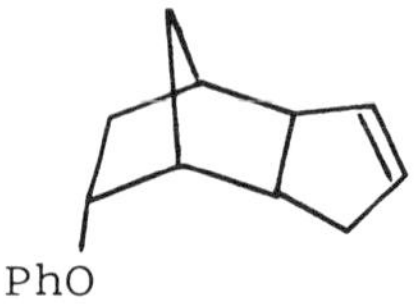

288

In 1974, the alkenylation of mono- and dimethoxybenzenes with 1,3-cyclohexadiene was studied by Nesterenko et al. [321]. 3-Cyclohexenyl derivatives were obtained in good yields using H_3PO_4 or BF_3-Et_2O catalysts.

IV. ALKYLATIONS WITH MISCELLANEOUS DI- AND POLYFUNCTIONAL REAGENTS

A. Stereospecific Alkylations with Optically Active Chloro Alcohols, Acids, and Esters

Stereospecific alkylations with dichloroalkanes were described earlier (Sec. I.A.6). Suga and co-workers also demonstrated stereospecific alkylations of benzene with optically active 3-chloro-1-butanol, 3-chlorobutanoic acid, and their esters in the presence of $AlCl_3$ [322]. All of these reactions were found to proceed in good optical yield with inversion of configuration at the chiral atom. Although the products were not racemized under the conditions of the reaction, the starting materials were to a considerable extent. Taking into account the optical purity of the starting material recovered before completion of the reaction, the net stereospecificity of each reaction was calculated to be about 90% except for the reaction of 3-chloro-1-butanol and its acetate. The high degree of stereospecificity was interpreted in terms of a mechanism involving the cyclic intermediates 289-292. The higher specificity of the reactions of the 3-chlorobutanoic acid and its ethyl ester was attributed to an en-

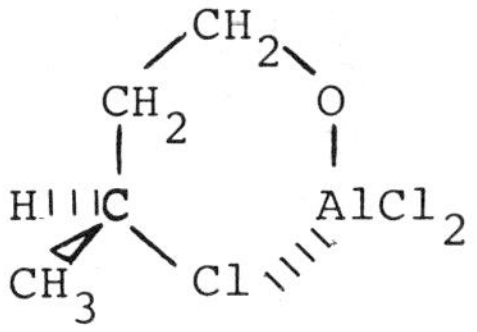

289

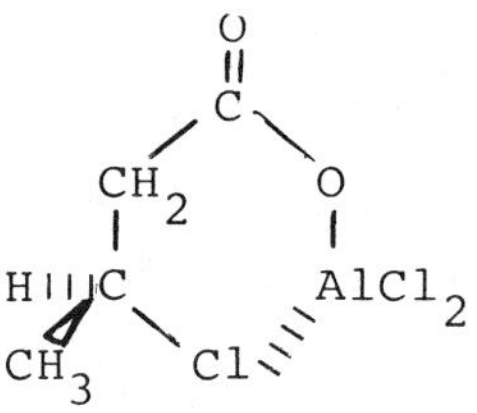

290

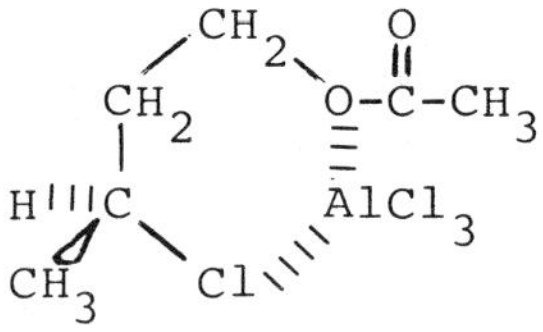

291

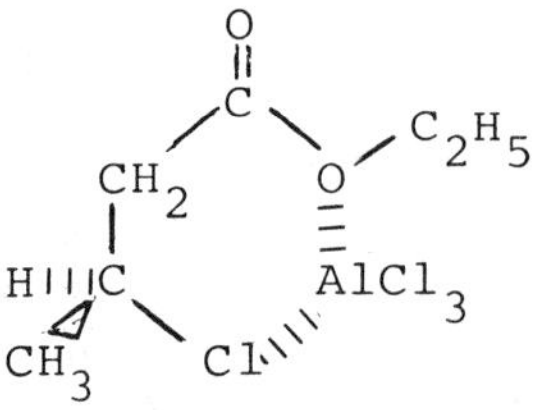

292

forced tightness of the ion pair of 290 and 292 owing to the electron-withdrawing effect of the carbonyl group.

B. Alkylations with Other Di- and Polyfunctional Reagents

In recent years numerous reports of alkylations of aromatics with di- and polyfunctional compounds have appeared. Some illustrative examples of various types will be given, as well as references to additional reports.

1. Unsaturated alcohols [323-325]

$$CH_2{=}CH{-}CH_2{-}C(R)(R'){-}OH + C_6H_5OH \xrightarrow[\substack{90-100^\circ \\ 10\ hr}]{H_3PO_4} HO{-}C_6H_4{-}C(R)(R'){-}CH_2{-}CH{=}CH_2$$

293a, $R=R'=CH_3$ [324]

293b, $R=CH_3$, $R'=C_2H_5$

293c, $R=R'=i\text{-}C_4H_9$

293d, $R=CH_3$, $R'=C_6H_{13}$

2. Unsaturated acids [326-331].

$$CH_3CH_2CH_2CH{=}CHCO_2H + C_6H_6 \xrightarrow{H_2SO_4} \gamma\text{-lactone} + \delta\text{-lactone} \xrightarrow{AlCl_3} Ph(C_5H_{10})CO_2H$$

$$CH_3CH_2CH_2CH{=}CHCO_2H + C_6H_6 \xrightarrow{AlCl_3} Ph(C_5H_{10})CO_2H$$

[329]

This area was reviewed by Glatz in 1974 [331].

3. Unsaturated and epoxy keto acids [332-334]

$$Br{-}C_6H_4{-}C(=O){-}CH{=}CH{-}CO_2H + p\text{-cymene} \xrightarrow[HCl]{AlCl_3} Br{-}C_6H_4{-}C(=O){-}CH_2{-}CH(Ar){-}CO_2H$$

[333]

[334]

4. Unsaturated acid chlorides [335-340]

X = F,Cl,Br

[340]

5. Unsaturated esters, ketones, and nitriles [341,342]

$CH_2=CH-CH(R)(R')$ + anisole $\xrightarrow[\text{hexane}]{AlCl_3}$ (major) + (minor)

[341]

294a, R=R'=$CO_2C_2H_5$

294b, R=H, R'=$COCH_3$

294c, R=H, R'=$CO_2C_2H_5$

The unusual amount of *ortho* product was attributed to association of the two reactants with the same Al atom so that the alkylating agent is delivered through a cyclic transition state.

6. Unsaturated ketones [337,341,343-347]

$CH_3CH{=}CHCH_2C(O)CH_3$ or $CH_2{=}CH{-}CH_2CH_2C(O)CH_3$ + C_6H_6 $\xrightarrow{AlCl_3}$ $CH_3{-}CH(C_6H_5){-}CH_2{-}CH_2C(O)CH_3$

[346]

$R{-}C_6H_4{-}C(O){-}CH{=}CH{-}C_6H_4{-}R'$ + ArH $\xrightarrow{AlCl_3}$

$R{-}C_6H_4{-}C(O){-}CH_2{-}CH(Ar){-}C_6H_4{-}R'$ (major) + R-substituted 3-(R'-phenyl)indan-1-one (minor)

[336, 344]

Other unsaturated ketones gave mixtures of phenylated alkanones, the composition of which depended on the position of the double bond relative to the carbonyl group in the starting material.

7. Unsaturated nitro compound [348]

$O_2N{-}C_6H_4{-}CH{=}CH{-}NO_2$ + C_6H_6 $\xrightarrow[\text{catalyst"}]{\text{"acid}}$ $O_2N{-}C_6H_4{-}CH(Ph){-}CH_2{-}NO_2$ (major) + $O_2N{-}C_6H_4{-}CH(Ph){-}CH_2{-}Ph$ (minor)

8. Halonitriles [349]

$X(CH_2)_nCN$ + C_6H_6 $\xrightarrow[6^\circ\text{, then reflux}]{AlCl_3}$ $Ph(CH_2)_nCN$

X = Cl, Br n = 3, 4

9. Ketonitrile [350]

$$C_6H_5COCH_2CN + C_6H_4(OH)R \xrightarrow[HCl]{AlCl_3} \left((C_6H_5)(HO(R)C_6H_3)C(OH)CH_2CN \right) \xrightarrow{-H_2O} (C_6H_5)(HO(R)C_6H_3)C{=}CH{-}CN$$

R = H, CH_3

10. Sulfolanyl sulfonates [351]

$$\text{3-}(OSO_2R)\text{-sulfolane} + ArH \xrightarrow{AlCl_3} \text{3-Ar-sulfolane}$$

Ar = C_6H_5, MeC_6H_4, $Me_2C_6H_3$, etc.

R = Me, Ph, p-MeC_6H_4, p-ClC_6H_4

11. Amino alcohols [352]

$$CH_3{-}N(CH_3)(CH_2)_nC(CH_3)_2{-}OH + ArH \xrightarrow[(PhNO_2)]{AlCl_3} CH_3{-}N(CH_3)(CH_2)_nC(CH_3)_2{-}Ar$$

n = 2-4 ArH = toluene, o-xylene, naphthalene, thiophene

12. Perchlorocyclobutenone [353]

$$\text{Perchlorocyclobutenone} + \text{anisole} \xrightarrow{AlCl_3} \mathbf{295}$$

Cl, Cl, O, Cl, Cl + OCH_3 (anisole) $\xrightarrow{AlCl_3}$ Cl, OH, OCH_3, Cl, Cl, CH_3O

<u>295</u>

$$\mathbf{295} \xrightarrow[HCl]{SnCl_4}$$

Cl, Cl, OCH_3, +, Cl, CH_3O $\quad SnCl_5^-$

13. Chloro alcohols, ethers, and esters [354,355]

$$R\text{-}CH_2\text{-}\underset{Cl}{CH}\text{-}CH_3 + C_6H_6 \xrightarrow{AlCl_3} R\text{-}CH_2\text{-}\underset{Ph}{CH}\text{-}CH_3 + \underset{\mathbf{296}}{PhCH_2CH_2CH_3}$$

$$+ \underset{\mathbf{297}}{Ph_2CHCH_2CH_3} + \underset{\mathbf{298}}{PhCH_2\text{-}\underset{Ph}{CH}\text{-}CH_3}$$

$R = OH,\ OCH_3$

[355]

Mechanisms are proposed to explain the unexpected products <u>296</u>-<u>298</u>.

REFERENCES

1. Friedel, C., and J. M. Crafts, Bull. Soc. Chim. Fr. (2), *41*, 322 (1884); J. Chem. Soc. Abstr., XLVI, 1312 (1884).
2. Lavaux, J., C. R. Acad. Sci., *139*, 976 (1904); J. Chem. Soc. Abstr., (I), 43 (1905).
3. Lavaux, J., C. R. Acad. Sci., *140*, 44 (1905); J. Chem. Soc. Abstr., (I), 125 (1905).

4. Lavaux, J., C. R. Acad. Sci., *152,* 1400 (1911), J. Chem. Soc. Abstr., *100* (I), 533 (1911).
5. Lavaux, J., and M. Lombard, Bull. Soc. Chim. Fr., (4), *7,* 913 (1910); J. Chem. Soc. Abstr., *98* (I), 747 (1910).
6. Friedel, C., and J. M. Crafts, Ann. Chim. Phys., (6), *11,* 263 (1887); J. Chem. Soc. Abstr., *52,* 1102 (1887).
7. Wenzel, F., and R. Kugel, Monatsh. Chem., *35,* 953 (1914); J. Chem. Soc. Abstr., *108,* (I), 514 (1915).
8. Wenzel, F., and G. Drada, Monatsh. Chem., *35,* 973 (1914); J. Chem. Soc. Abstr., *108* (I), 514 (1915).
9. Adam, P., C. R. Acad. Sci., *103,* 207 (1986); J. Chem. Soc. Abstr., *50,* 1033 (1886).
10. Bodroux, F., Bull. Soc. Chim. Fr., (3), *25,* 491 (1901); J. Chem. Soc. Abstr., *80* (I), 374 (1901).
11. Homer, A., J. Chem. Soc., *97,* 1141 (1910).
12. Silva, R. D., C. R. Acad. Sci., *89,* 606 (1897); J. Chem. Soc. Abstr., 259 (1880).
13. Silva, R. D., Bull. Soc. Chim. Fr., (2), *36,* 24 (1881); J. Chem. Soc. Abstr., 913 (1881).
14. Friedel, C., and M. Balsohn, Bull. Soc. Chim. Fr., *35,* 52 (1881); J. Chem. Soc. Abstr., 260 (1881).
15. Shiozaki, K., and M. Shiga, Jpn. Kokai 73-18, 270 (1973); Chem. Abstr., *79,* 18336e (1973).
16. Roux, L., Ann. Chim. Phys., *12,* 297 (1887).
17. Lespieau, R., Bull. Soc. Chim. Fr., *6,* 288 (1891).
18. Yura, S., and R. Oda, J. Soc. Chem. Ind. Jpn., *46,* 531 (1943); Chem. Abstr., *42,* 6348h (1948).
19. Yura, S., and T. Hashimoto, J. Soc. Chem. Ind. Jpn., *47,* 814 (1944); Chem. Abstr., *42,* 6347i (1948).
20. Yura, S., K. Sato, T. Koizumi, and R. Oda, J. Soc. Chem. Ind. Jpn., *44,* 622 (1941); Chem. Abstr., *42,* 2248e (1948).
21. Sisido, K., J. Soc. Chem. Ind. Jpn. Suppl. Bind., *44,* 463 (1941); Chem. Abstr., *45,* 1558i (1951).
22. Shinkle, S. D., A. E. Brooks, and G. H. Cady, Ind. Eng. Chem., *28,* 257 (1936).
23. Nicolescu, I. V., and M. Iovu, Ind. Plast. Mod. (Paris), *10*(10), 46 (1958); Chem. Abstr., *53,* 11273d (1959).
24. Kolesnikov, G. S., V. V. Khorskak, M. A. Andreeva, and A. T. Kitaigorodskii, Izv. Akad. Nauk SSSR, Div. Chem. Soc., *107* (1956); Chem. Abstr., *50,* 13814b (1956).
25. Dolgov, B. N., and N. A. Larin, J. Gen. Chem. USSR, *20,* 450 (1950); Chem. Abstr., *45,* 566f (1951); Chem. Zentralbl., 8325 (1953).
26. Roberts, R. M., and A. A. Khalaf, Unpublished results.
27. Anshutz, R., Justus Liebigs Ann. Chem., *235,* 329 (1886).
28. Lee, C. C., A. G. Forman, and A. Rosenthal, Can. J. Chem., *35,* 220 (1957).
29. McMahon, M. A., and S. C. Bunce, J. Org. Chem., *29,* 1515 (1964).
30. Khalaf, A. A., and R. M. Roberts, J. Org. Chem., *31,* 89 (1966).
31. Vartanyan, S. A., A. G. Vartanyan, and E. A. Araratyan, Arm. Khim. Zh., *25,* 948 (1972); Chem. Abstr., *78,* 83924p (1973).

32. Silva, R. D., Bull. Soc. Chim. Fr., (2), *41*, 448 (1884); J. Chem. Soc. Abstr., 1356 (1884).
33. Silva, R. D., C. R. Acad. Sci., *89*, 606 (1879); J. Chem..Soc. Abstr., 259 (1880).
34. Angeblis, A., and R. Anschütz, Chem. Ber., *17*, 165 (1884); J. Chem. Soc. Abstr., 753 (1884).
35. Anschütz, R., and E. Romig, Chem. Ber., *18*, 662 (1885); J. Chem. Soc. Abstr., *48*, 768 (1885).
36. Anschütz, R., Justus Liebigs Ann. Chem., *235*, 302 (1886).
37. Schmerling, L., R. W. Welch, and J. P. Luvisi, J. Am. Chem. Soc., *70*, 3636 (1957).
38. Lavaux, J., C. R. Acad. Sci., *141*, 354 (1905); J. Chem. Soc. Abstr., *88* (I), 698 (1905).
39. Silva, R. D., Jahresber. Fortschr. Chem., 379 (1879).
40. Ransley, D. L., J. Org. Chem., *31*, 3595 (1966).
41. Winstein, S., and K. G. Schreiber, J. Am. Chem. Soc., *74*, 2171 (1952).
42. Sabatier, P., and M. Murat, Ann. Chim., *4*, 286 (1915).
43. Goudet, H., and F. Schenker, Helv. Chim. Acta, *10*, 132 (1927); Brt. Chem. Abstr. A, 440 (1927).
44. Greene, R. N., Ph.D. dissertation, University of Texas at Austin, 1965.
45. Bodroux, F., C. R. Acad. Sci., *132*, 155 (1901); J. Chem. Soc. Abstr., *80* (I), 196 (1901).
46. Tsukervanik, I., and K. Yatsimirskii, J. Gen. Chem. USSR, *10*, 1075 (1940); Chem. Abstr., *35*, 3981 (1941).
47. Ransley, D. L., J. Org. Chem., *33*, 1517 (1968).
48. Heck, R., and S. Winstein, J. Am. Chem. Soc., *79*, 3105 (1957).
49. Roberts, R. M., Y. W. Han, C. H. Schmid, and D. A. Davis, J. Am. Chem. Soc., *81*, 640 (1959).
50. Khalaf, A. A., Ph. D. dissertation, University of Texas at Austin, 1965.
51. Khalaf, A. A., Diss. Abstr., XXVI, No. 2, 699 (1965).
52. Nenitzescu, C. D., and A. Glatz, Acad. Repub. Pop. Rom. Stud. Cercet. Chim., *7*, 505 (1959); Chem. Abstr., *54*, 19546c (1960).
53. Roberts, R. M., A. A. Khalaf, and R. N. greene, J. Am. Chem. Soc., *86*, 2846 (1964).
54. Sisido, K., and H. Nozaki, J. Am. Chem. Soc., *69*, 961 (1947).
55. Tsukervanik, I. P., and L. V. Bugrova, Zh. Obshch. Khim, *27*, 889 (1957); Chem. Abstr., *52*, 2827h (1958).
56. Shadmanov, K. M., Uzb. Khim. Zh., *7*(1), 57 (1963); Chem. Abstr., *59*, 2732g (1963).
57. Shadmanov, K. M., Dokl. Akad. Nauk Uzb. SSSR, No. 4, 40 (1959); Chem. Abstr., *54*, 10980d (1960).
58. Shadmanov, K. M., Dokl. Akad. Nauk Uzb. SSSR, No. 10, 42 (1960); Chem. Abstr., *56*, 15434a (1962).
59. Shadmanov, K. M., Dokl. Akad. Nauk Uzb. SSSR, No. 4, 42 (1961); Chem. Abstr., *60*, 14407f (1964).
60. Shadmanov, K. M., Uzb. Khim. Zh., No. 4, 70 (1961); Chem. Abstr., *56*, 5897 (9162).
61. Shadmanov, K. M., Dokl. Akad. Nauk Uzb. SSR, *20*(11), 26 (1963); Chem. Abstr., *61*, 4286f (1964).
62. Flanagan, Pat. W. K., and A. Shadan, Am. Chem. Soc., Div. Petrol. Chem., Prepr., *14*(3), A101 (1969); Chem. Abstr., *74*, 141246g (1971).

63. Krafft, F., Chem. Ber., *19*, 2982 (1886); J. Chem. Soc. Abstr., LII, 252 (1887).
64. Auger, V., Bull. Soc. Chim. Fr., (2), *47*, 49 (1887); J. Chem. Soc. Abstr., LII, 814 (1887).
65. Sisido, K., and H. Nozaki, J. Am. Chem. Soc., *70*, 1288 (1948).
66. Sisido, K., and H. Nozaki, J. Soc. Chem. Ind. Jpn., *48*, 35 (1945); Chem. Abstr., *42*, 6521d (1948).
67. Nozaki, H., M. Kawanisi, M. Okazaki, M. Yamae, Y. Nisikawa, T. Hisida, and K. Sisido, J. Org. Chem., *30*, 1303 (1965).
68. Roberts, R. M., G. P. Anderson, A. A. Khalaf, and Chow-Eng Low, J. Org. Chem., *36*, 3342 (1971).
69. Khalaf, A. A., and R. M. Roberts, J. Org. Chem., *37*, 4227 (1972).
70. Grunwald, E., J. Am. Chem. Soc., *73*, 5458 (1951).
71. Matsuda, S., Asahi Garasu Kogyo Gijutsu Shoreikai Kenkyu Hokoku, *22*, 197 (1973); Chem. Abstr., *81*, 77154h (1974).
72. Gelin, R., S. Gelin, and B. Chantegrel, C. R. Acad. Sci., Ser. C, *266*(11), 813 (1968); Chem. Abstr., *69*, 35589a (1968).
73. Gelin, R., B. Chantegrel, and S. Gelin, Bull. Soc. Chim. Fr., (11), 4136 (1969); Chem. Abstr., *72*, 54623k (1970).
74. Gelin, R., B. Chantegrel, and S. Gelin, C. R. Acad. Sci., Ser. C, *270*(12), 1123 (1970); Chem. Abstr., *73*, 14037m (1970).
75. Sisido, K., and H. Nozaki, J. Soc. Chem. Ind. Jpn., *47*, 516 (1944); Chem. Abstr., *48*, 2016 (1954).
76. Ranny, M., P. Kondelik, J. Pasek, and M. Zbirovsky, Tenside Deterg., *16*(1), 23 (1979); Chem. Abstr., *90*, 153772v (1979).
77. Masuda, S., M. Segi, T. Nakajima, and S. Suga, J. Chem. Soc., Chem. Commun., 86 (1980).
78. Somerville, W. T., and P. E. Spoerri, J. Am. Chem. Soc., *74*, 3803 (1952); *72*, 2185 (1950).
79. Bodroux, F., C. R. acad. Sci., *132*, 1333 (1901).
80. Khalaf, A. A., and R. M. Roberts, J. Org. Chem., *31*, 926 (1966).
81. Schmerling, L., R. W. Welch, and J. P. West, J. Am. Chem. Soc., *78*, 5406 (1956).
82. Bugrova, L. V., and I. P. Tsukervanik, Zh. Org. Khim., *1*(4), 714 (1965); J. Org. Chem. USSR, *1*(4), 714 (1965).
83. Bugrova, L. V., and I. P. Tsukervanik, Zh. Obshch. Khim., *32*, 3575 (1965); Chem. Abstr., *58*, 11242g (1963).
84. Khalaf, A. A., Rev. Chim. (Bucharest), *18*, 479 (1973); Chem. Abstr., *76*, 5064q (1973).
85. Bruson, H. A., and J. W. Kroeger, J. Am. Chem. Soc., *62*, 36 (1940).
86. Ransley, D. L., J. Org. Chem., *34*, 2618 (1969).
87. Shadmanov, K. M., Uzb. Khim. Zh., *6*(6), 49 (1962); Chem. Abstr., *59*, 3847h (1963).
88. (a) Gelin, R., B. Chantegrel, and S. Gelin, C. R. Acad. Sci., Ser. C, *270*(12), 1123 (1970); Chem. Abstr., *73*, 14037m (1970); (b) Gelin, R., and B. Chantegrel, Bull. Soc. Chim. Fr., 2627 (1971).
89. Krimm, H., H. J. Buysch, and H. Schnell, Ger. Offen. 2,055,968 (1972); Chem. Abstr., *77*, 48026e (1972).
90. Schmerling, L., J. P. Luvisi, and R. W. Welch, J. Am. Chem. Soc., *77*, 1774 (1955).
91. Schmerling, L., R. W. Welch, and J. P. West, J. Am. Chem. Soc., *79*, 2636 (1957).

92. Schmerling, L., J. P. Luvisi, and R. W. Welch, J. Am. Chem. Soc., *81*, 2718 (1959).
93. Friedel, C., and J. M. Crafts, Bull. Soc. Chim. Fr., (2), *28*, 50 (1877).
94. Scharz, H., Chem. Ber., *14*, 1516 (1881); J. Chem. Soc. Abstr. XL, 912 (1881).
95. Fischer, E., and O. Fischer, Chem. Ber., *14*, 1942 (1881); J. Chem. Soc. Abstr., 62 (1882).
96. Friedel, C., and J. M. Crafts, Bull. Soc. Chim. Fr., (2) *37*, 6 (1882); J. Chem. Soc. Abstr., 621 (1882).
97. Böeseken, J., Recl. Trav. Chim. Pays-Bas, *22*, 301 (1903); J. Chem. Soc. Abstr., *84* (I), 617 (1903).
98. Friedel, C., and C. Vincent, Bull. Soc. Chim. Fr., (2), *36*, 1 (1881).
99. Böeseken, J., Recl. Trav. Chim. Pays-Bas, *24*, 1 (1905); J. Chem. Soc. Abstr., *88* (I), 423 (1905).
100. Gomberg, M., and R. L. Jickling, J. Am. Chem. Soc., *37*, 2575 (1915).
101. Böeseken, J., Recl. Trav. Chim. Pays-Bas, *27*, 5 (1908); J. Chem. Soc. Abstr., *94* (I), 189 (1908).
102. Fischer, E., and O. Fischer, Pays-Bas, Justus Liebigs Ann. Chem., *194*, 242 (1878).
103. Friedel, C., and J. M. Crafts, Ann. Chim. Phys., (6), *1*, 449 (1884).
104. Meyer, V., Chem. Ber., *28*, 2276 (1895).
105. Norris, J. F., *Organic Syntheses*, Vol. 4, Wiley, New York, 1925, pp. 81-83.
106. Norris, J. F., Ind. Eng. Chem., *16*, 184 (1924).
107. Norris, J. F., and R. C. Young, J. Am. Chem. Soc., *46*, 2580 (1924).
108. Marvel, C. S., and W. N. Sperry, *Organic Syntheses*, Vol. 8, Wiley, New York, 1928 pp. 26-29.
109. Wilson, S. D., and Hsiao-Yun Huang, J. Chin. Chem. Soc., *4*, 142 (1936); Chem. Abstr., *30*, 8192 (1936).
110. Chernyakovskaya, K. A., G. S. Mironov, M. I. Farberov, M. I. Tyuleneva, and N. A. Rovnyagina, Uch. Zap. Yaroslav. Tekhnol. Inst., No. 13, 92 (1970); Chem. Abstr., *77*, 48006y (1972).
111. Elbs, K., and O. Wittich, Chem. Ber., *18*, 347 (1885); J. Chem. Soc. Abstr., 517 (1885).
112. Lavaux, J., C. R. Acad. Sci., *146*, 345 (1908); J. Chem. Soc. Abstr., *94* (I), 256 (1908); C. R. Acad. Sci., *139*, 976 (1904); J. Chem. Soc. Abstr., *88* (I), 43 (1905).
113. Honig, M., and F. Berger, Monatsh. Chem., *3*, 668 (1882); J. Chem. Soc. Abstr., 68 (1882).
114. Homer, A., J. Chem. Soc., *97*, 1141 (1910).
115. Gomberg, M., and O. W. Boedisch, J. Am. Chem. Soc., *23*, 177 (1901).
116. Tousley, N. E., and M. Gomberg, J. Am. Chem. Soc., *26*, 1516 (1904).
117. Gomberg, M., and J. D. Todd, J. Am. Chem. Soc., *39*, 2392 (1917).
118. Dolgov, B. N., N. T. Sorokina, and S. A. Cherkasov, J. Gen. Chem., USSR, *21*, 509 (1951); Chem. Abstr., *45*, 8464e (1951).
119. Beckert, W. F., and J. U. Lowe, Jr., J. Org. Chem., *32*, 582 (1967).

120. Drahowzal, F. A., in *Friedel-Crafts and Related Reactions*, Vol. 2, (G. A. Olah, ed.), Wiley-Interscience, New York, 1964, pp. 428, 435.
121. Brovko, V. V., V. A. Sokolenko, and G. G. Yakobson, Zh. Org. Chem., *10,* 300 (1974); Chem. Abstr., *80,* 120417q (1974).
122 Guseinov, M. M., T. A. Kambarova, Z. K. Mekhtieva, M. R. Mirzoeva, and G. A. Yur'eva, Azerb. Khim. Zh., No. 4, 102 (1972); Chem. Abstr., *79,* 52892t (1973).
123. McBee, E. T., E. P. Wesseler, D. L. Crain, R. Hurnaus, and T. Hodgens, J. Org. Chem., *37,* 683 (1972).
124. Hine, J., *Physical Organic Chemistry,* McGraw-Hill, New York, 1962, p. 357.
125. Ullman, F., and A. Munzhuber, Chem. Ber., *36,* 404 (1903).
126. Hart, H., and F. A. Cassis, J. Am. Chem. Soc., *76,* 1634 (1954).
127. Fonken, G. J., J. Org. Chem., *28,* 1909 (1963).
128. Gardeur, A., Bull. Acad. R. Belg., *34*(3), 920 (1898).
129. Kuntze-Fechner, M., Chem. Ber., *36,* 472 (1903).
130. Angeblis, A., and R. Anschutz, Chem. Ber., *17,* 167 (1884); J. Chem. Soc. Abstr., 753 (1884).
131. Anschütz, R., Justus Liebigs Ann. Chem., *235,* 333 (1886).
132. Korshak, V. V., and K. K. Samplavskaya, Sb. Statei Obshch. Khim., *2,* 1020 (1953); Chem. Abstr., *49,* 8207c (1955).
133. Anschütz, R., and H. Immendorff, Chem. Ber., *17,* 2816 (1884); J. Chem. Soc. Abstr., *48,* 269 (1885).
134. Lavaux, J., C. R. Acad. Sci., *146,* 345 (1908); J. Chem. Soc. Abstr., *94* (I), 256 (1908); C. R. Acad. Sci., *139,* 976 (1904); J. Chem. Soc. Abstr., *88* (I), 43 (1905); C. R. Acad. Sci., *146,* 135 (1908).
135. Homer, A., J. Chem. Soc., *97,* 1141 (1910).
136. Mouneyrat, A., Bull Soc. Chem., *19*(3), 554, 557 (1898); J. Chem. Soc. Abstr., *76* (I), 490 (1899).
137. Dolgov, B. N., and N. A. Larin, J. Gen. Chem. USSR, *20,* 450 (1950); Chem. Abstr., *45,* 566f (1951).
138. Schmerling, L., J. P. West, and R. W. Welch, J. Am. Chem. Soc., *80,* 576 (1958).
139. Peterson, P. E., C. Casey, E. V. P. Tao, A. Agtarap, and G. Thompson, J. Am. Chem. Soc., *87,* 5163 (1965).
140. Stevenson, G. W., and D. Williamson, J. Am. Chem. Soc., *80,* 5943 (1958).
141. Olah, G. A., S. J. Kuhn, and J. A. Olah, J. Chem. Soc., 2174 (1957).
142. Wohl, W., and E. Wertyporoch, Chem. Ber., *64,* 1357 (1931).
143. Burwell, R. L., Jr., and S. Archer, J. Am. Chem. Soc., *64,* 1032 (1942).
144. Olah, G. A., and S. J. Kuhn, J. Org. Chem., *29,* 2317 (1964).
145. Laubengeyer, A. W., and D. S. Sears, J. Am. Chem. Soc., *67,* 164 (1945).
146. Brown, H. C., and R. R. Holmes, J. Am. Chem. Soc., *78,* 2173 (1956).
147. Greenwood, N. N., and B. G. Perkins, J. Chem. Soc., 1141 (1960).
148. Müller, C., and P. Weyerstahl, Tetrahedron, *31,* 1787 (1975).
149. (a) Nerdel, F., and J. Buddrus, Tetrahedron Lett., 3197 (1965); (b) Buddrus, J., Chem. Ber., *101,* 4152 (1968).

150. Skattebol, L., and B. Boulettte, J. Org. Chem., *31*, 81 (1966).
151. Brovko, V. V., V. A. Sokolenko, and G. G. Yakobson, Zh. Org. Chem., *10*, 300 (1974); Chem. Abstr., *80*, 120417q (1974).
152. Sheppard, W., and C. Sharto, *Organic Fluorine Chemistry*, W. A. Benjamin, New York, 1969.
153. McBee, E. T., E. P. Wesseler, R. Hurnaus, and T. Hodgins, J. Org. Chem., *37*, 1100 (1972).
154. Tsukervanik, I. P., V. A. Ermokhina, and A. R. Absurasuleva, USSR Patent 351,822 (1972); Chem. Abstr., *78*, 29418p (1973).
155. Böeseken, J., and M. C. Bastet, Recl. Trav. Chim. Pays-Bas, *32*, 184 (1914).
156. Angelbis, A., and R. Anschütz, Chem. Ber., *17*, 167 (1884).
157. Anschütz, R., Justus Liebigs Ann. Chem., *235*, 299 (1886).
158. Davidson, J. M., and A. Lowy, J. Am. Chem. Soc., *51*, 2978 (1929).
159. Malinovskii, M. S., Zh. Obshch. Khim., *17*, 2235 (1947).
160. Hanriot, and Guilbert, C. R. Acad. Sci., *98*, 525 (1884).
161. Sisido, K., J. Soc. Chem. Jpn., Suppl. Bind., *45*, 169B (1942); Chem. Abstr., *44*, 8340 (1950).
162. Sisido, K., and T. Isida, J. Am. Chem. Soc., *70*, 1289 (1948).
163. Tsukervanik, I. P., and Kh. Yu. Yuldashev, Zh. Obshch. Khim., *31*, 858 (1961).
164. Yuldashev, Kh. Yu., and I. P. Tsukervanik, Uzb. Khim. Zh., *6*, 40 (1961); Chem. Abstr., *57*, 16443b (1962).
165. Roberts, R. M., and M. B. Abdel-Baset, J. Org. Chem., *41*, 1698 (1976).
166. Hoffman, A., J. Am. Chem. Soc., *51*, 2542 (1929).
167. Barnes, R. A., and B. D. Beitchman, J. Am. Chem. Soc., *76*, 5430 (1954).
168. Curtis, R. F., and K. O. Lewis, J. Chem. Soc., 418 (1962).
169. Gramenitskaia, V. N., G. I. Nikishin, and A. D. Petrov, Proc. Acad. Sci. USSR, *118*, 65 (1958).
170. Schmerling, L., U.S. Patent 2,485,017 (Oct. 18, 1949).
171. Schmerling, L., J. P. West, and R. W. Welch, J. Am. Chem. Soc., *80*, 576 (1958).
172. Belyaev, V. F., and A. I. Abrazhevich, Zh. Org. Khim., *3*, 910 (1967); J. Org. Chem. USSR, *3*, 876 (1967).
173. Belyaev, V. F., A. I. Abrazhevich, and V. P. Prokoproich, Zh. Org. Khim., 4, 1647 (1968); J. Org. Chem. USSR, *4*, 1583 (1968).
174. Belyaev, V. F., and L. A. Ivanova, Zh. Org. Khim., *4*, 1839 (1968); J. Org. Chem. USSR, *4*, 1777 (1968).
175. Kochetkov, N. K., and V. F. Belyaev, Zh. Org. Khim., *30*, 1495. (1960).
176. Belyaev, V. F., and A. I. Abrazhevich, Khim. Geterotsikl. Soedin., No. 2, 228 (1967); Chem. Abstr., *67*, 73466z (1967).
177. Belyaev, V. F., V. I. Grushevich, and A. I. Abrazhevich, Zh. Org. Khim., *7*, 610 (1971); Chem. Abstr., *75*, 5617q (1971).
178. Belyaev, V. F., Vestsi Akad. Nauk SSR, Ser. Khim. Navuk, No. 5, 190 (1969); Chem. Abstr., *72*, 43047 (1970).
179. Belyaev, V. F., Zh. Obshch. Khim., *34*(3), 861 (1964); Chem. Abstr., *60*, 15767c (1964).
180. Demole, E., Chem. Ber., *12*, 2245 (1879).
181. Korshak, V. V., and K. K. Samplavskaya, Zh. Obshch. Khim., *18*, 1470 (1948).

182. Guseinov, M. M., T. A. Kambarova, Z. K. Mekhtieva, and Z. G. Yarieva, Dokl. Akad. Nauk Az. SSR, *22*, 38 (1966); Chem. Abstr., *67*, 2822b (1967).
183. Guseinov, M. M., T. A. Kambarova, Z. K. Mekhtieva, and T. Enfendieva, Azerb. Khim. Zh., No. 5, 65 (1967); Chem. Abstr., *69*, 66996k (1968).
184. Guseinov, M. M., T. A. Kambarova, Z. K. Mekhtieva, and M. T. Dzhafarova, Dokl. Akad. Nauk Az. SSR, *25*(8), 37 (1969); Chem Abstr., *72*, 132175g (1970).
185. Dolgov, B. N., and N. A. Larin, J. Gen. Chem. USSR, *20*, 475 (1950).
186. Mouneyrat, A., Bull. Soc. Chim. Fr., (3), *19*, 557 (1898).
187. Bodrikov, I. V., and I. S. Okrokova, Tr. Khim. Tekhnol., No. 2, 179 (1967); Chem. Abstr., *70*, 67306t (1969).
188. Bodrikov, I. V., and I. S. Okrokova, Zh. Org. Khim., *4*(10), 1806 (1968); J. Org. Chem. USSR, *4*(10), 1742 (1968).
189. Bodrikov, I. V., and I. S. Okrokova, Zh. Org. Khim., *3*(9), 1706 (1967); J. Org. Chem. USSR, *3*(9), 1663 (1967).
190. Mamedaliev, Yu. G., R. A. Babakhanov, M. N. Magerramov, M. A. Salimov, and A. R. Musaeva, Azerb. Khim. Zh., No. 5, 3 (1963).
191. Bodrikov, I. V., I. S. Okrokova, and A. N. Egorochkin, Izv. Vyssh. Uch. Zaved., Khim. Tekhnol., *14*, 1362 (1971); Chem. Abstr. *76*, 13957m (1972).
192. Truffault, R., and Y. Monteils, Bull. Soc. Chem. Fr., (5), *6*, 726 (1939).
193. Truffault, R., C. R. Acad. Sci., *202*, 1286 (1936).
194. Truffault, R., and Y. Monteils, Bull. Soc. Chim. Fr., (5), *6*, 726
195. Mamedaliev, Yu. G., R. A. Babakhanov, and A. R. Musaeva, Azerb. Khim. Zh., Nos. 2-3 (1961); Chem. Abstr., *56*, 2357h (1962).
196. Mamedaliev, Yu. G., R. A. Babakhanov, and M. N. Magerramov, Dokl. Akad. Nauk Az. SSR, *18*, 25 (1962).
197. Magerranov, M. N., N. S. Sadykhov, and Sh. T. Akhmedov, Azerb. Khim. Zh., No. 2, 92 (1968); Chem. Abstr., *70*, 3376h (1969).
198. Petrov, A. D., O. M. Nefedov, and Yu. N. Ogibin, Izv. Akad. Nauk SSSR, Otd. Khim. Nauk, 1004 (1957).
199. Smith, W. T., Jr., and J. T. Sellas, *Organic Synthesis*, Vol. 32, Wiley, New York, 1952, p. 90.
200. Whitmore, F. C., C. A. Weisgarber, and A. C. Shabica, Jr., J. Am. Chem. Soc., *65*, 1469 (1943).
201. Smith, W. T., Jr., and J. T. Sellas, *Organic Synthesis*, Coll. Vol. 4, New York, 1963, p. 702.
202. Vdovtsova, E. A., and G. F. Fedorova, Zh. Org. Khim., *6*, 767 (1970); J. Org. Chem. USSR, *6*, 770 (1970); Chem. Abstr., *73*, 14356q (1970).
203. Monteils, Y., Bull. Soc. Chim. Fr., 637 (1951).
204. Sadykov, N. S., M. N. Megerramov, and Sh. T. Akhmedov, Zh. Org. Khim, *10*, 318 (1974); J. Org. Chem. USSR, *10*, 318 (1974).
205. Archer, S., J. D. Malkemus, and C. M. Suter, J. Am. Chem. Soc., *67*, 43 (1945).
206. Pines, H., B. Kvetinskas, and J. A. Vesely, J. Am. Chem. Soc., *72* 1568 (1950).
207. Kreuz, K. L., and R. T. Sanderson, U. S. Patent 2,529,298 (Nov. 7, 1950).
208. Smith, R. A., and J. B. Niederl, J. Am. Chem. Soc., *55*, 4151 (1933).

209. Vdovtsova, E. A., and E. N. Khar'yanov, Dokl. Akad. Nauk Uzb. SSR, No. 11, 43 (1960); Chem. Abstr., *61*, 2991e (1964).
210. Vdovtsova, E. A., Zh. Org. Khim., *1*, 2192 (1965); J. Org. Chem. USSR, *1*, 2233 (1965).
211. Vdovtsova, E. A., Zh. Org. Khim., *6*, 1238 (1970); J. Org. Chem. USSR, *6*, 1245 (1970).
212. Yanichkin, L. P., and E. A. Vdovtsova, Zh. Org. Khim., *3*(5), 883 (1967); J. Org. Chem. USSR, *3*(5) 848 (1967).
213. Vdovtsova, E. A., and A. G. Kakharov, Zh. Org. Khim., *3*(5), 889 (1967); J. Org. Chem. USSR, *3*(5), 854 (1967).
214. Vdovtsova, E. A. and G. F. Fedorova, Zh. Org. Khim. USSR, *6*(6), 1230; J. Org. Chem. USSR, *6*(6) 1238 (1970).
215. Vdovtsova, E. A., and G. F. Fedorova, Kh. Org. Khim., *6*(4), 767 (1970); J. Org. Chem. USSR, *6*(4), 770 (1970).
216. Vdovtsova, E. A., and A. G. Kakharov, Zh. Org. Khim., *6*, 1653 (1970); J. Org. Chem. USSR, *6*, 1663 (1970); Chem. Abstr., *73*, 109409y (1970).
217. Vdovtsova, E. A., and A. G. Kakharov, Zh. Org. Khim., *6*, 1664 (1970); J. Org. Chem. USSR, *6*, 1672 (1970); Chem. Abstr., *73*, 109410s (1970).
218. Vdovtsova, E. A., L. P. Yanichkin, and I. P. Tsukervanik, Zh. Org. Khim., 2(7), 1279 (1966); J. Org. Chem. USSR, *2*(7), 1277 (1966).
219. Petrov, A. D., V. N. Gramenitskaya, A. S. Lebedeva, and G. I. Nikishin, Neftekhimiya, *2*, 776 (1962).
220. Vdovtsova, E. A., and L. P. Yanichkin, Zh. Org. Khim., *5*(5), 947 (1969); Chem. Abstr., *71*, 38489v (1969).
221. Koncos, R. and B. S. Friedman, in *Friedel-Crafts and Related Reactions*, Vol. 1, (G. A. Olah, ed.), Wiley-Interscience, New York, 1964, p. 291.
222. Nenitzescu, C. D., and A. Isacescu, Chem. Ber., *66*, 1100 (1933).
223. Farkhadova, S. M., M. N. Magerramov, R. A. Babakhanov, and Sh. T. Akhmedov, Azerb. Khim. Zh., No. 4, 65 (1972); Chem. Abstr., *79*, 52900y (1973).
224. Wispek, P., and R. Zuber, Justus Liebigs Ann. Chem., *218*, 379 (1883).
225. Wispek, P., and R. Zuber, Bull. Soc. Chim. Fr., (2), *43*, 588 (1885).
226. Losev, I. P., O. V. Smirnova, and T. A. Pfeifer, J. Gen. Chem. USSR, *21*, 737 (1951).
227. Konowalow, M., and S. Dobrowolski, J. Russ. Phys.-Chem. Soc., *37*, 547 (1905); Chem. Zentralbl., *2*, 825 (1905).
228. Schukowsky, S., Bull. Soc. Chim. Fr., (3), *16*, 126 (1896).
229. Shukow, A., and P. J. Schestakow, J. Russ. Phys.-Chem. Soc., *35*, 1 (1903); Chem. Zentralbl., (I), 825 (1903).
230. Weston, A. W., U. S. Patent 2,654,791 (Oct. 6, 1953).
231. Zokhrabbekov, E. Z., M. N. Magerramov, E. N. Usubova, and Sh. T. Akhmedov, Azerb. Khim. Zh., No. 2, 70 (1973); Chem. Abstr., *80*, 108083v (1974).
232. Hennion, G. F., and R. A. Kurtz, J. Am. Chem. Soc., *65*, 1001 (1943).
233. Magerramov, M. N., D. A. Guserinov, Sh. T. Akhmedov, E. Z. Zokhrabbekova, and S. M. Farkhadova, Dokl. Akad. Nauk Az. SSR, *24*(2), 33 (1968); Chem. Abstr., *70*, 28507q (1969).

234. Magerramov, M. N., E. Z. Zokhrabbekova, E. N. Usubova, and Sh. T. Akhmedov, Azerb. Khim. Zh., No. 3, 71 (1972); Chem. Abstr., *79*, 52896x (1973).
235. Lyutfaliev, A. G., M. N. Magerramov, R. A. Khalilova, and Sh. R. Nagieva, Azerb. Khim. Zh., No. 1, 22 (1972); Chem. Abstr., 43100k (1973).
236. Schmerling, L., U. S. Patent 3,678,122 (1972); Chem. Abstr., *77*, 101088e (1972).
237. Tarnopol'sku, Yu. I., and L. I. Denisovich, Khim. Geterotsikl. Soedin., No. 2, 29 (1971); Chem. Abstr., *77*, 5257p (1972).
238. Tsukervanik, I. P., and K. Y. Yuldashev, Uzb. Khim. Zh., No. 6, 58 (1960).
239. Vdovtsova, E. A., and L. P. Yanichkin, Zh. Org. Chem., *5*(3), 510 (1969); J. Org. Chem. USSR, *5*(3), 497 (1969).
240. Vdovtsova, E. A., L. P. Yanichkin, and I. P. Tsukervanik, Zh. Org. Khim., 2(7), 1275 (1966); J. Org. Chem. USSR, *2*(7), 1272 (1966).
241. Vdovtsova, E. A., L. P. Yanichkin, and I. P. Tsukervanik, Dokl. Akad. Nauk Uzb. SSR, *19*(1), 19 (1962); Chem. Abstr., *59* 2683h (1963).
242. Magerramov, M. N., A. G. Lyutfaliev, Z. A. Sadgkhov, R. A. Khalilova, and Sh. T. Akhmedov, Azerb. Khim. Zh., No. 4, 118 (1971); Chem. Abstr., *77*, 126284d (1972).
243. Schmerling, L., U.S. Patent, 3,810,947 (1974); Chem. Abstr., *81*, 25323t (1974).
244. Khudaverdyan, G. A., L. G. Grigoryan, and V. O. Babayan, Sb. Nauch. Tr. Erevan, Arm. Gos. Pedagog., Inst. Khim., No. 1, 45 (1970); Chem. Abstr., *77*, 100950t (1972).
245. Willgerodt, C., and A. Genieser, J. Prakt. Chem., *37*(2), 371 (1888).
246. Kundiger, D. G., and H. Pledger, J. Am. Chem. Soc., *78*, 6098 (1956).
247. Kundiger, D. G., and H. Pledger, Jr., J. Am. Chem. Soc., *78*, 6101 (1956).
248. Sisido, K., and H. Nozaki, J. Am. Chem. Soc., *70*, 1609 (1948).
249. Mironov, V. F., N. S. Fedotov, and G. E. Goler, Zh. Obshch. Khim., *42*, 1173 (1972); Chem. Abstr., *77*, 101744r (1972).
250. Mironov, V. F., N. W. Fedotov, G. E. Evert, T. E. Latysehva, and V. N. Bochkarev, Zh. Obshch. Khim., *44*, 561 (1974); Chem. Abstr., *80*, 146229e (1974).
251. Barry, A. J., J. W. Gilkey, and D. E. Hook, in *Metal Organic Compounds*, Advances in Chemistry, Monograph 23, American Chemical Society, Washington, D. C., 1959.
252. Gilman, H., and G. E. Dunn, Chem. Rev., *52*, 77 (1953).
253. Russell, G. A., J. Am. Chem. Soc., *81*, 4831 (1959).
254. Olah, G. A., *Friedel-Crafts Chemistry*, Wiley, New York, 1973, pp. 73 and 74.
255. Azatyan, V. D., Dokl. Akad. Nauk SSSR, *59*, 901 (1948).
256. Bert, P., C. R. Acad. Sci., *213*, 619 (1941).
257. Mochida, I., K. Takeshita, M. Ohgai, T. Tanaka, and T. Seiyama, Chem. Lett., 357 (1975).
258. Simons, J. H., and S. Archer, J. Am. Chem. Soc., *61*, 1521 (1939).
259. Calcott, W. S., J. M. Tinker, and V. Weinmayer, J. Am. Chem. Soc., *61*, 1010 (1939).

260. Huston, R. C., and D. D. Sager, J. Am. Chem. Soc., *48*, 1955 (1926).
261. McKenna, J. F., and F. J. Sowa, J. Am. Chem. Soc., *59*, 470 (1937).
262. Ackermann, W., and A. Heesing, Chem. Ber., *108*, 1182 (1975); Chem. Abstr., *82*, 169691e (1975).
263. Sadykhov, N. S., M. N. Magerramov, and Sh. T. Akhmedov, Dokl. Akad. Nauk Az. SSR, *24*(10), 30 (1968); Chem. Abstr., *71*, 21762f (1969).
264. Yusifov, Ch. A., and M. N. Magerramov, D. A. Guseinov, Sh. T. Akhmedov, and R. A. Khalilova, Azerb. Khim. Zh., No. 4, *12* (1967); Chem. Abstr., *68*, 87037t (1968).
265. Akperov, O. G., and Sh. T. Akhmedov, Zh. Org. Khim., *4*(11), 1939 (1968); J. Org. Chem. USSR, *4*(11), 2000 (1968).
266. Niederl, J. B., R. A. Smith, and M. E. McGreal, J. Am. Chem. Soc., *53*, 3390 (1931).
267. Niederl, J. B., and E. A. Storch, J. Am. Chem. Soc., *55*, 4549 (1933).
268. Isagulyants, V. I., N. S. Gosalova, and N. A. Chursina, Zh. Org. Khim., *7*, 1960 (1971).
269. Kakhniashrili, A. I., D. S. Parkzhikiya, and D. Sh. Toramshvili, Soobshch. Acad. Nauk Gruz. SSR, *76*, 337 (1974); Chem. Abstr., *81*, 49365j (1974).
270. Kakhniashvili, A. I., and E. N. Chikovani, Soobshch. Akad. Nauk Gruz. SSR, *53*(1), 93 (1969); Chem. Abstr., *71*, 80841d (1969).
271. Kakhniashvili, A. I., and E. N. Chikovani, Soobshch. Akad. Nauk Gruz. SSR, *52*(2), 357 (1968); Chem. Abstr., *70*, 106143j (1969).
272. Sonawane, H. R., M. S. Wadia, and B. C. SubbaRoa, Indian J. Chem. *6*(4), 191 (1968); Chem. Abstr., *69*, 106077v (1968).
273. Cardillo, B., L. Merlini, and S. Servi, Tetrahedron Lett., No. 10, 945 (1972).
274. Nazarov, I. N., and A. I. Kakhniashvili, Zh. Org. Khim., *22*, 617 (1952).
275. Kakhniashvili, A. I., and D. S. Parkzhikiya, Tr. Tbilis. Gos. Univ., *80*, 147 (1962).
276. Kakhniashvili, A. I., D. S. Pardzhikiya, and M. L. Kantariya, Zh. Org. Khim., *33*, 667 (1963).
277. Kakhniashvili, A. I., and D. Ya. Bugianishirli, Zh. Org. Khim., *1*, 1043 (1965); J. Org. Chem. USSR, *1*, 1051 (1965).
278. Kakhniashvili, A. I., D. S. Pardzhikiya, D. Ya. Bugianishvili, and L. I. Dzhibladze, Soobshch. Akad. Nauk Gruz. SSR, *54*(3), 569 (1969); Chem. Abstr., *71*, 112534z (1969).
279. Ecke, G. G., and A. J. Kolka, U.S. Patent 2,831,898 (Apr. 22, 1958).
280. Matsuda, H., A. Yamamoto, N. Iwamoto, and S. Matsuda, J. Org. Chem., *43*, 4567 (1978).
281. Mamedaliev, Yu. G., M. M. Guseinov, D. E. Mishiev, P. A. Petrosyan, and M. A. Salimov, Azerb. Khim. Zh., No. 5, 19 (1962); Chem. Abstr., *59*, 2676e (1963).
282. Schmerling, L., U. S. Patent 2,848,512 (1958).
283. Wood, T. F., and J. Angiolini, Tetrahedron Lett., *1*, 1 (1963).
284. Guseinov, M. M., D. E. Mishiev, P. A. Petrosyan, and I. M. Aliev, Azerb. Khim. Zh., Nos. 1-2, 85 (1971); Chem. Abstr., *74*, 42026w (1971).
285. Smith, F. M., U.S. Patent 2,623,912 (Dec. 30, 1952).
286. Axe, W. N., U.S. Patent 2,404,120 (July 16, 1946).

287. Axe, W. N., U.S. Patent 2,430,660 (Nov. 11, 1947).
288. Axe, W. N., U.S. Patent 2,471,922 (May 31, 1949).
289. Bansal, R. C., J. R. Mattox, E. J. Eisenbraun, P. W. K. Flanagan, and A. B. Carel, J. Org. Chem., *31*, 2716 (1966).
290. Eisenbraun, E. J., R. C. Bansal, J. R. Mattox, and P. W. K. Flanagan, Am. Chem. Soc., Div. Petrol. Chem., Prepr. *10*, 157 (1965).
291. Eisenbraun, E. J., J. R. Mattox, R. C. Bansal, M. A. Wilhelm, P. W. K. Flanagan, A. B. Carel, R. E. Laramy, and M. C. Hamming, J. Org. Chem., *33*, 2000 (1968).
292. Roberts, R. M., and D. Shiengthong, J. Am. Chem. Soc., *86*, 2851 (1964).
293. Shadan, A., and P. W. Flanagan, Am. Chem. Soc., Div. Petrol. Chem., Prepr., *15*(3), B60 (1970).
294. Bader, A. R., and W. C. Bean, J. Am. Chem. Soc., *80*, 3073 (1958).
295. Vdovtsova, E. A., J. Org. Chem. USSR (Engl.), *1*, 2233 (1965).
296. Isagulyants, V. I., and V. P. Evstaf'ev, Zh. Org. Khim., *1*(1), 102 (1965); J. Org. Chem. USSR, *1*, 99 (1965).
297. Isagulyants, V. I., and V. P. Evstaf'ev, Zh. Prikl. Khim., *36*, 2064 (1963).
298. Vdovtsova, E. A., Zh. Org. Khim., *5*, 504 (1969); J. Org. Chem. USSR, *5*, 491 (1969).
299. Mishiev, D. E., M. M. Guseinov, A. A. Mekraliev, and A. R. Musaeva, Uch. Zap. Azerb. Gos. Univ. Ser. Khim. Nauk No. 3, 76 (1967); Chem. Abstr., *7*, 12721a (1969).
300. Srebrodol'skaya, I. I., S. V. Zavgorodnii, and D. A. Pisanenko, Zh. Org. Khim. USSR, *6*, 1444 (1970); J. Org. Chem. USSR, *6*, 1457 (1970).
301. Schaad, R. E., U. S. Patent 2,283,465 (May 19, 1942).
302. Proell, W., J. Org. Chem., *16*, 178 (1951); U. S. Patent 2,564,077 (Aug. 14, 1951).
303. Bader, A. R., J. Am. Chem. Soc., *79*, 6164 (1957).
303a. Guseinov, M. M., S. Yu. Mamedalieva, and P. A. Petrosyan, Azerb. Khim. Zh., No. 4, 25 (1970); Chem. Abstr., *75*, 19820n (1971).
304. Mishiev, D. E., and M. T. Agaev, Sb. Tr. Azerb. Inst. NeftiKhim., 72 (1971); Ref. Zh., Khim., Abstr. 15Zh247 (1972); Chem. Abstr., *78*, 71739u (1973).
305. Smith, L. I., H. E. Ungnade, J. R. Stevens, and C. C. Christman, J. Am. Chem. Soc., *61*, 2615 (1939).
306. Vdovtsova, E. A., and A. M. Romanikhin, Zh. Obshch. Khim., *31*, 479 (1961).
307. Vdovtsova, E. A., and S. V. Zavgorodnii, Dokl. Akad. Nauk SSSR, *113*, 590 (1957).
308. Vdovtsova, E. A., Zh. Obshch. Khim., *31*, 95 (1961).
309. Vdovtsova, E. A., and G. I. Popova, Zh. Obshch. Khim., *33*(6), 1870 (1963); Chem. Abstr., *59*, 9860b (1963).
310. Vdovtsova, E. A., and G. I. Popova, Zh. Vses. Khim. Ova., *5*, 470 (1960); Chem. Abstr. *55*, 3516b (1961).
311. Vdovtsova, E. A., Zh. Obshch. Khim., *31*, 102 (1961).
312. Vdovtsova, E. A., and R. M. Yagudaev, Uzb. Khim. Zh., *6*(6), 37 (1962); Chem. Abstr., *59*, 3804b (1963).
313. Vdovtsova, E. A., and M. A. Aleksyuk, Zh. Obshch. Khim., *32*, 1491 (1962); Chem. Abstr., *58*, 4451h (1963).
314. Yagudaev, M. R., and E. A. Vdovtsova, Zh. Obshch. Khim., *32*, 2184 (1962); Chem. Abstr., *58*, 8878d (1963).

315. Yagudaev, M. R., and E. A. Vdovtsova, Dokl. Akad. Nauk Uzb. SSR, *20*(9), 31 (1963); Chem. Abstr., *60*, 8645c (1964).
316. Vdovtsova, E. A., Zh. Obshch. Khim., *33*(12), 3911 (1963); Chem. Abstr., *60*, 9183b (1964).
317. Vdovtsova, E. A., and A. G. Kakharov, Zh. Org. Khim., *6*, 1653 (1970); J. Org. Chem. USSR, *6*, 1663 (1970); Chem. Abstr., *73*, 109409y (1970).
318. Vdovtsova, E. A.,,and A. G. Kakharov, Zh. Org. Khim., *6*, 1664 (1970); J. Org. Chem. USSR, *6*, 1672 (1970); Chem. Abstr., *73*, 109410s (1970).
319. Bader, A. R., J. Am. Chem. Soc., *75*, 5967 (1953).
320. Bruson, H. A., and T. W. Riener, J. Am. Chem. Soc., *68*, 8 (1946).
321. Nesterenko, S. A., D. A. Piasarenko, and G. V. Sandul, Zh. Org. Khim., *10*, 286 (1974);J. Org. Chem. USSR, *10*, 287 (1974), Chem. Abstr., *80*, 120435u (1974).
322. (a) Suga, S., T. Nakajima, Y. Nakamoto, and K. Matsumoto, Tetrahedron Lett., 3283 (1969); (b) Nakajima, T., S. Masuda, S. Nakashima, T. Kondo, Y. Nakamoto, and S. Suga, Bull. Chem. Soc. Jpn. *52*, 2377 (1979).
323. Quasim, C., H. R. Sonawane, M. S. Wadia, and N. L. Dutta, Indian J. Chem., *9*, 181 (1971); Chem. Abstr., *74*, 111230x (1971).
324. Kuchkarev, A. B., D. A. Khankhodtaeva, and Kh. Z. Babaeva, Tr. Tashk. Politekh. Inst., No. 64, 85 (1970); Ref. Zh., Khim., Abstr. 22Zh298 (1971); Chem. Abstr., *77*, 113947k (1972).
325. Kakhniashvili, A I., L. I. Dzhibladze, and D. Sh. Ioramashvili, Soobshch. Akad. Nauk Gruz. SSR, *65*, 601 (1972); Chem. Abstr., *77*, 19286j (1972).
326. Glatz, A. M., A. Razus, F. Badea, and C. D. Nenitzescu, Rev. Chim., (Bucharest), *16*, 1567 (1971).
327. Glatz, A. M., F. Badea, Z. Arvay, V. Dragutan, F. Chiraleu, and A. Razus, Rev. Chim (Bucharest), *18*, 2087 (1973); Chem. Abstr., *80*, 120413k (1974).
328. Four, P., Tetrahedron Lett., No. 59, 6143 (1968); Chem. Abstr., *70*, 57355u (1969).
329. Four, P., Bull. Soc. Chim. Fr., (10), 4032 (1972); Chem. Abstr., *78*, 42396z (1973).
330. Elsworth, J. F., and M. Lamchen, J. S. Afr. Chem. Inst., *25*, 31 (1972).
331. Glatz, A. M., Rev. Chim. (Bucharest), *19*, 455 (1974).
332. Sammour, A., and M. El Hashash, J. Prakt. Chem., *314*, 906 (1972); Chem. Abstr., *78*, 57954d (1973).
333. El Hashash, M. A., M. Y. El Kady, and M. M. Mohamed, Ind. J. Chem., Sect. B, 18B(2), 136 (1979).
334. El Kady, M., M. M. Mohamed, and M. A. El Hashash, Rev. Chim. (Bucharest), *24*, 1499 (1979).
335. Drusiani, A., and L. Plessi, Atti Mem. Accad. Patavina Sci., Lett. Arti, *83* (Pt. 2), 193 (1970-1971); Chem. Abstr., *78*, 124335t (1973).
336. El Khawaga, A. M., M. A. thesis, University of Assiut, Assiut, Egypt, 1975.
337. Darzens, G., C. R. Acad. Sci., *189*, 766 (1929).
338. Konowalow, M., J. Russ. Phys.-Chem. Soc., *27*, 457 (1895).
339. Simonis, H., and S. Denischweski, Chem. Ber., *59*, 2914 (1926).
340. Shotter, R. G., K. M. Johnston, and J. F. Jones, Tetrahedron, *34*, 741 (1978).

341. Kretchner, R. A., and M. B. McCloskey, J. Org. Chem., *37,* 1989 (1972).
342. El Hashash, M., and M. M. Mohamed, Pak. J. Sci. Ind. Res., *20,* 325 (1977); Chem. Abstr., *91,* 56722h (1979).
343. Ried, W. and D. P. Shäfer, Chem. Ber., *102,* 4193 (1969).
344. Sammour, A., and M. Elkasaby, J. Chem., UAR, *13,* 409 (1970); Chem. Abstr., *77,* 101261f (1972).
345. Sammour, A., and A. A. Hamed, Egypt. J. Chem., *16,* 101 (1973).
346. Ansell, M. F., and S. A. Mahmud, J. Chem. Soc., Perkin Trans. *1,* 2789 (1973).
347. Abdel Maksoud, A., G. Hosni, O. Hassan, and S. Shafik, Rev. Chim. (Bucharest), *23,* 1541 (1978).
348. Grebenyuk, A. D., Zh. Org. Khim., *5,* 1469 (1969); Chem. Abstr., *71,* 112525x (1969).
349. Butler, D. E., Tetrahedron Lett., No. 19, 1929 (1972); Chem. Abstr., *77,* 6158x (1972).
350. Kikumasa, S., and T. Amakasu, J. Org. Chem., *33,* 2446 (1968).
351. Benzmenova, T. E., and T. V. Lysukho, Khim. Geterotsikl. Soedin., *7,* 1481 (1971); Chem. Abstr., *77,* 5261k (1972).
352. Cope, A. C., and W. D. Burrows, J. Org. Chem., *31,* 3093 (1966).
353. Ried, W., and R. Lantzsch, Justus Liebigs Ann. Chem., *756,* 173 (1972); Chem. Abstr., *77,* 5023j (1972).
354. Ermokhina, V. A., D. U. Arifova, and A. R. Abdurasulev, Sb. Nauchn. Tr., Tashk. Gos. Univ. V. I. Lenina, *553,* 61 (1978); Chem. Abstr., *92,* 198047m (1980).
355. Matsuda, H., and H. Shimohara, Chem. Lett., 95 (1978); Chem. Abstr., *88,* 104798z (1978).

6

Friedel-Crafts Cyclialkylation Reactions

I. INTRODUCTION

The term *cyclialkylation* was coined by Bruson and Kroeger [1] to refer to Friedel-Crafts alkylations of aromatic substrates with difunctional alkylating agents. Later, this definition was considerably extended by Barclay [2] in his excellent review of cyclialkylation of aromatics to include all electrophilic ring closures on aromatic systems. Through various modifications, this general type of ring formation has been of great synthetic usefulness, e.g., the synthesis of quinoline derivatives by the Bischler-Napieralski [3], the Pictet-Spengler [4], and the Pomeranz-Fritsch [5] reactions. In Chap. 5, we described alkylations of arenes with di- and polyfunctional alkylating agents, and we noted that bicyclic and polycyclic products frequently are formed. In this chapter we deal primarily with cyclialkylations involving intramolecular ring closures of arylhaloalkanes, arylhydroxyalkanes, and arylalkenes. By definition, these are respectively alkyl halides, alkanols, and alkenes which bear an aromatic residue in the chain. Aside from these systems, related ones are considered briefly.

II. CYCLIALKYLATIONS OF ARYLHALOALKANES AND RELATED SYSTEMS

A. Cyclialkylations of Arylhaloalkanes

In 1912, von Braun and Deutsch [6] investigated the action of $AlCl_3$ on a nubmer of phenylalkyl chlorides having the general formula $Ph(CH_2)_nCl$ and indicated that intramolecular ring closure took place readily only when *n* was greater than 3. Thus $Ph(CH_2)_2Cl$ gave no cyclialkylation product and $Ph(CH_2)_3Cl$ gave only about 10% of what was believed to be indan, whereas $Ph(CH_2)_4Cl$ gave 60% of tetralin, and $Ph(CH_2)_5Cl$ gave a similar yield of a product claimed, at that time, to consist chiefly "not of the expected benzsuberane but of phenylcyclopentane."

In 1927, von Braun and Kuhn [7] confirmed the latter result for $Ph(CH_2)_5Cl$ and extended the previous work to include the action of $AlCl_3$ on 1-chloro-2-methyl-5-phenylpentane (1), 1-chloro-5-(*p*-tolyl)pentane (2), and 1-chloro-5-(*p*-xylyl)pentane (3). The obviously incorrect results shown in Eqs. (1) to (3) were reported.

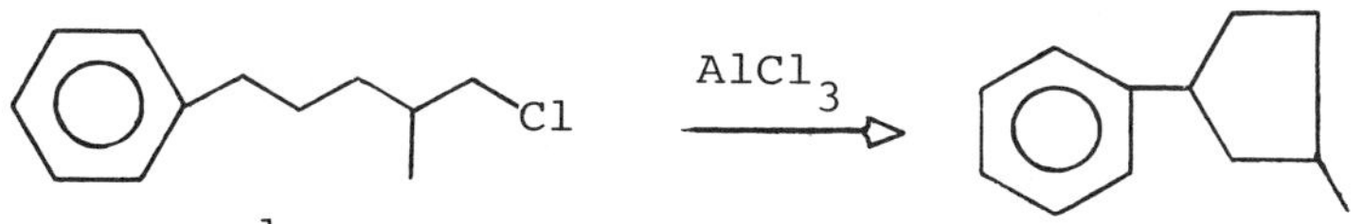

1 —$AlCl_3$→ (1)

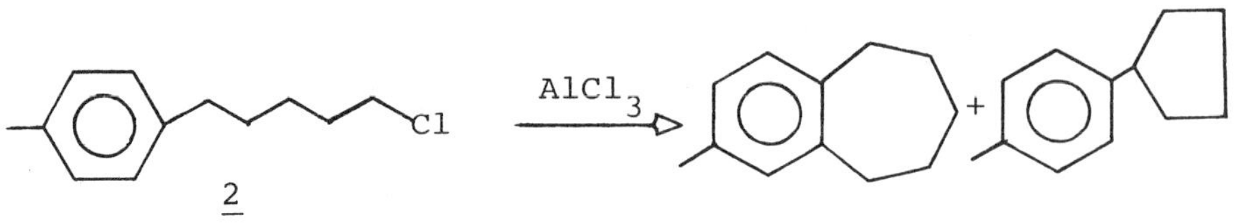

AlCl$_3$ (2)

2

AlCl$_3$ (3)

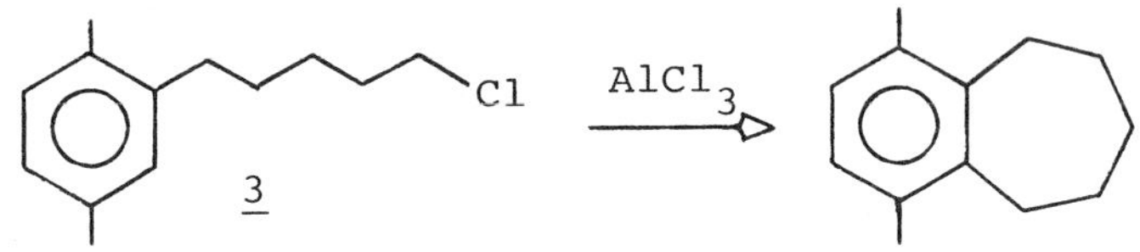

3

In 1956, Baddeley and Williamson [8,9] conducted two series of experiments primarily designed to gain information regarding the effect of proximity and electronic deactivation on competition between intramolecular ring closure and rearrangement processes. The results, which proved to be mostly misleading, are presented in Table 1.

From these results the authors concluded that intramolecular alkylations involving the formation of a five-, six-, or seven-membered ring were faster and less affected by isomerization of the alkylating moiety than were intermolecular alkylations. Moreover, by offering the types of transformation shown in scheme 1 (4 → 5 → 6 → 7) to account for the claimed production of 5-acetyl-2-methylindan (7) from 1-chloro-4-(*p*-acetylphenyl)butane (4), the authors also concluded that intramolecular alkylations provide facile primary attack even on a phenyl group which is deactivated by an acyl substituent.

It is to be noted that the foregoing generalizations of Baddeley and Williamson [8,9] were based on product identification by means of derivatives, a technique that is qualitative at best and certainly liable to failure in detecting minor products. Thus, although it is quite understandable that intramolecular alkylations to form cyclic products might occur more readily than the corresponding intermolecular Friedel-Crafts reactions (under similar conditions), certain generalizations regarding the extent or lack of accompanying rearrangements that were made were too broad and were, indeed, invalidated by more recent work. For example, with reference to Table 1, the reported closure of 1-chloro-5-phenylpentane to benzosuberane (entry 3) was shown to be in error by Barclay and co-workers in 1964 [10]. Instead, the product was 1-methyltetralin, resulting from primary-to-secondary rearrangement followed by ring closure. Also, the formation of indan from 1-chloro-3-phenylpropane (entry 1), of 1-ethylindan from 3-chloro-5-phenylpentane (entry 5), of only 1,1-dimethyltetralin from 2-chloro-2-methyl-4-phenylbutane (entry 6), and of 5-acetyl-2-methylindan from 1-chloro-4-(*p*-acetylphenyl)butane (entry 7) were shown to be incorrect by Khalaf and Roberts in 1966 [11]. 1-

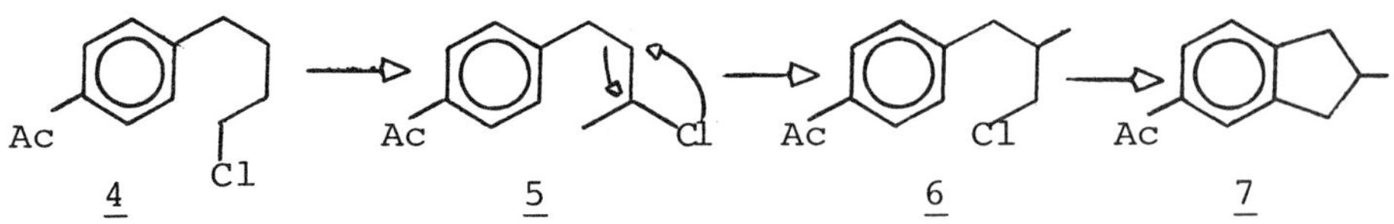

Ac Cl 4 → Ac Cl 5 → Ac Cl 6 → Ac 7

Scheme 1

Table 1. Results of Some Cyclialkylation Reactions as Reported by Baddeley and Co-workers

Entry	Halide treated	Catalyst	Product	Yield, %
1	$PhCh_2CH_2CH_2Cl$	$AlCl_3$	Indan	50
2	$PhCH_2CH_2CH_2CH_2Cl$	$AlCl_3$	Tetralin	50
3	$PhCH_2CH_2CH_2CH_2CH_2Cl$	$AlCl_3$	Benzosuberane	50
4	$PhCH_2CH_2CH_2CHClMe$	$AlCl_3$	1-Methyltetralin	40
5	$PhCH_2CH_2CHClEt$	$AlCl_3$	1-Ethylindan	40
6	$PhCH_2CH_2CH_2CMe_2Cl$	$AlCl_3$	1,1-Dimethyltetralin	—
7	p-$AcC_6H_4CH_2CH_2CH_2CH_2Cl$	$AlCl_3$-NaCl, 100°C	5-Acetyl-2-methylindan	60-70
8	p-$AcC_6H_4CH_2CH_2CHClMe$	$AlCl_3$-NaCl, 100°C	5-Acetyl-2-methylindan	60-70
9	p-$AcC_6H_4CH_2CHMeCH_2Cl$	$AlCl_3$-NaCl, 100°C	5-Acetyl-2-methylindan	60-70
10	p-$AcC_6H_4CHMeCH_2CH_2Cl$	$AlCl_3$-NaCl, 100°C	5-Acetyl-1-methylindan	60-70
11	p-$AcC_6H_4(CH_2)_5Cl$	$AlCl_3$-NaCl, 100°C	5-Acetyl-2-ethylindan	60-70
12	p-$AcC_6H_4CH_2CH_2CH_2CHClMe$	$AlCl_3$-NaCl, 100°C	5-Acetyl-2-ethylindan	60-70
13	p-$AcC_6H_4CH_2CH_2CHClEt$	$AlCl_3$-NaCl, 100°C	5-Acetyl-2-ethylindan	60-70
14	$PhCOCH_2CH_2CH_2Br$	$AlCl_3$-NaCl, 100°C	3-Methyl-1-indanone	—
15	$PhCOCH_2CH_2CHBrMe$	$AlCl_3$-NaCl, 100°C	4-Methyl-1-tetralone	—

Source: Refs. 8 and 9.

Table 2. Cyclialkylations Reported by Khalaf and Roberts in 1966

Compound cyclized	Catalyst[a]	Solvent[b]	Temp, °C	Time, hr	Product composition[c,d]
1-Chloro-3-phenylpropane	$AlCl_3$	CS_2	25	2.5	*n*-Propylbenzene (61), di-*n*-propylbenzene (39), residue (56)
	$AlCl_3$	CS_2	25-45	4.5	*n*-Propylbenzene (94), di-*n*-propylbenzene (6), residue (25)
	$AlCl_3$	SKB[c]	ca. 70	1.0	*n*-Propylbenzene (73), di-*n*-propylbenzene (27), indan (trace), residue (53)
1-Chloro-4-phenylbutane	$AlCl_3$	SKB	25	2.5	Tetralin (93), *n*-propylbenzene (7), residue (6)
2-Chloro-4-phenylbutane	$AlCl_3$	SKB	25	2.5	*n*-Butylbenzene (92), di-*n*-butylbenzene (8), residue (50)
	$AlCl_3$	SKB	70	1.0	*n*-Butylbenzene (77), di-*n*-butylbenzene (23), residue (41)
2-Chloro-2-methyl-4-phenylbutane	$AlCl_3$	SKB	0	2.5	1,1-Dimethylindan (25), isopentylbenzene (75), residue (47)
	$AlCl_3$	SKB	25	2.5	1,1-Dimethylindan (38), isopentylbenzene (61), *t*-pentylbenzene (1), residue (32).
	$AlCl_3$	SKB	70	1.0	1,1-Dimethylindan (47), isopentylbenzene (50), *t*-pentylbenzene (3), residue (38)
	$AlCl_3$	CS_2	25	0.5	1,1-Dimethylindan (31), isopentylbenzene (69)

	$FeCl_3$	CS_2	25	2.5	1,1-Dimethylindan (100), residue (54)
1-Chloro-2-methyl-4-phenylbutane	$AlCl_3$	SKB	25	2.5	1,1-Dimethylindan (4), 2-methyltetralin (33), 1-methyltetralin (ca. 1), isopentylbenzene (62), residue 27
	$AlCl_3$	SKB	70	1.0	1,1-Dimethylindan (8), 2-methyltetralin (52), 1-methyltetralin (ca. 6), isopentylbenzene (34), residue (23)
	$AlCl_3$	CS_2	25	2.5	1,1-Dimethylindan (2), 2-methyltetralin (65), 1-methyltetralin (ca. 2), isopentylbenzene (31), residue (27)
3-Chloro-5-phenylpentane	$AlCl_3$	SKB	25	2.5	1-Methyltetralin (98), n-$C_5H_{11}C_6H_5$ (1), residue (22)
	$AlCl_3$	SKB	70	1.0	1-Methyltetralin (99), n-$C_5H_{11}C_6H_5$ (1), residue (22)
2-Chloro-2-methyl-5-phenylpentane	$AlCl_3$	CS_2	25	2.5	1,1-Dimethyltetralin (35), isohexylbenzene (55), unidentified (10), residue (41)
	$AlCl_3$	SKB	70	1.0	1,1-Dimethyltetralin (25), isohexylbenzene (56), unidentified (19), residue (46)
	$FeCl_3$	CS_2	25	2.5	1,1-Dimethyltetralin (100), residue (17)
1-Chloro-3,4-diphenylbutane	$AlCl_3$	CS_2	0	3.0	2-Phenyltetralin (93), tetralin (7), 1-benzylindan (trace)

Table 2. (Continued)

Compound cyclized	Catalyst[a]	Solvent[b]	Temp, °C	Time, hr	Product composition[c,d]
	$AlCl_3$	CS_2	25	2.0	2-Phenyltetralin (15), tetralin (70), 1-benzylindan (3), unidentified (12)
	$AlCl_3$	SKB	70	1.0	2-Phenyltetralin (1), tetralin (51), unidentified (48)
	$FeCl_3$	CS_2	25	3.0	Unchanged starting material (100)
γ-Chlorobutyrophenone	$AlCl_3$	SKB	25	2.5	No cyclialkylation products
	$AlCl_3$-NaCl	—	100	1.0	3-Methyl-1-indanone (83), unidentified (17), residue (32)
1-Chloro-4-(*p*-acetylphenyl)-butane	$AlCl_3$-NaCl	—	100	1.0	5- and 6-Acetyl-1-methylindan isomers in roughly equal amounts (32), *p*-acetylbutylbenzene (12), 6-acetyltetralin (5), acetophenone (3), unidentified (48), residue (36)

[a]Molar ratio of $AlCl_3$ or $FeCl_3$ to phenylalkyl chloride was 0.50.
[b]SKB is Skellysolve B.
[c]Quantitative values from GLPC are relative.
[d]Residue is expressed in weight percent relative to starting material.
Source: Ref. 11.

Chloro-4-phenylbutane (entry 2) and γ-halobutyrophenone (entry 14), however, gave the reported products, but with some contamination. The corrected results for entries 1 to 7 and 14 as well as the results for some other related cyclialkylations studied by Khalaf and Roberts [11] in their 1966 paper are given in Table 2 and partly in Eqs. (4) to (10).

$AlCl_3$, 25°; CS_2, 2.5 hr → 61% + 39% (4)

$AlCl_3$, 25°; SKB,* 2.5 hr → 93% + 7% (5)

$AlCl_3$, 25°; SKB, 2.5 hr → 92% + 8% (6)

$AlCl_3$, 25°; SKB, 2.5 hr → 38% + 61% + 1%
$FeCl_3$, 25°; CS_2, 2.5 hr → 100% (7)

$AlCl_3$, 25°; CS_2, 2.5 hr → 35% + 2% (8)

8

*Skellysolve B

$AlCl_3$, 25°
CS_2, 2.5 hr
35% + 55%
+ unidentified
10%
Cl
$FeCl_3$, 25°
CS_2, 2.5 hr
100%
(9)

$AlCl_3$-NaCl
100°, 1 hr
+ unidentified
17%
O
C
83% O
Cl
$AlCl_3$, 25°
SKB, 2.25 hr
No cyclialkylation products
(10)

It can be seen from both Table 2 and Eqs. (4) to (10) that, in contrast to the older results of Baddeley and Williamson [8]:

1. 1-Chloro-3-phenylpropane (Eq. 4) gave only traces of indan; the major products were those formed by hydride transfer (*n*-propylbenzene), by disproportionation (di-*n*-propylbenzene), and by intermolecular polyalkylation (high-boiling residues).

2. 1-Chloro-4-phenylbutane (Eq. 5) gave tetralin which was contaminated with minor amounts of the hydride transfer product, *n*-butylbenzene. (Interestingly, Roth and von Auwers in 1915 [12] also recognized the production of *n*-butylbenzene in a similar reaction.)

3. The product of cyclization of 3-chloro-5-phenylpentane (8, Eq. 8) was not 1-ethylindan as reported previously, but 1-methyltetralin, whose formation was attributed to a secondary-to-secondary rearrangement prior to ring closure (Eq. 11). The possibility that 1-methyltetralin could have been

$-Cl^-$
$\sim H:^-$
Cl + +
8
(11)
$-H^+$

formed by isomerization of an initially formed 1-ethylindan was ruled out by demonstrating the stability of the latter to the reaction conditions.

4. With $AlCl_3$ as catalyst, 2-chloro-2-methyl-5-phenylpentane (Eq. 9) gave not only the reported 1,1-dimethyltetralin but also the hydride exchange product isopentylbenzene (mainly).

$$\mathbf{4} \xrightarrow[100°,\ 1\ hr]{AlCl_3\text{-}NaCl} \mathbf{9}\ (16\%) + \mathbf{10}\ (16\%) + \mathbf{11}\ (12\%) + \mathbf{12}\ (5\%) + \mathbf{13}\ (3\%) + \text{Unidentified}\ (48\%) \tag{12}$$

5. 1-Chloro-4-(*p*-acetylphenyl)butane (4) gave, among other unidentified products, the compounds shown in Eq. (12). None of the previously reported [8] 5-acetyl-2-methylindan was found in the mixture. In this case the formation of 6-acetyl-1-methylindan (9) was explained by a simple primary-to-secondary carbocation rearrangement followed by ring closure, while that of the isomeric 5-acetyl-1-methylindan (10) was attributed to subsequent isomerization of 9, a type of isomerization which is well known [13]. At the same time, *p*-acetyl-*n*-butylbenzene (11) and acetophenone (13) were assumed to result from hydride transfer and dehaloalkylation, respectively. The fact that the product of direct cyclialkylation, 6-acetyltetralin (12), was produced only in a minor amount, is in good accord with the assumption that straightforward primary closures can hardly be effected after the reactivity of the phenyl group as a nucleophile has been decreased by acetylation [3,11].

We shall see later that the recent results of Roberts and Khalaf agree closely with those obtained earlier by Bogert and co-workers [14,15] from the cyclialkylations of the corresponding phenylalkanols with sulfuric and polyphosphoric acids.

Examining the cyclialkylations of phenylalkyl halides above, it is quite clear that production of tetralins (six-membered ring formation) is the predominant reaction pathway. There is also a considerable amount of evidence from other methods of cyclization to show that six-membered ring formation, when possible, is the most favored in terms of both entropy and strain factors [16].

In an attempt to determine the nature of the various competing factors influencing cyclialkylation processes, Roberts and co-workers have investigated the cyclialkylation behaviors of a number of carefully designed mono- and diphenylalkyl chlorides. A study of the reaction of 1-chloro-2-methyl-4-phenylbutane (14) was chosen by Khalaf and Roberts [11] to provide more insight into the difference in behavior of 1-chloro-4-phenylbutane, which cyclizes without rearrangement (Eq. 5), and 3-chloro-5-phenylpentane (Eq. 8), which shows rearrangement. The following question was then asked: Is it because the cyclialkylation of a primary chloride is a concerted process, with no free primary carbocation intermediate, whereas the reaction of the secondary chloride involves a carbocation intermediate? When 1-chloro-2-

methyl-4-phenylbutane (14) was prepared and cyclized with $AlCl_3$ under various conditions, it gave the four products (15-18) shown in Eq. (13).

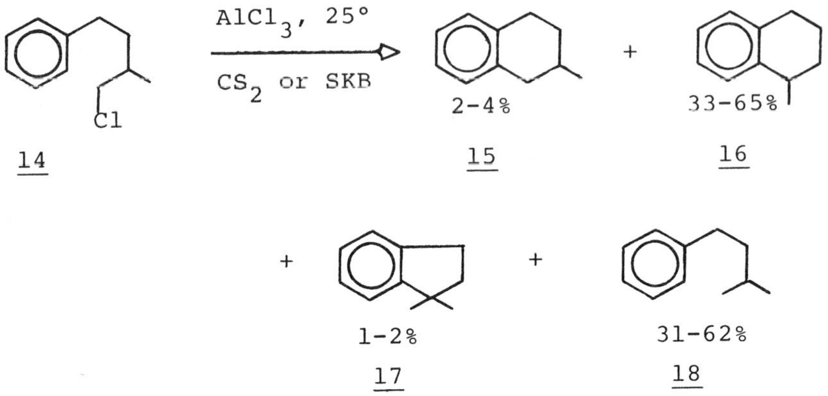

$AlCl_3$, 25°; CS_2 or SKB

14 → 15 (2-4%) + 16 (33-65%) + 17 (1-2%) + 18 (31-62%) (13)

The finding that the primary chloride 14 did give a small amount of rearrangement products (16 and 17) resulting from both methide and hydride

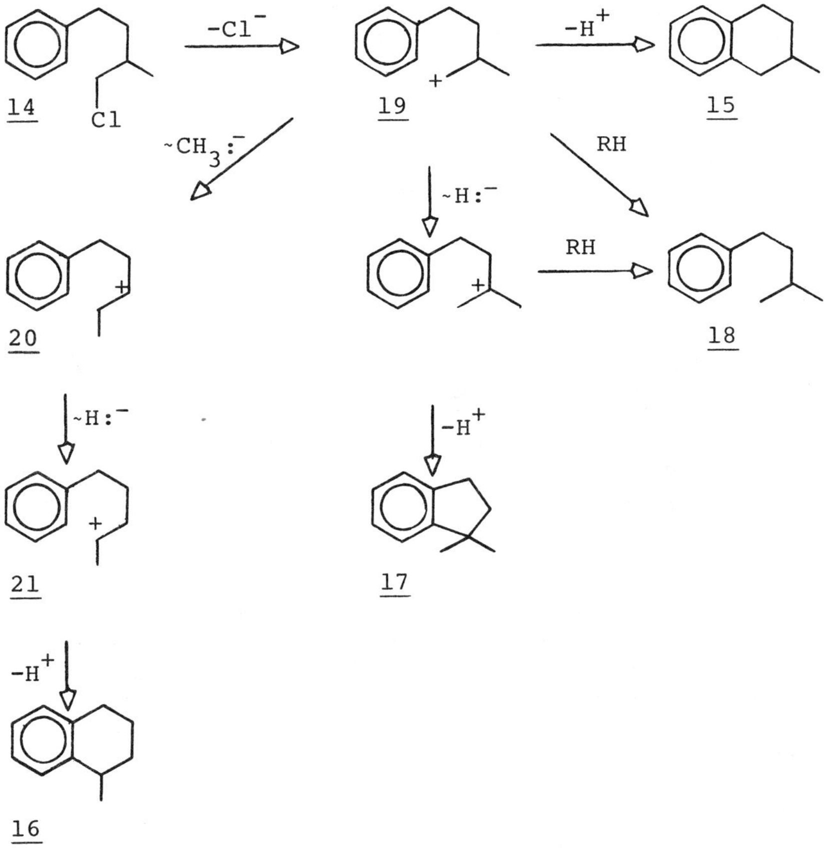

14 —($-Cl^-$)→ 19 —($-H^+$)→ 15

14 —($\sim CH_3:^-$)→ 20

19 —(RH)→ 18

19 —($\sim H:^-$)→ [cation] —(RH)→ 18

[cation] —($-H^+$)→ 17

20 —($\sim H:^-$)→ 21 —($-H^+$)→ 16

Scheme 2

transfers provided a partial answer for the question above. Just as in intermolecular alkylation, when the thermodynamic incentive is great enough (i.e., rearrangement of a primary carbocation to a tertiary one by a simple 1,2 hydride shift) rearrangement does occur. Scheme 2 was then suggested to rationalize the formation of compounds 15-18 from chloride 14 upon treatment with $AlCl_3$.

According to scheme 2, the formation of the rearrangement product 16 was visualized as taking place by a series of carbocation transformations (19 → 20 → 21 → 16), all preceding the ring closure step. More recently, however, this route was disqualified by Khalaf [17] on the ground that none of the rearrangement product 16 was obtained when the reaction of 14 was induced by the nonisomerizing $AlCl_3$-CH_3NO_2 catalyst. Under these conditions, the product consisted only of 15 (44%) and 17 (56%). Accordingly, 1-methyltetralin (16) was suggested to result from the isomerization of the initially formed 2-methyltetralin (15) by the steps shown in Eq. (14).

$$\underset{15}{\text{2-methyltetralin}} \underset{RH}{\overset{R^+}{\rightleftharpoons}} \text{(cation)} \underset{}{\overset{\sim CH_3:^-}{\rightleftharpoons}} \text{(cation)} \underset{R^+}{\overset{RH}{\rightleftharpoons}} \underset{16}{\text{1-methyltetralin}} \quad (14)$$

The latter alternative route for the formation of 16 gained experimental support from the fact that 15 and other similar methylated tetralins underwent partial rearrangement in the presence of $AlCl_3$ catalyst [17,18].

It became clear from the results described so far that rearrangements do accompany cyclialkylations with primary and secondary phenylalkyl chlorides [10,11]. However, there remained the possibility that these rearrangements could have been more extensive if it were not for the difference in the ease of forming tetralin rings compared with indan rings. A clear-cut test of this difference was provided by the synthesis of 1-chloro-3,4-diphenylbutane (22) and a study of its reactions with aluminum chloride. With reference to scheme 3, cyclialkylation of this molecule might occur in three different ways: (1) without rearrangement to form 2-phenyltetralin (23), (2) without rearrangement to form 1-benzylindan (24), or (3) with rearrangment to form 1-methyl-2-phenylindan (25). The fact that the almost-exclusive cyclialkylation products were 2-phenyltetralin and its dealkylation product, tetralin (Eq. 15), showed clearly that six-membered ring formation took precedence over five-membered ring formation, either directly from a primary chloride or via rearrangement to a secondary intermediate. The possibility that 1-benzylindan was formed first and subsequently rearranged to 1-phenyltetralin and 2-phenyltetralin was excluded by the fact that 1-benzylindan was proved to be stable to the reaction conditions and, also, no traces of 1-phenyltetralin were found in the reaction product mixture (see scheme 3 and Eq. 15).

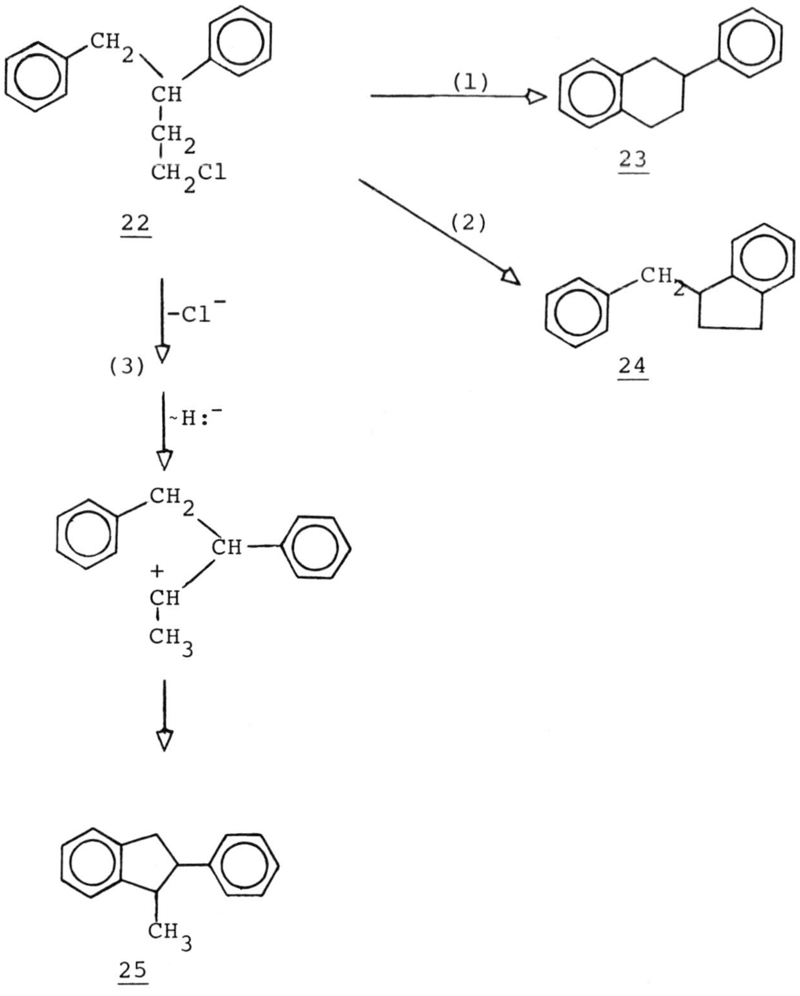

22 → (1) → 23

22 → (2) → 24

22 → (3) $-Cl^-$ → $\sim H:^-$ → 25

Scheme 3

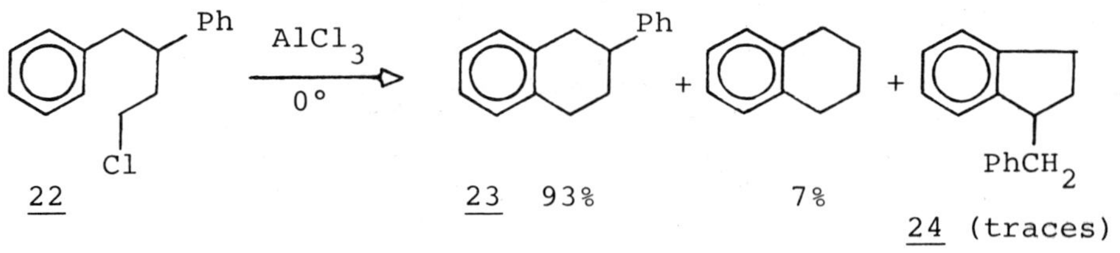

22 → $AlCl_3$, 0° → 23 93% + 7% + 24 (traces)

(15)

Since the preceding results for chloride 14 gave excellent evidence of the great preference for direct 1° closure to tetralin over rearranged 2° closure to indan, it was considered desirable by Roberts and co-workers [19] to test the extent of isomerization when direct 1° and rearranged 2° or 3° carbocations were given the equal opportunity of competing for closure to a tetralin ring. It was expected that a 1° → 2° or 1° → 3° cationic isomerization would now occur more extensively than before due to removal of the ring-size factor. The systems chosen to test these competing possibilities were 1-chloro-4,5-diphenylpentane (26a, for direct 1° versus rearranged 2° closure to tetralin) and 1-chloro-2-methyl-4,5-diphenylpentane (26b, for direct 1° versus rearranged 3° closure to tetralin). The various possible direct and rearranged cyclialkylations of these phenylalkyl chlorides are outlined in scheme 4. The numbers over the arrows refer to the ring size produced in the cyclialkylation step. The actual products formed from each of the two chlorides upon treatment with $AlCl_3$ in CS_2 at 25°C and Skellysolve B (SKB) at 25 and 70°C are shown in Table 3 and partly in Eqs. (16) and (17).

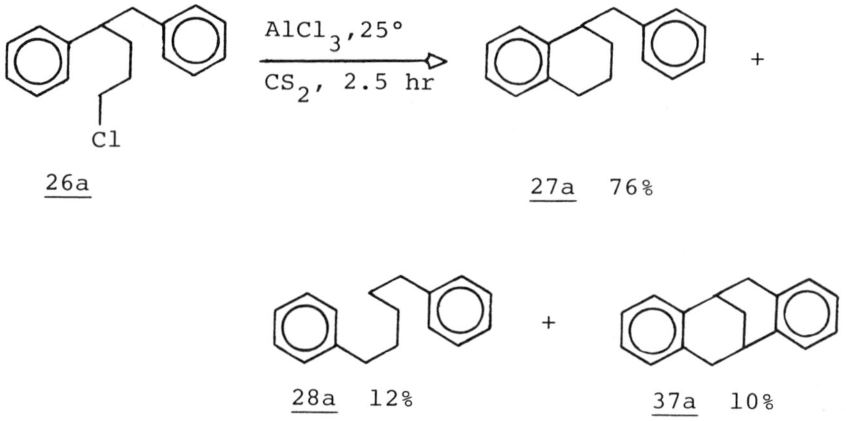

Cl

26a $\xrightarrow[CS_2,\ 2.5\ hr]{AlCl_3,\ 25°}$ 27a 76% + 28a 12% + 37a 10% (16)

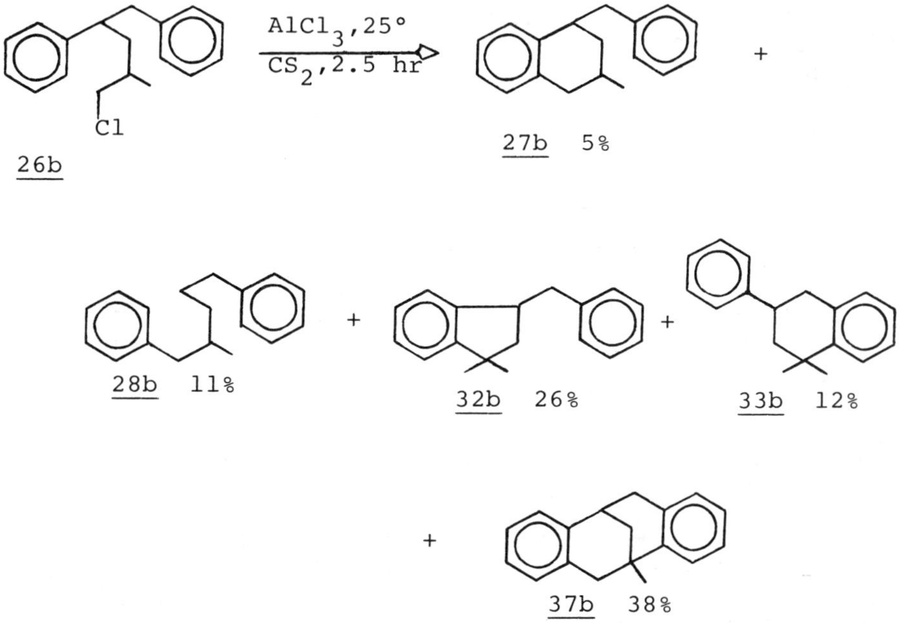

Cl

26b $\xrightarrow[CS_2,\ 2.5\ hr]{AlCl_3,\ 25°}$ 27b 5% + 28b 11% + 32b 26% + 33b 12% + 37b 38% (17)

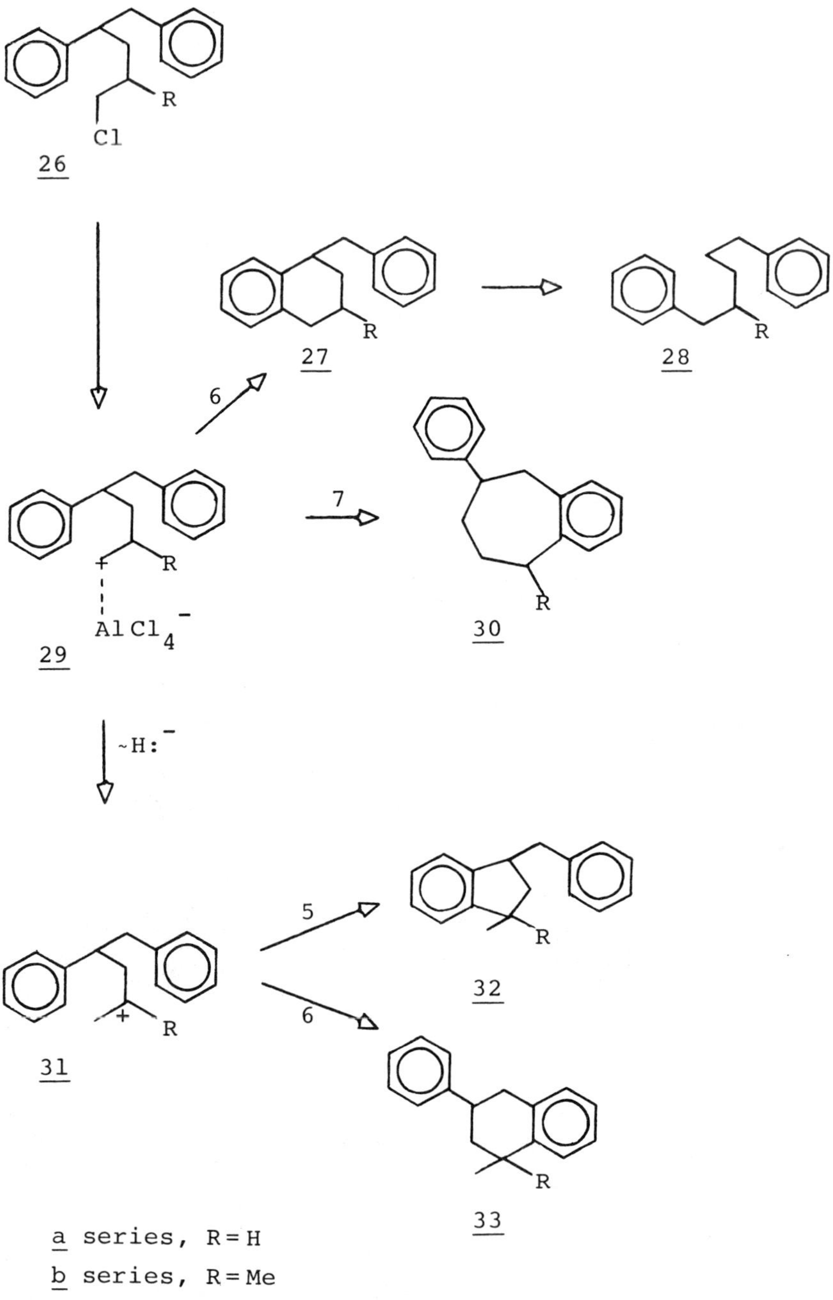
R
Cl
26
27
28
6
7
29
30
AlCl4
~H:
5
6
31
32
33
a series, R = H
b series, R = Me

Scheme 4

With reference to scheme 4, Table 3, and Eqs. (16) and (17), the treatment of 1-chloro-4,5-diphenylpentane (26a) with $AlCl_3$ in CS_2 at room temperature gave 1-benzyltetralin (27a) as the major product mixed with some 1,5-diphenylpentane (28a) and 2,3:6,7-dibenzobicyclo[3.3.1]nona-2,6-diene

Table 3. Products from Cyclialkylation and Bicyclialkylation of Diphenylalkyl Chlorides (26a and 26b) with Aluminum Chloride

Diphenylalkyl chlorides	Solvent	Temp., °C	Time hr.	Products,[a] %				
				27a	28a	32a	33a	37a
1-Chloro-4,5-diphenylpentane (26a)	SKB	25	2.5	47	36	0	0	15
	SKB	70	1.0	21	51	0	0	27
	CS_2	25	2.5	76	12	0	0	10
				27b	28b	32b	33b	37b
1-Chloro-2-methyl-4,5-diphenylpentane (26b)	SKB	25	2.5	9	10	16	19	39
	CS_2	25	2.5	5	11	26	12	38

[a]Relative amounts of products distilling in the diphenylalkane range; about 50% of reaction products were in the monophenylation range. The relative amounts were determined by GLPC; totals do not add up to 100% because small amounts of known and unidentified products are not included in the table.

(37a). Both 28a and 37a were suggested to result from the initially formed 27a; the former (28a) by dealkylation at the tertiary carbon followed by hydride exchange (27a → 34a → 35a → 28a) and the latter (37a) by hydride abstraction followed by bicyclialkylation (27a → 36a → 37a) (Eq. 18). When

(18)

a series, R = H

b series, R = Me

the reaction was carried out in petroleum ether at 70°C for 2.5 hr, the amount of 27a decreased and the amounts of 35a and 37a increased. No 2-phenylbenzosuberane (30a), 1-benzyl-3-methylindan (32a), or 1-methyl-3-phenyltetralin (33a) could be detected under any of the conditions employed, however.

The lack of formation of 30a showed, in another way, the preference for six-membered ring formation over seven-membered ring formation. Moreover, the lack of formation of 32a and 33a, which are expected to result form a 1° → 2° carbocation rearrangement prior to the cyclialkylation step, showed that the degree of participation of the nucleophile in a concerted displacement of the halide is greater than in analogous intermolecular reactions.

Treatment of 1-chloro-2-methyl-4,5-diphenylpentane (26b) with aluminum chloride in petroleum ether or carbon disulfide at room temperature gave a more complex mixture of products, including 1-benzyl-3-methyltetralin (27b), 1-benzyl-3,3-dimethylindan (32b), 1,1-dimethyl-3-phenyltetralin (33b), 2-methyl-1,5-diphenylpentane (28b), and 1-methyl-2,3:6,7-dibenzobicyclo[3.3.1]-nona-2,6-diene (37b) (Table 3 and Eq. 17). No 1-methyl-4-phenylbenzosuberane (30b) could be detected. The ratio fo 37b/27b was much higher than the ratio of 37a/27a produced under similar reaction conditions, reflecting the greater ease of abstracting a hydride ion from the tertiary C-3 carbon of 27b than from the corresponding secondary carbon of 27a. The formation of the rearranged cyclialkylation products 32b and 33b may also be attributed to the greater driving force for rearrangement of the intermediate primary complex 29 to the tertiary carbonium ion 31b, rather than to the secondary carbonium ion 31a. The more facile formation of a five-membered ring by an intermediate tertiary carbonium ion is also a factor; this has been noted before [11].

The intermediacy of 27a and 27b in the formation of 37a and 37b from 26a and 26b was confirmed by treating 27a and 27b separately with aluminum chloride. The major products, as expected, were as follows: from 27a, 37a, and 28a; from 27b, 37b, and 28b. No products corresponding to structures 32 and 33 were found.

The 1-chloro-2-methyl-4,5-diphenylpentane (26b) used as starting material in the cyclialkylations and bicyclialkylations reported here was a mixture of diastereomers. In order to examine the possibly different behaviors of the individual diastereomers toward cyclialkylation and bicyclialkylation [19,20], Khalaf and Roberts [21] obtained the pure diastereomeric alcohols (38a and 38b) and corresponding chlorides (39a and 39b) and subjected them to various cyclization conditions. The conditions employed and the results obtained are depicted in Table 4.

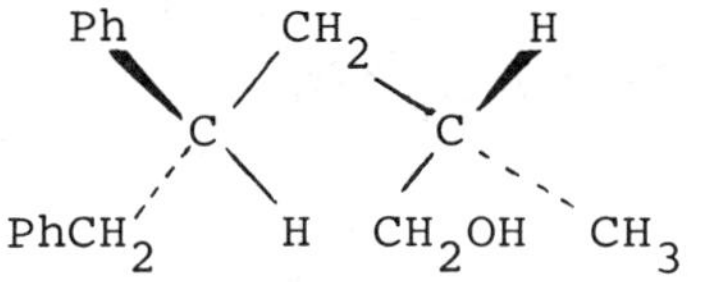

38a

a series, *R,S;S,R*-pair

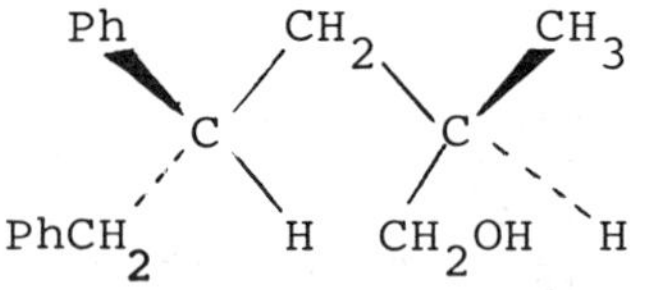

38b

b series, *R,R;S,S*-pair

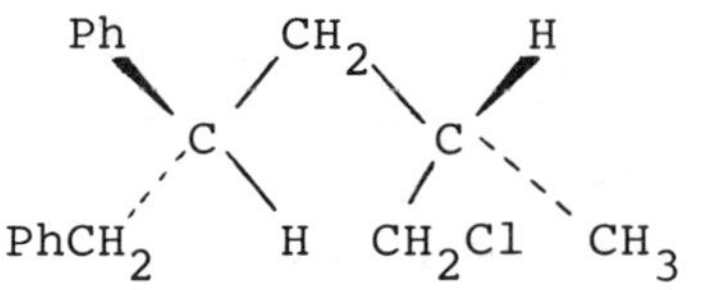

39a

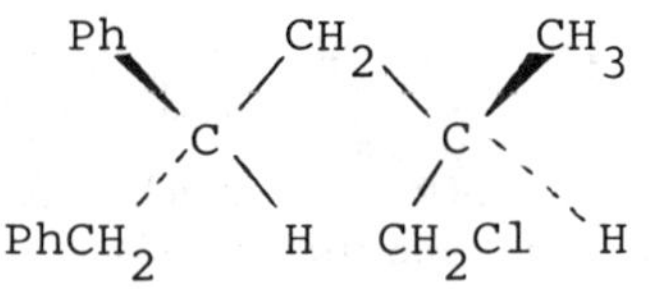

39b

Table 4. Products from Cyclialkylation and Bicyclialkylation of Diastereomeric 1-Chloro- and 1-Hydroxy-2-methyl-4,5-diphenylpentanes

Starting compound	Catalyst	Time, hr	Products, %,[a-c] 41	32b	28b	*trans*-27b	*cis*-27b	33b	37b	Starting compound
Chloride 39a	$AlCl_3$	2.5	1	10	4	43	12	16	10	
Chloride 39b	$AlCl_3$	2.5	2	21	4	3		29	38	
Chloride 39a	$AlCl_3$-CH_3NO_2	4		Trace		18	Trace	20		62
		9		Trace		20	Trace	30		50
		24		Trace		25	Trace	35		40
Chloride 39b	$AlCl_3$-CH_3NO_2	4		Trace		Tr	4	14		82
		9		Trace		Trace	7	22		71
		24		Trace		Trace	10	30		60
Alcohol 38a	H_3PO_4	0.25		7		31		62		
Alcohol 38b	H_3PO_4	0.25		5		6		85		

[a]Products are arranged in order of increasing GLPC retention times on a 16-ft X 0.125-in. DEGA column.
[b]Relative amounts of products distilling in the diphenylalkane range; about 15 to 20% of the reaction products were in the monophenylalkane range as a result of dephenylation. The "monophenylalkanes" produced consisted chiefly of 1,1-dimethylindan and 2-methyl-5-phenylhexane, with minor amounts of 1,3-dimethyltetralin.
[c]The relative amounts were determined by GLPC; totals do not add up to 100% because small amounts of unidentified products are not included in the table.

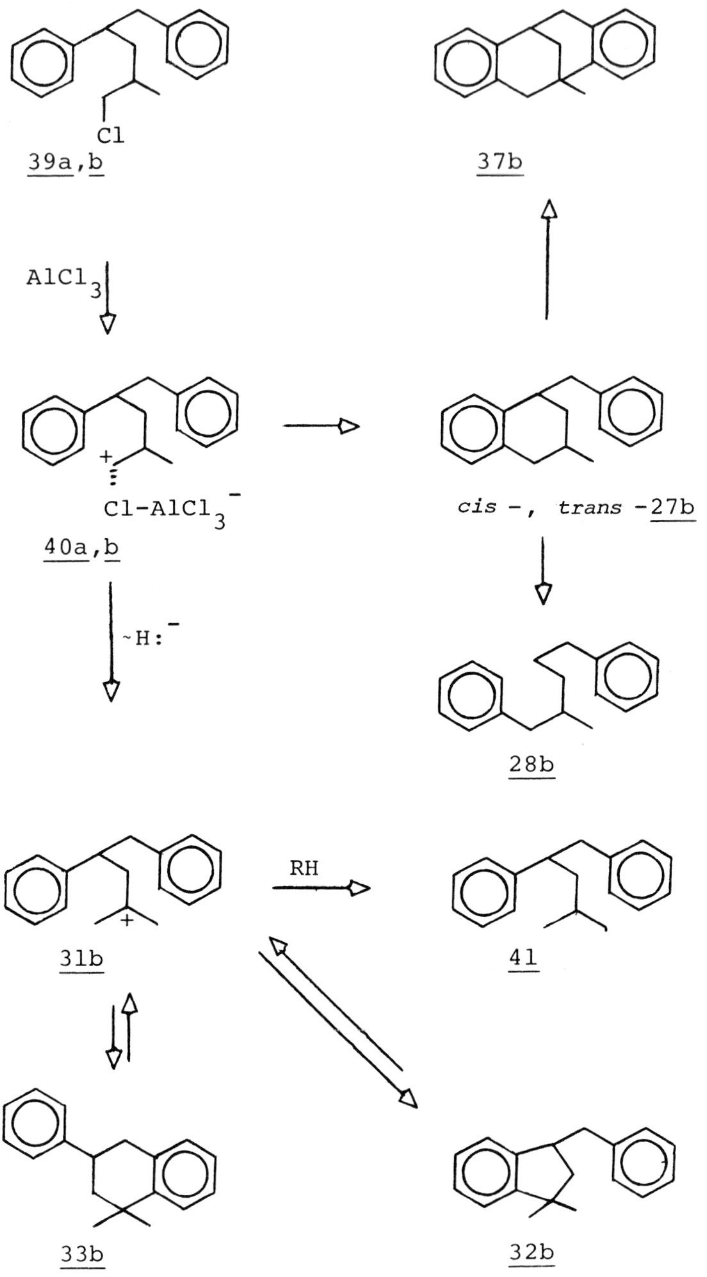

Cl
39a,b
37b
$AlCl_3$
$Cl{-}AlCl_3^-$
40a,b
cis –, *trans* –27b
~H:⁻
28b
RH
31b
41
33b
32b

Scheme 5a

By comparing the results of Table 4 with those of Table 3, it can be seen that, in accord with our former result, both chlorides 39a and 39b gave with $AlCl_3$ product mixtures composed of *cis*- and/or *trans*-1-benzyl-3-methyltetralin (27b, scheme 5a), 1,1-dimethyl-3-phenyltetralin (33b), 1-benzyl-3,3-dimethylindan (32b), 1-methyl-2,3:6,7-dibenzobicyclo[3.3.1]nona-2,6-diene

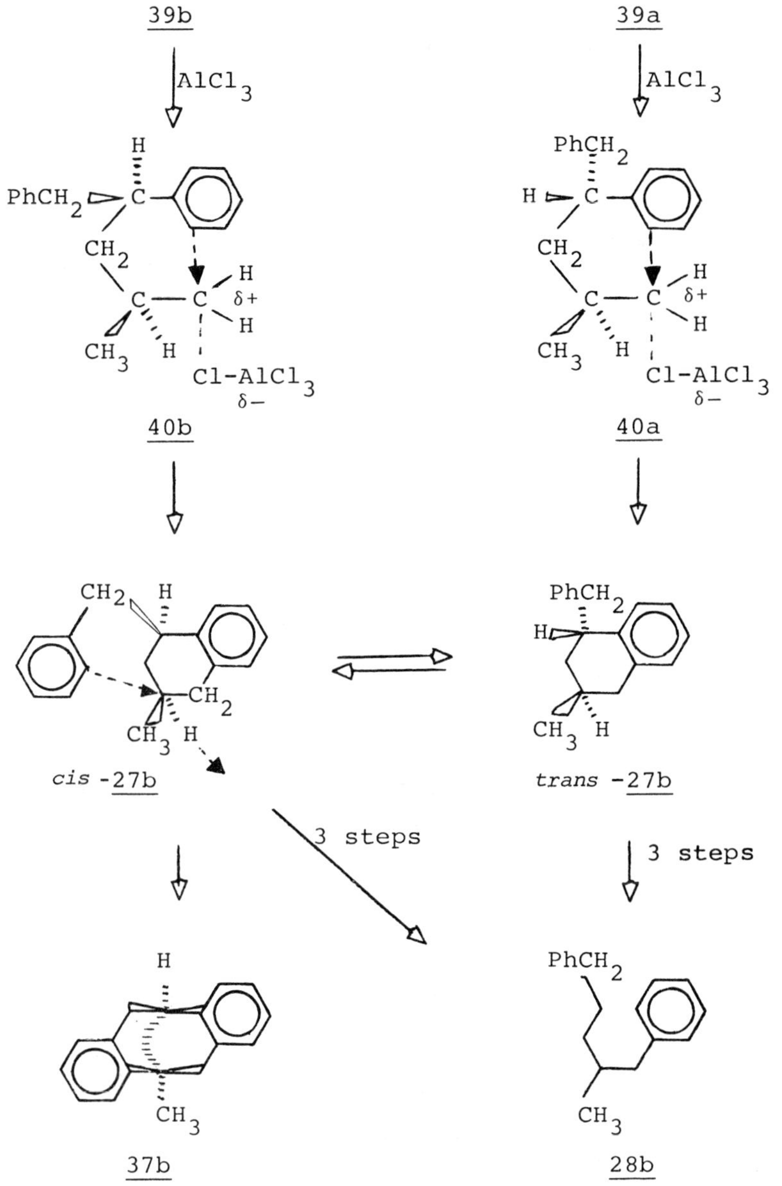

Scheme 5b

(37b), and 2-methyl-1,5-diphenylpentane (28b). An additional product found was 2-methyl-4,5-diphenylpentane (41). However, the relative yields of these components in the product mixture from chloride 39a were significantly different from those in the product mixture from chloride 39b.

In the first place, the yields of the tertiary cyclialkylation products 32b and 33b were twice as great from chloride 39b as from chloride 39a. This was explained by inspection of models of chlorides 39a and 39b (or their enantiomers) which indicated that 39a should cyclize with no difficulty to *trans*-1-benzyl-3-methyltetralin (*trans*-27b, scheme 5b), whereas 39b should give *cis*-1-benzyl-

3-methyltetralin (*cis*-27b), but with considerable steric opposition exerted by the developing 1,3 benzyl-methyl interaction. This steric hindrance to cyclialkylation by the primary complex 40b may allow rearrangement to the tertiary carbocation 31b to compete significantly, resulting in higher yields of the tertiary cyclialkylation products 32b and 33b from 39b than from 39a.

The other major product from 39b was 37b resulting from hydride abstraction at the tertiary C-3 carbon atom of *cis*-27b concerted with phenyl participation in the cyclialkylation. Apparently, most of the *cis*-1-benzyl-3-methyltetralin (*cis*-27b) formed by primary cyclilakylation of 40b (in competition with rearrangement to 31b as mentioned above) underwent facile bicyclialkylation; no *cis*-27b remained in the reaction mixture. By contrast, *trans*-1-benzyl-3-methyltetralin (*trans*-27b) is the major product (43%) from 39a. Although it is produced more easily by primary cyclialkylation than the *cis* isomer, it does not undergo bicyclialkylation readily. In *trans*-27b the C-3 hydrogen is on the same side of the tetralin ring as the C-1 benzyl group, so that the phenyl group cannot assist in the hydride abstraction. The 10% of 37b found in the reaction mixture from 39a probably comes from bicyclialkylation of *cis*-27b, which is produced by isomerization of *trans*-27b* [122].

In order to obtain additional information on the stereochemical course of cyclialkylation reactions, with less complication from side reactions such as dealkylations and bicyclialkylations, Khalaf and Roberts studied the cyclialkylation of chlorides 39a and 39b in the presence of nitromethane-moderated aluminum chloride, and the reactions of alcohols 38a and 38b in the presence of phosphoric acid. Moreover, they also followed the progress of the $AlCl_3$-CH_3NO_2-catalyzed reaction by analyzing samples taken from the reaction mixture after various time intervals. The results of these reactions are also included in Table 4.

Examination of these results indicates that they not only confirm the previous conclusions about the stereochemical behavior of chlorides 39a and 39b, but also provide two more interesting pieces of information that are in good agreement with the present theories. First, from a comparison of the rates of primary cyclialkylation (to form *trans*-27b and *cis*-27b) of chlorides 39a and 39b, determined after similar durations, it was judged that chloride 39a reacts roughly three times as fast as does chloride 39b. This difference in reactivity was explained simply in terms of the severe benzyl-methyl interactions experienced by chloride 39b upon cyclialkylation to *cis*-27b.

Second, the data from both the $AlCl_3$-CH_3NO_2- and H_3PO_4-catalyzed reactions indicated that 1,1-dimethyl-3-phenyltetralin (33b) is the chief tertiary-cyclialkylation product resulting form the closure of cation 31b. This is not unexpected on the basis of the fact that six-membered ring formation, when possible, is the most favored in terms of both entropy and strain factors [16]. Furthermore, this finding is in accord with the observation that 33b was the sole product obtained when 4,5-diphenyl-2-methyl-1-pentene was subjected to the action of 85% sulfuric acid [18].

No bicyclialkylation product 37b was found in any of the reactions of the chlorides 39a and 39b with $AlCl_3$-CH_3NO_2 or of the alcohols 38a and 38b with phosphoric acid. This is understandable in terms of the known fact that these catalysts are much poorer hydride-abstracting agents than unmodified $AlCl_3$, and it is the abstraction of hydride from the tertiary C-3 carbon in 27b that is essential to the bicyclialkylation.

---

*R. M. Roberts et al. [22] showed that *cis*-27b and *trans*-27b can be interconverted under the influence of $AlCl_3$.

The result above, together with the finding that in the $AlCl_3$-catalyzed reactions both chlorides 39a and 39b gave 32b and 33b in an apparent equilibrium ratio of about 1:1.2 to 1.4, directed attention to the possibility that the latter ratio may be the result of a secondary process involving the isomerization of 33b to 32b. To examine this possibility, Khalaf and Roberts investigated the behavior of both 32b and 33b in the presence of $AlCl_3$ under conditions comparable to those of the alkylation reactions. They found that both hydrocarbons gave isomerization mixtures in which the ratios of 33b to 32b were indeed very similar to those observed in the alkylation mixture, thus substantiating the proposition that 32b is formed mainly by intramolecular isomerization of 33b by $AlCl_3$.

The presence of 2-methyl-1,5-diphenylpentane (28b) among the products of reaction of 39a and/or 39b with $AlCl_3$ was indicated previously [19], and was also confirmed by the later results. Investigation of its mode of formation confirmed the proposition that it was formed by the dealkylation at the C-1 position of 1-benzyl-3-methytetralin (27b) (see Eq. 18). The small amount of the isomeric 2-methyl-4,5-diphenylpentane (41) detected in the reaction mixtures from 39a and 39b was assumed to come from hydride exchange with the intermediate 31b.

The question concerning competition between rearranged 3° closure to indan and direct 2° or 3° closure to tetralin was answered by Khalaf and Roberts [18] through an investigation of the cyclialkylations of the 3-methyl-5-phenyl system 42 and the 2,3-dimethyl-5-phenyl system 43. The results

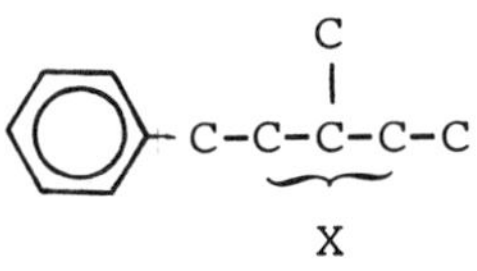

42 (X = Cl or OH)

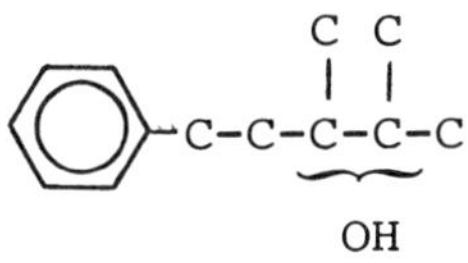

43

showed that, under nonisomerizing conditions in the presence of $AlCl_3$-CH_3NO_2 or H_2SO_4 as the catalyst, compounds 42 produced essentially a mixture of 1,2-dimethyltetralin (44) (mainly) and 1-ethyl-1-methylindan (45), showing the preference for secondary six-ring formation over tertiary five-ring formation (Eq. 19). Under similar conditions compounds 43 gave only 1,1,2-trimethyl-

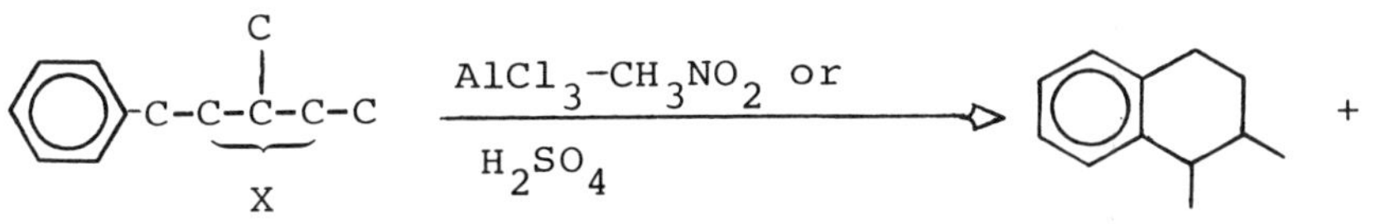

$$\text{42 (X = Cl or OH)} \xrightarrow[H_2SO_4]{AlCl_3-CH_3NO_2 \text{ or}} \text{44 (70-80\%)} + \text{45 (20-30\%)} + \text{46 (1-3\%)} \tag{19}$$

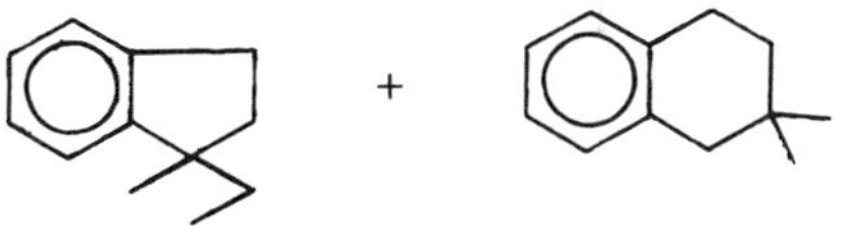

Table 5. Cyclialkylation of 3-Chloro-3-methyl-5-phenylpentane (48) and 4-Chloro-3-methyl-5-phenylpentane (49) at 25°,

Phenylalkyl chloride	Catalyst,[a] solvent[b]	Time, hr	Yield,[c] %	Product composition, %[d]				
				1,2-Dimethyltetralin (44)		1-Ethyl-1-methylindan (45)	2,2-Dimethyltetralin (46)	Unidentified
				cis	*trans*			
$PhCH_2CH_2C(CH_3)(Cl)CH_2CH_3$ (48)	$AlCl_3$-CH_3NO_2, petroleum ether	4	78	7	71	16	3	3
	$AlCl_3$, CS_2	4	70	20	27	15	32	6
	$AlCl_3$, petroleum ether	4	76	8	34	21	24	9
	$AlCl_3$ (0.5 mol), petroleum ether	2	—	13	5	9	57	16
		4	57	8	11	8	59	14
$PhCH_2CH(Cl)CH(CH_3)CH_2CH_3$ (49)	$AlCl_3$, petroleum ether	4	55	4	3	30	47	16
	$AlCl_3$, CS_2	4	60	16	25	24	25	10
	$AlCl_3$-CH_3NO_2	4	50	12	63	19	2	4

[a]Unless specified otherwise, the molar ratio of $AlCl_3$ to substrate was 0.1.
[b]In all cases a ratio of solvent (ml) to substrate (g) of ca. 4.5 was used.
[c]Total yields were calculated by GLPC using 2-phenylhexane as internal standard.
[d]Calculated by integration of GLPC recordings.
Source: Ref. 18.

tetralin (47), showing that tertiary six-ring formation takes precedence over tertiary five-ring formation (Eq. 20).

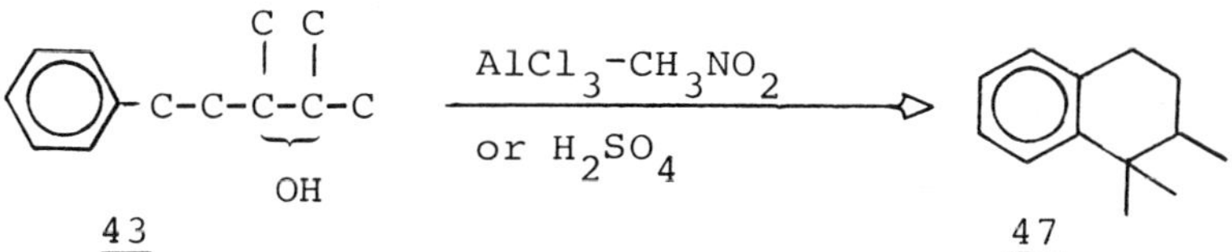

$$\text{C}_6\text{H}_5\text{-C-C-C(C)(OH)-C(C)-C} \ \xrightarrow[\text{or } H_2SO_4]{AlCl_3\text{-}CH_3NO_2} \ \textbf{47} \quad (20)$$

43 47

The conditions and results of cyclialkylation of the two isomeric chlorides 48 and 49, which belong to system 42, are presented in Table 5.

Examination of the data included in the table shows that the cyclization of the isomeric chlorides 48 and 49 gave similar product mixtures, whose compositions were determined by the type of catalyst employed. Thus, while the mild catalyst $AlCl_3$-CH_3NO_2 produced mixtures consisting of *cis*- and *trans*-1,2-dimethyltetralin (44) and 1-ethyl-1-methylindan (45) with little or none of the isomeric 2,2-dimethyltetralin (46) present, the strong catalyst $AlCl_3$ produced mixtures in which the latter isomer predominated. The results from cyclialkylation of chlorides 48 and 49 were explained in terms of the carbocation processes outlined in scheme 6.

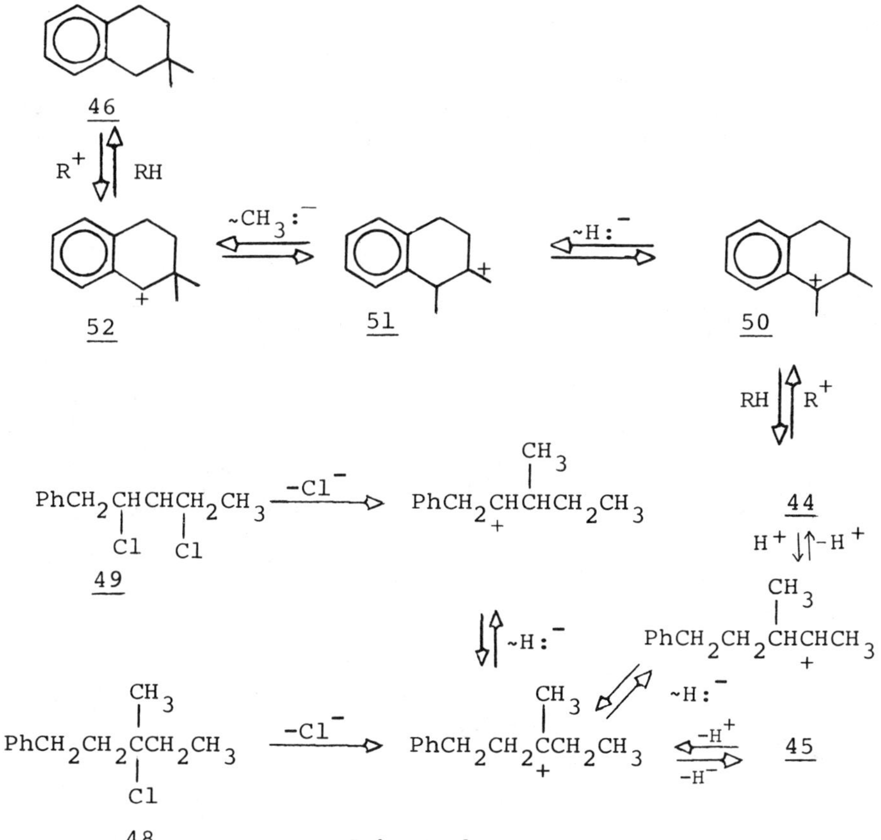

46

R^+ RH

52 ⇌ ($\sim CH_3:^-$) 51 ⇌ ($\sim H:^-$) 50

RH R^+

44

$PhCH_2CHClCHClCH_2CH_3 \xrightarrow{-Cl^-} PhCH_2\overset{+}{C}HCH(CH_3)CH_2CH_3$

49

H^+ ⇅ $-H^+$

$\sim H:^-$ ⇅

$PhCH_2CH_2CH(CH_3)\overset{+}{C}HCH_3$

$\sim H:^-$

$PhCH_2CH_2CCl(CH_3)CH_2CH_3 \xrightarrow{-Cl^-} PhCH_2CH_2\overset{+}{C}(CH_3)CH_2CH_3 \underset{-H^-}{\overset{-H^+}{\rightleftharpoons}}$ 45

48

Scheme 6

According to scheme 6, the formation of 2,2-dimethyltetralin (46) when $AlCl_3$ was used as cyclialkylation catalyst was attributed to the known [23] ability of $AlCl_3$ to abstract hydride ions and thus produce from 44 intermediate ion 50, which, by undergoing successive hydride and methyl shifts, gives carbocation 52; the latter then abstracts a hydride ion to produce 46. Credibility of the route 44 → 50 → 51 → 52 → 46 is found in the following observations: (1) the predominant formation of 46 only when $AlCl_3$ was used as catalyst; (2) the enhanced formation of 46 with increased amount of $AlCl_3$ (see results with chloride 48); and (3) the rapid rearrangement of 1,2-dimethyltetralin (44) to 2,2-dimethyltetralin (46) in the presence of $AlCl_3$, but not in the presence of the weaker catalyst $AlCl_3$-CH_3NO_2.

It was pointed out before that the claimed closure of 1-chloro-5-phenylpentane (53) to benzosuberane was disproved by Barclay and co-workers [10], who found the product to be 1-methyltetralin (56), assumed to result via the route 53 → 54 → 55 → 56 (Eq. 21). From that it was concluded that seven-

$AlCl_3$, $-Cl^-$; $\sim H:^-$; $-H^+$; Cl; $AlCl_4^-$ (21)

53 54 55 56

membered rings cannot form by primary cyclialkylation under Friedel-Crafts conditions. To find out whether or not benzosuberanes can be prepared by Friedel-Crafts cyclialkylations with secondary or tertiary derivatives, and to establish the stage(s) at which the rearrangement to tetralins occurs, Barclay and co-workers [10] performed a number of cyclialkylation experiments whose conditions and results are summarized in Eqs. (22) to (25). The

57 + $FeCl_3$ → (CH_3NO_2, 0°, 4 hr; 89.2% yield) (22)

58 → ($AlCl_3$, CS_2, −10°, 7 hr; 75% yield); HO-analog → (HF, CCl_4, 10°, 1 hr; 100% yield) (23)

59 + $AlCl_3$ → (CS_2, 0°, 1 hr; 93.3% yield) (24)

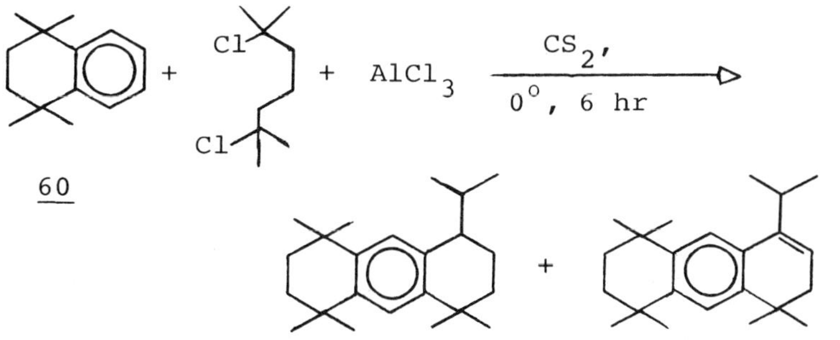

$$CS_2,\ 0^\circ,\ 6\ hr \qquad AlCl_3 \tag{25}$$

60

60% yield 12% yield

results indicated by Eq. (25) disproved an earlier claim by Colonge and Lagier [24] that a benzosuberane derivative was produced from 60 by this reaction.

In view of the results above it was established that (1) as with alkylations by primary derivatives, those by secondary or tertiary derivatives also proceed with accompanying rearrangement to form alkyltetralins instead of alkylbenzosuberanes; and (2) the formation of tetralins must take place through rearrangement in the side chain *prior* to cyclization, most probably, by the isomerization of initially formed carbocations 61 to the rearranged carbocations 62 *prior* to ring closure to the products (scheme 7). Understandably,

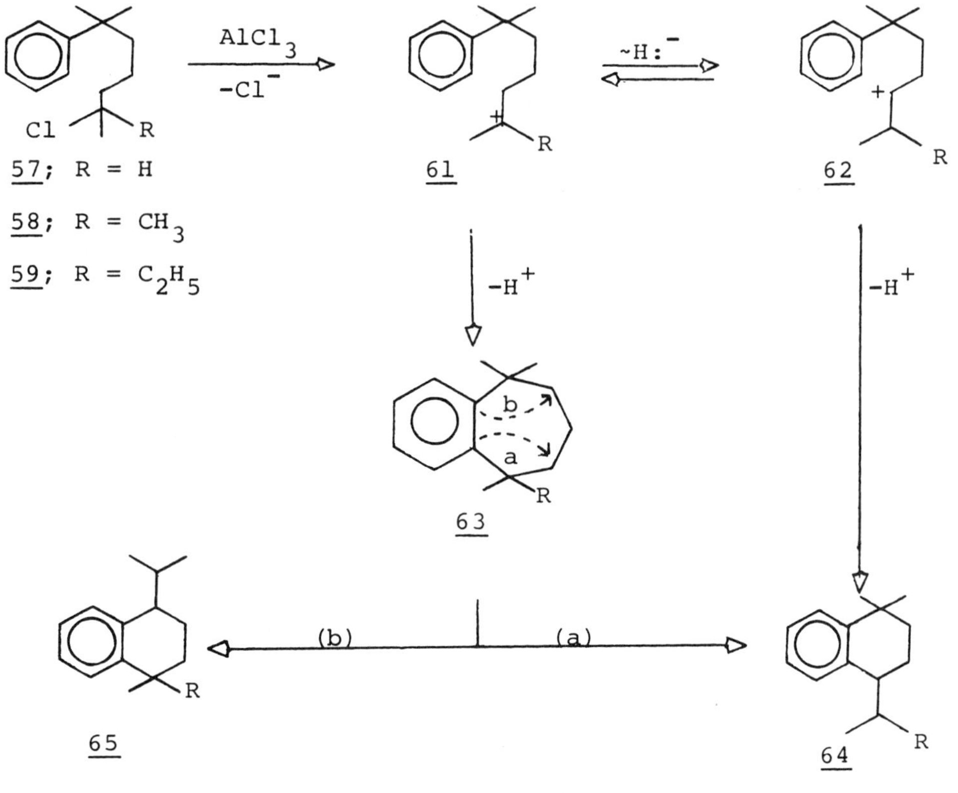

57; R = H

58; R = CH_3

59; R = C_2H_5

$AlCl_3$, $-Cl^-$; $\sim H:^-$; $-H^+$

61 62 63 64 65

(a) (b)

Scheme 7

such a mechanism would be expected to produce only one tetralin isomer (64) from each of the starting phenylalkyl chlorides 57-59. If the rearrangement occurred by initial formation of an *ortho*-diquaternary benzosuberane (63) and subsequent ring contraction (either via a dealkylation-rearrangement-realkylation sequence or via hydride abstraction followed by a 1,2 aryl shift, one would expect equal proportions of two tetralin derivatives, one (64) from route (a) and another (65) from route (2).

As pointed out by Barclay et al. [10], the mechanism of scheme 7, established primarily for the cyclization of chlorides 57-59, cannot be generalized to apply to other similar systems in the absence of a supporting evidence. For example, 2,5,5-trimethyl-6-phenyl-2-hexene (66) has been shown by Hart [25] to cyclize in the presence of boron trifluoride etherate to yield not only 2,2-dimethyl-4-isopropyltetralin (67) but also some tetramethylbenzosuberane (68, Eq. 26). Likewise, it has been claimed by Tilak et al. [26] that the

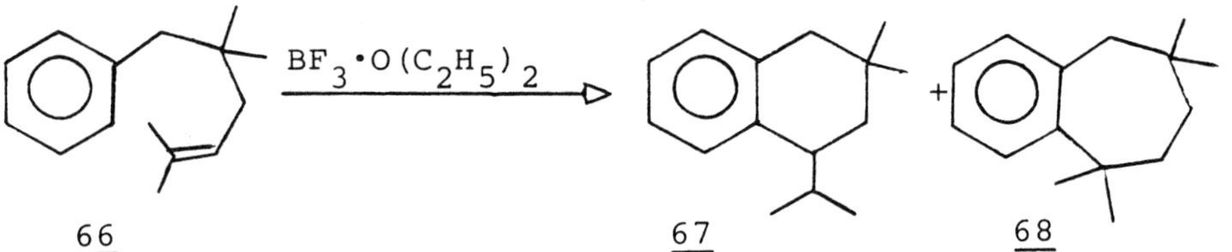

$BF_3 \cdot O(C_2H_5)_2$ (26)

66 67 68

cyclization of 2-methyl-5-phenylamino-3-pentanol (69) and of 2,2-dimethyl-5-phenylamino-3-pentanol (70) with 70% perchloric acid at 60°C produced 5,5-

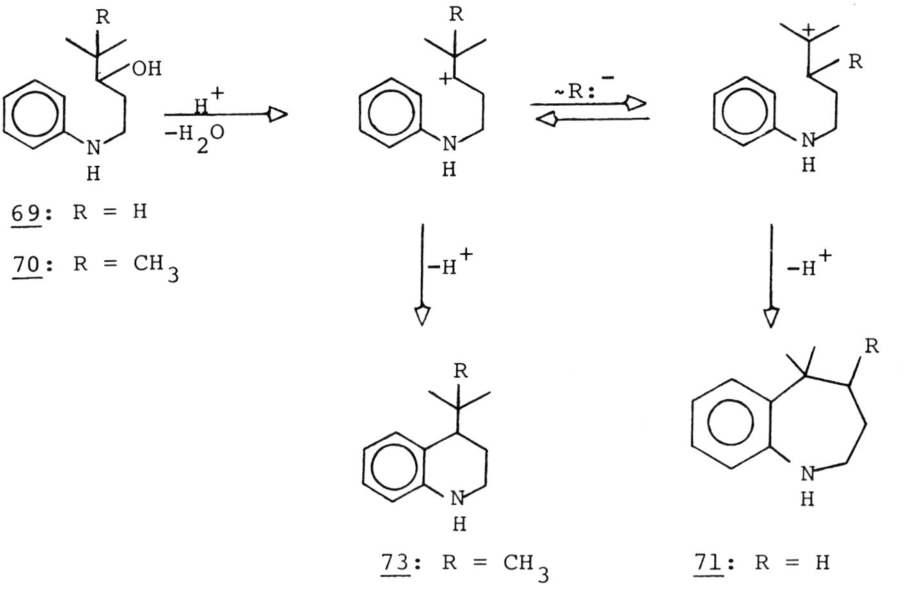

R OH H^+ $-H_2O$ ~R:⁻ $-H^+$ $-H^+$

69: R = H

70: R = CH_3

73: R = CH_3

71: R = H

72: R = CH_3

Scheme 8

dimethyl-2,3,4,5-tetrahydro-1H-1-benzazepine (71) from the former and 4,5,5-trimethyl-2,3,4,5,-tetrahydro-1H-1-benzazepine (72) in high yield along with 4-*t*-butyl-1,2,3,4-tetrahydroquinoline (73), from the latter. These products and the steps involved in their formation are shown in scheme 8.

The results of Eq. (26) and scheme 8 illustrate that, in some cases, 3° closure to a seven-ring system may compete successfully with 2° closure to a six-ring system.

In continuation of the efforts aimed at the understanding of Friedel-Crafts cyclialkylation, Barclay and Sanford [27,28] have utilized deuterium labeling of phenylalkyl chlorides as a means of providing information on the mechanism of rearrangement and hydride exchange accompanying these reactions. In one paper [27], these authors described cyclialkylations of 4,4-d_2- (74), 3,3-d_2- (75) and 2,2-d_2-1-chloro-5-phenylpentane (76), and 2,2-d_2-1-chloro-4-phenylbutane (77) under a variety of reaction conditions. A summary of conditions and cyclization products from these reactions is given in Table 6.

The assignment of 3,3-d_2- (78), 2,2-d_2- (79), and of 80 and 81 as the only labeled methyltetralins obtained from the cyclialkylation of 4,4-d_2- (74), 3,3-d_2- (75), and 2,2-d_2-1-chloro-5-phenylpentane (76), respectively, proved that benzosuberane could not form even as an intermediate in the pathways leading from starting materials to products. For example, if 2,2-d_2-benzosuberane (83, scheme 9) was involved as intermediate in the transformation of 74-78, one should get not only 78 but also 84 as a result of ring contraction via routes (a) and (b). Accordingly, the rearrangement leading to 1-methyltetralin must have taken place in the side chain via path 1 (scheme 9) *prior* to ring closure.

As to the cyclialkylation of 2,2-d_2-1-chloro-4-phenylbutane (77), the major product was found to be 2,2-d_2-tetralin (82) by NMR analysis. This, together with the findings by Roberts and co-workers [19] that treatment of

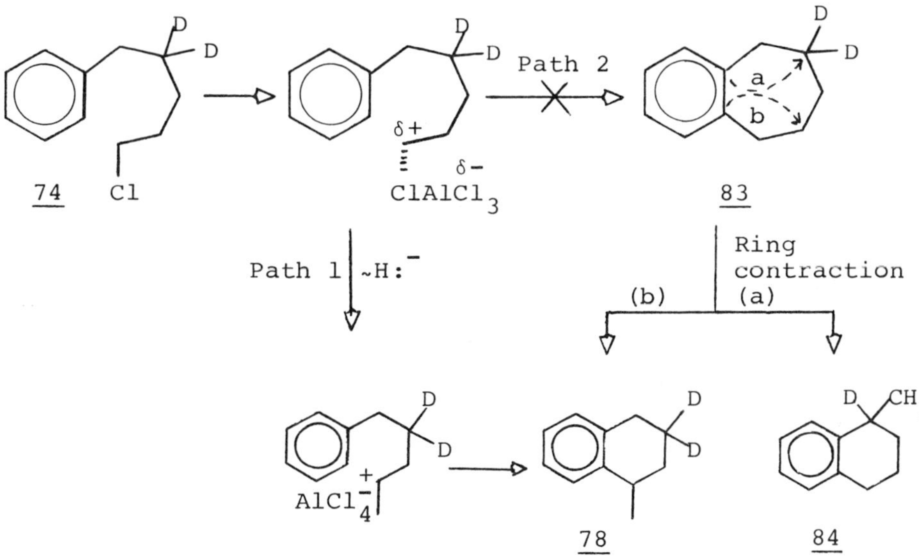

74 Cl

D D

δ+ δ− ClAlCl$_3$

Path 2

83

a b

Path 1 ~H:$^-$

Ring contraction (a) (b)

$AlCl_4^-$ +

78

84

D CH_2D

Scheme 9

Table 6. Cyclialkylations of Dideuterated Phenylalkyl Chlorides at 25°C[a-c]

Compound (mmol)	Catalyst (mmol)	Temp., °C	Time, hr	Cyclialkylation products
$C_6H_5CH_2CD_2CH_2CH_2CH_2Cl$ 74	$AlCl_3$ (6)	25	2.5	78
$C_6H_5CH_2CH_2CD_2CH_2CH_2Cl$ 75	$AlCl_3$ (9)	25	3.0	79
$C_6H_5CH_2CH_2CH_2CD_2CH_2Cl$ 76	$AlCl_3$	25	3.0	80 + 81 (CH_2D)
$C_6H_5CH_2CH_2CD_2CH_2Cl$ 77	$AlCl_3$ (4.5)	25	1.5	82

[a]The label was located by NMR and mass spectroscopy.
[b]Petroleum ether (bp 60 to 80°C) was used as solvent.
[c]Various by-products due to hydride ion exchange and dehydrogenation also formed, but are not included in the table.

1-chloro-4,5-diphenylpentane (26a) with $AlCl_3$ produced 1-benzyltetralin (27a) and no benzosuberane, suggests that the cyclialkylations of primary chlorides to tetralins occur by a concerted assisted mechanism (Eqs. 27 and 28).

$$77 \; (Cl:AlCl_3) \xrightarrow{-AlCl_4^-} [\text{arenium ion}] \xrightarrow{-H^+} 82 \qquad (27)$$

[Eq. (28): 26a (CH2Ph-substituted tetralin precursor with Cl:AlCl3) —$-AlCl_4^-$→ arenium ion intermediate —$-H^+$→ 27a] (28)

Further support for the assisted pathway of Eqs. (27) and (28) was given by Khalaf [17] through a study of inter- or intramolecular reactions of some β-, γ-, and δ-arylhaloalkanes in the presence of $AlCl_3$-CH_3NO_2 as catalyst. The results demonstrated that the ionization of arylalkyl halides under Friedel-Crafts conditions are assisted, to a varying degree, by aryl participations depending on the location of the aryl group relative to the halogen atom. The observed sequence was found to be β-aryl > δ-aryl > γ-aryl. Thus under similar conditions compounds 14 and 85-87 were found to react with $AlCl_3$-CH_3NO_2 in the following order:

Ph-C(C)(C)CCl (85) > Ph-CCC(C)CCl (14) > PhCCCCCl (86) > p-$CH_3C_6H_4$C(C)(C)CCCl (87)

The fact that the rate of reaction of 1-chloro-2-methyl-4-phenylbutane (14) was several times faster than that of 1-chloro-4-phenylbutane (86) was attributed to assisted removal of the chloride ion by both β-methyl and δ-phenyl in the case of 14, but only by δ-phenyl in the case of chloride 86.

The latter findings of Khalaf are in good accord with earlier reports of aryl participation during the solvolysis of arylalkyl esters [27,29,30]. While it was found that a phenylbutyl chlorosulfite ester underwent assisted direct cyclization to tetralin (25%) [27], there was no significant aryl participation in the solvolysis of the 5-aryl-1-pentyl series [29].

In another paper, Barclay and Sanford [28] undertook a study of the effect of varying substrate (phenylalkyl chloride) and reaction conditions on the ratio of cyclialkylation to hydride exchange and other competing reactions. The conditions and products of cyclialkylation of the various phenylalkyl chlorides studied (53, 86, and 88-92) are outlined in Table 7.

From the table, it is evident that the main cyclization product from 1-chloro-5-phenylpentane (53), under varying conditions of catalyst, solvent, and temperature, was 1-methyltetralin (56). The hydride exchange product *n*-pentylbenzene (93) was formed in each case, even with the milder catalyst $FeCl_3$-CH_3NO_2 (entry 4). A small yield of 1-methyl-3,4-dihydronaphthalene (94) was formed in both petroleum ether and carbon disulfide solvents and a significant amount of methylnaphthalenes was formed in dichloromethane solvent. The pathway which shows the interrelationships between the various compounds in the cyclialkylation of phenylpentyl chloride (53) is given in scheme 10. The production of *n*-pentylbenzene (93) and 1-methylnaphthalene (95) from 1-methyltetralin (56) upon treatment with $AlCl_3$, together with the

Table 7. Barclay and Sanford's Cyclialkylations of Phenylalkyl Chlorides in the Presence of $AlCl_3$[a]

Entry	Phenylalkyl chloride	Solvent	Temp, °C	Time, hr	Products, (*No.*) (%)[b]
1	$C_6H_5(CH_2)_5Cl$ (53)	PE[c]	Reflux	0.6	1-Methyltetralin (56) (42), n-$C_5H_{11}C_6H_5$ (93) (23), 1-methyl-3,4-dihydronaphthalene (94) (5), residue (27)
2	$C_6H_5(CH_2)_5Cl$ (53)	CS_2	25	50	(56) (43), (93) (34), (94) (3), residue (20)
3	$C_6H_5(CH_2)_5Cl$ (53)	CH_2Cl_2	0	1	(56) (17), (93) (3.3), methylnaphthalene (95) (26), residue (54)
4	$C_6H_5(CH_2)_5Cl$ (53)	CH_3NO_2[d]	25	72	(56) (64), starting chloride (53) (34)
5	$C_6H_5(CH_2)_4Cl$ (86)	PE	25	1	Tetralin (56), n-$C_4H_9C_6H_5$ (12) naphthalene (17), residue (17)
6	$C_6H_5C(CH_3)_2(CH_2)_3Cl$ (88)	PE	25	2	1,1-Dimethyltetralin (34), $C_6H_5C(CH_3)_2$-$CH_2CH_2CH_3$ (3.5), i-$C_6H_{13}C_6H_5$ (32)
7	$C_6H_5(CH_2)_3Cl$ (89)	PE	Reflux	1.2	n-$C_3H_7C_6H_5$ (48), indan (trace), residue (52)
8	$C_6H_5CH_2CH_2CD_2Cl$ (90)	PE	Reflux	1	$C_6H_5CH_2CH_2CD_2H$
9	$C_6H_5CH_2CD_2CH_2Cl$ (91)	PE	Reflux	1	$C_6H_5CH_2CHDCH_2D = C_6H_5CH_2CHDCH_2D$
10	$C_6H_5C(CH_3)_2CH_2CH_2Cl$ (92)	PE (bp 30-40°C)	Reflux	0.2	2-Methyl-3-phenylbutane (mainly) + t-$C_5H_{11}C_6H_5$ (22), compounds mass 158 (58) residue (21)

[a]An equimolar ratio of aluminum chloride and phenylalkyl chloride was used unless otherwise specified.
[b]Percentages are relative amounts by vapor-phase chromatographic analyses. Products were identified by vapor-phase chromatographic internal standards and nuclear magnetic resonance analyses. The "residues" refer to long retention-time components on the vapor-phase chromatograph which were not identified.
[c]Petroleum ether (bp 60 to 80°C), unless otherwise specified.
[d]Ferric chloride (0.50 molar ratio) was used.
Source: Ref. 28.

Table 8. Effect of Time on the Product of Reaction of $AlCl_3$ with 1-Chloro-5-phenylpentane (53)[a]

Time, min	Composition of product, %		
	Chloride	1-Methyltetralin	*n*-Pentylbenzene
20	73.6	24.4	Trace
30	59.6	40.2	0.31
75	0.0	92.0	7.9
1350	0.0	58.4	41.6

[a]An equimolar ratio of phenylpentyl chloride and $AlCl_3$ in petroleum ether at room temperature.
Source: Ref. 28.

observed increase in the ratio *n*-pentylbenzene/1-methyltetralin by time during cyclialkylation of 1-chloro-5-phenylpentane (53) (Table 8), suggested that most or all of the pentylbenzene produced from 53 was formed by dealkylation and hydride exchange on 1-methyltetralin rather than by hydride exchange *prior* to cyclialkylation.

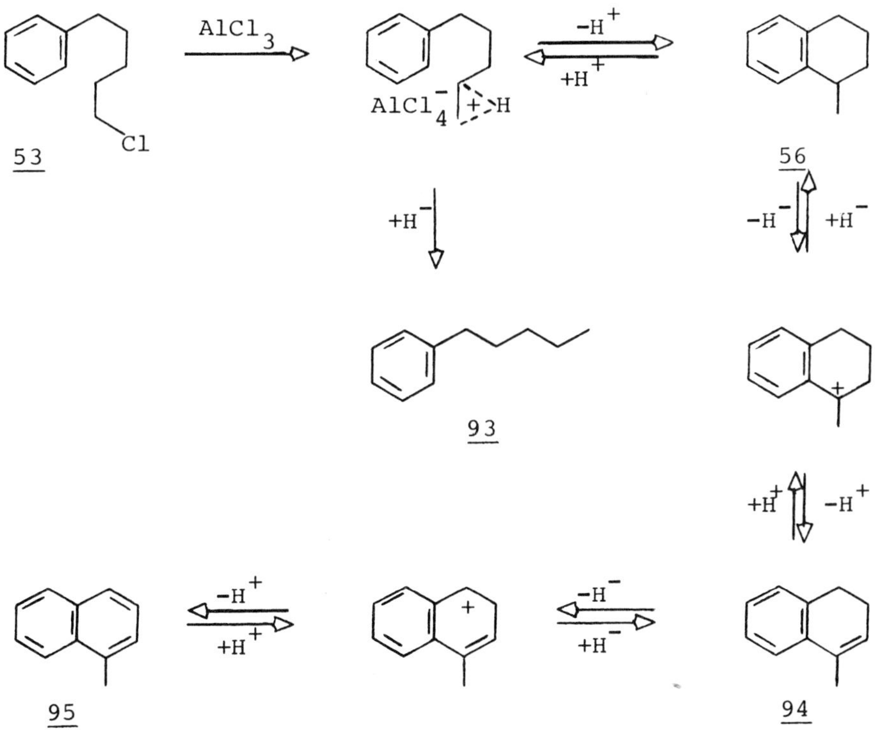

Scheme 10

In line with results from chloride 53, the cyclialkylation of 1-chloro-4-phenylbutane (86) yielded tetralin, *n*-butylbenzene, and naphthalene and that of 1-chloro-4-methyl-4-phenylbutane (88) yielded 1,1-dimethyltetralin (96), 2-methyl-5-phenylpentane (79), and a small amount of 2-methyl-2-phenylpentane (98). The fact that treatment of 1,1-dimethyltetralin (96) with $AlCl_3$ yielded hydrocarbon 97 and none of the isomeric 98 indicated that hydrocarbon 97 resulted from 96 after cyclialkylation, while hydrocarbon 98 resulted from hydride transfer preceding alkylation. This is illustrated in Eq. (29).

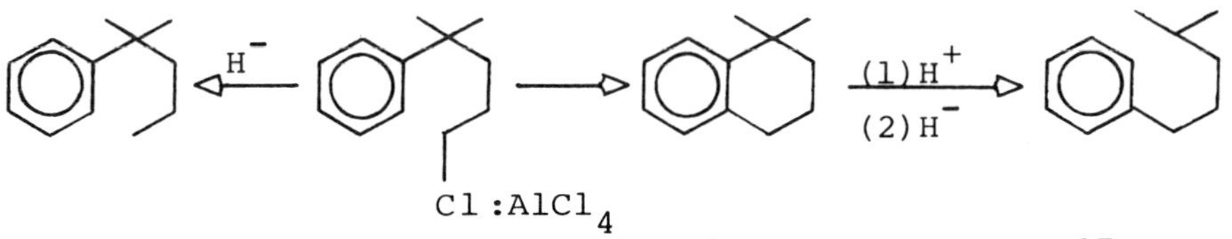

98 $\xleftarrow{H^-}$ 88 ($Cl:AlCl_4$) $\longrightarrow$ 96 $\xrightarrow[(2)\ H^-]{(1)\ H^+}$ 97 (29)

Barclay and Sanford [28] also studied the hydride transfer reactions in cyclialkylation of 1-chloro-3-phenylpropane (89, entry 7) and 1-chloro-3-methyl-3-phenylbutane (92, entry 10) by $AlCl_3$. In agreement with Khalaf and Roberts [11], these authors showed that the products from phenylpropyl chloride were *n*-propylbenzene (mainly), di-*n*-propylbenzene, and polymers formed by hydride transfer. Treatment of this chloride containing deuterium labels at C-1 (90, entry 8) and C-2 (91, entry 9) showed that the hydride ion rearranged from the carbon adjacent to that bearing the halogen as shown in Eq. (30).

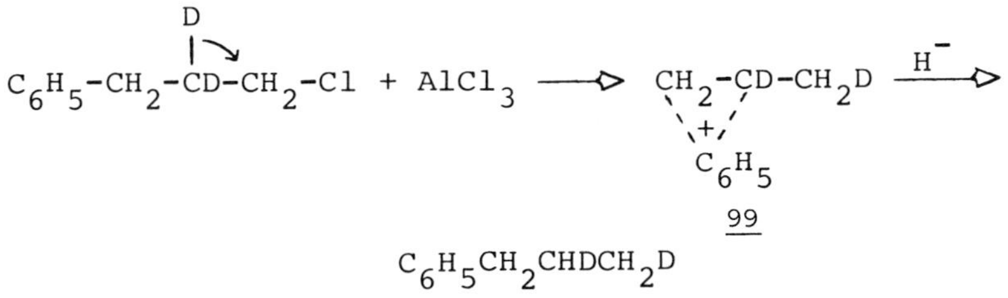

$$C_6H_5\text{-}CH_2\text{-}CD\text{-}CH_2\text{-}Cl + AlCl_3 \longrightarrow CH_2\text{-}CD\text{-}CH_2D\ (\text{bridged by } {}^{+}C_6H_5,\ \mathbf{99}) \xrightarrow{H^-} C_6H_5CH_2CHDCH_2D \qquad (30)$$

The proposed phenonium ion intermediate (99) was related to that proposed earlier by Douglass and Roberts [31] to explain the equilibration of ^{14}C between the α and β positons of propylbenzene.

Upon treatment with $AlCl_3$, 1-chloro-3-methyl-3-phenylbutane (92) yielded a complex mixture including the hydride exchange products *t*-pentylbenzene (100) and 2-methyl-2-phenylbutane (101) besides two other unidentified components of higher molecular weight (entry 10, Table 7). These results were rationalized in terms of the steps outlined in scheme 11.

More recently (1974), Khalaf [32] investigated the cyclialkylation behavior of the structurally related 1-chloro-3-methyl-3-*p*-tolylbutane (102) in the presence of $AlCl_3$ catalyst and CS_2 or petroleum ether solvent. Because of the presence of the ring methyl group, the products obtained from chloride 102 were far more complex than those obtained previously from 1-chloro-3-methyl-3-phenylbutane (92). These products are depicted in Table 9 and the steps believed to be responsible for their formation are represented in scheme 12.

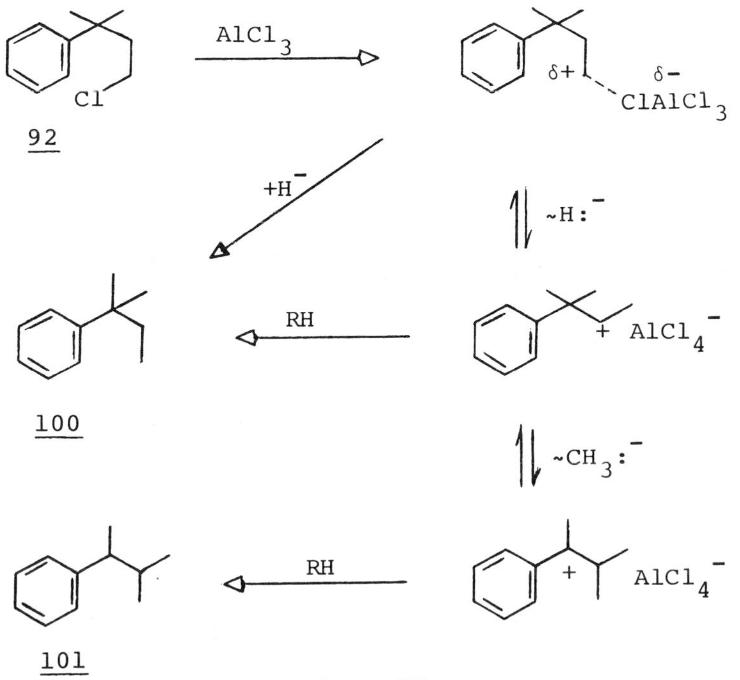

92

$AlCl_3$

δ+ δ− $ClAlCl_3$

$+H^-$

$\sim H:^-$

RH

$AlCl_4^-$

100

$\sim CH_3:^-$

RH

$AlCl_4^-$

101

Scheme 11

Table 9. Different Products Formed in the Reaction of 1-Chloro-3-methyl-3-*p*-tolylbutane with $AlCl_3$ in CS_2 and Petroleum Ether (60 to 80°C) at Room Temperature[a]

Product[b]	Relative percent[c] in: CS_2	Petroleum ether
Toluene	19	13
Xylene	2	3
2-Methyl-4-*m*-tolylbutane (110)	1	4
2-Methyl-3-*p*-tolylbutane (117)	31	35
2-Methyl-2-*p*-tolylbutane (112)	37	29
1,1,5-Trimethylindan (105)	2	2
1,1,7-Trimethylindan (109)	Traces	1
Higher unidentified products	8	10

[a]In all the reactions a ratio of tolylalkyl chloride (mol) to $AlCl_3$ (mol) to solvent (ml) of 0.2:0.002:20 was used.
[b]Products are listed in order of increasing GLPC retention time on a 16-ft DEGA column operated at 140°C with nitrogen as carrier gas.
[c]The percentage composition of various products was calculated by integration of GLPC recordings of the crude mixtures. So, the quantitative values are relative amounts of products and not the yields.

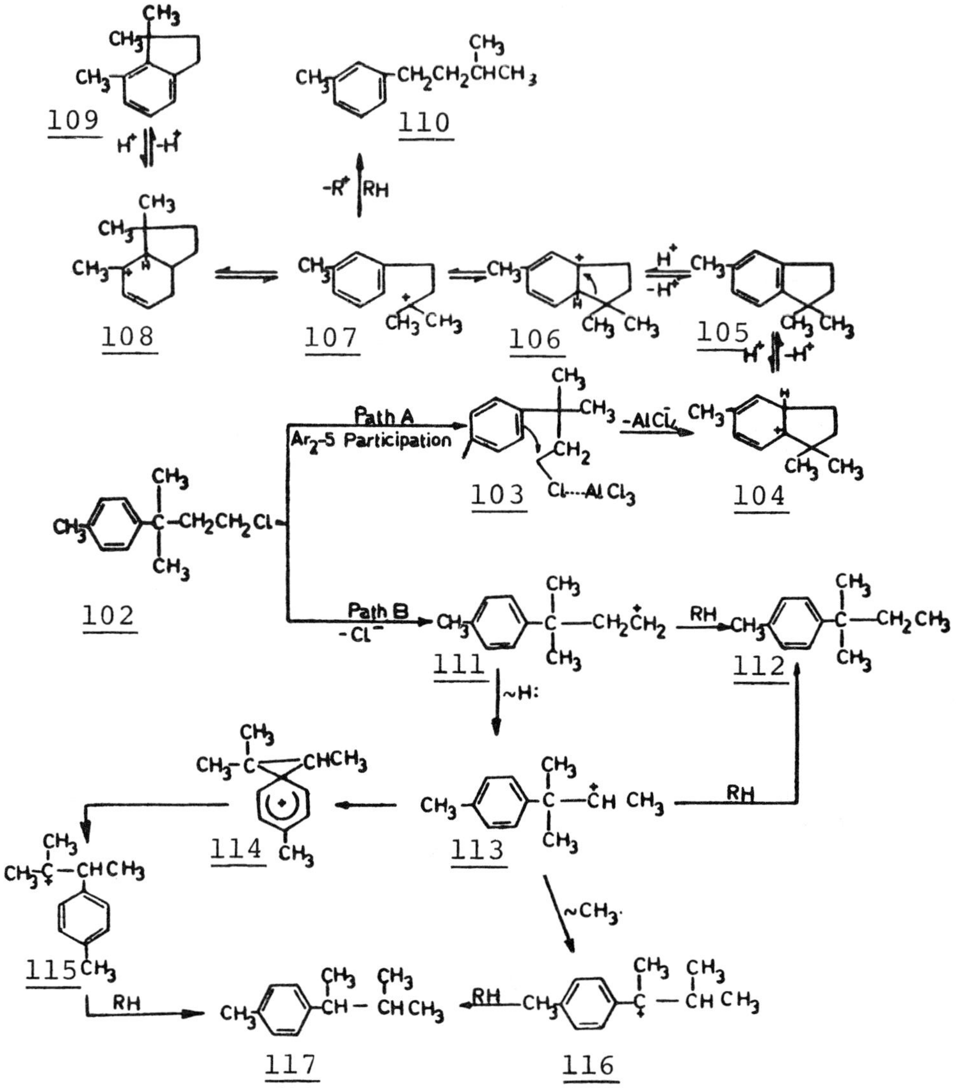

Scheme 12

Mechanism for the formation of different products in the reaction of 1-chloro-3-methyl-3-*p*-tolylbutane (102) with $AlCl_3$

We have seen from the previous discussion that cyclialkylations induced by the strong catalyst $AlCl_3$ are usually complicated by accompanying rearrangement, hydride transfer, and ring opening by dealkylation. We have also noted that these complicating reactions could be eliminated or greatly minimized by replacing $AlCl_3$ with milder catalysts such as $AlCl_3$-CH_3NO_2 or $FeCl_3$. For example, if we go back to Eqs. (7) and (9), in which $PhCH_2CH_2C(CH_3)_2Cl$ (118) and $PhCH_2CH_2CH_2C(CH_3)_2Cl$ (119) were cyclized in the presence of $AlCl_3$ and $FeCl_3$ catalysts, we find that with $AlCl_3$, chlorides 118 and 119 yielded multicomponent mixtures in which the ring closure products 1,1-dimethylindan and 1,1-dimethyltetralin constituted only 35 to 38%, whereas the ring opening products isopentylbenzene and isohexylbenzene constituted

Table 10. Cyclialkylation Reactions of Secondary and Tertiary Phenylalkyl Chlorides in the Presence of $AlCl_3$–CH_3NO_2 and $FeCl_3$ Catalysts at 25°C[a]

Cyclized chloride	Catalyst	Solvent	Product composition, %	Yield, %
2-Chloro-2-methyl-4-phenylbutane	$AlCl_3$-CH_3NO_2	SKB[b]	1,1-Dimethylindan (100)	79
	$AlCl_3$-CH_3NO_2	CS_2	1,1-Dimethylindan (100)	60
	$FeCl_3$	SKB	1,1-Dimethylindan (100)	65
2-Chloro-2-methyl-4-(*o*-tolyl)butane	$AlCl_3$-CH_3NO_2	SKB	1,1,4-Trimethylindan (100)	75
	$FeCl_3$	SKB	1,1,4-Trimethylindan (100)	73
2-Chloro-2-methyl-4-(*m*-tolyl)butane	$AlCl_3$-CH_3NO_2	SKB	1,1,5-Trimethylindan (56) and 1,1,7-trimethylindan (44)	70
	$FeCl_3$	SKB	1,1,5-Trimethylindan (60) and 1,1,7-trimethylindan (40)	65
2-Chloro-2-methyl-4-(*p*-tolyl)butane	$AlCl_3$-CH_3NO_2	SKB	1,1,6-Trimethylindan (100)	72
	$FeCl_3$	SKB	1,1,6-Trimethylindan (100)	68
	$AlCl_3$	SKB	1,1,6-Trimethylindan (30) and isopentylbenzene (70)	--
2-Chloro-2-methyl-4-phenylpentane	$AlCl_3$-CH_3NO_2	SKB	1,1-Dimethyltetralin (100)	74
	$FeCl_3$	SKB	1,1-Dimethyltetralin (100)	76
2-Chloro-2-methyl-6-phenylhexane	$AlCl_3$-CH_3NO_2	SKB	1-Isopropyltetralin (100)	68
	$FeCl_3$	SKB	1-Isopropyltetralin (100)	65

Table 10. (continued)

Cyclized chloride	Catalyst	Solvent	Product composition, %	Yield, %
2-Chloro-5-phenylpentane	$AlCl_3$-CH_3NO_2	SKB	1-Methyltetralin (100)	83
	$FeCl_3$	SKB	1-Methyltetralin (100)	76
3-Chloro-5-phenylpentane	$AlCl_3$-CH_3NO_2	SKB	1-Methyltetralin (100)	75
	$FeCl_3$	SKB	1-Methyltetralin (100)	70

[a] In all experiments the reactants were as follows: arylalkyl chloride (0.02 mol); $AlCl_3$ or $FeCl_3$ (0.002 mol); solvent (50 ml). In reactions catalyzed by $AlCl_3$-CH_3NO_2, the $AlCl_3$ catalyst was dissolved in 0.02 mol of CH_3NO_2 before use.
[b] SKB is Skellysolve B.
Source: Ref. 33.

55 to 61%. On the other hand, when $FeCl_3$ was used as catalyst, chlorides 118 and 119 gave exclusively the cyclialkylation products 1,1-dimethylindan and 1,1-dimethyltetralin, respectively, in good yields.

With these findings in mind, Khalaf [33,34] extended the $AlCl_3$-CH_3NO_2- and $FeCl_3$-catalyzed cyclialkylations to include other phenylalkyl chlorides. The aim was to improve the synthetic utility of cyclialkylation reactions by controlling their course in such a way as to optimize the yield of the ring closure products and to eliminate or minimize side products resulting from complicating secondary processes. The results of these investigations are compiled in Table 10. Examination of these results demonstrates that $AlCl_3$-CH_3NO_2- or $FeCl_3$-catalyzed cyclialkylations of secondary and tertiary phenylalkyl chlorides proceed without complications to give high yields of pure cyclization products.

In summing up the reliable Friedel-Crafts cyclialkylation results of this section, we would like to emphasize the following conclusions:

1. Rearrangements do accompany intramolecular as well as intermolecular Friedel-Crafts alkylations; primary, secondary, and tertiary phenylalkyl chlorides have been shown to rearrange during cyclialkylation.

2. The extent of rearrangement in intramolecular alkylations may be considerably less than in intermolecular alkylations, but only if a six-membered ring is produced directly.

3. The formation of a five- or a seven-membered ring is not so favorable, and competing processes such as rearrangement, hydride transfer, and intermolecular polyalkylation will intervene in reactions that otherwise would form these ring systems. This can be seen by examining the results for 1-chloro-3-phenylpropane (Eq. 4) and 2-chloro-4-phenylbutane (Eq. 6).

4. The formation of five-membered rings is facilitated in phenylalkyl halides by having the halogen on a tertiary carbon in the γ-position or having the possibility of rearrangement to produce a positive charge at that position. This is illustrated by the results for 2-chloro-2-methyl-4-phenylbutane (Eq. 7) and 1-chloro-2-methyl-4-phenylbutane (Eq. 13).

5. Rearrangements of the side chain are not so extensive as has been reported [8] in cyclialkylations of acetylphenylalkyl halides. This is evident from the case of 1-chloro-4-(*p*-acetylphenyl)butane (Eq. 12).

6. Cyclialkylations of tertiary phenylalkyl chlorides to indans or tetralins having tertiary cycloalkyl groups attached to the aromatic ring may be accomplished without extensive hydride exchange to produce alkylbenzenes only with a catalyst less active than aluminum chloride, such as $FeCl_3$ or $AlCl_3$-CH_3NO_2. This is best illustrated by the cyclialkylation results of Table 10.

7. Secondary or tertiary closure to tetralins is favored over tertiary closure to indans. This is illustrated by the results of Eqs. (19) and (20).

8. Closure at secondary carbon to tetralin is much more favored than closure at a tertiary carbon to benzosuberane. This is illustrated by the results of Eqs. (23) to (25) as well as by the preferred closure of 2-chloro-2-methyl-6-phenylhexane to 1,1-dimethyltetralin (Table 10).

9. When two aromatic nuclei are present in the same molecule with one of them at a position suitable for direct closure to a six-membered ring, cyclialkylations occur at rates faster than those of competing 1,2 hydride shifts, provided that retardations are not encountered. This is illustrated by the cyclialkylation results of 1-chloro-4,5-diphenylpentane (Eq. 16) and the diastereomers of 1-chloro-2-methyl-4,5-diphenylpentane (39a and 39b, Table 4).

B. Cyclialkylations of Systems Related to Arylhaloalkanes

Several examples of intramolecular alkylations of arylhalo esters, amides, amines, ketones, and ethers have been reported. In the following we shall consider briefly some of these examples. However, it should be noted that most of the results in this area were obtained before the advent of modern analytical techniques and were based exclusively on outdated methods of structural identification. Because of this, and to some extent to the under-developed state of knowledge at that time, the results of these cyclization reactions should be taken with considerable reservation.

1. Halo esters

Early in this century, von Auwers and co-workers reported that aryl esters of α-halopropionic (120) [35] and α-halobutyric (120b) [36] acids underwent ring closure, probably via preliminary Fries rearrangement, to β-halo hydroxyketones (121), giving the corresponding substituted indanones (122a and 122b) (Eq. 31).

$$\text{C}_6\text{H}_5\text{–O–C(=O)–CH(Br)CH}_2\text{R}\ (\mathbf{120}) \xrightarrow{AlCl_3} \left[o\text{-HO–C}_6\text{H}_4\text{–COCH(Br)CH}_2\text{R}\ (\mathbf{121}) \right] \xrightarrow{-HBr} \mathbf{122} \quad (31)$$

120, 122 — a: R = H; b: R = CH_3

Aryl esters of β-halopropionic acid have also been claimed to yield substituted indanones upon treatment with $AlCl_3$ [37,38]; the intermediacy of hydroxyketones was proposed (e.g., Eq. 32).

$$p\text{-CH}_3\text{C}_6\text{H}_4\text{–O–CO–CH}_2\text{CH}_2\text{Br} \xrightarrow{AlCl_3} \text{(OH)(CH}_3\text{)C}_6\text{H}_3\text{–COCH}_2\text{CH}_2\text{Br} \xrightarrow{AlCl_3} \text{indanone (OH, CH}_3\text{)} + \text{HBr} \quad (32)$$

2. Halo amides

Ring closure of α-chloroacetanilide [39], β-chloropropionanilide, and γ chlorobutyranilide [40] have been reported to occur readily in the presence of $AlCl_3$, giving good yields of oxindole and 2-ketoterahydroquinolines. These are shown in Eqs. (33) to (35), respectively.

$\xrightarrow{AlCl_3}$ + HCl (33)

$\xrightarrow{AlCl_3}$ + HCl (34)

95%

$\xrightarrow{AlCl_3}$ + HCl (35)

Other aryl and aralkyl amides of chloroacetyl chloride [39,41-43a] and β-chloropropionyl chloride [40,43] were also reported to react analogously in the presence of $AlCl_3$ to yield the respective lactams. Four of the examples reported are presented in Eqs. (36) to (38a).

$\xrightarrow{AlCl_3}$ (36) [40]

($R = CH_3$ or C_2H_5)

$\xrightarrow[\Delta]{AlCl_3}$ (37) [43]

$\xrightarrow{AlCl_3}$ (38) [43a]

123 124 (not formed)

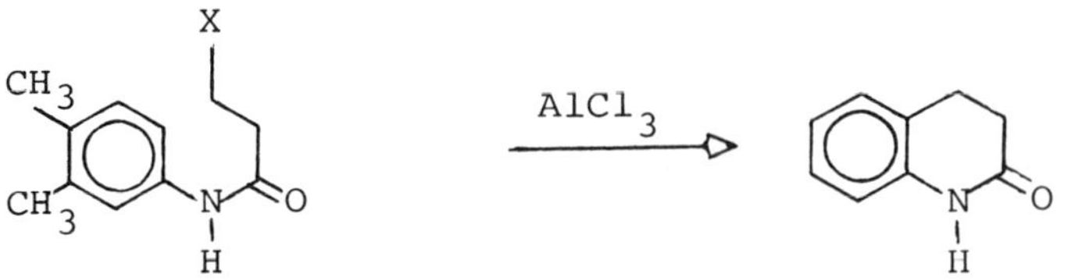

(38a) [44]

On the basis of the results of Eqs. (36) and (37), it was expected that the cyclization of α-chloro-N-(β-phenethyl)acetanilide (Eq. 38) would yield a mixture of the oxindole 123 and the benzazepinone 124 as a result of five- and seven-membered ring closures, respectively. Far from expectation, only oxindole 123 was obtained, illustrating the preference for five- over seven-membered ring formation. Further manifestation of this preference was also observed by Khalaf and Roberts [18] during the cyclialkylation of 2-methyl-4,6-diphenyl-2-hexanol and the cycliacylation of 3,5-diphenylpentanoyl chloride (see Sec. II).

The cyclialkylation of halo amides was likewise applied to both N,N'-*bis*(β-chloropropionyl) and N,N'-*bis*(β-chlorobutyryl) derivatives of phenylenediamines [40]. Examples of these applications are shown in Eqs. (39) and (40).

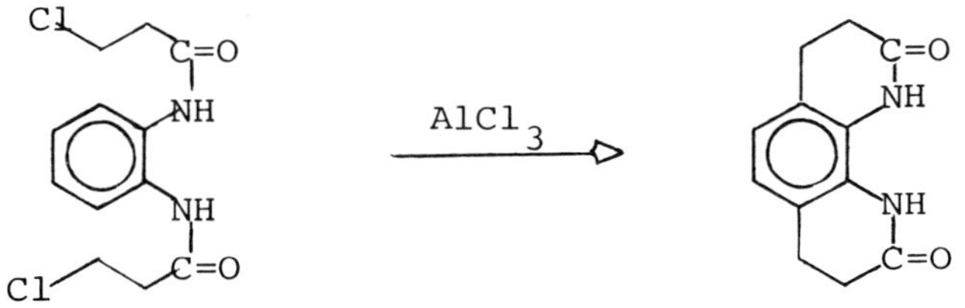

(39)

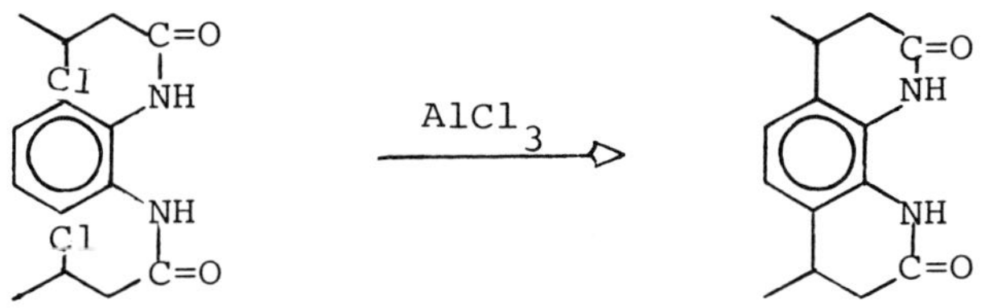

(40)

3. Halo amines

Both haloalkyl aryl amines and N-haloaryl alkyl amines underwent Friedel-Crafts cyclization [45,46]. An example of the $AlCl_3$-induced cyclization of haloalkyl aryl amines can be illustrated by the reported [45] 1° closure of 125 to 126 (Eq. 41).

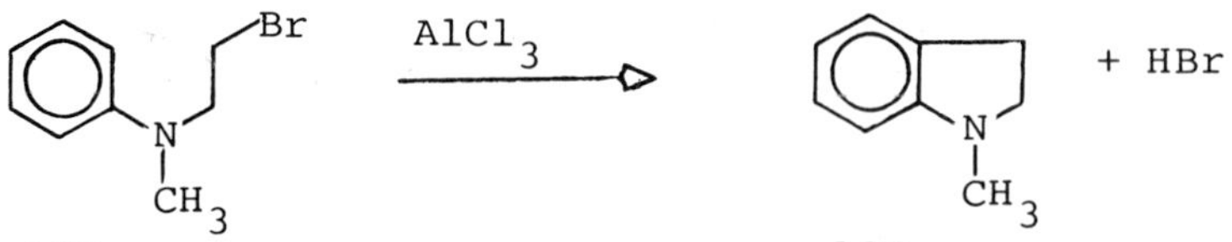

$$125 \xrightarrow{AlCl_3} 126 + HBr \qquad (41)$$

An illustration of the cyclization of N-halo amines is shown in the reported [46] cyclization of methyl(3-phenylpropyl)-N-chloramine (127) and methyl(2-phenylethyl)-N-chloramine (129) to give N-methyltetrahydroquinoline (128) and N-methylindoline (130), respectively, by treatment with sulfuric acid in the presence of powdered ferrous sulfate (Eqs. 42 and 43).

127 $\xrightarrow[\text{81\% yield}]{FeSO_4\text{-dilute } H_2SO_4}$ 128 (42)

129 $\xrightarrow[\text{27\% yield}]{FeSO_4\text{-conc. } H_2SO_4}$ 130 (43)

4. Haloalkyl aryl ketones

Under special conditions, haloalkyl aryl ketones can be cyclized into cyclic ketones, i.e., indanones or tetralones. Understandably, the conditions have to be somewhat drastic to overcome the deactivation of the aromatic ring by the electron-attracting carbonyl group. Thus upon treatment with $AlCl_3$-NaCl at 100°C. γ-bromo [8] and γ-chlorobutyrophenone [11] underwent rearrangement and cyclization to 3-methyl-1-indanone (Eq. 44).

$$C_6H_5\text{-}C(=O)\text{-}C\text{-}C\text{-}C\text{-}X \xrightarrow[100^\circ,\ 1\text{ hr}]{AlCl_3\text{-NaCl}} \text{3-methyl-1-indanone} + \text{Unidentified} \quad (44)$$

X = Cl or Br (mainly) (some)

Under similar conditions, γ-bromovalerophenone gave 4-methyl-1-tetralone (Eq. 45) [8].

$$C_6H_5\text{-}C(=O)\text{-}C\text{-}C\text{-}C(Br)\text{-}C \xrightarrow[100^\circ,\ 1\text{ hr}]{AlCl_3\text{-NaCl}} \text{4-methyl-1-tetralone} \quad (45)$$

In 1927-1931, Mayer and co-workers [47,48], described the preparation of various types of 1-indanones from β-haloalkyl aryl ketones (or the correspondingly dehydrohalogenated α,β-unsaturated ketones) by ring closure with acids such as H_2SO_4 or $AlCl_3$. Some of their results are shown in Eq. (46).

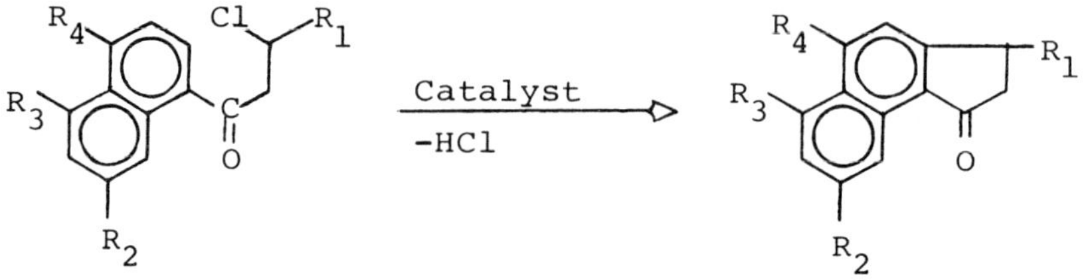

Catalyst
-HCl
(46)

$R_1 = R_2 = R_3 = R_4 = H$

$R_1 = R_2 = R_3 = H, R_4 = CH_3$

$R_1 = R_4 = CH_3, R_2 = R_3 = H$

$R_1 = R_2 = R_3 = H, R_4 = Cl$

$R_1 = R_4 = H, R_2 = R_3 = CH_3$

In the same years, Mayer et al. [49,50] also reported that indanones can alternatively be obtained via intermolecular reaction between arenes and β-chloropropionyl chloride in the presence of $AlCl_3$ (Eq. 47).

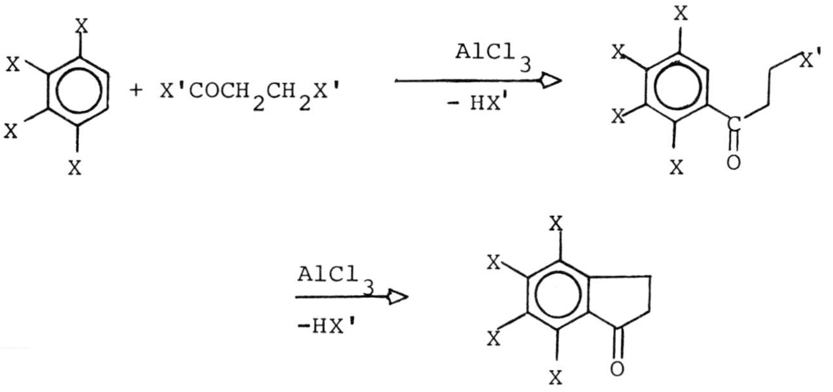

+ $X'COCH_2CH_2X'$ $\xrightarrow[-HX']{AlCl_3}$ $\xrightarrow[-HX']{AlCl_3}$ (47)

In the latter general equation X' stands always for halogen but X may stand for hydrogen, halogen, or one of the groups hydroxyl, hydroxylalkyl, or alkyl, or two adjacent Xs may be a ring linked to the benzene nucleus, in which formula, however, at least one X signifies a substituent other than hydrogen [49].

Cyclialkylations involving other β-haloalkyl [51] as well as α,β-dihaloalkyl [52] aryl ketones have also been reported. Two examples are shown in Eqs. (48) [51] and (49) [52].

$COCH_2CH_2Cl$ $\xrightarrow[\text{4 hr, steam bath}]{H_2SO_4}$ (48) [51]

$COCH(Br)CBr(CH_3)_2$ $\xrightarrow[50^\circ]{AlCl_3}$ Br (49) [52]

Aside from the cyclizations of β- and γ-haloalkyl aryl ketones above, several cyclialkylations of α-haloalkyl aryl ketones have been described. Until 1956, the only reference to the cyclialkylation of an α-bromoalkyl aryl ketone was in the work of Kishner [53], who reported in 1914 that α-bromoisobutyrophenone yielded 2-methyl-1-indanone upon treatment with $AlCl_3$ (Eq. 50).

Br $\xrightarrow{AlCl_3}$ (50)

In 1956, Layer and MacGregor [54] investigated the cyclialkylation of a number of α-bromoalkyl aryl ketones in the presence of $AlCl_3$ under a variety of reaction conditions. They found that ketones 131a-131c in which the bromine atom was attached to a tertiary carbon atom, could readily be

R O Br R_2 R_1 R_3 $\xrightarrow{AlCl_3}$ R O R_2 R_1 R_3 (51)

131 132

a: $R = R_1 = R_2 = CH_3$; $R_3 = H$

b: $R = R_1 = R_3 = CH_3$; $R_2 = C_2H_5$

c: $R = R_1 = H$; $R_2 + R_3 = CH_2CH_2CH_2$

d: $R = R_1 = CH_3$; $R_2 = R_3 = H$

e: $R = R_1 = R_3 = CH_3$; $R_2 = H$

f: $R = R_1 = R_2 = R_3 = H$

cyclized to the corresponding 1-indanones 132a-132c upon treatment with $AlCl_3$ in refluxing CS_2, whereas ketones 131d-131f, in which the bromine atom was attached to a secondary carbon atom, required the more severe conditions of heating with $AlCl_3$ in the absence of solvent at 150°C. Moreover, when α-bromopropiophenone and α-bromobutyrophenone were treated with $AlCl_3$ in refluxing ligroin at 125°C, they gave 60 and 64% of the hydride exchange products propiophenone and butyrophenone, respectively.

In the same investigation, Layer and MacGregor explored the effect of structural variations on the ease of ring closure. For determining the effects of nuclear substituents on the ease of cyclization, compounds 133 were prepared and subjected to cyclialkylation conditions (Eq. 52). Upon treat-

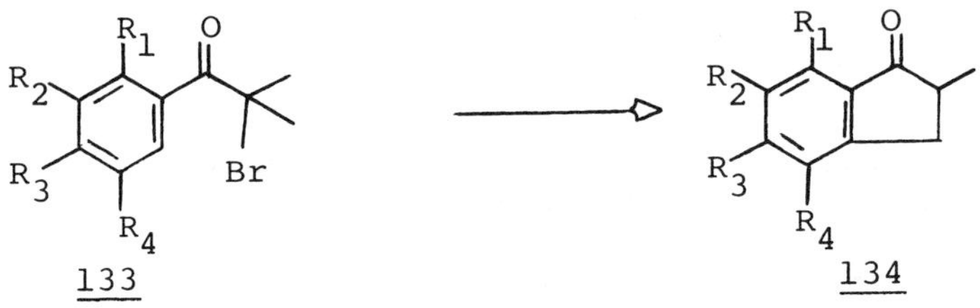

133 → 134 (52)

a: $R_1 = R_2 = R_4 = H$; $R_3 = CH_3$

b: $R_1 = R_3 = CH_3$; $R_2 = R_4 = H$

c: $R_1 = R_4 = CH_3$; $R_2 = R_3 = H$

d: $R_1 = R_2 = R_4 = H$; $R_3 = Cl$

e: $R_1 = Cl$; $R_2 = R_3 = R_4 = H$

f: $R_1 = R_3 = R_4 = H$; $R_3 = NO_2$

ment with $AlCl_3$ in refluxing CS_2, compounds 133a-133e yielded the corresponding 1-indanones 134a-134e while compound 133f gave a large amount of polymer along with some unreacted α-bromoketone. From this it was concluded that α-bromoisobutyrophenones bearing methyl group(s) in the aromatic nucleus can be made to cyclize regardless of the position(s) of the methyl substituent(s) and that *o,p*-directing groups, even deactivating ones, will allow ring closure.

As to the effect of α and β substituents on the ease of intramolecular ring closure, it was concluded on the basis of the then available experimental observations that (1) ring closure was facilitated by the presence of α-alkyl groups and retarded by β-alkyl groups and (2) in compounds with both α and β substitutents, the activating effect of the α-alkyl groups overcame the deactivating effect of the β-alkyl group.

The unexpected facility with which some α- and β-halo aralkyl ketones were claimed to close to 1-indanones [47-54], added to the long-standing uncertainty about the mechanistic nature of these closures [2,47-54,55], inspired some workers to restudy this segment of Friedel-Crafts chemistry with modern techniques. In 1976, Pines and Douglas [56] conducted a detailed study of the $AlCl_3$-catalyzed cyclialkylations of 2- and 3-bromo- and 2- and 3-chloro-4'-fluoro-2-methylpropiophenones 135b, 135c, 136b, and 136c.

135

136

a: R = H; X = Br

b: R = F; X = Br

c: R = F; X = Cl

This study included investigation by in situ ^{13}C NMR spectroscopy as well as by more conventional techniques involving chromatographic analysis of reaction mixtures. The results obtained from the in situ observation of the conversion of α-bromo ketone 135b into the 2-methyl-1-indanone 138 established, for the first time, that the cyclization proceeded through the methacrylophenone 137 (Eq. 53).

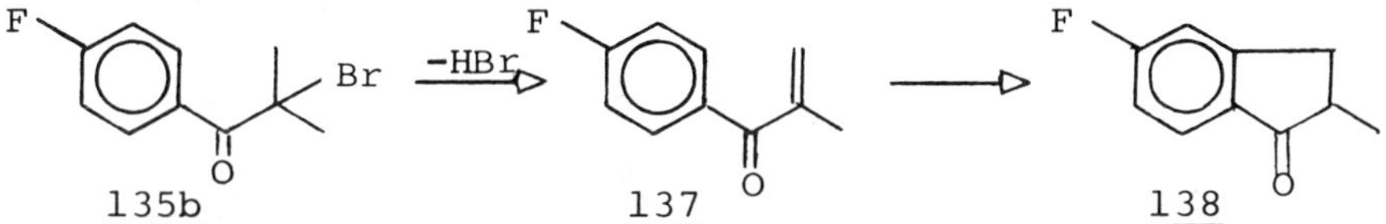

135b —(−HBr)→ 137 → 138 (53)

The cyclialkylation behavior of the isomeric β-halo aralkyl ketones 136b and 136c showed significant differences from that of the isomeric α-halo counterparts (135b and 135c). Not only were the cyclization rates much slower, but also the products were more complex, consisting of the 2-methylindanone (138), its rearranged 3-methyl isomer (141), and the oxonium ion 142. The formation of 138, 141, and 142 was accounted for in terms of the carbocation transformations shown in scheme 13.

According to scheme 13, ionization of 136c gives the primary carbocation 139, which by direct closure provides the major product of the reaction, indanone 138. Rearrangement of 139 by consecutive methyl and hydride shifts gives carbocation 140. The latter can either close to the 3-methylindanone 141 (path a) or undergo a concerted process (path b) whereby hydride migration accompanies the formation of the O → C donor bond to give the oxonium ion 142. However, due to the involvement of primary carbocations, scheme 13 was proposed with much reservation.

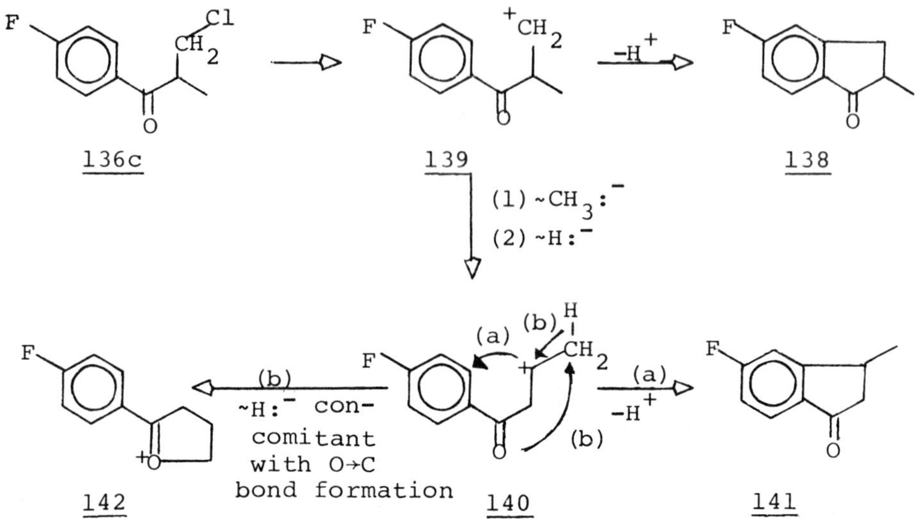

136c → 139 → ($-H^+$) 138

139 → (1) ~CH_3:$^-$ (2) ~H:$^-$ → 140

140 → (a) $-H^+$ → 141

140 → (b) ~H:$^-$ concomitant with O→C bond formation → 142

Scheme 13

In 1978, Pines and Douglas [57] extended their study to include the $AlCl_3$- and $AlCl_3$-CH_3NO_2-catalyzed alkylation of ^2H- and ^{13}C-labeled 3-chloro-4'-fluoro-2-methylpropiophenone (136c-^{13}C and 136c-D). Their aim was to facilitate choice between the various alternative cyclialkylation mechanisms of β-halo

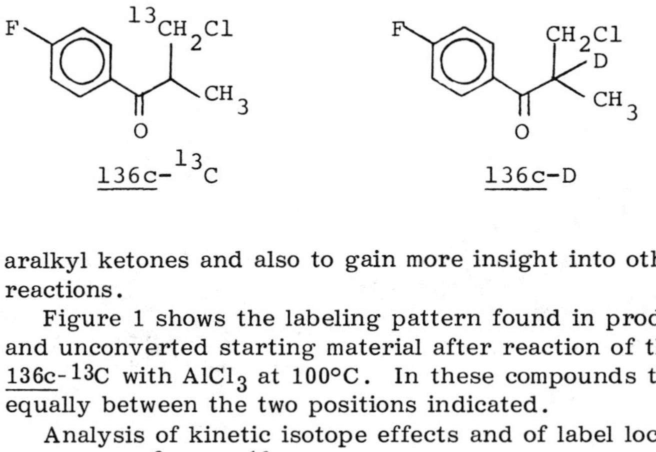

136c-^{13}C 136c-D

aralkyl ketones and also to gain more insight into other aspects of these reactions.

Figure 1 shows the labeling pattern found in products 138, 141, and 142 and unconverted starting material after reaction of the ^{13}C-labeled chloride 136c-^{13}C with $AlCl_3$ at 100°C. In these compounds the label was shared equally between the two positions indicated.

Analysis of kinetic isotope effects and of label location in the products as revealed in ^2H and ^{13}C spectra allowed definition of the major pathways involved. In cyclization to 2-methylindanone (138), an isotope rate effect (k_H/k_D = 2.5) supported ionization concerted with C_2-H (or C_2-D) migration as the rate-determining step. Scheme 13a was thus proposed to account for this and also for the observed presence of deuterium in both the methyl and methylene groups, but not in the methine group.

As to the skeletally rearranged products 141 and 142, the data obtained supported their formation via initial methyl migration and not acyl migration (Schemes 14 and 15).

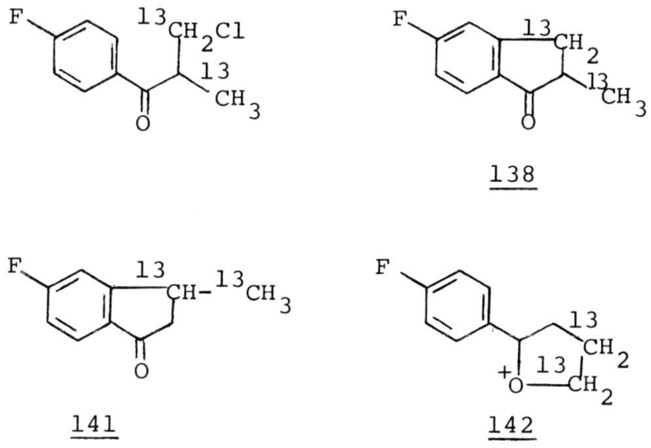

138

141

142

Figure 1

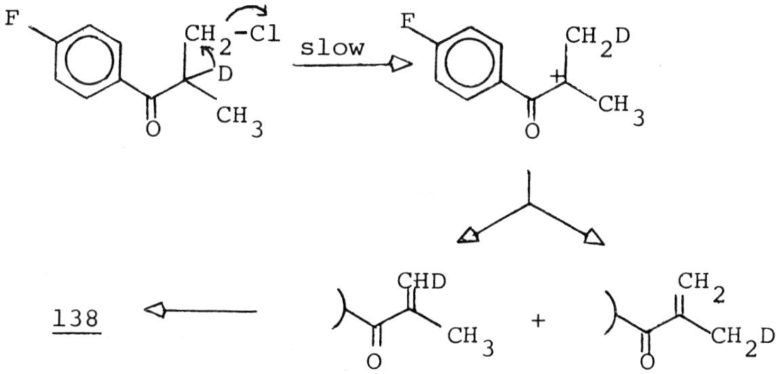

Scheme 13a

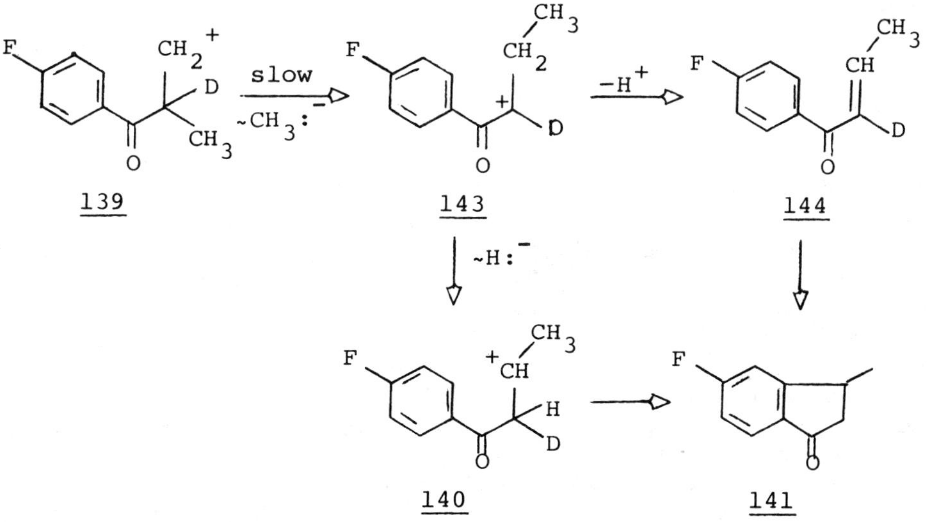

139

143

144

140

141

Scheme 14

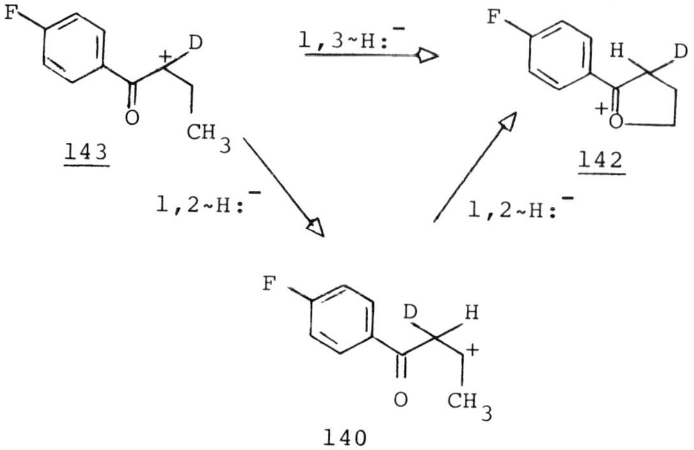

Scheme 15

According to scheme 14, the conversion of 143-141 could proceed via either 140 or 144. However, the characterization of 144 in the mixture [56] suggested some preference for this route.

The cyclialkylation of 3-bromo- (136b) and 3-chloro-4'-fluoro-2-methylpropiophenone (136c) was also studied in the presence of H_2SO_4 [56] and $AlCl_3$-CH_3NO_2 [57] catalysts. With H_2SO_4, 136b underwent facile cyclization to 2-methyl-5-fluoroindanone (138), giving 92% yield after treatment at 50°C for 4 hr. The behavior of the β-bromo analog 136c in H_2SO_4 was somewhat different. In this case, facile oxidation of Br^- to Br_2 in H_2SO_4 intruded upon the simplicity of the reaction of 136b leading to brominated products (e.g., 145) (Eq. 54).

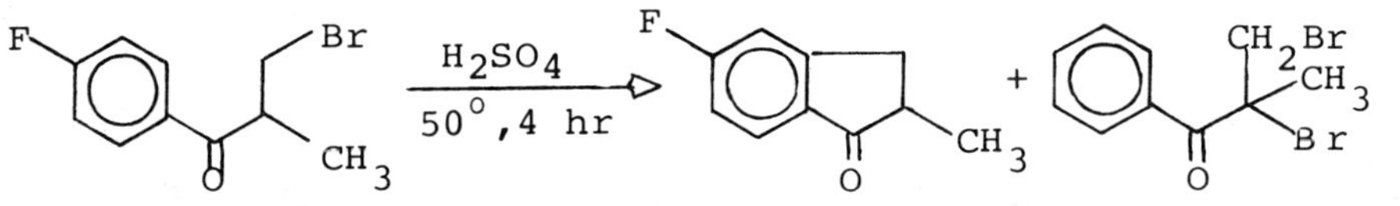

$$\textbf{136b} \xrightarrow[50^\circ, 4\ hr]{H_2SO_4} \textbf{138}\ 65\% + \textbf{145}\ 34\% \quad (54)$$

The cyclialkylation of 136c with $AlCl_3$-CH_3NO_2 showed characteristics similar to those observed with H_2SO_4. For example, it proceeded at a milder temperature than required for $AlCl_3$ alone, neither skeletal nor isotopic rearrangement was observed in the recovered starting material or in the product, but the reaction was accompanied by subsequent conversion of indanone 138 to chloroindanone 146 and isocoumarin 147 (scheme 16). The similar behavior of H_2SO_4 and $AlCl_3$-CH_3NO_2 in catalyzing the closure of β-halo aralkyl ketones was attributed to their protic nature [58], which supports product formation through an enolization-initiated mechanism.

In other recent development, Khalaf and co-workers [59] have undertaken a systematic reinvestigation of the work of Mayer et al. The preliminary results obtained by the application of modern analytical tools proved that the products from these reactions are not as simple as previously claimed. Be-

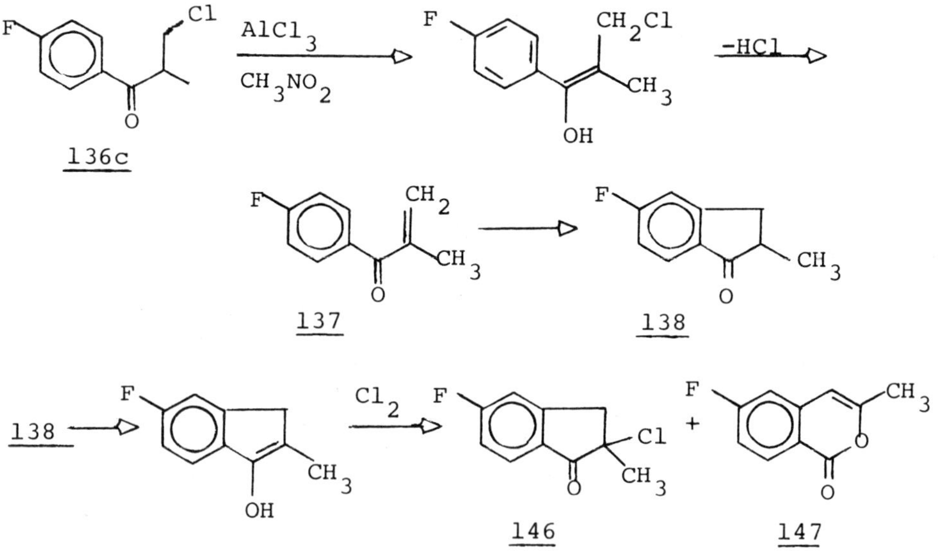

136c $\xrightarrow[CH_3NO_2]{AlCl_3}$ $\xrightarrow{-HCl}$

137 ⟶ 138

138 ⟶ $\xrightarrow{Cl_2}$ 146 + 147

Scheme 16

sides ring closure products, those resulting from hydride transfer and *inter-* and *intra*molecular acyl migration were also characterized in the resulting mixtures.

5. Haloalkyl ethers

Both aryl and arylalkyl haloalkyl ethers have been cyclized to cyclic ethers. For example, chroman and its 6-methyl, 6-bromo, and 6-chloro derivatives (149) were obtained upon treatment of ethers 148 with $SnCl_4$ (Eq. 55) [60].

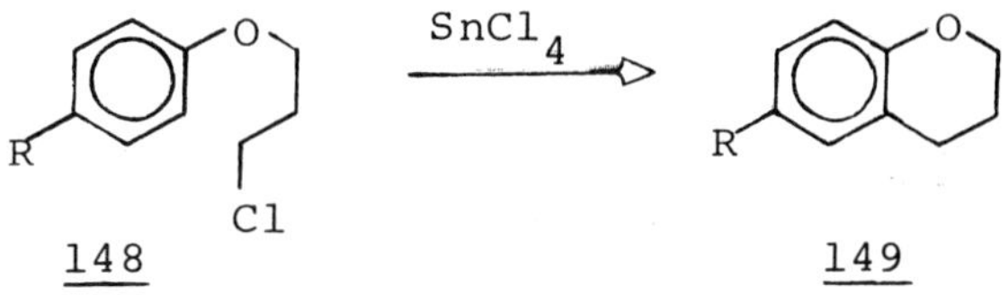

148 $\xrightarrow{SnCl_4}$ 149 (55)

$R = H, CH_3, Br, Cl$

Similarly, isochroman and its 7-methyl and 5,6-benzo derivatives were isolated in good yields (85 to 91.5%) when compounds 150 and 151, respectively, were treated with $AlCl_3$ (Eqs. 56 and 57) [61].

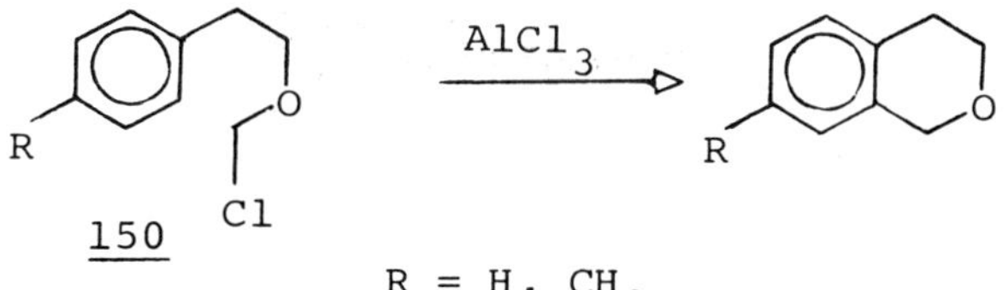

150 $\xrightarrow{AlCl_3}$ (56)

$R = H, CH_3$

$$\xrightarrow{AlCl_3} \quad (57)$$

151

III. CYCLIALKYLATIONS OF ARYLALKANOLS, ARYLALKENES, AND RELATED SYSTEMS

This section is devoted primarily to cyclialkylations involving intramolecular, alkylations of arylalkanols and arylalkenes. Other related systems such as hydroxyaralkyl and aralkenyl ketones, ethers, amines, and amides will be briefly considered. Moreover, to complete the picture of the ring closure processes, we have also included some pertinent intermolecular reactions which have been used alternatively to form cyclic derivatives. However, in the case of indans, for which intermolecular reactions have been extensively employed, these will be described under a separate heading.

A. Cyclialkylations Yielding Bicyclic Systems

1. Indan and tetralin derivatives

a. Production of Indans and Tetralins by Intramolecular Alkylations

Starting in 1932, Bogert and co-workers [14,15,62-73] undertook a systematic study to evaluate the various factors affecting intramolecular ring closure of arylalkanols and arylalkenes to indan and tetralin derivatives. Their results, together with other related ones published prior to 1969, are summarized in Table 11. This table also includes the cyclialkylation results of Colonge and Pichat [74] and Mukherji and co-workers [75-77], who developed the method outlined in Eq. (58) for the synthesis of substituted tetralin derivatives.

R_1, R_2, R_3, R_4 + O $\xrightarrow{AlCl_3}$ R_1, R_2, R_3, R_4, O (58)

$\xrightarrow[\text{or Grignard } (R_5MgX)]{\text{Reduction}}$ R_1, R_2, R_3, R_4, OH, R_5 $\xrightarrow{H_2SO_4}$ R_1, R_2, R_3, R_4, R_5

Classified according to final products, the cyclialkylation reactions incorporated in Table 11 have been divided into four categories: those giving mainly polymers, those giving indans, those giving tetralins, and those giving both indans and tetralins.

Table 11. Cyclialkylations of Arylhydroxyalkanes to Indans and Tetralins Prior to 1969

Cyclized compounds	Catalyst	Yield, %	Product	Ref.
(a) Cyclialkylations Producing Polymers				
$PhCH(OH)CH_2CH_2CH_2CH_3$	H_2SO_4	—	Polymer	67
$PhCH_2C(CH_3)(OH)CH_2CH_3$	H_2SO_4	—	Polymer	65
$PhCH_2CH_2CH_2OH$	H_3PO_4	—	Polymer	65
$PhCH_2CH_2CH(OH)CH_3$	H_3PO_4	—	Polymer	65
$PhCH_2CH(CH_3)CH(OH)CH_3$	H_2SO_4	—	Polymer	65
$PhCH_2CH_2CH = CH_2$	H_2SO_4	—	Polymer	65
(b) Cyclialkylations Producing Indans				
$PhCH_2CH(OH)CH(CH_3)CH_3$	H_2SO_4	55	1,1-Dimethylindan	65
$PhCH_2C(CH_3)(OH)CH(CH_3)CH_3$	H_2SO_4	90	1,1,2-Trimethylindan	65
$PhC(CH_3)_2CH_2CH_2OH$	H_3PO_4	15	1,1-Dimethylindan	65
$PhCH_2CH_2C(CH_3)(OH)CH_3$	H_2SO_4	65	1,1-Dimethylindan	65
$PhC(CH_3)_2CH_2CH(OH)CH_3$	H_2SO_4	35	1,1,3-Trimethylindan	65
$PhC(CH_3)_2CH_2C(CH_3)(OH)CH_3$	H_2SO_4	85	1,1,3,3-Tetramethylindan	65
p-$CH_3C_6H_4C(CH_3)_2CH_2C(OH)(CH_3)_2$	H_2SO_4	75	1,1,3,3,5-Pentamethylindan	78

Table 11. (continued)

Cyclized compounds	Catalyst	Yield, %	Product	Ref.
p-$CH_3C_6H_4C(CH_3)_2CH_2C(OH)(CH_3)(C_2H_5)$	H_2SO_4	74	1,1,3,5-Tetramethyl-3-ethylindan	78
p-$CH_3C_6H_4C(CH_3)_2CH(CH_3)C(OH)(CH_3)_2$	H_2SO_4	40	1,1,2,3,3,5-Hexamethylindan	78
p-$CH_3C_6H_4C(CH_3)_2CH_2C(OH)(CH_3)(C_3H_7)$	H_2SO_4	64	1,1,3,5-Tetramethyl-3-propylindan	79
p-$CH_3C_6H_4C(CH_3)_2CH_2C(OH)(CH_3)$ (*iso*-C_4H_9)	H_2SO_4	18	1,1,3,5-Tetramethyl-3-isobutylindan	79
p-$CH_3C_6H_4CH(CH_3)CH(CH_3)C(OH)(CH_3)_2$	H_2SO_4	68	1,2,3,3,5-Pentamethylindan	78
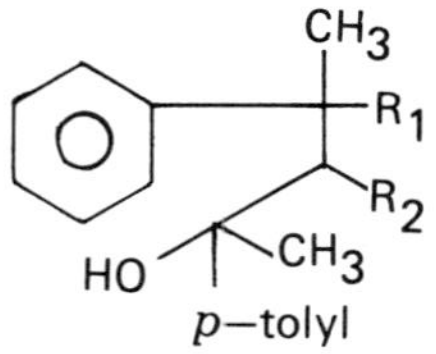 CH_3, R_1, R_2, HO, CH_3, p-tolyl			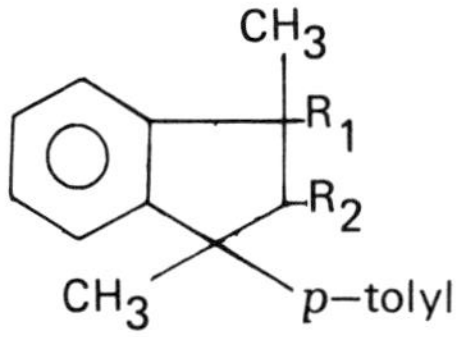 CH_3, R_1, R_2, CH_3, p-tolyl	
$R_1 = CH_3$, $R_2 = H$	HF	54	$R_1 = CH_3$, $R_2 = H$	80
$R_1 = C_2H_5$, $R_2 = CH_3$	HF	65	$R_1 = C_2H_5$, $R_2 = CH_3$	81
3,5-$(CH_3)_2C_6H_3C(CH_3)_2CH_2C(OH)(CH_3)_2$	AcOH-H_2SO_4	—	1,1,3,3,4,6-Hexamethylindan	82

3,4-$CH_2O_2C_6H_3CH_2CH_2CH_2C(OH)(CH_3)_2$	H_2SO_4	70	1,1-Dimethyl-5,6-methylene-dioxyindan	68
3,4-$CH_2O_2C_6H_3CH_2CH(CH_3)C(OH)(CH_3)_2$	H_2SO_4	65	1,1,2-Trimethyl-5,6-methylene-dioxyindan	68
$PhCH_2CH = C(CH_3)CH_3$	H_2SO_4	—	1,1-Dimethylindan	65
(c) Cyclialkylations Producing Tetralins				
$PhCH_2CH(OH)CH_2CH_2CH_3$	H_2SO_4	35	1-Methyltetralin	67
$PhCH_2CH_2CH(OH)CH_2CH_3$	H_2SO_4	60	1-Methyltetralin	67
$PhCH_2CH_2CH(OH)CH(CH_3)CH_3$	H_2SO_4	—	1,1-Dimethyltetralin	66
p-$CH_3OC_6H_4CH_2CH_2CH(OH)CH_2CH(CH_3)_2$	H_3PO_4	86	1-Isopropyl-7-methoxytetralin	76
m-$CH_3C_6H_4CH_2CH_2C(OH)(CH_3)CH(CH_3)_2$	H_2SO_4	83	1,1,2,6-Tetramethyltetralin (irene)	70
$PhCH_2CH_2CH_2CH_2OH$	H_3PO_4	55	Tetralin	65
$PhCH_2CH_2CH_2CH(OH)CH_3$	H_2SO_4	70	1-Methyltetralin	67
$PhCH_2CH_2CH_2C(CH_3)(OH)CH_3$	H_2SO_4	—	1,1-Dimethyltetralin	66
$PhCH_2CH_2CH_2C(OH)(CH_3)(C_6H_5)$	H_2SO_4	5	1-Methyl-1-phenyltetralin	83
p-$CH_3OC_6H_4CH_2CH_2CH_2CH(OH)CH(CH_3)_2$	H_3PO_4	86	1-Isopropyl-7-methoxytetralin	84
$PhCH(CH_3)CH_2CH_2C(OH)(CH_3)_2$	H_2SO_4	68	1,1,4-Trimethyltetralin	76
$PhCH_2CH_2CH(COOH)CH(OH)CH_3$	H_2SO_4	37	*trans*-1-Methyl-2-carboxytetralin	85
$PhCH(CH_3)CH_2CH_2CH(OH)CH_3$	H_2SO_4	68	1,4-Dimethyltetralin	74

Table 11. (continued)

Cyclized compounds	Catalyst	Yield, %	Product	Ref.
p-$CH_3C_6H_4CH(CH_3)CH_2CH_2CH(OH)CH_3$	H_2SO_4	53	1,4,6-Trimethyltetralin	74
2,5-$(CH_3)_2C_6H_4CH(CH_3)CH_2CH_2CH(OH)CH_3$	H_2SO_4	43	1,4,5,8-Tetramethyltetralin	74
2,4-$(CH_3)_2C_6H_4CH(CH_3)CH_2CH_2CH(OH)CH_3$	H_2SO_4	68	1,4,5,7-Tetramethyltetralin	74
p-$CH_3C_6H_4CH(CH_3)CH_2CH_2C(OH)(CH_3)_2$	H_2SO_4	79	1,1,4,7-Tetramethyltetralin	74
2,5-$(CH_3)_2C_6H_4CH(CH_3)CH_2CH_2C(OH)(CH_3)_2$	H_2SO_4	88	1,1,4,5,8-Pentamethyltetralin	74
2,4-$(CH_3)_2C_6H_4CH(CH_3)CH_2CH_2C(OH)(CH_3)_2$	H_2SO_4	85	1,1,4,5,7-Pentamethyltetralin	74
3,4-$(CH_3)_2C_6H_4CH(CH_3)CH_2CH_2CH(OH)CH_3$	H_2SO_4	—	1,4,6,7-Tetramethyltetralin	86
p-$CH_3OC_6H_4CH(CH_3)CH_2CH_2CH(OH)CH_3$	H_2SO_4	69.5	1,4-Dimethyl-6-methoxytetralin	75
p-$(CH_3)_2CHC_6H_4CH(CH_3)CH_2CH_2CH(OH)CH_3$	H_2SO_4	71	1,4-Dimethyl-6-isopropyltetralin	77
2-CH_3-5-$(CH_3)_2CHC_6H_4CH(CH_3)CH_2CH_2CH$-$(OH)CH_3$	H_2SO_4	72	1,4,8-Trimethyl-5-isopropyltetralin	77
$PhCH(CH_3)CH_2CH_2C(OH)(CH_3)_2$	H_2SO_4	84	1,1,4-Trimethyltetralin	74

p-$CH_3OC_6H_4CH_2CH_2CH_2CH_2C(OH)(CH_3)_2$	H_3PO_4	65	1-Isopropyl-7-methoxytetralin	76
$PhCH_2CH_2CH_2CH_2CH_2OH$	H_3PO_4	—	1-Methyltetralin	67
$PhCH_2CH_2CH_2CH_2CH(OH)CH_3$	H_2SO_4	55	1-Ethyltetralin	67
$PhCH_2CH_2CH_2CH_2C(CH_3)(OH)CH_3$	H_2SO_4	35	1-Isopropyltetralin	67
$PhCH_2CH_2CH_2CH_2CH_2CH(OH)CH_3$	H_2SO_4	45	1-*n*-Propyltetralin	67
$PhCH_2CH_2CH = C(CH_3)_2$	H_2SO_4	—	1,1-Dimethyltetralin	66
$PhCH_2CH_2CH_2CH = CH_2$	H_2SO_4	75	1-Methyltetralin	67
$PhCH_2CH_2CH_2CH_2CH = CH_2$	H_2SO_4	65	1-Ethyltetralin	67
(d) Cyclialkylations Producing Mixtures of Indans and Tetralins				
$PhCH_2CH(OH)CH(CH_3)CH_2CH_3$	H_2SO_4	13	1,2-Dimethyltetralin and	67
		37	1-Methyl-1-ethylindan	
$PhCH_2CH_2C(CH_3)(OH)CH_2CH_3$	H_2SO_4	69	1,2-Dimethyltetralin and	67
		18	1-Methyl-1-ethylindan	

Table 12. Cyclidehydrations Reported by Khalaf and Roberts

Cyclized compound	Catalyst	Temp, °C	Yield, %	Product composition, %[a]	Ref.
		A. Primary Alcohols			
$PhCH_2CH_2CH_2OH$	85% H_3PO_4	230-240	45	Indan (trace), *n*-propylbenzene, (11), three isomeric phenylpropenes (89)	87
	$AlCl_3$-CH_3NO_2	25	—	Starting material	89
$PhC(CH_3)_2CH_2CH_2OH$	85% H_3PO_4	230-240	85	1,1-Dimethylindan (18), 2-methyl-3-phenyl-2-butene (82)	87
	$AlCl_3$-CH_3NO_2	25	—	Starting material	89
p-$CH_3C_6H_4C(CH_3)_2CH_2CH_2OH$	85% H_3PO_4	230-240	24	*p*-*t*-Pentylbenzene (trace), 2-methyl-3-(*p*-tolyl)butane (6), 2-methyl-3-(*p*-tolyl)-2-butene (33), 1,1,6-trimethylindan (13), 1,1,4-trimethylindan (32), 1,1,5-trimethylindan (10), 1,1,7-trimethylindan (3), and unidentified (2)	88
	$AlCl_3$-CH_3NO_2	25	—	Starting material	89
$PhCH_2CH_2CH_2CH_2OH$	85% H_3PO_4	230-240	61	Tetralin (80), three lower-boiling unidentified (20)	87
	$AlCl_3$-CH_3NO_2	25	—	Starting material	89
$PhCH_2CH_2CH(CH_3)CH_2OH$	85% H_3PO_4	230-240	84	1,1-Dimethylindan (40), 2-methyltetralin (32), 1-methyltetralin (14), unidentified (14)	87

$PhCH_2CH_2CH_2CH_2CH_2OH$	$AlCl_3$-CH_3NO_2	25	—	Starting material	89
$PhCH_2CH(Ph)CH_2CH(CH_3)$-CH_2OH	$AlCl_3$-CH_3NO_2	25	—	Starting material	89
		Secondary Alcohols			
$PhCH_2CH_2CH(OH)Ph$	90% H_2SO_4	10-25	—	1-Phenylindan (34), 1-phenylindene (1), bimolecular condensation products (50), unidentified (14)	87
$PhCH_2CH_2CH(OH)CH_3$	85% H_3PO_4	230-240	75	Mixture of three isomeric phenylbutenes (100)	87
	$AlCl_3$-CH_3NO_2	25	—	High polymers	89
$PhCH_2CH_2CH_2CH(OH)CH_3$	$AlCl_3$-CH_3NO_2	25	80	1-Methyltetralin (100)	89
$PhCH_2CH_2CH(OH)CH_2CH_3$	$AlCl_3$-CH_3NO_2	25	76	1-Methyltetralin (100)	87
$PhCH_2CH(OH)CH_2CH_2CH_3$	90% H_2SO_4	10-25	55	1-Methyltetralin (97), unidentified (3)	87
	$AlCl_3$-CH_3NO_2	25	50	1-Methyltetralin (100)	89
		Tertiary Alcohols			
$PhCH_2CH_2C(OH)(CH_3)_2$	85% H_2SO_4	10-25	57	1,1-Dimethylindan (100)	87
	$AlCl_3$-CH_3NO_2	25	82	1,1-Dimethylindan (100)	89

Table 12. (continued)

Cyclized compound	Catalyst	Temp, °C	Yield, %	Product composition, %[a]	Ref.
o-$CH_3C_6H_4CH_2CH_2C(OH)(CH_3)_2$	85% H_2SO_4	10-25	40	1,1,4-Trimethylindan (100)	88
	$AlCl_3$-CH_3NO_2	25	76	1,1,4-Trimethylindan (100)	89
p-$CH_3C_6H_4CH_2CH_2C(OH)(CH_3)_2$	85% H_2SO_4	10-25	50	1,1,6-Trimethylindan (100)	88
	$AlCl_3$-CH_3NO_2	25	69	1,1,6-Trimethylindan (100)	89
m-$CH_3C_6H_4CH_2CH_2C(OH)(CH_3)_2$	85% H_2SO_4	10-25	15	1,1,5- (68) and 1,1,7-Trimethylindan (32)	88
	85% H_3PO_4	230	50	1,1,5- (53) and 1,1,7-Trimethylindan (47)	
	$AlCl_3$-CH_3NO_2	25	74	1,1,5- (57) and 1,1,7-Trimethylindan (43)	89
$PhCH_2CH_2CH_2C(OH)(CH_3)_2$	85% H_2SO_4	10-25	83	1,1-Dimethyltetralin (100)	87
	$AlCl_3$-CH_3NO_2	25	50	1,1-Dimethyltetralin (100)	89
$PhCH_2CH_2CH_2CH_2C(OH)(CH_3)_2$	$AlCl_3$-CH_3NO_2	25	65	1-Isopropyltetralin (100)	89
$PhCH_2CH(Ph)CH_2C(OH)(CH_3)_2$	$AlCl_3$-CH_3NO_2	25	82	1,1-Dimethyl-3-(β-phenylethyl)indan (100)	89

[a]Percentage composition as determined by GLPC.
Source: Refs. 87-89.

It is to be noted that the cyclization results incorporated in Table 11 were identified by unreliable conventional means, including elemental analyses, comparison of physical constants, and characterizations of oxidation products. The fact that such early identification methods are vulnerable to introduction of errors inspired Khalaf and Roberts [18,21,87-89] to conduct some new studies on cyclidehydration reactions in which modern instrumental methods of separation and identification were employed. These authors not only reinvestigated some of the cyclidehydrations reported earlier, but also examined some new ones especially designed to provide more insight into the mechanisms of cyclidehydration. Some of these author's results are compiled in Table 12 and a summary of their interpretations follows.

In comparing the early results of Table 11 with the newer ones of Table 12, some differences can be recognized which are probably due to the improved identification techniques used by the more recent workers. For example, 3-phenyl-1-propanol (152) and 4-phenyl-1-butanol were reported by Bogert and Davidson [65] (Table 11) to yield polymer and pure tetralin, respectively, upon treatment with phosphoric acid at high temperature. Khalaf and Roberts [87], however, showed the product from the former to contain a trace of indan and a little *n*-propylbenzene, but to consist mainly of the three isomeric phenylpropenes expected to result from normal dehydration (Eq. 59). The product from the latter alcohol was found to consist not only

$$\text{C}_6\text{H}_5\text{-C-C-C-OH} \;(\underline{152}) \longrightarrow \text{indan (trace)} + \text{Ph-C-C-C}\;(11\%) + \text{Ph-C=C-C} + \text{Ph-C=C-C} + \text{PhC-C=C}\;(89\%) \qquad (59)$$

of tetralin (80%), but also of three lower-boiling unidentified products (20%).

Again, 3-methyl-3-phenyl-1-butanol (153) was reported by Bogert and Davidson [65] to yield a mixture of hydrocarbons, suggested on the basis of oxidative degradation to be a mixture of the cyclidehydration product, 1,1-dimethylindan and rearranged normal dehydration products having either of the skeletal structures $C_6H_5C(CH_3)CCCH_3$ or $C_6H_5CCC(CH_3)_2$. When this reaction was repeated by Khalaf and Roberts [87], they found this product to be a mixture of 82% 2-methyl-3-phenyl-2-butene (154) and 18% 1,1-dimethylindan (17) (Eq. 60). The routes proposed to account for the production of 154 and 17 from 153 are shown in Eqs. (61) and (62), respectively.

$$\text{C}_6\text{H}_5\text{-C(C)(C)-C-C-OH}\;(\underline{153}) \xrightarrow[230\text{-}240^\circ]{H_3PO_4} \text{C}_6\text{H}_5\text{-C(C)=C(C)-C}\;(\underline{154}) + \text{1,1-dimethylindan}\;(\underline{17}) \qquad (60)$$

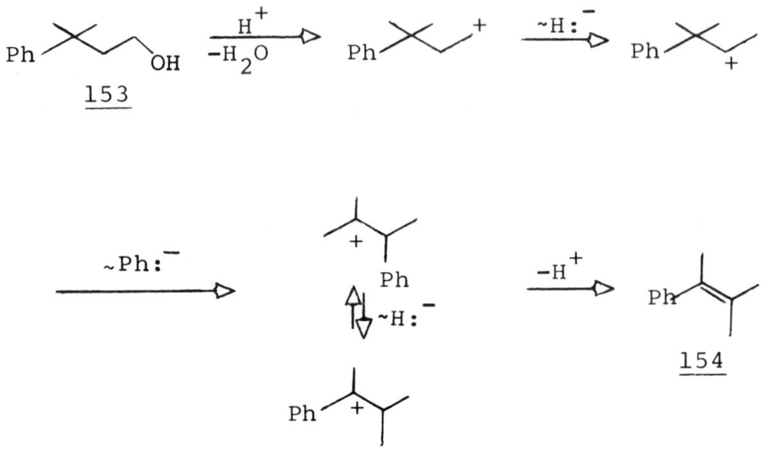

(61)

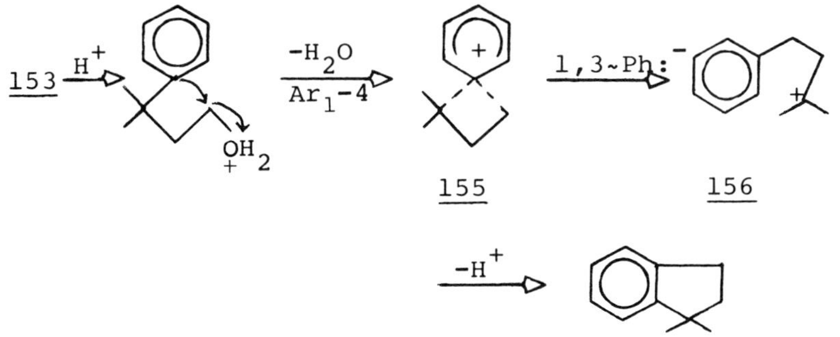

(62)

According to the mechanism of Eq. (62), the cyclization of 153 to 1,1-dimethylindan (17) involves an Ar_1-4 participation and a 1,3-phenyl shift via the bridged carbocation 155 to yield a tertiary carbocation intermediate (156) which is capable of cyclizing to indan. On the basis of this mechanism, the failure of the nonmethylated 3-phenyl-1-propanol (152) to undergo similar closure to indan was rationalized by proposing that a parallel Ar_1-4 participation and 1,3-phenyl migration, if they occurred, would produce another identical carbocation intermediate which again is incapable of closing to an indan.

Although the mechanism of Eq. (62) offered a plausible rationale for the role of *gem*-dimethyls in promoting primary closures to indans, there was no available experimental data to support it or to distinguish it form a similar likely mechanism that may involve Ar_1-5 rather than Ar_1-4 type participation. In an effort to gain more insight into the mechanism of acid-catalyzed cyclidehydrations of primary alcohols to indans, Khalaf and Roberts [88] investigated the phosphoric acid catalyzed dehydration of the methyl-labeled 3-methyl-3-*p*-tolyl-1-butanol (157). On dehydration, this alcohol gave a complex hydrocarbon mixture comprising the acyclic components *p*-*t*-pentyltoluene (112), 2-methyl-3-*p*-tolylbutane (117), and 2-methyl-3-*p*-tolyl-2-butene (158) and the cyclic components 1,1,4- (159), 1,1,5- (105), 1,1,6- (160), and 1,1,7-trimethylindan (109). These components and their relative percentages are shown in Eq. (63).

H_3PO_4, 240–250°

157 → 112 (trace) + 117 7% + 158 33% + 159 32% + 105 10% + 160 13% + 109 3% (63)

The finding that all of the trimethylindans produced were stable to the reaction conditions indicated that they must have resulted directly from the alcohol by rearrangements occurring *prior* to their formation. Since the mechanisms of Eqs. (61) and (62) could not account for all of the products given by 3-methyl-3-*p*-tolyl-1-butanol (157), expanded mechanisms were thus proposed. Of these, the mechanism leading to the formation of the normal acyclic dehydration products 112, 117, and 158 was essentially that used before to account for their formation from the corresponding chloride (102) in the presence of $AlCl_3$ (scheme 14). In terms of this mechanism, aralkene 158 could result from deprotonation of either or both of the two carbocations (115 and 116) leading to 117. Turning to cyclidehydration, the formation of the four isomeric trimethylindans 105, 109, 159, and 160 was best explained in terms of the steps outlined in scheme 17.

In scheme 17, the formation of 104 by Ar_2-5 participation followed by direct deprotonation gives 105. 1,1,7-Trimethylindan 109 is also produced from 104, but by ring opening to give the tertiary carbocation 107, followed by a recyclization *ortho* to the ring methyl group. Although some 105 may arise via 107 as an intermediate, the direct deprotonation of 104 was considered to be the major route for its production, because a higher proportion of 105-109 was obtained from 157 than in the case of cyclidehydration of the isomeric 2-methyl-4-*m*-tolyl-2-butanol with either sulfuric or phosphoric acid, a reaction which should involve the ion 107 as a common intermediate for the formation of 105 and 109. The 1,1,6-trimethylindan (160), which was produced in an amount about equal to the sum of 105 and 109, was suggested to form via Ar_1-4 participation to produce 161, followed by ring opening to give the tertiary carbocation 162, which can cyclize to 160 via 163. The finding of 1,1,4-trimethylindan (159) as the major trimethylindan isomer was accounted for in terms of the route 163 → 164 → 165 → 166 → 159. The stabilization of the intermediate carbocations 164 and 166 by the electron release of the ring methyl group enhances these rearrangements. It was concluded on the basis of the results above that the closure of primary alcohols to five-membered ring products is a rather complex process involving anchimerically assisted

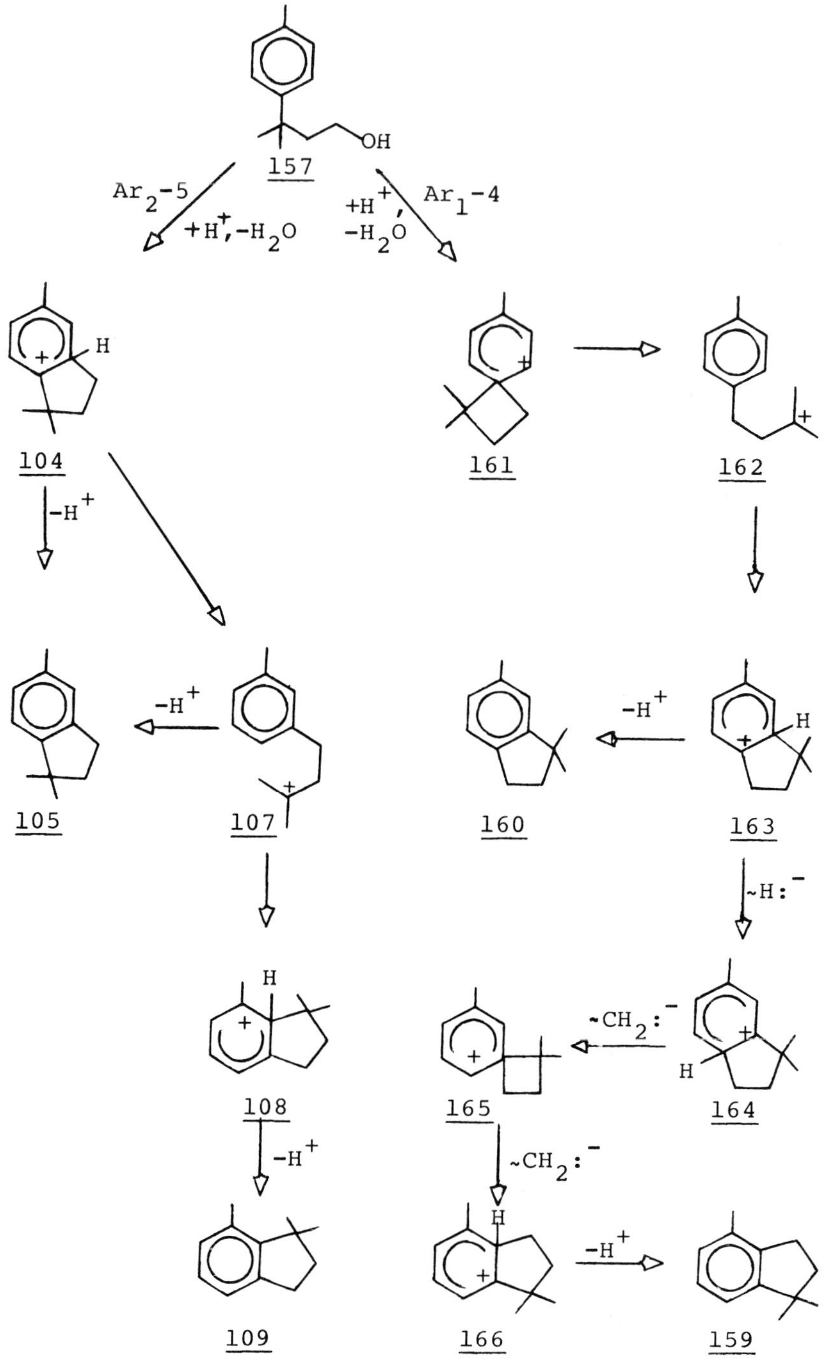

157

Ar_2-5 $+H^+, -H_2O$

$+H^+, -H_2O$ Ar_1-4

104 H

161

162

$-H^+$

105

$-H^+$

107

160

$-H^+$

163 H

$\sim H:^-$

108 H

165

$\sim CH_2:^-$

164 H

$-H^+$

$\sim CH_2:^-$

109

166 H

$-H^+$

159

Scheme 17

ionization through both Ar_1-4 and Ar_1-5 participations in combination with the usual Wagner-Meerwein shifts.

In comparing the results of the H_3PO_4-catalyzed reactions of primary γ-arylalkyl alcohols 152, 153, and 157 with those of the $AlCl_3$-catalyzed reactions of their corresponding γ-arylalkyl chlorides, several significant differences appear. To pinpoint these differences and rationalize them, it seems profitable to present a summary of the recent literature results obtained from the reactions of both the alcohols and the chlorides under various Friedel-Crafts conditions. These results are depicted in Table 13.

With reference to Table 13, the most striking differences between the results of the treatment of primary γ-arylalkyl alcohols with H_3PO_4 and of their corresponding chlorides with $AlCl_3$ are as follows:

1. The total yields of monophenylated hydrocarbons are usually higher with alcohols than with chlorides.
2. Under the conditions employed, the relative yields of the cyclized components are usually higher in product mixtures from alcohols than from the corresponding chlorides.
3. In contrast to the products from alcohols, which consisted mainly of unsaturated components, those from the corresponding chlorides were completely devoid of unsaturates.
4. The formation of cyclic components was shown to occur mostly by Ar_1-4 participation with hydroxyl as the leaving group, but exclusively by Ar_2-5 participation with chloride as the leaving group. This is clear from the fact that the overall ratio of Ar_1-4 to Ar_2-5 participation gave products estimated to be in the ratio 3.5:1 in the case of 3-methyl-3-*p*-tolylbutan-1-ol, but 0:1 in the case of the corresponding chloride.

It can therefore be concluded that the product distribution in the reaction of primary γ-arylalkyl derivatives under Friedel-Crafts conditions varies over a wide range, depending not only on the nature of the leaving group but also on other reaction variables, such as temperature, time, solvent, and ratio of reactants.

Going back to the data of Table 12, it seems of interest to elaborate on the reaction of 2-methyl-4-phenyl-1-butanol (167), as it provides some information about the cyclidehydration mechanism. According to Khalaf and Roberts [87], treatment of this alcohol with phosphoric acid gave not only the direct cyclidehydration product, 2-methyltetralin, but also the rearrangement products, 1,1-dimethylindan and 1-methyltetralin (Eq. 64). The points of similarity between the cyclidehydration of 167 with phosphoric acid and the cyclidehydro-

167 (OH) $\xrightarrow[230-240^{\circ}]{H_3PO_4}$ 32% + 40% + 14% + Unidentified 14% (64)

Table 13. Results of the Reactions of Primary γ-Arylalkyl Alcohols and Chlorides Under Friedel-Crafts Conditions

	Reaction conditions						Product type,[c] %			
								Open chain		
γ-Arylalkyl derivative	Time, hr	Catalyst	Solvent	Temp, °C	Yield, %	Product composition[a,b] (%)	Cyclic indans	Saturated	Unsaturated	Ref.
$PhCH_2CH_2CH_2OH$	0.25	H_3PO_4	None	230-40	46	Indan (trace) *n*-propylbenzene (11), three isomeric phenylpropenes (89)	Trace	11	89	88
$PhCH_2CH_2CH_2Cl$	2.5	$AlCl_3$	CS_2	25	—	*n*-Propylbenzene (61), di-*n*-propylbenzenes (39)	Nil	100	Nil	11
	4.5	$AlCl_3$	CS_2	25-45	—	*n*-Propylbenzene (94), di-*n*-propylbenzenes (6)	Nil	100	Nil	11
	1.0	$AlCl_3$	Pet. ether (80-60°C)	70	—	*n*-Propylbenzene (73), di-*n*-propylbenzenes (27), indan (trace)	Nil	100	Nil	11
	1.2	$AlCl_3$	Pet. ether (80-60°C)	Reflux	—	Indan (trace), *n*-propylbenzene (100)	Trace	100	Nil	28
$Ph{-}C(CH_3)_2{-}CH_2CH_2OH$	0.25	H_3PO_4	None	230-40	85	1,1-Dimethylindan (18), 2-methyl-3-phenyl-2-butene (82)	18	Nil	82	88
$Ph{-}C(CH_3)_2{-}CH_2CH_2Cl$	0.2	$AlCl_3$	Pet. ether (30-60°C)	Reflux	—	2-Methyl-3-phenylbutane (mainly) + *t*-pentylbenzene (27), compounds with mass 158 (73)	Nil	100	Nil	28

$p-CH_3C_6H_4-C(CH_3)_2-CH_2CH_2OH$	0.25	H_3PO_4	None	230-40	27	*p-t*-Pentyltoluene (trace), 2-methyl-3-*p*-tolylbutane (6), 2-methyl-3-*p*-tolyl-2-butene (33), 1,1,6-trimethylindan (13), 1,1,4-trimethylindan (32), 1,1,5-trimethylindan (10), unidentified (3), 1,1,7-trimethylindan (3)	58	6	34	88
$p-CH_3C_6H_4-C(CH_3)_2-CH_2CH_2Cl$	2.5	$AlCl_3$	CS	25	7	Toluene (19), xylene (2), 2-methyl-4-*m*-tolylbutane (1), 2-methyl-3-*p*-tolylbutane (31), 2-methyl-2-*p*-tolylbutane (37), 1,1,5-trimethylindan (2), 1,1,7-trimethylindan (trace), unidentified (8)	2	98	Nil	32
	2.5	$AlCl_3$	Pet. ether (60-80°C)	25	10	Toluene (13), xylenes (3), 2-methyl-4-*m*-tolylbutane (4), 2-methyl-3-*p*-tolylbutane (35), 2-methyl-2-*p*-tolylbutane (35), 2-methyl-2-*p*-tolylbutane (29), 1,1,5-trimethylindan (5), 1,1,7-trimethylindan (1), unidentified (10)	6	94	Nil	32

[a]Relative percentage of each component in the mixture as determined by integration of GLPC recordings.
[b]To facilitate comparison, high-boiling products and residues are not included.
[c]Relative percentage of the product types (cyclic as well as open-chain saturated and unsaturated) in the mixture as calculated from GLPC data.

chlorination of the corresponding chloride (14) with aluminum chloride (Eq. 13 and scheme 2) suggested that similar cyclialkylation pathways must have been followed in both cases. Thus scheme 2 was also used to rationalize the results of Eq. (64). The fact that the major component from the chloride was the hydride exchange product isopentylbenzene, which was not produced from the phenylalkanol, suggested that it is formed by dealkylation and hydride exchange on 1,1-dimethylindan under the influence of $AlCl_3$ rather than by hydride exchange prior to ring closure. The stability of 1,1-dimethylindan to the action of phosphoric acid under cyclidehydration conditions and the intramolecular dealkylation of this compound upon treatment with $AlCl_3$ under cyclidehydrochlorination conditions provides additional support for this view.

Parallel to the behavior of 167, the structurally related 2-methyl-4,5-diphenyl-1-pentanol [21] gave a mixture of the direct cyclidehydration product *trans*-1-benzyl-3-methyltetralin (27b) and the rearrangement cyclidehydration products, 1,1-dimethyl-3-benzylindan (32b) and 1,1-dimethyl-3-phenyltetralin (33b) (Eq. 65).

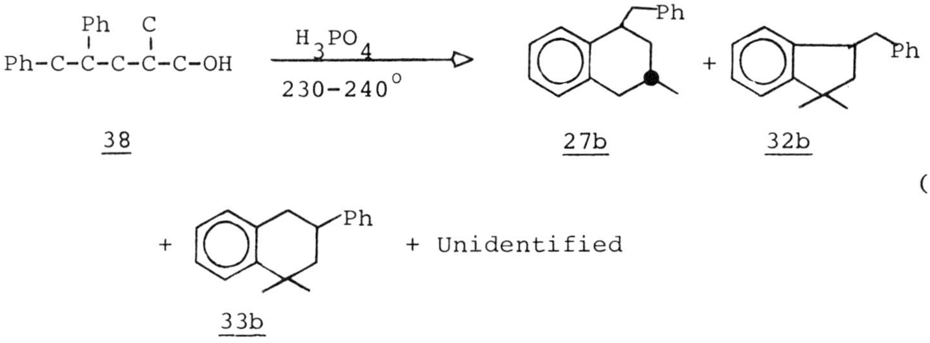

Ph C
Ph-C-C-C-C-C-OH $\xrightarrow[230\text{-}240^\circ]{H_3PO_4}$ 27b + 32b + 33b + Unidentified (65)

38

	27b	32b	33b	Unident.
R,S; S,R-pair (38a)	31%	7%	62%	-
R,R; S,S-pair (38b)	6%	5%	85%	4%

As can be seen from Eq. (65), only the *R,S:S,R* pair of diastereomers of the alcohol (38a) was able to undergo facile 1° closure to *trans*-1-benzyl-3-methyltetralin. Similar closure of the *R,R:S,S* isomer (38b) to *cis*-1-benzyl-3-methyltetralin was retarded by the developing steric repulsion exerted by the 1,3 benzyl-methyl interactions in this case (see schemes 5a,b).

In 1972, Khalaf and Roberts [18] extended their investigation of cyclialkylations to include the cyclization of phenylalkanols 168-174, diphenylalkanenes 175 and 176, diphenylalkanols 177 and 178, and acid chloride 179. A summary of the results of these reactions is given in Tables 14 and 15.

Examination of the data of Table 14 shows that the cyclization of the isomeric alcohols 168-171 gave similar product mixtures whose compositions were determined by the type of catalyst employed. Thus, while the nonisomerizing catalysts H_2SO_4 and $AlCl_3$-CH_3NO_2 produced mixtures consisting of 1-ethyl-1-methylindan (45) and *cis*- and *trans*-1,2-dimethyltetralin (44) with little or none of the isomeric 2,2-dimethyltetralin (46), the isomerizing catalyst $AlCl_3$ produced mixtures consisting predominantly of the latter isomer. The

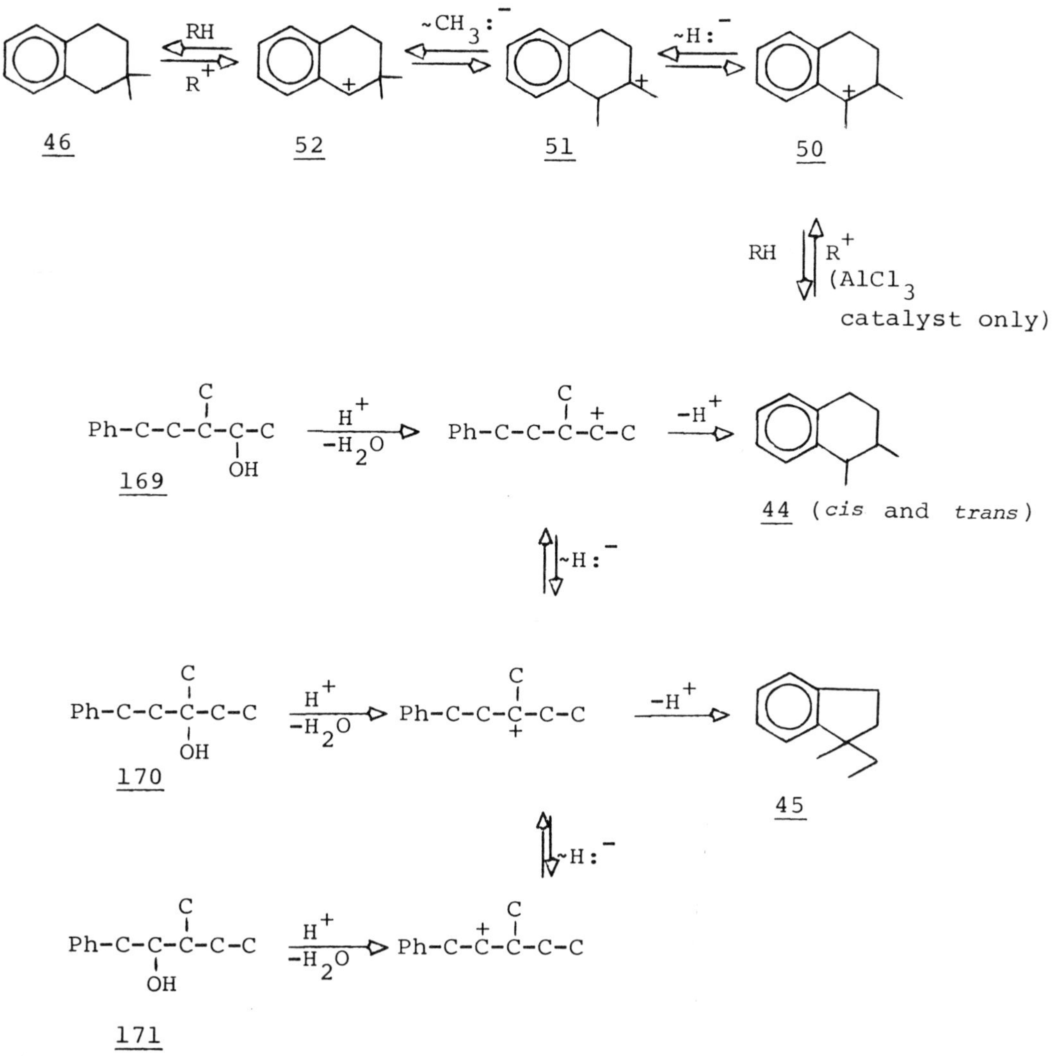

46 ⇌(RH / R^+) 52 ⇌($\sim CH_3:^-$) 51 ⇌($\sim H:^-$) 50

50 ⇌(RH / R^+ ($AlCl_3$ catalyst only)) 44

Ph-C-C-C(C)-C(OH)-C (169) →(H^+, $-H_2O$) Ph-C-C-C(C)-$\overset{+}{C}$-C →($-H^+$) 44 (*cis* and *trans*)

⇅ $\sim H:^-$

Ph-C-C-C(C)(OH)-C-C (170) →(H^+, $-H_2O$) Ph-C-C-$\overset{+}{C}$(C)-C-C →($-H^+$) 45

⇅ $\sim H:^-$

Ph-C-C(OH)-C(C)-C-C (171) →(H^+, $-H_2O$) Ph-C-$\overset{+}{C}$-C(C)-C-C

Scheme 18

results for cyclialkylations of compounds 169-171 were reasonably explained in terms of carbocation processes parallel to those used to explain the results from the corresponding chlorides (see scheme 6). These steps are shown in scheme 18.

The results from the cyclialkylation of 2,2-dimethyl-4-phenyl-1-butanol (168) deserves special comment. The study of this alcohol has been of particular interest not only because of its structural relationship to compounds 169-171, but also because of its illustration of a unique case of intramolecular alkylation with a primary neopentyl-type system. In contrast to intermolecular alkylations with neopentyl derivatives, which invariably proceeded with complete rearrangement (see Chap. 4), the intramolecular alkylation of 168 with phosphoric acid proceeded with partial direct closure to give a mixture of the rearrangement cyclidehydration products 1,2-dimethyltetralin (44) and

Table 14. Friedel-Crafts Cyclialkylations of Some Isomeric Phenylhexyl Alcohols and Their Corresponding Chlorides[a]

Cyclized compound	Catalyst (mol)	Time, hr	Total yield, %
C | Ph—C—C—C—C—OH | C 168	H_3PO_4		—
	$AlCl_3$ (1.2)		—
	$AlCl_3$ (1.2)- CH_3NO_2 (10)	4	—
C | Ph—C—C—C—C—C | OH 169	H_2SO_4	3	75
	$AlCl_3$ (1.2)- CH_3NO_2 (10)	1	—
		8	70
	$AlCl_3$ (1.2)	14	66
C | Ph—C—C—C—C—C | OH 170	H_2SO_4	3	73
	$AlCl_3$ (1.2)- CH_3NO_2 (10)	2	—
		8	—
		20	75
	$AlCl_3$ (1.2)	10	
		20	57
C | Ph—C—C—C—C—C | OH 171	H_2SO_4	3	28
	$AlCl_3$ (1.2)- CH_3NO_2 (10)	2	
		8	75
		20	—
	$AlCl_3$ (1.2)	10	—
		20	31

[a]All reactions were carried out at room temperature except with H_3PO_4, for which the temperature was 230 to 240°C.

Source: Ref. 18.

Products, %				
		1,2-Dimethyl-tetralin		
1-Ethyl-1-methylindan	2,2-Dimethyl-tetralin	*cis*	*trans*	Unidentified
24	18	6	25	7
No cyclization				
31	—	15	54	—
21	2	7	56	14
20	14	15	35	16
27	38	15	2	18
34	Trace	10	56	—
19	Trace	11	70	—
19	Trace	11	70	—
20	Trace	9	71	—
22	59	10	7	2
23	54	11	5	7
36	—	16	47	—
16		12	72	
19		10	71	
17	—	13	70	
20	51	—	9	—
20	51	18	5	5

Table 15. Friedel-Crafts Cyclialkylations of Some Mono- and Diphenylated Alkanols and Alkenes[a,b]

Cyclized compound formula	Catalyst	Time, hr	Total yield, %	Product composition, %
Ph—C—C—C(C)—C(C)(OH)—C 172	H_2SO_4	3	60	1,1,2-Trimethyltetralin (100)
	$AlCl_3$-CH_3NO_2	3	70	1,1,2-Trimethyltetralin (98) and 1,2,2-trimethyltetralin (2)
	$AlCl_3$	3	50	1,1,2-Trimethyltetralin (17), 1,2,2-trimethyltetralin (50), and two unidentified compounds (32)
Ph—C—C—C(C)(OH)—C(C)—C 173	H_2SO_4	3	58	1,1,2-Trimethyltetralin (100)
	$AlCl_3$-CH_3NO_2	3	65	1,1,2-Trimethyltetralin (100)
	$AlCl_3$	3	55	1,1,2-Trimethyltetralin (10), 1,2,2-trimethyltetralin (80), and other unidentified compounds (10)
Ph—C—C—C—C—C(C)(OH)—C 174	H_2SO_4	3	50	1-Isopropyltetralin (100)
	$AlCl_3$-CH_3NO_2	2	65	1-Isopropyltetralin (100)[c]
Ph—C—C(Ph)—C—C(C)=C 175	H_2SO_4	3	65	1,1-Dimethyl-3-phenyltetralin (100)

Compound	Catalyst	Time	Yield	Products
Ph—C—C(Ph)—C—C=C—C **176**	H_2SO_4	2.5	60	1-Benzyl-4-methyltetralin (87) and 1-ethyl-3-phenyltetralin (13)
Ph—C—C—C(Ph)—C—C(C)(OH)—C **177**	H_2SO_4	3	85	1,1,-Dimethyl-3-(β-phenylethyl)-indan (100)
	$AlCl_3$-CH_3NO_2	2.25	82	Same product (100)[c]
Ph—C—C—C—C—C(OH)—Ph **178**	H_2SO_4	3	10	1-Phenylbenzosuberane (100)
	$AlCl_3$-CH_3NO_2	3	15	1-Phenylbenzosuberane (100)
	$AlCl_3$	3	—	Tetralin (3), 1-methyltetralin (1), benzosuberane (6), 1-phenylbenzosuberane (54), and 1-benzyltetralin (36)
Ph—C—C—C(Ph)—C—COCl **179**	$AlCl_3$	2	70	3-(β-Phenylethyl)-1-indanone (100)
	$AlCl_3$-CH_3NO_2	1	75	Same product (100)

[a]Skellysolve B was used as solvent in reactions catalyzed by $AlCl_3$ and $AlCl_3$-CH_3NO_2.
[b]The cyclized compound, $AlCl_3$ and CH_3NO_2 (if used) were used in the ratio 1:1.2:10.
[c]These two reactions were reported in Ref. 89.
Source: Ref. 18.

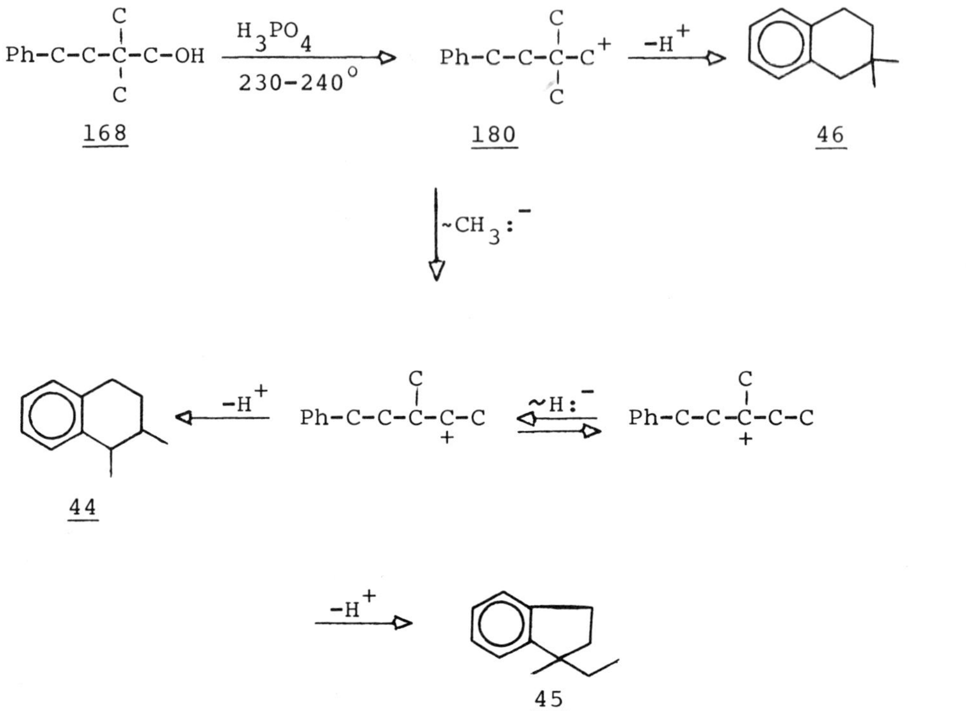

Ph-C-C-C(C)(C)-C-OH ($\underline{168}$) $\xrightarrow[230-240^\circ]{H_3PO_4}$ Ph-C-C-C(C)(C)-C$^+$ ($\underline{180}$) $\xrightarrow{-H^+}$ $\underline{46}$

$\downarrow$ ~CH_3:$^-$

$\underline{44}$ $\xleftarrow{-H^+}$ Ph-C-C-C(C)-C$^+$-C $\underset{\longrightarrow}{\overset{\sim H:^-}{\longleftarrow}}$ Ph-C-C-C$^+$(C)-C-C

$\xrightarrow{-H^+}$ $\underline{45}$

Scheme 19

1-ethyl-1-methylindan (45) and the direct cyclidehydration product 2,2-dimethyltetralin (46) (scheme 19). The possibility that 46 was formed by subsequent rearrangement of either 44 or 45 rather than by direct chlosure via 180 (or its equivalent) was excluded on the ground that both 44 and 45 were shown to be stable to the reaction conditions. These results from 168 gave more evidence in confirmation of the view that intramolecular alkylations proceed with participation by aryl and/or alkyl groups and hence are considerably faster and less prone to accompanying rearrangements than the corresponding intermolecular reactions.

Based on the results of Table 14, Khalaf and Roberts [18] concluded that 2° closure to a tetralin is favored over 3° closure to indan either directly or via a tertiary-to-secondary rearrangement. Meanwhile, from the cyclialkylations of the monophenylalkanols 172 and 173 and the diphenylalkene 175 (Table 15) these authors concluded that the 3° closure to a tetralin takes precedence over 3° closure to an indan either directly or via a tertiary-to-tertiary rearrangement (scheme 20 and Eq. 66). Justification of this conclusion is found in the absence of 1-methyl-1-isopropylindan (181) from the

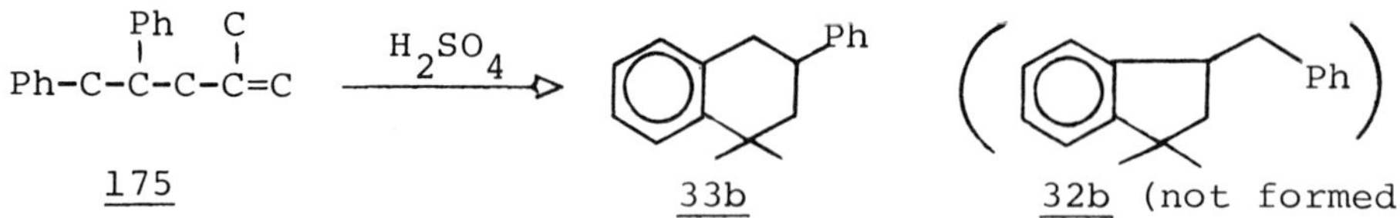

Ph-C-C(Ph)-C-C(C)=C ($\underline{175}$) $\xrightarrow{H_2SO_4}$ $\underline{33b}$ ($\underline{32b}$ (not formed)) (66)

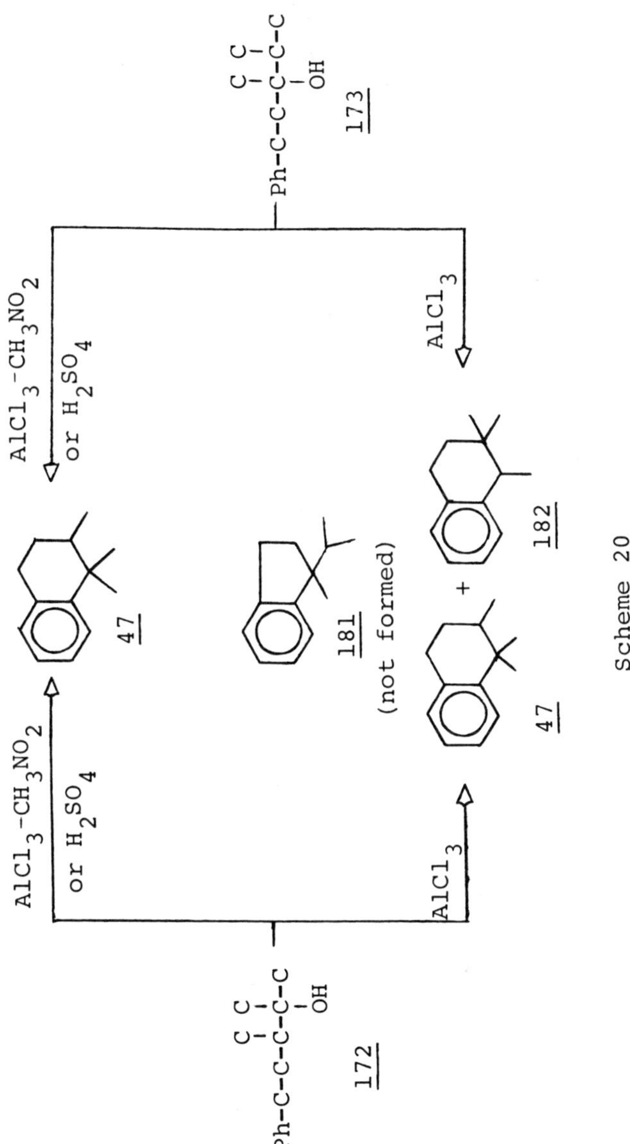

Scheme 20

Table 16. Cyclidehydrations of Some Isomeric Dimethylarylhexanols

Starting alcohol	Catalyst	Time, min	Total yield, %	Product composition, %					
				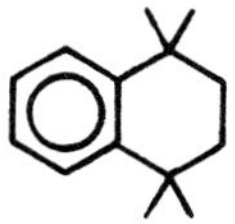 192	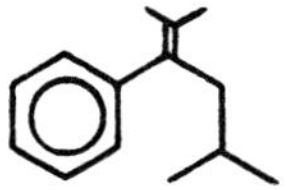 188	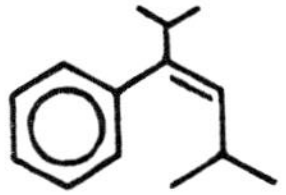 193	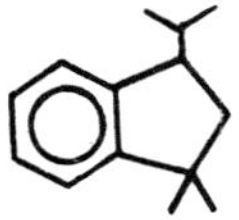 194	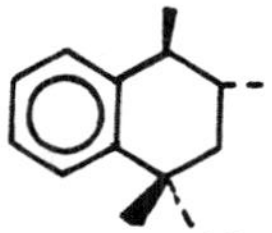 *trans*-195	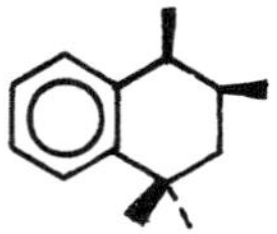 *cis*-195
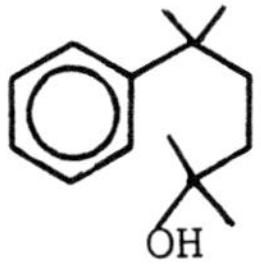 183	Conc. H_2SO_4	60	75	100	-	-	-	-	-
	$H_2SO_4 \cdot 1H_2O$	60	88	100	-	-	-	-	-
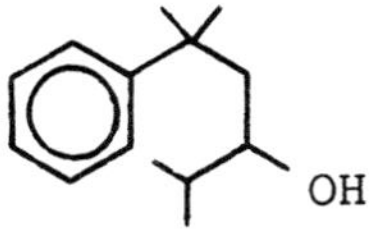 184	Conc. H_2SO_4	30	63	90	-	-	-	-	-
	$H_2SO_4 \cdot 1H_2O$	60	73	90	-	-	-	-	-

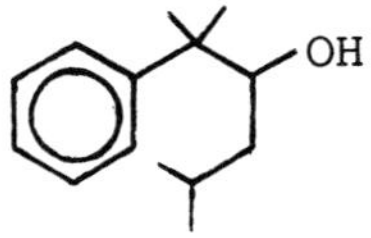 185	Conc. H_2SO_4	15	86	-	0.9	-	74	12	13
	$H_2SO_4 \cdot 1H_2O$	60	92	-	66.4	2.7	21	4.5	5.1
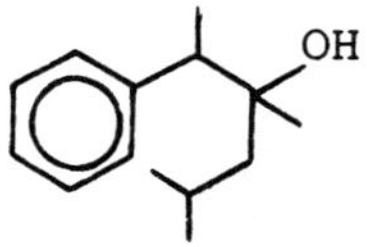 186	Conc. H_2SO_4	10	68	-	-	-	7.3	42	50
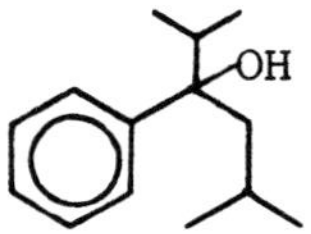 187	Conc. H_2SO_4	15	79	-	1.4	-	68	10.4	12.2
		30	87	-	83	2.9	5.8	2.7	
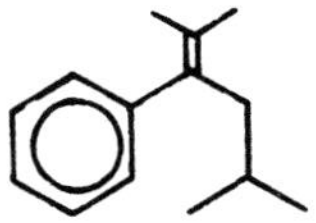 188	Conc. H_2SO_4	10	56	-	0.8	-	74	25	

Source: Ref. 90.

products of 172 and 173 and of 1,1-dimethyl-1-benzylindan (32b) from the products of 175. The formation of 182 from both 172 and 173 when $AlCl_3$ was used as a cyclialkylation catalyst was attributed to rearrangement of the primarily formed 1,1,2-trimethylindan (47).

In a related development, Giovannini and Brandenberger [90] investigated the cyclialkylation of dimethylphenylhexanols 183 to 187, dimethylphenylhexene 188, and the dimethylanisylhexanols 189-191 in the presence of concentrated H_2SO_4 or $H_2SO_4 \cdot 1H_2O$ catalyst. The conditions and products (192-199) of these investigations are summarized in Table 16 and in Eqs. (67) to (69).

CH_3O OH 189 — Conc. H_2SO_4, 30 min, 51% yield → CH_3O 196

CH_3O OH 190 — $H_2SO_4 \cdot 1H_2O$, 90 min, 82% yield → 196 (67)

189 — $H_2SO_4 \cdot 1H_2O$, 120 min, 57% yield → OCH_3 197 65% + CH_3O 196 35% (68)

OCH_3 OH 191 — Conc. H_2SO_4, 25 min or $H_2SO_4 \cdot 1H_2O$, 90 min, 82-85% yield → H_3CO 198 75-78%

+ OCH_3 *trans*-199 19% + OCH_3 *cis*-199 2.1-3.4% (69)

The results of Table 16 and Eqs. (67) to (69), while further confirming the facility with which 3° closures to indans and/or tetralins usually occur, also further demonstrated that such closures can be preceded by multiple carbocation rearrangements. For example, whereas the results from alcohols 184, 186, 187, and 190 are explicable in terms of simple 1,2 hydride shifts, those from alcohols 185 (Table 16) and 191 (Eq. 69) require additional methyl and aryl shifts.

In a continuation of their effort to explore the vairous factors governing the course of ring closure reactions, especially in regard to seven- versus five- and six-membered ring formation, Khalaf and Roberts [18,89] investigated the cyclization behavior of compounds 174 and 176-179 (Table 15). Their aim was to test the following competing possibilities: (1) 2° closure to a seven versus 2° closure to a five- and/or six-membered ring in 176; (2) 3° closure to a seven- versus 2° closure to a five- and/or six-membered ring in 174; (3) 3° closure to a seven- versus 3° closure to a five- and/or 2° closure to a six-membered ring in 177; (4) benzylic 2° closure to a seven- versus ordinary 2° closure to a five- and/or six-membered ring in 178, and (5) closure to a seven- versus closure to a five-membered ring in 179.

The results of cyclialkylation of compounds 174 and 176 to 179 disclosed some interesting facts about ring closure reactions. For example, upon treatment with H_2SO_4, 5,6-diphenyl-2-hexene (176) gave a product consisting of 87% 1-benzyl-4-methyltetralin (201) and 13% of the isomeric 1-ethyl-3-phenyltetralin (203) (Eq. 70). None of the products expected from 2° closure to a five- or to a six-membered ring were formed.

$$PhCH_2CH(Ph)CH_2CH{=}CHCH_3 \ (\mathbf{176}) \xrightarrow{H^+} PhCH_2CH(Ph)CH_2CH_2\overset{+}{C}HCH_3 \ (\mathbf{200}) \xrightarrow{-H^+} \mathbf{201}\ (87\%)$$

$$PhCH_2CH(Ph)CH_2CH{=}CHCH_3 \ (\mathbf{176}) \xrightarrow{H^+} PhCH_2CH(Ph)CH_2\overset{+}{C}HCH_2CH_3 \ (\mathbf{202}) \xrightarrow{-H^+} \mathbf{203}\ (13\%)$$

(70)

Since the enhanced production of 201 could hardly be attributed to the small differences in stabilities expected on the basis of the numbers of hyperconjugated hydrogen in cations 200 and 202 (5H in 200 versus 4 in 202), it was rather attributed to steric reasons. In that respect, the developing 1,4 interactions between the benzyl and the methyl groups in going from 200 to 201 are much more favorable than the 1,3 interactions between the ethyl and the phenyl groups in going from 202 to 203.

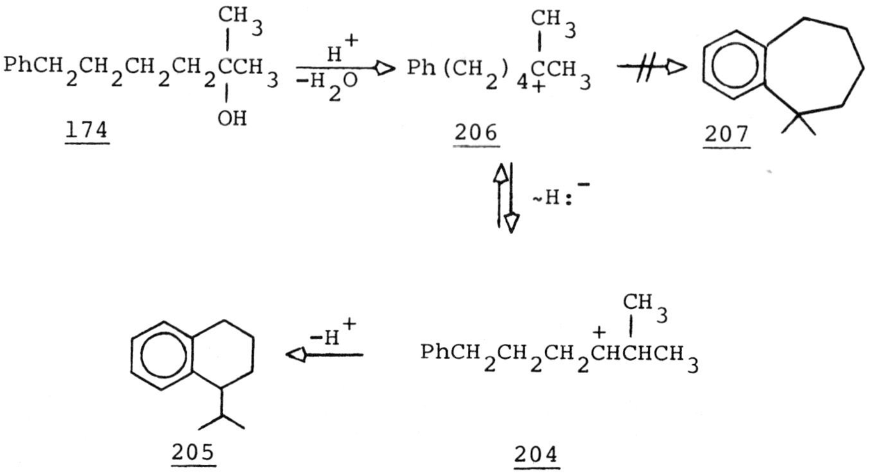

$PhCH_2CH_2CH_2CH_2C(CH_3)(OH)CH_3$ (174) $\xrightarrow[-H_2O]{H^+}$ $Ph(CH_2)_4\overset{+}{C}(CH_3)CH_3$ (206) ⇸ 207

206 ⇌ (~H:⁻) $PhCH_2CH_2CH_2\overset{+}{C}HCH(CH_3)CH_3$ (204) $\xrightarrow{-H^+}$ 205

Scheme 21

$PhCH_2CH_2CH(Ph)CH_2C(CH_3)(OH)CH_3$ (177) $\xrightarrow[-H_2O]{H^+}$ $PhCH_2CH_2CH(Ph)CH_2\overset{+}{C}(CH_3)CH_3$ (208)

208 $\xrightarrow{-H^+}$ 211 (CH_2CH_2Ph)

208 ⇸ 212 (Ph)

208 ⇌ (~H:⁻) $PhCH_2CH_2CH(Ph)\overset{+}{C}HCH(CH_3)CH_3$ (209) ⇸ 210 (Ph)

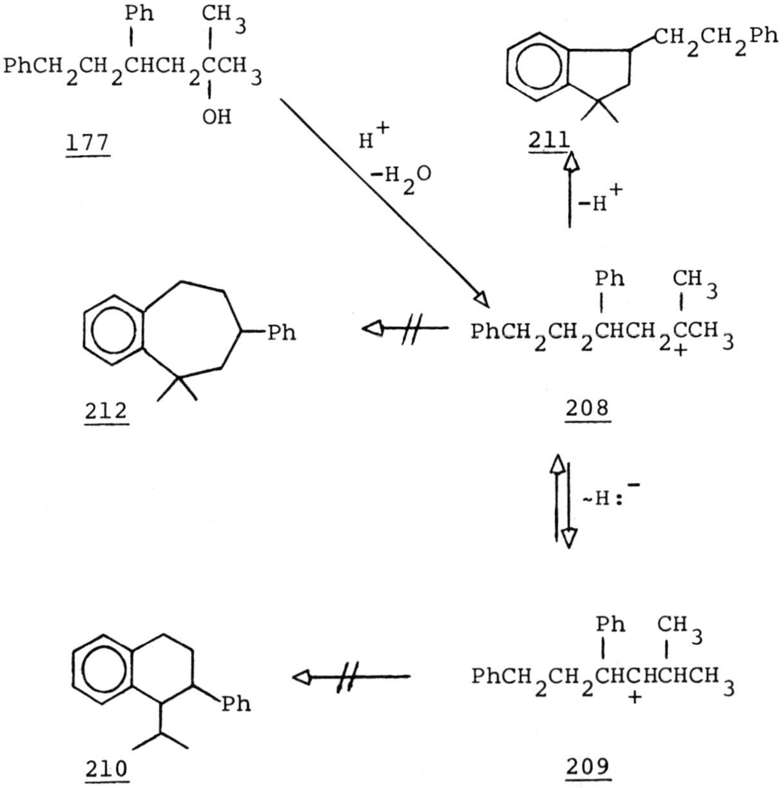

Scheme 22

In line with results from diphenylhexene 176, the treatment of 2-methyl-6-phenyl-1-hexanol (174) with either H_2SO_4 or $AlCl_3$-CH_3NO_2 gave the rearranged product 1-isopropyltetralin (205) through the secondary carbocation 204, rather than the direct cyclization product 207 through the tertiary cation 206 (scheme 21) [18,89]. This fact, which was formerly noted by other workers [10,67], has been attributed to the greater driving force for six- than for seven-membered ring formation, in spite of the required tertiary-to-secondary carbocation rearrangement. In other words, the energy required to overcome the strain in going from a six- to a seven-membered ring intermediate is more than that required to overcome the barrier in going from a tertiary to a secondary cation.

In light of the foregoing results with alcohol 174, it was expected that the H_2SO_4-catalyzed reaction of 2-methyl-4,6-diphenyl-2-hexanol (177) would yield a mixture of 1-isopropyl-2-phenyltetralin (210) and 1,1-dimethyl-3-(β-phenylethyl)indan (211). Far from expectations, however, treatment of alcohol 177 with either H_2SO_4 or $AlCl_3$ gave exclusively 211 and none of 210 (scheme 22). Explanation for this irregular behavior of alcohol 177 was given in terms of steric effects. Thus, whereas the developing 1,2 interactions between the bulky phenyl and isopropyl groups would strongly inhibit closure of the rearranged secondary carbocation 209 to 210, the tertiary carbocation 208 would encounter much less steric repulsion to give the trisubstituted indan 211. None of 212 was produced, illustrating again the failure of 3° closure to a seven-membered ring.

In contrast to ordinary secondary and tertiary carbocation intermediates, secondary benzylic carbocations have been shown to cyclize to seven- and even larger-membered ring systems. For example, when the secondary benzylic alcohol 213 was treated with H_2SO_4 or $AlCl_3$-CH_3NO_2, it gave a 10 to 15% yield of the seven-membered ring product 1-phenylbenzosuberane (214) (Eq. 71) [18]. Treatment of 213 with the stronger catalyst $AlCl_3$ resulted

$$Ph(CH_2)_4CH(OH)Ph \xrightarrow{H_2SO_4} Ph(CH_2)_4\overset{+}{C}HPh \xrightarrow{-H^+} \text{214} \quad (71)$$

213 214 10-15%

in a more complex product mixture consisting mainly of 214. Furthermore, it was reported that 215 gave 216 upon treatment with either H_2SO_4 or H_3PO_4

$$\text{215} \xrightarrow{H_2SO_4} \text{216}$$

215 (CH$_2$Ph, OH) 216

[91] and that 217 gave 218 upon treatment with polyphosphoric acid at 120 to 130°C for 15 min [92], although an eight-membered ring is formed in the latter case (Eq. 72).

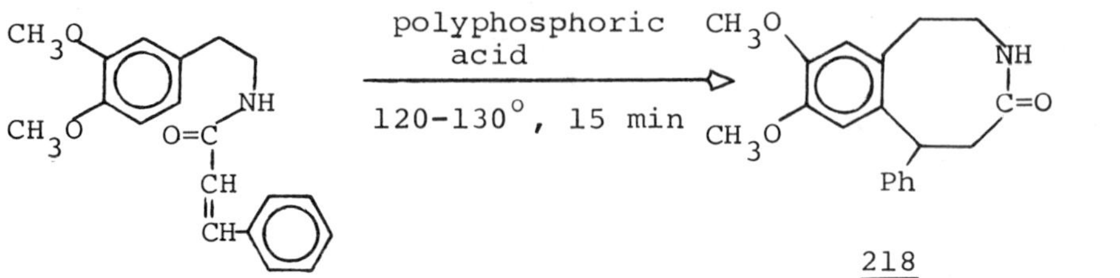

$$\text{217} \xrightarrow[\text{120-130°, 15 min}]{\text{polyphosphoric acid}} \text{218} \qquad (72)$$

So far, the results from compounds 174 (scheme 21) and from 177 (scheme 22) have demonstrated that 2° closure to tetralin or 3° closure to indan is favored over 3° closure to benzosuberane. To provide more insight into five- versus seven-membered ring formation, Khalaf and Roberts [18] examined the cyclialkylation of 3,5-diphenylpentanoyl chloride (219). In view of the widely applicable closures of 3-aryl- [11,55,56,88,93-104] and 5-aryl [55,93,105-112] alkanoic acids or acid chlorides to form five- and seven-membered cyclic ketones, respectively, it was expected that 219 would give a mixture of 220 and 221 (Eq. 73). Far from expectations, when 219 was treated with either $AlCl_3$ or $AlCl_3$-CH_3NO_2 catalysts it gave only the five-membered ring product, 3-(β-phenylethyl)-1-indanone (220); none of the benzosuberone 221 was found.

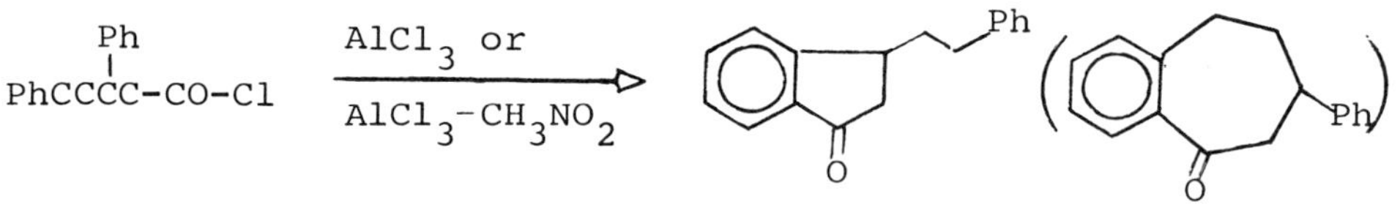

$$\text{PhCCC(Ph)C-CO-Cl (219)} \xrightarrow[AlCl_3-CH_3NO_2]{AlCl_3 \text{ or}} \text{220} \quad (\text{221, not formed}) \qquad (73)$$

Taken together, the results from compounds 174, 177, and 219 prove beyond doubt that five-membered ring closure, whenever possible, takes precedence over seven-membered ring formation in both cyclialkylation and cycliacylation.

In reviewing these results on the cyclialkylations of arylalkanols and arylalkenes, it becomes clear that they are entirely compatible with the corrected results of cyclizations of structurally related arylhaloalkanes to indans and tetralins. In fact, the conclusions drawn previously on the basis of cyclialkylations of arylhaloalkanes can also apply equally well to cyclialkylations of arylalkanols and arylalkenes. However, it may be useful to emphasize the following points:

1. Intramolecular ring closure of arylalkanols and arylalkenes to form a new ring may only occur if the hydroxyl group (or the double bond) is located at position 2 or higher with respect to the aryl position in the chain. With the hydroxyl (or the double bond) located at position 1, mainly polymeric materials are produced upon acid treatment (see Table 11).

2. The data obtained under normal Friedel-Crafts conditions showed clearly that primary, secondary, and tertiary arylalkanols (or related alkenes) are capable of cyclizing to tetralins, but only secondary benzylic and tertiary arylalkanols (or related alkenes) are capable of cyclizing to indans, and

only secondary benzylic arylalkanols (or related alkenes) are capable of cyclizing to benzosuberane and larger rings (see Tables 11 to 16).

3. Under forcing Friedel-Crafts conditions or in deamination reactions, deviation from point 2 may occur. For example: (a) 3-Phenyl-1-propanol (152, Eq. 59) [87], 3-methyl-3-phenyl-1-butanol (153, Eq. 60) [87], and 3-methyl-3-*p*-tolyl-1-butanol (157, Eq. 63) [88] underwent detectable primary closures to indans, but only upon heating with H_3PO_4 at 230 to 240°C. (b) 4-Phenyl-1-butene showed, among other reactions, some cyclization to methylindan and methylindenes, but only upon treatment with metal oxide catalysts at elevated temperatures [113]. (c) The primary carbocation (223) generated from 9-(δ-aminobutyl)triptycene (222) by deamination with sodium nitrite in acetic acid gave, among products 226, 20% of the direct 1°-closure product 1,9-butanotriptycene (224) and 5% of the rearranged 2°-closure product 1,9-(γ-methylpropano)triptycene (225) (Eq. 74) [112].

$(CH_2)_4NH_2$ — $NaNO_2$ / CH_3COOH → [$(CH_2)_3\overset{+}{C}H_2$] →

222 223

(74)

224 20% + 225 5% (CH_3) + noncyclized products 226 68%

4. Primary closure to indan is enhanced only in systems where the alkanol residue contains a *gem*-dimethyl or a keto group attached directly to the aromatic ring or when the aromatic ring is activated by electron-releasing groups (see Table 13). With reference to Table 13, the latter effect is illustrated by the finding that a better yield of cyclic products was obtained from 3-methyl-3-*p*-tolyl-1-butanol than from 3-phenyl-1-propanol; in the former alcohol, the ring is activated by the methyl groups.

5. The yields of cyclized products are dependent on a number of factors, which include the position of the hydroxyl group (or double bond) with respect to the aryl group, the stability of the intermediate carbocation(s) involved, the size of the ring produced, and the nucleophilicity of the aromatic center at which closure is taking place. In general, the yields increased as the OH approached position 4 with respect to the aromatic, and in all the examples studied, tertiary closure to tetralins gave the best yields.

6. The weak catalyst $AlCl_3$-CH_3NO_2 can catalyze cyclidehydrations of secondary and tertiary phenylalkanols but not of primary alkanols, which were

recovered unchanged after similar treatment with the catalyst at room temperature [89] (see Table 12). Compared to other catalysts such as $AlCl_3$ or H_2SO_4, the $AlCl_3$-CH_3NO_2 is superior, as it gives better yields of much cleaner products.

7. In general, five-membered ring closure, whenever possible, takes precedence over seven-membered ring formation in both cyclialkylation and cycliacylation. Taken together, the previous cyclialkylation results confirm the conclusion that both seven- and four-membered rings close at about the same rate [114].

8. Steric factors play a determining role in directing the course of ring closure reactions. For example, they have been held responsible for the retarded closure of *R,R:S,S*-2-methyl-4,5-diphenyl-1-pentanol (38b, Eq. 65) to *trans*-1-benzyl-3-methyltetralin, the closure of 5,6-diphenyl-2-hexene (175, Eq. 71) to 1-benzyl-4-methyltetralin (mainly), and the exclusive closure of 2-methyl-4,6-diphenyl-2-hexanol (177, scheme 22) to the more strained five-membered 1,1-dimethyl-3-(β-phenylethyl)indan (211).

9. The results with arylalkanols gave more evidence in support of the view that intramolecular alkylations are anchimerically assisted and hence are considerably faster and less prone to accompanying rearrangements than the corresponding intermolecular reactions (e.g., see scheme 19) [11,18,21,32, 87,88].

b. *Production of Indans and Tetralins by Intermolecular Alkylations*

Besides their preparation by intramolecular alkylations, indans have also been obtained by a number of methods involving intermolecular alkylations. These comprised (1) dimerization of styrenes, (2) reaction of styrenes with alkenes, and (3) reactions of arenes with alkenes and alkyl halides via hydride transfer. Since the early literature on these reactions was extensively reviewed by Barclay [2], we shall present only a short account of this subject, placing special emphasis on recent advances.

(i) Dimerization of styrenes. The cyclic dimerization of styrenes in the presence of acid catalysts constitutes one of the most applicable procedures for the formation of indans. Depending on nature of substrate, type of catalyst, and reaction conditions, indans 233 (scheme 23), accompanied by varying amounts of hydroanthracenes 232 [115], unsaturated dimers 230 and 234 [116,117,118], and polymers [116], were found to be formed by this method. The formation of these products has been suggested to occur by the general mechanism formulated in scheme 23. In this mechanism, the starting styrene 227 is converted by the acid into the monomeric cation 228 which, upon attack by another molecule of 227, yields either the dimeric cation 229 (addition) or the dimeric alkene 230 (substitution). The latter can undergo protonation to 231 followed by ring closure to 232 while the former can undergo one or more of the following reactions: (1) intramolecular cyclization to form 233, (2) deprotonation to form 234 or (3) cationic polymerization to form higher polymers.

Some of the indans (233) prepared by the dimerization reaction above are shown in Table 17.

Interestingly, it was recognized by Muller et al. [115,127-132] and Baker et al. [133,134] and their co-workers in the 1950s that the cyclization step proceeds with a high degree of stereospecificity, such that the resulting 1-

Scheme 23

Table 17. Indans 223 Synthesized by Dimerization of Styrenes

Compound	R_1	R_2	R_3	R_4	Ref.
233a	H	H	H	H	116,119
233b	H	H	CH_3	H	120-123
233c	H	H	H	CH_3	115,124
233d	H	H	C_2H_5	H	118
	H	H	CH_3	CH_3	118
233e	H	CH_3	CH_3	H	120
233f	H	COOH	CH_3	H	120
233g	H	NH_2	CH_3	H	120
233h	H	OCH_3	CH_3	H	120
233i	H	OCH_3	H	CH_3	115
233j	OCH_3	OH	H	CH_3	115
233k	OCH_3	OCH_3	H	CH_3	115
233l	OCH_2Ph	OCH_2Ph	H	CH_3	115
233m	OCH_3	OC_2H_5	H	CH_3	115
233n	OH	OCH_3	H	CH_3	115
233o	$R_1 + R_2 = OCH_2O$		H	CH_3	115
233p	H	H	C_6H_5	H	126

[a]Indan 233b was also obtained when α-phenylisobutanoyl chloride, $PhC(CH_3)_2COCl$, was decarbonylated to the corresponding carbocation 228 by refluxing with $SnCl_4$ in CS_2 for 23 hr [135].

[b]The cyclic product from these two reactions was incorrectly assigned as 1,3-diethyl-1-methyl-3-phenylindan [125], but was corrected by later workers [118]. See forthcoming discussion.

ethyl-2-methyl-3-arylindans 233c and 233i-233o formed in only one stereoisomeric modification, the sterically favored *trans-trans* 235.

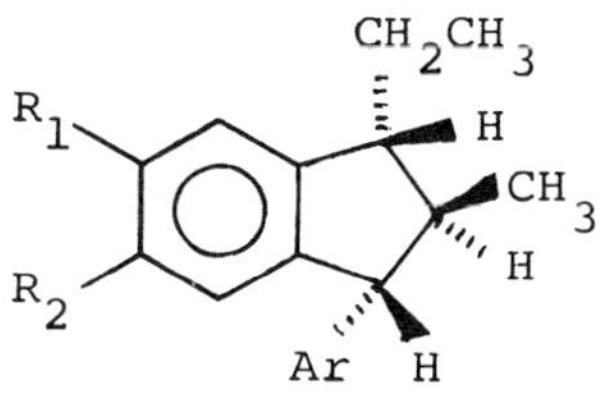

R_1, R_2, CH_2CH_3, H, CH_3, H, Ar, H

235

The history of the dimerization of the isomeric α-ethyl- and α,β-dimethylstyrenes (235 and 236, respectively) requires brief mention. In 1958, both isomers were claimed by Overberger et al. [125] to yield similar product mixtures consisting of 1,3-diethyl-1-methyl-3-phenylindan (237) and an unsaturated dimer. This claim seemed to be inconsistent with scheme 23, on the basis of which the expected product would be 1-ethyl-3-phenyl-1,2,3-trimethylindan (233d).

$PhC(C_2H_5)=CH_2$ 235; $PhC(CH_3)=CHCH_3$ 236; Ph 237; Ph 233d

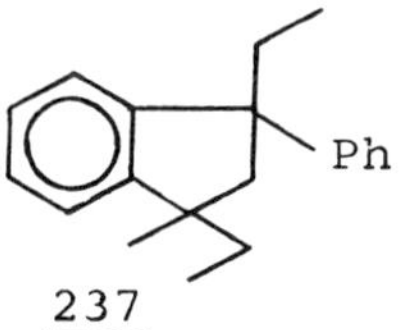

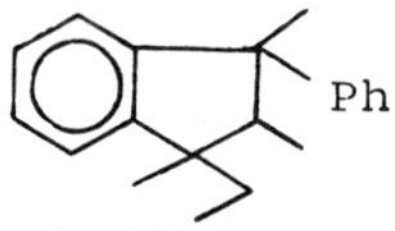

In fact, a more recent reinvestigation by Khalaf and Roberts [118] proved this to be the case, as treatment of α,β-dimethylstyrene with H_2SO_4 gave 233d as the cyclic dimer. Moreover, the latter compound was also obtained as a major product during the attempted alkylation of benzene with 2-chloro-2-phenylbutane in the presence of either $FeCl_3$ or $AlCl_3$-CH_3NO_2 catalysts. The formation of 233d in this case was also explained in terms of the general mechanism of scheme 23, with the exception that the formation of the initial carbocation corresponding to 228 involved the loss of a chloride ion.

The results of the acid-catalyzed dimerization of *m*-isopropenyltoluene (238) remains confusing. In 1958, Petropoulos and Fisher [120] reported the product to be a mixture of the two possible cyclic dimers, 239 and 240, resulting from closure *ortho* and *para* to the ring methyl group (Eq. 75). However, these authors claimed on the basis of inductive effects that closure at the *ortho* position was favored over the *para* position by a factor of 2. This

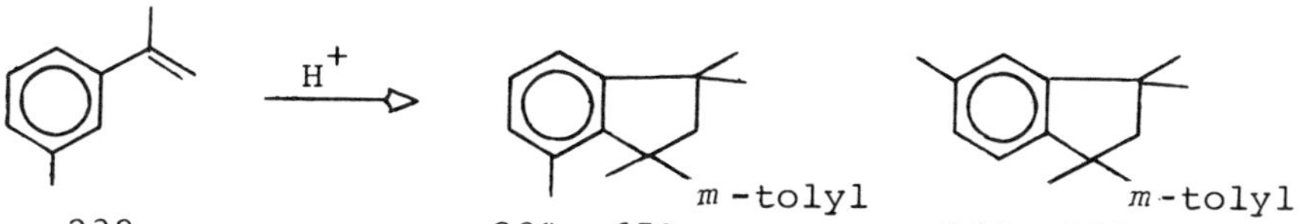

238 $\xrightarrow{H^+}$ 239 (*m*-tolyl) 65% + 240 (*m*-tolyl) 35% (75)

preferred formation of 239 seems to be in doubt, as it contradicts many substantiated findings, among which are the following: (1) the well-recognized steric opposition to tertiary alkylation *ortho* to a methyl group (see Chaps. 3 and 4); (2) when *m*-isopropenyltoluene (238) was treated with diisobutylene (241), the ratio of 242:(243 + 244) in the product was found to be 1.4 (Eq. 76) [136]; and (3) a similar cyclialkylation of 2-methyl-4-*m*-tolyl-2-butanol

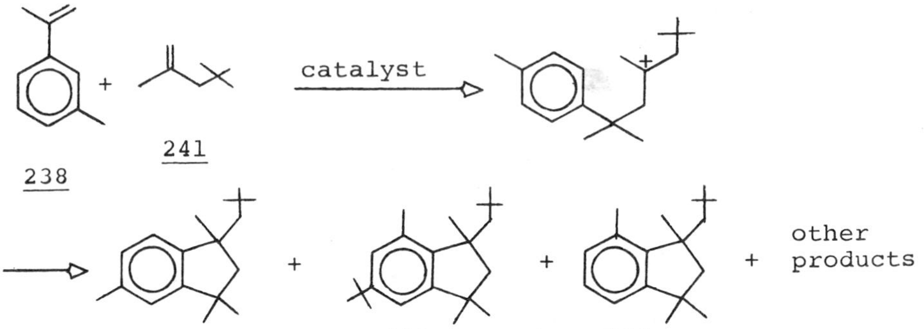

238 + 241 —catalyst→ → 242 + 243 + 244 + other products (76)

(245) under various conditions always gave more of 246 than of 247 (Eq. 77).

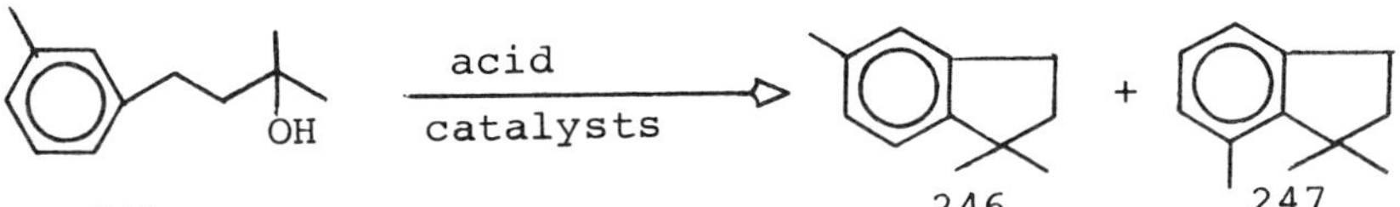

245 —acid catalysts→ 246 (mainly) + 247 (77)

In line with the results from styrene derivatives, other structurally related isopropenylarenes gave analogous products upon treatment with acid catalysts. For example, in 1966, Paquette and Phillips [137] found that treatment of β-isopropenylnaphthalene (248) with polyphosphoric acid gave a cyclic dimer to which they assigned structure 249. This product was assumed to form by a mechanism similar to that postulated in scheme 24.

More recently, Horspool et al. [138] studied the acid-catalyzed dimerization of ferrocenylethylenes and their corresponding carbinols. Using 2-ferrocenylpropene (250) as substrate and 90% HCOOH (or 40% H_2SO_4) as catalyst they obtained an 85% yield of a mixture of the hetero- and homoannularly cyclized dimers 252 and 253, respectively (scheme 25). The assignment of structures 252 and 253a to the dimers was based on a number of grounds, including lack of unsaturation, interpretation of mass and NMR spectra, comparison with the related cyclizations of ferrocenylalkanoic acids [139,140], and consideration of steric effects. In fact, steric reasons were of prime importance in deciding between 253a and 253b, in which the ferrocenyl substitutent is *exo* (253a) or *endo* (253b) to the other iron atom. It appeared more likely that the bulky ferrocenyl group would be *exo* during the cyclization step, hence structure 253a was preferred.

The lack of formation of the unsaturated dimer 254 in the reaction above suggested that, in this case, the intramolecular ring closure reactions of ion 251 to 252 and 253 were much faster than its deprotonation to 254.

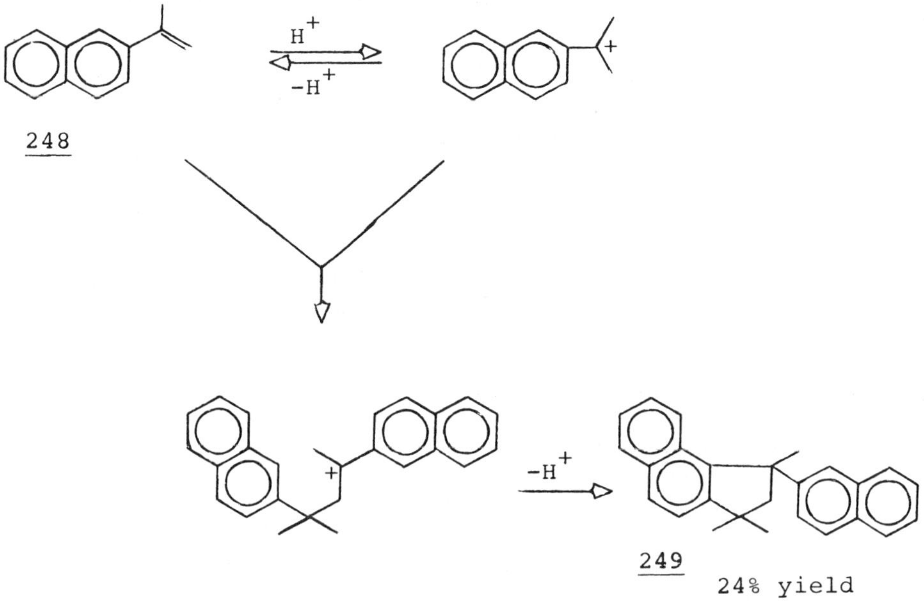

248

H^+ / $-H^+$

$-H^+$

249 24% yield

Scheme 24

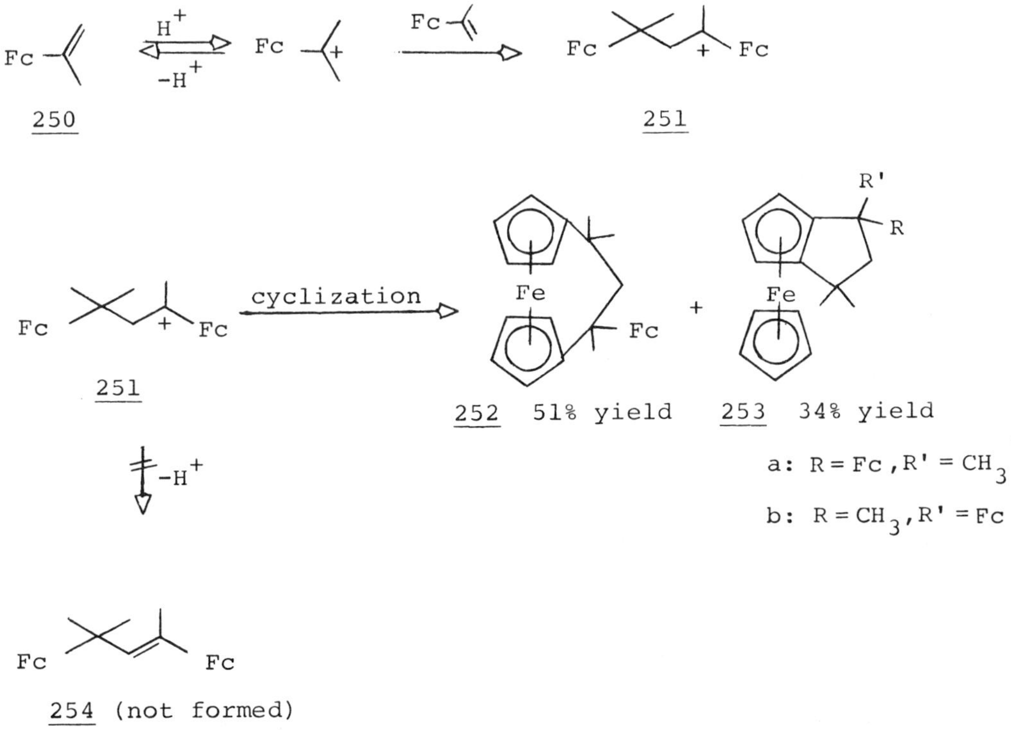

Fc

H^+ / $-H^+$

Fc

Fc

Fc Fc

250

251

Fc Fc

251

cyclization

Fe Fc

Fe R' R

252 51% yield 253 34% yield

a: R = Fc, R' = CH_3

b: R = CH_3, R' = Fc

$-H^+$

Fc Fc

254 (not formed)

Scheme 25

Besides their formation as major products in the acid-catalyzed dimerization of styrene-like alkenylarenes, indans were also produced as minor byproducts in reactions involving carbocations of general formula $Ar\overset{+}{C}(R)CH_2R'$ as possible intermediates. For example, Roberts and Abdel-Baset [141] found that the reaction of 2-bromopropene (255) with benzene and Al_2Br_6 gave 2,2-diphenylpropane (258a) as major product and 1,1,3-trimethyl-3-phenylindan (259a), 1,1,3-trimethylindene (260a), and 3,3,3',3'-trimethyl-1,1'-spirobiindan (263a) as minor products (scheme 26, R = H). As in the reaction with benzene, the major product of the reaction of 2-bromopropene with toluene was 2,2-di-*p*-tolylpropane (258b), and the other corresponding minor products, 1,1,3,5-tetramethyl-3-*p*-tolylindan (259b) and 1,1,3,5-tetramethylindene (260b), were observed. An additional product was also observed which was shown to be 1,1,3,7-tetramethylindene (262b). The formation of all of these products was rationalized in terms of carbocation transformations shown in scheme 26.

As can be seen from scheme 26, the key intermediates to all of the resulting products are the 2-arylpropene (256) and the 2-aryl-2-propyl carbocation (257) produced from it by proton addition. In fact, 2-*p*-tolylpropene (256b) was characterized among the products of reaction of 2-bromopropene with toluene and $AlCl_3$.

With its structure finally established and its mode of formation fitting logically into scheme 26, the long and checkered history of the tetramethyl-1,1'-spirobiindan (263a) is worthy of comment. This compound was first reported in 1929 by Hoffman [142] as a hydrocarbon of unknown structure from the reaction of 4-methyl-4-phenyl-2-pentanone with anhydrous $ZnCl_2$. The structure 264 was later assigned to the hydrocarbon by Barnes and Beitchman [143] on the basis of their investigations of similar reactions. This formula was accepted as correct for the products obtained from (1) treatment of the "saturated dimer of α-methylstyrene" [1,1,3-trimethyl-3-phenylindan (259a)] with $AlCl_3$ at 100°C [121], (2) reaction of methylacetylene with benzene and $AlCl_3$ [144], (3) treatment of α-methylstyrene (256a) with $AlCl_3$ [144], and (4) reaction of cumene with nitrosonium hexafluorophosphate [145].

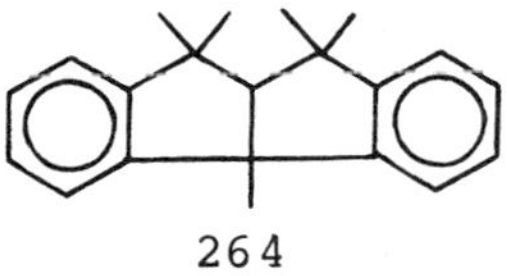

264

However, in 1962 Curtis and Lewis [146] demonstrated by NMR studies that the product obtained from all of these reactions was indeed the 1,1'-spirobiindan of structure 263a. In 1964, Barclay and Chapman [147] came to the same conclusion, and they proposed a mechanism for the formation of 263a from 1,1,3-trimethylindene (260a) and the carbocation 257a corresponding to the last three steps of scheme 26. This mechanism whose key step is the isomerization of 260a to 261a was supported by a demonstration of the production of 263a from 260a and α-methylstyrene (256a) in sulfuric acid and of two other new polyalkyl-1,1'-spirobiindans by analogous reactions. Notably, the homologous 3,3,6,3',3',6'-hexamethyl-1,1'-spirobiindan (263b) isolated from the reaction of 2-bromopropene with toluene and Al_2Br_6 had also been identified incorrectly earlier [143] as a homolog of 264, but the correct

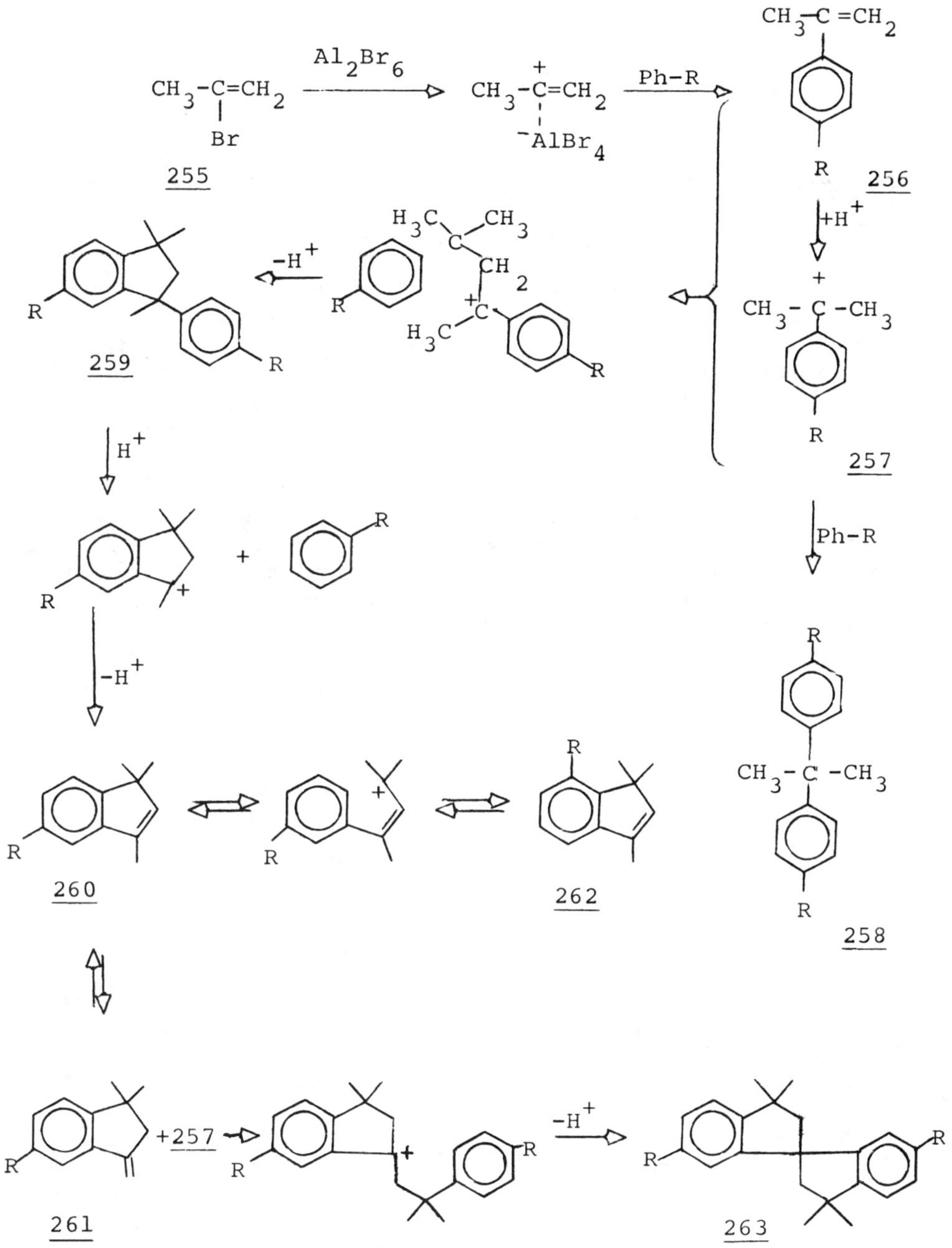

a: R = H

b: R = CH_3

Scheme 26

structure was established at the same time as that of 263a [146]. In view of our current knowledge of carbocation chemistry, the formation of 1,1'-spirobiindans 263 in each of the reactions reported in Refs. 142 to 145 can also reasonably be correlated in terms of the mechanisms outlined in scheme 26.

It was pointed out before that the products of the acid-catalyzed reactions of styrenes are determined primarily by both reaction conditions and nature of reactants. Elaborating on this, the following observations may be made:

1. From a study of the dimerization of styrene in the presence of aqueous sulfuric acid under various conditions, Rosen [116] found that the formation of the saturated indan dimer could generally be enchanced by increasing reaction time and by adopting a procedure in which solutions of low acid concentration are used during the early stages of the reaction and solutions of relatively higher concentration during later stages.

2. By reviewing the acid-catalyzed dimerizations of ring-substituted styrenes, Muller and Karczag-Willhelms [131] noted that isopropyenylphenols and isopenylphenol ethers that are alkoxyl or hydroxyl substituted in the 3 or 5 position of the aromatic moiety yield the cyclic dimer with great ease, so that the unsaturated dimers are not observed. An explanation for this was given by assuming that an alkoxyl or hydroxyl group placed *meta* to the isopropenyl side chain should increase the electron density at the carbons *ortho* to the side chain in the transition state, leading to facile ring closure of the dimeric carbocation (265, Fig. 2).

3. Styrenes of the 1,1-di(*p*-alkoxyphenyl)ethylene type (266, R = CH_3 or C_2H_5) have been shown [117] to dimerize in the presence of Brönsted

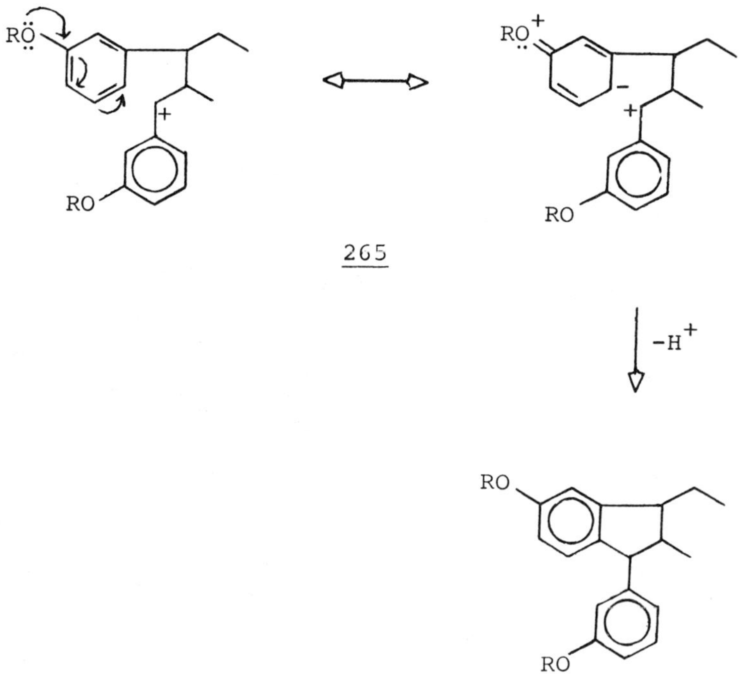

RÖ RO RO RO

265

$-H^+$

RO RO

Figure 2

catalysts such as HCl, HBr, Cl_3CCOOH, or $AcOH$-H_2SO_4 to give only dimers of the unsaturated open-chain type (267, Eq. 78).

$$(p\text{-}ROC_6H_4)C{=}CH_2 \xrightarrow{H^+} Ar_2CMeCH{=}CAr_2 \quad (78)$$

266 → 267

$R = CH_3, C_2H_5$

4. Indans 268 (R = H or Me) of high purity, useful as intermediates for dyes and pesticides, were prepared in increased yields by the acid-catalyzed dimerization of PhRC = CH_2 in the presence of inorganic oxidizing agents such as MoO_3, $K_3Fe(CN)_6$, HNO_3, $NaClO_3$, HgO, etc., and, optionally, polymerization inhibitors such as phenothiazine or phenylthiourea [148,149].

268

Thus the dimerization of styrene in the presence of both $NaNO_3$ and 98% H_3PO_4 at 50 to 60°C gave 80% of 268 (R = H) as compared with 60% of the same product in the absence of $NaNO_3$.

5. The dimerization of styrenes with acids was practically suppressed by the presence of activated aromatics such as phenols or alkylbenzenes; instead, alkylation of the aromatic by the styrene took place to produce 1,1-diarylethanes [141,150-161]. Illustrative reactions are shown in Eqs. (79) to (81).

$$PhCH{=}CH_2 + \text{cumene} \xrightarrow[3\text{ hr},\ 150^\circ]{\text{acid clay}} \text{(isopropylphenyl)(Ph)ethane} \quad (79)\ [154]$$

$$PhCH{=}CH_2 + \text{phenol} \xrightarrow[50^\circ,\ 2\text{ hr};\ 96\%\text{ yield}]{H_3PO_4} o\text{-(1-phenylethyl)phenol} + p\text{-(1-phenylethyl)phenol} \quad (80)\ [151]$$

1 mol, 3 mol; products 33.5% (ortho), 66.5% (para)

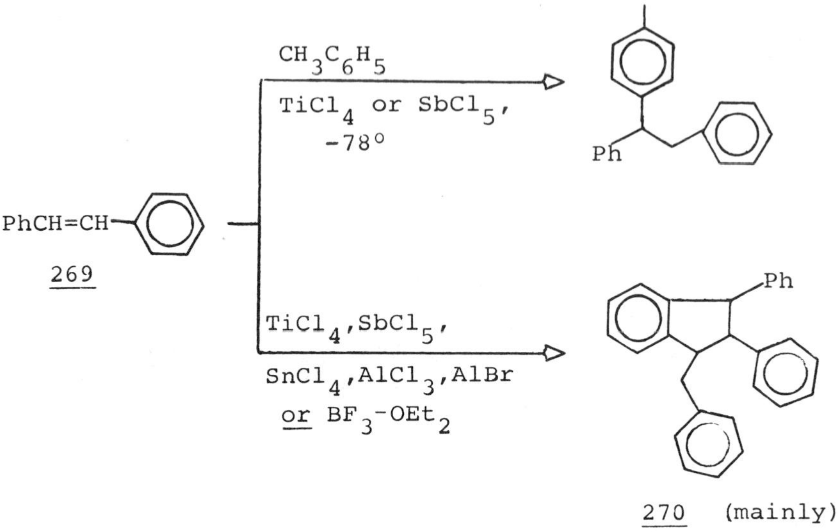

PhCH=CH–Ph (269) → ($CH_3C_6H_5$; $TiCl_4$ or $SbCl_5$, $-78°$) → [product]; → ($TiCl_4$, $SbCl_5$, $SnCl_4$, $AlCl_3$, AlBr or BF_3-OEt_2) → 270 (mainly)

(81) [153]

6. The acid-catalyzed dimerization of a binary mixture of styrenes results in the formation of all of the theoretically possible indan dimers. Thus the chlorophenylindans 271 (R = 2-, 3-, or 4-Cl; R' = H) and chloro analogs 271 (R' = 5-, 6-, or 7-Cl) together with 271 (R = R' = H_ were all obtained by reaction of PhCH = CH_2 with the ClC_6H_4CH = CH_2 in the presence of 89 to 100% H_3PO_4 at 30 to 45°C in $ClC_6H_4C_2H_5$ [162].

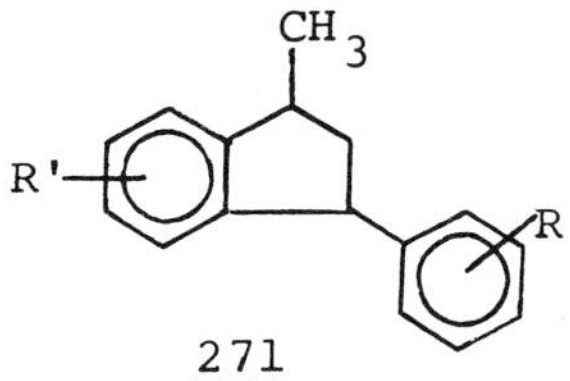

271

(ii) Reaction of styrenes with alkenes. The reaction of styrene and its derivatives with alkenes in the presence of H_2SO_4-AcOH or BF_3-$O(CH_3)_2$ was shown by Polak and co-workers [163,164] to provide an accessible route for the preparation of polyalkylindans. A general formulation of the reaction and its probable mechanism is given in scheme 27.

A large number of alkyl-substituted indans 273 were prepared by this reaction [163,164].

In a more recent modification of the preceding reaction, triphenylmethyl cation (274) was used in place of styrene. Thus 1,1-dimethyl-3,3-diphenylindan (275) was prepared by addition of isobutylene to a solution of either $Ph_3C^+ClO_4^-$ or $Ph_3C^+BF_4^-$ in organic solvents [165]. This reaction, which represents a direct intermolecular addition of a carbocation to an alkene to form a new ring, was suggested to proceed as shown in scheme 28.

Notably, indan 275 was also shown to result from reactions of salts of the triphenylmethyl cations with *t*-butyl ethers such as methyl *t*-butyl ether and

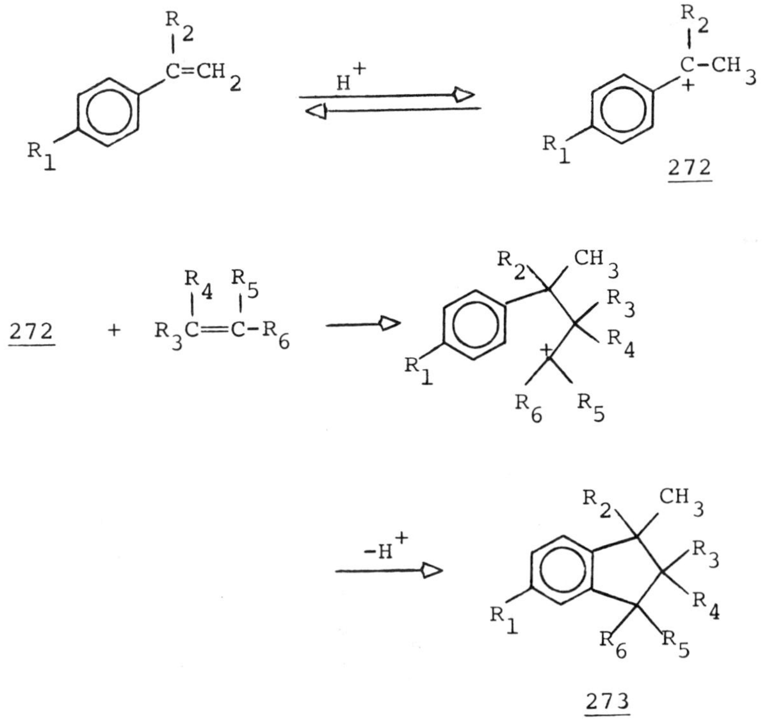

272

272 + $R_3C{=}C(R_4)\text{-}C(R_5)\text{-}R_6$

273

Scheme 27

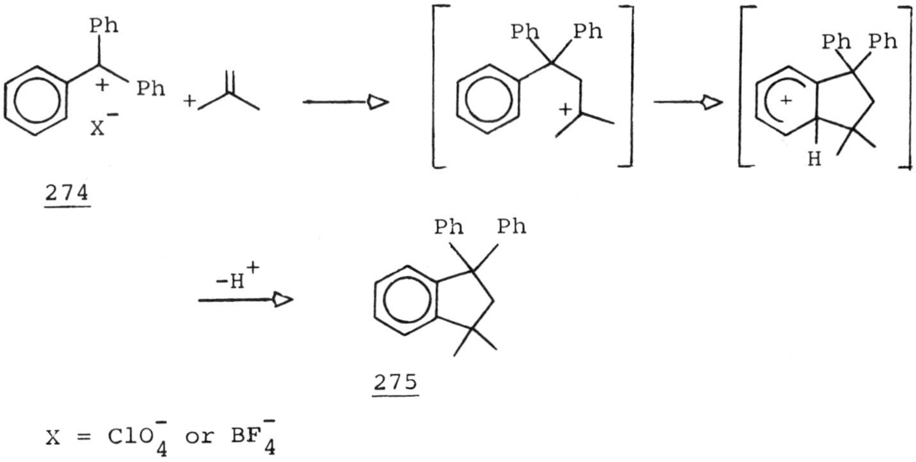

274

275

$X = ClO_4^-$ or BF_4^-

Scheme 28

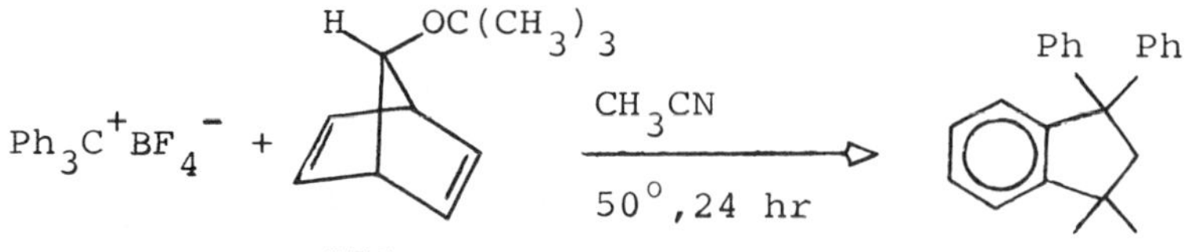

$Ph_3C^+BF_4^-$ + 276 $\xrightarrow[50^\circ, 24\ hr]{CH_3CN}$ 275 (82)

7-*t*-butoxynorbornadiene (276, Eq. 82) [165]. In this case, indan 275 was presumed to form by reaction of triphenylmethyl cation with isobutylene generated from the *t*-butyl ethers under the conditions employed. Preliminary experiments showed that indans similar to 275 could not be obtained when isobutylene was replaced by either *cis*-2-butene or 2,3-dimethyl-2-butene.

(iii) Reactions of arenes with alkenes and alkyl halides via hydride transfer. Cyclialkylations initiated by hydride ion abstractions were first reported by Ipatieff et al. [166] in 1948. At that time, these authors observed that 1,3,3,6-tetramethyl-1-*p*-tolylindan (233e) was also produced during the acid-catalyzed alkylation of *p*-cymene with branched alkenes such as dihydrolimonene, methylcyclohexene, and trimethylethylene. These authors rationalized their observations in terms of the mechanism shown in scheme 29 using trimethylethylene (277) as the alkene.

As evident from scheme 29, the key step is the generation of the 2-*p*-tolyl-2-propyl cation 257b from *p*-cymene via transfer of its tertiary α hydrogen to the tertiary aliphatic carbocation generated in the medium. This 2-*p*-tolyl-2-propyl cation then adds to the 2-*p*-tolylpropene (256b) produced from it by deprotonation to give the dimeric carbocation 278, followed by proton loss to 233e. However, as we shall see shortly, the 2-*p*-tolyl-2-propyl cation 257b may also add to the alkene present in the medium to give other cyclic derivatives.

In later years, the hydride transfer cyclialkylation method received more attention and was extended to include other arenes and other carbocation precursors. This led to the preparation of a variety of indans, hydrindacenes,

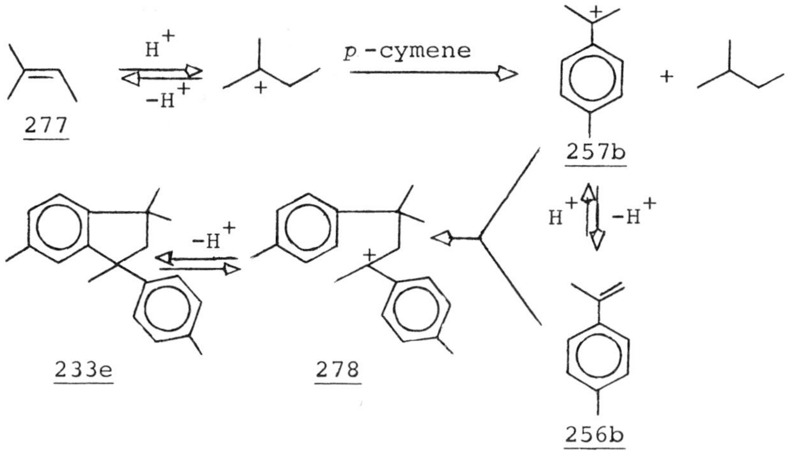

277 (H^+ / $-H^+$) → *p*-cymene → 257b + ; 257b (H^+ / $-H^+$) 256b; 278 ($-H^+$) 233e

Scheme 29

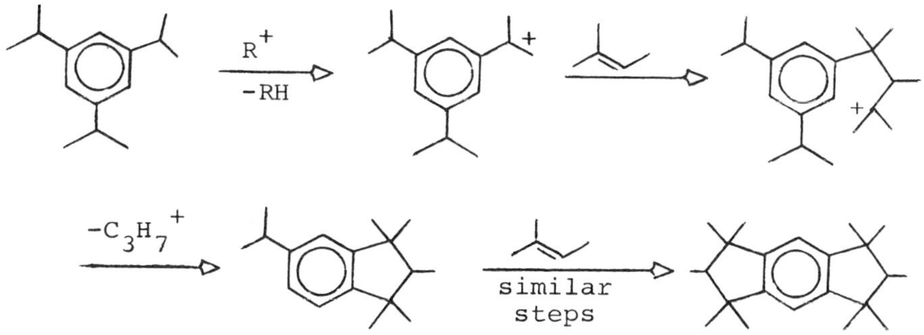

R^+ / $-RH$

$-C_3H_7^+$

similar steps

281

Scheme 30

tetralins, and octahydroanthracenes [167,168]. For example, under Friedel-Crafts conditions, Barclay and co-workers obtained octamethyloctahydroanthracene (279) from 1,3,5-tri-*t*-butylbenzene and *t*-butyl chloride [168], octamethylhydrindacene (280) from 1,3,5-*t*-butylbenzene and isopropyl chloride or from 1,3,5-triisopropylbenzene and isobutylene in the presence of isopropyl chloride [167], decamethylhydrindacene (281) from triisopropylbenzene and 2-methyl-2-butene [167], and 1,1,3-trimethyl-5,6-diisopropyl-3-neopentylindan (282) from triisopropylbenzene and 2-chloro-2,4,4-trimethylpentane or 2,4,4-trimethyl-1-pentene [167]. These results were explained

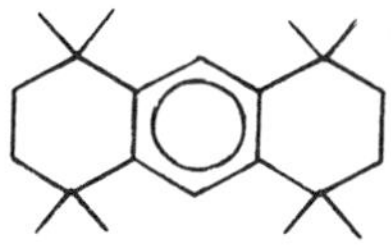

279 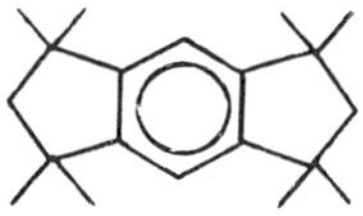 280 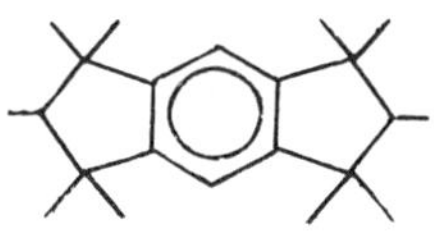 281 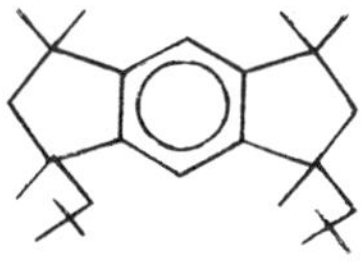 282

in terms of multistep rearrangement-dealkylation-alkylation-hydride transfer reactions. To illustrate, the mechanism suggested to explain the formation of decamethylhydrindacene (281) from triisopropylbenzene and 2-methyl-2-butene is shown in scheme 30.

In a more recent development, Eisenbraun and co-workers [136] conducted a thorough study of the reactions of *o*-, *m*-, and *p*-cymenes and iso-, diiso-, and triisobutylene in the presence of various Brönsted and Lewis acid catalysts. The products obtained consisted of complex mixtures of cyclialkylation and/or alkylation hydrocarbons. *o*-Cymene formed only alkylation products with no products due to cyclialkylation. In this case, the alkylation products consisted mainly of the two *t*-butyl-*o*-cymenes 283 and 284, mixed with at least six minor products (Eq. 83).

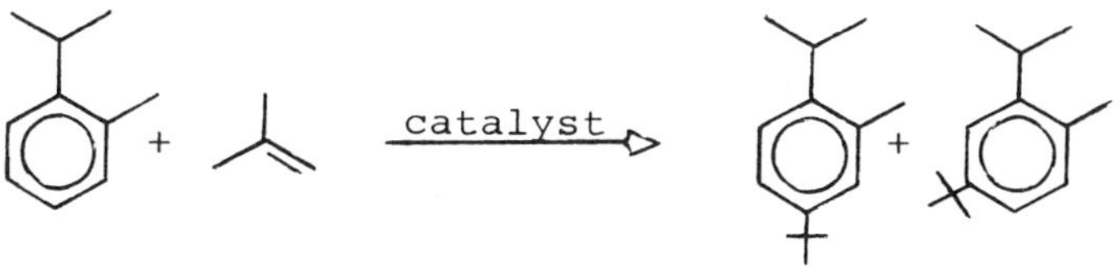

catalyst → 283 + 284 + minor products ca. 3% (83)

With *p*-cymene (Eq. 84), eight components were identified as cyclialkylation

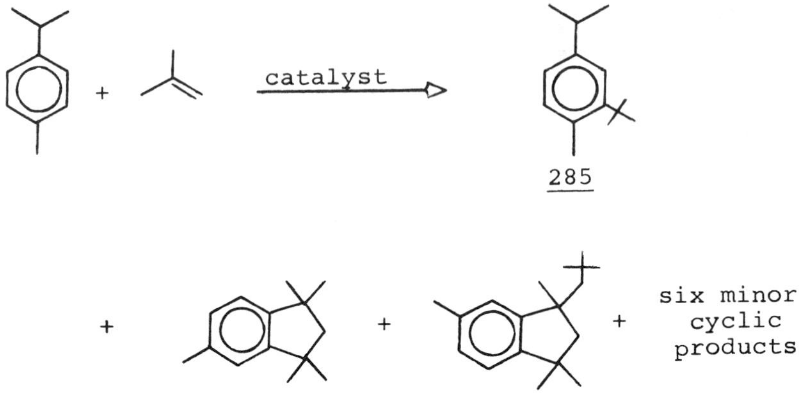

catalyst

285

(84)

286

287

(major)

\+ six minor cyclic products

products but only one, 2-*t*-butyl-4-isopropyltoluene (285), a minor product, was due to alkylation. In this case the two most dominant cyclic products were 1,1,3,3,5-tetramethylindan (286) and 1,1,3,5-tetramethyl-3-neopentylindan (287), the minor cyclic products contained both indan and tetralin derivatives. As expected, *m*-cymene occupied an intermediate position, providing essentially the alkylation product 3-*t*-butyl-5-isopropyltoluene (288) mixed with minor amounts of other products, including the three indans 289-291 (Eq. 85).

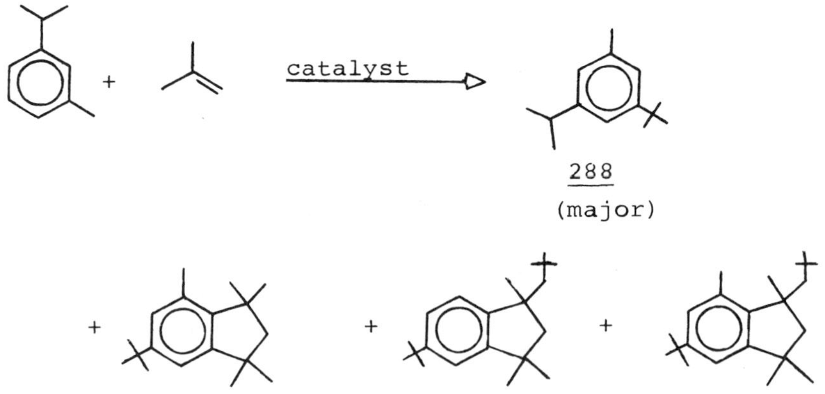

catalyst

288

(major)

(85)

289

290

291

\+ two minor alkylation products

Apart from the findings above, the work of Eisenbraun and co-workers illustrated the following:

1. For alkylation, the order is o- $>$ m- $\gg$ p-cymene, and for cyclialkylation is o- $\ll$ m- $<$ p-cymene.

2. In the cyclialkylation products from *m*-cymene, the point of ring closure (*ortho* or *para* position to the methyl group) was strongly influenced by alkyl groups located at the 5 position of *m*-cymene. Thus, whereas some of 290 was found in the *m*-cymene-isobutylene reaction, none of the isomeric 292 could be detected. This supports the conclusion that steric factors are important in determining the course of cyclialkylation.

290 292

3. With sulfuric acid catalyst, the best results were obtained with a reaction temperature below 10°C. Also, within limits, the quantity of sulfuric acid used had little effect on the products formed.
4. Anhydrous hydrogen fluoride was effective in catalyzing the cyclialkylation reaction.
5. The products of these reactions were obtained not only from alkylation and cyclialkylation of the starting alkene, but also from alkenes produced in the reaction system via polymerization, rearrangement, and fragmentation.

2. Production of indenes by intra- and intermolecular reactions

Like indans and tetralins, indenes have been obtained by methods involving both intra- and intermolecular cyclialkylations. Starting with the former, Heindel et al. [169] found that diaryldialkyl pinacols (293), in which the aryl nucleus is electronically activated by substituents such as NH_2, $N(CH_3)_2$, or OH underwent 1° closures to 2-aryl-3-methylindenes (294) in the presence of acid catalysts (Eq. 86).

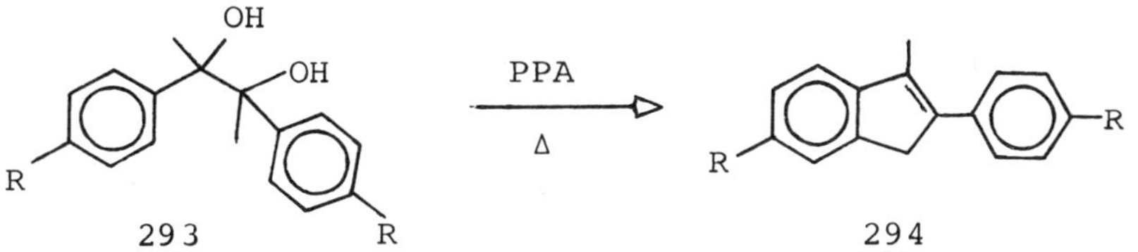

OH OH R R 293 — PPA, Δ → R R 294 (86)

(R = NH_2, $N(CH_3)_2$, OH)

Indenes have also been prepared in good yields by intramolecular cyclizations involving allylic-type alcohols. The recent examples shown in Eqs. (87) and (88) are illustrative [170].

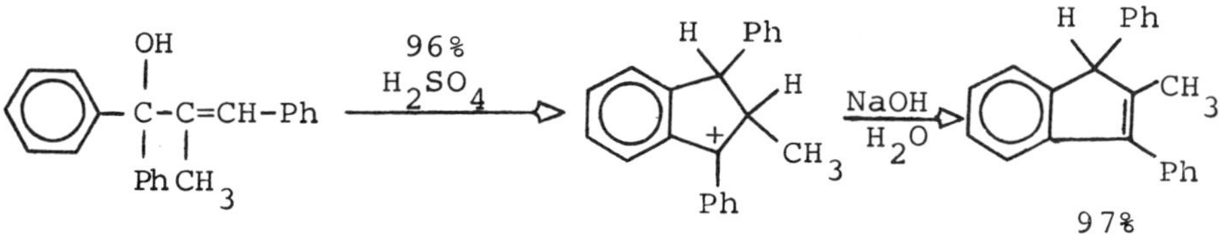

OH C-C=CH-Ph Ph CH_3 — 96% H_2SO_4 → H Ph H + CH_3 Ph — NaOH, H_2O → H Ph CH_3 Ph 97% (87)

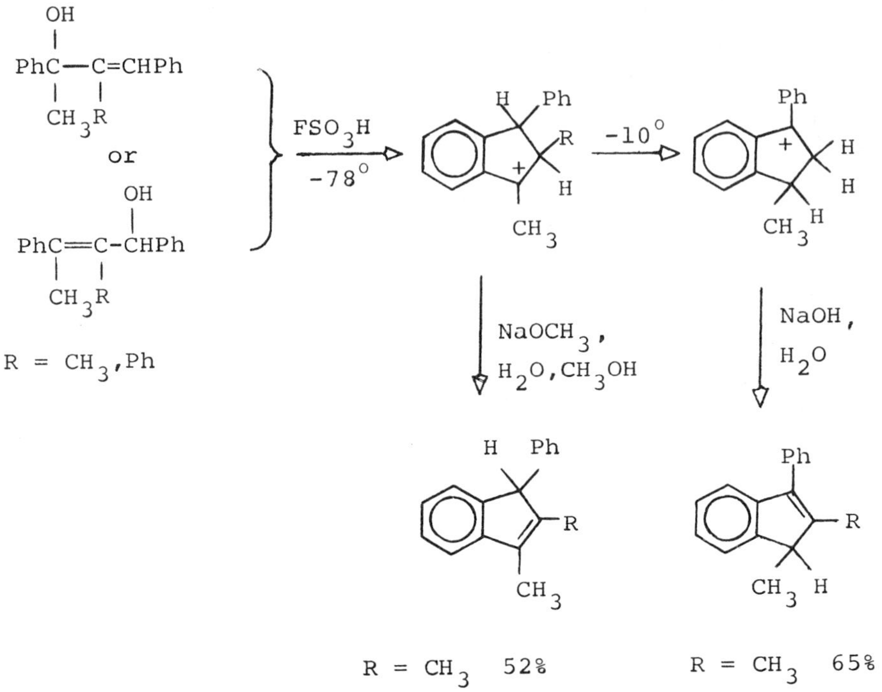
OH
PhC—C=CHPh
CH3 R
or
OH
PhC=C-CHPh
CH3 R
R = CH3, Ph
FSO3H
-78°
H Ph
R
H
CH3
-10°
Ph
H
H
CH3 H
NaOCH3,
H2O, CH3OH
NaOH,
H2O
H Ph
R
CH3
Ph
R
CH3 H
R = CH3 52%
R = CH3 65%

(88)

Other reported intramolecular closures to indene derivatives involve the cyclizations of benzaldesoxybenzoin (295) with PCl_5 in chlorinated solvents (Eq. 89) [171] and of β-aralkyl carbonyl compounds with $AlCl_3$ (Eq. 90) [172].

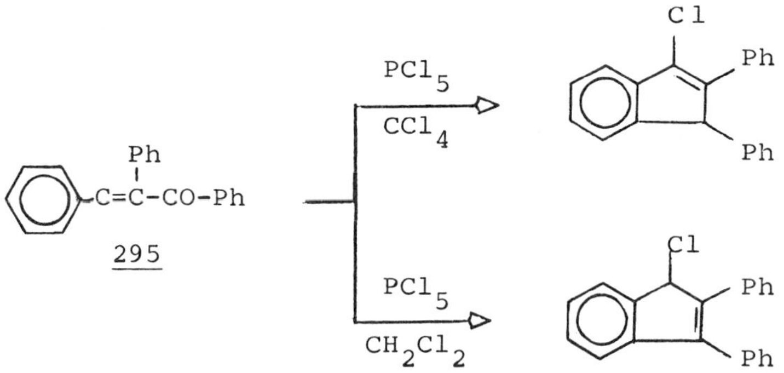
Ph
C=C-CO-Ph
295
PCl5
CCl4
Cl
Ph
Ph
PCl5
CH2Cl2
Cl
Ph
Ph

(89)

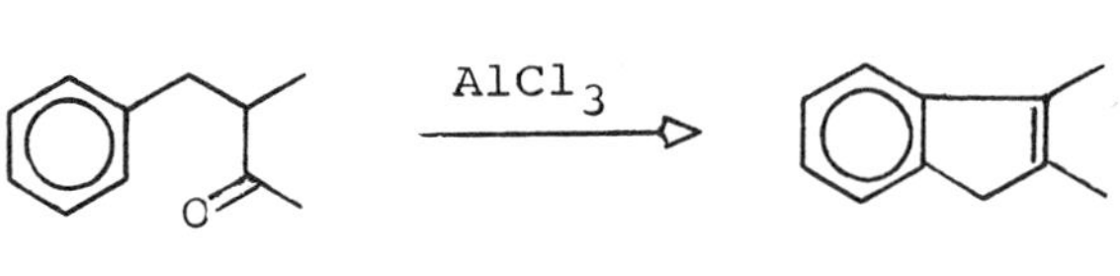
AlCl3

(90)

As to intermolecular cyclizations, indenes have been obtained by a number of unusual reactions. One of these, which was developed by Buddrus and Nerdel in 1965 [173] and extended by Skattebol and Boulette in 1966 [174], involved the $AlCl_3$- or $FeCl_3$-catalyzed alkylation of arenes with *gem*-dihalocyclopropanes possessing a quaternary (or a phenylated) carbon atom. Some of the reported examples are depicted in Table 18.

Table 18. Production of Indenes by Alkylation of Arenes with *gem*-Dihalocyclopropanes in the Presence of $AlCl_3$ at 25°

Arene	Dihalocyclopropane	Catalyst	Yield, %	Product composition, %	Ref.
Benzene	1,1-Dichloro-2,2-dimethyl-cyclopropane	$AlCl_3$	66	1,2-Dimethylindene (89%) and 1,1-dimethyl-2-phenylindane (11%)	173
Benzene	1,1-Dibromo-2,2-dimethyl-cyclopropane	$AlCl_3$	56	2,2-Dimethylindene (100%)	174
Benzene	1,1-Dibromotrimethylcyclo-propane	$AlCl_3$	54	1,2,3-Trimethylindene (100%)	174
Benzene	1,1-Dichlorotrimethylcyclo-propane	$AlCl_3$	63	1,2,3-Trimethylindene (55%) and 1,1,3-trimethylindene (45%)	173
Benzene	1,1-Dichlorotetramethyl-cyclopropane	$AlCl_3$	48	1,1,2,3-Tetramethylindene (100%)	173
Benzene	1,1-Dibromotetramethyl-cyclopropane	$AlCl_3$	80	1,1,2,3-Tetramethylindene (100%)	174
Benzene	1,1-Dichloro-2-phenylcyclo-propane	$AlCl_3$	24	2,3-Diphenylindene (100%)	173
Benzene	1,1-Dibromo-2-phenylcyclo-propane	$AlCl_3$	25	3-Phenylindene (100%)	174
Benzene	1,1-Dichloro-2-methyl-2-phenyl-cyclopropane	$AlCl_3$	53	2-Phenyl-3-methylindene	173
Toluene	1,1-Dibromotetramethyl-cyclopropane	$AlCl_3$	81	1,1,2,3,5-Pentamethylindene (70%) and 1,1,2,3,6-penta-methylindene (30%)	174

Table 18. (continued)

Arene	Dihalocyclopropane	Catalyst	Yield, %	Product composition, %	Ref.
o-Xylene	1,1-Dichloro-2,2-dimethyl-cyclopropane	$AlCl_3$	76	2,3,5,6-Tetramethylindene (76%) + 2 isomers (28% + 9%)	173
o-Xylene	1,1-Dibromotetramethylcyclo-propane	$AlCl_3$	79	1,1,2,3,5,6-Hexamethylindene (60%) and 1,1,2,3,4,5- + 1,1,2,3,6,7-hexamethylindene (40%)	174
m-Xylene	1,1-Dichloro-2,2-dimethyl-cyclopropane	$AlCl_3$	47	2,3,5,7-Tetramethylindene (100%)	173
p-Xylene	1,1-Dichloro-2,2-dimethyl-cyclopropane	$AlCl_3$	67	2,3,4,7-Tetramethylindene (100%)	173
1,4-Diethyl-benzene	1,1-Dichloro-2,2-dimethyl-cyclopropane	$AlCl_3$	43	2,3-Dimethyl-4,7-diethylindene	173
1,2,3,5-Tetra-methylben-zene	1,1-Dichloro-2,2-dimethyl cyclopropane	$AlCl_3$	32	2,3,4,5,6,7-Hexamethylindene (100%)	173
p-Cymene	1,1-Dichloro-2,2-dimethyl-cyclopropane	$AlCl_3$	28	2,3-Dimethyl-4-isopropyl-7-methylindene	173
Chlorobenzene	1,1-Dichloro-2,2-dimethyl-cyclopropane	$AlCl_3$	75	Two isomeric 2,3-dimethyl-chloroindenes in a 77:23 percent ratio	173

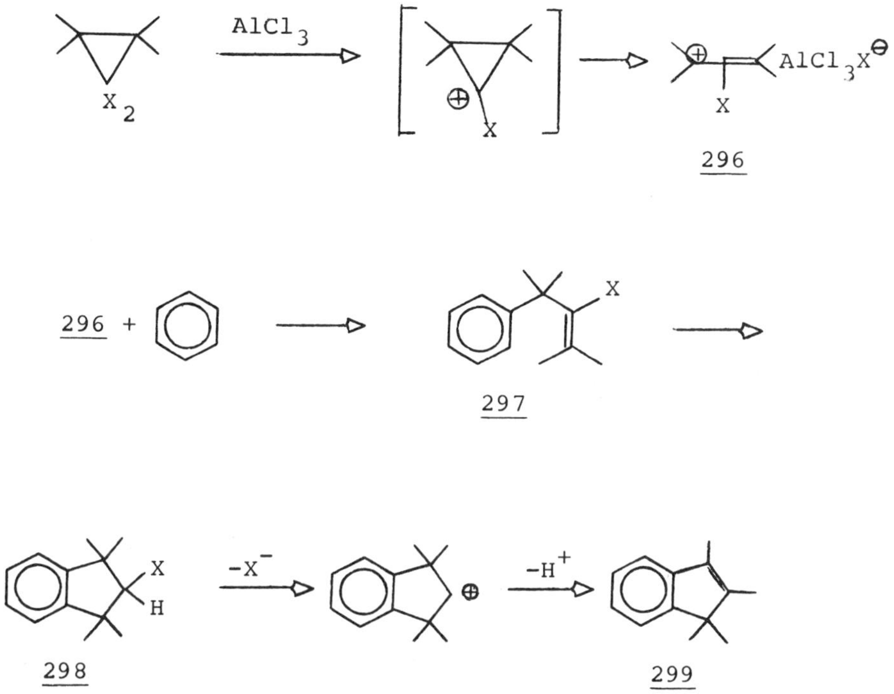

296

297

298 299

Scheme 31

In examining the results of Table 18 it is to be noted that mixtures of indene isomers were isolated in some cases and that the reaction of 1,2,3,5-tetramethylbenzene with 1,1-dichloro-2,2-dimethylcyclopropane (last column) involved migration of methyl groups to yield 2,3,4,5,6,7-hexamethylindene. Altogether the results of Table 18 were rationalized by Skattebol and Boulette [174] in terms of a unified mechanism which is illustrated in scheme 31 by the reaction between 1,1-dichlorotetramethylcyclopropane and benzene in the presence of $AlCl_3$.

As in reactions of cyclopropane and its derivatives under Friedel-Crafts conditions [175], the mechanism of scheme 31 assumes initial opening of the cyclopropane ring by the catalyst to an allylic ion 296 followed by attack on the arene. This attack leads primarily to the vinylic halide 297, which undergoes closure to 298, then dehydrohalogenation to the final indene product 299.

In addition to the reactions above, indenes were also obtained as sole, major, or minor products in several other intermolecular Friedel-Crafts reactions. For example, indenes were obtained as minor products from the reaction of arenes with 2-bromopropene and $AlCl_3$ or Al_2Br_6 (see scheme 26) [141]; as major products from the reaction of mesitylene with isopropyl bromide and $AlCl_3$ (Eq. 91) [176]; and as sole products from the reactions of (1) benzene with 1,1-dichloro-2-phenylcyclopropane (300) in the presence of $AlCl_3$ catalyst (Eq. 92) [177], (2) phenylperfluoropropylene (301) and $AlCl_3$ in acetyl chloride (Eq. 93) [178], (3) benzene with styryl ketones 302 in the presence of acid catalysts (Eq. 94) [179], and (4) diphenylmethyl chloride (303a) with diphenylacetylene in the presence of acid catalyst (scheme 32) [179a].

$\xrightarrow{AlCl_3}$ (major) + (minor) (91) [176]

300 + PhH $\xrightarrow{AlCl_3}$ (major) + ... (92) [177]

301 $\xrightarrow[CH_3COCl]{AlCl_3}$ (93) [178]

302 $\xrightarrow{\text{Friedel-Crafts catalysts}}$ (94) [179]

(R = H, CH_3; R' = CH_3, C_2H_5, OH)

In scheme 32, it is to be noted that the preferred path was the intramolecular cyclization (path b) in the case of diphenylmethyl chloride (303a) but the reaction with the chloride ion (path a) in the case of benzyl chloride (303b). The competition between the two paths has been related to the relative reactivity of linear vinyl cations toward internal and external nucleophiles [180].

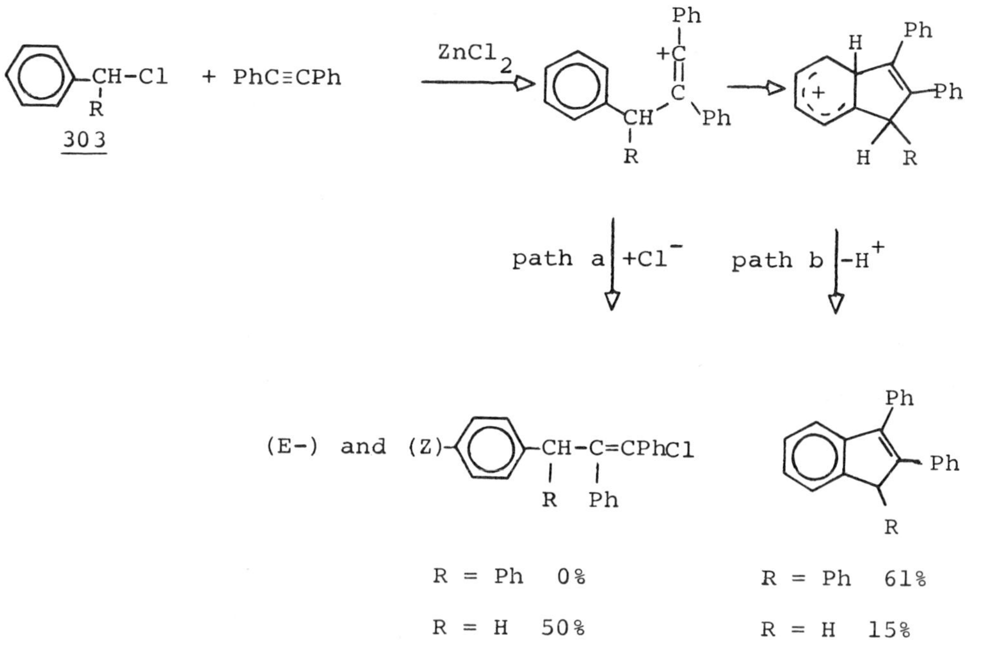

303

path a +Cl^- path b $-H^+$

(E-) and (Z)-C_6H_4-CH(R)-C(Ph)=CPhCl

R = Ph 0%

R = H 50%

R = Ph 61%

R = H 15%

303a: R = Ph

303b: R = H

Scheme 32 [179a]

3. Production of indanones and tetralones by intra and intermolecular reactions

In the literature, several indanones and tetralones were reported to form by the intramolecular cyclization of ketonic precursors such as alkenyl [47,52,82, 135,181-207a], hydroxylalkyl [188,208,209], and haloalkyl [8,47-51,53,56,57, 181,200,210-217] aryl ketones. However, examination of the early literature results and discussions [2,11,87,218] on this subject reveals a considerable number of discrepancies and contradictions. For example, α, β-unsaturated ketones of general formula 304 have been claimed to undergo facile five-membered ring closures to 1-indanones of general formula 305 upon treatment with acid catalysts [82,181-204]. The conditions employed to effect these cyclialkylations were diverse and included $AlCl_3$ in CS_2 at room temperature

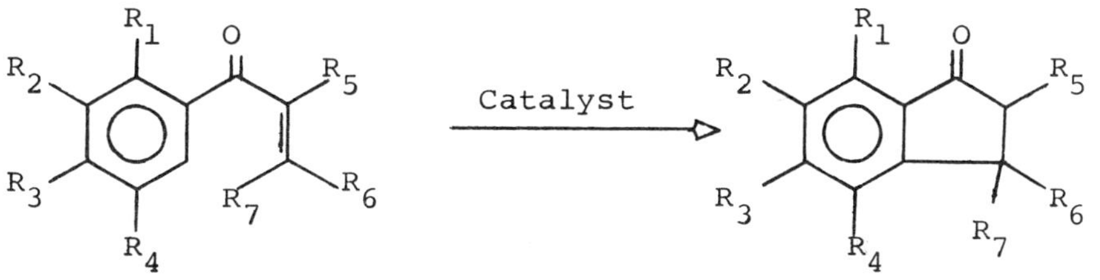

304 —Catalyst→ 305

[184], $AlCl_3$ in refluxing CS_2 [184,186], $AlCl_3$ in benzene at steam-bath temperature [191], excess $AlCl_3$-HCl in refluxing CS_2 [183,190], H_3PO_4-HCOOH [194,203], and $AlCl_3$-NaCl at 100 to 180°C [192,194,197,204].

Added to direct closures, α,β-unsaturated aralkenyl ketones have also been proposed as reaction intermediates in the acid-catalyzed formation of 1-indanones from α-halo aralkyl ketones (306, Eq. 95) [181,200], α-alkyl-β-hydroxypropiophenones (307, Eq. 96) [188,208,209], and from mixtures of arenes with α,β-unsaturated acid chloride (Eq. 97) [183,186,187,191,202].

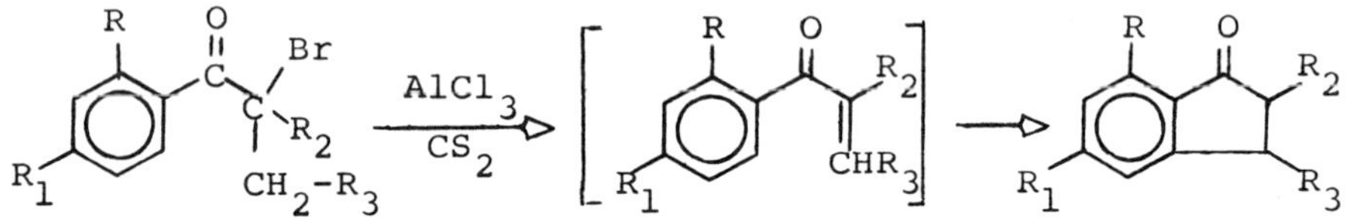

(95)

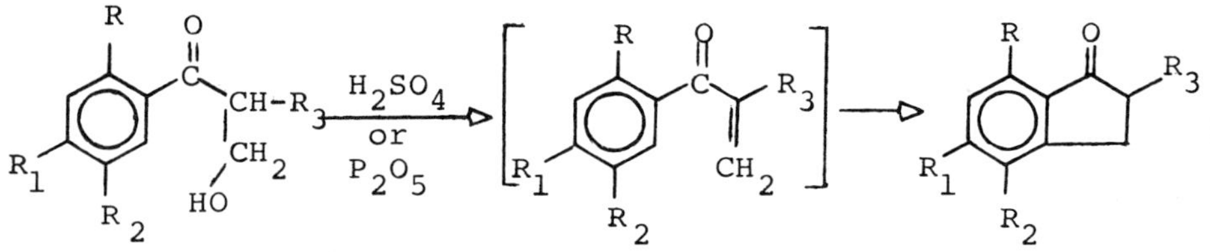

(96)

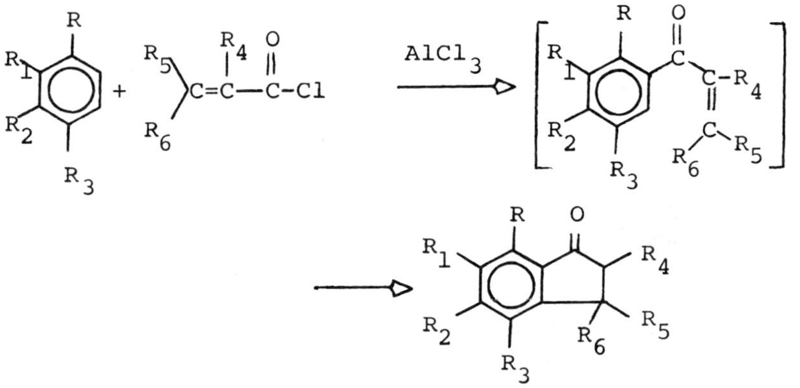

(97)

In contrast with the preceding claims, Colonge and Grimaud [189] reported that unsaturated aromatic ketones of the general formula 308 failed to cyclize to 1-tetralones 309 upon treatment with acid catalysts (Eq. 98).

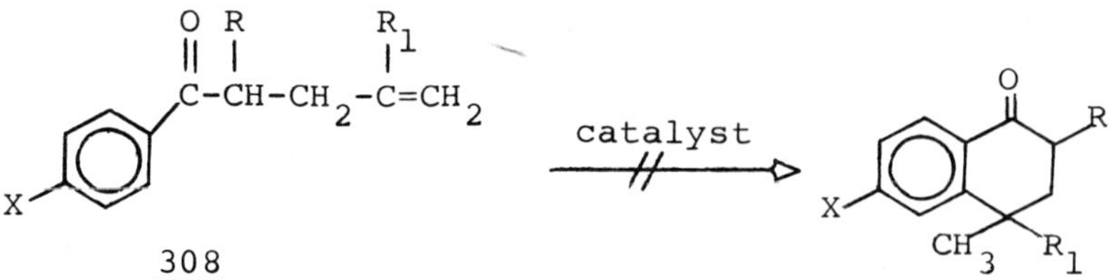

(98)

$X = Cl, CH_3 \quad R = CH_3; R_1 = H, CH_3$

In the literature confusion also exists in regard to the effect of nuclear substituents on the ease of closure of aromatic ketones. On one hand, von Auwers and Risse [181] and Smith and his collaborators [52,82] concluded that methyl substituents placed *ortho* and *para* to the closure site (i.e., R_2 and $R_4 = CH_3$ in formula 304) promote the closure, whereas a methyl substituent placed *meta* to the site (i.e., $R_3 = CH_3$ in formula 304) inhibits the closure. On the other hand, Layer and MacGregor [200] suggested that the preceding conclusion was misleading and that both α,β-unsaturated and α-bromo aromatic ketones will ring-close, regardless of the positions of the nuclear substituents.

In recent years, a few studies have been aimed at the resolution of some of the above-mentioned literature contradictions. Two of these studies were carried out by Pines and Douglas [56,57], who analyzed in detail some aspects of the cyclialkylation reactions of α- and β- haloaralkylphenones. Their results, which we presented in detail earlier in connection with cyclialkylations of haloalkyl aryl ketones, proved that cyclialkylations of haloaralkylphenones to 1-indanones proceeded via an elimination mechanism (e.g., Eq. 53) with $AlCl_3$ catalyst but via an enolization-initiated mechanism with $AlCl_3$-CH_3NO_2 or H_2SO_4 catalyst (see scheme 16).

Very recently (1981), Khalaf and El-Khawaga [219] tackled another aspect of the problem. Inspired by the well-known fact that the reactivity of aromatic carbonyl compounds toward electrophilic reagents is low [8,11,213,220-222], they reevaluated the previously indicated readiness with which cyclialkylations to 1-indanones and 1-tetralones were claimed to occur. To do this, they studied the cyclialkylation behavior of some model ketones in which the carbonyl function and the aromatic nucleus at which closure is expected to occur were (1) directly attached to each other as in 310-312, (2) conjugated with each other as in 313, and (3) isolated from each other as in 314 and 315.

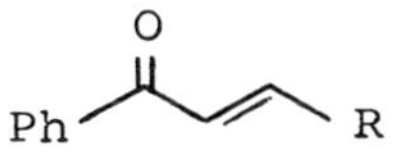

310 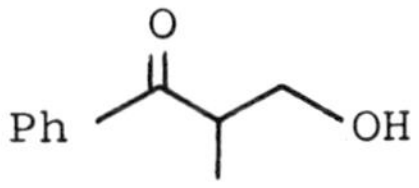 311 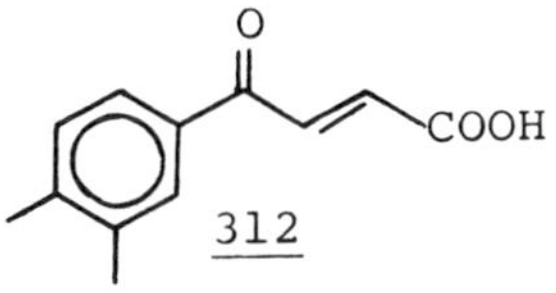 312

a: R = phenyl; R'=H

b: R = *p*-anisyl; R'=H

c: R = methyl; R'=H

d: R = R'=CH_3

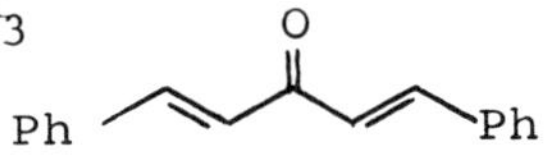

313 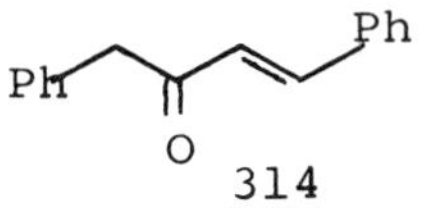 314 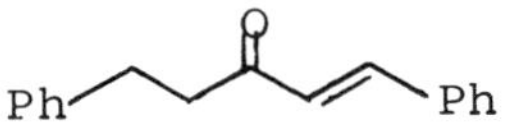 315

Some of the experiments carried out were repetitions of earlier ones to examine their validity by the help of modern techniques. The results obtained, while confirming some of the earlier reports, were in sharp contrast to several others. For example:

1. In contrast to earlier claims [186], both ethylideneacetophenone (310c) and isopropylideneacetophenone (310d) failed to cyclize to 3-methylindanone and 3,3-dimethylindanone, respectively, upon treatment with $AlCl_3$ in CS_2 at room temperature.

2. In the literature [192], β-(3,5-dimethylbenzoyl)acrylic acid (312) was reported to yield 87% of 3-carboxyl-4,7-dimethyl-1-indanone (316) upon treatment with $AlCl_3$-NaCl catalyst at 135°C. In the present work,

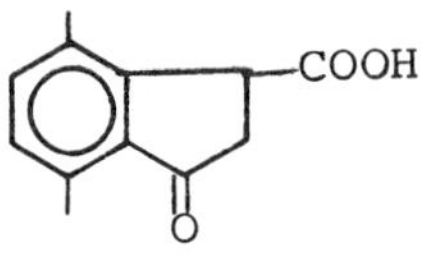

COOH

O

316

treatment of 312 with similar amounts of catalyst but at 100°C for 2 hr gave none of the previously reported product; only unreacted starting material could be recovered.
3. The primary alcohol 2-benzoyl-1-propanol (311) was reported by Fuson et al. [208] and by Colonge and Weinstein [209] to give 67 and 81% yields, respectively, of 2-methyl-1-indanone upon treatment with concentrated H_2SO_4. In the recent work, treatment of 311 under the specified literature conditions gave only a sulfur-containing resinous material of unidentified nature.
4. In agreement with Fine and Stern [223], the benzyl styryl ketone 314 cyclized easily to 4-phenyl-2-tetralone (317) upon treatment with $AlCl_3$-HCl in CS_2 at ambient temperature (Eq. 99).

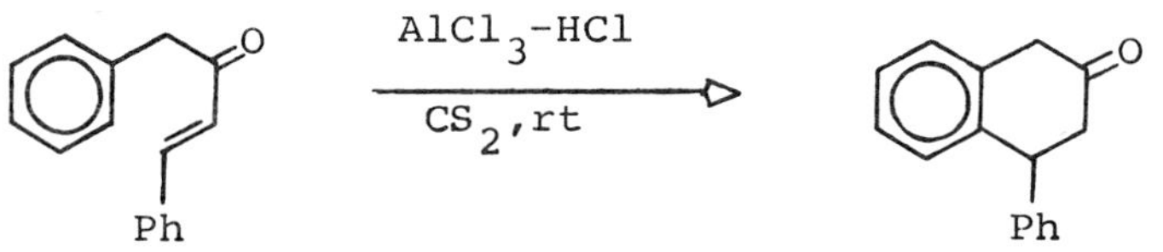

$AlCl_3$-HCl / CS_2, rt (99)

314 (O, Ph) → 317 (O, Ph) 41% yield

This recent work of Khalaf and El-Khawaga demonstrated clearly that ketones in which the deactivating carbonyl group was directly attached to or conjugated with the aromatic nucleus at which ring closure is expected to occur (i.e., compounds 310-313) showed severe resistance to the ring closure. In fact, attempts to enforce these closures by treatment with 96% H_2SO_4 at 25, 70, or 100°C; with $AlCl_3$ or $AlCl_3$-HCl in CS_2, or in petroleum ether (bp 60 to 80°C) at 25°C, or at reflux temperature; and with $AlCl_3$-NaCl at 108°C were all unsuccessful. However, when the carbonyl group was isolated from the ring by one or more CH_2 groups as in 314 (Eq. 99) or in 318 (Eq. 100), ring closures were observed.

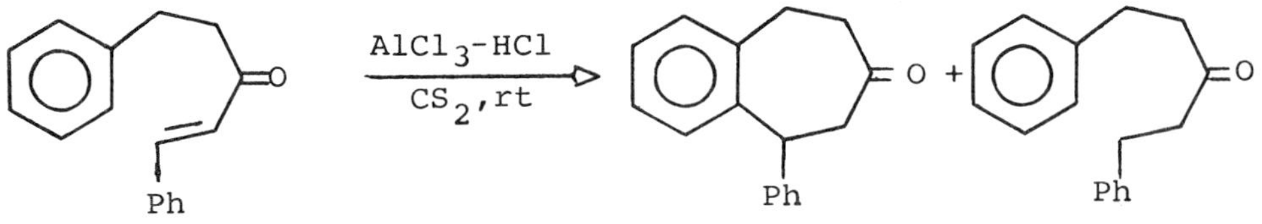

$AlCl_3$-HCl / CS_2, rt

318 (O, Ph) → (O, Ph) + (O, Ph) (100)

+ unidentified

Based on these and other substantiated findings, the latter authors concluded that the nucleophilicity of the aromatic nucleus upon which the attack occurs plays a determining role in Friedel-Crafts cyclizations, as in other electrophilic substitution reactions. Accordingly, the readiness with which intramolecular closures are believed to occur on carbonyl-deactivated rings calls for reassessment.

Support for this view can readily be found in the following pertinent observations:

1. According to recent reports [11,56,57,59], the cyclialkylation of haloketones 135a, 136a [56], 319 and 320 could be effected only under strenuous conditions such as heating with excess $AlCl_3$-NaCl catalyst at temperatures higher than 100°C.

135a 136a 319 320

2. Although intramolecular reactions are known to be faster than intermolecular ones, reactions of benzene with dibasic acid chlorides 321-325 gave intermolecular but no intramolecular acylation end products. The latter process is undoubtedly retarded by the carbonyl deactivation effect.

$$\underset{\mathbf{321}}{CH_2COCl\text{–}CH_2COCl} \xrightarrow{PhH,\ AlCl_3} PhCOCH_2CH_2COCl + HCl \quad [219,224]$$

$$\underset{\mathbf{321}}{CH_2COCl\text{–}CH_2COCl} \xrightarrow{2PhH,\ AlCl_3} PhCOCH_2CH_2COPh + 2HCl \quad [219]$$

$$\underset{\mathbf{322}}{CHXCOCl\text{–}CHXCOCl} + 2PhH \xrightarrow{AlCl_3} PhCOCHXCHXCOPh + 2HCl \quad [225]$$

$$\underset{\mathbf{323}}{CH_2CH_2COCl\text{–}CH_2CH_2COCl} + 2PhH \xrightarrow{AlCl_3} CH_2CH_2COPh\text{–}CH_2CH_2COPh + 2HCl \quad [226]$$

$$\underset{\mathbf{324}}{ClCOCH_2SO_2Cl} + PhH \xrightarrow{AlCl_3} PhCOCH_2SO_2Cl \quad [227]$$

325 + 2PhH $\xrightarrow{AlCl_3}$ [228]

3. Careful reinvestigation of the reaction of crotonyl and β-methylcrotonyl chlorides (326a and 326b) with benzene and $AlCl_3$ showed the products to be 3-methylindanone (328a) and crotonophenone (329a) from the former and 3,3-dimethylindanone (328b) and β-methylcrotonophenone (329b) form the latter [219]. Since it has been demonstrated that indanones 328 could be formed

326 + PhH $\xrightarrow{AlCl_3}$ alkylation (a) → 327 $\xrightarrow{AlCl_3}$ 328; acylation (b) → 329

a: $R_1 = CH_3$, $R_2 = H$

b: $R_1 = R_2 = CH_3$

from chlorides 327 but not from ketones 329 upon treatment with $AlCl_3$ under the conditions employed, the alkylation path (a) was assigned for the formation of indanones 328 and the competing acylation path (b) for the formation of unsaturated ketones 329. Again, the survival of 329 under the conditions employed was undoubtedly the result of carbonyl retardation to ring closure. Parallel results were also reported for the reaction of cinnamoyl chloride with benzene and $AlCl_3$ [229].

4. Ketones such as 314 (Eq. 99) [219], 318 (Eq. 100) [219], and 330 (Eq. 101) [230], in which the carbonyl group was not conjugated with the ring, underwent facile closure to the corresponding cyclic ketones.

330 $\xrightarrow{\text{acid catalyst}}$ (101)

Indanones and tetralones have also been prepared by intermolecular Friedel-Crafts cyclizations. An early example is the reported preparation of polyalkyl-2-tetralones by the reaction of arenes with 2,2,5,5-tetramethyltetrahydrofuranone and $AlCl_3$ [231]. By using benzene, toluene, and naphthalene in this synthesis, it was possible to obtain 1,1,4,4-tetramethyl-2-tetralone (331), 1,1,4,4,7-pentamethyl-2-tetralone (332), and 2-keto-1,1,4,4-tetramethyl-1,2,3,4-tetrahydroanthracene (333), respectively [135]. However, due to accompanying rearrangements, considerable amounts of other products also resulted, which amounted to about 50% of the product in the reaction with benzene.

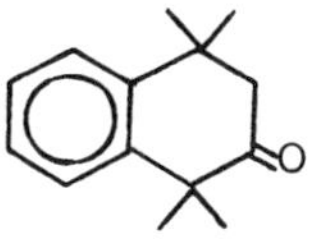

331 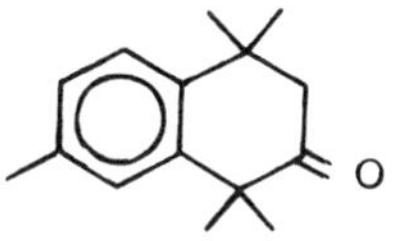 332 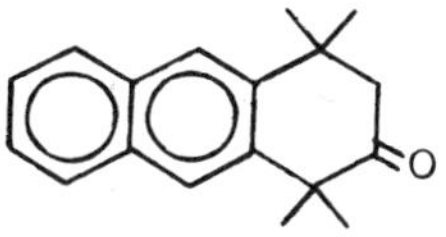 333

As illustrated by the reaction with toluene in scheme 33, the formation of 331-333 was explained by cleavage *a* of 334, remote from the electropositive carbonyl carbon, followed by alkylation to yield intermediate 335, then cyclialkylation to the final product.

In a better procedure, developed by Burckhalter and Campbell [232] for the preparation of 2-tetralone (Eq. 102), and extended by Barclay and co-

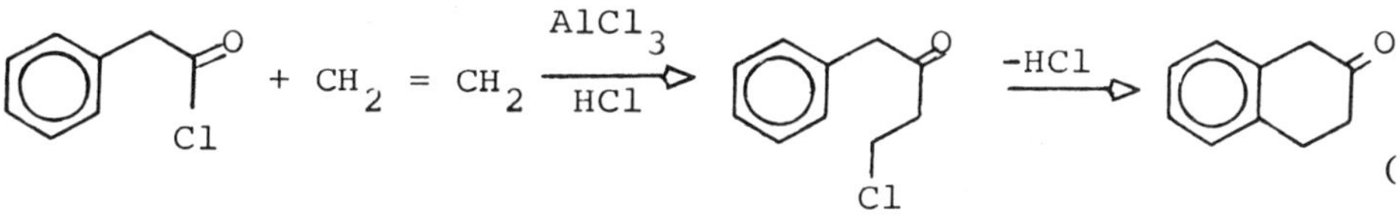

$AlCl_3$, HCl; + $CH_2=CH_2$; Cl; -HCl; Cl; O

(102)

workers [135], polyalkyl-2-tetralones were prepared by a reaction involving

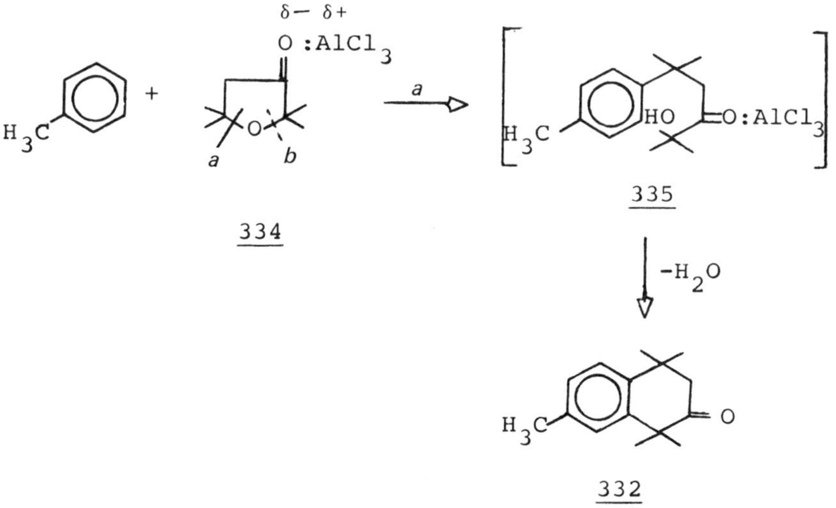

δ− δ+ O :$AlCl_3$; H_3C; + ; *a*; *b*; *a*; HO; O:$AlCl_3$; H_3C; 335; 334; $-H_2O$; H_3C; O; 332

Scheme 33

336 + $(CH_3)_2C{=}CH_2$ $\xrightarrow[-HCl]{SnCl_4}$ 337 $\xrightarrow[BF_3\text{-}OEt_2]{SnCl_4 \text{ or}}$ 338 (103)

336

a: $R_1 = R_2 = CH_3$

b: 336a + $(CH_3)_2C{=}CHCH_3$

c: $R_1 = R_2 = CH_3$; 4-CH_3

d: $R_1 = R_2 = C_2H_5$

e: $R_1 = R_2 = CH_3$; 5,6-benz

337

a $R_1 = R_2 = CH_3$

b $R_1 = R_2 = CH_3$; 5,6-benz

338

a: $R_1 = R_2 = CH_3$

b: $R_1 = R_2 = CH_3$; 3-CH_3

c: $R_1 = R_2 = CH_3$; 6-CH_3

d: $R_1 = R_2 = C_2H_5$

e: $R_1 = R_2 = CH_3$; 7,8-benz

acylation of appropriately branched alkenes by α,α-dialkylarylacyl chlorides followed by Friedel-Crafts cyclization of the resulting unsaturated ketone [233a]. Some examples of the utilization of this synthetic procedure are summarized in Eq. (103).

When attempts were made to use 1,1-diphenylethylene as the alkene in this reaction (Eq. 103) with α-phenylisobutanoyl chloride (336a), the only products identified were the hydrocarbons, 1,1,3-triphenyl-3-methylindan, and 1,1,3-trimethyl-3-phenylindan. The former was attributed to dimerization of the starting alkene and the latter to dimerization of the 2-phenylisopropyl cation, $Ph\overset{+}{C}(CH_3)_2$, resulting from decarbonylation of 336a.

A similar facile procedure for the preparation of β-substituted α-indanones was developed by Bruson and Plant [177]. In 1967, these authors found that carbon monoxide interacts with benzene, $AlCl_3$, and certain polyfunctional halides having at least three carbon atoms in an aliphatic chain to give β-substituted α-indanones by a new type of cyclization reaction. The reaction occurs at room temperature and atmospheric pressure with 1 mol of $AlCl_3$ required per mole of polyfunctional halide employed. Using this method, 2-methyl-2-phenylindanone (339) was obtained in varying yields by reaction of benzene, CO and $AlCl_3$ with either propylene dichloride, allyl chloride, 1,2,2- or 1,1,2-trichloropropane, 2,3-dichloro-1-propene, 1,2-dichloro-1-propene, 1-chloro-1-propene, or (β-chloropropyl)benzene; the best yield was obtained with 1,2,2-trichloropropane (Eq. 104). Similarly, 2,2-dimethylindanone (342) was obtained in good yields from reaction of methallyl chloride, isocrotyl

Table 19. Production of 2,2-Dimethyl-1-indanone by Reaction of Various Aliphatic Halides with PhH, CO and $AlCl_3$

Aliphatic halide		Yield of pure 2,2-dimethylindanone-1, %
$CH_2 = C(CH_3)CH_2Cl$		46-48.5
$(CH_3)_2C = CHCl$		65-68
$(CH_3)_2CClCH_2Cl$		74.5
$CH_2 = CClCH_2CH_3$ $CH_3CCl = CHCH_3$	mixture	48.5
$ClCH_2CHClCH_2CH_3$		61
$CH_3CHClCH_2CH_3$		69.5
$CH_3CHBrCHBrCH_3$		55
$ClCH_2CH_2CHClCH_3$		14
$CH_3CCl_2CH_2CH_3$		63.5

Source: Ref. 177.

chloride, isobutylene dichloride, isobutylene dibromide, 2-chloro-1-butene, 2-chloro-2-butene, or 1,2-, 2,3-, 1,3- and 2,2-dichlorobutane with benzene, CO, and $AlCl_3$ (Table 19). Dimethylindanone (342) was also obtained in even better yields by reaction of neophyl chloride (340) or 2-chloro-2-methyl-3-phenylpropane (341) with CO and $AlCl_3$ (Eq. 105). Also prepared by analogous methods were 2,3-dimethyl-2-phenylindanone (343, Eq. 106), 2,2,3-trimethylindanone (344; Eq. 107), 2,2,3,3-tetramethylindanone (345, Eq. 108), and 10-methyl-1,2,3,4,10,11-hexahydrofluorenone (346, Eq. 109).

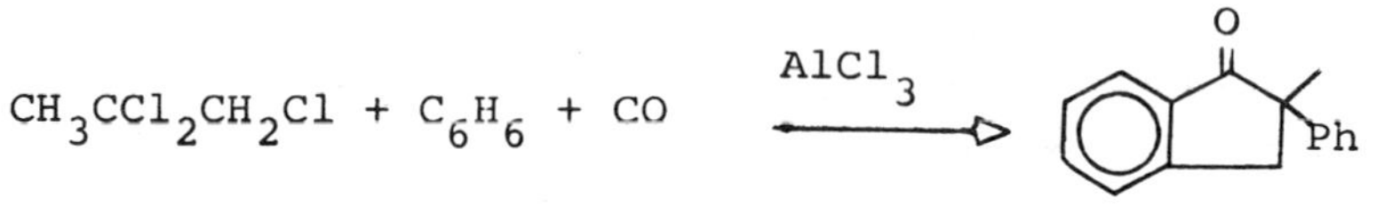

$$CH_3CCl_2CH_2Cl + C_6H_6 + CO \xrightarrow{AlCl_3} \text{339} \quad 58\text{-}60\% \tag{104}$$

$$\text{340} \xrightarrow[81\%\ \text{yield}]{CO + AlCl_3} \text{342} \xleftarrow[60\%\ \text{yield}]{CO + AlCl_3} \text{341} \tag{105}$$

The fact that both the chlorides of Eq. (105) gave the same indanone was attributed to an isomerization of neophyl chloride to 2-chloro-2-methyl-3-phenylpropane under the influence of $AlCl_3$ *prior* to reaction with carbon monoxide to give the product.

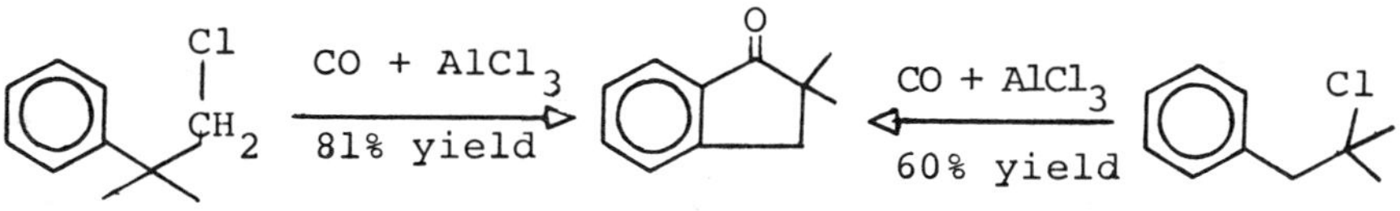

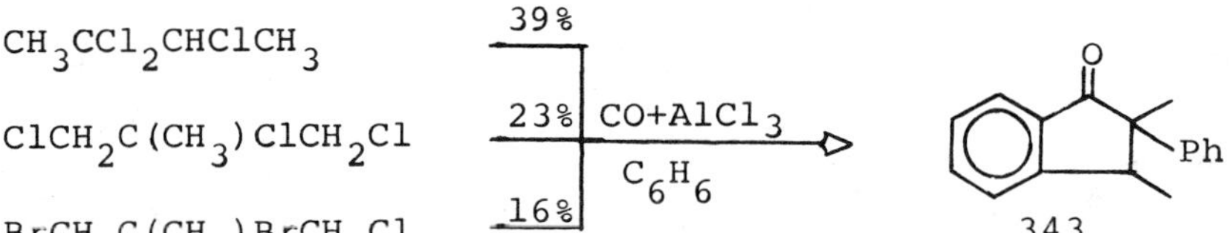

$$\left.\begin{array}{ll} CH_3CCl_2CHClCH_3 & 39\% \\ ClCH_2C(CH_3)ClCH_2Cl & 23\% \\ BrCH_2C(CH_3)BrCH_2Cl & 16\% \end{array}\right\} \xrightarrow[C_6H_6]{CO+AlCl_3} \text{343} \tag{106}$$

$$CH_3\text{-}C(CH_3)(Br)\text{—}CH(Br)CH_3 + C_6H_6 + CO \xrightarrow{AlCl_3} \text{344} \quad 60\% \tag{107}$$

$$\left.\begin{array}{ll} (CH_3)_2CBrCBr(CH_3)_2 & 64\% \\ (CH_3)_3CCCl_2CH_3 & 30\% \\ (CH_3)_3CCl{=}CH_2 & 40\% \\ (CH_3)_3CCHBrCH_2Br & 54\% \end{array}\right\} \xrightarrow{C_6H_6 + AlCl_3 + CO} \text{345} \tag{108}$$

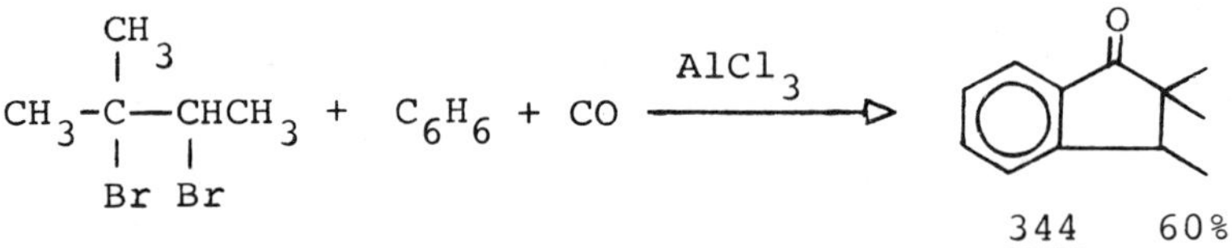

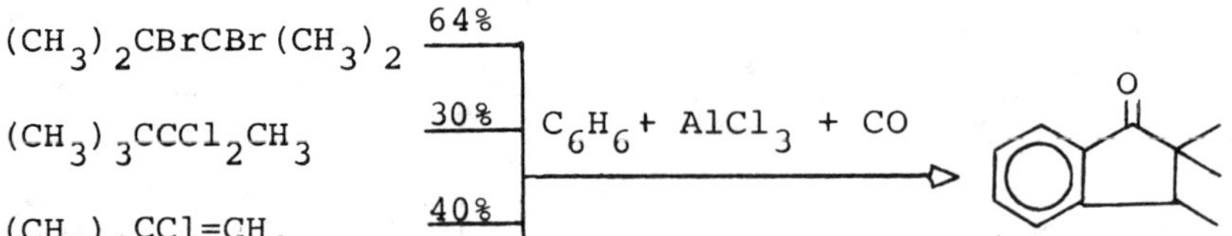

$$\text{C}_6\text{H}_6 + \text{1-chloro-2-methylcyclohexene} + \text{CO} \xrightarrow{\text{AlCl}_3} \underline{246} \qquad (109)$$

As to mechanism, the reactions above were suggested to proceed via initial alkylation of benzene by the polyfunctional halide followed by reaction with formyl chloride (formed in situ) to give an acyl halide, then closure to the product. These steps are illustrated in Eq. (110) by the formation of 2,2-dimethylindanone (342) from the reaction of benzene with methallyl chloride, CO, and $AlCl_3$. The possibility that the reaction above could have occurred

$$\text{PhH} + \text{ClCH}_2\text{C(CH}_3\text{)=CH}_2 \xrightarrow{\text{AlCl}_3} \text{PhCH}_2\text{C(CH}_3\text{)=CH}_2 \xrightarrow[\text{(HCOCl)}]{\text{HCl + CO}}$$

$$\text{PhCH}_2\text{C(CH}_3\text{)}_2\text{-COCl} \xrightarrow{\text{-HCl}} \underline{342} \qquad (110)$$

by initial combination of the organic halide with CO to yield an acyl halide ($RX + CO \rightarrow RCOX$), followed by further reaction with benzene, was excluded on the ground that no carboxylic acids were formed upon hydrolysis of the reaction mixtures, even in the absence of benzene.

In recent years, a number of unconventional procedures have been advanced for the preparation of tetralones by Friedel-Crafts cyclialkylation reactions. In one such reaction, 4-phenyl-2-tetralone (317) was obtained by the action of concentrated hydrochloric acid on 2-(N,N-dimethylamino)-1,4-diphenyl-1,4-butanediol (347) (Eq. 111) [223,224]. The reaction was visualized as involving initial cyclidehydration to tetralol 348 followed by dehydration to enamine 349, then hydrolysis to the 2-tetralone 317 [234].

$$\underline{347} \xrightarrow[\text{-H}_2\text{O}]{\text{H}^+} \underline{348} \xrightarrow{\text{-H}_2\text{O}} \qquad (111)$$

$$\underline{349} \xrightarrow[\text{H}_2\text{O}]{\text{H}^+} \underline{317}$$

In another procedure, 4,4-dialkyl-1-tetralones were obtained in sufficiently good yields by the action of HF-SbF_5 on alkyl phenyl ketones at 40 to 50°C [235]. The results obtained are shown in Eqs. (112) to (114).

HF-SbF_5, 3 hr, 40° — 68% (112)

HF-SbF_5, 6 hr, 40° — 95% (113)

HF-SbF_5, 3 hr, 50°, 95% yield — 82% (113a)

Ph ... + Ph ... 18%

HF-SbF_5 — ~H:$^-$ — (114)

350

~CH_3:$^-$

The results of Eqs. (112) to (114) were accounted for in terms of initial protolysis of the tertiary C-H bond with subsequent ring closure of the resulting carbocation intermediates. Three features of the reactions above should be emphasized: (1) Like other cycliakylations induced by ordinary acid catalysts, reactions induced by superacids also showed preference for six- over seven-membered ring formation. (2) Intramolecular ring closure on carbonyl-deactivated rings is considerably enhanced by superacid catalysis. We have seen in Chap. 3 that intermolecular alkylation of acetophenone with propyl and butyl halides was also facilitated in superacid media. (3) Rearrangements are more extensive with superacids than with ordinary acids. For example, in the reaction of 350 with HF-SbF_5 both methyl and hydride shifts took place *prior* to ring closure (Eq. 114), whereas in the closures of the structurally related 2-chloro- (Table 10) or 2-hydroxy-2-methyl-4-phenylhexane (Table 11 and scheme 25), only a hydride shift was observed, yielding isopropyltetralin [10,18,33,67,89] (Eq. 115).

catalyst → $\sim H:^-$ → $-H^+$ →

X = Cl or OH

(115)

The reason for this difference might be due to some retardation to closure on the part of ketones which allows further rearrangements to occur.

4. Production of heterobicyclic systems by intramolecular reactions

Friedel-Crafts cyclialkylation reactions of aryl-substituted alkanols and alkenes have been used to obtain various heterocyclic systems. These include cyclic ethers, amines, amides, sulfides, sulfones, and other related heterocyclics. In the following, we present some of the reported examples, emphasizing mainly recent ones.

Starting with cyclic ethers, it may be stated that cyclialkylation reactions have long been utilized for their synthesis. In fact, a well-publicized standard synthesis of chromans, especially those substituted in the aromatic ring, involved cyclidehydration of the corresponding 3-aryloxy-1-propanol by boiling with P_2O_5 in benzene. This procedure, which was originally developed by Rindfusz in 1919 [236], was attempted in 1966 by Borowitz and Williams [237] for the cyclization of 4- and 2-*t*-butylphenyl α-hydroxypropyl ethers (351 and 353, respectively). Whereas the former gave the expected 6-*t*-butylchroman (352, Eq. 116) in good yield, the latter gave a mixture of products which contained 8-*t*-butylchroman (354) only as a minor component, 6-*t*-butylchroman (352) as the major component, chroman (355) and *t*-butylbenzene (356) (Eq. 117). The formation of this complex mixture was attributed to the acid-catalyzed reorientation and disproportionation of the highly mobile *t*-butyl group.

351 $\xrightarrow[C_6H_6]{P_2O_5}$ 352 75% (116)

353 $\xrightarrow[C_6H_6]{P_2O_5}$ 354 + 352 + 355 + 356 (117)

Recently (1980), Mendelson et al. [238] reported an efficient synthesis of biologically active tetrahydroquinolines of type 358 from N-(2-hydroxyethyl)benzylamines (357, X = OH) or N-(2-haloethyl)benzylamines (357, X = halogen) (Eq. 118) via intramolecular alkylation by fusion with AlX_3-NH_4X. The results obtained are summarized in Table 20.

357 $\xrightarrow[185^\circ]{AlCl_3/NH_4Cl}$ 358 (118)

Table 20. Synthesis of Tetrahydroisoquinolines from Benzylamines (357)

X	R_1	R_2	R_3	Yield, %
OH	Cl	H	H	60
Cl	H	Ph	H	77
OH	Br	H	H	59
Cl	Cl	Cl	Cl	51
OH	H	Cl	Cl	41[a]

[a]Plus 22% of 5,6 isomer from same starting material.
Source: Ref. 238.

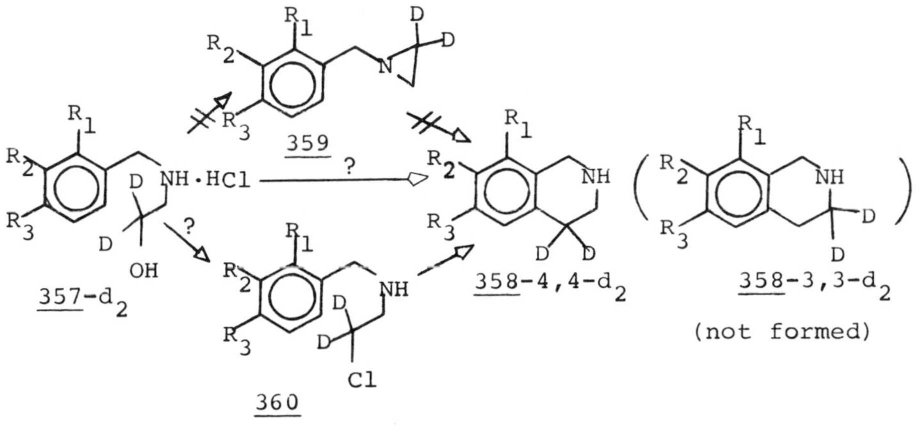

Scheme 34

As indicated by the authors, the preceding method offers several advantages: "It is a simple, direct, one-pot preparation of tetrahydroquinolines in high yield, which is readily applied to large scale work." Based on the finding that reaction of labeled amino alcohols (357-d_2) gave a product consisting entirely of 358-4,4-d_2 with no 358-3,3-d_2, it was possible to rule out the possible intermediacy of aziridines (359) in these reactions (scheme 34). However, the available results did not solve the question of whether or not the closure involved direct displacement of the OH in 357 or of the chloride 360 that could be produced from it by the action of $AlCl_3$.

Another interesting application of cyclialkylation reactions, is their use to prepare the five- and six-membered cyclic amides, oxindoles, and dihydrocarbostyrils [44,239-259]. Starting in 1951, Petyunin and his collaborators [239-255] undertook an extensive study of the conversion of N-arylamides of benzylic acid and derivatives (361) to diaryloxindoles (363) by cyclidehydration with sulfuric acid in the presence of acetic acid (Eq. 119).

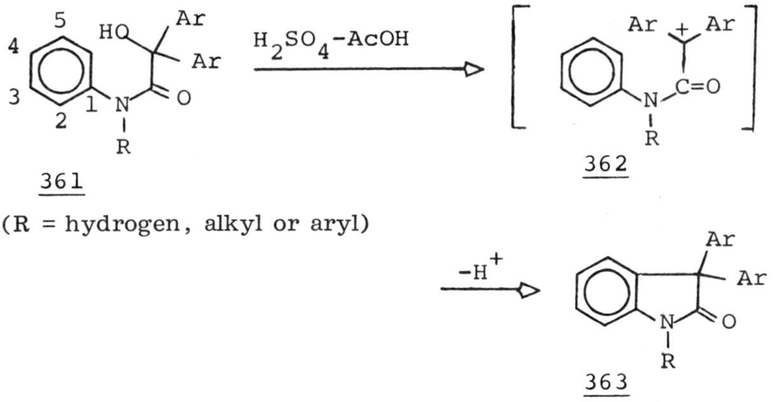

(R = hydrogen, alkyl or aryl) (119)

As suggested [247], the reaction proceeded through formation of a tertiary carbocation intermediate (362) followed by electrophilic attack on the aromatic nucleus.

Similarly, dihydrocarbostyrils of general formula 365 were obtained by the $AlCl_3$-catalyzed cyclialkylations of the arylides of α, β-unsaturated acids having general formula 364 (Eq. 120) [44, 256-258].

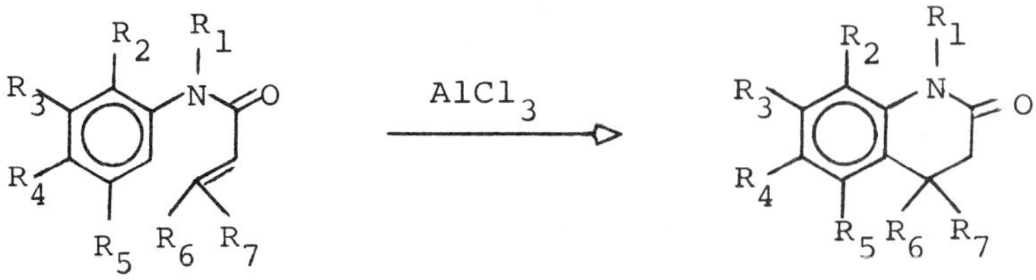

$$364 \xrightarrow{AlCl_3} 365 \quad (120)$$

364

365

Using this procedure, several dihydrocarbostyrils were obtained in which both R_6 and R_7, as well as one of the other R substituents in 365, were methyls [44, 256-258]. However, when the reaction was applied to N-aryl-cinnamide of general formula 366, the reported products were carbostyrils 367 with $AlCl_3$ catalyst [258], but dihydrocarbostyrils (368) with polyphosphoric acid catalyst [259].

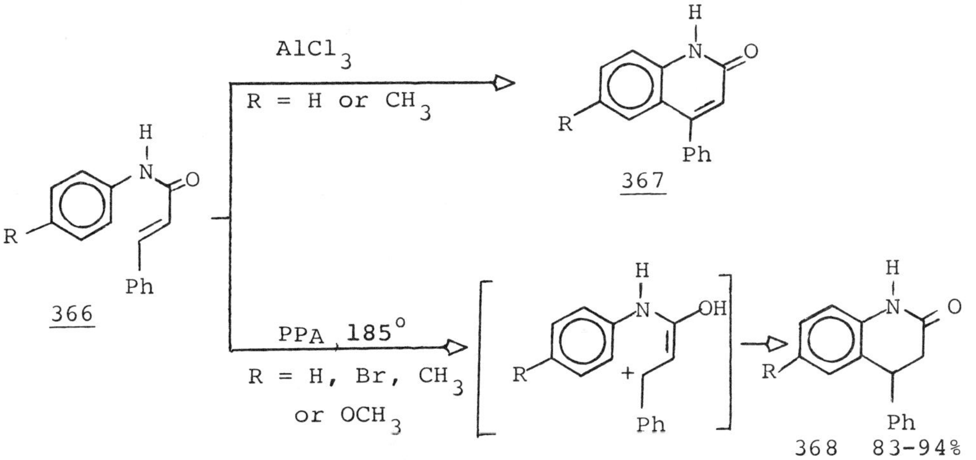

$AlCl_3$, R = H or CH_3 → 367 [258]

PPA, 185°, R = H, Br, CH_3 or OCH_3 → [intermediate] → 368 83-94% [259]

366

367

368 83-94%

Harmon et al. [92] investigated the cyclialkylation behavior of a number of β-phenylethylamide derivatives of general formula 369. In the presence of $POCl_3$ catalyst, these unsaturated amides underwent cyclidehydration to 1-styryl-3,4-dihydroisoquinolines 370 through the Bischler-Napieralski reaction (Eq. 121, Table 21). No products resulting from reaction at the double bond could be found.

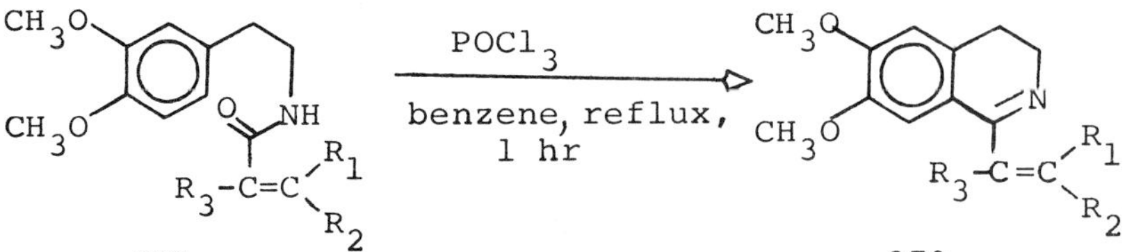

$$369 \xrightarrow[\text{benzene, reflux, 1 hr}]{POCl_3} 370 \quad (121)$$

369

370

Table 21. Results of Cyclidehydration of β-Phenylethylamides 369 to 1-Styryl-3,4-dihydroisoquinolines 370

R_1	R_2	R_3	Yield, %
Ph	Ph	H	78
Ph	*p*-tolyl	H	92
p-tolyl	*p*-tolyl	H	97
p-$C_2H_5C_6H_4$	*p*-$C_2H_5C_6H_4$	H	86
p-ClC_6H_4	*p*-ClC_6H_4	H	54
CH_3	Ph	H	43
H	Ph	CH_3	58
CH_3	CH_3	H	58

In contrast to the behavior of amides 369, the corresponding cinnamic acid amide (217, Eq. 72) failed to undergo the expected cyclidehydration in the presence of $POCl_3$. However, in analogy with the behavior of N-phenylcinnamides 366 [259] indicated above and as shown schematically in Eq. (72), when 217 was heated in polyphosphoric acid it underwent eight-membered ring closure to 8,9-dimethoxy-6-phenyl-3-benzazocin-4-one 218 [92].

To test the generality of cyclization to seven-, eight-, and nine-membered rings, Harmon et al. attempted the cyclization of substituted benzyl-, phenylethyl-, and phenylpropylacrylamides with polyphosphoric acid. However, all these attempts were unsuccessful, as only tars and/or starting materials were isolated from the reactions.

Friedel-Crafts cyclizations of aralkenyl sulfones have similarly been utilized for the synthesis of cyclic sulfones. In 1949, Backer and Dost [260] reported several examples, some of which are shown in Eqs. (122) to (124).

R; SO_2; H_2SO_4; R; SO_2; 70-80% (122)

(R = H or NO_2)

SO_2; H_2SO_4; SO_2 (123)

$$\text{PhSO}_2\text{CH}_2\text{CH=CH}_2 \xrightarrow{BF_3 \text{ in } H_2SO_4} \text{(2-methyl-2,3-dihydrobenzothiophene 1,1-dioxide)} \quad 28\% \tag{124}$$

The results of Eq. (124) seems unlikely in view of previous discussions, as it involves secondary closure on a deactivated benzene nucleus to give a five-membered ring compound.

In 1965, Loev and Kormendy [261], examined the cyclization of ethylenesulfonanilide (371a) and styrenesulfonanilide (371b) in the presence of $AlCl_3$ and/or polyphosphoric acid catalysts. Whereas the former compound failed to cyclize to dihydrosulfostyril (372a), the suitably activated 371b underwent smooth cyclization to give 4-phenyl-3,4-dihydrosulfostyril (372b). These results are shown in Eq. (125).

$$\text{C}_6\text{H}_5\text{-HNSO}_2\text{CH=CHR} \ (\mathbf{371}) \xrightarrow[\text{or PPA}]{AlCl_3} \mathbf{372} \tag{125}$$

a: R = H

b: R = Ph

In addition to the foregoing systematic applications of Friedel-Crafts cyclialkylations to the synthesis of heterocycles, the following examples were also encountered (Eqs. 126 to 129).

$$(\text{2-C}_{10}\text{H}_7\text{CH}_2\text{CH}_2\text{S-})_2 \xrightarrow[C_6H_6]{AlBr_3} \text{(naphthothiophene, dihydro)} \ 35\% + \text{2-C}_{10}\text{H}_7\text{CH}_2\text{CH}_2\text{SH} \ \text{(minor)} \tag{126}$$

[262]

$$\text{C}_6\text{H}_5\text{CH}_2\text{CH}_2\text{SeCl} \xrightarrow{AlCl_3} \text{2,3-dihydrobenzoselenophene} \tag{127}$$

[263]

$$\text{R-C}_6\text{H}_4\text{-NHNHC(=S)-NH}_2 \xrightarrow{PPA} \text{R-(2-aminobenzothiazole)} \tag{128}$$

[264]

R = H, CH_3, Cl, OCH_3

Yield % =85 65 70 14

$$\xrightarrow[20^{\circ}]{H_2SO_4} \quad \xrightarrow[20^{\circ}]{H_2SO_4}$$

(129)
[265]

B. Cyclialkylations Yielding Polycyclic Systems

In the preceding section we considered some of the Friedel-Crafts cyclialkylations used to produce indan and tetralin derivatives. Friedel-Crafts cyclialkylations have also been used to produce higher condensed ring systems, i.e., systems with more than two condensed rings, which we shall refer to here as *polycyclic compounds* or *polycycles*. Polycycles prepared in this manner have long served as intermediates in the preparation of steroids [266,267], alkaloids [268-276], and carcinogenic polycyclic hydrocarbons [277].

Three main routes have been developed for the production of polycycles by Friedel-Crafts cyclialkylations. The first of these routes involves multiple intermolecular alkylations of aromatics with difunctional alkylating agents and can be exemplified by the reactions shown in Eqs. (130) to (138).

$$+ \quad (CH_3)_2C(X)CH_2CH_2C(X)(CH_3)_2 \xrightarrow{AlCl_3} \quad + \quad (130)$$

X = Cl [1,278], OH [279]

$$+ \quad \xrightarrow{AlCl_3} \quad (131)$$

[1]

$$+ \quad \xrightarrow{AlCl_3} \quad (132)$$

[1]

CH_3–C(Cl)(CH_3)–$(CH_2)_3$–CHCl–CH(CH_3)...

(133) [10]

+ $Cl(CH_2)_4Cl$ $\xrightarrow{AlCl_3}$

(134) [280]

+ $Cl(CH_2)_4Cl$ $\xrightarrow{AlCl_3}$

(135) [281]

+ (2 × CH_2Br) $\xrightarrow{AlBr_3}$

(136) [282]

$HOCH_2$ CH_2OH $\xrightarrow[H_2SO_4,\ 25°,\ 20\ hr]{}$

20%

(137) [283]

OH; $BF_3 \cdot OEt_2$; 21 hr, 21% conversion; OH; OH

39.6% 13.2%

(138) [284]

O; OH; other products

4.3% 14.9%

26.1%

The second route involves multiple intramolecular ring closures of aryl-substituted unsaturated long-chain alcohols, acids, acid chlorides, and ethers in the presence of acid catalysts. This route was used repeatedly by Ansell and co-workers [277,285-287] for the preparation of polynuclear hydroaromatic hydrocarbons and polycyclic ketones. Examples illustrating the utility of multiple intramolecular ring closures are shown in Eqs. (139) to (143).

OH; H_2SO_4 or PPA

(139) [277]

O; C-X; X = OH; PPA; HO-C(=O); 73%; O

X = Cl; $AlCl_3$; O; O; 17%

(140) [285]

S; $AlCl_3$ or BF_3 etherate; S

(141) [288]

trans-isomer

$\xrightarrow[CH_2Cl_2]{BF_3\ etherate}$

CH_3 CH_2OH 24% + CH_3 CH_2OH 9% (142) [289]

+ CH_3 OH 3% + Other products

$\xrightarrow[-2H_2O]{HF}$ (143) [290]

$(R = CH_3, C_6H_5)$

The stereochemical course of these multiple ring closures was shown to be dependent on both the nature of the cyclized compound and reaction variables such as type of catalyst and temperature. This dependence is well elucidated by the following double cyclization results obtained for the isomeric arylalkenols 373 (Eq. 144) and 375 (Eq. 145). Isomer 373 was shown by Fetizon and Delobelle [291] to yield the *trans* isomer (*trans*-374) upon treatment with oxalic acid at 140°C, but the *cis* isomer (*cis*-374) upon treatment with polyphosphoric acid at 100°. Isomer 375, however, was shown by Ansell and Gadsky [286] to yield the *trans* isomer (*trans*-374) as the major product of reaction with polyphosphoric acid. Under conditions similar to those used

cis-374 $\xleftarrow[100^\circ]{PPA}$ 373 $\xrightarrow[140^\circ]{(COOH)_2}$ *trans*-374 (144) [291]

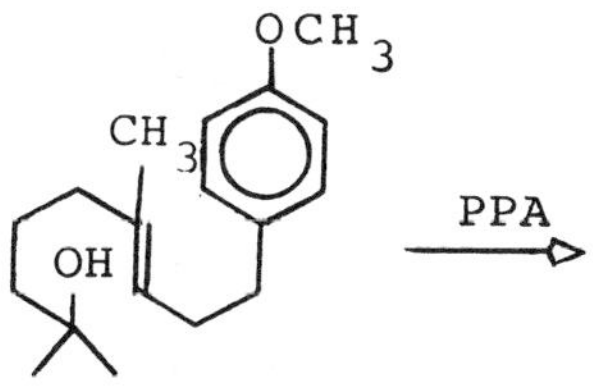

375 —PPA→ *trans*-374 (mainly) + *cis*-374 (145) [286]

for cyclizing 375, arylalkenol 376 gave a mixture of the *cis* and *trans* isomers of 377 (Eq. 146).

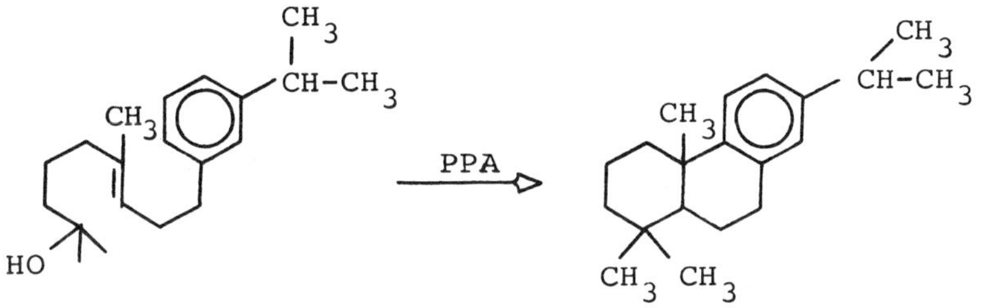

376 —PPA→ 377 (*cis* + *trans*) (146) [282]

The third and most useful route for the preparation of polycycles by Friedel-Crafts cyclialkylation involves the cyclization of aralkylcycloalkanols and aralkylcycloalkenes in the presence of acid catalysts. In fact, this route forms the basis of the convenient synthetic methods adopted by numerous groups of workers for the synthesis of hydrophenathrene derivatives [71-73,292-314]. As represented generally in Eq. (147), all these methods depend fundamen-

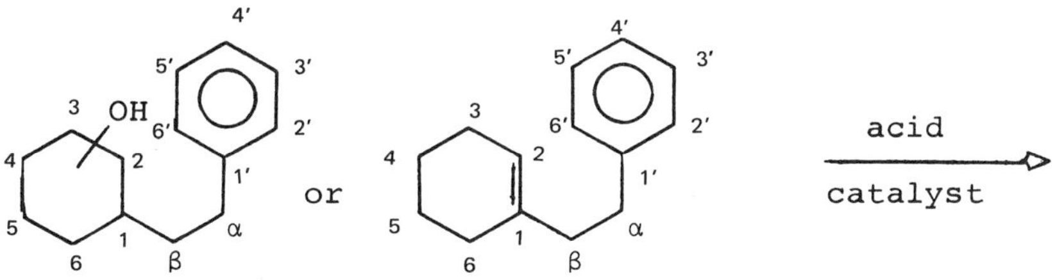

378 or 379 —acid catalyst→ *cis*-380 *trans*-380 381 (147)

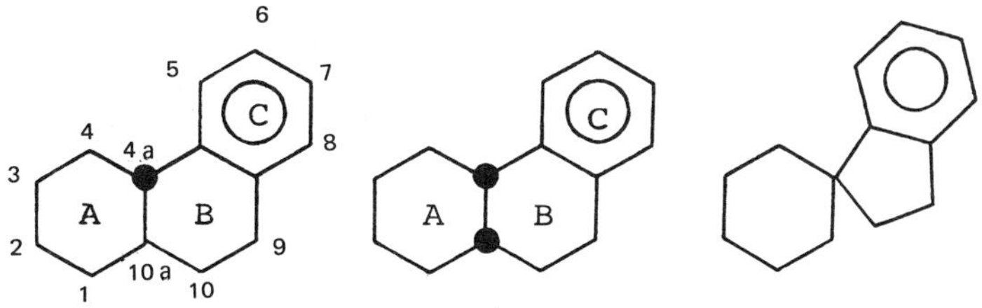

tally on the facile six-membered ring closures of suitably substituted β-arylethylcyclohexanols (378) or β-arylethylcyclohexenes (379) to octahydrophenanthrenes (380).

These octahydrophenanthrenes, which may have *cis* and/or *trans* fusion of rings A and B, were shown to be mixed in some cases, with varying amounts of spiranes having the parent structure 381. The formation of one or more of these isomeric products was shown by Barnes and co-workers [306-310] to depend on various factors, including reaction conditions, catalyst type, and both structural and electronic nature of the cyclized substrate. Generally, under normal cyclization conditions, alcohols (or corresponding alkenes) similar to 382 and 383 yielded hydrophenanthrenes, whereas those similar to 384 and 385 yielded mixtures of hydrophenanthrenes and spiranes. Moreover, the relative yields of spiranes and *cis*-hydrophenanthrenes increased with increasing nucleophilicity of the aromatic nucleus at the closure site.

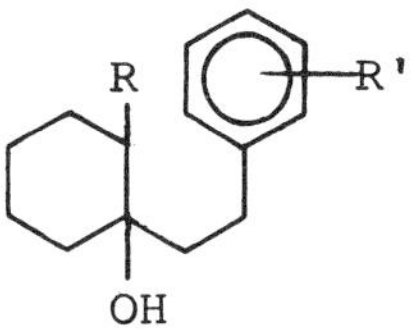

382: R = H

384: R = CH_3

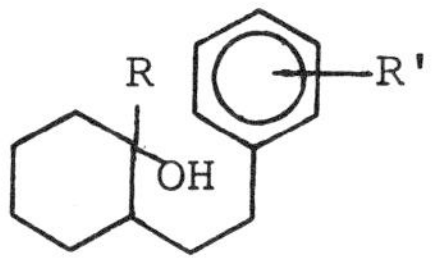

383: R = H

385: R = CH_3

Based on these and other related results, Barnes [308] invoked a mechanism in which various possible cationic intermediates are in competition with each other. For example, he speculated that the cyclizations of alcohols 382 and 383 in which R = H, might involve one or more of four intermediates: oxonium ion 386, bridged ion 387, tertiary ion 388, and secondary ion 389. With a moderately reactive group such as phenyl (382 and 383, R = H) cycliza-

386

387

388

389

tion proceeded mainly via the bridged ion 387 to yield *cis*-octahydrophenanthrene [71,309]. With a more reactive aryl gorup such as *m*-methoxyphenyl (382 and 383, R = *m*-OCH_3) the cyclization proceeded largely via the tertiary carbocation 388 to yield spirane [295]. Finally, with unreactive aryl groups such as *o*-chlorophenyl (382 and 383, R = *o*-Cl), the cyclization proceeded via the higher-energy secondary carbocation 389 to yield *trans*-octahydrophenanthrene [307].

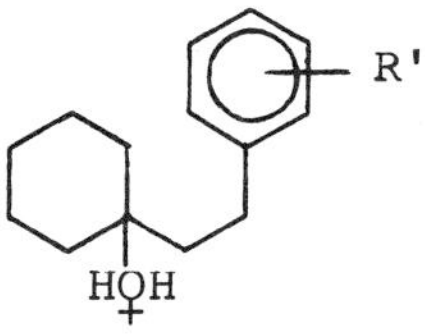

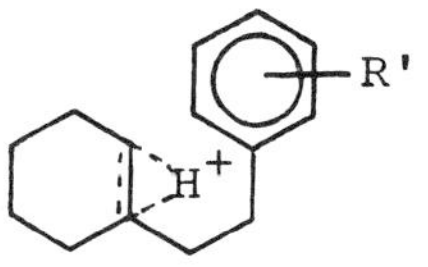

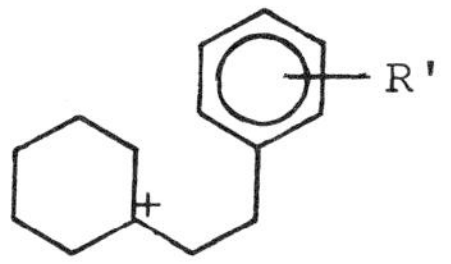

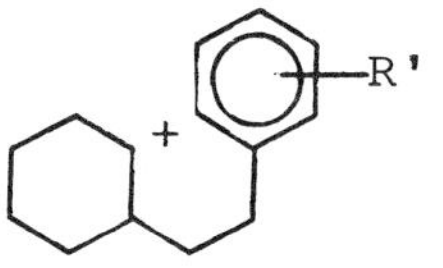

To gain more support for these speculations, Barnes and Olin [310] investigated the cyclization of the optically active alcohol *trans*-2-(2-phenylethyl) cyclohexanol (390). Their finding that the product was the optically active *cis*-octahydrophenanthrene (391, Eq. 148) ruled out the alkene [71,293] and secondary carbocation 389 (R = H) as possible intermediates in this cyclization. Thus it was concluded that the formation of the *cis* isomer must have taken place either by a displacement mechanism or via a bridged ion.

OH $\xrightarrow{H_2SO_4}$ (148)

390 Optically active

391 Optically active, *cis*-fusion

More recently [315], the result of Eq. (148) has alternatively been explained by a consideration of the conformation of the intermediate carbocation in the following manner. If the intermediate is regarded as having the conformation depicted in structure 392 in which the parent cyclohexyl ring

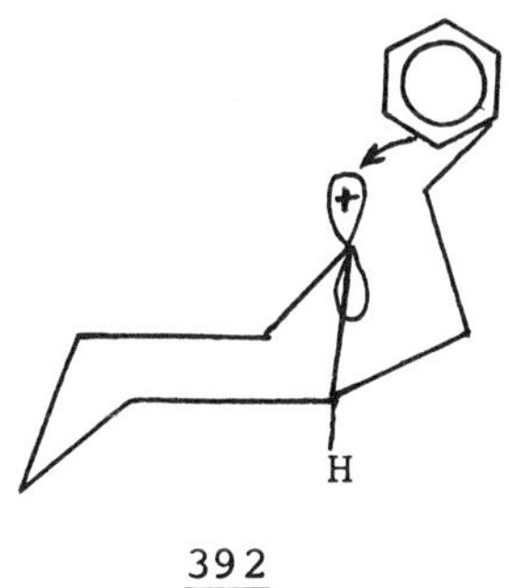

392

has the chair conformation and the bulky substituent prefers the equatorial position, then an easy overlap of the empty *p* orbital of the secondary carbocation of the cyclohexyl ring is obvious. This will therefore result in the observed *cis* fusion. On the other hand, if the intermediate 393 is invoked, in order to produce a *trans* fusion, it is obvious that overlap of the *p* orbital and the carbocation center is not facilitated, so *trans* fusion is not observed. Advantageously, the foregoing treatment, which emphasizes the conformational

H

H

393

requirements for the bond formation via a carbocation, can explain the conversion of 390 to 391 without invoking either a displacement mechanism or a hydrogen-bridged intermediate. Explanations invoking conformationally fixed cations have also been suggested by Goldsmith and Phillips [289] to account for the stereochemistry of the products obtained from the *trans* isomer used as starting material in Eq. (142).

Stereospecific cyclialkylations of aralkylcycloalkanols and related alkenes and alkenones have proved to be potentially useful in approaching the syntheses of some complex organic molecules. Over the years, many examples demonstrating the utility of this method have accumulated in the organic literature [316-329]. Many of these have been reviewed by Barclay [2]. Three more recent examples are described in the following three equations [149 to 151).

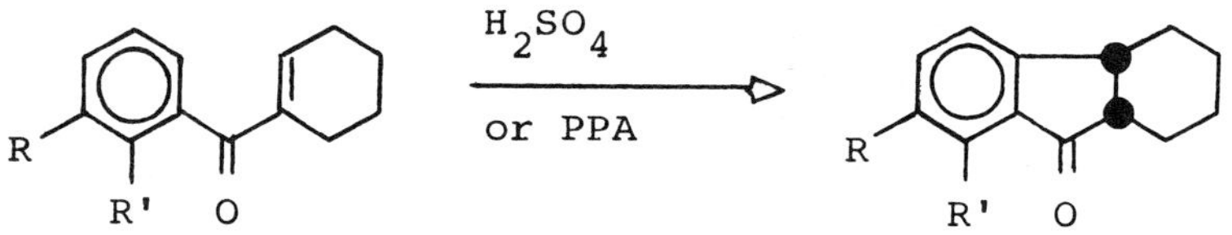

H_2SO_4 or PPA

(149) [206, 322 323, 324]

R = R' = H; H_2SO_4 catalyst [206]

R = H, R' = CH_3; H_2SO_4 catalyst [322]

R = OCH_3, R' = H; PPA [323] or H_2SO_4 catalyst [324]

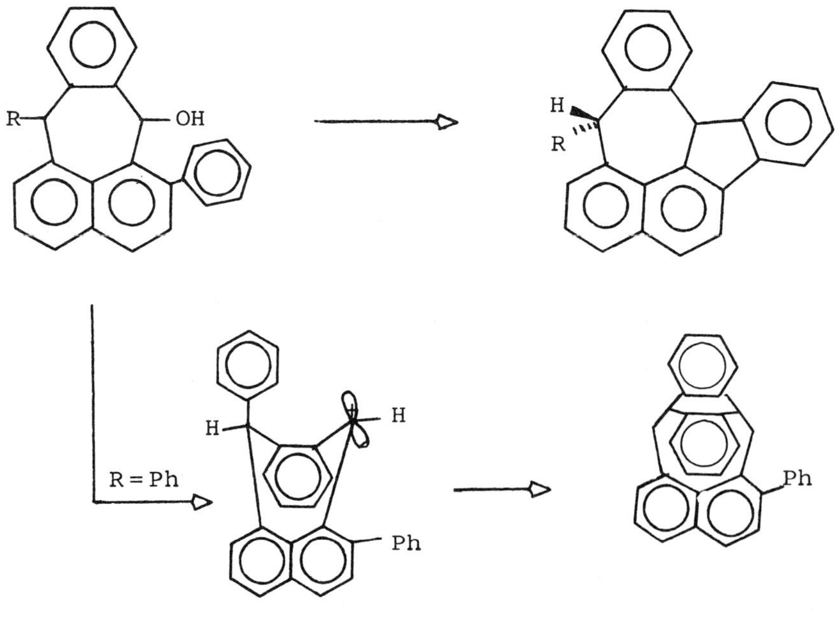

OH

R = Ph

Ph

(150) [327, 328]

R = CH_3, C_2H_5, i-C_3H_7

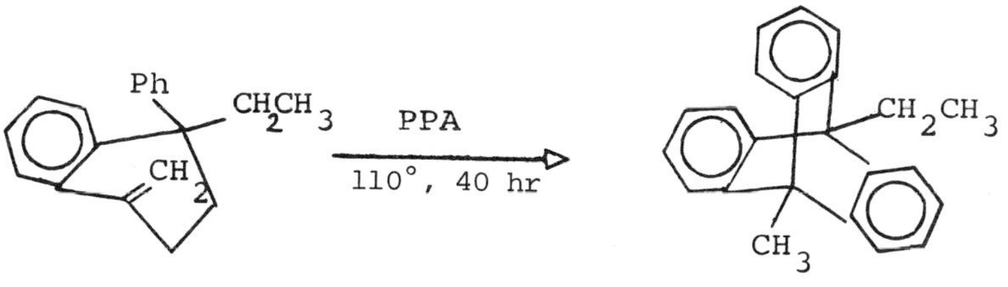

PPA, 110°, 40 hr

(151) [329]

9-Phenyl-9-ethyl-10-methylene-9,10-dihydroanthracene

80%

9-Ethyl-10-methyltriptycene

In 1968 it was reported [330] that reflux of 394 in concentrated HBr brought about hydrolysis of the methoxy groups to produce the indole analog of norlaudanosoline, which cyclized under the conditions being utilized for the hydrolysis to give 395 (Eq. 152).

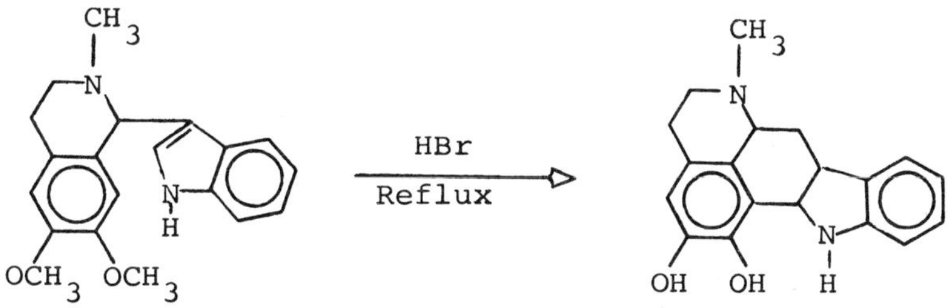

HBr, Reflux

(152)

394 395

Thompson [331] described an interesting route for the synthesis of triptindan derivatives which invoked a double polyphosphoric acid-catalyzed cyclization of the dibenzylindanones 396. Thus treatment of 396a with polyphosphoric acid for 30 min at 100°C provided 400a and 400b in a combined yield of 96% and in a ratio of 3:2. Similar treatment of 396b yielded 400c and 400d in a combined yield of 96% and in the ratio 7:1. The reaction sequence shown in scheme 35 was suggested to account for these transformations.

Low and Roberts [91] prepared a number of bridged polycyclic dibenzobicyclononadienes by acid-catalyzed cyclidehydration of benzyltetralols. According to these workers, treatment of 4-benzyl-1-tetralol (401) with sulfuric or phosphoric acid under a variety of conditions resulted in direct (nonrearranged) cyclization via 402 to 2:3,6:7-dibenzobicyclo[3.2.2]nona-2,6-diene (403), wherein a seven-membered rigid ring system was produced (Eq. 153).

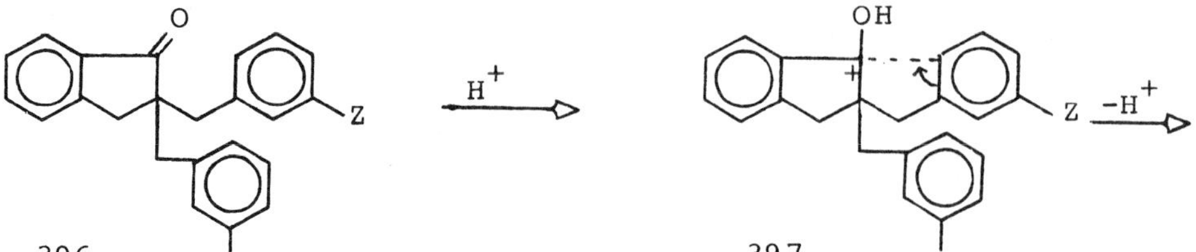

396

a: Y = Z = OMe
b: Y = H; Z = OMe

397

a: Y = Z = OMe
b: Y = H; Z = OMe

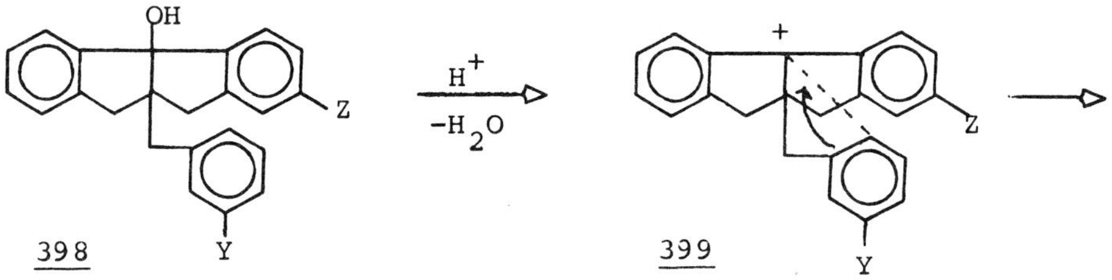

398

a: Y = Z = OMe
b: Y = H; Z = OMe

399

a: Y = Z = OMe
b: Y = H; Z = OMe

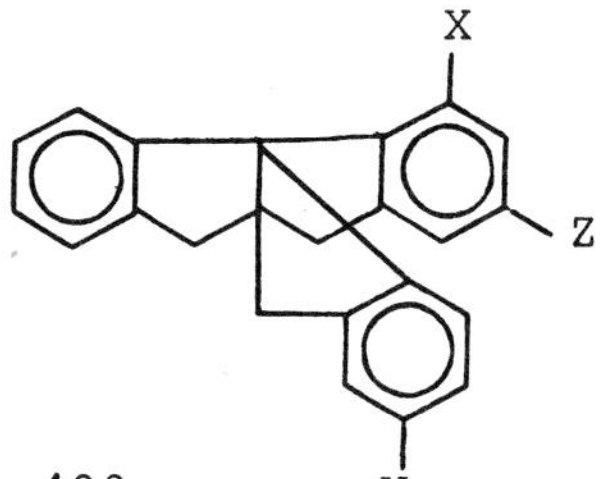

400

a: X = H; Y = Z = OMe
b: X = Y = OMe; Z = H
c: X = Y = H; Z = OMe
d: X = OMe; Y = Z = H

Scheme 35

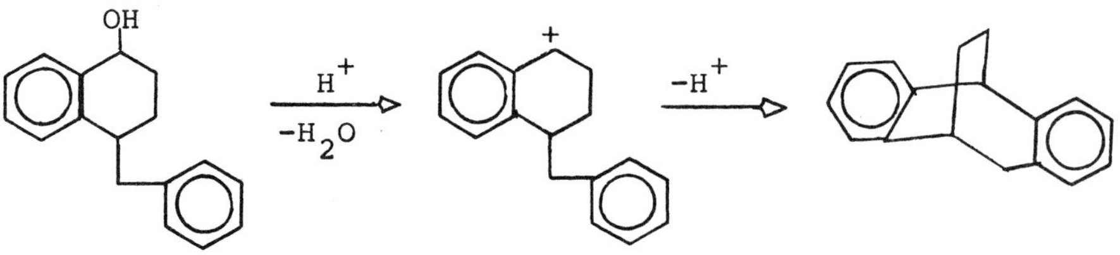

401 402 403 (153)

Similar treatments of 4-benzyl-2-methyl-1-tetralol (404) and 4-benzyl-2,2-dimethyl-1-tetralol (408), however, yielded the rearrangement products, 1-methyl-2:3,6:6-dibenzobicyclo[3.3.1]nona-2,6-diene (407) and 1,8-dimethyl-2:3,6:7-dibenzobicyclo[3.3.1]nona-2,6-diene (411), as shown in Eqs. (154) and (155), respectively.

OH

H^+ / $-H_2O$; $\sim H:^-$

404 405

(154)

$-H^+$

406 407

OH

H^+ / $-H_2O$; $\sim CH_3:^-$

408 409

(155)

$-H^+$

410 411

The differences in behavior between the homologous benzyltetralols 401, 404, and 408 were attributed to a combination of factors, including carbocation stability, ring strain, and steric interactions. Thus the unfavorable rearrangement of carbocation 402, having the charge in benzylic position, to another in which the charge would otherwise be in an ordinary secondary position, resulted in direct closure to the seven-membered ring system 403. On the other hand, the not unfavorable rearrangement of the secondary benzylic carbocation 405 to the tertiary carbocation 406, coupled with the favorable formation of a six-membered ring, accounts for the conversion of 404 to 407. Similarly, the favorable rearrangement of 409 to 410, coupled with six-membered ring formation, accounts partly for the behavior of 408. In addition, direct cyclization in the case of 409 is unfavorable because of the steric hindrance imposed by the *gem*-dimethyl groups, which makes it difficult for the benzylic carbon to approach the reaction site.

It was also noted by Roberts and co-workers [22], that dibenzobicyclononadienes such as those resulting from the cyclization of benzyltetralols can be formed (together with other products) by the action of $AlCl_3$ on benzyltetralins. Equations (156) to (160) summarize the results of treatment of a number of benzyltetralins with $AlCl_3$.

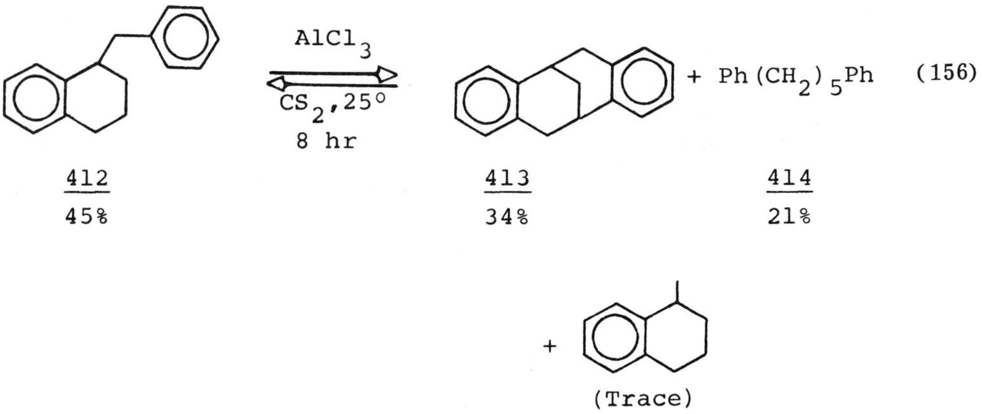

$AlCl_3$, CS_2, 25°, 8 hr

412 45% ⇌ 413 34% + $Ph(CH_2)_5Ph$ 414 21% + (Trace) (156)

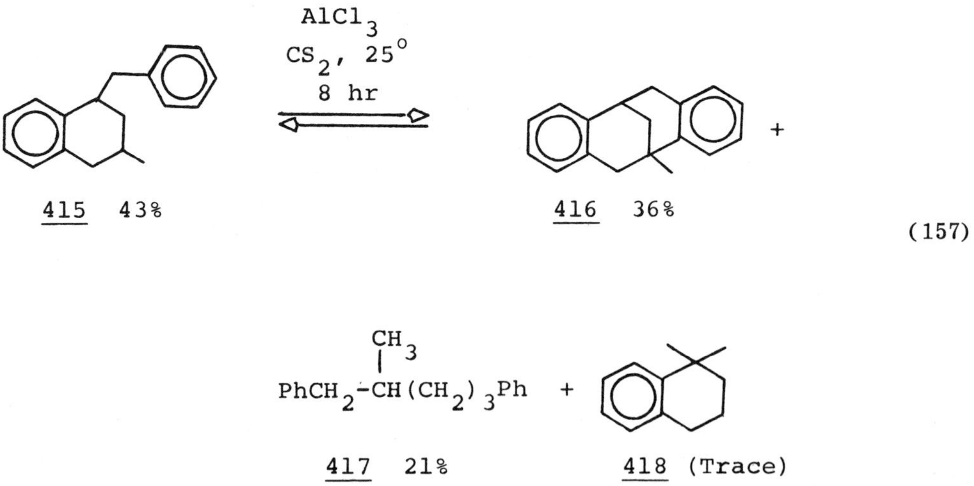

$AlCl_3$, CS_2, 25°, 8 hr

415 43% ⇌ 416 36% + (157)

$PhCH_2-CH(CH_3)(CH_2)_3Ph$ 417 21% + 418 (Trace)

$AlCl_3$, CS_2, 25°, 10 hr

419 → 415 and/or 419 15%; 416 53%; 417 32%; 418 (Trace) (158)

$AlCl_3$, HCl, CS_2, 25°, 48 hr

420 30% → 421 24% + Unidentified 46% (159)

$AlCl_3$, CS_2, 25°, 18 hr

422 21% → 421 30% + Other products 49% (160)

In these reactions, the major products from 1-benzyltetralin (Eq. 156), 1-benzyl-3-methyltetralin (Eq. 157), and 1-benzyl-1-methyltetralin (Eq. 158) were accounted for in terms of scheme 36. According to this scheme, the ring-cleaved products are formed via protonation of the tetralin benzene ring, followed by cleavage of the phenyl-C_1 bond (path A), whereas the bridged bicyclization products are formed via initial hydride abstraction from the alicyclic ring, followed by cyclialkylation of the benzyl ring (path B). The conversions of Eqs. (159) and (160) were accounted for in terms of scheme 37.

The results discussed so far in this section have demonstrated clearly that six-membered ring formation is, as in the case of bicyclic systems, also favored over the formation of smaller or larger rings in the synthesis of polycyclic compounds. Further confirmation of this fact can be found in the cyclizations of related systems. For example, cyclization of benzylcyclohexanols (415) in which the hydroxyl group was placed at position α, 1, 2, or 3 took place at position 3, yielding the bicyclononane system 424 (Eq. 161) [332,333].

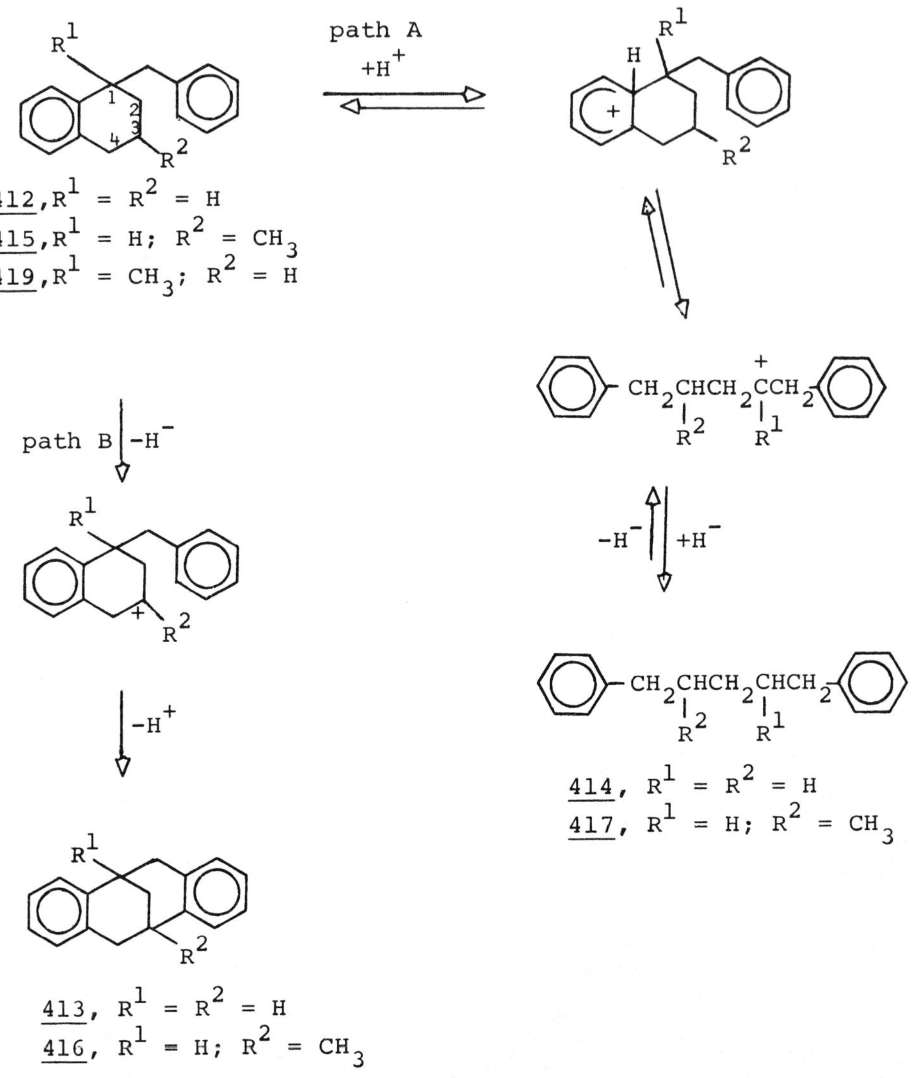

Scheme 36

(161)

423 424

Analogously, the $AlCl_3$-catalyzed cyclization of the alkenes resulting from dehydration of benzylcyclopentanols having the hydroxyl group at position

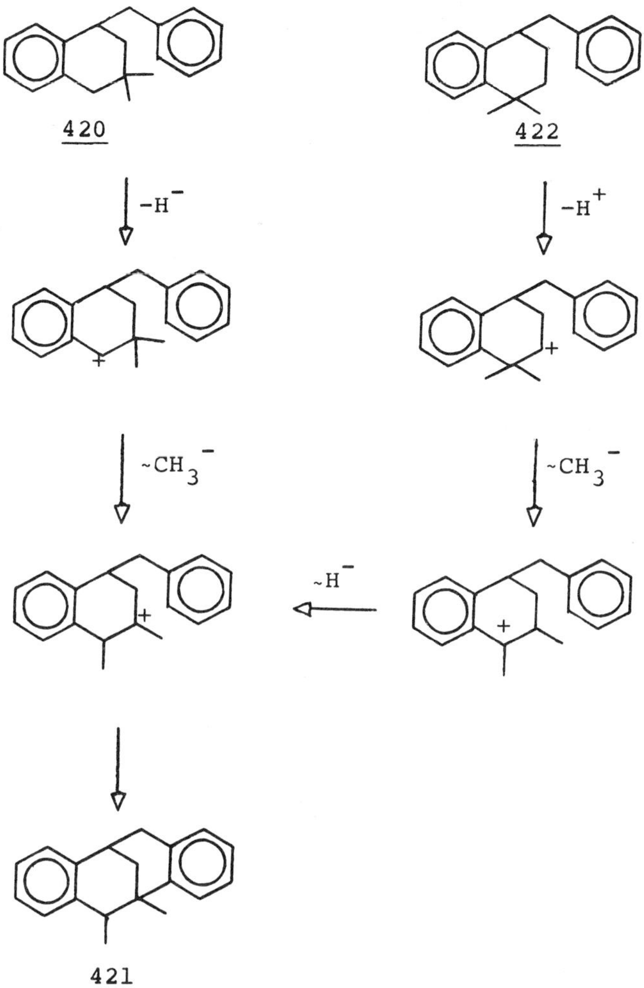

Scheme 37

1 or 2 (425) gave, in all cases, 1,3-*endo*-ethylene-1,2,3,4-tetrahydronaphthalene (426) (Eq. 162) [334].

$$\underset{\mathbf{425}}{\text{(1-benzyl-1-methylcyclopentan-2-ol; positions } \alpha, 1, 2, 3\text{; OH)}} \xrightarrow[\text{(2) } AlCl_3]{\text{(1) } P_2O_5-H_2O} \underset{\mathbf{426}}{} \qquad (162)$$

Further confirmation of the preference for six-membered ring formation in routes to polycyclics can be found in the results of Eqs. (163) to (166), in which direct 2° or 3° closure to a six-membered ring system was always favored over rearranged 3° or 2° closure to smaller or larger ring systems.

(163) [335]

($R = H, CH_3, n\text{-}C_4H_9$)

H_2SO_4

(164) [336]

$R_1 = R_2 = CH_3$

$R_1 = CH_3, R_2 = C_2H_5$

$R_1 = R_2 = C_2H_5$

$R_1 — R_2$ = cyclopentyl

(165) [337]

($R = H, CH_3$)

$AlCl_3, CS_2$ reflux

(166) [337a]

H_3PO_4 $\Delta, -H_2O$ $-H^+$

(167) [338]

However, in accordance with simpler systems, tertiary closures to five-membered rings (Eqs. 168 to 170) and secondary benzylic closures to either five- or seven-membered ring systems (Eqs. 171 and 172, respectively) were observed in cases where competing closures to six-membered ring systems are actually lacking.

R, R', CH_3-C-OH, CH_3 — CH_3COOH-HCl / sealed tube, 200°, 24 hr → R, R' (168) [339]

R = R' = H

R = OH, R' = CH_3

CH_3, OCH_3, CH_3, CH_3-C-OH, CH_3 — CH_3COOH-HCl / sealed tube, 187°, 24 hr → (169) [340]

Amberlyst-15 / toluene, 30 min, 110° → (170) [341]

Ph, Ph, CH-Ph, OH — 97% HCOOH → Ph, Ph Ph (171) [342]

Ph, CH_2, CH_3, CD_2, CH-OH — 97% HCOOH / steam bath, 5 hr → CD_2CH_3 (172) [343]

Analogous to these preparations of polycyclic derivatives, a number of substituted fluorenes were obtained by reactions involving di- and triarylcarbocation precursors [344]. An early example is the reported closure of diphenyl-α-naphthylcarbinol (427) to 9-phenyl-1,2-benzfluorene (428) in almost quantitative yield upon refluxing in trifluoroacetic acid (Eq. 173) [345]. In

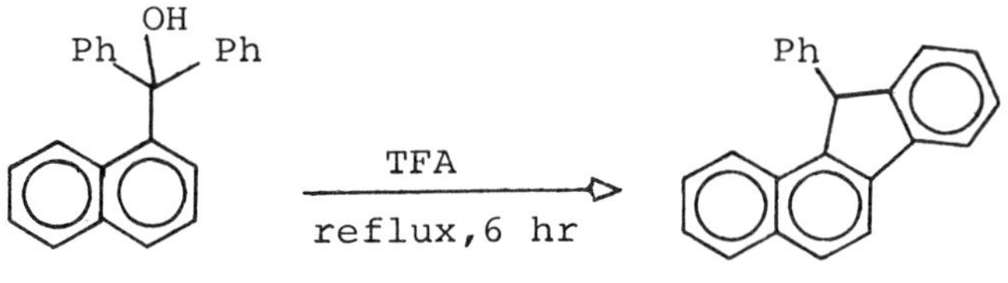

TFA, reflux, 6 hr (173)

427 428

more recent work, Lansbury and co-workers [346] found that the 7,12-dihydropleiadenes 429 and 431 underwent rapid cyclizations to the 1,12-(*o*-phenylene)-7,12-dihydropleiadenes 430 and 432 upon treatment with HCOOH or trifluoroacetic acid catalyst (Eq. 174). This finding was followed by the observations that (1) cyclization of 429 in TFA-d gave 430, which was essentially devoid of

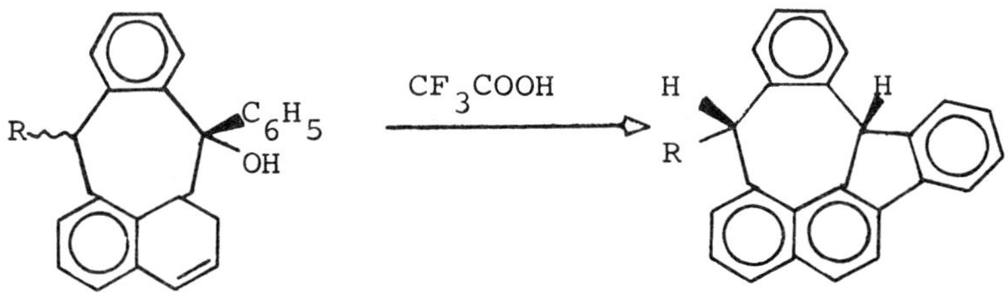

CF_3COOH (174)

429: R = H 430: R = H

431: R = CH_3 432: R = CH_3

deuterium; (2) cyclization of 7-(phenyl-d_5)-7-hydroxy-7,12-dihydropleiadene in TFA gave 430, again with no detectable C_{12}-D; and (3) cyclization of specifically deuterated 429 (429-d_2) in TFA gave 430-d_2 with ≥96% deuterium retention (Eq. 175) [347]. Based on these deuterium tracer results, it was con-

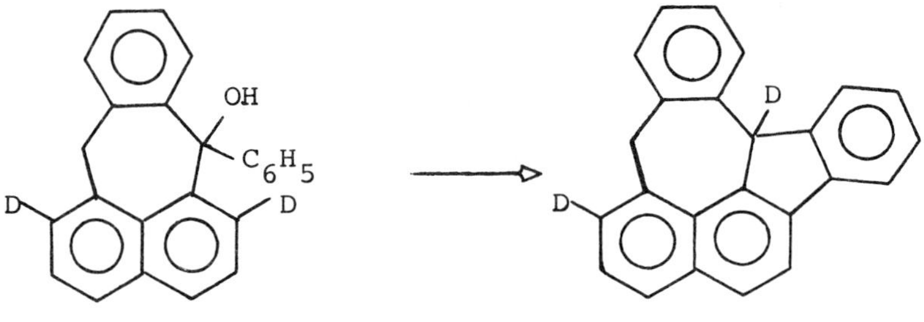

(175)

429-d_2 430-d_2

cluded that the reaction is overwhelmingly intramolecular; that isofluorene 435 is the intermediate intervening *en route* to 430; and that the conversion of 435 to 430 involves a 1,3-hydride shift. Accordingly, the route 429 → 433 → 434 → 435 → 430 of scheme 38 was suggested for this reaction, thus ruling out the intervention of 436.

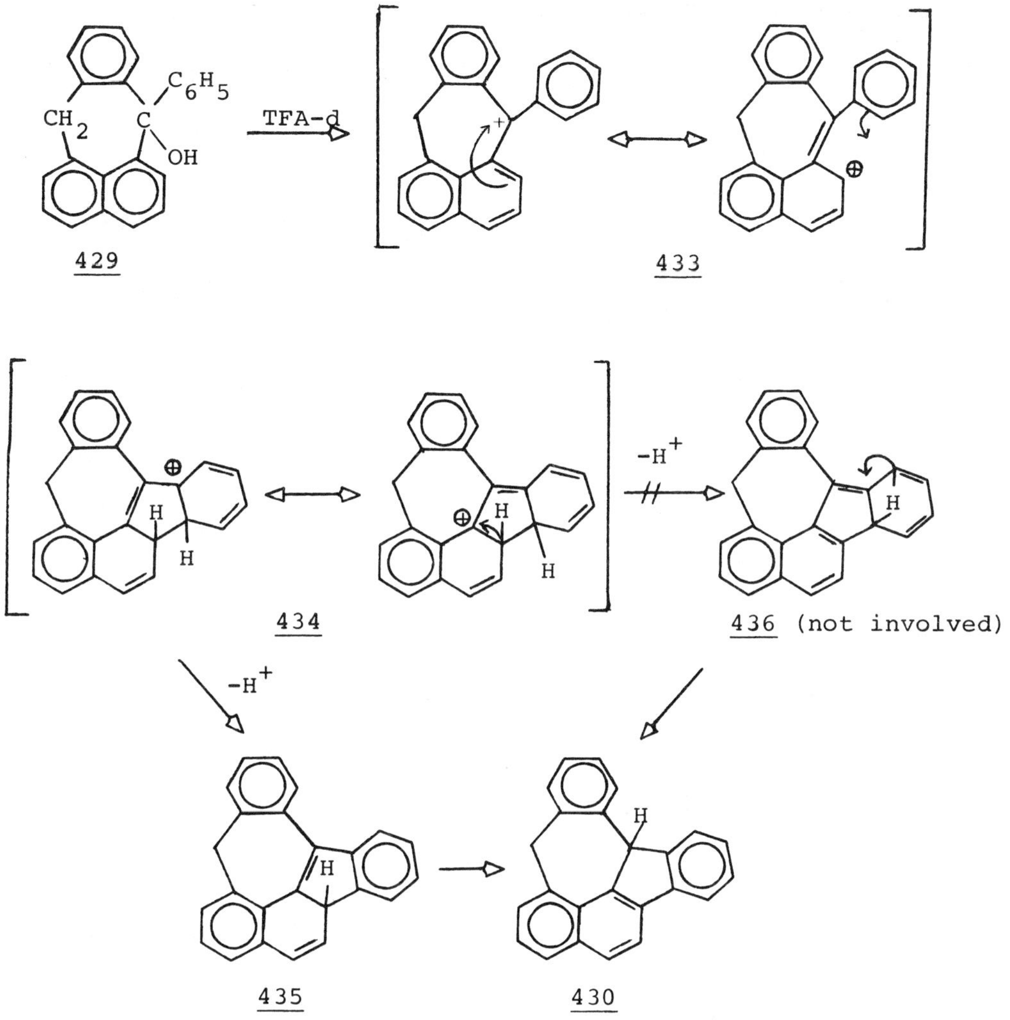

429 TFA-d 433

434 -H+ 436 (not involved)

435 430

Scheme 38

Considering the utilization of cyclialkylation reactions for the preparation of naturally occurring bridged heteropolycyclics, one may cite the reported preparation of isopavine (438) by the action of concentrated sulfuric acid on acetal 437 (Eq. 176) [348].

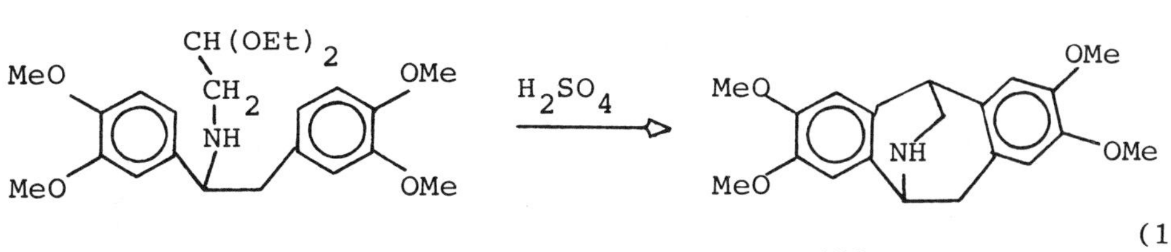

437 $\xrightarrow{H_2SO_4}$ 438 (176)

However, the acetals 439 and 440, although structurally similar to 437, reacted only at one site to give papaverine 441 (Eq. 177) [349]; no analogs of 438 resulting from double-ring closures could be found.

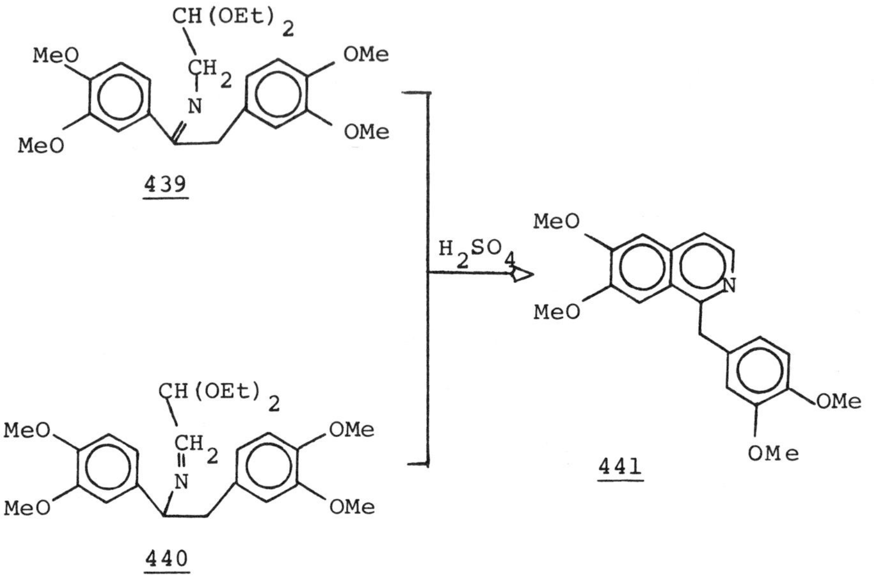

(177)

One step in a recent synthesis of ^{13}C-labeled dibenzo[*a,i*]pyrene involved an aluminum chloride-catalyzed cyclization to form a polycyclic diketone (Eq. 178) [350].

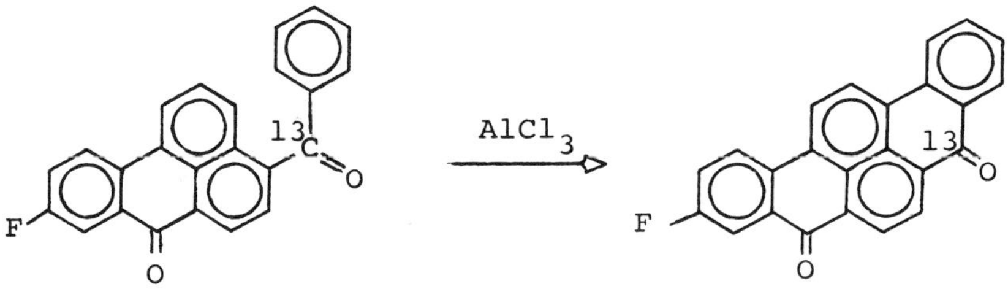

(178)

In concluding this chapter, we can state that Friedel-Crafts cyclialkylations are very important, not only because they offer facile routes for the synthesis of many otherwise difficult to obtain bi- and polycyclic derivatives, but also because they have disclosed much interesting theoretical information regarding the importance of electronic, steric, and ring-size effects in governing cyclialkylation reactions.

REFERENCES

1. Bruson, H., and J. Kroeger, J. Am. Chem. Soc., *62*, 36 (1940).
2. Barclay, L. R. C., in *Friedel-Crafts and Related Reactions*, Vol. 2, (G. A. Olah, ed.), Wiley-Interscience, New York, 1964, Chap. 22.
3. For a review see: Whaley, W. M., and T. R. Govindachari, in *Organic Reactions*, Vol. 6, Wiley, New York, 1960, p. 74.
4. For a review see: Whaley, W. M., and T. R. Govindachari, in Ref. 3, p. 151.
5. For a review see: Gensler, W. J., in Ref. 3, p. 191.
6. Von Braun, J., and H. Deutsch, Chem. Ber., *45*, 1267 (1912); Chem. Abstr., *6*, 2605 (1912).
7. Von Braun, J., and M. Kuhn, Chem. Ber., *60B*, 2557 (1927); Chem. Abstr., *22*, 1146 (1928).
8. Baddeley, G., and R. Williamson, J. Chem. Soc., 4647 (1956).
9. Baddeley, G., and W. Pickles, J. Chem. Soc., 2855 (1957).
10. Barclay, L. R. C., B. A. Ginn, and C. E. Milligan, Can. J. Chem., *42*, 579 (1964).
11. Khalaf, A. A., and R. M. Roberts, J. Org. Chem., *31*, 89 (1966).
12. Roth, W. A., and K. von Auwers, Justus Liebigs Ann. Chem., *407*, 159 (1915).
13. For some recent references to inter- and intramolecular acyl transfers, see: (a) Marx, J. M., J. C. Argyle, and L. R. Norman, J. Am. Chem. Soc., *96*, 2121 (1974); (b) Kagan, J., D. A. Agdeppa, Jr., S. P. Singh, D. A. Mayers, C. Bogajian, C. Poorker, and B. E. Firth, J. Am. Chem. Soc., *98*, 4581 (1976); (c) Domagala, J. M., R. D. Bach, and J. Wemple, J. Amer. Chem. Soc., *98*, 1975 (1977); (d) Andreou, A. D., P. H. Gore, and D. F. C. Morris, J. Chem. Soc., Commun., 271 (1978); (e) Agranat, I., Y. Bentor, and Yu-Shan Shih, J. Am. Chem. Soc., *99*, 7068 (1977); (f) Andreou, A. D., R. V. Bulbulian, and P. H. Gore, J. Chem. Res. Synop., 225 (1980); Chem. Abstr., *93*, 238978v (1980).
14. Bogert, M. T., and D. Davidson, J. Am. Chem. Soc., *56*, 185 (1934).
15. Bogert, M. T., D. Davidson, and P. M. Apfelbaum, J. Am. Chem. Soc., *56*, 959 (1934).
16. Eliel, E. L., *Stereochemistry of Carbon Compounds*, McGraw-Hill, New York, 1964, p. 198.
17. Khalaf, A. A., Rev. Chim. (Bucharest), *19*, 1361 (1974).
18. Khalaf, A. A., and R. M. Roberts, J. Org. Chem., *37*, 4227 (1972).
19. Roberts, R. M., G. P. Anderson, Jr., A. A. Khalaf, and Chow-Eng Low, J. Org. Chem., *36*, 3342 (1971).
20. Anderson, G. P., Jr., Ph.D. dissertation, University of Texas at Austin, 1967, pp. 134-149.
21. Khalaf, A. A., and R. M. Roberts, J. Org. Chem., *38*, 1388 (1973).
22. Roberts, R. M., K. H. Bantel, and Chow-Eng Low, J. Org. Chem., *38*, 1903 (1973).
23. Khalaf, A. A., and R. M. Roberts, J. Org. Chem., *35*, 3717 (1970).
24. Colonge, J., and A. Lagier, C. R. Acad. Sci., *225*, 1160 (1947).
25. Hart, H., private communication with L. R. C. Barclay, cited in Ref. 10.
26. Tilak, B. D., V. N. Gogte, and T. Ravindranathan, Indian J. Chem., *7*, 24 (1969).
27. Barclay, L. R. C., and E. C. Sanford, Can. J. Chem., *46*, 3315 (1968).
28. Barclay, L. R. C., and E. C. Sanford, Can. J. Chem., *46*, 3325 (1968).

29. Corey, E. J., and C. K. Savers, J. Am. Chem. Soc., *79*, 248 (1957).
30. Heck, R., and S. Winstein, J. Am. Chem. Soc., *79*, 3105, 3114 (1957).
31. Douglass, J. E., and R. M. Roberts, J. Org. Chem., *28*, 1225, 1229 (1963).
32. Khalaf, A. A., Indian J. Chem., *12*, 476 (1974).
33. Khalaf, A. A., Rev. Chim. (Bucharest), *18*, 297 (1973).
34. Khalaf, A. A., unpublished data.
35. Von Auwers, K., and E. Hillinger, Chem. Ber., *49*, 2410 (1916).
36. Von Auwers, K., Justus Liebigs Ann. Chem., *439*, 132 (1924).
37. Mayer, F., and L. van Zutphen, Chem. Ber., *57*, 200 (1924).
38. Krollpfeiffer, F., and H. Schultze, Chem. Ber., *57*, 600 (1924).
39. Stolle, R., R. Bergdoll, M. Luther, A. Auerhahn, and W. Wacker, J. Prakt. Chem., *128*, 1 (1930); Chem. Abstr., *25*, 293 (1931).
40. Mayer, F., L. van Zutphen, and H. Phillips, Chem. Ber., *60*, 858 (1927).
41. Wahl, A., and V. Livovschi, Bull. Soc. Chim. Fr., 5(5), 653 (1938); Chem. Abstr., *32*, 7037 (1938).
42. Stolle, R., Chem. Ber., *47*, 2120 (1914).
43. Orito, K., and T. Matsuzaki, Hokkaido Daigaku Kogakubu Kenkyu Hokoku, 41 (1979).
43a. Mathew, K. K., and K. N. Menon, Proc. Indian Acad. Sci., *29A* (1949); Chem. Abstr., *44*, 3998 (1950).
44. Reist, E. J., H. P. Hamlow, I. G. Junga, R. M. Silverstein, and B. R. Baker, J. Org. Chem., *25*, 1368 (1960).
45. Von Braun, J., J. Heider, and E. Muller, Chem. Ber., *50*, 1637 (1917); J. Chem. Soc. Abstr., *114*, 107 (1918).
46. Minisci, F., R. Galli, and M. Perchinunno, Org. Prep. Proc., *1*, 77 (1969).
47. Mayer, F., and P. Muller, Chem. Ber., *60*, 2278 (1927); Chem. Abstr., *22*, 417 (1928).
48. Mayer, F., Ger. Patent 513,204 (1926); Chem. Abstr., *25*, 1260 (1931).
49. Mayer, F., K. Billing, K. Horst, and K. Schirmacher, U. S. Patent 1,754,031 (1930); Chem. Abstr., *24*, 2469 (1930).
50. Mayer, F., Ger. Patent 515,110 (1927); Chem. Abstr., *25*, 1841 (1931).
51. Arnold, R. T., and E. Rondestvedt, J. Am. Chem. Soc., *67*, 1265 (1945).
52. Smith, L. I., and W. W. Prichard, J. Am. Chem. Soc., *62*, 772 (1940).
53. Kishner, N., J. Russ. Phys.-Chem. Soc., *46*, 1411 (1914).
54. Layer, R. W., and L. R. MacGregor, J. Org. Chem., *21*, 1120 (1956).
55. Gore, P. H., in *Friedel-Crafts and Related Reactions*, Vol. 3, (G. A. Olah, ed.), Wiley-Interscience, New York, 1964, Chap. 31.
56. Pines, S. H., and A. W. Douglas, J. Am. Chem. Soc., *98*, 8119 (1976).
57. Pines, S. H., and A. W. Douglas, J. Org. Chem., *43*, 3126 (1978).
58. Paul, R. C., R. Kaushal, and S. S. Pahil, J. Indian Chem. Soc., *44*, 995 (1967), and references therein.
59. Khalaf, A. A., and co-workers, research in progress.
60. Maitte, P., Ann. Chim. (Paris), *9*, 431 (1954); Chem. Abstr., *49*, 6929 (1955).
61. Colonge, J., and P. Boisde, Bull. Soc. Chim. Fr., 1337 (1956).
62. Bogert, M. T., Science, *76*, 475 (1932).
63. Bogert, M. T., Science, *77*, 197 (1933).

64. Bogert, M. T., Science, *77,* 289 (1933).
65. Bogert, M. T., and D. Davidson, J. Am. Chem. Soc., *56,* 185 (1934).
66. Bogert, M. T., D. Davidson, and P. M. Apfelbaum, J. Am. Chem. Soc., *56,* 959 (1934).
67. Roblin, R. O., Jr., D. Davidson, and M. T. Bogert, J. Am. Chem. Soc., *57,* 151 (1935).
68. Oratt, R. M., and M. T. Bogert, J. Am. Chem. Soc., *58,* 2055, 2057 (1936).
69. Prince, D., D. Davidson, and M. T. Bogert, J. Org. Chem., *2,* 540 (1937).
70. Bogert, M. T., and P. M. Apfelbaum, J. Am. Chem. Soc., *60,* 930 (1938).
71. Periman, D., D. Davidson, and M. T. Bogert, J. Org. Chem., *1,* 288, 300 (1936).
72. Periman, D., and M. T. Bogert, J. Am. Chem. Soc., *59,* 2534 (1937).
73. Papa, D., D. Periman, and M. T. Bogert, J. Am. Chem. Soc., *60,* 319 (1938).
74. Colonge, J., and L. Bichat, Bull. Soc. Chim. Fr., 853 (1949).
75. Vig, O. P., N. K. Maheswary, and S. M. Mukherji, Res. Bull. Panjab. Univ., No. 34, 107 (1953); J. Indian Chem. Soc., *34,* 9 (1957).
76. Bardhan, J. C., and D. N. Mukherji, J. Chem. Soc., 4629 (1956).
77. Mukherji, S. M., O. P. Vig., N. K. Maheswary, and S. S. Sandahu, J. Indian Chem. Soc., *34,* 1 (1957).
78. Weber, S. H., D. B. Spoelstra, and E. H. Polak, Recl. Trav. Chim. Pays-Bas, *74,* 1179 (1955).
79. Weber, S. H., J. Stofberg, D. B. Spoelstra, and R. J. C. Kleipool, Recl. Trav. Chim. Pays-Bas, *75,* 1433 (1956).
80. Ipatieff, V. N., H. Pines, and R. C. Olberg, J. Am. Chem. Soc., *70,* 2123 (1948).
81. Pines, H., D. R. Strehlau, and V. N. Ipatieff, J. Am. Chem. Soc., *72,* 1563 (1950).
82. Smith, L. I., and L. J. Spillane, J. Am. Chem. Soc., *65,* 202 (1943).
83. Adkins, H., and J. W. Davis, J. Am. Chem. Soc., *71,* 2955 (1949).
84. Kloetzel, M. C., J. Am. Chem. Soc., *62,* 3405 (1940).
85. Julia, S., and Y. Bonnet, Bull. Soc. Chim. Fr., 419 (1959); Chem. Abstr., *54,* 24578 (1960).
86. Vig, O. P., and S. S. Sandhu, Sci. Cult. (Calcutta), *19,* 331 (1953).
87. Khalaf, A. A., and R. M. Roberts, J. Org. Chem., *34,* 3571 (1969).
88. Khalaf, A. A., and R. M. Roberts, J. Org. Chem., *36,* 1040 (1971).
89. Khalaf, A. A., Rev. Chim. (Bucharest), *19,* 1373 (1974).
90. Giovannini, E., and K. Brandenberger, Helv. Chim. Acta, *56,* 1775 (1973); Chem. Abstr., *80,* 59763f (1974).
91. Low, Chow-Eng, and R. M. Roberts, J. Org. Chem., *38,* 1909 (1973).
92. Harmon, R. E., B. L. Jensen, S. K. Gupta, and J. D. Nelson, J. Org. Chem., *35,* 825 (1970).
93. Peto, A. G., in *Friedel-Crafts and Related Reactions,* Vol. 3, (G. A. Olah, ed.), Wiley-Interscience, New York, 1964, Chap. 34.
94. Drusiani, A., and L. Plessi, Atti Mem. Accad. Patavina Sci., Lett. Arti, *83,* 193 (1970-1971); Chem. Abstr., *78,* 124335t (1973).
95. Johnston, K. M., and R. G. Shotter, Tetrahedron, *30,* 4059 (1974).
96. Brouwer, D. M., J. A. Van Doorn, A. A. Kiffen, and P. A. Kramer, Recl. Trav. Chim. Pays-Bas, *93,* 189 (1974); Chem. Abstr., *82,* 42695d (1974).

97. Mavoungou-Gomes, L., and P. Coffin, C. R. Acad. Sci., Ser. C, *282,* 429 (1976); Chem. Abstr., *85,* 5437n (1976).
98. Hashimoto, I., and R. Takatsuka, Bull. Chem. Soc. Jpn., *50,* 2495 (1977).
99. Jaysukhlal, R. M., M. S. Kamath, and S. Y. Dike, J. Chem. Soc., Perkin Trans. *1,* 2089 (1977).
100. Palmer, M. H., and D. S. Leitch, Tetrahedron, *34,* 1015 (1978).
101. Migachev, G. I., L. V. Eremenko, Ya. G. Urman, A. Kh. Buki, and K. M. Dyumaev, J. Org. Chem. USSR, *15,* 1332 (1979).
102. Shen, T. -Y., R. B. Greenwald, H. Jones, B. O. Linn, and E. B Witzel, U. S. Patent 3,956,363 (1976); Chem. Abstr., *85,* 123658h (1976).
103. Shen, T. -Y., and J. Jones, U.S. Patent 3,954,852 (1976); Chem. Abstr., *85,* 123659j (1976).
104. Okaniwa, M., Y. Murase, H. Ito, and T. Goto, Jpn. Kokai 76-65, 742 (1976); Chem. Abstr., *85,* 123661d (1976).
105. Anaell, M. F., and M. H. Palmer, Q. Rev. Chem. Soc., *18,* 211 (1964).
106. Johnson, W. S., Acc. Chem. Res., *1,* 1 (1968).
107. Groves, J. K., Chem. Soc., Rev., *1,* 73 (1972).
108. Bhargava, S. S., and G. S. Saharia, Indian J. Chem., *13,* 1100 (1975).
109. Cole, J. P., and G. Saint-Ruff, Bull. Soc. Chim. Fr., 2249 (1975); Chem. Abstr., *85,* 5571b (1975).
110. Nagasaka, T., and S. Ohki, Chem. Pharm. Bull., *25,* 3023 (1977); Chem. Abstr., *88,* 105048s (1978).
111. Ninomiya, I., and T. Naito, Heterocycles, *10,* 237 (1978); Chem. Abstr., *91,* 5138q (1979).
112. Brunovlenskaya, I. I., A. A. Kolontsov, and V. R. Skvarchenko, J. Org. Chem. USSR, *15,* 1341 (1979).
113. Csicsery, S. M., J. Catal., *12,* 183 (1968).
114. Winnik, M. A., Chem. Rev., *81,* 491 (1981).
115. Muller, A., J. Org. Chem., *17,* 1077 (1952), and references reported therein.
116. Rosen, M. J., J. Org. Chem., *18,* 1701 (1953).
117. Tadros, W., A. F. Sakla, and S. E. Abdou, J. Chem. Soc., Perkin Trans. *1,* 2839 (1972); Chem. Abstr., *78,* 29337m (1973).
118. Khalaf, A. A., and R. M. Roberts, J. Org. Chem., *31,* 926 (1966).
119. Corson, B. B., J. Dorsky, J. E. Nickels, W. M. Kutz, and H. I. Thayer, J. Org. Chem., *19,* 17 (1954).
120. Petropoulos, J. C., and J. J. Fisher, J. Am. Chem. Soc., *80,* 1938 (1958).
121. Adams, L. M. R., J. Lee, and F. T. Wadsworth, J. Org. Chem., *24,* 1186 (1959); Wadsworth, F. T., and L. M. Adams, U.S. Patent 2,953,609 (1960); Chem. Abstr., *55,* 5448c (1961).
122. Benkeser, R. A., J. Hooz, T. V. Liston, and A. E. Trevillyan, J. Am. Chem. Soc., *85,* 3984 (1963).
123. Greene, R. N., Ph.D. dissertation, University of Texas at Austin, 1965, p. 59.
124. Gibson, T. L., Ph.D. dissertation, University of Texas at Austin, 1972, p. 75.
125. Overberger, C. G., E. M. Pearce, and D. Taner, J. Am. Chem. Soc., *80,* 1761 (1958).

126. Bergmann, E., and H. Weiss, Justus Liebigs Ann. Chem., *480*, 49 (1930).
127. Muller, A., L. Toldy, G. Halmi, and M. Meszaros, J. Org. Chem., *16*, 481 (1951).
128. Muller, A., M. Meszaros, M. Lempert-Streter, and I. Szara, J. Org. Chem., *16*, 1003 (1951).
129. Muller, A., M. Meszaros, K. Kormendy, and A. Kussman, J. Org. Chem., *17*, 787 (1952).
130. Muller, A., and K. Kormendy, J. Org. Chem., *18*, 1237 (1953).
131. Muller, A., and A. Karczag-Wilhelms, Chem. Ber., *87*, 1727 (1954).
132. Muller, A., M. Meszaros, and A. Karczag-Wilhems, Chem. Ber., *87*, 1735 (1954).
133. Baker, W., J. A. Godsell, J. F. McOmie, and T. L. V. Ulbright, J. Chem. Soc., 4058 (1953).
134. Baker, W., C. N. Haksar, J. F. W. McOmie, and T. L. V. Ulbright, J. Chem. Soc., 4310 (1952).
135. Barclay, L. R. C., K. L. Adams, H. M. Foote, E. C. Sandford, and R. H. Young, Can. J. Chem., *48*, 2763 (1970).
136. Boone, D. E., E. J. Eisenbraun, P. W. K. Flanagan, and R. D. Grigsky, J. Org. Chem., *36*, 2042 (1971), and earlier references therein.
137. Paquette, L. A., and T. R. Phillips, J. Org. Chem., *30*, 3883 (1965).
138. Horspool, W. M., R. G. Sutherland, and J. R. Sutton, Can. J. Chem. *48*, 3542 (1970).
139. Rosenblum, M., A. K. Banerjee, N. Danieli, R. Fish, and V. Schlatter, J. Am. Chem. Soc., *85*, 316 (1963).
140. Traylor, T. G., and J. C. Ware, J. Am. Chem. Soc., *89*, 2304 (1967).
141. Roberts, R. M., and M. B. Abdel-Baset, J. Org. Chem., *41*, 1698 (1976).
142. Hoffman, A., J. Am. Chem. Soc., *51*, 2542 (1929).
143. Barnes, R. A., and B. D. Beitchman, J. Am. Chem. Soc., *76*, 5430 (1954).
144. Tsukervanik, I. P., and Kh. Yu. Yuldashev, Zh. Obshch. Khim., *33*, 3497 (1963); Chem. Abstr., *60*, 7635a (1964).
145. Olah, G. A., and N. Friedman, J. Am. Chem. Soc., *88*, 5330 (1966).
146. Curtis, R. F., and K. O. Lewis, J. Chem. Soc., 418 (1962).
147. Barclay, L. R. C., and R. A. Chapman, Can. J. Chem., *42*, 25 (1964).
148. Armburst, H., H. J. Sturm, G. Kilpper, W. Koehler, and H. G. Schecker, Ger. Offen. 2,029,026 (1970); Chem. Abstr., *76*, 72307x (1972).
149. Armbrust, H., G. Kilpper, W. Koehler, H. J. Quadbeck-Seeger, H. G. Schecker, and H. J. Sturm, Ger. Offen. 2,101,089 (1972); Chem. Abstr., *77*, 164307w (1972).
150. Kokhanova, I. V., T. A. Rednikova, S. P. Starkov, F. M. Egidis, A. S. Taranenko, and K. A. Zolotareva, Zh. Org. Khim., *1*, 648 (1965); J. Org. Chem. USSR, *1*, 648 (1965).
151. Paushkin, Ya. M., G. M. Mamedaliev, L. G. Sementsova, and M. V. Kurashev, Neftekhimiya, *9*, 842 (1969).
152. Gadzhibalaev, A. A., and M. I. Arkhipov, Izv. Vyssh. Uch. Zaved., Khim. Khim. Tekhnol., *14*, 472 (1971); Chem. Abstr., *75*, 19841v (1971).

153. Heublein, G., and O. Barth, Z. Chem., *12*, 19 (1972); Chem. Abstr., *76*, 126485z (1972).
154. Miki, H., and H. Hasui, Jpn. Kokai 74-31,652 (1974); Chem. Abstr., *81*, 25316t (1974).
155. Nametkin, N. S., T. G. Veretyakhina, M. V. Kurashev, and G. M. Mamedaliev, Izv. Akad. Nauk SSSR, Ser. Khim., No. 10, 2268 (1975).
156. Yamao, T., and S. Shiga, Jpn. Kokai 75-04,049 (1975); Chem. Abstr., *84*, 43575t (1976).
157. Hasegawa, H., and T. Higashimura, Polym. J., *12*, 407 (1980); Chem. Abstr., *94*, 156425u (1980).
158. Yuldashev, Kh. Yu., and I. P. Tsukervanik, Zh. Obshch. Khim., *32*, 1293 (1962); Chem. Abstr., *58*, 2386 (1963).
159. Furukawa, Y., Tadashi, M., and T. Ishikawa, Jpn. Kokai 75-12, 063 (1975); Chem. Abstr., *82*, 155970h (1975).
160. Furukawa, Y., T. Matsumoto, and T. Ishikawa, Jpn. Kokai 75-12, 064 (1975); Chem. Abstr., *82*, 155971j (1975).
161. Ibrahim, Y. A. and M. M. Eid, Indian J. Chem., *13*, 1098 (1975).
162. Eilingsfeld, H., K. G. Baur, M. Patsch, R. Platz, G. H. Schecker, and M. Fischer, Ger. Offen. 2,256,702 (1974); Chem. Abstr., *81*, 49463q (1974).
163. Benz, G., and E. H. Polak, U.S. Patent 2,851,501 (1952).
164. Polak's Frutal Work, Inc., U.S. Patents 796,129 and 796,130 (1954).
165. Richey, H. G., Jr., R. K. Lustgarten, and J. M. Richey, J. Org. Chem., *33*, 4543 (1968).
166. Ipatieff, V. N., H. Pines, and R. C. Olberg, J. Am. Chem. Soc., *70*, 2123 (1948).
167. Barclay, L. R. C., J. W. Hilchie, A. H. Gray, and (in part) N. D. Hall, Can. J. Chem., *38*, 94 (1960).
168. Barclay, L. R. C., and E. E. Betts, J. Am. Chem. Soc., *77*, 5735 (1955).
169. Heindel, N. D., S. M. Lemke, and W. A. Mosher, J. Org. Chem., *31*, 2680 (1966).
170. Miller, W. G., and C. U. Pittman, Jr., J. Org. Chem., *39*, 1955 (1974), and references therein.
171. Newman, M. S., and G. Kaugars, J. Org. Chem., *30*, 3105 (1965).
172. Colonge, J., and J. Chambion, C. R. Acad. Sci., *224*, 128 (1947).
173. (a) Buddrus, J., and F. Nerdel, Tetrahedron Lett., 3197 (1965); (b) Buddrus, J., Chem. Ber., *101*, 4152 (1968); Chem. Abstr., *70*, 28677v (1969).
174. Skattebol, L., and B. Boulette, J. Org. Chem., *31*, 81 (1966).
175. (a) Grosse, A. V., and V. N. Ipatieff, J. Org. Chem., *2*, 447 (1937); (b) Pines, H., W. D. Huntsman, and V. N. Ipatieff, J. Am. Chem. Soc., *73*, 4343 (1951); (c) Pinnick, H. W., S. P. Brow, E. A. McLean, and L. W. Zoller, III. J. Org. Chem., *46*, 3758 (1981); (d) Khosrovi, M., I. Partchamazad, and M. Fakhrai, Tetrahedron Lett., 2619, 4017 (1975).
176. Gambacorta, A., R. Nicoletti, and A. Oratore, Chim. Ind. (Milan), *54*, 997 (1972); Chem. Abstr., *78*, 124328t (1973).
177. Bruson, H., and H. L. Plant, J. Org. Chem., *32*, 3356 (1967).
178. Failkov, Yu. A., A. P. Sevast'yan, and Y. M. Yagupol'ski, J. Org. Chem. USSR, *15*, 1121 (1979).
179. Vecchionacci, J. P., J. C. Canevet, and Y. Graff, Bull. Soc. Chim. Fr., 7-8(2), 1683 (1974); Chem. Abstr., *82*, 86168p (1975).

179a. Marcuzzi, F., and G. Melloni, Tetrahedron Lett., 2771 (1975).

180. (a) Richey, H. G., and J. M. Richey, in *Carbonium Ions*, Vol. 2, (G. A. Olah and P. v. R. Schleyer, eds), Wiley-Interscience, New York, 1970, p. 899; (b) Hanack, M., Acc. Chem. Res., *3*, 209 (1970); (c) Modena, G., and U. Tonellato, Adv. Phys. Org. Chem., *9*, 185 (1971).

181. Von Auwers, K., and E. Risse, Ann. Chim. (Paris), *502*, 282 (1933).

182. Cohen, A., and J. W. Cook, J. Chem. Soc., 1570 (1935).

183. House, H. O., V. Paragamian, R. S. Ro, and D. J. Wluka, J. Am. Chem. Soc., *82*, 1452 (1960).

184. Koelsch, C. F., and J. A. Anthes, J. Org. Chem., *6*, 558 (1941).

185. Gutsche, C. D., and W. S. Johnson, J. Am. Chem. Soc., *68*, 2239 (1946).

186. Colonge, J., and J. Chambion, Bull. Soc. Chim. Fr., 999 (1947).

187. Blattner, P. A., A. Furst, and K. Jirasek, Helv. Chim. Acta, *30*, 1320 (1947); Chem. Abstr., *42*, 1256 (1948).

188. Burckhalter, J. H., and R. C. Fuson, J. Am. Chem. Soc., *70*, 4184; see Ref. 2, p. 826.

189. Colonge, J., and E. Grimaud, Bull. Soc. Chim. Fr., 857 (1951).

190. Smith, L. I., and P. N. Gordon, J. Am. Chem. Soc., *73*, 3747 (1951).

191. Baker, W., and P. G. Jones, J. Chem. Soc., 787 (1951).

192. Baddeley, G., G. Holt, and S. M. Maker, J. Chem. Soc., 3289 (1952).

193. Johnson, W. S., D. K. Banerjee, W. P. Schneider, C. D. Gutsche, W. E. Shelberg, and L. J. Chinn, J. Am. Chem. Soc., *74*, 2832 (1952).

194. Baddeley, G., G. Holt, S. M. Maker, and M. G. Ivinson, J. Chem. Soc., 3605 (1952).

195. Braude, E. A., and W. F. Forbes, J. Chem. Soc., 2208 (1953).

196. Baddeley, G., S. M. Maker, and M. G. Ivinson, J. Chem. Soc., 3969 (1953).

197. Baddeley, G., and R. Williamson, J. Chem. Soc., 2120 (1953).

198. Khastgir, H., S. M. Mukherji, and B. K. Bhattacharyya, J. Indian. Chem. Soc., *31*, 351 (1954).

199. Parham, W. E., E. L. Wheller, and R. M. Dodson, J. Am. Chem. Soc., *77*, 1166 (1956).

200. Layer, R. W., and I. R. MacGregor, J. Org. Chem., *21*, 1120 (1956).

201. Bergmann, E. D., Bull. Res. Council Isr, *5A*, 150 (1956); Chem. Abstr., *50*, 15493 (1956).

202. Granger, R., M. Corbier, J. Vinas, and P. Nau, C. R. Acad. Sci., *244*, 1048 (1957); Chem. Abstr., *51*, 12048 (1957).

203. Bergmann, E. D., and R. Ikan, J. Am. Chem. Soc., *80*, 5803 (1958).

204. Mahar, S. M., H. F. Bassilios, and A. Y. Salem, J. Chem. Soc., 2437 (1958).

205. Gilman, H., and J. F. Nelson, Recl. Trav. Chim. Pays-Bas, *55*, 518 (1936).

206. House, H. O., V. Paragamian, R. S. Ro, and D. J. Wluka, J. Am. Chem. Soc., *82*, 1457 (1960), and other references therein.

207. Sammour, A., and M. Elkasaby, J. Chem. UAR, *13*, 409 (1970); Chem. Abstr., *77*, 101261f (1972).

207a. Fieser, L. F., and E. B. Hershberg, J. Am. Chem. Soc., *61*, 1272 (1939).

208. Fuson, R. C., W. E. Ross, and C. H. McKeever, J. Am. Chem. Soc., *60*, 2935 (1938).

209. Colonge, J., and G. Weinstein, Bull. Soc. Chim. Fr., 462 (1952); Chem. Abstr., *47,* 3287 (1953).
210. Zakharkin, L., Izv. Akad. Nauk SSSR, Otd. Kim. Nauk, 313, (1956); Chem. Abstr., *50,* 15492 (1956).
211. (a) Buu-Hoi, N. P., and R. Royer, Bull. Soc. Chim. Fr., 812 (1947); Chem. Abstr., *42,* 3382 (1948); (b) Marchant, A., J. Chem. Soc., 3325 (1957).
212. Hart, R. T., and R. F. Tebbe, J. Am. Chem. Soc., *72,* 3286 (1950).
213. Granger, R., M. Corbier, J. Vinas, and P. Nau, C. R. Acad. Sci., *244,* 1048 (1957); Chem. Abstr., *51,* 12048 (1957).
214. Martens, H., and G. Hoonaert, Synth. Commun., *2,* 147 (1972).
215. Martens, H., and G. Hoornaert, Tetrahedron Lett., 1821 (1970).
216. Pines, S. H., and A. W. Douglas, Tetrahedron Lett., 1955 (1976).
217. Murrary, R. J., and N. H. Cromwell, J. Org. Chem., *41,* 3540 (1976).
218. Olah, G. A., *Friedel-Crafts Chemistry,* Wiley, New York, 1973, pp. 104, 114-115, 435-439.
219. Khalaf, A. A., and A. M. El-Khawaga, Rev. Chim. (Bucharest), *26,* 739 (1981).
220. Gore, P. H., and J. A. Hoskins, J. Chem. Soc., C, 517 (1970).
221. Gore, P. H., in *The Chemistry of Acyl Halides,* (S. Patai, ed.), Wiley-Interscience New York, 1972, Chap. 5, pp. 138-141, 157-168, and references therein.
222. March, J., *Advanced Organic Chemistry: Reactions, Mechanisms and Structures,* McGraw-Hill, New York, 1968, pp. 406-415, and references therein.
223. Fine, S. A., and R. L. Stern, J. Org. Chem., *32,* 4132 (1967).
224. Lutz, R. E., and A. W. Winne, J. Am. Chem. Soc., *56,* 445 (1934).
225. Lutz, R. E., J. Am. Chem. Soc., *49,* 1106 (1927).
226. (a) Etaix, L., Ann. Chim. Phys., *9,* 372 (1896); (b) Fuson, R. C., and J. T. Walker, Org. Synth., *13,* 32 (1933).
227. Olah, G. A., unpublished results, cited in Ref. 218, p. 431.
228. (a) Cobb, P. H., Am. Chem. J., *35,* 486 (1906); (b) Freud, M., and K. Fleischer, Ann. Chim. (Paris), *411,* 14 (1916); (c) List, R., and M. Stein, Chem. Ber., *31,* 1648 (1898).
229. (a) Konowalow, M., J. Russ. Phys.-Chem. Soc., *27,* 457 (1895); (b) Simonis, H., and S. Danischewski, Chem. Ber., *59,* 2914 (1926).
230. Oikawa, Y., and O. Yonemitsu, Heterocycles, *8,* 307 (1977); Chem. Abstr., *88,* 105051n (1978).
231. Bruson, H. A., F. W. Grant, and E. Bobko, J. Am. Chem. Soc., *80,* 3633 (1958).
232. Burckhalter, J. H., and J. R. Campbell, J. Org. Chem., *26,* 4232 (1961).
233. For a review of similar reactions, see: Nenitzescu, C. D., and A. T. Balaban in *Friedel-Crafts and Related Reactions,* Vol. 3, (G. A. Olah, ed.), Wiley-Interscience, New York, 1964, p. 1033.
234. Fine, S. A., and R. L. Stern, J. Org. Chem., *35,* 1857 (1970).
235. Yoneda, N., Y. Takahashi, and A. Suzuki, Chem. Lett., 231 (1978).
236. Rindfusz, R. E., J. Am. Chem. Soc., *41,* 667 (1919); (b) Rindfusz, R. E., P. M. Gennings, and V. L. Harnack, J. Am. Chem. Soc., *42,* 157 (1920).
237. Borowitz, I. J., and G. J. Williams, J. Org. Chem., *31,* 603 (1966).

238. Mendelson, W. L., C. B. Spainhour, Jr., S. S. Jones, B. L. Lam, and K. L. Wert, Tetrahedron Lett., *21,* 1393 (1980).
239. Petyunin, P. A., and I. S. Berdinsky, J. Gen. Chem. USSR, *21,* 1859 (1951); Chem. Abstr., *46,* 6638 (1952).
240. Petyunin, P. A., and I. S. Berdinsky, J. Gen. Chem. USSR, *21,* 1877 (1951).
241. Petyunin, P. A., and I. S. Berdinsky, J. Gen. Chem. USSR, *21,* 2249 (1951).
242. Petyunin, P. A., J. Gen. Chem. USSR, *21,* 2455 (1951).
243. Petyunin, P. A., J. Gen. Chem. USSR, *21,* 1853 (1951); Chem. Abstr., *46,* 6638 (1952).
244. Petyunin, P. A., J. Gen. Chem. USSR, *22,* 237 (1952).
245. Petyunin, P. A., J. Gen. Chem. USSR, *22,* 359 (1952).
246. Petyunin, P. A., J. Gen. Chem. USSR, *22,* 761 (1952).
247. Petyunin, P. A., J. Gen. Chem. USSR, *22,* 1029 (1952).
248. Petyunin, P. A., J. Gen. Chem. USSR, *22,* 697 (1952); Chem. Abstr., *47,* 5385 (1953).
249. Petyunin, P. A., and A. S. Pesis, J. Gen. Chem. USSR, *22,* 1187 (1952); Chem. Abstr., *47,* 7490 (1953).
250. Petyunin, P. A., and V. S. Shklyaev, Zh. Obshch. Khim., *23,* 1364 (1953); Chem. Abstr., *48,* 161 (1954).
251. Petyunin, P. A., and V. S. Skhlyaev, Zh. Obshch. Khim., *23,* 853 (1953); Chem. Abstr., *48,* 4439 (1954).
252. Petyunin, P. A., V. S. Shklyaev, and I. S. Berdinsky, Zh. Obshch. Khim., *24,* 1078 (1954); Chem. Abstr., *49,* 8888 (1955).
253. Petyunin, P. A., and V. S. Shklyaev, J. Gen. Chem. USSR, *27,* 805 (1957); Chem. Abstr., *52,* 17229i (1958).
254. Petyunin, P. A., I. S. Berdinsky, and N. G. Panferova, Zh. Obshch. Khim., *27,* 1901 (1957); Chem. Abstr., *52,* 4647h (1958).
255. Petyunin, P. A., V. S. Shklyaev, and A. S. Pesis, J. Gen. Chem. USSR, *27,* 1628 (1957); Chem. Abstr., *52,* 3763h (1958).
256. Smith, L. I., and W. W. Prichard, J. Am. Chem. Soc., *62,* 778 (1940).
257. Colonge, J., and R. Chambard, Bull. Soc. Chim. Fr., 982 (1953).
258. Reist, E. J., H. P. Hamlow, I. G. Junga, R. M. Silverstein, and B. R. Baker, J. Org. Chem., *25,* 1455 (1960).
259. Conley, R. T., and W. N. Knopka, J. Org. Chem., *29,* 496 (1964).
260. Backer, H. J., and N. Dost, Recl. Trav. Chim. Pays-Bas, *68,* 1143 (1949); Chem. Abstr., *44,* 6857 (1950).
261. Loev, B., and M. F. Kormendy, J. Org. Chem., *30,* 3163 (1965).
262. Campaigne, E., and B. G. Heaton, Chem. Ind., 96 (1962); Chem. Abstr., *57,* 4621d (1962).
263. Thibaut, P., L. Christiaena, and M. Reson, C. R. Acad. Sci. Ser. C, *28,* 937 (1975); Chem. Abstr., *84,* 121591m (1975).
264. Kost, A. N., N. Yu. Lebedenko, and L. A. Sviridova, J. Org. Chem. USSR, *12,* 2374 (1976).
265. Shabarov, Yu. S., E. V. Pisanova, and L. G. Saginova, J. Org. Chem. USSR, *17,* 555 (1981).
266. Kon, G. A. R., J. Chem. Soc., 1081 (1933); Harper, S. H., G. A. R. Kon, and F. C. J. Ruzika, J. Chem. Soc., 124 (1934); Gamble, D. J. C., and G. A. R. Kon, J. Chem. Soc., 443, 644 (1935); Kon, G. A. R., and E. S. Narracott, J. Chem. Soc., 672 (1938).
267. Linstead, R. P., Ann. Rep., Prog. Chem. (Chem. Soc. Lond.), *33,* 312 (1936); Springall, H. D., Ann. Rep. Prog. Chem. (Chem. Soc. Lond.), *36,* 288 (1939).

268. Grewe, R., Chem. Ber., *72*, 1314 (1939).
269. Grewe, R., Chem. Ber., *76B*, 1072 (1943).
270. Grewe, R., and A. Mondon, Chem. Ber., *81*, 279 (1948).
271. Sugasawa, S., and S. Saito, Jpn. Anal, 5532, 5533 (1958); Chem. Abstr., *53*, 18061, 18062 (1959).
272. Grewe, R., A. Mondon, and E. Nolte, Ann. Chim. (Paris), *564*, 161 (1949).
273. Rodd, E. H., *Chemistry of Carbon Compounds*, Vol. 4c, Elsevier, New York, 1960.
274. Manske, R. H. F., and H. L. Holmes, *The Alkaloids*, Vol. 2, Academic Press, New York, 1952.
275. May, E. L., and E. M. Fry, J. Org. Chem., *22*, 1366 (1957).
276. Eddy, N. B., J. G. Murphy, and E. L. May, J. Org. Chem., *22*, 1370 (1957).
277. Ansell, M. F., and M. E. Selleck, J. Chem. Soc., 1238 (1956).
278. Barclay, L. R. C., A. H. Gray, and C. E. Milligan, Can. J. Chem., *39*, 870 (1961).
279. Labunskii, I. P., and I. P. Tsukervanik, Dokl. Akad. Nauk SSSR, *80*, 369 (1951); Chem. Abstr., *46*, 5022b (1952).
280. Hausigh, D., and G. Koelling, Chem. Ber., *101*, 469 (1968).
281. Hausigh, D., Chem. Ber., *101*, 473 (1968).
282. Hausigh, D., Chem. Ber., *103*, 659 (1970).
283. Sato, T., M. Waka-ayashi, and K. Hata, Bull. Chem. Soc. Jpn. *43*, 3632 (1970); Chem. Abstr., *74*, 42023t (1971).
284. Balquist, J. M., and E. R. Degginger, J. Org. Chem., *22* 3345 (1971).
285. Ansell, M. F., and S. S. Brown, J. Chem. Soc., 3956 (1958).
286. Ansell, M. F., and B. Gadsby, J. Chem. Soc., 2994 (1959).
287. Ansell, M. F., and J. W. Ducker, J. Chem. Soc., 206 (1961).
288. Canonne, and J. Gourier, C. R. Acad. Sci., Ser. C, *268*(26), 1319 (1969); Chem. Abstr., *71*, 61106b (1969).
289. Goldsmith, D. J., and F. Phillips, J. Am. Chem. Soc., *91*, 5862 (1969).
290. Barclay, L. R. C., and R. A. Chapman, Can. J. Chem., *43*, 1754 (1965).
291. Fetizon, M., and J. Delobelle, C. R. Acad. Sci., *246*, 2774 (1958); Chem. Abstr., *52*, 20092 (1958).
292. Bardhan, J. C., and S. C. Sengupta, J. Chem. Soc., 2520, 2798 (1932).
293. Bardhan, J. C., and R. N. Adhya, J. Chem. Soc., 260 (1956).
294. Bardhan, J. C., and R. N. Adhya, and K. C. Bhattacharyya, J. Chem. Soc., 1346 (1956).
295. Cook, J. W., C. L. Hewett, and A. M. Robinson, J. Chem. Soc., 168 (1939).
296. Colonge, J., and F. Collomb, Bull. Soc. Chim. Fr., *18*, 285 (1951).
297. Mukherji, S. M., V. S. Gaind, and P. N. Rao, J. Org. Chem., *19*, 328 (1954).
298. Vig, O. P., S. V. Kessar, V. P. Kubba, and S. M. Mukherji, J. Indian Chem. Soc., *32*, 697 (1955).
299. Gaind, V. S., R. P. Gandhi, I. C. Lakhumna, and S. M. Mukherji, J. Indian Chem. Soc., *33*, 1 (1956).
300. Gaind, V. S., M. L. Vashisht, and S. M. Mukherji, J. Indian Chem. Soc., *33*, 697 (1956).

301. Gandhi, R. P., K. Chander, O. P. Vig, and S. M. Mukherji, J. Indian Chem. Soc., *34,* 163 (1957).
302. Mukherji, S. M., and N. K. Bhattacharyya, J. Org. Chem., *17,* 1202 (1952).
303. Barnes, R. A., and L. Gordon, J. Am. Chem. Soc., *71,* 2644 (1949).
304. Barnes, R. A., H. P. Hirschler, and B. R. Bluestein, J. Am. Chem. Soc., *74,* 32 (1952).
305. Barnes, R. A., and R. T. Gottesman, J. Am. Chem. Soc., *74,* 35 (1952).
306. Barnes, R. A., H. P. Hirschler, and B. R. Bluestein, J. Am. Chem. Soc., *74,* 4091 (1952).
307. Barnes, R. A., and M. D. Konort, J. Am. Chem. Soc., *75,* 303 (1953).
308. Barnes, R. A., J. Am. Chem. Soc., *75,* 3004 (1953).
309. Barnes, R. A., and M. T. Beachem, J. Am. Chem. Soc., *77,* 5388 (1955).
310. Barnes, R. A., and A. D. Olin, J. Am. Chem. Soc., *78,* 3830 (1956).
311. Slater, S. N., J. Chem. Soc., 68 (1941).
312. Renfrow, W. B., A. Renfrow, E. Shoun, and C. A. Sears, J. Am. Chem. Soc., *73,* 317 (1951).
313. Church, R. F., R. E. Ireland, and J. A. Marshall, Tetrahedron Lett., No. 17, 1 (1960).
314. Phillips, D. D., and A. W. Johnson, J. Am. Chem. Soc., *77,* 5977 (1955).
315. Low, Chow-Eng, Ph.D. dissertation, University of Texas at Austin, 1970.
316. Haworth, R. D., and R. L. Barker, J. Chem. Soc., 1299 (1939).
317. Haworth, R. D., and B. P. Moore, J. Chem. Soc., 633 (1946).
318. King, F. E., T. J. King, and J. G. Topliss, Chem. Ind., 118 (1956).
319. Ghatak, U. R., Tetrahedron Lett., 19 (1959).
320. Barltrop, J. A., and N. A. J. Rogers, J. Chem. Soc., 2566 (1958).
321. Saha, N. N., B. K. Ganguly, and P. C. Dutta, J. Am. Chem. Soc., *81,* 3670 (1959).
322. House, H. O., V. Paragamian, and D. J. Wluka, J. Am. Chem. Soc., *82,* 2561 (1960).
323. Dauben, W. G., and J. W. Collette, J. Am. Chem. Soc., *81,* 967 (1959).
324. House, H. O., T. M. Bare, and W. E. Hanners, J. Org. Chem., *34,* 2209 (1969).
325. Church, R. F., R. E. Ireland, and J. A. Marshall, Tetrahedron Lett., No. 17, 1 (1960).
326. Hodges, R., and R. A. Raphael, J. Chem. Soc., 50 (1960).
327. Lansbury, P. T., J. F. Bieron, and J. A. Lacher, J. Am. Chem. Soc., *88,* 1482 (1966).
328. Lansbury, P. T., J. B. Bieber, F. D. Saeva, and K. R. Fountain, J. Am. Chem. Soc., *91,* 399 (1969).
329. Walborsky, H. M., and T. Bohnert, J. Org. Chem., *33,* 3934 (1968).
330. Morrison, G. C., R. O. Waite, and J. Shavel, Jr., J. Org. Chem., *33,* 1663 (1968).
331. Thompson, H. W., J. Org. Chem., *33,* 621 (1968).

332. Cook, J. W., and C. L. Hewett, J. Chem. Soc., 62 (1936).
333. Challis, A. A. L., and G. R. Clemo, J. Chem. Soc., 1692 (1947).
334. Groves, L. H., and G. A. Swan, J. Chem. Soc., *71,* 2965 (1949).
335. Adkins, H., and G. F. Hager, J. Am. Chem. Soc., *71,* 2965 (1949).
336. Adkins, H., and D. C. England, J. Am. Chem. Soc., *71,* 2958 (1949).
337. Vig, O. P., R. P. Gandi, and R. K. Gulati, J. Indian Chem. Soc., *34,* 281 (1957).
337a. Canonne, P., and A. Regnault, Tetrahedron Lett., 243 (1969); Chem. Abstr., *70,* 87376a (1969).
338. Finger, C., and M. Zander, Chem. Ber., *103,* 1001 (1970).
339. Anchel, M., and A. H. Blatt, J. Am. Chem. Soc., *63,* 1948 (1941).
340. McPhee, W. D., and F. J. Ball, J. Am. Chem. Soc., *66,* 1636 (1944).
341. Burnham, J. W., R. G. Melton, E. J. Eisenbraun, G. W. Keen, and M. C. Hamming, J. Org. Chem., *38,* 2783 (1973).
342. Lewis, G. E., J. Org. Chem., *31,* 749 (1966).
343. Lansbury, P. T., J. A. Lacher, and F. D. Saeve, J. Am. Chem. Soc., *89,* 4361 (1967).
344. For reviews of earlier literature, see: (a) Ref. 2, pp. 843-847; (b) Bergmann, F., and S. Israelashvili, J. Am. Chem. Soc., *68,* 354 (1946).
345. (a) Guyot, A., and A. Kovache, C. R. Acad. Sci., *155,* 838 (1912); (b) S. T. Bowden, W. L. Clarke, and W. E. Harris, J. Chem. Soc., 874 (1940).
346. Lansbury, P. T., J. F. Bieron, and A. J. Lacher, J. Am. Chem. Soc., *88,* 1482 (1966); (b) Lansbury, P. T., J. B. Bieber, F. D. Saeva, and K. R. Fountain, J. Am. Chem. Soc., *91,* 399 (1969).
347. Lansbury, P. T., and K. R. Fountain, J. Am. Chem. Soc., *90,* 6544 (1968).
348. (a) Waldmann, E., and C. Chwala, Justus Liebigs Ann. Chem., *609,* 125 (1957); (b) Battersby, A. R., and D. A. Yeowell, J. Chem. Soc., 1988 (1958).
349. (a) Fritsch, P., Justus Liebigs Ann. Chem., *329,* (1903); (b) Schlittler, E., and J. Muller, Helv. Chim. Acta, *31,* 914 (1948).
350. (a) Sardella, D. J., H. A. Mariani, P. Mahathalang, and E. Boger, J. Labelled Comp. Radiopharm., *16,* 633 (1979); (b) Sardella, D. J., P. Mahathelang, H. A. Mariani, and E. Boger, J. Org. Chem., *45,* 2064 (1980).

7

Transalkylations and Reorientations of Arenes Induced by Friedel-Crafts Catalysts

I. EARLY REPORTS OF TRANSALKYLATIONS AND REORIENTATIONS OF METHYL-ETHYL-, AND *n*-PROPLYBENZENE

The disproportionation of an arene is that type of reaction in which an alkylbenzene is converted into a mixture of benzene and dialkylbenzene; higher alkylbenzenes may also be produced by further reaction of the dialkylbenzene with benzene and alkylbenzene. *Transalkylation* is a term which may be used synonymously with disproportionation, but is better used for the more general transfer of alkyl groups between aromatic rings. *Reorientation* is the term we shall use for reactions such as the conversion of *p*-xylene to *m*-xylene and/or *o*-xylene, to distinguish these reactions in which changes in the orientation of two or more side chains on an aromatic ring occur from other types of molecular rearrangements.

The first disproportionations were described by Anschütz in 1886 [1]. Toluene was heated with aluminum chloride at reflux and was found to be converted into a mixture of benzene and xylenes, said to be mainly the *meta* and *para* isomers. Similarly, ethylbenzene was found to produce benzene and a mixture of *m*- and *p*-diethylbenzenes. *m*-Xylene gave benzene, *p*-xylene, and tri- and tetramethylbenzenes, reported to be the 1,3,5-, 1,2,4-, and 1,3,4,6-isomers, respectively. It should be noted that the products from *m*-xylene are indicative of both disproportionation and reorientation reactions. We shall see that in later work it has been demonstrated that reorientation of xylenes can be produced without any accompanying disproportionation.

A few years later, Heise and Töhl reported the disproportionation of *n*-propylbenzene by heating it with aluminum chloride at 100°C, stating that both the recovered propylbenzene and the dipropylbenzene contained only *n*-propyl groups [2]. Their experimental evidence for the structure of the propyl groups was less than convincing.

It was early noted that Friedel-Crafts alkylations and disproportionations often produced a high proportion of *m*- and 1,3,5-oriented alkylated arenes, especially when aluminum chloride catalysts were used in large amounts. These results were generally explained in terms of either initial formation of 1,2,4-trialkylbenzenes followed by transalkylation, as shown in Eq. (1).

$$\text{1,2,4-}C_6H_3R_3 + C_6H_6 \longrightarrow \text{1,3-}C_6H_4R_2 + C_6H_5R \qquad (1)$$

or in terms of isomerization of initially formed 1,2- and/or 1,4-dialkylbenzenes, as shown in eq. (2) [3]. Actually, we shall see that both of these processes

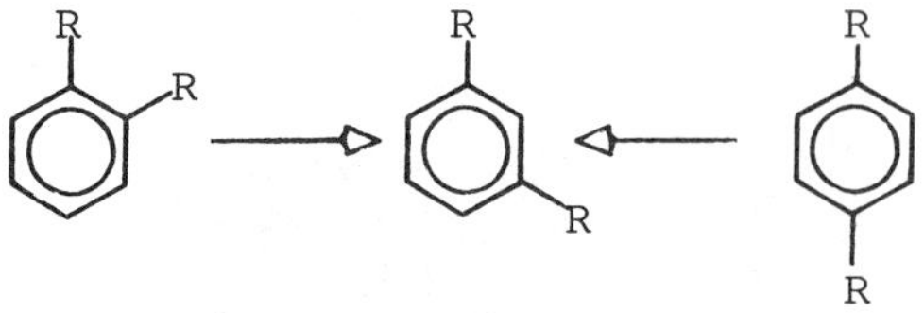

(7.2)

may be operative, depending on the nature of the alkyl groups (cf. Sec. III B).

In 1935, Baddeley and Kenner presented convincing evidence that the products of disproportionation of *p*-di-*n*-propylbenzene by aluminum chloride were *n*-propylbenzene, *m*-di-*n*-propylbenzene, and 1,3,5-*n*-propylbenzene [4]. Nightingale and her co-workers [5] studied the reorientation of 1,3-dimethyl-4-alkylbenzenes to 1,3-dimethyl-5-alkylbenzenes in the presence of aluminum chloride and reported that the migrating alkyl groups (*n*-propyl and *n*-butyl) underwent rearrangement [5(a),(b)]. However, this result was later shown to be in error [5(c),(d)], the alkyl groups actually being found to have the same structure after the reorientation. McCaulay and Lien described the disproportionation of *n*-propylbenzene with hydrogen fluoride-boron trifluoride; they found no rearrangement of *n*-propyl groups to isopropyl groups when the reactions were carried out at temperatures between 5 and 25°C [6]. Kinney and Hamilton found no rearrangement in the disproportionation of *n*-butylbenzene by aluminum chloride at 100°C [7]. The primary interest in all of these experiments was in whether or not the *n*-propyl and *n*-butyl groups would undergo rearrangement in the course of their transalkylations and reorientations, and the finding that they did not was somewhat surprising in view of the known rearrangement of these groups accompanying alkylations with the corresponding alkyl halides.

II. TRANSALKYLATIONS AND REORIENTATIONS OF METHYLARENES

A. Reorientations

Methylarenes are unique in that they may undergo reorientations without any observable transalkylations. This has been demonstrated by several investigators. Baddeley et al. reported that a mixture of *p*-xylene and aluminum bromide kept at room temperature for 1 day gave no change in the hydrocarbon, but in the presence of hydrogen bromide, reorientation occurred to produce a mixture of the three xylene isomers (in a ratio of ca. 67:27:2, *m*/*p*/*o*), with no appreciable formation of transalkylation products [8]. They observed that under the same conditions ethylbenzene and propylbenzene gave disproprotionation products. The mechanism of the reorientation of the methyl group was described as a "Wagner-Meerwein rearrangement." In the same year (1952), McCaulay and Lien reported experiments in which rate studies were made of the reactions of xylenes, trimethylbenzenes, and tetramethylbenzenes catalyzed by hydrogen fluoride and boron trifluoride [9]. *o*- and *p*-Xylenes were found to isomerize to *m*-xylene at low temperature (3°C), with no accompanying disproportionation. The amount of catalyst was found to have a pronounced effect on the composition of the isomerized equilibrium mixture. At low BF_3 concentrations the *m*-xylene content of 60% was in agreement with the thermodynamically calculated value of 57%, whereas at high concentrations the *m*-xylene content approached 100%. This result was attributed to the greater stability of the *m*-oriented σ complex, in which the added proton is *ortho* or *ortho* and *para* to both methyl groups of the xylene.

Owing to the value of *p*-xylene as a starting material for the production of terephthalic acid, which is used in the synthesis of polyester fibers and films, it was of interest to show that *m*-xylene could be isomerized to *p*-xylene in an economically feasible way. Boedeker and Erner [9a] explored this possibility and demonstrated that vapor-phase isomerization at 515°C over a synthetic silica-alumina catalyst would convert a feedstock consisting of *o*-, *m*-, and *p*-xylene in a mole percent ratio of 5:84:9 (plus 2 mol% ethylbenzene) into a mixture of xylenes having the mole percent ratio of 7:60:21, respectively (plus 11% ethylbenzene), with only 1% of toluene produced by disproportionation.

Brown and Jungk studied the reorientation of methyl-, ethyl-, and isopropyltoluenes by Al_2Br_6-HBr [10]. They reported that infrared analysis showed that *o*-xylene isomerized to *m*-xylene without the formation of any detectable amount of *p*-xylene, and likewise, the isomerization of *p*-xylene proceeded directly to *m*-xylene without any observable formation of the *ortho* isomer. They commented on the much faster rates of reorientation of *p*-ethyl- and *p*-isopropyltoluenes.

The definitive work establishing the intramolecular nature of the reorientation of methyl groups on aromatic rings induced by Friedel-Crafts catalysts was that of Steinberg and Sixma using ^{14}C-labeled toluene [11]. Toluene is unchanged by treatment with Al_2Br_6-HBr at 0°C for 8 days [12], indicating that the methyl group has little or no tendency to migrate *inter*molecularly. Steinberg and Sixma synthesized toluene-1-^{14}C and subjected it to treatment with Al_2Br_6-HBr at 35°C. After various periods of time the recovered toluene was degraded in such a way that radioactivity distribution in the benzene nucleus could be determined [13]. When the reaction was allowed to continue for a long period of time, the activity was found to be statistically distributed over the ring carbon atoms. Using short reaction times, the activity distributions were found to be in good agreement with the theoretical values calculated on the assumption that the migration goes by a series of consecutive 1,2 shifts. The activity at carbon atom 1 diminished progressively toward one-sixth of the original value. The activity at carbon atom 2 went through a maximum value and thereafter approached asymptotically the value for statistical distribution, whereas the activities at the *meta* and *para* positions showed a steady increase toward the same value.

More recently, rate constants were measured for the reorientation of *p*-cresol to *m*-cresol and of 3,4-xylenol to 3,5-xylenol in HF-BF_3 [13a]. The reorientation rates were substantially lower than those of methylbenzenes. This difference was ascribed to the thermodynamic unfavorability of benzenonium ions with a hydroxyl group in the *meta* position.

B. Transalkylations of Toluene, Xylenes, and Higher Methylarenes

Although it is clear from the research that has been described in the preceding paragraphs that reorientation of xylenes can be produced without any significant disproportionation, many examples of disproportionations of toluene, xylenes, and higher-methylated benzenes have been reported, most of them coming from industrial applications. Benzene and xylenes are more valuable industrial starting materials than toluene, so the activity in this area of investigation is understandable.

Many different catlaysts have been tested for such disproportionations; a few examples will be mentioned here. Zeolite-X-type catalysts containing Ca, Cd, Mn, La, and Ce cations were studied at 300 to 500°C and found to have high activity for both reorientation of *o*-xylene and disproportionation to give toluene and trimethylbenzenes, but a conventional Al_2O_3 catalyst promoted reorientation selectively [14]. Four zeolite-X-type catalysts prepared by ion exchange with La and Ce chlorides were found to catalyze reorientation of xylenes at lower temperatures than amorphous SiO_2-Al_2O_3 catalysts, but they gave larger amounts of by-products [15]. In transalkylations between toluene and trimethylbenzenes using zeolite-X and zeolite-Y catalysts at 440 to 450°C, the Y-type catalysts proved to be more effective [16]. NaScY and NaMgHY zeolites were reported to catalyze both dealkylation and disproportionation of toluene to benzene and xylenes at 370 to 600°C. The higher temperature gave higher yields of benzene and lower yields of xylenes [17]. Transalkylations between toluene and penta- and hexamethylbenzenes to give *p*-xylene selectively under N, Ar, and H was catalyzed by $AlCl_3$, $GaCl_3$, and a NaCe Y-zeolite. Also used were $AlCl_3$-CH_3NO_2, $AlCl_3$-H_2O, and $AlCl_3$-PhOMe [18]. Fluorinated alumina and silica catalysts were used [19,20], and combinations of Al, Si, Ti, and B oxides with metal fluorides, as well as synthetic zeolites [21], for toluene and xylene disproportionations. Sodium complexes with anthracene, phenanthrene, and chrysene were also used to catalyze the disproportionation of toluene to benzene and xylene [22].

Recently, a superacidic perfluorinated resinsulfonic acid catalyst (DuPont's) Nafion-H) was tested for reorientation and disproportionation of methylbenzenes [23]. It was found to effect reorientation of trimethyl- and tetramethylbenzenes under relatively mild conditions (193°C) at atmospheric pressure with short contact times (4 to 5 sec). Reorientation was said to be a purely intramolecular process and was faster than intermolecular methyl transfer (disproportionation). Penta- and hexamethylbenzenes transmethylated benzene and toluene readily.

To our knowledge, there is only one report of disproportionation of methylarenes taking place more readily than reorientation of the methyl groups, although Olah and Kaspi [23] refer to two such reports. On examination of the English translation of one of these, the article by Delone et al. [24], one finds that the Russian workers *did* observe reorientation of mesitylene and pseudocumene when they were disproportionated in the liquid phase by an aluminosilicate catalyst to xylenes and tetramethylbenzenes. The second report quoted by Olah and Kaspi is that of Molchanova et al. [25], and the original article is not generally available. The *Chemical Abstracts* abstract states that "Intermolecular processes involving disproportionation of Me groups proceeded quicker than isomerization among $Me_3C_6H_3$, $Me_4C_6H_2$, and $Me_4C_6H_2$ mixtures." The catalyst and temperature are not specified in the abstract.

III. HIGHER ALKYLARENES: INTRA- AND INTERMOLECULAR MECHANISMS OF TRANSALKYLATIONS AND REORIENTATIONS

A. Ethylbenzene

As we reported in a preceding section, *n*-propyl- and *n*-butylbenzenes were shown to undergo reorientation and disproportionation reactions without internal rearrangement of the side chains. With the availability of ^{14}C as a label for organic molecules, it became possible to determine whether or not the simplest alkyl group that could undergo rearrangement (i.e., ethyl), would do so during disproportionation of ethylbenzene. This problem was addressed by Roberts et al. [26], who synthesized ethyl-β-^{14}C-benzene and heated it with aluminum chloride (0.3 mole per mole of hydrocarbon) under reflux for 2.5 hrs. The recovered ethylbenzene and the diethylbenzenes produced (mainly *meta* and *para* isomers) were oxidized to benzoic and phthalic acids, which were found to contain none of the radioactive label. Thus the disproportionation took place with no isomerization of the ^{14}C-labeled ethyl group (scheme 1).

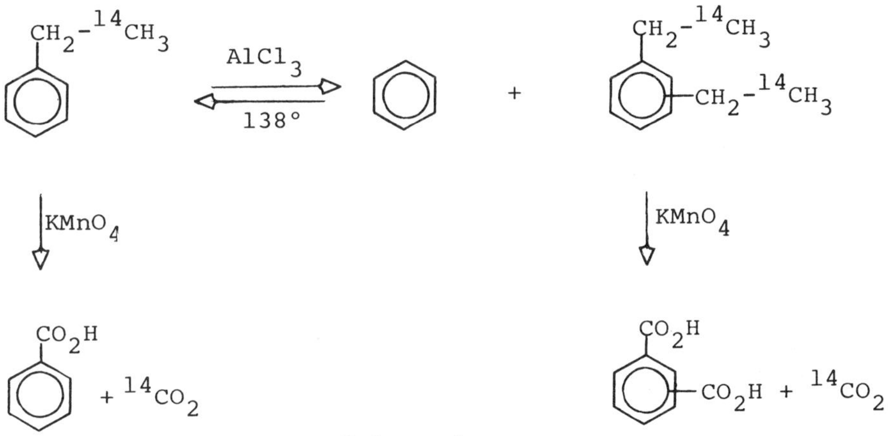

$CH_2-^{14}CH_3$ (on benzene) $\xrightleftharpoons[138°]{AlCl_3}$ benzene + $CH_2-^{14}CH_3$ / $-CH_2-^{14}CH_3$ (diethylbenzene)

$\downarrow KMnO_4$ $\qquad\qquad$ $\downarrow KMnO_4$

CO_2H (benzoic acid) $+ {}^{14}CO_2$ $\qquad\qquad$ CO_2H / $-CO_2H$ (phthalic acid) $+ {}^{14}CO_2$

Scheme 1

B. Methyl-, Ethyl-, Isopropyl-, and *t*-Butyltoluenes

In 1959-1960, R. H. Allen and co-workers published reports of studies of the kinetics of "three-compound equilibrations," in which reorientations of *ortho*, *meta*, and *para* isomers of xylenes, ethyltoluenes, isopropyltoluenes, and *t*-butyltoluenes by $AlCl_3$-HCl in toluene solution were examined [27]. From the measured rates and equilibrium constants for the reorientations of xylenes [27(b)] and of ethyltoluene [27(c)] and isopropyltoluene [27(a)] isomers, they concluded that the reorientation of xylene isomers takes place entirely by an intramolecular mechanism, the reorientation of ethyltoluene takes place mainly by an intramolecular mechanism, the reorientation of isopropyltoluene takes place mainly by an intermolecular mechanism, and the reorientation of *t*-butyltoluene takes place entirely by an intermolecular mechanism. As evidence for an intermolecular mechanism for the reorientation of *t*-butyltoluene, they reported [27(d)] the transfer of *t*-butyl groups from *p*-*t*-butyltoluene to *o*-xylene under conditions that were too mild to isomerize *p*-*t*-butyltoluene to *m*-*t*-butyltoluene to an appreciable extent (Eq. 3). The intermolecular reorientation of *neat* *p*-*t*-butyltoluene should not be possible because of steric hindrance, and indeed Allen found that treatment of pure *p*-*t*-butyltoluene with $AlCl_3$-HCl did not produce reorientation. The reorien-

Table 1. Mechanism of Alkyltoluene Reorientations

Alkyltoluene	Intramolecular (%)	Intermolecular (%)
Xylene	100	0
Ethyltoluene	>84	<16
i-Propyltoluene	>14	<86
t-Butyltoluene	0	100

Source: Data from Ref. 27(d).

$AlCl_3/CH_3NO_2$ (fast) (slow) (3)

tation of *p*-*t*-butyltoluene to *m*-*t*-butyltoluene in toluene solution was said to proceed by intermolecular transalkylation of toluene. In summary of his work, Allen presented the data in Table 1.

Olah and coworkers investigated the reorientation of the three *t*-butyltoluene isomers with water-promoted aluminum chloride (a heterogeneous system) and with aluminum chloride in nitromethane solution (homogeneous) [28]. Analysis of reaction mixtures was by means of gas chromatography. The equilibrium mixture obtained starting with *ortho*, *meta*, or *para* isomers contained ca. 64% *m*- and 36% *p*-*t*-butyltoluene (no *ortho* isomer). In the formation of the equilibrium mixture from the *ortho* isomer, the concentration of the *para* isomer built up first, went through a maximum, and then decreased. Olah suggested that a π-type intermediate might be involved, with the *t*-butyl group never completely detached from the aromatic ring.

C. Diethyl-, Diisopropyl-, and Di-*t*-butylbenzenes

The reorientations of diethyl-, diisopropyl-, and di-*t*-butyl benzene isomers were also investigated by Olah and co-workers [29]. Starting with any one of the diethylbenzene isomers in reaction with water-promoted aluminum chloride, an equilibrium mixture containing ca. 3% *ortho*, 69% *meta*, and 28% *para* isomer was obtained [29(a)]. Disproportionation products (mainly ethylbenzene and 1,3,5-triethylbenzene) were always observed. In the isomerization of *o*-diethylbenzene, the *para* isomer appeared only after a significant amount of *meta* isomer had formed, which was taken to indicate a predominance of an intramolecular 1,2-shift mechanism of reorientation, in agreement with the conclusion of Allen [27(c)] regarding the reorientation of ethyltoluenes. The reactions of the diisopropylbenzenes were similar to those of the diethylbenzenes [29(b)], and the mechanism of reorientation was concluded to be predominantly an intramolecular 1,2 shift, although substantial disproportionation was observed. The equilibrium mixture obtained from each of the three isomers was ca. 62% *m*- and 38% *p*-diethylbenzene, with no *ortho* isomer present.

The reorientations of *o*-, *m*-, and *p*-di-*t*-butylbenzene by water-promoted aluminum chloride were investigated in carbon disulfide solution at 25°C [29(c)]. The isomerization of *o*-di-*t*-butylbenzene proceeded rapidly to yield a composition of 28% *meta* and 72% *para* isomer, a ratio that stayed relatively constant for some time and then changed slowly to the equilibrium composition of 52% *meta* and 48% *para* isomer. Starting with either *m*- or *p*-di-*t*-butylbenzene gave (more slowly) the same final equilibrium concentration of 52% *meta* and 48% *para* isomer, with no *ortho* isomer present. Disproportionation products (mainly *t*-butylbenzene and 1,3,5-tri-*t*-butylbenzene) amounted, at the time equilibrium was reached, to ca. 50% of the reaction mixture. To explain the dynamics of the reorientation of the *ortho* isomer, Olah again invoked a π-type intermediate for a fast intramolecular rearrangement of the *ortho* to the *para* isomer, followed by a much slower 1,2-shift leading to the *meta* isomer. This was essentially the same mechanism as that proposed for the reorientation of *t*-butyltoluenes [16].

A paper published two years later, however, cast doubt on an intramolecular 1,2-shift of a *t*-butyl group as the major pathway for such reorientations. Myhre et al. [30], in considering the reported synthesis of 1,3,5-tri-*t*-butylbenzene by alkylation of *p*-di-*t*-butylbenzene with *t*-butyl chloride and $AlCl_3$ [31], suggested that the reaction involved a reorientation of *p*-di-*t*-butylbenzene to the *meta* isomer by a dealkylation-realkylation mechanism, followed by alkylation to give the 1,3,5-tri-*t*-butylbenzene (scheme 2). Evidence to substantiate this view was obtained by preparing *p*-di-*t*-butylbenzene-β-^3H and showing that a loss of ca. 80% of the ^3H occurred in conversion of the ^3H-labeled di-*t*-butylbenzene to the tri-*t*-butylbenzene.

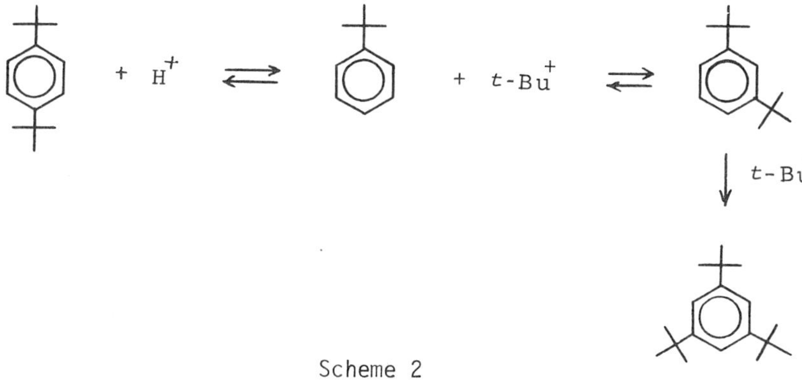

+ H^+ ⇌ + $t\text{-}Bu^+$ ⇌

↓ $t\text{-}Bu^+$

Scheme 2

D. *sec*-Butyl- and *sec*-Pentylbenzenes

As we have noted previously, the disproportionations of ethyl-, *n*-propyl-, and *n*-butylbenzenes have been demonstrated to take place without any isomerization of the alkyl side chains. *sec*-Butylbenzene was also found to undergo disproportionation by treatment with HF-BF_3 without side chain rearrangement [6]. However, when either 2- or 3-phenylpentane was treated with $AlCl_3$ at 25°C, the rapid disproportionation produced besides benzene a mixture of 2- and 3-phenylpentane isomers [32]. The equilibrium proportion of these isomers was 71:29 (2-/3-). The disproportionation was initially faster than the interconversion of 2- and 3-phenylpentane. When mixtures of *p*-di-3-pentylbenzene and benzene were treated with $AlCl_3$ at 25°C, reorientation to *m*-di-pentylbenzene and transalkylation (disproportionation) took place at comparable rates. Optically active 2-phenylpentane was found to be almost completely racemized by $AlCl_3$ in experiments in which disproportionation had proceeded to the extent of only 1.5%. Burwell and Shields thus concluded that transalkylation did not contribute significantly to the racemization, which was viewed as a hydride ion transfer chain process. The reorientations, disproportionations, and side-chain isomerizations were all considered to involve secondary carbocation intermediates.

In view of the lack of rearrangement of *n*-propyl - and *n*-butyl groups during disproportionation, McCaulay and Lien [6] proposed an S_N2-type mechanism. However, in their study, Brown and Smoot [12] found relative rates of disproportionation of alkylbenzenes by Al_2Br_6-HBr to be: Me, 10^{-7}; Et, 1; *i*-Pr, 10^2, and the difference between the rates of toluene and ethylbenzene were considered to be too large to be compatible with a direct displacement mechanism. Hence Brown and Smoot proposed a rapid equilibrium to a "localized π-complex" intermediate in a displacement reaction. However, this mechanism was invalidated by the finding that ethyltoluene does not undergo intramolecular reorientation much faster than intermolecular disproportionation [27(c)], as would seem to be demanded by the localized π-complex mechanism.

E. Diarylalkane Intermediates for Transalkylations of Primary Alkylbenzenes

In 1964 there was still no satisfactory rationale for the disproportionation reactions of primary alkylarenes, until Streitwieser and Reif reported an elegant study in which optically active ethylbenzene-α-^{2}H-ring-^{14}C was utilized [33]. The kinetics of transalkylation of this molecule in benzene at 50°C with gallium bromide-hydrogen bromide catalyst were determined in homogeneous medium. Recovered ethylbenzene was examined for radioactivity, optical rotation, and deuterium content. Loss of ^{14}C activity and optical activity were found to occur at equal rates, and the scrambling of deuterium was only a little slower. The only plausible explanation consistent with these results is a mechanism in which the three processes whose rates were measured have the same rate-determining step, as outlined in scheme 3.

The first step, abstraction of hydride ion from the α position by a cation, is postulated as the rate-determining step. The authors speculated that the cation might arise from protonation of a small amount of styrene present as an impurity in the ethylbenzene. Since a cation is regenerated in the fourth step, only minute amounts would be required to initiate the chain reaction. A necessary corollary of this mechanism is that 1,1-diphenylethane should cleave readily under the conditions of the reaction, and such a cleavage was demonstrated by the rapid conversion of 1,1-di-*p*-tolylethane in benzene

$$*C_6H_5-\underset{D}{\overset{H}{C}}-CH_3 + R^+ \longrightarrow *C_6H_5-\overset{+}{C}\begin{smallmatrix}H(D)\\CH_3\end{smallmatrix} + RH \quad (4)$$

$$*C_6H_5-\overset{+}{C}\begin{smallmatrix}H(D)\\CH_3\end{smallmatrix} + C_6H_5C_2H_5 \longrightarrow *C_6H_5-\underset{CH_3}{\overset{H(D)}{C}}-C_6H_4C_2H_5 + H^+ \quad (5)$$

$$*C_6H_5-\underset{CH_3}{\overset{H(D)}{C}}-C_6H_4C_2H_5 \longrightarrow *C_6H_6 + \begin{smallmatrix}(D)H\\CH_3\end{smallmatrix}\overset{+}{C}-C_6H_4-C_2H_5 \quad (6)$$

$$\begin{smallmatrix}(D)H\\CH_3\end{smallmatrix}\overset{+}{C}-C_6H_4-C_2H_5 \longrightarrow CH_3-\underset{H}{\overset{H(D)}{C}}-C_6H_4-C_2H_5 + R^+ \quad (7)$$

Scheme 3

containing gallium bromide-hydrogen bromide into 1,1-diphenylethane and toluene.

Evidence for the intermediate formation of 1,1-diarylalkanes in disproportionation reactions of primary alkylbenzenes had already been demonstrated in a less dramatic way by Pines and Arrigo [34]. They found that ethyl groups were transferred from one ethylxylene molecule to another in the presence of hydrogen fluoride and an alkene such as 4-methylcyclohexene (Eq. 8). The mechanism they proposed was substantially the same as that of Streitwieser and Reif. The later work demonstrated that high concentrations of a hydrogen ion acceptor such as 4-methylcyclohexene is not required, and

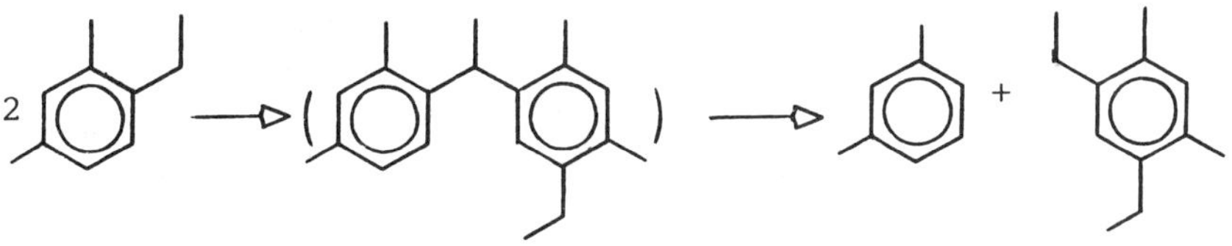

(8)

the experimental data of three different types is well accommodated by the mechanism. Streitwieser and Reif go on to note that some other seemingly anomalous data can be understood in terms of a mechanism of disproportionation that involves hydride abstraction from a benzylic carbon with the subsequent formation of a 1,1-diarylalkane intermediate. The low relative rate of disproportionation of toluene [12] may be attributed to the fact that hydride abstraction produces a benzyl cation that is primary, in contrast to the secondary benzyl cations that are produced from primary alkylbenzenes such as ethyl-, *n*-propyl, and *n*-butylbenzene. The failure of neopentylbenzene to undergo disproportionation with hydrogen fluoride under conditions that give rapid disproportionation of ethylbenzene [6] is understandable, since the phenyl-*t*-butylcarbinyl cation is severely destabilized by steric hindrance to coplanarity.

F. Comparative Ease of Transalkylations of Primary and Secondary Alkylarenes

An observation by Roberts and Shiengthong [35] may appropriately be mentioned here. They found that although an isopropyl group was transferred from isopropylmesitylene to benzene in the presence of $AlCl_3$, there was no transalkylation between *n*-propylmesitylene and benzene, or between *n*-propylbenzene and mesitylene. Presumably the transition state for the formation of the 1,1-diarylpropane from the latter two pairs of reactants is too crowded. The transalkylation of the (secondary) isopropyl group can take place by a different mechanism, probably via a secondary propyl cation intermediate, by analogy to the behavior of the secondary pentylbenzenes described by Burwell and Shields [32] and mentioned earlier.

The greater ease of intermolecular transfer of secondary than of primary alkyl groups may be put to practical advantage. When a mixture of primary and secondary dodecylcumenes, produced by alkylation of cumene (isopropylbenzene) with *n*-dodecyl bromide, was treated with Al_2Br_6-HBr in a large excess of benzene, selective transfer of secondary groups took place to give a mixture of *n*-dodecylbenzene, cumene, and *sec*-dodecylbenzenes [36]. The proportion of *n*-dodecylbenzne produced in this way by transalkylation was higher than that obtainable by direct alkylation of benzene by *n*-dodecyl bromide.

Disproportionations involving isopropylbenzenes for practical reasons have also been reported by several groups of workers. Mazonski and co-workers investigated the transalkylation of triisopropylbenzene (structure not specified) with benzene using $AlCl_3$ catalyst [37]. Strohmeyer studied the effects of using complexes of various polymethylbenzenes with $AlCl_3$ and HCl as catalysts for the disproportionation and reorientation of 1,2,4-triisopropylbenzene [38]. Shimada compared the proportions of *p*-ethyl- and *p*-isopropyltoluene obtained by transalkylations between ethyl and isopropylalkylbenzenes and toluene with those obtained by alkylation of toluene with ethyl and isopropyl bromides, using $AlCl_3$ and $AlCl_3$-CH_3NO_2 catalysts [39]. The higher ratios of para isomers from the transalkylations were attributed to steric effects in bimolecular reactions. Optimum conditions were determined for the disproportionation of cumene to *m*- and *p*-diisopropylbenzene using an Al_2O_3 catalyst modified with BF_3 [40]. The recovery of *m*- and *p*-diisopropylbenzene from the "bottom fraction" produced in the synthesis of cumene from benzene and propene, by treating the bottom fraction with propene and $AlCl_3$-HCl was described in a patent [41].

Similar studies of the relative rates of transalkylation of different alkyl groups were made by Japanese and Russian chemists. In the Japanese work [42], Hamanaka et al. determined the migratory tendencies of Me, Et, *i*-Pr, and *t*-Bu groups in transalkylations catalyzed by $AlCl_3$ at 4°C, and they reported the sequence *t*-Bu > *i*-Pr > Et > Me (Me did not undergo transalkylation under their conditions). The sequence of the alkylbenzenes for acceptability of the migrating alkyl groups was said to be PhMe > PhEt > Ph*i*-Pr > Ph*t*-Bu. Lipovich et al. carried out transalkylations between alkylbenzenes and toluene [43a] using Al_2Br_6 as catalyst in *n*-hexane solution at 40°C. The order of relative rates of transalkylation to toluene was found to be *t*-Bu ≫ *i*-Pr > *sec*-Bu > Et > Pr > *n*-Bu > *n*-C_5H_{11} > *n*-C_6H_{13}. In a second paper, Lipovich et al. determined relative rates of transalkylation of the *n*-alkylbenzenes to benzene, by using ^{14}C-labeled benzene [43(b)], with Al_2Br_6 catalyst in *n*-hexane at 40°C. The order of the C_2 to C_6 groups was the same as for the transfer to toluene, and the difference in the rates of ethyl and *n*-hexyl was only a factor of 10.

In a more recent continuation of this work [43(c)], the substrate selectivity (k_t/k_b) for transalkylation of toluene and benzene-1-^{14}C by various alkylbenzenes (ArR, R = Et, Pr, *i*-Pr, *sec*-Bu, *c*-Hex) was determined and found to be small, as was also the positional (*o,m,p*) selectivity in toluene. These results were said to favor a transition state of the π-type for the transalkylations.

Chenets and co-workers also reported a kinetic study of transalkylation of benzene and toluene by *p*-dialkylbenzenes, in which one of the alkyl groups was always ethyl [43(d)]. The transfer rate of the ethyl group was greater than that of any of the other alkyl groups (propyl, butyl, isopropyl, and *sec*-butyl), whereas in earlier work [43(a)] the ethyl group was found to occupy an intermediate position among the other alkyl groups. The increased rates of transalkylation of the *p*-ethylalkylbenzenes was attributed to the electron-donating properties of the alkyl group in the *para* position to the migrating group. The transition state for the transalkylation reaction was considered to be of the π-complex type.

G. Reorientations and Transalkylations of Ethylarenes

1. Aluminum bromide-catalyzed reactions

Although the elegant work of Streitwieser and Reif appeared to give the solution to the problem of the disproportionation of primary alkylbenzenes without side-chain isomerization, the mechanism of *reorientation* of dialkylbenzenes, except for that of the methylarenes, remained uncertain. More light on the mechanism of reorientation of ethyl groups was sought by Unseren and Wolf [44(a)], by using a technique like that of Steinberg and Sixma [11] for the methyl group. Ethylbenzene-1-^{14}C was synthesized and treated with Al_2Br_6-HBr at 0°C. After a short period of time (17 min), the reaction mixture was quenched, and recovered ethylbenzene was degraded in such a way that the radioactivity distribution in the benzene ring could be determined. Although experimental errors were admittedly large, it was clearly shown that the bulk of the rearranged ^{14}C activity was in the *meta* and *para* positions, which was said to be incompatible with the localized π complex of Brown and Smoot [12]. Unseren and Wolf also reported one experiment in which a doubly labeled ethylbenzene molecule, ethyl-β-^{14}C-benzene-1-^{14}C (ethylbenzene-1,8-^{14}C), was treated with Al_2Br_6-HBr, and they reported that their "results on rearrangement in the ethyl group were in complete accord with the published results of Roberts" [26]. In a second publication from the same laboratory [44(b)], Moore and Wolf extended the earlier work with experiments in which ethylbenzene-1-^{14}C was treated with Al_2Br_6-HBr for various periods of time, and they developed improved degradation procedures to give more accurate distributions of the isotope in the ring of the recovered ethylbenzene. The distributions from four samples withdrawn from the same reaction mixture after different reaction periods are shown in Table 2. Moore and Wolf concluded that their data did not allow them to choose any one of several possible combinations of 1,2 shifts of ethyl groups and intermolecular transalkylations as the precise mechanism of isotopic reorientation, but they emphasized the fact that the small amount of radioactivity in the *ortho* position showed that a 1,2 shift of an ethyl group in ethylbenzene is slow compared to disproportionation. This result appears to be at odds with the conclusions of Allen [27(d)] and Olah [29(a)] about the mechanism of reorientation of ethyltoluene and diethylbenzene.

Table 2. Reorientation of ^{14}C in Ethylbenzene-1-^{14}C by Al_2Br_6-HBr

Time (min):	0	7.5	9.3	13.3	23
	98	74	72	72	68
	0.6	3.2	4.0	4.0	4.6
	0.3	3.6	4.7	4.7	5.4
	0.8	18.8	20.2	20.2	22.0

The isotopic distributions found by Moore and Wolf in all of their reaction mixtures were high in the *para* position compared to the *meta* and *ortho* positions, but there was not a great difference between the distribution found after the shortest reaction period (7.5 min) and the longest reaction period (23 min). This led one of the present authors to wonder if a pseudoequilibrium composition had not been produced even during the shortest reaction period, and if more meaningful kinetic data might be obtained during even shorter reaction periods. The tremendous experimental simplification intrinsic to the use of ^{13}C-labeled substrates monitored by NMR and mass spectrometry (which became available after the work of Moore and Wolf) appeared to offer the possibility of yielding definitive new information about the mechanisms of reorientation and disproportionation of ethylbenzene and its homologs. For example, with this technique it should be possible to follow the isotopic reorientation of ethylbenzene-1-^{13}C almost continuously from the first few seconds after it is put in contact with a catalyst, by taking small samples of the reaction mixture at frequent time intervals and, after appropriate workup, analyzing them by ^{13}C NMR spectrometry. Just such an approach was made by Roberts and Roengsumran [45]. They first tested the technique by essentially repeating the experiments of Steinberg and Sixma [13], but using ^{13}C-labeled toluene instead of ^{14}C-labeled toluene. The results were in good agreement with the earlier work. A slightly larger amount of disproportionation was observed, especially when a reaction was carried out at 50°C, but the rates of increase in the ^{13}C content in the *meta* and *para* positions of the toluene ring were unchanged even after 10% of xylenes and 10% of trimethylbenzenes had formed, showing that the intermolecular transalkylation of toluene was too slow to play an appreciable part in the isotopic reorientation process.

The ethylbenzene-1-^{13}C was treated with Al_2Br_6-HBr in 1,2,4-trichlorobenzene solution at 10°C. The first increase in the ^{13}C content occurred in the *para* position after about 30 min. (The reaction was slower in 1,2,4-trichlorobenzene solution than when no solvent was used by Moore and Wolf.) After 2 hr, the isotopic content in the *para* position had tripled, while there was no increase in the *ortho* and *meta* positions. As the reaction course was followed for 5 hr, the ^{13}C content began slowly to increase in the *ortho* and *meta* positions as well, and after 20 hr the ^{13}C content was near the equilibrium proportion, or: C_1 35%; $C_{2,6}$ 16%; $C_{3,5}$ 18%; C_4 31%. These results not only agreed with those of Moore and Wolf, but they also excluded any possibility that a rapid 1,2-shift mechanism plays an important role in the isotopic reorientation of C_1-labeled ethylbenzene.

Further to confirm an intermolecular transalkylation process as the major mechanism for the reorientation of an ethyl group, an experiment was performed in which simultaneous reaction of ethylbenzene molecules labeled, separately, in the aromatic ring and in the side chain was effected. Ethylbenzene-8-^{2}H was prepared and mixed with ethylbenzene-1-^{13}C, and the mixture was treated with Al_2Br_6-HBr. The increase in the ^{13}C content in the *para* position of the ethylbenzene and the appearance of doubly labeled molecules, as detected by the m/e M + 2 ions in the mass spectrum, occurred at almost precisely the same time, 15 min after the addition of catalyst and when about 6% of *p*-diethylbenzene was detected. The amount of doubly labeled molecules expected on the basis of equilibration by intermolecular transalkylation was detected at approximately the same time that the equilibrium distribution of ^{13}C was found in the ethylbenzene ring. This result was interpreted as good evidence for an intermolecular transalkylation-dealkylation mechanism as the major process leading to the reorientation of ^{13}C in the ethylbenzene ring, since, if an intramolecular 1,2-shift of the ethyl group took place to an appreciable extent, the ^{13}C distribution in the ring would reach equilibrium before the maximum amount of doubly labeled ethylbenzene was formed.

Roberts and Roengsumran also prepared *n*-propylbenzene-1-^{13}C and studied its reorientation and disproportionation [45]. The results were very similar to those from ethylbenzene-1-^{13}C, indicating that a similar intermolecular mechanism was predominantly responsible for reorientation of the *n*-propyl group.

2. Zeolite-catalyzed reactions

The reactions of the three isomeric diethylbenzenes catalyzed by a type Y zeolite catalyst were studied by Bolton et al. [46]. The catalyst was prepared by partial multivalent cation exchange (40% Ce^{3+}) and partial decationization (50%) of a type Y zeolite with a SiO_2/Al_2O_3 molar ratio of 5.0. At 100°C, 2 mol % of *p*-diethylbenzene underwent transalkylation to benzene and triethylbenzene, but no reorientation occurred during a 24-hr reaction time. At 150°C for 16 hr, 30 mol % of the *p*-diethylbenzene underwent transalkylation, and 10% reorientation occurred. The triethylbenzenes formed were the 1,2,4- and 1,3,5-isomers, with the former predominating in the early stages of reaction. Similar results were obtained starting with *o*- and *m*-diethylbenzene. The equilibrium product distribution at 170°C consisted of ca. 50 mol % transalkylate: the diethylbenzene fraction contained ca. 5% ortho, 62% meta, and 33% para isomers and the triethylbenzene fraction consisted of 31% 1,2,4- and 69% 1,3,5-isomers. Bolton et al. stated that the reorientations of the diethylbenzenes are best explained in terms of transalkylations via 1,2,4-triethylbenzene as an intermediate. Seeking further support for this theory, they made a test in which the isomerization of the *o*-diethylbenzene was carried out in the presence of added ethylbenzene and triethylbenzene, assuming that if transalkylation was responsible for reorientation of the diethylbenzene, an increase in the rate of this isomerization should occur, whereas if the mechanism involved 1,2-shifts, a decrease in rate due to dilution of the reactant should occur. As expected, after 4 hr the reorientation of the *o*-diethylbenzene was found to have proceeded at a faster rate in the reaction mixture containing the added transalkylation components than in the reaction mixture of pure *o*-diethylbenzene and catalyst. They concluded that 1,2,4-triethylbenzene is formed as an intermediate from each of the three diethylbenzene isomers by a Streitwieser-Pines [33,34] intermolecular mechanism. They emphasized the fact that 1,2,4-triethylbenzene has not been isolated or recognized as a possible intermediate previously, and they suggested that this was probably because the catalysts used most often, ($AlBr_3$-HBr, $HF-BF_3$, etc.), are ones that form stable complexes with 1,3,5-trialkylarenes and thus favor them in an equilibrium mixture, whereas the zeolite catalyst does not form such complexes.

A study of the "dealkylation and isomerization" of dialkylbenzenes, including diethylbenzene, was reported by Matsumoto et al. [47]. The catalyst used was silica-alumina poisoned with NaOH, and they operated at 200 to 500°C, keeping conversion below 15%. No products were said to be detected except "those from isomerization and dealkylation." The "isomerization" was apparently what we are calling reorientation, which they state was "reasonably considered to take place by an intramolecular 1,2-shift mechanism." It is difficult to determine what they mean by "dealkylation" of the alkylbenzenes, or how they measured it. Their major interest was in measuring rate constants, activation energies, and Hammett linear free-energy relationships for the reactions of methyl-, ethyl-, isopropyl-, and *t*-butylbenzenes. They refer to the different effects of alkyl groups on "isomerization and dealkylation" in terms of "the stability of an alkyl carbonium ion in activated complexes . . . interacting with a benzene ring," without differentiating between a methyl group and a *t*-butyl group.

3. Reactions catalyzed by a perfluorinated resinsulfonic acid

Nafion-H, a "superacidic" perfluorinated resinsulfonic acid developed by DuPont was assessed as a catalyst for transalkylations and reorientations of alkylbenzenes by Olah and Kaspi [48]. From the reaction of the diethylbenzene isomers at 170°C, they found an equilibrium distribution of diethyl- and triethylbenzene isomers very similar to that from the zeolite-Y catalyst of Bolton et al. [44], but Olah and Kaspi conclude that the reorientation of diethylbenzene by Nafion-H occurs by both intermolecular transalkylation and intramolecular 1,2-shift, in contrast to the exclusive intermolecular mechanism in the zeolite-Y-catalyzed reaction.

Disproportionation of *n*-propylbenzene by the Nafion-H catalyst produced benzene, *m*-di-*n*-propylbenzene, and *p*-di-*n*-propylbenzene. No *o*-di-*n*-propylbenzene was produced; steric hindrance probably precludes formation of detectable amounts of this isomer. No isopropylarenes were detected in the *n*-propylbenzene disproportionation. Disproportionation of *n*-propylbenzene by $AlCl_3$ or $AlBr_3$ is known to produce only 2 to 5% of isopropylbenzene [49].

Transalkylations and reorientations of *m*- and *p*-diisopropylbenzene by the Nafion-H catalyst were rapid, and produced almost the same proportion of products. The only triisopropylbenzene detected was the 1,3,5-isomer, as was true of the $AlCl_3$-catalyzed reaction [29(b)]. The reorientation of the diisopropylbenzenes was assumed to take place mainly by an intermolecular mechanism.

H. Transalkylations and Reorientations of Diarylethanes

The reactions of some 1,2-diarylethanes in the presence of $AlBr_3$-HBr were studied by Kunichika et al. [50]. The *ortho, ortho*-, *meta,meta*-, and *para, para*-isomers of 1,2-ditolylethane, after treatment with $AlBr_3$-HBr at 50°C for 2 to 3 hr, gave identical mixtures containing 42% *meta,meta*- and 35% *meta, para*-1,2-ditolylethane, with smaller maounts of all other reorientation isomers, including only a trace of the *ortho, ortho*-isomer [50(a)]. Similar reaction of

$$\text{CH}_3\text{C}_6\text{H}_4\text{CH}_2\text{CH}_2\text{C}_6\text{H}_4\text{CH}_3 \xrightarrow[\text{HBr}]{\text{AlBr}_3} \text{(3,3')}\ 42\% + \text{(3,4')}\ 35\% + \text{(2,3')}\ 12\% + \text{smaller amounts of other isomers}$$

1,2-diphenylethane in toluene gave, in a very slow reaction, 1-phenyl-2-tolylethane and a mixture of isomers of 1,2-ditolylethane in which the ratio of isomers was the same throughout the reaction. These results indicated that the reorientation of the methyl groups occurred by a fast intramolecular mechanism and the transalkylation of the 2-arylethyl group was much slower, as might be expected on the basis of the relative ease of reorientations and transalkylations of methyl- and ethylbenzenes, considering the phenylethyl group to be similar to an unsubstituted ethyl group in its reactivity. A second publication from the same laboratory [50(b)] described similar results from the reaction of 1-phenyl-2-tolylethanes. No reaction occurred in nitromethane solution, and only slow reaction in carbon disulfide or cyclohexane solution. The methyl reorientation was described as occurring by a 1,2-shift and the slower transalkylation via hydride abstraction in the rate-determining step, in accordance with the mechanism of Streitwieser and Reif [33].

I. Intermolecular Transfer of *t*-Butyl and Benzyl Groups Between Arenes and Phenols

In a previous section we described the facile transfer of a *t*-butyl group from *p*-*t*-butyltoluene to *o*-xylene [27(d)]. When a similar reaction between *t*-butylbenzene and *p*-xylene was attempted using $AlCl_3$ as catalyst, a complex mixture of products was obtained which did not include any *t*-butyl-*p*-xylene, but consisted mostly of disproportionation products from both starting materials [51]. Apparently, steric hindrance made attachment of a *t*-butyl group adjacent to a methyl group difficult, and the strong catalyst led to many side reactions. In 1978, Tashiro and co-workers reported a preparation of 2-*t*-butyl-*p*-xylene in satisfactory yields (40 to 60%) by transalkylation between *p*-xylene and 2-*t*-butyl-*p*-cresol or 2,6-di-*t*-butyl-*p*-cresol using $AlCl_3$-CH_3NO_2 as catalyst (Eq. 9) [52].

$$\text{2-}t\text{-butyl-6-R-}p\text{-cresol} + p\text{-xylene} \xrightarrow[\text{CH}_3\text{NO}_2]{\text{AlCl}_3} \text{2-R-}p\text{-cresol} + 2\text{-}t\text{-butyl-}p\text{-xylene} \qquad (9)$$

R = H or t-Bu

Very recently, Tashiro et al. [52a] reported a convenient synthesis of dibenzofurans which was facilitated by the use of *t*-butyl groups, which could later be removed by transalkylation (Eqs. 9a and 9b). The presence of the

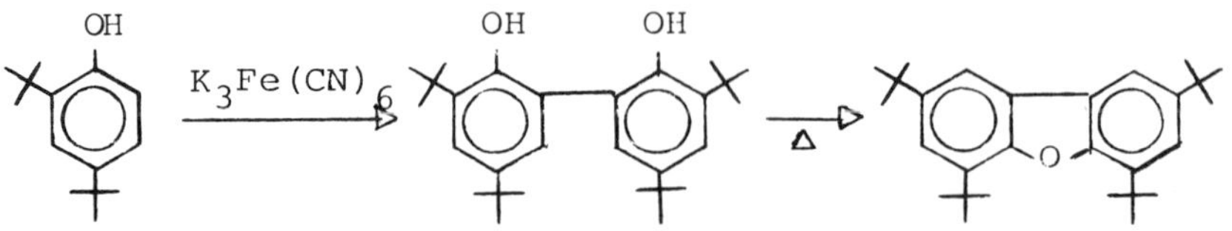

$K_3Fe(CN)_6$ (9a)

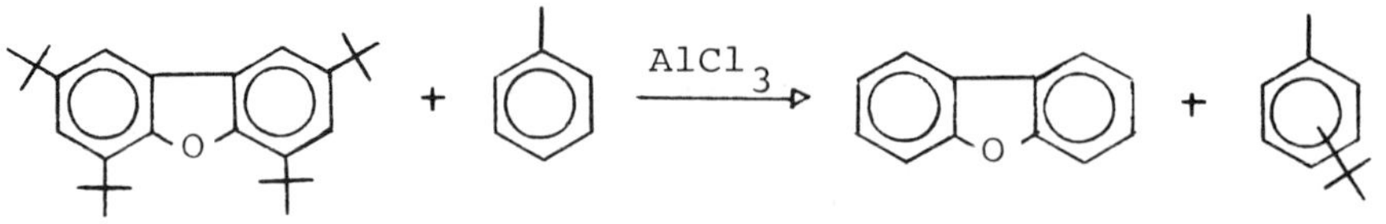

$AlCl_3$ (9b)

t-butyl groups in the commercially available starting material ensures *o-o* coupling.

There were earlier reports of facile exchanges of *t*-butyl groups between an aromatic hydrocarbon and a phenol, or between two phenols. For example, *t*-butylphenol isomers were shown to undergo reorientation by a transalkylation mechanism when heated at 100 to 200°C with a type Y zeolite catalyst [53]. The mechanism was said to be similar to that proposed by the same authors for the reorientation of diethylbenzenes [46]. The equilibrium distribution of the *t*-butylphenols at 200°C was approximately 1% *ortho*, 72% *meta*, and 25% *para*.

Itoh and co-workers reported transfer of the *t*-butyl group from *t*-butylbenzene to phenol in 50% yield (the catalyst was not specified in the abstract) [54]. Two studies were made on the equilibria established between *o-t*-butylphenols and other phenols having free *ortho* positions, of the type shown in Eq. (10). In the first study [55(a)], $Al(OPh)_3$ was used as catalyst, and

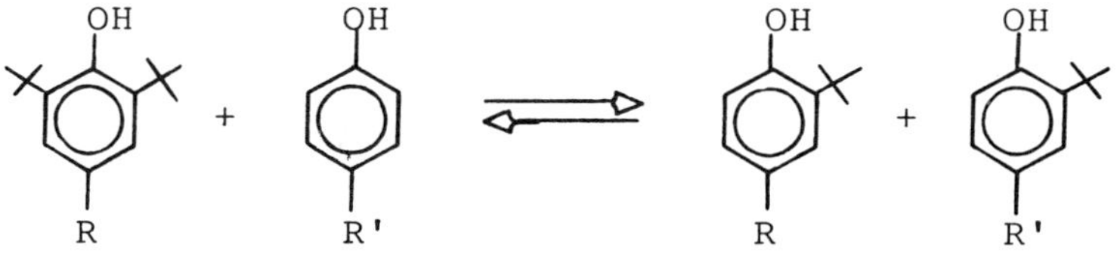

(10)

in the second [55(b), *p*-toluenesulfonic acid. Some reorientations were observed when the 3-, 4-, and 5-positions were unsubstituted.

In 1977 the intermolecular transfer of benzyl groups under the influence of Friedel-Crafts catalysts was demonstrated [56]. For example, treatment of compound 1 with $AlCl_3$-CH_3NO_2 in toluene at −33°C for 1 hr gave a mixture of products, among which were 2, 3, and *p-t*-butyltoluene (Eq. 11). The

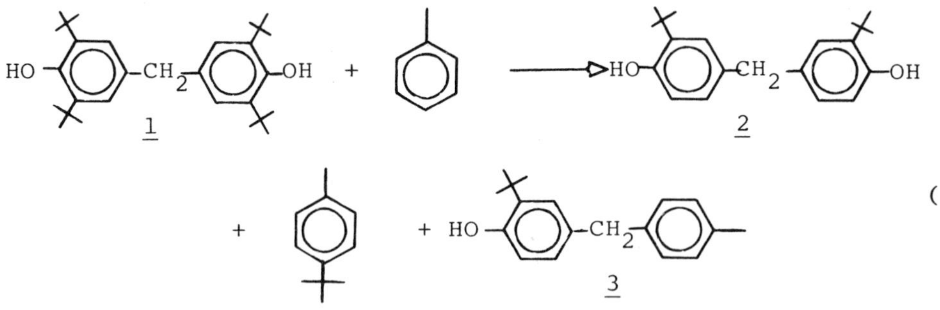

(11)

production of 2 and *p-t*-butyltoluene involves transbutylation. The formation of 3 represents a *transbenzylation*, as well; that is, after loss of one *t*-butyl group, the 4-hydroxy-3-*t*-butylbenzyl group has been transferred to a toluene molecule in the *para* position. The intermediates and/or transition states were depicted as shown in scheme 4. The "π-outer" complexes had been described earlier by Olah et al. [57]. Some of the other products identified by GC-MS were those differing in the orientation of the methyl group of 3. The authors emphasized the fact that the mild catalyst ($AlCl_3$-CH_3NO_2) which was effective for the transalkylations of these phenolic compounds is known to be ineffective for transbenzylations and reorientations of diphenylmethanes.

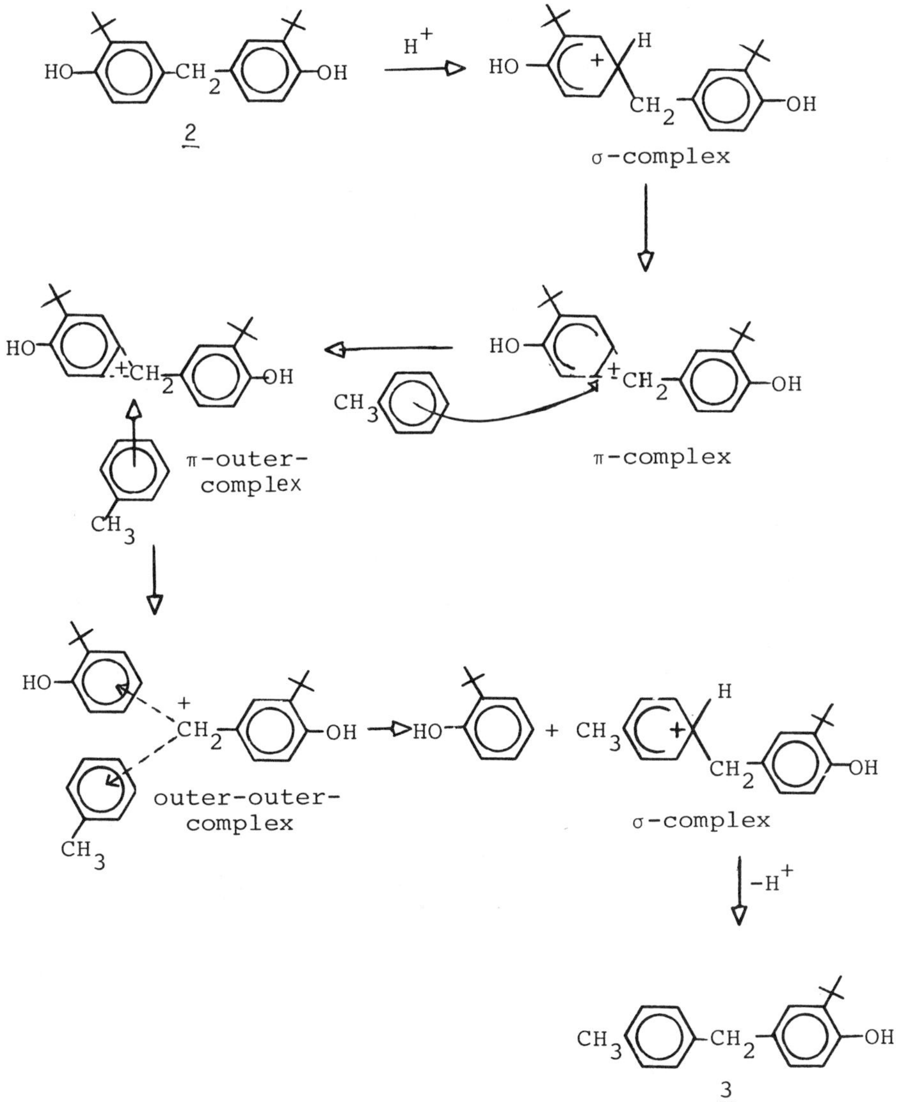

HO–C₆H₃(t-Bu)–CH₂–C₆H₃(t-Bu)–OH (2) $\xrightarrow{H^+}$ σ-complex → π-complex (+ $CH_3C_6H_5$) → π-outer-complex → outer-outer-complex → HO–C₆H₄(t-Bu) + σ-complex $\xrightarrow{-H^+}$ CH_3–C₆H₄–CH₂–C₆H₃(t-Bu)–OH (3)

Scheme 4

In another paper from the same laboratory, Tashiro and co-workers examined the possible transalkylations of *t*-butyl derivatives of diphenylmethanes and diphenylethanes, catalyzed by Lewis acids [58]. Reaction of compound 4 (Eq. 12) gave 5 and 6, which are products of trans-*t*-butylation. Reaction of compound 7 gave not only the products of trans-*t*-butylation, 8 and 9, but also 10 and 11, which come from transbenzylation (Eq. 13). Reaction of 12 gave a mixture of products indicative of mainly transbenzylations. When the milder catalyst $TiCl_4$ was used with 12, the transbenzylation products 13 and 14 were found as the only products, and in good yield (Eq. 14). The different reaction courses taken by compounds 4, 7, and 12 were ascribed to the different degrees of steric crowdedness at the methylene position. In

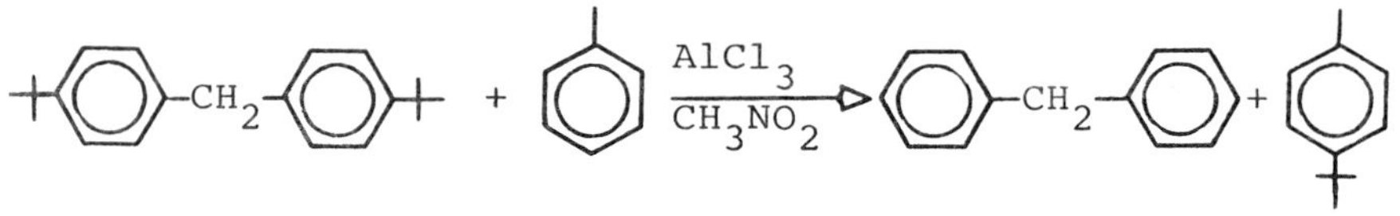

$AlCl_3$ / CH_3NO_2 (12)

4 5 6

$AlCl_3$ / CH_3NO_2

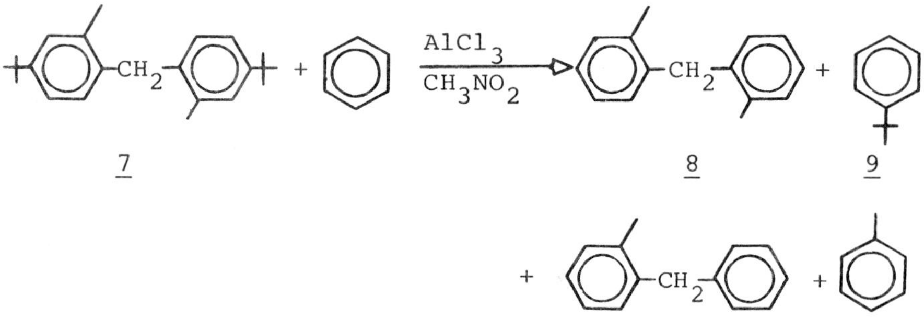

7 8 9

(13)

10 11

$TiCl_4$

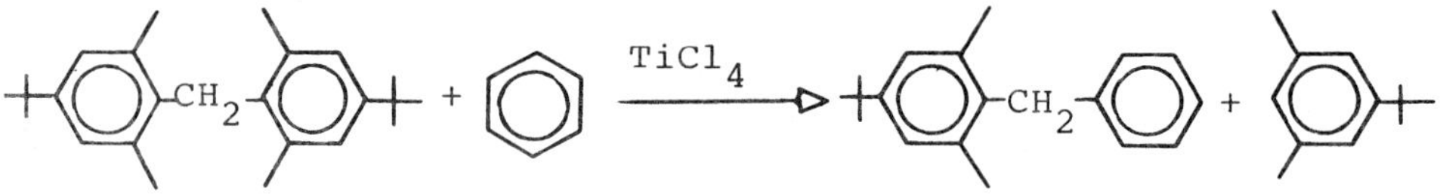

12 13 14

(14)

contrast to the behavior of the dihydroxy diphenylmethanes, the diphenylethane derivatives (15) underwent no transaralkylation (transfer of arylethyl groups) with $AlCl_3$-CH_3NO_2 or $TiCl_4$ catalysts, but trans-*t*-butylation occurred readily (Eq. 15).

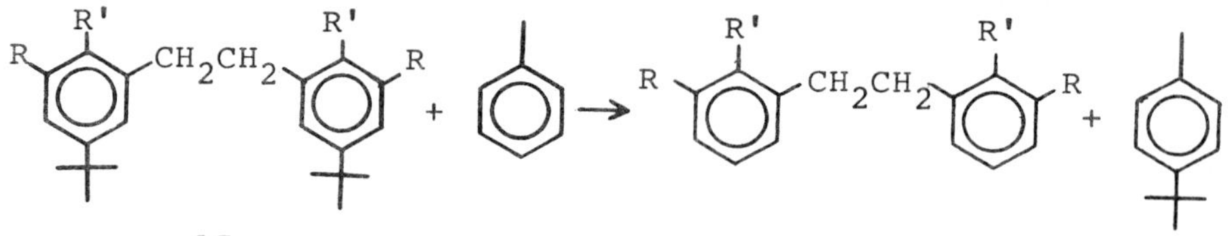

15 16 (15)

R,R' = H, CH_3, OCH_3, Cl, etc.

The lack of transaralkylation in the case of the diphenylethanes is understandable in terms of the mechanism proposed in the earlier paper [56]. The π complexes proposed as intermediates and depicted in scheme 3 have the nature of benzyl cations; the corresponding intermediates required for an analogous mechanism for transaralkylation of diphenylethanes would have the nature of β-aryethyl cations, which would be much less stable than benzyl-type cations.

On the basis of the results of their studies of the transalkylations of the diphenylmethane and diphenylethane derivatives, Tashiro and co-workers suggested that the *t*-butyl group might be used as a positional protective group for the synthesis of certain diphenylethanes.

J. Reorientations and Disproportionations of Alkylnaphthalenes

An investigation of the reactions of alkylnaphthalenes with $AlCl_3$ was carried out by K. Y. Zee-Cheng in the laboratory of one of the authors of this monograph, but the full details of his results were not published at the time of completion [59]. They will be presented here for comparison with published results from other laboratories which followed.

α-Methylnaphthalene did not undergo either reorientation to β-methylnaphthalene or disproportionation when treated with $AlCl_3$ (0.1 mol per mole of hydrocarbon) at 25°C for 24 hr. α-Ethylnaphthalene, under the same treatment, gave both reorientation and disproportionation. The α/β isomer ratio produced was 23:77, and about 10 mol% of naphthalene and diethylnaphthalene were formed. α-*n*-Propyl and α-*n*-butylnaphthalenes also underwent reorientation to the β isomers, at a rate about one-half that of α-ethylnaphthalene, and the rate of α-isobutylnaphthalene reorientation was much slower still. No side-chain rearrangement of the *n*-propyl, *n*-butyl, and isobutyl groups accompanied these reorientations. The α/β isomer ratios observed at the end of 24-hr reaction times at 25°C were as follows, but they may not represent equilibrium values:

	α	β
n-Propyl	28	72
n-Butyl	35	65
Isobutyl	67	33

α-Isopropyl- and α-*sec*-butylnaphthalenes rearranged to the β isomers at a rate approximately 100 times faster than that of α-ethylnaphthalene. The α/β ratios after 5 hr at 25°C were isopropyl, 7:93 and *sec*-butyl, 4:96. We were especially interested in the result from heating α-*sec*-butylnaphthalene with $AlCl_3$ at 100°C for 0.5 hr, because it was known that *sec*-butylbenzene under similar reaction conditions is isomerized to isobutylbenzene (cf. Chap. 8). There was no side-chain rearrangement of the *sec*-butylnaphthalene, however, only reorientation to a 4:96 ratio of α- to β-*sec*-butylnaphthalene. A supplementary experiment was carried out in which an equimolar amount of naphthalene was added to *sec*-butylbenzene and then the mixture of hydrocarbons was heated with $AlCl_3$. There was only a 4% rearrangement of the *sec*-butylbenzene to isobutylbenzene, whereas in the absence of naphthalene, conversion to the pseudoequilibrium proportion of *sec*-butylbenzene:isobutylbenzene of 1:2 would be expected, as described in Chap. 8, where an explanation of this behavior will be given. The same equilibrium proportions of α- and β-isopropylnaphthalene and α- and β-*sec*-butylnaphthalene were produced by treating the β isomers with $AlCl_3$.

Zee-Cheng did no experiments with α-*t*-butylnaphthalene, but he showed that β-*t*-butylnaphthalene underwent disproportionation with $AlCl_3$ at 25°C with no detectable production of the α isomer, and no dealkylation to isobutane (as is observed with *t*-butylbenzene; cf. Chap. 8).

Martan et al. investigated the dependence of alkylnaphthalene isomer distribution on catalyst, solvent, temperature, and extent of conversion, using the reaction of 2-butene with naphthalene as the test system [60]. They reported that $AlCl_3$ gave an isomer ratio that was dependent on extent of conversion, because of reorientation of the α isomer to the thermodynamically more stable β isomer [59]. With H_2SO_4 as catalyst (10% in CCl_4), a constant ratio of α and β isomers was found up to 90% conversion, indicating that no reorientation accompanied the alkylation. (The products were referred to in their table as isobutylnaphthalenes, but this must be a mistake, judging from the finding of Zee-Cheng that alkylation with *sec*-butyl chloride and $AlCl_3$ gave *only* α- and β-*sec*-butylnaphthalene [59].) An unusual temperature dependence was observed. At 10 and −10°C the α/β ratio was 60:40, but at −5°C it was 80:20. No explanation for this result was offered.

In 1976, George and Judith Olah reported a study of the alkylation of naphthalene with methyl, ethyl, isotropyl, and *t*-butyl halides, using $AlCl_3$ and $SnCl_4$ as catalysts [61]. In the alkylations with $AlCl_3$ in benzene or carbon disulfide solution, the α/β isomer ratio stayed high and quite constant for methylation and ethylation, showed relatively slow variation in isopropylation, and changed rapidly with conversion only in the case of *t*-butylation (even at 0°C). To ascertain the effect of reorientation on the observed isomer distributions in the alkylation products, they investigated separately the reaction of each of the pure α-alkylnaphthalenes with $AlCl_3$ in CS_2 solution.

The progress of the reorientations was followed by capillary gas chromatography. Table 3 summarizes data which show that the order of rate of reorientation of the alkyl groups from the α to the β position of naphthalene was t-Bu > i-Pr > Et > Me, and gives the equilibrium α/β isomer ratios. As to the mechanism of the α-to-β reorientation, the authors speculate that the alkyl shifts can take place in "arenium ion-type intermediates," and they further state; "As methyl and ethyl group migration takes place predominantly

Table 3. Relative Reorientation Rates and Equilibrium Ratios of Alkylnaphthalenes

	Temp. (°C)	Time[a] (min)	α-/β-
Me	46	25	25:75
Et	46	2.5	9.5:90.5
i-Pr	0	13	1.5:98.5
t-Bu	0	6	0:100

[a]Approximate time required to reach a 1:1 ratio of α/β isomers, starting with α-alkylnaphthalene.

via intramolecular processes, these can be less suppressed than those of isopropyl or *t*-butyl groups, which tend to migrate intermolecularly." Perhaps this statement should be qualified with respect to the intramolecular nature of an ethyl shift in view of the work of Wolf and co-workers [44] and Roberts and Roengsumran [45].

The Olahs' work showed clearly the ease with which kinetically controlled alkylations can be affected by thermodynamically controlled reorientations. In the case of naphthalene, predominantly kinetic reactions give high α-/β-alkylnaphthalene isomer ratios initially, but the β isomer may be the major one isolated, owing to a secondary reorientation reaction which is controlled by the greater thermodynamic stability of the β isomer. In comparing the equilibrium α/β isomer ratios of the Olahs' with the ratio for Et, *n*-Pr, *n*-Bu, and isobutyl observed by Zee-Cheng (which were higher), it seems probable that Zee-Cheng's reorientation reactions had not reached equilibrium.

The reorientations of dimethylnaphthalenes in HF-BF_3 were investigated by Suld and Stuart [62]. They found that the proportions of the isomers were dependent on the amount of BF_3 in the catalyst medium, as is true in the dimethylbenzene (xylene) system [9]. The unique feature of the reorientations of the dimethylnaphthalenes was the finding that there were three discrete sets of isomer groups among which reorientations occurred, but there was no crossover between the three sets. These sets of isomers were (1) 2,6-, 1,6-, and 1,5-dimethylnaphthalene; (2) 2,7-, 1,7-, and 1,8-dimethylnaphthalene; and (3) 2,3-, 1,3-, and 1,4-dimethylnaphthalene. It is obvious that the reorientations in each of the three sets occurred by 1,2-shifts of methyl groups between the 1-(α-) and 2-(β-) positions, and that such shifts did not occur between the 2,3- or 1,8-positions The authors state that their results indicate there must be a migrational barrier between the adjacent 2,3- or β,β' positions, as well as between the two rings of the naphthalene nucleus. They cite a similar conclusion reached by Russian workers [63] on the basis of their results from an isomerization study of 1-methyl-1-^{14}C-naphthalene. Suld and Stuart rationalized this migrational barrier in terms of the stability of the intermediate σ complex (17) required for the 1,2 methyl shift between the two β positions, which does not have the resonance stabilization of an intact benzene ring as the intermediates for all of the observed reorientations do. Similar reasoning applies to the intramolecular migration of the methyl groups from one ring to another, as the consecutive methyl shifts would involve another quinonoid structure, such as 18.

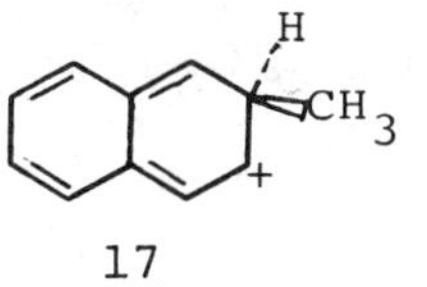

17 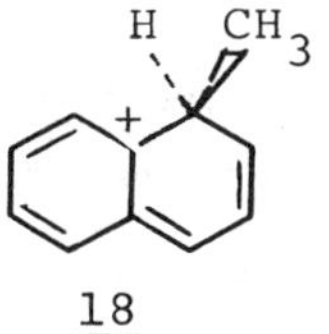 18

Some interesting reorientations and disproportionations of polymethylnaphthalenes induced by trifluoroacetic acid were described by Oku and Uyzen [64]. They found that α,β-methyl migrations occurred smoothly in seven polymethylnaphthalenes with methyl substituents in peri positions and with at least one adjacent β position unsubstituted. For example, 19 and 20, which in turn gave 21 (Eq. 16); 22 gave 23, which further gave a mixture of 24

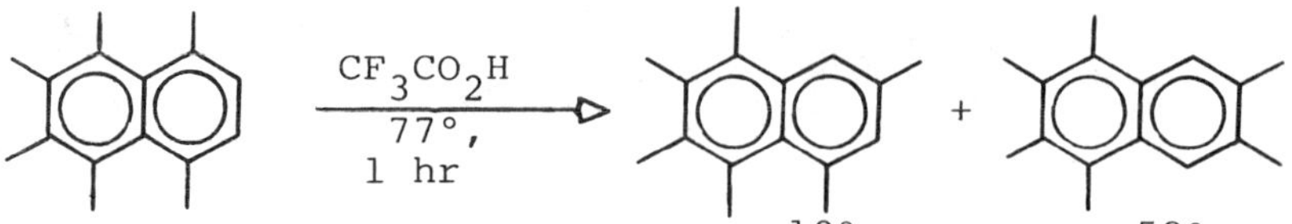

$$19 \xrightarrow[\text{77°, 1 hr}]{CF_3CO_2H} 20 \; (19\%) + 21 \; (58\%) \qquad (16)$$

and 25 (Eq. 17). Other examples were the conversion of 26 to 27 and 28 to 29 (Eqs. 18 and 19).

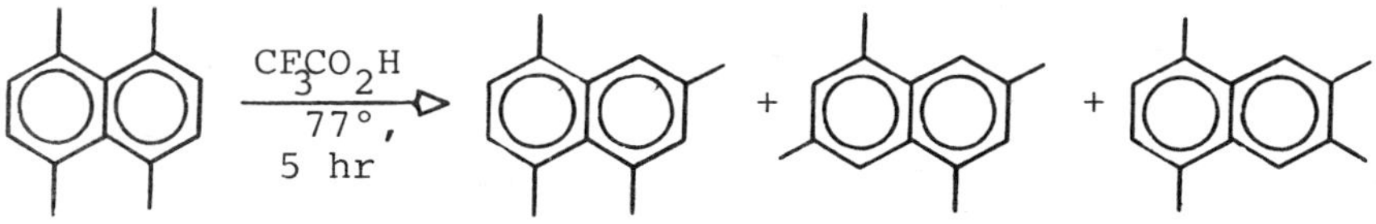

$$22 \xrightarrow[\text{77°, 5 hr}]{CF_3CO_2H} 23 \; (76\%) + 24 \; (12\%) + 25 \; (1\%) \qquad (17)$$

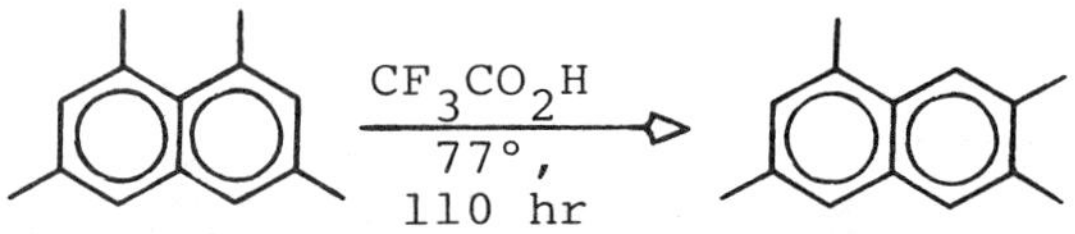

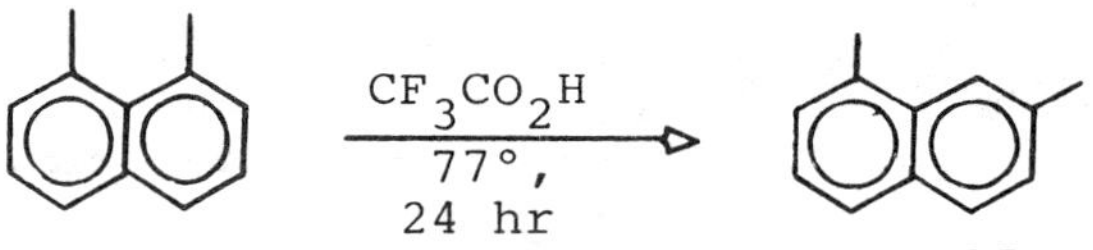

$$26 \xrightarrow[\text{77°, 110 hr}]{CF_3CO_2H} 27 \; (93\%) \qquad (18)$$

$$28 \xrightarrow[\text{77°, 24 hr}]{CF_3CO_2H} 29 \; (15\%) \qquad (19)$$

Other acidic media than CF_3CO_2H were also tried for the reorientations. Hydrogen chloride in acetic acid was found to induce reorientations with relatively low yields of by-products, and more effectively when $AlCl_3$, BF_3, or $Zn(CN)_2$ were added, but none of these combinations was as effective as CF_3CO_2H with respect to yields, reaction time, and simplicity of products.

For the methyl reorientations, the authors propose a mechanism involving intramolecular, 1,2-shifts of methyl groups via σ-complex intermediates, and they state that "the migrating force mainly derives from the strain release of peri interaction, and the carbocation stability from both the strain release and the electronic effects of methyl substitution."

In compounds without peri position methyl groups, little reorientation occurred; instead, intermolecular transalkylations took place at slow rates and with low yields. For example, heating 1,2,3,4-tetramethylnaphthalene (30)

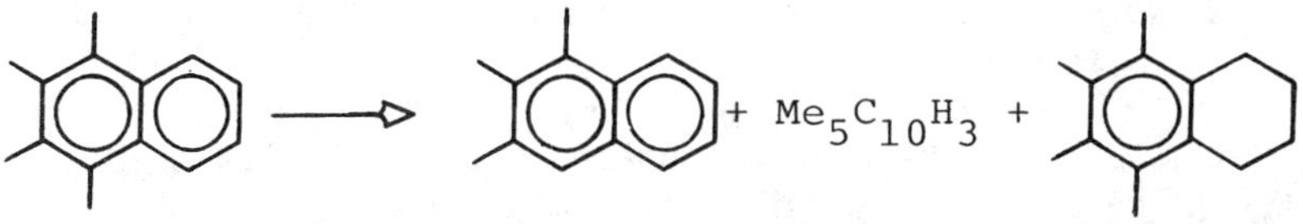

$$30 \longrightarrow 31 \; (13\%) + Me_5C_{10}H_3 \; 32 \; (1\%) + 33 \; (4\%) \qquad (20)$$

for 110 hr in refluxing CF_3CO_2H gave 31-33 in the amounts shown in Eq. (20), besides unchanged 30 (80%). The transalkylation of the methyl group was proposed to occur via a benzyl-type carbocation (35), which might result from an intermolecular hydride ion transfer producing a dihydronaphthalene (34) (Eq. 21). A second sequence of proton and hydride addition to 34 would ac-

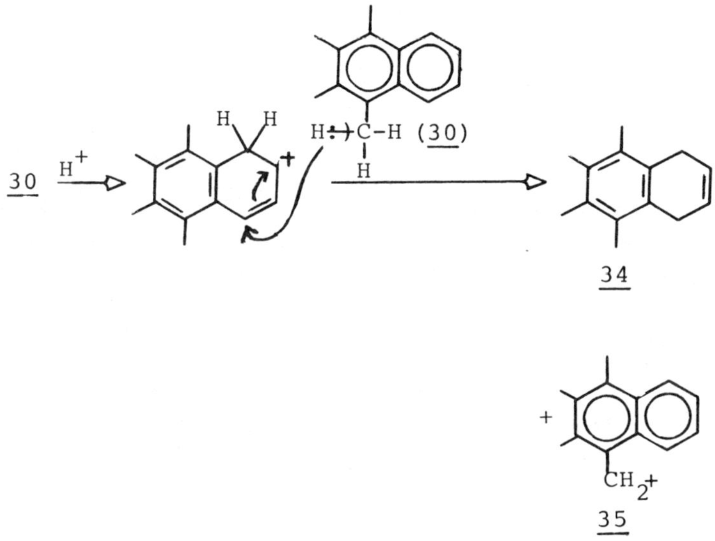

(21)

count for the formation of the tetrahydronaphthalene (33), and the carbocation 35 could alkylate another molecule of 30 and subsequently produce the trimethylnaphthalene (31) and pentamethylnaphthalene (32). Such a mechanism of intermolecular transalkylation of a methyl group is, of course, of the type proposed by Streitwieser and Reif [33] and Roberts et al. [65] for transalkylations of ethyl- and *n*-propylbenzenes, respectively. There is good precedent for the formation of benzyl-type cations from *p*-xylene and the subsequent production of diarylmethane derivatives (Eq. 22), even with a mild catalyst

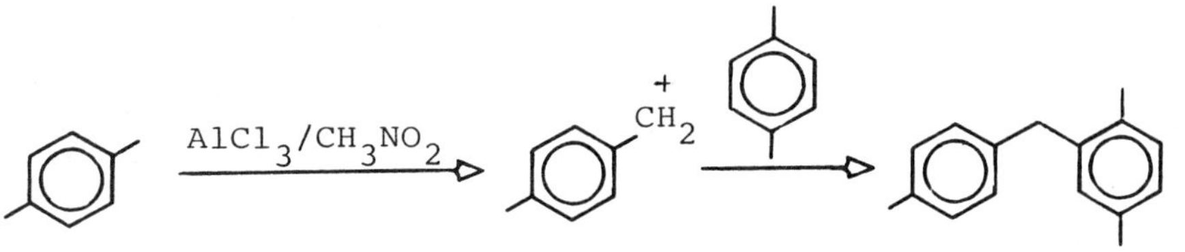

(22)

such as $AlCl_3$-CH_3NO_2 [66], so it is possible that the mechanism proposed by Oku and Yuzen for the disproportionations of the polymethylnaphthalenes may hold in CF_3CO_2H solution.

Reduction of naphthalenes to tetralins has previously also been observed to occur in the presence of acid catalysts. Treatment of α-methylnaphthalene in benzene with $AlCl_3$-H_2O catalyst at 80°C was found to produce besides β-methylnaphthalene small amounts of methyltetralins [67].

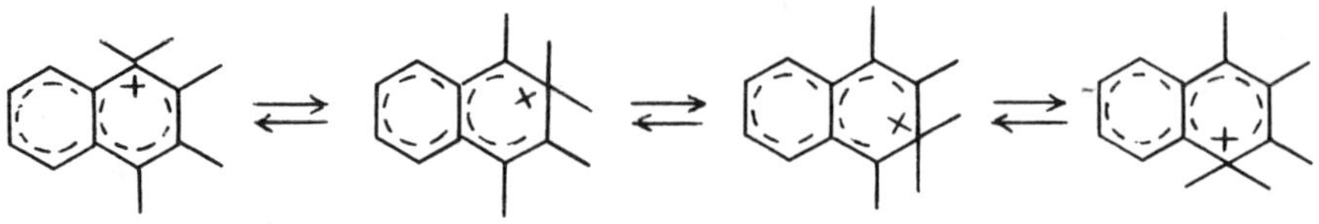

Scheme 5

A degenerate reorientation of the 1,1,2,3,4-pentamethylnaphthalenonium ion, which is realized by three successive 1,2 shifts of a methyl group (scheme 5), has been demonstrated to occur in strong protonic acids (e.g., CF_3SO_3H) [67a]. The degenerate ion was produced by solution of appropriate alkenes and alcohols in acid systems, and the reorientations were observed by means of dynamic PMR spectroscopy. CD_3-labeled molecules were utilized in some of the experiments.

K. Reorientations of Hydrindacenes

s-Hydrindacene (46) and *as*-hydrindacene (37) may be interconverted easily by HF-BF_3 or HF-BCl_3 [68]. Mixtures containing less than the equilibrium

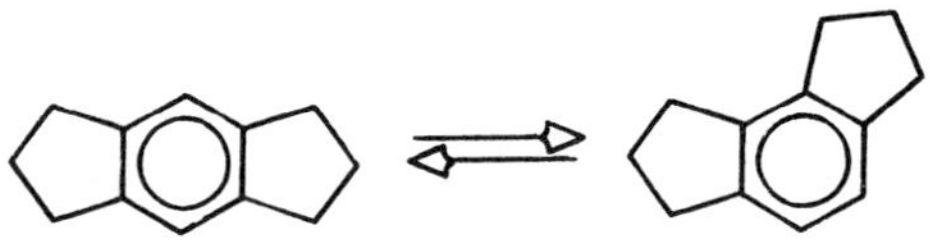

36 37

amount (36/37 = ca. 1/4) of either 36 or 37 can be brought toward the equilibrium composition by heating in HF-BF_3 (or HF-BCl_3) at 40 to 80°C and, after removal of the catalyst, the isomers may be separated by distillation. The procedure was of patent interest because the hydrindacenes can be oxidized to benzenetetracarboxylic acids, or their anhydrides, which are valuable as monomers for polymer manufacture.

L. Transalkylations and Reorientations of Bromotoluenes

As we have seen, there has been considerable interest in the mechanism of reorientation of alkylbenzenes, and there is experimental evidence that in some cases it proceeds by an intramolecular 1,2-shift, in others by an intermolecular transalkylation, and in some, quite possibly by the concurrent operation of both mechanisms.

The reactions of halotoluenes with Al_2Br_6-H_2O were examined by Olah and Meyer in 1962 [69]. They found that *o*- and *p*-bromotoluene isomerized (reoriented) by a rapid intermolecular transbromination, followed by a slower 1,2-shift of bromine. *m*-Bromotoluene was said to isomerize by "an apparent

1,2-shift." But they added: "The present evidence, however, does not permit differentiation between the intramolecular (1,2-shift) and the intermolecular mechanism." The reorientations of the bromotoluenes were always accompanied by disproportionation. Varying amounts of toluene and dibromotoluenes were found, indicating that the bromo substituent moves and not the methyl groups.

A few years later, Dutch workers undertook to clarify the ambiquity of the reorientation of *m*-bromotoluene. They did this by treating the bromotoluenes with Al_2Br_6-HBr in the presence of ^{14}C-labeled toluene and determining the radioactivity of the separated *o*-, *m*-, and *p*-bromotoluenes in the reaction mixtures after various time intervals [70]. Their results confirmed Olah's conclusions that *o*- and *p*-bromotoluene undergo reorientation by an intermolecular mechanism, and they obtained evidence which they interpreted to mean that *m*-bromotoluene does indeed reorient to the *o* and *p* isomers by a 1,2-shift of bromine in σ-complex intermediates.

REFERENCES

1. Anschütz, R., Justus Liebigs Ann. Chem., *235*, 177 (1886).
2. Heise, R., and A. Töhl, Justus Liebigs Ann. Chem., *270*, 155 (1892).
3. Price, C. C., in *Organic Reactions*, Vol. 3, Wiley, New York, 1946 Chap. 1.
4. Baddeley, G., and J. Kenner, J. Chem. Soc., 303 (1935).
5. (a) Nightingale, D., and L. I. Smith, J. Am. Chem. Soc., *61*, 101 (1939); (b) Nightingale, D., and B. Carton, J. Am. Chem. Soc., *62*, 280 (1940); (c) Nightingale, D., and J. M. Shackelford, J. Am. Chem. Soc. *76*, 5767 (1954); (d) Nightingale, D., and J. M. Shackelford, J. Am. Chem. Soc. *78*, 1225 (1956).
6. McCaulay, D. A., and A. P. Lien, J. Am. Chem. Soc., *75*,2411 (1953).
7. Kinney, R. E., and L. A. Hamilton, J. Am. Chem. Soc., *76*, 786 (1954).
8. Baddeley, G., G. Holt, and D. Voss, J. Chem. Soc., 100 (1952).
9. McCaulay, D. A., and A. P. Lien, J. Am. Chem. Soc., *74*, 6246 (1952).

9a. Boedeker, E. R., and W. E. Erner, J. Am. Chem. Soc., *76*, 3591 (1954).

10. Brown, H. C., and H. Jungk, J. Am. Chem. Soc., *77*, 5579 (1955).
11. Steinberg, H., and F. L. J. Sixma, Recl. Trav. Chim. Pays-Bas, *81*, 185 (1962).
12. Brown, H. C., and C. R. Smoot, J. Am. Soc., *78*, 2176 (1956).
13. Steinberg, H., and F. L. J. Sixma, Recl. Trav. Chim. Pays-Bas, *79* 679 (1960).

13a. Buraev, V. I., I. S. Isaev, and V. A. Koptyug, J. Org. Chem. USSR, *15*, 697 (1979).

14. Matsumoto, H., and Y. Morita, Kogyo Kagaku Zasshi, *70*(10), 1674 (1967); Chem. Abstr., *68*, 77861 (1968).
15. Parasiewicz-Kaczmarska, J., Zesz. Nauk. Uniw. Jagiellon., Pr. Chem., No. 17, 173 (1972); Chem. Abstr., *78*, 29338 (1973).
16. Chojnacki, R., D. Kruszkowa, J. Obloj, J. M. Berak, Z. Lisicki, W. Tecza, and M. Popowicz, Przem. Chem., *53*(6), 338 (1974); Chem. Abstr., *81*, 91148 (1974).
17. Gryaznova, Z. V., I. V. Nicolescu, G. V. Tsitsishvili, and K. A. Baskun'yan, Neftekhimiya, *17*(4), 531 (1977).

18. DeKoninck, A., I. Aguirre, and C. Marcilly, Fr. Patent 2,291,957 (1976); Chem. Abstr., *86*, 120959 (1977).
19. Sampson, R., A. L. Crowther, J. K. January, and I. J. S. Lake, Ger. Offen. 2,353,327 (May 2, 1974); Chem. Abstr., *81*25317 (1974).
20. Matsumura, T., S. Hayashi, D. Ogawa, M. Sato, S. Otani, T. Iwamura, K. Hashiguchi, T. Handa, and T. Miyata, Jpn. Patent 72-20,211 (June 8, 1972); Chem. Abstr., *77*, 61497 (1972).
21. Miyoshi, K., and M. Shigeyasu, *Maruzen Sekiyu Giho*, No. 17, 39 (1972); Chem. Abstr., *79*, 5063 (1973).
22. Tsuchiya, S., Y. Saito, T. Murakami, H. Iwata, and T. Annaka, Jpn. Kokai 74-80,024 (Aug. 2, 1974); Chem. Abstr., *81*, 169269 (1974).
23. Olah, G. A., and J. Kaspi, Nouv. J. Chimie, *2*, 581 (1978).
24. Delone, I. O., L. Z. Osityanskaya, and A. A. Petrov, Neftekhimiya, *2*(2), 189 (1962); Petrol. Chem. USSR, *2*, 123 (1963); Chem. Abstr., *58*, 5583 (1963).
25. Molchanova, V. V., M. P. Tsvetkova, V. Skobeleva, and V. A. Valiulina, Tr. Inst. Khim., Ural. Nauchn, Tsentr., Akad. Nauk USSR, *28*, 17 (1974); Chem. Abstr., *83*, 9336 (1975).
26. Roberts, R. M., G. A. Ropp, and O. K. Neville, J. Am. Chem. Soc., *77*, 1764 (1955). This work was actually done in Oak Ridge in the summer of 1951, before the publication of the results of disproportionations of *n*-propyl- and *n*-butylbenzene without rearrangement (Refs. 6 and 7).
27. (a) Allen, R. H., T. Alfrey, Jr., and L. D. Yats, J. Am. Chem. Soc., *81*, 42 (1959); (b) Allen, R. H., and L. D. Yats, J. Am. Chem. Soc., *81*, 5289 (1959); (c) Allen, R. H., J. D. Yats, and D. S. Erley, J. Am. Chem. Soc., *82*, 4853 (1960); (d) R. H. Allen, J. Am. Chem. Soc., *82*, 4856 (1960).
28. Olah, G. A., M. W. Meyer, and N. A. Overchuk, J. Org. Chem., *29*, 2310 (1964).
29. (a) Olah, G. A., M. W. Meyer, and N. A. Overchuk, J. Org. Chem., *29*, 2313 (1964); (b) Olah, G. A., M. W. Meyer, and N. A. Overchuk, J. Org. Chem., *29*, 2315 (1964); (c) Olah, G. A., C. G. Carlson, and J. C. Lapierre, J. Org. Chem., *29*, 2687 (1964).
30. Myhre, P. C., T. Rieger, and J. T. Stone, J. Org. Chem., *31*, 3425 (1966).
31. (a) Bartlett, P. D., M. Roha, and R. M. Stiles, J. Am. Chem. Soc., *76* 2349 (1954); (b) Barclay, L. R. C., and E. E. Betts, Can. J. Chem., *33*, 672 (1955).
32. Burwell, R. L., Jr., and A. D. Shields, J. Am. Chem. Soc., *77*, 2766 (1955).
33. Streitwieser, A., Jr., and L. Reif, J. Am. Chem. Soc., *86*, 1988 (1964).
34. Pines, H., and J. T. Arrigo, J. Am. Chem. Soc., *80*, 4369 (1958).
35. Roberts, R. M., and D. Shiengthong, J. Am. Chem. Soc., *86*, 2851 (1964).
36. (a) Sharman, S. H., J. Am. Chem. Soc., *84*, 2951 (1962); (b) Sharman, S. H., U.S. Patent 3,235,616 (Feb. 15, 1966); Chem. Abstr., *64*, 17479 (1966).
37. Mazonski, T., D. Gasztych, and A. Piatek. Zesz. Nauk. Politech. Slask, Chem., No. 13, 49-55 (1963); Chem. Abstr., *62*, 13064 (1965).
38. Strohmeyer, M., Erdoel Kohle, Erdgas, Petrochem., 1969, 22(9), 519; Chem. Abstr., *71*, 123739 (1969).
39. (a) Shimada, K., Nippon Kagaku Zasshi, *90*, 1041 (1969); Chem. Abstr., *72*, 42991 (1970); (b) Shimada, K., Nippon Kogaku Kaishi, 616 (1972); Chem. Abstr., *76*, 152852 (1972).

40. Kozorezov, Y. I., and A. N. Kuleshova, Neftepererab. Neftekhim. (Moscow), No. 11, 33 (1973); Chem. Abstr., *80*, 70451 (1974).
41. Yanagihara, T., and K. Fukahori, Jpn. Kokai 74-54,336 (May 27, 1974); Chem. Abstr., *81*, 169266 (1974).
42. Hamanaka, S., H. Yoshida, and M. Ogawa, Technol. Rep. Kansai Univ., *10*, 51 (1969); Chem. Abstr., *72*, 11778 (1970).
43. (a) Lipovich, V. G., A. V. Vysotskii, and V. V. Chenets, Neftekhimya, *11*(5), 656 (1971); Chem. Abstr., *76*, 13565 (1972); (b) Lipovich, V. G., A. V. Vysotskii, and V. V. Chenets, Neftekhimya, *11*(6), 838 (1971); Petrol. Chem. USSR, *11*(4) 237 (1971); (c) Chenets, V. V., O. R. Sergeeva, T. V. Chizhevskaya, and V. G. Lipovitch, J. Org. Chem. USSR, *15*, 475 (1979); (d) Chenets, V. V., O. R. Sergeeva, T. V. Chizhevskaya, T. I. Glebco, and V. G. Lipovitch, J. Org. Chem, USSR, *15*, 480 (1979).
44. Ünseren, E., and A. P. Wolf, J. Org. Chem., 27, 1509 (1962); (b) Moore, G. G., and A. P. Wolf, J. Org. Chem., *31*, 1106 (1966).
45. Roberts, R. M., and S. Roengsumran, J. Org. Chem., *46*, 3689 (1981).
46. Bolton, A. P., M. A. Lanewala, and P. E. Pickett, J. Org. Chem., *33*, 1513 (1968).
47. Matsumoto, H., J.-I. Take, and Y. Yoneda, J. Catal., *19*, 113 (1970).
48. Olah, G. A., and J. Kaspi, Nouv. J. Chim., *2*, 585 (1978).
49. (a) Roberts, R. M., and S. G. Brandenberger, J. Am. Chem. Soc., *79*, 5484 (1957); (b) Roberts, R. M., and J. E. Douglass, J. Org. Chem., *28*, 1225 (1963).
50. (a) Kunichika, S., S. Oka, T. Sugiyama, C. Inoue, and M. Ichii, Nippon Kagaku Zasshi, *92*(6), 539 (1971); Chem. Abstr., *76*, 58676 (1972); (b) Kunichika, S., S. Oka, T. Sugiyama, and M. Ichii, Nippon Kagaku Zasshi, *92*(9), 801 (1971); Chem. Abstr., *76*, 71771 (1972).
51. Hamanaka, S., S. Saigan, K. Itoh, and M. Ogawa, Kogyo Kagaku Zasshi, *70*(5), 695 (1967); Chem. Abstr., *67*, 116652 (1967).
52. Tashiro, M., T. Yamato, and G. Fukata, J. Org. Chem., *43*, 743 (1978).
52a. Tashiro, M., H. Yoshiya, and G. Fukata, J. Org. Chem., *46*, 3784 (1981).
53. Bolton, A. P., M. A. Lanewala, and P. E. Pickert, J. Org. Chem., *33*, 3415 (1968).
54. Itoh, K., Kadokawa, Y., S. Hamanaka, and M. Ogawa, Technol. Rep. Kansai Univ., *9*, 59 (1967); Chem. Abstr., *70*, 96306 (1969).
55. (a) Willemse, F. R. J., J. Wolters, and E. C. Kooijman, Recl. Trav. Chim. Pays-Bas, *90*, 5 (1971); (b) Willemse, F. R. J., J. Wolters, and E. C. Kooijman, Recl. Trav. Chim. Pays-Bas, *90*, 14 (1971).
56. Tashiro, M., and G. Fukata, J. Org. Chem., *42*, 1208 (1977).
57. Olah, G. A., S. Kobayashi, and M. Tashiro, J. Am. Chem. Soc., *94*, 7448 (1972).
58. Tashiro, M., T. Yamato, and G. Fukata, J. Org. Chem., *43*, 1413 (1978).
59. Zee-Cheng, K. Y., Ph.D. dissertation, University of Texas at Austin, Diss. Abstr., *26*(4), 1925 (1965); Chem. Abstr., *64*, 3436 (1966).
60. Martan, M., J. Manassen, and D. Vofsi, Chem. Ind. (Lond.), 434 (1970).
61. Olah, G. A., and J. A. Olah, J. Am. Chem. Soc., *98*, 1839 (1976).
62. Suld, G., and A. P. Stuart, J. Org. Chem., *29*, 2939 (1964).
63. Vorozhtsov, N. N., and V. A. Koptyug, J. Gen. Chem. USSR, *30*, 1014 (1960).

64. Oku, A., and Y. Yuzen, J. Org. Chem., *26,* 3850 (1975).
65. Roberts, R. M., A. A. Khalaf, and R. N. Greene, J. Am. Chem. Soc., *86,* 2846 (1964).
66. Friedman, B. S., F. L. Morritz, C. J. Morrisey, and R. Koncos, J. Am. Chem. Soc., *80,* 5867 (1958).
67. Abdel-Baset, M. B., Ph.D. dissertation, University of Texas at Austin, 1973, p. 48.
67a. Morozov, S. V., M. M. Shakirov, V. G. Shubin, and V. A. Koptyug, J. Org. Chem. USSR, *15,* 685 (1979).
68. Bushick, R. D., U.S. Patent 3,637,882 (Jan. 25, 1972); Chem. Abstr., *76,* 99401 (1972).
69. Olah, G. A., and M. W. Meyer, J. Org. Chem., *27,* 3464 (1962).
70. deValois, P. J., M. P. van Albada, and J. U. Veenland, Tetrahedron, *24,* 1835 (1968).

8

Rearrangements, Dealkylations, and Fragmentations of Arenes Induced by Friedel-Crafts Catalysts

It is almost funny to find out that the so much investigated field of hydrocarbon isomerization is still hiding new rearrangements to be discovered.

(C. D. Nenitzescu in a letter to R. M. Roberts, 1958)

I. REARRANGEMENTS OF ARENES

Although the Friedel-Crafts alkylation reaction has been widely used as a synthetic procedure, there are inherent problems which have complicated its practical applications. In Chaps. 2 to 5 we have discussed the rearrangements of alkylating agents which may occur previous to or simultaneous with the alkylation process. In Chap. 7 we described the reorientations and transalkylations which also often accompany alkylation reactions. These aspects of Friedel-Crafts alkylations have been recognized for many years, and indeed some examples were described by Friedel and Crafts themselves in their earliest papers. More recently, however, an additional complication was discovered, the isomerization of the side chain of an alkylbenzene *after* attachment to the aromatic ring. This new development has been called the *alkylbenzene rearrangement* [1].

A. Secondary Alkylbenzenes Produced by Alkylation with Tertiary Alkyl Chlorides

The earliest example of the alkylbenzene rearrangement was reported by Schmerling and West in 1954 [2], but the acceptance of it as such came much later, after considerable controversy. Schmerling and West described alkylations of benzene with *tertiary* pentyl and hexyl chlorides, and they found that when aluminum chloride was used as catalyst, the major products were the *secondary* alkylbenzenes 2 and 4, as shown in Eqs. (1) and (2):

$$CH_3\text{-}CH_2\text{-}C(CH_3)(Cl)\text{-}CH_3 + C_6H_6 \xrightarrow[0^\circ]{AlCl_3} CH_3CH_2\text{-}C(CH_3)(C_6H_5)\text{-}CH_3 + CH_3\text{-}CH(CH_3)\text{-}CH(C_6H_5)CH_3 \quad (1)$$

1 (15%) 2 (85%)

$$CH_3\text{-}CH(CH_3)\text{-}CCl(CH_3)\text{-}CH_3 + C_6H_6 \xrightarrow[0^\circ]{AlCl_3} \underset{\underline{3}\ (10\%)}{CH_3\text{-}CH(CH_3)\text{-}C(CH_3)(C_6H_5)\text{-}CH_3} + \underset{\underline{4}\ (90\%)}{CH_3C(CH_3)_2\text{-}CH(C_6H_5)\text{-}CH_3} \quad (2)$$

When BF_3, $FeCl_3$, or $AlCl_3$-CH_3NO_2 were used as catalysts, the only products were 1 and 3, respectively—the expected products with side chains of retained constitution [2,3]. The explanation offered by Schmerling and West for the production of the *secondary* alkylbenzenes involved initial alkylation to produce the expected *tertiary* alkylbenzenes, followed by their rearrangement to the secondary isomers. The mechanism suggested for the pentyl system was as shown in Eqs. (3) to (7).

$$\underset{\underline{1}}{CH_3\text{-}CH_2\text{-}C(CH_3)(C_6H_5)\text{-}CH_3} + \underset{\underline{5}}{CH_3\text{-}CH_2\text{-}\overset{\oplus}{C}(CH_3)\text{-}CH_3} \longrightarrow \underset{\underline{6}}{CH_3\text{-}\overset{\oplus}{C}H\text{-}C(CH_3)(C_6H_5)\text{-}CH_3} + \underset{\underline{7}}{CH_3\text{-}CH_2\text{-}CH(CH_3)\text{-}CH_3} \quad (3)$$

$$\underline{6} \xrightarrow{\sim C_6H_5:} \underset{\underline{8}}{CH_3\text{-}CH(C_6H_5)\text{-}\overset{\oplus}{C}(CH_3)\text{-}CH_3} \quad (4)$$

$$\underline{8} \xrightarrow{\sim H:} \underline{9} \quad (5)$$

$$\underline{6} \xrightarrow{\sim CH_3:} \underset{\underline{9}}{CH_3\text{-}\overset{\oplus}{C}(C_6H_5)\text{-}CH(CH_3)\text{-}CH_3} \quad (6)$$

$$\underline{9} + \underline{1} \longrightarrow \underline{2} + \underline{6} \quad (7)$$

This explanation was not generally accepted, however. Several authors preferred to consider the formation of the secondary alkylbenzenes as being the result of initial rearrangement of the tertiary alkyl chlorides to secondary alkyl chlorides, followed by secondary alkylation, which was considered to be faster and less reversible than tertiary alkylation [3-5], e.g., Eqs. (8) and (9).

$$CH_3-CH_2-\underset{Cl}{\overset{CH_3}{C}}-CH_3 \xrightleftharpoons{AlCl_3} CH_3-\underset{Cl}{CH}-\overset{CH_3}{CH}-CH_3 \quad (8)$$

$$CH_3-\underset{Cl}{CH}-\overset{CH_3}{CH}-CH_3 + C_6H_6 \xrightarrow{AlCl_3} CH_3-\underset{C_6H_5}{CH}-\overset{CH_3}{CH}-CH_3 \quad (9)$$

2

Thus for many years it remained uncertain whether the rearrangement that produced the tertiary alkylbenzenes occurred before or after the side chains became attached to the benzene ring.

B. The *n*-Propylbenzene Isotopic Rearrangement

In Chap. 7 we mentioned the demonstration of the disproportionation of ethyl-β-^{14}C-benzene [6], *n*-propylbenzene [7], and *n*-butylbenzene [8] with no accompanying rearrangement of the side chains. However, before the papers on *n*-propyl- and *n*-butylbenzene appeared, Roberts and Brandenberger had embarked on a reexamination of the earliest report of disproportionation of *n*-propylbenzene in which Heise and Töhl [9] claimed that no side-chain rearrangement accompanied the disproportionation by $AlCl_3$ at 100°C. The experimental evidence of these early workers for the retention of the straight chain in the disproportionation products and the recovered propylbenzene was less than convincing, whereas in the 1950s there were several techniques that could be used easily to distinguish between *n*-propylbenzene and isopropylbenzene. Either gas chromatography or infrared spectrometry would have been entirely satisfactory, but because of the previous experience of Roberts with ^{14}C-labeled ethylbenzene, he chose to apply the isotopic tracer technique to the propylbenzene test. As it turned out, this was a serendipitous choice, because if the products of the reaction had been examined by means of gas chromatography and/or infrared spectrometry, these analyses would have only confirmed that Heise and Töhl were correct in their 1892 report that the *n*-propyl groups were unchanged in both the dipropylbenzene produced and the propylbenzene recovered from their experiment. However, when Roberts and Brandenberger [10] treated *n*-propyl-β-^{14}C-benzene with $AlCl_3$ at 100°C for 6 hr, separated the propyl- and dipropylbenzene fractions, and oxidized them to benzoic and phthalic acids, these were found to contain radioactivity corresponding to about 30% rearrangement of the isotope from the β-C to the α-C of the side chains. When the infrared spectrum of the propylbenzene fraction showed it to be almost pure (at least 96%) *n*-propylbenzene, Roberts and Brandenberger realized they had turned up something quite unusual.

More extensive degradation studies revealed that *no* ^{14}C became incorporated into the γ-C of the side chains. Further experiments gave the following additional information:

1. The ^{14}C could be equilibrated to a 50:50 distribution between the α- and β-C's of the *n*-propyl side chain, with no ^{14}C appearing in the γ-C, starting with either *n*-propyl-β-^{14}C-benzene or *n*-propyl-α-^{14}C-benzene [11].
2. The equilibration could be accomplished either by successive treatment of neat *n*-propyl-α- or β-^{14}C-benzene with fresh $AlCl_3$ at 100°C [11], or by a single treatment with the catalyst in benzene solution (6 mol of benzene per mole of *n*-propyl-^{14}C-benzene) at ca. 83°C [12].
3. Disproportionation of *n*-propyl-α- or β-^{14}C-benzene was produced readily by $AlCl_3$ at room temperatures, but without isotopic rearrangement.
4. The extent of rearrangement to isopropylbenzene reached only about 5% as a maximum. By using isotope-dilution technique, it was shown that the isotopic distribution between the α- and β-carbons of the isopropylbenzene was the reverse of that in the *n*-propylbenzene before a 50:50 distribution was reached in both [13].
5. Water was found to be an effective cocatalyst with both $AlCl_3$ and Al_2Br_6; Al_2Br_6-HBr was also an effective catalyst system for the isotopic rearrangement [11b].

C. Butylbenzene Rearrangements

Following the discovery of the isotopic rearrangement of *n*-propyl-α- or β-^{14}C-benzene, Roberts and his co-workers began investigations of the possibly similar rearrangement reactions of butyl- and pentylbenzenes. *n*-Butyl-α-^{14}C-benzene was synthesized and treated with $AlCl_3$ under the same conditions used with the lower homolog. Oxidative degradation of the recovered butylbenzene as before and radioassay of the benzoic acid showed a much smaller amount of isotopic rearrangement (ca. 5%) than *n*-propyl-α-^{14}C-benzene gave, and no rearrangement to *sec*-butylbenzene [14]. (A more extensive comparison of the behavior of isotopically labeled *n*-propyl- and *n*-butylbenzene was made several years later using ^{13}C as the isotope; the results from this study will be described in Sec. I.F.)

The next butylbenzene to be examined by Roberts' group was isobutylbenzene, which also gave a surprise, because it was isomerized by $AlCl_3$ not to *t*-butylbenzene, but to *sec*-butylbenzene. This rearrangement could be followed by infrared spectrometry, as the infrared spectra of the four isomeric butylbenzenes are quite distinctive, and it was seen that the same equilibrium mixture of isobutylbenzene and *sec*-butylbenzene, in a 2:1 ratio, was produced from either isomer by heating with $AlCl_3$ at 50 to 100°C [15]. The discovery of the rearrangement of *sec*-butylbenzene to isobutylbenzene was made independently and almost simultaneously by Nenitzescu's group in Romani [16]. This finding of the interconversion of *sec*-butylbenzene and isobutylbenzene by $AlCl_3$ was of some practical importance, because these two isomers have boiling points that differ by only 0.5°C. Whereas *n*-butylbenzene and *t*-butylbenzene can be separated from each other and from the other two isomers by distillation, and easily by gas chromatography, *sec*-butylbenzene and isobutylbenzene are difficult to separate even by gas chromatography. Thus it appeared likely that alkylations carried out with *sec*-butyl derivatives in the presence of aluminum chloride catalyst might well result in a mixture of *sec*-butyl- and isobutylarenes which would not be recognized as such. Also, since *n*-butyl derivatives are known to produce (mainly) *sec*-butyl alkylates (as well as some *n*-butyl), an alkylation carried out with an *n*-butyl derivative might well give an alkylate containing *n*-butyl-, *sec*-butyl-, and isobutyl side chains. These possibilities were quickly substantiated by alkylation experiments [17] (see Sec. 3.II.D for details). It should be noted, however, that the rearrangement of *sec*-butylbenzene to isobutylbenzene in the presence of aluminum chloride is not significant below 50°C, and since alkylations by *n*-butyl and *sec*-butyl derivatives can be carried out readily at room temperature, this complicating rearrangement can be avoided. However, there has been a tendency in the past for Friedel-Crafts alkylations to be carried out in refluxing benzene (or other arene). Products from such procedures involving *n*-butyl or *sec*-butyl alkylating agents should thus be considered to be of questionable identity and purity.

D. Pentylbenzene Rearrangements

New evidence bearing on the rearrangement of *t*-pentylbenzene (1) to 2-methyl-3-phenylbutane (2) by aluminum chloride was reported almost simultaneously from Nenitzescu's [16] and Roberts' [18] laboratories. Both groups confirmed that 1 is indeed isomerized to 2, and they found, further, that neopentylben-

CH_3-C(CH_3)(C_6H_5)-CH_2CH_3 (1) ⇄ CH_3-CH(CH_3)-CH(C_6H_5)-CH_3 (2) ⇄ CH_3-C(CH_3)(CH_3)-CH_2-C_6H_5 (10) (10)

zene (10) is produced by extended heating of either 1 or 2 with aluminum chloride (Eq. 10). Later a more detailed study of the effect of aluminum chloride on all of the pentylbenzene isomers was reported by Roberts and Han [19]. 2-Phenylpentane (11) and 3-phenylpentane (12) were found to be rapidly interconverted, while both were more slowly isomerized to 2-methyl-1-phenylbutane (13) by heating with $AlCl_3$-H_2) in benzene solution at 80 to 82°C. The rearrangement of 11 and 12 to 13 was not so nearly irreversible as that of 1 and/or 2 to 10; long heating of 13 with $AlCl_3$ produced small amounts of 11 and 12. Burwell and Shields described the interconversion of 11 and 12 by $AlCl_3$ in 1953, but they found no further isomerization of these isomers by this catalyst [20]. The relationship of the interconversion of these two secondary pentylbenzenes to racemization and transalkylation reactions is discussed in a later section.

Isopentylbenzene (3-methyl-1-phenylbutane, 14) was found to be unchanged

CH_3-CH(C_6H_5)-CH_2-CH_2-CH_3 (11) ⇄ CH_3CH_2-CH(C_6H_5)-CH_2-CH_3 (12) ⇄ C_6H_5-CH_2-CH(CH_3)-CH_2-CH_3 (13)

after 24 hr of heating with $AlCl_3$-H_2O, and *n*-pentylbenzene (15) underwent about 9% rearrangement to 14 under these conditions.

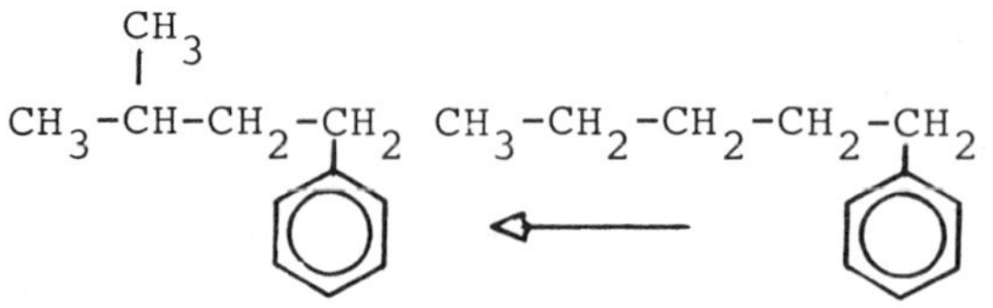

CH_3-CH(-CH_3)-CH_2-CH_2-Ph ← CH_3-CH_2-CH_2-CH_2-CH_2-Ph

14 15

E. Other Alkylbenzene Rearrangements

Because of their formal resemblance to the *sec*-butylbenzene/isobutylbenzene and 2-phenylpentane/3-phenylpentane/2-methyl-1-phenylbutane systems, cyclohexylbenzene (16) and cyclopentylmethylbenzene (17) were examined with respect to their behavior on heating with aluminum chloride [21,22].

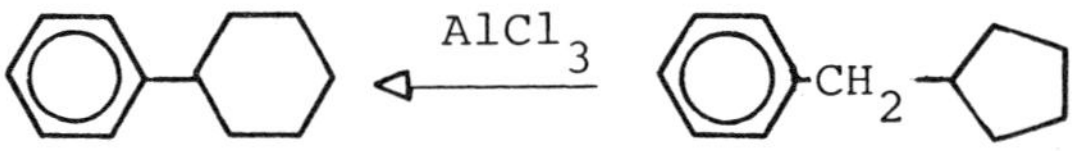

16 ← ($AlCl_3$) 17 (Ph-CH_2-cyclopentyl)

Cyclohexylbenzene (16) was found to undergo disproportionation, but no rearrangement on heating with aluminum chloride. When 17 was tested similarly, it was found to rearrange rapidly and irreversibly to 16. A ^{14}C-labeled sample of 17 was synthesized (17a, cyclopentyl-1-^{14}C-methylbenzene) and treated

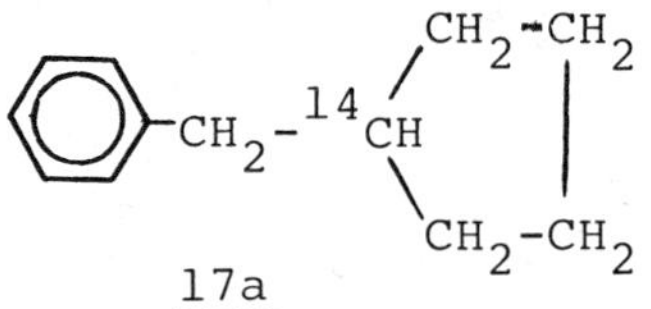

Ph-CH_2-^{14}CH (CH_2-CH_2-CH_2-CH_2 ring)

17a

with aluminum chloride; it was found that under conditions such that only 7% conversion to 16 had occurred, the ^{14}C had become evenly distributed around the cyclohexane ring in 16. This indicated that the equilibration of the ^{14}C in the cyclohexane ring, which may occur by 1,2 shifts of phenyl and is thus analogous to the interconversion of 2- and 3-phenylpentane, is a much faster process than the ring-expansion rearrangement of the cyclopentylmethyl-to-cyclohexyl group.

A Russian group also studied the rearrangement of ^{14}C in the cyclohexane ring of cyclohexylbenzene, and compared the rate of this rearrangement to the rate of transalkylation of cyclohexylbenzene [23]. The first rate was measured in the reaction of cyclohexyl-1-^{14}C-benzene (16a) in benzene, catalyzed by $AlCl_3$ (Eq. 11), by taking samples after various time periods and

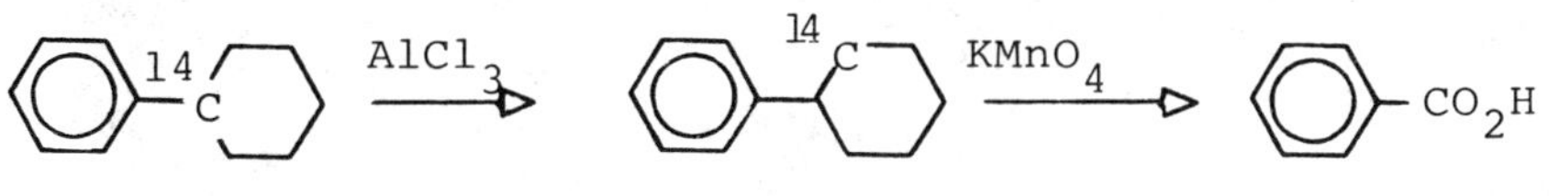

16a →($AlCl_3$) (etc.) →($KMnO_4$) Ph-CO_2H (11)

determining the decrease in the radioactivity of the benzoic acid obtained by oxidation. The rate of transalkylation was measured in the reaction of cyclohexylbenzene with benzene-1-6-^{14}C (Eq. 12), by determining the decrease in the radioactivity of the benzene with time. Both reactions were carried

$$\text{C}_6\text{H}_5\text{-C}_6\text{H}_{11} + {}^{14}\text{C}_6\text{H}_6 \xrightarrow{AlCl_3} {}^{14}\text{C}_6\text{H}_5\text{-C}_6\text{H}_{11} + \text{C}_6\text{H}_6 \qquad (12)$$

out at 20°C. Although the rates of rearrangement and transalkylation were of the same order of magnitude, the rate of transalkylation was slightly higher, especially in the initial stages of the reactions. In fact, some transalkylation occurred before any rearrangement of ^{14}C in the cyclohexane ring took place. This led the authors to suggest at 1,1-diphenylcyclohexane intermediate for the transalkylation, which is in accord with the Pines-Streitwieser mechanism for transalkylation [24,25] (cf. Sec. I.F and Sec. 7.III.E). When the $AlCl_3$/hydrocarbons ratio was lowered, the rate of rearrangement was decreased significantly more than the rate of transalkylation. No explanation of this effect was offered. For the isotopic rearrangement in the cyclohexane ring, the authors suggested two mechanisms, one involving phenyl shifts around the cyclohexane ring, the other hydride shifts in the secondary cyclohexyl cation, which they showed being produced as an intermediate in the transalkylation. Thus they apparently envisioned the transalkylation as occurring by two different routes, with the initial process going via a 1,1-diphenylcyclohexane intermediate (no ^{14}C rearrangement allowed), followed by a later process in which the cyclohexyl cation is formed, allowing ^{14}C rearrangement through hydride shifts, with the further possibility that 1,2 shifts of phenyl groups may contribute to the rearrangement.

F. Mechanisms of Alkylbenzene Rearrangements

1. Isotopic *n*-propylbenzene and *sec*-butyl-/isobutylbenzene rearrangements

It may have been noted that there is some similarity between the rearrangement of *t*-pentylbenzene to 2-methyl-3-phenylbutane and the rearrangement of *sec*-butylbenzene to isobutylbenzene. In the pentyl system the alkyl side chain goes from tertiary to secondary attachment to benzene, and in the butyl system, from secondary to primary. [Which is exactly opposite to the generalization applied to alkylation (too generally, as we mentioned in Chap. 1!); that is, a primary alkyl halide yields a secondary or tertiary alkylbenzene.]

However, one of the two mechanisms proposed for the pentylbenzene rearrangement [3-5] can be ruled out for the butylbenzene rearrangement (Eqs. 13 to 17). Dealkylation of *sec*-butylbenzene to a secondary carbocation, followed by rearrangement to the isobutyl carbocation and realkylation (Eq. 15), is untenable, because isobutyl derivatives are known to rearrange completely to give *t*-butyl alkylates (Eqs. 16 and 17). A more reasonable mechanism is one analogous to the one proposed by Schmerling and West [2] (Eqs. 18 to 20).

$$C_6H_5\text{-}CH(CH_3)\text{-}CH_2\text{-}CH_3 + H^{\oplus} \longrightarrow C_6H_6 + CH_3\text{-}\overset{\oplus}{C}H\text{-}CH_2CH_3 \quad (13)$$

$$\downarrow \sim CH_3: \quad (14)$$

$$C_6H_5\text{-}CH_2\text{-}CH(CH_3)\text{-}CH_3 \;\not\longleftarrow\; C_6H_6 + CH_3\text{-}CH(CH_3)\text{-}CH_2^{\oplus} \quad (15)$$

$$\downarrow \sim H: \quad (16)$$

$$C_6H_5\text{-}C(CH_3)_2\text{-}CH_3 \longleftarrow C_6H_6 + CH_3\text{-}\overset{\oplus}{C}(CH_3)\text{-}CH_3 \quad (17)$$

$$C_6H_5\text{-}CH(CH_3)\text{-}CH_2\text{-}CH_3 + R^{\oplus} \rightleftharpoons \underset{\underline{18}}{C_6H_5\text{-}CH(CH_3)\text{-}\overset{\oplus}{C}H\text{-}CH_3} + RH \quad (18)$$

$$\underline{18} \underset{}{\overset{\sim CH_3}{\rightleftharpoons}} \underset{\underline{19}}{C_6H_5\text{-}\overset{\oplus}{C}H\text{-}CH(CH_3)\text{-}CH_3} \quad (19)$$

$$\underline{19} + RH \rightleftharpoons C_6H_5\text{-}CH_2\text{-}CH(CH_3)\text{-}CH_3 + R^{\oplus} \quad (20)$$

A mechanism similar to this may be written for the *n*-propylbenzene isotopic rearrangement, but it has the difficulty that a primary carbocation must be included as an intermediate. It should be noted, too, that there is no reasonable pathway from *sec*-butylbenzene or isobutylbenzene to *t*-butylbenzene that does not include a primary carbocation intermediate, which probably accounts for the absence of significant amounts of *t*-butylbenzene.

A clue to another plausible mechanism for the *sec*-butylbenzene/isobutylbenzene rearrangement which is also applicable to the *n*-propylbenzene isotopic rearrangement came out of studies of *dealkylation* reactions of alkylbenzenes. When alkylbenzenes are heated above 50°C with aluminum chloride, some alkane corresponding in carbon number to the alkyl side chain is always produced. For example, both butane and isobutane are produced from *sec*-butylbenzene or isobutylbenzene. Among the products of a reaction in which *sec*-butylbenzene was heated with aluminum chloride at 100°C for 3 hr was found crystalline *meso*-2,3-diphenylbutane [26]. This compound was recognized as a by-product of the dealkylation reaction (see Sec. II.A).

Careful investigation, principally by means of gas chromatography, showed that diphenylalkanes are always present in reaction mixtures resulting from heating alkylbenzenes with aluminum chloride. 1,2-Diphenylpropane was identified among the products of reactions in which *n*-propyl-α-^{14}C-benzene was isomerized to *n*-propyl-β-^{14}C-benzene [12]. In 1958, Pines and Arrigo proposed that 1,1-diarylethanes were intermediates in the disproportionation of ethylxylenes in the presence of hydrogen fluoride and methylcyclohexene [24]. A little later, Streitwieser and Reif arrived at a similar mechanism, based on work which showed that added alkene is not required as a hydrogen acceptor when a gallium bromide-hydrogen bromide catalyst system is employed in the disproportionation of ethylbenzene [25]. The latter authors remarked that the facile alkylation-dealkylation demonstrated in their work suggested a mechanism for the isotopic *n*-propylbenzene rearrangement, i.e., one involving diphenylpropane intermediates.

Roberts et al. had arrived at the same conclusion independently at about the same time, and they outlined a more complete mechanism (scheme 1). [12]. This mechanism is analogous to the bimolecular process proposed by Karabatsos and Vane [27] to rationalize the isotopic isomerization of *t*-pentyl chloride by aluminum chloride without resorting to primary carbocation ion intermediates. As can be seen in scheme 1, the 1,2 shift of the methyl group can occur in the diphenylpropane system without producing a primary carbocation ion, whereas it cannot in a monomolecular system, analogous to the mechanism for the butylbenzenes given earlier (Eqs. 18 to 20). It is noteworthy that although equilibration of ^{14}C between the α- and β-C's of the side chain can proceed via secondary carbocations, according to the mechanism of scheme 1, there are no pathways to *n*-propyl-γ-^{14}C-benzene or to isopropylbenzene which do not involve primary carbocation intermediates.

Implicit in the mechanism of scheme 1 is the interconversion of the isotopically isomeric diphenylpropanes under the conditions of the *n*-propylbenzene rearrangement. This was tested experimentally, using both 1,2-diphenylpropane-2-^{14}C and 1,1-diphenylpropane-1-^{14}C as starting materials [28]. The expected equilibration of ^{14}C between the two positions in the diphenylpropanes was observed; furthermore, the *n*-propylbenzene produced was found to have the ^{14}C evenly distributed between the α- and β-C's, with none in the γ-C.

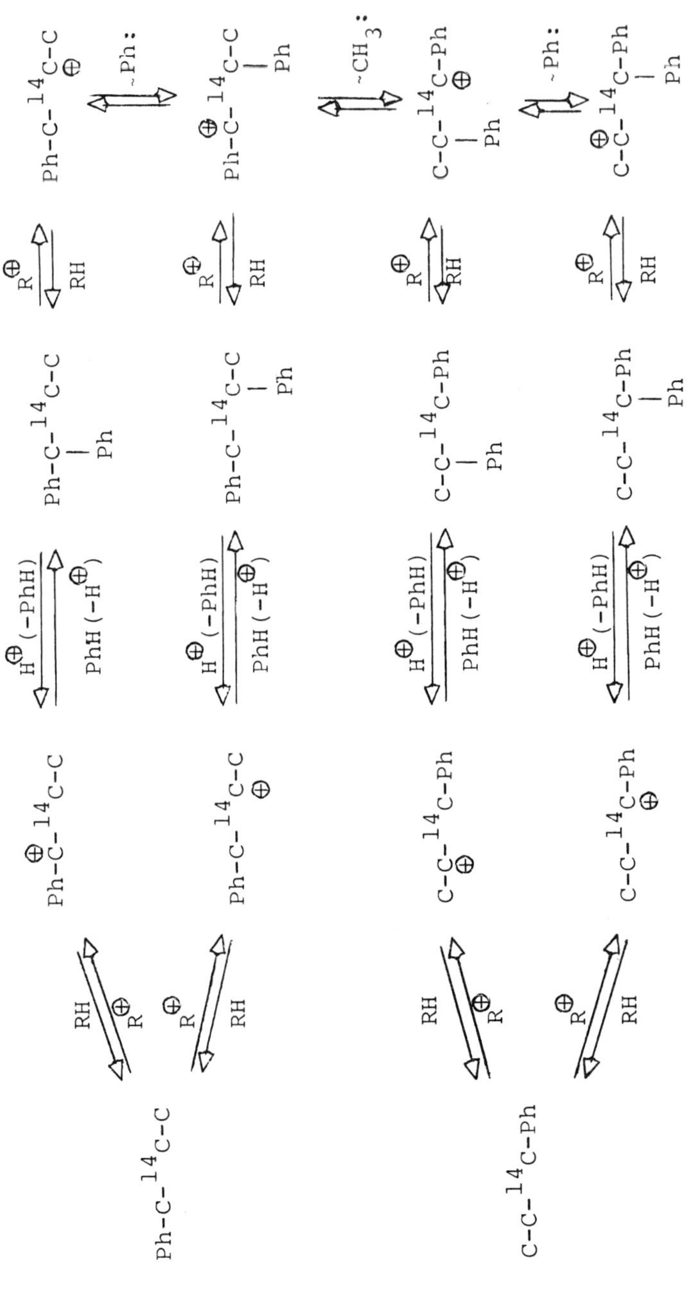

$Ph-C-{}^{14}C-C \xrightleftharpoons[R^{\oplus}]{RH} Ph-\overset{\oplus}{C}-{}^{14}C-C \xrightleftharpoons[PhH(-H^{\oplus})]{H^{\oplus}(-PhH)} Ph-C(Ph)-{}^{14}C-C \xrightleftharpoons[RH]{R^{\oplus}} Ph-C-{}^{14}\overset{\oplus}{C}-C$

$Ph-C-{}^{14}C-C \xrightleftharpoons[RH]{R^{\oplus}} Ph-C-{}^{14}\overset{\oplus}{C}-C \xrightleftharpoons[PhH(-H^{\oplus})]{H^{\oplus}(-PhH)} Ph-C-{}^{14}C(Ph)-C \xrightleftharpoons[RH]{R^{\oplus}} Ph-\overset{\oplus}{C}-{}^{14}C(Ph)-C$

$Ph-C-{}^{14}\overset{\oplus}{C}-C \xrightleftharpoons{\sim Ph:} Ph-\overset{\oplus}{C}-{}^{14}C(Ph)-C \xrightleftharpoons{\sim CH_3:} C-C(Ph)-{}^{14}\overset{\oplus}{C}-Ph \xrightleftharpoons{\sim Ph:} C-\overset{\oplus}{C}-{}^{14}C(Ph)-Ph$

$C-C-{}^{14}C-Ph \xrightleftharpoons[R^{\oplus}]{RH} C-\overset{\oplus}{C}-{}^{14}C-Ph \xrightleftharpoons[PhH(-H^{\oplus})]{H^{\oplus}(-PhH)} C-C(Ph)-{}^{14}C-Ph \xrightleftharpoons[RH]{R^{\oplus}} C-C(Ph)-{}^{14}\overset{\oplus}{C}-Ph$

$C-C-{}^{14}C-Ph \xrightleftharpoons[RH]{R^{\oplus}} C-C-{}^{14}\overset{\oplus}{C}-Ph \xrightleftharpoons[PhH(-H^{\oplus})]{H^{\oplus}(-PhH)} C-C-{}^{14}C(Ph)-Ph \xrightleftharpoons[RH]{R^{\oplus}} C-\overset{\oplus}{C}-{}^{14}C(Ph)-Ph$

Scheme 1

At this point it is appropriate to consider some of the other observed details of the rearrangement that were described earlier (Sec. I.B) without explanation. The failure to produce complete isotopic equilibration between the α and β positions in some of the experiments is undoubtedly due to deactivation of the catalyst, probably mainly by the di-, tri-, and higher-propylbenzenes produced by transalkylations. This rationale is supported by the fact that repeated treatment of partially rearranged samples of *n*-propyl-^{14}C-benzene, which were separated from the benzene and other disproportionation products, with fresh $AlCl_3$ did produce complete equilibration of the isotope between the α and β positions [11]. Also, the complete equilibration can be accomplished by one catalyst treatment in benzene solution [12], because the disproportionation to polypropylbenzenes is inhibited by the high concentration of benzene. It is well known that the strength of an arene as a Lewis base, and hence its ability to complex with a Lewis acid like $AlCl_3$, increases with the number of alkyl substituents on the benzene ring.

The effectiveness of water as a cocatalyst is thought to be due to the formation of the strong protonic acid $H^+(AlCl_3OH)^-$. Nenitzescu et al. [29] demonstrated the production of a small amount of molecular hydrogen from reaction of water-activated aluminum chloride with a hydrocarbon. This may be the way in which the first hydride abstraction from *n*-propylbenzene occurs; after the reaction is initiated by the catalyst, the carbocations produced by hydride donation will propagate a chain reaction (scheme 1).

We mentioned in Sec. 7.I the demonstration by Nightingale and co-workers of the reorientation of 1,3-dimethyl-4-*n*-propylbenzene to 1,3-dimethyl-5-*n*-propylbenzene without rearrangement of the *n*-propyl group [30]. This reaction was reexamined with the modification of using ^{14}C-labeled *n*-propyl groups, so that a possible isotopic rearrangement might be observed even though no constitutional rearrangement to isopropyl derivatives occurred; a similar test was made of the reorientation of *p*-*n*-propyl-^{14}C-toluene [31]. The lack of constitutional rearrangement of the propyl group was confirmed, and the isotopic rearrangements accompanying the reorientations were found to be less than that observed for *n*-propyl-^{14}C-benzene. In fact it was found in one experiment at room temperature that reorientation of 1,3-dimethyl-4-*n*-propylbenzene to 1,3-dimethyl-5-*n*-propylbenzene was 60% complete without any isotopic rearrangement. It may be recalled that isotopically labeled *n*-propylbenzene undergoes disproportionation without isotopic rearrangement at low temperatures. Thus it has been demonstrated that both disproportionation and reorientation of *n*-propyl groups may occur by a mechanism or mechanisms which are independent of the isotopic rearrangement. The Pines-Streitwieser [24,25] mechanism can account for both disproportionation and reorientation without isotopic rearrangement. It is also possible that the reorientation of the propyl group occurs by an intramolecular 1,2 shift, but this seems less likely on the basis of the study of the reorientation of ethylbenzene-1-^{13}C (cf. Chap. 7).

Although the reactions were not carried out under identical experimental conditions, the relative extents of isotopic rearrangements under similar conditions of time, temperature, and catalyst concentration were calculated to be for *n*-propyl-^{14}C-benzene, *p*-*n*-propyl-^{14}C-toluene, and 1,3-dimethyl-4-*n*-propyl-^{14}C-benzene, approximately 30, 17, and 5%, respectively. The relative extents of rearrangement are in the order expected on the basis of greater deactivation of the catalyst by the additional methyl substituents in the propyltoluene and propylxylene.

As we mentioned above, the mechanism of scheme 1 provides a rationale for the equilibration of the isotope between the α and β positions but not the γ position of the side chain, and for the occurrence of the isotope rearrangement in the side chain without significant isomerization to isopropyl groups, on the basis that this mechanism provides routes for the observed isotopic rearrangement that avoid the incursion of primary carbocations, whereas the other isomerizations would require these higher-energy intermediates. It is interesting to note that the finding of the reverse distribution of the isotope in the α and β positions of the small amount of isopropylbenzene produced can also be rationalized in terms of the diphenylpropane intermediates of scheme 1. Starting with *n*-propyl-β-^{14}C-benzene, if cleavage of the first-formed 1,2-diphenylpropane occurs at the original α-C-phenyl bond and before the methyl shift, the isopropylbenzene produced from the resulting *primary* carbocation intermediate will have the ^{14}C in the α position (scheme 1A).

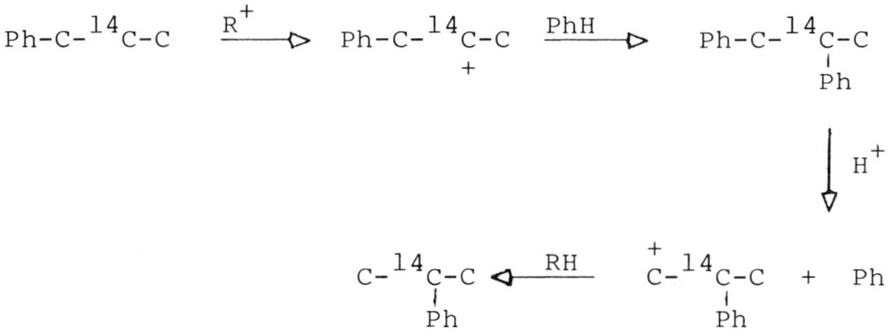

$$\text{Ph-C-}^{14}\text{C-C} \xrightarrow{R^+} \text{Ph-C-}\overset{+}{^{14}\text{C}}\text{-C} \xrightarrow{\text{PhH}} \text{Ph-C-}^{14}\underset{\text{Ph}}{\text{C}}\text{-C} \xrightarrow{H^+} \overset{+}{\text{C}}\text{-}^{14}\underset{\text{Ph}}{\text{C}}\text{-C} + \text{Ph} \xrightarrow{\text{RH}} \text{C-}^{14}\underset{\text{Ph}}{\text{C}}\text{-C}$$

Scheme 1A

Shortly after Roberts et al. proposed the mechanism of scheme 1 for the isotopic *n*-propylbenzene rearrangement, an alternative mechanism (scheme 2) was proposed by Farcasiu [32], in which diphenylhexyl cations (as contrasted with diphenyl*propyl* cations) are produced, rearranged, and undergo fission in a way similar to the "bimolecular mechanism" proposed by Karabatsos for the isotopic rearrangements of ^{13}C- and ^{14}C-labeled *t*-pentyl cations [27]. Farcasiu suggested that a mechanism in which diphenylhexyl cations intervened rationalized some facts that are not explained by scheme 1, such as (1) the identification of small amounts of butylbenzenes in a reaction mixture produced from 1,2-diphenylpropane and $AlCl_3$ [12] and, (2) detection of 1-methyl-1,3-diethyl-3-phenylindan [33] in a reaction mixture produced from *sec*-butylbenzene and $AlCl_3$ [12]. However, Farcasiu acknowledged that the possible operation of the diphenylhexyl cation mechanism does not preclude "the other possible intermediates (as diphenylpropane in the isomerization of propylbenzene, and even the simple monomeric cations in the case of *sec*-butylbenzene . . .)." In a footnote he states: "A conclusive argument for the diphenylhexyl cation-intermediate mechanism would be the finding of dilabeled and unlabeled molecules in the isomerization product of an α- (or β-) isotope labeled *n*-propylbenzene by the mass spectrum."

In 1971, Roberts and Gibson reported the results from a test just such as that sugested by Farcasiu [34]. *n*-Propyl-benzene-α-^{13}C-benzene (samples of 32% and 17% ^{13}C enrichment) was synthesized and treated with aluminum chloride in the presence of benzene and water in the molar ratio 1.0:0.5:6.0:0.1, respectively, at reflux for 7 hr. After the usual decomposition with

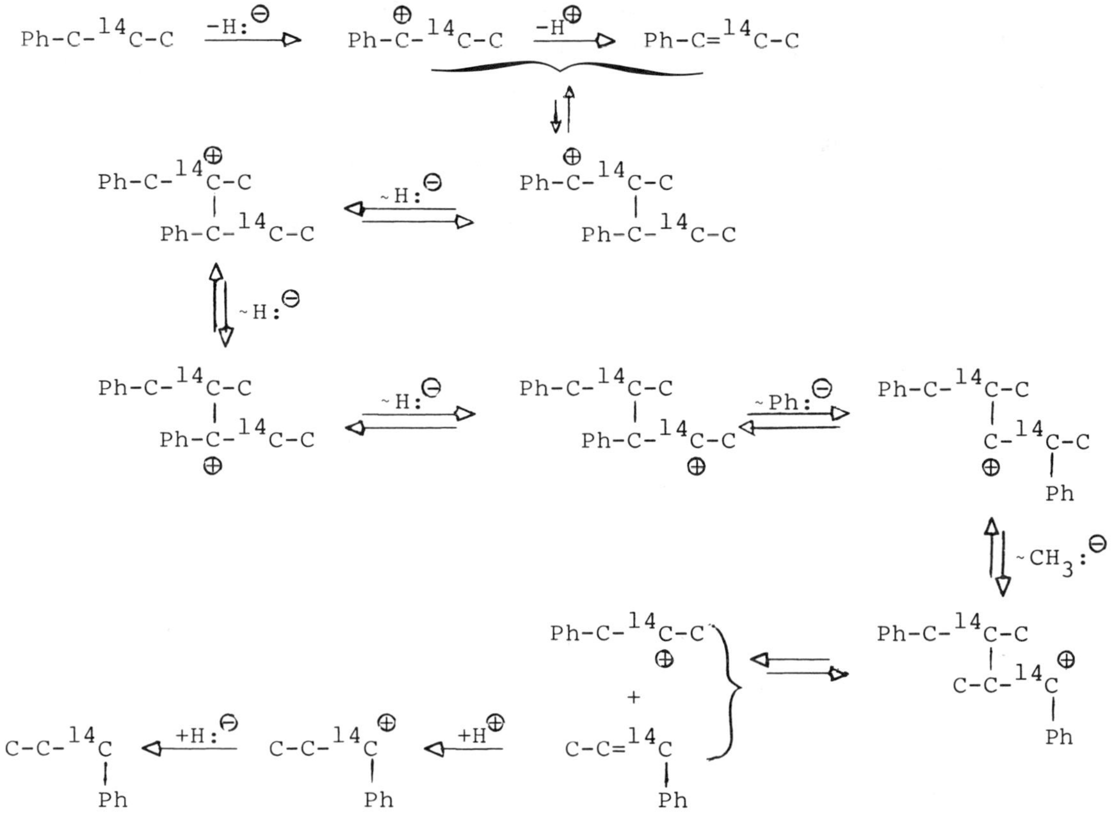

```
Ph-C-14C-C  --(-H:⊖)-->  Ph-C⊕-14C-C  --(-H⊕)-->  Ph-C=14C-C

Ph-C-14C⊕-C                      ⊕
    |          <--(~H:⊖)-->  Ph-C-14C-C
Ph-C-14C-C                       |
                             Ph-C-14C-C
   ↕ ~H:⊖

Ph-C-14C-C                 Ph-C-14C-C                 Ph-C-14C-C
    |       <--(~H:⊖)-->       |        <--(~Ph:⊖)-->     |
Ph-C-14C-C                 Ph-C-14C-C                   C-14C-C
   ⊕                            ⊕                       ⊕    |
                                                             Ph
                                                      ↕ ~CH3:⊖

                          Ph-C-14C-C                  Ph-C-14C-C
                                  ⊕                       |      ⊕
                                 +      <-->         C-C-14C
                                                             |
C-C-14C  <--(+H:⊖)--  C-C-14C⊕  <--(+H⊕)--  C-C=14C           Ph
     |                     |                     |
     Ph                    Ph                    Ph
```

Scheme 2

water, n-propylbenzene was recovered in 60-67% yield by preparative gas chromatography. The rearrangement of the ^{13}C from the α to the β position of the side chain was demonstrated by both mass spectrometry and ^{13}C NMR analysis. The products of two experiments which were found to have undergone almost complete isotopic rearrangement were examined by mass spectrometry for the presence of unlabeled and dilabeled n-propylbenzene molecules. None were found. This was the expected result if the isotopic rearrangement occurred via diphenylpropane intermediates, but if the rearrangement proceeded via diphenylhexyl cations, as Farcasiu suggested, one would expect to find scrambling of the isotope within the dimeric cations, with the result that their fission would produce some unlabeled and some dilabeled molecules, as well as singly labeled ones.

Although this study of ^{13}C-labeled n-propylbenzene by mass spectrometry apparently excludes diphenylhexyl cations as a major pathway for the isotopic rearrangement, and leaves the diphenylpropane system of scheme 1 as a likely mechanism, it is still quite possible that diphenylhexyl cations are intermediates in the formation of the butylbenzenes and the substituted indan that Farcasiu mentioned. They may also be responsible for the small amount of isopropylbenzene observed; as Farcasiu indicated, it is possible for isopropylbenzene to be produced from diphenylhexyl cations without the involvement of any primary carbocations.

2. Comparison of isotopic *n*-propylbenzene and *n*-butylbenzene rearrangements

With the increased availability of ^{13}C for labeling organic molecules and the development of convenient ^{13}C NMR and mass spectrometric techniques for analyzing ^{13}C-labeled compounds, it became attractive to Roberts and his co-workers to reexamine the earlier tests of an isotopic rearrangement of *n*-butylbenzene, especially as the early experiments were done before the optimum conditions for the *n*-propylbenzene rearrangement were worked out.

When *n*-butyl-α-^{13}C-benzene was prepared and treated with $AlCl_3$-H_2O in benzene under conditions known to give complete equilibration of the isotope between the α and β positions of the *propyl* side chain, *complete lack* of isotopic rearrangement of the *n*-butyl-α-^{13}C-benzene was determined by ^1H NMR analysis [35]. Even more surprising, when an equimolar mixture of *n*-propyl-α-^{13}C-benzene and *n*-butyl-α-^{13}C-benzene were heated with $AlCl_3$-H_2O or Al_2Br_6-H_2O in benzene, neither alkylbenzene gave any isotopic rearrangement with the usual concentration of catalyst. When the concentration of catalyst was made much higher, isotopic rearrangement of the propylbenzene occurred, but not enough of the butylbenzene could be recovered to check for rearrangement. It was converted into a mixture of disproportionation and other products, one of which was 1-methylindan. Further tests indicated that the higher disproportionation products from *n*-butylbenzene were mainly responsible for the inhibitory effect on the isotopic rearrangement of *n*-propylbenzene, but it was not clear why they should have a stronger effect than the disproportionation products from *n*-propylbenzene.

In considering the effect the additional methylene group of *n*-butylbenzene might have on the tendency of this alkylbenzene to exchange α- and β-carbon atoms by a mechanism of the scheme 1 type, two factors can be discerned. One is the requirement of a 1,2 shift of an ethyl group instead of a methyl group. Although it appeared unlikely that the difference in migratory aptitude of methyl and ethyl groups is large enough to be responsible for the wide discrepancy in the behavior of *n*-propyl- and *n*-butylbenzene [36], an experimental test was made by preparing 1,2-diphenyl-2-^{13}C-butane and treating it with $AlCl_3$-H_2O in benzene. In contrast to 1,2-diphenyl-2-^{14}C-propane, in which equilibration of the isotope occurred at room temperature and which gave isotopically equilibrated *n*-propylbenzene [28], the diphenylbutane was found to have undergone only 8% rearrangement after an hour at reflux temperature, and the *n*-butylbenzene produced from it exhibited the same small amount of rearrangement.

The other significant difference between the two homologous alkylbenzenes has to do with the hydride abstraction in the first step of scheme 1. In the case of both hydrocarbons, hydride abstraction from the β carbon opens the way for production of a 1,2-diphenylalkane intermediate, which is essential for allowing the 1,2-shift of a methyl or ethyl group without producing a primary carbocation. In the case of *n*-butylbenzene, hydride abstraction may take place at either the β or γ carbon, producing a secondary carbocation in either event, and the γ cation can lead to 1,3-diphenylbutane and substituted indans and indenes (scheme 3). Such a pathway is, of course, not possible for *n*-propylbenzene except via a *primary* carbocation. Analysis of the product mixture from *n*-butylbenzene, $AlCl_3$-H_2O, and benzene at reflux for 1 hr showed the presence of compounds <u>21</u>-<u>24</u>, all of which can be derived from the γ cation, <u>20</u>.

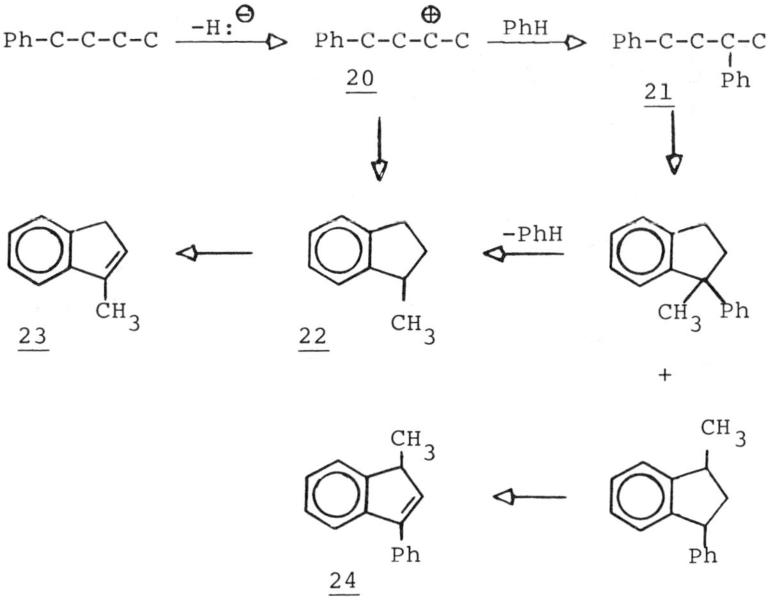

$$\text{Ph-C-C-C-C} \xrightarrow{-\text{H:}^{\ominus}} \underset{\underline{20}}{\text{Ph-C-C-}\overset{\oplus}{\text{C}}\text{-C}} \xrightarrow{\text{PhH}} \underset{\underline{21}}{\text{Ph-C-C-}\underset{\text{Ph}}{\text{C}}\text{-C}}$$

CH3 23 ← CH3 22 ←(−PhH) CH3 Ph

\+

CH3 Ph 24 ← CH3 Ph

Scheme 3

3. Competing reactions of secondary alkylbenzenes

The transalkylation aspects of the investigation by Burwell and Shields [20] of the reactions of 2- and 3-phenylpentane with $AlCl_3$ were described in Sec. 7.III.D. It is appropriate here to restate the fact that they found rapid rearrangement (at 25°C) of each of these isomers to an equilibrium mixture of 71% 2- and 29% 3-phenylpentane. They examined the mixture of these two isomers by infrared spectroscopy for evidence of *t*-pentylbenzene and 2-methyl-3-phenylbutane and found none. They did report the formation of up to 7% of the latter isomer when methanesulfonic acid was used as catalyst at 150°C, but it is more likely that this isomer was actually 2-methyl-1-phenylbutane, since this is the isomer found by Roberts and Han [19] to be produced by $AlCl_3$ catalyst. Burwell and Shields also prepared optically active 2-phenylpentane and found it to be racemized by $AlCl_3$ at 25°C so rapidly that the racemization was 99% complete when disproportionation had proceeded to the extent of only 1.5%. The interconversion of 2- and 3-phenylpentane proceeded at approximately the same rate as their disproportionation, and the isomerization was said to occur during the transfer of the secondary pentyl group as a cation. However, in one experiment they observed "considerable disproportionation" with only 1% rearrangement of the 2- to the 3-phenylpentane; hence rearrangement does not occur every time a pentyl group is transferred from one ring to another, and thus the two processes may be independent of one another.

Olson described the conversion of 2-phenyldodecane to an equilibrium mixture of 2-, 3-, 4-, 5-, and 6-phenyldodecanes by $AlCl_3$ at 50°C [37].

In connection with a systematic study concerned primarily with the dealkylation reaction of secondary and tertiary alkylbenzenes (which is described further in Sec. II). Hamanaka and co-workers observed some rearrangements of secondary and tertiary pentyl-, hexyl-, and heptylbenzenes under mild conditions [38]. The reactions were carried out at low temperatures (0 to 30°C for secondary alkylbenzenes and −3 to 5°C for tertiary alkylbenzenes) with 0.2 mol $AlCl_3$ per mole of hydrocarbon. Identification of products was entirely by gas chromatography and only relative yields were reported. With 2- and 3-phenylpentane and 2- and 3-phenylhexane they found that migration of the phenyl group to the other secondary position occurred rapidly at low temperature, followed by disproportionation, as judged by the appearance of benzene. Very little dealkylation occurred, even at 30°C. With the five secondary alkylbenzenes derived by substitution of phenyl on a secondary carbon atom of isohexane (25 and 26) and isoheptane (27-29), they reported only a little rearrangement at 30°C of 26 to 25, of 28 to 27 and 29, and of 29

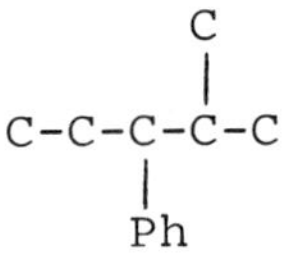

```
      C                  C
      |                  |
C-C-C-C-C          C-C-C-C-C
    |                |
    Ph               Ph

   25                 26
```

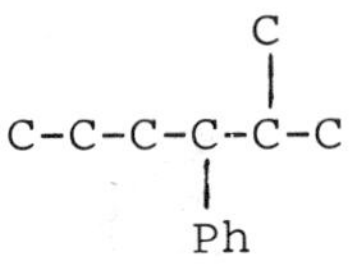

```
        C                  C                  C
        |                  |                  |
C-C-C-C-C-C        C-C-C-C-C-C        C-C-C-C-C-C
      |                |                |
      Ph               Ph               Ph

    27                 28                 29
```

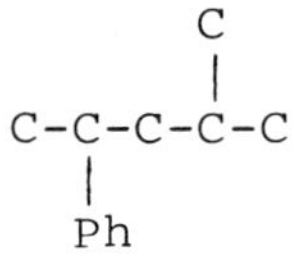

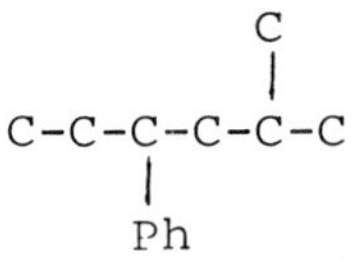

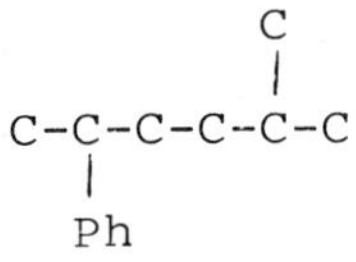

to 28. The tertiary alkylbenzenes 2-methyl-2-phenylpentane (69) and 2-methyl-2-phenylhexane (70) were found to undergo transalkylation and dealkylation at −3 to 5°C. The transalkylation was said to be the faster of these two reactions, and rearrangement was said to be the slowest of the three types of reaction. Actually, judging from the results of Roberts and Han [19] and those of Roberts et al. [26] described in Sec. II, it is likely that all of the branched secondary alkylbenzenes (25-29) rearranged, at least partially, to the tertiary alkylbenzenes by phenyl shifts, and the tertiary alkylbenzenes were dealkylated rapidly at 30°C, so that they were not found as products of the reactions of the secondary alkylbenzenes. The tertiary hexylbenzene was said to rearrange partially to 25 and the tertiary heptylbenzene to 27, which is analogous to the rearrangement of *t*-pentylbenzene (1) to 3-methyl-

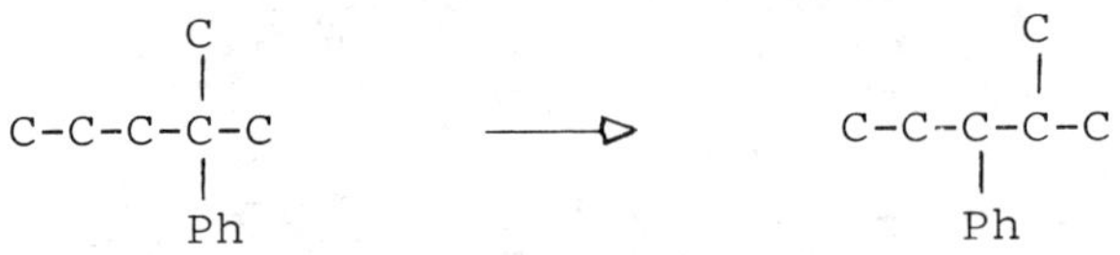

```
      C                    C
      |                    |
C-C-C-C-C    ---->   C-C-C-C-C
      |                  |
      Ph                 Ph
     69                   25

        C                    C
        |                    |
C-C-C-C-C-C  ---->   C-C-C-C-C-C
        |                  |
        Ph                 Ph
       70                   27
```

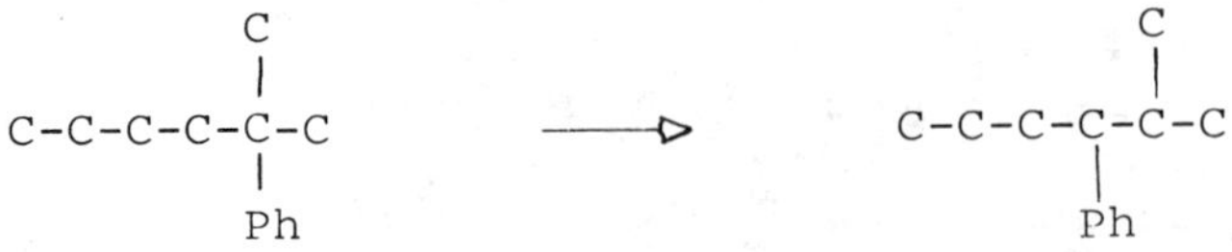

2-phenylpentane (2) [16,18], which was mentioned earlier and is discussed later (cf. scheme 6).

With the aim of obtaining information of the type apparently desired by the Hamanaka group [38], but utilizing a simple secondary alkylbenzene as substrate, Roberts et al. prepared 2-phenyl-2-^{14}C-butane [39]. They then sought to establish the conditions under which the ^{14}C-labeled *sec*-butylbenzene would react with Friedel-Crafts catalysts to undergo (1) disproportionation, (2) rearrangement to isobutylbenzene, and (3) isotopic rearrangement (i.e., 2-phenyl-2-^{14}C-benzene $\rightleftarrows$ 2-phenyl-3-^{14}C-benzene).

As described in Chap. 7, *sec*-butylbenzene (2-phenylbutane) has been shown to undergo disproportionation by $AlCl_3$ [15] or HF-BF_3 [7] at 25°C without side-chain rearrangement.

Roberts et al. found that disproportionation of the labeled *sec*-butylbenzene occurred in 15 min even at −14°C with either $AlCl_3$-H_2O or Al_2Br_6 catalysts, with no isotopic rearrangement. At 0°C for 60 min with $AlCl_3$-H_2O, isotopic rearrangement occurred to produce a mixture of 21% 2-phenyl-3-^{14}C-butane and 79% 2-phenyl-2-^{14}C-butane. At 25°C for 60 min with $AlCl_3$-H_2O, complete isotopic equilibration occurred. No rearrangement to isobutylbenzene or other isomers occurred in any of these experiments.

It is interesting to consider the possible reaction pathways by which 2-phenyl-2-^{14}C-butane may be converted into four different kinds of products (30i, 30e, 33, and 35, scheme 4). Isomerization to the isotopic isomer 25i and to the constitutional isomer 33 may both involve the secondary carbocation 31. On the basis of stability of the carbocations produced, competition between the phenyl and methyl shifts would be expected to be in favor of the latter, since the charge on 32 may be delocalized by the benzene ring. The experimental observation that the isotopic isomerization (30 $\rightleftarrows$ 30i) reaches equilibrium under conditions that produce no *sec*-butylbenzene-isobutylbenzene rearrangement (30 $\rightleftarrows$ 33) seems to suggest that the phenyl group participates in the 1,2 shift in a way that is more important than its function in the stabilization of the ion 32. (Evidence for phenyl participation in reactions of phenylalkyl halides will be mentioned in later sections.) The fact that the formation of 35 and benzene by disproportionation is an even more facile reaction than the isotopic isomerization may be rationalized, however, on the basis that the tertiary benzylic cation 34 is formed much more readily that the aliphatic secondary cation 31, and the disproportionation proceeds by the Pines-Streitwieser mechanism [24,25]. Burwell and Shields [20] found that disproportionation was slightly faster than rearrangement of 2- and 3-phenylpentane, but the difference in rates was apparently much smaller than in the case of 2-phenylbutane. The data from both the phenylbutane and phenylpentane work are not quantitative enough to justify anything more than the speculation as to a reason for the difference in the relative rates of the competing reactions of the homologs. One is tempted, however, to consider that the lower activation energy for disproportionation of a *sec*-alkylbenzene than for the nonbranching rearrangement of the phenyl group is evidence that the disproportionation does take place by a Pines-Streitwieser mechanism involving diphenylalkane intermediates rather than by cleavage to form secondary carbocations which transfer to another aromatic ring.

Although Roberts et al. did not use optically active 2-phenylbutane, and hence could not observe the interconversion of the enantiomers indicated by the formulas 30 and 30e, the extremely rapid racemization of optically active 2-phenylpentane compared to its rearrangement to 3-phenylpentane [20] is consistent with scheme 4 in that a more stable cation homologous to 34 would

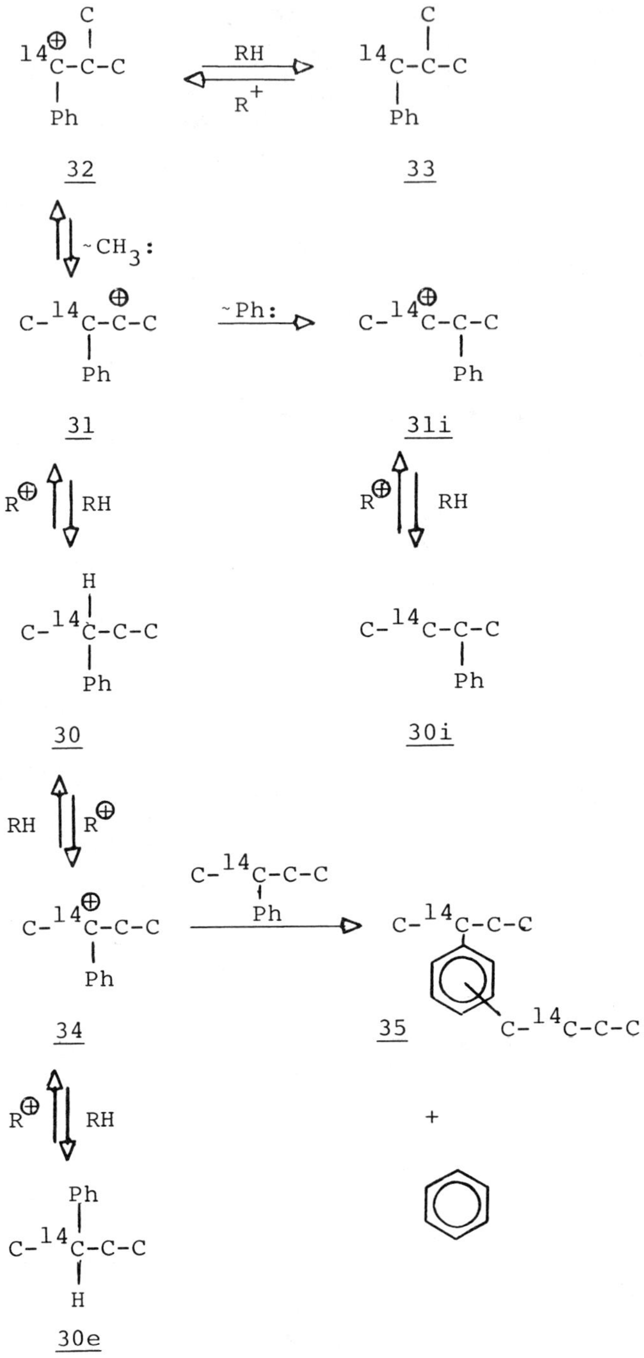

$^{14}\overset{\oplus}{C}$–C(C)–C (Ph) **32** ⇌ (RH / R^+) ^{14}C–C(C)–C (Ph) **33**

32 ⇌ (~CH_3:) C–^{14}C–$\overset{\oplus}{C}$–C (Ph) **31** —(~Ph:)→ C–$^{14}\overset{\oplus}{C}$–C(Ph)–C **31i**

31 ⇌ ($R^{\oplus}$ / RH) C–^{14}C(H)(Ph)–C–C **30**

31i ⇌ ($R^{\oplus}$ / RH) C–^{14}C–C(Ph)–C **30i**

30 ⇌ (RH / $R^{\oplus}$) C–$^{14}\overset{\oplus}{C}$(Ph)–C–C **34**

34 + C–^{14}C(Ph)–C–C → C–^{14}C–C(Ar)–C, Ar = C_6H_4–C–^{14}C–C–C **35** + benzene

34 ⇌ ($R^{\oplus}$ / RH) C–^{14}C(Ph)(H)–C–C **30e**

Scheme 4

be involved in the racemization whereas the less stable homolog of 31 is required for the rearrangement.

In 1968, Itoh and co-workers reported a study of the reactions of *sec*-butylbenzene induced by $AlCl_3$ [40]. They carried out reactions at 20, 40, 50, and 70°C and analyzed the products after various time intervals by capillary gas chromatography, and infrared and ^{1}H-NMR spectrometry. The products first formed at 20°C were benzene and *m*- and *p*-di-*sec*-butylbenzene; 1,3,5-tri-*sec*-butylbenzene appeared 30 min later and, after another 120 min, small amounts of di- and tributylbenzenes having one isobutyl group per molecule. Traces of *n*-butane and *meso*-2,3-diphenylbutane were detected after 4 hr. The same products were identified in the reaction mixtures produced at 40 and 50°C, but an apparent equilibrium composition was reached in 120 min at 40°C and in 60 min at 50°C, whereas 180 min had been required at 20°C. At 70°C all of the disproportionation products were detected much sooner, and diisobutylbenzene, 1,3,5-diisobutyl-*sec*-butylbenzene, 1,3,5-triisobutylbenzene, and both 2,2- and 2,3-diphenylbutane were identified among the products. None of the products had butyl groups in ortho positions.

These results were said to indicate the following reaction sequence: transalkylation > side-chain rearrangement > dealkylation. It is clearly apparent that dealkylation is much faster than rearrangement and dealkylation, but the reported data do not indicate much difference in the rates of the latter two processes. Perhaps the most unusual finding of these workers was that the side-chain rearrangement of *sec*-butyl groups to isobutyl groups appeared progressively in the *meta* di- and the tributylbenzenes, and was never seen in the monobutylbenzene or *para* dibutylbenzene, even at 70°C. They comment that this behavior must be related to the relative order of basicity of the alkylbenzenes, e.g., mesitylene > *m*-xylene > *p*-xylene > toluene, but they do not offer any explanation of the relationship of the basicity to the ease of rearrangement. They do say that they favor as a mechanism for the rearrangement an early scheme suggested by Roberts and co-workers [15] over a later modification [19], because it better illustrates how the rearrangement can proceed without the formation of any butane or diphenylbutanes, and how *m*-di- and tri-*sec*-butylbenzene are isomerized faster than mono- and *p*-di-*sec*-butylbenzene. The present authors of this monograph do not understand how these results of the Japanese workers are better accommodated by the earlier mechanism. One can actually make a good case for the effect of the greater nucleophilicity of the more-substituted aryl groups in the phenyl participation involved in the rearrangement according to the later mechanism.

With respect to the occurrence of rearrangement without dealkylation, we also do not see any basis for choice of the earlier mechanism of rearrangement, as the alkyl carbocations proposed as intermediates in the dealkylation process presumably are of higher energy than the intermediate ions proposed in either of the mechanisms of rearrangement.

Ogawa et al. outline a dealkylation mechanism for the formation of butane and diphenylbutanes which is identical with that suggested by Roberts et al. in 1963 [26]. This topic is discussed in detail in Sec. II.A).

In a second paper [41], Ogawa and co-workers extended their study to the rearrangement of di- and tri-*sec*-butylbenzenes by $AlCl_3$. They concluded that the rearrangement of *sec*-butyl to isobutyl groups was an intramolecular process occurring in a complex of the alkylbenzene with a proton (36) formed from $AlCl_3$ and H_2O (Eq. 21). This conclusion was based on the

$$\text{C}_6\text{H}_5\text{R} + \text{AlCl}_3 + \text{H}_2\text{O} \longrightarrow \text{C}_6\text{H}_5\text{R} \longrightarrow \text{H}^+ \; [\text{AlCl}_3\text{OH}]^- \qquad (21)$$

36

fact that they observed no rearrangement in 1,3,5-tri-*sec*-butylbenzene when *anhydrous* $AlCl_3$ was used at 50°C or at 80°C after 2 hr. No transalkylation products were obtained at 50°C, and only a small amount at 80°C.

A more probable explanation of the activating effect of water on aluminum chloride is that which was mentioned in Sec. I.F.1. Nenitzescu and co-workers [29] demonstrated that the cocatalyst produced, $H^+(AlCl_3OH)^-$, is a strong hydride abstractor. Roberts and co-workers in 1959 demonstrated the importance of water in promoting the aluminum chloride-catalyzed rearrangement of *sec*-butylbenzene [15]. The mechanisms of Eqs. (18) to (20) and scheme 1 are initiated by hydride abstraction from the α or β carbon of a side chain to give the phenylalkyl cations which are intermediates in the rearrangements of propyl- and butylbenzene, and also in their transalkylations by the Pines-Streitwieser mechanism.

One additional brief study of the *sec*-butylbenzene/isobutylbenzene rearrangement was reported recently. Because of the interesting results obtained by Brouwer and Hogeveen [36] and others with fluoroantimonic acid, the potential of this catalyst for the *sec*-butylbenzene/isobutylbenzene rearrangement was investigated. Roberts and Chen [42] found that whereas $AlCl_3$-H_2O produced no rearrangement in 24 hr at 25°C and only 10% in 6 hr at 50°C [15], HF-SbF_5 produced 25% rearrangement in 1 hr at 25°C and complete equilibration to the 1:2 ratio of isomers in 1 hr at 50°C. The greater effectiveness of HF-SbF_5 in isomerizing *sec*-butylbenzene was attributed to the well-known ability of the SbF_6^- counterion to stabilize carbocations such as those which are involved as intermediates in this alkylbenzene rearrangement.

4. Further studies of pentylbenzene rearrangements

As we mentioned earlier, there was a lack of agreement on the explanation of the formation of 2-methyl-3-phenylbutane by alkylation of benzene with *t*-pentyl chloride and unmodified $AlCl_3$ catalyst [2], whereas *t*-pentylbenzene is obtained as the sole or major product when milder catalysts such as BF_3 [2], $ZrCl_4$ [2], $FeCl_3$ [43], and $AlCl_3$-CH_3NO_2 [3] are used. This uncertainty about the mechanism for the formation of the rearranged product persisted for many years. The interpretation of the catalyst effect given by Schmerling and West [2] was that $AlCl_3$ is known to be an active catalyst for isomerization of alkanes, whereas the weaker catalysts are not, and the rearrangement of the side chain (Eqs. 3 to 7) was considered to be analogous to alkane isomerization. The proponents of the other mechanism [3-5] considered that the rearrangement resulted from the rearrangement of the *t*-pentyl cation to the *sec*-pentyl cation, which, being less stable, is more reactive and would alkylate benzene faster and less reversibly than the *t*-pentyl cation.

Khalaf and Roberts decided that one of the major assumptions on which the second mechanism rests could be tested experimentally, namely, the reversibility of *t*-pentylation [44]. If it could be demonstrated that dealkylation of *t*-pentylbenzene takes place readily, even with a weak catalyst, this mechanism would be discredited. Consider scheme 5. The mechanism proposed

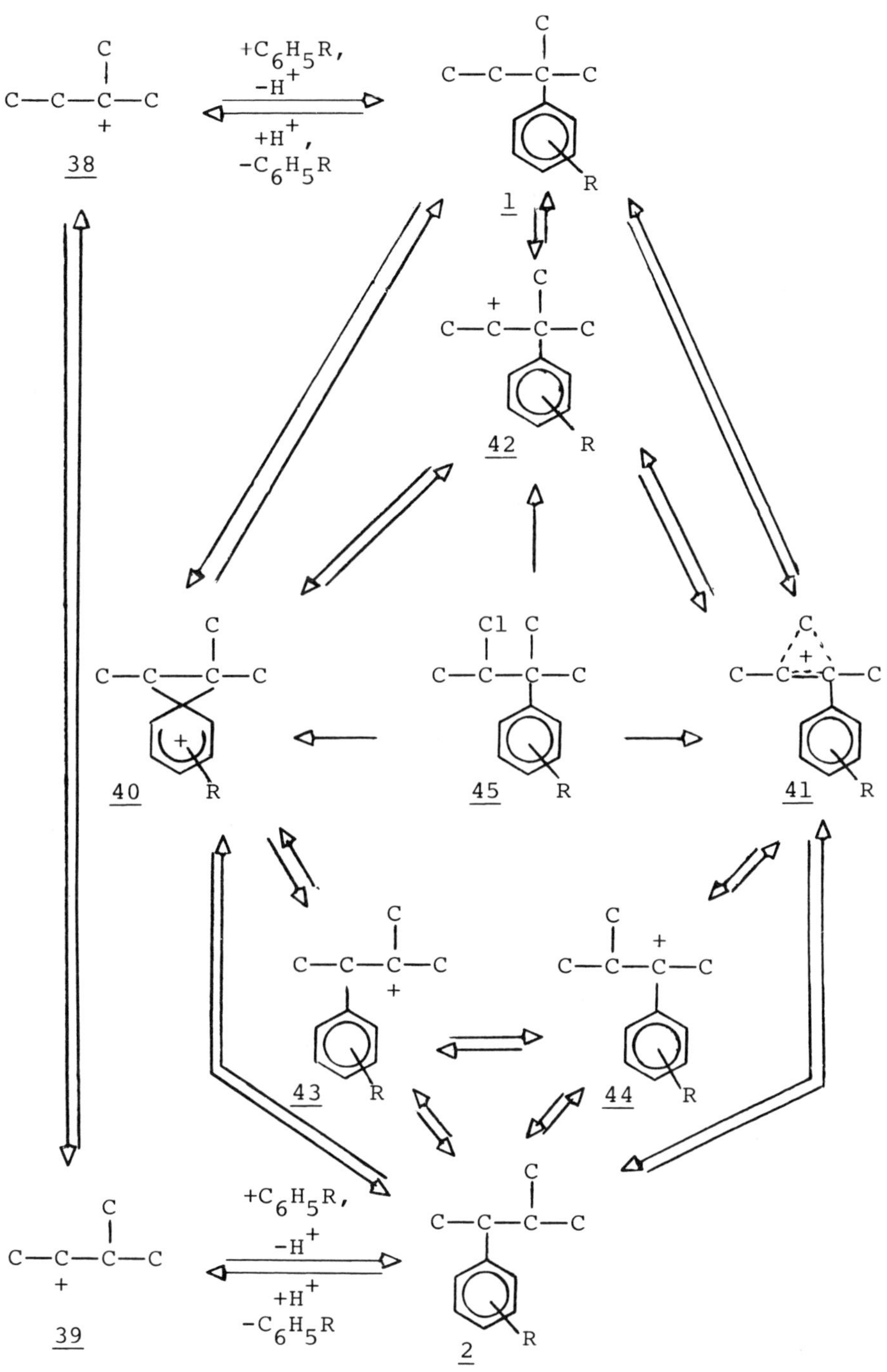

1a, R=H; 1b, R=*p*-CH_3; 1c, R=*m*-CH_3; etc.

Scheme 5

by Schmerling and West follows the sequence 1a → 42a → 43a → 2a; it may be referred to as *intra*molecular, since the rearranging side chain never becomes separated from the aromatic ring. We have added the possibility that hydride abstraction may be concerted with phenyl or methyl participation, in which case the routes may be more direct, 1a → 40a → 2a or 1a → 41a → 2a. We demonstrated that dealkylation of 1a does indeed take place readily in the presence of $AlCl_3$-CH_3NO_4, by carrying out transalkylations between 1a and toluene [44]. The *t*-pentyl group was also transferred from *p*-*t*-pentyltoluene (1b) to benzene. These transalkylations took place with no significant internal rearrangement of the *t*-pentyl side chain. A transalkylation between 2-methyl-3-*p*-tolylbutane (2b) and benzene was also carried out. It was interesting to note that the 2-methyl-3-*p*-tolylbutane underwent no reorientation to the *m* isomer in this experiment, indicating that the transalkylation is much faster than reorientation of the secondary alkyl group.

The demonstration of facile transfers of *t*-pentyl groups between benzene and toluene molecules in the presence of $AlCl_3$-CH_3NO_2, with little or no isomerization of the pentyl group, makes it appear unlikely that the effect of unmodified $AlCl_3$ in producing rearrangement of 1a to 2a is simply to increase the rate of dealkylation of 1a. More probably, the effect is to open up another mechanistic route from 1a to 2a by virtue of the hydride abstracting capability of $AlCl_3$. One of our experiments showed that, although equilibrium was reached in transfer of *t*-pentyl groups between benzene and toluene in about 1 min, only 3% of rearrangement of the side chain occurred. We had demonstrated previously that the addition of an alkyl halide to $AlCl_3$ augments its hydride abstracting capability [12]. When the transalkylations between 1a and toluene were repeated using $AlCl_3$-CH_3NO_2, with the addition of a molar equivalent of isopropyl chloride, extensive rearrangement of the *t*-pentyl side chain occurred. It was noteworthy that 14% rearrangement of 1a to 2a occurred before significant *meta-para* equilibration of the *t*-pentyltoluenes took place. Apparently the carbocation source provided by the isopropyl chloride in conjunction with even the moderated catalyst produces a medium capable of initiating and sustaining the hydride exchanges required for the intramolecular mechanism, so that the rearrangement of 1a to 2a becomes faster even than transalkylation and reorientation.

Consistent with this theory, too, was the finding that 2-chloro-3-methyl-3-phenylbutane (45a) produced a mixture of 1a and 2a upon treatment with $AlCl_3$-CH_3NO_2 and methylcyclohexane, which serves as a hydride donor. In this same paper [44] two examples were offered illustrating the driving force of phenyl participation in hydride abstraction processes. On the basis of these examples and of the results from the transalkylation experiments, Roberts and Khalaf concluded that the rearrangement of *t*-pentylbenzene (1a) to 2-methyl-3-phenylbutane (2a) induced by $AlCl_3$ takes place by a hydride abstraction process concerted with phenyl participation, producing a phenonium ion intermediate (40a). The rearrangement is completed by a second hydride abstration by the phenonium ion from the side chain of another molecule of hydrocarbon, so that a chain reaction is set up. This mechanism is similar to those described for the *sec*-butylbenzene/isobutylbenzene and isotopic *n*-propylbenzene rearrangements described earlier, in that they all involve hydride exchanges. The *n*-propylbenzene rearrangement is unique in requiring an alkylation-dealkylation step to provide a diphenylpropane intermediate in which the 1,2 methyl shift can take place without the incursion of primary carbocations.

In an earlier section, we mentioned briefly the observation by Nenitzescu et al. [16] and Roberts and Han [18] that prolonged heating of *t*-pentylbenzene or 2-methyl-3-phenylpentane with $AlCl_3$ produces neopentylbenzene. This additional rearrangement was rationalized by Nenitzescu and Roberts in quite similar formulations (scheme 6). Nenitzescu's mechanism proceeded from 2 to 10 via intermediates 44a, 43a, and 46. Roberts' mechanism postulated the same route and intermediates, but included the possibility that the phenonium ion 40a may be a part of the scheme. Nenitzescu emphasized the fact that the rearrangement of 1 → 2 → 10 is similar to that of *sec*-butylbenzene → isobutylbenzene in the sense that the sequence is in the direction tertiary → secondary → primary alkylbenzene in both cases, and he suggested that it is related to the order of thermodynamic stability of the three types of alkylbenzenes. Roberts and Han suggested that a steric factor may be involved in the fact that 10 is formed so much more slowly than 2 from 1. The carbocation 46 may not have much stabilization by resonance with the structure 47 because of steric interference to coplanarity between the *t*-butyl group and the *ortho* hydrogen.

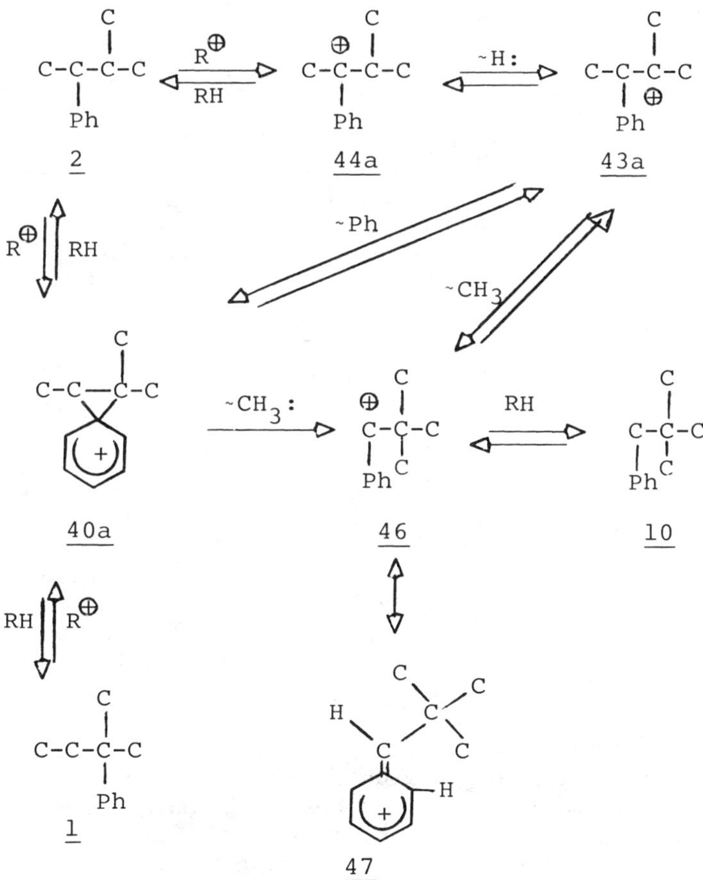

Scheme 6

The rearrangement of 2-phenylpentane (11) and 3-phenylpentane (12) to 2-methyl-1-phenylbutane (13) by $AlCl_3$, discovered by Roberts and Han [18, 19[, was rationalized as shown in scheme 7. The conversion of 11 and 12

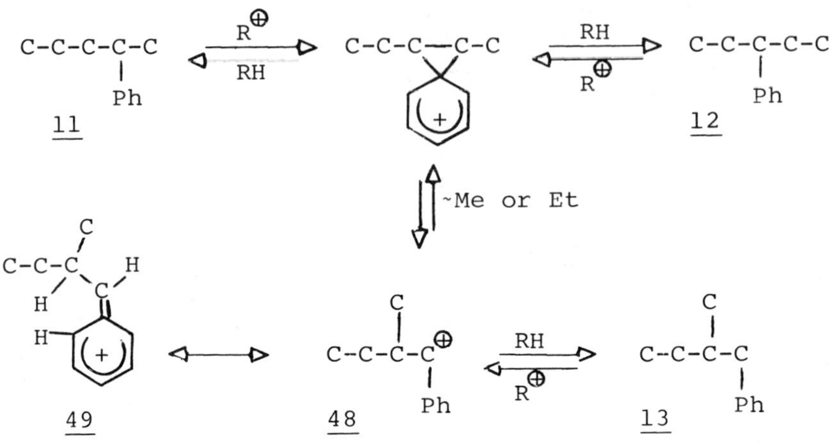

Scheme 7

to 13 is not as irreversible as the conversion of 1 and 2 to 10. This may be attributed to the lesser steric inhibition of the resonance stabilization of cation 48 by contributions from structures such as 49.

In connection with a series of dealkylation studies which are described in Sec. II, Roberts and co-workers considered the possibility of rearrangements of higher linear secondary alkylbenzenes according to Eq. (21a), by a mechanism analogous to that of scheme 7. As we have seen, *sec*-butylben-

$$CH_3(CH_2)_nCH_2\underset{\substack{|\\ Ph}}{CH}\text{-}CH_3 \xrightarrow{AlCl_3} CH_3(CH_2)_n\overset{\substack{CH_3\\ |}}{CH}\text{-}\underset{\substack{|\\ Ph}}{CH} \qquad (21a)$$

zene (n = 0 in Eq. 21a) rearranges to an apparent equilibrium mixture of *sec*-butyl- and isobutylbenzene containing ca. 67% of isobutylbenzene [15]. 2-Phenylpentane (n =1) gives mixtures of 2- and 3-phenylpentane and 2-methyl-1-phenylbutane containing up to 35% of the last isomer [19]. 2-Phenylhexane (n =2) gave a complex mixture of products, as will be described in Sec. II. The hexylbenzenes recovered (ca. 10%) in one experiment consisted of 2-phenylhexane, 3-phenylhexane, and 2-methyl-1-phenylpentane in ratio 61%:30%:9% [45]. No traces of 2-methyl-1-phenylhexane or of 2-methyl-1-phenylheptane could be detected among the products from heating 2-phenylheptane [46] and 2-phenyloctane [47] with aluminum halides. This decline in the yields of the 2-methyl-1-phenylalkanes in going up the homologous scale from 1-phenylbutane can plausibly be attributed to the conversion of the initially formed 2-methyl-1-phenylalkanes, or their carbocation precursors, to other products, including indan and tetralin derivatives when the side chain is longer than four carbon atoms.

This section on alkylbenzene rearrangements was introduced by a description of the report of Schmerling and West that alkylation of benzene by *t*-

pentyl chloride in the presence of unmodified aluminum chloride gave the secondary alkylbenzene 2-methyl-3-phenylpentane as the major product. The explanation they gave for this finding was initial formation of the expected product, *t*-pentylbenzene, followed by rearrangement of this initial product to the isomeric secondary pentylbenzene. This rationale has been supported by considerable additional experimental evidence and is now generally accepted.

It has now been shown that alkylation of benzene by two other pentyl halides, when carried out with unmodified aluminum chloride as catalyst, gives 2-methyl-3-phenylbutane as the major product and a minor amount of *t*-pentylbenzene, whereas when a milder catalyst ($AlCl_3$-CH_3NO_2) is employed, *t*-pentylbenzene is the almost exclusive product [48]. The pentyl halides are 1-chloro-2-methylbutane (50) and 2-chloro-3-methylbutane (55). The product mixtures were almost the same from either pentyl chloride; none of the primary alkylbenzene (52) was obtained. These results are rationalized in scheme 8. Using $AlCl_3$-CH_3NO_2 catalyst, the primary carbocation or complex (51) formed from 50 rearranges so rapidly to the tertiary carbocation (54) that none of the primary alkylbenzene 52 is produced. The secondary cation (56) formed from 55 also rearranges to the tertiary cation, so that *t*-pentylbenzene (1) is produced from either pentyl chloride. With unmodified $AlCl_3$ catalyst, the sequence of events is probably the same, but now rearrangement of 1 to 2 occurs because of the ability of the strong catalyst to abstract a hydride ion from the β-carbon of 1, with phenyl participation, to produce the phenonium ion intermediate 40a. Whether or not any of the 2 is produced

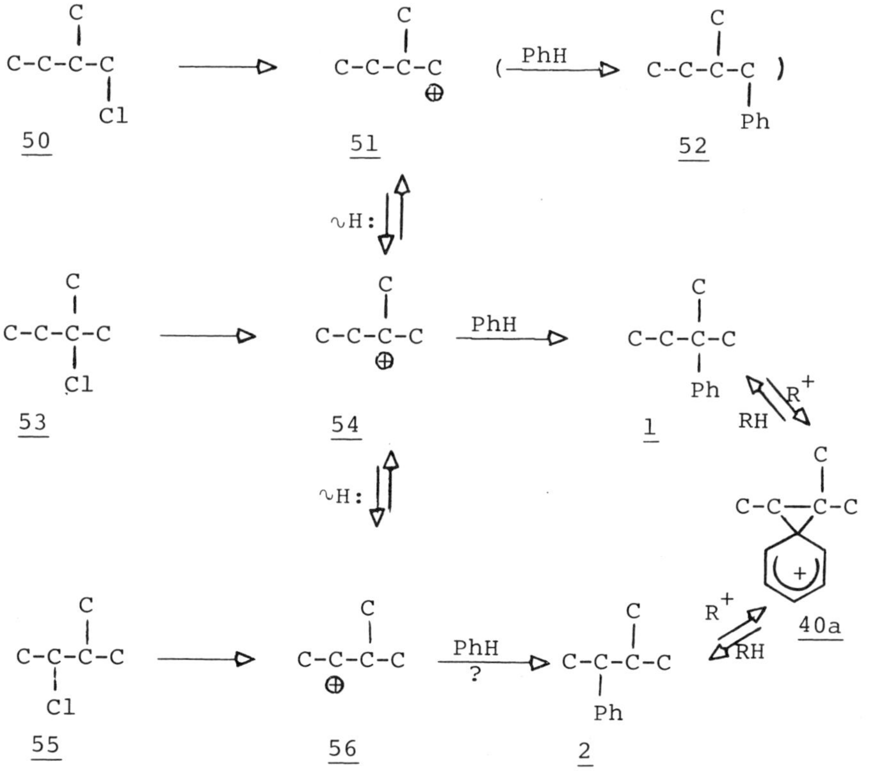

```
     C                       C                           C
     |                       |                           |
C-C-C-C       ----->    C-C-C-C     ( --PhH-->     C-C-C-C  )
       |                      ⊕                            |
       Cl                                                  Ph
 50                      51                          52

                        ~H: ⇅

     C                       C                           C
     |                       |                           |
C-C-C-C       ----->    C-C-C-C      --PhH-->      C-C-C-C
     |                       ⊕                           |
     Cl                                                  Ph   ⇅ R+ / RH
 53                      54                           1
                                                                   C
                        ~H: ⇅                                      |
                                                              C-C——C-C
                                                                 \/
                                                               (benzenium +)
      C                      C                           C    R+ / RH ⇅
      |                      |                           |        40a
C-C-C-C       ----->    C-C-C-C      --PhH-->      C-C-C-C
  |                     ⊕              ?             |
  Cl                                                 Ph
 55                      56                           2
```

Scheme 8

by direct alkylation by the secondary carbocation 56 is uncertain. However, when the alkylation of *p-xylene* by the primary pentyl chloride (50) was examined, the only alkylxylene formed was 2-methyl-3-*p*-xylylbutane (57), the same product that had been obtained previously by alkylation of *p*-xylene

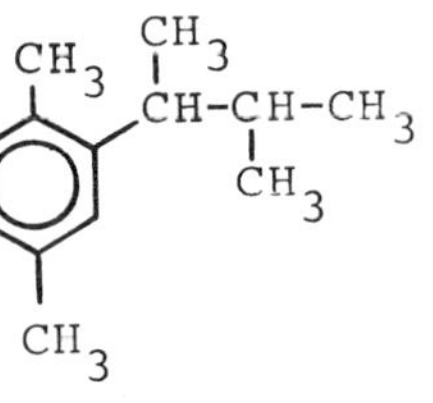

CH_3 $CH(CH_3)$-$CH(CH_3)$-CH_3

CH_3

57

with *t*-pentyl chloride [3,49]. Apparently, the steric requirements of the *t*-pentyl cation (54) are so great that it does not easily become attached ortho to a methyl group, but the effective size of the secondary cation 56 is smaller so that it does alkylate adjacent to the methyl group in *p*-xylene. This result demonstrates the interesting carbocation rearrangement sequence primary → tertiary → secondary.

II. DEALKYLATIONS OF ARENES

The dealkylation reaction of an alkylbenzene induced by a Friedel-Crafts catalyst is characterized by the production of an alkane corresponding in carbon number to the original alkyl side chain. One of the earliest examples of dealkylation was evidenced by the observation of a saturated gas evolved when *p*-cymene was heated with aluminum chloride at 150°C, and the detection of toluene among the liquid products of this reaction [50]. The gas was later identified as propane and various disproportionation products were found in addition to toluene, the liquid product of the dealkylation [51]. Similarly, heating isopropylbenzene with $AlCl_3$-HCl at 100°C was found to produce propane, but *n*-propylbenzene and *n*-butylbenzene subjected to identical treatment were reported to give no alkanes [9]. In 1937 Ipatieff and Pines made a study of the reactions of a series of alkylbenzenes when heated at 65 to 80°C with a hydrogen donor (cyclohexane or decalin) in the presence of aluminum chloride [52]. They found that the ease of dealkylation to yield the alkanes was in the order of the side chains *t*-butyl > *sec*-butyl > isopropyl. Ethyl and methyl groups were not cleaved. McCaulay and Lien observed the production of isobutane from *t*-butylbenzene when it was warmed to 45°C with HF in BF_3; disproportionation without any dealkylation occurred at 0°C [7]. Sharman observed the formation of "branched saturated hydrocarbons corresponding in carbon content to the original side chain" from the secondary dodecylbenzenes produced by alkylation of benzene with 1-bromododecane and Al_2Br_6 [53]. The dealkylation was said to occur at a measurable rate at 6°C and very rapidly at 35°C. Since the primary dodecylbenzene did not undergo the dealkylation, the percentage of the primary isomer was found to increase in the reaction mixture, owing to the loss of the secondary dodecylbenzene isomers by dealkylation.

A. Dealkylations at 100°C with $AlCl_3$-H_2O

In 1963, Roberts et al. reported a systematic investigation of the dealkylation of alkylbenzenes, a study that was made feasible by modern instrumental methods available by the time, particularly gas chromatography [26]. At the outset of describing this research, we should reiterate the fact that it is usually difficult or impossible to study individually a single type of reaction of an alkylbenzene with a Friedel-Crafts catalyst. The catalyst and experimental conditions required to produce dealkylation of an alkylbenzene are strong enough to produce several other reactions, such as disproportionation, reorientation, rearrangement of a side chain, and fragmentation. All of these except fragmentation have been discussed in previous chapters or sections. Fragmentation is the term we use to describe the reaction of an alkylbenzene to produce alkanes with fewer carbon atoms that the original alkyl side chain, and also alkylbenzenes with fewer carbon atoms than the original hydrocarbon. Although fragmentation products were found as well as dealkylation products by Roberts et al. we shall concentrate on the dealkylation reaction in this section and consider fragmentation processes in Sec. III.

"Dealkylation" might be interpreted as meaning true reversal of Friedel-Crafts alkylation, with the production of benzene and an alkyl halide or alkene from an alkylbenzene. However, under ordinary alkylating conditions the thermodynamic equilibrium is so far on the side of alkylation that the reaction is irreversible for all practical purposes. Nevertheless, the apparent reversibility of some cyclialkylations (Chap. 6) tells us that the process is partially reversible to an unstable intermediate, probably a carbocation, which can rearrange before recyclizing.

The evolution of propane from isopropylarenes suggests that a transient isopropyl cation may abstract a hydride ion from some source in a fast step that diverts it from the equilibrium. This hypothesis is supported by the results of the experiments of Ipatieff and Pines [52], in which cycloalkanes served as the hydride donors. The observed order of rate and ease of dealkylation these workers found (*t*-butyl > *sec*-butyl > isopropyl >> ethyl, methyl) is, of course, in line with the stability of the carbocation which may be produced by cleavage of the side chain from the benzene ring.

The experiments of Roberts et al. were carried out on all of the alkylbenzene homologs and isomers having from one to five carbon atoms in the side chain. Their results confirmed the finding of Ipatieff and Pines that the order of susceptibility of alkylbenzenes toward alkane formations in the presence of $AlCl_3$-H_2O was in the order of tertiary > secondary > primary > methyl side chains, and they further demonstrated that no added source of hydride ions is required. It appears that the side chains of some alkylbenzenes can serve equally as well as a cycloalkanes as a hydride donor since they hold hydrogens in secondary, tertiary, and/or benzylic positions. Examination of the data in Table 1 shows that the yields of alkanes produced from the alkylbenzenes by heating with $AlCl_3$-H_2O at 100°C for 3 hr are related to the primary, secondary, or tertiary nature of the side chain, as we have mentioned.

The gas produced from both *sec*-butyl- and isobutylbenzene during a 3-hr period had about the same ratio of butane and isobutane, ca. 2:1, but in some samples of gas taken after shorter heating periods, the ratios were quite different. A more extensive study of the change in composition with time showed that the earliest gas samples from either *sec*-butyl- or isobutylbenzene were rich in isobutane. This finding led Roberts et al. to suggest that when

Table 1. Dealkylation Products of Ethyl-, Propyl-, Butyl-, and Pentylbenzenes[a]

Starting material, $R-C_6H_5$	Alkanes produced	
	Mole %[b]	Identity (ratio, %)
Ethyl	7	Ethane
n-Propyl	7	Propane
Isopropyl	18	Propane
n-Butyl	4	Butane/isobutane (36:64)
Isobutyl	15	Butane/isobutane (70:30)
sec-Butyl	24	Butane/isobutane (70:30)
t-Butyl	72	Isobutane
n-Pentyl	14	Pentane/isopentane (14:81)[c]
Isopentyl	13	Pentane/isopentane (6:80)[d]
2-Methylbutyl	19	Pentane/isopentane (11:74)[e]
1-Methylbutyl	20	Pentane/isopentane (4:96)
1-Ethylpropyl	27	Pentane/isopentane (4:96)
Neopentyl	18	Isopentane (83)[f]
1,2-Dimethylpropyl	40	Isopentane (73)[g]
t-Pentyl	56	Isopentane (72)[h]

[a]Reactions at 100°C, 3 hr; reactants: $R-C_6H_5$, 0.1 mol; $AlCl_3$, 0.03 mol; H_2O, 0.01 mol.
[b]Yield based on starting material.
[c]Also isobutane, 5%.
[d]Also isobutane, 8%; butane, 2%; propane, 4%.
[e]Also isobutane, 5%; butane, 6%; propane, 4%.
[f]Also isobutane, 15%; butane, 1%; propane, 1%.
[g]Also isobutane, 27%.
[h]Also isobutane, 28%.

the alkylbenzene is added to the fresh, hot catalyst, both isomers are isomerized to *t*-butylbenzene, which is immediately dealkylated to isobutane. The rapid rate of gas evolution was of short duration and within a few minutes butane became the major component, the ratio of butane to isobutane then representing the relative rates of dealkylation of *sec*-butyl- and isobutylbenzene. A different interpretation of the initial production of isobutane from both butylbenzene isomers, and of isopentane from all of the pentylbenzenes will be given a little later (Sec. II.D.1).

The rates of dealkylation of all of the alkylbenzenes decreased with time and gas evolution usually stopped in 3 hr or less. That this could be atttributed to deactivation of the catalyst by the polyalkylbenzenes formed by disproportionation was demonstrated by (1) adding fresh catalyst, which reinitiated gas evolution; (2) adding pentamethylbenzene, which stopped gas evolution; and (3) by adding pentamethylbenzene at the beginning, in which case no gas evolution occurred under the otherwise identical conditions of catalyst and heating that produced dealkylation.

As we mentioned earlier, the facile dealkylation of alkylbenzenes in the absence of an added hydride donor suggested that the side chains themselves may serve as hydride donors. Support for this theory came from the fortui-

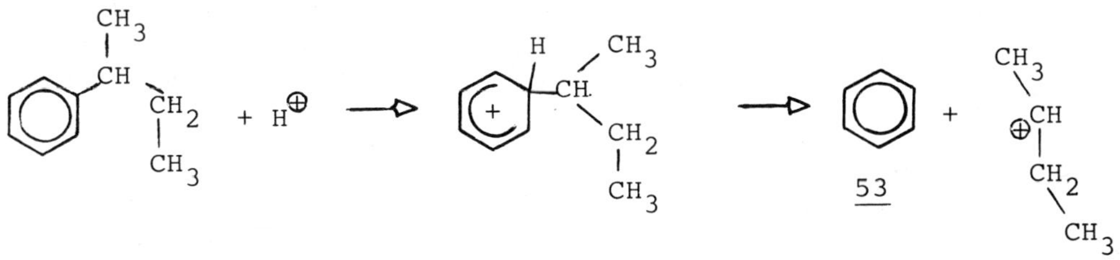

CH3 CH CH2 CH3 + H⊕ → H CH3 CH CH2 CH3 → 53 + CH3 ⊕CH CH2 CH3

(22)

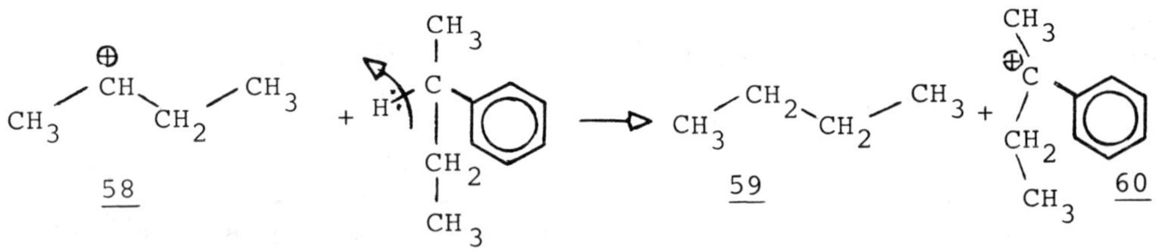

58 CH3 CH⊕ CH2 CH3 + H: CH3 C CH2 CH3 → 59 CH3 CH2 CH2 CH3 + CH3 ⊕C CH2 CH3 60

(23)

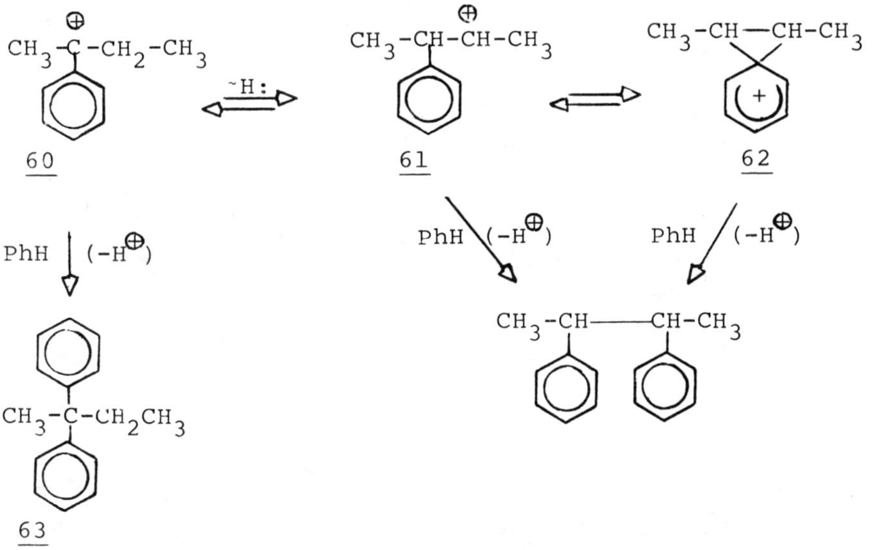

60 → ~H: → 61 → 62

(24)

PhH (-H⊕) → 63

PhH (-H⊕) PhH (-H⊕) → CH3-CH—CH-CH3

(25)

Scheme 9

tous isolation and identification of *meso*-2,3-diphenylbutane from one of the reaction mixtures produced from the reaction of *sec*-butylbenzene and $AlCl_3$-H_2O. Scheme 9 illustrates a possible mechanism for the dealkylation of *sec*-butylbenzene. In this scheme, the function of a second molecule of *sec*-butylbenzene as a hydride donor is shown in Eq. (23), and in Eq. (24) are depicted some of the possible rearrangements of the benzyl cation (60) which results from the hydride donation. Equations (25) illustrate two likely fates of the ions 60-62 in alkylating benzene (present owing to concurrent disproportionation as well as dealkylation), *sec*-butylbenzene, or other arene molecules. Other possible fates of these ions include ejection of a proton to produce substituted styrenes which (mainly) may polymerize. McCauley and Lien [7] reported spectroscopic (ultraviolet) evidence for the presence of substituted styrenes in the high-boiling residue from a dealkylation of *t*-butylbenzene with HF-BF_3 at 45°C. Besides *meso*-2,3-diphenylbutane, which could be isolated because of its unusually high melting point and low solubility in the reaction medium, Roberts et al. were able to demonstrate the presence of racemic 2,3-diphenylbutane, 2,2-diphenylbutane (63 in scheme 9), and 1,2-diphenyl-2-methylpropane in reaction mixtures from *sec*-butyl- and isobutylbenzene by gas chromatographic analysis [12]. We have mentioned these same diphenylbutanes in Sec. I as possible intermediates in the rearrangement of *sec*-butyl- and isobutylbenzenes, and we shall encounter them again when describing fragmentation reactions in Sec. III.

The extremely facile and extensive dealkylation of *t*-butylbenzene deserves special mention. The first step of the reaction, analogous to Eq. (22) for the *sec*-butylbenzene, should take place readily since the *tertiary* butyl cation is produced (Eq. 26). The second step (Eq. 27), if it is analogous to Eq. (23), requires the donation of a hydride ion from a methyl group, apparently resulting in the formation of a primary carbocation (64), which would be a high-energy intermediate. However, as we have mentioned before, there is

$$C_6H_5\text{-}C(CH_3)_3 \xrightarrow{H^{\oplus}} [H(C_6H_5^{+})\text{-}C(CH_3)_3] \longrightarrow C_6H_6 + (CH_3)_3C^{\oplus} \tag{26}$$

$$(CH_3)_3C^{\oplus} + H\text{-}CH_2\text{-}C(CH_3)_2\text{-}C_6H_5 \longrightarrow (CH_3)_3CH + {}^{\oplus}CH_2\text{-}C(CH_3)_2\text{-}C_6H_5 \ (\underline{64}) \tag{27}$$

evidence that hydride transfer from a carbon atom *beta* to a phenyl group may take place with phenyl participation [54], so that the second step may be more like Eq. (28), followed by the alkylation step (Eq. 29) to produce

$$(CH_3)_3C^{\oplus} + C_6H_5\text{-}CH_2\text{-}C(CH_3)_2\text{-}H \longrightarrow (CH_3)_3CH + \underline{65} \longrightarrow \underline{66} \quad (28)$$

$$\underline{66} + C_6H_6 \xrightarrow{-H^{\oplus}} C_6H_5CH_2\text{-}C(CH_3)_2\text{-}C_6H_5\ (\underline{67}) \quad (29)$$

1,2-diphenyl-2-methylpropane (67). This compound, which was identified in trace amount by gas chromatography among the products from reaction of *t*-butylbenzene with $AlCl_3$, may serve as a good additional hydride donor. Abstraction of the two methylene hydrogens in the benzyl position should be easy, and one can account for a 75 mol % yield of isobutane by the stoichiometry of Eq. (29a). The tetraphenylbutane (68) may undergo dealkylation

$$4Ph\text{-}C(CH_3)_2\text{-}CH_3 \longrightarrow 3CH_3\text{-}CH(CH_3)\text{-}CH_3 + Ph_3C\text{-}C(CH_3)(Ph)\text{-}CH_3\ (\underline{68}) \quad (29a)$$

and fragmentation and/or comprise a part of the almost coal-like residue from the reaction.

The addition of decalin, which has two tertiary hydrogens and thus is a potential hydride donor, was found to have little effect on the yield of propane from treatment of *n*-propyl- and isopropylbenzene with $AlCl_3$-HCl for 3 hr at 50°C, but its addition to *t*-butylbenzene increased the yield of isobutane from 27% to 88% [55]. This led Miethchen and Ewald to state as a generalization that the dealkylation of alkylbenzenes by $AlCl_3$-HCl is increased by adding decalin only if the starting alkylbenzene contains no C-H bonds in benzyl positions having stronger hydride-donor properties than the tertiary, secondary, and primary C-H bonds of alkanes and cycloalkanes.

B. Dealkylations at 40°C and Below by $AlCl_3$

A few years after this work was reported by Roberts et al., Ogawa and coworkers at Kansai University in Japan published a series of articles on dealkylation, rearrangement, and transalkylation reactions of alkylbenzenes. Some of the results from one of these papers were described in Sec. I in regard to rearrangements of the pentyl-, hexyl-, and heptylbenzenes studied [38]. In an earlier paper, the dealkylations of all of the four butyl- and eight pentylbenzene isomers by $AlCl_3$ were investigated [56]. This paper duplicated the Roberts et al. work in many respects, without comparing the results or acknowledging the identity or similarity of the mechanisms proposed, although the earlier paper was referenced. The only unique aspect of the research was the mild conditions employed. The reactions were carried out at 30 to 40°C in closed test tubes with the alkylbenzene, $AlCl_3$, and hexane in molar ratios of 1:0.2:0.5, for from 1 min (for *t*-butyl- and *t*-pentylbenzene) to 15 or 20 hr for the primary alkylbenzenes. Analysis was entirely by gas chromatography. Yields of alkanes were expressed relative to benzene, produced by transalkylation in all cases and, presumably, by dealkylation to the extent in which it occurred. The following results were reported:

1. The rates of transalkylation were, in the case of all the butyl- and pentylbenzenes, faster than the rates of dealkylation, as determined by the observed benzene and alkane peaks on the GC traces after various reaction times.
2. The rates of dealkylation were in the order tertiary-RPh > secondary-RPh > primary-RPh (as reported by Roberts et al. [26]). Amounts of alkanes produced from primary alkylbenzenes were apparently *very* small.
3. The alkanes produced were (a) the same in structure as in the original side chain in the case of all of the primary alkylbenzenes, except neopentylbenzene, which gave isopentane; (b) isopentane from all of the secondary pentylbenzenes and from *t*-pentylbenzene, and isobutane from *t*-butylbenzene; and (c) *n*-butane from *sec*-butylbenzene.

The mechanism proposed for the dealkylation of the secondary and tertiary alkylbenzenes was essentially that of Roberts et al., as outlined earlier in this section. For the primary alkylbenzenes, the Japanese authors suggest an analogous mechanism, in which the hydride abstraction to form the alkane takes place by attack of a "polarized alkylbenzene" on the side chain of another alkylbenzene molecule.

The unusual production of *n*-butane from *sec*-butylbenzene (in view of the production of isopentane from *sec*-pentylbenzenes) and of isopentane (rather than neopentane) from neopentylbenzene were emphasized, but no explanation was offered.

We will suggest at this time an explanation for the latter observation, and the first will be treated a little further on. In the paper by Roberts and Han [19] mentioned earlier, treatment of neopentylbenzene with $AlCl_3$-H_2O in benzene at 80 to 82°C for 24 hr was reported to give a 33% recovery of neopentylbenzene, and 2% of 2-methyl-3-phenylbutane was detected. This indicated that the conversion of 2-methyl-3-phenylbutane (2 in scheme 6) to neopentylbenzene (10) is not completely irreversible, and it was also demonstrated [19] that *t*-pentylbenzene (1) was produced from 2 on heating with $AlCl_3$-H_2O. Thus it is probable that the isopentane obtained from neopentylbenzene actually comes from the dealkylation of both 2-methyl-3-phenylbutane and *t*-pentylbenzene, produced in small amounts by rearrangement of the original neopentylbenzene.

The Kansai group extended their investigation to the secondary and tertiary hexyl- and heptylbenzenes [38]. The aspects of this work related to rearrangements of these alkylbenzenes were discussed in Sec. I. We mentioned there that they observed little isomerization among the branched secondary alkylbenzenes 25-29 probably because these were also converted to the *tertiary*

```
        C                          C
        |                          |
C-C-C-C-C                    C-C-C-C-C
    |                          |
    Ph                         Ph

        25                         26
```

```
         C                 C                 C
         |                 |                 |
C-C-C-C-C-C       C-C-C-C-C-C       C-C-C-C-C-C
     |                 |                 |
     Ph                Ph                Ph

  27                28                  29
```

hexyl- and heptylbenzenes (69 and 70), which underwent rapid dealkylation.

```
       C                      C
       |                      |
C-C-C-C-C              C-C-C-C-C-C
       |                      |
       Ph                     Ph
  69                     70
```

The authors reported that for compounds 25-29, dealkylation was faster than transalkylation, based on the earlier and greater production of alkanes than of benzene. They attributed this to two processes: (1) the alkylation of the benzene, produced in the dealkylation step, by the phenylalkyl cation, produced by hydride abstraction, to produce a diphenylalkane (see scheme 9, the mechanism proposed by Roberts et al. [26]); and (2) the formation of indans and tetralins by cyclialkylation of the tertiary phenylalkyl cations produced from compounds 26, 28, and 29. They did not identify any indans and tetralins, but a substituted tetralin (72) had previously been reported

```
 C     C                     C     C
 |     |       AlCl3         |     |
C-C-C-C-C-C  ---------->   C-C-C-C-C-C   +   PhH
 |             20°
 Ph
                              +
  71                    [tetralin structure]
                              72
```

by some of these same workers from a reaction of 2,5-dimethyl-2-phenylhexane (71) with $AlCl_3$ at 20°C [57]. Indans have since been found among the products from *n*-butylbenzene [35]; cf. scheme 3.

Dealkylation of the tertiary alkylbenzenes 69 and 70 was much faster than that of any of the secondary alkylbenzenes (observable in an hour at −3°C), but disproportionation was still faster. A small amount of rearrangement of 69 to 25 and of 70 to 27 was observed, as might be expected (cf. Sec. I).

C. Effect of Dealkylations on the Distribution of Products of Alkylations

One of the long-standing mysteries of Friedel-Crafts chemistry has been the early reports that alkylations with certain primary alcohols take place without rearrangement when aluminum chloride is used as catalyst, whereas rearrangements expected of carbocation intermediates occur when "milder" catalysts such as sulfuric acid and boron trifluoride are employed [58-61]. For example, neopentyl alcohol was reported to yield neopentylbenzene with $AlCl_3$ catalyst [59], but *t*-pentylbenzene with H_2SO_4 [58] or BF_3 [61] catalyst.

It has now been shown that the detection of the so-called "unrearranged product" in this case is actually the end result of multiple rearrangements. The first rearrangement is one of the alkylating agent, before attachment to the benzene ring so that the initial product is *t*-pentylbenzene (cf. Chaps. 2 and 3). This hydrocarbon is isomerized by $AlCl_3$, first to 2-methyl-3-phenylbutane and then to neopentylbenzene, as we described in Sec. I [16,18,19]. Both H_2SO_4 and $AlCl_3$ isomerize the alkylating agent, but only the latter is capable of isomerizing *t*-pentylbenzene.

Other primary alcohols that have been reported to alkylate benzene without rearrangement when $AlCl_3$ is used are *n*-propyl alcohol [58], cyclobutylcarbinol, cyclopentylcarbinol, and cyclohexylcarbinol [60]. The report about the last three compounds has been shown to be incorrect; they all do give alkylation products of rearranged structure [62], as described in Chap. 4. The report about *n*-propyl alcohol is also incorrect, but for a different reason, one which is related to dealkylation, so it will be described in detail.

n-Propyl alcohol was first reported to give only isopropylbenzene when 80% H_2SO_4 was used as catalyst [58]; more recently, with $AlCl_3$ catalyst it was found to give about equal amounts of *n*-propyl- and isopropylbenzene [63]. Their earlier observation of the greater ease of dealkylation of isopropylbenzene than of *n*-propylbenzene [26] suggested to Roberts and co-workers an explanation for the results of alkylation with *n*-propyl alcohol and $AlCl_3$. The conditions employed by Ipatieff et al. [58] (110 to 120°C, 1 hr) and Nield [63] (80°C, 6.5 hr) were quite severe, and it seemed possible that the initial major product was isopropylbenzene, as expected, but that the prolonged heating with $AlCl_3$ decomposed it by dealkylation more rapidly than *n*-propylbenzene. Thus the relative amount of the latter isomer became increasingly larger until it was the only isomer detected by the qualitative identification technique employed in the early work, preparation of a derivative [58].

This explanation was confirmed by the results obtained by Roberts et al. in 1969 [62] and presented in Table 2. In reaction mixtures stirred at room temperature for 24 to 60 hr, isopropylbenzene was the major product (74 to 77%). When the reaction mixtures were heated, the relative amount of isopropylbenzene decreased and that of *n*-propylbenzene increased until, in the second experiment after 7 hr, *n*-propylbenzene was the major alkylation prod-

Table 2. Alkylation of Benzene with 1-Propanol and Aluminum Chloride

	Experiment 1			Experiment 2		
	S1[a]	S2[b]	FRM[c]	S1[a]	S2[b]	FRM[c]
Time at 25°C (hr)	60	60	60	24	24	24
Time at 80°C (hr)	—	1	8	—	1	6
Composition[d] (%)						
Me-Ph	—	1	11	—	—	7
Et-Ph	—	3	28	—	2	23
i-Pr-Ph	77	46	29	74	64	29
n-Pr-Ph	23	51	29	26	34	38
s/*i*-Bu-Ph[e]	—	—	2	—	—	2

[a]First sample from reaction mixture.
[b]Second sample from reaction mixture.
[c]Final reaction mixture.
[d]Determined from relative GLPC peak areas.
[e]*sec*-Butyl- and isobutylbenzene were not resolved by the GLPC column (SE-30 silicone gum rubber).

uct. Gas evolved during the course of the heating, collected and analyzed by gas chromatography (GLPC), was found to consist mainly of propane, although an appreciable amount of isobutane was produced during the second and third hours of heating. Isobutane was previously observed as a gaseous product when 1,3-diphenylpropane was heated with $AlCl_3$ [12]. The detection of this alkane, as well as that of *sec*-butylbenzene and/or isobutylbenzene among the liquid products (Table 2), is indicative of the complex sequence of dealkylations, carbocation additions, rearrangements, fissions, realkylations, and further dealkylations that must occur when arenes are heated with $AlCl_3$. Significant amounts of the fragmentation products ethylbenzene and toluene were also found in the liquid reaction mixture.

The study of the alkylation of benzene with *n*-propyl alcohol illustrates the possible effect of dealkylation as a side reaction on the distribution of products that may be obtained when one of the products is more susceptible to dealkylation than the other. This kind of selectivity is probably not restricted to alkylations with alcohols, but occurs whenever alkylations are carried out under strenuous conditions of temperature and time, expecially with the stronger Freidel-Crafts catalysts. For example, although the difference in the proportion of *n*-propyl- and isopropylbenzene obtained by alkylation with *n*-propyl chloride at low temperature (e.g., −6°C) and higher temperature (e.g., 80°C) is not as great [17] as first reported [58], it might be found to be somewhat greater if a correction were made for the amount of isopropylbenzene lost by dealkylation at the higher temperature. Similarly, the small amounts of isopropylbenzene detected in the reaction mixtures from the isotopic rearrangement of *n*-propyl-^{14} or ^{13}C-benzene (Sec. I), and of *t*-butylbenzene found in *sec*-butylbenzene-isobutylbenzene rearrangement reaction mixtures may not represent accurately the amounts of these isomers actually produced by rearrangements.

Table 3. Gaseous Dealkylation Products from Secondary Butyl- and Pentylbenzenes[a]

	Run						
	1	2	3	4	5	6	7
Alkylbenzene	2-PB[b]	2-PB[b]	2-PP[c]	2-PP[c]	2-PP[c]	3-PP[d]	3-PP[d]
Solvent (mol)		MCH[e] (0.6)			MCH[e] (1.2)		MCH[e] (1.2)
Reaction temp. (°C)	100	100	100	30	80	80	80
Mole percent dealkylation[f]	31	48	21	1	8	14	7
Alkane distribution[g] (%)							
n-C_4H_{10}	72	89					
i-C_4H_{10}	27	11	Trace				
n-C_5H_{12}			4	Trace	16	3	19
i-C_5H_{12}	1		96	100	84	97	81

[a]Mole ratio of reactants, R-C_6H_5/$AlCl_3$/H_2O = 1.0:0.33:0.11; reaction time, 3 hr.
[b]2-Phenylbutane.
[c]2-Phenylpentane.
[d]3-Phenylpentane.
[e]Methylcyclohexane.
[f]Mole percent conversion to alkanes based on alkylbenzene and calculated from volume of gas collected.
[g]GLC analysis.

Table 4. Liquid Dealkylation and Fragmentation Products from Secondary Butyl- and Pentylbenzenes[a]

	Run					
	1	3	4	5	6	7
Alkylbenzene	2-PB	2-PP	2-PP	2-PP	3-PP	3-PP
Product total[b] (%)	27	21	68	17	38	15
C_6H_5-R[c] (mol %)						
Me	4	2				Trace
Et	13	12		1	1	2
n-Pr	3	1				
i-Pr	5	10		1	1	1
sec-Bu	26			Trace	Trace	Trace
i-Bu	49	Trace		Trace	Trace	Trace
n-Bu				4		2
2-Pe[d]		37	56	42	53	46
3-Pe		11	33	10	12	12
2-MB[e]		27	11	42	33	37

[a]Conditions, reactants, and run numbers are the same as in Table 3.
[b]Weight percent of original alkylbenzene of products; bp 80 to 220°C.
[c]Mole percent of alkylbenzenes from GLC analysis.
[d]2-Pentyl.
[e]2-Methyl-1-butyl.

D. Relationships Between the Structure of the Side Chain and the Alkane Formed by Dealkylation

1. *sec*-Butyl-, pentyl-, and hexylbenzenes

a. sec-Butyl- and pentylbenzenes

Many years after the publication of the paper on the dealkylation of propyl-, butyl-, and pentylbenzenes [26], while reviewing some of the results, Roberts became intrigued with the difference in the proportions of normal and branched alkanes produced by 2-phenylpentane and 2-phenylbutane (*sec*-butylbenzene). The major product from 2-phenylbutane was *n*-butane (70%), whereas 2-phenylpentane gave only 4% *n*-pentane and 96% isopentane. Some of the experiments were repeated to check the validity of these results, and research was extended to higher alkylbenzenes as well [45]. The results are summarized in Tables 3 and 4. Runs 1 and 3 duplicated experiments reported in the 1963 paper, and the results were the same as before. In the earlier study, the proportion of *n*-butane and isobutane produced from both *sec*-butyl- and isobutylbenzene with short reaction times at 100°C was found to be high in isobutane. The explanation offered was that both butylbenzenes were isomerized initially by the active catalyst to *t*-butylbenzene, which was immediately dealkylated, producing isobutane. A similar explanation for the almost exclusive formation of isopentane from both 2- and 3-phenylpentane (runs 3 and 6, Table 3) is not satisfactory, since (1) there was no change in the ratio of isopentane to *n*-pentane with time or with different temperatures (cf. run 4 at 30°C); (2) no *t*-pentylbenzene could be detected in any reaction mixtures, and there was evidence that some of it could survive the reaction conditions (cf. [19]); and (3) the only pentylbenzene isomers that were detected were 2- and 3-

phenylpentane and 2-methyl-3-phenylbutane (Table 4), as expected from earlier work [19].

These results demanded a new explanation for the very different behavior of the two similar secondary alkylbenzenes with regard to the alkanes produced by dealkylation. It occurred to Roberts that there was an analogy in the greater susceptibility of *n*-pentane than *n*-butane toward isomerization by HF-SbF_5, which had been noted by Brouwer and Oelderik [64]. Also, Kramer [65] had demonstrated that the secondary carbocation produced from 2-chlorobutane in HSO_3FSbF_5 could be trapped by a hydride donor before rearrangement to the tertiary cation much more successfully than the corresponding secondary carbocation from 2-chloropentane; i.e., the tendency of the secondary *pentyl* carbocation to rearrange was much greater than that of the butyl cation. The principle proposed by Brouwer and Oelderik to rationalize their observed differences in alkane isomerizations, and which was applied by Kramer to explain his results, was that carbocation isomerizations such as these, which involve a change in the degree of branching, pass through protonated cyclopropane intermediates. The significant difference in the behavior of butyl and pentyl systems can be rationalized in terms of the ability of the pentyl intermediates thus to avoid primary carbocation character, whereas butyl intermediates cannot.

It seemed to Roberts that this principle might also be applied to explain the difference in the amount of branched alkanes produced from dealkylation of butyl- and pentylbenzenes. Referring to scheme 10, one may see that in

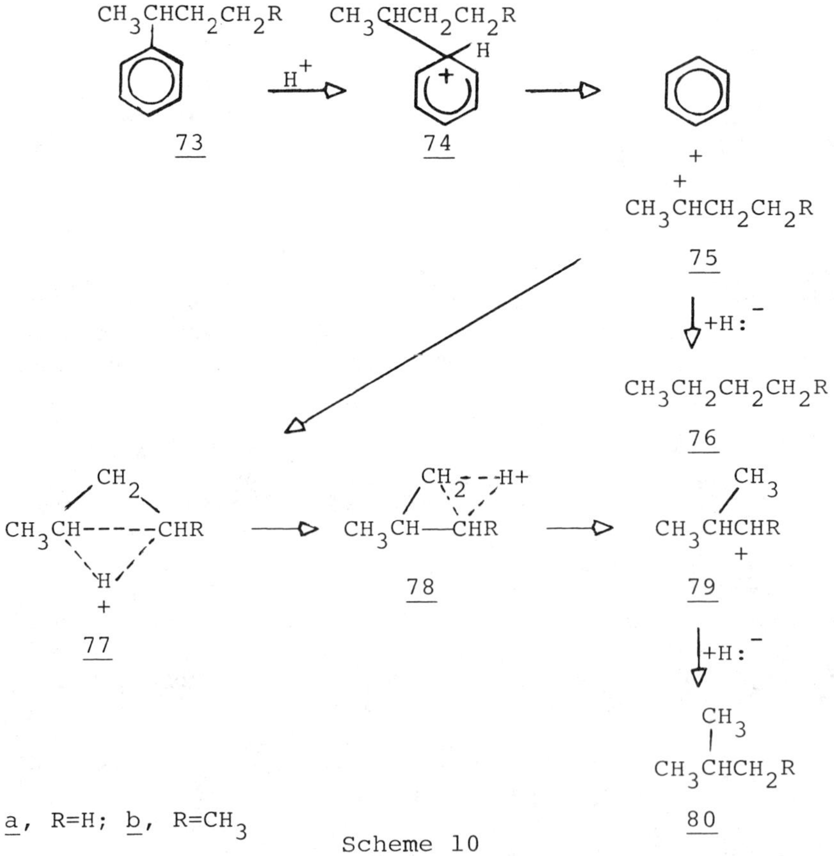

$CH_3CH(C_6H_5)CH_2CH_2R$ (73) $\xrightarrow{H^+}$ 74 (ring-protonated σ-complex, $CH_3CHCH_2CH_2R$ and H on the same ring carbon) $\longrightarrow$ C_6H_6 + $CH_3\overset{+}{C}HCH_2CH_2R$ (75)

75 $\xrightarrow{+H:^-}$ $CH_3CH_2CH_2CH_2R$ (76)

75 $\longrightarrow$ 77 (edge-protonated cyclopropane: CH_3CH- - - - -CHR bridged by CH_2 and H^+) $\longrightarrow$ 78 (CH_3CH—CHR ring with CH_2- - -H^+) $\longrightarrow$ $CH_3C(CH_3)\overset{+}{C}HR$ (79)

79 $\xrightarrow{+H:^-}$ $CH_3CH(CH_3)CH_2R$ (80)

a, R=H; b, R=CH_3

Scheme 10

the case of 2-phenylbutane (73a), the carbocation intermediate 79a formed by the opening of the protonated cyclopropane intermediate 78a would have primary carbocation character, whereas the corresponding intermediate (79b) from 2-phenylpentane (73b) would have secondary carbocation character.* Assuming that the capture of the initial secondary carbocation, 75, by a hydride ion is competitive with its rearrangement to 79 via 77 and 78, increasing the concentration of potential hydride donors in the reaction medium should increase the proportion of *n*-alkane to isoalkane.

In run 2 (Table 3) it may be seen that addition of methylcyclopentane to 2-phenylbutane increased the proportion of *n*-butane from 72% to 89% in the mixture of butane isomers produced. Similarly, the proportion of *n*-pentane from 2-phenylpentane was increased from 4% (run 3) to 16% (run 5) by the addition of methylcyclohexane, and a slightly larger increase (from 3% to 19%) was produced from 3-phenylpentane (runs 6 and 7).

b. sec-Hexylbenzenes

(i) Reaction of *sec*-Hexylbenzenes with Al_2Br_6. Results from dealklyations of three secondary hexylbenzenes are presented in Table 5. Heating either 2- or 3-phenylhexane with Al_2Br_6 at 80°C gave 2-methylpentane, 3-methylpentane, and *n*-hexane in ratios of 50-52:34–38:10–60%, respectively (runs 1 and 5). The addition of HBr increased the yield of dealkylation without changing the proportion of alkane isomers (run 2). When the reaction was carried out in benzene solution, however, none of the unbranched alkane isomer could be detected (runs 3 and 6). This is puzzling at first, but when one considers the source of the hydrogen which traps the secondary hexyl cation before rearrangement to convert it to *n*-hexane, a reasonable explanation emerges. It is the 2- or 3-phenylhexane itself which can donate a tertiary hydrogen from its side chain. When the reaction is carried out in benzene solution, the concentration of this hydride donor is much reduced by the dilution in benzene, and the secondary hexyl cation rearranges before it can acquire a hydrogen. This assumption is supported by the results from experiments in which an additional hydride donor, methylcyclohexane or methylcyclopentane, was added as well as benzene (i.e., run 4); in all such experiments *n*-hexane was observed, and the overall yield of dealkylation was also increased.

The results from the studies of dealkylation of 3-methyl-2-phenylpentane also fit this picture. When the reaction was carried out in cyclohexane solution (run 8), the ratio of 3-methylpentane to 2-methylpentane was 31:69, essentially the equilibrium proportion of these isomers. However, when the solvent was methylcyclopentane (run 7), the ratio of isomeric methylpentanes was reversed, owing to the ability of the better hydride donor to trap some of the 3-methyl-2-pentyl cations before the methyl shift occurred.

(ii) Reaction of *sec*-Hexylbenzenes with $HF\text{-}SbF_5$. Brouwer and Oelderik [65] found that treatment with fluoroantimonic acid produced all of the hexane isomers from 2-methylpentane. The formation of 3-methylpentane was the fastest

---

*To explain the observed initial formation of isobutane from *sec*-butylbenzene, one may reasonably suggest that when the hydrocarbon is added to the catalyst at 100°C, the high concentration of the active catalyst and the exothermicity of the reaction provide enough energy to open the ring to the primary carbocation intermediate. When isobutylbenzene is the starting material, for the same reasons there may be enough energy initially to cleave the isobutyl group directly.

Table 5. Dealkylation of Secondary Hexylbenzenes

Run	Hexylbenzene	Catalysts	Solvents	Temp. (°C)	Time (hr)	Dealkylation[a] (mol %)	Distribution of alkanes (%)		
							2-MP[b]	3-MP[c]	H[d]
1	2-PH[e] (0.1)[f]	Al_2Br_6 (0.033)		80	3	19	53	37	10
2		Al_2Br_6 (0.033) HBr (0.02)		80	3	38	56	37	7
3		Al_2Br_6 (0.033) HBr (0.059)	PhH (4.9)	35	1	52	65	35	
4		Al_2Br_6 (0.023) HBr (0.065)	PhH (4.9) MCH[g] (1.2)	70	2	70	59	27	14
5	3-PH[h] (0.1)	Al_2Br_6 (0.033)		80	2		50	34	16
6	3-PH (0.1)	Al_2Br_6 (0.033)	PhH (4.9)	80	2	51	67	33	
7	3-M-2-PP[i] (0.025)	Al_2Br_6 (0.008)	MCP[j] (0.3)	35	2	33	30	70	
8		Al_2Br_6 (0.004)	CH[k] (0.357)	35	2	24	69	31	

[a]Based on hexylbenzenes and calculated from GLC analysis with an internal standard.
[b]2-Methylpentane.
[c]3-Methylpentane.
[d]*n*-Hexane.
[e]2-Phenylhexane.
[f]Numbers in parentheses are molar quantities.
[g]Methylcyclohexane.
[h]3-Phenylhexane.
[i]3-Methyl-2-phenylpentane.
[j]Methylcyclopentane.
[k]Cyclohexane.

reaction, followed by the conversion of these two isomers to 2,3-dimethylbutane, while the isomerizations to *n*-hexane and 2,2-dimethylbutane were the slowest reactions. Although Roberts et al. found neither of the doubly branched hexanes as dealkylation products from reaction of 2- or 3-phenylhexane with aluminum bromide, they thought they might be produced in the presence of fluoroantimonic acid. In benzene solution at 80°C, only 2- and 3-methylpentane were produced, in about the same proportion as with aluminum bromide catalyst. However, when the fluoroantimonic acid was not diluted by benzene, all five isomeric hexanes were produced, even at 35°C. Addition of methylcyclohexane gave a higher yield of dealkylation products and increased the proportion of *n*-hexane and 2,2-dimethylbutane.

(iii) Rearrangement and Cyclialkylation Products from *sec*-Hexylbenzenes. The high-boiling aromatic components of the reaction mixtures were also examined, with the aim of gaining more insight into the mechanism of the dealkylation process and its relationship to other competing or conjugated processes. In the earlier study of dealkylations [26] diphenylalkanes were identified as the probable by-products resulting from the hydride donation by the alkylbenzenes required to produce the alkanes. In support of this theory, the isolation of *meso*-2,3-diphenylbutane from an aluminum chloride-catalyzed dealkylation of *sec*-butylbenzene was cited. The formation of this compound is reasonably explained in terms of the intermediate carbocation 81, which results from hydride donation to a butyl cation by *sec*-butylbenzene (the source of the H^-: shown in scheme 10); this ion then alkylates benzene to give the diphenylbutane.

$$\underset{\underset{\underline{81}}{}}{CH_3\underset{|\atop Ph}{C}H\overset{+}{C}HCH_3} + C_6H_6 \longrightarrow CH_3\underset{|\atop Ph}{C}H\text{-}\underset{|\atop Ph}{C}HCH_3 + H^+$$

In the case of higher secondary alkylbenzenes such as pentyl- and hexylbenzenes, there is a favored alternative fate for the carbocation intermediate analogous to 81. For example, if hydride donation occurs from a secondary carbon in the 3 or 4 position of the side chain, intramolecular alkylation (cyclialkylation) may take precedence over the intermolecular alkylation, which would produce diphenylalkanes. Consider the sequence of reactions in scheme 11. By the sequence of Eqs. (30) + (31) + (32), two molecules of hexylbenzene may be converted into one molecule each of hexane, benzene, and 1-ethyl-3-methylindan, and by the sequence of Eqs. (30) + (31) + (33) into hexane, benzene, and 1,4-dimethyltetralin. It is significant that these indan and tetralin derivatives are major components of the high-boiling aromatic reaction products.

Another observation may be made about the products from the hexylbenzenes. This concerns the *absence* of branched chain hexylbenzene isomers such as 3-methyl-2-phenylpentane among the products. If the 2- and 3-methylpentanes came from dealkylation of branched-chain hexylbenzenes produced by rearrangement prior to dealkylation, one would expect to find some of these branched-chain hexylbenzenes among the reaction products. The only one detected was 2-methyl-1-phenylpentane, and it was a very minor component (cf. Sec. I.F.4). These results are analogous to those from the pentylbenzenes, and they lend further support to the theory that the branched alkanes result from subsequent rearrangement of the straight chain secondary carboca-

$$CH_3CH(Ph)(CH_2)_3CH_3 + H^+ \longrightarrow PhH + CH_3\overset{+}{C}H(CH_2)_3CH_3 \quad (30)$$

$$CH_3\overset{+}{C}H(CH_2)_3CH_3 + CH_3CH(Ph)(CH_2)_3CH_3 \longrightarrow C_6H_{14} + \underset{\underline{82}}{(PhC_6H_{12})^+} \quad (31)$$

$$(PhC_6H_{12})^+ = \underset{\underline{82a}}{CH_3\overset{+}{C}(Ph)(CH_2)_3CH_3}, \quad \underset{\underline{82b}}{CH_3CH(Ph)\overset{+}{C}HCH(CH_2)_2CH_3},$$

$$\underset{\underline{82c}}{CH_3CH(Ph)CH_2\overset{+}{C}HCH_2CH_3} \text{ or } \underset{\underline{82d}}{CH_3CH(Ph)(CH_2)_2\overset{+}{C}HCH_3}$$

$$\underset{\underline{82c}}{PhCH(CH_3)CH_2\overset{+}{C}HCH_2CH_3} \longrightarrow \text{1-methyl-3-ethylindan} + H^+ \quad (32)$$

$$\underset{\underline{82d}}{PhCH(CH_3)CH_2CH_2\overset{+}{C}HCH_3} \longrightarrow \text{1,4-dimethyltetralin} + H^+ \quad (33)$$

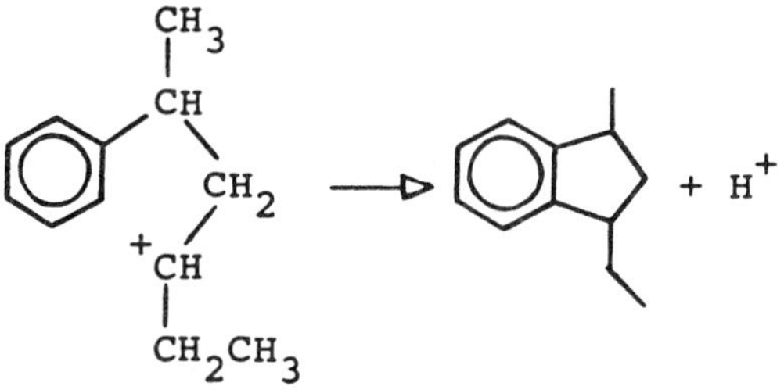

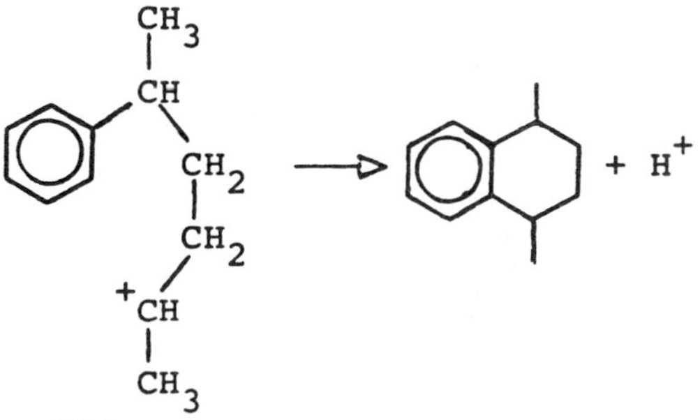

Scheme 11

tions produced by initial cleavage from the aromatic ring. The wide difference in the proportions of normal and branched alkanes produced from *sec*-butylbenzene on the one hand and from secondary pentyl- and hexylbenzenes on the other hand is nicely rationalized in terms of protonated cyclopropane intermediates for the rearrangement of the straight-chain secondary pentyl and hexyl carbocations to branched isomers, which then capture hydride ions to become branched alkanes. Since the corresponding protonated cyclopropane intermediate from *sec*-butyl cation cannot open to a branched isomer without acquiring primary carbocation character, much less isobutane than *n*-butane is formed.

2. *sec*-Heptylbenzenes

Going up the homologous series by one more unit, Roberts and Elrod examined the dealkylation of three *sec*-heptylbenzenes, 2-, 3-, and 4-phenylheptane [46]. As in the case of the pentyl- and hexylbenzenes, the major dealkylation products were branched alkanes, but more isomers were observed. Seven of the nine isomeric heptanes were produced by treatment with Al_2Br_6-HBr, and all nine were identified as products of the reaction with HF-SbF_5.

The major isomers from all of the dealkylations were 2- and 3-methylhexane, in approximately equal amounts, apparently. (Because of overlap in GC retention times among some of the isomeric heptanes, it was difficult to get a quantitative measure of the proportions of these two isomers.) The next most abundant branched heptanes were 2,3- and 2,4-dimethylpentanes, with the former predominating. The amount of unbranched heptane found depended on the availability of a hydride donor in the reaction mixture, as in the case of the *sec*-butyl-, pentyl-, and hexylbenzenes as reported in the preceding section. When the reaction was carried out with 2-phenylheptane and Al_2Br_6-HBr in the absence of inert solvent and added hydride donor, the proportion of unbranched heptane produced was 3 to 6%. When 1,4-dimethylcyclohexane (an effective hydride donor) was added, the proportion of heptane increased to 17 to 24%. Similar results were obtained with 3- and 4-phenylheptane; this was not surprising since the rate of isomerization of the three phenylheptanes to an equilibrium mixture containing all three is probably of the same order of magnitude as the rate of dealkylation.

The number and proportions of the heptane isomers produced under various conditions were considered by Roberts and Elrod in terms of mechanisms involving fission of the secondary heptyl cation from the benzene ring, followed by hydride exchanges and rearrangements among simple secondary and tertiary carbocations, as well as protonated cyclopropane intermediates. The experimental results could be rationalized with some success on the basis of the relative stability of these intermediate carbocations of various types. One such interesting example was the large effect that the addition of the hydride donor, 1,4-dimethylcyclohexane, had on the ratio of the unbranched heptane isomers to the singly branched isomers, and the neglible effect it had on the ratio of singly branched to doubly branched isomers. This was attributed to the effectiveness of the 1,4-dimethylcyclohexane in donating its hydrogen to a secondary carbocation, wherein it becomes a tertiary cation, and to a much lesser effectiveness in exchanging hydrogen with a tertiary carbocation intermediate. This is reminiscent of the generalization of Miethchen and Ewald [55] about the effectiveness of decalin in promoting dealkylations of alkylbenzenes (Sec. II.A).

Careful examination of the arenes in the reaction mixtures indicated the presence of no branched-chain isomers of the 2-, 3-, and 4-phenylheptanes that were found in the approximate ratio 1:0.6:0.2 when any one of the three isomers was the starting material treated with Al_2Br_6-HBr or HF-SbF_5. The possible presence of 2-methyl-1-phenylhexane was checked especially carefully because of the known rearrangements of the lower homologs to the corresponding 2-methyl-1-phenylalkanes (cf. Sec. I.F.4.).

E. Identification and Utilization of the *t*-Butyl Cation Produced by Dealkylation of *t*-Butylarenes

Evidence for the formation of *t*-butyl cation in equilibrium with *t*-butyl arenes by means of Friedel-Crafts catalysts has been obtained in ways other than by the observation of the evolution of isobutane. In 1962, Friedman and Cotton reported that treatment of *t*-butylbenzene with carbon monoxide under pressure in hydrogen fluoride solution at 30°C gave a moderate yield of pivalic (trimethylacetic) acid, which they explained in terms of Eqs. (34) and (35)

$$\text{Ph-C(C)}_2\text{-C} + H^+ \rightleftharpoons \text{PhH} + \text{C-C}^{\oplus}(\text{C})_2 \qquad (34)$$

$$\text{C-C}^{\oplus}(\text{C})_2 + \text{CO} \longrightarrow \text{C-C(C)}_2\text{-C}^{\oplus}{=}\text{O} \qquad (35)$$

[66]. A year later Knight et al. extended this work, using boron trifluoride as catalyst at 25°C [67]. *t*-Butylbenzene and 1,3-dimethyl-5-*t*-butylbenzene gave excellent yields of pivalic acid from carbon monoxide, and *m*-*t*-pentyltoluene and 1,3-dimethyl-5-*t*-butylbenzene gave 2,2-dimethylbutanoic acid. Benzene, toluene, and *m*-xylene were isolated in high yields as the dealkylated arenes. Both Friedman and Cotton and Knight and co-workers reported that dealkylation of cumene (isopropylbenzene) failed completely under the conditions used with *t*-butylbenzene. Phosphoric acid, sulfuric acid, and methanesulfonic acid were ineffective for the dealkylation-carbonylation of *t*-butylbenzene [67].

Direct evidence of the formation of the *t*-butyl cation was obtained by Brouwer [68] and Olah et al. [69] by the observation of its ^1H NMR spectrum when it was produced from *t*-butylbenzene in superacid solutions at low temperatures. Even the presence of electron-withdrawing and bulky substituents on the *t*-butylbenzene ring did not prevent the cleavage of the *t*-butyl cation at −30°C by Hf-SbF_5-SO_2ClF [70]. In all cases the *t*-butyl cation was formed as evidenced by its ^1H NMR singlet absorption at δ 4.0 to 4.2. The behavior of *t*-butylbenzene in another strong acid system, HF-TaF_5, was examined by Farcasiu [71], who found that although the alkylbenzene is protonated, it did not form any significant amount of *t*-butyl cations between −60 and −10°C. However, a dealkylation-realkylation equilibrium was established, as was indicated by partial disproportionation to benzene, *m*-di-*t*-butylbenzene, and 1,3,5-tri-*t*-butylbenzene, and by trapping of carbon monoxide to form pivaloyl cations, a portion of which subsequently reacted with benzene to form pivalophenone (Eqs. 34 to 36).

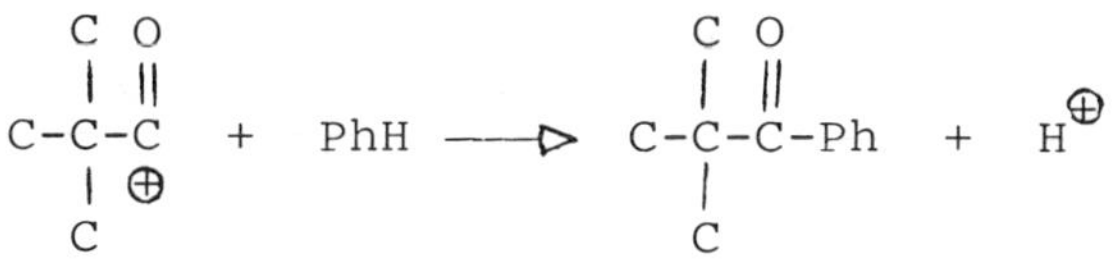

```
  C O                       C O
  | ||                      | ||
C-C-C   +   PhH  ---->   C-C-C-Ph   +   H⊕          (36)
  | ⊕                       |
  C                         C
```

Very recently, the possibility of using such a reaction to synthetic advantage has been reported. Farcasiu and Schlosberg [72] found that *p*-di-*t*-butylbenzene reacted with carbon monoxide in the presence of $AlCl_3$-HCl at room temperature to produce *p*-*t*-butylpivalophenone (83, Eq. 37). The in-

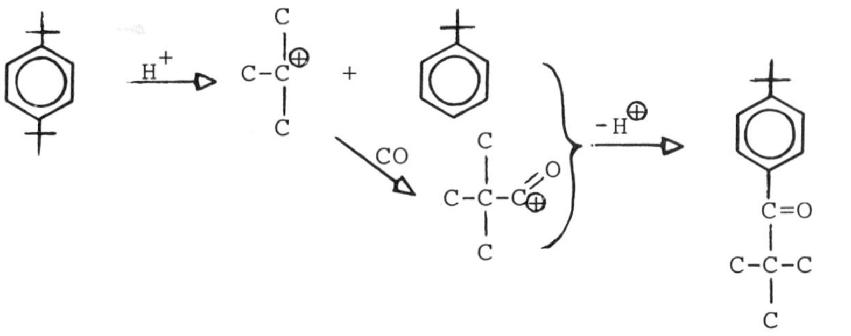

(37)

83

sertion of CO into a ring was demonstrated by treatment of 1,1-dimethylindan in HF-TaF_5 solution at $-20°C$ with carbon monoxide (Eq. 38).

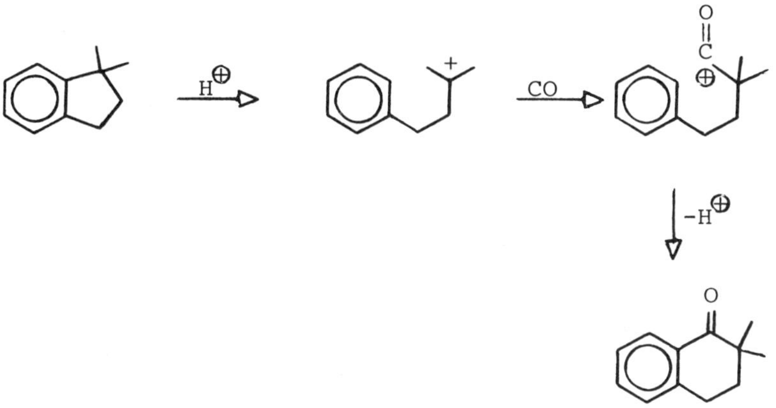

(38)

F. Dealkylations with Miscellaneous Catalysts

Many catalytic reactions of hydrocarbons are carried out industrially, particularly in petroleum and petrochemical processing. Although most of the catalysts used would not be considered typical "Friedel-Crafts catalysts," some of the reactions are closely related to those produced by aluminum chloride, so we shall include a few examples of these.

The reactions of several pentylbenzenes over silica-alumina catalysts at over 400°C were studied by Dimitrov and co-workers at the University of Sofia, Bulgaria [73-75]. 1-Phenylpentane gave isopentane, benzene, 2-methylnaphthalene, toluene, and ethylbenzene [73]. 2-Methylnaphthalene was also obtained from 3-phenylpentane [74]; 1-ethylindan and 2-methyltetralin were proposed as intermediates, with dehydrogenation of the latter producing the 2-methylnaphthalene. *p*-Di-*n*-pentylbenzene was the substrate for the third study [75(a)]. A 40-component mixture was obtained, 16 of which were gases, including H_2 and C_1-C_7 hydrocarbons. The liquids included C_6 to C_{10} alkanes, C_1 to C_5 alkylbenzenes, indan, tetralin, and methylnaphthalenes. *p*-Di-*n*-butylbenzene was similarly studied [75(b)]; the effect of substituting H^+ and different metals in the zeolite catalysts on competing dealkylation, isomerization, and fragmentation processes was determined.

Benzene is a more valuable petrochemical starting material than toluene, and often more toluene is produced than needed, so that the excess toluene is converted into benzene, some by disproportionation (with concurrent formation of xylenes, which are also more valuable than toluene), but mainly by "hydrodealkylation" (Eq. 39) and steam dealkylation (Eq. 40) [76]. Hydrodealkylation has been estimated to provide about 29% of the benzene produced currently. Typical processes utilize Ni and other catalysts at temperatures of 680 to 720°C under a pressure of 570 psi. Steam dealkylations have the advantage of producing hydrogen rather than using it; Ni-Cr_2O_3 and Ni-Al_2O_3 catalysts may be employed at temperatures between 320 and 360°C [76].

$$C_6H_5CH_3 + H_2 \xrightarrow{\text{catalyst}} C_6H_6 + CH_4 \quad (39)$$

$$C_6H_5CH_3 + 2H_2O \xrightarrow{\text{catalyst}} C_6H_6 + CO_2 + 3H_2 \quad (40)$$

Among the many reports that have appeared in the primary literature, the following are representative. Toluene and xylene were dealkylated over Pd, Pt, and/or Rh catalysts supported on Al_2O_3, optionally containing Ni or Co oxides [77]. The effects of methylcyclohexane and other H-donor compounds on yields of benzene from toluene using a "D-26" catalyst were studied [78]. Alkylbenzenes, especially toluene, were dealkylated by steam at 400 to 750°C and 1 to 40 atm over catalysts containing 0.15 to 0.6% noble metals, preferably Rh, Pt, and/or Pd, and ca. 2% K_2CO_3 on a Cr_2O_3-Al_2O_3 support [79]. Ethylbenzene, *m*-xylene, and cumene were selectively dealkylated with steam over a Ni-BeO catalyst (20%) Ni) with <10% decomposition. Temperature and space velocity had little effect on the decomposition, whereas the steam/alkylbenzene ratio greatly affected the decomposition [80]. The selective dealkylation of alkylaromatic C_8 + hydrocarbons by hydrogenation over a mordenite (zeolite) catalyst containing Co, Ni, Pt, Pd, or Ag was described [81].

A"reforming fraction" was dealkylated at 475°C and 46.2 atm; the changes produced in the complex mixture were detailed.

Csicsery described very comprehensive studies of the catalytic reactions of *n*-butylbenzene [82(a)] and of *n*-pentylbenzene and 2-phenylpentane [82(b)] in the presence of hydrogen. The catalysts employed were Pt-SiO_2, Pt-SiO_2-Al_2O_3, Pt-Al_2O_3, and SiO_2-Al_2O_3 and reaction temperatures were 316 to 482°C. All of the basic reforming reactions were observed, including "cracking" (dealkylation and fragmentation) rearrangement, cyclization, and dehydrogenation. Some of the reactions were considered to be "acid-catalyzed," and carbocation intermediates were proposed; others were considered to be nonionic, occurring in molecules absorbed on the surface of the catalyst.

III. FRAGMENTATIONS OF ARENES

A. Propyl-, Butyl-, and Pentylbenzenes

The development of sensitive and convenient instruments for analyzing complex mixtures led to the recognition of another type of reaction of alkylbenzenes with Friedel-Crafts catalysts, in addition to those that have been discussed so far in this monograph. This is fragmentation, which was mentioned in the section on dealkylations, because fragmentations frequently accompany dealkylations, especially at higher temperatures. Whereas dealkylation refers to the production of an alkane corresponding, in the number of carbon atoms, to the side chain of an alkylbenzene, fragmentation describes the formation of alkanes with fewer carbon atoms than the side chain and alkylbenzenes with smaller side chains than the original alkylbenzene.

Only a few scattered observations of such reactions were noted before gas chromatography made it easy to demonstrate that almost a complete spectrum of lower alkylbenzenes and alkanes is produced by heating an alkylbenzene with four or more carbon atoms in the side chain with $AlCl_3$-H_2O at 100°C [26]. Alkylation of benzene with *t*-butyl alcohol and $AlCl_3$ at 80 to 95°C was reported in 1939 to give toluene, ethylbenzene, and isopropylbenzene, but no *t*-butylbenzene [83]. Ethylbenzene was identified among the products of alkylation of benzene with *n*-propyl ether at 80°C with $AlCl_3$ as catalyst [84], and also in the mixture produced by heating isopropylbenzene with $AlCl_3$-H_2O in benzene under reflux [16].

Roberts et al. made a systematic investigation of the gaseous and liquid hydrocarbons produced by heating ethylbenzene, *n*-propyl- and isopropylbenzene, all of the butylbenzene isomers, and all of the pentylbenzene isomers with $AlCl_3$-H_2O [26]. The dealkylation results were presented in Table 1. The complete results of dealkylation, fragmentation, disproportionation, and rearrangement (of both alkanes and alkylbenzenes) reactions are presented in Tables 6 to 8. The difficulty of studying the reactions separately may be illustrated by the behavior of *t*-butylbenzene. When it was heated with $AlCl_3$-H_2O at only 50°C for 1 hr, 17 mol % of isobutane was evolved, and the liquid reaction mixture was found to contain toluene, ethylbenzene, isopropylbenzene, *sec*-butyl- and isobutylbenzene, and a small amount of recovered *t*-butylbenzene. The disproportionations that occur in all cases have an effect on the recovery of the alkylbenzenes which is superimposed on the effects of the dealkylation and fragmentation reactions. For example, note that only 30% of ethylbenzene was recovered, although very little dealkylation and no fragmentation of this hydrocarbon occurred.

Table 6. Dealkylation and Fragmentation Products of Ethyl-, Propyl-, and Butylbenzenes[a]

Starting material $R-C_6H_5$	Gaseous products		Liquid mixture, $R-C_6H_5$[d]						
						Pr (%)[f]		Bu (%)[f]	
	Percent[b]	Alkane[c]	Total (%)	Me (%)[f]	Et (%)[f]	*n*–	*i*–	*n*–	*sec*/*i*[g]
Et	7	C_2H_6	30	Trace	100				
n-Pr	7	C_3H_8	27	5	4	87	4		
i-Pr	18	C_3H_8[h]	17	2	22	Trace	76		
n-Bu	4	*n*/*i*-C_4H_{10}[i]	37		0.5	0.5		97	2
i-Bu	15	*n*/*i*-C_4H_{10}[j]	20	18	10	2	5		65
sec-Bu	24	*n*/*i*-C_4H_{10}[j]	19	14	6	3	3		74
t-Bu	72	*i*-C_4H_{10}	3	25	56	5	12		2[k]

[a]Reactions at 100°C for 3 hr; reactants: $R-C_6H_5$, 0.10 mol; $AlCl_3$, 0.033 mol; H_2O, 0.017 mol.
[b]Mole percent conversion to alkane based on alkylbenzene, calculated from volume of gas collected.
[c]Analyzed by GLPC.
[d]Boiling range 81 to 172°C; benzene was produced in all experiments.
[e]Weight percent of original alkylbenzene; residue (bp > 172°C) was ca. 30 wt %.
[f]Percentage of each alkylbenzene in the liquid mixture.
[g]Not resolved by GLPC.
[h]Traces of a hexane were found.
[i]*n*/*i* = 36:64.
[j]*n*/*i* = 70:30.
[k]No *t*-Butylbenzene was recovered.

A significant clue to the mechanism of the acid-catalyzed fragmentation reactions was the observation that no methane or ethane was ever found among the gaseous products from any of the alkylbenzenes higher than ethylbenzene, and tests showed that they would have been detected by the gas chromatographic analysis had they been formed. Assuming some kind of mechanism involving carbocation intermediates, a plausible explanation for the absence of methane and ethane lies in the fact that propane and butanes may be formed via secondary and tertiary carbocation intermediates, but not methane and ethane. However, fragmentation products are found which obviously involve loss of one- and two-carbon units from the original alkylbenzene; i.e., propylbenzene yields toluene and ethylbenzene (Table 6). Hence a mechanism must be sought that can accommodate the preferential formation of three-carbon and higher alkanes as well as loss of one- and two-carbon units from the side chains. This is not as difficult as it might at first appear, and it is interesting that the fortuitous isolation and identification of *meso*-2,3-diphenylbutane by Fonken provided support for a mechanism of fragmentation as well as for the mechanism of dealkylation described in Sec. II [26].

Roberts, Baylis, and Fonken suggested that the C-C bond cleavages that produce the fragmentation products occur mainly in diphenylalkane molecules such as 2,3-diphenylbutane [26]. These cleavages can occur so as to produce

Table 7. Gaseous Dealkylation and Fragmentation Products from Pentylbenzenes[a]

	Gaseous products					
			C_4H_{10} (%)[c]		C_5H_{12} (%)[c]	
R-C_6H_5	Total (%)[b]	C_2H_5 (%)[c]	*n*	*i*	*n*	*i*
n-Pe	14			5	14	81
i-Pe	13	4	2	8	6	80
2-MeBu[d]	19	4	6	5	11	74
1-MeBu[e]	20			Trace	4	96
1-EtPr[f]	27				3	97
neo-Pe[g]	18	1	1	15		83
1,2-di-MePr[h]	40			27		73
t-Pe	56			28		72

[a]Reactions at 100°C for 3 hr; reactants; alkylbenzene, 0.10 mol; $AlCl_3$, 0.033 mol; H_2O; 0.011 mol.
[b]Mole percent conversion to alkane based on alkylbenzene, calculated from volume of gas collected.
[c]Percentage of each alkane in the gaseous mixture.
[d](2-Methylbutyl)-benzene.
[e](1-Methylbutyl)-benzene.
[f](1-Ethylpropyl)-benzene.
[g]Neopentylbenzene.
[h](1,2-Dimethylpropyl)-benzene.

more stable carbocation intermediates than would be obtained by cleavages in simple alkylbenzene molecules. For example, the formation of ethylbenzene from *sec*-butylbenzene may involve the initial formation of a diphenylbutane according to scheme 9 (Sec. II.A), and the diphenylbutane may than undergo fission according to scheme 12. Further support for this mechanism was afforded by the treatment of pure *meso*-2,3-diphenylbutane with $AlCl_3$-H_2O at 70°C for 1 hr. The liquid products in the 100 to 200°C boiling range were found to be *sec*-butyl-/isobutylbenzene (50%), ethylbenzene (45%), isopropylbenzene (3%), and *n*-propylbenzene (2%). The production of toluene and ethylbenzene from *n*-propylbenzene can be explained in terms of an analogous mechanism in which 1,2-diphenylpropane is an intermediate. We have mentioned in Sec. I of this chapter that diphenylpropanes were identified in reaction mixtures resulting from heating isotopically labeled *n*-propylbenzene [12], and it is thought that these same diphenylpropanes serve also as intermediates in the isotopic rearrangement of *n*-propylbenzene.

We shall not attempt to outline here detailed mechanisms for the formation of all of the observed fragmentation products; many of them can be reasonably explained by similar sequences of alkylation, rearrangement, cleavage, and hydride exchange. It is quite possible that direct cleavage in an alkylbenzene molecule may occur when a secondary or tertiary alkyl carbocation will result. The predominance of isobutane as a fragmentation product from neopentylbenzene and the two isomeric pentylbenzenes which rearrange to neopentyl ben-

Table 8. Liquid Dealkylation and Fragmentation Products from Pentylbenzenes[a]

Starting material $R-C_6H_5$	Liquid mixture, $R-C_6H_5$[b] Total (%)[c]	Me (%)[d]	Et (%)[d]	Pr (%)[d] *n*	Pr (%)[d] *i*	Bu (%)[d] *n*	Bu (%)[d] *sec/i*	Pe (%)[d] Original	Pe (%)[d] Others	
n-Pe	35	5	5	4		4	5	70	7[e]	
i-Pe	28	4	10		8		6	72		
2-MeBu[f]	28	1	4	3				92		
1-MeBu[h]	16	1	8	8			4	43	24[f]	12[g]
1-EtPr[g]	20	Trace	6	5				18	50[h]	20[f]
neo-Pe[j]	41	6	2	2	2		2	85	1[i]	
1,2-di-MePr[i]	13	1	4		8		4	5	78[j]	
t-Pe	5	1	10	2	18		10		50[j]	9[i]

[a]Conditions and reactants same as in Table 7.
[b]Boiling range, 100 to 200°C.
[c]Weight % of original alkylbenzene; residue (bp > 200°C) was ca. 30 wt %.
[d]Percentage of each alkylbenzene in the liquid mixture.
[e]Isopentylbenzene.
[f](2-Methylbutyl)-benzene.
[g](1-Ethylpropyl)-benzene.
[h](1-Methylbutyl)-benzene.
[i](1,2-Dimethylpropyl)-benzene.
[j]Neopentylbenzene.

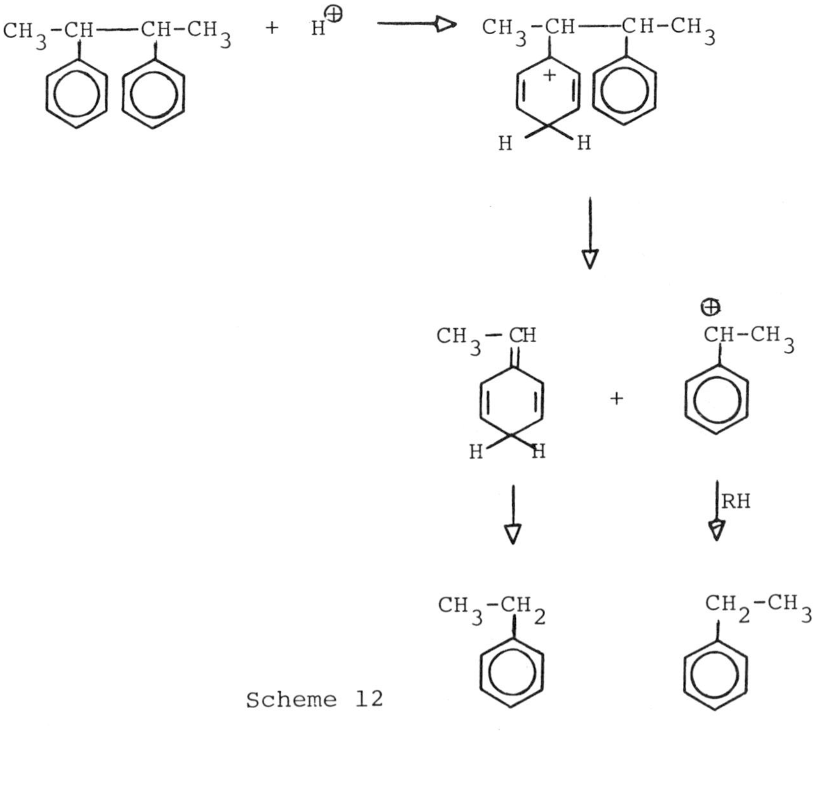

Scheme 12

zene (cf. Tables 6 and 7) may be indicative of direct cleavage to toluene and *t*-butyl cation (scheme 13).

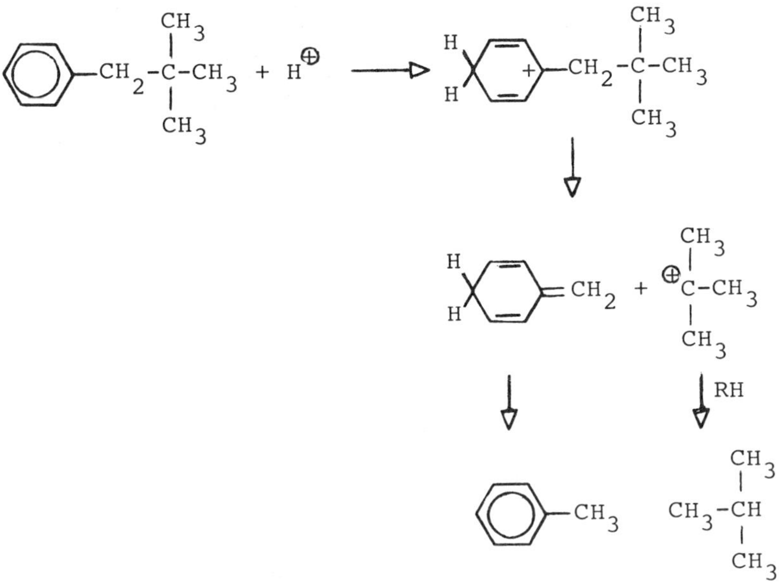

$$C_6H_5\text{-}CH_2\text{-}C(CH_3)_2\text{-}CH_3 + H^{\oplus} \longrightarrow [H_2C_6H_5^{+}]\text{-}CH_2\text{-}C(CH_3)_2\text{-}CH_3$$

$$\downarrow$$

$$H_2C_6H_4{=}CH_2 + {}^{\oplus}C(CH_3)_2\text{-}CH_3$$

$$\downarrow \qquad\qquad \downarrow RH$$

$$C_6H_5\text{-}CH_3 \qquad CH_3\text{-}CH(CH_3)\text{-}CH_3$$

Scheme 13

After the fortuitous isolation of *meso*-2,3-diphenylbutane from a dealkylation-fragmentation reaction in which *sec*-butylbenzene was the starting material [26], Roberts et al. examined reaction mixtures from other alkylbenzenes by gas chromatography, and they were able to demonstrate the presence of a number of other diphenylalkanes [12]. From the reaction of *sec*-butylbenzene, besides *meso*-2,3-diphenylbutane, the racemic (±) isomers were detected, as well as 2,2-diphenylbutane and 1,2-diphenyl-2-methylpropane. From the reaction of isobutylbenzene, (±)- and *meso*-2,3-diphenylbutane and 1,2-diphenyl-2-methylpropane were obtained; from *t*-butylbenzene, 1,2-diphenyl-2-methylpropane and 1,1-diphenyl-2-methylpropane were obtained; and from *n*-butylbenzene, 1,2- and 1,3-diphenylbutane were obtained.

On the assumption that a good source of carbocations (which should act as hydride abstractors) should augment the production of diphenylalkanes from alkylbenzenes, a number of butylbenzenes were treated with *t*-butyl chloride and $AlCl_3$ in benzene solution. *sec*-Butylbenzene and isobutylbenzene were converted to diphenylbutanes in significantly greater amounts (40 to 50 mol % at 25°C) than *t*-butyl- and *n*-butylbenzene, in keeping with the greater extent of fragmentation of the first two isomers. The diphenylbutanes produced in these experiments were the same ones found among the products of fragmentations carried out at higher temperatures in the absence of *t*-butyl chloride.

An obvious test of an intermediate is to ascertain whether or not it gives the same products as its assumed precursor under the same experimental conditions. Several of the pure diphenylbutane isomers were synthesized and treated with $AlCl_3$ under the same conditions that produce fragmentations of butylbenzenes. The results from four of these tests are shown in Table 9. All of the alkylbenzenes expected from fragmentation were produced, and in approximately the same proportions as from the butylbenzenes (compare Table 6), with the exception that the ratio of ethylbenzene to toluene was

Table 9. Fragmentation and Dealkylation Products from Diphenylbutanes[a]

	Alkylbenzenes, R-C_6H_5[b]						
Diphenylbutanes	Total (%)[c]	Me (%)[d]	Et (%)[d]	*n*-Pr (%)[d]	*i*-Pr (%)[d]	*sec*-Bu (%)[d,e]	*i*-Bu (%)[d,e]
C—C(Ph)—C(Ph)—C	12	3	23	2	4		68
C—C(Ph)(Ph)—C—C	16	11	26	1	4	28	30
Ph—C(C)(C)—C—Ph	21	11	18	2	8	22	38
Ph—C(Ph)—C(C)—C	24	4	16	2	9	25	44

[a]Reactions at 100°C, 3 hr; reactants: $(Ph)_2C_4H_8$, 0.10 mol; $AlCl_3$, 0.066 mol; H_2O, 0.033 mol.
[b]Boiling range 81 to 176°C; benzene was produced in all experiments.
[c]Weight percent of original diphenylbutane.
[d]Percentage of each alkylbenzene in the mixture from GLPC analysis.
[e]Distinction between *sec*-Bu and *i*-Bu was by infrared.
[f]Experiment of G. J. Fonken on *meso* isomer; slightly smaller amount of $AlCl_3$ used; temperature 70°C, 3 hr.
[g]Isomeric ratio not determined.

higher in the products from the diphenylbutanes. This may be attributed to the fact that a major source of ethylbenzene is the cleavage of (±)- and *meso*-2,3-diphenylbutane, which can also be produced by isomerization of the other three diphenylbutanes (as was demonstrated in other experiments). Some direct cleavage of isobutylbenzene probably occurs (i.e., not via a diphenylbutane intermediate) without producing ethylbenzene, e.g., scheme 14. No propane was found to be produced from *sec*-butyl- or isobutylbenzene. Possibly the reactive secondary propyl cation reacts faster with benzene or other arene than with hydride donors, whereas the more stable *t*-butyl cation chooses hydride exchange over alkylation (scheme 13). Since the rate of isomerization of *sec*-butylbenzene to isobutylbenzene is of the same order as its rate of fragmentation, it is reasonable that these two butylbenzene isomers

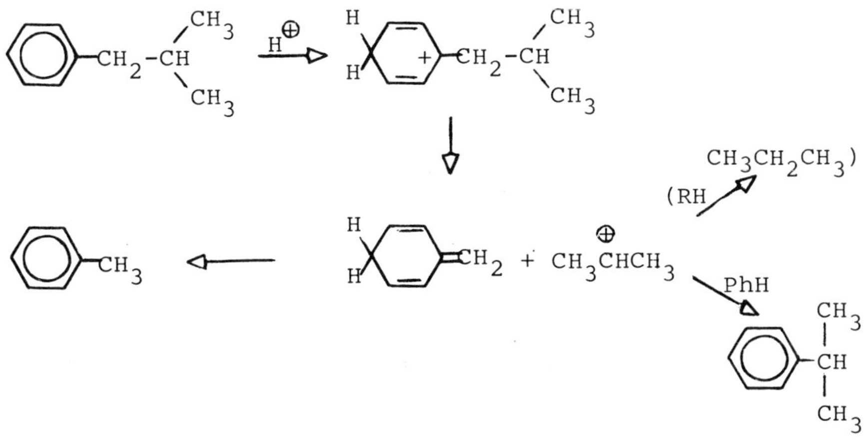

$$C_6H_5CH_2CH(CH_3)_2 \xrightarrow{H^{\oplus}} [\text{protonated arenium ion}] \rightarrow \text{methylenecyclohexadiene} + CH_3\overset{\oplus}{C}HCH_3 \rightarrow C_6H_5CH_3$$

$CH_3\overset{\oplus}{C}HCH_3 \xrightarrow{RH} CH_3CH_2CH_3$; $CH_3\overset{\oplus}{C}HCH_3 \xrightarrow{PhH} C_6H_5CH(CH_3)_2$

Scheme 14

give about the same ratio of ethylbenzene and toluene by fragmentation, and that this ratio is lower than that produced by the fragmentation of the dibutylbenzene isomers.

B. Diphenylbutanes: Further Investigations of Rearrangements, Dealkylations, and Fragmentations

We have already discussed some experiments in which diphenylbutanes were tested as intermediates in fragmentation reactions of butylbenzenes [12]. This research was extended by examining the reaction mixtures from a series of diphenylbutane isomers heated with different amounts of $AlCl_3$-H_2O for various periods of time in order to determine changes in the composition of the product mixtures throughout the course of the reactions [85-87].

During the preparation of the pure diphenylbutane isomers required for this work, some interesting observations were made [85]. The alkylation of benzene with neophyl chloride (84) using $AlCl_3$ catalyst gave not only 1,2-diphenyl-2-methylpropane (85), but also 1,1-diphenyl-2-methylpropane (86) and both *meso*- and (±)-2,3-diphenylbutane (87 and 88) (Eq. 41). With the milder catalysts $AlCl_3$-CH_3NO_2 or $FeCl_3$, only the first two isomers (85 and

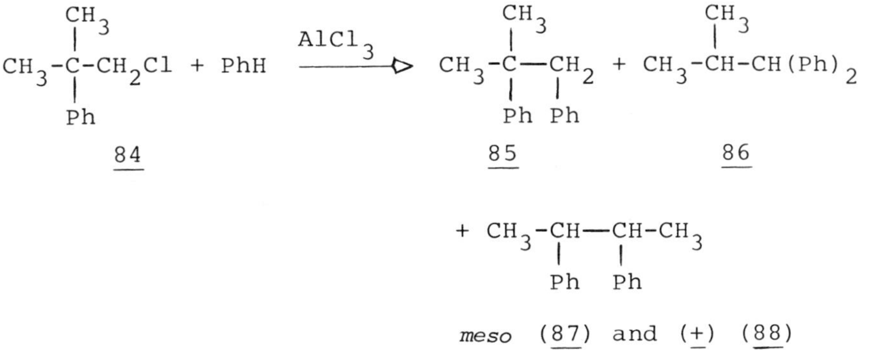

$$\underset{\underline{84}}{CH_3C(CH_3)(Ph)CH_2Cl} + PhH \xrightarrow{AlCl_3} \underset{\underline{85}}{CH_3C(CH_3)(Ph)CH_2Ph} + \underset{\underline{86}}{CH_3CH(CH_3)CH(Ph)_2} \tag{41}$$

$$+ CH_3CH(Ph)CH(Ph)CH_3$$

meso (87) and (±) (88)

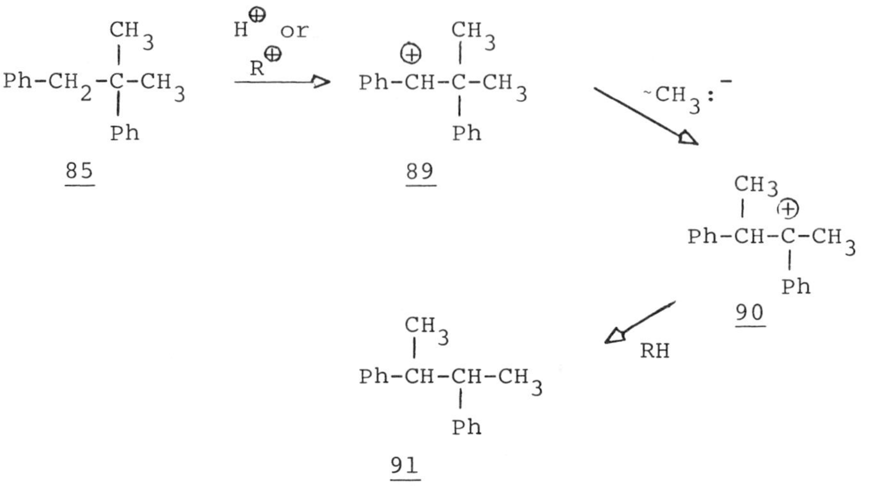

$$\underset{\underline{85}}{\mathrm{Ph{-}CH_2{-}C(CH_3)(Ph){-}CH_3}} \xrightarrow[R^{\oplus}]{H^{\oplus}\ \mathrm{or}} \underset{\underline{89}}{\mathrm{Ph{-}\overset{\oplus}{C}H{-}C(CH_3)(Ph){-}CH_3}} \xrightarrow{\sim CH_3:^-} \underset{\underline{90}}{\mathrm{Ph{-}CH(CH_3){-}\overset{\oplus}{C}(Ph){-}CH_3}} \xrightarrow{RH} \underset{\underline{91}}{\mathrm{Ph{-}CH(CH_3){-}CH(Ph){-}CH_3}}$$

Scheme 15

86) were obtained. The formation of 85 and 86 (in ca. 2:1 ratio) is the result of a rearrangement accompanying the alkylation, and as such is discussed in Chap. 3. The formation of 87 and 88 when the unmodified $AlCl_3$ catalyst is used can be rationalized as another example of an "alkylbenzene rearrangement." The well-known capability of $AlCl_3$ to abstract hydride ions comes into play and the carbocation 89 is produced, leading to rearrangement (scheme 15). Apparently, the weaker catalysts fail to initiate the rearrangement by the hydride abstraction.

The dealkylation and fragmentation reactions of 85 and 86 with $AlCl_3$-H_2O at 100°C were described in detail in 1973 [86(b)]. Comparison of the product mixtures at earlier stages (from 5 min) with those present after a 3 hr reaction period (Table 8) leads to the following observations:

1. Similar but nonidentical spectra of products were given by 85 and 86 at all times, indicating that the rate of interconversion of 85 and 86 is of the same order but not much faster than the rates of dealkylation and fragmentation of the two isomers.
2. Dealkylation exceeded fragmentation in the early stages. In the first 30 min only 15% and 6% of liquid fragmentation products were formed from 85 and 86, respectively, whereas after 180 min these values were 39% and 31%.
3. Isobutane was the major gaseous product at all times from both 85 and 86, with *n*-butane and isopentane as minor components. After 30 min there was more isopentane than *n*-butane formed from both diphenylbutanes. Isopentane comes from unsymmetrical fission of isooctanes which are produced by bimolecular reaction of isobutyl cations and isobutylene. (The three-carbon fission fragment must be converted mostly into isopropylbenzene.) No more than traces of ethane, and a maximum of 4% of propane was observed at any time.
4. Doubling the proportion of $AlCl_3$ to diphenylbutane increased the extent of fragmentation, as represented by the toluene, ethylbenzene, and propylbenzenes produced, much more than the amount of dealkylation.

Similar studies were made on the reactions of 1,1-, 1,2-, 1,3-, and 2,2-diphenylbutane (92-95) with $AlCl_3$ [86(a),87]. The results may be outlined and interpreted as follows:

```
                                                      Ph
                                                      |
Ph-C-C-C-C     Ph-C-C-C-C     Ph-C-C-C-C     C-C-C-C
   |                |                  |          |
   Ph               Ph                 Ph         Ph

   92               93                 94         95
```

1. At 25 to 70°C in benzene solution, 92 and 93 underwent dealkylation to *n*-butylbenzene, and isomerization occurred in the sequence 92 → 93 → 94 [86(a)].

2. When these diphenylbutanes were heated with $AlCl_3$-H_2O at 100°C for periods of from 5 min to 3 hr [87], the products from 92-94 were found to be similar, as might be expected, owing to the facile rearrangements that had been demonstrated. The first reaction from these three was clearly dealkylation to produce *n*-butylbenzene (95-98%).

3. Compound 95, on the other hand, gave *sec*-butylbenzene initially (85%), with isobutylbenzene in minor amounts which increased with time; 95 does not rearrange to 92, 93, or 94 easily, but it does rearrange to 2,3-diphenylbutane by a simple 1,2-phenyl shift. Loss of a phenyl group from a secondary carbon atom from either 95 or 2,3-diphenylbutane produces *sec*-butylbenzene.

4. The major alkane from dealkylation of all four diphenylbutanes was isobutane, with the exception of one experiment in which a double amount of catalyst was used with 92. In this one experiment the ratio was reversed to 70:30 *n*-butane/isobutane. With this exception, the ratio of isobutane to *n*-butane was almost always higher in the gases evolved from diphenylbutanes than in those from the butylbenzenes. No explanation for this difference was given.

5. There were very small amounts of fragmentation products from 92-94 with the usual amount of catalyst (0.3 M equivalent). *n*-Propylbenzene appeared first from 92, *n*-propylbenzene and ethylbenzene from 93, and isopropylbenzene and ethylbenzene from 94. Ethylbenzene was produced in largest amount from 95, probably coming via its rearrangement product, 2,3-diphenylbutane, as outlined earlier in scheme 12. The similarity in the pattern of fragmentation products from 2,2-and 2,3-diphenylbutane can be seen in Table 9. Ethane was reportedly produced by 92 and 94; *cis*- and *trans*-2-butene were also reportedly produced from 94. Isopentane was found in all of the reaction mixtures, usually in larger amount than *n*-butane; its mode of formation was described in the preceding section.

To summarize the significance of these studies of the diphenylbutanes: the results provide deductive evidence for the intervention of these compounds as intermediates in the rearrangements, dealkylations, and fragmentations of alkylbenzenes induced by aluminum chloride and other strong catalysts of the Friedel-Crafts type. Most of the observed reactions can be rationalized on the basis that the favored routes to products are those that involve secondary, tertiary, and/or benzyl carbocations as intermediates.

C. Miscellaneous Fragmentations and Cracking Reactions

As we mentioned at the end of the preceding section, many catalytic reactions of hydrocarbons, including arenes, are involved in petroleum processing. Some of these that are related to "cracking" and fragmentation will be listed here, although the catalysts and conditions are quite different from those of typical Friedel-Crafts reactions.

The results of Dimitrov and co-workers [73-75] that could be interpreted as dealkylations were described in the preceding section. These workers also reported products which resulted from fragmentations. Toluene and ethylbenzene were among the products from *n*-pentylbenzene passed over Al_2O_3-SiO_2 at 410 to 490°C [73]; 3-phenylpentane [74], and *p*-di-*n*-pentylbenzene [75(a)] gave many fragmentation products (cf. the preceding section for details). *p*-Di-*n*-butylbenzene was used as the substrate in studies of the effects of varying composition of the zeolite catalysts [75(b)]. The addition of Na was said to give more fragmentation in the side chains, whereas the substitution of HX decreased fragmentation and gave more isomerization and dealkylation.

Rudenko and Rodicheva [88] reported that in the cracking of cumene (isopropylbenzene) over aluminosilicates at 250 to 750°C, as the temperature was raised, the detection of propene and ethylbenzene decreased and the formation of benzene, xylene, and cymene increased.

The work of Csicsery [82] on butyl- and pentylbenzene reactions induced by a variety of catalysts was described in Sec. II.F.

In their report on the study of Nafion-H as a catalyst for reorientation and disproportionation reactions, Olah and Kaspi mentioned that some fragmentation of isopropyl to ethyl groups occurred in the reactions of cumene and diisopropylbenzenes with "fresh catalyst" at 170°C [89].

D. "Retro-fragmentations"

If a Friedel-Crafts fragmentation reaction is one in which smaller arenes are produced from larger arenes, the reverse of this process might be called a "retro-fragmentation." Reactions of this type on a minor scale have actually been reported from time to time, beginning in 1883 when Friedel and Crafts themselves described the first example [90-95]. Friedel and Crafts did not indicate the extent of the reaction, but they identified toluene, ethylbenzene, and biphenyl as products from benzene heated at 200°C with 0.2 mol of $AlCl_3$ [90]. In the most recent report, Siskin and Porcelli describe heating benzene with HF-TaF_5 and hydrogen at 200°C, and they reported that toluene and ethylbenzene were produced in yields of 36% and 40%, respectively, with about 40% conversion of benzene [95]. If their interpretation of the mechanism of formation of these products is correct, the overall "retro-fragmentation" may actually have some of the earmarks of a real fragmentation. They, like Friedel and Crafts, found biphenyl among the products, some of which were polyphenyls and more highly condensed aromatics, and they believe the condensations by which these arenes are formed provide the hydrogen required for the production of toluene and ethylbenzene. However, the yields of toluene and ethylbenzene were higher when the reaction was carried out in the presence of added hydrogen. The source of the methyl and ethyl groups was suggested to be through hydrogenation of benzene to cyclohexane, followed by its hydrogenolysis to hexane, which then undergoes acid catalyzed cracking (fragmentation!) to carbocations, which alkylate benzene (scheme 16). Toluene was found to give ethylbenzene and cumene under similar reaction conditions.

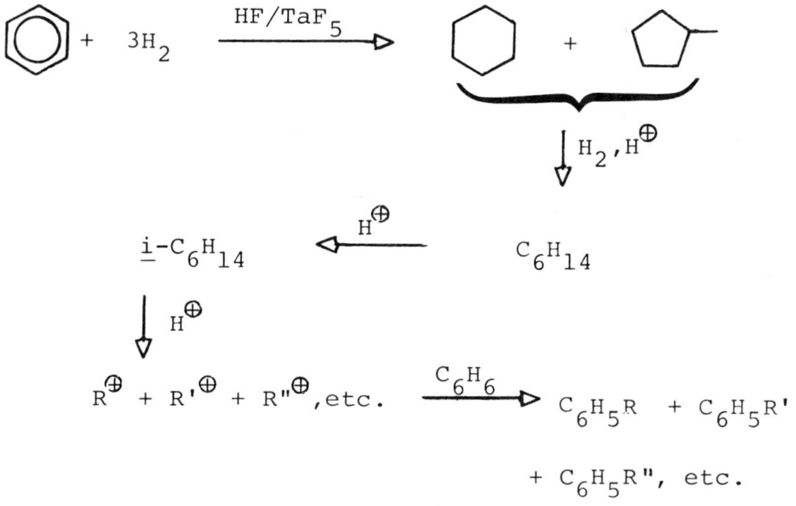

$$\text{C}_6\text{H}_6 + 3\text{H}_2 \xrightarrow{\text{HF/TaF}_5} \text{cyclohexane} + \text{methylcyclopentane}$$

$$\downarrow \text{H}_2, \text{H}^{\oplus}$$

$$\underline{\text{i}}\text{-C}_6\text{H}_{14} \xleftarrow{\text{H}^{\oplus}} \text{C}_6\text{H}_{14}$$

$$\downarrow \text{H}^{\oplus}$$

$$\text{R}^{\oplus} + \text{R}'^{\oplus} + \text{R}''^{\oplus}, \text{etc.} \xrightarrow{\text{C}_6\text{H}_6} \text{C}_6\text{H}_5\text{R} + \text{C}_6\text{H}_5\text{R}' + \text{C}_6\text{H}_5\text{R}'', \text{etc.}$$

Scheme 16

E. Confirmation of Diphenylalkanes as Intermediates in Fragmentations from Experiments with ^{14}C-Labeled Molecules

In this section we conclude the discussion of fragmentations of arenes induced by Friedel-Crafts catalysts with descriptions of two essentially serendipitous observations which tend to confirm the function of diphenylalkanes as intermediates in these reactions.

While pursuing his research primarily concerned with elucidating the mechanism of the isotopic *n*-propylbenzene rearrangement [12], Robin Greene used isotope-dilution technique to determine the isotopic distribution in the small amount of ethylbenzene produced from 1,2-diphenylpropane-2-^{14}C which had been treated with $AlCl_3$-H_2O at room temperature [28(a)]. Degradation and radioassay of recovered 1,2-diphenylpropane had shown that the ^{14}C had become completely equilibrated between the 1- and 2-positions of the diphenylpropane. The *n*-propylbenzene produced by dealkylation (ca. 20%) of the diphenylpropane had also been found to have the ^{14}C evenly distributed between the α and β positions of the side chain. (See scheme 1 and the discussion accompanying it in Sec. I.) The ethylbenzene isolated from the same reaction mixture was diluted with ordinary ethylbenzene. Half of the sample was radioassayed directly, the other half was degraded to benzoic acid, which was then radioassayed. The ethylbenzene and benzoic acid had the same ^{14}C molecular activity, thus all of the ^{14}C resided in the α position of the ethylbenzene side chain.

Now, if the ethylbenzene had come from direct fragmentation of *n*-propylbenzene, in which the isotope was known to be equally distributed between the α and β positions of the side chain, the ethylbenzene would have been expected to have ^{14}C in both positions (Eq. 42). But if the fragmentation

$$\text{Ph-}^{14}\text{CH}_2\text{-}^{14}\text{CH}_2\text{-CH}_3 \longrightarrow \text{Ph-}^{14}\text{CH}_2\text{-}^{14}\text{CH}_3 \qquad (42)$$

$^{14}CH_2-^{14}CH-CH_3$
Ph
Ph
$H^{\oplus}$
$H^{\oplus}$
$^{14}CH_2-^{14}CH-CH_3$
Ph
H
H
$^{14}CH_2-^{14}CH-CH_3$
Ph
H
H
$^{14}CH_2$
H
H
$^{14}CH_3$
RH
$^{14}CH_2^{\oplus}$
Ph
+
+
$^{14}CH-CH_3$
H
H
$^{14}CH_2^{\oplus}-CH_3$
Ph
RH
$^{14}CH_2-CH_3$
(0)
$^{14}CO_2H$

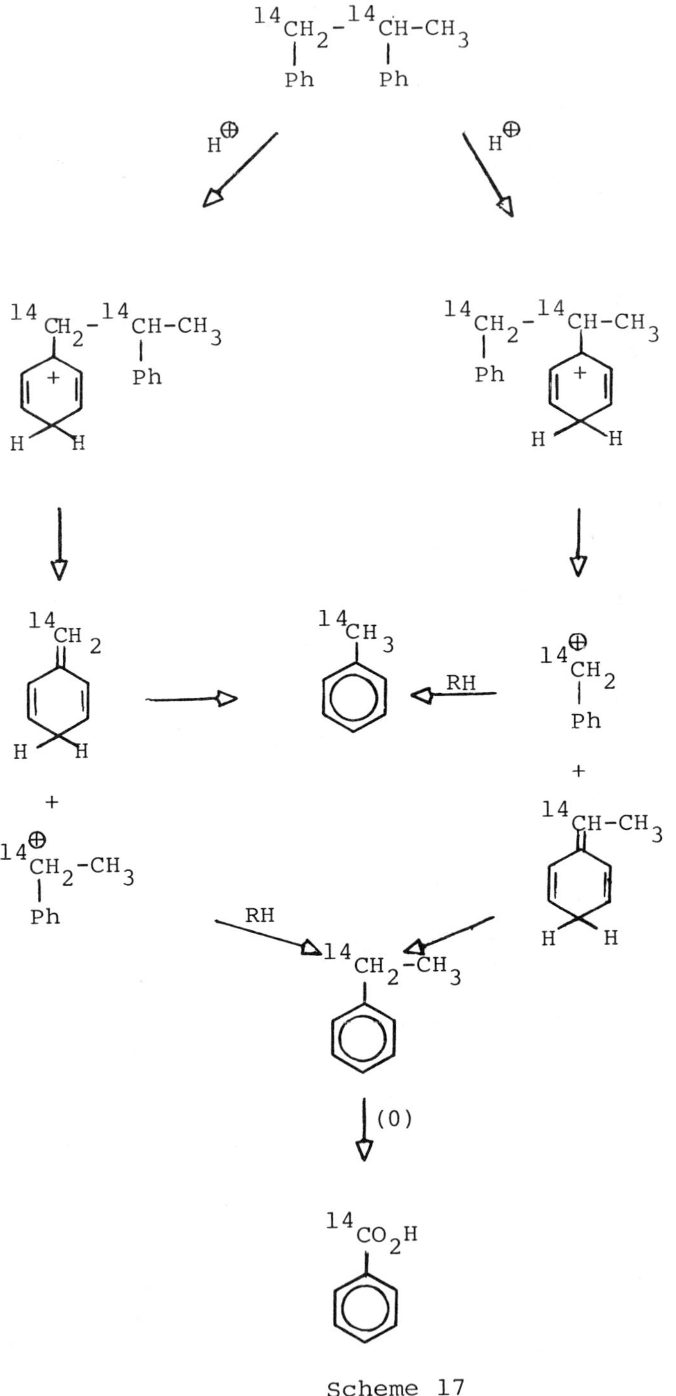

Scheme 17

occurs in the diphenylpropane by the type of mechanism proposed earlier (scheme 12), the results would be those observed, as outlined in scheme 17.

The second serendipitous observation was one made by C. C. Lee and co-workers, which was also somewhat off the mainstream of their planned research project [96]. Lee and his group were interested in determining whether isotopic rearrangement of ethyl-2-^{14}C iodide might occur preceding attachment to the benzene ring if the alkylation was carried out in an inert solvent (cf. Chap. 3). The experimental results indicated that rearrangement preceding alkylation did not occur, but the ethylbenzene-2-^{14}C initially produced at lower temperatures did undergo up to 20% rearrangement to ethylbenzene-1-^{14}C when the reaction mixture was heated at higher temperatures for longer times. In the course of separation of the ethylbenzene from various experiments by gas chromatography, peaks corresponding to benzene and toluene were observed. The benzene was assumed to come from disproportionation and the toluene from fragmentation of ethylbenzene. When the toluene collected from several experiments was radioassayed and its molecular activity was compared with that of the ethylbenzene, it was found to be about 0.5 in all cases. Lee recognized that this would be the expected result if the toluene arose from cleavage of 1,2-diphenyl-1-^{14}C-ethane [97), which could be derived from either ethylbenzene-2-^{14}C (96) or ethylbenzene-1-^{14}C (98) (scheme 18). Thus the intermediate formation and cleavage of 1,2-diphenyl-

$^{14}CH_3-CH_2I$ + PhH ⟶ $Ph-CH_2-^{14}CH_3$ (96)

$CH_3-^{14}CH_2-Ph$ (98) ⇄ $Ph-CH_2-^{14}CH_2-Ph$ (97) (← $Ph-CH(Ph)-^{14}CH_3$)

H^+ / H^+

$H_2C{=}C_6H_5{=}CH_2$ (H, H) + $^{14}\overset{\oplus}{C}H_2-Ph$ $Ph\overset{\oplus}{C}H_2$ + $^{14}CH_2{=}C_6H_4H_2$

RH / RH

$PhCH_3$ $^{14}CH_3-Ph$ $PhCH_3$ $^{14}CH_3-Ph$

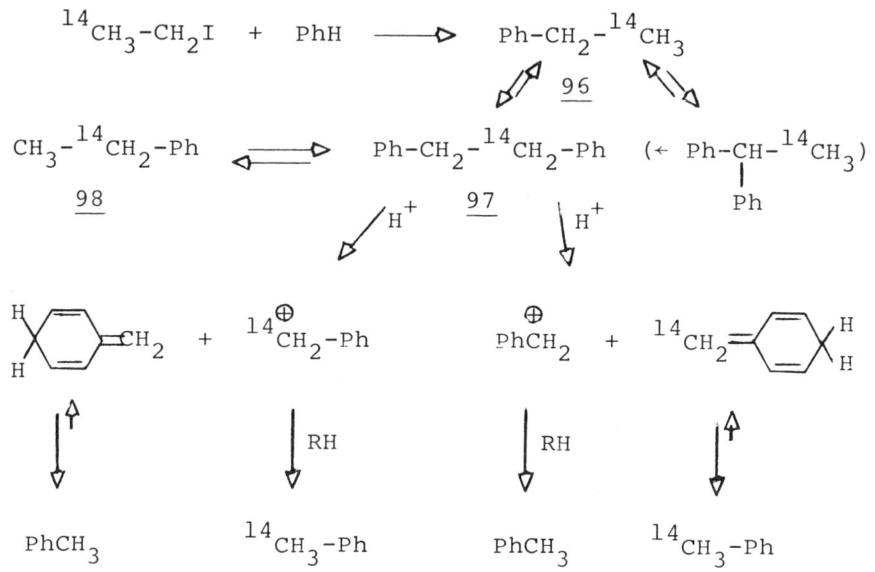

Scheme 18

ethane was seen as a rationale for both the observed isotopic rearrangement of ethylbenzene at high temperature and for the fragmentation that produced toluene, which correlates very well with the theories of Roberts and co-workers about analogous reactions of propylbenzenes and butylbenzenes.

REFERENCES

1. Roberts, R. M., Intra-Sci. Chem Rep., *6*, 89 (1972).
2. Schmerling, L., and J. P. West, J. Am. Chem. Soc., *76*, 1917 (1954).
3. Friedman, B. S., F. L. Morritz, C. J. Morrissey, and R. Koncos, J. Am. Chem. Soc., *80*, 5867 (1958).
4. Baddeley, G., Q. Rev. (Lond.), *8*, 355 (1954).
5. Gould, E. S., *Mechanism and Structure in Organic Chemistry*, Henry Holt, New York, 1959, p. 450.
6. Roberts, R. M., G. A. Ropp, and O. K. Neville, J. Am. Chem. Soc., *77*, 1764 (1955).
7. McCauley, D. A., and A. P. Lien, J. Am. Chem. Soc., *75*, 2411 (1953).
8. Kinney, R. E., and L. A. Hamilton, J. Am. Chem. Soc., *76*, 786 (1954).
9. Heise, R., and A. Töhl, Justus Liebigs Ann. Chem., *270*, 155 (1892).
10. (a) Roberts, R. M., and S. G. Brandenberger, Chem. Ind. (Lond.), 227 (1955); (b) Roberts, R. M., and S. G. Brandenberger, J. Am. Chem. Soc., *79*, 5484 (1957).
11. (a) Roberts, R. M., and J. E. Douglass, Chem. Ind. (Lond.), 1557 (1958); (b) Roberts, R. M., and J. E. Douglass, J. Org. Chem., *28*, 1225 (1963).
12. Roberts, R. M., A. A. Khalaf, and R. N. Greene, J. Am. Chem. Soc., *86*, 2846 (1964).
13. (a) Douglass, J. E., and R. M. Roberts, Chem. Ind., 926 (1959); (b) Douglass, J. E., and R. M. Roberts, J. Org. Chem., *28*, 1229 (1963).
14. Roberts, R. M., S. G. Brandenberger, and S. G. Panayides, J. Am. Chem. Soc., *80*, 2507 (1958).
15. Roberts, R. M., Y. W. Han, C. H. Schmid, and D. A. Davis, J. Am. Chem. Soc., *81*, 640 (1959).
16. Nenitzescu, C. D., I. Necsoiu, A. Glatz, and M. Zalman, Chem. Ber., *92*, 10 (1959).
17. Roberts, R. M., and D. Shiengthong, J. Am. Chem. Soc., *82*, 732 (1960).
18. Roberts, R. M., and Y. W. Han, Tetrahedron Lett., No. 6, 5 (1959).
19. Roberts, R. M., and Y. W. Han, J. Am. Chem. Soc., *85*, 1168 (1963).
20. Burwell, R. L., Jr., and A. D. Shields, J. Am. Chem. Soc., *77*, 2766 (1955).
21. Han, Y. W., Ph.D. dissertation, University of Texas at Austin, 1960.
22. Lin, Y.-T., Ph.D. dissertation, University of Texas at Austin, 1966.
23. Lipovich, V. G., V. V. Chenets, A. I. Rudenkov, and I. V. Kalechits, J. Org. Chem. USSR, *4*, 285 (1968).
24. Pines, H., and J. T. Arrigo, J. Am. Chem. Soc., *80*, 4369 (1958).
25. Streitwieser, A., and L. Reif, J. Am. Chem. Soc., *86*, 1988 (1964).
26. Roberts, R. M., E. K. Baylis, and G. J. Fonken, J. Am. Chem. Soc., *85*, 3454 (1963).
27. (a) Karabatsos, G. J., and F. M. Vane, J. Am. Chem. Soc., *85*,729 (1963); (b) Karabatsos, G. J., and F. M. Vane, J. Am. Chem. Soc., *85*, 79 (1963).
28. (a) Greene, R. N., Ph.D. dissertation, University of Texas at Austin, 1965; (b) Roberts, R. M., and R. N. Greene, Acta Cient. Venez. *15*(6), 251 (1965).
29. Nenitzescu, C. D., M. Avram, and E. Sliam, Bull. Soc. Chim. Fr., 1266 (1955).
30. Nightingale, D., and J. M. Shackelford, J. Am. Chem. Soc., *78*, 1225 (1956).
31. Roberts, R. M., A. A. Khalaf, and J. E. Douglass, J. Org. Chem., *29*, 1511 (1964).

32. Farcasiu, D., Rev. Chim. (Bucharest), *10*, 457 (1965).
33. This compound was later shown actually to be 1-ethyl-1,2,3-trimethyl-3-phenylindan; Khalaf, A. A., and R. M. Roberts, J. Org. Chem., *31*, 926 (1966).
34. Roberts, R. M., and T. L. Gibson, J. Am. Chem. Soc., *93*, 7340 (1971).
35. Roberts, R. M., T. L. Gibson, and M. B. Abdel-Baset, J. Org. Chem., *42*, 3018 (1977).
36. Brouwer, D. M., and H. Hogeveen [Prog. Phys. Org. Chem., *9*, 213 (1972)] reported no great difference in rate between methyl and ethyl shifts in superacid medium.
37. Olson, A. C., Ind. Eng. Chem., *52*, 833 (1960).
38. Hamanaka, S., T. Kimura, K. Itoh, and M. Ogawa, Kogyo Kagaku Zasshi, *72*, 1305 (1969); Chem. Abstr., *71*, 101426 (1969).
39. Roberts, R. M., G. P. Anderson, Jr., and N. L. Doss, J. Org. Chem., *33*, 4259 (1968).
40. Itoh, K., G. Yamada, S. Hamanaka, and M. Ogawa, Bull. Chem. Soc. Jpn., *41*, 2504 (1968).
41. Okami, Y., S. Hamanaka, K. Itoh, and M. Ogawa, Kogyo Kagaku Zasshi, *73*, 1729 (1970); Chem. Abstr., *74*, 12717 (1971).
42. Roberts, R. M., and H. H. Chen, Rev. Chim. (Bucharest), *25*, 687 (1980).
43. Inatome, M., K. W. Greenlee, J. M. Derfer, and C. E. Boord, J. Am. Chem. Soc., *74*, 292 (1952).
44. Khalaf, A. A., and R. M. Roberts, J. Org. Chem., *35*, 3717 (1970).
45. Roberts, R. M., H. H. Chen, and B. Junhasavasdikul, J. Org. Chem., *43*, 2977 (1978).
46. Roberts, R. M., and L. W. Elrod, Rev. Chim. (Bucharest), *27*, 761 (1982).
47. Roberts, R. M., and M. R. M. Ismail, unpublished data, 1977; M. R. M. Ismail, Ph.D. dissertation, University of Texas at Austin, Austin, 1977.
48. Roberts, R. M., and S. E. McGuire, J. Org. Chem., *35*, 102 (1970).
49. Schmerling, L., J. P. Luvisi, and R. J. Welch, J. Am. Chem. Soc., *81*, 2718 (1959).
50. Anschütz, R., Justus Liebigs Ann. Chem., *235*, 177 (1886).
51. Schorger, A. W., J. Am. Chem. Soc., *39*, 2671 (1917).
52. Ipatieff, V. N., and H. Pines, J. Am. Chem. Soc., *59*, 56 (1937).
53. Sharman, S. H., J. Am. Chem. Soc., *84*, 2945 (1962).
54. Khalaf, A. A., and R. M. Roberts, J. Org. Chem., *31*, 926 (1966).
55. Miethchen, R., and K. Ewald, Z. Chem., *14*, 240 (1974).
56. Itoh, K., R. Kurashige, S. Hamanaka, and M. Ogawa, Kogyo Kagaku Zasshi, *70*, 918 (1967); Chem. Abstr., *67*, 116653 (1967).
57. Itoh, K., R. Kurashige, S. Hamanaka, and M. Ogawa, Kogyo Kagaku Zasshi, *71*, 1570 (1968); Chem. Abstr., *70*, 47145 (1969).
58. Ipatieff, V. N., H. Pines, and L. Schmerling, J. Org. Chem., *5*, 253 (1940).
59. Pines, H., L. Schmerling, and V. N. Ipatieff, J. Am. Chem. Soc., *62*, 2901 (1940).
60. Huston, R. C., and K. Goodemoot, J. Am. Chem. Soc., *56*, 2432 (1934).
61. Streitwieser, A., Jr., D. P. Stevenson, and W. D. Schaeffer, J. Am. Chem. Soc., *81*, 1110 (1959).
62. Roberts, R. M., Y.-T. Lin, and G. P. Anderson, Jr., Tetrahedron, *25*, 4173 (1969).
63. Nield, P. G., J. Chem. Soc., 2278 (1964); 712 (1966).
64. Brouwer, D. M., and J. M. Oelderik, Recl. Trav. Chim, Pays-Bas, *87*, 721 (1968).

65. Kramer, G. M., J. Am. Chem. Soc., *91*, 4819 (1969).
66. Friedman, B. S., and S. M. Cotton, J. Org. Chem., *27*, 481 (1962).
67. Knight, H. M., J. T. Kelly, and J. R. King, J. Org. Chem., *28*, 1218 (1963).
68. Brouwer, D. M., Recl. Trav. Chim. Pays-Bas, *87*, 210 (1968).
69. Olah, G. A., R. H. Schlosberg, R. D. Porter, Y. K. Mo, D. P. Kelly, and G. D. Mateescu, J. Am. Chem. Soc., *94*, 2034 (1972).
70. Olah, G. A., and Y. K. Mo, J. Org. Chem., *38*, 3221 (1973).
71. Farcasiu, D., J. Org. Chem., *44*, 2103 (1979).
72. Farcasiu, D., and R. H. Schlosberg, J. Org. Chem., *47*, 151 (1982).
73. Dimitrov, Khr., and R. Pelova, God. Sofii, Univ., Khim, Fak., *57*, 149 (1962-1963) (publ. 1964); Chem. Abstr., *63*, 9776 (1965).
74. Dimitrov, Khr., and P. Ignatiev, J. Catal., *7*, 103 (1967).
75. (a) Dimitrov, Khr., and Tsv. Bezukhanova, God. Sofii, Univ., Khim, Fak., *64*, 411 (1969-1970) (publ. 1972); Chem. Abstr., *78*, 147637 (1973); (b) Dimitrov, Khr., Tsv. Bezukhanova, M. D. Mukareva, and Kh. M. Georgiev, Dokl. Bolg. Akad. Nauk, *27*, 55 (1974).
76. Hatch, L. F., and S. Matar, *From Hydrocarbons to Petrochemicals*, Gulf Publishing Co., Houston, 1981, Chap. 10.
77. Maslyanskii, G. N., G. L. Rabinovich, N. Kh. Avtonomova, and K. L. Brisker, Ger. Patent 1,793,129 (Nov. 4, 1971); Chem. Abstr., *76*, 45905 (1972).
78. Safaev, A. S., and R. I. Radyuk, Katal. Pererab. Uglevodorodnogo Syr'ya, *4*, 40 (1970); Chem. Abstr., *76*, 126486 (1972).
79. Kochloefl, K., Ger. Offen., 2,357,405 (Aug. 1, 1974); Chem. Abstr., *81*, 120170 (1974).
80. Tsuchiya, M., A. Igarashi, and Y. Ogino, Sekiyu Gakkai Shi, *13*, 643 (1970); Chem. Abstr., *73*, 120,208 (1970).
81. British Petroleum, Fr. Demande 2,010,151 (Feb. 13, 1970); Chem. Abstr., *73*, 55790 (1970).
82. (a) Csicsery, S. M., J. Catal., *9*, 336 (1967); (b) Csicsery, S. M., J. Catal., *15*, 111 (1969).
83. Norris, J. F., and B. M. Sturgis, J. Am. Chem. Soc., *61*, 1413 (1939).
84. Searles, S., J. Am. Chem. Soc., *76*, 2313 (1954).
85. Khalaf, A. A., Rev. Chim. (Bucharest), *18*, 470 (1973).
86. (a) Roberts, R. M., and A. A. Khalaf, Rev. Chim. (Bucharest), *18*, 1929 (1973); (b) Roberts, R. M., and A. A. Khalaf, Rev. Chim. (Bucharest), *18*, 1937 (1973).
87. Khalaf, A. A., and R. M. Roberts, Rev. Chim. (Bucharest), *19*, 1351 (1974).
88. Rudenko, A. P., and M. F. Rodicheva, Vestn, Mosk, Univ., Khim., *12*, 705 (1071); Chem. Abstr., *76*, 59079 (1972).
89. Olah, G. A., and J. Kaspi, Nouv. J. Chimie, *2*, 585 (1978).
90. Friedel, C., and J. M. Crafts, Bull. Soc. Chim. Fr., *39*, 195 (1893).
91. Wertyporoch, E., and H. Sagal, Chem. Ber., *66*, 1306 (1933).
92. Ipatieff, V. N., and V. I. Komarewsky, J. Am. Chem. Soc. *56*, 1926 (1934).
93. Ipatieff, V. N., and G. S. Monroe, J. Am. Chem. Soc., *69*, 710 (1947).
94. Kazanskii, B. A., M. Rozengart, and Z. F. Kuznetsova, Dokl. Acad. Nauk SSSR, *126*, 571 (1959).
95. Siskin, M., and J. Porcelli, J. Am. Chem. Soc., *96*, 3640 (1974); U.S. Patent 3,728,411, (Apr. 17, 1973); Chem. Abstr., *78*, 159, 170 (1973).
96. Lee, C. C., M. C. Hamblin, and J. F. Uthe, Can. J. Chem., *42*, 1771 (1964).

Index

B

C

D

E

F

I

K

R